MECHANICS OF STRUCTURES AND MATERIALS

PROCEEDINGS OF THE 16TH AUSTRALASIAN CONFERENCE
ON THE MECHANICS OF STRUCTURES AND MATERIALS
SYDNEY / NEW SOUTH WALES / AUSTRALIA / 8-10 DECEMBER 1999

Mechanics of Structures and Materials

Edited by

Mark A. Bradford

School of Civil and Environmental Engineering, The University of New South Wales, Sydney, N.S.W., Australia

Russell Q. Bridge

School of Civic Engineering and Environment, University of Western Sydney, Nepean, N.S.W., Australia

Stephen J. Foster

School of Civil and Environmental Engineering, The University of New South Wales, Sydney, N.S.W., Australia

A.A. BALKEMA / ROTTERDAM / BROOKFIELD / 1999

Cover: Olympic sculptures; AMP Tower, Sydney.
Photo supplied by Hyder (Australia)

The texts of the various papers in this volume were set individually by typists under the supervision of each of the authors concerned.

Published by
A.A. Balkema, P.O. Box 1675, 3000 BR Rotterdam, Netherlands
Fax: +31.10.413.5947; E-mail: balkema@balkema.nl; Internet site: www.balkema.nl

A.A. Balkema Publishers, Old Post Road, Brookfield, VT 05036-9704, USA
Fax: 802.276.3837; E-mail: info@ashgate.com

ISBN 90 5809 107 4

Printed in the Netherlands

Mechanics of Structures and Materials, Bradford, Bridge & Foster (eds)
© 1999 Balkema, Rotterdam, ISBN 90 5809 107 4

Table of contents

Composite structures

Dynamic analyses of structures

Optimisation and reliability

Mechanics of Structures and Materials, Bradford, Bridge & Foster (eds)

Preface

The Australasian Conference on the Mechanics of Structures and Materials (ACMSM) is a biannual symposium for the exchange of ideas for practitioners and researchers in the broad areas of structural, material and foundation engineering. The first ACMSM was held at The University of New South Wales in 1967, under the chairmanship of the late Prof. F.S.Shaw. It is therefore appropriate that the sixteenth ACMSM, which is the final in the series for the twentieth century, returns again to The University of New South Wales. The conference was jointly organised and run by Civil Engineering staff members from The University of New South Wales and The University of Western Sydney, Nepean. A pleasing feature of ACMSM symposia is that they provide a forum for the exchange of ideas, not only between established academics, researchers and practitioners, but also between younger research students and engineers. The evolution of a strong and widespread research culture within the Australasian region has necessitated a frequent forum for young researchers to present the results of their findings, and to interact with more senior established research scholars and practising engineers.

The papers in this volume cover a very broad spectrum of topics in structural mechanics, including computational and fracture mechanics, advances in design and construction methodologies, steel, concrete and composite structures, foundation engineering, dynamic analysis of structures, composites and new materials, and environmental loadings. As well as papers from Australia and New Zealand, submissions from Canada, USA, Japan, UK, Iran, China and Singapore appear in this volume. Three keynote speakers, Dr A.J.Davids, Dr A.Currie and Prof. S.Valliappan, presented overviews from consulting and academia of the latest advances in structural and materials technology.

Each paper was subjected to a critical and rigorous review process, by experts within the field in which the paper was written. The editors are appreciative of the help of the reviewers in maintaining the high standard of assessment of each paper.

Mechanics of Structures and Materials, Bradford, Bridge & Foster (eds)
© 1999 Balkema, Rotterdam, ISBN 90 5809 107 4

Acknowledgements

The successful organisation of a conference requires financial support as well as individual effort from a number of organisations and people. The conference organising committee has been fortunate to gain both sponsorship and personal support. The secretarial support provided by Ms Betty Wong of the University of New South Wales is especially acknowledged with thanks.

Sponsorship has been provided by
Compumod Pty Ltd,
Hyder Consulting (Australia) Pty Ltd,
The University of New South Wales,
University of Western Sydney

The papers were reviewed by
P. Ansourian
M.A. Bradford
R.Q. Bridge
J.W. Butterworth
M.J. Clarke
C. Clifton
A. Deeks
S.J. Foster
S. Fragomeni
R.I. Gilbert
N. Gowripalan
M.C. Griffith
M.N.S. Hadi
N. Haritos
Y-C. Loo
M. Mahendran
R.E. Melchers
P. Mendis
D.J. Oehlers
M.D. O'Shea
J. Petrolito
Y-L. Pi
K.J.R. Rasmussen
M.G. Stewart
F. Tin-Loi
B. Uy
G.M. Van Erp
R.F. Warner
M. Xie

Mechanics of Structures and Materials, Bradford, Bridge & Foster (eds)
© 1999 Balkema, Rotterdam, ISBN 90 5809 107 4

Organisation

The 16th ACMSM Conference Committee

Professor Mark A. Bradford (Chairman)
Dr Stephen J. Foster (Secretary)
Professor R. Ian Gilbert
Associate Professor Francis Tin-Loi
Dr Mario M. Attard
Dr Brian Uy

School of Civil and Environmental Engineering, The University of New South Wales

Professor Russell Q. Bridge (Co-Chairman)
Dr E. Stefan Bernard
Dr Yang Xiang

School of Civic Engineering and Environment, University of Western Sydney, Nepean

Keynote papers

Mechanics of Structures and Materials, Bradford, Bridge & Foster (eds)
 ISBN 90 5809 107 4

Postcard from a busy consulting practice

A. Davids
Hyder Consulting, St. Leonards, Sydney, N.S.W., Australia

ABSTRACT: The real marvel of designing and building structures these days is the attitude of our society. They have now reached the point where they no longer worry about asking whether something is technically possible, they simply assume that it is possible. This is a double-edged sword for engineers. We outwardly accept the invitation with a casual ease that masks an inner searching for confidence, such searching being flashed sometimes in whispers between engineering brothers on massive or complex projects.

In these circumstances, raw human nature emerges. Structural engineers being basically boys who never grew up and still delight at the physics of the real world accept this illusion without reluctance, and go about pulling technical rabbits from a hat whilst hoping that the audience can't see their smile too. Such is the life of an engineer in today's consulting practice armed with contemporary technology and a wealthy patron with attitude.

This postcard from a busy consulting practice jots down some notes on the way in which design engineers go about their craft and the roles which computational technology, the research of structural mechanics and intuition play in that pursuit. The observations are illustrated with some recent projects undertaken by the Hyder group including very tall buildings, tuned mass dampers, bridge collapse investigations, a long span dome and other items a well dressed society demands.

Mechanics of Structures and Materials, Bradford, Bridge & Foster (eds)
© 1999 Balkema, Rotterdam, ISBN 90 5809 107 4

Developments in computational methods for structures and materials

A.Currie
Compumod Pty Limited

ABSTRACT: Structural analysis has come a long way since the early days of computers when Livesley's matrix methods were promoted in the 1960s to solve frame structures. We can even go back to the mid-1600s for Robert Hooke's elastic springs, Euler in the 1700s for elastic curves, followed by the likes of Navier, Young, Cauchy, Poisson, Lame, Saint-Venant and Kirchhoff in the 1800s and then Mohr and, of course Timoshenko, for the development of elasticity theory. Zienkiewicz was in the right place at the right time to show, around 1970, how the computational finite element method could be used to solve the analytical equations for structures of arbitrary geometry.

We are still limited by computer storage and speed but can now tackle non-linear structural mechanics at the micro-structure materials level and also solve for a complete structure with stage construction, changing contact plus non-linear material and geometric response to a load history. Loadings can be static, harmonic, transient and even coupled problems can be solved. The definition of a "structure" is very context sensitive with manufacturing processes, paper reinforced cement, human bone/tendon/tissue, foams and many other problems all able to be solved using the same computational software. In fact the question can be asked, when is a structure a fluid and when is a fluid a structure?

This paper gives a brief history on where we have come from and a current update on what can be achieved today. Case histories from civil, structural, transportation, biomechanics and manufacturing fields are given to illustrate actual use of computational solutions. The use of computer based data acquisition and signal analysis is also discussed to show that computational analyses are both valid and also based on a realistic model of the operating environment.

Mechanics of Structures and Materials, Bradford, Bridge & Foster (eds)

Developments in computational mechanics applied to the behaviour of materials and structures

S.Valliappan
School of Civil and Environmental Engineering, The University of New South Wales, Sydney, N.S.W., Australia

ABSTRACT: The developments in computational mechanics in Australia during the past three decades, related to the behaviour of materials and structures will be presented in this lecture.

These developments include non-linear behaviour of materials, fracture mechanics, damage mechanics, seismic analysis of structures, soil-structure interaction, micro-mechanics, composite materials, contact problems, fuzzy logic and optimisation. The solutions for the above-mentioned problems using finite elements, infinite elements and boundary elements will be illustrated through typical examples.

Finally, some thoughts on new as well as the extension of present approaches related to the mechanics of structures and materials for the next millennium are provided.

Computational and fracture mechanics

Mechanics of Structures and Materials, Bradford, Bridge & Foster (eds)
© 1999 Balkema, Rotterdam, ISBN 90 5809 107 4

Fracture simulation using a discrete triangular element

Mario M. Attard & Francis Tin-Loi
School of Civil and Environmental Engineering, The University of New South Wales, Sydney, N.S.W., Australia

ABSTRACT: A triangular element constructed from constant strain triangles with nodes along its sides is developed for the simulation of fracture. Fracture is captured through a constitutive softening-fracture law at the interface nodes. The material within the triangular unit remains elastic. The plastic displacement at the interface nodes represents the relative crack opening displacement, which are related to the conjugate inter-nodal force by the appropriate softening relationship. The path-dependent softening behaviour is solved in non-holonomic rate form based on a quasi-prescribed displacement formulation. At each event in the loading history all solutions are obtained and the critical path with the minimum second order work is adopted. All solutions are obtained using an enumerative procedure based on the observation that the solution of an LCP can be found by an exhaustive exploration of a binary tree. The classical three point bending and two notch tensile tests are simulated and compared with other published results.

1 INTRODUCTION

Over the last decade there has been intense research interest in the Modeling of fracture in quasi-brittle materials. Most of the investigations have been limited to Mode I failure (tension). The various methods proposed can be broken into two main classes: the discrete crack methods and those based on a continuum Model with smeared cracking. The formidable difficulties in providing a numerical simulation of the fracturing process within quasi-brittle material lie with the fact that the fracturing zone consists of discontinuous cracks and shear bands where damage is localized. Since the crack opening is discontinuous, it cannot easily be handled by continuum Models. The principal defect of the classical plastic continuum Models is that they do not incorporate an internal length scale or higher-order continuum structure and, therefore, the governing equations under quasi-static loading lose ellipticity, (Li and Cescotto, 1996).

Strain localization can give rise to a multiplicity of solutions to the softening path, Bolzon et al. (1997). Generalised continuum theories such as gradient continuum, Cosserat continuum and rate-dependent continuum have been developed in response to these problems (Li and Cescotto 1996). The discrete crack methods, on the other hand, take account of cracking by defining a boundary to the finite element mesh along the crack path, allowing the crack to open between continuum elements that remain linear elastic. Consideration of the fracture process zone ahead of the crack tip is also essential and can be incorporated using the fictitious crack Model of Hillerborg. The crack tip is extended over a fictitious length over which the opening normal to the crack is related to the normal traction by a softening law having an enclosed area equal to the fracture energy. Bolzon et al. (1995, 1997) used a novel approach where the discrete-crack Model is formulated as a linear complementarity problem (LCP) and solutions are sought using mathematical programming algorithms. One of the advantages of this approach is that all possible equilibrium paths could theoretically be followed.

The discrete-crack methods often require remeshing with the crack path determined by the trajectory of the principal tensile stress in Mode 1, unless the crack profile is known before hand as in the three point bending test. A challenge to the discrete-crack approach is to correctly predicting crack branching and the interaction between the developing cracks. Recently, Carol and Prat (1997) presented an interface element approach for Modeling normal/shear cracking in quasi-brittle materials that can be used in either in the context of discrete crack analysis or smeared crack analysis.

Lattice type Models such as those proposed by Van Mier et al. (1995) and Schlangen (1995), attempt to mimic the response of concrete at the meso-level. A concrete continuum is Modeled using a triangular lattice composed of beam elements connected at nodes. The heterogeneous material properties of concrete are incorporated by distinguishing between regions representing the aggregate, mortar and aggregate-mortar interface. To simulate the fracture process and growth of cracking, a step by step linear elastic analysis is carried out. Subject to a fracture law, a beam element is removed and the analysis continued. Some of the major drawbacks of the lattice Models are that if a crack closes there is no prevention of overlap, there is no reconnection of elements in contact and there is no account taken of friction when crack surfaces in contact move across one another. Because of this, the present lattice Models have limited success in simulating concrete under compressive loading. Models, which represent concrete as an assemblage of particles, have also been proposed in Zubelewicz and Bazant (1987), Bazant et al. (1990), Beranek and Hobbelman (1995). These particle type Models attempt to more closely mimic the structure of concrete. Usually the particles are assumed to be rigid or linear elastic and the fracture/softening effects are concentrated in the interface contact zone. Figure 1 shows a particle representation of a material using sphere particles. The particle representation only approximates a continuum depending on the density of voids. The particle and lattice Models can both be classified as "interface" type Models.

A discrete element Model is proposed in this paper which is a particle/interface type formulation. This paper summarises some of the results with respect to Mode 1 cracking. The formulation is based on a linear complementarity formulation as in Bolzon et al. (1995, 1997) and uses a enumerative mathematical programming algorithm to obtain all possible solutions to an incremental non-holonomic rate formulation. A constitutive law based on the fictitious crack Model is used. The path with the minimum second order work is chosen as the critical path. The advantage of this formulation is that, as with the discrete crack Models, there is no length scale required and hence the softening response is only marginally affected by the mesh used. The discrete Models are not mesh sensitive but the discretization used may of course, as in conventional finite element Modeling, affect the accuracy of the solution obtained.

2 DISCRETE TRIANGULAR ELEMENT

The basic unit in the formulation is a triangle formed by assembling nine constant strain triangles and condensing out the freedoms at the vertices (see Figure 2(a)). There are two nodes on each of the three sides/interfaces of the triangle unit. The position of the interface nodes is set at $L_i/2n$ from the vertices, where L_I is half the interface length and n an integer (refer to Figure 2(a)). The material within the triangular unit is assumed to remain linear elastic. The generalised

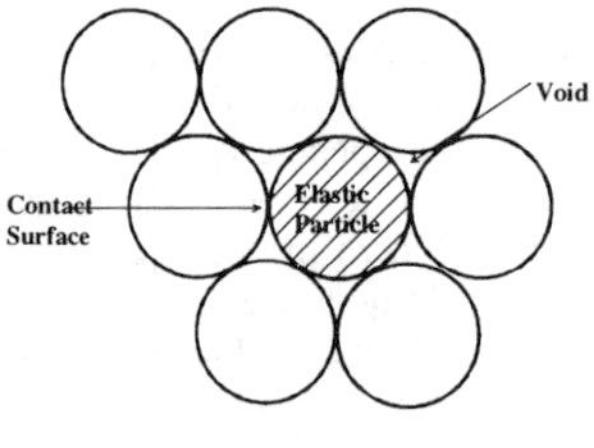

Figure 1 Sphere particles in contact.

displacements at the interface nodes are the displacements normal and tangential to the interface, which are assumed to consist of an elastic and plastic component. The conjugate generalised forces are taken as the normal and shear forces at the interface nodes (see Figure 2(a)). A material is Modeled by connecting these triangular units at the interface nodes. The Model only approximates a continuum since the vertices are not connected (see Figure 2(b)) mimicking the voids and contact surfaces that would exist in a particle representation (see Figure 1).

3 NON-HOLONOMIC RATE FORMULATION AND SOLUTION PROCEDURE

Paucity of space precludes a detailed description of the mathematical programming formulation used and the algorithmic details of the solution process; the interested reader is referred to Bolzon et al. (1997) for details. Briefly, a special mathematical programming problem known as a "linear complementarity problem" is set up from the equilibrium, compatibility and (softening) constitutive relations pertaining to the suitably discretized Model. In conventional finite element fashion, concatenation of appropriate elemental vectors and block-diagonalization of elemental matrices are carried out.

Following the classical plasticity formulation described in Maier (1970), the yield surface for a body is taken as a piece-wise linear assemblage of the yield surfaces for the generalised stresses. In the present context, the generalised stresses are the interface normal and shear forces at the connecting nodes. In classical continuum formulations, the yield surface is defined in terms of the stresses at a point. Here, the yield surface is in terms of the interfaces forces. General yield surfaces for particle type formulations exist. Beranek and Hobbelman (1995) modeled concrete as an assemblage of equal spheres and were able to determine a failure criterion for the contact layer for the spheres. The yield surface is a hyperbola with the strength in tension the same in shear, which they noted is just as Coulomb had suggested. For the present paper only Mode 1 cracking is considere. An associative flow rule has been assumed. It should be noted that the present formulation can also be extended to more general yield surfaces and an non-associative flow rule.

For simplicity, without undue loss of generality, the constitutive law is a linear softening law (see Figure 3) having a fracture energy determined in pure tension. In Figure 3, w is the crack-

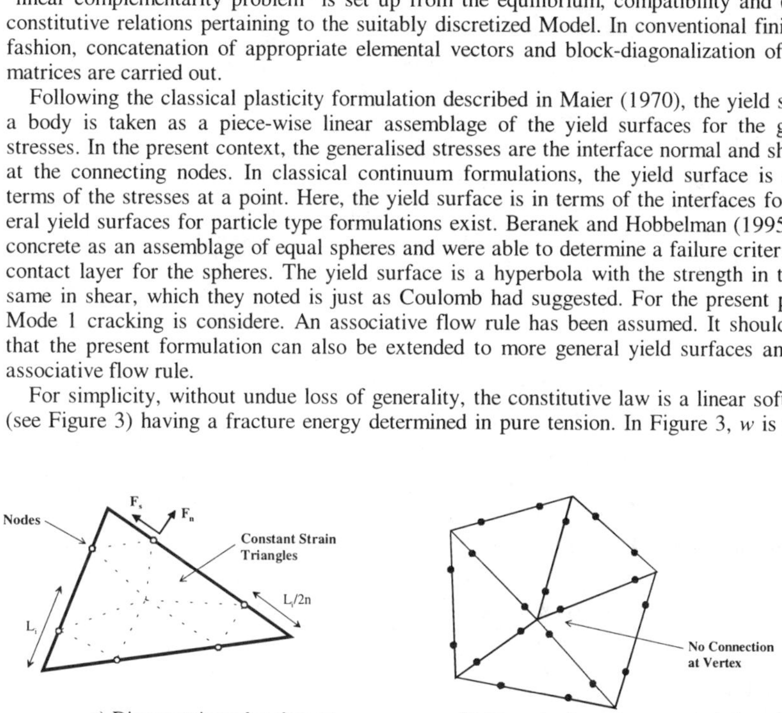

a) Discrete triangular element

b) Discrete elements connected along interfaces

Figure 2 Finite element model.

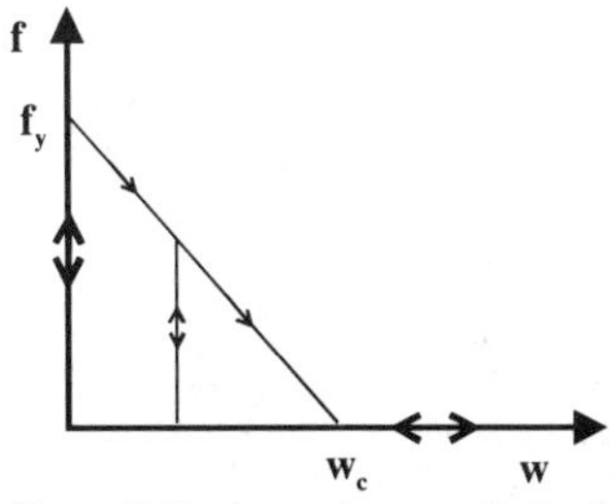

Figure 3 Crack opening constitutive law.

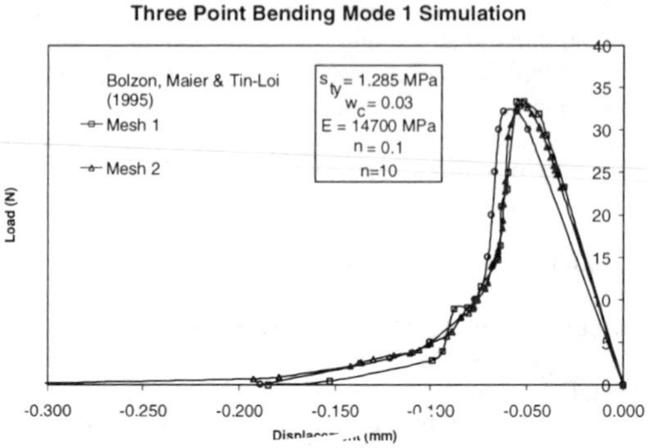

Figure 4 Simulation of three point bending test.

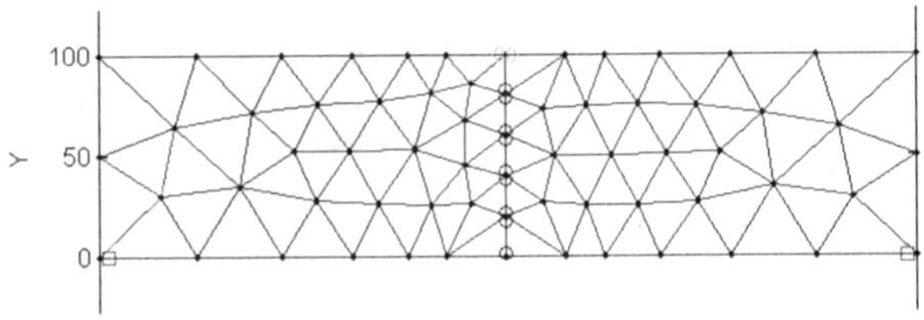

Figure 5 Three point bending simulation - Mesh 1.

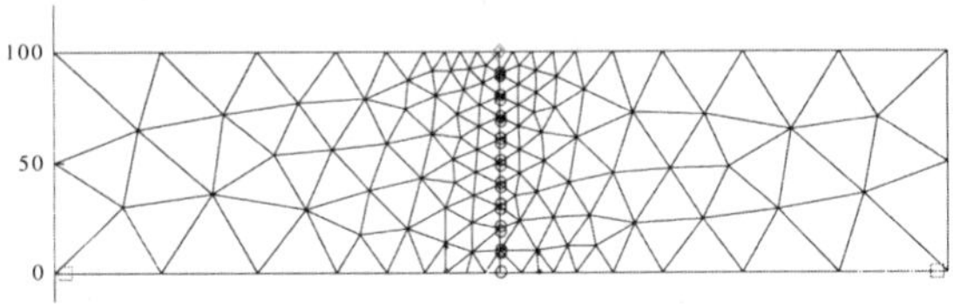

Figure 6 Three point bending simulation - Mesh 2.

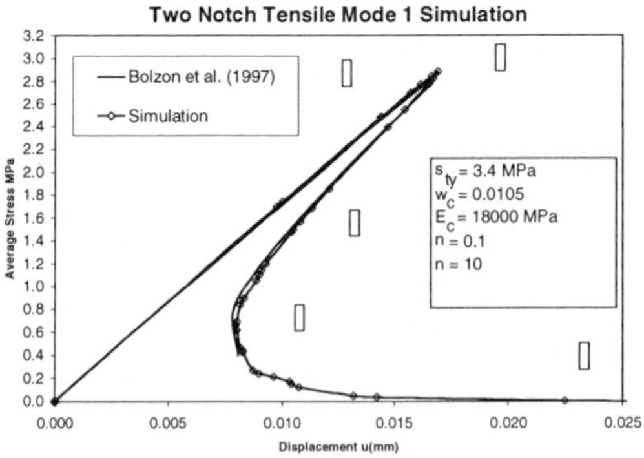

Figure 7 Simulation of the two notch tensile test.

ing opening displacement, f is the generalised stress and f_y the yield value. The critical crack opening under pure tension is denoted by w_c and the softening slope is denoted by h. No allowance has been made for the energy dissipated due to friction and interlocking since only Mode 1 is considered here.

Crucial to the formulation is an appropriate description of the softening constitutive laws in a so-called "complementarity" format, involving the orthogonality of two sign-constrained vectors. This, of necessity, is carried out in a rate form to account for the nonholonomic (path-dependent) nature of the fracture processes.

In particular, for a generic "node" and with reference to a single linear branch softening law (Figure 3) for simplicity, we can describe the incremental processes that can arise from a known (current situation) as follows:

$$\text{(a)} \quad \dot{w}=0,\ \dot{f}\ unconstrained,\ if\ 0\leq w\leq w_c,\quad \phi=f-f_y<0$$

$$\text{(b)} \quad \dot{\phi}\leq 0,\ \dot{\phi}.\dot{w}=0,\quad if\ 0<w\leq w_c,\quad \phi=f-(f_y-hw)=0$$

$$\text{(c)} \quad \dot{w}\ unconstrained,\ \dot{f}=0,\ if\ w_c<w$$

The first relation refers to a stress point in an "elastic" condition. That is, it describes the movement of a point either on the initial elastic branch or one which has elastically unloaded from the softening curve. The second condition assumes that the current stress point is "active" (i.e. is on the softening branch). The evolution of that point can follow either of two paths. It can elastically unload or can continue yielding down the softening curve. In both cases, the complementarity condition $\dot{\phi}.\dot{w}$ is thus satisfied. The third condition refers to the fully cracked case and is self-evident.

The adopted solution algorithm essentially follows the one used by Bolzon et al. (1997) in that only those stress points that are "active" (i.e. belong to the fracture process zone or "craze") are monitored. This active set is continually updated through (exact) identification of such critical events as new crack activation and/or elastic unloading from a softening branch. For this purpose, associated linear (in view of the adopted piecewise linear laws) complementarity problems need to be solved. The challenge involved is to capture the possible multiplicity of solutions expected to arise in view of the softening-related bifurcation phenomena. For the current work, in view of the tractable size of the linear complementarity problems involved, an enumerative code developed specifically for this purpose (Bolzon et al. 1995) has been adopted. All solutions to the LCP are obtained. The solution, which provides the minimum second order work, is taken as the critical solution. This is seen as very important when investigating softening and especially for interacting and/or branching cracks.

4 EXAMPLES

4.1 *Three Point Bending*

The first example is that of an idealised three point bending test taken from Bolzon et al. (1995). The simply supported plain concrete beam, is loaded by a concentrated load at mid-span. The length of the beam is 400 mm, while the width and depth are 100 mm. The elastic modulus of the concrete is taken as E_c=14700 MPa, the Poisson's ratio ν=0.1, the tensile strength of the concrete σ_{ty}=1.285 MPa, and the critical crack opening displacement w_c=0.03 mm. The discrete triangular elements used in the simulation, adopted an n=10. Two meshes were used, both were generated using a Delaunay triangularization algorthim and were prescribed a straight boundary at the mid-span. No notch was used. A comparison between the results of Bolzon et al. (1995) and the simulations for the load versus the load point deflection is shown in Figure 4. Both meshes show good agreement with the results of Bolzon et al. (1995). Figures 5 and 6 show the the yielding zones in the two meshes at the end of the softening branch of the load deflection curve. The only regions to reach the critical crack opening displacement are along the straight boundary. The circles in Figures 5 and 6 indicate the points that have reached the critical crack opening displacement. There was only one solution to the LCP at all the events along the load deflection curve indicating that there was no branching of the equilibrium path except for the elastic unloading path. Both the coarse and finer mesh give similar softening responses.

4.2 *Two Notch Tensile*

The second example is taken from Bolzon et al. (1997). The two notch tensile has been studied extensively in the literature and is a benchmark test for formulations used in the analysis of brittle fracture. Multiple solutions to the equilibrium path exist and are related to whether the end platens are allowed to rotate or are fixed. The material parameters chosen in Bolzon et al. were selected to produce a severe snap-back response. The length of the tensile specimen was taken as 250 mm, the depth was 60 mm, the elastic modulus of the concrete E_c=18000 MPa, the Poisson's ratio ν=0.2, the tensile strength of the concrete σ_{ty}=3.4 MPa, and the critical crack opening displacement w_c=0.0105 mm. Two notches 5 mm deep on each side at mid-span were used. The simulations here used an n=10. A mesh generated using Delaunay triangularization was used with a prescribed straight boundary between V-shaped notches at the center of the specimen. An axial displacement at one of the load points was used as a control variable. Figure 7 compares the simulation with the results of Bolzon et al. (1997) where excellent agreement is shown.

5 CONCLUSIONS

A discrete triangular element constructed from constant strain triangles with nodes along its sides has been developed for the simulation of brittle fracture. Fracture is captured through a constitutive softening-fracture law at the interface nodes. A linear softening law with a single softening branch was adopted. Since the softening law is formulated in terms of the forces at the interface nodes, no internal length scale or higher-order continuum structure was required. The path-dependent softening behaviour was solved in non-holonomic rate form. At each event in the loading history all solutions to the LCP were obtained and the critical equilibrium path with the minimum second order work was adopted. Two examples involving Mode 1 brittle fracture were presented and compared to known solutions. The formulation compared well and demonstrated its robustness by capturing the severe snap-back solution of the two notch tensile problem. Current work is aimed at applying the same methodology to the solution of problems with unknown, possibly interacting crack itineraries.

REFERENCES

Bazant, Z.P., Tabbara, M.R., Kazemi, M.T. and Pijaudier-Cabot, G., 1990, "Random Particle Model for Fracture of Aggregate or Fibre Composites", ASCE Journal of Engineering Mechanics, Vol. 116, No. 8, August, pp. 1686-1705.

Beranek, W.J. and Hobbelman, G.J., 1995, "2D and 3D-Modelling of Concrete as an Assemblage of Spheres, Revaluation of the Failure Criterion", Fracture Mechanics of Concrete Structures, Proceeding FRAMCOS-2, edited by Folker H. Wittman, Freiburg, pp. 965-978.

Bolzon, G., Maier, G. and Tin-Loi, F., 1997, "On Multiplicity of Solutions in Quasi-Brittle Fracture Computations", Computational Mechanics, Vol. 19, pp. 511-516.

Bolzon, G., Maier, G. and Tin-Loi, F., 1995, " Holonomic and Nonholonomic Simulations of Quasi-Brittle Fracture: A Comparative Study of Mathematical Programming Approaches", Fracture Mechanics of Concrete Structures, Proceedings FRAMCOS 2 edited by Folker H. Wittman, Frelburg, pp. 885-898.

Carol, I. and Prat, P.C., 1997, "Normal/Shear Cracking Model: Application to Discrete Crack Analysis", Journal of Engineering Mechanics, Vol. 123, No. 8, August, pp. 765-773.

Li, X. and Cescotto, S., 1996, "Finite Element Method for Gradient Plasticity at Large Strains", International Journal for Numerical Methods in Engineering, Vol. 39, pp. 619-633.

Maier, G., 1970, "A Matrix Structural Theory of Piecewise-Linear Elastoplasticity with Interacting Yield Planes", Meccanica, Vol. 5, pp. 54-66.

Schlangen, E., 1995, "Computational Aspects of Fracture Simulations with Lattice Models", Fracture Mechanics of Concrete Structures, Proceedings FRAMCOS-2, edited by Folker H. Wittmann, pp 913-928.

Van Mier, J.G.M., Schlangen, E., Vervuurt, A and Van Vliet, M.R.A., 1995. "Damage Analysis of Brittle Disordered Materials: Concrete and Rock", Mechanical Behaviour of Materials, Edited by A. Bakker, Seventh International Conference on Mechanical Behaviour of Materials – ICM7, Delft University Press, Delft, The Netherlands, pp 101-126.

Zubelewicz, A., and Bazant, Z.P., 1987, "Interface Element Modelling of Fracture in Aggregate Composites", ASCE Journal of Engineering Mechanics, Vol. 113, No. 11, November pp. 1619-1630.

Mechanics of Structures and Materials, Bradford, Bridge & Foster (eds)
© 1999 Balkema, Rotterdam, ISBN 90 5809 107 4

Finite element method for shakedown analysis of pavements

S.H.Shiau & H.S.Yu
Department of Civil, Surveying and Environmental Engineering, University of Newcastle, N.S.W., Australia

ABSTRACT: This paper presents a lower bound formulation for shakedown analysis of pavements using finite elements and linear programming. Both the elastic stress and the residual stress fields required in the shakedown analysis are assumed to be linear across the media and modelled by displacement and stress-based finite elements respectively. The proposed formulation is first verified with a homogeneous isotropic half space. The variation of shakedown limits with different material properties for layered pavements are then investigated. Finally, the design charts for layered pavement are also presented.

1. INTRODUCTION

The response of a layered pavement to repeated surface loading from moving wheels is such that it may gradually fail by the accumulation of plastic strain resulting in a form of rutting and surface cracks. Alternatively, the pavement may behave as a purely elastic structure after the first few loading cycles and after which no further permanent strains will be accumulated in the pavement. If the latter case occurs, the pavement is said to have 'shaken down'. The load under which the shakedown can occur is termed as the "Shakedown Load" (or "Shakedown Limit"). In other words, when the magnitude of the repeated load is greater than the shakedown load, the pavement will eventually fail due to excessive plastic deformation.

Since pavement structures frequently operate beyond the elastic limit (Sharp and Booker, 1984), there is a need to develop a plastic design method based on the concept of shakedown theorem for the pavement design. As direct calculation of the exact shakedown load is difficult, it is usual to estimate the best lower or upper bounds to the shakedown limits. The lower bound approach to the shakedown limit can be achieved by using the Melan's static shakedown theorem with an optimisation procedure. As the lower bound shakedown loads are conservative estimates of the true shakedown load, they are therefore more useful in the design of pavements.

The followings are objectives of the present paper: (1) to implement the lower bound shakedown theorem using finite elements and linear programming; (2) to compare the present numerical shakedown loads with the existing numerical results; (3) to perform a parametric study for shakedown of layered pavements; and (4) to present the charts for design of layered pavements based on the shakedown concept.

2. PLANE STRAIN PAVEMENT MODEL

Following Sharp and Booker (1984), a trapezoidal traffic load distribution is assumed in a vertical plane along the travel direction. It is further assumed that the resulting deformation is plane strain by replacing the wheel load as an infinite wide roller (Figure 1). With these two assumptions, it can be easily shown that both the permanent deformation and residual stress distribution are uniform over any horizontal plane and vary only with the depth z. Thus, the only non-zero residual stress is σ_x^r. The normal stresses (P_V) and longitudinal shear stresses (P_H) are related by the parameter $\mu = P_H/P_V$. This coefficient of surface friction μ is assumed to be constant so that the longitudinal shear stresses are also trapezoidal (Figure 2).

3. FINITE ELEMENT FORMULATION

Melan's static shakedown theorem states that "If the combination of a time independent, self-equilibrated residual stress field σ_{ij}^r and the elastic stresses $\lambda\sigma_{ij}^e$ can be found which doesn't violate the yield condition anywhere in the region, then the material will shakedown". Supposing that the elastic stresses are proportional to a load factor λ, the total stresses are therefore

$$\sigma_{ij}^t = \lambda\sigma_{ij}^e + \sigma_{ij}^r \tag{1}$$

where λ is the shakedown load factor, σ_{ij}^e are the elastic stresses and σ_{ij}^r are the residual stresses.

In the present formulation, both elastic stresses and residual stresses are assumed to be linearly distributed across the continuum by making use of the displacement and stress finite elements respectively. The formulation also allows for discontinuities of residual stresses at the edge of each triangle. By insisting that both the total stresses and the residual stresses do not violate the linearised Mohr-Coulomb yield condition in the mesh, the lower bound shakedown limit loads are then obtained as a solution to a large linear programming problem: the maximisation of the shakedown load factor λ subjected to the constraints on stresses due to: (1) element equilibrium; (2) discontinuity equilibrium; (3) stress boundary condition; and (4) yield criterion.

The elastic stresses are calculated by making use of the displacement finite element code, SNAC, developed at the University of Newcastle. A 6-noded triangular element is utilised so that the elastic stress field is linearly distributed across each element. As shown in Figure 3(a), this plane strain quadratic triangle has six nodes and two degrees of freedom at each node. Using the isoparametric formulation, the quadratic functions are used to model both the displacements and the geometry so that the resulted stresses are linear functions. Since the triangles used do not have curved sides, a three-point integration rule is sufficient to evaluate the element stiffness exactly.

The displacement finite element formulation is well known and will not be repeated here. The stress finite element formulation used in this paper is a modified version of the earlier work of Yu and Hossain (1998) and will be discussed briefly here.

The stress element used in the study is shown in Figure 3(b). The variation of the residual stresses is assumed to be linear and varies through an element according to

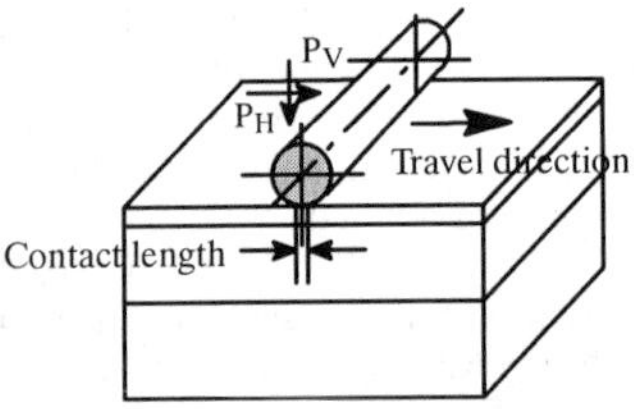

Figure 1. Idealized pavement model

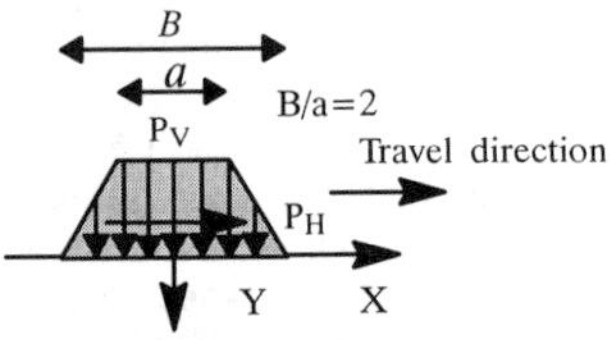

Figure 2. Trapezoidal load distribution for plane strain model

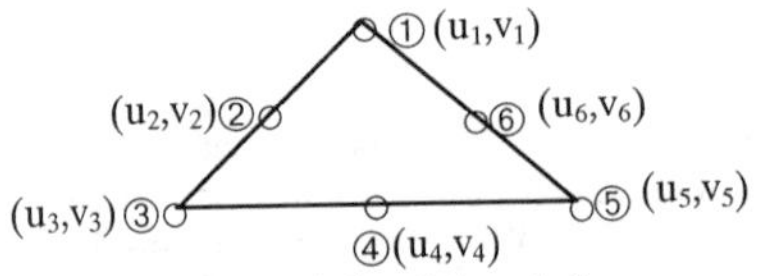

Figure 3(a). A 6-noded quadratic displacement triangle

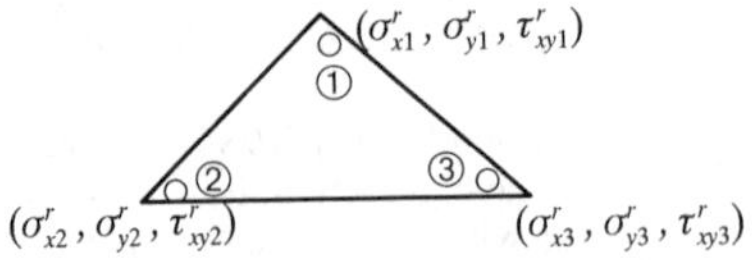

Figure 3(b). A 3-noded linear residual stress triangle

$$\sigma_x^r = \sum_{i=1}^{3} N_i \sigma_{xi}^r \; ; \; \sigma_y^r = \sum_{i=1}^{3} N_i \sigma_{yi}^r \; ; \; \tau_{xy}^r = \sum_{i=1}^{3} N_i \tau_{xyi}^r \tag{2}$$

where σ_{xi}^r, σ_{yi}^r and τ_{xyi}^r are the nodal residual stresses and N_i are linear shape functions.

Statically admissible stress discontinuities are permitted at shared edges between adjacent stress triangles. Unlike the usual form of the displacement finite elements, each node is unique to a particular element and more than one node share the same coordinates (Figure 4). If E denotes the number of triangles in the mesh, then there are $3E$ nodes and $9E$ unknown residual stresses.

The Mohr-Coulomb yield criterion for plane strain condition can be approximated as a linear function of the unknown stresses and details of this linearization can be found in Sloan (1987) and Yu and Hossain (1998). The linearized yield surface must be internal to the Mohr-Coulomb yield circle to ensure that the obtained solutions are rigorous lower bounds. In each element, both the elastic and residual stress variations are linearly distributed across the mesh. The yield conditions in terms of both residual and total stresses will be satisfied at any point within an element provided that the yield criterion is enforced at corner nodes. As a result, the condition of not violating the yield criterion in the mesh can be replaced by the following inequality constraints: (a) yield criterion at corner nodes by the residual stresses (b) yield criterion at corner nodes by the total stresses. This differs from the formulation by Yu and Hossain (1998) in which the total stresses need to be enforced at several sampling points within each element.

The task of the lower bound shakedown analysis is to seek a residual stress distribution that would maximise the load factor λ. The problem can then be stated as

Minimise $-\lambda$;

Subject to $\mathbf{A}_1 \mathbf{x} = \mathbf{B}_1$ and $\mathbf{A}_2 \mathbf{x} \le \mathbf{B}_2$.

where $\mathbf{A}_1$ is the matrix of equality constraints, $\mathbf{A}_2$ is the matrix of yield constraints, $\mathbf{B}_1$ and $\mathbf{B}_2$ are the respective vectors containing strength properties. The modified active set strategy developed

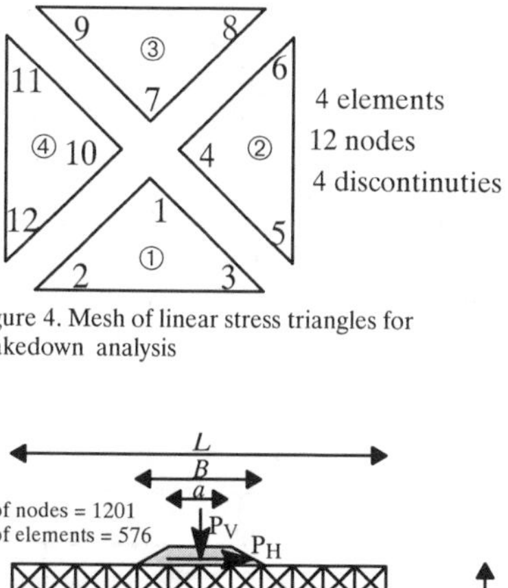

Figure 4. Mesh of linear stress triangles for shakedown analysis

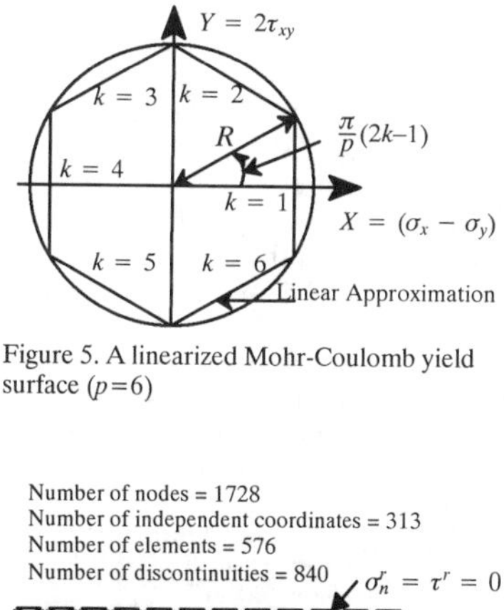

Figure 5. A linearized Mohr-Coulomb yield surface (p=6)

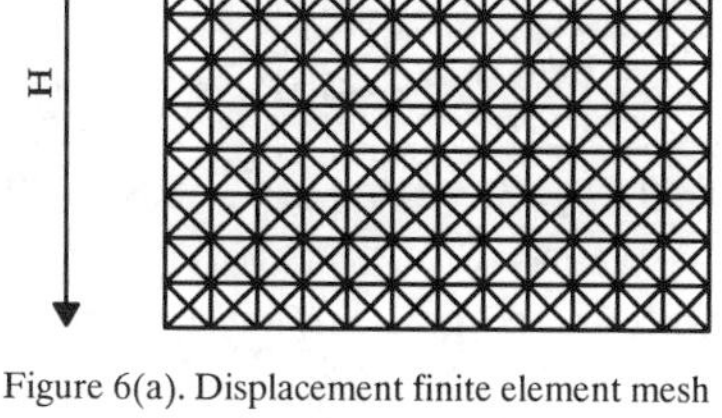

Figure 6(a). Displacement finite element mesh

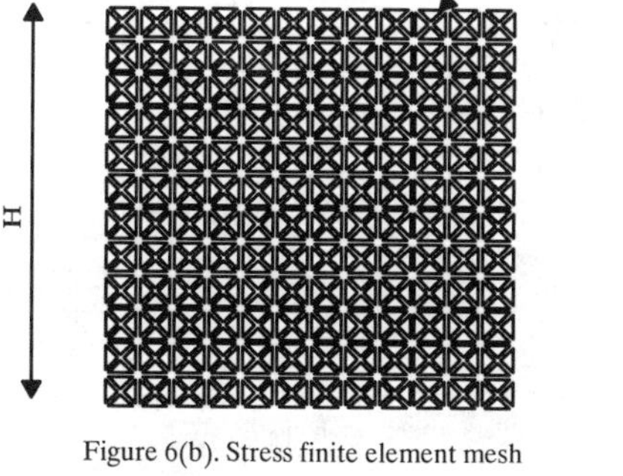

Figure 6(b). Stress finite element mesh

by Sloan (1998) is employed to solve the above linear programming problem. A comprehensive discussion of the theory and implementation of the active set algorithm can be found in Best and Ritter (1985).

4. APPLICATION

4.1 *Introduction*

A dimensionless shakedown limit is best defined by $\lambda P_V/c$, where P_V is the vertical pressure at which the elastic stresses are calculated and c is the cohesion of the basecourse. Since the elastic stresses are evaluated based on the application of a unit pressure in the finite element calculations, the dimensionless shakedown limit $\lambda P_V/c$ reduces to λ/c as $P_v = 1$. λ is the shakedown load factor that is to be maximised.

A sensitivity analysis suggests that a mesh with the dimension of L/B=3 and H/B=3 would be sufficient to accurately model the infinite half space. A further study on the effect of the number of elements in the mesh indicates that it is adequate to use a mesh with a total of 576 elements in all the calculations. The typical finite element meshes used for both elastic stress field and residual stress field are presented in Figure 6(a) and (b) respectively.

4.2 *Verification of the numerical formulation*

For the purpose of verification, only the vertical loading is considered in this section. The results of dimensionless shakedown limits with the variation of soil internal friction angle from 0° to 30° are presented in Figure 7. It is shown that the value of $\lambda P_V/c$ in this paper is very close to that obtained by Sharp and Booker (1984) and this is particularly true for the case of a purely cohesive clay which gives a value of 3.696.

As shown in Figure 8, the dimensionless shakedown limit decreases dramatically with the increase in the coefficient of surface friction μ. This is likely due to the existence of high elastic shear stresses in the top layer which is possibly resulted in a type of shear failure of that layer when the value of μ is high. In this Figure, it is also found that the results from this formulation yield smaller values than that in Sharp and Booker (1984) at higher frictional angle of soil. It is believed that the present formulation would improve the solution accuracy due to the use of a large number of finite elements for modelling both elastic and residual stress fields.

Figure 9 shows the distribution of horizontal residual stresses with the depth for the case of $\phi = 0$, $\mu = 0$. The value of σ_x^r/c reaches a maximum at D/B=0.21 with $\sigma_x^r = 0.9c$. This value converged to zero at a depth of 0.8D/B. These results are in good agreement with the experimental data presented in Radovsky and Murashina (1996). Figure 10 presents the principal stress vectors beneath the loaded area. As shown, only σ_x^r exists in the media. In such an analysis, residual stresses vary with the depth only and are uniform over any horizontal plane.

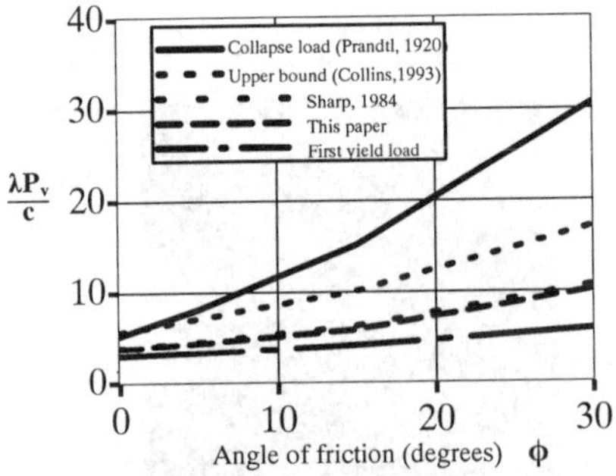

Figure 7. Effect of internal friction angle upon dimensionless shakedown limits

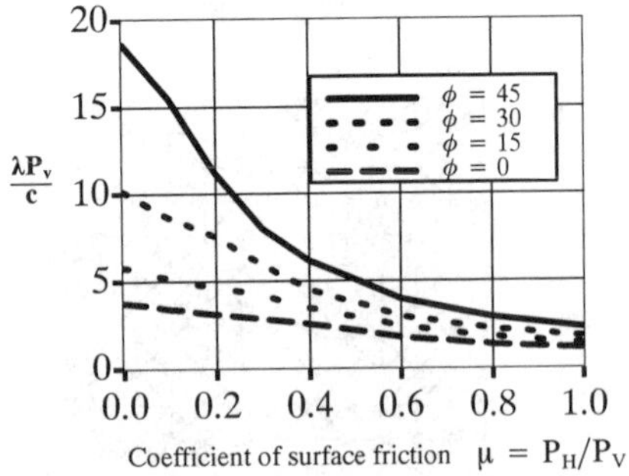

Figure 8. Effect of Coefficient of surface friction upon dimensionless shakedown limits

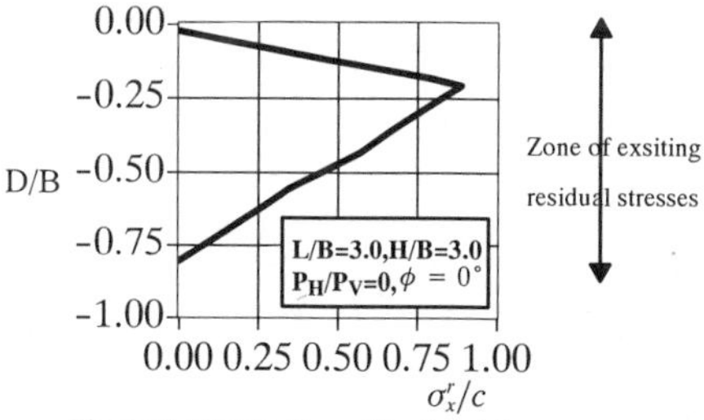

Figure 9. Distribution of horizontal residual stresses with D/B

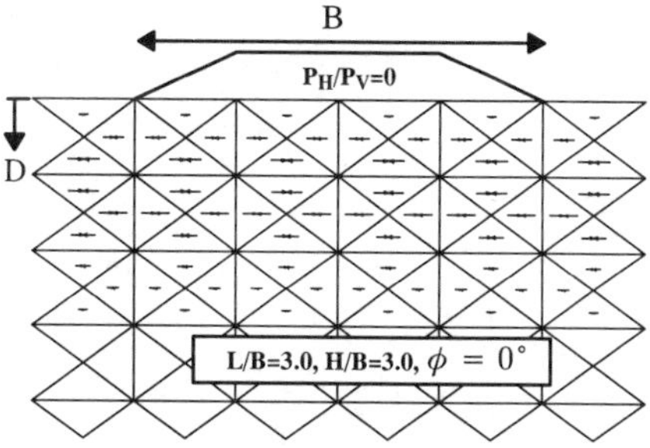

Figure 10. Principal residual stress vectors beneath the loaded area

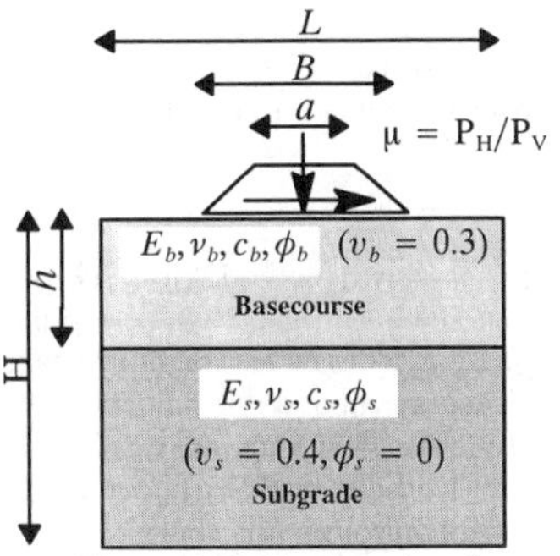

Figure 11. Problem notation of two layer pavement

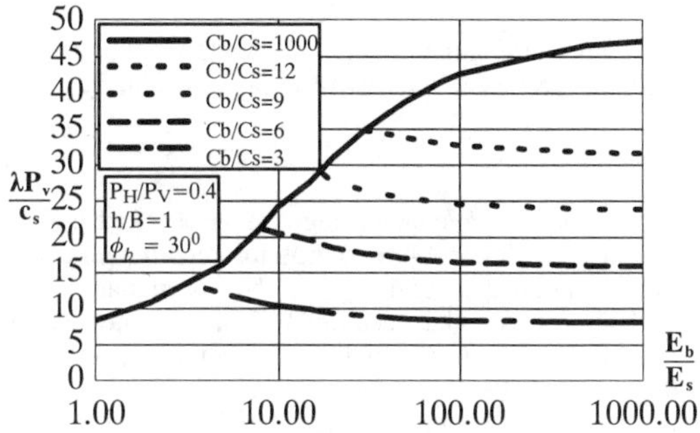

Figure12. Influence of E_b/E_s and C_b/C_s on dimensionless shakedown limits

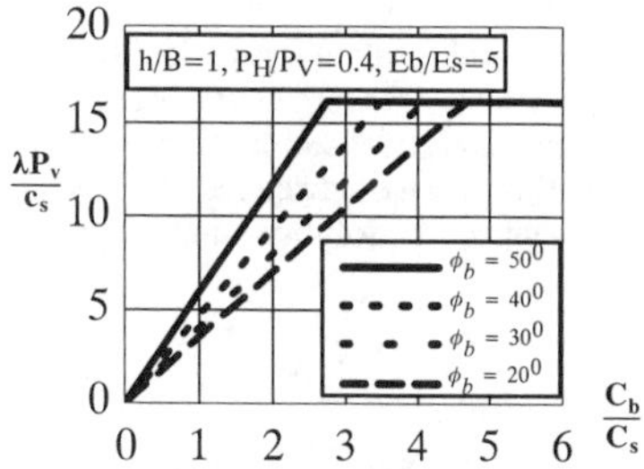

Figure13. Influence of Cb/Cs and internal friction angle (Eb/Es=5)

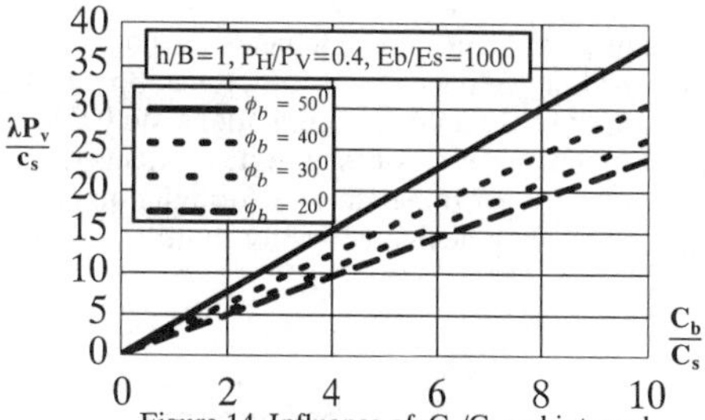

Figure 14. Influence of C_b/C_s and internal friction angle (Eb/Es=1000)

4.3 *Parametric study on shakedown of two layered pavements*

Figure 11 shows the problem notation of a two layered pavement. A purely cohesive subgrade ($\phi_s = 0$) is considered in this section. The influence of relative stiffness ratio (E_b/E_s) on the dimensionless shakedown limit for different values of relative strength ratio (C_b/C_s) is shown in Figure 12. It is clear that, at a given value of relative strength ratio, there exists an optimum relative stiffness ratio at which the resistance to incremental collapse is maximised. Further increase in relative stiffness ratio does not contribute to an increase in shakedown limit.

As shown in Figure 13, it is important to note that for a lower value of E_b/E_s the dimensionless shakedown limit ceases to increase at a particular C_b/C_s for different basecourse friction angles. This may indicate a transfer of failure mode from the bottom of basecourse (tensile failure) to the compressive failure at the top of subgrade when the value of E_b/E_s is low. Thus, further increase

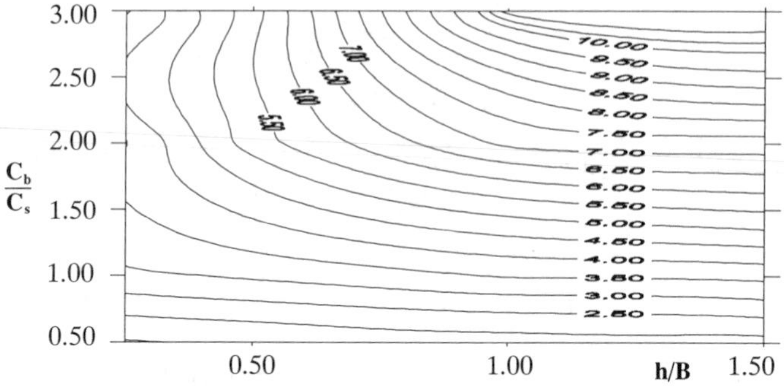

Figure15. Contours of dimensionless shakedown limits as an example chart for the thickness design of two layer pavement using shakedown analysis. $E_b/E_s=3$, $P_H/P_V=0.4$, $\phi_b = 20°, \phi_s = 0°$

in basecourse strength will not improve the shakedown capacity. In the case that the value of E_b/E_s is high, an increase in C_b/C_s will improve the shakedown capacity as shown in Figure 14.

4.4 *Design charts for two layered pavements*

Results of current pavement design can be checked against the shakedown method through Figure 15. For a given layered pavement with known parameters, shakedown limits can be determined from this chart and compare against the design load. The design load has to be less than the shakedown load to ensure that accumulation of permanent strains will not occur and this pavement will eventually behave elastically even though initial permanent strain has been previously occurred. In the case that the design load is much less than the shakedown load, the design of pavement is conservative and the adjustment of basecourse thickness is possible through this chart.

5. CONCLUSIONS

A formulation using finite elements for both elastic stress field and residual stress field has been developed in this paper to perform lower bound shakedown analysis of layered structures. Unlike other existing shakedown formulations, both the elastic and residual stress fields are modelled by displacement and stress-based finite elements respectively in this paper. Selected results presented in this paper suggest that the new formulation gives accurate shakedown solutions and can be used easily to develop design charts for multilayered pavements.

REFERENCES

Best, M.J. & Ritter. K. (1985) *Linear programming: active set analysis and computer programs*, Prentice-Hall, New Jersey.

Collins, I.F., Wang, A.P. and Saunders, L.R.(1993) "Shakedown in layered pavements under moving surface loads," *Int. J. for Numerical and Analytical Methods in Geomechanics*, Vol. 17, pp. 165-174.

Prandtl, L. (1920) "Uber die harte plastischer korper", *Gottinger Nachricten, Math. Phys.* K1, pp74-85.

Radovsky, B.S. and Murashina, N.A. (1996) Shakedown of Subgrade Soil Under Repeated loading, *Transportation Research Record*, n 1547, pp 82-88.

Sharp, R.W. & Booker, J.R. (1984) Shakedown of Pavements Under Moving Surface Loads, *Journal of Transportation Engineering*, Vol 110, pp 1-14.

Sloan, S.W. (1988) Lower Bound Limit Analysis Using Finite Elements and Linear Programming, *Int. J. of Numerical and Analytical Methods in Geomechanics*, Vol 12, pp 61-77.

Yu, H.S. and Hossain, M.Z. (1998) Lower Bound Shakedown Analysis of Layered Pavements Using Discontinuous Stress Fields, *Computer Methods in Applied Mechanics and Engineering*, Vol 167, pp 209-222.

Mechanics of Structures and Materials, Bradford, Bridge & Foster (eds)
© 1999 Balkema, Rotterdam, ISBN 90 5809 107 4

Computational mechanics – Limit analysis of unreinforced masonry shear walls

D.J. Sutcliffe, H.S. Yu & A.W. Page
Department of Civil, Surveying and Environmental Engineering, University of Newcastle, N.S.W., Australia

ABSTRACT: This paper describes a new numerical method for lower bound limit analysis of unreinforced masonry shear walls. Two materials which display similar properties to unreinforced masonry, and which have already been successfully modelled using the lower bound technique are jointed rock and reinforced earth. While these materials are treated as homogenous and anisotropic, the masonry composite must be treated as an inhomogeneous, anisotropic medium. The overall behavior of the masonry is controlled by the mechanical properties of the unit (brick/block) and the discontinuities or joints, as well as the relative position and orientation of the joint sets. Several examples are analyzed to illustrate the effectiveness of the proposed numerical procedure for computing lower bound solutions for the in-plane behavior of unreinforced masonry walls.

1 INTRODUCTION

It is well recognized that the mortar joints, or more precisely the unit/mortar interfaces, in a masonry composite often have a much lower strength than that of the intact unit or mortar. The presence of these joints creates planes of weakness along which failures may initiate and propagate resulting in masonry displaying distinct directional properties. The overall properties of the masonry composite are determined then by the properties of the intact materials (units and mortar) and the strength and orientation of the unit/mortar interfaces. Other factors which may influence the strength and stiffness of the masonry composite include anisotropy of the units, unit size and aspect ratio, joint dimension, joint orientation, relative position of head and bed joints, properties of the units and mortar, properties of the unit/mortar bond, and workmanship. The presence of all these factors makes numerical simulation of masonry assemblages very difficult.

The practical application of the displacement finite element method, which has proved to be useful in predicting the collapse behavior of unreinforced masonry walls, is somewhat limited by the need for extensive testing in order to provide an accurate description of the masonry response. The lower bound theorem is an alternative approach which can be used to predict the in-plane collapse load of an unreinforced masonry wall. Yu and Sloan (1994) applied the lower bound theorem to problems involving the ultimate bearing capacity of footings on jointed rock. In their formulation, Yu and Sloan pointed out that if the joint sets are reasonably constant in orientation and closely spaced, then the rock mass could be considered as anisotropic but homogenous. Further to this assumption, it is necessary that the joints must be continuous. While jointed rock and unreinforced masonry have similar properties, the assumption of continuous joints may not be applied to the masonry material (eg. the vertical head joints). The result of this is that the masonry material must now be treated as anisotropic and inhomogeneous.

The purpose of this paper is to briefly outline an extension of Yu and Sloan's (1994) formulation for application to an inhomogeneous, anisotropic material and use this formulation to calculate the lower bound collapse load of some masonry wall tests reported in the literature. A simplified micro-model, which is similar to that of Lourenco (1996) and Lourenco and Rots (1997), is adopted in this paper. The intact brick units are assumed to obey a Mohr-Coulomb failure criterion, while the behavior of the mortar and the two interfaces is lumped into an 'average' single failure surface based on the 'cap' model presented by Lourenco (1996). Suitable linear approximations of these yield surfaces ensures that the application of the lower bound theorem results in a linear programming problem, a major advantage of which is the ease with which complex loading and geometry can be dealt with.

2 FAILURE CRITERION FOR UNREINFORCED MASONRY

The intact brick unit material is assumed to be isotropic, homogenous and obey the Mohr-Coulomb failure condition. Assuming tensile stresses are positive, the Mohr-Coulomb criterion for plane problems can be expressed as follows.

$$F_b = (\sigma_x - \sigma_y)^2 + (2\tau_{xy})^2 - (2c\cos\phi - (\sigma_x + \sigma_y)\sin\phi)^2 = 0 \tag{1}$$

where σ_x, σ_y are the normal stresses in the horizontal and vertical directions respectively, τ_{xy} is the shear stress, and c, ϕ denote the cohesion and friction angle of the brick unit material.

A linear approximation to the spherical cap model proposed by Lourenco (1996) is used for the joint failure criterion. Figure 1 shows the Cap model proposed by Lourenco and the linear approximation utilized in this paper. Lourenco's failure model includes a tension cut off for Mode I failure (i.e. tensile failure of the unit/mortar interface) a Coulomb friction envelope for Mode II failure (i.e. shear failure of the unit/mortar interface) and a spherical cap model for compressive failure. Three linear Mohr-Coulomb failure surfaces are used in order to provide a suitable linear approximation to the yield criterion.

The Mohr-Coulomb failure surface may be expressed in terms of normal and shear stresses as follows:

$$F_i = |\tau| - c_i + \sigma_n \tan\phi_i = 0 \tag{2}$$

where τ is the shear stress and σ_n is the normal stress acting on the joint. This failure criterion can be expressed in terms of σ_x, σ_y and τ_{xy} using the following stress transformation:

$$\sigma_n = \sin^2\theta_i\,\sigma_x + \cos^2\theta_i\,\sigma_y - \sin 2\theta_i\,\tau_{xy} \tag{3}$$

$$\tau = -\frac{1}{2}\sin 2\theta_i\sigma_x + \frac{1}{2}\sin 2\theta_i\sigma_y + \cos 2\theta_i\tau_{xy} \tag{4}$$

where θ_i is the angle of the joint set i from the horizontal axis (positive anticlockwise).

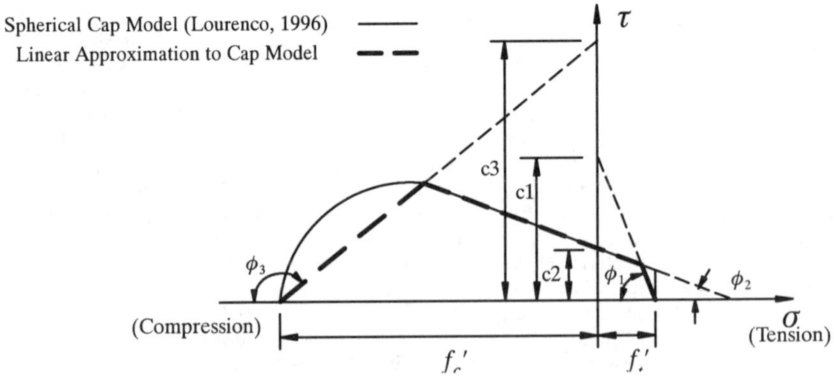

Figure 1. Spherical cap model (Lourenco, 1996) and linear approximation to cap model

3 THE LOWER BOUND THEOREM

The lower bound theorem states that the collapse load obtained from any statically admissible stress field will give a lower bound of the true collapse load. A statically admissible stress field is one which a) satisfies the equations of equilibrium, b) satisfies the stress boundary conditions and c) does not violate the yield criterion.

In order to model the stress field in the masonry material, a three noded triangular stress element is used (Sloan, 1988). Each node is associated with three unknown stresses, σ_x, σ_y, and τ_{xy} with the variation of stresses throughout each element assumed to be linear. Unlike the elements used in displacement finite element analysis, several nodes may share the same coordinate with each node associated with only one element. Statically admissible stress discontinuities are allowed to occur at all edges between adjacent triangles.

3.1 *Yield Condition*

As described earlier, the effect of the joints in the masonry are included in the analysis by using 3 separate linear failure surfaces, one for the brick unit and one each for the head and bed joints. In this way, the masonry composite is represented as a inhomogeneous, anisotropic material. For the masonry composite, the complete failure conditions are expressed by equations (1) and (2). By imposing the constraints $F_b \le 0$ and $F_i \le 0$, we ensure the yield conditions are satisfied. It is readily seen that for a joint set *i* the requirement that $F_i \le 0$ results in six linear constraints on the nodal stresses (one for each linear branch of the failure surface. If the inequality constraints $F_b \le 0$ is applied directly however, non-linear constraints result since the function F_b is quadratic in the unknown stresses. Since the lower bound theorem is to be formulated as a linear programing problem, it is necessary to approximate (1) by a yield condition which is a linear function of the unknown stress variables. In order for the solution to be a rigorous lower bound the linear approximation to the failure surface must lie inside the generalized Mohr-Coulomb failure surface.

If *p* is the number of sides used to approximate the yield function (1) then the linearized yield function can be shown to be (Sloan, 1988):

$$A_k\sigma_x + B_k\sigma_y + C_k\tau_{xy} \le E, \quad k = 1,2,\ldots,p \tag{5}$$

where

$$A_k = \cos\left(\frac{2\pi k}{p}\right) + \sin\phi\cos\left(\frac{\pi}{p}\right); \quad B_k = -\cos\left(\frac{2\pi k}{p}\right) + \sin\phi\cos\left(\frac{\pi}{p}\right)$$

$$C_k = 2\sin\left(\frac{2\pi k}{p}\right); \quad E = 2c\cos\phi\cos\left(\frac{\pi}{p}\right)$$

Each linear branch of the joint failure surface results in the following two linear constraints on the unknown stresses:

$$A_k\sigma_x + B_k\sigma_y + C_k\tau_{xy} \le E, \quad k = p+2i-1,\ p+2i, \quad j = 1,2,3 \tag{6}$$

where

$$A_{p+2i-1} = \sin^2\theta_i\tan\phi_{ij} - \frac{1}{2}\sin 2\theta_i; \quad B_{p+2i-1} = \frac{1}{2}\sin 2\theta_i + \cos^2\theta_i\tan\phi_{ij}$$

$$C_{p+2i-1} = \cos 2\theta_i - \sin 2\theta_i\tan\phi_{ij}$$

$$A_{p+2i} = \sin^2\theta_i\tan\phi_{ij} + \frac{1}{2}\sin 2\theta_i; \quad B_{p+2i} = -\frac{1}{2}\sin 2\theta_i + \cos^2\theta_i\tan\phi_{ij}$$

$$C_{p+2i} = -\cos 2\theta_i - \sin 2\theta_i\tan\phi_{ij}$$

$$E = c_{ij}$$

and j represents the branch number (i.e. tension cut-off, shear sliding and compression ‘cap’).

It can be easily proved that by using a linearized failure surface it is sufficient to enforce the linear constraints (5) and (6) at each nodal point to ensure that the stresses satisfy the yield conditions throughout the mesh.

The above yield conditions give rise to inequality constraints on the nodal stresses which may be expressed in matrix form as:

$$\boldsymbol{a}_1 \boldsymbol{x}_1 \leq \boldsymbol{b}_1 \tag{7}$$

where $\boldsymbol{x}_1$ is a vector of the nodal stresses, $\boldsymbol{b}_1 = [E]^T$ and $\boldsymbol{a}_1$ is a function of A, B and C above.

3.2 *Objective Function*

The problem of finding a statically admissible stress field which maximizes the collapse load may be expressed as

$$\begin{aligned} &\textit{Minimize} && -\boldsymbol{C}^T\boldsymbol{X} \\ &\textit{Subject to} && \boldsymbol{A}_1\boldsymbol{X} = \boldsymbol{B}_1 \\ & && \boldsymbol{A}_2\boldsymbol{X} = \boldsymbol{B}_2 \end{aligned} \tag{8}$$

where $\boldsymbol{A}_1, \boldsymbol{B}_1$ represent the coefficients due to the equilibrium and stress boundary conditions; $\boldsymbol{A}_2$, $\boldsymbol{B}_2$ represent the coefficients for the yield conditions; $\boldsymbol{C}$ is the vector of objective function coefficients and $\boldsymbol{X}$ is the global vector of unknown stresses. An active set algorithm is used to solve the above linear programming problem, the details of which can be found in Sloan (1988).

4 NUMERICAL EXAMPLES

4.1 *Deep Beam*

As shown in Figure 2, the first problem to be analyzed is the capacity of a masonry panel acting as a deep beam. Page (1978) tested a masonry deep beam with dimensions 757mm x 457mm, rigidly supported at each end over a 188mm length. The panel was made of half scale, pressed solid clay bricks 122mm x 37mm x 54mm, with 5mm thick mortar joints (1:1:6 cement:lime:sand by volume). The lower bound mesh used in the analysis is shown in Fig. 3.

The failure surface for the brick is a simple Mohr-Coulomb failure criterion while the joint failure surface is similar to the tri-linear failure surface in Figure 1. Basic values of c, ϕ, f_c' and f_t'

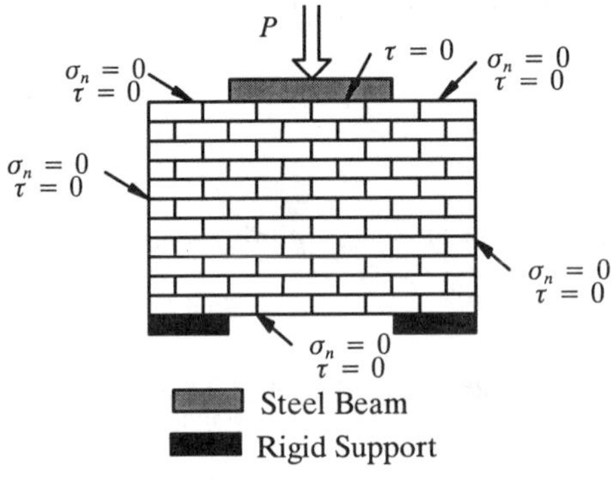

Figure 2. Deep beam test, Page (1978) with imposed boundary conditions for lower bound calculations.

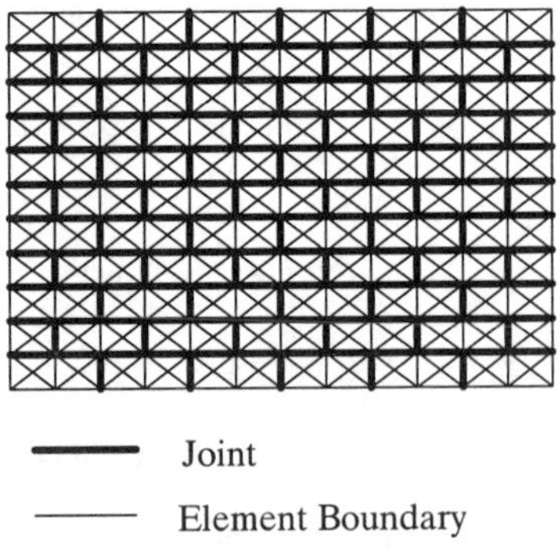

Figure 3. Lower bound mesh

were taken from Lourenco (1996).

The slope of the 'cap' was determined by taking a straight line from the value of f_c' on the normal stress axis to the point of intersection of the straight line representing mode II failure and the spherical cap assumed by Lourenco (1996). If a straight line is drawn from the intersection of the spherical cap with the mode II failure line to the value of f_c' on the normal stress axis then $\phi_3 \approx 154°$

The lower bound on the collapse load using the parameters from Lourenco is 105.2 kN which is only around 4% lower than the experimental collapse load of 109.2 kN. A sensitivity analysis was performed on effects of the strength parameters c and ϕ, which control the unit failure surface, and the parameters ϕ_1 and ϕ_3, which control the tension and compression branches of the joint failure surfaces. It was found that the two most influential parameters were ϕ_3 and c. The parameter ϕ_3 controls the orientation of the compression cap, and given that the main failure mode reported by Page (1978) was crushing of the masonry composite, this result is expected. The cohesion value (c) of the unit governs the transition of tension failure from occurring, mainly in the relatively weak joints (at low values of c) to tensile failure in both the units and the joints (at low values of c).

4.2 *Shear Wall*

Tests on masonry shear walls carried out by Raijmakers and Vermeltfoot (1992) and Vermeltfoot and Raijmakers (1993) were simulated. The walls have dimensions 990mm x 1000mm and are composed of wire-cut solid clay bricks with dimensions 210mm x 52mm x 100mm, with 10mm thick mortar joints (1:2:9 cement:lime:sand by volume). A lower bound mesh similar to that shown in Figure 3 is used, with the test configuration and boundary conditions shown in Figure 4.

Similar shape failure surfaces as those used in the previous example were adopted for this analysis, with typical strength parameters taken again from Lourenco (1996). The analysis was performed at three different levels of normal load.

A comparison of the lower bound collapse load with the experimental collapse shows that at high values of normal compression (2.12 MPa) the lower bound collapse load is 82.9 kN which compares well with the experimental value of approximately 97 kN. However, as the level of compression is reduced, the accuracy of the lower bound result tends to decline. For example, when a compressive load of 1.21 MPa is applied the lower bound result of 57.5 kN is 20% lower than experimental collapse load of approximately 72 kN, while at a compressive load of 0.3 MPa the lower bound result is 26.5 kN while the experimental collapse load is approximately 50 kN, a difference of approximately 47%.

Again a sensitivity analysis was performed on the effects of the strength properties controlling the failure surface. The effect of the cohesion value (c) of the unit was significant under all normal load levels, with the collapse load increasing as the cohesion increased. This is largely due to the fact that as c increases failure is transferred from both the joints and bricks to the 'weak' joints alone. Under low levels of normal load the strength parameter ϕ_1, which controls the slope of the joint tension cut-off, was found to be most significant. This was expected given the primary failure mode is the direct tensile failure of the joints. As the normal load increases the significance of ϕ_1 tends to reduce while the influence of ϕ_3 becomes more important. This is because of the change of failure

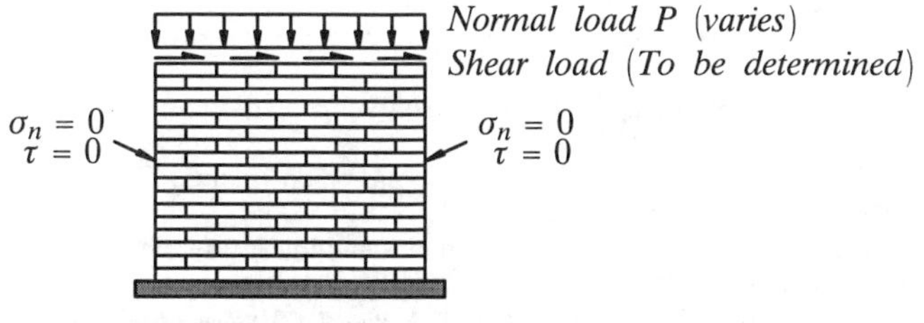

Figure 4. Shear wall test configuration as modelled by Lourenco (1996) with boundary conditions used in lower bound calculations.

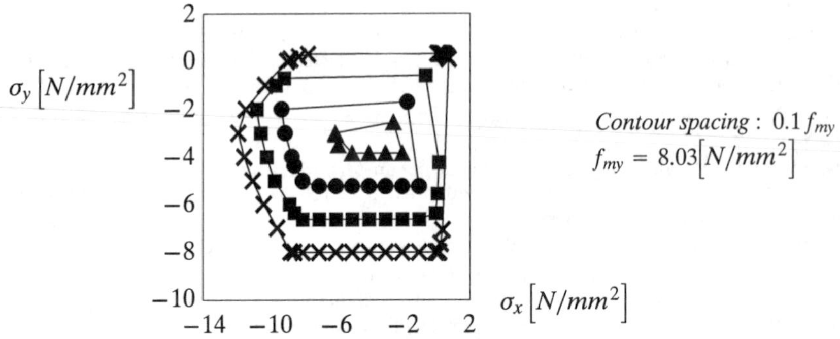

Figure 5. Composite yield criterion from lower bound solution.

mode from a tensile failure of the joints to a compression failure of the masonry composite.

4.3 *Biaxial panel tests*

The most complete set of strength data of biaxially loaded masonry is given by Page (1981) and Page (1983). 102 panels with dimensions 360mm x 360mm x 50mm composed of half scale solid clay brick masonry were tested. The material parameters used in the lower bound approach were again taken from Lourenco (1996).

A plot of the composite yield criterion as calculated from the lower bound results is shown in Figure 5. These results compare well with the previously reported experimental composite failure surface of Page (1981, 1983) and the composite yield surface calculated by Lourenco (1996) using a displacement finite element formulation.

5 CONCLUSIONS

A general finite element formulation of the lower bound theorem for the in-plane strength of an inhomogeneous, anisotropic material has been developed. The formulation has been used successfully to determine the in-plane collapse load of unreinforced masonry. By including separate failure surfaces for the units and the joints the model allows for all the failure modes of masonry to be taken into account. Four numerical examples presented in this paper suggest that the proposed numerical procedure is useful in obtaining results for the collapse load of unreinforced masonry structures for in-plane behavior.

6 REFERENCES

Lourenco P.B. 1996. Computational strategies for masonry structures. *PhD Dissertation, Civil Engineering Department, Delft University of Technology.*

Lourenco P.B. and Rots J.G. 1997. Multisurface interface model for analysis of masonry structures. *Journal of Engineering Mechanics,* 123(7):650-668.

Page A.W. 1978. Finite element model for masonry. *Journal of the Structural Division, ASCE,* 104(8):1267-1285.

Page A.W. 1981. The biaxial compressive strength of brick masonry. *Proceedings of the Institution of Civil Engineers, Part 2,* 71:893-906.

Page A.W. 1983. The strength of brick masonry under biaxial compression-tension. *The International Journal of Masonry Construction,* 3(1):26-31.

Sloan S.W. 1988. Lower bound limit analysis using finite elements and linear programming. *International Journal for Analytical Methods in Geomechanics,* 12:61-77.

Yu H.S. and Sloan S.W. 1994. Bearing capacity of jointed rock. *Computer Methods in Advanced Geomechanics, Siriwardane and Zaman (eds), Rotterdam.* 2403-2408.

Yu H.S. and Sloan S.W. 1997. Finite element limit analysis of reinforced soils. *Computers and Structures,* 63(3):567-577.

Mechanics of Structures and Materials, Bradford, Bridge & Foster (eds)
© 1999 Balkema, Rotterdam, ISBN 90 5809 107 4

FE modelling of confined eccentrically loaded columns subject to second order effects

M.Ghazi, S.J.Foster & M.M.Attard
School of Civil and Environmental Engineering, The University of New South Wales, Sydney, N.S.W., Australia

ABSTRACT: A numerical model is developed to examine the effect of core confinement on the behaviour of eccentrically loaded columns. The 3-D model uses the Microplane-M4 formulation with second order effects included using a total Lagrangian formulation. Comparisons with experimental results are presented and shown to model satisfactorily the load versus lateral displacement. Further development is required to include the effects of cover spalling into the formulation.

1 INTRODUCTION

The level of the ductility of a reinforced concrete structure is a major parameter contributing to the overall safety of that structure. For a reinforced concrete column, ductility is directly tied to the deformation of the column beyond the peak load (Attard and Foster, 1995). As experiments have shown, the cover portion of a reinforced concrete column spalls away from the body when the applied load approaches the peak load. Thus the behaviour of a tied reinforced concrete column beyond the peak load is dependent on the behaviour of the effectively confined core. Numerical modelling of concrete columns requires a 3-D formulation due to the effect of stresses acting orthogonally to the directions of the main stresses (Kotsovos and Pavlovic, 1995). This 3-D formulation should be able to correctly incorporate the relatively large lateral deflection we expect in the post peak phase of eccentrically loaded columns. In this paper a total Lagrangian 3-D finite element model has been developed to investigate the behaviour of eccentrically loaded concrete columns. The constitutive relationships are developed using the Microplane formulation of Bazant et al. (1999).

2 CONSTITUTIVE LAW

Concrete is a heterogeneous material with a complex physical behaviour that is closely related to the grain size and shape, and the physical characteristics of its constituents. Concrete may be seen as a brittle material under tensile stress where it fractures with little deformation beyond the peak stress. Concrete may also be treated as a plastic material under compressive stresses where it undergoes greater deformation after attaining the peak stress. However, unlike plastic materials such as mild steel, softening of the concrete matrix follows the peak stress.

The irreversible deformation of concrete under compressive stresses justifies the use of the theory of plasticity. However, as this deformation is induced by internal micro-cracking and not the flow of the material; and as the behaviour of concrete is fundamentally different from that of genuinely plastic materials, major modifications have been incorporated into the classical plasticity theory to fit with the observed behaviour of concrete (Chen, 1982).

Whether using the modified plasticity theory for compressive stresses, or fracture theory for tensile stresses, or a combination of both, these traditional approaches use tensorial stress-strain relationships to calculate the macro stress tensor from the macro strain tensor. In reality, concrete is not a simple homogenous material; it is fundamentally a composite structure with progressive micro cracking under loading, which is translated into inelastic and anisotropic behaviour. These anisotropic properties add to the complexity of the modelling in a way that the suitability of traditional formulations that have evolved from the analysis of simple homogenous materials becomes questionable.

The Microplane method is a different approach that describes the complicated inelastic properties of concrete, not globally but individually for planes at various orientations at a point within the material (Bazant, 1984). In this approach continuity is maintained by constraining the microscopic deformations on the microplanes to the macroscopic deformations of the point. The stresses on the microplanes are explicitly determined from the stress-strain relationships that have been developed for a generic microplane. The stresses are then combined using the principle of virtual work to get the macro stress tensor at the point.

The Microplane method has the potential to capture the inelastic deformations of the material by using the stress and strain components on the microplanes and considerably simplifies the constitutive formulations. In this paper, the latest version of the Microplane model (model M4) has been used with full details of the formulation given in Bazant et al. (1999). In the Microplane model, components of the micro-stresses are taken normal and tangential to each microplane with the normal component further split into deviatoric and volumetric components. The governing relationships between micro-stresses and micro-strains are given by:

$$d\sigma_V = E_V\, d\varepsilon_V, \quad d\sigma_D = E_D\, d\varepsilon_D\ , \quad d\sigma_M = E_T\, d\varepsilon_M\ , \quad d\sigma_L = E_T\, d\varepsilon_L \tag{1}$$

$$E_V = E/(1\text{-}2\nu)\ , \quad E_D = 5E/[(2+3\mu)(1+\nu)]\ , \quad E_T = \mu\, E_D \tag{2}$$

where E is the modulus of elasticity of concrete; ν is the Poisson's ratio, E_V, E_D and E_T are volumetric, deviatoric and tangential elasticity moduli, respectively; σ_V and σ_D are volumetric and deviatoric stresses; σ_M and σ_L are tangential stresses along orthogonal axes on the microplane; ε_V, ε_D , ε_M and ε_L are corresponding strains, and μ is the ratio of the tangential modulus to the deviatoric modulus.

In the M4 Microplane model, as for the M3 Microplane model (Bazant et al., 1996), the micro stress-strain relationship remains proportional until a stress boundary is reached. When the stress boundary is reached the governing equations are given by:

$$\sigma^b_N = E\, K_1 C_1\, exp[\, - < \varepsilon_N - C_1 C_2 K_1 > / (K_1 C_3 + < C_4\, \sigma_V / \varepsilon_V >)] \tag{3}$$

$$\sigma^b_D = E\, K_1 C_5 / \{\, 1 + [\, < \varepsilon_D - C_5 C_6 K_1 > / (K_1 C_{20}\, C_7)\,]^2 \} \quad \text{for } \sigma_D > 0 \tag{4}$$

$$\sigma^b_D = E\, K_1 C_8 / \{\, 1 + [\, < - \varepsilon_D - C_8 C_9 K_1 > / (K_1\, C_7)\,]^2 \} \quad \text{for } \sigma_D < 0 \tag{5}$$

$$\sigma^b_T = E_T K_1\, K_2 C_{10} < - \sigma_N + \sigma^0_N > / \ [\, E_T K_1\, K_2 + C_{10} < - \sigma_N + \sigma^0_N >] \tag{6}$$

$$\sigma^0_N = E_T K_1\, C_{11} / (\, 1 + C_{12}\, |\varepsilon_V|\,) \tag{7}$$

$$\sigma^b_V = E_V\, K_1 C_{13} / [\, 1 + (C_{14} / K_1) < \varepsilon_V - C_{13} C_{15}\, K_1 >] \quad \text{for } \sigma_V > 0 \tag{8}$$

$$\sigma^b_V = -\, E\, K_1 K_3\, exp\, (\, - \varepsilon_V / K_1\, K_4) \quad \text{for } \sigma_V < 0 \tag{9}$$

where b stands for boundary, $\sigma_N = \sigma_V + \sigma_D$ and $\varepsilon_N = \varepsilon_V + \varepsilon_D$. Equations 1 to 9 with other relationships governing unloading and reloading form the basis of the stress-strain relationships for the M4 Microplane model. In Equations 3 to 9 the parameters C_1 to C_{20} are fixed dimensionless parameters calibrated against standard experimental tests. Parameters K_1 to K_4 are adjustable parameters used to define the concrete strength, the volumetric and the hydrostatic properties and confinement properties.

3 COMPARISON WITH COLUMN TESTS

In this paper, eccentrically loaded 150 mm square columns of Attard and Foster (1994) are modelled. The in-situ concrete strength is taken as 36 MPa being 90 percent of the measured cylinder strength (40 MPa). The fixed model parameters used to define the material model, parameters C_1 to C_{20} in Equations 3 to 9 were taken as $C_1 = 0.73$, $C_2 = 2.76$, $C_3 = 1.5$ and $C_7 = 104$, with the remainder taken as given by Bazant et al. (1999b).

Figure 1 shows the application of the model developed in this study on the simulation of the uniaxial compression response of a 36 MPa concrete brick element. The element was modelled with the adjustable parameters $K_1 = 110 \times 10^{-6}$, $K_2 = 160$, $K_3 = 10$ and $K_4 = 150$, with the elastic modulus at the level of the micro model was taken as $E = 38.3$ GPa and with $\nu = 0.18$. At the macro level the initial modulus of elasticity is $E_c = 29.2$ GPa. Figure 2a shows the triaxial stress-strain behaviour for a single 8-node brick element subject to confining pressures and in Figure 2b the peak loads of the confined concrete are compared with the triaxial failure criterion of Ottosen (1997). Figures 1 and 2 show the model represents well the response of the concrete both in uniaxial and triaxial compression.

The column contains 4% longitudinal reinforcement (8 Y12) and 6.3 mm diameter tie bars at 30 mm spacings. The cover to the outside of the ties was 10 mm. The yield stress and modulus of elasticity of the longitudinal bars are $f_y = 480$ MPa and $E_s = 195$ GPa, and for the tie bars

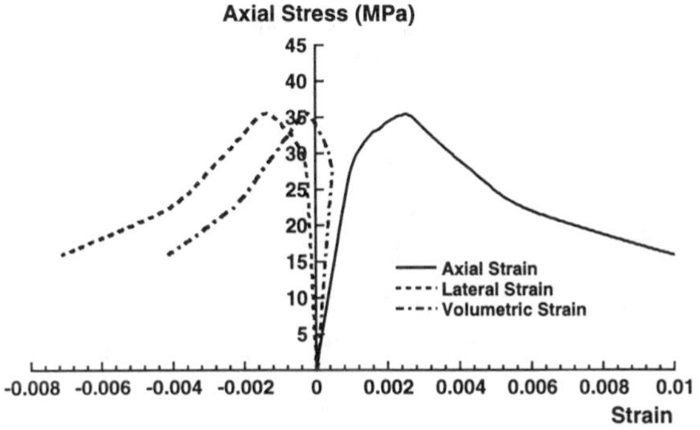

Figure 1. Stress versus strain curve for a single brick element in uniaxial compression.

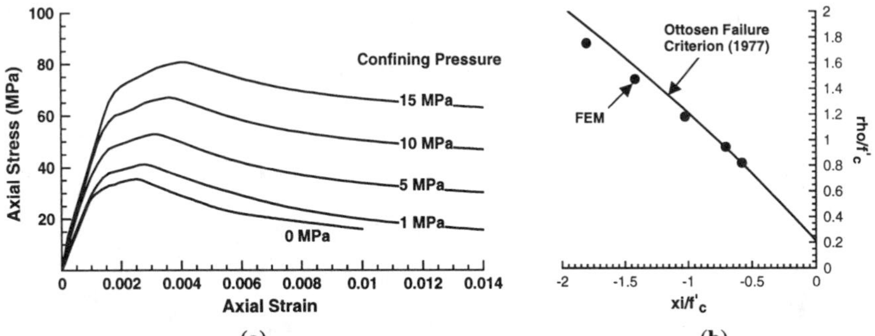

(a) (b)

Figure 2. Response for a single concrete element in triaxial compression; a) stress verse strain and, b) peak loads versus the Ottosen (1977) failure criterion.

are $f_y = 360\,\text{MPa}$ and $E_s = 210\,\text{GPa}$. The columns were loaded with initial eccentricities of 8 mm, 20 mm and 50 mm and are marked as 4L8-30, 4L20-30 and 4L50-30, respectively.

With the column doubly symmetric, only half the length of the column and half the width was modelled. Concrete was modelled using 8-node isoparametric solid elements using 2 x 2 x 2 Gaussian integration. Longitudinal and tie bars were modelled with 2-node truss elements. A linear elastic, perfectly plastic, stress–strain relationship was used for the steel reinforcement and the Microplane-M4 model used for the concrete. The FE mesh used for the analysis is shown in Figure 3.

The results of the numerical modelling are shown in Figure 4. The loads versus lateral displacement show a similar trend to that of the experimental data. In addition to eccentricities of 8 mm, 20 mm and 50 mm, the columns were also analysed for initial eccentricities of 0 mm, 3 mm, 14 mm, 30 mm 75 mm and 150 mm. The results are plotted in Figure 5 together with the experimental data and the axial force-moment interaction diagram derived using the rectangular stress block defined in AS3600 (1994).

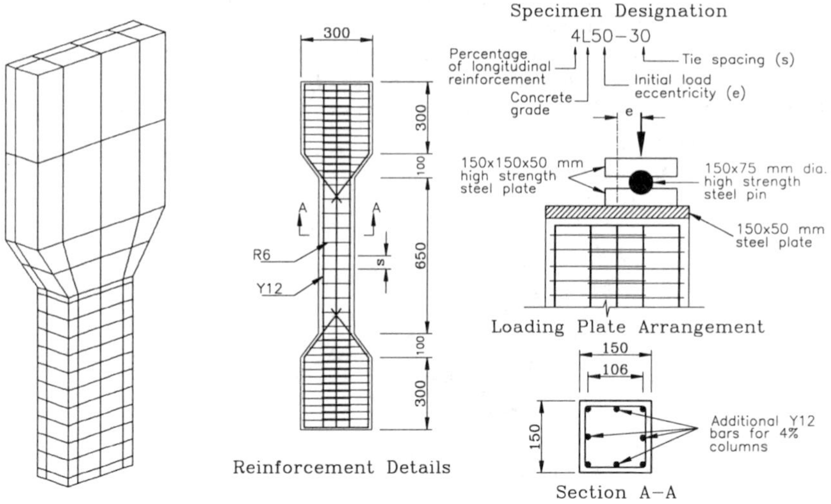

Figure 3. FE mesh and experimental details of 150 mm square columns.

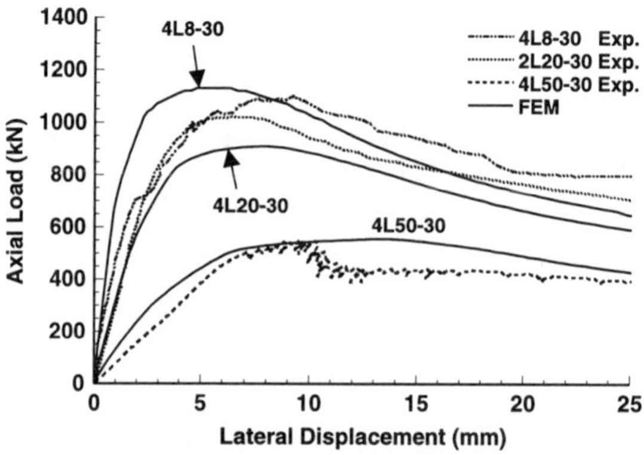

Figure 4. Analytical simulation of eccentrically loaded column tests of Attard and Foster (1995).

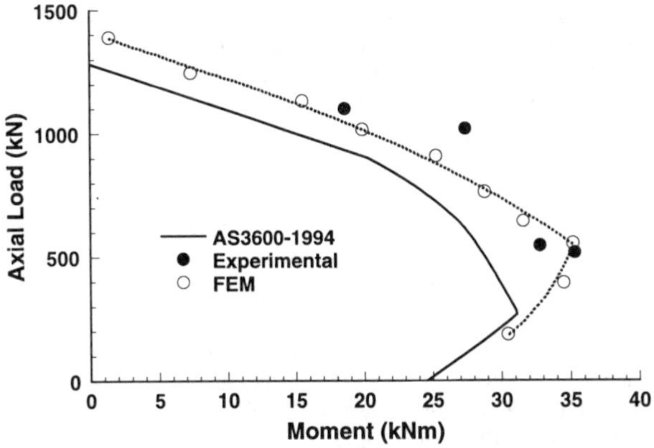

Figure 5. Comparison of FE and experimental data with the interaction diagram of AS3600 (1994)

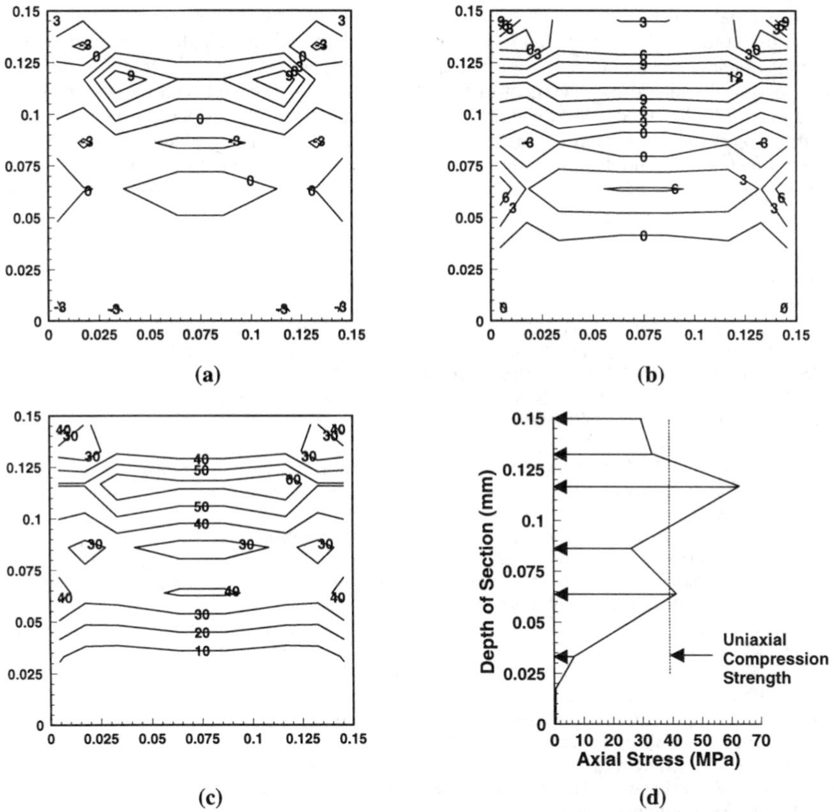

Figure 6. Principal stress contours, at the peak load, for the mid-height of the 150 mm square column loaded with an initial eccentricity of 20 mm: a) major principal stress; b) intermediate principal stress and; c) minor principal stress; d) axial stress through the middle of the section. Compression plotted as positive.

Figure 5 shows that the interaction curve obtained from the numerical model compares well with the experimental data. The trend of the experimental and FE data compares well to the interaction diagram obtained by AS3600 (refer Figure 5) but give consistently higher failure moments for any given axial load, particularly when the section fails in a primary compression mode. In Figure 6, the principal stresses are plotted for the FE model of column 4L20-30 at a plane through the mid-length of the column. The figure shows that the compression zone is in a triaxial stress state. At the peak load the maximum major (σ_1) and intermediate (σ_2) principal stresses are 12 MPa and 14 MPa, respectively. At the middle of the section in the compression zone, $\sigma_1 = 6$ MPa and $\sigma_2 = 14$ MPa. The confinement provided by the tie steel has lead to a maximum minor compression stress of $\sigma_3 = 64$ MPa, significantly greater than the uniaxial strength of the concrete of 39 MPa.

At the squash point (that is M=0) on Figure 5, the confinement provided by the ties gives an increased capacity over that predicted by AS3600. However, at this point of the FE model development, cover spalling has not been included (refer Foster, 1999). With the inclusion of the effects of cover spalling it is expected that the FE squash load predictions will be reduced.

4 CONLUSIONS

A 3-dimensional total Lagrangian FE model has been developed using the Microplane M4 formulation of Bazant et al. (1999). A 150 mm square column was modelled with various initial eccentricities and compared to the experimental data. The analyses showed that the trend of the applied load to the lateral displacement of the FE model compares well with the experimental observations. The FE model shows that confinement can have a considerable effect on the strength of eccentrically loaded columns and explains the higher loads obtained from the experimental tests than is predicted by the interaction model of AS3600 (1994). At this point in the development of the FE formulation the effect of cover spalling has not been included and further investigation is required on this aspect.

REFERENCES

AS3600. 1994. Concrete structures code. Sydney: Standards Association of Australia.

Attard, M. M., & Foster, S. J., 1995. Ductility of high strength concrete columns. *UNICIV Report R-344*, Sydney, School of Civil Engineering, The University of New South Wales.

Attard, M. M., & Setunge, S. 1994. The stress – strain relationship of confined and unconfined normal and high strength concretes. *UNICIV Report R-341*, Sydney: School of Civil Engineering, The University of New South Wales.

Bazant, Z. P., Caner, C. F., Carol, I., Adley, M. D., and Akers, S. A. 1999a. Microplane model M4 for concrete: I. Formulation with work - conjugate deviatoric stress. *ASCE Journal of Structural Engineering* (in press).

Bazant, Z. P., Caner, C. F., Carol, I., and Adley, M. D. 1999b. Microplane model M4 for concrete: II. Algorithm, Calibration and application. *ASCE, Journal of Structural Engineering* (in press).

Bazant, Z. P., & Gambarova, P. G. 1984. Crack shear in concrete: crack band Microplane model. *ASCE Journal of Structural Engineering*, 110(9): 2015-2035.

Bazant, Z. P., Xiang, Y. & Prat, P. C., 1996. Microplane model for concrete. I: stress-strain boundaries and finite strain. *ASCE, Journal of Engineering Mechanics*, 122(3): 245-254.

Chen, W. F., 1982. *Plasticity in reinforced concrete.* McGraw-Hill

Dahl, K. B. 1992. Uniaxial stress – strain curves for normal and high strength concrete. *Research Report Series No. 282.* Copenhagen: Department of Structural Engineering, Technical University of Denmark.

Foster, S. J. 1999. The fallacy of early cover spalling in HSC columns. 5th International Symposium on *Utilization of High Strength/High Performance Concrete.* Sandefjord, Norway: 282-291.

Kotsovos, M. D., & Pavlovic, M. N. 1995. *Structural Concrete: finite element analysis for limit state design.* Thomas Telford Publications London.

Ottosen, N.S. 1977. A failure criterion for concrete. *ASCE, Journal of Engineering Mechanics,* and 103 (EM4): 527-535.

Mechanics of Structures and Materials, Bradford, Bridge & Foster (eds)

A study of the convergence of the scaled boundary finite element method

A.J. Deeks & C. Costello
Department of Civil and Resource Engineering, University of Western Australia, Nedlands, W.A., Australia

ABSTRACT: The scaled boundary finite element method is a boundary element method which does not require a fundamental solution or integration of singular functions. The solutions are exact in the radial direction (relative to a similarity centre), and converge in a finite element sense in the tangential direction(s). The method can be used to solve both bounded and unbounded problems, and problems of static and elasto-dynamics, amongst others. This paper provides a simple overview of the method, and then applies it to the solution of two simple plane stress problems. The rate of convergence as the number of degrees of freedom are increased is compared with standard finite element analysis. The method is found to be more accurate than the standard finite element approach for a given number of degrees of freedom.

1 INTRODUCTION

The scaled boundary finite element method (originally called the consistent infinitesimal finite element cell method) is a recent development by Wolf and Song (1994, 1996, 1998). Inspired by Dasgupta's cloning algorithm (Dasgupta, 1982), it only requires the boundary of a model to be discretised. However, unlike the boundary element method, it does not require a fundamental solution. The solutions are exact in the radial direction (relative to a similarity centre), and converge in a finite element sense in the tangential direction(s). The method can be used to solve both bounded and unbounded problems, and problems of statics and elasto-dynamics, amongst others.

The equations governing the scaled boundary finite element method have been presented in full by Wolf and Song (1998). The derivation of the equations and the process required to obtain a solution are not as transparent as the well-known finite element method, or even the boundary element method. However, the method has significant advantages over both these methods, in particular reducing the spatial dimension by one while not requiring fundamental solutions or the integration of singular functions.

This paper provides a brief overview of the scaled boundary finite element method. In the discussion of the solution process, the equations involved are deliberately simplified to illustrate the salient points. For a complete derivation of the full equations, the reader should refer to Wolf and Song (1996) or Wolf and Song (1998).

The paper then reports the results of a study in which both the scaled boundary finite element method and the standard finite element method were applied to two problems, and the number of degrees of freedom varied. The rates of convergence of the two methods were examined in detail.

2 OVERVIEW OF THE SCALED BOUNDARY FINITE ELEMENT METHOD

Considering a plane stress problem, the finite element method works by weakening the governing set of differential equations in both directions, obtaining a linear set of simultaneous equations, where the number of equations is equal to the number of nodal degrees of freedom. Boundary conditions are applied at the nodal degrees of freedom, either as nodal forces or as prescribed displacements.

The scaled boundary method defines a ξ–η coordinate system based on a 'similarity centre' within the solution domain. The ξ coordinate is directed from the similarity centre towards the boundary, and has unit value at each point on the boundary. The η coordinate runs around the boundary of the domain. Each value of ξ defines a scaled boundary, around which the η coordinate also runs. This coordinate system is called the scaled boundary coordinate system. The scaled boundary coordinates are illustrated in Figure 1.

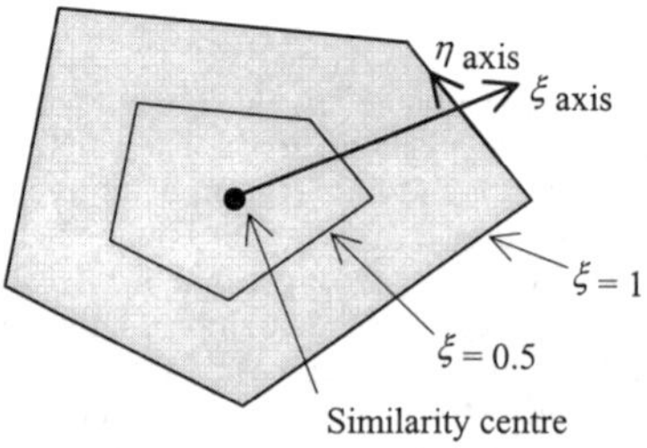

Figure 1 : Definition of the scaled boundary coordinates.

The governing differential equations are re-written in terms of the scaled boundary coordinates. These equations are then weakened in the η direction by introducing one-dimensional finite elements around the boundary. If $\{u(\xi)\}$ represents a vector of displacement functions at the nodal η coordinates (so each function $u(\xi)$ represents the variation of a displacement component along a straight line drawn from the similarity centre to a node), the weakened set of equations (without boundary conditions) can be written in the form:

$$\xi^2 [A] \{u(\xi)\}_{,\xi\xi} + \xi ([A]+[B]^{\mathrm{T}}-[B]) \{u(\xi)\}_{,\xi} - [C] \{u(\xi)\} = \{0\} \tag{1}$$

where $[A]$, $[B]$, and $[C]$ are square matrices of coefficients depending only on the locations of the nodes and the properties of the elements, and independent of both ξ and η. This equation represents a set of linear second-order ordinary differential equations in ξ.

The resultant nodal force functions $\{R(\xi)\}$ (found by truncating the domain at ξ and integrating the stress resultants with the shape functions at this truncated boundary) can also be related to the displacements.

$$\{R(\xi)\} = \xi [A] \{u(\xi)\}_{,\xi} + [B]^{\mathrm{T}} \{u(\xi)\} \tag{2}$$

These two sets of equations can be combined to form a single set of first-order ordinary differential equations by placing the nodal displacement and nodal force functions in a single vector $\{X(\xi)\}$.

$$\xi \{X(\xi)\}_{,\xi} = [Z] \{X(\xi)\} \quad \text{where} \quad \{X(\xi)\} = \begin{Bmatrix} \{u(\xi)\} \\ \{R(\xi)\} \end{Bmatrix} \tag{3}$$

The coefficient matrix $[Z]$ is a combination of $[A]$, $[B]$, and $[C]$. Solutions to these equations can be found in the form $\{X(\xi)\} = \xi^{\lambda} \{\phi\}$, where both λ and the vector $\{\phi\}$ consist of constant coefficients. Substituting this solution into Eqn. 3 leads to an eigenvalue problem, which can be solved using conventional techniques. Note that the weakened differential equations have been solved exactly in the ξ direction.

$$\lambda \{\phi\} = [Z] \{\phi\} \tag{4}$$

The general solution to the weakened governing differential equations is a linear combination of the modes of deformation represented by the eigenvalues and eigenvectors obtained from this equation. These deformation modes are orthogonal, and form a basis for the solution space. For each mode the eigenvalue λ specifies how the deformation varies in the ξ direction, while the first n values in the eigenvector represent the movement at the nodal degrees of freedom, and the second n values represent the nodal forces required on the boundary to produce this deformation, where n represents the number of nodal degrees of freedom on the boundary.

Solution of the eigenvalue problem actually gives n modes of deformation for the bounded domain, and another n modes of deformation for the unbounded domain. Taking just the modes for the bounded domain leaves n vectors, each of size $2n$. These vectors can be written as the columns of a matrix, and the upper section (containing the n rows of nodal displacements) can be designated $[\Phi_u]$, while the lower section (containing the corresponding n rows of nodal forces) is designated $[\Phi_R]$. Any vector of nodal boundary displacements $\{u\}$ can be represented by a linear combination of the deformation modes. This is written as $[\Phi_u]\{u^*\} = \{u\}$, where $\{u^*\}$ are coefficients indicating the contribution of each mode. It follows immediately that $\{u^*\} = [\Phi_u]^{-1}\{u\}$. The nodal forces required to cause the boundary displacements $\{u\}$ can then be found by multiplying the nodal forces for each mode by the corresponding modal coefficient, and summing.

$$\{R\} = [\Phi_R]\{u^*\} = [\Phi_R][\Phi_u]^{-1}\{u\} = [K]\{u\} \tag{5}$$

Consequently, once the deformation modes have been found, the static stiffness matrix can be found using $[K] = [\Phi_R][\Phi_u]^{-1}$. Any particular load case can then be solved in the usual finite element manner.

The modes computed for a square region discretised with a single quadratic boundary element along each side are illustrated in Figure 2. To prevent singularity of [Z] a small value had to be added to the diagonal terms. This allowed the eigenvalue procedure to find the two translational modes and the rigid body rotation.

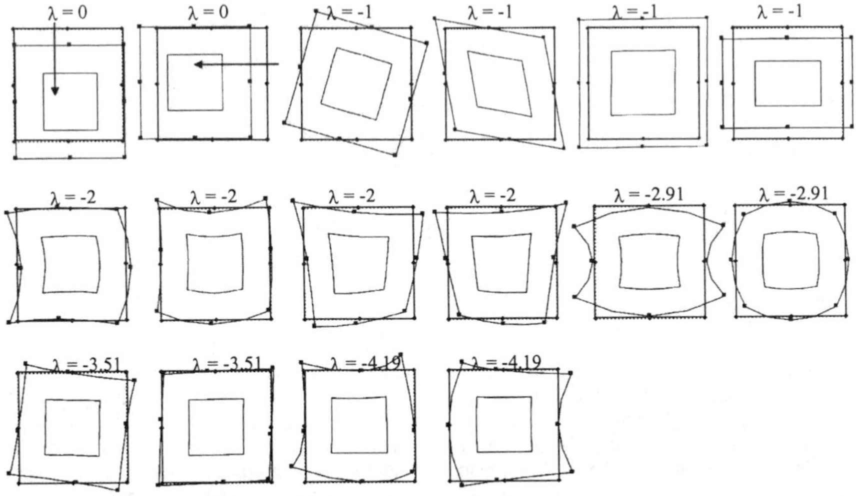

Figure 2. Mode shapes for a square domain with 16 nodal degrees of freedom.

3 FIRST EXAMPLE – CANTILEVER WITH A POINT LOAD

A deep cantilever of rectangular cross-section with an aspect ratio of 1/3 was analysed. The problem was treated as one of plane stress. The point load at the end of the member was applied as a uniform shear over the depth. This arrangement is illustrated in Figure 3. A Poisson's ratio of 0.2 was used.

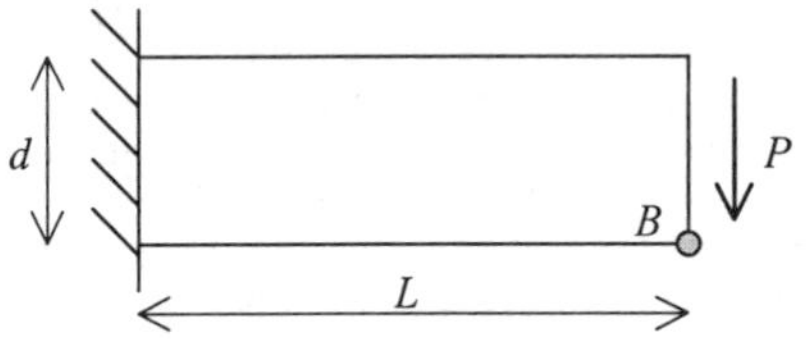

Figure 3. Diagram of first example.

The rates of convergence of the scaled boundary finite element method and the standard finite element method were compared by solving the problem using several different levels of discretisation. The initial mesh was constructed by using one element over the depth and two elements over the length of the cantilever. Subsequent meshes were obtained by congruent subdivision of the initial mesh. The dimensionless vertical deflection (Δ^*) was measured at the point marked B on Figure 3. The actual deflection (Δ) is related to the dimensionless deflection by Equation 6.

$$\Delta = \Delta^* \frac{PL^3}{EI} \tag{6}$$

The displacements obtained using the two methods for meshes with various numbers of degrees of freedom are illustrated in Figure 4. Both methods show rapid convergence, although the scaled boundary finite element method requires significantly less degrees of freedom to obtain the same level of accuracy as the standard finite element method.

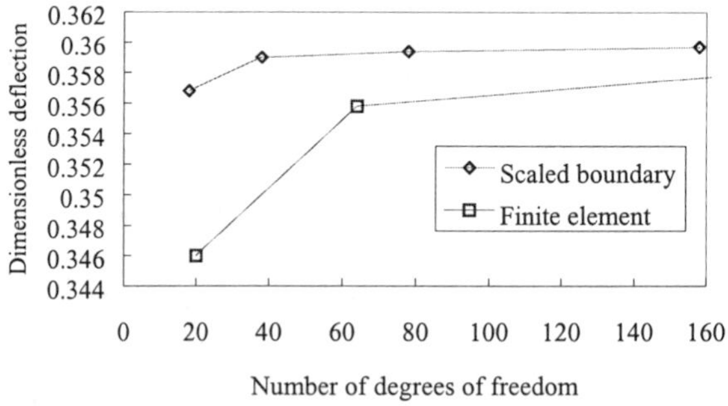

Figure 4. Convergence of deflections in Example 1.

An 'exact' solution was obtained using an extremely fine finite element model (over 6000 d.o.f). This solution was used to compute normalised errors. These errors were plotted against number of degrees of freedom on a logarithmic plot (Figure 5). This figure shows that the slope of the graphs is the same for each method. This is a consequence of both methods using quadratic elements. However, the scaled boundary finite element method graph is significantly lower than the finite element graph. This indicates that, although the scaled boundary technique requires significantly less degrees of freedom to obtain a given level of accuracy, it retains the same convergence characteristics as the finite element method.

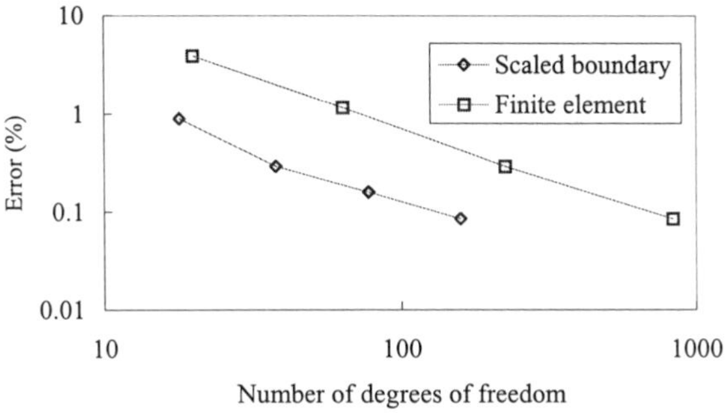

Figure 5. Rate of convergence of normalised error in Example 1.

4 SECOND EXAMPLE – SQUARE PLATE WITH A SQUARE HOLE

A square plate with a square central hole subjected to a uniform in-plane stress along one axis was analysed. The hole was taken to have a width of 1/6 of the plate width. Poisson's ratio was set at 0.25. Advantage was taken of the symmetry of the problem, and only one quarter of the plate was modelled, as shown in Figure 6. The point designated by O in Figure 6 was taken as the similarity centre in the scaled boundary finite element method.

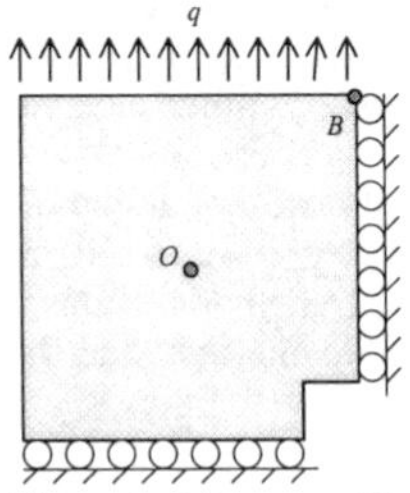

Figure 6. Diagram of Example 2.

Displacement was recorded at the point indicated as B on Figure 6. The results were non-dimensionlised using the 'exact' displacement computed using a very fine finite element model (in excess of 20 000 degrees of freedom). The convergence of the displacements is shown in Figure 7, and the rate of convergence of the normalised errors in Figure 8. Both of these figures show the same trends as Example 1.

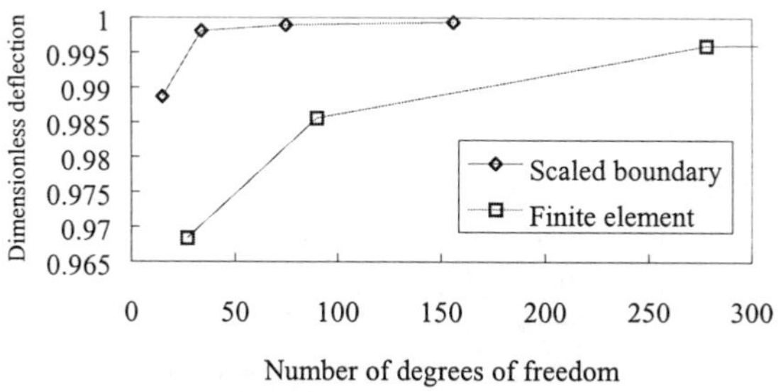

Figure 7. Convergence of deflection in Example 2.

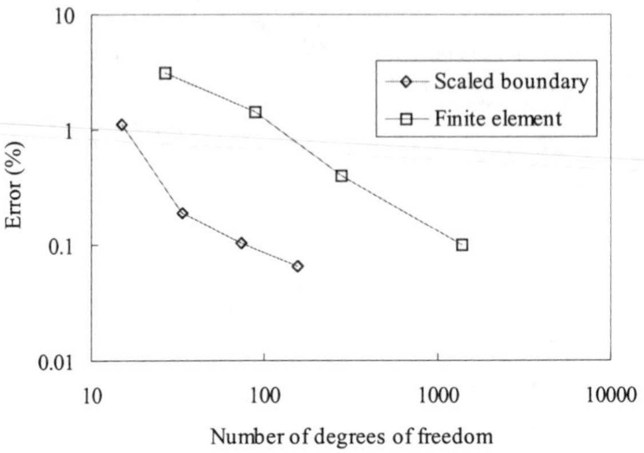

Figure 8. Rate of convergence of normalised error in Example 2.

5 CONCLUSIONS

The scaled boundary finite element method element method is a relatively new technique which reduces the dimension of the spatial discretisation of a solution domain by one. The governing differential equations are solved exactly in the radial direction from a chosen similarity centre. This paper briefly detailed the theory behind the method, and then applied it to two example problems, comparing the rate of convergence of the solutions with that obtained by traditional finite element techniques. These problems showed that the scaled boundary finite element method significantly reduces the number of degrees of freedom required to obtain a solution of a particular level of accuracy. The rate of convergence, interpreted as the slope of a graph plotting the logarithm of an error norm against the logarithm of the number of degrees of freedom, remained approximately equal, as quadratic elements were employed in both methods. Consequently, the scaled boundary finite element method is seen to offer a valuable alternative to the finite element method.

REFERENCES

Dasgupta, G. 1982. A finite element formulation for unbounded homogeneous continua. *Journal of Applied Mechanics* 49:136-140.

Wolf, J.P. and Song, C. 1994. Dynamic-stiffness matrix in time domain of unbounded medium by infinitesimal finite-element cell method. *Earthquake Engineering and Structural Dynamics* 23:1181-1198.

Wolf, J.P. and Song, C. 1996. *Finite Element Modelling of Unbounded Media*. England:John Wiley & Sons Ltd.

Wolf, J.P. and Song, C. 1998. The scaled boundary finite-element method – A fundamental-solution-less boundary element method. *Proc. of the Fourth World Cong. on Comp. Mech.* Buenos Aires, Argentina.

Mechanics of Structures and Materials, Bradford, Bridge & Foster (eds)
 ISBN 90 5809 107 4

BE mathematical programming approach for identifying quasibrittle fracture parameters

F.Tin-Loi & N.Que
School of Civil and Environmental Engineering, The University of New South Wales, Sydney, N.S.W., Australia

ABSTRACT: This paper deals with the identification of quasibrittle fracture parameters that characterize Hillerborg's discrete cohesive crack model. Starting from a boundary element (BE) space-discretized model of the standard three-point bending (3PB) test, we formulate this task as an inverse problem, requiring the minimization of a suitable norm of some error function characterizing the discrepancy between experimental and theoretical values of reactions corresponding to suitably imposed displacements. This particular nonconvex optimization problem is difficult to solve due to the presence of "complementarity" constraints, involving the orthogonality of two sign-constrained vectors. A smoothing algorithm, designed to cater for the nondifferentiability of the complementarity constraints, is proposed. Application is illustrated by means of an example concerning actual test data for a composite polymeric material.

1 INTRODUCTION

The discrete cohesive crack model, popularized by Hillerborg (1991), is widely used to numerically simulate fracture processes in structures made of quasibrittle or concrete-like material. This model is attractive in view of its simplicity and potential for the desired predictive accuracy. A disadvantage, however, is that it involves, even in the case of mode I fracture that prevails in crack propagation processes, a softening constitutive law that cannot be directly obtained experimentally. In particular, this softening law relates displacement discontinuities (crack widths) across the locus of potential cracks to corresponding cohesive tractions, while standard tests measure such parameters as fracture energy (RILEM 3PB test) and material tensile strength (pure tensile test and Brazilian test).

In this paper, we consider the inverse problem of identifying the relevant material properties characterizing mode I cohesive crack interface laws on the basis of standard 3PB tests. We focus, without loss of generality, on a particular idealization of the four-parameter constitutive law consisting of two linear softening branches — as recently recommended by RILEM for concrete-like materials and which has been found to correlate well with extensive laboratory experiments (Wittmann & Hu 1991).

The idea of identifying quasibrittle fracture parameters from the load-displacement responses of 3PB tests is not new (see e.g. Elices & Planas 1995, Nanakorn & Horii 1996, Bolzon *et al.* 1997). Particularly promising in our view is the approach of Bolzon *et al.* (1997). In essence, the parameter identification problem is cast as a specially challenging optimization problem involving nonconvex "complementarity" (a term describing the orthogonality of two sign-constrained vectors) constraints. In this paper, we extend the work of Bolzon *et al.* (1997) by proposing an alternative, arguably more robust scheme, for solving this optimization problem.

2 COHESIVE CRACK MODEL AND STATE PROBLEM

Consider the 3PB cementitious specimen shown in Figure 1(a). Crack propagation in the expected mode I behavior can be assumed to occur along the known crack locus Γ_c (on the vertical axis of symmetry) of possible displacement discontinuity w. The RILEM recommended two-branch law, relating normal tractions t across the crack interfaces to w, is shown in Figure 1(b). Since the controlled displacement u is monotonically increasing, local unloading (from a softening branch) is unlikely to occur so that the interface law can be assumed to be holonomic (path-independent).

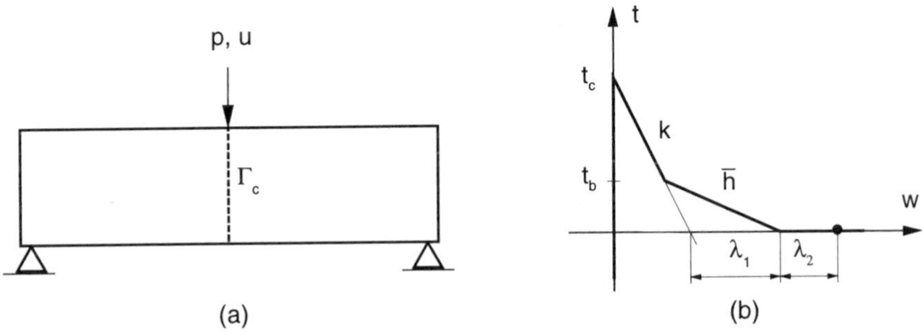

Figure 1. Problem definition (a) 3PB specimen, (b) cohesive law.

The cohesive crack model can be conceived as a union of cracks ($t=0, w>0$), process zone or craze ($t \neq 0, w \neq 0$) and undamaged material ($w=0$). A mathematical description of the two-branch piecewise linear model, applicable to any point i on the crack locus, can be conveniently given as (Bolzon *et al.* 1994) the following linear complementarity system:

$$\mathbf{f}^i = t_c \mathbf{v}_1^i - t_b \mathbf{v}_2^i - [k\mathbf{M}_1^i + h\mathbf{M}_2^i]\,\mathbf{z}^i - \mathbf{N}^i t^i \geq \mathbf{0}, \qquad \mathbf{z}^i \geq \mathbf{0}, \qquad \mathbf{f}^{iT}\mathbf{z}^i = 0 \tag{1}$$

where t_c = tensile strength; t_b = breakpoint strength; k = initial (positive) slope; h = (positive) parameter that relates k to the (positive) slope $\bar{h} = hk/(h+k)$ of the second branch; and where we have set

$$\mathbf{f}^i = \begin{bmatrix} f_1 \\ f_2 \\ f_3 \end{bmatrix}^i, \; \mathbf{z}^i = \begin{bmatrix} \lambda_1 \\ \lambda_2 \\ w \end{bmatrix}^i, \; \mathbf{v}_1^i = \begin{bmatrix} 1 \\ 1 \\ 1 \end{bmatrix}, \; \mathbf{v}_2^i = \begin{bmatrix} 1 \\ 0 \\ 0 \end{bmatrix}, \; \mathbf{N}^i = \begin{bmatrix} 0 \\ 0 \\ 1 \end{bmatrix}, \; \mathbf{M}_1^i = \begin{bmatrix} -1 & -1 & 1 \\ -1 & -1 & 1 \\ -1 & -1 & 1 \end{bmatrix}, \; \mathbf{M}_2^i = \begin{bmatrix} -1 & 0 & 0 \\ 0 & 0 & 0 \\ 0 & 0 & 0 \end{bmatrix}$$

More general nonlinear laws, which can be specialized to such well-known models as the power softening laws (Foote *et al.* 1986), can be elegantly expressed in the same, albeit nonlinear, complementarity format (Tin-Loi & Ferris 1997).

Further, the assumption of nonlinearity only in the crack locus and linear elasticity elsewhere in the domain allows us to construct, after suitable discretization by finite or boundary element, the following system of linear equations for those nodes pertaining only to the potential crack interface:

$$\mathbf{t} = \mathbf{t}^e + \mathbf{Z}\mathbf{w} \tag{2}$$

where, for n nodes, $\mathbf{t} \in \Re^n$ is a vector of total interface tractions conceived as being made up of (a) an elastic component $\mathbf{t}^e \in \Re^n$ caused by the external action (consisting in our case of an imposed displacement u) in the absence of displacement discontinuities and (b) of tractions due to the displacement discontinuities $\mathbf{w} \in \Re^n$ in the otherwise unloaded body. $\mathbf{Z} \in \Re^{n \times n}$ is a matrix of influence coefficients reflecting geometric and elastic properties of the specimen. Analytical expressions for $\mathbf{Z}$ are rarely possible so that these influence coefficients need to be calculated from a suitably space-discretized structure. Moreover, as Eq. (2) involves variables on the boundary and at the assumed interface only, a BE approach is more suitable than a FE method which would involve domain variables as well. Motivated primarily by simplicity of implementation, we have adopted a multizone, direct collocation BE approach involving isoparametric quadratic three-node boundary elements. Details regarding the associated BE model and the construction of $\mathbf{Z}$ are given in Tin-Loi & Li (1999).

Finally, assuming that Eq. (1) applies pointwise to all interface nodes $i = 1 \ldots m$, combination of Eq. (2) with the collected nodal softening constitutive laws leads to the following (mixed) complementarity problem:

$$\begin{aligned} &\mathbf{t} = \mathbf{t}^e + \mathbf{Z}\mathbf{w} \\ &\mathbf{f} = t_c \mathbf{v}_1 - t_b \mathbf{v}_2 - [k\mathbf{M}_1 + h\mathbf{M}_2]\,\mathbf{z} - \mathbf{N}\,\mathbf{t} \geq \mathbf{0}, \quad \mathbf{z} \geq \mathbf{0}, \quad \mathbf{f}^T\mathbf{z} = 0 \end{aligned} \tag{3}$$

where, as is conventional with FE or BE formalism, extension of Eq. (1) to the whole structure has been carried out by a simple reinterpretation of the quantities involved as new indexless symbols. In particular, elemental vectors and matrices are "assembled" as appropriate concatenated vectors and block-diagonal matrices, respectively. For example, $\mathbf{f}^T \equiv [\mathbf{f}^{1T} \ldots \mathbf{f}^{nT}]$, $\mathbf{v}_1^T \equiv [\mathbf{v}_1^{1T} \ldots \mathbf{v}_1^{nT}]$, $\mathbf{M}_1 \equiv \mathrm{diag}(\mathbf{M}_1^1, \ldots, \mathbf{M}_1^n)$, $\mathbf{N} \equiv \mathrm{diag}(\mathbf{N}^1, \ldots, \mathbf{N}^n)$, etc.

Solution of the mixed complementarity problem (3), it must be noted, is not guaranteed by any of known methods in view of the presence of softening (Tin-Loi & Ferris 1996). However, we have had considerable success with the *de facto* industry standard mixed complementarity solver PATH (Dirkse & Ferris 1995), especially from within the GAMS (Brooke *et al.* 1998) modeling environment.

For any imposed displacement u, it would then be possible to compute the corresponding "reaction" p from a formula such as

$$p = p^e + \mathbf{r}^T\mathbf{w} \tag{4}$$

where p^e is the pure elastic response corresponding to imposed displacement u, and $\mathbf{r} \in \Re^n$ is a known vector, easily obtainable as a by-product of the computational model.

3 INVERSE PROBLEM

The inverse or parameter identification problem can now be briefly described. It is assumed that a number of pairs of readings (p_j^m, u_j^m) have been measured during a 3PB test where $j \in \mathrm{J}$ represents a measurement. Let us also assume that we need to identify the four parameters related to the two-branch piecewise linear law (1).

If we now denote by a superscript c values that would be obtained from the structural model for the same loading, then a natural measure of the discrepancy (or error) ω between measured and theoretical reactions is provided by a suitable norm of the difference between p_j^m and p_j^c, $\forall j \in \mathrm{J}$. The identification problem obviously requires the *global* minimum of ω subject to Eqs. (3) and (4), and other constraints such as any known bounds (based on engineering knowledge of the material) on the four parameters.

The identification problem can be formally stated as the following constrained optimization problem, assuming a two-norm for the error function:

$$\min_{\forall j \in J} \quad \omega \equiv \sum_j (p_j^m - p_j^c)^2$$

subject to:

$$\mathbf{t}_j = \mathbf{t}_j^e + \mathbf{Z}\mathbf{w}_j \tag{5}$$

$$\mathbf{f}_j = t_c \mathbf{v}_1 - t_b \mathbf{v}_2 - [k\mathbf{M}_1 + h\mathbf{M}_2]\,\mathbf{z}_j - \mathbf{N}\,\mathbf{t}_j \geq \mathbf{0}, \quad \mathbf{z}_j \geq \mathbf{0}, \quad \mathbf{f}_j^T \mathbf{z}_j = 0$$

$$p_j^c = p_j^e + \mathbf{r}^T \mathbf{w}_j$$

$$t_c \geq t_b$$

The last constraint is obvious and we have omitted, for brevity, any bounds on the parameters.

The optimization problem given by Eq. (5) is a special case of the challenging class of so-called Mathematical Programs with Equilibrium Constraints or MPEC (Luo *et al.* 1996) for which the equilibrium constraints are complementarity conditions. The latter distinguish an MPEC from a standard nonlinear program, and impart a combinatorial flavor to the problem. In spite of extensive research in the area, there are currently no known algorithms guaranteed to solve this problem. The various algorithms which have been proposed are categorized by the way they handle the complementarity constraints. Motivated by recent work within the mathematical programming community, we propose in the following a scheme involving "smoothing" of the complementarity constraints.

4 SMOOTHING ALGORITHM

We first note that it is difficult to directly solve the MPEC in the form given in (5) as a nonlinear program. As is well-known (Luo *et al.* 1996), certain conditions (constraint qualifications) do not hold at any feasible point of this smooth nonlinear programming problem. This is essentially synonymous with numerical instability.

An idea, first introduced by Facchinei *et al.* (1996) is to replace the complementarity conditions such as $\mathbf{x}^T \mathbf{y} = 0$ (where $\mathbf{y}$ is a vector-valued function of $\mathbf{x}$) by the equations

$$\varphi_\mu(x_i, y_i) = 0, \quad \forall i \tag{6}$$

The function $\varphi_\mu(a,b)$ has the property that $\varphi_\mu(a,b) = 0$ if and only if $a \geq 0$, $b \geq 0$, $ab = \mu$. The parametrization φ_μ is in effect a "smoothing" of the mapping $\varphi_{\mu=0}$. That is, φ_μ is differentiable if the scalar μ is nonzero. Moreover, the most important advantage of introducing a smoothing parameter is that the reformulation of the MPEC as a nonlinear program is more likely to satisfy constraint qualifications and feasibility for all μ, leading to improved solvability.

Many such functions have been proposed in the literature, particularly in relation to the solution of complementarity problems. In this paper, we use the following smoothing of the so-called Fischer function (Facchinei *et al.* 1996):

$$\varphi_\mu(a,b) = \sqrt{a^2 + b^2 + 2\mu} - (a+b) \tag{7}$$

Our approach is simple: we replace the complementarity conditions with Fischer functions and attempt to solve a series of nonlinear programs for decreasing values of μ until complementarity is satisfied to within a preset tolerance. The initial value of μ and its decrease for each major iteration is of course problem dependent. Too small a decrease will lead to slow convergence but possibly a better solution. Also, too large an initial value may not provide a solution to the nonlinear program while too small a value may lead to a much more ill-conditioned initial nonlinear program.

A smoothing approach involving nonlinear programming subproblems is eminently viable when a modeling system such as GAMS is used. Access to industry tested and proven nonlinear programming codes such as GAMS/CONOPT (Drud 1994) is then possible. GAMS also provides a powerful facility to "warm start" each subsequent iteration.

5 EXAMPLE

The methodology briefly described in this paper has been used to identify the fracture properties of a polymeric composite made of an epoxy matrix with silicon micro-spherical hollow inclusions (Bolzon *et al.* 1997). Known properties are: E = 3707 MPa, ν = 0.39. The 3PB test was carried out on a 107.1×30 mm notched specimen, 15 mm thick and 15 mm notch depth. Although, horizontal displacements of 30 points in a regular 5×6 grid were monitored through laser interferometry and are available, we have only used the recorded $p-u$ points.

We developed a standard two-zone collocation boundary element model of the specimen, with 15 quadratic elements (31 nodes) on the interface, in order to obtain necessary quantities such as **Z** and **r**. These, together with the recorded experimental values, were input as data into a GAMS model we set up for the inverse problem given by the smoothed version of Eq. (5). GAMS/CONOPT2 was the adopted nonlinear programming solver.

Assuming a two-branch softening law of the form given by Eq. (1), we solved the smoothed MPEC using arbitrarily chosen starting values of $t_c = 1$, $t_b = 1$, $k = 200$, $h = 200$. Parameter μ was reduced by a factor of 10 at each major iteration.

The results are shown in Figure 2, in the form of predicted $p-u$ curves versus experimental points (+). Note that we used only 8 (out of a total of 48) experimental points ($*$) for the identification process. As is evident, reasonably good predictions have been obtained.

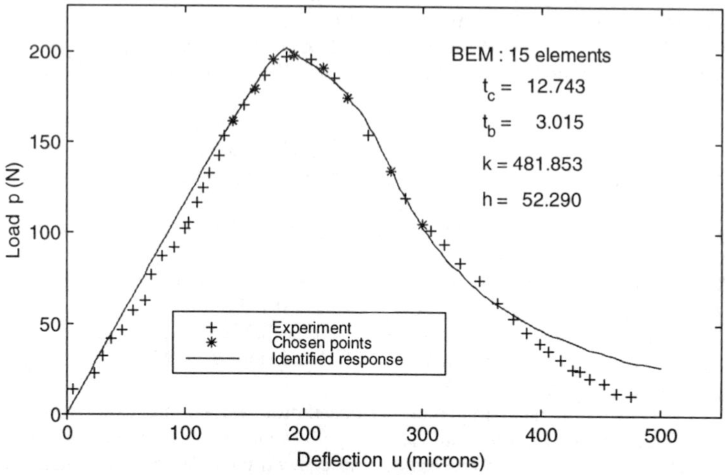

Figure 2. Experimental versus identified behavior.

6 CONCLUSIONS

In this paper, we describe an optimization-based inverse scheme suitable for identifying key quasibrittle fracture parameters characterizing softening laws, with special reference to the commonly adopted two-branch piecewise linear law. The formulation of the inverse problem leads to an MPEC which is then solved as a series of standard (smoothed) nonlinear programs. A realistic example concerning an actual test on a polymeric composite is given to illustrate suitability of the method. Current work is aimed at enhancing the robustness of the method, to develop new algorithms and to find the best (global) minimum.

ACKNOWLEDGMENTS

This work was supported by the Australian Research Council. We would also like to thank Professor Michael Ferris for helpful discussions regarding GAMS and MPECs, and Professor Giulio Maier and Dr. Gabriella Bolzon for kindly providing us with the experimental data.

REFERENCES

Bolzon, G., Ghilotti, D. & Maier, G. 1997. Parameter identification of the cohesive crack model. In: *Material Identification Using Mixed Numerical Experimental Methods*. eds. H. Sol & C.W.J. Oomens, Kluwer Academic Publishers, Dordrecht, 213-222.

Bolzon, G., Maier, G. & Novati, G. 1994. Some aspects of quasi-brittle fracture analysis as a linear complementarity problem. In: *Fracture and Damage in Quasibrittle Structures*. eds. Z.P. Bazant, Z. Bittnar, M. Jirasek & J. Mazars, E&FN Spon, London, 159-174.

Brooke, A., Kendrick, D., Meeraus, A. & Raman, R. 1998. *GAMS: A User's Guide*. Gams Development Corporation, Washington, DC 20007.

Dirkse, S.P. & Ferris, M.C. 1995. The PATH solver: a non-monotone stabilization scheme for mixed complementarity problems. *Optimization Methods & Software* 5: 123-156.

Drud, A. 1994. CONOPT - a large-scale GRG code. *ORSA Journal on Computing* 6: 207-216.

Elices, M. & Planas, J. 1995. Numerical modeling and determination of fracture mechanics parameters: Hillerborg type models. In: *Fracture Mechanics of Concrete Structures*. ed. F.H. Wittmann, Aedificatio Publishers, Freiburg, 1611-1620.

Facchinei, F., Jiang, H. & Qi, L. 1996. A smoothing method for mathematical programs with equilibrium constraints. Applied Mathematics Report AMR96/15, School of Mathematics, University of New South Wales.

Foote, R.M.L., Mai, Y.W. & Cotterel, B. 1986. Crack growth resistance curves in strain-softening materials. *Journal of the Mechanics and Physics of Solids* 34: 593-607.

Hillerborg, A. 1991. Application of the fictitious crack model to different types of materials. *International Journal of Fracture* 51: 95-102.

Luo, Z.Q., Pang, J.S. & Ralph, D. 1996. *Mathematical Programs with Equilibrium Constraints*. Cambridge University Press, Cambridge.

Nanakorn, P. & Horii, H. 1996. Back analysis of tension-softening relationship of concrete. *Journal of Materials, Concrete Structures and Pavements, JSCE* 32: 265-275.

Tin-Loi, F. & Ferris, M.C. 1997. Holonomic analysis of quasibrittle fracture with nonlinear softening. In: *Advances in Fracture Research*. eds. B.L. Karihaloo, Y.W. Mai, M.I. Ripley & R.O. Ritchie, Pergamon, Oxford, 2183-2190.

Tin-Loi, F. & Li, H. 1999. Numerical simulations of quasibrittle fracture processes using the discrete cohesive crack model. *International Journal of Mechanical Sciences* (forthcoming).

Wittmann, F.H. & Hu, X. 1991. Fracture process zone in cementitious materials. *International Journal of Fracture* 51: 3-18.

Mechanics of Structures and Materials, Bradford, Bridge & Foster (eds)
© 1999 Balkema, Rotterdam, ISBN 90 5809 107 4

Study on normal flow rule and radial flow rule

K.Zama, T.Nishimura & K.Miura
Department of Mechanical Engineering, College of Science and Technology, Nihon University, Tokyo, Japan

ABSTRACT: The normal flow rule as a plastic flow rule, suggested by R.Hill and D.C.Drucker, is widely applied to many plastic analyses. However, there is little experimental verification concerning the normal flow rule. In this study, some experiments are performed to investigate the plastic flow rule. The experimental results suggest that the actual plastic strain proceeds parallel to the current stress vector, that is, the radial flow rule. Here the experimental study on the plastic flow is mentioned, and the suggestions by Hill and Drucker are discussed.

1 INTRODUCTION

The constitutive equation of plastic deformation should be defined by the incremental plastic strain and the associated stress, because the plastic deformation depends on the loading history. It is widely believed without the experimental verification that the increment of plastic strain is induced on the basis of the normal flow rule, which is suggested by R.Hill (1950) or D.C.Drucker (1951). The normal flow rule is characterized by that the produced plastic strain increment is directed to the outward normal direction on the yield curve. On the other hand, Prandle and Reuss assert that the deviatoric plastic strain increment is proportional to the current deviatoric stress, that is, the radial flow rule. However, no one pays attentions to Prandle-Reuss assertion recently. Moreover there are few experimental studies on the plastic flow rule.

We have attempted to investigate the plastic flow rule experimentally for a decade. The experimental study on Mises' material is trivial, because the normal direction cannot be distinguished from the radial direction on Mises' yield criterion,. It is known that annealed mild steel has a yield plateau, and within the yield plateau it behaves as an elastic and perfectly plastic material and also yields in accordance with Tresca's yield criterion. Hence, mild steel is adopted in our experiments. In order to perform the experiment under the multi-axial loading path, a thin walled circular specimen is subjected to the axial force and the twisting moment simultaneously. The experimental results concerning the plastic flow rule are mentioned in this paper.

2 EXPERIMENTAL METHOD

Annealed mild steel, tested in the experiment, has a yield plateau, which spreads over the eight times length of the elastic limit strain. Annealed mild steel can be regarded as an elastic and perfectly plastic material and obeys Tresca's yield criterion within the yield plateau. Since a usual hardening material may transform the configuration of yield surface with the increasing plastic strain because of the strain hardening, the configuration of the current yield surface is

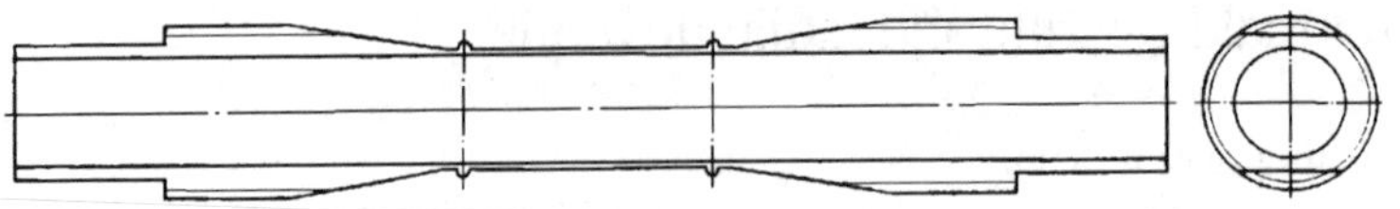

Outside diameter	Inside diameter	Gauge length
23 [mm]	20 [mm]	50 [mm]

Figure 1 Specimen

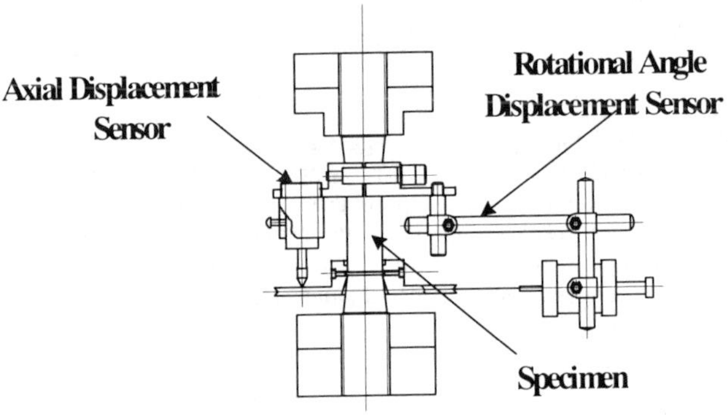

Figure 2 Experimental Apparatus

indeterminate. Hence the hardening material is improper for investigating the plastic flow rule. Therefore annealed mild steel is useful for the experimental verification of the plastic flow rule.

The originally designed experimental apparatus, which can apply tensile force and torsional moment simultaneously to the thin-walled circular specimen (Fig.1), is prepared for the experiment. Thus, the proceeding direction of a plastic strain increment is investigated experimentally under multi-axial loading paths. And the following three types of experiment are carried out.

1) Proportional loading, where the ratios of shearing stress to normal stress is kept constant through the entire loading path.
2) Tensile loading under a constant shearing stress.
3) General loading.

Plastic deformation on the yield plateau is not induced uniformly but irregularly along the specimen. A plastically deformed portion appears at random along the specimen and spreads over the specimen one by one. Hence, it is hard to measure the plastic strain by a strain gage exactly. In the experiment, the mean plastic strain over the specimen is measured by the displacement sensors directly attached to the specimen (Fig.2).

3 EXPERIMENTAL RESULT

The results of several experiments are shown in (Figs 3-5). The abscissa denotes the non-dimensional deviatoric tensile stress σ_X/σ_Y and strain $3G(\varepsilon_X - \sigma_X/(9K))/\sigma_Y$, and the ordinate denotes the non-dimensional shearing stress $\sqrt{3}\tau_{XY}/\sigma_Y$ and strain $\sqrt{3}G\gamma_{XY}/\sigma_Y$, where σ_Y is the yielding normal stress, and **G** and **K** denote the modulus of rigidity and the bulk modulus, respectively. The ellipse in the figures is the Tresca's yield curve, and the solid and dashed line designates the radial and normal direction, respectively.

Figure 3. shows the experimental result on the proportional loading. The strain varies in

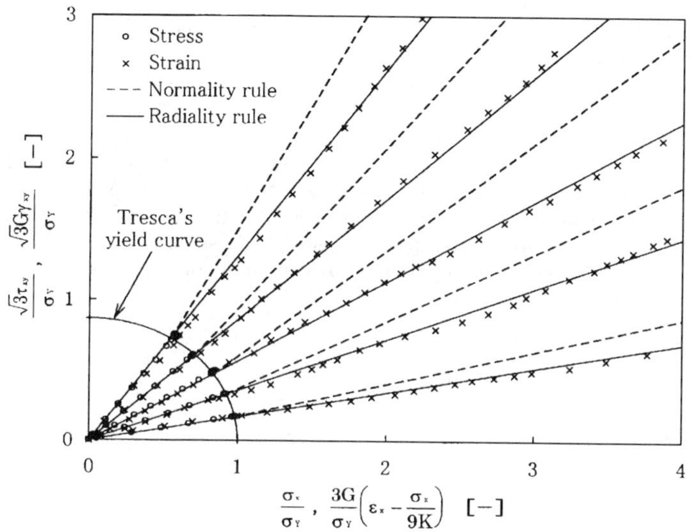

Figure 3 Proportional Loading

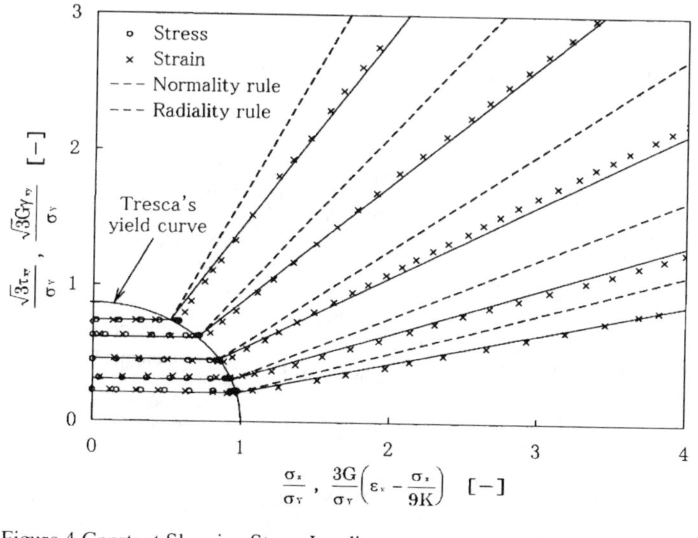

Figure 4 Constant Shearing Stress Loading

company with the stress within the elastic limit, that is, the inside region of the Tresca's yield curve. After the stress reaches the Tresca's yield curve, the plastic strain proceeds to the radial direction and the stress remains on the Tresca's yield curve. It is obvious from the figure that the actual plastic strain proceeds along the solid line, that is, the radial flow rule is reasonable rather than the normal flow rule. Figure 4. shows the deformational behavior by the tensile loading under a constant shearing stress. After the stress reaches the yield curve, the proceeding direction of the plastic strain changes certainly into the radial direction.

It should be noted that the stress remains on the yield curve without moving while the strain proceeds to the radial direction. Consequently, the deformation, after arriving at the yield curve, is pure plastic because of no variation of stress. It is imagined from Figure 3. and Figure 4. that the proceeding direction of the plastic strain increment depends on not the loading path but the current stress state on the yield curve.

Figure 5. shows the experimental result of stress variation (△) against the given arbitrary deformation (□) beyond the yield curve. It is recognized from the figure that the stress moves along the yield curve corresponding to the given deformation. Since the stress variation along the yield curve implies the elastic strain increment, the pure plastic strain increment can be obtained by subtracting the associate elastic strain increment from the given total strain increment.

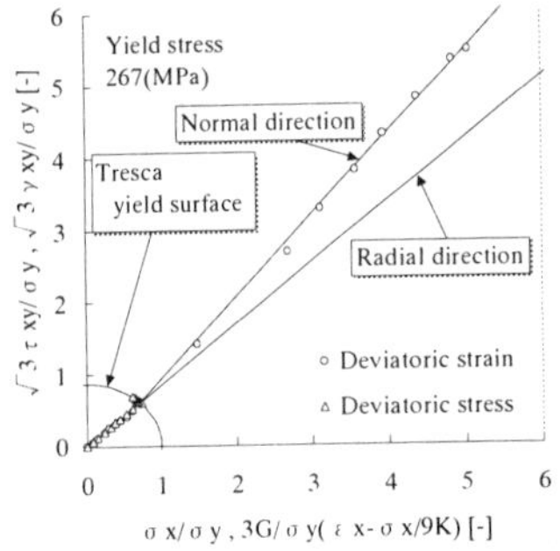

Figure 5 General Loading Path

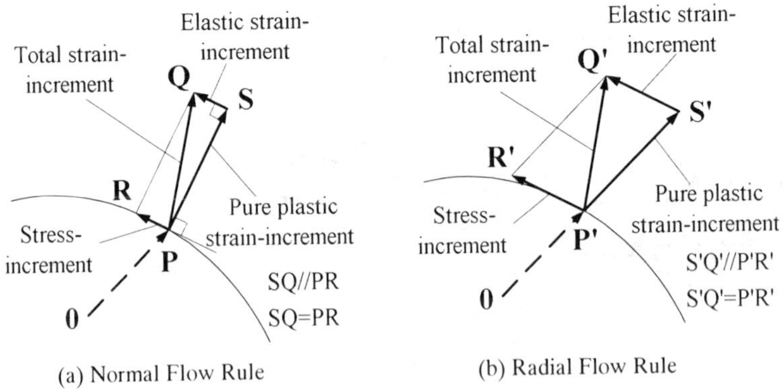

Figure 6 Strain Increment and Stress Increment

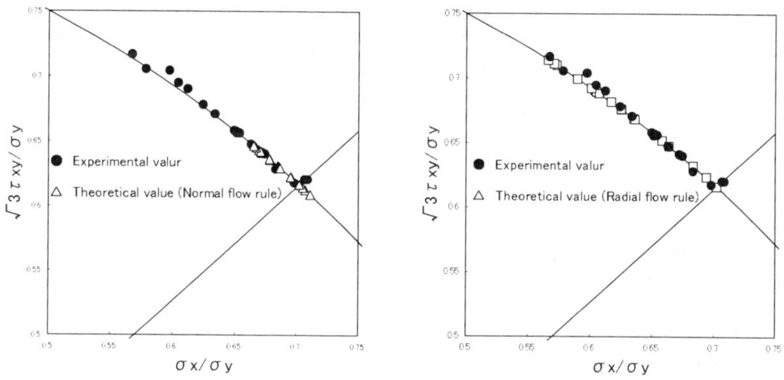

(a) Normal Flow Rule (b) Radial Flow Rule

Figure 7 Comparison of Experimental Value and Theoretical Value

Since the deviatoric elastic strain vector multiplied by 2**G** coincides with the deviatoric stress vector for an isotropic material, the stress increment is equal to the strain increment multiplied by 2**G**. Hence the total strain increment is resolved into the plastic strain increment and the pure elastic strain increment, which is parallel to the tangential direction on the yield curve. Accordingly, although the tangential component to the yield curve is always the elastic strain increment, the different plastic strain increment is resolved by the individual plastic flow rule. The methods of resolution are shown in Figure 6. The orthogonal resolution is applied to the normal flow rule (Fig.6(a)), on the contrary, the oblique resolution must be carried out for the radial flow rule (Fig.6(b)).

Figure 5. is expanded into Figure 7. in order to explain in detail. The experimentally measured stress variation is represented by "●" in Figure 7. The calculated stress variation in accordance with the normal flow rule is plotted by "△" in Figure 7(a), and also the stress variation resolved by the radial flow rule is shown by "◇" in Figure 7(b). It is obvious from the figures that the actual stress variation can be explained well by radial flow rule rather than the normal flow rule.

4 DISCUSSION

The experimental results obtained by our investigation seem to assert that the radial flow rule is reasonable rather than the normal flow rule. Here it is attempted to discuss logically the Hill's suggestion and the Drucker's postulation.

4.1 *Discussion on Hill's suggestion*

Hill suggests as follows: He introduces the concept of the plastic potential surface, which is normal to the increment of plastic strain. And he states that if the plastic potential surface coincides with the yield surface, the proceeding direction of the plastic strain increment is normal to the yield surface, and consequently, the variational principle and uniqueness theorem can be formulated. His suggestion is correct logically, but he has no idea to verify the inevitability on the coincidence of the both surfaces.

4.2 *Discussion on Drucker's postulation*

Drucker's Postulation is as follows: He affirms that the work hardening phenomenon is stable. Concerning uni-axial loading conditions, he mentions that the scalar product of the stress increment and the plastic strain increment must be positive for the stable phenomenon. He also states that "it is not possible to draw a simple picture to depict work-hardening for general paths of loading" (Drucker 1951). Next, he cites the loading cycle that the external agency is added to an initially stressed material and removed, and he states that "if plastic deformation has occurred in the cycle, the net work performed by the external agency over the complete cycle is positive". As a result of it, he suggests the normal flow rule as the plastic flow.

Let discuss the plastic deformation on an elastic and perfectly plastic material. Here the plastic deformation with no elastic strain increment is defined as the pure plastic deformation, and the deformation along the yield surface without plastic strain increment is defined as the neutral deformation. The general deformation on loading process is composed of the neutral (pure elastic) and the pure plastic deformation which dose not induce the stress increment. It should be noted that the neutral deformation is independent from the pure plastic deformation, because the neutral deformation never induces the plastic strain increment. The multiplication of the increment of pure plastic strain and the increment of elastic strain caused by the neutral deformation dose not imply the strain energy. For the hardening material, the pure plastic deformation produces the increment of stress by the strain hardening. However the stress variation caused by strain hardening should be distinguished from the stress variation by the neutral deformation, which is parallel to the tangential direction of the yield surface. Here the pure plastic deformation for hardening materials is defined such that the proceeding direction of plastic strain (not yet determined) coincides with that of the stress increment by the strain hardening. Hence the general plastic deformation is represented by the superposition of the pure plastic and the neutral deformation.

It is imagined that the Drucker's postulation lacks the concept of the pure plastic and the neutral deformation. The pure plastic deformation is actually stable, even if the plastic strain increment dose not proceed toward the normal direction, because the plastic strain increment has the same direction of the stress increment by strain hardening. It is no matter to say, the neutral deformation is always stable. It is incredible that the unstable phenomenon is realized by the superposition of two independent stable deformations, that is, the pure plastic deformation and the neutral deformation.

On the contrary, the radial flow rule is suggested by Prandle and Reuss. The proceeding direction of plastic strain increment coincides with the proceeding direction of stress increment on the radial flow rule.

5 CONCLUSION

The normal flow rule, which has been generally adopted as the plastic flow, is not reasonable because the experimental results give the evidence for the correctness of the radial flow. It is mentioned the incompleteness of Hill and Drucker's suggestion in this paper.

REFERENCE

Hill, R. 1950. The mathematical theory of plasticity. London: Oxford at the Clarendon Press

Drucker, D.C. 1951. Some implication of work hardening and ideal plasticity: *Quaterly of Applied Mathematics* no.4 vol.7: 411-418.

Mechanics of Structures and Materials, Bradford, Bridge & Foster (eds)

 ISBN 90 5809 107 4

A finite element cloning technique for transient flow problems in unbounded domains

N. Khalili, W. Wang & S. Valliappan
School of Civil and Environmental Engineering, The University of New South Wales, Sydney, N.S.W., Australia

ABSTRACT: The concept of cloning is applied to transient groundwater flow problems in infinite media. The property matrices of the infinite flow region are obtained using a two-cell cloning system. The accuracy and the efficiency of the proposed model is demonstrated through numerical modelling of a pumping well located at the surface of a half plane. The results show that the proposed finite-element cloning technique can be effectively used to simulate transient flow problems in infinite media.

1 INTRODUCTION

Groundwater flow problems often involve unbounded domains in one or more dimensions. The common engineering approach in numerical simulation of these problems is to truncate the remote boundary at a large distance from the zone of interest and then impose free or fixed boundary conditions. The imposition of such boundaries can lead to spurious solutions, particularly if the truncation occurs too close to the zone of interest.

In this paper, the concept of finite-element cloning, first introduced by Dasgupta (1982), is applied to groundwater flow in infinite media. In this approach an expression for the property matrix of the infinite media is obtained as a function of the property matrix of a cloning cell of finite elements. The approach is based on the assumption that adding a finite number of elements to an infinite domain will not change the property matrix of the domain. The main advantage of this approach is that it is algebraic in nature and makes no appeal to physical reasoning of the problem. Thus, it can be applied to a range of engineering problems, including those for which the fundamental solution does not exist in a closed form. In this paper, a two cell cloning approach is proposed to obtain the property matrices of transient flow problems in infinite media. The accuracy and the efficiency of the approach are demonstrated through a numerical example involving pumping well in 2D semi-infinite media. In general, it is shown that the finite-element cloning technique can be effectively used to simulate transient flow problems in infinite media.

2 FUNDAMENTALS OF THE CLONING ALGORITHM

2.1 *One cell cloning*

The fundamental idea of one-cell cloning is illustrated in Figure 1. In this figure, an unbounded domain with a characteristic length of r_i (interior boundary) is decomposed into a cell of finite elements and an unbounded medium with the characteristic length r_e (exterior boundary).

According to the finite element formulation, the total head-fluid flux relationship for the cell of finite elements can be written as,

$$\begin{bmatrix} [K_{ii}] & [K_{ie}] \\ [K_{ei}] & [K_{ee}] \end{bmatrix} \begin{Bmatrix} \{h_i\} \\ \{h_e\} \end{Bmatrix} = \begin{Bmatrix} \{q_i\} \\ \{q_e\} \end{Bmatrix} \tag{1}$$

where, $\{q\}$ and $\{h\}$ are the nodal flux and total head vectors; and $[K]$ is the property matrix of the cell composed of finite elements. Indices i and e refer to the interior and the exterior boundaries, respectively.

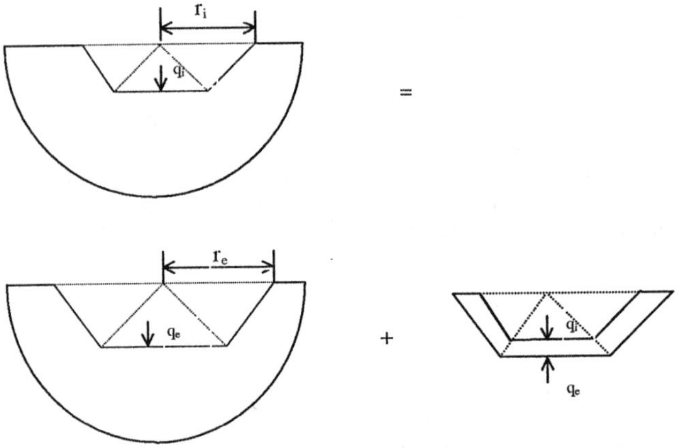

Figure 1: Fundamental concept of one-cell cloning

The matrix equations for the interior and exterior unbounded domains may be formulated as,

$$[K_i^{\infty}]\{h_i\} = \{q_i\} \tag{2}$$

$$[K_e^{\infty}]\{h_e\} = -\{q_e\} \tag{3}$$

Applying a set of independent total head vectors at the interior boundary, $[h_i]$, and eliminating $\{q_i\}$ and $\{q_e\}$ from equations (1) to (3) yield (Wolf and Song, 1994; Khalili et al, 1996),

$$[K_i^{\infty}][h_i] = [K_{ii}][h_i] + [K_{ie}][h_e] \tag{4}$$

$$-[K_e^{\infty}][h_e] = [K_{ei}][h_i] + [K_{ee}][h_e] \tag{5}$$

Setting $[h_i] = [I]$ (where $[I]$ is the unit matrix), and combining equations (4) and (5) yield,

$$[K_{ie}][h_e]^2 + [G][h_e] + [K_{ei}] = 0 \tag{6}$$

where,

$$[G]=[K_{ii}]+[K_{ee}]-[K_i^{\infty}]+[K_e^{\infty}] \tag{7}$$

According to the basic cloning assumption that adding a finite number of elements to an infinite domain does not change the property matrix of the domain, we have,

$$[K_i^{\infty}]=[K_e^{\infty}] \quad \text{and thus} \quad [G]=[K_{ii}]+[K_{ee}] \tag{8}$$

Now, the quadratic matrix equation (6) can be solved for $[h_e]$, using the eigenvalue decomposition technique[10]. Let λ_i and $\{\phi_i\}$ be an eigenvalue and eigenvector pair such that:

$$[h_e]\{\phi_i\}=\lambda_i\{\phi_i\} \tag{9}$$

Then, post-multiplication of (6) by $\{\phi_i\}$ yields the following eigenvalue problem,

$$\left(\lambda_i^2[K_{ie}]+\lambda_i[G]+[K_{ei}]\right)\{\phi_i\}=0 \tag{10}$$

It can be proven that if λ_i is an eigenvalue, $\mu_i=1/\lambda_i$ will also be an eigenvalue. Since $[G]$ is symmetric and $[K_{ei}]=[K_{ie}]^T$, the complete set of eigenvalues of (10) can be partitioned into two parts: $[\Lambda_1]$, with $|\lambda_i|\leq 1$; and $[\Lambda_2]$, with $|\lambda_i|\geq 1$. However, only $|\lambda_i|\leq 1$ should be chosen according to Sommerfeld condition. Having obtained the eigenvalues $[\Lambda_1]$ and the corresponding eigenvector matrix $[\Phi]$, $[h_e]$ can be obtained from equation (9) as,

$$[h_e]=[\Phi][\Lambda_1][\Phi]^{-1} \tag{11}$$

Substituting $[h_e]$ into (4), $[K_i^{\infty}]$ is obtained.

2.2 *Two cell cloning*

In one cell cloning, the property matrices of the interior and exterior boundaries are set equal $[K_i^{\infty}]=[K_e^{\infty}]$ or $[K_i^{\infty}]-[K_e^{\infty}]=0$. This of course is only valid in steady state problems, satisfying the conditions of similarity with respect to both the geometry and material properties of the unbounded domains. If the conditions of similarity are not satisfied or the governing equations involve dissipative terms such as the mass matrix, this assumption will lead to incorrect results, particularly in transient flow problems in which the long term response of the system does not approach a steady state condition.

To alleviate this, a two cell cloning approach, which can be applied to the general case of $[K_i^{\infty}]\neq[K_e^{\infty}]$, is proposed involving the following basic steps:

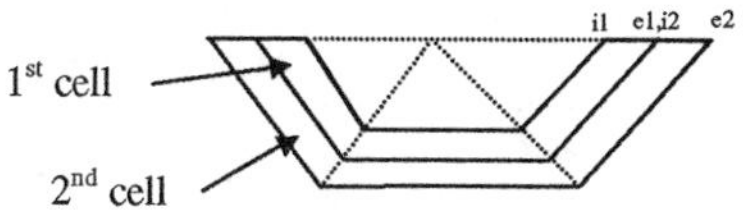

Figure 2: A two-cell cloning system

1. For the given dimensions of the geometry (Figure 2), calculate the property matrices $[K_{ii}]$, $[K_{ie}]$, $[K_{ei}]$, and $[K_{ee}]$ for the cloning cells 1 and 2.
2. Assuming $\left[K_i^{\infty}\right]-\left[K_e^{\infty}\right]=0$, solve the quadratic equation (6) for $\{h_e\}$ and obtain the property matrix $\left[K_i^{\infty}\right]$ at the interior boundaries of cells 1 and 2 using equation (4).
3. Obtain a revised estimate of $\left[K_{i1}^{\infty}\right]-\left[K_{e1}^{\infty}\right]$ for cell 1, taking $\left[K_{e1}^{\infty}\right]=\left[K_{i2}^{\infty}\right]$.
4. Repeat Step 2 for cell 1, using the revised $\left[K_{i1}^{\infty}\right]-\left[K_{e1}^{\infty}\right]$.

Notice that the above approach is based on the assumption that the error involved in calculating $\left[K_{i1}^{\infty}\right]$ and $\left[K_{i2}^{\infty}\right]$ in Step 2 will be of the same order of magnitude for cells 1 and 2. This is reasonable given the similarity of the cloning cells. The approach can be readily extended to three or four cloning cells, for further accuracy of the results. However, as will be shown later, a two-cell cloning system will provide numerical results of sufficient accuracy for most practical problems.

3 APPLICATION TO TRANSIENT FLOW

For the case of two dimensional transient flow, the governing differential equation is expressed as,

$$k_x \frac{\partial^2 h}{\partial x^2} + k_y \frac{\partial^2 h}{\partial y^2} = S \frac{\partial h}{\partial t} \tag{12}$$

in which, h is the total head; k_x and k_y are the permeabilities in the x and y directions; S is the storativity; and t is time. In the Laplace domain, equation (13) can be conveniently written as,

$$k_x \frac{\partial^2 \bar{h}}{\partial x^2} + k_y \frac{\partial^2 \bar{h}}{\partial y^2} = Ss\bar{h} \tag{13}$$

in which, $\bar{h}$ is the Laplace transform of h defined as,

$$\bar{h} = \int_0^{\infty} h e^{-st}\, dt \tag{14}$$

Using the Galerkin weighted residual approach, the discretised form of (12) can be written as,

$$[H]\{\bar{h}\} = \{\bar{q}\} \tag{15}$$

in which,

$$[H] = [K] + Ss[M] \tag{16}$$

where, $[K]$ is the property matrix, $[M]$ is mass the matrix and $\{\bar{q}\}$ is the flux vector in the Laplace domain.

According to (15), the total head-fluid flux relationship for the cell of finite elements can be written as,

$$\begin{bmatrix} [H_{ii}] & [H_{ie}] \\ [H_{ei}] & [H_{ee}] \end{bmatrix} \begin{Bmatrix} \{\bar{h}_i\} \\ \{\bar{h}_e\} \end{Bmatrix} = \begin{Bmatrix} \{\bar{q}_i\} \\ \{\bar{q}_e\} \end{Bmatrix} \tag{17}$$

Now, following the procedure described in Section 2, $\left[H_i^{\infty}\right]$ can determined. Having obtained the numerical results in the Laplace domain, it will then be a simple matter to convert them into time domain using a technique similar to that proposed by Talbot (1979).

To demonstrate the application of the technique a pumping well located at the surface of a half plane is considered, Figure 3.

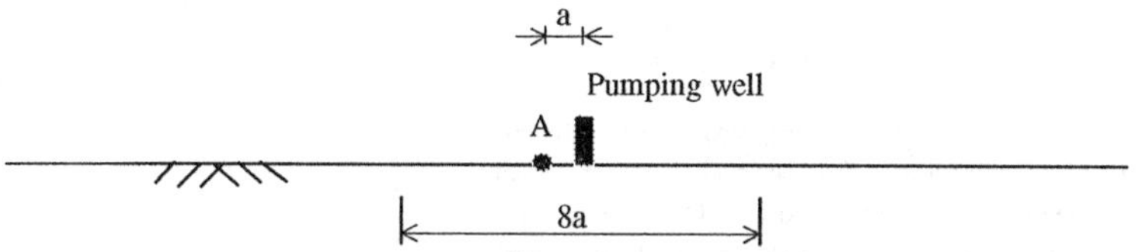

Figure 3: Pumping test problem in 2D semi-infinite medium

The finite element mesh used in the analysis is shown in Figure 4. The near field is modelled using 32 finite elements, and the infinite domain is modelled using two cells of finite elements each consisting of 16 elements. Both similar and non-similar cloning cells were used in the analysis. The problem was also analysed using conventional finite element technique by imposing arbitrarily terminated boundaries located at sufficiently large distances from the near field area of the system (4000a x 4000a). To investigate the effectiveness of the proposed infinite element, analyses were also performed using conventional finite elements with fixed or free total head conditions imposed at the truncated near field boundary. The results of the analyses in terms of dimensionless time (kt / Sa^2) versus dimensionless drawdown ($4\pi ks / Q$) at the observation point, A, are shown in Figure 5.

As can be observed from Figure 4, there is a very good agreement between the results obtained using the finite element two-cell cloning mesh and the large mesh. However, when the second cloning cell is removed or artificial free and fixed boundaries are introduced at the near field boundary, the accuracy of the numerical results is significantly affected. It is interesting to note that non-similar cloning cell also produces results similar to those of the cloning cell with similarity. This is significant and implies that satisfying the conditions of similarity may not be necessary in dealing with ground water flow problems in infinite media.

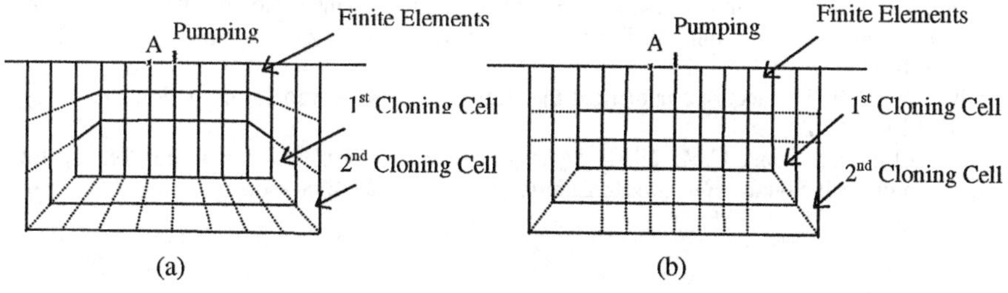

Figure 4: Finite element mesh and the cloning cell: a) with similarity, b) without similarity

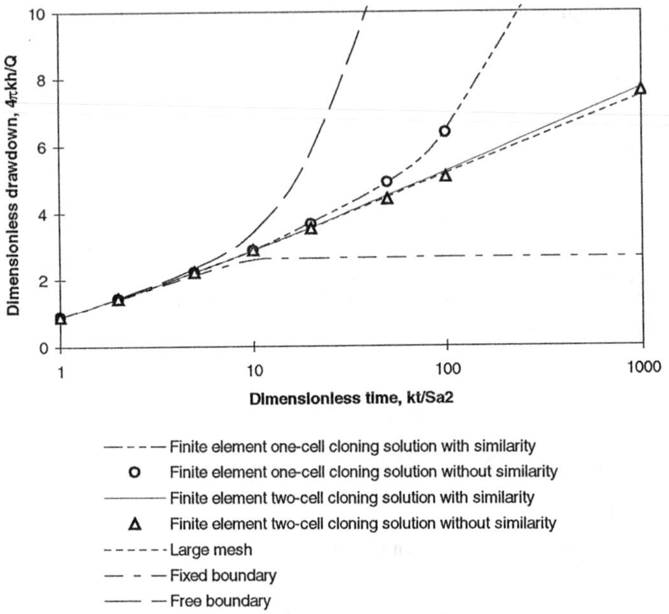

Figure 5: Dimensionless time versus dimensionless drawdown

4 CONCLUSIONS

In this paper, the finite-element cloning method has been used to simulate the groundwater flow problem in infinite media. The property matrices of the infinite flow region have been obtained using a two-cell cloning system for transient problems.

Through a numerical example involving pumping well in 2D semi-infinite media, it is shown that the finite-element cloning technique can be efficiently used to simulate ground flow problems in infinite media. It is also shown that, for the case considered, satisfying the conditions of similarity may not be necessary for ensuring the accuracy of the numerical results.

5 REFERENCES

Dasgupta, G.G. 1982. A finite element formulation for unbounded homogeneous continua. *J. Appl. Mech. ASME*, 49, 136-140.

Khalili, N., Wang, W. and Valliappan, S. 1996. A finite element cloning method for ground water flow problems in infinite media. *Proc. 3rd Asian-Pacific Conf. on Comp. Mech.,* Korea, 3, 1849-1857.

Talbot, A. 1979. The accurate numerical integration of Laplace transforms. *J. Ins. Math. Appl.* 23, 97-120.

Wolf, G.J.P. and Song, C.M. 1994. Dynamic stiffness matrix of unbounded soil by finite element multi-cell cloning. *Earthquake Engng. Struc. Dyn.*, 23, 233-250.

Mechanics of Structures and Materials, Bradford, Bridge & Foster (eds)
© 1999 Balkema, Rotterdam, ISBN 90 5809 107 4

Joint stiffness estimation of body structure using neural network

A.Okabe & N.Tomioka
College of Science and Technology, Nihon University, Tokyo, Japan

ABSTRACT: In this paper, an application of hierarchical neural networks to joint stiffness estimation of automobile body structure is described. We deal with simple joint structures, which are composed of thin walled box beams as typical models of actual body structure. Plate thickness, sectional dimension, length of partition, and position of flange, are considered as the design parameters on joint stiffness. The joint stiffness is expressed with a joint stiffness matrix. The design parameters and joint stiffness are used as input and output data for a neural network, respectively. The sample data of some design parameters vs. joint stiffness are calculated by the finite element method as the training data sets for hierarchical neural network. The error-back-propagation neural network is trained using the sample data. Finally, it was found that the values of joint stiffness could be estimated from the design parameters by the trained neural network.

1 INTRODUCTION

The automobile body structures consist of spot-welded thin-walled box beams with complex sectional contour. Most body connections are formed of overlapping sheet metals by spot welds and of necessity are not rigid perfectly. Their flexible characteristic (called joint stiffness after this) contributes one of the key factors dominating the following basic characteristics, a strength, a vibration, a collision, and a stiffness of a whole body structure of the automobile. And in case of carrying out structural analysis of automobile body structure by a finite element method, it is well known that we have to introduce the effect of the joint stiffness into modeling for the analysis. Thus it must be required to estimate the value of joint stiffness quantitatively in the stage of designing body structure.

However, in general, it is hard to make estimation of joint stiffness quantitatively. Because the jointed parts have many spot welded parts, service holes and box beams with complex contour. If we will try to obtain the value of joint stiffness as exactly as possible, the detailed analyses by both experimental method and finite element method would have to be carried out. Hard labor and a lot of times to do the analyses may be required. If the quick and accurate method giving the value of the joint stiffness is developed, it must be useful in designing body structure (Okabe et al. 1998).

In present paper, we proposed the new method, which estimates the value of joint stiffness to the design parameters using hierarchical neural network. A set of design parameters should determine one joint structure. We deal with L-shape jointed beam as simple joint structures. The joint stiffness is expressed with a joint stiffness matrix. And we consider the sectional dimension as the design parameters.

2 MODEL OF JOINTED BOX BEAMS AND DEFINING THE JOINT STIFFNESS

In this study, jointed part uniting two beams as shown in Figure 1 is employed as the typical models simplified the actual body connections. The joint stiffness of jointed part was defined as follows. The number of beams combined in jointed box beams is 2, and the longitudinal direction of beam, the sectional dimension and the direction of principal axis are assumed to be decided arbitrarily. It is assumed that the elastic deformation of the jointed part is allowed only against moments about three orthogonal axes.

On the above assumptions, the jointed part may be considered the elastic body consisted of the 2 nodal points with same coordinates. The relationships between moments $[\mathbb{M}]$ applied to 2 nodal points and rotations $[\Theta]$ can be represented by the stiffness matrix and it defines the joint stiffness. It is here called the joint stiffness matrix (Shimomaki et al. 1990), and can be represented by following equation (1).

$$\begin{bmatrix}\mathbb{M}_1\\ \mathbb{M}_2\end{bmatrix}=\begin{bmatrix}\mathbb{K}_{11} & \mathbb{K}_{12}\\ \mathbb{K}_{21} & \mathbb{K}_{22}\end{bmatrix}\begin{bmatrix}\Theta_1\\ \Theta_2\end{bmatrix} \tag{1}$$

$$[\mathbb{M}_i]=[M_{ix} \quad M_{iy} \quad M_{iz}]^{\mathrm{T}}\ (i=1,2),\quad [\Theta_i]=[\theta_{ix} \quad \theta_{iy} \quad \theta_{iz}]^{\mathrm{T}}\ (i=1,2)$$

where T = transpose of a matrix, $[\mathbb{K}_{ij}]$ = square matrix of order 3, (i , j = 1, 2) (Nm/rad.), $[\mathbb{M}_i]$, $[\Theta_i]$ = moment vector applied to section i and the corresponding rotation vector, (i = 1, 2), $[M_{ik}]$, $[\theta_{ik}]$ = component vectors of $[M_i]$ and $[\theta_i]$, (i = 1, 2, k = x, y, z).

If the jointed part is rotated rigidly ($\Theta_1=\Theta_2\neq 0$), moment is zero. Thus the following relationship exists.

$$\mathbb{K}_{i1}+\mathbb{K}_{i2}=\mathbb{O}\ (i=1,2)$$

And the joint stiffness matrix in equation (1) is symmetric by reciprocal theorem.

$$[\mathbb{K}_{ij}]=[\mathbb{K}_{ji}]^{\mathrm{T}}\ (i,j=1,2,\ \mathrm{T}:\text{ transpose of a matrix})$$

Therefore the number of independent components is one in the four components of the matrix of equation (1). If the matrix $[\mathbb{K}_{22}]$ is obtained, the other partial joint stiffness matrices are as follows.

$$[\mathbb{K}_{11}]=-[\mathbb{K}_{12}]=-[\mathbb{K}_{21}]^{\mathrm{T}}=[\mathbb{K}_{22}]^{\mathrm{T}} \tag{2}$$

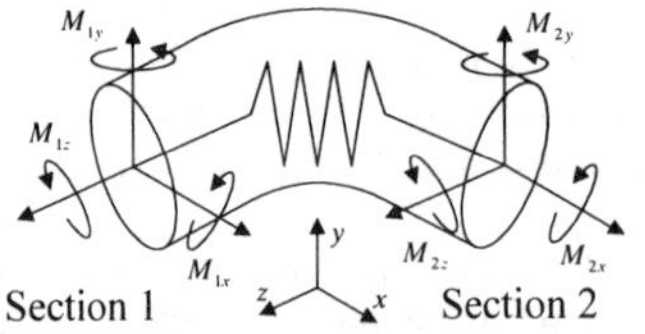

Figure 1. Jointed part.

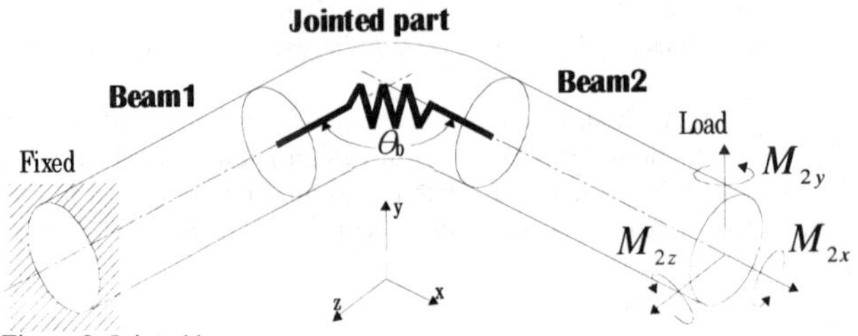

Figure 2. Jointed beam.

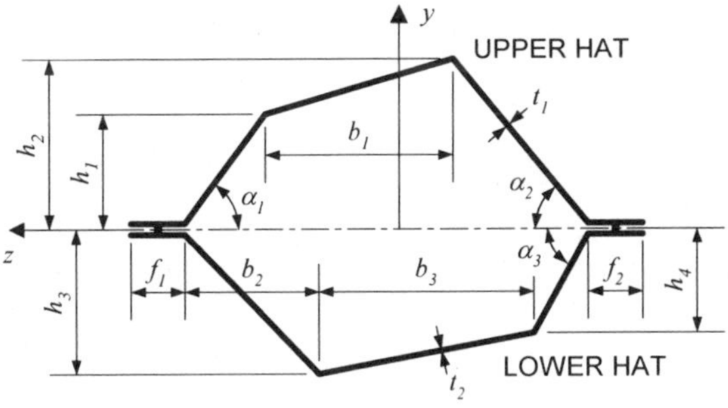

the design parameters	
$h_1\ h_2\ h_3\ h_4$	4
$b_1\ b_2\ b_3$	3
$f_1 f_2$	2
$\alpha_1\ \alpha_2\ \alpha_3$	3
$t_1\ t_2$	2
	14

Figure 3. The section scale (the design parameter).

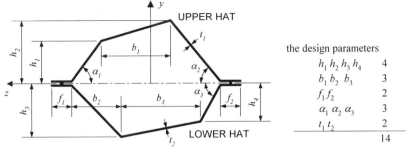

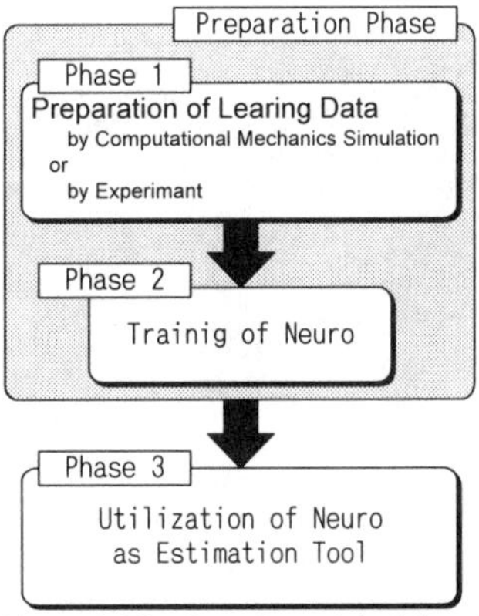

Figure 4. The flow chart of the joint stiffness estimation method.

Table 1. The values of the design parameters.

Parameters	Length (mm)	Step (mm)
b_1	40~60	
b_2	5~20	
b_3	45~55	5
h_1	10~40	
h_2	20~40	
h_3	15~35	

The jointed beam with the jointed part as shown in Figure 1 is considered (Figure 2). About the part except jointed part (called the general part after this), the shape of its section is identical with the end section of jointed part, the length of its part is more 3 times of the maximum sectional dimension of the end part of jointed part. About the rectangular coordinate axis, y-axis and z-axis are taken in the directions of the sectional principal axis of the externally loaded beam (beam2), x-axis is taken in the direction of the length of beam2. Thus $\left[\mathbb{K}_{22}\right]$ in the equation (1) corresponds to the joint stiffness of jointed beam as shown in Figure 2.

3 DESIGN PARAMETER

The physical design parameters must be independent and form a complete set, i.e., only one joint design should correspond to a set of values of these parameters. There are many ways to select physical design parameters. However, a good set of parameters should include only important parameters, i.e., parameters affect significantly the joint performance (Nikolaidis et al. 1992).

In this paper, the design parameters are considered only sectional dimensions. Figure 3 show the end of section 2 in Figure 1. This section is composed of two hat members jointed by the spot welding. The jointed part has the uniform section as shown in Figure 3. Thus, about the jointed part as shown in Figure 1, the structure of jointed part is determined by 14 independent physical parameters as shown in Figure 3.

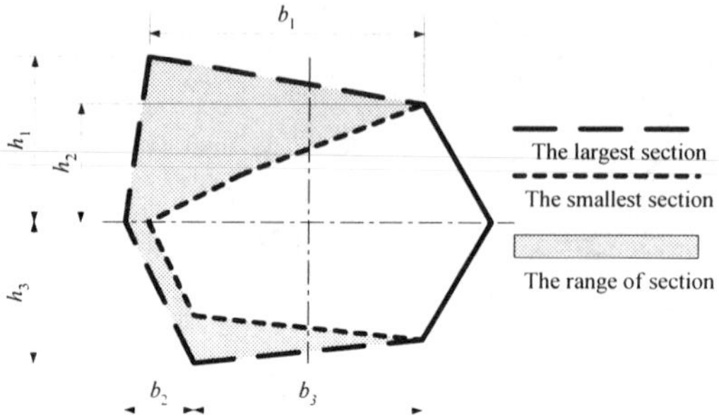

Figure 5. The range of the jointed part section.

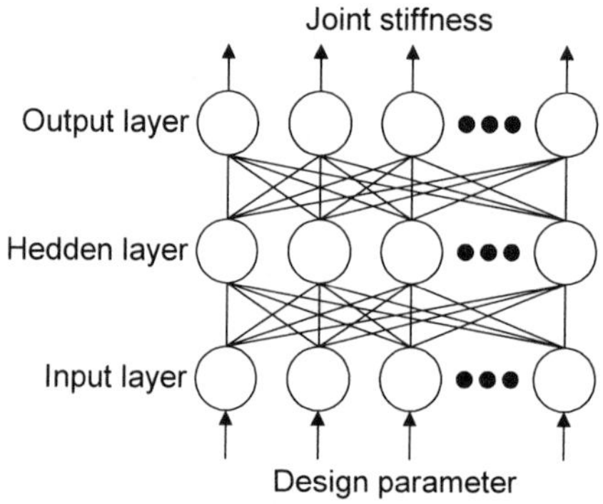

Figure 6. Neural network model.

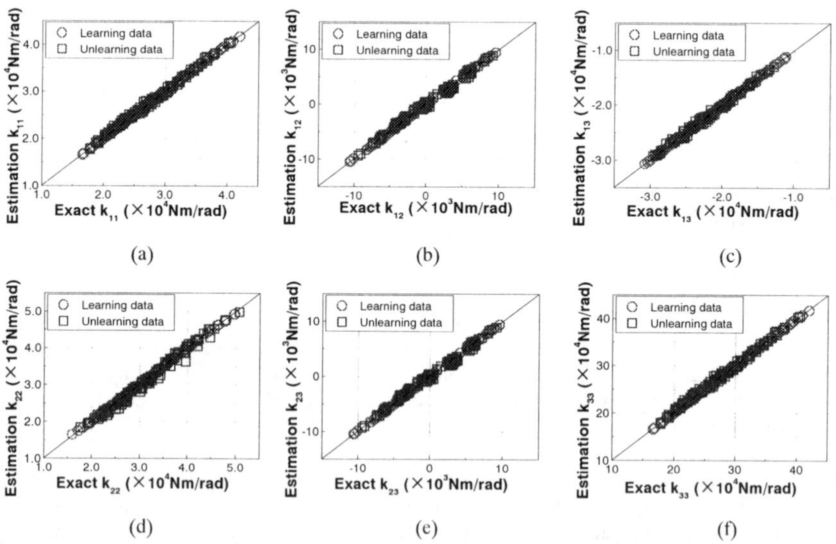

Figure 7. Comparison of estimated data and exacts.

4 JOINT STIFFNESS ESTIMATION

The basic idea of the method that we propose in this paper shows in Figure 4. Each piecewise relation between a set of the design parameters and a set of the joint stiffness is called here "learning data". At first, some learning data are prepared by computational mechanics simulation or by experiment. Next, the neural network is trained using these some learning data. Finally, if a set of the design parameters is given to the trained neural network, it may estimate a set of the joint stiffness.

We chose h_1, h_2, h_3, b_1, b_2, b_3, which are greatly influenced the joint stiffness, as the design parameters (Figure 5), and the other parameters were constant. And the two flange width f_1, f_2 are zero.

The values of the design parameters show Table 1. The range of the jointed part section shows in Figure 5. The number of learning data is 108. The sample data sets are prepared to examine the general ability of trained neural network. Here, the sample data sets are called "unlearning data". The number of unlearning data is 392.

In this paper, the hierarchical neural network model as shown in Figure 6 was employed, the study algorithm is the back-propagation algorithm (D.E.Rumelhart et al. 1986). The input data were the design parameters h_1, h_2, h_3, b_1, b_2, b_3. The output data were the values of the independent elements in the joint stiffness matrix. To decide how many units are in hidden layer, we tried to perform numerical calculations by changing the number of units in hidden layer. As a result, 20 were chosen as the number of units in hidden layer.

Figure 7 shows the some comparison between value of the joint stiffness matrix estimated by trained neural network and the exact. The training frequency was 100,000. As shown in Figure 7, the estimated values of not only the data used training but also not used were in good agree-

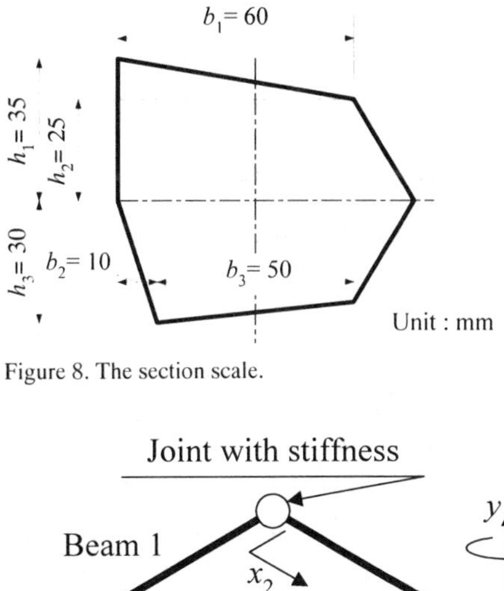

Figure 8. The section scale.

Table 2. Values of estimated data and exact data.

	Exact (Nm/rad)	Estimation (Nm/rad)
k_{11}	3.923×10^4	3.947×10^4
k_{12}	-7.440×10^3	-7.525×10^3
k_{13}	-3.019×10^4	-3.024×10^4
k_{22}	4.625×10^4	4.593×10^4
k_{23}	-7.440×10^3	-7.526×10^3
k_{33}	3.922×10^4	3.947×10^4

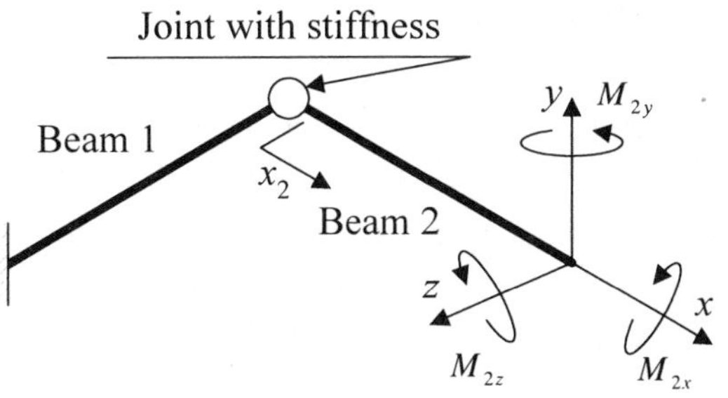

Figure 9. The model to obtain deformation by joint stiffness matrix.

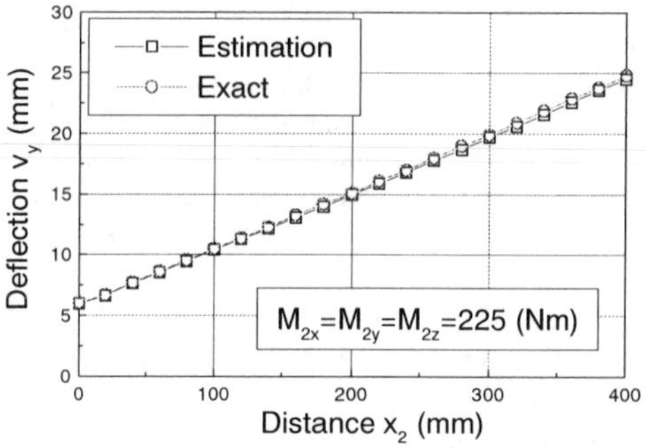

Figure 10. The distribution of deflection of beam 2.

ment with the exact values. It was founded that the relation between the design parameter and the joint stiffness could be constructed well.

About the jointed part with the section as shown in Figure 8, the estimated values of the joint stiffness matrix by using the trained neural network show Table 2. The estimated values were in good agreement with the exact values.

When these values of the joint stiffness are given, the deflection of each beam in the jointed box beam can be calculated as follows. The L-shape model as shown in figure 2 may be considered the structure that two beam (beam 1, beam 2) are jointed by a nodal joint with flexible rigidity defined by equation (1), as shown in figure 9

When the end of beam 2 with the section shown in Figure 8 was loaded $M_{2x} = M_{2y} = M_{2z}$ =225 (Nm) as shown in Figure 9, the distribution of deflection of beam 2 is shown in Figure 10.

5 CONCLUSIONS

The new method to estimate the joint stiffness by using the trained neural network was proposed. The conclusions obtained from this work were as following.

It is found that the relation between the joint stiffness and the design parameter can be constructed well by using a hierarchical neural network.

REFERENCES

Okabe,A. Tmioka,N. 1998. Joint stiffness estimation of body structure using neural network - jointed part composed of 2 beams -. *Modeling and simulation based engineering vol. 1*, 588-593

Shimomaki,K. 1990. Joint stiffness of body structure —part 1 its evaluation method—. *Transactions of JSAE No.43*, 138-142 (Japanese)

Nikolaidis,E. Zhu,M. Jasuja,S.C. 1992. Translators for design guidance of joints in automotive structures. *FISITA Inter national Congress, Total Vehicle Dynamics, vol.2, Technical papers, No.925236*, 179-187

Rumelhart,D.E. McClelland,J.L. and the PDP Research Group, 1986. *Parallel Distributed Processing*, MIT Press

Advances in design and construction methodologies

Mechanics of Structures and Materials, Bradford, Bridge & Foster (eds)
© 1999 Balkema, Rotterdam, ISBN 90 5809 107 4

A successful concrete repair work for clarifiers in Bandar Imam water supply project

M. Damghani
Department of Civil Engineering, Tehran-Boston Consulting Engineers, Iran

ABSTRACT: In 1986, various parts of a water supply plant in the Iranian town of Bandar-Imam were tested before their utilisation was started. The plant had been designed to deliver 450,000 cubic meters of water per day. The utilisation of the plant had, however, to be halted as the water leakage from the walls and bases of clarifiers No. 4 and No. 5 were found to exceed the permissible limit. The walls were on the brink of overturning, and the water stop tapes in the expansion joints were split. This paper reports on the investigation of the causes of the failure and on the successful repair operation.

1 INTRODUCTION

In 1986, various parts of a water supply plant in the Iranian town of Bandar-Imam were tested before their utilisation was started. The plant had been designed to deliver 450,000 cubic meters of water per day. The utilisation of the plant had, however, to be halted as the water leakage from the walls and bases of clarifiers No. 4 and No. 5 were found to exceed the permissible limit. The walls were on the brink of overturning, and the water stop tapes in the expansion joints were split. The water outlet of the plant was reduced to 40% of the target value, which the plant had been designed for. The clarifiers were cylindrical with a cross sectional diameter of 54 meter each. Their 5 meter-high walls were composed of 20 separate section. Some of these sections were tilted causing the failure of the scraper.

Heading an investigation team, the authors of this paper were assigned to find the cause(s) of the problem and specify a method for repairing and waterproofing the clarifiers. At first, all design and construction specification and documents were studied. The investigation showed that the problem was not due to a faulty design but to improperly fixing the water stop tapes and hastily filling the units with water.

Kot-Amir water treatment plant is located beside Karon River, near the city of Ahwaz in the south of Iran. The ground of the plant is mostly silty-clay and underground water table is high. Hence, for preliminary consolidation, a four-metre layer of soil was first filled. This layer was later replaced with a well-graded and compacted layer of soil of 1.8 metres thick.

Five turbo circular clarifiers stand with 54 meters inner diameter on the ground. During the utilisation of the plant differential settlement appeared in some wall segments of clarifiers No. 4 and No. 5, due to their water leakage. Thus, a study to find the causes and to design a repair method was initiated.

Today, it is 9 years after the successful completion of the repair work, and since then, there have been no problems in "“Bandar-Imam” water supply system.

Table 1. Solid mechanics properties.

Kind of soil	Depth sample (m)	Special gravity (T/m^3)	ϕ(0)	C (T/m^2)	E (T/m^2)	μ
Gravel with silty clay	-20<D<-10	1.9	32	0	2000	0.3
silty clay	-10<D<-5	1.9	20	2	1000	0.4
clay silty	-5<D<0	2.0	28	1	700	0.3
Filled layer	0<D<1.8	2.0	30	2	900	0.3

2 CONDITIONS AND DOCUMENTS

Each clarifier consists of 20 separate wall and base segments. They are joined to each other by expansion joints with water stop tapes and shown that each segment is stable against water, weight and earthquake loads. It resisted strongly against sliding and overturning in normal loading situation, by the shear key element, but it was necessary to investigate the conditions after water leakage and make a mathematical interaction model between the structure and soil.

For control the behaviour of soil under service load and new conditions, normal tests were applied to 20 meters depth. Soil mechanics parameters under the clarifiers have been determined by these test results (Table 1).

2.1 *Soil stability control*

A finite element plastic analysis program for axisymmetric models (PLAXIS) was used. Results have shown that only few points had plastic displacement. But, even these points did not reach to rapture limit under service loads. Due to this calculation the maximum settlements were: 8.7 cm in the centre of the reservoir 7.4 cm under its wall, and 8.5 cm in the front of its wall base. Also wall horizontal displacement was limited to 2.6 cm. Even, with applying of harsh conditions, results were hopeful.

2.2 *Structural stability*

The structure has been modified with a structural program for standard loading and steel concrete strength assumption (fy=280 MPa & fc = 25 MPa). The calculation results have shown that there were no problems in the structure members, but horizontal displacement will increase to a limit of 5.5 cm by the 2.5-cm soil settlement.

3 PRINCIPALS AND BASIC CONCEPT OF THE REPAIR METHOD

The results of the investigation have shown that it new similar accident does not occur, the cylindrical wall structure will be stable. So, it was necessary to control the horizontal displacement limit during the service life of the plant. Hence, the points below, were decided as the principals and basic concept of the repair method:

i. Design a structural system for the horizontal displacement limitation.
ii. Clarifiers should be returned to its circular shape with exact 54 metre inner diameter, which was necessary for mechanical equipment operation.
iii. Repair the reservoir watertight system.
iv. A concrete mix design, which should be compatible with concrete and also protection for new steel bars, were necessary.
v. A control system should be considered for measuring of the water leakage and settlement, during utilisation and operation period.
vi. Economical study of the method.

4 DESIGN AND DETAILS

Due to the mentioned studies, a repair method was designed for repairing the clarifiers:

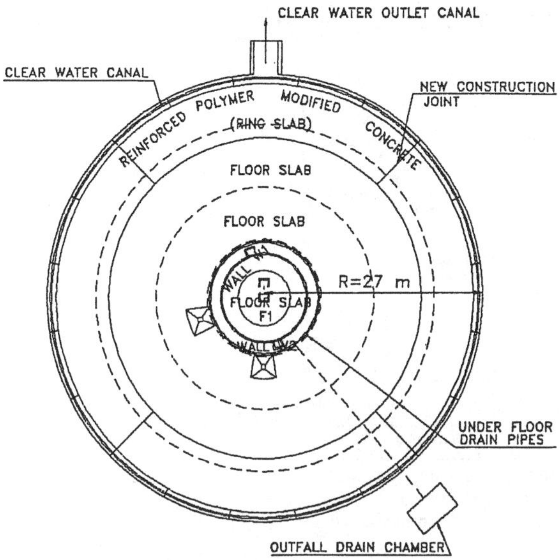

Figure1. Plan view of the clarifier.

(a) A reinforced concrete ring in clear water canal was designed for suffering hoop tension (refer Figure 2). This ring has 14 steel bars with 26mm diameter each and 3000 kg/cm^2 field point strength. The bars are located in polymer modified concrete layer to protect against corrosion.

The mix design of this concrete satisfies low permeability and low electrical conductivity, shown in Table 2.

(b) A new reinforced concrete layer has been designed with a width of 525 cm and 25-cm thickness as a ring in the structural system. It also protects repaired watertight system.

(c) A polymer modified concrete layer with a 10 cm thickness used in the other parts of the base floor to protect the watertight system.

(d) Every clarifier had a drainage system for collecting water leakage. This system consists of three circular raw and a radial pipe drainage which run to a out fall chamber. After repair, it was so important to do control the water leakage. Hence, it was programmed to check and repair this system, before any concreting.

Table 2 – New concrete layer mix properties

Materials	Quantities
Portland Cement (type V)	450 kg/m^3
Fine aggregates (natural sands)	900 kg/m^3
Course aggregates (max. size 6mm)	850 kg/m^3
Water	1485 kg/m^3
Polymer (styrene butadiene rubber)	80 kg/m^3

5 EXECUTION PROCEDURES

In this stage the flowing steps were undertaken:

(a) All the damaged expansion joints have been found by passing coloured water through joints.

(b) The drainage system have been checked and repaired.

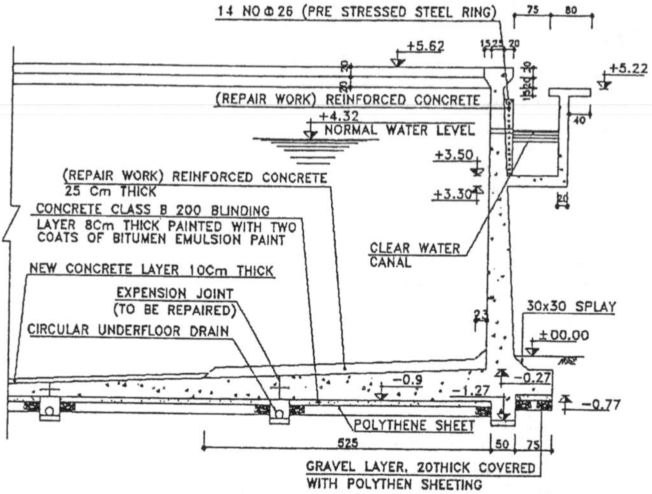

Figure 2. Typical section through the clarifier tank for remedial works.

(c) The steel bars of reinforced ring were prestressed during installation in the clear water canal. This was necessary for interaction between tow structural system. It was also important to remove some wall segments against their previous horizontal displacement and was done by a special tensile technology and measuring strain gauge and surveying operation.
(d) All the steel bars were coated with epoxy paint.
(e) Polymer modified concrete were used in rings and base floor.
(f) The clarifiers were tested with water in five stages and measuring settlements and water leakage in each stage.

6 RESULTS & RECOMMENDATIONS

It is 9 years after successful completion of the repair work, and since then, there have been no water leakage or any structural problem in the clarifiers. Also any durability problem due to cracks or steel bar corrosion have not been reported.

The following points could be recommended for repairing these systems:
(i) It is necessary to do soil and hydraulic tests for brief study and during design and execution.
(ii) Repair system should be a compatible structure with the previous system in materials and behaviour.
(iii) Training the execution crew in the different stages of repair is necessary.
(iv) After repair, test and commissioning should be done carefully and settlement and water leakage in each stage considered.

REFERENCES

L.A. Kuhlmann. 1990. Styrene – Butadine latex modified concrete. *ACI,* Cconcrete International.
Tehran – Boston Engineer's documents and reports. 1990-1995.
Choi, O.C., Hong, G.S., Hong Y.K. & Y.S. Shin. 1995. Proceedings of the International conference Consec, 95, Sapporo, Japan.
Tamai, M. & Yamaguchi, K. 1997. Proceedings of the second East Asia symposium on polymers in concrete, Japan.

Mechanics of Structures and Materials, Bradford, Bridge & Foster (eds)
© 1999 Balkema, Rotterdam, ISBN 90 5809 107 4

Performance of open girder timber floors

H.R. Milner & C.Y. Adam
Monash Engineering Timber Centre, Monash University, Melbourne, Vic., Australia

ABSTRACT: This paper is concerned with the static and dynamic performance of open girder timber floor systems. With this method of floor construction, flooring is laid directly on the open girders that, in turn, span directly between supports. It is traditional to design such floors under distributed and point/patch loading using two dimensional analysis/design software. For the former load, design is straightforward but for point and patch loading determination of lateral distribution factors presents problems. There also exists a problem with analysing the dynamic characteristics of open-girder floors. The evaluation of both static and dynamic response are addressed herein.

1 INTRODUCTION

"Open girder" is the term applied in this paper to parallel chord, truss-like floor support members; see Figure 1. Such members have considerable longitudinal strength and stiffness and therefore directly support flooring. This contrasts with more conventional timber floor construction that involves a two-layer support system of bearers and joists.

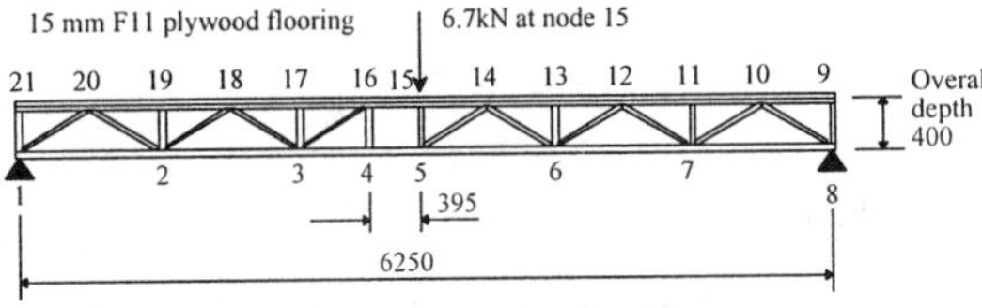

Figure 1 Typical open girder floor system directly supporting the flooring.

In the design of such systems, it is the analysis of patch patch loads that cause the most difficulty. Industry uses two dimensional analysis/design software that relies on the use of a lateral distribution factors known in AS1720.1 as g_{41} and g_{42} (point loads) or g_{43} (patch loads). The expressions, were developed for a system of n_C joists, see Figure 2, crossing a system of bearers. The relevant expressions are provided in Table 1.

Table 1 AS1720.1 expressions for computing grid lateral load distribution factors.

Applicability	AS1720.1 equation used to compute g value	
Point loads at mid-span	$g_{41} = 0.883 - 0.34\log(R + 0.44)$	(1)
Point loads anywhere within middle half of beam	$g_{42} = 0.2\log(1/R) + 0.95$	(2)
Patch loads of width s anywhere within middle half of beam	$g_{43} = 0.15\log(R) + 0.75$	(3)

where $R = \frac{E_B I_B}{L^3} \frac{s^3}{n_C E_C I_C}$, $E_B I_B$, $E_C I_C$ = bending stiffness of bearers and joists respectively, L, s = span and spacing of the bearers. For continuous flooring $R = \frac{E_B I_B}{L^4} \frac{s^3}{E_C I_C}$.

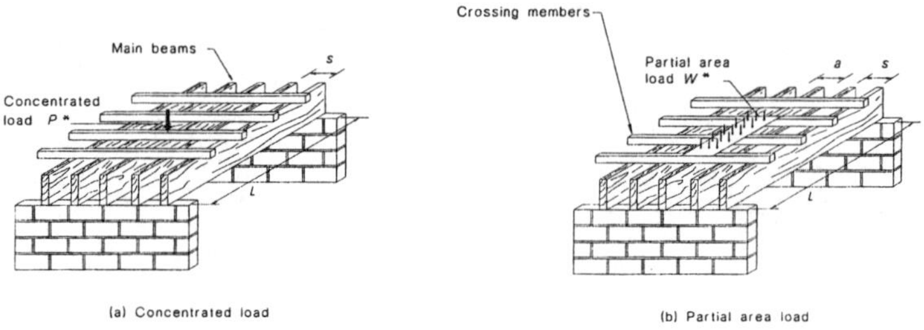

Figure 2 Diagram extracted from AS1720.1 showing definitions used in conjunction with Table 1 formulae.

In applying such formulas to lateral distribution of static point/patch loads in open girder floor systems, there is an immediate problem in that $E_B I_B$ is not a property normally associated with a truss. The situation is complicated by the fact that it is a common practice to provide a Vierendeel bay "for the plumber" and this can significantly affect deflection performance.

A further complication concerns floor dynamics. In a conventional floor, the most widely accepted method for evaluating floor performance is provided by Ohlsson (1984). His methodology is incorporated into Euro Code 5 and, in a modified form, in other standards. The Ohlsson criterion requires knowledge of the fundamental frequency, usually computed from an $E_B I_B$ value but it can also be obtained from finite element truss modeling with much inconvenience. It is highly desirable that an approximate design method be available to determine the fundamental frequency based on 2D analysis of the form

$$f_1 = \frac{\pi}{2}\sqrt{\frac{E_B I_B / s}{\rho L^4}} \tag{4}$$

2 OPEN GIRDER "BENDING" STIFFNESS

If standard formulas are to be used in conjunction with the lateral distribution factors of Table 1, it is clear that a methodology must be found to convert truss performance to an equivalent beam performance. Consider the typical parallel chord truss of Figure 1 with F17 Mountain Ash (*e regnans*) upper and lower chords of size 90 x 45 in the top ($E = 19800MPa$) and bottom ($E = 16600MPa$) chords. The webs details are shown in Figure 1, $E = 6900MPa$.

2.1 *Beam Approximation*

The simplest approximate analysis of a parallel chord, open girder as a beam would liken it to an I-beam having 90 x 45 flanges and a very thin web with infinite resistance to shear deformation. If the flanges were of size $b \times d$ and the overall member depth centre to centre of these members was D, then , assuming composite action with the flooring

$$(E_B I_B)_b = E_1 b_1 t_1^3/3 + E_2 b_2 t_2 (D - t_2/2)^2 + 12 + E_3 b_3 t_{3e} (D + t_2/2)^2 + EI_{0\,floor} - (EA)\bar{y}^2 \tag{5}$$

where the details are given in the Appendix. Under point load P applied as shown in Figure 23the deflection may be computed according to the usual beam bending theory by

$$\delta_{b,f} = \left(PL^3/48(E_B I_B)_b\right)\left(3a/L - 4a^3/L^3\right) = PL^3/48(E_B I_B)_b \quad \text{for mid-span loading} \tag{6a}$$

Equation 6a only allows for pure chord bending effects – no web shear deformation. Shear deformation is estimated at $0.00004L$ (L in *mm*) of the bending deformation leading to

$$\delta_{b,f+w} = (1 + 0.00004L)PL^3/48(E_B I_B) \tag{6b}$$

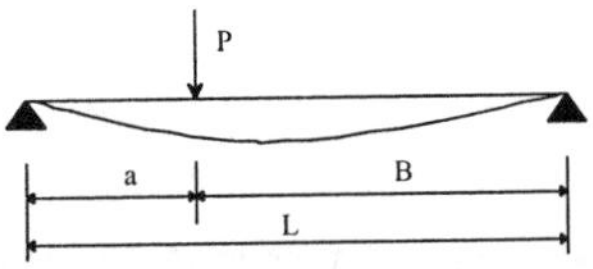

Figure 3 Beam deflection under point load

Where the member also contains a Vierendeel bay additional deformation arises. Suppose that the Vierendeel bay is of length L_1 and the adjacent back bays of length L_2. Let the shear be V. Assuming that a point of contraflexure occurs at mid-bay, Figure 4, the Vierendeel deflection contribution is given approximately by

$$\delta_{b,V} = \frac{V}{12(EI_f)}\left[L_1^3 + 2L_1^2 L_2\right] \tag{7}$$

where (EI_f) = sum of the bending stiffness of the upper and lower chord flanges about their own centroidal axes. The final deflection is given as

$$\delta_b = \delta_{b,f+w} + \delta_{b,V} \tag{8}$$

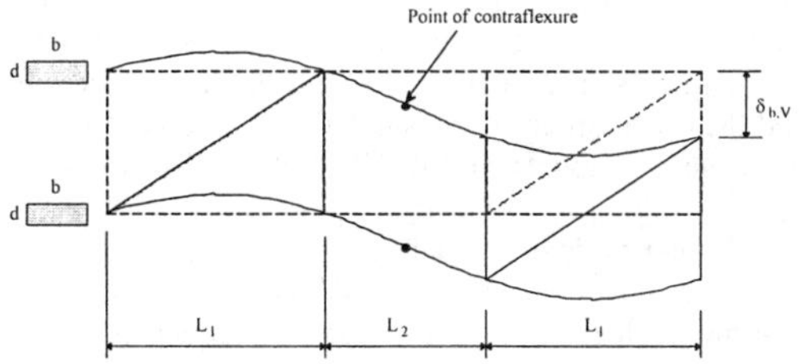

Figure 4 Model used to evaluate the effect of a Vierendeel bay.

2.2 *Rigid frame analysis*

The girder can also be analysed as a rigid jointed two-dimensional frame, current industry practice, with pin releases at the ends of all chord members with the following variants.

1. With deflection contributions from flange, web and Vierendeel bay deformation, ie, the usual analysis, $\Rightarrow \delta_{rf,f+w+V}$,

2. With deflection contributions from flange and Vierendeel bay deformations but with web shear deformations suppressed $\Rightarrow \delta_{rf,f+V}$
3. With deflection contributions from flange deformation only, ie, assign very high EA/L stiffness to the webs and use a diagonal member in the Vierendeel bay $\Rightarrow \delta_{rf,f}$

Using a rigid frame computer analysis and the value $\delta_{rf,f+w+V}$ derived therefrom, in conjunction with equation 6a, it is possible to compute an $(E_B I_B)_{rf}$ value using

$$(E_B I_B)_{rf} = \frac{PL^3}{48\delta_{rf,f+w+V}} \tag{9}$$

2.3 *Comparison of beam and rigid frame analyses*

The results of Table 2 were obtained for the case of a 6.7 *kN* load applied at node 15 for the truss of Figure 1. Both analyses can be taken with and without composite action (between the flooring and the top chord) included simply by adjusting the model parameters.

Table 2 Contributions to deflections from various deformation sources and comparison with beam approximation in which composite action is ignored, ie, $E_3 = 0$.

	Beam approximation				2D rigid frame analysis			
	$\delta_{b,f}$	$\delta_{b,f+w}$	$\delta_{b,V}$	$\delta_{b,f+w+V}$	$\delta_{rf,f}$	$\delta_{rf,f+w}$	$\delta_{rf,V}$	$\delta_{b,f+w+V}$
Non-composite	7.33	9.16	1.64	10.8	6.39	8.00	1.21	9.38
Composite	6.23	7.79	1.30	9.09	6.37	7.98	0.87	8.84

NOTE: $\delta_w = \delta_{f+w+V} - \delta_{f+V}$, $\delta_V = \delta_{f+V} - \delta_f$ etc.

3 THREE AND TWO DIMENSIONAL ANALYSES – COMPARISON WITH EXPERIMENTS

Finite element modeling and experimental observations are available for a floor of length 6.25 *m* and width 2.4 *m*; see Figure 5. The finite element model represented the open girders using beam elements for the upper and lower chords and truss elements for the web elements. The flooring was modeled using shell elements connected to the top chord by short beam elements. Altering the stiffness of these elements can be used to control the degree of composite action between the flooring and the open girder. The results given in Table 3 were obtained from the finite element and experimental models. In the 3D model full composite action was invoked.

The experimental model was tested with both a static dead load applied through a patch of 100 x 100 *mm* and dynamically using a calibrated hammer technique. With the dynamic test, the force input and acceleration response are both measured. The natural frequencies and modal damping ratios were then extracted using Fast Fourier transform techniques.

Table 3 Finite element and test model results. g_{42} values are based on 8.84 *mm* composite or 9.38 *mm* non-composite finite element model analysis.

	Test Results	3-D FEM	Composite 2D Model	Non-composite 2D Model
Fundamental Frequency (*Hz*)	20.5	20.4	23.8	22.1
Deflection all deformation	5.48	5.48	8.84	9.38
g_{42}	na	-	0.62	0.58

The agreement between the finite element modeling and the experimental results is remarkable and reflects the care that went into collecting the raw data. The modulus of elasticity of the chords of all trusses was measured prior to their being used to fabricate the truss and was thus known with considerable precision.

It is clear that the static performance of the open girders within the grid under patch/point loads differs from its performance when loaded in isolation. This is due mainly to the lateral distribution of load to adjacent members but is also caused by differences in flooring composite action behaviour. It was felt initially that 2D rigid frame mathematical models that took account of composite action would be more consistent with short-term loads and dynamic response but this is not borne out by the results. On the basis of results to date, lateral load distribution as predicted by a g_{42} style factor is consistent with the non-composite 2D model but the non-composite 2D model more closely predicts the fundamental frequency.

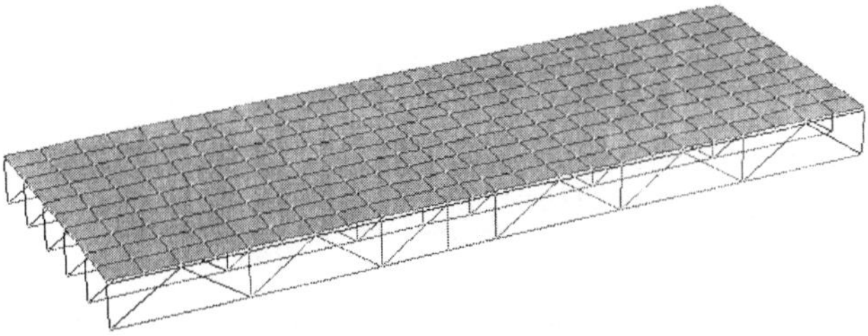

Figure 5 3D finite element model of open girder floor.

The determination of the lateral distribution factor for a point load can be based on the following procedure.

1. Compute the deflection using a two-dimensional, rigid frame analysis. $\Rightarrow \delta_{b+w+V}$
2. Compute $E_B I_B = \dfrac{PL^3}{48\delta_{rf,b+w+V}}\left(\dfrac{3a}{L} - \dfrac{4a^3}{L^3}\right) = \dfrac{PL^3}{48\delta_{b+w+V}}$ if $a = L/2$ (11)
3. Compute the g_{42} factor as given below and adjust the two dimensional deflections.

As an alternative, $\delta_{rf,b+w+V}$ of Step 1 can be replaced by the beam deflection $\delta_{b,b+w+V}$ given by equation 8 with composite action taken into account.

The recommended g_{42}, based on preliminary studies, is given by

$$g_{42} = 0.868 - 0.22\log(1.1/R + 0.25) \quad (12)$$

The dynamic analysis method proposed by Ohlsson (1984) relies heavily on knowledge of the fundamental frequency and damping ratios. Damping ratios appear to be of the order of 2% - 3% for open girder floor systems but the data is limited at the time of writing. Fundamental frequencies can be computed from finite element analysis directly or obtained from equation 4 with $E_B I_B$ = girder stiffness obtained from equation 11. As an alternative, $\delta_{rf,b+w+V}$ of Step 1 can be replaced by the beam deflection $\delta_{b,b+w+V}$ given equation 8 with composite action ignored. The floor mass is computed as indicated in Table A1 of the Appendix.

4 CONCLUSIONS

- The prospects for devising two-dimensional approximate analyses of floor grid systems that describe their response to static point/patch loads and dynamic excitation within a grid appear to be excellent.
- Where estimates of g_{42} are required these can be obtained from a two-dimensional frame or beam analysis using the procedures described in Section 5. If using the beam approximation, composite action should be taken into account.
- Where the fundamental frequency is being computed, either by rigid frame analysis or by the beam approximations, composite action should be ignored.

5 REFERENCES

Ohlsson, S 1984. Springiness and human-induced floor vibrations, *Swedish Council for Building Research*, Report BFR T20:1984

Euro Code 5 ENV 1995-1-1: 1993.

AS 3623 – 1993 Domestic metal framing.

AS 1684 – 1998 Draft Domestic construction – Performance criteria and verification.

Kalkert. RE, Dolan, JD, Woeste, FE 1993. The current status of analysis and design for annoying wooden floor vibrations, *Wood and Fibre Science*, 25(3):305-314.

Yang, J, Paevere, P, Pham, L 1993. Evaluating dynamic performance of Australian domestic floors, *13th Australasian Conference on the Mechanics of Structures and Materials*, University of Wollongong.

6 APPENDIX

$$(EA) = E_1 b_1 t_1 + E_2 b_2 t_2 + E_3 b_3 t_{3e}$$
$$(EAy) = E_1 b_1 t_1^2/2 + E_2 b_2 t_2 (D - t_2/2) + E_3 b_3 t_{3e} (D + t_2/2)$$
$$\bar{y} = (EAy)/(EA)$$
$$(EI) = E_1 b_1 t_1^3/3 + E_2 b_2 t_2 (D - t_2/2)^2 + 12 + E_3 b_3 t_{3e} (D + t_2/2)^2 + EI_{0\,floor} \_ (EA)\bar{y}^2$$

where E_1, E_2, E_3, b_1, b_2, b_3, t_1, t_2, t_3 = Young's modulus, width, thickness of lower chord, upper chord and flooring respectively, t_{3e} = effective thickness of flooring parallel to the open girder (eg with plywood only parallel plies), $I_{0\,floor}$ = second moment area of flooring about its own centroidal axis.

Table A1 Typical computation of flooring and open girder mass.

Top Chord	$0.09 \times 0.045 \times 650 \times 6.3$	$16.6kg$
Bottom Chord	$0.09 \times 0.045 \times 650 \times 6.3$	$16.6kg$
Web diagonals	$0.09 \times 0.045 \times 550 \times (11 \times 0.659)$	$16.1kg$
Web verticals	$0.09 \times 0.045 \times 550 \times (8 \times 0.355)$	$6.3kg$
	Total girder mass (kg)	$55.6kg$
	Girder mass / m^2 $= 55.6/(0.45 \times 6.3)$	$19.6\,kg/m^2$
Flooring (take as 15 mm plywood)	0.015×550	$8.3\,kg/m^2$
	Total floor mass / m^2	$27.9\,kg/m^2$

Mechanics of Structures and Materials, Bradford, Bridge & Foster (eds)
© 1999 Balkema, Rotterdam, ISBN 90 5809 107 4

Performance of particleboard walls as deep beams

J.A.Taylor & H.R.Milner
Department of Civil Engineering, Monash University, Melbourne, Vic., Australia

ABSTRACT: This paper describes the physical and mathematical modelling, of a 9 *m* span x 2.7 *m* deep wall beam constructed with particleboard and timber. In total, three models were developed. Because of laboratory space limitations the first was a truncation of the full size prototype wall. It was constructed to the design specification and successfully tested to 5 times design load. A second, one third scale, model was built to validate the first model and its test method. The deformations observed on these models were geometrically similar. Finally, a finite element model was used to provide further insight into the behaviour of the deep wall beam. The behaviour of all three models matched one another reasonably well.

1 INTRODUCTION

A study undertaken at Monash University is aimed at the development of a long span wall beam supporting roof and floor loads over a span of 9m. A prototype Vierendeel beam, 2.7*m* in overall height and spanning 9*m*, was designed to support a limit states permanent load of 10*kN/m* as specified in AS1170.1, SAA Loading Code. It consisted of three 1.8*m* wide particleboard diaphragms sandwiched between twin timber flanges with two 1.8*m* openings formed either side of the central diaphragm. According to the provisions of AS1720.1, Appendix D, a prototype test load is given by equation D2 in the form

$$Q_E = \frac{k_2 k_{26} k_{27} k_{28}}{k_1} Q^* \tag{1}$$

where $k_2 = 1$, k_{26}, k_{27}, k_{28} are statistical parameters, and k_1 is a duration of load factor not available in AS1720.1 for particleboard. Following an extensive literature search it was determined that the short term test load corresponding to a strength limits state permanent load of 10*kN/m* was 50*kN/m*. The critical reference is due to Hunt (1976).

Because a test involving a 9*m* long wall subjected to a total load of 45*t* could not be accommodated in the laboratory, modelling became necessary. The first model, see Fig 1, was a truncated, full size 2.7*m* long section of the prototype subjected to the heaviest shear force, ie, the section adjacent to a support. The test load was resisted and the model recovered fully, but it was considered prudent to validate the results. The method of loading the model, while correctly dividing the shear equally between the two flanges crossing the open panel, failed to apply any axial forces to them. Any attempt to do so would have posed considerable difficulty in ensuring that they remained truly axial with the flanges rotating and deflecting appreciably under the high loading applied during the test. Accordingly, a second model of one-third scale, $3m$ long by $0.9m$ high was built and tested. The deformations obtained from this model, when scaled up, agreed very closely with those observed in the truncated model. With a view to structural design and to a better understanding of its structural behaviour, a finite element of the wall was also developed. The results obtained from all three models were sufficiently consistent to conclude that each could represent a full size wall beam.

2 THE PROTOTYPE

The prototype wall beam, designed in accordance with AS1720.1, consisted of a pair of 290 × 45 MGP 12 machine stress graded Radiata pine members that formed each top and bottom flange and two 900mm wide × 19mm thick flooring grade particleboard panels sandwiched between them to make each of the three $1.8m$ wide diaphragms. Stiffeners were also made from the same timber. All timber to particleboard connections were effected with pneumatically driven $\phi 3.1mm$ diameter wire-drawn nails. Half their number were driven through the near flange to pass through the particleboard and penetrate the far member of the pair so that they acted in double shear, the other half were driven similarly from the opposite side. The joint between the flange and the diaphragm edge stiffeners was reinforced at each diaphragm corner with a pair of steel plates nailed to the face of the timber with $\phi 2.8mm$ diameter nails in single shear.

Characteristic properties of the materials employed are given in Table 1.

Table 1 Characteristic properties used to design the prototype wall beam.

Flanges MGP 12	Particleboard Flooring Grade	Nails – Wire Drawn	Steel Reinforcing Nail-on plates
$f_t^{'} = 15MPa$	$f_t^{'} = 9MPa$	$2280N/mm$ for $\phi 3.1$	$1.2mm$ x $75mm$
$f_s^{'} = 6.5MPa$	$f_s^{'} = 9MPa$	$2040N/mm$ for $\phi 2.8$	
$f_c^{'} = 29MPa$	$f_b^{'} = 20MPa$		
$E = 12700MPa$	$E = 4000MPa$		$E = 200{,}000MPa$

3 TRUNCATED FULL SIZE MODEL

A $2.7m$ long full height section of the prototype described above was fabricated in the field, transported to the structures laboratory, and mounted horizontally on a reaction frame

Figure 1 Truncated full size model.

Figure 2 One third scale model.

secured by friction grip bolting to anchorages in the strong floor; see Figure 1. Perimeter deformations under load were recorded on a data logger.

4 ONE THIRD SCALE MODEL

To validate the truncated full-size model, a 3*m* long, 0.9*m* high model was made in the laboratory; see Figure 2. The pair of flange members was made from the same MGP 12 timber with each of the pair reduced in width to 18*mm* and in depth to 125*mm* so that the combined stiffness was one third that of full-size. Particleboard one-third the thickness of the prototype was not available, and hardboard that had a similar Young's modulus and Poisson's ratio as the prototype was substituted. After scaling down, the model's diaphragms were considered to represent the buckling tendency and strength of the prototype.

The flanges were connected to the diaphragms by $\phi 2$ *mm* nails spaced uniformly to develop a stiffness per unit length of connection that proportionally matched the prototype's nailing stiffness. The reinforcement at the diaphragm corners was also connected with $\phi 2$ *mm* nails. The model was subjected to 5 times design load and subsequently recovered fully. Perimeter deformations and loads were again recorded on a data logger.

5 THE FINITE ELEMENT MODEL

A finite element model of the one-third scale wall beam was assembled using the 1998 version of the STRAND 6.17 computer program. A 25*mm* mesh was regarded as sufficiently fine to provide the information sought and required the face dimensions of the timber components in the plane of the stress plates to be modified to fit 25*mm* multiples. This, in turn, required adjustments to the thickness of the plate elements so that they matched the stiffness of the respective model components.

Three plane stress plates, each subdivided with a 25*mm* square mesh, were separately formed to represent the wall beam. One was fitted to the particleboard web, another to the timber flanges and web stiffening elements, and a third to the steel reinforcing located on the timber surfaces at each outer diaphragm corner; see Figure 3. The plates of the timber elements were laid over the web elements and displaced 5*mm* horizontally and vertically from the latter. These plates were then connected to each other with truss elements at their common nodes to transfer the shear between the two plates. The plate elements representing the steel reinforcement were also displaced 5*mm* each way from the underlying timber plates to facilitate similar connections; see Figure 4.

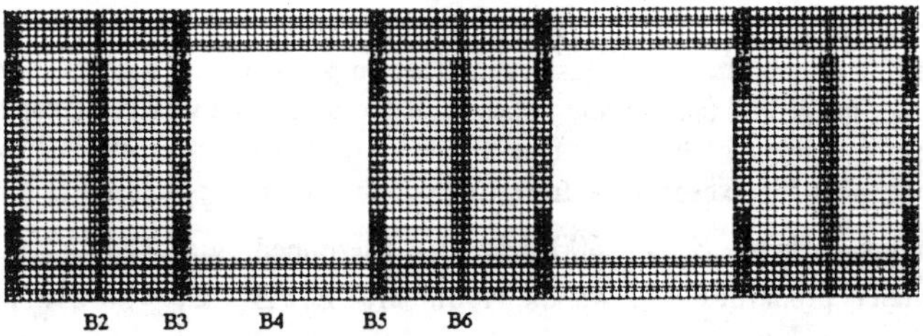

Figure 3 Finite element model

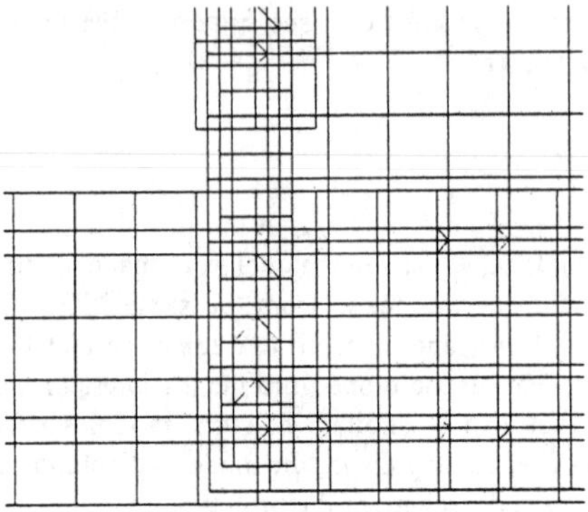

Figure 4 Finite element model showing the overlay pattern and nail representation.

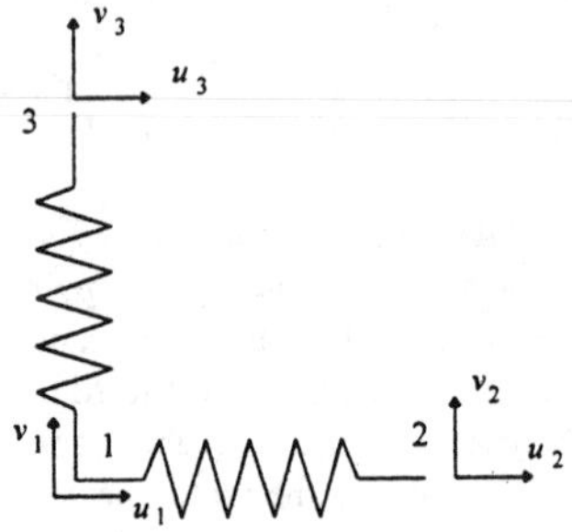

Figure 5 Nail modelling detail.

5.1 *Timber properties*

Timber is a strongly orthotropic material as evidenced from the data published in AS1720.1 which, typically, gives $G = E/15 = E/2(1+\nu)$. To be isotropic, there is an implied Poisson's ratio of 6.5, which is illogical. A full treatment requires a statement of timber properties relative to the longitudinal axis (L direction), the radial (R) direction and the tangential (T) direction. Assuming that the timber is quarter sawn, it is the L-R direction that is significant, (if back sawn the L-T direction becomes significant), and the constitutive equation for timber in compliance form - strain as a function of stress - is given by

$$\begin{bmatrix} \varepsilon_{LL} \\ \varepsilon_{RR} \\ \varepsilon_{LR} \end{bmatrix} = \begin{bmatrix} 1/E_L & -\nu_{RL}/E_R & 0 \\ -\nu_{LR}/E_L & 1/E_R & 0 \\ 0 & 0 & 1/G_{LR} \end{bmatrix} \begin{bmatrix} \sigma_{LL} \\ \sigma_{RR} \\ \sigma_{LR} \end{bmatrix} \tag{2}$$

or, in stiffness form, as

$$\begin{bmatrix} \sigma_{LL} \\ \sigma_{RR} \\ \sigma_{LR} \end{bmatrix} = \begin{bmatrix} E_L E_R/(E_R - E_L\nu_{RL}^2) & E_L E_R \nu_{LR}/(E_R - E_L\nu_{RL}^2) & 0 \\ E_L E_R \nu_{RL}/(E_R - E_L\nu_{RL}^2) & E_L E_R/(E_L - E_R\nu_{LR}^2) & 0 \\ 0 & 0 & 1/G_L \end{bmatrix} \begin{bmatrix} \varepsilon_{LL} \\ \varepsilon_{RR} \\ \varepsilon_{LR} \end{bmatrix} \tag{3}$$

This relationship is ostensibly unsymmetrical and, indeed, test data show that an unsymmetrical constitutive relationship should be used. As this renders all standard finite element software useless, it is usual to ignore this feature that, in practice, is not especially significant. Typical values quoted in the literature show that the assumption of orthotropy holds reasonably well; see Table 2. Because piece-to-piece coefficients of variation in mechanical properties of timber can exceed 30%, finessing of the values is not justified.

The flange bending stiffness $E_L = 12600MPa$ was determined by a bending test. Nominal ratios for timber properties, (Dinwoodie 1981) are $E_L : E_R : E_T = 20:1.4:1$, $G_{LR} : G_{LT} : G_{RT} = 10:9:1$, $E_L : G_{LR} = 15:1$, which gave values that were adopted for the model of $E_R = 880MPa$, $\nu_{RL}/E_R = 40 \times 10^{-6} mm^2/N = \nu_{LR}/E_L$.

Table 2 Typical compliance values for wood elastic properties. Units $mm^2/N \times 10^{-6}$.

Species	L, R		L, T		R, T	
	ν_{LT}/E_L	ν_{TL}/E_T	ν_{LR}/E_L	ν_{RL}/E_R	ν_{RT}/E_R	ν_{TR}/E_T
Aspen, quaking	56.0	74.4	41.8	84.3	1513	1856
Beech	33.1	33.1	37.0	39.0	320	330
Spruce, Engelmann	28.0	41.0	23.9	33.1	640	680

5.2 *Hardboard properties*

The manufacturing of both hardboard and particleboard does not impose any orthotropic character on its planar elastic properties so that it may be treated as isotropic. An elastic modulus was measured as 5500*MPa*. Typically Poisson's ratio was taken as 0.4.

5.3 *Nail properties*

The nails were represented as truss elements inter-connecting nodes in the overlaid system of plane stress elements representing the flange and web; see Figure 4.

Given that the direction of the load on the nail is unknown, it is necessary to represent each nail by two springs at 90° to one another, each having the nail stiffness measured by test or estimated from AS1720.1 formulas. The nail modulus is not influenced by the direction of the load with respect to the timber grain and was determined by test as 1440*N/mm* for 2.0*mm* diameter nails.

Let the nail in the flange be at node 1 and linked to nodes 2 and 3 in the diaphragm; see Figure 5. With the two springs at each connection represented by truss elements at $\theta = 0$, $\theta = \pi/2$, the respective stiffness relationships are

$$\begin{Bmatrix} F_{1x} \\ F_{1y} \\ F_{2x} \\ F_{2y} \end{Bmatrix} = k \begin{bmatrix} 1 & 0 & -1 & 0 \\ 0 & 0 & 0 & 0 \\ -1 & 0 & 1 & 0 \\ 0 & 0 & 0 & 0 \end{bmatrix} \begin{Bmatrix} u_1 \\ v_1 \\ u_2 \\ v_2 \end{Bmatrix} \qquad \begin{Bmatrix} F_{1x} \\ F_{1y} \\ F_{3x} \\ F_{3y} \end{Bmatrix} = k \begin{bmatrix} 0 & 0 & 0 & 0 \\ 0 & 1 & 0 & -1 \\ 0 & 0 & 0 & 0 \\ 0 & -1 & 0 & 1 \end{bmatrix} \begin{Bmatrix} u_1 \\ v_1 \\ u_3 \\ v_3 \end{Bmatrix}$$

After element assembly and setting u_2, v_2, u_3, v_3 $= 0$ these become

$$\begin{Bmatrix} F_{1x} \\ F_{1y} \end{Bmatrix} = k \begin{bmatrix} 1 & 0 \\ 0 & 1 \end{bmatrix} \begin{Bmatrix} u_1 \\ v_1 \end{Bmatrix} \tag{4}$$

Suppose that movement U occurs at angle θ to the x direction, then $F_{1x} = Uk\cos\theta$, $F_{1y} = Uk\sin\theta$ and $F = \sqrt{F_{1x}^2 + F_{1y}^2} = Uk\sqrt{\sin^2\theta + \cos^2\theta} = Uk$ as required for nail stiffness to be independent of load direction. To facilitate plate interconnections in the model, the two truss elements were placed at 45° to the longitudinal axis of the individual timber component, that is, at right angles to each other and nominally at 45° to the direction of grain.

A load of 16*kN/m* was applied to this model, double the limits state serviceability load as specified by AS1170.1. The increase took account of the anticipated long term deflection.

Table3 Comparative long term deflections (*mm*) from tests and finite element analysis.

Location	B2	B3	B4	B5	B6
Truncated full size model		10	18		
One third scale model	1.7	3.4	6.0	8.1	8.7
Finite element model	1.2	2.8	6.7	9.7	10.3

6 RESULTS

When the deflections were measured under the long term serviceability load of 16 *kN/m*, agreement between the truncated full-size model and the one-third scale model was excellent with deflections in the latter nearly exactly three times those of the former; see Table 3. The results from the finite element model did not match those obtained from the one-third-scale model as closely as the writers would have wished; see Table 3. In an effort to determine a reason for this, the model's sensitivity to nail stiffness and component elasticity was examined.

A very high nail stiffness was introduced. Nail test data is collected using specimens in which there is an 0.8*mm* gap between the pieces being joined. This deliberately introduces a degree of conservatism into structural design by reflecting the effect of wood shrinkage. When pieces are tightly nailed together, as in the model, considerable friction develops across the joined pieces and, in theory at least, it is possible to have no relative movement until the static friction is overcome.

Some uncertainty exists in relation to the timber flange stiffness and various moduli of elasticity (MOE) were inserted. It was observed that the deflections across the open Vierendeel panel, were higher in the finite element model than observed experimentally. The reasons are unknown but may be related to the length over which the MOE is measured. It is not uncommon in timber to have a higher MOE over a short length although this does not seem to have been the case.

7 CONCLUSIONS

The results from all models form a consistent pattern.

Analysis of the finite element model shows that deflection levels are very sensitive to nail stiffness. For example, deflections increased some 60% without nail slip being prevented by the assumed high static friction. Sensitivity to variations in elastic properties was not as pronounced. When either the flange modulus or the web modulus was increased 50% above the measured value, deflections decreased around 20%.

A finite element model that assumes isotropy in the timber components is sufficiently accurate for structural design purposes.

8 REFERENCES

Dinwoodie, JM, 1981 *Timber its nature and behaviour*, Van Nostrand Reinhold.

Hunt, DG, 1976 *Rupture tests of wood chipboard under long-term loading*. J.Inst.Wood Sci. Vol.7.

Standards Australia 1989 AS1170.1 SAA Loading code Part 1: Dead and live loads and loading combinations. Standards Association of Australia, Sydney, Australia.

Standards Australia 1997 AS1720.1 Timber structures code Part 1; Design methods. Standards Association of Australia, Sydney, Australia.

Mechanics of Structures and Materials, Bradford, Bridge & Foster (eds)
 ISBN 90 5809 107 4

Plasterboard peaking and cracking under timber roof trusses

H.R. Milner & C.Y. Adam
Monash Engineering Timber Centre, Monash University, Melbourne, Vic., Australia

ABSTRACT: This paper is concerned with an experimental investigation into the phenomenon of plasterboard peaking, the unsightly undulations and cracks which develop in ceilings as a result of moisture movement. It is shown that the peaking is caused by the differential expansion / contraction rates that occur under seasonal moisture change of plasterboard and timber. The differential movement is accommodated by a combination of sheet stressing and joint movement. Where the joints are formed by flexible joining tapes the accommodation is almost entirely within the joint with high joint movements. Where more rigid joints are formed, for example with back blocking, more sheet stressing and less joint movement arises. The construction of joints with tapes on the underside only leads to high joint movement and the unsightly rotations characteristics of peaking.

1 INTRODUCTION

Plasterboard peaking refers to the phenomenon in plasterboard ceilings of developing undulations which either protrude downwards into the room space (peaking) or protrude upwards into the roof space (inverse peaking) and display a tendency to crack. The matter became of particular concern in Australia recently when the phenomenon reportedly occurred in a number of houses. As a result, housing industry and timber industry associations as well as plasterboard manufacturers commissioned Monash University to investigate the phenomenon by constructing and observing the performance of a full size model roof and ceiling system; see Figure 1.

2 CAUSE OF PEAKING – A QUALITATIVE DESCRIPTION

2.1 *Material Related Factors*

At the commencement of the project, there was considerable speculation about the causes of plasterboard peaking with many conflicting statements extant. Ultimately, it was established that the phenomenon was triggered by the differential rates at which timber and plasterboard expand and contract with moisture sorption and desorption. The longitudinal expansion of timber is actually slightly higher than that of plasterboard resulting in a tendency to inverse peaking during periods of high humidity and a tendency towards peaking during periods of drying. Had the shrinkage rate of plasterboard been higher than that of timber it would simply result in inverse peaking during periods of drying.

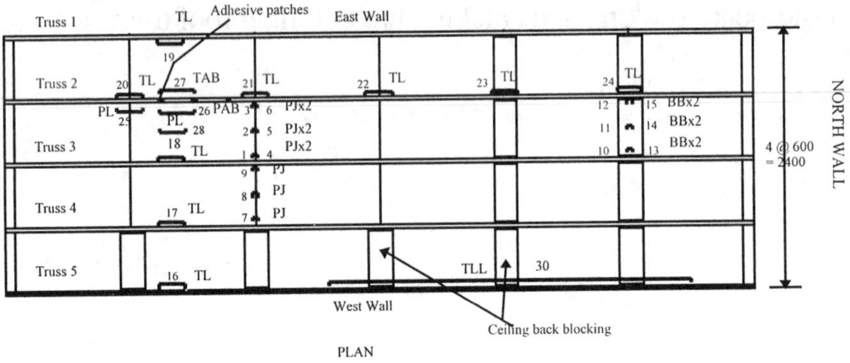

Figure 1. Locations of measurement points (displacements).

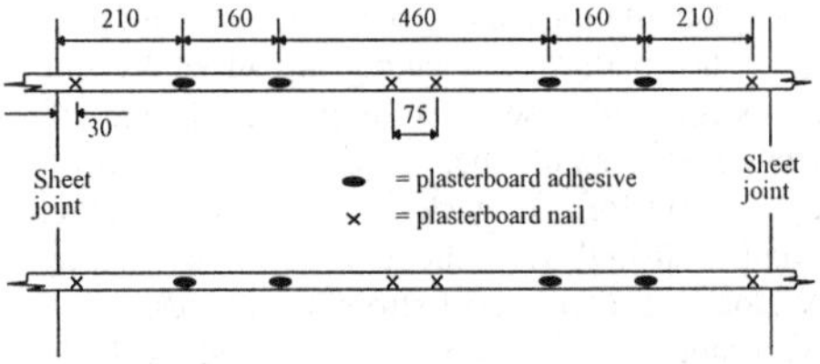

Figure 2. Plasterboard nails and glue patches positions.

2.2 *Building Practices Factors*

The accepted practice in the industry is to rigidly attach plasterboard ceilings directly to the under-side of the roof truss lower chord; no ceiling battens are used. The plasterboard sheets are typically 6*m* in length x 1.2*m* and oriented with the joints running at right angles to the truss. These sheets are attached by nailing and glue patching spaced at approximately 200*mm* centres; see Figure 2. The sheet joints are formed by placing an adhesive tape across the joint on the underside only and then plastering over it. This practice contributes significantly to plasterboard peaking by causing joint rotations. Some builders use a back blocking technique whereby a piece of plasterboard is also glued on the topside of the joint; this practice almost completely eliminates the joint rotations and thus the objectionable undulations observed in plasterboard ceilings but involves additional labour costs.

3 TEST PROGRAM

The test model is illustrated in Figure 1. It consists of truncated girder trusses of span 7350*mm*, sitting on stud walls. The ceiling and all walls were lined with plasterboard such that a room space was created which could be sealed from the external environment in the test chamber.

The model was erected in a 10*m* x 8*m* electronically controlled environment chamber and then subjected the test model to the set of conditions detailed in Table 1. Estimates of moisture conditions induced in the timber are available from AFRDI (1997).

The nail-plated truss itself was constructed from F5 Radiata Pine to a design and manufacture by a local (Melbourne) fabricator. The timber was all F5 Radiata Pine.

Table 1 Chamber test conditions.

Description	RH (%)	Temperature (°C)	Time held at this condition (days)
Hot and dry (commenced 98/2/2)	20	40	12
Hot and moist	90	40	13
Cool and moist	90	15	18
Cool and moist	90	10	7
Cool and moist but room interior heated to 23°C	90	10	7

3.1 *Construction Details*

- The plasterboard was 10*mm* thick ceiling board (Unispan).
- The trusses were erected on the walls and tile battens were nailed to the trusses.
- The plasterboard ceiling was next fixed in accordance with the details shown in Figure 2. All joints were set at this time and back-blocking installed at the locations shown in Figure 1.
- All displacement transducers were next installed and connected to the data logger. These were located at the positions shown in Figure 1.
- The tiles were laid and the plasterboard walls completed

3.2 *Measurements*

The displacement transducers were resistive type linear position sensors. All transducers were individually calibrated using a drum micrometer mounted in a special rig.

Moisture contents of the timber and plasterboard were monitored by the weighing of specimens hanging freely in the chamber. While such weighing provide data on moisture contents during the course of testing it did provide data at the end of the tests when oven drying was completed. Care was needed with plasterboard drying as temperatures in excess of 45°C cause chemical changes. This temperature is quite low compared with the oven drying temperatures used with wood of more than 100°C and thus separate drying chambers are required.

3.3 *Chamber Control*

The control system for the environment chamber maintained its own log but an independent monitoring system was maintained. The sensors for the drying chamber were located in the chamber's outlet duct but a calibration factor was used in the computer program to convert the electrical signal from these sensors to RH and temperature data. This factor was adjusted so that it corresponded with the independent readings from sensors located just above the test ceiling. These readings were checked against dry-wet bulb temperature measurements.

4 RESULTS

4.1 *Wood and Plasterboard Moisture Content*

The moisture conditions achieved in the wood and plasterboard can be observed in Figures 3 and 4. The levels are within the range of expectations published in the literature [1,2]. It is important to know that the two materials are achieving the expected moisture levels at different humidity conditions since moisture drives peaking response.

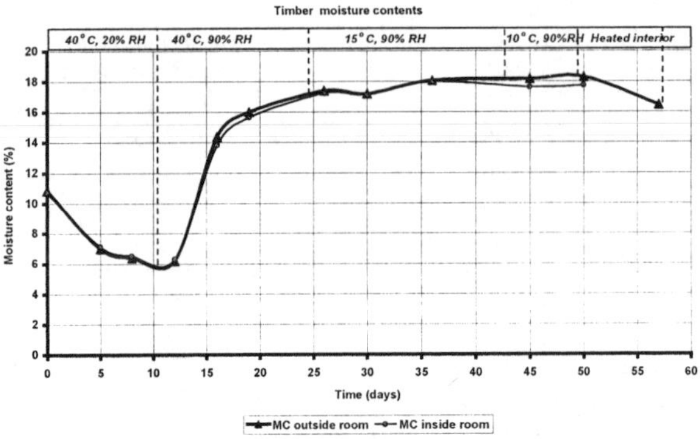

Figure 3. Timber moisture content

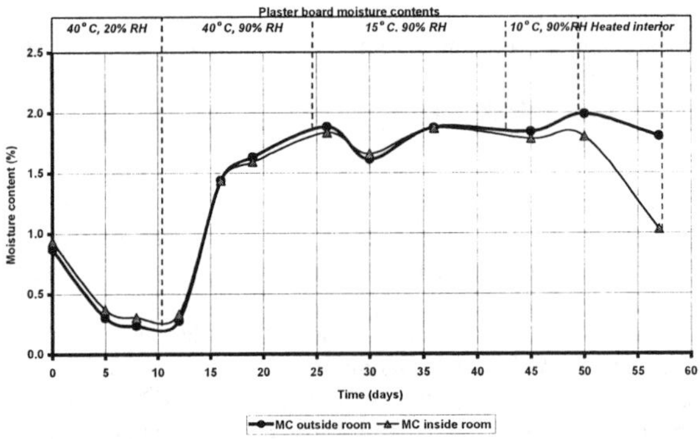

Figure 4. Plasterboard moisture content.

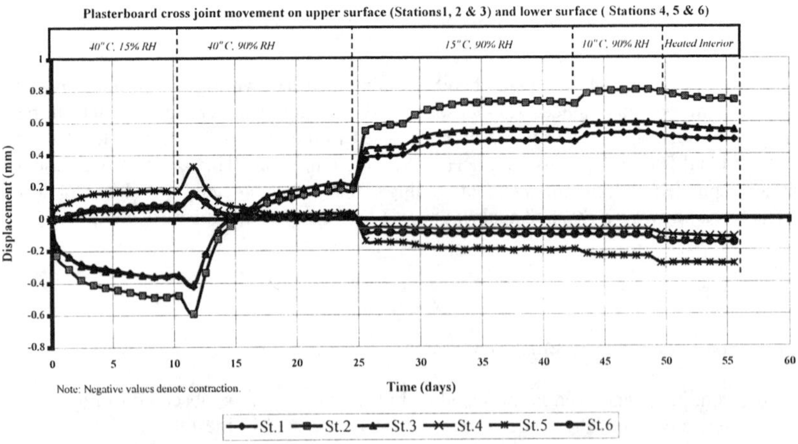

Figure 5. A joint without back blocking

4.2 *Horizontal Cross Joint Movements*

The horizontal cross joint movements are given in Figure 5 for a joint without back blocking, Figure 6 for a joint with back blocking.

In Figure 5 the data plots fall into two distinct groups that move in contrariwise patterns. Stations 1, 2 and 3 data apply to the upper surface implying closure in the early stages and stations 4, 5 and 6 data to the lower surface implying opening in the early stages. The

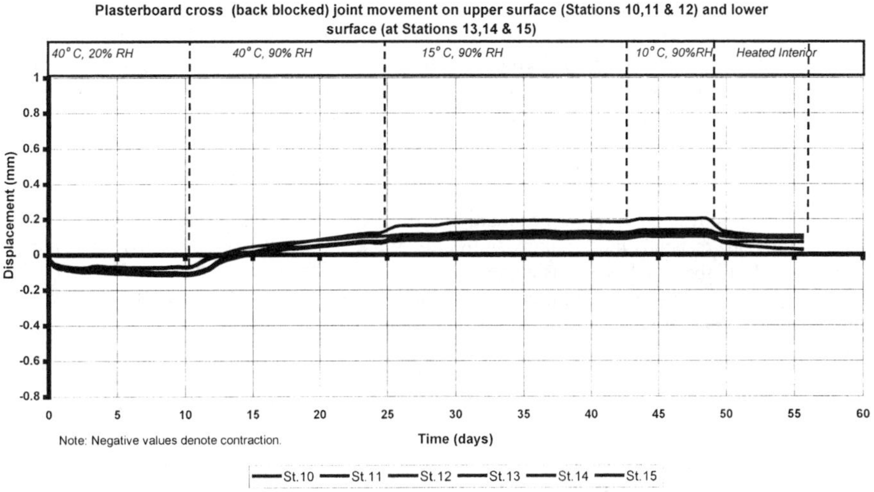

Figure 6. A joint with back blocking

Figure 7. The inverse peaking phenomenon

separation of the two data sets is a measure of the magnitude of peaking displacement that takes place. From days 0 to 15 (materials drying and shrinking) peaking displacements occur and, subsequently, inverse peaking is evident (materials moistening). Visual evidence of inverse peaking was also available in the photograph of Figure 7, taken around day 45 when inverse peaking reached maximum levels.

In Figure 6, plotted to exactly the same scale as Figure 5, not only are the joint movements smaller but the movements on the upper and lower surfaces move in sympathy implying that almost no joint rotations take place when back blocking is used. It is also evident that the mean movements in the back blocked joint are smaller than the movements in the joint without back blocking. This is attributed to the action of the back blocks that stretch the plasterboard sheet between joints and require less accommodation of the differential in movement between timber and plasterboard within the joint. When a tape only is used all such movement differential must be accommodated within the joint tape which has neither the strength nor stiffness to stretch the sheet.

4.3 *Movements within the Timber and Plasterboard*

Overall timber movements tend to be slightly higher than those of the plasterboard, although the plasterboard responds more rapidly in the initial stages to any change of environmental conditions. In timber it takes a considerable length of time for any environmental changes to affect moisture levels in the central core of the timber away from any exposed surfaces. On the other hand, plasterboard expansion seems to be affected by moisture uptake in the outer cardboard layers that tend to exchange moisture quite rapidly.
The difference in movements of timber and plasterboard that occurs once moisture take up is complete is the principal driving force behind the tendency towards peaking and inverse peaking behaviour.

5 CONCLUSIONS

- Peaking and inverse peaking are the result of differential shrink-swell rates of plasterboard and timber. It does not matter which of them expands most, it is the differential which causes peaking. The longitudinal expansion rates of timber and plasterboard are both quite small.
- The practice of forming joints by a tape on the underside only of joints tends to exacerbate inverse peaking action. It tends to restrain the joint eccentrically and thereby induce a joint moment and consequent rotation characteristic of peaking undulations.
- Back blocking counteracts both inverse peaking rotations and reduces nett joint movement by progressively stretching the plasterboard sheet between joints.
- The use of back blocking will increase building costs and is unlikely to be accepted as a general practice. Other jointing techniques can be considered which will ameliorate the problem.

6 REFERENCES

Australian Furnishing Research and Development Institute Limited 1997. Australian Timber Seasoning Manual, *Third Ed.*

Reinforced and prestressed concrete structures

Mechanics of Structures and Materials, Bradford, Bridge & Foster (eds)
© 1999 Balkema, Rotterdam, ISBN 90 5809 107 4

Spacing and width of flexural cracks in reinforced concrete

S. Fragomeni, R. Piyasena & Y.C. Loo
School of Engineering, Griffith University, Southport, Qld, Australia

ABSTRACT: The effects of various variables on the spacing and width of cracks are investigated and the results are presented. For this investigation, spacing and width of cracks are calculated using the results of finite element analysis on various flexural members having different material and sectional properties. The calculated values of crack spacing and crack width are then used in a parametric study to develop new prediction formulas. The new formulas are verified for accuracy by comparing the predicted crack spacing and crack width values with those measured by other researchers.

1 INTRODUCTION

Limiting crack width under service loads is an important consideration in the design of reinforced concrete flexural members. For this purpose, designers may use simplified rules specified by various authorities including Standards Association of Australia (1994) and ACI Committee 318 (1995) in distributing tension reinforcement. These rules are based on certain crack width prediction formulas proposed by various investigators.

A comparison of various crack width prediction procedures indicates that there is no general agreement among different investigators on the relative significance of the variables affecting crack width (Nawy-1968, Gergely and Lutz-1968, Kaar-1968). A major reason for this is that in each experiment only a certain group of variables, amongst the large number involved in cracking of flexural members, are considered.

In this paper, crack spacing and crack width within constant moment regions of flexural members are determined using concrete tensile stresses evaluated by the finite element method. Effects of various variables are investigated by calculating crack spacing and crack width of flexural members having different material and sectional properties. These values are then used in a parametric study to develop simplified prediction formulas. The proposed formulas are verified by comparing the predicted average crack spacing and maximum crack width with those measured by other investigators.

2 DETERMINATION OF CRACK SPACING AND CRACK WIDTH

The average crack spacing of a flexural member is based on the maximum crack spacing for the particular load level.

Piyasena (1999) found that the measured average crack spacing S_{ave} in a constant moment region of a flexural member varies between the upper limit S_{max} and the lower limit $0.5S_{max}$ where S_{max} is the maximum crack spacing corresponding to the particular load level. The predicted average crack spacing $S_{ave\text{-}pred}$ can be taken as the mean of these two limits as

$$S_{ave\text{-}pred}=0.75S_{max} \tag{1}$$

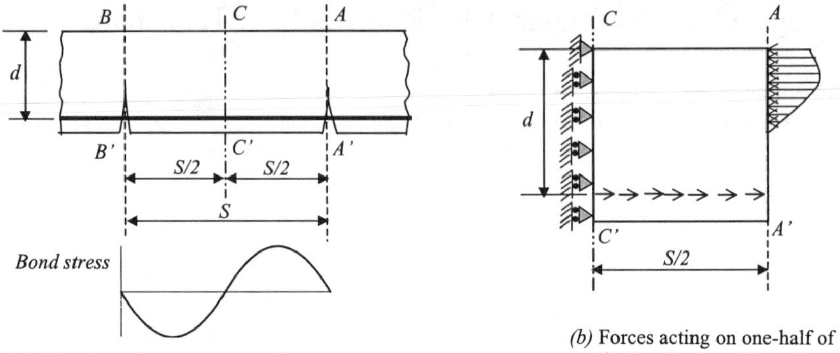

(a) A free-body between two cracks in a constant moment region and the assumed bond stress distribution

(b) Forces acting on one-half of the free-body and boundary conditions

Figure 1. Assumed bond stress distribution and forces acting on a free-body between two cracks

The maximum crack spacing S_{max} was taken as the distance between two adjacent cracks that produces a concrete tensile stress equal to the flexural strength of concrete. For the evaluation of concrete tensile stress, a free body of concrete block bounded by the top and bottom faces of the member and located between two adjacent cracks in a constant moment region was analysed (see Figure 1). Note that due to symmetry, only one half of the concrete block needs to be analysed. This analysis was carried out by the finite element method using two-dimensional plane stress rectangular elements and the standard software package STRAND6 (1993). The crack width was taken as the difference in steel and concrete extensions evaluated at the crack. More details may be found elsewhere (Piyasena-1999).

3 EFFECTS OF VARIOUS VARIABLES ON SPACING AND WIDTH OF CRACKS

The effects of a number of variables on the spacing and width of cracks were investigated. These variables are: concrete strength f'_c, steel stress f_s at the crack, effective depth d, width of the member b, number of bars m, clear concrete cover measured from the outside surface of the steel bar c_o and bar diameter ϕ. A large number of data points were produced by calculating the crack spacing and crack width values changing only one of the above variables at a time. The results are summarised below.

(a) Concrete strength f'_c has no significant effect on crack spacing and crack width.
(b) An increase in steel stress f_s at the crack decreases the crack spacing while it increases the crack width.
(c) Crack spacing and crack width remain unchanged when the effective depth d is varied if $d \geq 300mm$. For members with $d < 300mm$, spacing and width of cracks will increase with the effective depth.
(d) An increase in the width of the member b will increase crack spacing and crack width.
(e) An increase in the number of bars m will decrease crack spacing and crack width.
(f) If both b and m are changed in such a way that the ratio b/m is constant, the resulting crack spacings and crack widths remain unchanged. An increase in the effective width per one bar (b/m) results an increase in both crack spacing and crack width.
(g) An increase in the clear concrete cover c_o will increase the crack spacing and crack width.
(h) Effects of the bar diameter ϕ on the crack spacing and crack width are negligibly small, if other variables are kept constant.
(i) An increase in the bar diameter ϕ by reducing the number of bars m to have a constant steel area will increase both crack spacing and crack width.

4 DEVELOPMENT OF NEW PREDICTION FORMULAS

According to the results presented in Section 3, the most important variables affecting crack spacing and crack width are the steel stress f_s at the crack, effective width of the member per one bar (b/m), clear concrete cover c_o, bar diameter ϕ and the effective depth d (only if d>300mm). Instead of using the clear concrete cover c_o as a parameter, the non-dimensional ratio $c'=c_o/\phi$ is considered more appropriate, because in many practical situations the concrete cover is related to the bar diameter ϕ. For convenience, the effective width of the member per one bar is also divided by ϕ to be used as another non-dimensional parameter $e=b/(m\phi)$.

The maximum crack spacing S_{max} and the resulting maximum crack width $W_{s,max}$ at reinforcement level were calculated for flexural members assuming various values for the above parameters, using the procedure described in Section 2. The results obtained were then used in a semi-regression analysis to develop the following new prediction formulas.

4.1 *Maximum crack spacing*

The following formula is proposed to determine the maximum crack spacing S_{max} (in *mm*).

$$S_{max} = \alpha S_f\left(\frac{200}{f_s} + 0.5\right) \qquad (2a)$$

where the steel stress f_s at the crack is in *MPa* and

$$S_f = 8e' + (0.65 + 0.35e')(\phi - 10) \qquad \phi > 10mm \qquad (2b)$$

$$S_f = 8e' \qquad \phi \leq 10mm \qquad (2c)$$

$$e' = (0.8+0.2c')e \qquad (2d)$$

$$\alpha = 1.1 - \frac{e'}{6 + 0.22d} \leq 1 \qquad d<300mm \qquad (2e)$$

$$\alpha = 1 \qquad d \geq 300mm \qquad (2f)$$

4.2 *Maximum crack width*

The following formula is proposed to determine the maximum crack width $W_{s,max}$ (in *mm*) at reinforcement level.

$$W_{s,mas} = 2\beta W_f\left[1 - \frac{400 - f_s}{400 + 500W_f}\right] \qquad (3a)$$

where

$$W_f = 0.007\{e' + 0.2\phi(1 + 0.4e')\} \qquad (3b)$$

$$\beta = 1.05 - \frac{2e'}{d} \leq 1 \qquad d<300mm \qquad (3c)$$

$$\beta = 1 \qquad d \geq 300mm \qquad (3d)$$

The resulting crack width $W_{t,max}$ at the tension face of the member is then determined using the following equation.

$$W_{t,max} = W_{s,max}\{1.67(h/d) - 0.67\}. \tag{3e}$$

In the derivation of Eq. (3*e*), it is assumed that the two concrete faces of the crack are plane. A further assumption made in Eq. (3*e*) is that the ratio of the depth of compression zone to the effective depth of the member is equal to 0.4. More details are found elsewhere (Piyasena-1999).

5 VERIFICATION OF NEW PREDICTION FORMULAS

The accuracy of the proposed crack spacing and crack width prediction formulas is verified by comparing predicted values with those measured by other investigators. Figure 2 shows the comparison of the average crack spacing predicted using Eqs. (1) and (2), and those measured by Clark (1956), Chi and Kirstein (1958) and Stewart (1997) on 72 flexural members, at various load levels. This figure, which contains 420 data points, indicates that most of the measured average crack spacings lie between the upper limit S_{max} and lower limit 0.5 S_{max}, as predicted in Section 2.

Figure 3(*a*) shows the comparison of the maximum crack width predicted by the proposed method (Eq. 3) and the measured values for the same 72 flexural members mentioned above. A similar comparison based on the Gergely and Lutz (1968) prediction procedure which is adopted by ACI Committee 318 (1995) is shown in Figure 3(*b*). Inspection of these two figures, each containing 420 data points, indicates that both methods can predict the maximum crack width with sufficient accuracy. It should be noted however that approximately 50% of the measured crack widths are underpredicted using the Gergely and Lutz (1968) prediction procedure, whereas only 30% are underpredicted using the proposed method.

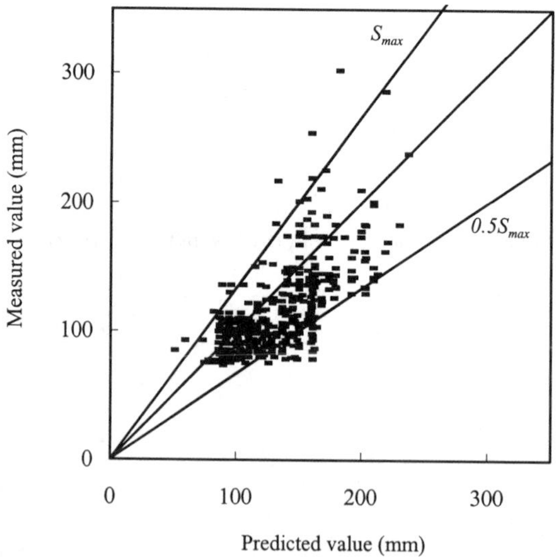

Figure 2. Comparison of predicted and measured average crack spacing

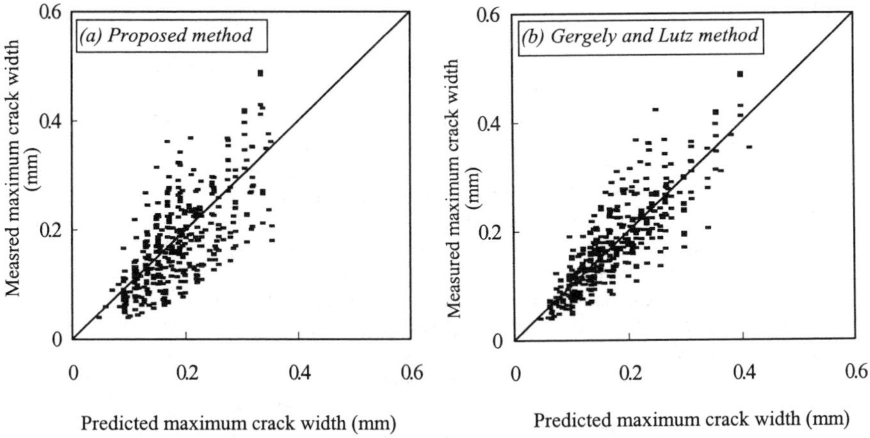

Figure 3. Comparison of predicted and measured maximum crack width

6 EXAMPLE

To demonstrate the proposed procedure, the average crack spacing and the maximum crack width are calculated for Beam No. 6-9.5-7-2 tested by Clark (1956). The properties of the beam are: overall depth h=152.4mm, effective depth d=128.5mm, member width b=241.3mm, bar diameter ϕ=22.2mm, clear concrete cover c_o=12.8mm, number of bars m=2. The crack spacing and crack width are calculated at a steel stress f_s=206MPa.

Solution: $c'=c_o/\phi=0.57$; $e=b/(m\phi)=5.43$; and $e'=(0.8+0.2c')e=4.97$.

Average crack spacing:
Eq. (2e): α=0.955; Eq. (2b): S_f=73.9mm; Eq. (2a): S_{max}=103mm.
Eq. (1): $S_{ave\text{-}pred}$=78mm (Measured average crack spacing=85mm).

Maximum crack width:
Eq. (3c): β=0.973; Eq. (3b): W_f=0.128mm; Eq. (3a): $W_{s,max}$=0.145mm
Eq. (3e): $W_{t,max}$=0.190mm (Measured maximum crack width=0.185mm).
Predicted maximum crack width using the Gergely and Lutz (1968) procedure=0.140mm.

7 CONCLUSION

The effects of various variables on the spacing and width of cracks were investigated for flexural members having different material and sectional properties, and the results presented. These calculated crack spacing and crack width values were based on the concrete tensile stresses and displacements determined by the finite element method. A large number of crack spacing and crack width values evaluated by this method were used in a parametric study to develop new prediction formulas. A comparison of predicted and measured values indicates that the accuracy of the proposed formulas is adequate.

8 REFERENCES

ACI Committee 318 1995. *Building Code Requirements for Reinforced Concrete (ACI 318-95)*. Detriot: American Concrete Institute.

Chi, M. & A. F. Kirstein 1958. Flexural Cracks in Reinforced Concrete Beams. *ACI Journal Proceedings* 54(10): 865-878. Detriot: American Concrete Institute.

Clark, A. P. 1956. Cracking in Reinforced Concrete Flexural Members. *ACI Journal Proceedings* 52(8): 851-862. Detriot: American Concrete Institute.

Gergely, P. & L. A. Lutz 1968. Maximum crack width in Reinforced Concrete Flexural Members. *Causes, Mechanism, and Control of Cracking Concrete, SP-20*: 87-117. Detriot: American Concrete Institute.

Kaar, H. P. 1968. An Approach to the Control of Cracking in Reinforced Concrete. *Causes, Mechanism, and Control of Cracking Concrete, SP-20*: 141-157. Detriot: American Concrete Institute.

Nawy, E. G. 1968. Crack Control in Reinforced Concrete Structures. *Journal of the American Concrete Institute Proceedings* 65(10): 825-836. Detriot: American Concrete Institute.

Piyasena, R. 1999. *Cracking and Deflection Analysis of Reinforced Concrete Flexural Members, PhD Thesis*. School of Engineering: Griffith University, Gold Coast Campus, Australia.

Standards Association of Australia 1994. *Concrete Structures (AS 3600)*. Sydney: Standards Association.

Stewart R. A. 1997. *Crack Widths and Cracking Characteristics of Simply Supported and Continuous Reinforced Concrete Box Beams, B. Eng Thesis*. School of Engineering: Griffith University, Gold Coast Campus, Australia.

STRAND6 1993. *Finite Element Analysis System*. Sydney: G+D Computing Pty Ltd, NSW, Australia.

Mechanics of Structures and Materials, Bradford, Bridge & Foster (eds)
© 1999 Balkema, Rotterdam, ISBN 90 5809 107 4

Moment redistribution and rotation capacity in prestressed concrete structures

M. Rebentrost, K.W. Wong & R.F. Warner
Department of Civil and Environmental Engineering, University of Adelaide, S.A., Australia

ABSTRACT: Preliminary results are presented from a study of secondary moments, moment redistribution and rotation capacity in indeterminate, continuous, prestressed concrete members. The purpose of this ongoing work is to check the adequacy of the relevant design provisions in the current Australian design standard, AS 3600, and propose modifications as necessary. Available test data for fully bonded continuous prestressed beams, published over the last fifty years, have been reviewed and used to check the adequacy of the relevant AS 3600 clauses. Additional computer simulations were undertaken to augment the test data. The results suggest that the present clauses are conservative in dealing with members in which concrete crushing governs the rotation capacity, but that additional clauses may be needed to deal with the possibility of fracture of the tensile reinforcement and the prestressing tendon.

1 INTRODUCTION

The current design provisions in AS 3600 for moment redistribution in prestressed concrete structures are a simple extrapolation of those originally derived for reinforced concrete. The applicable requirements are as follows:

- secondary (hyperstatic) moments must be treated as load effects, with a load factor of unity to be used when checking the strength of cross-sections in bending;
- redistribution of the design moments of up to 30 per cent is allowed, provided: (a) the secondary moments are included in the redistribution; and (b) the neutral axis parameter k_u is limited to the values specified for reinforced concrete.

Only one research study known to the authors has investigated the adequacy of these clauses. Wyche *et al.* (1992) concluded from a limited study that the provisions were adequate. However, the current clauses assume that failure and rotation-capacity depend on crushing of the compressive concrete, and this is the case investigated by Wyche *et al.* The possibilities of fracture of the steel reinforcement and fracture of the prestressing tendon at high overload were ignored. With the proposed introduction of new high-strength reinforcing steel with limited uniform elongation, premature steel fracture has recently become a matter of concern in the design of reinforced concrete structures (Patrick *et al*, 1997). Clearly, the possibility of fracture of the reinforcing steel in prestressed concrete construction also needs to be considered, as well as the further possibility of fracture of the prestressing tendon.

In order to evaluate the adequacy of the current AS 3600 provisions, a review was undertaken of laboratory tests of fully bonded prestressed concrete continuous beams that have been published in the last 50 years. The aim was to assemble as much data as possible, and use it to check the AS 3600 design provisions. Computer simulation studies were also undertaken to supplement the test data.

2 EVALUATION OF AS 3600 PROVISIONS USING TEST DATA

2.1 *Test data for continuous prestressed concrete beams*

Although many tests have been conducted and reported in the technical literature, much of the published information was found to be unusable. In many cases there was incomplete reporting of the properties of the materials, the test specimens, or the test results. From 27 separate publications, involving more than 200 beam tests, only eight publications were found to be useable. The results of 64 two-span and 42 three-span beams containing bonded tendons were finally selected for this study. All two-span members had a single point load applied in one or both spans. In the case of the three-span beams, either one or two point loads were applied in the central span, while the outer spans were unloaded. The span to depth ratio ranged from 10 to 20, and the mode of failure was reported as flexure in all cases. The concrete compressive strength was between 30 and 50 MPa for the beams. The sections were rectangular except for four beams, which had T-sections. Most of the test beams had values of the neutral axis parameter k_u well above 0.4. According to AS 3600, these beams would be classified as non-ductile, and no moment redistribution would be allowed in their design.

2.2 *Definition of moment redistribution*

Moment redistribution is a design concept, which is based on elastic analysis. A clear physical definition is required for the purpose of evaluating test data. The majority of the beams were symmetric over two spans, with a single central point load in each span. In such a beam, the static moment in a span is $M_O = PL/4$, where P is the point load and L is the span. The elastic moments at the central support A, M_A, and at the mid-span C under the point load, M_C, can be expressed in terms of the static moment as follows:

$$M_A = \frac{6}{32} PL = \frac{6}{8} M_0 \quad \text{and} \quad M_C = \frac{5}{32} PL = \frac{5}{8} M_0$$

If the peak load, $P_{u.test}$, and the moments at A, $M_{A.test}$, and C, $M_{C.test}$, are determined in the beam test, then the moment redistribution at A, $\beta_{A.test}$, and at C, $\beta_{C.test}$, can be defined in terms of the static moment, $M_{0.test} = P_{u.test} L/4$, as follows:

$$\beta_{A.test} = \frac{M_{A.test} - \frac{6}{8} M_{0.test}}{\frac{6}{8} M_{0,test}} \quad \text{and} \quad \beta_{C.test} = \frac{M_{C.test} - \frac{5}{8} M_{0.test}}{\frac{5}{8} M_{0.test}}$$

A positive $\beta_{A.test}$ implies a reduction in the negative support moment; this is accompanied by an increase in span moment, ie a positive $\beta_{C.test}$. Similarly, an increase in the negative support moment implies a negative $\beta_{A.test}$, and a decrease in positive span moment, with a negative $\beta_{C.test}$. Similar expressions can be derived for moment redistributions in other test set-ups.

2.3 *Observed and permissible moment redistribution*

The moments in an indeterminate test beam can be obtained experimentally by measuring the support reactions, as well as the applied loads. The moment redistribution can then be calculated using the definitions given above. In Fig 1 the moment redistribution at the support is plotted as a function of the maximum neutral axis parameter, k_u, for the test beams.

The dashed line in Fig 1 shows the maximum moment redistribution allowed by AS 3600 for varying k_u. Although k_u for most beams is much higher than the upper limit of 0.4 set by AS 3600, very large moment redistributions were achieved. This suggests that the present design provisions may be extremely conservative.

The five points in Fig 1 with k_u values less than 0.4 come from tests by Hawkins *et al* (1962). Of the three beams with points within the AS 3600 region, only one achieved full plastic redistribution; the other two failed, for unreported reasons, before the full section capacities were reached. Further tests are clearly needed, with k_u less than 0.4, to check the AS 3600 requirements.

2.4 *Observed and predicted load capacity*

The load capacities of the test beams were calculated using two different methods. A fully non-linear computer analysis was carried out, using mean material properties. In a second simplified method, the ultimate moments at the sections of maximum moment were calculated using mean values of the material properties. For the many cases with $k_u > 0.4$, a reduced moment capacity M_{ud} was calculated in accordance with AS 3600. This moment applies to an "equivalent" section which only contains sufficient tensile steel to produce a value of $k_u = 0.4$. The collapse load was then estimated using an elastic moment distribution and the governing moment capacity, with the capacity reduction factor, ϕ, set equal to unity.

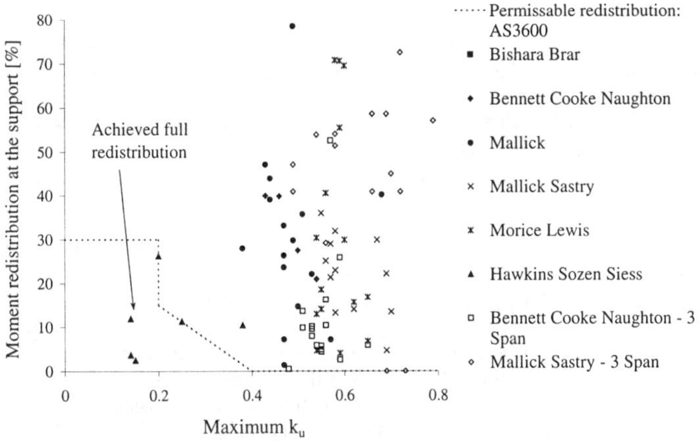

Figure 1– Moment redistribution attained at the support against critical k_u

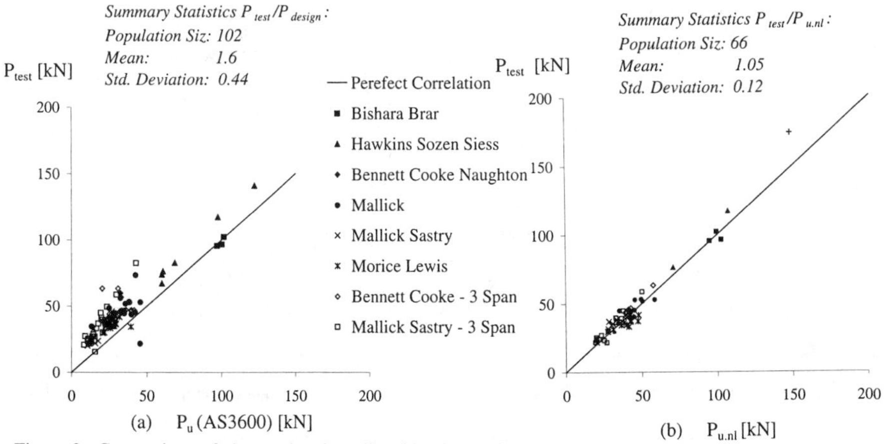

Figure 2– Comparison of observed and predicted load capacity

The results of the simplified analysis, presented in Fig 2a, show that the use of M_{ud} significantly underestimates the load carrying capacity. In Fig 2b the results of the non-linear analysis show good correlation, and confirm the accuracy of this particular method of analysis.

2.5 *Observed and predicted rotation capacities*

In many European Standards, including CEB, DIN and EC2, rotation capacity is used to measure local structural ductility. In AS 3600, the neutral axis parameter k_u provides a simplified, approximate measure of rotation capacity. In a small number of tests the rotations in the peak-moment regions were recorded and published. In Fig 3 the recorded rotations at the support and at the load point are compared with the rotations predicted by the non-linear analysis.
In considering these results it must be remembered that the moment-rotation curve is horizontal at peak moment, so that both experimental and calculated rotations are prone to very large errors. However, the calculated values are generally less than the observed ones. This may be due to underestimations of the compressive concrete strain and the plastic hinge length.

3 COMPUTER SIMULATION STUDIES

To supplement the very limited test data, simulation studies of continuous beams were undertaken. Owing to space limitations, the results of only six two-span T-beam simulations are presented here. The calculations were carried out using a computer program that takes full account of non-linear behaviour of the prestressed structure at all stages of loading (Wong and Warner, 1998). The calculations for Fig 2b above were undertaken with this program.

3.1 *Computer simulations*

All beams had the same dimensions and the same prestressing details (see Fig 4). The basic beam is adapted from one used by Lin and Thornton (1972). The prestressing cable produces large positive secondary moments throughout the beam, with a maximum value of 1867 kNm at the interior support. Varying amounts of reinforcement were placed in the peak moment regions in the different beams to induce both positive and negative moment redistribution at collapse. All beams had a mean concrete strength of 56.8 MPa and the tendon consisted of 21-12.7mm prestressing strands (f_{pym} = 1850 MPa). Additional steel reinforcement was assumed to be of Grade400Y. The details of the beams are given in Table 1, and the main results in Table 2.

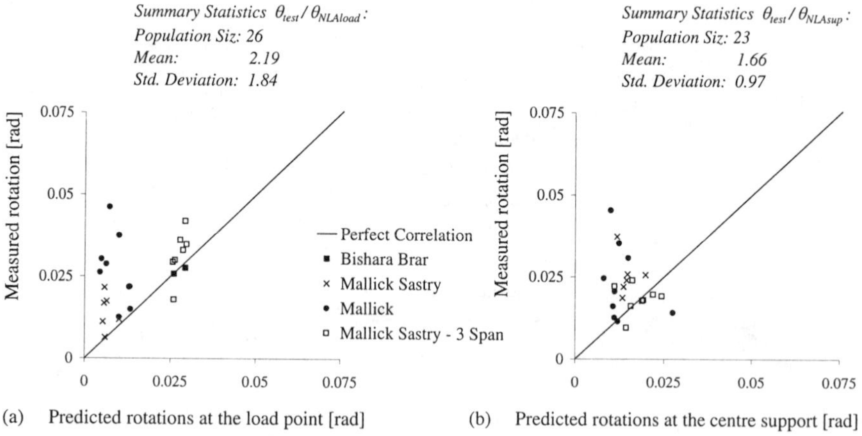

Figure 3 – Predicted and observed total rotation

Table 1 - Details of beams

Beam No.	A_{stA}	A_{scA} [mm^2]	A_{stC}	M_{UA}	M_{UC} [kNm]	k_{uA}	k_{uC}	M_2/M^*_{sup}	β_A [%]
Beam 1	0	0	0	3513	3990	0.38	0.04	-0.41	-3.3
Beam 2	1800	4100	0	4460	3990	0.30	0.04	-0.38	-15.0
Beam 3	0	0	3666	3510	5630	0.38	0.05	-0.32	30.0
Beam 4	0	0	6000	3510	6660	0.38	0.06	-0.39	42.0
Beam 5	6000	0	0	5330	3990	0.61	0.04	-0.39	-9.7
Beam 6	6000	6000	0	6090	3990	0.40	0.04	-0.50	-30.0

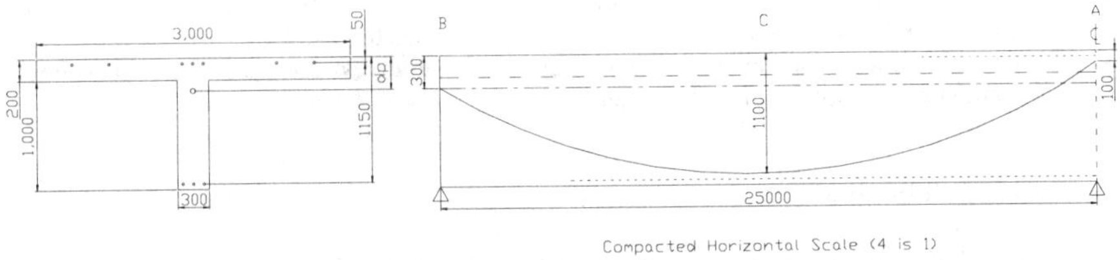

Figure 4 - Details of beams

Table 2 - Results of non-linear beam analysis

Beam No.	w^*	$w_{u.nl}$ [kN]	$w_{u.pl}$	ε_p at ult	ε_s at ult [%]
Beam 1	58.8	73.6	75.6	1.62	-
Beam 2	63.7	78.9	81.5	1.69	1.11
Beam 3	75.5	95.8	100.1	1.82	0.28
Beam 4	86.1	107.6	115.6	1.83	0.23
Beam 5	61.3	84.5	86.1	0.96	0.31
Beam 6	72.0	87.5	91.5	1.28	0.74

3.2 Simulation results

The maximum values for the neutral axis parameter k_u always occurred at the interior support, and varied between 0.38 and 0.61. According to AS 3600, the allowable design moment redistribution would therefore be negligible for all beams. In fact, the redistributions indicated by the non-linear analysis were substantial. In Table 2, values of the redistribution parameter $\beta_{A.test}$ (previously defined) are as high as 40 per cent. Negative values occurred when tensile reinforcement was placed at the support to carry additional negative moment.

The load capacity of each beam was calculated assuming simple plastic theory ($w_{u.pl}$) and using the full non-linear analysis ($w_{u.nl}$). The factored design load w^* was also calculated using the AS 3600 provisions and an elastic moment distribution, but with $\phi = 0.8$. The results in Table 2 are very consistent. For all beams $w_{u.nl}$ is only slightly smaller than $w_{u.pl}$, but significantly higher than w^*. This indicates quite large moment redistributions, even though the beams are non-ductile according to AS 3600. The ductility demand is greatest for Beams 3 and 4, because the positive mid-span moment capacity has been augmented by tensile steel. Even with a k_u value of 0.38, Beam 3 achieved a ratio of $w_{u.nl}/w_{u.pl}$ of 0.96 and Beam 4, a ratio of 0.93. It is noteworthy that the positive redistribution acts together with the large positive secondary moment to create a condition of very high ductility demand. Nevertheless, these beams almost reached their plastic capacities before failure occurred due to concrete softening. Values of the tensile strains in the tendon and reinforcing steel at beam collapse, given in Table 2, are much less than the fracture values. This suggests that, for these beams at least, steel fracture is unlikely.

4 CONCLUDING REMARKS

An unexpected result of this study was that very few of the many published beam tests were useable. In view of the high costs of experimental work, this is very disappointing.

In regard to the moment redistribution Clauses in AS 3600 for the design of continuous prestressed concrete flexural members, the following conclusions follow from this study:

- The AS 3600 limits on allowable moment redistribution are very conservative.
- Both the tests and the simulations showed that significant moment redistribution can occur for k_u values much higher than 0.4. The test data did not show any clear trend of achievable moment redistribution with critical k_u.
- For the beams analysed, the steel strains at collapse were well below the fracture strains, but further studies are needed to allow reliable conclusions to be drawn.
- Further experimental and analytical studies are needed to investigate the collapse behaviour of ductile beams with k_u values less than 0.4.

5 ACKNOWLEDGEMENT

Financial support through a large Australian Research Council grant is gratefully acknowledged. The ongoing support of SRIA and BHP, in particular Mark Patrick and Mark Turner, is also acknowledged.

6 REFERENCES

Bennett, E. W., Cooke, L. P. & Naughton, L. P.,1967, *Deformation of Continuous Prestressed Concrete Beams and its Effect on the Ultimate Load*, Proceedings Institute of Civil Engineers, 37: 57-74, London, England

Bishara, A. G. & Brar, S., 1974, *Rotational Capacity of Prestressed Concrete Beams*, ASCE Journal of Structural Engineering, 100(ST9): 1883-1895

Hawkins, N. M., Sozen, M. A. & Siess, C. P., 1964, *Behavior of Continuous Prestressed Concrete Beams*, Proceedings of the International Symposium: Flexural Mechanics of Reinforced Concrete, ASCE/ACI: 259-294, Miami

Kenyon, J. M. & Warner, R. F., (1993), *Refined Analysis of Non-Linear Behavior of Concrete Structures*, Australian Civil Engineering Transactions, CE35(3)

Lin, T. Y. & Thornton, K., (1972), *Secondary Moments and Moment Redistribution in Continuous Prestressed Concrete Beams*, PCI 42(1):8-20

Lin, T. Y., 1955, *Strength of Continuous Prestressed Concrete Beams Under Static and Repeated Loads*, ACI Proceedings, 26 (10): 1037-1059

Mallick, S. K. & Sastry, M. K. L. N., 1966, *Redistribution of Moments in Prestressed Concrete Continuous Beams*, Magazine of Concrete Research, 18 (57): 207-220

Mallick, S. K., 1962, *Redistribution of Moments in Two-Span Prestressed Concrete Beams*, Magazine of Concrete Research, 14 (42):171-183

Morice, P. B. & Lewis, H. E., 1955 (published 1958), *The Ultimate Strength of Two-Span Continuous Prestressed Concrete Beams as affected by Tendon Transformation and Un-tensioned Steel*, 2nd Congress of the Federation Internationale de la Preconstraints, Session IIIa (5): 561-584, Amsterdam

Patrick, M., Akbarshahi, E. & Warner, R. F., 1997, *Ductility limits for the design of concrete structures containing high-strength, low-elongation steel reinforcement*, Concrete Institute of Australia, Concrete 97

Standards Australia, 1994, *AS3600-1994 Concrete Structures*, Australia, Australian Standards

Wong, K. W. & Warner, R. F., 1998, *Collapse Load Analysis of Prestressed Concrete Structures*, Departmental Research Report, University of Adelaide, R162

Wyche, P. J., Uren, J. G. & Reynolds, G. C.(1992), *Interaction between Prestressed Secondary Moments, Moment Redistribution, and Ductility - A treatise on the Australian Concrete Code*, Journal of the ACI Structural Division, 89(1):57-70

Mechanics of Structures and Materials, Bradford, Bridge & Foster (eds)
© 1999 Balkema, Rotterdam, ISBN 90 5809 107 4

Modelling of high-moment plastification regions in concrete structures

R.J.Gravina & R.F.Warner
Department of Civil and Environmental Engineering, University of Adelaide, S.A., Australia

ABSTRACT: A review of recent research into the rotation capacity of reinforced concrete flexural members shows renewed interest in the topic has been generated by the introduction, in various countries, of high-strength reinforcing steel with limited uniform elongation. This has caused concern over the possibility of steel fracture in high-moment regions. Also the methods of analysis presently available do not give an accurate or adequate treatment of flexural deformations and deformation capacity in high moment regions. This paper proposes a method of analysis, which can trace the local flexural deformations in a beam as the moment increases up to, and beyond, the moment capacity. The analysis predicts the progressive formation of tensile cracks, the initial yielding of the tensile steel, and the subsequent spread of the yielded region to create a plastic hinge. The method is being developed to create an improved finite element for the non-linear analysis of concrete structures.

1 INTRODUCTION

The collapse of an indeterminate concrete structure usually occurs only after large plastic deformations occur in one or more peak moment regions. In such a region, yielding of the tensile steel reinforcement first occurs in one cracked section and then progressively spreads outwards on either side of the crack for a significant distance, and eventually may encompass a number of adjacent cracks. The length of yielding greatly influences the rotation capacity of the hinge region, and hence the ability of the structural system to redistribute internal moments. Without a spread of steel yield through a significant finite length, the structure would collapse prematurely and suddenly.

The progressive increase in plastic deformation in the hinge region, while the moment remains almost constant, allows the moments elsewhere to increase and form additional plastic hinges and, eventually, a plastic collapse mechanism in the system. Such ductile behaviour at high overload is highly desirable, and code clauses are designed to achieve such behaviour. However, brittle collapse can intervene if premature failure occurs locally in a high-moment region, for example because of steel fracture, or shear failure, or anchorage failure.

Many factors affect the rotation capacity of a hinge region, including the properties of the component materials and the amount of tensile steel in the section. The local steel-concrete bond properties also play a vital role in determining the spread of steel yield and hence the rotation capacity.

In the non-linear computer analysis of concrete structures, the plastification regions are often treated in a simplified manner by assuming a constant value for the hinge length, and undertaking a smeared analysis for a "typical" cross-section within the hinge (Wong *et al.* 1988). Simplifying assumptions are made, notably that perfect bond exists between the steel and concrete, strains are linearly distributed across the section, and an "effective" stress-strain relation applies for the concrete in tension. The effective tensile stress-strain curve allows indirectly for local bond breakdown and tension stiffening between adjacent tensile cracks.

While smeared approaches can give satisfactory results, especially when the high-moment regions are ductile, a more accurate analysis of the local region is needed to take proper account of progressive bond breakdown and the spread of the yield region. The limited accuracy of the smeared approach can be seen in Figs. 1 and 2 below, where analytic predictions of rotation capacity are compared with test data. Although the scatter is slightly reduced by including tension stiffening in the analysis, the correlation is not good and an improved approach is clearly desirable. Figures 1 and 2 were prepared using the results from forty-four beam tests, conducted by Bigaj and Walraven (1993), Bosco and Debernardi (1993), Calvi *et al.* (1993), and Eibl and Buhler (1993). The beams chosen for the comparison were all simply supported reinforced concrete beams loaded at mid-span by a single point load. In the tests, important parameters were varied, such as the member size, reinforcement ratio and steel ductility, and bond properties.

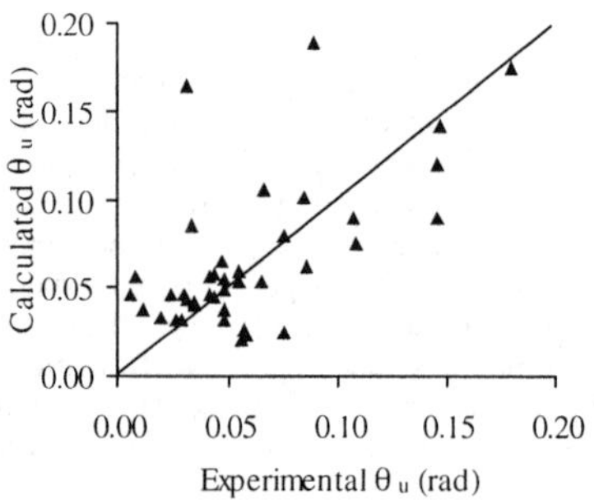

Figure 1. Smeared model-no tension stiffening

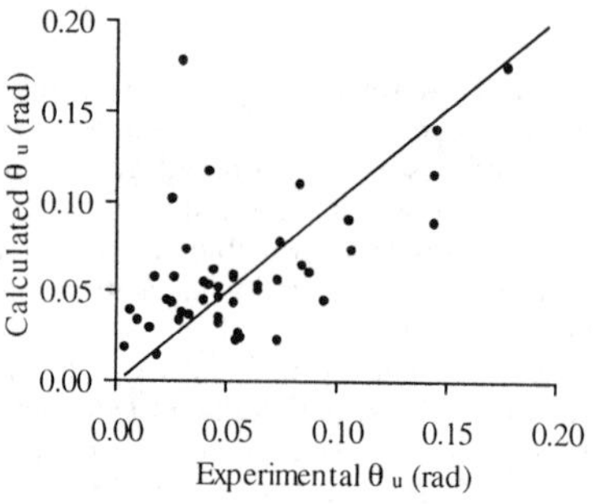

Figure 2. Smeared model-with tension stiffening

2 DEFORMATION CAPACITY MODELS FOR FLEXURAL MEMBERS

In deformation models proposed recently by a number of researchers, the bond-slip behaviour at the steel-concrete interface has played a key role (Langer and Eligehausen (1987); Eligehausen and Li (1990); Sigrist and Marti (1993), Cosenza *et al.* (1991); Beeby (1997)). Most of the methods are very similar in overall approach, and vary only in the detailed development. Attention is therefore restricted here to two representative methods, namely the Stuttgart model of Langer and Eligehausen (1987) and the Zürich tension-chord model of Sigrist and Marti (1993) and Sigrist (1995).

The method developed by the Stuttgart group determines the local strain distribution along the tensile steel reinforcement in a high-moment plastification region, using bond-slip relations proposed by Ciampi *et al* (1981). The rotation in the region is obtained by integration of strains. The method can take account of inclined shear cracking as well as vertical tensile cracking. Variation in steel stress between cracks is calculated using a numerical integration procedure similar to the one proposed by Bachmann (1967), together with appropriate bond-slip relations.

In the Zürich approach, a composite steel-concrete tension chord model allows the deformations along the tensile steel reinforcement to be calculated. A stepped, rigid-plastic relation between bond stress and slip provides the basis of the analysis. Although this bond relation is a crude simplification of the complex real behaviour it takes account of the main parameters and leads to reasonable predictions. The main factors influencing the deformation capacity of a flexural region are shown to be: the strain hardening modulus and ultimate strain of the steel, the bar size, and the bond properties including bond degradation in the post-yield range. Analytical expressions for the average steel strain are used in a non-linear analysis of the member.

In order to investigate the applicability of the Stuttgart and Zürich models, Figs. 3 and 4 were constructed using the same beam test data that were used in Figs. 1 and 2. While the correlation between experimental and calculated rotation capacities is slightly better than in Figs. 3 and 4, further improvement is obviously desirable.

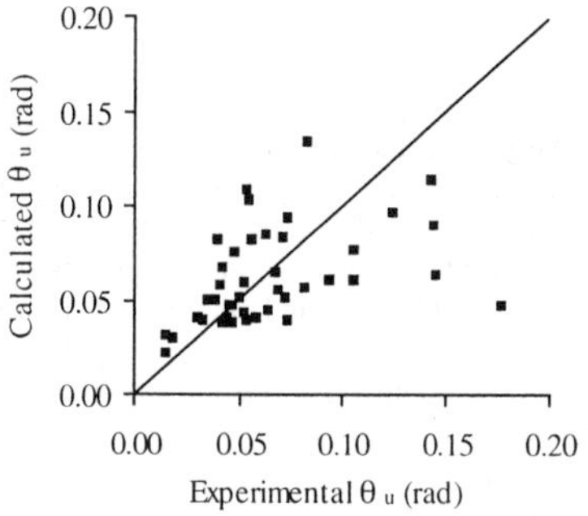

Figure 3. Stuttgart Model

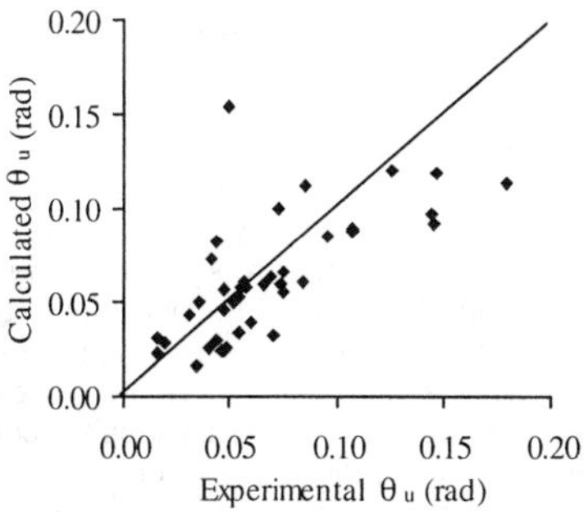

Figure 4. Zurich Model

A disadvantage of both the Stuttgart and Zürich approaches is that an average crack spacing has to be assumed before the analysis can be undertaken. Furthermore, the calculations proceed for a progressively increasing moment in the section, so that the rotation capacity is associated with the peak moment. The possibility of post-peak softening is not considered. As indicated with Figs. 1 to 4, the correlation with test data does not give accurate or adequate treatment of plastification regions of reinforced concrete elements

3 ANALYSIS OF LOCAL STEEL STRAINS, STRESSES AND DEFORMATIONS IN A POTENTIAL HINGE REGION

In this section, an outline is given of a new approach for a more basic method of analysis. The method predicts progressive cracking and crack spacing and subsequent changes in crack spacing as the applied moment increases. This allows the local distribution of strain and stress to be determined in the steel between adjacent cracks in high moment plastification regions. The extent of the plastic hinge region can thus be obtained. The approach is similar in concept to one used be Ferretti (1995) in a doctoral dissertation. However, the method being proposed is more general, and allows bar pullout and anchorage failure to be detected.

3.1 *Details of analysis method*

To illustrate the method, it is applied here to the full span of a beam with a single central point load. It is assumed that the shear span is so large that only flexural cracks develop. The right hand half of the beam is shown in Fig. 5. The analysis proceeds numerically by incrementing the strain in the tensile steel at mid-span. A simple tensile concrete stress criterion is used to determine when the first crack forms at mid-span, and when and where subsequent cracks appear along the span. The local variation in steel strain and stress around each crack is determined using bond-slip relations and a numerical integration procedure similar to the one adopted by Bachmann (1967). An iterative calculation determines the conditions after each increment in strain is applied at mid-span.

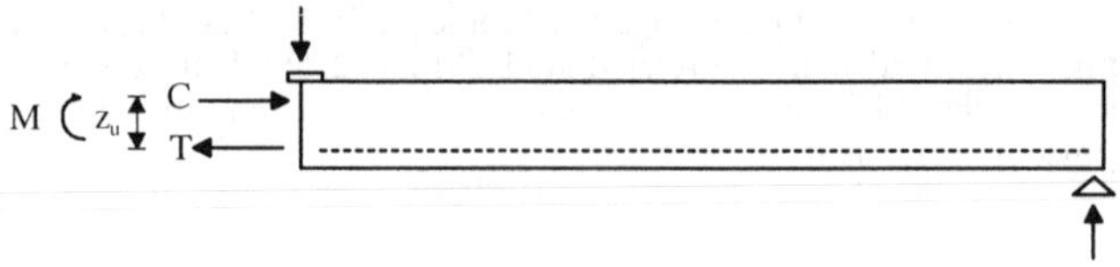

Figure 5. Beam with central point load for analysis

With increasing load, new cracks appear progressively at points further and further away from the mid-span. However, there is also a build up in tensile stress in the steel and concrete between adjacent cracks until secondary cracks appear between adjacent primary cracks. At each step, the steel stress distribution is therefore recalculated between all adjacent cracks. At overload, the extent of the plastified region around the original crack at mid-span is determined, together with the rotation in this plastified region.

The final failure of the beam can occur in various ways, including:

- destruction of the concrete in the compressive zone;
- fracture of the tensile steel;
- anchorage failure and pullout of the tensile steel bars.

The calculations are continued until the uniform elongation of the steel reinforcement is reached and steel fracture occurs. Anchorage failure, which can intervene if the bars are curtailed, is detected from the bond-slip calculations.

3.2 *Numerical Example*

To illustrate how the procedure works, it is applied here to a beam with a rectangular section, 300 mm by 600 mm, with tensile reinforcement at a depth of 550 mm. The span is 4,800 mm. The area of reinforcement is 0.5 per cent, with a yield strength of 550MPa, and an ultimate strength of 660 MPa and uniform elongation of 10 per cent. The properties of the steel correspond to a 500MPa grade hot-rolled deformed bar produced in Australia. The concrete strength is 30MPa. An elastic, strain-hardening stress-strain relation is used for the tensile steel (Fig. 6), with the bond–slip relationship model proposed by Ciampi *et al.* (1981) (Fig. 7).

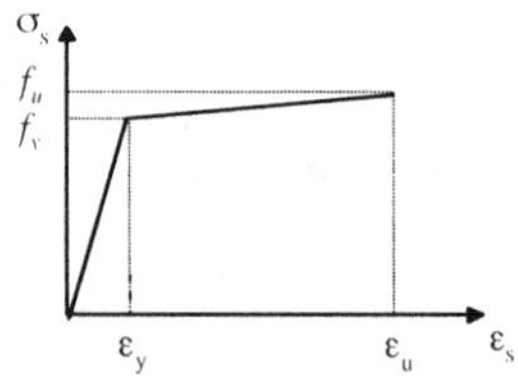

Figure 6. Steel stress strain relationship

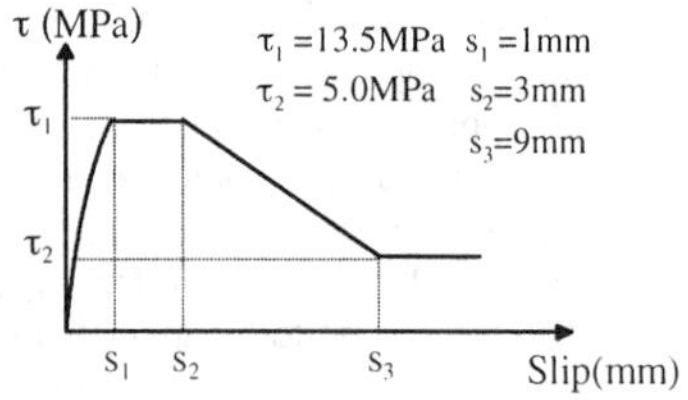

Figure 7. Bond slip

Figures 8 to 11 show some results of the analysis. The bond breakdown length following first cracking was found to be 210mm and the second crack formed at a distance of 215mm from the first crack (Fig. 8). With increasing load, new cracks formed further out into the span; however, a check showed that new secondary cracks also appeared between the first and second cracks, and, at a high stage of overload, even between the first primary crack and the first secondary crack. The yielded hinge region at mid-span was nevertheless restricted to a length of about 645 mm, which contained the first three primary cracks (Fig. 9). This is fairly close to the depth of the section, ie 600mm. In the smeared analysis, it has often been assumed that the hinge length is equal to the beam depth (Warner and Yeo, 1986).

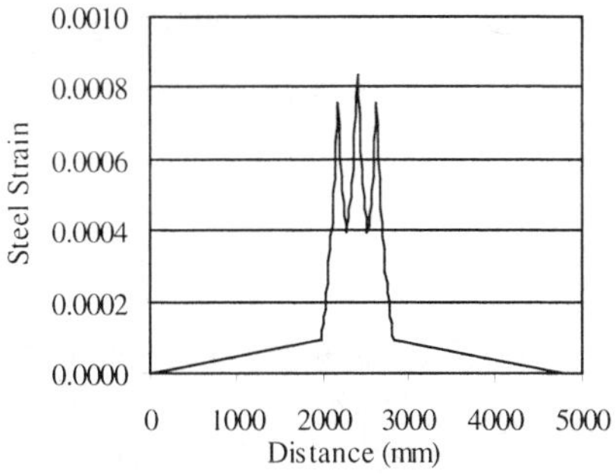

Figure 8. Strain distribution 1st and 2nd cracks formed

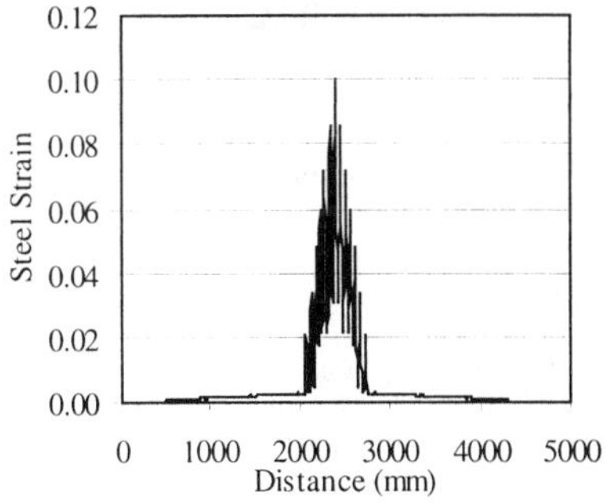

Figure 9. Strain distribution at ultimate

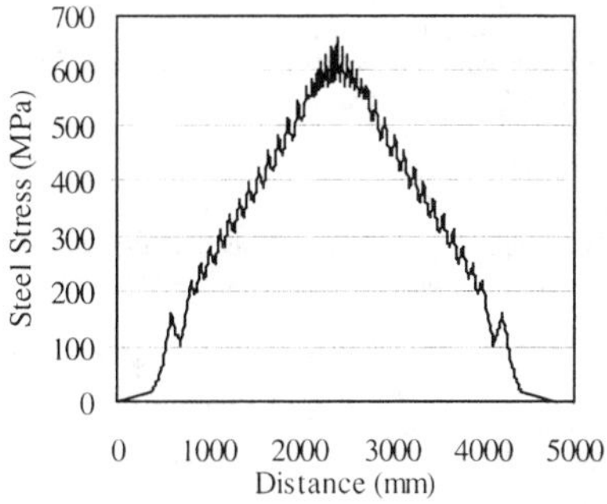

Figure 10. Steel stress distribution at ultimate

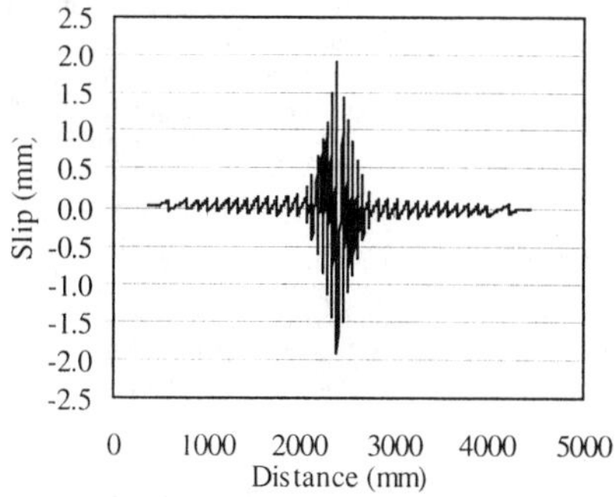

Figure 11. Slip distribution at ultimate

In Figs.10 and 11 an uncracked region can be seen at each end of the beam. Regularly spaced cracks then occur until, in the high-moment plastified region close to the load, the cracks become closely spaced. At peak load a local maximum slip value of 2mm is reached at the first primary crack.

The bond-slip relationship adopted in the analysis clearly has a great influence on the bond breakdown length, and hence on the crack spacing and the length of the plastified region. For comparison purposes, calculations were also made using poor bond conditions with the maximum bond stress reduced from 13.5Mpa to 5MPa. This approximately doubled the bond breakdown length to 450mm. It is clear that the bond properties, the bar diameter, the deformation pattern on the bars, and concrete tensile strength have a decisive effect on local behaviour, including crack spacing, the size of the plastification region, and hence also rotation capacity. The post-yield shape of the stress-strain curve for the reinforcement is also important. An extensive parametric study is currently being undertaken in order to quantify these effects.

4 CONCLUDING REMARKS

The method of analysis proposed in this paper allows the local stresses and strains and deformations along a flexural member to be determined at each stage of loading. It also provides a detailed history of crack formation. Further refinements are at present being undertaken to improve the treatment of the local variations in stress and strain in the compressive concrete above the neutral axis.

5 ACKNOWLEDGEMENTS

The Steel Reinforcement Institute of Australia (SRIA) supports this research project through an APA Industry Award. The authors are grateful for this financial support, and also for detailed technical advice provided by Mr Mark Turner and Dr Mark Patrick.

6 REFERENCES

Bachmann, H. (1967). *Zur plastizitätstheoretischen Berechnung statisch unbestimmter Stahlbetonbalken.* PhD Thesis, ETH Zürich.

Beeby, A. W. (1997). *Ductility in reinforced concrete.* The Structural Engineer, 75(18):311-318.

Bigaj, A., & Walraven, J. C. (1993). *Size effect on rotational capacity of plastic hinges in reinforced concrete beams.* Ductility Requirements for Structural Concrete-Reinforcement Progress report of Task Group 2.2, Comite Euro International du Beton, Lausanne, p7-24.

Bosco, C., & Debernardi, P. G. (1993). *Influence of some basic parameters on the plastic rotation of reinforced concrete elements.* Ductility Requirements for Structural Concrete-Reinforcement Progress report of Task Group 2.2, Comite Euro International du Beton, Lausanne, p25-44.

Calvi, G. M., Cantu, E., Macchi, G., & Magenes, G. (1993). *Rotation capacity of R.C. slabs as a function of steel properties.* Ductility Requirements for Structural Concrete-Reinforcement Progress report of Task Group 2.2, Comite Euro International du Beton, Lausanne.

Ciampi, V., Eligehausen R.,Bertero, V., & Popov, E. (1981). *Analytical model for deformed bar bond under generalised excitations.* IABSE Report, 33-34:53-67.

Cosenza, E., Greco, C., & Manfredi, G. (1991). *La valutazione teorica di spostamenti e rotazioni in fase anelastica negli elementi monodimensionali in cemento armato.* Serie IX, Volume II - Fascicolo 3, Atti Della Accademia Nazionale Dei Lincei.

Eligehausen, R., & Langer, P. (1987). *Rotation capacity of plastic hinges and allowable degree of moment re-distribution.* Univesrsitat Stuttgart, paper presented to CEB TG VII/5 Ductility requirements of reinforcing steel, Com II 'Structural Analysis' and Com VII 'Reinforcing Steel'.

Ferretti, D., (1995). *Sul comportomento a breve termine di elementi inflessi in conglomerato cementizio armato.* PHD Thesis, Politecnico di Torino, Turin, Italy.

Langer, P. (1987). *Verdrehfähigkeit plastifizieter Tragwerksbereiche im Stahlbetonbau*, University of Stuttgart, Stuttgart.

Sigrist, V., & Marti, P. (1993). *Versuche zum Verformungsvermögen von Stahlbetonträgern*, Institut für Baustatik und Konstruktion, ETH, Zürich.

Sigrist, V. (1995). *Zum Verformungsvermögen von Stahlbetonträgern*, Institut für Baustatik und Konstruktion, ETH, Zürich.

Eibl, J., & Buhler, B. (1993). *Rotational behaviour of reinforced concrete slabs.* Ductility Requirements for Structural Concrete-Reinforcement, Progress report of Task Group 2.2, Comite Euro International du Beton, Lausanne, p25-44.

Warner, R. F., & Yeo, M. F. (1986). *Collapse behaviour of concrete structures with limited ductility.* Research Report R-61, The University of Adelaide, Department of Civil and Environmental Engineering, Adelaide.

Wong, K. W., Yeo, M. F., & Warner, R. F. (1988). *Non-linear behaviour of reinforced concrete frames*, Civil Engineering Transactions, CE30(2); 57-65.

Mechanics of Structures and Materials, Bradford, Bridge & Foster (eds)
© 1999 Balkema, Rotterdam, ISBN 90 5809 107 4

Reinforced concrete columns in combined tension and bending

K.N.Baker
Department of Civil and Resource Engineering, University of Western Australia, Perth, W.A., Australia

M.T.Sanders
Halpern Glick Maunsell Pty Limited, Perth, W.A., Australia

ABSTRACT: An investigation of symmetrical columns subject to combined tension and biaxial bending using assumptions commonly adopted for compression columns, reveals that under small eccentricity abrupt curvature transition and premature failure may occur. A limited testing programme appears to support the theoretical analysis. Precautionary design measures are suggested.

1 INTRODUCTION

The duty of most reinforced concrete columns and struts is to sustain combined axial compression and bending. Some columns, either for intended duty, or under actions induced by extreme events, are subjected to tension, and this action typically is eccentric. Tension piles, tension hangers and ties, and earthquake-induced tension in columns, are typical examples.

The ultimate strength of columns in compression and bending has been exhaustively examined and reported. Both rigorous and simplified methods of analysis are readily available, and use basic assumptions of the type: a) plane sections remain plane throughout all stages of bending, b) tensile strength of concrete is ignored, c) accredited stress/strain relationships are used for both concrete and steel. By comparison, the literature pertaining to tension-columns is sparse. Extrapolation of the procedure adopted for compression-columns provides an attractive course for the analyst, and is frequently adopted. However such solutions may lead to over confident prediction of ultimate strength, in some cases of low eccentricity of loading. Identifying the conditions under which this may be true is the purpose of this paper.

2 TENSION IN LINEAR/PLASTIC SECTIONS

The performance of a prismatic, rectangular column, constructed of linear-elastic/plastic material, subject to uni-axial eccentric tension, is first used to identify some relevant principles. Consider the section performance of a column of section b.D, subject to an increasing axial tension N, at a fixed eccentricity e. See Figure 1a. Prior to first yield, and for sufficiently low eccentricity ratio e/D, a virtual neutral axis occurs, that is, the axis lies outside the section.

Subsequent to the onset of yield, the neutral axis migrates progressively until maximum load is achieved, and the final neutral axis always lies within the section. The ultimate strength for e may be represented by the simple parabolic interaction:

$$(N_u / N_{uot})^2 + (N_u.e / M_{uo}) = 1. \qquad (1)$$

where N_u and $N_u.e$ are applied axial and moment actions respectively, and
N_{uot} and M_{uo} are ultimate strengths in pure tension and pure moment respectively.

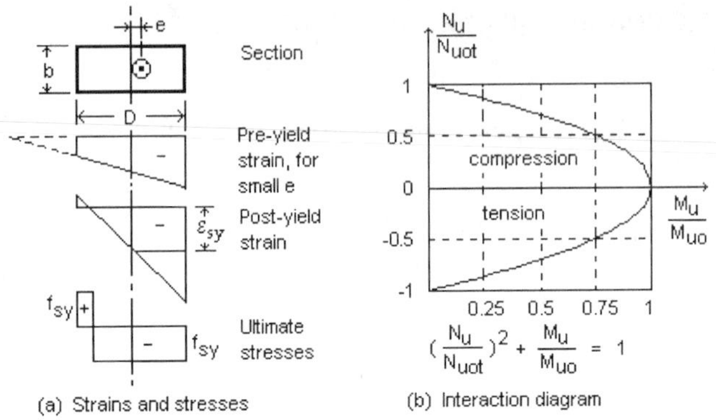

Figure 1. Rectangular linear elastic/plastic section under eccentric tension.

This is symmetrically analogous to compression loading, and can be shown as the second quadrant of a non-dimensional interaction diagram as in Figure 1b. Highly ductile performance is demonstrated for all e/D.

The following observations apply to the rectangular section, and to a wide range of less regular sections:

1. Migration of the neutral axis occurs after first yield;
2. A compression zone is required, even for small values of e; and
3. The 'web' area contributes usefully to ultimate performance.

3 REINFORCED CONCRETE TENSION COLUMN.

3.1 *Rectangular section with no side-face reinforcement.*

Consider first a symmetrical, rectangular tension-column, of cross section b.D, in which reinforcement of area A_1 is provided to each 'main' face as shown in Figure 2, the cage dimension being D'. The yield strength and yield strain of the reinforcement are given by f_{sy} and ε_y respectively. The section is subjected to a tensile force of N, at eccentricity e. For sufficiently low eccentricity (small e/D'), cracking of the entire section occurs when the tensile strain capacity of the concrete is reached, and the pre-yield neutral axis lies outside the section as shown. The section resistance immediately prior to yield of the tension face steel (N_y), and the corresponding section curvature (κ_y), are given by Figure 2a:

$$N_y = A_1 f_{sy} / (0.5 + e/D') \quad \text{and} \tag{2}$$

$$\kappa_y = (1 - C)\, \varepsilon_y / D', \text{ where } C = (0.5 - e/D') / (0.5 + e/D') \tag{3}$$

Immediately at yield of the tension face steel, and with no change to the section resistance, a new strain configuration is assumed by the section, as shown in Figure 2 (b). Strain in the tension face steel increases abruptly to ε_a, and the section curvature becomes

$$\kappa'_y = 2\, C\, \varepsilon_y / (D - D') \tag{4}$$

The change of curvature is potentially large. For example, for D'/D = 0.8 and e = 0.1D', the increase is from $\kappa_y = 0.333\ \varepsilon_y / D'$ to $\kappa'_y = 5.33\ \varepsilon_y / D'$, a 15-fold increase. Impact caused by this transition must be considered. Firstly, the compression face cover, typically only partially confined, may burst under a steadily increasing load. Secondly, either the fracture tensile strain, or the post-yield softening region of the steel, may be reached. In either case, the full tensile resistance of the section may not be available. Some mitigation is possible due to the real stress-strain characteristic of the steel. In addition, the provision of side-face reinforcement will assist, as discussed below.

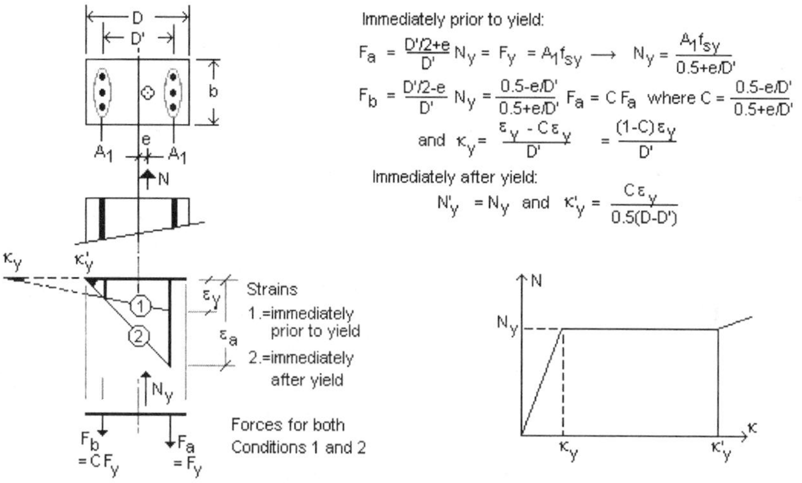
D
D'
b
e
A1
A1
N
κy
κ'y
εy
εa
Strains
1.=immediately prior to yield
2.=immediately after yield
Ny
Forces for both Conditions 1 and 2
Fb = CFy
Fa = Fy
Immediately prior to yield:
Fa = (D'/2+e)/D' Ny = Fy = A1fsy ⟶ Ny = A1fsy / (0.5+e/D')
Fb = (D'/2-e)/D' Ny = (0.5-e/D')/(0.5+e/D') Fa = C Fa where C = (0.5-e/D')/(0.5+e/D')
and κy = (εy - C εy)/D' = (1-C) εy / D'
Immediately after yield:
N'y = Ny and κ'y = C εy / 0.5(D-D')
N
Ny
κ
κy
κ'y

(a) Section performance before and after first yield.

(b) Abrupt curvature transition.

Figure 2. Concrete tension column, no side face reinforcement.

3.2 *General rectangular section, with all faces reinforced.*

Figure 3 shows a general rectangular section in which the reinforcement is distributed symmetrically between the main and side faces, and idealised as uniform strips of areas A_1 and A_2 respectively. The section is subjected to tension N at eccentricity e. The limiting value of e for tension across the full section, prior to first field, is

$$(e / D')_{lim} = [(A_2 + 3A_1) / (A_1 + A_2)]\,[D' / (6D)] \tag{5}$$

When $e / D' < (e / D')_{lim}$, N_y and κ_y are given by

$$N_y = f_{sy} / [1 / (2(A_1 + A_2)) + 3e / (D'(A_2 + 3A_1))] \text{ and} \tag{6}$$

$$\kappa_y = (1 - C)\, e_y / D' \tag{7}$$

where $C = [1 / (2(A_1 + A_2)) - 3e / (D'(A_2 + 3A_1))] / [1 / (2(A_1 + A_2)) + 3e / (D'(A_2 + 3A_1))]$ (8)

As the applied load is increased above N_y, stresses increase in the side face reinforcement, and migration of the neutral axis towards the section is retarded. The section resistance when the neutral axis just reaches the compression face depends on the geometry of the section and the degree to which the reinforcement has reached yield. The resistance N'_y is determined by simultaneous solution of the equations:

$$N'_y = A_1 f_{sy}\,[1 + (D - D') / (D + D' - 2x)] + 2A_2 f_{sy}\,[1 - (D' - x)^2 / (D + D' - 2x] \text{ and} \tag{9}$$

$$N'_y e = A_1 f_{sy}\,[1 - (D - D') / (D + D' - 2x)]\, D'/ 2 + 2A_2 f_{sy}\,(D' - x)^2\,(D' + 2x)/ 6 /(D + D'- 2x) \tag{10}$$

The corresponding curvature is given by:

$$\kappa'_y = 2\,\varepsilon_a / (D + D') \tag{11}$$

Parametric methods have been used to solve equations (9) and (10), and to develop the N vs κ diagram for the full loading range to ultimate. Typical non-dimensionalised solutions $0 < N < N'_y$ are shown in Figure 4, for e/D' values 0.05, 0.1 and 0.2.

By comparison with Figure 2, it is seen that a smoother transition of curvature occurs as a result of the inclusion of side face reinforcement.

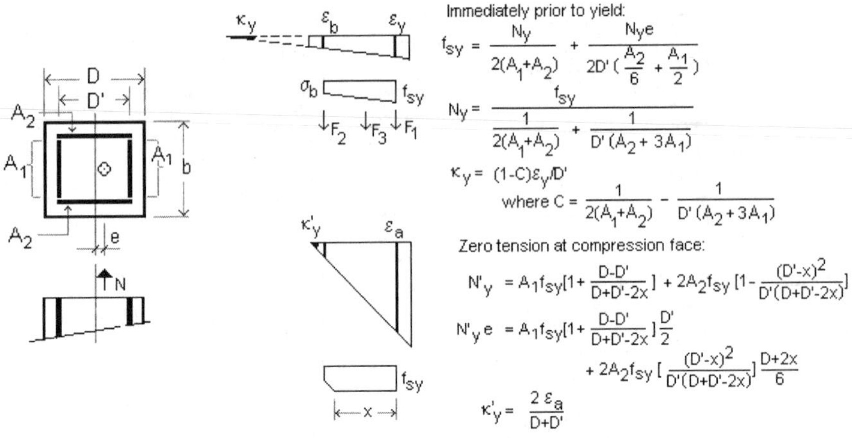

Figure 3. Reinforced concrete tension column, all faces reinforced.

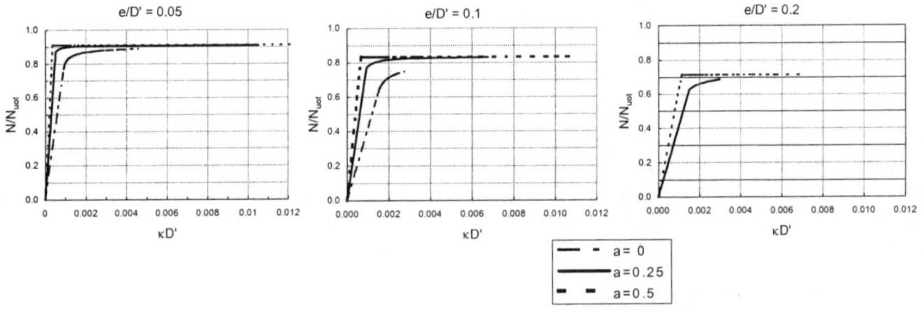

Figure 4. N/N_{uot} vs $\kappa D'$ up to N'_y, with $a = A_1 / A_s$

3.3 *Section ductility with low eccentricity.*

When the eccentricity of loading is low, a steep strain gradient is induced in the section, and the strain in the tension face steel becomes very large as ultimate load is approached. If the post-yield softening strain of this steel is reached, then some loss of expected ultimate strength may occur. The critical neutral axis depth d'_n for this condition is readily shown to be

$$d'_n = \tfrac{1}{2} (D - D') . (\varepsilon_{cu} / \varepsilon_{su}) \tag{12}$$

where ε_{cu} = ultimate strain capacity of extreme fibre concrete in compression, and
ε_{su} = post-yield softening strain of reinforcement, corresponding to uniform elongation.

Since for such cases all the rebar is probably yielding, the eccentricity e' corresponding to this case is given by

$$N_{ut} . e' = F_c (D / 2 - \gamma d_n / 2) \tag{13}$$

where F_c = force in the compression block,
γ = appropriate neutral axis factor for a rectangular stress block approximation, and
d_n = depth to the neutral axis from the compression face.

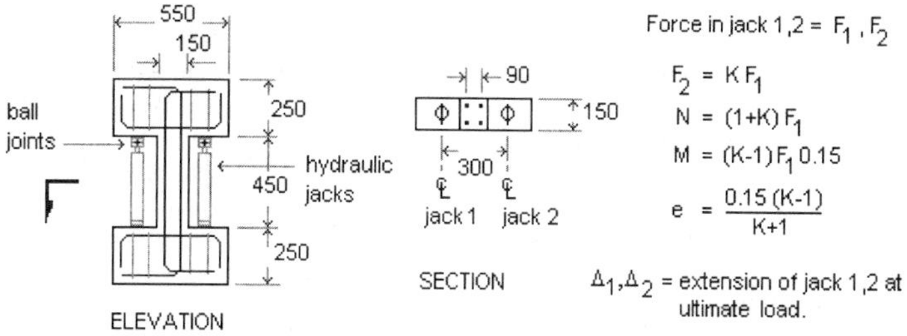

Figure 5. Test apparatus.

The critical eccentricity e' is therefore dominantly dependent on the ratio of F_c to F_s - F_c, where F_s is the sum of the forces in the tensile reinforcement. In the absence of reliable data on the post-yield softening of the rebar being considered, a lower bound estimate of e' may be made by treating ε_{su} as a strength limiting criterion. For normal concrete and steel proportions likely to be used in tension columns, the ratio e/D' varies in the approximate range 0.05 to 0.2 for rebar of 5% uniform elongation.

Accordingly, caution should be observed in each specific case to check whether the strength of a tension column, loaded at low eccentricity, is governed by post-yield softening of the tension face steel. Testing is currently proceeding to check this postulate.

4 TEST PROGRAMME.

As part of a tension - column test programme, a limited set of tests was conducted on square, reinforced concrete sections, subjected to an eccentric tensile load. The test specimens comprised a test section of 150x150 mm, with monolithically cast hammer-head caps, as shown diagrammatically in Figure 5. The test section was reinforced with 4-Y12 (Grade 400) deformed bars, to achieve a reinforcement ratio of 0.02. Load was applied through two hydraulic jacks, placed centrally and at 300 mm centres, the required eccentricity being achieved by selecting the ratio of jacking forces. Various eccentricities were tested and duplicate tests were carried out. Extension of the left and right hand sides of the test length were observed with each load increment. Loading continued until peak load was achieved.

Extracts from selected test results are shown in Table 1, in which Δ_1 and Δ_2 are total specimen deformations of compression and tension faces at peak load, respectively. For all series, transverse displacement of the specimen occurred as the section axis moved towards the axis of load application. Failure was concentrated at column ends. For Series 1 tests (e/D' = 26 / 90 = 0.29) an abrupt, but small, change of curvature occurred as yielding of the tensile face steel became apparent. For Series 2 and 3, a smoother transition of curvature occurred. Testing is proceeding (May 1999) to quantify the curvature transitions.

Table 1.

Series	e (mm)	N_u test (kN)	Δ_1 (mm)	Δ_2 (mm)	N_u calc (kN)
1	26	191.9	-0.16	14.4	157
2	50	146.4	-0.12	9.0	115
3	150	83.8	-0.77	6.5	62

5 DISCUSSION AND CONCLUSIONS.

For compression columns, design practice commonly requires that a minimum eccentricity be adopted for the sizing of members. See for example AS3600 - 1994, Section 10. This eccentricity accounts for construction tolerances and material non-uniformity. A similar approach to tension columns, and for the same reasons, should be considered. It would be prudent for designers to consider the effects of curvature transition, and low eccentricity ductility effects for each application. Where no reliable redundant load paths are available, premature failure might occur.

Conclusions drawn from this study are:

1. Some unexpected hazards exist in the sizing of tension columns where low eccentricity of loading applies.
2. Tension columns should be sized for a minimum eccentricity, that is for a simultaneous bending moment estimated as the product of the design axial load and the minimum eccentricity. A value of e_{min} of 0.1 D' is tentatively proposed.
3. Side face reinforcement should be provided to tension columns wherever possible. In practical terms, this may mean the use of at least eight bars, distributed around the faces of the member.

REFERENCES

Standards Australia 1994. *AS3600-1994 Australian Standard Concrete Structures.*

Mechanics of Structures and Materials, Bradford, Bridge & Foster (eds)
© 1999 Balkema, Rotterdam, ISBN 90 5809 107 4

Finite element modelling of HSC squat walls in shear

S.J. Foster
School of Civil and Environmental Engineering, The University of New South Wales, Sydney, N.S.W., Australia

B.V. Rangan
School of Civil Engineering, Curtin University of Technology, Perth, W.A., Australia

ABSTRACT: In this paper a three-dimensional finite element model is developed to study the behaviour of reinforced concrete shear walls. The concrete is modelled using non-linear orthotropic 20-node isoparametric brick elements, the reinforcing steel using 3-node elasto-plastic bar elements. Four high strength concrete shear wall panels were analysed using the proposed formulation with a good correlation observed between the finite element and the experimental results.

1 INTRODUCTION

Many multi-storey structures have reinforced concrete cores that are designed to carry either the whole or part of the lateral loading resulting from wind shear or seismic events, as well as the dead load and live loads. As with columns in the lower storey of multi-storey buildings, economic benefits can be gained by reducing the core wall thickness through the use of high strength concrete. To take full advantage of the use of high strength concrete, a more detailed assessment is required to ensure adequate ductility. This paper presents a finite element formulation for the analysis of reinforced concrete walls. The formulation is based on an orthotropic constitutive model for the concrete and elastic-plastic model for the reinforcing steel. The FE model takes into consideration the effects of cracking, crushing and the loss of compressive strength due to transverse tension fields and other biaxial and triaxial effects. Experimental corroboration is undertaken for a number of panels tested by Gupta and Rangan (1996).

2 3D MATERIAL MODELLING

The reinforcing steel is modelled using 3-node isoparamentric bar elements with a bi-lieaer elasto-plastic relationship between stresses and strains. The concrete is modelled using twenty node isoparametric plane stress elements with a 3 x 3 x 3 Gauss quadrature being used to formulate the element stiffness matrix. The element can be in any of four states in any of the three principal stress directions at any point in the analysis; that is undamaged, cracked, crushed, or both cracked and crushed.

The concrete element is formulated using the principle of equivalent uniaxial strains, as outlined by Darwin and Pecknold (1977), but with the simplification that the equivalent uniaxial strain is approximated by the linear system such that

$$\begin{Bmatrix} \varepsilon_{1u} \\ \varepsilon_{2u} \\ \varepsilon_{3u} \end{Bmatrix} = \frac{1}{(1+\mu)(1-2\mu)} \begin{bmatrix} (1-\mu) & \mu & \mu \\ \mu & (1-\mu) & \mu \\ \mu & \mu & (1-\mu) \end{bmatrix} \begin{Bmatrix} \varepsilon_1 \\ \varepsilon_2 \\ \varepsilon_3 \end{Bmatrix} \tag{1}$$

where ε_i is the strain in the ith principal direction (i=1, 2, 3), ε_{iu} is the equivalent uniaxial strain in the ith principal direction and μ is Poisson's ratio.

The equivalent uniaxial strain model was established in order to predict the multi-axial behaviour of concrete by subtracting the Poisson's ratio effect and allowing the use of uniaxial stress-strain curves. The equivalent uniaxial strain, ε_{iu}, can be thought of as the strain that would exist in one direction when the stress is zero in the other. The stress-strain relationship is expressed as

$$\{\sigma\} = \mathbf{D}_{iu}\{\varepsilon_{iu}\} \tag{2}$$

where $\mathbf{D}_{iu}$ is the material elasticity matrix in the material coordinate system and is given by the diagonal matrix

$$\mathbf{D}_{iu} = \lceil E_1 \quad E_2 \quad E_3 \rfloor \tag{3}$$

and where E_i (i=1, 2, 3) are the secant moduli in the principal stress directions.

The stress versus strain relationship for the concrete is shown in Figure 1. The CEB-FIP (1993) model is used to describe the base curve for the ascending branch of the compression stress-strain response. That is

$$f_c = \nu f_{cp}\left[\frac{n\eta - \eta^2}{1+(n\eta-2)\eta}\right] \quad \text{..........} \quad \text{with } \eta = \varepsilon_{iu}/\varepsilon_o \tag{4}$$

where f_c is the stress in the concrete corresponding to the equivalent uniaxial strain ε_{iu}; f_{cp} is the strength of the in-situ concrete; $n = E_o/E_s$ where E_o is the initial elastic modulus and E_s is the secant modulus corresponding to the strain at peak stress $\left(E_s = \gamma_1 f_{cp}/\varepsilon_o\right)$; ε_o is the strain corresponding to the peak stress νf_{cp} and; ν is the modified compression field factor to account for the influence of the transverse strains on the concrete uniaxial compressive strength or the efficiency factor. The descending branch of the compression stress-strain curve and the tension relationships are modelled using the linear relationships shown in Figure 1.

Vecchio and Collins showed that the peak compressive stress, νf_{cp}, is a function not only of the principal compressive strain but also of the coexisting principal tensile strain, with concrete subject to high tensile strains being softer and weaker than for the uniaxial compressive state.

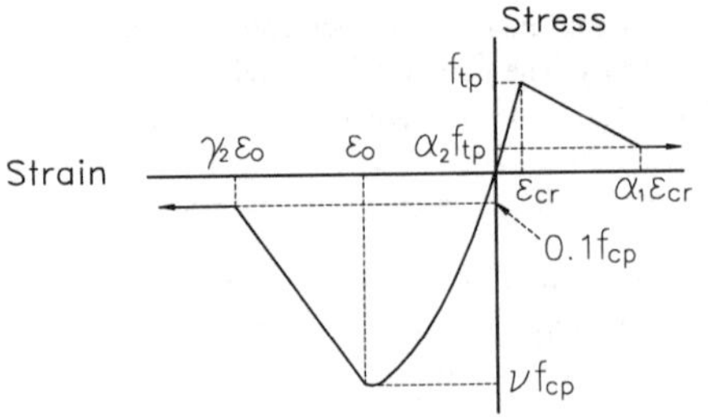

Figure 1. Stress-strain relationship used to model the concrete.

The modified compression field factor of Vecchio and Collins (1993) is given by

$$\nu = \frac{1}{1.0 - K_c K_f} \leq 1.0 \tag{5}$$

where

$$K_c = 0.35(-\bar{\varepsilon}_t / \varepsilon_o - 0.28)^{0.80} \geq 0.0 \tag{6a}$$

$$K_f = 0.1825\sqrt{f_{cp}} \geq 1.0 \tag{6b}$$

For panels subject to two-dimensional stress fields $\bar{\varepsilon}_t = \varepsilon_1$, where ε_1 is the tension strain transverse to the direction of the minor principal stress. However, the writers are not aware of any research on the effect of transverse tension strains, on compression strength, in three-dimensional stress fields. In this study Equation 5 is adopted with

$$\bar{\varepsilon}_t = \sqrt{\varepsilon_{1u}^2 + \varepsilon_{2u}^2} \quad \text{............... for Compression - Tension - Tension} \tag{7a}$$

$$\bar{\varepsilon}_t = \varepsilon_{1u} \quad \text{............... for Compression - Compression - Tension} \tag{7b}$$

On obtaining the principal stress vector given by Equation 2 the stresses are transformed into global the coordinate system and the out of balance forces calculated. Convergence was monitored using a displacement norm with a solution tolerance of 0.1 percent.

3 NUMERICAL EXAMPLES

Wall panels S4, S5, S6 and S7 of Gupta and Rangan (1996) were modelled using 2D membrane elements and using the 3D finite element formulation presented above. The 2D model uses a non-linear formulation based on equivalent uniaxial strains with the concrete modelled as a non-linear, orthotropic, material. The model is a rotating crack model and takes into account concrete cracking and crushing, the Modified Compression Field theory and biaxial compression strengthening effects. Details of the 2D FE formulation are given in Foster (1992) and Foster et al. (1996) but with the uniaxial concrete stress-strain law given by Equation 4, above. The reinforcing steel was modelled using elastic-plastic stress-strain laws to match that of the experiments.

Details of the panels are shown in Figure 2 and details of the reinforcing bar material properties are given in Table 1. The 2D mesh (shown in Figure 3a) consisted of 30 by 8-node concrete elements for the web and flanges and 71 by 3-node bar elements for the reinforcing steel. The stiff top and bottom boundary elements were taken as linear elastic. The 3D mesh (shown in Figure 3b) consisted of 40 non-linear 20-node isoparametric brick elements to model the flanges and the web, 96 linear-elastic brick elements to model the top and bottom plates and 119 by 3-node bar elements to model the reinforcing steel. One half the panel only is modelled due to symmetry. The in-situ concrete compressive strengths (f_{cp}) were taken as 0.85 times the mean cylinder strength (f_{cm}) obtained in the laboratory and are given in Table 2 together with the tension strength and elastic moduli of the concrete used in the FE modelling. A concrete strain at peak stress of $\varepsilon_o = -0.0025$ was adopted and the material modelling parameters α_1, α_2 and γ_2 were taken as 3.0, 0.0 and 3.0, respectively.

The walls were subject to a constant vertical loading and increasing lateral load to failure. The axial thrusts applied to each specimen are given in Table 2

A comparison between the experimental and numerical failure loads, given in Table 3, shows that both the 2D and 3D models give a good prediction of the failure loads. However, comparisons of the load versus lateral displacement responses, shown in Figure 4, show that the

Table 1. Reinforcing bar material properties.

Bar	W5	W7.1	W8	W10	W12.5
Area (mm2)	19.6	39.6	50.2	78.5	122.5
Yield Strength (MPa)	578	545	533	529	531

Table 2. Details of experimental walls.

Panel	f_{cm} (MPa)	f_{cp} (MPa)	f_{tp} (MPa)	E_o (GPa)	Applied Axial Thrust (kN)
S4	75	64	4.2	35	0
S5	73	62	4.2	35	610
S6	71	60	4.2	35	1230
S7	71	60	4.2	35	610

Table 3. Comparison of experimental and numerical failure loads.

Panel	Exp. Failure Load (kN)	2D FEM Failure Load (kN)	2D FEM / Exp.	3D FEM Failure Load (kN)	3D FEM / Exp.
S4	600	650	1.08	645	1.08
S5	790	835	1.06	803	1.02
S6	970	993	1.02	910	0.94
S7	800	816	1.02	831	1.04

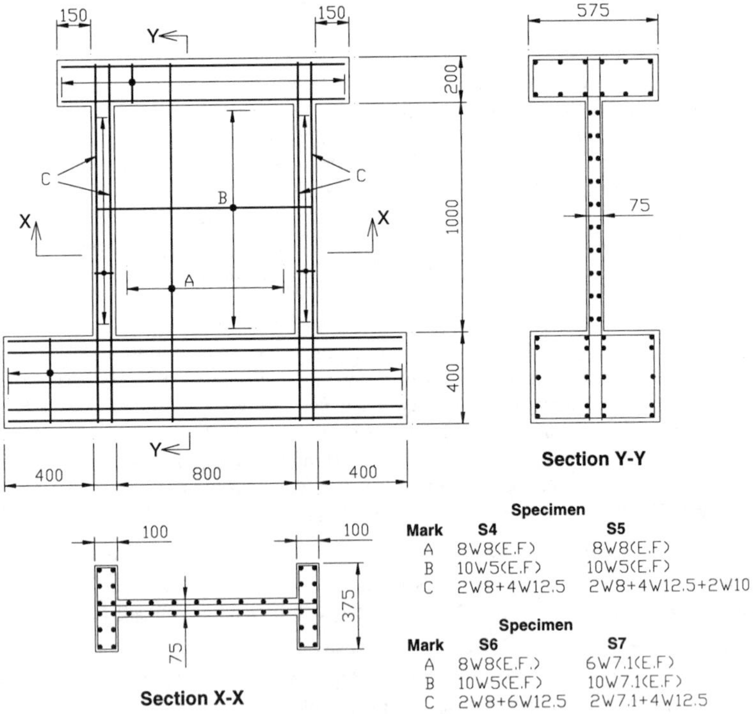

Figure 2. Details of experimental shear walls tested by Gupta and Rangan (1996).

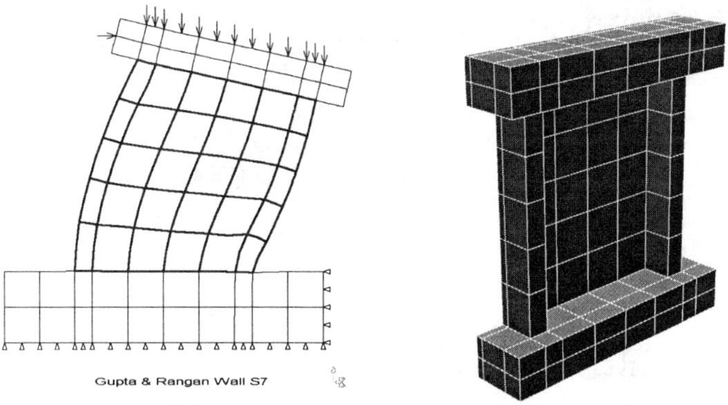

(a) (b)

Figure 3. Finite element meshes for Gupta and Rangan (1996) wall panels; a) 2D model showing deflected shape of panel S7 at failure (50x displacement magnification); b) 3D model.

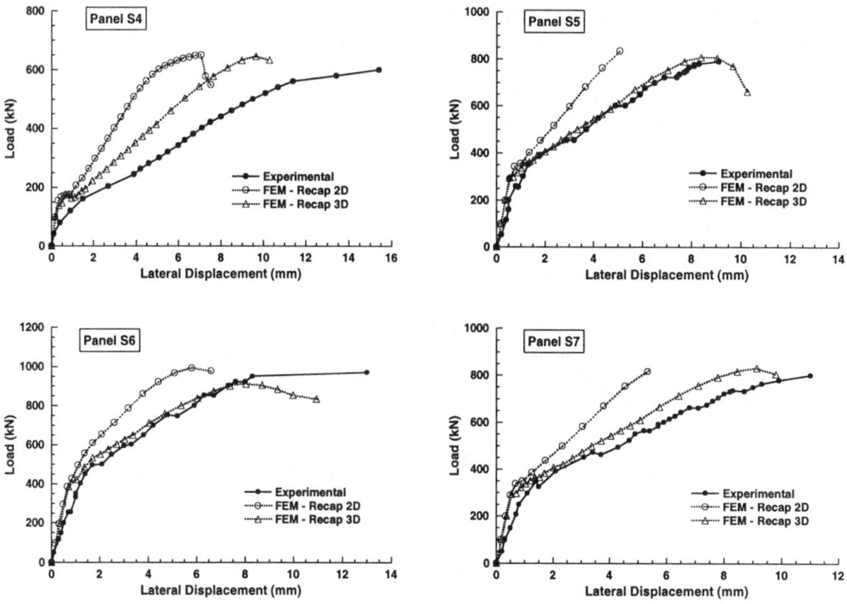

Figure 4. Load versus lateral displacement for Gupta and Rangan (1996) wall panels S4, S5, S6 and S7.

2D model under predicts the lateral drift of the walls, possibly due to an over calculation of the stiffness contribution of the flanges (edge elements) to the lateral displacement of the walls. The results of the 3D model show good correlations to the experimental data both for the failure loads and the load versus lateral displacement response.

Stress contours at failure are plotted for panel S7 in Figure 5a. The figure shows a wide diagonal compression band through the web of the wall with the maximum compression stresses in the web of the order of 35 MPa. The major principal strains at failure for panel S7 is plotted

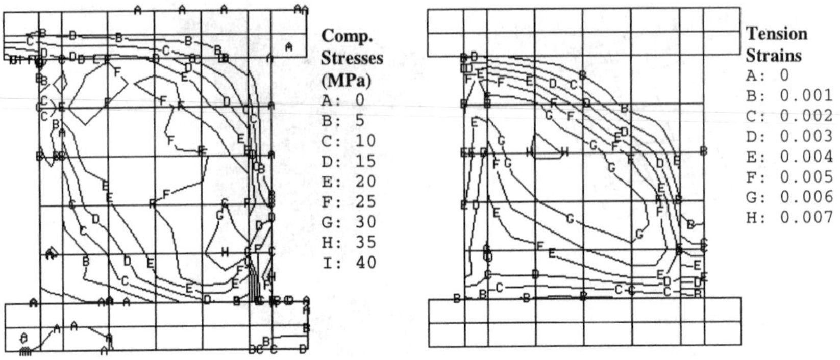

(a) **(b)**

Figure 5. FE results at failure for Gupta and Rangan (1996) panel S7, a) minor principal stresses and, b) major principal strains.

in Figure 5b. At the peak load the principal tension strains in the web are of the order of 6000 $\mu\varepsilon$. With this level of transverse tension strain in the web, the effective strength of the concrete (given by Equations 5 to 7) is 36 MPa. Thus at failure the capacity of the web was exhausted. No yielding of the horizontal or vertical web reinforcement occurred but at failure the tension flange steel had yielded at the base of the wall.

4 CONCLUSIONS

In this paper a three dimensional finite element model is developed for the analysis of reinforced concrete shear walls. The model was used to analyse four high strength concrete shear wall panels tested by Gupta and Rangan (1996) with the FE results compared to the experimental data and with the predicted load versus displacement response using 2D plane stress elements. The failure loads predicted by the 2D finite element model compared well with the experimental data, however, the load versus displacement response of the 2D model was stiffer that that of the experimental data. The 3D finite element results compared well with that of the experimental data for both the failure load and the load versus displacement data. The difference between the two and three-dimensional models is the contribution of the flanges to the in-plane stiffness of the wall and shows the importance of modelling the behaviour of shear walls in three-dimensional space.

REFERENCES

CEB-FIP Model Code. 1993. Comite Euro-International du Beton. Thomas Telford.

Darwin, D., & Pecknold D.A. 1977. Nonlinear biaxial stress-strain law for concrete. *ASCE, Journal of the Engineering Mechanics Division,* 103(EM2): pp 229-241.

Foster, S.J. 1992. An application of the arc length method involving concrete cracking. *International Journal for Numerical Methods in Engineering* 33(2): 269-285.

Foster, S.J., Budiono, B., & Gilbert, R.I. 1996. Rotating crack finite element model for reinforced concrete structures. *International Journal of Computers & Structures* 58(1): 43-50.

Gupta, A., & Rangan, V. 1996. Studies on reinforced concrete structural walls. *Research Report No. 2/96.* School of Civil Engineering, Curtin University of Technology, Perth, Australia.

Vecchio F.J., & Collins M.P. 1993. Compression response of cracked reinforced concrete. *ASCE, Journal of Structural Engineering,* 119(12): pp 3590-3610.

Mechanics of Structures and Materials, Bradford, Bridge & Foster (eds)

 ISBN 90 5809 107 4

Biaxial bending of high strength concrete columns

A.S. Bajaj & P. Mendis
Department of Civil and Environmental Engineering, University of Melbourne, Vic., Australia

ABSTRACT: This paper presents the details of a project to investigate the behaviour of high strength concrete columns under biaxial bending. The numerical results of a case study are presented to show the differences in failure surfaces obtained with the conventional rectangular stress block and the revised blocks suitable for high strength concrete. Finally the moment curvature relationship for a typical column tested during the current experimental program is presented. A comparison of the theoretical results with experimental results is also presented.

1 INTRODUCTION

Applications of high strength concrete (HSC) in structures are increasing day by day. The use of HSC is very common in heavily loaded lower storey columns in buildings. Biaxial bending occurs in corner columns of these buildings. In recent years, columns subjected to biaxial bending are often encountered and designed as building and bridge systems become simpler. However, to a large extent practice has preceded theory with constitutive equations simply being extrapolated to higher strengths. The research at the University of Melbourne and elsewhere has shown that simply extrapolating the constitutive equations for normal strength concrete to HSC could either be unsafe or uneconomical.

A comprehensive literature survey and other work done at the University of Melbourne have shown the inadequacy of the design rules in this area. This paper describes the Ph.D. research project at the University of Melbourne to formulate design rules for these columns. The important findings up to now are also presented.

2 BIAXIAL INTERACTION EQUATION

In AS 3600, a horizontal plane through the failure surface is approximated by an interaction formula given by:

$$\left(\frac{M_x}{\phi M_{ux}}\right)^{\alpha_n} + \left(\frac{M_y}{\phi M_{uy}}\right)^{\alpha_n} = 1.0 \tag{1}$$

In Eq. (1), the design strengths in uniaxial bending about the principal axes x and y are $\phi\ M_{ux}$ and $\phi\ M_{uy}$, respectively, for a design axial force N*, while M^*_x and M^*_y are the design moments, and α_n varies with N* and is given by:

$$\alpha_n = 0.7 + 1.7\left(N^*/\ 0.6N_{uo}\right) \text{ with } 1.0 < \alpha_n < 2.0 \tag{2}$$

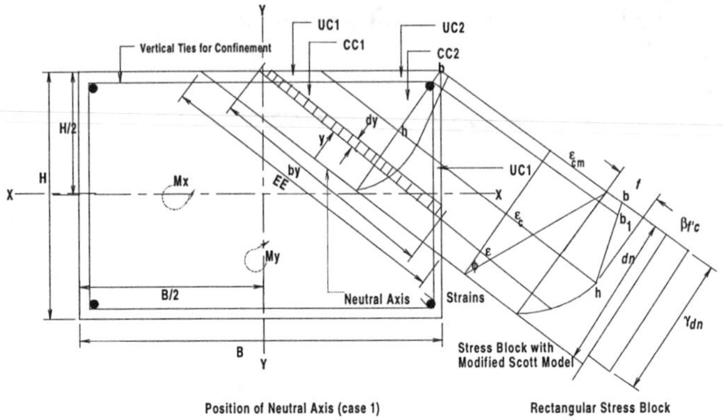

Figure 1. Stress-Strain Distribution for Concrete with Modified Scott Model and Rectangular Stress Block

The format of Eq. (1) was originally proposed by Bresler (1960), although he did not specify values for α_n. The value of α_n given by Eq. (2) has been adopted from the British Standard BS 8110 (1985). A revised value of α_n has been derived from the failure surfaces of high-strength concrete columns with different dimensions, reinforcement arrangements, etc. The three dimensional failure surfaces incorporate the interaction diagrams for α between 0^0 to 90^0 These failure surfaces have been used to calibrate the formula given in the Australian Concrete Structures code, AS 3600 (1994), to cover high-strength concrete. These results have been presented elsewhere (Bajaj and Mendis 1999).

3 RECTANGULAR STRESS BLOCK FOR NORMAL AND HIGH STRENGTH CONCRETE

Presently most codes for concrete structures define the rectangular stress block concrete (AS 3600 - 1994) as given below and is mainly used for Normal Strength Concrete (NSC) with concrete compressive strength $f'c$ up to 50 MPa. The normal conventional rectangular stress block (ORSB) has been defined by two parameters $\beta = 0.85$ as shown in Figure 1, a horizontal constant (which is to determine the intensity of the stress block) and γ (the ratio of the depth of the stress block to the depth of the neutral axis d_n from the top) is given below:

$$\gamma = 0.85 - 0.007\left(f'c - 28\right), \text{ where } f'c \text{ is in MPa.} \quad (3)$$

The Modified Scott Model shown in Figure1 is explained in section 4.

The research projects conducted at The University of Melbourne and other universities around the world have shown that the shape of the stress-strain curve for high-strength concrete is different. Therefore the currently used parameters for an equivalent rectangular stress block are not applicable to high strength concrete (Pendyala and Mendis 1997).

The revised stress block (RSB-1) suggested by Pendyala and Mendis (1997) is given below:

$$\gamma = \left[0.85 - 0.007(f'c - 28)\right] \text{ within limits 0.65 and 0.85 for } f\ c \leq 60\ MPa \quad (4a)$$

$$\gamma = \left[0.65 - 0.00125(f'c - 60)\right],\ 60 < f'c < 100\ MPa \quad (4b)$$

$$\beta = 0.85 \text{ for } f'c \le 60\ MPa \qquad (5a)$$

$$\beta = 0.85 - 0.0025\left(f'c - 60\right) \text{ for } 60 < f'c \le 100\ MPa \qquad (5b)$$

The above mentioned stress-block is a combination of rectangular and triangular stress blocks. More details are given by Pendyala and Mendis (1997).

For comparison, the proposed revised stress block (RSB-2) for normal and high strength concrete up to 100 *MPa* based on experimental testing of uniaxial eccentrically loaded columns by Ibrahim and Macgregor (1997) is also considered (Eq. 6(a) and 6(b)).

$$\gamma = 0.95 - \frac{f'c}{400} \ge 0.70,\ f'c \text{ in } MPa \qquad (6a)$$

$$\beta = 0.85 - \frac{f'c}{800} \ge 0.725,\ f'c \text{ in } MPa \qquad (6b)$$

4 THEORETICAL STRESS-STRAIN MODEL

For establishing design and moment curvature relationships, the Modified Scott Model (Pendyala and Mendis 1997) has been recommended as the most suitable model to predict the full range stress-strain behaviour of concrete strength ($f'c$) between 25 *MPa* and 100 *MPa*. A computer program has been developed to incorporate the full-integrated equations for all the five possible cases of neutral axis for biaxial bending. The concrete compressive force, P_C, can be calculated as the integral sum of the two stages of stress variation multiplied by area varying in width and depth directions. This force is defined in Eq. (7). Figure 1 shows the first case of neutral axis with Modified Scott Model used to denote the stress-strain relationship for confined concrete. These cases are described in detail elsewhere (Bajaj and Mendis 1999).

Total concrete compressive force, P_C

Unconfined compressive force in stages 1 & 2 = $P_{UC} = UC1 + UC2$

Confined compressive force in stages 1 & 2 = $P_{CC} = CC1 + CC2$

$$\text{Therefore, } P_C = P_{UC} + P_{CC} \qquad (7)$$

5 EXPERIMENTAL WORK

A total of 10, 150x150 columns with concrete compressive strengths ($f'c$) of 50 and 100 *MPa* representing normal-strength and high-strength concrete respectively were cast and tested in this study. The geometry of the specimens and the test set-up is shown in Figure 3. Two loading brackets (300x300) were being provided at each column end to assist with the application of biaxially eccentric loads. First six specimens were tested with biaxial eccentricities $\left(e_x, e_y\right)$ of (27,100), (57,100) and (100,100) corresponding to α values of 15 or 75, 30 or 60 and 45 degrees $\left(\alpha = \tan^{-1} e_y / e_x\right)$. Results for these tests are shown in Table1.

Continuous loading was maintained at a low displacement rate of 0.1 mm per minute with Material Testing System (MTS) machine and 500 kN actuator capacity as shown in Figure 2, with linear variable displacement transducers (LVDT^S) and electrical strain gauges mounted closed to the centre of the columns. Six displacement transducers were used to measure the curvature and deflections at the mid-height of the columns in both x and y directions. All data were collected and processed using a micro-computer.

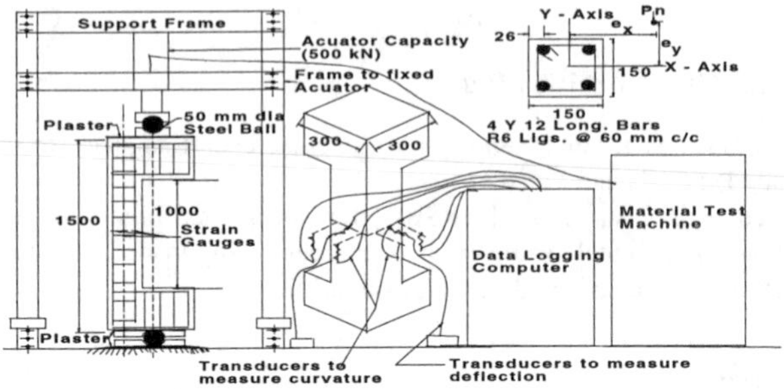

Figure 2 Test set-up (Not to scale)

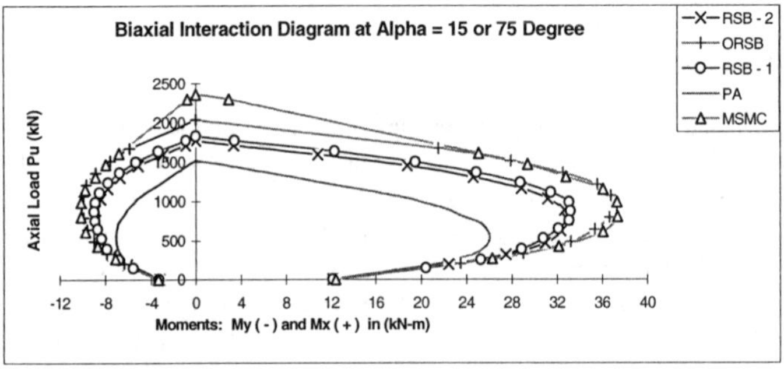

Figure 3 Interaction diagrams for square column with $f_c^{'} = 98\ MPa\ and\ \alpha = 75^0$

6 CASE STUDY

In biaxial bending, the inclination and depth of neutral axis will vary depending upon the magnitude of applied axial load or moments or both. A computer program (Ehsani 1986) has been modified to incorporate the revised rectangular stress blocks. It uses simple shifting of neutral axis with respect to height and angle of inclination of the neutral axis and biaxial moments.

As an example, the column cross-section used in the experimental program (Figure 2) with same amount of reinforcement has been analysed In this Figure 2, the moments about axes x and y are M_x and M_y and nominal axial load is P_n acting at a point such that eccentricity, $e_x = M_y / P_n$ and $e_y = M_x / P_n$. Therefore angle of rotation also defined as angle of eccentricity, $\alpha = \tan^{-1} e_y / e_x$. When the angle of eccentricity is either 0^0 or 90^0, the corresponding moments will be uniaxial bending moments. The design of a column under an axial load associated with bending moments about both axes is tedious. The ultimate load and moments for certain values of eccentricities is influenced by such factors as dimensions of cross section, percentage of reinforcement, number and arrangement of bars, yield stress of steel, compressive strength of concrete and cover to the steel reinforcement. The square columns with $f'c = 56.8, 98, 50.6$ and $90\ MPa$ have been analysed by ORSB, RSB-1, RSB-2, MSMC and with proposed approach (PA) based on Modified Scott Model. Proposed approach

(PA) is to limit the maximum cylinder stress $(f'c)$ in concrete for the tested column specimen stress to $C_1 f'c$, where C_1 is defined in Eq. (8). Figure 3 shows a typical interaction diagram for the square column respectively with PA, MSMC, old and revised rectangular stress blocks. As seen from the diagram, the rectangular stress blocks overestimate the moment capacity of these biaxially loaded columns. There is a need to redefine a separate position factor in the "Revised Stress Block" for high strength concrete biaxial loaded columns. This will be attempted during the Ph.D. project. In proposed approach, maximum concrete strength, f, (as shown in stress block of Figure 1), has been taken as $C_1 f'c$ instead of $\beta C_1 f'c$.

$$C_1 = 1.1 - 0.005\, f'c \leq 0.70, \text{ where } f'c \text{ is in } MPa \qquad (8)$$

7 ANALYSIS OF RESULTS

The numerical and experimental analysis are useful to derive the complete load-moment curvature relationship, including ascending and descending behaviour for a short reinforced concrete column subjected to biaxial bending. The theoretical results are obtained by further modifying the existing Modified Scott Model as mentioned earlier. A computer program has been developed using theoretical stress-strain model (Figure 1). The computer program provides failure load capacity along with curvatures with respect to both principal axes at given eccentricities and maximum concrete strain. The computer analysis ignores effects of creep, shrinkage and deflections along both axes.

Tests were continued beyond the maximum load and were terminated when the pinned ends could no longer rotate or when the concrete spalled off outside due to brittle cracks in high strength concrete columns. The experimental curvatures and deflections were measured by LVDT^S as shown in Figure 2. The experimental values of M_x and M_y were computed using the experimental axial load values obtained from the load cell measurements, and the load

Table 1 -Comparison of Computed Nominal Failure Load Capacity with Test Loads

Particulars of Columns Tested (150 X 150 mm) with Clear Cover = 20 mm									
(Average Cylinder Strength)	Tested Eccentricity in mm		Failure Loads — Maximum Concrete Strain $(\varepsilon_{cm} = 0.003)$						
f'_c in MPa	e_x	e_y	ORSB (kN)	RSB-1 (kN)	RSB-2 (kN)	MSMC (kN)	Proposed Approach (kN)	Experi-mental (kN)
1 56.8	27	100	208.2	208.2	209.7	217.5	185.7	184.1
2 98	27	100	264.8	250.8	251.6	259	222	221.2
3 64.9	57	100	177.2	175.5	176.9	180.7	157.9	172
4 96	57	100	213.6	197.5	201.4	215	175.1	190
5 50.6	100	100	117.8	117.8	120.4	122.3	107.7	111
6 90	100	100	146	137.2	143.4	151.8	128.8	136

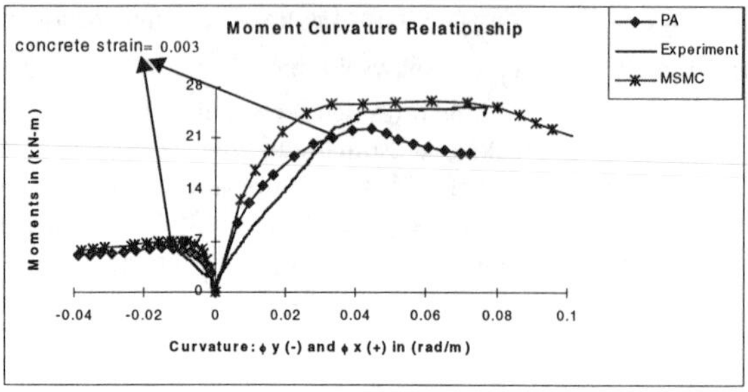

Figure 4 Comparison of moment curvature curves for square column with $f'c = 98MPa$ and $\alpha = 75^0$

eccentricities were corrected for the mid height deflections of the column. The maximum concrete compressive stresses were determined using 76.2 X 152.4 mm cylinders. These are reported in Table 1. Figure 4 shows moment curvature $(M - \phi)$ curves for a typical specimen (No. 2 of Table1), tested at eccentricities $e_x = 27\,mm$ and $e_y = 100\,mm$. As seen from Table 1 and Figure 4, good agreement is obtained between the experimental strengths and the analytical values calculated using the proposed approach at maximum concrete strain of 0.003.

8 CONCLUDING REMARKS

1) With significant differences in material behaviour compared to low strength concrete, important differences have appeared in concrete columns with concrete compressive strength higher than 50 MPa, subjected to biaxial bending.

2) Different failure surfaces are obtained with the proposed approach and rectangular stress blocks suitable for normal and high strength concrete. For the columns tested in the experimental program, the suggested rectangular stress blocks overestimate the load and moment capacity of these types of columns. Position factor in the rectangular stress block must be redefined for these types of columns.

3) In this project, design constitute equations for biaxial bending of high-strength concrete columns will be developed.

4) More testing for multiple bar columns is recommended.

REFERENCES

AS 3600 1994. Australian Standard Concrete Structures Code. *Standard Australia:*93-102.NSW.

Bajaj, A.S. & Mendis, P.A. 1999. Interaction Exponent For Biaxial Bending of High Strength Concrete Columns, Submitted for Publication, *ACI Structural Journal, Detroit.*

Bresler, B. 1960. Design Criteria for Reinforced Columns under Axial Load and Biaxial Bending, *Journal Of The American Concrete Institute:* 481-490.

Ehsani, R.M. 1986. CAD for Columns, *Concrete International:* 43-47.

Hisham, H.H. Ibrahim, and Macgregor, J.G. 1997. Modification of the ACI Rectangular Stress Block for High-Strength Concrete, *ACI Structural Journal:* 40-48.

Pendyala, R.S. & Mendis, P.A. 1997. A Rectangular Stress Block For High Strength Concrete, *Civil Engineering Transactions, IE Australia:*135:142

Mechanics of Structures and Materials, Bradford, Bridge & Foster (eds)
© 1999 Balkema, Rotterdam, ISBN 90 5809 107 4

An experimental study of the long-term behaviour of reinforced concrete flat slabs under sustained service loads

R.I.Gilbert
School of Civil and Environmental Engineering, The University of New South Wales, Sydney, N.S.W., Australia

ABSTRACT: An experimental program of long-term testing of large-scale reinforced concrete flat slab structures is described and the results from the first series of tests on two continuous flat slab specimens are presented. Each specimen was subjected to sustained service loads for periods up to 220 days and the deflection, extent of cracking and column loads were monitored throughout. The measured long-term deflection is many times the initial short-term deflection, due primarily to the loss of stiffness associated with time-dependent cracking under the combined influences of transverse load and drying shrinkage. This effect is not accounted for in the current code approaches for deflection calculation and control, and the test results will be used to suggest improvements to the current design procedures for the serviceability of flat slabs.

1 INTRODUCTION

The design of a reinforced concrete structure is complicated by the difficulties involved in estimating the service load behaviour. The deflection and extent of cracking in reinforced concrete flexural members depend primarily on the non-linear and inelastic properties of concrete and, as such, are difficult to predict with confidence. The problem is particularly difficult in the case of slabs, which are typically thin in relation to the their spans and are therefore deflection sensitive. It is stiffness rather than strength that usually governs the design of slabs, particularly in the cases of flat slabs and flat plates.

In most concrete codes (including AS3600-1994), there are two basic approaches for deflection control. The first is deflection control by the calculation of deflection. The second is deflection control by the satisfaction of a minimum depth requirement or a maximum span to depth ratio. Code procedures for the calculation of the final deflection of a slab are necessarily design-oriented and simple to use, involving crude approximations of the complex effects of cracking, tension stiffening, concrete creep and shrinkage and the load history. Unfortunately, in most code procedures and in most practical situations, the effects of cracking in a slab are not adequately accounted for, particularly the loss of stiffness resulting from time-dependent cracking caused by restraint to the load-independent shrinkage and temperature deformations.

The final deflection of a slab depends very much on the extent of initial cracking which, in turn, depends on the construction procedure (shoring and re-shoring), the amount of early shrinkage, the temperature gradients in the first few weeks after casting, the degree of curing and so on. Many of these parameters are, to a large extent, out of the control of the designer. In field measurements of the deflection of many identical flat slab panels (Sbarounis, 1984, Jokinen and Scanlon, 1985), large variability was reported. Deflections of identical panels after one year differed by over 100 percent in some cases. These differences can be attributed to the different conditions (both in terms of load and environment) that existed in the first few weeks after the casting of each slab.

Over the past several years, numerous cases have been reported, in Australia and elsewhere, of flat slabs for which the calculated deflection is far less than the actual deflection. Many of these slabs complied with AS3600's serviceability requirements, but still deflected excessively. Evidently, the deflection calculation procedures embodied in the code do not adequately model the in-service behaviour of slabs.

Surprisingly, no laboratory controlled long-term measurements of deflection and crack propagation in large-scale reinforced concrete slabs under sustained service loads have been reported in the literature to date. This lack of reliable experimental data has hindered both analytical research and the development of reliable design procedures for serviceability. In this paper, an extensive experimental program of long-term testing of large-scale flat slab structures, which is currently underway at the University of New South Wales, is described and the results are presented for the first series of tests. These bench-mark experimental results include the measured time-varying material properties, slab deflection, extent of cracking and column loads and will be used to develop and evaluate alternative design procedures.

The need for a more reliable deflection calculation procedure has been exacerbated by the introduction in recent years of higher strength reinforcing steels. The use of higher strength steel results in less steel being required for strength and, consequently, less stiffness being available after cracking, leading to greater deflection and crack widths under service loads. The design for serviceability will increasingly assume a more prominent role in the design of slabs. Designers will need to pay more attention to the specification of both an appropriate concrete mix, particularly with regard to the creep and shrinkage characteristics of the mix, and a suitable construction procedure, involving acceptably long stripping times, adequate propping, effective curing procedures and rigorous on-site supervision.

2 EXPERIMENTAL PROGRAM

A three-year experimental program to measure the time-dependent in-service behaviour of reinforced concrete flat slabs commenced at the University of New South Wales in 1998. The work is funded by the Australian Research Council and will eventually involve the testing of eight large-scale flat slab structures. At the time of writing this paper, two slab specimens had been tested under sustained service loads for periods up to 220 days. These tests are described and the results are presented herein.

2.1 *Slab specimens and test set-up*

Two 100 mm thick reinforced concrete flat plates (designated S1 and S2) have been constructed and tested. Each slab is continuous over two spans in two orthogonal directions and is supported on nine 200 mm x 200 mm x 1250 mm long columns below the slab. The dimensions of each slab and the top and bottom reinforcement layouts are identical and are shown in plan in Figure 1. The slab reinforcement consists of 10 mm deformed bars (Y10) and the clear concrete cover to the outer layer of bars is 15 mm at both the top and bottom surfaces of the slab. The slabs were cast and tested approximately 3 months apart and the concrete for each specimen had significantly different properties and deformation characteristics.

The base of each column for specimen S1 was pinned with all exterior columns mounted on roller supports to eliminate support restraint to drying shrinkage in each direction. The base of each exterior column in Specimen S2 was poured monolithically with a 700 x 700 x 300 mm pad footing fixed to the laboratory floor to prevent translation and thereby provide support restraint to shrinkage in the slab (which is more typical of the conditions in practical slabs).

Both slabs were cast and initially moist cured for nine days at which time the formwork was removed and the slab back-propped to the laboratory floor. At age 14 days, each slab was subjected to a uniformly distributed load applied via concrete blocks carefully constructed and arranged to ensure uniform loading and uninhibited air flow over both the top and bottom surfaces of the slab. The props were removed and testing commenced. Slab S1 was initially loaded with superimposed load of 3.15 kPa, in addition to its self-weight (2.40 kPa). The load was sustained from first loading (age 14 days) to age 169 days, when the superimposed load was increased to 6.25 kPa. The load was then held constant for the remainder of the test. Slab S2 was initially subjected to a superimposed load of 3.32 kPa, in addition to its self-weight, and the load was held constant throughout the test. Figure 2 shows S1 soon after first loading.

Bottom Reinforcement:
B1 - Y10 bars @ 220mm ctrs

Top Reinforcement:
T1 - Y10 bars @ 250mm ctrs
T2 - Y10 bars @ 140mm ctrs

E-W bars placed 1st and last
N-S bars placed 2nd and 3rd

Slab thickness = 100 mm
Concrete cover = 15 mm

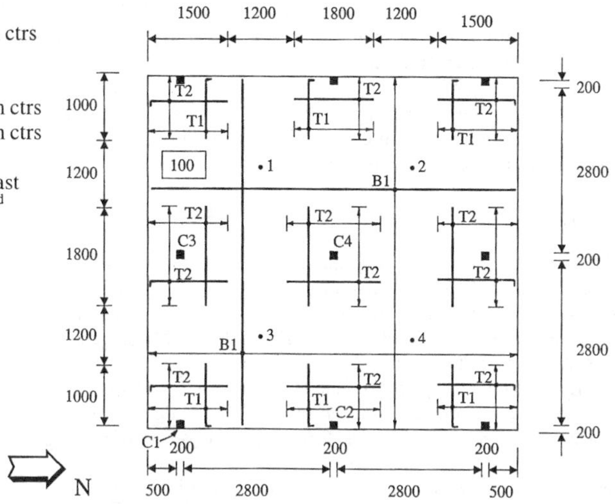

Figure 1 Plan of slab specimens and reinforcement layout (all dimensions in mm).

2.2 *Test measurements*

Throughout the duration of the tests, deflections were measured at the midpanel of each bay (points 1 to 4 in Figure 1), at the midspan on each column line and on the column lines at the cantilevered slab edge. Concrete strains on the slab edges at various locations were recorded and the vertical reaction at the base of columns C1, C2, C3 and C4 were measured using load cells. The extent of cracking on both the top and bottom surfaces of each slab was also monitored and the time dependent changes in cracking were recorded.

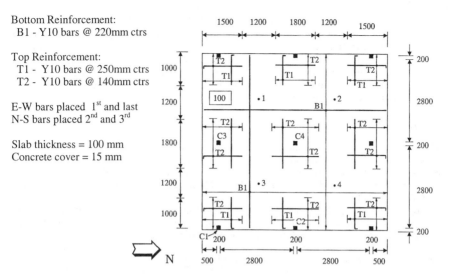

Figure 2 Specimen S1 under load.

Concurrent with the slab tests, material properties were also measured on companion specimens. The compressive strength and elastic modulus of concrete were measured at various ages on 150mm and 100mm diameter cylinders, and the flexural tensile strength was measured on prisms. Creep of concrete was measured on concrete cylinders loaded at 14 days mounted in standard creep rigs. Shrinkage strains in the concrete were recorded from 600 x 600 x 100 mm thick slab specimens (with edges sealed to ensure that drying only took place at the top and bottom surfaces of the specimen).

2.3 *Test results*

The measured compressive strength, elastic modulus and flexural tensile strength for the concrete at age 14 days (first loading) are, respectively, 34.5 MPa, 30020 MPa and 4.39 MPa for S1 and 29.0 MPa, 29100 MPa and 2.72 MPa for S2. At age 28 days and 100 days the compressive strength had increased to 37.9 MPa and 41.0 MPa, respectively, for S1 and 33.9 MPa and 35.1 MPa, respectively, for S2.

The measured creep coefficient and shrinkage strains are given in Table 1, together with the average midpanel deflections. For each slab, the deflections measured at points 1 to 4 (Figure 1) were within 8% of each other throughout the period of loading.

Table 1. Midpanel deflection, creep coefficient and shrinkage strain

	Slab S1			Slab 2		
Age (days)	Avge Midpanel Deflection (mm)	Shrinkage Strain ($x10^{-6}$)	Creep Coefficient	Avge Midpanel Deflection (mm)	Shrinkage Strain ($x10^{-6}$)	Creep Coefficient
14	2.62	124	0	2.84	302	0
20	3.85	153	0.65	4.58	409	0.84
28	4.63	228	0.89	6.16	529	1.29
40	5.19	270	1.27	7.67	569	1.56
56	5.84	361	1.47	8.45	592	1.92
80	6.34	386	1.64	9.94	738	2.39
120	7.15	442	1.84	10.95	780	2.56
169	7.60	489	2.05	-	-	-
169	8.68	489	2.05	-	-	-
190	9.10	548	2.04	-	-	-
220	9.14	555	2.07	-	-	-

The column reactions in S1 measured immediately after first loading (age 14 days) and at age 50 days were: for C1, 15.9 kN and 13.0 kN; for C2, 29.7 kN and 28.8 kN; for C3, 40.7 kN and 41.1 kN; and for C4, 63.7 kN and 65.0 kN. For S2, the central column reaction (C4) increased from 72.4 kN immediately after first loading (age 14 days) to 77.6 kN at age 50 days.

At first loading, no cracking was observed in S1, although very fine flexural cracks occurred within the first two weeks under load on the top surface of the slab and radiating from the interior column C4. These cracks increased in width with time, but remained quite serviceable with a maximum recorded crack width of less than 0.15mm. When the superimposed load was doubled at age 169 days, the existing cracks extended and widened but no new cracks developed. In specimen S2, flexural cracks formed at first loading on the top surface of the slab over columns C2, C3 and C4 and additional cracking occurred on the top surface at all columns with time. Crack widths increased with time throughout the tests, but the maximum crack width was less than 0.25mm. No cracks were observed on the bottom surface of either specimen at any stage. Crack patterns on the top surface of each slab at various ages are shown in Figure 3.

2.4 *Discussion of results*

By treating the imposed load as a typical service load, both slabs were designed to satisfy the strength and serviceability requirements of AS 3600. Both slabs did in fact behave in a serviceable manner, although slab S2 with midpanel deflections exceeding span/300 after just four months under load may well eventually suffer excessive deflection.

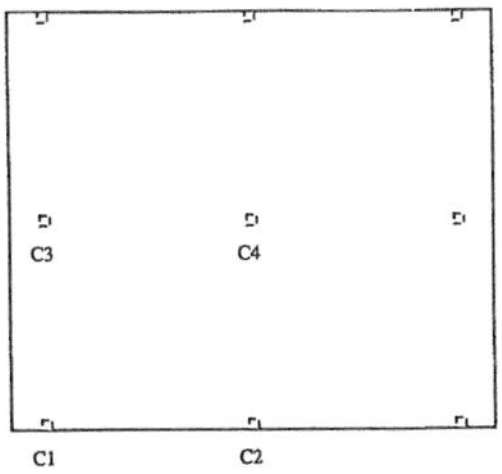

(a) Slab S1 (age 14 days)

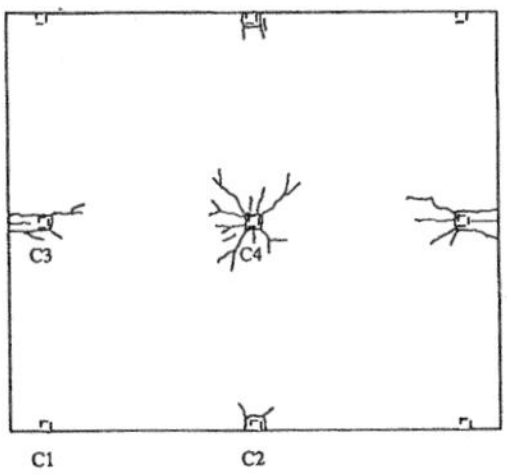

(b) Slab S1 (age 120 days)

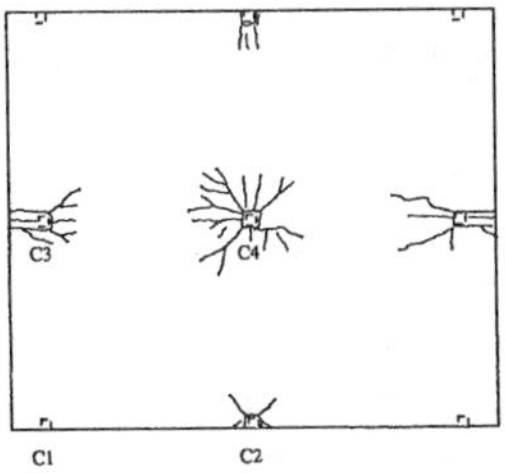

(c) Slab S1 (age 220 days)

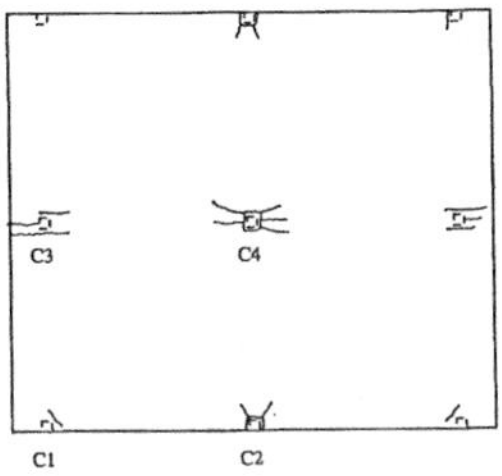

(d) Slab S2 (age 14 days)

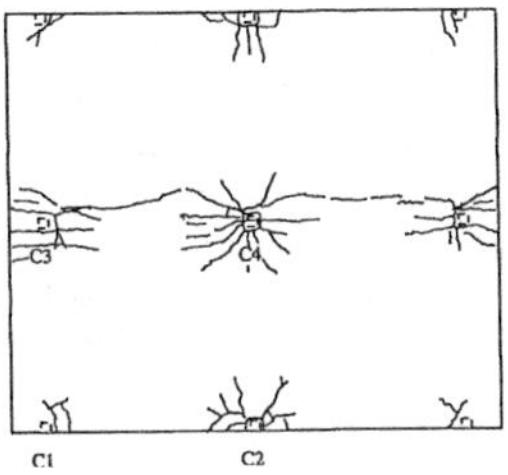

(e) Slab S2 (age 120 days)

Figure 3 Crack patterns on top surface of slabs.

As can be seen in Table 1, the concrete in S2 had significantly higher creep coefficient and shrinkage strain than that in S1. Experience suggests that the deformation characteristics of the concrete in S2 is more typical of 25-32 MPa normal-class concretes supplied in Sydney and the recorded creep coefficient and shrinkage strain for S1 were unusually low. Nevertheless, the time-dependent to instantaneous deflection ratio after just four months (120 days) is 1.73 for S1 and 2.86 for S2. This suggests that the long-term deflection multiplier k_{cs} in AS 3600 (which has a maximum value at time infinity of 2.0) is unrealistically low.

The significantly larger long-term deflections and the more extensive distribution of cracking in S2 (compared to S1) is due to many factors including the lower concrete tensile strength, the higher shrinkage strains, the higher creep coefficient and the increased restraint to shrinkage provided by the footing pads under each exterior column. Further experimentation is currently underway to ascertain the relative significance of each of these factors.

It is of interest to note that the column reactions C1 and C2 decreased with time and the reactions at C3 and C4 increased. This effectively means that the peak negative moment over the interior column (in the E-W direction) increased with time and the positive span moments decreased. This is primarily due to shrinkage and depends to a large extent on the reinforcement layout. The increase in negative moments with time is entirely consistent with the time-dependent increase in flexural cracking observed on the top surface of both slabs. The time-dependent reduction in the positive span moments has lead to the observation that the slab soffits remained uncracked throughout. This is typical of observations on flat slabs structures suffering serviceability problems (Gilbert, 1998a), where frequently the slab soffit is free from cracking while the top surface is extensively and excessively cracked.

The results presented in Table 1 also highlight the relatively small part the actual superimposed load plays in the final deflection of a slab. When the superimposed load on slab S1 was doubled at age 169 days, the additional instantaneous deflection was 1.08 mm and, over the next 50 days, the change in deflection was just 0.46 mm. Despite the large difference in applied load, the deflection of S1 at age 220 days was significantly less than the deflection of S2, which was carrying about one half the sustained load for only 120 days. The critical factors affecting slab deflection are the tensile strength of concrete, the creep and shrinkage characteristics of the concrete, and the environmental conditions, in addition to the load history and slab geometry.

3 COMMENTS ON DEFLECTION CALCULATION PROCEDURES

Current design approaches for the calculation of the deflection of flat slabs have recently been shown to be inadequate, Gilbert (1998b). They fail to adequately account for the loss of stiffness due to cracking, in particular time-dependent cracking resulting from shrinkage. The tests described here confirm the significance of time-dependent cracking and its influence on long-term deflection.

In general, slabs are lightly reinforced and lose a large percentage of their stiffness when cracking occurs. Shrinkage (and, in practical slabs, temperature changes) usually results in a continuing, gradual expansion of the cracked region in flat slabs, with a resultant gradual reduction in slab stiffness with time. This is evident in slabs S1 and S2 (Figure 3). Consequently, the ratio of final deflection to short-term deflection in flat slabs is usually greater than 5. In this study, the ratio after just four months (120 days) was 2.73 for S1 and 3.86 for S2.

Recently, several alternative methods were proposed to include the influence of shrinkage induced cracking in deflection calculations, Gilbert (1998b). The experimental data generated in the present study will be used to evaluate and calibrate these design proposals.

4 CONCLUSIONS

The time-dependent in-service behaviour of two large-scale flat slab structures has been presented. The deflection, the extent of cracking and the column loads were monitored with time, together with the strength and deformation characteristics of the concrete. Significant time-dependent cracking was observed and the long-term to short-term deflection ratios were significantly greater than those predicted using the current design procedures for deflection calculation. The tests described here are part of an on-going experimental study in which eight flat slabs will eventually be tested.

5 ACKNOWLEDGEMENTS

The study is being funded by the Australian Research Council and this support is gratefully acknowledged. The author would also like to thank Mr Patrick Zou, Mr X. Guo and the staff of the Heavy Structures Laboratory at UNSW for their assistance in the experimental work.

REFERENCES

AS3600-1994. *Australian Standard for Concrete Structures*. Standards Australia, Sydney.

Gilbert, R.I. 1998a. *Case History – A Capital Crunch – The Silverton Building*. ASCE seminar. *Protecting your R's – risk, responsibility and retribution*. Sydney. 13pp.

Gilbert, R.I. 1998b. Deflection calculation and control for reinforced concrete structures. *Proceedings ASCE- 1998. Australasian Structural Engineering Conf.*. Auckland: 269-274.

Jokinen, E.P. & Scanlon, A. 1985. Field measured two-way slab deflections. *CSCE Annual Conference*. Saskatoon. Canada. 16pp.

Sbarounis, J.A. 1984. Multi-storey flat plate buildings-measured and computed one-year deflections. *Concrete International*. 6(8): 31-35.

Mechanics of Structures and Materials, Bradford, Bridge & Foster (eds)
© 1999 Balkema, Rotterdam, ISBN 90 5809 107 4

Theoretical assessment of flat slab bridges

A. Hira, N. Haritos & P. Mendis
Department of Civil and Environmental Engineering, University of Melbourne, Vic., Australia

ABSTRACT: The methodology currently applied by road authorities for assessing the load carrying capacity of their flat slab bridge stock is to identify peak load demands (bending moments and shear forces) using a continuous beam model in combination with an "effective width" criterion. Such an approach does not account for the influence of the kerb edge conditions and the three-dimensional character of the load distribution, which can be accounted for using Finite Element Analysis models. This paper summarises the findings of both analytical and experimental studies performed on Barr Creek Bridge and highlights the benefits to be gained in assessing the performance of typical flat slab bridges using more refined modelling techniques.

1 INTRODUCTION

The Load Capacity Factor (LCF) for a RC flat slab bridge can be expressed as:

$$\text{LCF} = \frac{\text{Residual Moment Capacity}}{\text{Live Load Moment Demand}} \quad (1)$$

where the Residual Moment Capacity is estimated from the appropriately factored flexural section capacity of the deck, less the factored Dead Load Moment Demand. The Live Load Moment Demand in this equation includes a factor for the Dynamic Load Allowance on the Live Load Moment. The lowest value of LCF for any bridge that is acceptable, under currently adopted guidelines in Australia, is a value of 2.0.

Traditionally, the evaluation of the lowest LCF for a flat slab bridge by VicRoads, (the Road Authority in the State of Victoria, Australia), has been based upon the identification of the most critical factored Live Load Moment Demand from an adopted design vehicle configuration, identified from an elastic "continuous beam model". This model is based upon assumed articulation conditions of the bridge over its supports and at its ends and incorporates a "design lane width" approach that does not take into account the influence on bridge behaviour posed by the kerb beams in the deck. The resultant estimates of the most critical factored Live Load Moment Demand, when using such an approach, would therefore tend to be conservative when compared with corresponding estimates from a more realistic structural modelling of bridge behaviour, such as can be obtained from a Finite Element Analysis (FEA) model of the bridge.

Studies performed by the authors on Barr Creek Bridge, (both analytical and experimental), have demonstrated the relative importance of the role played by the kerb beams in the structural performance of the bridge deck at load levels associated with serviceability and beyond.

This paper highlights the benefits to be gained in the load capacity assessment of flat slab bridges when using such more realistic FEA modelling procedures, drawing upon the experience gained from the Barr Creek Bridge studies.

2 DESCRIPTION OF BARR CREEK BRIDGE

2.1 *General*

Barr Creek Bridge was designed as a RC bridge of four relatively short continuous spans, (see Figure 1 for a schematic of the bridge that details its geometry) and was constructed in 1939. The bridge was designed with its deck to be cast-in-place with the retaining abutments at both ends with the intent of providing rotational fixity but otherwise contained design features considered typical of the RC slab-on-pier bridges of its era.

The bridge was removed from general service in 1971 due to extensive degradation of its substructure and was subsequently made available to the VicRoads' corporate R & D program, "Assessment of Load Capacity of RC Flat Slab Bridges", with the intention that it be instrumented and load tested to collapse prior to its scheduled demolition. The tests were intended to provide comprehensive observations of the in-situ behaviour of the bridge when subjected to a range of loading conditions to the elastic limit and subsequently to failure. Such observations could then assist The University of Melbourne with the development of analytical tools that could be used to better determine the performance of these bridges than would be possible using currently adopted methods.

Barr Creek Bridge was deemed by VicRoads to be deficient in negative moment capacity at positions corresponding to slab top-steel reinforcement curtailment points in the internal spans. Their estimate of the LCF using the traditional approach of a continuous beam model, an "effective width" equal to the lane width and assumed material properties, was only 1.60.

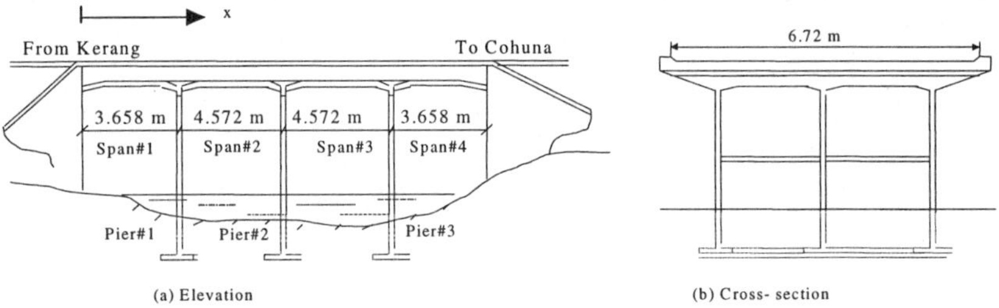

Figure 1. Schematic of geometric features of Barr Creek Bridge

2.2 *Experimental Program*

The University of Melbourne suggested the performance of a dynamic test program on Barr Creek Bridge, prior to the execution of a comprehensive static test program involving a number of test load case configurations where loading was taken to serviceability levels and beyond, (Haritos et al 1999).

The purpose of the dynamic testing was to establish the in-situ low load behavioural characteristics of the bridge from identification of its dynamic modes of vibration via a Simplified Experimental Modal Analysis, (SEMA).

The SEMA results clearly identified that the condition of support of the bridge deck at the two abutment ends of the bridge was virtually "pinned" (as opposed to "fixed"as assumed in the original design) and that the effective Young's modulus of the concrete, based upon gross section properties in a STRAND6 Finite Element Analysis (FEA) model of the bridge, was approximately 20 GPa. This value was substantially lower than the approximate value of 28 GPa obtained from the concrete cylinder tests, inferring that the actual section properties applicable to the deck would be substantially below gross sectional values. (Inspection of the deck during the performance of both the dynamic and static testing programs, and during the course of demolition of the bridge, indicated substantial delamination of the concrete cover to the top steel in the slab which could largely account for this "softening").

3 FINITE ELEMENT ANALYSIS

3.1 *FEA model*

Several modelling options were investigated in the selection of the FEA mesh for Barr Creek Bridge using the STRAND6 program.

These options included:

- the use of plate elements with varying thickness to model the haunched and flat slab regions of the deck of the bridge
- the use of local thickening in the plate elements along the deck edges to simulate the kerb beams and, as an alternative, the introduction of "offset beam elements" to model these kerbs
- modelling of the piers as a sequence of beams connected integrally with the deck, and alternatively, treating the piers as simple supports providing vertical restraint only to the deck as a Boundary Condition in the model

Variations in the FEA model features associated with particular choices in the alternative modelling options listed above generally led to only minor differences in predicted results when applying these models. By far the most critical modelling parameters were the Boundary Conditions of support at the abutment ends and the effective Young's modulus of the concrete. Since both of these parameters were able to be identified accurately from the performance of the SEMA tests, any further "tuning" of the FEA model was considered to be of minor significance

3.2 *Elastic behaviour*

The "tuned" FEA model version obtained from the dynamic testing was used to compare the modelled and observed stiffness of the bridge at serviceability load levels. Figure 2 compares the load/deflection behaviour for Load Case 1 of the static test program (single Load Position mid-span of Span 1 representing twin axle loading from a T44 truck up to the total axle serviceability level load of 250 kN), by way of example of this comparison. Results presented in the figure are seen to compare very favourably providing confidence in the use of the FEA model as a predictive tool of the behaviour of the bridge.

Some of the key results of the sensitivity study are presented in Figure 3 where it is seen that significant reductions in level of Live Load Moment Demand in the slab at the critical section (corresponding to the curtailment point of the negative reinforcement in span 2, designated by x = 4.60m) can be achieved by considering the influence of kerb beam stiffness. The study identified moment reductions of 15-20% at this critical location when incorporating a 400mm deep by 350mm wide kerb compared with a deck configuration without a kerb, which was attributed to the three-dimensional load distribution effects that can be realised from FEA modelling. In ad-

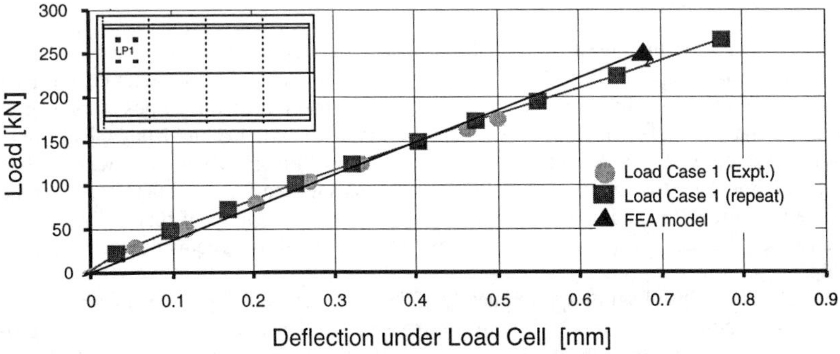

Figure 2. Comparison of load-deflection behaviour in Span 1 for Load Case 1

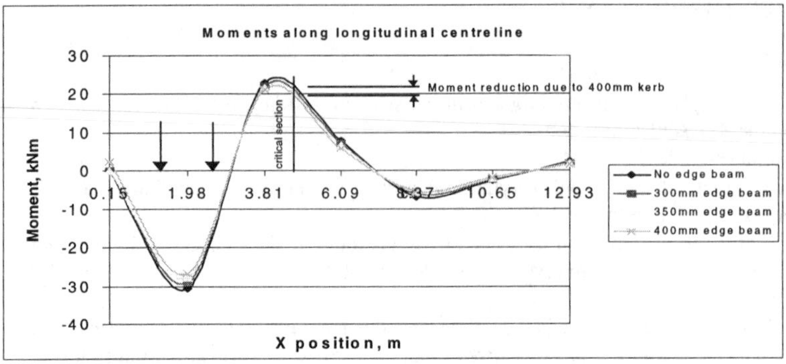

Figure 3. Influence of kerb stiffness on flexural demand along bridge centre-line

dition to this benefit the FEA also confirmed that the full width of the bridge deck, including the kerbs (i.e. 3.70m per design lane) is effective in resisting the flexural demand. This is significantly greater than the 2.92m equivalent width recommended by NAASRA (1976).

The FEA model was consequently used to perform sensitivity studies of the role that the kerb beam plays in controlling the level and distribution of bending moments (positive and negative) under different loading scenarios. The load scenarios chosen included searching for the T44 truck configuration considered most critical in terms of Live Load Moment Demand on the bridge for the purpose of performing a revised LCF assessment.

3.3 *LCF determination*

To maximise the benefits of using advanced analyses such as FEA, for load capacity assessment of the Barr Creek Bridge, an accurate assessment of the flexural capacity throughout the bridge length is essential. A yield stress of 280MPa was adopted based on tensile strength test results of 10 reinforcement specimen extracted from the bridge. This strength is significantly greater than the 230MPa adopted for the preliminary assessment prior to the ultimate load testing of the bridge.

By application of FEA, a LCF of 2.24 was achieved for the critical T-44 vehicle configuration satisfying the present code requirements. This higher capacity was realised due to the ability of FEA to better model the lateral distribution of the forces to the kerbs particularly in the regions of the unloaded spans which dictated the load capacity rating due to inadequate negative flexural capacity.

3.4 *Post -elastic behaviour*

An integral objective of the investigation was to carry out a theoretical assessment of the post elastic behaviour to collapse of Barr Creek Bridge and compare the predictions with the ultimate test results. Load case 6 of the static load test program consisting of two load positions (centre of each design lane at mid-span of Span 4), representing a twin axle loading from a T44 truck, was used for the comparison. A staged linear elastic FEA with incremental loads was used as the analytical tool. The stiffness of each finite element was adjusted for each stage based on the accumulated flexural stress-state of the previous stages. For the purposes of this study the moment curvature relationships for each element type was simplified to an equivalent tri-linear relationship.

The predicted collapse load of approximately 1375kN per design lane (corresponding to 5.5 times the T-44 loading) compared well with the measured collapse load of 1500kN. The progressive onset of yielding and subsequent plasticity predicted by the model compared well with the observed collapse mechanism. The study revealed the significance of the participation of

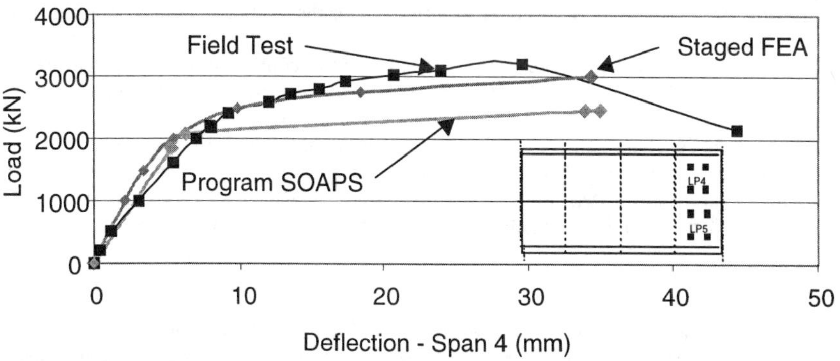

Figure 4. Comparisons of load-deflection behaviour in span 4 for Load Case 6.

edge kerb beams in attracting loads in the post-elastic range. The final mechanism predicted a positive hinge in span 4 near the abutment at location of the curtailment point of the positive reinforcement followed by a negative hinge in span 3 near pier 3 at location of the curtailment point of the negative reinforcement and a final positive hinge at pier 2.

The FEA is expected to overestimate the collapse load capacity due to the assumption that all sections maintain the ultimate flexural capacity through plastic rotation with no allowance for plastic softening. The overestimation would be further enhanced due to the element stiffness being based on the stress-state at the beginning of each load increment. The onset of membrane action at high deflections may be sufficient to counteract this tendency and account for the higher observed collapse load.

Despite its approximate nature, the staged FEA was found to be a useful qualitative tool to predict the behaviour of the bridge deck in the post-elastic range to collapse. Favourable comparisons of the load/deflection behaviour between predicted values and observed values, presented in Figure 4, gave further confidence in using this method.

4 PLASTIC DESIGN METHODS

Generally, the fundamental philosophy adopted for load capacity assessment of existing bridges has been to evaluate the ultimate limit state, since the serviceability limit state criteria has generally been satisfied by default for structures maintained in service. The ultimate limit state criterion is commonly evaluated by ensuring that the critical flexural demand at all locations, based on elastic methods, is less than the flexural capacity at the corresponding location. This is generally, a conservative approach since attaining the section capacity locally in a structure is used as the definition for "failure" of the structure as a whole. The consequences of such a "failure condition" are likely to be small in most structures that exhibit a high degree of redundancy. (Middleton 1995) This implies that a better approach would be to use collapse as the criterion for assessing bridges. Based on this premise, plastic methods to evaluate LCFs were included in the scope of our study.

4.1 *Yield Line methods*

Yield line methods are not widely adopted in assessing bridge capacities due to a number of perceived difficulties. In particular this method is an upper-bound technique therefore there is always a degree of uncertainty that the critical mode has not been found. The Yield Line Method was used to predict collapse load capacity of the Barr Creek Bridge for Load Case 6. The results of this method, being an upper bound solution, formed the basis for evaluating the design loads for the testing frame and the specification for the loading equipment for the static load test.

The application of this technique in practice is limited to configurations where the mechanism, defined by the yield lines, falls within a single span. However this technique can be applied to multi-span continuous slab systems by providing full fixity at the continuous supports restricting the collapse mechanism to a single span. The yield line pattern investigated for Barr Creek Bridge was a box-type pattern located over the full width of span 4. The calculated yield-line collapse load for this pattern was 1420kN per design lane which compared favourably with the observed collapse load of 1500kN for Load Case 6 of the static load test.

4.2 *Continuous beam model*

The yield line method of investigation to determine the LCF becomes extremely difficult to apply when multiple spans are loaded. A non-linear two dimensional equivalent beam analysis model is more effective. The computer program SOAPS (Second Order Analysis with Plastic Softening) was used to perform a step by step elastic-plastic full-range analysis for Load Case 6 and for the critical T-44 vehicle configuration. The moment-curvature relationships adopted for the model are identical to those used in the staged FEA outlined in the previous section.

The collapse load predicted by this method was 1100kN per design lane compared with the observed value of 1500kN. This discrepancy can be attributed to a number of factors including the effects of the kerb, variation in moment capacity and spread of yielding. Figure 4 presented earlier in this paper also compares the load-deflection predictions from SOAPS with the observed values.

The SOAPS software was applied to assess the collapse behaviour of Barr Creek Bridge when subjected to the critical T-44 vehicle configuration as used for the FEA. The LCFs achieved using SOAPS method was in the order of 3.15, 3.22 and 3.25 for formation of the first three hinges respectively which are well above the acceptable value of 2.0.

5 CONCLUDING REMARKS

This paper has investigated several methods of determining the LCF for Barr Creek Bridge which failed to comply with present code requirements when assessed using traditional simplified methods. FEA, through its superior modelling capabilities and ability to incorporate features such as the edge kerb conditions and slab haunching have given a higher LCF, in excess of 2.0 thereby demonstrating acceptable capacity.

The study has also revealed the importance of behavioural tests and material tests to establish realistic FEA model and to allow accurate section capacity determination.

The study has shown that the application of plastic methods of analysis, when compared with traditional strength limit state design approach based on elastic analysis, can lead to significant enhancement in the LCF for existing bridges. The University of Melbourne is currently investigating the use of advanced methods, including non-linear FEA, and plastic analyses to load rate flat slab bridges and to better check the serviceability limit state. The research team will also be embarking on a comprehensive testing program on laboratory models to verify the slab behaviour in the post-elastic range.

6 REFERENCES

Haritos, N., Hira, A. & Mendis, P. 1999. Theoretical Analysis of Barr Creek Bridge, *Proc. Intl. Conf. on Mechanics of Structures, Materials and Systems*, Feb., Uni. of Wollongong, 67-74.

Middleton, C.R., 1995. Concrete Bridge Assessment using Plastic Collapse Analysis, Proc. 14th Australasian Conf. On the Mechanics of Structures and Materials, Tas., Dec., 95, pp 422-427.

Mechanics of Structures and Materials, Bradford, Bridge & Foster (eds)
© 1999 Balkema, Rotterdam, ISBN 90 5809 107 4

Transmission of high strength concrete column loads through slabs

J. Portella & P.A. Mendis
Department of Civil and Environmental Engineering, University of Melbourne, Vic., Australia
D. Baweja
CSR Limited, Sydney, N.S.W., Australia

ABSTRACT: Design and construction of reinforced concrete buildings often leads to the use of high strength concrete columns with intervening layers of normal strength concrete slabs. Past investigators have found that under some circumstances the slab layer may significantly decrease the load carrying capacity of the column. This paper summarises past research and presents the latest developments in this area. Present guidelines for column-slab joints in AS3600 are shown to be unconservative in some instances when compared to overseas code requirements and the latest research findings. Further study currently being undertaken at The University of Melbourne is outlined.

1 INTRODUCTION

In present day construction of multistorey reinforced concrete buildings, considerable economy may be achieved by designing the columns with high-strength concrete and the floor slabs with concrete of normal strength. The favoured construction technique for such buildings is to cast the columns up to the soffit of the slab they will support and then to cast a continuous slab. The columns of the next storey are then cast, resulting in a layer of slab concrete intersecting the high-strength concrete columns at each floor level. The axial load on a column must therefore transverse a layer of weaker floor concrete and under some circumstances this layer may cause a decrease in the load carrying capacity of the column. A design dilemma arises as to what compressive strength should be used in the design of the column; should it be the column concrete strength, the floor concrete strength, or some intermediate value?

Current design practices in Australia and overseas with respect to the transmission of column loads through slabs consist of three approaches. The first approach, commonly referred to as "puddling" or "mushrooming", involves placing column concrete within the joint region and allowing it to extend at least 600mm beyond the face of the column. Design calculations are then based simply on the cylinder strength of the column concrete. This method demands a high level of supervision and extensive planning as two different grades of concrete are required to cast the slab and they need to be well integrated. No studies have yet been reported on the effects of puddling and integration of concretes on the performance of slabs (Kayani, 1992). The second approach is to provide adequate longitudinal and lateral reinforcement to the column based on the lower-strength slab concrete. However this approach can only be used when the column concrete strength does not exceed the slab concrete strength by more than a factor of 1.4 (AS3600, 1994, ACI 318, 1995). The third method of dealing with the transmission of high strength concrete column loads through slabs is to design the column using an "effective" concrete strength which may be greater than the slab concrete strength, but limited by the column concrete strength. The value of effective concrete strength of the column has been found by past researchers to be predominantly dependant on the ratio between the column concrete strength and slab concrete strength, the aspect ratio of the joint (the ratio of slab thickness to column side dimension) and the degree of lateral confinement offered to the joint region by the surrounding slab which will differ for each of the three types of columns: interior, edge and corner. Based on these research findings, a number of codes have put forward recommendations on how to calculate the effective strength of a column with an intervening layer of floor concrete. The Australian Concrete Structures code, AS3600 also provides some guidance in the area, but as will be demonstrated later in this paper, it can be unconservative and is in urgent need of revision.

2 PAST RESEARCH

2.1 *General*

A relatively limited amount of research has been conducted on the transmission of high strength concrete column loads through normal strength slabs. The common approach that investigators have taken to the problem has been to treat the slab layer as part of the column and to make use of the design equation for the cross-sectional capacity of a column under concentric load which is given in AS3600, ACI and CSA design standards as

$$P_o = 0.85 f'_c (A_g - A_{st}) + f_y A_{st} \quad (1)$$

where f'_c is the column concrete strength, A_g is the gross area of the column cross section, A_{st} is the area of longitudinal steel and f_y is the yield strength of the steel. If the terms of this equation are rearranged and f'_c is replaced with f'_{ce}, the effective strength of the column, then we have

$$f'_{ce} = \frac{P_{max} - f_y A_{st}}{0.85(A_g - A_{st})} \quad (2)$$

where P_{max} is the maximum axial load carried by the column. In this way, an effective strength can be calculated for a column specimen.

Figures 2.1(a) through to 2.1(c) show typical interior, edge and corner slab-column connections which have been subject to testing in the past. Figure 2.1(d) depicts what is termed a "sandwich" column and it is often used to simulate the behaviour of a corner column.

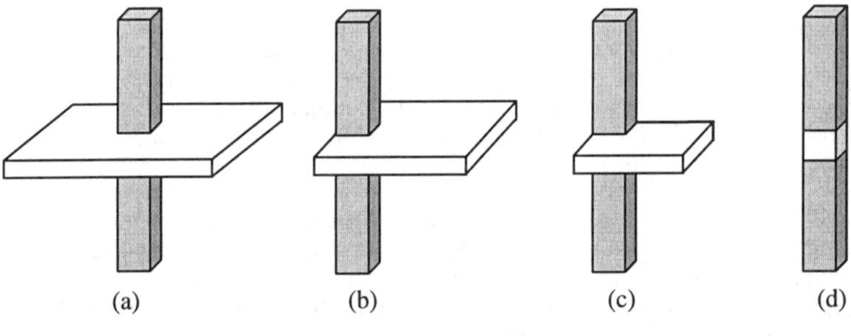

Figure 2.1. (a) Interior Column Specimen. (b) Edge Column Specimen. (c) Corner Column Specimen. (d) Sandwich Column Specimen.

2.2 *A summary of past literature*

Bianchini et al. (1960) were the first investigators to conduct a comprehensive series of tests on the behaviour of column-slab joints. The tests consisted of 11 interior, 9 edge, 9 corner and 4 sandwich column specimens, all of which were subject to concentric column loads. Bianchini et al. concluded that for interior columns, provided the ratio of column to slab strength, f'_{cc}/f'_{cs}, does not exceed 1.5, then the effective strength is simply equal to the column concrete strength. If this ratio is exceeded however, Bianchini et al. proposed that only 75% of the column concrete can be considered to be effective. For edge and corner columns, the limiting value was found to be 1.4. Bianchini et al suggested that for edge and corner columns, increasing the column strength beyond 1.4 times the slab strength bears no additional gain in effective strength.

Gamble and Klinar (1991) tested 6 interior and 6 edge column-slab specimens and applied concentric axial loads to all of them. The tests extended both the range of concrete strengths and column to slab strength ratios. The design guidelines they established are given in Table 1.0.

Kayani (1992) tested 2 edge and 4 sandwich columns, applying concentric axial loads similar to other investigators, but adding lateral reinforcement to the joint region of two of the specimens. Kayani's relationship for column effective strength is shown in Table 1.0, along with the findings of other researchers. Kayani's experiment results suggested that the addition of hoops in the joint region resulted in an increase in ductility of the column, but not necessarily an increase in effective strength.

An extensive study was conducted by Ospina and Alexander (1997) through the testing of 20 interior, 6 edge and 4 sandwich column specimens. While investigating the influence of the column strength to slab strength ratio, Ospina and Alexander introduced two other variables into the tests; the slab thickness to column side dimension ratio, h/c, and slab loading. It was established that the smaller the h/c ratio for a column-slab joint, the more significant is the lateral restraint offered from the upper and lower column ends to the joint region and thus higher effective strengths are achievable. For large h/c ratios, the restraint is not as effective thus resulting in a reduced effective strength of the overall column. Table 1.0 shows the resultant design recommendations which Ospina and Alexander developed. With h/c equal to 1, the equation they propose for interior columns matches that given by the Canadian Standards, while with h/c equal to 1/3, it matches the equations given by the American Code.

2.3 *Code Recommendations*

Based on the test results of Bianchini et al, the American Concrete Institute Code (ACI 318-95) has defined equations for the effective strength of interior, edge and corner columns and these are given in Table 2.0. The suggestions provided by the Canadian Standards (CSA A23.3-94) are also presented here. Clause 10.8 of AS3600 offers recommendations on the transmission of column loads through floors as presented in Table 2.0 also. It is important to note that AS3600 adopts a critical f'_{cc}/f'_{cs} ratio of 2, which exceeds the more commonly accepted value amongst past investigators and other codes of 1.4.

Table 1.0. Design provisions given by past investigators

	Interior Columns
Gamble & Klinar 1991	$f'_{cc}/f'_{cs} \le 1.4$; $f'_{ce} = f'_{cc}$ & $f'_{cc}/f'_{cs} \ge 1.4$; $f'_{ce} = 0.47 f'_{cc} + 0.67 f'_{cs}$
Kayani 1992	$f'_{ce} = 2.0\lambda_G \dfrac{f'_{cc} f'_{cs}}{f'_{cc} + f'_{cs}}$ with $\lambda_G = 1.25$
Ospina & Alexander 1997	$f'_{ce} = \dfrac{0.25}{h/c} f'_{cc} + \left(1.4 - \dfrac{0.35}{h/c}\right) f'_{cs} \le f'_{cc}$ and $h/c \ge 1/3$
	Edge Columns
Gamble & Klinar 1991	$f'_{cc}/f'_{cs} \le 1.4$; $f'_{ce} = f'_{cc}$ & $f'_{cc}/f'_{cs} \ge 1.4$; $f'_{ce} = 0.32 f'_{cc} + 0.85 f'_{cs}$
Kayani 1992	$f'_{ce} = 2.0\lambda_G \dfrac{f'_{cc} f'_{cs}}{f'_{cc} + f'_{cs}}$ with $\lambda_G = 1.00$
Ospina & Alexander 1997	$f'_{cc}/f'_{cs} \le 1.4$; $f'_{ce} = f'_{cc}$ & $f'_{cc}/f'_{cs} \ge 1.4$; $f'_{ce} = 1.4 f'_{cs}$
	Corner Columns
Kayani 1992	$f'_{ce} = 2.0\lambda_G \dfrac{f'_{cc} f'_{cs}}{f'_{cc} + f'_{cs}}$ with $\lambda_G = 0.90$
Ospina & Alexander 1997	$f'_{cc}/f'_{cs} \le 1.2$; $f'_{ce} = f'_{cc}$ & $f'_{cc}/f'_{cs} \ge 1.2$; $f'_{ce} = 1.2 f'_{cs}$

Table 2.0. Code design provisions

	Interior Columns
ACI 318-95	$f'_{cc}/f'_{cs} \le 1.4$; $f'_{ce} = f'_{cc}$ & $f'_{cc}/f'_{cs} \ge 1.4$; $f'_{ce} = 0.75f'_{cc} + 0.35f'_{cs}$
CSA A23.3 - 94	$f'_{cc}/f'_{cs} \le 1.4$; $f'_{ce} = f'_{cc}$ & $f'_{cc}/f'_{cs} \ge 1.4$; $f'_{ce} = 0.25f'_{cc} + 1.05f'_{cs}$
AS3600, 1994	$f'_{cc}/f'_{cs} \le 2.0$; $f'_{ce} = f'_{cc}$
	Edge Columns
ACI 318-95 & CSA A23.3-94	$f'_{cc}/f'_{cs} \le 1.4$; $f'_{ce} = f'_{cc}$ & $f'_{cc}/f'_{cs} \ge 1.4$; $f'_{ce} = 1.4f'_{cs}$
AS3600, 1994	$f'_{cc}/f'_{cs} \le 2.0$; $f'_{ce} = f'_{cc}$ & $h/c \le 1/2$
	Corner Columns
ACI 318-95	$f'_{cc}/f'_{cs} \le 1.4$; $f'_{ce} = f'_{cc}$ & $f'_{cc}/f'_{cs} \ge 1.4$; $f'_{ce} = 1.4f'_{cs}$
CSA A23.3 - 94	$f'_{cc}/f'_{cs} \le 1.0$; $f'_{ce} = f'_{cc}$ & $f'_{cc}/f'_{cs} \ge 1.0$; $f'_{ce} = f'_{cs}$
AS3600, 1994	$f'_{cc}/f'_{cs} \le 2.0$; $f'_{ce} = f'_{cc}$ & $h/c \le 1/4$

3 COMPARISON OF AVAILABLE EQUATIONS TO CALCULATE THE EFFECTIVE STRENGTH OF A COLUMN-SLAB JOINT

Figures 3.1 to 3.3 indicate how the available equations for effective strength of a column-slab joint compare. Figure 3.1 shows that the ACI 318-95 code provision grossly overestimates the effective strength of interior columns. Gamble and Klinar (1991) and Kayani (1992) are also unsafe for high h/c ratios. The AS3600 provision may be very unconservative for f'_{cc}/f'_{cs} ratios between 1.4 and 3.5, in particular when the h/c ratio is greater than 0.3. The CSA A23.3-94 recommendation is on the other hand unduly conservative for low values of h/c. Ospina and Alexander appear to predict the behaviour of interior column-slab joints most adequately, with data points falling below the design equations in all cases.

Figure 3.2 for edge columns indicates that the critical value of 1.4 for the f'_{cc}/f'_{cs} ratio is a fair recommendation, with only one data point falling below the design line. The AS3600 equation is only valid for h/c ratios less than or equal to 0.5 for which no test data is currently available, however it is conceivable that it may be unconservative for edge columns also.

Figure 3.3 contains data points from available corner column (CC) and sandwich column (SC) tests. It shows that ACI 318-95 once again overestimates the effective strength of this time the corner column-slab joints. Ospina and Alexander provide the best fitting prediction, while

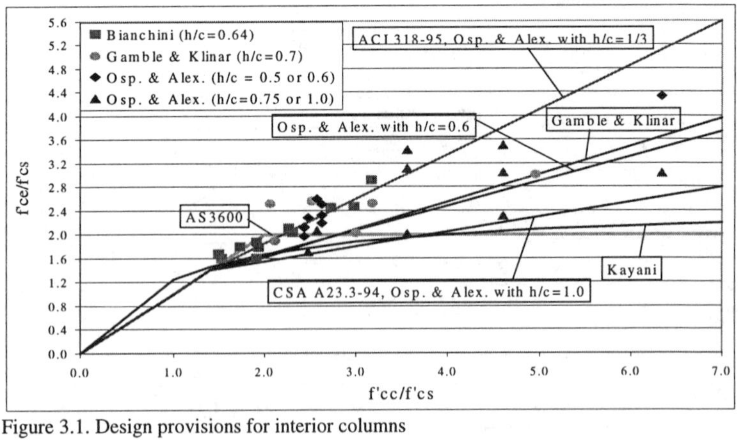

Figure 3.1. Design provisions for interior columns

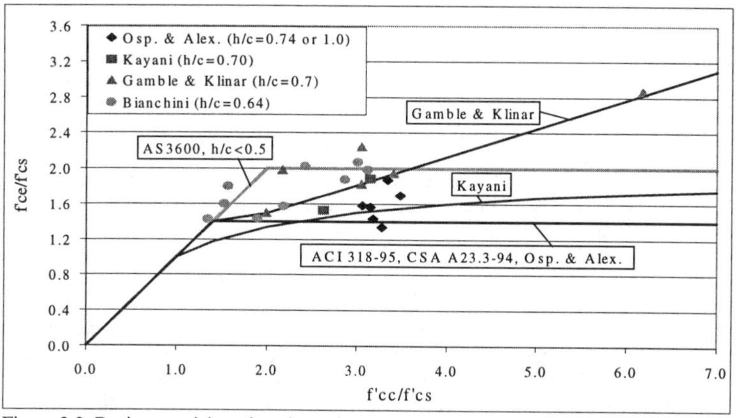

Figure 3.2. Design provisions for edge columns

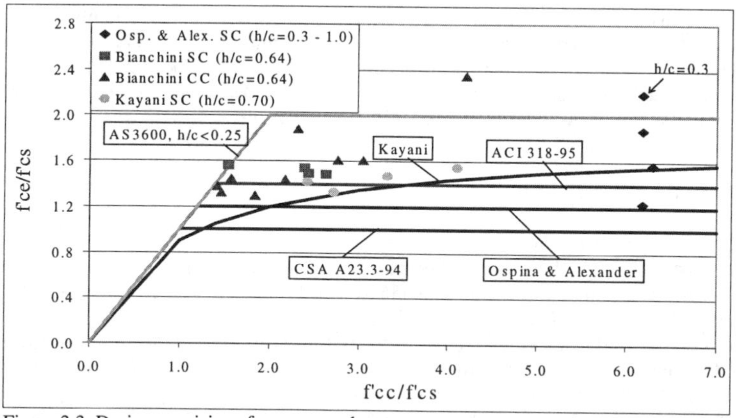

Figure 3.3. Design provisions for corner columns

CSA A23.3-94 appear to underestimate the effective strength of the corner columns on all occasions. The AS3600 recommendation is restricted to corner columns with h/c values less than or equal to 0.25 which is a significantly lower value than most values used in past investigations, hence it is difficult to predicate on its adequacy.

Based on all available data, the predictions of Ospina and Alexander interpret column-slab joint behaviour most accurately. In keeping with the latest research findings, AS3600 should move towards adopting Ospina and Alexander's recommendations.

4 FURTHER RESEARCH BEING UNDERTAKEN AT THE UNIVERSITY OF MELBOURNE

A relatively minor amount of research has been performed up to date and many opportunities exist for new and improved developments. Researchers at The University of Melbourne, in conjunction with CSR Ltd., are currently working towards this aim. The two main objectives of the research are firstly, to consider the effect of an eccentric axial load on the column and secondly, to assess the influence of adding lateral reinforcement in the column-slab joint. The investigations will involve both experimental work and theoretical work using finite element modeling. The tests will be limited to sandwich columns.

4.1 *Effect of an Eccentric Axial Load on the Column*

Past investigators have always applied concentric axial loads to column specimens. If one was to consider an eccentrically applied column load on the other hand, then this would induce flexural action such that part of the column experiences an increased vertical compression, while the other part experiences a reduced compression or for large eccentricities, possibly tension. The beneficial influence of having a small h/c ratio relies on axial compression over the entire joint section, hence it can be envisaged that the effect of the h/c ratio will be diminished with an eccentric load. If this is the case, then Ospina and Alexander's equations may actually be unconservative for eccentrically loaded columns.

4.2 *Effect of Lateral Reinforcement in the Column-Slab Joint*

The prospect of laterally confining the slab concrete in the joint region to improve its strength characteristics has been virtually ignored up to now, with the exception of Gamble and Klinar who tested one interior column specimen with spiral reinforcement and found that it lead to an increase in effective strength and Kayani, who tested two sandwich columns with ties in the joint region and concluded that rectangular ties do not improve the effective strength of the column. Kayani argued that the addition of ties in the joint region causes spalling of the cover concrete and thus a loss in the overall load carrying capacity of the column. Much research has been done however and many models formulated to show that rectangular ties do actually have a positive impact on the effective strength of a column, as well as functioning to improve ductility behaviour (Mander et al., 1988, Setunge, 1992). In light of this and the fact that only 3 tests have investigated the effect of lateral confining steel on column-slab joints, the proposed research at The University of Melbourne should provide valuable results.

5 CONCLUSIONS

This paper has presented the findings of past researchers on the transmission of high strength concrete column loads through slabs and commented on their adequacy. The guidelines set by Australian Standards for Concrete Structures, AS3600, can be unconservative, especially for interior columns at f'_{cc}/f'_{cs} values between 1.4 and 3.5 and therefore are in need of revision to reflect the latest research findings. Ospina and Alexander currently provide the most adequate equations to determine the effective strength of column-slab joints for interior, edge and corner columns. Researchers at The University of Melbourne are currently investigating the impact of eccentric column loading which has not been considered up to now. Also, the effects of providing lateral reinforcement in the joint region to improve its performance shall be further studied.

REFERENCES

Bianchini, A.C., Woods, R.E., Kesler, C.E. 1960. Effect of floor concrete strength on column strength. *ACI Journal* 31(11): 1149-1169.

Committee BD/2, Concrete Structures 1994. *Australian Standards for Concrete Structures (AS3600).* North Sydney: Standards Australia.

Gamble, W.L., Klinar, J.D. 1991. Tests of high strength concrete columns with intervening floor slabs. *ASCE Journal of Structural Engineering* 117(5): 1462-1476.

Kayani, M.K. 1992. Load transfer from high strength concrete columns through lower strength concrete slabs. *Ph.D. Thesis*. Department of Civil Engineering, University of Illinois, Urbana-Champaign.

Mander, J.B., Priestly, M.J.N, Park, R. 1988. Theoretical Stress-Strain Model for Confined Concrete. *ASCE Journal of Structural Engineering* 114(8): 1804-1826.

Ospina, C.E., Alexander, S.D.B. 1997. Transmission of high-strength concrete column loads through concrete slabs. *Structural Engineering Report No. 214*. Department of Civil Engineering, University of Alberta, Edmenton.

Setunge, S., Attard M.M., Darvall, P. 1992. Ultimate Strength Criterion for very high strength concrete subjected to triaxial loadings. *Civil Engineering Research Report No. 1/92*. Monash University, Melbourne.

Mechanics of Structures and Materials, Bradford, Bridge & Foster (eds)
© 1999 Balkema, Rotterdam, ISBN 90 5809 107 4

Influence of boundary conditions on punching shear behaviour of flat plate-column connections

Y.R. Khwaounjoo, S.J. Foster & R.I. Gilbert
School of Civil and Environmental Engineering, The University of New South Wales, Sydney, N.S.W., Australia

ABSTRACT: Boundary conditions are an important factor in the performance of flat plate structures in experimental studies and in practice. In this study the response of slab-column connections under vertical load is investigated using finite element modelling for various boundary conditions. A numerical analysis was undertaken for a slab-column connection tested by Lim and Rangan (1995) to investigate the possible effect of boundary conditions on the results obtained from the experimental investigation. The results of the study show that the transfer of unbalanced moment to the column is highly sensitive to the boundary conditions.

1 INTRODUCTION

Flat plate and flat slab floor systems are widely used in multistorey buildings as they offer the advantage of simplicity in construction and lower floor to floor heights. Many investigations have been undertaken to study the behaviour of flat plate floor systems, with most studies undertaken using scaled laboratory specimens. Many different types of boundary condition have been used in these tests. Linear elastic finite element studies have also been used to investigate the behaviour of slab-column connections using layered plate elements. However, the behaviour of slab-column connections is a complex three-dimensional problem and linear analyses give limited insight into the behaviour of these connections.

Experimental studies of the behaviour of slab-column connections (Zaghlool, 1971, Zaghlool and de Paiva, 1973, Robertson and Durrani, 1990, Foutch et al., 1990, Lim and Rangan, 1995, Elgabry and Ghali, 1996, and others) have been conducted using different boundary conditions for the slabs. There appears to be no general consensus on the experimental boundary conditions necessary to simulate the behaviour of multi-bay flat-plate column systems. In this study the effect of boundary conditions on the behaviour of flat plate-column connections is investigated using materially non-linear finite element brick elements and the finite element program DIANA.

2 MATERIAL MODELS

The concrete is modelled using 20-node 3D brick elements and the steel reinforcement with embedded bar elements. The uniaxial stress-strain curve used for modelling the concrete is shown in Figure 1. Confinement effects, if any, on the compressive behaviour are taken into account using the Drucker-Prager yield criterion (shown in Figure 2) with the angle of internal friction $\phi = 30$ degrees. The uniaxial compressive response is scaled to match the peak load corresponding to the strength of the confined concrete.

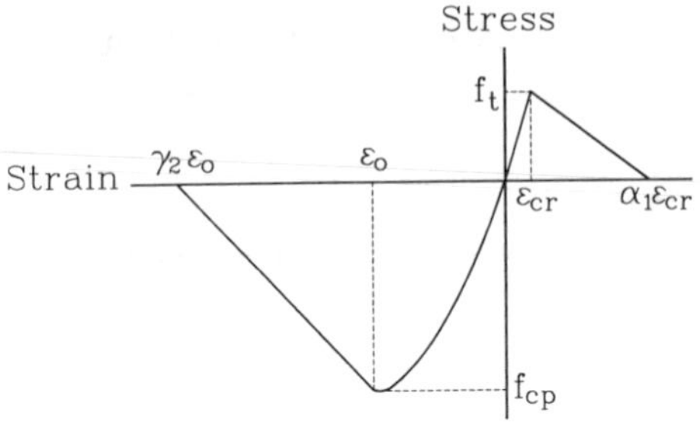

Figure 1. Stress-strain curve for concrete in uniaxial tension and compression.

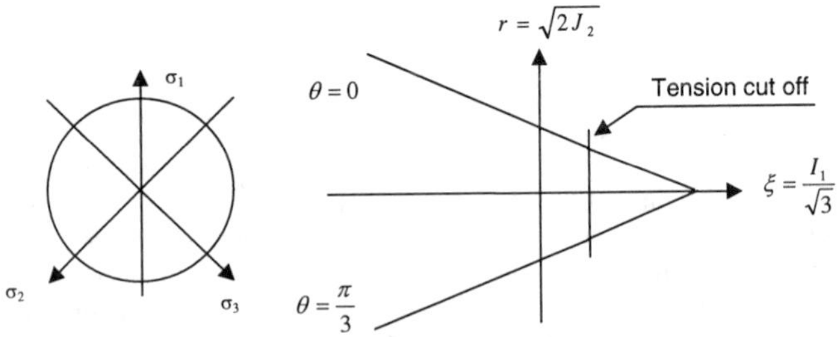

Figure 2. Drucker-Prager yield condition with tension cut-off.

For concrete in tension a linear tension cut-off model is used with smeared cracking added to the plasticity model discussed above. A crack occurs if the major principal tensile stress exceeds the triaxial tensile strength of the concrete. After cracking a linear tension-softening model is used, as shown in Figure 1. The slope of the descending branch is adjusted to allow for tension stiffening (van Mier, 1987) with the strain at the point of zero stress ε_u^{cr} given by

$$\varepsilon_u^{cr} = \alpha_1 \varepsilon_{cr} = 0.5 f_y / E_s \tag{1}$$

and with the cracking strain taken as

$$\varepsilon_{cr} = f_t / E_c \tag{2}$$

where E_c = elastic modulus of concrete, f_t = tensile strength of the concrete, f_y = yield stress of reinforcing steel and E_s = modulus of elasticity of steel. In the cracked concrete the shear stiffness due to aggregate interlocking is modelled using a shear retention factor of $\beta = 0.2$.

The reinforcing steel is modelled using an elasto-plastic stress-strain law. Perfect bond is assumed between reinforcing bars and the surrounding concrete elements.

3 FINITE ELEMENT MODEL

To demonstrate the effect of the boundary condition on the behaviour of flat plate-column connections, specimen one of Lim and Rangan (1995) was modelled using the finite element program DIANA. The model consisted of a 4950 mm by 3170 mm slab with a 250 mm by 250 mm column at the middle of one of its longer edges (as shown in Figure 3) and with the base of the column resting on a pinned support. The slab edge adjacent to the column is free while the opposite edge and the side edges rested on I-beams. The slab was restrained against lifting at the corners by placing a beam across the top corner of the slab. However no direct connection was made from the supports to the slab and, thus, the slab was free to slip horizontally against the corner restraints and the boundaries. The equilibrating horizontal forces for the horizontal reaction measured at the base of column could then only have come from friction of the slab against the I-beams, or from restraint developed at corners where the slab was prevented from lifting, or both.

The slab-column connection was modelled using 504 by 20-node isoparametric brick elements. The reinforcing steel was modelled using 78 embedded bar elements. One half of the specimen was modelled due to symmetry (see Figures 4 and 5). To model the boundary friction the corners of the slab (marked A and B in Figure 4) were restrained in the x, y, and z directions and the fixity of the column in x was replaced with a spring of variable stiffness. The stiffness of the spring at the base of the column was then tuned to simulate different boundary conditions.

The recorded mean cylinder strength at the time of testing was 25 MPa. In the FE modelling, the in-situ strength of the concrete was taken as $f_{cp} = 0.9 f_{cm}$, that is with $f_{cp} = 22.5$ MPa. The concrete tensile strength was taken as $f_t = 1.7$ MPa and the initial modulus of elasticity of the concrete as $E_0 = 26$ GPa. For the reinforcing steel in the slab the yield strength was $f_y = 520$ MPa and the elastic modulus $E_s = 170$ GPa. In the column the yield strength of steel was $f_y = 410$ MPa and with an elastic modulus of $E_s = 200$ GPa.

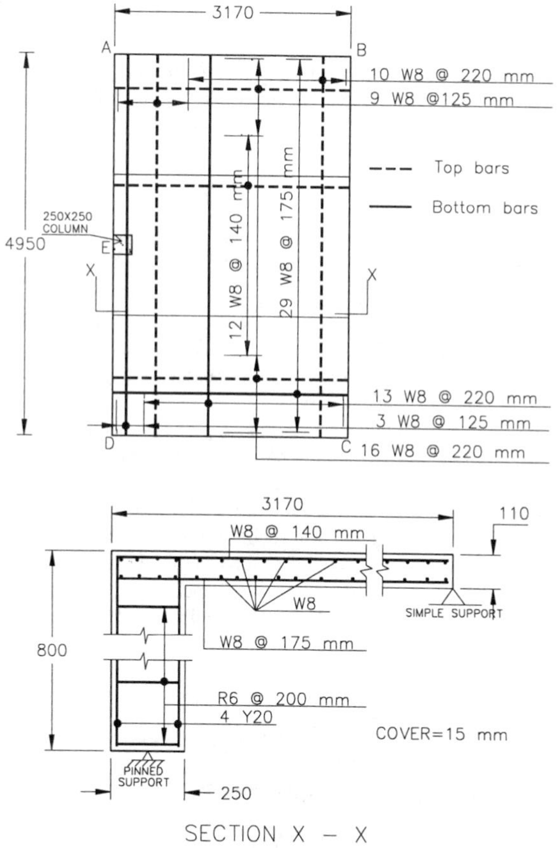

Figure 3. Detailing of Lim and Rangan (1995) specimen one used for the FE analysis.

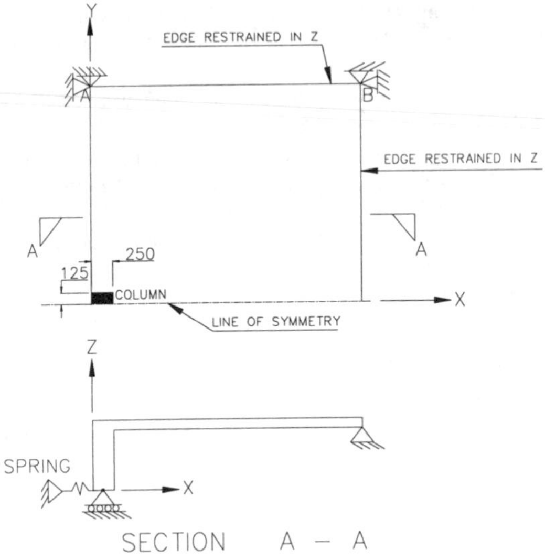

Figure 4. Boundary conditions for the FE model.

The spring at the base of the column (in the x direction) was modelled using stiffnesses of $K = 0$ N/mm (free), $K = 800$ N/mm, $K = 2400$ N/mm and $K = \infty$ (fully restrained). The stiffness $K = 800$ N/mm was selected to match the measured horizontal load in the column under elastic conditions.

4 NUMERICAL RESULTS

The calculated failure loads from each of the analyses are shown in Table 1. The best correlation with the experimental result is when the spring stiffness $K = 2400$ N/mm. With the increase of stiffness, the load carrying capacity of the system is found to decrease. The transfer of load through the connection depends on the combined effect of shear and moment interaction. With the increase of spring stiffness the bending moment transferred to the connection increases causing a reduction in overall load carrying capacity.

The mid-span deflections for the finite element analyses and the experimental slab are shown in Figure 6. A reasonable correlation is seen between the experimental data and the FE model for the slab. Figure 6 also shows that the slab deflections are not sensitive to the spring stiffness. The main variance between the numerical deflections and those recorded in the laboratory is in the pre-cracking elastic range with negligible deflections being measured in the laboratory.

The vertical reaction at the base of the column is plotted against the total slab load in Figure 7. The figure shows that the vertical reaction in the column and hence, the shear transferred to the slab is not sensitive to the spring stiffness. The best results are for stiffness of $K = 2400$ N/mm.

Table 1 – Failure loads for different base spring stiffnesses.

Stiffness, K (N/mm)	Failure Load (kPa)	Mode of Failure
0	39	Flexure
800	32	Flexure/Punching Shear
2400	25	Punching Shear
∞	19	Punching Shear
Experimental	22	Punching Shear

The greatest sensitivity to the boundary conditions is seen in the development of base shear at the column (refer Figure 8). With K = 0 N/mm, that is with no lateral restraint at the base of the column, no unbalanced moment occurs at the column-slab joint. With infinite spring stiffness the horizontal reaction and, consequently the unbalanced moment at the connection, is significantly greater than that measured in the laboratory. The best match is with K = 800 N/mm. Thus the influence of boundary conditions is clearly demonstrated.

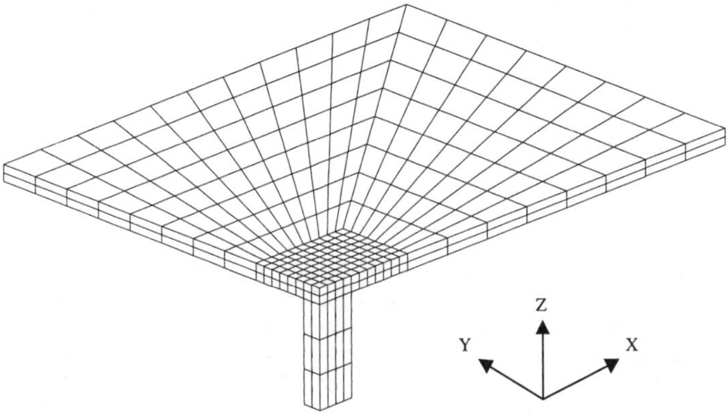

Figure 5. Finite element mesh for the Lim and Rangan (1995) specimen.

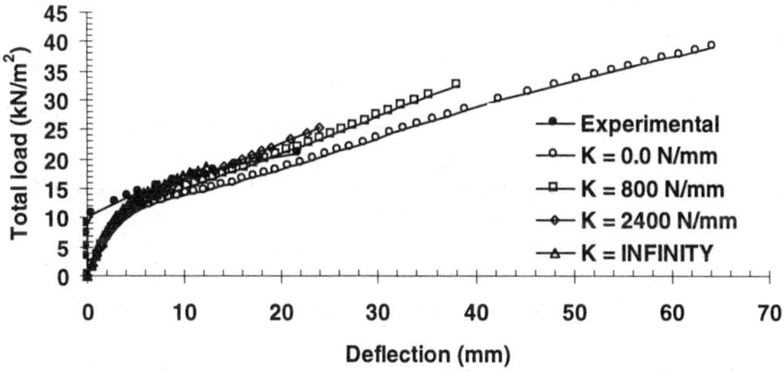

Figure 6. Load–deflection curves with different lateral restraints and experimental results.

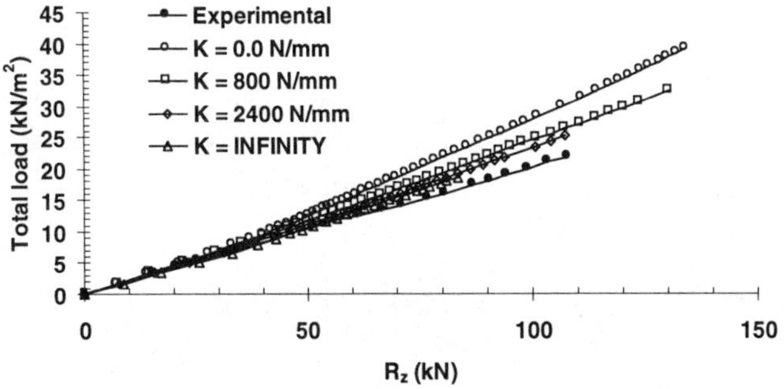

Figure 7. Load-vertical column reaction for different restraints and experimental results.

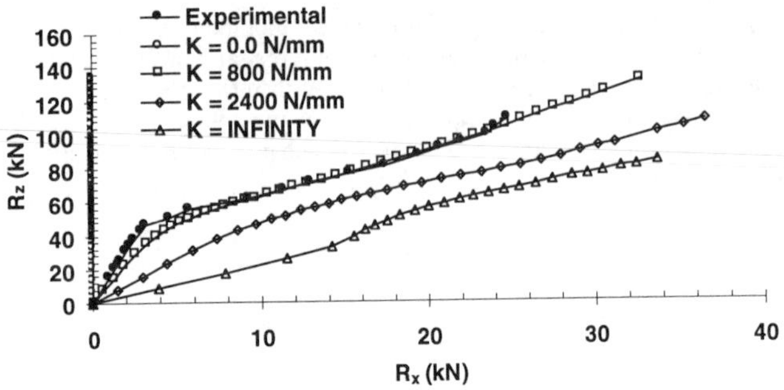

Figure 8. Vertical-horizontal reactions at column for different restraints and experimental results.

5 CONCLUSIONS

The boundary conditions have a significant effect on the moment transferred from the slab to the column in flat plate and flat slab structures. This in turn affects the capacity of the joint and puts into question how well the experimental model simulates the real structures. This study also showed that the slab deflections and the transfer of vertical load to the column are not significantly influenced by the boundary conditions. Further investigations are needed to determine what effect this may have on design practice.

REFERENCES

Elgabry, A. A. & Ghali, A. 1996. Moment transfer by shear in slab-column connections. *ACI Structural Journal.* 93(2) *March-April*: 187-196.

Foutch D. A., Gamble, W. L., & Sunidja, H. 1990. Tests of post-tensioned concrete slab-edge column connections. *ACI Structural Journal* 87(2) *March-April*: 167-179.

Lim, F. K., & Rangan, B. V. 1995. Studies on concrete slabs with stud shear reinforcement in vicinity of edge and corner columns. *ACI Structural Journal.* 92(5) *Sept.-Oct.:* 515-525.

Robertson, I. N., & Durrani, A. J. 1990. Seismic response of connections in indeterminate flat-slab subassemblies. *Department of Civil Engineering Rice University Houston Texas Report NO. 41 July*: 266.

TNO Building and Construction Research 1996. DIANA. 1996: Displacement analyzer Finite element code. *TNO Building and Construction Research 2600AA Delft Netherlands.*

van Mier, J. G. M. 1987. Examples of non-linear analysis of reinforced concrete structures with DIANA. *HERON.* 32(3): 147.

Zaghlool, E. R. F. 1971. Strength and behavior of corner and edge column-slab connections in reinforced concrete flat plates. PhD thesis. *Department of Civil Engineering. University of Calgary.* Alberta: 366.

Zaghlool, E., R. F. & de Paiva, H. A. R. 1973. Tests of flat-plate corner column-slab connections. *Proceedings ASCE.* 99(ST3) *March:* 551-572.

Mechanics of Structures and Materials, Bradford, Bridge & Foster (eds)
© 1999 Balkema, Rotterdam, ISBN 90 5809 107 4

The influence of concrete material parameters on the FE modelling of punching shear in flat plate-column connections

Y.R.Khwaounjoo, S.J.Foster & R.I.Gilbert
School of Civil and Environmental Engineering, The University of New South Wales, Sydney, N.S.W., Australia

ABSTRACT: In a companion paper, finite element modelling was used to show that the results of experimental tests on slab-column connections, under vertical load, are significantly influenced by the boundary conditions. The experimental boundary conditions influence the moment-shear ratio in the connection, the development of strain in the reinforcing steels and, consequently, the mode of failure. In this paper sensitivity analyses are presented on the material model parameters required to define the concrete stress-strain law and the shear retention used in the finite element models. It is concluded that the FE modelling of punching type problems is sensitive to the material properties assumed to describe tension stress-strain law for the concrete. In addition to the sensitivity analyses, the finite element and the experimental data are compared with design code formulae. Whilst the code formulae are generally conservative, both the experimental and the numerical data correlate poorly with code predictions.

1 INTRODUCTION

In a companion paper (Khwaounjoo et al., 1999) a finite element investigation was undertaken on a punching shear experiment of Lim and Rangan (1995). In that paper it was shown that the behaviour of slab-column connections depends on the degree of restraint imposed at the boundaries. In this paper the sensitivity of the finite element results to the assumptions made in defining the tension and shear response of the concrete post-cracking are examined.

Different design codes use different criteria for calculating the behaviour and strength of flat slab-column connections. In all cases, however, empirical or semi-empirical models are used. In this paper the results of the finite element analyses, together with experimental data, are compared to that calculated using various design codes.

2 FINITE ELEMENT MODEL

To demonstrate the effect of boundary conditions on the behaviour of flat plate-column connections, specimen one of Lim and Rangan (1995) was modelled using the finite element program DIANA. The model consisted of a 4950 mm by 3170 mm slab with a 250 mm by 250 mm column at the middle of one of its longer edges. The concrete was modelled using 504 by 20-node isoparametric brick elements and the reinforcing steel with 78 embedded bar elements. One half of the specimen was modelled due to symmetry using the FE mesh shown in Figure 1. Full details of the finite element model are given in the companion paper (Khwaounjoo et al., 1999).

In the FE modelling of the slab, the in-situ strength of the concrete was taken as, $f_{cp} = 22.5$ MPa , being 90 percent of the measured mean cylinder strength. The concrete tensile strength and the initial modulus of elasticity of the concrete were taken as $f_t = 1.7$ MPa and $E_0 = 26$ GPa , respectively. To simulate boundary friction a spring was placed at the base of the column in the x-direction (Khwaounjoo et al., 1999). In this paper the spring stiffness was held constant at $K = 800$ N/mm and is selected to match the vertical load distribution observed in the experimental study. All other details, including those of the reinforcing steel are given in (Khwaounjoo et al., 1999).

3 MATERIAL MODELS

The concrete is modelled using 20-node 3D brick elements and the steel reinforcement with embedded bar elements. The uniaxial stress-strain curve used for modelling the concrete is shown in Figure 2. In tension, the concrete is modelled using a bilinear stress-strain relationship using the smeared cracking approach. A crack arises if the major principal tensile stress exceeds the triaxial tensile strength of the concrete. After cracking a linear tension-softening model is used with the slope of the descending branch adjusted to allow for tension stiffening (van Mier, 1987). The strain at the point of zero stress ε_u^{cr} (refer Figure 2) is given by

$$\varepsilon_u^{cr} = \alpha_1 \varepsilon_{cr} = 0.5 f_y / E_s \tag{1}$$

and with the cracking strain taken as

$$\varepsilon_{cr} = f_t / E_c \tag{2}$$

where E_c = elastic modulus of concrete, f_t = tensile strength of the concrete, f_y = yield stress of reinforcing steel and E_s = modulus of elasticity of steel. After cracking, it is well acknowledged that the shear stiffness is reduced, but not to zero due to the influence of aggregate interlock and dowel effects. After cracking, the diagonal terms of the local stiffness matrix are

$$[k] = \lceil E_{nn} \quad E_{ss} \quad E_{st} \quad \beta G_{ns} \quad \beta G_{st} \quad \beta G_{nt} \rfloor \tag{3}$$

where E_{ij} and G_{ij} are tangent modulus and shear modulus, respectively, with reference to fixed principal *n, s, t* axes; *n* referring to the direction normal to the crack, *s* and *t* referring to the directions tangential to the crack and β is a shear retention factor.

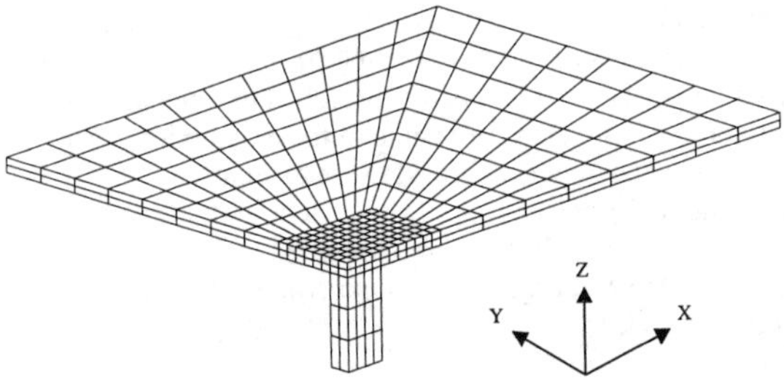

Figure 1. Finite element mesh for the Lim and Rangan (1995) specimen 1.

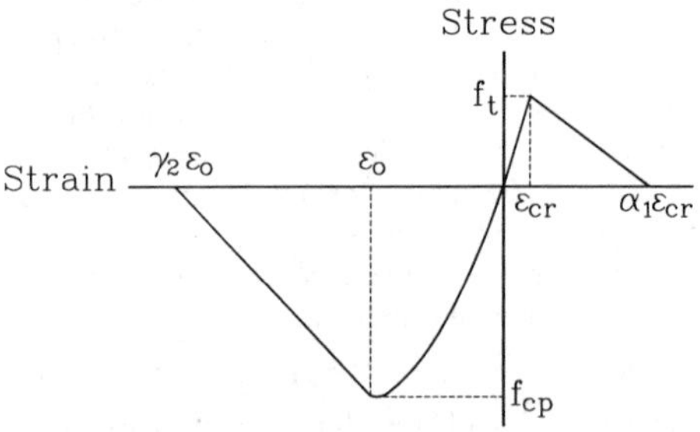

Figure 2. Stress-strain curve for concrete in uniaxial tension or compression.

4 SENSITIVITY ANALYSES

In the modelling of punching type problems the behaviour of the cracked concrete is important in obtaining accurate and reliable results for the failure load and failure mechanism. In this respect, the parameters α_1 and β are of fundamental importance. Following are the results of sensitivity analyses for the concrete tension parameters α_1 and β.

4.1 *Sensitivity to* α_1

With the development of cracks, the stiffness of the concrete reduces. This reduction and its rate is dependent on ε_u^{cr} (Figure 2) as represented by the tension softening model (van Mier, 1987). The ultimate uniformly distributed load on the slab predicted from the FE analyses with $\beta = 0.2$ and for different values of the parameter α_1 are given in Table 1 and shown in Figures 3 and 4. The parameter α_1 has a significant influence on the measured deflections and on the failure load.

Table 1.Ultimate load for different values of α_1.

Value of α_1	Ultimate Load (kPa)
0	21.7
5	23.5
15	32.4
25	34.4
Experimental	**22.0**

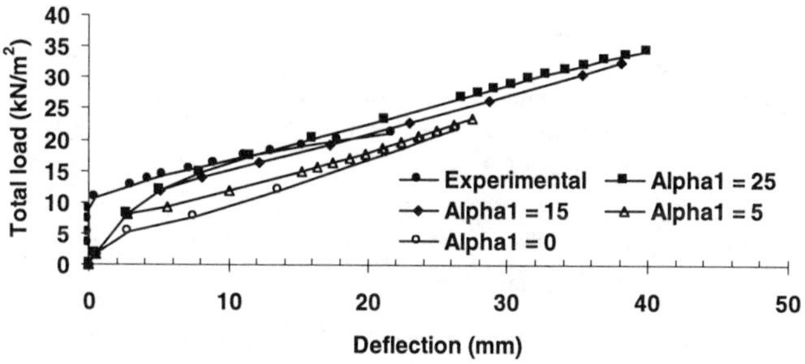

Figure 3. Load versus deflection curves for the Lim and Rangan (1995) slab with different values of α_1.

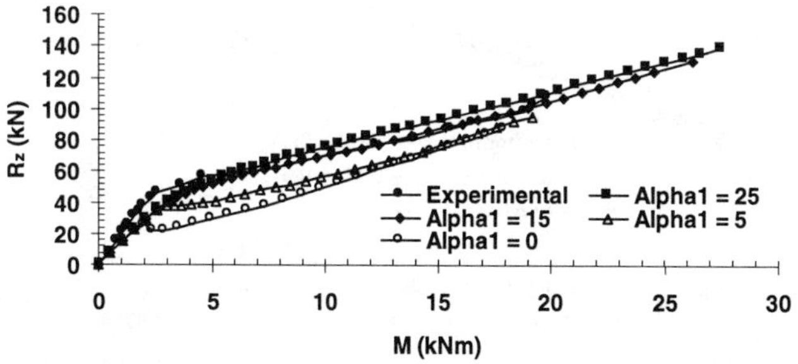

Figure 4. Shear versus moment at the column slab connection for the Lim and Rangan (1995) slab with different values of α_1.

4.2 *Sensitivity to β*

For fully cracked plain concrete the shear stiffness becomes small after the onset of cracking. However, in reinforced concrete where the interaction of the reinforcement crossing the cracks and the influence of aggregate interlock is significant, the shear stiffness of the concrete may be significant. Although the effect of shear stiffness of cracked concrete is acknowledged as important in the FE modelling community, no consensus is found amongst researchers on a satisfactory mathematical model.

The effect of the shear retention factor β on the deflections, the load distribution and the failure load are given in Table 2 and in Figures 5 and 6 for a constant value of $\alpha_1 = 15$. The analyses show that the distribution of load to the supports and the slab deflections are insensitive to β, however, the failure load is significantly influenced by the shear retention factor β selected.

Table 2.Ultimate load for different values of β.

Value of β	Ultimate Load (kN)
0.01	20.0
0.2	32.4
0.5	40.5
0.9	45.7
Experimental	**22.0**

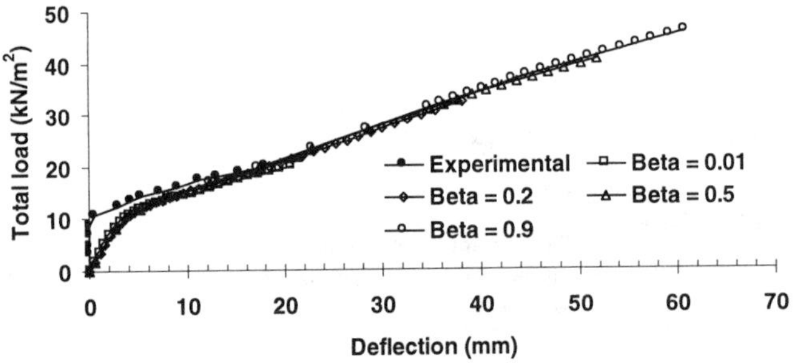

Figure 5. Load versus deflection curves for the Lim and Rangan (1995) slab-column connection with different values of β.

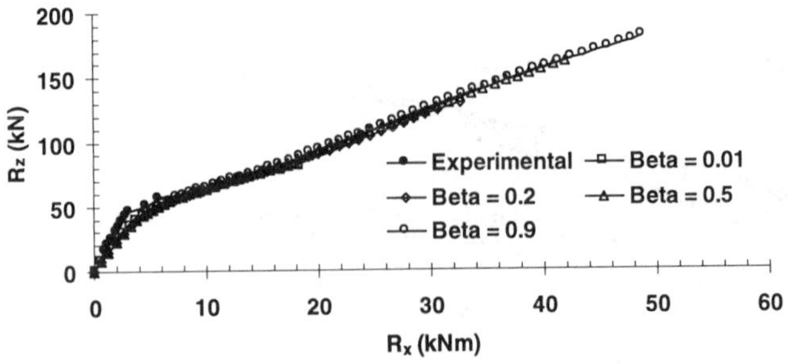

Figure 6. Vertical-horizontal reactions at the column for the Lim and Rangan (1995) slab-column connection with different values of β.

5 COMPARISON WITH DESIGN CODES

Design codes use empirical or semi-empirical models to calculate the capacity of flat slab-column connections. The main differences between codes are the concrete shear strength and the position of critical shear perimeter. In AS3600 (1994), ACI318 (1995) and CSA-A23.3 (1994) the critical shear perimeter is taken at a distance of $d/2$ from the face of the column, where d is the effective depth of the slab. In BS 8110 (1997) and Eurocode 2-(1992), the critical shear perimeter is taken at $1.5d$ from the face of the column and in CEB-FIP (1990) at $2d$ from the face of the column. This creates a large variation in the critical shear perimeters causing the design shear stress to be relatively smaller for the British and European codes compared to that of the Australian and North American codes. This large difference is offset, however, by the variation of sectional shear capacities given in the respective codes of practice.

The results of the FE modelling are compared with design code predictions in Figure 7, together with available experimental data. The interaction curves shown in Figure 7 for the AS, ACI and CEB models are plotted in the non-dimensional form of V_u/V_{uo} versus M_u/M_{uo}, where V_u is the shear capacity of the section for the ultimate moment M_u and V_{uo} and M_{uo} are the shear capacity for zero moment and the moment capacity for zero shear, respectively. In the Eurocode 2 model, the effect of moment transferred into the column is taken into consideration by increasing the design shear. In this case, the applied shear per unit length V_{sd} is plotted against the design shear resistance per unit length of critical perimeter V_{rd1}.

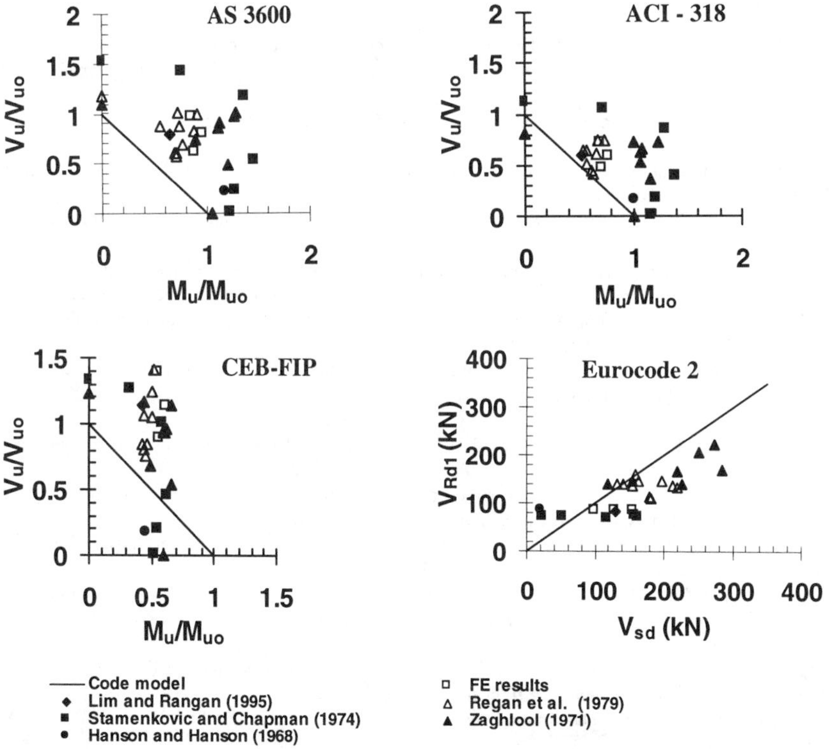

Figure 7. Comparison of FE and test results AS 3600 (1994), ACI 318 (1995), CEB-FIP (1990) and Eurocode 2 (1992).

Comparing the results of Figure 7 shows that AS 3600 (1994) is slightly more conservative than other codes. However, some non-conservative results are seen when comparing the CEB and Eurocode models to the experimental data. Finally, the large scatter of data for all code models suggests that the design models may not reflect well the mechanics of shear and moment transfer from flat plates and flat slabs to the supporting columns.

6 CONCLUSIONS

This paper is divided into two parts, a study of the effect of the tension and shear material parameters on the results of FE analysis of punching type problems and a comparison of FE results with experimental data and design code models. The failure load, the displacement of the slab and the shear to moment ratio at the junction of the column and the slab were found to be significantly affected by the slope of the descending branch of the tension stress-strain curve for the concrete. In the case of post cracking shear retention, the load versus displacement and the shear to moment ratio are not significantly influenced, but the failure load varied up to double that of the experimental result, depending on the value of β selected.

In the second part of the paper, FE and experimental data are compared to design code models. It was found that the design models do not represent well the data, but for the bulk of the test data are conservative. The high scatter of data suggests further investigations are needed to more fully understand the mechanics of load transfer from flat plate slabs to their supporting columns.

REFERENCES

ACI 318M-95 & ACI 318RM-95. Building code requirements for structural concrete (ACI 318M-95) and commentary (ACI 318RM-95). *American Concrete Institute. P. O. Box 9094. Farmington Hills. MI 48333:* 371.

AS3600-1994. Australian Standard for Concrete Structures. *Standard Association of Australia. Sydney*: 154.

British Standards Institution. 1997. Code of practice for design and construction (BS 8110: Part I). *British Standard Institution.* London: 126.

CEB-FIP Model Code for Concrete Structures 1990. *Comite Euro-International du Beton/Federation International de la Precontrainte. Lausanne.* Switzerland.

CSA 23.3-94. Design of Concrete Structures. Canadian Standards Association. Rexdale. Canada: 220.

European Committee for Strandardization. 1992. Eurocode 2: Design of concrete structures – Part I: General rules and rules for buildings (DD ENV 1992-1-1). European Committee for Standardization. Central Secretariat: rue de Stassart 36. B-1050 Brussels: 253.

Hanson, N. W., & Hanson, J. M. 1968. Shear and moment transfer between concrete slabs and columns. *Journal of Portland Cement Association Research and Development Laboratories.Illinois.* 10(1) *January*: 2-16.

Khwaounjoo, Y. R. Foster, S. J. & Gilbert, R. I. 1999. Influence of boundary conditions on punching shear behaviour of flat plate-column connections. ACMSM16. Sydney. December.

Lim, F. K. & Rangan, B. V. 1995. Studies on concrete slabs with stud shear reinforcement in vicinity of edge and corner columns. *ACI Structural Journal.* 92(5) *Sept.-Oct.*: 515-525.

Regan, P. E., Walker, P. R. & Zakaria, K. A. A. 1979. Tests of reinforced concrete flat slabs. *CIRIA Project No. RP 220. Polytechnic of Central London UK*: 217.

Stamenkovic, A. & Chapman, J. C. 1974. Local strength at column heads in flat slabs subjected to a combined vertical and horizontal loadings. *Proceedings. Institution of Civil Engineers. London.Part* 2. 57. *June:* 205-232.

TNO Building & Construction Research 1996. DIANA. 1996: Displacement analyzer Finite element code *TNO Building and Construction Research. 2600AA Delft. Netherlands.*

van Mier, J. G. M. 1987. Examples of non-linear analysis of reinforced concrete structures with DIANA. *HERON.* 32(3): 147.

Zaghlool, E. R. F. 1971. Strength and behavior of corner and edge column-slab connections in reinforced concrete flat plates. PhD thesis. *Department of Civil Engineering. University of Calgary.* Alberta: 366.

Mechanics of Structures and Materials, Bradford, Bridge & Foster (eds)
© 1999 Balkema, Rotterdam, ISBN 90 5809 107 4

High strength concrete deep beams with different web reinforcements

K.H.Tan
School of Civil and Structural Engineering, Nanyang Technological University, Singapore

L.W.Weng
Development Resources Pte Limited (A subsidiary of Singapore Power Limited), Singapore

ABSTRACT: In this paper, eighteen high strength concrete deep beams with concrete cube strengths f_{cu} generally exceeding 80 MPa were tested to failure under top symmetric two-point loading. The specimens were organised into three series of six beams each. The first beam in each series had no web reinforcement and it served as a control beam. The other five incorporated different web reinforcement details. Based on the obtained ultimate shear strength of the control beam, the web steel contribution to shear resistance for different types of web reinforcement is obtained. The paper gives an account of the diagonal crack development, the crack patterns and the failure modes of the specimens.

1 INTRODUCTION

Unlike flexural failure, shear failure is characteristically brittle (Tan et al. 1995) and occurs with little warning. Failures associated with high strength concrete can even be explosive (Ahmad et. al 1986, Elzanty et al. 1986, Roller & Russell 1990, Sarsam et al. 1992, Tan et al. 1997). Up to now, most design codes (ACI 1992, BS8110:1985, CIRIA 1984) on deep beams are based on test beams made of medium or low-strength concrete ($f_{cu} \leq 40$ MPa). (The authors have proposed a revision to the CIRIA equation (Tan et al. 1998) which covers the design of both short and deep beams). Although there are some literature on normal strength deep beams with web reinforcement (Kong 1970, Kong 1971, Smith & Vantsiotis 1982), there is relatively less information on high strength deep beams. This paper describes the effects of different types of web reinforcement on high strength concrete (HSC) deep beams.

2 EXPERIMENTAL PROGRAMME

2.1 *Details of beam specimens*

The testing program consisted of three series of 6 rectangular beams of 500 mm height and 110 mm width. The geometric details of the specimens are given in Weng (1995). All the beams had longitudinal main steel content ρ of 2.58%, consisting of four 200 mm diameter high strength deformed bars with yield strength of 498.9 MPa. Within each series, six specimens with different web reinforcements (Fig. 1) were tested to failure.

In the specimen notation, the series number is given first, followed by the type of web reinforcement, and then the a/h ratio. For example, I-1/0.75 refers to a beam with web reinforcement pattern of Type 1 and an a/h ratio of 0.75. In particular, the vertical web reinforcements on either side of Type 2 and Type 6 specimens are different. The shear spans are designated as the "N" shear span and the "S" shear span, respectively. The "N" span was provided with mild steel stirrups of f_{yv} = 353.2 MPa while the "S" span with high strength deformed stirrups of f_{yv} = 446.7 MPa.

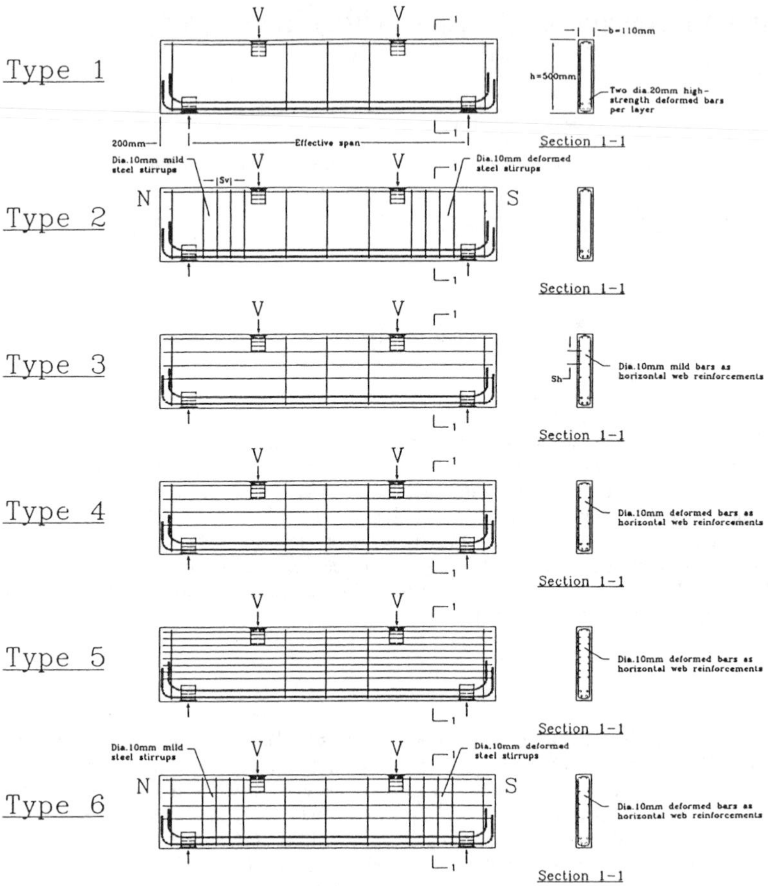

Figure 1 Different types of web reinforcement

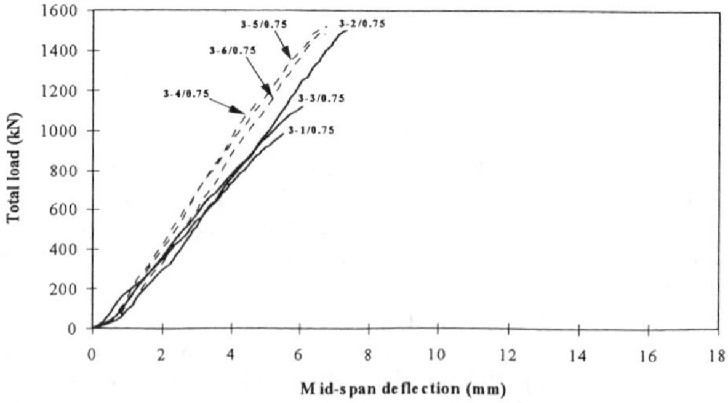

Figure 2 Total load versus mid-span deflection for Series I beams

The web reinforcement details are classified into six types, as follows:

Type 1: This specimen with no web reinforcement in either shear span served as a *control beam*.

Type 2: Vertical web steel consisting of lower strength plain bars was placed in the "N" shear span whereas vertical web steel of high strength deformed bars was placed in the "S" shear span.

Type 3: Horizontal web reinforcement consisting of mild steel bars was provided, giving ρ_h = 1.59%.

Type 4: Horizontal web reinforcement was identical to Type 3 specimen, except that high strength deformed bars were used instead of lower strength plain bars.

Type 5: Horizontal web steel ratio using high strength deformed bars was doubled to ρ_h = 3.17%.

Type 6: Here, Type 2 vertical web steel and Type 4 horizontal web steel were combined together to investigate the effect of an orthogonal web reinforcement.

2.2 *Materials*

In the concrete mix design ordinary portland cement was used, with an aggregate to cement ratio of 2.90 and water to cement ratio of 0.27. Due to congestion in reinforcement, 10 mm chippings were used as aggregates. Water-reducing plasticising admixture and silica fumes were added to give the high strength performance. The concrete compressive strengths f_{cu} and f_c' were obtained from four 100 mm concrete cubes and four 150 mm concrete cylinders respectively.

2.3 *Test Procedure*

All the beams were tested to failure under two-point symmetric top loading. Initial loading was applied at increments of 20 kN until the first crack occurred. Subsequently, the load increments were increased to 40 kN.

3 TEST RESULTS AND DISCUSSIONS

3.1 *Deflections*

The mid-span deflection curves for all six beams in Series I are shown in Figure 2. Beams with horizontal web reinforcement show greater bending stiffness. The control beam I-1/0.75 has the lowest stiffness, followed by Beam I-2/0.75 with ony mild steel stirrups. Thus, from the test results, providing stirrups alone does not seem to increase the beam stiffness significantly.

3.2 *Crack widths*

In all the specimens, the first crack was a flexural crack of width between 0.02 – 0.04 mm when it first initiated from the beam soffit in the mid-span region. Compared to the diagonal cracks, the flexural cracks were much narrower in width although in some cases, the crack widths exceeded the serviceability of 0.3 mm (BS 8110:1985) when they first occurred.

Diagonal crack development was the fastest in Type 1 control beam with no web reinforcement. It was also observed that diagonal cracks in Type 3 (ρ_h = 1.59%) and Type 4 (ρ_h = 3.17%) specimens widened faster than those in Type 2 (ρ_v = 2.86%) or Type 6 beams. At failure, all the beams had diagonal crack widths exceeded the serviceability limit.

3.3 *Diagonal cracking strength, serviceability strength and ultimate strength*

The diagonal cracking strength V_{cr} remains more or less constant within the same series, relatively undisturbed by the type of web reinforcement. It is also observed that V_{cr} decreases with increasing *a/h*.

In this context, the serviceability load V_{ser} is defined as the load corresponding to an observed crack width of 0.30 mm. The greatest V_{ser} is associated with Type 6 orthogonal web reinforce-

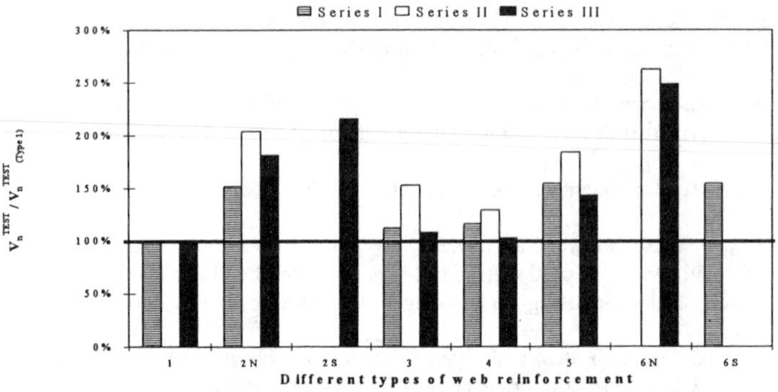

Figure 3 Ultimate shear strengths of Series I, II and III specimens

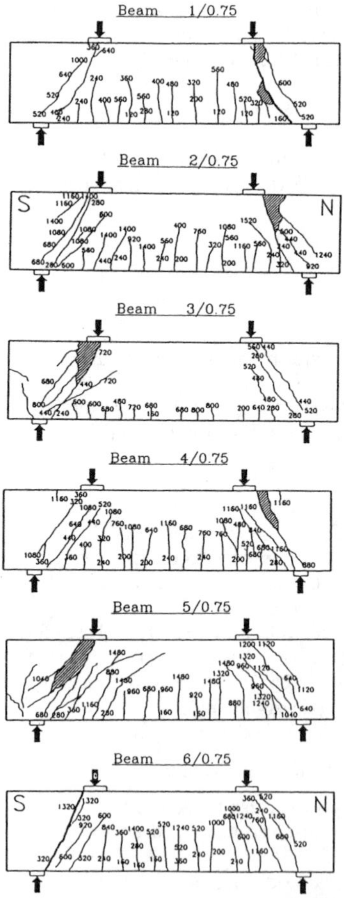

Figure 4 Crack patterns of Series I specimens

ment, whereas the least V_{ser} with Type 1 control beams. This confirms that orthogonal web reinforcement offers the most effective restraint on diagonal crack development of HSC deep beams.

Figure 3 shows the percentage ratios of the ultimate strengths (V_n^{TEST}) of different beams to that of Type 1 beam ($V_{n(Type1)}^{TEST}$) within each series; for the three control specimens, $V_n^{TEST}/V_{n(Type1)}^{TEST}$ = 100 %. This comparison is valid as within each series there is no significant variation in the concrete strengths. The term $V_{n(Type1)}^{TEST}$ represents the pure concrete contribution to shear strength V_c. To obtain the web steel contribution to shear strength V_s, $V_{n(Type1)}^{TEST}$ is subtracted from V_n^{TEST} of each beam in a series. Clearly, from Figure 3, the web steel contribution associated with Type 6 orthogonal web reinforcement is the highest, followed on by the Type 2 vertical web steel and the Type 5 heavy horizontal web steel. Thus, the effective contributions of different types of web reinforcements can be ranked in a descending order as follows: Type 6 > Type 2 > Type 5 > Type 3 ≈ Type 4. The failure loads of specimens with high-strength horizontal web reinforcement (Type 4) were about the same as those of beams with smooth horizontal bars (Type 3). But beams with double the horizontal web reinforcement ratio (Type 5) achieved significantly greater load as compared with either Type 3 or Type 4 specimens. However, for *a/h* ≥ 1.00, Type 5 specimens (ρ_h = 2.58%) with high strength deformed horizontal bars, had smaller V_s as compared with Type 2 beams (ρ_v = 1.43%) with low strength smooth bars, showing that the vertical web reinforcement was more effective. Thus, it can be inferred that the effect of horizontal web reinforcement on shear strength diminishes when *a/h* exceeds 1.00.

There was a total of six specimens with two different types of web reinforcement in both shear spans. The initial objective was to compare the shear contributions from high strength deformed bars and plain mild steel bars. In the six tests, after the first shear span had failed, the beams were externally clamped with steel yokes and re-loaded to failure. Unfortunately, only Beam III-2/1.50 was successfully re-tested. In Series I beams, apart from Beam I-6/0.75, all the other five specimens failed first in the "N" spans which were reinforced with lower strength vertical web steel, implying the shear strength contribution of high strength deformed bars is greater than that of mild steel bars.

4 MODES OF FAILURE

From the strain readings on main tension steel (Weng 1995), clearly, the reinforcement had not yielded. Thus, three failure modes are identified, i.e. *crushing of strut failure, diagonal-splitting failure* and *shear-compression failure*. In the crushing of strut failure, there normally exist more than one inclined cracks. The concrete portion between the inclined cracks is in compression, forming a concrete compression strut, which crushes under high compression. This mode of failure is brittle and sudden. An equally brittle failure mode is the shear-compression mode in which after the appearance of the inclined crack, the concrete portion above the upper end of this crack experiences high compression. When the inclined crack further propagates upwards, the concrete above the crack fails by crushing, accompanied by a loud noise. The diagonal splitting mode is a less brittle mode in comparison, characterised by a critical diagonal crack joining the outside edge of the bearing block at loading point and the inside edge of the bearing block at support point. No explosive sound was heard for this mode of failure, which was akin to the splitting failure of a concrete cylinder in a tensile splitting test.

The crack patterns at failure of Series I specimens are shown in Figure 4. For Series I specimens, Beams I-1/0.75 and I-6/0.75 failed in the diagonal-splitting mode. The failure was not as explosive as the crushing of strut mode experienced by Beams I-2/0.75, I-3/0.75 and I-5/0.75 in which loud noise could be heard. An equally abrupt failure was that of Beam I-4/0.75, which failed in shear-compression mode.

5 CONCLUSIONS

From the study of 18 HSC deep beams with six different web reinforcement details, the following conclusions are made:

1. It is evident that web reinforcement can play an important role for HSC deep beams. The most favourable pattern is the Type 6 orthogonal web reinforcement; it is the most effective in in-

creasing the beam stiffness, restricting the diagonal crack width development (thereby increasing the serviceability load) and in increasing the ultimate shear resistance.

2. For deep beams with $a/h \geq 1.00$ (equivalent to $a/d \geq 1.13$), the vertical web steel has greater effect on restraining the diagonal crack width and increasing the ultimate shear resistance of HSC deep beams than the horizontal web steel of the same steel ratio.

3. It is also confirmed that the web steel contribution of high strength deformed bars is significantly greater than that of lower strength plain mild steel bars.

ACKNOWLEDGMENT

The Nanyang Technological University, Singapore provided the funding for this research through the Applied Research Project No. 9/91.

REFERENCES

ACI Committee 318 1992. Building Code Requirements for Reinforced Concrete (ACI 318-89) and Commentary — ACI 318R-89. *American Concrete Institute*, Detroit.

Ahmad, S.H., Khaloo, A.R. & Poveda, A. 1986. Shear Capacity of Reinforced High-Strength Concrete Beams. *ACI Journal*. V.83, No.2, Mar.-Apr. 1986, pp.297-305.

BS8110: 1985. Structural Use of Concrete – Part 1: Code of Practice for Design and Construction. *British Standards Institution*. London.

Elzanaty, A.H., Nilson, A.H. & Slate, F.O. 1986. Shear Capacity of Reinforced Concrete Beams Using High-strength Concrete. *ACI Journal*. V.83, No.2, Mar.-Apr. 1986, pp. 290-296.

CIRIA Guide-2 1984. The Design of Deep Beams in Reinforced Concrete. *Construction Industry Research and Information Association*, London, Jan. 1977 (Reprinted 1984), 131pp.

Kong, F.K. & Robins, P.J. 1971. Web Reinforcement Effects on Lightweight Concrete Deep Beams. *ACI Journal*, V.68, No.7, July 1971, pp. 514-520.

Kong, F.K., Robins, P.J. & Cole, D.F. 1970. Web Reinforcement Effects on Deep Beams. *ACI Journal*. V.67, No.12, Dec.1970, pp.1010-1017.

Roller, J.J. & Russell, H.G. 1990. Shear Strength of High-strength Concrete Beams with Web Reinforcement. *ACI Structural Journal*. Vol.87, No.2, Mar.-Apr. 1990, pp. 191-198.

Sarsam, K.F. & Al-Musawi, J.M.S. 1992. Shear Design of High- and Normal Strength Concrete Beams with Web Reinforcement. *ACI Structural Journal*. Vol.89, No.6, Nov.-Dec. 1992, pp. 658-664.

Smith, K.N. & Vantsiotis, A.S. 1982. Shear Strength of Deep Beams. *ACI Journal*. V.79, No.3, May-June 1982, pp.201-213.

Tan, K.H., Kong, F.K., Teng, S. & Guan, L. 1995. High Strength Concrete Deep Beams with Effective Span and Shear Span Variations. *ACI Journal*. V. 92, No. 4, July-Aug 1995, pp. 395 - 405.

Tan, K.H., Teng, S., Kong, F.K. & Lu, H.Y. 1997. Main Tension Steel in High Strength Concrete Deep and Short Beams. *ACI Structural Journal*, Vol. 94, No. 6, Nov-Dec, 1997.

Tan, K.H., Kong, F.K. & Weng, L.W. 1998. High Strength Reinforced Concrete Deep and Short Beams: Shear Design Equations in North American and UK Practice. *ACI Structural Journal*. Vol. 95, No. 3, May-June 98.

Weng, L.W. 1995. High Strength Concrete Deep Beams with Different Web Reinforcements under Combined Loading. *MEng* Thesis, Nayang Technological University, August, 1995.

Mechanics of Structures and Materials, Bradford, Bridge & Foster (eds)
© 1999 Balkema, Rotterdam, ISBN 90 5809 107 4

Strut-and-tie model for large reinforced deep beams

K.H.Tan
School of Civil and Structural Engineering, Nanyang Technological University, Singapore
H.Y.Lu
L&M Concrete Specialists Pte Limited, Singapore

ABSTRACT: The strut-and-tie model presented here is used to predict the ultimate shear strengths of reinforced deep beams. In addition to 233 deep beam data collected from past research works, predictions from the proposed model are compared with two studies of deep beams tested in-house, viz. Study I comprising 22 reinforced beams with different main steel ratios, and Study II of 12 large reinforced beams with overall height ranging from 500 to 1750 mm.

The strut-and-tie model incorporated with the appropriate failure criteria for the nodal zones and the members, gives very good shear strength predictions of deep beams, that is, the accuracy does not dip as member sizes increase. Derivations for the model are given and they can be implemented in a personal computer. More importantly, the suggested approach is rational and the failure criteria explicit. The model can also be extended as a useful design tool as it can indicate the "weakest link" in the truss.

1 INTRODUCTION

The "sectional" approach for the shear design of deep beams is adopted in the current design practices (Tan et al. 1998), for example, the ACI 318 Code (1995) and the UK CIRIA Guide-2 (1984). These two methods are based on the assumption that the shear resistance comprises two parts, viz. the concrete contribution V_c and the steel contribution V_s. Though easily understood and conveniently used, the above assumption ignores the interplay between the concrete and steel. Besides, studies (Tan & Lu 1999a) have shown that the conservatism in these code predictions erodes with increasing member size. Thus, it is necessary to develop an approach that not only provides accurate predictions of shear strength and does not suffer from the size effect.

In a strut and tie model, concrete is modeled as compression struts, while steel is modeled as tension ties. At the nodal zones, concrete strength is subjected to the maximum nodal stress limits (Canadian CSA Code 1994). Although there are a number of studies on strut-and-tie modeling of deep beams (Ramirez & Breen 1991, Schlaich et al. 1987, Foster & Gilbert 1996), very few papers relate to the investigation of *size effect* using such an approach. To establish confidence in this approach, the model predictions are compared with test results from Study I (Tan et al. 1997c) and II (Tan & Lu 1999a), and test results from others (Tan & Lu 1999b).

2 STRUT-AND-TIE MODEL

2.1 *Basic assumptions*

The proposed strut-and-tie model is shown in Figure 1, taking advantage of symmetry. Top loading is applied at Node 4; provision for bottom loading is allowed for at Node 5, in which case, a tie member between Node 4 and 5 (Tie 4) is required. The horizontal force at Node 1 represents the prestress force, acting at the level of main tension tie. For reinforced beams, this horizontal force is simply zero.

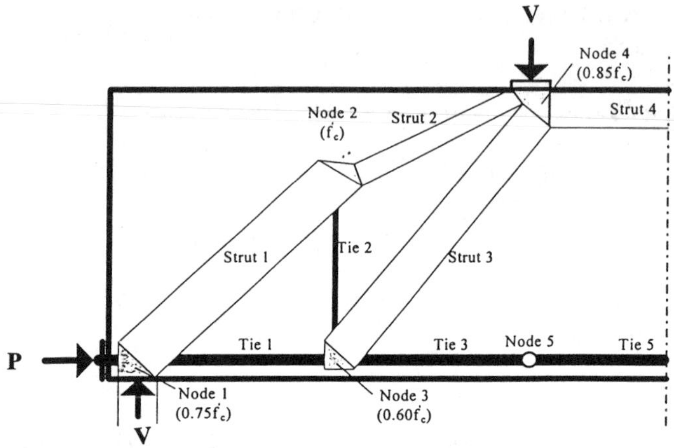

Figure 1 Strut and tie model for deep beams

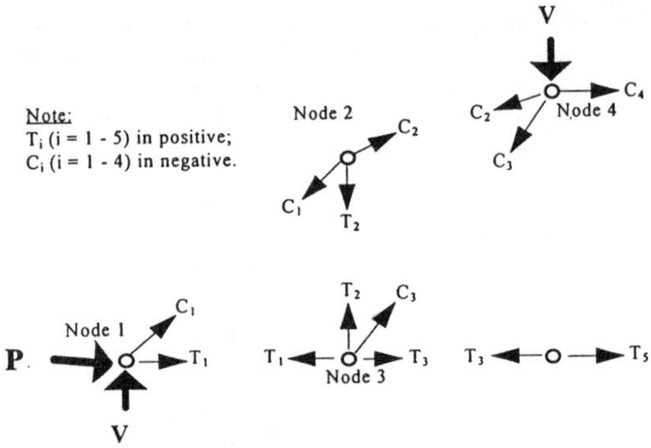

Figure 2 Equilibrium of forces at the nodes

In the analysis, the following assumptions are made:

1. Inclined uniaxial struts idealized the compressive stress trajectories along the path joining the loading and support points (Fig. 1). Similarly, the "ties" are under uniaxial tension. The struts and ties are connected at the nodes, which are biaxially stressed and do not resist any moment.
2. The cross sectional area of each member is constant along the member length. This is different from the truss model in the CSA Code where two ends of an inclined strut member can have different dimensions.
3. The effect of horizontal web reinforcement is ignored in the proposed strut-and-tie model for simplicity. However, vertical web reinforcement is represented by a vertical link at the mid-shear-span (Tie 2 in Figure 1) as its influence on the ultimate capacity is more dominant than horizontal web reinforcement (Tan et al. 1997b). This is an improvement over the current CSA model that does not consider the beneficial effect of web reinforcement.
4. At the junction of the inclined strut and the main tie, the compressive stress in the nodal zone cannot exceed the effective compressive strength of concrete f_{ce}.

2.2 *Compressive and tensile stress limits*

All the nodes in Figure 1 are represented by idealized pins and are subjected to the CSA (1994) code restrictions on maximum concrete compressive stresses (Tan et al. 1997a). These restrictions at the nodal zones are also clearly indicated in Figure 1.

The main longitudinal reinforcement A_s is represented by Tie 1, Tie 3 and Tie 5 located at the effective depth d of the beam. Since considerable tensile straining is experienced at Node 1, the effective compressive stress f_{ce} in that nodal zone is reduced in accordance with the equation proposed by Collins & Mitchell (1986):

$$f_{ce} = \frac{f_c'}{0.8 + 170\varepsilon_1} \leq 0.85 f_c' \qquad (1)$$

in which
f_c' is the cylinder compressive strength of concrete;
ε_1 is the principal tensile strain crossing the strut and is given by $\varepsilon_1 = \varepsilon_s + (\varepsilon_s + 0.002)/\tan^2\alpha_s$;
ε_s is the tensile strain in the main tie and $\alpha_s = \alpha_1$.

For all the tie members except Tie 2, the maximum tensile stress f_s is limited by:

$$f_s \leq f_y \qquad (2)$$

in which f_y is the yield strength of the main longitudinal reinforcement.

The load-carrying capacity of Tie 2 comprises the strength of vertical links and the tensile strength of concrete, that is, it acts like a "reinforced concrete" tie. However, if there is no vertical link within the shear span, then Tie 2 becomes a "concrete tie". To calculate the strength of this concrete tie, a critical crack is assumed to extend between the loading and the support points and the capacity of Tie 2 is given by:

$$T_{2max} = 0.5\left(f_t b\sqrt{a^2 + h^2} \right)\frac{a}{\sqrt{a^2 + h^2}} = 0.5 f_t ab \qquad (3)$$

where f_t is the cylinder splitting tensile strength of concrete; a is the length of shear-span; h is the overall height of a beam; b is the beam width.

A factor of 0.5 is introduced to take account that only portion of the concrete at the upper location along the critical crack contributes effectively towards the shear resistance.

2.3 *Calculation of angles*

The length of a member between Node i and Node j is denoted as l_{ij}. In Figure 1, Node 3 is located at the mid-length of l_{15}. Besides, the length of Tie 2 (l_{23} in Fig. 2) is taken to be a function of the truss height l_{45}, assumed as $l_{23} = k\, l_{45}$, where k is defined as the ***strut factor***. The angles α_1, α_2 and α_3 can be expressed in terms of the strut factor k, shear span a and truss height l_{45}, as follows:

$$\tan\alpha_1 = l_{23}/l_{13} = (kl_{45})/(0.5a) = (2kl_{45})/a$$
$$\tan\alpha_2 = (l_{45} - l_{23})/l_{35} = (1-k)l_{45}/(0.5a) = 2(1-k)l_{45}/a \qquad (6)$$
$$\tan\alpha_3 = l_{45}/l_{35} = (2l_{45})/a$$

Employing a linear correlation between the limits of k and a/d, the following formulas for the strut factor k are suggested:

$$k = \begin{cases} 0.5 & \text{when } a/d \leq 0.5 \\ 0.4 + 0.2a/d & \text{when } 0.5 < a/d < 2.5 \\ 0.9 & \text{when } a/d \geq 2.5 \end{cases} \qquad (7)$$

2.4 *Calculation of internal forces*

From nodal equilibrium of forces in Figure 2, the following expressions can be determined:

For strut members:

$$C_1 = -V/(\sin\alpha_1)$$

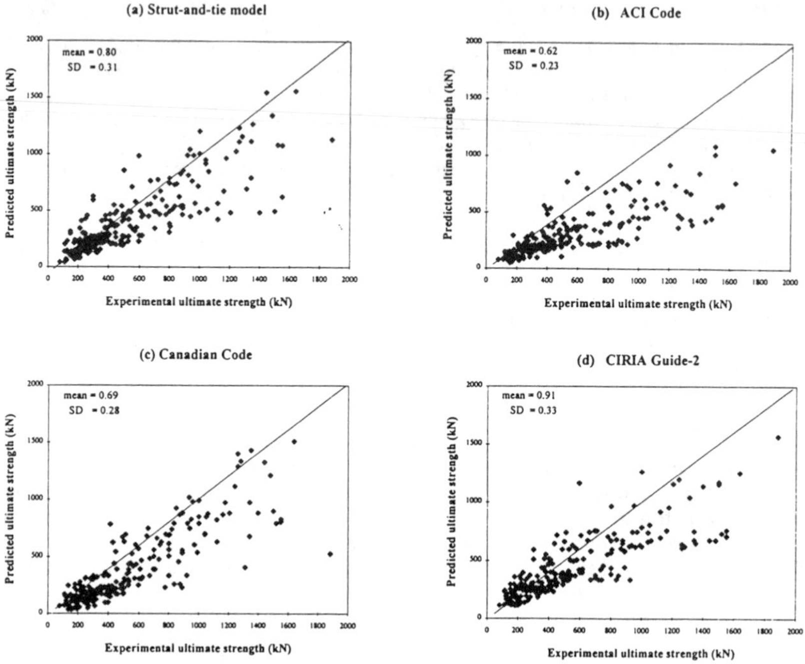

Figure 3 Ultimate strength predictions for 267 deep beams

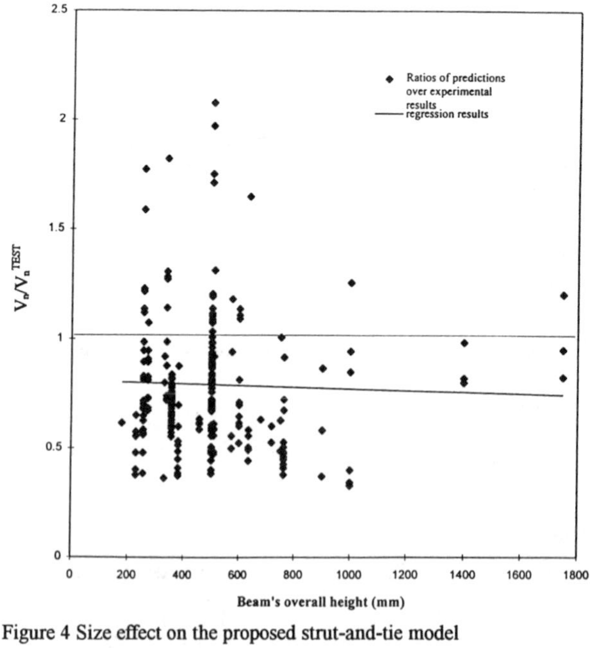

Figure 4 Size effect on the proposed strut-and-tie model

$C_2 = - V/(\tan\alpha_1 \cos\alpha_2)$

$C_3 = - V(1\text{-}\tan\alpha_2/\tan\alpha_1)/\sin\alpha_3$

$$C_4 = -V\{(\tan\alpha_1)^{-1} + (\tan\alpha_3)^{-1}(1\text{-}\tan\alpha_2/\tan\alpha_1)\} \tag{8}$$

<u>*For tie members:*</u>

$T_1 = V/\tan\alpha_1 - P$

$T_2 = V(1\text{-}\tan\alpha_2/\tan\alpha_1)$

$T_3 = V\{(\tan\alpha_1)^{-1}+(\tan\alpha_3)^{-1}(1\text{-}\tan\alpha_2/\tan\alpha_1)\}\text{-}P$

$T_4 = V_2$

$$T_5 = V\{(\tan\alpha_1)^{-1}+(\tan\alpha_3)^{-1}(1\text{-}\tan\alpha_2/\tan\alpha_1)\}\text{-}P \tag{9}$$

where $C_1, C_2, C_3, C_4 \leq 0$, $T_1, T_2, T_3, T_4, T_5 \geq 0$ and $0 < \alpha_1, \alpha_2, \alpha_3 < \pi/2$.

2.5 *Dimensioning truss elements*

The respective widths of struts 1 to 4 are denoted by W_1 to W_4. All the struts maintain the same thickness as the beam width b. The expressions for the respective strut widths W_1 to W_4 are derived as follows:

$$W_1 = a_1 \cos\alpha_1 + b_1 \sin\alpha_1 \tag{10a}$$

$$W_2 = (b_2 - W_3 \sin \alpha_3) \sin \alpha_2 + (W_4 - W_3 \cos \alpha_3) \cos \alpha_2 \tag{10b}$$

$$W_3 = W_4 \cos \alpha_3 \tag{10c}$$

where a_1 is the vertical dimension of Node 1; b_1 is the stiff bearing length of the support area at Node 1; b_2 is the stiff bearing length of the loading area at Node 4 (see Fig. 1).

To calculate W_1 to W_3, the width of Strut 4, W_4, is required. Substituting the relevant stress limits for C_4 and T_1 in Figure 1, W_4 can be determined:

$$W_4 = -(\frac{0.75}{0.85})\frac{C_4}{T_1}a_1 = \left(\frac{1}{1.13}\right)\left[1 + (\tan\alpha_3)^{-1}(\tan\alpha_1 - \tan\alpha_2)\right]a_1 \tag{12}$$

The truss height l_{45} is related to a_1 by the following equation:

$$l_{45} = h - \frac{1}{2}a_1 - \frac{1}{2}W_4 \tag{13}$$

Thus, the value of l_{45} that satisfies the external equilibrium of the truss model is given by:

$$l_{45} = -\frac{V}{C_4}a \tag{14}$$

in which a is the horizontal distance between Node 1 and Node 5.

3 COMPARISON OF TEST RESULTS WITH PREDICTIONS

Predictions from the strut-and-tie model are compared with test results from a total of 267 *reinforced* beams comprising 233 data (Tan & Lu 1999b). Among the test data, 22 beams in Study I with different main steel ratios (Tan et al. 1997c) and 12 large beams in Study II (Tan & Lu 1999a) were conducted in-house. Test results are also compared with predictions from the ACI 318-95 Code, the Canadian CSA 94 Code and the CIRIA Guide-2. The comparison is shown in Figure 3, in which the x-axis represents the experimental strength and the y-axis represents the predicted strength. The degree of strength overestimation can be assessed from the points above the diagonal line. Clearly, the ACI Code is the most conservative with a–mean value of 0.62, while CIRIA has the highest mean of 0.91. The proposed strut-and-tie model has an intermediate mean value of 0.80 with a competitive SD of 0.31.

The 267 test data are re-analyzed for the influence of size effect in Figure 4; the overall height h ranges from 178 to 1750 mm in the x-axis. By employing a least squares analysis on 267 values of V_n/V_n^{TEST}, an almost flat line is obtained with a gradient of negative 3.64×10^{-5}, thus indicating that the prediction accuracy is not affected by the size effect.

4. CONCLUSIONS

The proposed strut-and-tie model gives accurate and consistent predictions of shear strengths of reinforced concrete deep beams. Unlike the ACI code and the CIRIA guide, the prediction accuracy for a wide range of member sizes is not affected by the size effect. With suitable safety factors in-built into the stress and force limits, the model can also serve as a design tool.

REFERENCES

ACI Committee 318 1995. Building Code Requirements for Reinforced Concrete (ACI 318-95) and Commentary - ACI 318R-95. *American Concrete Institute*, Detroit, Michigan.

Alshegeir A. & Ramirez, J. A. 1992. Strut-Tie Approach in Pretensioned Deep Beams. *ACI Structural Journal*. Detroit, S 29, No. 89, May-June 1992, pp 296-304.

Canadian Standards Association 1994. Design of Concrete Structure for Buildings CAN3-A23.3-M94. *Canadian Standards Association*. Toronto, Canada.

CIRIA Guide 2 reprinted in 1984. The design of deep beams in reinforced concrete. *CIRIA Publication*. pp. 37-38.

Collins, M.P. & Mitchell, D. 1984. A Rational Approach to Shear Design — The 1984 Canadian Code Provisions. *ACI Journal*. V.83, No.6, Nov.-Dec. 1986, pp. 925-933.

Foster, S.J. and Gilbert R.I. 1996. The Design of Nonflexural Members with Normal and High-Strength Concretes. *ACI Structural Journal*. V. 93, No.1, Jan-Feb. 1996, pp. 3 – 10.

Ramirez, J.A. & Breen, J.E. 1991. Evaluation of a Modified Truss-Model Approach for Beams in Shear. *ACI Structural Journal*. Vol.88, No.5, Sept.-Oct. 1991, pp.562-571.

Schlaich, J., Schoefer, K. & Jennewein, M. 1987. Toward a Consistent Design of Structural Concrete. *PCI Journal*. Vol.32, No.3, May-June 1987, pp. 74-150.

Tan, K.H., Kong, F.K. & Weng, L.W., 1998. High Strength Reinforced Concrete Deep and Short Beams: Shear Design Equations in North American and UK Practice. *ACI Structural Journal*. Vol. 95, No. 3, May-June 98.

Tan, K.H., Weng, L.W. & Teng, S. 1997a. A Strut-and-tie Model for Deep Beams Subjected to Combined Top-and-bottom Loading. *The Structural Engineer*. Vol. 75, No. 13, 1 July 1997.

Tan, K.H., Kong, F.K., Teng, S. & Weng, L.W. 1997b. Effect of Web Reinforcement on High Strength Concrete Deep Beams. *ACI Structural Journal*, Vol. 94, No. 5, Sept-Oct, 1997.

Tan, K.H., Teng, S., Kong, F.K. & Lu, H.Y. 1997c. Main Tension Steel in High Strength Concrete Deep and Short Beams. *ACI Structural Journal*. Vol. 94, No. 6, Nov-Dec, 1997.

Tan, K.H. and Lu, H.Y. 1999a. Shear Behavior of Large RC Deep Beams and Code Comparisons. Accepted for publication in *ACI Structural Journal*.

Tan, K.H. & Lu, H.Y., 1999b. Strut-and-tie model for large reinforced and prestressed deep beams. Submitted to the *ACI Structural Journal* for publication.

Mechanics of Structures and Materials, Bradford, Bridge & Foster (eds)
© 1999 Balkema, Rotterdam, ISBN 90 5809 107 4

Using helices to enhance the performance of high strength concrete beams

M.N.S. Hadi & L.C. Schmidt
Department of Civil, Mining and Environmental Engineering, University of Wollongong, N.S.W., Australia

ABSTRACT: Beams made of high strength concrete (HSC) cannot utilise their full potential strength due to ductility provisions in the current design standards. However, the full strength potential and ductility can be achieved, provided certain modification techniques are adopted. This paper presents one new way of improving the structural performance of reinforced HSC beams. Exploration into the effects on strength and ductility of high strength reinforced concrete beams, through an increase in tensile reinforcement, is reviewed. The use of helical reinforcement in the compression zone of beams in order to improve ductility is proposed, in which different geometric configurations are explored. Beams of 1.2 metres span were subjected to standard flexural loading, with an emphasis placed on the mid span deflection and the plastic curvature capacity of the beams.

1 INTRODUCTION

Current developments of the construction industry have led to the continual improvement of construction materials. However, continual improvements in a material's strength capacity are burdened by a decrease in ductility. Theoretically, to increase the ductility will allow new materials to be used, as well as harness the full potential flexural strength of reinforced concrete beams. Improved ductility through the incorporation of helical reinforcement located in the compression region is investigated herein. Helical reinforcement has been successfully used to improve the ductility of high strength concrete columns, but is yet to be implemented in beams. The helix must successfully confine the inner core in order for substantial ductility improvements. In this work, different helical configurations are investigated, in order to determine the full implications of their use.

2 HIGH STRENGTH CONCRETE CHARACTERISTICS

High strength concrete has become increasingly popular in recent years. The Australian Standard (AS 3600) recognises the use of concrete up to 50 MPa, although a modification allowing the use of 60 MPa is expected in the near future. Recently, concretes of 70, 80 and 90 MPa have been used on a small number of projects throughout Australia, whilst in other countries, compressive strengths up to 130 MPa have been used (Webb, 1993). As a result of its highly brittle nature, its immediate use throughout the industry is restricted by limitations placed by the Australian Standard. A full understanding into its behaviour is required before it can be successfully incorporated into present and future structures.

The high compressive strength of high strength concrete can be obtained from a low water to cement ratio and the use of potential high strength cements, such as silica fume (Cement and Concrete Association of Australia and National Ready Mixed Concrete Association of Australasia, 1992).

Silica fume is a by-product from the production of ferrosilicate alloys in an arc furnace. The resulting powder is extremely fine (surface area of approximately 20 000 m^2/kg) and highly reactive (Cement and Concrete Association of Australia and National Ready Mixed Concrete Association of Australasia, 1992). The fine cementitous material will completely react, filling any small voids and providing a highly dense material.

Incorporation of a low water-cement ratio (≈ 0.25) and an increase in water demand from the silica fume, results in a highly unworkable mix. To improve the workability a superplasticiser is incorporated. The superplasticiser increases the slump of the mix, but does not affect the composition or characteristics of the concrete. However, the period of high workability is limited to around an hour.

3 DUCTILITY INVESTIGATIONS – HELICAL REINFORCEMENT

Lateral reinforcement will provide confinement to the concrete, and hence increase the compressive strength of the member. This effect is achieved by resisting lateral expansion, due to the Poisson effect upon loading, and thus acts on the concrete with a lateral compressive force. This allows the core of the member to be subjected to a multi-axial compression. In this state, both the deformation capacity and strength of the concrete are enhanced (Pessiki and Pieroni, 1997). A helical configuration will provide a sufficiently even distribution of lateral forces to the enclosed core of concrete.

Experiments preformed by Pessiki and Pieroni (1997) looked at the axial loading of large-scale spirally reinforced concrete columns. During the experimentation, it was observed that large cracks occurred along the length of the concrete, and nearing the ultimate loading capacity the concrete cover split and buckled and fell away. The closely spaced reinforcement physically separated from the concrete cover from the core, causing the early failure of the cover. It was then suggested that a decrease in spiral pitch was required for the high strength column to achieve the same ductility as that of normal strength columns. Foster et al. (1998) studied spalling of column covers and proposed a mathematical model to represent the concrete in the cover area. Results of the developed model have comparable values to the experimental results.

Helical reinforcement is used mainly in columns, however it may be incorporated into other applications to improve ductility or compressive strength. To be effective, the helical reinforcement must be placed in compression and the pitch must be small enough to confine the concrete core. Herein, the exploration of helical reinforcement in the compressive region of the reinforced concrete beams is considered, so as to determine whether the helix will increase the strength and ductility of the beams. Single and double helices were used.

The helical pitch size is a major influence on the effectiveness of the confinement. If the concrete is sufficiently confined, only the cover will spall near the ultimate loading capacity. However, if the pitch is increased the concrete core may not be effectively confined, causing some of the core to shed with the concrete cover. The reduction in cross sectional area affects the ultimate load capacity.

4 EXPERIMENTAL PROGRAM

Twelve beams were subjected to standard flexural loading. Six different beams were tested and each beam was duplicated in order to observe the typical behaviour. Table 1 shows a summary of the beams' composition.

Throughout this research a plain beam (PL) is classified as a reinforced concrete beam that does not contain any helical reinforcement, while SH and DH indicate a single helix and double helix, respectively. The notation of UR (Under reinforced) and OR (Over reinforced) are used to indicate respective design specifications. The pitches are alternated between 15 mm (P1) and 45 mm (P2). Modifications to a design mix obtained from Van and Montgomery (1998) were used with a targeted characteristic strength of 90 MPa, but the compressive strength obtained was approximately 25% lower than the intended design strength. The beams' dimensions and reinforcement placement are shown in Figure 1 (a) through (c). Cover for all reinforcement was 20 mm.

The reinforcing cage was constructed through the use of wire ties. Deformed steel bars located in the compression region of the beam were not considered in resisting flexural loads, but were used to support the shear stirrups and helical reinforcement. The helical reinforcement was coiled by cold working straight R6 steel around a hollow tube. To obtain the desired pitch, the helices were pulled longitudinally and cut to the desired length.

Superplasticiser was added to all the mixes to obtain a normal structural slump. The beams were then quickly poured and compacted through the use of an internal vibrator. All beams were left in their moulds until adequate strength was obtained, and were cured by covering with wet hessian bags. To obtain a comparative concrete strength, the test cylinders were cured in the same manner.

All beams were tested in an INSTRON 8500 (500 kN capacity) in the Wollongong University Laboratory. Strain gauges were used to monitor strains in the midspan longitudinal steel and helical reinforcement. The INSTRON has an internal deflection monitor, however a deflection gauge was still situated at midspan to measure vertical deflection. Demec gauges were approximately placed at the location of the longitudinal steel and 25 mm from the top of the cross section at midspan. The rate of loading was 0.01 mm/s, with readings taken in 5 kN increments. A plotter was used to graph the load versus the deflection of the beam during loading. Loading of the beam was ceased upon failure, or due to the inability to continue effective loading.

5 EXPERIMENTAL RESULTS

The loading of the beams incorporating a 15 mm pitch had to be stopped as the corner of the loading beam came in contact with the test beam. The helically reinforced beams experienced an initial failure of the compression region. The concrete cover failed and crumbled, yet loading

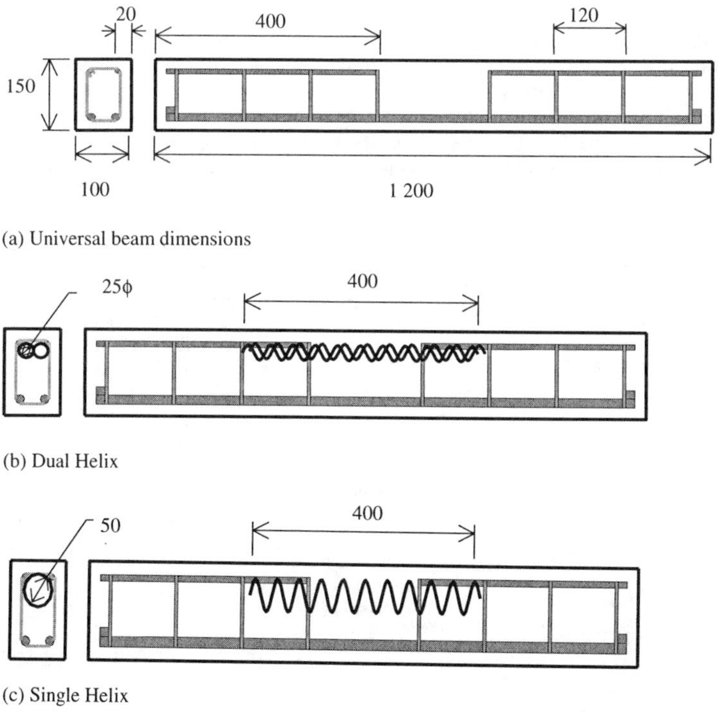

Figure 1 – Beam dimensions (mm)

was still maintained after an initial drop in load capacity. Inspection of the beams after loading showed that the concrete was effectively confined in the 15 mm pitch, while the 45 mm pitch furnished a less effective confinement. Figure 2 presents the load deflection curves for the beams tested.

The following tables provide a summary of the experimental results. Comparisons of strength (Table 2) and deflection (Table 3) are made between plain beams and similar beams incorporating the helical reinforcement.

6 EXPERIMENTAL ANALYSIS

The application of helical reinforcement made a substantial difference in the flexural behaviour, by significantly increasing the beams' ductility. The flexural strengths of the helical beams were increased as observed in Table 3. Except for the single helix with a 15 mm pitch, all helical beams experienced a reduction in strength in the order of 5% when compared with the over reinforced beams. This was due to early failure of the concrete cover, lowering the compressive resistant area of concrete.

On the other hand, the ductility was significantly increased through the use of helical reinforcement. In the case of the 15 mm pitch, the ductility was increased by a factor of at least 240%. The actual increase in ductility is not known, as the final failure of the beams was not observed. The helical beams with 45 mm pitch experienced a 300% increase in ductility; however, high flexural resistance was not sustained.

Comparisons of the single helix and dual helices yielded similar results. In the experiments performed no significant difference in strength or ductility was observed between the two sets of results. As the amount of steel required for the single helix was less than that of the dual helices, the single helix would present the more economical solution.

Both the plain and helical beams, behaved in an identical manner prior to the ultimate load. Where the plain beams deflected and then ultimately failed, the helical beams shed their cover and continued to deflect at a substantial load. However, the small plateau achieved at maximum loading was less for the helical beams. This was due to the concrete cover being separated by the presence of the helical reinforcement.

Once the concrete cover was shed, the flexural capacity of the beam was decreased. As the beams were quite small, compared with that used in current structures, the concrete cover of 20 mm was a significant portion of the total area. This left a small amount of concrete to resist all compressive forces. The reduction of flexural strength would not be as predominant in larger sized beams.

The variation in helical pitch played an important role in behaviour past the primary failure of the concrete cover. The 15 mm pitch was successful in confining the concrete. After failure of the concrete cover, the loading resistance of the beam stabilised and the beam continued to deflect effectively at constant load. A slight increase in loading was observed as the beam further deflected to 50 mm. The final result could not be observed, as the deflected beam came in contact with the loading beam.

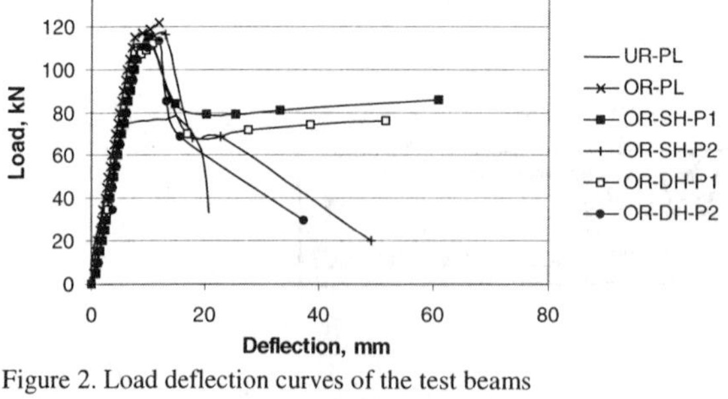

Figure 2. Load deflection curves of the test beams

Table 1. Summary of beam composition

Specimen	Concrete strength (MPa)	Tensile steel	Helical reinforcement Steel Bar Size	Diameter D (mm)	Type	Diameter D (mm)	Pitch P (mm)	D / P
UR – PL	70	2Y12	R6	6	-	-	-	-
OR – PL	70	2Y16	R6	6	-	-	-	-
OR – SH – P1	70	2Y16	R6	6	Single	50	15	3.3
OR – SH – P2	70	2Y16	R6	6	Single	50	45	1.1
OR – DH – P1	70	2Y16	R6	6	Dual	25	15	1.7
OR – DH – P2	70	2Y16	R6	6	Dual	25	45	0.6

Table 2. Strength comparison of helical beams

Beam Type	Tensile Reinforcement	Helix Number	Diameter (mm)	Pitch (mm)	Yield Strength[1] (kN)	Yield Ratio[2]	Maximum Strength (kN)	Maximum Ratio[2]
UR – PL	2Y12	-	-	-	72	1	82	1
OR – PL	2Y16	-	-	-	114	1.58	120	1.46
OR – SH - P1	2Y16	1	50	15	123	1.71	119	1.45
OR – SH - P2	2Y16	1	50	45	111	1.54	115	1.40
OR – DH - P1	2Y16	2	25	15	107	1.49	112	1.37
OR – DH - P2	2Y16	2	25	45	108	1.50	113	1.38

1 – The yield point is defined as the point where linear deflection transforms to non-linear
2 – UR-PL used as reference value

Table 3. Deflection comparisons of helical beams

Beam Type	Yield Strain Deflection (mm)	Yield Ratio[1]	Maximum Deflection (mm)	Deflection Ratio	Ductility[2] (mm)	Ductility Ratio[1]
UR-PL	6.0	1	18.6	1	12.7	1
OR - PL	8.3	1.38	14.3	0.77	6	0.47
OR – SH - P1	7.9	1.32	56.1	3.02	48.2	3.80
OR – SH - P2	8.3	1.38	38.6	2.08	30.3	2.39
OR – DH - P1	8.2	1.36	60.1	3.23	51.9	4.09
OR – DH - P2	8.5	1.42	38.1	2.05	29.6	2.33

1 – UR-PL used as reference value
2 – Ductility is defined as the distance measured from the yield point to failure

The 45 mm pitch did not effectively confine the inner core of concrete, consequently the improvement in ductility was significantly less than that for the 15 mm pitch. After failure of the concrete cover the loading resistance was not stable, and continued to decrease whilst deflecting. Plateaus at particular strengths were observed in discrete intervals, but did not remain for long.

According to strain gauge readings, the helical reinforcement did not experience large loading until the cover of the beams failed. Once the cover had spalled off the beam, the confined concrete resisted the majority of the compressive force.

The incorporation of helical reinforcement of an appropriate pitch significantly increases the ductility of a reinforced concrete beam. If an effective pitch is obtained for complete concrete confinement, the beams will continue to deflect and resist loading at constant load. The improvement in ductility allows an increase in strength through increasing the tensile reinforcement. This new approach differs from the conservative methods of AS3600, and offers to utilise the full potential of high strength reinforced concrete beams.

7 CONCLUSIONS

Based on this research, the following observations can be made:

- Due to the size of the beams cross sections, a significant cross-sectional area of compressive concrete cover existed and subsequently a significant fall in load capacity occurred when the ultimate load was reached.
- The reduced ductility, due to the increase in tensile steel and the use of high strength concrete was overcome through the use of helical reinforcement in the compression region of the beam.
- The use of helical reinforcement was effective due to the lateral confinement of the concrete. The use of a single helix of 6 mm diameter and with a helix diameter/pitch ratio (D/p) of 3.3 effectively confined the concrete so as to allow a large deflection at constant load to occur.
- An increased helical pitch size, with D/p=1.1, was not as effective in confining the concrete.
- The results from this research are encouraging. To improve further the effects of helical reinforcement on beam ductility, additional investigations into helical configuration and structural behaviour are required.

ACKNOWLEDGMENT

The authors wish to acknowledge the suggestion of Dr G.D. Base, formerly Reader of the University of Melbourne, for highlighting the potential advantages of the use of helical reinforcement in concrete beams. The financial support of the Mechanics of Structures, Materials and Systems Research Group, University of Wollongong is highly appreciated.

REFERENCES

Cement and Concrete Association of Australia and National Ready Mixed Concrete Association of Australasia 1992, *High-Strength Concrete*.

Foster, S.J., J.Liu & S.A. Sheikh 1998. Cover spalling in HSC columns loaded in concentric compression. *J. St. Eng. ASCE*. 124(12):1434-1437. Dec.

Pessiki, S. and Pieroni, A. 1997, Axial behaviour of large-scale spirally-reinforced high-strength concrete columns, *ACI Structural Journal*, 94(3): 304-314 May-June.

Van, B. and Montgomery, D. 1998, High-Performance concrete containing limestone modified cements, *International Conference on HPHSC*, Perth, Australia 701-713.

Webb, J. 1993, High-Strength concrete: economics, design and ductility, *Concrete International*, 15(1): 27-32 Jan.

Mechanics of Structures and Materials, Bradford, Bridge & Foster (eds)
© 1999 Balkema, Rotterdam, ISBN 90 5809 107 4

Flexural crack control for reinforced concrete beams and slabs: An evaluation of design procedures

R.I.Gilbert
School of Civil and Environmental Engineering, The University of New South Wales, Sydney, N.S.W., Australia

ABSTRACT: The crack control provisions of AS3600-1994 are currently under review. For the control of flexural cracking, AS3600 requires only the satisfaction of a minimum bar spacing requirement and, often, this does not guarantee acceptably narrow cracks. The advent of high strength reinforcing steels will inevitably lead to higher steel stresses under in-service conditions, thereby exacerbating the problem of crack control. In this paper, the current flexural crack control provisions of AS3600 are presented and the crack width calculation procedures in several of the major international concrete codes, including ACI 318 and Eurocode 2, are assessed. A parametric evaluation of the various code approaches is undertaken to determine the relative importance in each model of some of the factors that affect the final crack spacing and crack width. Some deficiencies in the existing approaches are exposed and an alternative model for the calculation of the width of flexural cracks is proposed.

1 INTRODUCTION

In reinforced concrete structures, cracking is inevitable. Excessively wide cracks are aesthetically unacceptable and may lead to corrosion of reinforcement and durability problems. Excessive cracking is the most common cause of damage in concrete structures and results in a huge annual cost to the construction industry.

There are many possible causes of cracking in concrete structures, including corrosion of reinforcement, plastic shrinkage, expansive chemical reactions within the concrete (such as alkali-silica reaction), restrained deformations (such as shrinkage or temperature movements, or foundation settlement or rotation); and external loads. Of these, only cracking resulting from restrained deformation and/or external loads can be effectively controlled by reinforcement. This paper is concerned with the control of *flexural cracking* in reinforced concrete beams and slabs by the provision of adequate amounts of tensile reinforcement.

In the current Australian Standard for Concrete Structures, AS3600-1994, the control of flexural cracking is deemed to be satisfactory, providing the designer satisfies certain detailing requirements. These involve maximum limits on the centre-to-centre spacing of tensile reinforcement bars and on the concrete cover to the side face or soffit of a beam. These procedures over-simplify the problem and do not always ensure adequate control of flexural cracking, particularly when the stress in the tensile steel at service loads exceeds about 250 MPa.

With the current move to higher strength reinforcing steels (characteristic strengths of 500 MPa and above), there is an urgent need to review the crack-control design rules in AS3600 for reinforced concrete beams and slabs. The existing design rules for reinforced concrete flexural elements are intended for use in the design of elements containing 400 MPa bars and are sometimes unconservative. They are unlikely to be satisfactory if higher strength steels are used. In such elements, the stress in the higher strength steel under service loads is higher than it would be if conventional steel is used due to the reduction in the steel area required for strength.

Standards Australia has established a Working Group to investigate and revise the crack control provisions of the current Australian Standard, to incorporate recent developments and to accommodate the use of high of high strength reinforcing steels. The Working Group is chaired by the writer and is currently undertaking an investigation to gain a better understanding of the factors that affect the spacing and width of cracks in reinforced concrete elements and to develop rational and reliable design-oriented procedures for crack control.

In an earlier paper, Gilbert et al. (1999) presented the current crack control provisions of AS3600 and compared them with the corresponding provisions in several of the major international concrete codes, including BS8110, ACI318 and Eurocode 2. A limited parametric evaluation of the various code approaches was also undertaken to determine the relative importance in each model of such factors as steel area, steel stress, bar diameter, and bar spacing on the final crack spacing and crack width. The applicability of each model was assessed by comparison with some local crack width measurements and problems were identified with each of the code models. The approach in BS8110 was dismissed as being unreliable and was found to significantly underestimate flexural crack widths.

In this paper, further assessments are made of the flexural crack width calculation procedures in Eurocode 2 and ACI318, and a new approach is proposed for consideration for inclusion in the next revision of AS3600.

2 FLEXURAL CRACK WIDTH CALCULATION PROCEDURES

2.1 *General*

Cracking should not lead to serviceability problems. It should not impair the proper functioning of the structure or cause its appearance to be unacceptable. As well as spoiling the appearance of a concrete structure, wide cracks may also allow the ingress of water and salts, which may lead to waterproofing and corrosion problems. To avoid excessive cracking adequate quantities of well-anchored reinforcement must be located wherever tensile stresses exist in the concrete. In other words, good detailing of reinforcement is the key to crack control.

Acceptable crack widths depend on the environment and the exposure. Crack widths exceeding about 0.3 or 0.4 mm are noticeable and aesthetically poor. Crack widths exceeding 0.1 or 0.2 mm may not be acceptable in aggressive environments, such as exposure to seawater or spray or de-icing salts.

For exposure classifications A1, A2 and B1, it may be assumed that limiting the maximum design crack width to about 0.3 mm under the quasi-permanent combination of loads will generally be satisfactory for reinforced concrete members in buildings with respect to appearance and durability. For exposure Classification A1, crack width has no influence on durability and the limit could be relaxed if this was acceptable for other reasons. Special crack limitation measures may be necessary for members subjected to Exposure Classifications B2 and C.

2.2 *Provisions of AS 3600-1994*

The Standard gives specific detailing rules as the means of controlling flexural cracking in reinforced concrete beams and slabs. For beams, flexural cracking is deemed to be controlled if the centre to centre spacing of bars near the tensile face does not exceed 200 mm and the distance from the side or soffit of the beam to the centre of the nearest longitudinal bar is not greater than 100 mm. In addition, where the overall beam depth exceeds 750 mm, longitudinal reinforcement consisting of Y12 bars at 200 mm centres or Y16 bars at 300 mm centres is required in each side face. It is of interest to note that the Standard does not place any limitation on the depth of top tensile reinforcement in negative moment regions.

For slabs, flexural cracking is deemed to be controlled if the centre to centre spacing of bars in each direction does not exceed the lesser of 2.5 times the slab thickness or 500 mm. No limit is placed on the distance of the tensile bars from the top or bottom surface of the slab.

The commentary to the Standard suggests that the flexural crack widths can be calculated as an alternative procedure for controlling cracking and nominates the Gergely-Lutz equation adopted by ACI318 and the method given in BS8110:Part 2 as suitable procedures.

2.3 *Provisions of Eurocode 2 (EC2)*

EC2 specifies a minimum area of reinforcement A_s required within the tensile zone given by

$$A_s = k_c k f_{ct.eff} A_{ct} / \sigma_s \tag{1}$$

where A_{ct} is the area of concrete in that part of the section which is in tension just before formation of the first crack; $f_{ct.eff}$ is the tensile strength of the concrete at the time when the cracks are expected to first occur (a value of 3.0 MPa is suggested when the time of first cracking is uncertain); σ_s is the maximum stress permitted in the reinforcement after formation of the crack and depends on the crack width requirements (but should not exceed f_{sy}); k_c is a coefficient which takes into account the nature of the stress distribution prior to cracking and equals 0.4 for bending without normal compressive force; and k allows for non-uniform self-equilibrating stresses and may be taken as 0.8 when the non-uniform stress is caused by shrinkage or temperature (or 0.5 when the overall depth of the section exceeds 800 mm).

The design crack width, w_k, may be calculated from

$$w_k = \beta s_{rm} \varepsilon_{sm} \tag{2}$$

where s_{rm} is the average final crack spacing; β is a coefficient relating the average crack width to the design value and equals 1.7 for load induced flexural cracking; and ε_{sm} is the mean strain allowing for the effects of tension stiffening, shrinkage etc. and may be taken as

$$\varepsilon_{sm} = (\sigma_s/E_s)[\, 1 - \beta_1\beta_2\, (\sigma_{sr}/\sigma_s)^2\,] \tag{3}$$

where σ_s is the stress in the tension steel calculated on the basis of a cracked section; σ_{sr} is the stress in the tension steel calculated on the basis of a cracked section under the loading conditions causing first cracking; β_1 depends on the bond properties of the bars and equals 1.0 for high bond bars and 0.5 for plain bars; and β_2 accounts for the duration of loading and equals 1.0 for a single, short-term loading and 0.5 for a sustained load or for many cycles of loading.

The average final spacing of flexural cracks, s_{rm} (in mm), can be calculated from

$$s_{rm} = 50 + 0.25\, k_1\, k_2\, d_b / \rho_r \tag{4}$$

where d_b is the bar size (or average bar size in the section) in mm; k_1 accounts for the bond properties of the bar and, for flexural cracking, $k_1 = 0.8$ for high bond bars and $k_1 = 1.6$ for plain bars; k_2 depends on the strain distribution and equals 0.5 for bending; and ρ_r is the effective reinforcement ratio, $A_s / A_{c.eff}$ where A_s is the area of reinforcement contained within the effective tension area, $A_{c.eff}$. The effective tension area is the area of concrete surrounding the tension steel of depth equal to 2.5 times the distance from the tension face of the section to the centroid of the reinforcement, but not greater than $\frac{1}{3}$ of the depth of the tensile zone of the cracked section, in the case of slabs.

2.4 *Provisions of ACI 318*

When f_{sy} exceeds 300 MPa, cross-sections of maximum positive and negative moments must be proportioned so that the quantity z given by

$$z = f_s \sqrt[3]{(d_c A)} \tag{5}$$

does not exceed 30 kN/mm for interior exposure and 25 kN/mm for exterior exposure. The calculated stress in the reinforcement at service loads f_s is calculated on the basis of the cracked section. The term d_c is the distance from the nearest surface to the centre of the bar (clear cover plus half the bar diameter) and A is an effective area of concrete in the tension zone associated with each reinforcing bar. This area has the same centroid as the tensile reinforcement and is bounded by the surfaces of the cross-section and a straight line parallel to the neutral axis.

According to ACI318, Eqn 5 will provide a reinforcement distribution that will control cracking. The equation is developed from the expression proposed by Gergely and Lutz (1968):

$$w = 1.1\times10^{-5} f_s \sqrt[3]{d_c A} \frac{(D-x)}{(d-x)} \tag{6}$$

where x is the neutral axis depth on the cracked section; D is the overall depth of the section; and d is the effective depth to the tensile steel. By setting limitations on the crack width of 0.4 mm for interior exposure and 0.33 mm for exterior exposure and by taking $(D-x)/(d-x) = 1.2$ for beams, Eqn 6 can be rearranged to give the limits on z in Eqn 5. For slabs, the ratio $(D-x)/(d-x)$ is typically about 1.35 and, accordingly, the limits on z in Eqn 5 should be reduced by 1.2/1.35.

3 EVALUATION OF EXISTING PROCEDURES

Gilbert et al. (1999) undertook a comparative study of the existing crack width calculation procedures and compared calculated crack widths to measured crack widths obtained from flexural tests on one-way slabs and T-beams. They concluded that both the EC 2 and ACI 318 procedures generally underestimate final crack widths, particularly in slabs with widely spaced tensile reinforcement bars, and that neither approach adequately models the increase in flexural crack width that occurs with time due to shrinkage.

To illustrate these deficiencies, consider a 200 mm thick reinforced concrete floor slab, singly reinforced and containing a reinforcement ratio $p=A_{st}/bd = 0.006$. The concrete cover is 20 mm, the elastic modulus and flexural tensile strength of concrete are $E_c = 25000$ MPa and $f_t = 3.0$ MPa, respectively. Table 1 contains the calculated crack widths for a range of tensile bar diameters and spacings, both short-term and long-term (after shrinkage).

Table 1. Calculated crack widths in a 200 mm thick slab (using EC 2 and ACI 318).

d_b (mm)	d (mm)	A_{st} (mm²/m)	s (mm)	Crack widths (mm)											
				$\sigma_s = 200$ MPa				$\sigma_s = 250$ MPa				$\sigma_s = 300$ MPa			
				Short-term		Long-term		Short-term		Long-term		Short-term		Long-term	
				EC2	ACI	EC2	ACI	EC2	ACI	EC2	ACI	EC2	ACI	EC2	ACI
12	174	1044	108	.118	.140	.151	.140	.177	.174	.205	.174	.233	.209	.255	.209
16	172	1032	195	.136	.181	.178	.181	.208	.226	.242	.226	.275	.272	.303	.272
20	170	1020	308	.154	.224	.206	.224	.239	.280	.280	.280	.317	.336	.351	.336
24	168	1008	449	.171	.269	.232	.269	.269	.337	.319	.337	.360	.404	.401	.404
28	166	996	618	.187	.317	.260	.317	.299	.396	.358	.396	.402	.475	.451	.475

The ACI procedure may be assumed to calculate final or long-term crack widths since it is time-independent and it provides reasonable agreement with the EC2 long-term calculations (when $\beta_2 = 0.5$ in Eqn 3). However, neither approach depends on the magnitude of concrete shrinkage and, according to EC2, the increase in crack width with time is only about 10% at steel stresses of 300 MPa. This increase is unreasonably low. Observation of reinforced concrete structures indicates that under normal in-service conditions, flexural crack widths often increase substantially with time, particularly when shrinkage is high.

It should be noted that EC2 limits the spacing of the tensile bars in a slab to the lesser of 1.5D and 350mm, so that the last two steel layouts in Table 1 are not permitted by the code. Nevertheless, the crack width calculations for these cases give worrying results. Experience suggests that these crack widths are far less than would normally be expected to occur at these steel spacings.

Most practising engineers will have had experience with excessive flexural crack widths in some slabs, with observed crack widths in excess of 1 mm not uncommon. If the procedures considered here are to be believed, it is difficult to understand how cracks of this width are possible. Clearly, there are problems with these crack width calculation procedures, particularly at high steel stress levels and for widely spaced bars.

Next consider a rectangular beam section 400 mm wide and containing a single layer of tensile bars at $d = 400$ mm. Calculated crack widths are given in Table 2 for a range of rein-

Table 2. Final (long-term) crack widths for beam (b = 400 mm, d = 400 mm).

d_b (mm)	No. of bars N	A_{st} (mm^2)	Crack widths (mm) Cover = 25 mm Steel stress, σ_s (MPa) 200 EC2	200 ACI	250 EC2	250 ACI	300 EC2	300 ACI	Cover = 50 mm Steel stress, σ_s (MPa) 200 EC2	200 ACI	250 EC2	250 ACI	300 EC2	300 ACI
20	2	620	.183	.193	.271	.241	.353	.289	.206	.296	.325	.370	.435	.444
20	3	930	.180	.169	.240	.212	.297	.254	.208	.261	.283	.326	.354	.392
20	4	1240	.165	.155	.213	.193	.261	.232	.186	.239	.244	.299	.299	.359
24	2	900	.210	.202	.282	.253	.351	.304	.238	.307	.326	.384	.410	.461
24	3	1350	.182	.178	.234	.223	.285	.267	.200	.271	.260	.339	.318	.407
24	4	1800	.161	.163	.205	.204	.248	.244	.173	.249	.221	.311	.268	.374
28	2	1240	.213	.212	.276	.265	.338	.318	.232	.318	.303	.398	.373	.477
28	3	1860	.177	.187	.225	.234	.272	.281	.185	.282	.236	.353	.287	.423
32	2	1600	.211	.222	.270	.278	.328	.333	.220	.330	.282	.412	.344	.495

forcement layouts, steel stresses and concrete covers. For each beam, the covers from the soffit and the side face to the nearest bar are the same. As before, E_c = 25000 MPa and f_t = 3.0 MPa. As expected, the crack width increases when the concrete cover, c, is increased. This increase is more pronounced in the ACI method, where the calculated crack widths increase typically by about 50% for this beam when the cover is increased from 25 to 50 mm. By comparison, EC2 predicts a corresponding increase of between 4% and 20%. Field observations support the ACI procedure in this regard.

When c = 25 mm, the final crack widths predicted by EC2 and ACI are generally in close agreement, with EC2 predicting slightly wider cracks than ACI at high in-service stress levels. However, when c = 50 mm, ACI predicts wider cracks (by up to 50%). For most of the reinforcement layouts considered in Table 2, when c = 25 mm, calculated crack widths are less than 0.3 mm at steel stress levels up to and exceeding 300 MPa. Even for the beams containing just two bars (spaced over 300 mm apart), ACI indicates acceptable crack widths at stresses up to 300 MPa, with EC2 only slightly more conservative.

4 PROPOSED PROCEDURE

From parametric studies, such as those outlined above, and comparisons with crack width measurements, an alternative approach is here proposed. It is similar to the EC2 model, but modified to include shrinkage shortening of the intact concrete between the cracks in the tensile zone and to more realistically represent the increase in crack width if the cover is increased.

The design crack width, w, may be calculated from

$$w = \beta_m \; s_{rm} \; (\varepsilon_{sm} + \varepsilon_{sc.t}) \tag{7}$$

As for the EC2 method, s_{rm} is the average final crack spacing given by Eqn 4 and ε_{sm} is the mean tensile steel strain under the sustained service loads allowing for the effects of tension stiffening, shrinkage etc. and may be calculated using Eqn 3. The coefficient β_m is a coefficient relating the average crack width to the design value and may be taken as β_m = 1.0 +0.025c ≥ 1.7, where c is the distance from the concrete surface to the nearest longitudinal reinforcing bar.

$\varepsilon_{sc.t}$ is the shrinkage induced shortening of the intact concrete at the tensile steel level between the cracks. For short-term crack width calculations $\varepsilon_{sc.t}$ is zero. Using the age-adjusted effective modulus method and a shrinkage analysis of a singly reinforced concrete section, see Gilbert (1988), it can be shown that

$$\varepsilon_{sc.t} = \varepsilon_{sc} \,/\, (1 + 3\, p \,\overline{n}) \tag{8}$$

where p is the tensile reinforcement ratio for the section (A_{st}/bd); $\overline{n}$ is the age-adjusted modular ratio (E_s/E_{ef}); E_{ef} is the age-adjusted effective modulus for concrete ($E_{ef} = E_c/(1+0.8\phi_c)$); and ε_{sc} and ϕ_c are final long-term values of shrinkage strain and creep coefficient, respectively.

5 EVALUATION OF PROPOSED PROCEDURE

The proposed procedure addresses the previously mentioned deficiencies in both the ACI and EC2 procedures and more accurately agrees with laboratory and field measurements of crack widths. In Tables 3 and 4, crack widths calculated using the proposed procedure are presented for the slab and beam examples considered earlier and may be compared directly to the ACI and EC2 predictions in Tables 1 and 2, respectively. In each case, $\varepsilon_{sc} = 0.0006$ and $\phi_c = 3.0$.

It should be pointed out that the steel stress under *sustained* service loads is usually less than 200 MPa for beams and slabs designed using 400 MPa steel. The range of steel stresses in Tables 3 and 4 are more typical of situations in which 500 MPa steel is used.

Table 3. Calculated final crack widths in a 200 mm thick slab - proposed procedure.

d_b (mm)	d (mm)	A_{st} (mm^2/m)	s (mm)	Crack widths (mm) σ_s (MPa) 200	250	300
12	174	1044	108	.226	.279	.330
16	172	1032	195	.267	.331	.392
20	170	1020	308	.309	.384	.455
24	168	1008	449	.352	.438	.519
28	166	996	618	.394	.492	.585

Table 4. Final crack widths for beam (b = 400 mm, d = 400 mm) - proposed procedure.

d_b mm	N	A_{st} mm^2	Crack width (mm) Cover = 25 mm σ_s (MPa) 200	250	300	Cover = 50 mm σ_s (MPa) 200	250	300
20	2	620	.309	.397	.479	.488	.646	.791
20	3	930	.267	.326	.384	.414	.513	.607
20	4	1240	.231	.280	.327	.349	.425	.498
24	2	900	.314	.386	.455	.480	.596	.707
24	3	1350	.251	.304	.355	.369	.449	.526
24	4	1800	.214	.258	.301	.304	.367	.430
28	2	1240	.299	.362	.424	.434	.529	.621
28	3	1860	.234	.281	.329	.325	.393	.459
32	2	1600	.285	.344	.402	.394	.477	.558

In general, the calculated crack widths are larger than those predicted by either of the existing procedures, but unlike them, the proposed model will signal serviceability problems to the structural designer in most situations where excessive crack widths are likely.

6 CONCLUSIONS

The existing procedures for the calculation of flexural crack widths in Eurocode 2 and ACI318 have been found to underestimate actual crack widths, particularly in members with widely spaced tensile reinforcement bars. Neither approach adequately models the increase in flexural crack width that occurs with time due to shrinkage. An alternative approach has been proposed that provides more realistic predictions of long-term crack width and more accurately models the effect on crack width of such factors as concrete cover, bar size and bar spacing. The proposed model is suitable for routine use in design and is recommended for inclusion in AS3600.

REFERENCES

ACI318-95. *Building code requirements for reinforced concrete.* American Concrete Institute. Committee 318. Detroit.

AS3600-1994. *Australian Standard for Concrete Structures.* Standards Australia. Sydney.

Eurocode 2. 1992. *Design of concrete structures.* DD ENV-1992-1-1. British Standards Inst.

Gergely,P. & Lutz, L.A. 1968. Maximum crack width in reinforced concrete flexural members. *ACI SP-20.* Detroit: 87-117.

Gilbert, R.I. 1988. *Time Effects in Concrete Structures.* Amsterdam. Elsevier.

Gilbert, R.I., Patrick, M. & Adams, J.C. 1999. Evaluation of crack control design rules for reinforced concrete beams and slabs. *Concrete 99. 19th Bienniel Conf of CIA.* Sydney: 21-29.

Mechanics of Structures and Materials, Bradford, Bridge & Foster (eds)
© 1999 Balkema, Rotterdam, ISBN 90 5809 107 4

Prestressed concrete design – Beyond load-balancing

J.L.van der Molen
Department of Civil and Environmental Engineering, University of Melbourne, Vic., Australia

ABSTRACT: When T.Y. Lin proposed his load-balancing concept there were at the time significant differences in the practice of prestressed concrete design, when compared with the present. Designs were based (mainly) on working stress rather than strength and partially prestressed concrete was not in general use. The paper suggests that more comprehensive design methods, which take account of advances in design requirements and construction technology since 1962, may be more appropriate. In these methods load-balancing techniques certainly have a place, be it is a less prominent one. A unified design and analysis system is proposed, capable of rapid iteration through the use of spreadsheets.

1 INTRODUCTION

1.1 Historical

T.Y. Lin proposed his load-balancing method at a meeting in Sydney in 1961 (Lin, 1961). An extended version of the paper (Lin, 1963a) appeared in the ACI Journal. This paper became the basis for a new Chapter in the second edition of Lin's book (Lin, 1963b).

It is well to note the historical context of these publications:

- At the time, partially prestressed concrete was not in general use and, while it was being used occasionally, a coherent design and analysis procedure had not yet been developed,
- Designs were done on a working stress basis with, sometimes but not always, a check for ultimate load capacity,
- T.Y. Lin was at the time engaged in developing design methods for two-way prestressed concrete slabs, using unbonded tendons, as well as using the load-balancing concept in the design of shells.

Contrasting this with the present situation, we have the development of the concept of "Structural Concrete" (Breen, Bruggeling, 1991), embodying all concrete with passive (non-stressed) and/or active (stressed) reinforcement, the transition from one to the other being seen as a continuum. This principle has been embodied in the Australian Standard for Concrete Structures, AS 3600 since 1988, when the Concrete Structures Code (AS 1480) and the Prestressed Concrete Code (AS 1481) were amalgamated. It is also embodied in the FIP Recommendations 1996 (FIP, 1996).

At present, designs are based on the ultimate limit state, having regard to serviceability requirements, listed in AS 3600 as ".....deflection, lateral drift, cracking and vibration, as appropriate....".

These are two paradigm shifts which took place after the introduction of load-balancing. This raises the question if the load-balancing design concept which Lin lists as a design method replacing either the working load or the strength method, is still as relevant now as it was before these shifts took place.

1.2 Design and Analysis

In the design of a structural concrete member we have to make a number of assumptions at the outset. Taking a beam as example, we have as unknowns the sizes b and D, as well as the active and passive reinforcement areas, A_p and A_{st}. The beam defined by these parameters has to withstand the ultimate actions M*, V* and (sometimes) N*. Obviously, there is an infinite number of combinations of member sizes and reinforcement areas which will satisfy this condition. Some of these will infringe serviceability requirements, but still a vast number of combinations will be found suitable.

It follows that the designer in practice makes assumptions regarding one or more of the design parameters and then determines the magnitude of the remaining ones using well-known design principles based on structural mechanics. He then may do a check analysis to ascertain the overall suitability of the design, including serviceability limit states, architectural considerations, constructability and economy.

It should be noted that the preliminary computations will generally be based on ultimate strength. However, in the load-balancing approach one determines the tendon force (i.e. A_p) from serviceability considerations: the deflection under the balanced load is set at zero. This may very well result in a suitable tendon size and profile from a strength point of view, but this is not necessarily so. Moreover, the active reinforcement is found in isolation. In structural concrete with prestressing there is an interplay between the amount of active reinforcement and that of the passive reinforcement. The load-balancing procedure does not allow for this.

It would seem that there is room for a more holistic approach to the design problem, so that advantage may be taken of the interconnection between the various design parameters. This would enable the designer quickly to evaluate the results of a number of different scenarios and would improve the quality of the final choice.

Finally, Bruggeling (1987) makes the point that all structural concrete should contain a basic amount of passive reinforcement. In load-balancing one does not take this reinforcement into account: it may, or may not, be provided.

2 DESIGN PROCEDURE

2.1 Theoretical Background

The design procedure presented here is based unequivocally on reinforced concrete strength design theory by extending it to take account of both active and passive reinforcement, making it applicable to structural concrete.

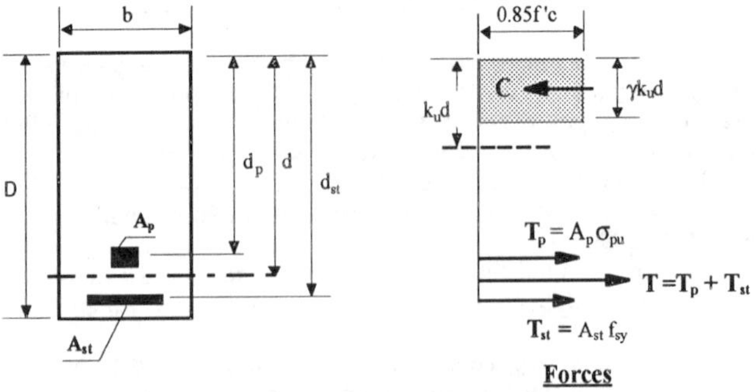

Figure 1 - Prestressed concrete beam - internal forces

Referring to Fig. 1, we have:

$$d = \frac{A_p \sigma_{pu} d_p + A_{st} f_{sy} d_{st}}{A_p \sigma_{pu} + A_{st} f_{sy}}, \tag{1}$$

d being the depth to the centre of the forces acting on both passive and active reinforcement in the ultimate condition. The ultimate moment capacity, φM_u, is now:

$$\varphi M_u = \varphi d\left(1 - \frac{\gamma k_u}{2}\right)\left(A_p \sigma_{pu} + A_{st} f_{sy}\right) \tag{2}$$

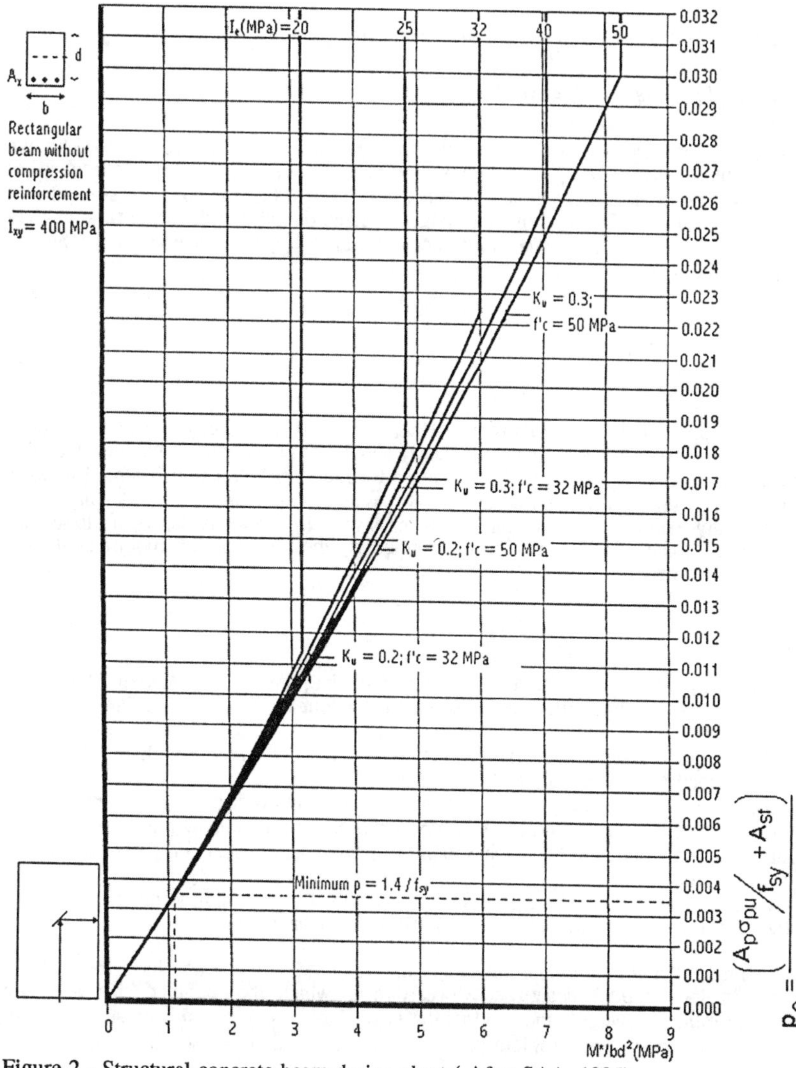

Figure 2 - Structural concrete beam design chart (After SAA, 1995)

Introducing the equivalent reinforcement index, q_{eq}, then:

$$q_{eq} = \left(\frac{A_p \sigma_{pu} / f_{sy} + A_{st}}{bd} \right) \left(\frac{f_{sy}}{f'_c} \right) \tag{3a}$$

and:

$$k_u = \frac{1}{0.85\gamma} q_{eq} \tag{3b}$$

Substituting eq. (3a) and (3b) in eq. (2) and rearranging yields:

$$\varphi M_u = \varphi f'_c\, q_{eq}\left(1 - 0.6 q_{eq}\right) bd^2 = Kbd^2 \tag{4}$$

It will be seen that eq. (4) is exactly in the form which is the basis of the well-known Chart 3.1 of the Concrete Design Handbook (1995). This chart has been reproduced as Fig. 2. It has been marked up to indicate k_u-values of 0.2 and 0.3 for concrete strengths of 32 and 50 MPa. This enables designers to judge expected ductility at the same time as they obtain values for *b*, *d* and p_e. The reinforcement ratio A_{st}/bd is now replaced by the equivalent reinforcement ratio, p_e.

$$p_e = \frac{\left(A_p \sigma_{pu} / f_{sy} + A_{st} \right)}{bd} \tag{5}$$

The above approach is due to Fairhurst (1990). Eq. (5), containing the areas of both passive (A_{st}) and active (A_p) reinforcement, clearly shows the continuum of choices available to the designer. Setting $A_p = 0$ reduces the beam to a reinforced concrete beam. Likewise, $A_{st} = 0$ produces a fully prestressed beam without passive reinforcement. The designer may now choose any combination of active and passive reinforcement between these two extremes. The choice will generally be determined by the serviceability requirements for the particular member.

2.2 *Application*

Referring to Fig. 3, we see there are a number of salient points in the Moment-Deflection diagram, each of which may influence the service behaviour of the member. They are:

- M_o, the moment under which the deflection is zero. Obtained by determining the "balanced" load (Warner et al, 1998).
- M_{cr}, the moment which causes the member to crack in the first instance. Repetition of this condition will cause the cracks to open at the decompression moment, M_{dec}. It is recommended to work with M_{dec} rather than M_{cr}, as the latter occurs only once in the life of the structure.
- M_{ser}, together with Δ_{ser}, the moment under service load (either long-term or short-term, depending on the circumstances) and the deflection under this load.

It is assumed that the ultimate moment conditions M* and φM_u have been fully considered in the original design activities, during which the member size and the reinforcement ratio, p_e, were found and that the next phase is concerned with serviceability issues. More succinctly, the problem is to determine A_{st} and A_p so as to satisfy any

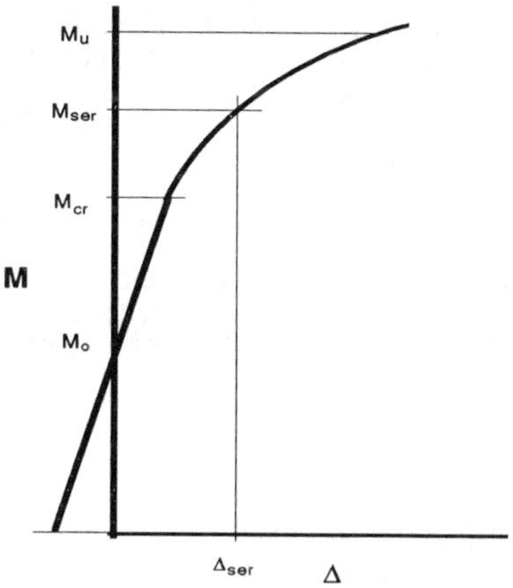

Figure 3 - Typical Moment-Defection Diagram

serviceability conditions which may obtain, while simultaneously satisfying the ultimate strength condition expressed in Eq. (5), i.e. p_e .

Because of the freedom of choice which exists at this point it is to be expected that different authorities should make different recommendations. Also, it is the serviceability limit state which is the determining factor, therefore the difficulty is not really how to apportion the active and passive reinforcement, but how to define the serviceability limit state.

Often, the serviceability limit state appears to contain no critical elements and the whole problem seems rather trivial. In this case, A_p is often determined by using data obtained from experience. The FIP Recommendations (1996) provide ranges of data based on average prestress, load balancing, and total mass of prestressing steel per unit surface area of the structure. The ranges are rather wide, eg. for load balancing a range is given of 0.6 - 1.0 total equivalent load due to prestress divided by total permanent loads.

It should be stressed however, that critical conditions affecting serviceability are often overlooked, not given proper weight, or unknown at the time of design. The designer could start by asking the basic question regarding the magnitude of the load under which cracking would be acceptable. Answering this question would yield the magnitude of M_{dec} in relation to M_{ser} , and so the magnitude of the prestressing force. Does the situation change in extremes of temperature, is there a possibility of the element being externally restrained, what do we know about the actual magnitude of shrinkage- and creep strains, what local forces are generated by horizontal and vertical curvature of the tendons? All these questions need at least some consideration.

Bruggeling (1986, 1987) in a series of articles mounts an eloquent argument for providing an amount of basic passive reinforcement in the first place and supplementing this with active and/or passive reinforcement so as to meet ultimate strength requirements. The amount of active reinforcement is then bngoverned by the serviceability requirements of cracking (occurrence and crack widths) and deflection. He bases his argument on a number of observed and reported failures in bridges and buildings. Some of these failures are described as the spalling of large areas of concrete from the bottom flanges of box girder bridges, an event which appears to have occurred recently in a new bridge in Melbourne.

3 CONCLUSION

Our industry has moved from the era when reinforced and prestressed concrete were considered different construction materials with a different design discipline, to the present situation where they are both recognised as structural concrete, having a unified design discipline irrespective of the degree of prestressing employed. Consequently, the designer has more freedom to determine the desired degree of prestress. This gives him/her the opportunity carefully to consider the serviceability requirements of the element, specially crack formation and deflection, and to determine the required degree of prestress accordingly.

It would seem that load balancing is not the only, nor even the best, method of tackling this particular problem.

In addition, there is circumstantial evidence that every so often the service behaviour of structural concrete members is less than satisfactory. It is suggested that more attention to the detailing of both passive and active reinforcement may go a long way towards improving this situation.

4 REFERENCES

Breen, J.E. 1991, Why structural concrete?, *IABSE Reports,* V. 62, *Colloquium Structural Concrete*:15-26. Zürich:IABSE.

Bruggeling, A.S.G., 1986, 1987, Constructief beton, een nieuwe aanpak (Structural concrete, a new approach), *Cement.* 1986, No. 7, 1987, No. 1, 2, 3, 4, 6, 7, 9, 10, 12, 1998, No. 1, 2, 5, 6. (in Dutch).

Bruggeling, A.S.G., 1987, Structural concrete: Science into practice, *Heron,* V.32, No. 2, 1987, Ed: Stevin Laboratory, Department of Civil Engineering, Delft University of Technology, Delft, Holland

Bruggeling, A.S.G., 1991, An engineering model for structural concrete, *IABSE Reports,* V. 62, *Colloquium Structural Concrete*:27-36. Zürich:IABSE.

Fairhurst, L., *Design and analysis of concrete structures,*Sydney:McGraw-Hill Book Co.,

FIP, 1996, FIP Recommendations 1996, *Practical design of structural concrete,* FIP Congress, Amsterdam, May 1998. London:Telford.

Lin, T.Y., 1961, A new concept for prestressed concrete, *Constructional Review,* V. 34, No. 9, Sept. 1961:36-52

Lin T.Y., 1963a, Load-balancing method for design and analysis of prestressed concrete structures, *Journ. ACI,* V. 60, No. 6, June 1963:719-742

Lin, T.Y., 1963b, *Design of Prestressed Concrete Structures,* 2nd Ed., New York: John Whiley & Sons, Inc.

SAA, 1995, *Concrete Design Handbook,* SAA HB71-1995, (C&CAA T38), ISBN 0 947 132 73 2:Sydney: SAA.

Warner, R.F., Rangan, B.V., Hall, A.S. & Faulkes, K.A., 1998, *Concrete Structures*: 129-132: Melbourne: Addison Wesley Longman Pty Ltd.

Mechanics of Structures and Materials, Bradford, Bridge & Foster (eds)
© 1999 Balkema, Rotterdam, ISBN 90 5809 107 4

Developments in displacement prediction methodology for reinforced concrete building structures

M. Edwards, J. Wilson, N. Lam & G. Hutchinson
Department of Civil and Environmental Engineering, University of Melbourne, Vic., Australia

ABSTRACT: In recent years there has been a significant shift in the focus of seismic research from the assessment of notional seismic design forces to the prediction of the structural displacements induced. Whilst progress has been made in the implementation of displacement based principles, a number of uncertainties related to displacement prediction remain. This paper describes research undertaken at the University of Melbourne directed at the prediction of structural displacement demand associated with Australian seismicity. As part of these efforts, rock site displacement response spectra (DRS) corresponding with design level Australian seismicity have been derived using stochastically based seismological modelling techniques. The DRS, in turn, have been modified for site soil flexibility. Finally, as part of the displacement prediction process, the substitute structure method described by Shibata and Sozen (1976) has been examined for idealising inelastic structural response. Utilising an ensemble of concrete sub-assemblage test data, this approach is being developed to permit displacement demand to be taken directly from elastic DRS.

1 INTRODUCTION

In the conventional seismic design of structures, the structural effects of an earthquake are represented by equivalent horizontal inertia forces. This is termed the Forced Based Method (FBM) of seismic design and forms the basis of essentially all seismic loadings codes. It is acknowledged, however, that earthquake ground motions induce displacements in a structure, rather than externally apply forces. Correspondingly, it has also been observed that the survival capacity of a structure is more a function of its displacement capability rather than its initial yield strength. Performance based structural evaluation and design, as presented in the recently published FEMA 273 document (1997), address this by focussing on structural performance at a target ultimate displacement, rather than on initial elastic strength and stiffness. The inelastic structural displacement demand is assessed and then compared to the structure's displacement capability. To successfully implement procedures of this type, a method is needed for accurately predicting inelastic displacements.

Two basic approaches to displacement prediction have been taken by researchers to date. Some have directed their efforts at addressing the inaccuracies of using the equal displacement observation in the manner presently codified. Corrections have been proposed to enable designers to continue to design by the familiar force based methodology while making more accurate predictions of inelastic displacement. The second approach has been to make use of the substitute structure concept described by Shibata and Sozen (1976). In this method the seismically induced displacements are evaluated using the structure's inelastic properties, rather than an initial elastic stiffness and damping level. This approach is central to the design process described by Priestley and Calvi (1997) in which displacement prediction by the substitute structure approach is an integral part.

The research underway at the University of Melbourne has been directed at evaluating the merits of the latter approach for intraplate regions such as Australia. Essential to this is the derivation of representative displacement response spectra for Australia, despite the paucity of strong motion records. Using magnitude (M) and epicentral distance (R) combinations representing design level Australian seismicity, seismological modelling has been carried out to derive DRS for rock sites. To modify the rock site spectra for site soil flexibility, a method has been developed based on a simple idealisation of site soils. Finally, the representation of a yielding multi-degree-of-freedom (MDOF) system with an elastic single-degree-of-freedom (SDOF) system is being examined with particular attention given to the assignment of damping.

The objectives of this paper are to:-

1) Summarise the results of the research effort undertaken towards producing accurate displacement response spectra for Australian seismicity.

2) Report on the initial observations made on assessing representative damping for the substitute structure approach.

2 ROCK SITE DISPLACEMENT RESPONSE SPECTRA

Strong motion data is very limited for intraplate regions of the world, and Australia is no exception. For this reason a stochastic approach was adopted to define displacement response spectra appropriate for Australian rock sites. Using a stochastically based seismological model and M-R combinations appropriate for Australian conditions (Table 1) numerous artificial accelerograms were generated. The accelerograms derived were then used to carry out a series of time history analyses. The spectral values from these were, in turn, used to calibrate simple expressions for defining generalised DRS for rock sites. Finally, the generalised DRS were modified for site flexibility. Each component of this work is described below.

2.1 *Seismological Model of Intraplate Seismicity*

Artificial accelerograms were generated using the seismological model developed by Atkinson (1993), Boore, and others. The model expresses the Fourier amplitude spectrum of earthquake ground motion as the product of a source function and various path modification functions. The method for generating accelerograms and a detailed description of the factors applied to the source function have been presented elsewhere by the authors (Lam, Wilson and Hutchinson, 1998). To account for the effect of the random assignment of phase angles to the individual frequency components of the ground motion, 18 ground motions were generated for each combination of parameters and the resultant spectral values averaged. Atkinson's Eastern North American (ENA) source model along with a Western North American (WNA) crustal model were taken as representative of intraplate Australia.

2.2 *Displacement Response Spectrum Model*

The displacement spectrum is assumed to be bi-linear and defined by the smaller of equations (1a) and (1b) :-

$S_d(T) = S_V \times T / 2\pi$ (1a)

$S_d(T) = S_D$ (1b)

Table 1. Predicted S_D and S_V from model

Epicentral Distance R [km]	10	20	30	50	70
Earthquake Magnitude M	5.0	5.5	6.0	6.5	7.0
S_D (ENA crust)	9 (6)	8 (8)	14 (13)	18 (17)	24 (22)
S_V (ENA crust)	110 (102)	78 (82)	95 (92)	95 (79)	104 (93)
S_D (WNA crust)	17 (11)	13 (13)	22 (21)	27 (25)	31 (30)
S_V (WNA crust)	188 (181)	129 (145)	152 (153)	138 (124)	135 (126)

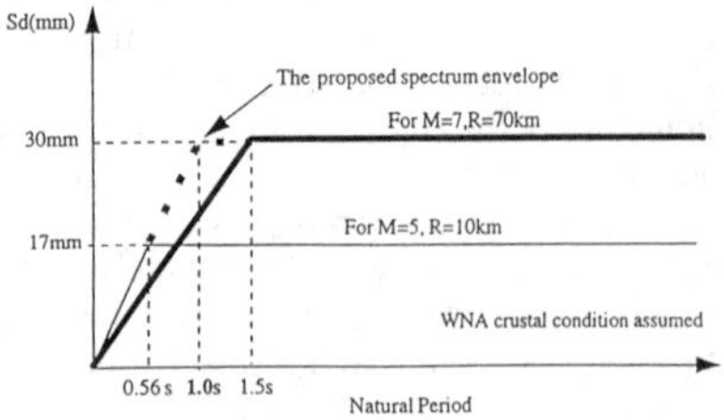

Figure 1. DRS for 500yr RP rock site event

By curve fitting the time history analysis spectral data obtained from the seismological model simulations, the peak spectral velocity, S_V, and the peak spectral displacement, S_D, have been approximated. Details of this work are described elsewhere (Lam, Wilson, Edwards et al, 1998). Spectral values predicted by the model are presented in Table 1. The accuracy of the parametric curve fitting can be assessed by comparing the time history values in brackets (the average result from each ensemble' of 18 accelerograms) with those from the model.

2.3 *Displacement Response Spectra for Rock Sites*

The displacement response spectra obtained from the extreme M-R combinations are presented in Figure 1, representing the displacement demand of a 500 year return period rock site earthquake. From the figure it can be seen that larger more distant events cause the greater displacement demand for medium to long period structures. However, even for the more distant events, the displacement demand for a rock site is very small but consistent with empirical data from studies by Somerville et al (1998).

2.4 *Displacement Response Spectra Modification For Site Flexibility*

Site flexibility has a marked affect on both the amplitude and frequency characteristics of earthquake ground motion. Site flexibility effects can be modelled by carrying out a one-dimensional non-linear shear wave analysis (e.g. using the program "SHAKE"). However, shear wave analyses of this type are not commonly carried out in practice due to the perceived complexity. For the purposes of producing DRS, a simpler method to SHAKE has been devised, taking into account the effects of the natural period of the site. The simplified method represents a resonating soil column by an imaginary shear building exhibiting a side-sway mode of vibration in response to basement rock excitation. The idealisation is presented in Figure 2 and the procedural steps described elsewhere (Lam et al, 1997). The result obtained from this method were subsequently favourably compared with the results from more rigorous SHAKE analyses

Figure 3 presents the M=7, R=70km, 500yr return period rock site DRS modified using the simplified method for a range of site flexibilities. As can be seen from the modified spectra, site flexibility does have a marked effect on the displacement demands. However, the modified demands are modest suggesting that Australian frame type construction will accommodate the earthquake displacement demands without excessive degradation. Further work is needed in this area before any firm conclusions can be drawn.

3 SUBSTITUTE STRUCTURE IDEALISATION OF DAMPING

At maximum seismic response a typical MDOF structure softens due to inelastic action. The effective structural stiffness reduces and the cyclic energy absorption increases. In the substitute structure approach the inelastic structure is represented by an equivalent elastically responding SDOF system to which a reduced secant stiffness and an accentuated equivalent viscous damping

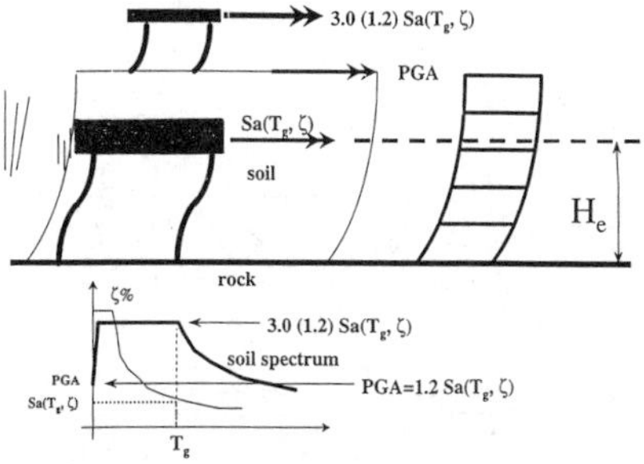

Figure 2. Soil idealisation for soft soil sites

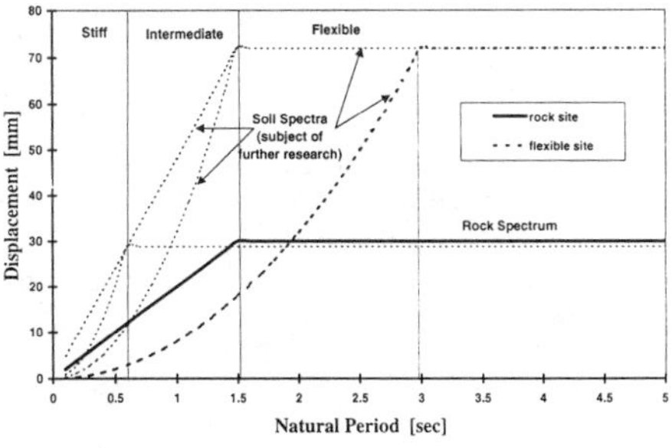

Figure 3. DRS for 500yr RP event with various site flexibilities.

are assigned. The key parameter of the assigned damping level has been the subject of research to date, with damping examined separately either side of flexural yield.

3.1 *Area Method Damping Assessment*

The "Area Method" illustrated in Figure 4 has been commonly used to assign equivalent viscous damping (EVD) levels to hysteretic energy absorption mechanisms. The theoretical basis for the equation presented in the figure is that an equivalent viscously damped SDOF oscillator at steady state resonant response should absorb the same energy per cycle as the real yielding system. As the theoretical assumptions do not match the complex time history of real hysteretic systems the damping values assigned by this method are only empirical reference values that will need some correction. In this research the area method has been used to assign damping levels, but the values will be calibrated to maximise the displacement prediction accuracy.

3.2 *Pre-yield Damping*

The assumption that a system is elastic up to yield is not strictly correct. For a reinforced concrete frame system, the framing members crack well before yield. The progressive appearance of system damping has been examined using the results from a series of exterior beam/column joint sub-assemblages tested at the University of Melbourne (Corvetti, 1994 ; Stehle et al, 1998). The Area Method damping levels for applied moments ranging from uncracked up to yield were evaluated and are presented in Figure 5. Stable hysteretic response is observed at very low moment amplitudes and appears to be relatively constant over the full pre-yield amplitude range. From the results the following observations are made:-

1) For Special Moment Resisting Frames (SMRF), an Area Method reference value of approximately 7% EVD appears to be appropriate for systems at the point of yield.
2) Damping level differences between Ordinary Moment Resisting Frames (OMRF) and SMRF are small. This is an expected result as detailing influences post yield behaviour, but has little influence prior to yield or bar anchorage loss.

3.3 *Post Yield Damping*

For post yield structural behaviour, damping can no longer be treated as uniformly distributed throughout the frame members. The localised hinges formed in structural frame systems during inelastic response dominate the energy absorption exhibited. For this reason an understanding of system damping requires an understanding of the behaviour of the hinges formed. The range of hinge types that can form in a frame system have been identified and experimental sub-assemblage test data back analysed to quantify the damping characteristics of each.

3.3.1 *Hinge Model*

The length of column or beam hinges where inelastic post yield behaviour occurs was arbitrarily assigned and total hinge region rotation selected as the parameter for determining hinge ductility demand. To assign hinge rotation an adaptation of the moment-curvature model proposed by Mander (1984) was used. Shear deformations were determined using the model proposed by

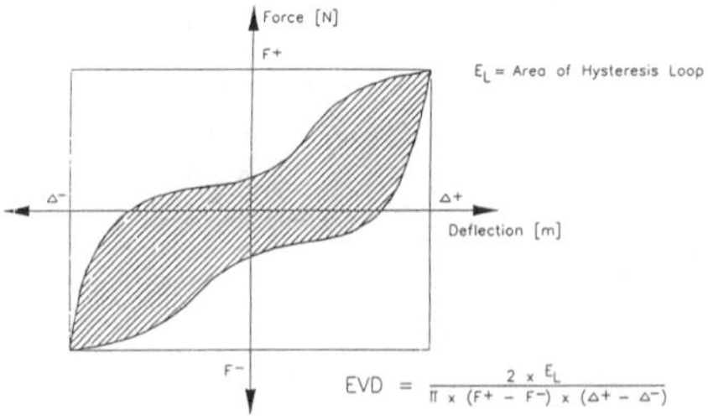

Figure 4. Area Method for assigning equivalent viscous damping to yielding systems

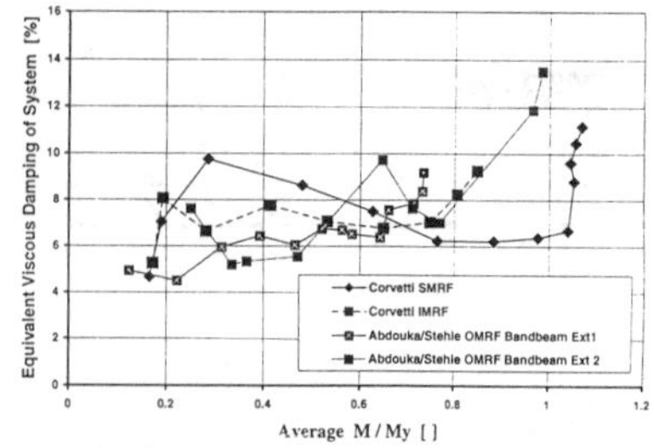

Figure 5. Sub-yield equivalent viscous damping

Park and Paulay (1975), and rotation of the beam at the column face due to bar extension was determined using the bond stress model proposed by Alsiwat and Saatcioglu (1992) . The deformation components are summarised in Figure 6 and the accuracy of its yield deflection predictions for column specimens is shown in Figure 7.

For assessing the damping, the first cycle hysteresis loop at each displacement test amplitude was used to provide the reference damping value for the hinge zone, thereby ignoring hinge degradation with subsequent cycles. Beam-column joints were assumed rigid, all energy absorption in the sub-assemblage was assumed to take place in the hinge, and the strain energy associated with the framing members beyond the hinge was subtracted from the observed system behaviour to give hinge behaviour.

3.3.2 *Beam Hinge Behaviour*

Typical results for SMRF beam hinges are presented in Figure 8. Pre-yield damping is also shown for ductilities less than unity with a sharp increase in damping observed following the yielding of the flexural steel. Where the aspect ratio of the beam is reduced, shear mechanisms play a more significant role. This corresponds with a pinching of the hysteresis loops, a more gradual increase in damping with ductility demand, and a lower plateau damping level. The plateau damping value for the higher aspect ratio beams was found to be 26%, with 23% noted for those beams more dominated by shear.

3.3.3 *Column Hinge Behaviour*

Typical results for SMRF column hinges are presented in Figure 9. Column behaviour was found to be greatly affected by axial load. With increasing axial load, increased quantities of confining steel in the hinge region is provided in the form of closed ties. The ties provide a hori-

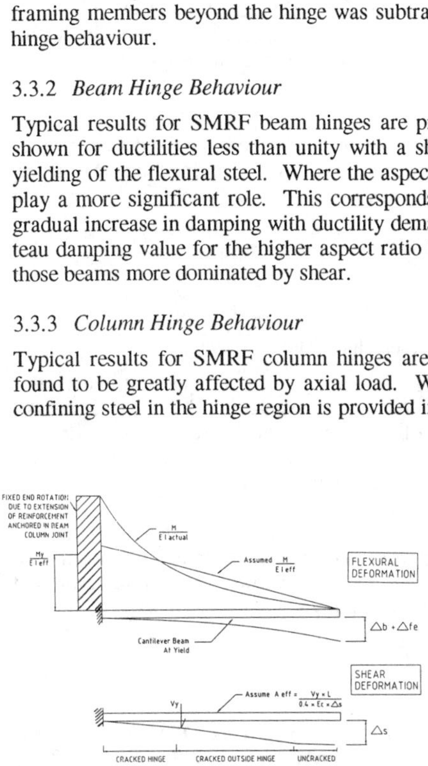

Figure 6. Cantilever beam deformation components

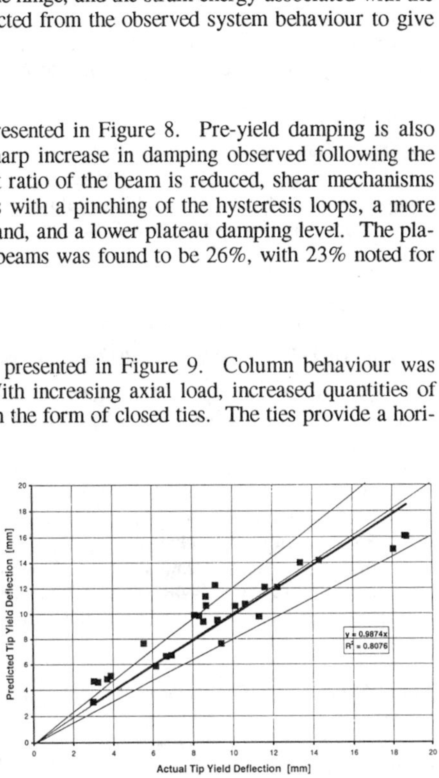

Figure 7. Analytical versus experimental yield deflections for columns

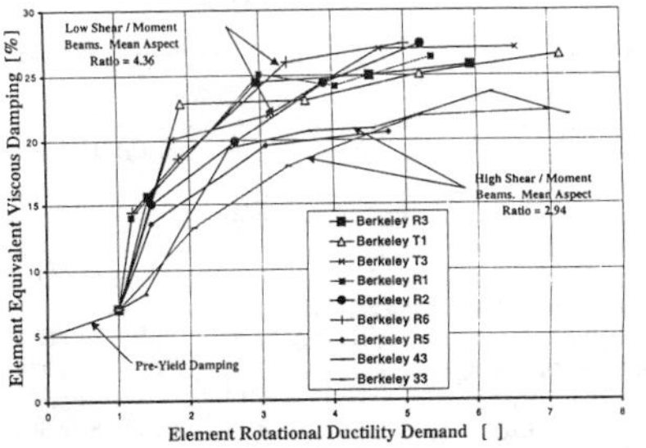

Figure 8. SMRF beam hinge damping versus rotational ductility demand

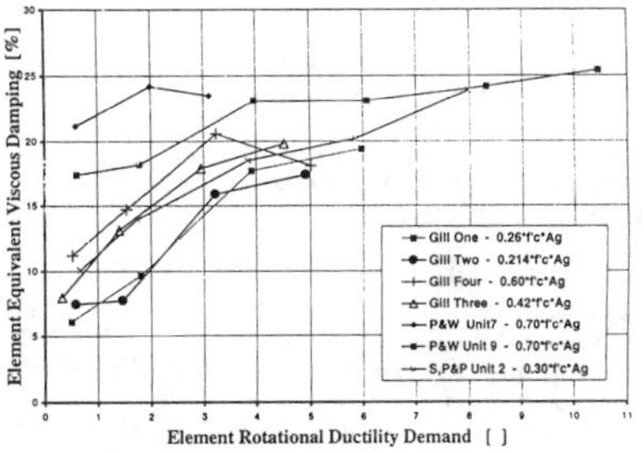

Figure 9. SMRF column hinge damping versus rotational ductility demand

zontal confining pressure on the core concrete permitting large hinge rotations with limited loss of load carrying capacity. Comparing Figure 9 to Figure 8, it can be seen that for low axial loads column hinge behaviour is similar to that of beams. However, column hinge damping levels climb more gradually with increasing ductility demand due, in part, to sequential yielding of the appreciable quantities of column side face reinforcement. In contrast, all beam reinforcement typically yields together.

At higher axial loads damping levels were found to increase markedly. For columns with high axial load the concrete reaches a compressive yield strain of 0.002 before the tension steel yields. Hence, prior to compression "yield", and thereafter, inelastic cyclic crushing of the concrete is taking place that may contribute to this feature. The influence of hinge region confining pressure furnished by column ties on this damping phenomenon is currently being investigated.

3.3.4 *System Behaviour*

The hinge behaviour presented in 3.3.2 and 3.3.3 above will form the basis for determining the damped hysteretic behaviour for SMRF frame systems. By utilising an inelastic time history analysis package it will be possible to calibrate an Area Method based damping curve for frame types to optimise displacement prediction accuracy.

4 SUMMARY

The following observations are made based on the research to date:-

- Stochastic modelling of accelerograms using a seismological model appears to be a valuable tool in deriving DRS representative of Australian seismicity. The displacement demands were found to be small for all site flexibilities, though further research is needed.
- The area method for assigning damping levels to yielding systems may provide useful reference values for the substitute structure method. Damping calibration work using time history analyses is proposed to optimise displacement prediction accuracy.
- SMRF beam hinges characteristically show a sharp increase in damping following yielding of reinforcement. Beam shear was found to reduce beam hinge damping in beams having low aspect ratios. SMRF column hinge damping is typically lower than that in beams and highly influenced by axial load and/or the degree of hinge region confinement.

5 ACKNOWLEDGEMENT

The work reported in this paper forms part of a project investigating the DBM funded by the Australian Research Council (Large Grant No. A89701689 and Small Grant No. 589711277)

6 REFERENCES

Alsiwat, J. M. ; Saatcioglu, M. ; 1992 ; "Reinforcement Anchorage Slip Under Monotonic Loading", *ASCE Journal of Struct. Eng., Vol 118, No.9.*

Atkinson, G. , 1993 ; "Earthquake Source Spectra in Eastern North America", *Bulletin of the Seismological Society of America, Vol.83, pp1778-1798.*

Corvetti, G. M. ; 1994 ; "Assessment of Reinforced Concrete External Beam-Column Joints Designed to the Australian Earthquake Standard", *Master of Engineering Science Thesis, Dept of Civil and Environmental Engineering, University of Melbourne.*

Federal Emergency Management Agency, 1997 , "NEHRP Guidelines for the Seismic Rehabilitation of Buildings", *FEMA 273.*

Lam, N. T. K.; Wilson, J. L. ; Edwards, M. R. ; Hutchinson, G. L. ; 1998 ; "A Displacement Based Prediction of the Seismic Hazard for Australia", *Australian Earthquake Engineering Society Seminar, Perth.*

Lam, N. T. K.; Wilson, J. L. ; Hutchinson, G. L. ; 1998 ; "Development of Intraplate Response Spectra for Bedrock in Australia", *Proceedings of the 1998 Technical Conference of the New Zealand National Society for Earthquake Engineering, Wairakei, pp137-144.*

Lam, N. T. K.; Wilson, J. L. ; Hutchinson, G. L. ; 1997 ; "Introduction to a New Procedure to Construct Site Response Spectrum", *Proceedings of ACMSM Conf., Melbourne, pp345-350.*

Mander, J. B. ; 1984 ; "Seismic Design of Bridge Piers", *Report No. 84-2, University of Canterbury, Christchurch, New Zealand.*

Park, R. ; Paulay, T. ; 1975 ; *"Reinforced Concrete Structures"*, Wiley.
Priestley, M. J. N. ; Calvi, G. M. ; 1997 ; "Concepts and Procedures for Direct Displacement Based Design of Reinforced Concrete Structures", *Workshop on Seismic Design procedures for the 21st Century.*
Shibata, A. ; Sozen, M. A. ; 1976 ; "Substitute-Structure Method for Seismic Design in R/C", *Journal of the Structural Division, ASCE, ST1, pp1-18.*
Somerville, M. ; McCue, K. ; Sinadinovski, C. ; 1998 ; "Response Spectra Recommended for Australia" , *Proceedings of the Australasian Structural Engineering Conference, Auckland, pp439-444.*
Stehle, J. S. ; Abdouka, K. ; Mendis, P. ; Goldsworthy, H. ; 1998 ; "Performance of a Reinforced Concrete Subassemblage With the Wide Beam Under Lateral Loading", *Proc of Australian Struct. Eng. Conference, Auckland.*

Mechanics of Structures and Materials, Bradford, Bridge & Foster (eds)
© 1999 Balkema, Rotterdam, ISBN 90 5809 107 4

Deformability of prestressed beams with AFRP and CFRP tendons

N.Gowripalan, X.W.Zou & R.I.Gilbert
School of Civil and Environmental Engineering, The University of New South Wales, Sydney, N.S.W., Australia

ABSTRACT: A rational basis for a combined factor for assessing the deformability of pretensioned beams with AFRP and CFRP tendons is proposed. Slow repeated load tests carried out on 18 pretensioned beams with concrete compressive strengths in the range of 40-100 MPa were used to verify the proposed method. These deformability factors are calculated for pretensioned beams containing FRP tendons and conventional steel tendons, and compared. Beams with FRP tendons reached sufficiently large deflections and curvatures immediately before failure, similar to those with steel tendons. A combined factor defined in terms of deflections and moment carrying capacities seems to be a useful deformability index. A minimum design limit of 30 for this factor is proposed to give sufficient warning of failure for pretensioned beams with FRP tendons. All beams tested and investigated satisfied this limit.

1 INTRODUCTION

In recent years, the use of fibre reinforced polymer (FRP) reinforcement in concrete members has emerged as one of the most promising developments in the construction industry, particularly in prestressed concrete applications. The most promising types of FRP tendons currently available include glass fibre based tendons (GFRP), aramid fibre based tendons (AFRP), such as Braided tendon, 'Parafil'(Burgoyne 1985) and 'Arapree'(Gerritse 1987), and carbon fibre based tendons such as Leadline (CFRP 1991) and CFCC. A typical cross-section of these tendons (other than Parafil) consists of about 40-70% fibres embedded in an epoxy, polyester or vinyl ester resin matrix. Typical size of these tendons range from 5mm diameter bars to 20mm diameter strands.

Existing ductility indices, including the models proposed by Naaman and Jeong (1995) (for all types of prestressing material), Mufti et al (1996) (for FRP reinforcement) and Abdelrahman et al (1995), are reviewed here and a new ductility index is proposed.

Ductility factors have been commonly expressed in terms of the various parameters related to deformation, such as displacements, strains, curvatures and rotations (Bachmann (1992), Park(1992)) and all relate ultimate deformation to deformation at first yield of the tendon. These ductility factors are not appropriate for elements containing FRP tendons, which are essentially linear-elastic with no yield point. Naaman and Jeong (1995) proposed a ductility index (μ_e) which is based on the ratio of the total energy (E_{tot}) to the elastic energy (E_{ela}) at the failure state of a beam. The elastic energy can be quantified by unloading the beam near the failure load and the ductility index is given by

$$\mu_e = 0.5\,(E_{tot}/E_{ela} + 1) \tag{1}$$

This definition is equally applicable for beams containing steel or FRP tendons. A minimum limit of between 3 and 4 is proposed for this factor for design purposes. More recently, Mufti et al (1996) proposed an overall ductility factor for reinforced concrete beams containing FRP and compared it with steel reinforced concrete beams. The overall factor is defined as the product of a deformability factor and a strength factor.

Overall factor = Deformability factor x Strength factor (2)

The deformability factor is the ratio of the curvature at ultimate to the curvature when the extreme fibre compressive strain $\varepsilon_c = 0.001$ and the strength factor is the ratio of the moment at ultimate to the moment when $\varepsilon_c = 0.001$.

Although this has been developed for reinforced concrete beams, similar concepts are applicable to prestressed concrete beams. Abdelrahman et al (1995) also proposed an equivalent ductility factor for prestressed beams with FRP tendons. This is defined as the ratio of the deflection at the ultimate state, Δ_u, to an equivalent deflection (Δ_l) of an uncracked beam at the same load level.

$$\mu_\Delta = \Delta_u / \Delta_l \quad (3)$$

The authors pointed out that this equation, when used for a prestressed concrete beam with steel tendons, might overestimate the ductility of the beam. A combined factor based on moment and rotation as proposed by Mufti et al (1996) may be a suitable factor. However, for pretensioned beams, the cracking moment seems to be a predictable, well-defined point on the load-deflection curve. Also the value of ultimate moment for these beams can be obtained quite accurately. Hence, a strength factor defined as the ratio of the ultimate moment to the cracking moment may be useful indicator in the consideration of ductility.

Rotations can be either directly measured or calculated (from concrete strains or deflections). However, the measured rotation at mid span can vary for a given moment, depending on the position of the first crack and subsequent cracks. The energy can easily be computed as the area under the load-deflection curve and, in this case, only load and deflection measurements are required. Also the beam can be unloaded at a suitable point near failure to obtain the elastic energy component. Hence, the energy factor (total energy/elastic energy) near failure can easily be calculated. In pretensioned beams containing steel tendons, even after the commencement of plastic deformation in the prestressing steel, a major proportion of the deflection is recoverable.

After considering a combination of factors for the overall ductility factor for FRP pretensioned beams, it is proposed to use a deformability index F_{se} which combines both a deflection factor and a moment factor. This combined index is defined as

$$F_{se} = (\Delta_u/\Delta_{cr}) \times (M_u/M_{cr}) \quad (4)$$

and can be easily calculated using existing theories for the computation of Δ_u, Δ_{cr}, M_u, and M_{cr}, separately. For the calculation of Δ_u, a bilinear load-deflection curve with a reduced stiffness after cracking can be used. The usefulness of the above index for design of pretensioned beams with FRP tendons is verified with experiments.

2 EXPERIMENTAL PROGRAMME

Eighteen simply-supported beams, all with the same rectangular cross section, 150 mm wide and 300 mm deep, and with spans of either 6000 mm or 3000mm, were tested. Five beams were pretensioned with AFRP (Arapree) tendons, seven with CFRP tendons (Leadline), and six with steel strands. All tendons were placed with a constant eccentricity of 85 mm. No steel stirrups or longitudinal steel reinforcement were used in the beams so that the effect of additional steel reinforcement was totally eliminated. Commercially available ready mixed concrete mixes, with characteristic compressive strengths at 28 days of either 40 or 80 MPa, were used. The initial prestress at transfer was 83.8 kN for the 40 MPa beams and 119.2 kN for the 80 MPa beams. Beams with AFRP tendons were designated as A40-80-S (A-Aramid tendon, 40-Concrete Strength (MPa), 80- Prestressing force (kN), S-Short term testing), whereas beams with CFRP tendons were designated as C40-80-S (C-Carbon tendon). Beams with AFRP tendons, tested after long-term loading, were designated as A40-80-L (L-Long-term loading). The number, diameter and type of tendon for each beam are given in Table 1.

The beams were cast in a 8 m long prestressing bed. A load cell was used to monitor the force in each tendon during jacking and up to the transfer of prestress. The two sides of the moulds of the beams were removed at the age of 5 days. Demec points were then attached to the side surface of the beam at mid-span and at various positions over the depth of the beam for the measurement of concrete strains. At the age of 9 days, the prestressing force was transferred to the concrete by cutting the tendons suddenly at one end. The beams were simply supported after the transfer of prestress.

At the age of about 28 days, the beams were subjected to seven cycles of loading, unloading and reloading under either two equal symmetrically placed point loads or a uniformly distributed load (designated by U in Table 1). The unloading part of the cycle was carried out at first cracking , at mid-span deflections of 15, 25, 40, 60 and 100 mm and near failure. The beams were finally loaded to failure. Deflections at mid-span were measured using LVDTs linked to a computer. The concrete strains at mid-span were measured using a demec gauge and LVDTs. Slippage of the tendons at the ends of each beam was monitored by dial gauges.

3 RESULTS

A typical load-deflection curve of a beam prestressed with CFRP, under slow cyclic loads, is shown in Figure 1. Beams with both AFRP and CFRP had a linear load-deflection relationship before cracking, with a very small increase of deflection when the load was increased. The load-deflection behaviour before cracking can be modelled using the gross cross-section stiffness(E_cI_g).

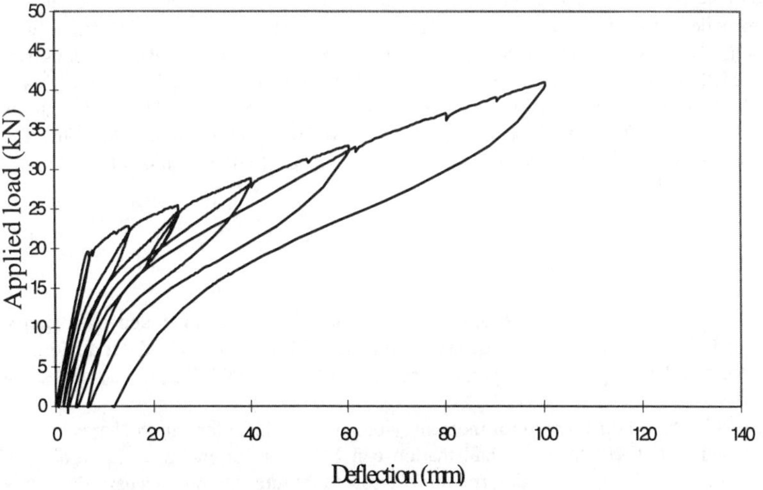

Figure 1 Load-deflection relationship of 80 MPa beam with CFRP tendons (C80-120-S)

Table 1 Details and deformability indices of beams with AFRP, CFRP and Steel Tendons

Beam .Designation	No. of tend -ons	Diam. of tendons (mm)	Span (m)	Δ_u/L	Δ_u/Δ_{cr}	M_u/M_{cr}	Eqn 1: $0.5(E_{tot}/E_{ela}+1)$	Δ_u/Δ_l	Eqn 4: $(\Delta_u M_u/\Delta_{cr} M_{cr})$
A40-80-S	4	8	3	1/44	17.4	2.5	1.37	9.0	43.7
S40-80-S	2	7.8	3	1/35	31.7	1.9	3.10	22.1	59.5
A80-80-S	4	8	3	1/56	15.9	2.1	1.35	9.0	33.3
S80-80-S	2	7.8	3	1/32	28.4	1.5	4.24	19.9	41.5
A40-80-L	4	8	3	1/44	(19.5)	(2.2)	1.46	(5.2)	(42.0)
S40-80-L	2	7.8	3	1/36	(32.8)	(1.8)	2.72	(14.2)	(59.0)
A80-80-L	4	8	3	1/59	(20.1)	(2.6)	1.42	(5.1)	(52.2)
S80-80-L	2	7.8	3	1/32	(36.2)	(1.9)	3.44	(12.5)	(68.8)
A80-120-L	3	8	6	1/52	(19.4)	(2.2)	1.46	(4.5)	(42.7)
S80-120-L	3	7.8	6	1/31	(31.0)	2	(1.45)	(5.1)	(62.0)
C80-120-S	2	8	6	1/46	21.0	2.44	1.32	10.9	50.9
C80-80-L-C	2	8	6	1/35	(24.0)	(2.7)	1.30	-	(64.8)
C40-80-S	2	8	6	1/44	24.3	2.77	1.34	11.4	67.8
C40-80-L-U	2	8	6	1/39	24.1	2.54	1.32	10.2	61.2
C40-80-L-C	2	8	6	1/36	(29.0)	(2.9)	1.24	-	(84.1)
C40-120-L-C	2	8	6	1/40	(26.2)	(3.0)	1.29	-	(78.6)
C80-120-L-U	2	8	6	1/42	17.2	2.45	1.31	8.5	42.2
S80-120-L-U	2	9.3	6	1/42	21.1	1.89	1.81	12	39.8

For values within ( ), Δ_{cr} and M_{cr} are based on a concrete tensile strength of zero, as these beams were pre-cracked due to sustained load testing.

The beams cracked at or close to the load level predicted using the current code of practice methods (such as ACI 318 and AS 3600), and the different deformability indices of beams with AFRP, CFRP and steel tendons are compared in Table 1. The beams with AFRP and CFRP tendons showed a reduced stiffness after cracking (I_{efl}) but still followed a linear load-deflection relationship. The residual deflections after the beams were unloaded, even from levels close to the failure moments, were less than 15% of the deflection prior to unloading.

The ratio of deflection at failure to deflection at first cracking, $\Delta_u/\Delta_{cr,}$ for all beams with AFRP tendons was greater than 15, while the same ratio for beams with steel tendons was greater than 25. The ratio of $M_u/M_{cr,}$ for beams with AFRP (without sustained loads) was greater than 2, while those for beams with steel tendons ranged from 1.8 to 2.0. The deformability index, F_{se}, was above 30 for AFRP beams and above 40 for beams with steel tendons. The results of beams with CFRP tendons showed similar trends to the results of the beams containing AFRP tendons.

4 DISCUSSION

The current research shows that the load-deflection behaviour can be modelled as a simple bilinear relationship for beams prestressed with AFRP and CFRP tendons. Abdelraman et al (1995) also reported that prestressed beams with CFRP showed a bilinear load-deflection relationship.

A ductility index based on strength (or moment) alone ignores the deformation characteristics of the beam and is not satisfactory. Deformation can be elastic or inelastic. According to Naaman and Jeong (1995), large deformations prior to failure do not necessarily imply acceptable ductility. Most of the energy stored in the beam near failure can be elastic energy. Hence it is useful to compute the components of both total energy and elastic energy

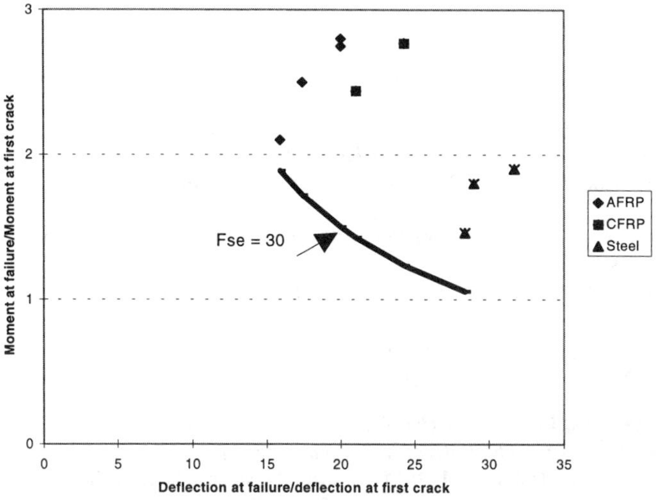

Figure 2 Comparison of test results with a limiting deformability index $F_{se} = 30$.

near failure. The factor proposed by Naaman and Jeong, however, varied from 1.24 to 1.46 for the beams tested here with AFRP or CFRP tendons and this range is not large enough to satisfactorily differentiate beams which may or may not have adequate deformability. For sufficient ductility, Naaman and Jeong suggested this factor should have values in the range of 3 to 4. Clearly, by this definition, these beams are not ductile. Yet all beams deflected by more than Span/60 at failure and exhibited considerable deformability.

Minimum limits on the ratio of deflection at failure to deflection at service loads and on the ratio of moment carrying capacity at failure to moment at service loads can ensure sufficient deformability of a beam prior to failure. The ratio of deflection at failure to span of the beams with AFRP tendons was about 35% lower than that of the beams with steel tendons. The same ratio for beams with CFRP tendons was 20% less than that of the beams with steel tendons. The beams with FRP tendons had lower deflections at failure, as expected, but the ratio of the moment carrying capacity at failure to the cracking moment was higher.

The deformability index proposed in Eqn 4, F_{se}, is essentially a measure of the elastic energy ratio at failure to that at first cracking. A graph comparing the experimental results with a limiting deformability index $F_{se} = 30$ is presented in Figure 2.

5 CONCLUSIONS

1. For all the beams tested in this study and pretensioned with AFRP and CFRP tendons, the ratio of Δ_u/Δ_{cr} exceeded 15. The same ratio for beams with steel tendons was in excess of 20. With a minimum limit of 15, large deformations near failure were recorded.
2. The ratio of M_u/M_{cr}, for beams with AFRP and CFRP exceeded 1.5 for all beams tested. For beams with steel tendons the above ratio was in excess of 1.3.
3. For pretensioned beams with AFRP and CFRP, a minimum limit of the deformability index F_{se} $(=\Delta_u/\Delta_{cr} \times M_u/M_{cr})$ of 30 appears reasonable to give sufficient deformation and moment carrying capacity prior to failure. For identical beams with steel tendons, a minimum of 40 for the overall factor was achieved.

REFERENCES

Abdelrahman,A.A., Tadros,G. and Rizkalla,S.H.1995. Test model for the first Canadian smart highway bridge, *ACI Structural Journal*, Vol.92, No.4. July, pp 451-458.

ACI Committee 440 1996. *State-of-the-Art Report on Fiber Reinforced Plastic Reinforcement for concrete structures,* ACI 440R-96, American Concrete Institute,USA, 68pp.

Bachmann 1992. Dynamische Belastingen (V) Earthquake Actions on Structures, *Constructief Ontwerp.*

Burgoyne, C.J. and Chambers, J.J. 1985. Prestressing with Parafil tendons, *Concrete (London),* Vol. 19, No. 10, October, pp12-15.

CFRP 1991. Carbon fibre reinforced plastic tendons, *Brochure, P.S. Corporation,* Saitama, 330 Japan, 4pp.

Gerritse,A. 1987. Prestressed concrete structures with Arapree, *Proceedings, IABSE Symposium,* Paris, 8pp.

Gilbert, R.I. and Gowripalan, N. 1993. Long-term behaviour of prestressed concrete beams using non-metallic tendons, *Proceedings CONCRETE 2000, International Conference on Economic and Durable Construction, University of Dundee,* Scotland, pp 255-264.

Mufti, A.A., Newhook,J.P. and Tadros,G. 1996. Deformability versus ductility in concrete beams with FRP reinforcement, *Proceedings of the Second International Conference on Advanced Composite Materials in Bridges and Structures,* Montreal, Canada, pp189-199.

Naaman, A.E. and Jeong, S.M. 1995. Structural ductility of concrete beams with FRP tendons, *Non-metallic (FRP) reinforcement for concrete structures,* Edited by Taerwe, E & FN Spon, London, pp379-401.

Park,R. 1992. Capacity design of ductile RC building structures for earthquake resistance, *The Structural Engineer,* Vol.70, No. 16/18, pp279-289.

Mechanics of Structures and Materials, Bradford, Bridge & Foster (eds)
© 1999 Balkema, Rotterdam, ISBN 90 5809 107 4

Behaviour of fibre reinforced high strength concrete columns

S.J. Foster & M.M. Attard
School of Civil and Environmental Engineering, The University of New South Wales, Sydney, N.S.W., Australia

ABSTRACT: This study consisted of the testing of 12 by 200 mm square high strength concrete columns cast with 70 MPa fibre reinforced concrete. The columns were tested in eccentric compression with initial loading eccentricities from 0 mm to 30mm. The study showed that the introduction of steel fibres into the mix design arrested the early spalling of the cover and increased the load capacity and the ductility of the columns over that of comparable non-fibre reinforced specimens.

1 INTRODUCTION

High strength concrete has been used in many lower storey columns of high-rise buildings, as well as low and mid-rise buildings, bridges and foundation piles. High strength concrete out performs conventional strength concrete in terms of strength, durability, and modulus of elasticity as well as in may other material properties. However, the advantages of using high strength concrete on columns predominantly loaded in compression is off-set by what has been termed "early cover spalling" (Cusson and Paultre, 1994, Foster and Attard, 1997). In fact the term "early cover spalling" is quite misleading as it implies an event that occurs when it should not, or at least, one that is unexpected (Foster, 1999a). Foster et al. (1998) showed that the load at which cover spalling begins is an inevitable consequence of the placement of tie steel within the column, and is independent of the concrete strength. Foster et al. (1998) postulated that the reasons for "early cover spalling" not being observed in earlier studies on conventional strength concrete was due to the effect being disguised by the increase in strength due to confinement, and by normal experimental variability. The event only became noticeable in the experimental data when high strength columns were tested with conventional tie detailing arrangements, giving relatively lower increases in the columns core strength due to confinement. Previous research has shown that increases in the strength and ductility of conventional strength columns can be significantly improved by providing an effectively confined core. The increase in strength and ductility being a function of the concrete cover, concrete strength, distribution of longitudinal reinforcement and the configuration, yield strength and spacing of the tie or spiral reinforcement (Sheikh and Uzumeri, 1980, Yong et al., 1998, Saatcioglu and Razvi, 1993, Cusson and Paultre, 1994). In tests on concentrically loaded HSC columns it is generally observed that the cover concrete spalls away from the section at a load lower than the axial load capacity calculated using current building codes. After separation of the cover concrete from the section, the load drops by 10 to 15 per cent (Cusson and Paultre, 1994). After separation of the cover concrete, the confining reinforcement becomes active and the load on the column can again be increased to a second peak load corresponding to the strength of the confined core. While the second peak load can be higher than the first peak load it requires a very high effective confinement ratio with generally greater than 4 percent of tie reinforcement to concrete by volume.

Tests by Paultre et al. (1996) have shown that the use of steel fibre in the mix design can improve the strength and ductility of HSC columns. In this paper 12, eccentrically loaded 200-mm square HSC, columns were tested to study the effect of adding steel fibres into the concrete mix design. On the strength and ductility of HSC columns of interest is the ability of the fibre reinforcement to prevent the cover spalling away from the section and to assess improvements in ductility afforded by the inclusion of the steel fibres.

2 EXPERIMENTAL PROGRAMME

2.1 *Testing Arrangements*

Twelve columns cast with high strength fibre reinforced concrete were tested in two series (Series G and S) with six columns in each series. The columns are designated as 2MF-XX-YYYZ where XX is the nominal initial eccentricity, YYY the spacing of the ties and Z is the series identifier. Details of the reinforcement arrangements and specimen dimensions are given in Figure 1. The cover for all specimens was 15 mm and the dimension "A" shown in Figures 1 and 2 was 140 mm.

The eccentrically loaded specimens were 200-mm square over the full 900-mm test region. The nominally concentrically loaded specimens (specimens 2MF0-YYY) contained a 5 mm block-out at the outside of the cover region over a length of 100 mm located at the mid-length of the column (refer Figure 2). Thus, for these specimens the critical cross-section was 190 mm square with a cover of 10 mm in the middle 100-mm of the column. The purpose of the reduced cross-section was to give a weakening in the column to ensure that failure was initiated in the gauged region.

2.2 *Materials*

A local ready-mix concrete with a maximum aggregate size of 10mm was used in this study. The concrete mixes contained 2 percent by weight of 35-mm, end hooked, steel fibres and were cast with nominally 70 MPa concrete. Cylinders of 100-mm diameter by 200 mm high were tested to obtain the compressive strengths. The cylinders were saw cut at the unformed end to remove extraneous fibres which protruded from the cylinder end and were capped at both ends. The concrete strengths for each series are given in Table 1.

The specimens contained 8 longitudinal bars of 12-mm diameter (Y-grade) and had a diamond arrangement for the tie steel, which consisted of 6-mm diameter ties (W-grade) spaced at either 50 or 100 mm centres. The tie reinforcement was bent to shape commercially with 135-degree end hooks used for stress development. The properties of the reinforcing steel used in the project are given in Table 2.

Table 1. Mean cylinder compressive strengths of the concrete.

Test Series	Days after casting	f_{cm} (MPa)	Std Dev. (MPa)	No. of Cylinders
Series G	70	67	2.3	10
Series S	56	73	3.5	9

Table 2. Material properties of the reinforcing steel.

Bar Designation	Bar Diameter (mm)	f_{sy} (MPa)	E_o (MPa)
Y12	12.2	450	200
W6	6.1	470	220

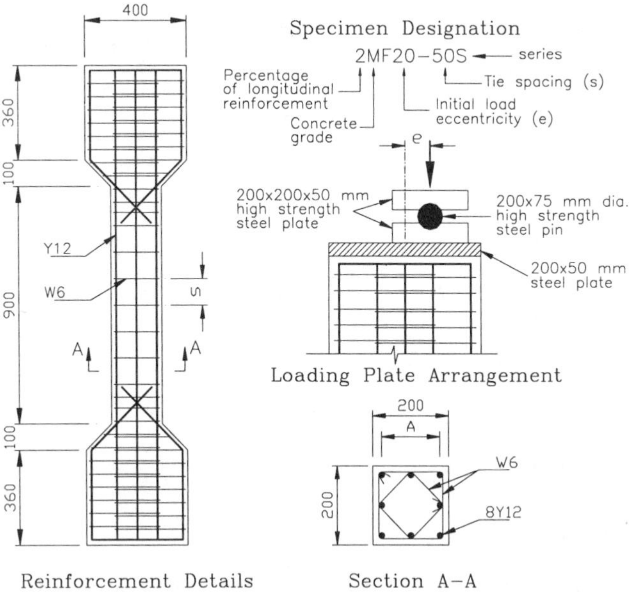

Figure 1. Test specimen dimensions and reinforcing details

2.3 *Instrumentation and Testing Procedure*

The columns were placed in the testing rig with 50mm thick steel plates placed at each end of the specimen to distribute the applied load. For the eccentrically loaded specimens the steel loading pins were positioned and the load cell placed between the pins and the hydraulic jack. The loading pins were located to the desired nominal eccentricity. For the concentric columns (specimens 2MF0-YYYZ) the loading pins were removed from the test set-up.

Six millimetre diameter lugs were welded to the outside of the longitudinal bars, on one face of each specimen, before the concrete was placed. Fifteen millimetre square polystyrene block-outs were placed around the lugs to be removed after the hardening of the concrete. The lugs were spaced at 100 mm each side of the mid-length of the specimen (as shown in Figure 2) giving a gauge length of 200 mm. Strains and curvatures at the column mid-length were calculated from the change in displacements measured between the lugs using LVDT's. A LVDT was also placed at the mid-length of the columns to measure the lateral deflection. Selected columns had strain gauges attached to the 4 legs of the outer tie and the 4 legs of the inner tie located at the mid-length of the column. Loads were applied using a 6000 kN capacity hydraulic jack and measured using a 5000 kN capacity electronic load cell. Figure 2 shows details of the instrumentation used.

The specimens were tested horizontally in a stiff testing frame with the lateral displacement closely monitored. An initial load of approximately 100 kN was applied and the external packing holding the specimen in place was removed and the gauges zeroed.

The true initial loading eccentricities were calculated after the completion of the test and were determined from the measured response of LVDT's 1 and 2 for the elastic stress conditions (refer Table 3). The second order eccentricity, caused by the lateral displacement of the specimens, were measured using LVDT 3 (refer Figure 2) and the total eccentricity given in Table 3 is taken as the sum of the calculated initial eccentricity plus the lateral displacement of the columns.

The loading was applied continuously at a slow rate with each test taking between 45 minutes and 90 minutes to complete. The stiffness of the testing rig enabled the descending curves to be obtained for all the specimens tested.

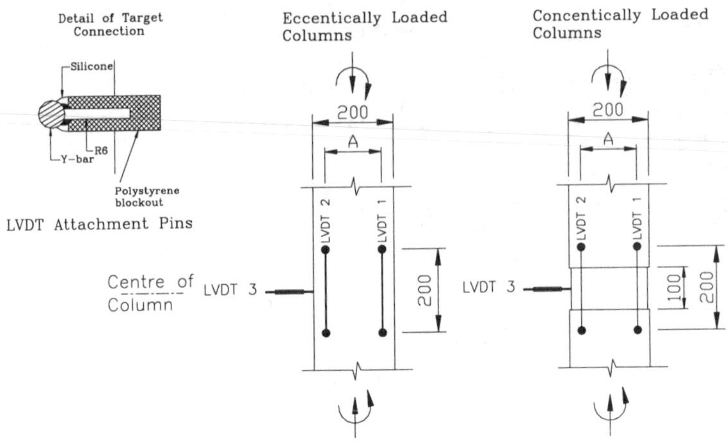

Figure 2. Arrangement of testing instrumentation.

3 ANALYSIS OF TEST RESULTS

All specimens tested failed in the gauged region with the exception of specimen 2MF30-100Ss, which failed adjacent to the end haunch. The results of the tests are given in Table 3 and are plotted in Figure 3 with the interaction diagram proposed by Foster (1998) for high strength concrete columns.

The squash load (N_{uo}) shown in Figure 3 is calculated by

$$N_{uo} = k_3 k_4 A_c + A_{st} f_{sy} \tag{1}$$

where A_c is the area of the concrete, A_{st} and f_{sy} the area and yield strength of longitudinal steel, respectively, k_3 is the in-situ strength factor and k_4 a cover spalling factor. In Figure 3 the in-situ strength factor was taken as k_3=0.9.

In all of the specimens tested the cover remained intact throughout the test, well beyond the peak load. The level of ductility as represented by the I_{10} ductility factors (as given by Foster and Attard, 1997) are given in Table 3. The I_{10} was measured on the load versus strain parameter curves, where the strain parameter is given by

$$\xi = \varepsilon_{av} + \kappa e \tag{2}$$

and where ε_{av} is the average strain across the section, κ is the curvature across the section and e is the total loading eccentricity. A value of I_{10}=1 represents a perfectly elastic-brittle response and I_{10}=10 a perfectly elastic-plastic response. The I_{10} values obtained for the 12 columns tested in this study are compared in Figure 4 with those for columns tested by Razvi and Saatcioglu (1996) and with the design equation proposed by Foster (1999b). In Figure 4, the effective confinement parameter is given by

$$k_e \rho_s f_{yt} / f_c' \tag{3}$$

where k_e is a confinement effectiveness factor (refer Foster et al., 1998), ρ_s is the volumetric ratio of the ties, f_{ty} is the yield strength of the ties and f'_c the concrete cylinder strength.

Figure 5 shows a comparison for the load versus axial strain results for column 2MF0-100G with the results of a 60 MPa 250-mm square column tested by Razvi and Saatcioglu (1996), column CS-24. The fibre column gave a similar performance to the Razvi and Saatcioglu specimen but with an effective confinement parameter less than 1/2 of that of the Razvi and Saatcioglu specimen. Figures 4 and 5 show that the fibre reinforced HSC columns have a greatly superior performance than columns cast with HSC without fibres in the concrete mix.

Table 3. Failure loads and loading eccentricities

Specimen	Maximum Axial Load (kN)	Initial eccentricity (mm) *	Total Eccentricity at the Peak Load (mm) #	Moment at the Peak Axial Load (kNm)	I_{10}
2MF0-50S	2612	4.9	7.4	19	8.0
2MF0-100S	2367	7.0	12	28	9.3
2MF5-50S	2534	8.0	9.9	25	9.6
2MF5-100S	2184	8.0	11	24	9.6
2MF30-50S	2123	24	33	71	8.4
2MF30-100S	1998	30	36	72	-
2MF0-50G	2356	9.5	12	27	8.7
2MF0-100G	2188	3.3	7.8	17	9.5
2MF10-50G	2450	7.6	10	25	9.0
2MF10-100G	2469	10	14	34	9.0
2MF20-50G	2058	21	32	66	8.4
2MF20-100G	2262	14	20	45	8.4

Notes: * Calculated from the initial response of the specimen under elastic stress conditions.
Initial eccentricity plus deflection at the mid-length of the column.

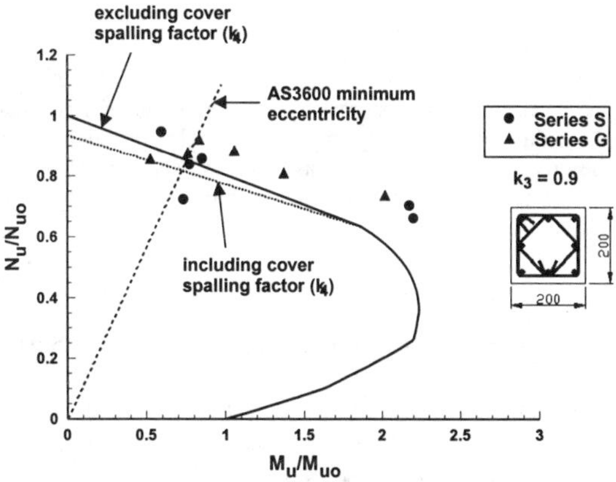

Figure 3. Comparison of the experimental data with the interaction diagram of Foster (1998).

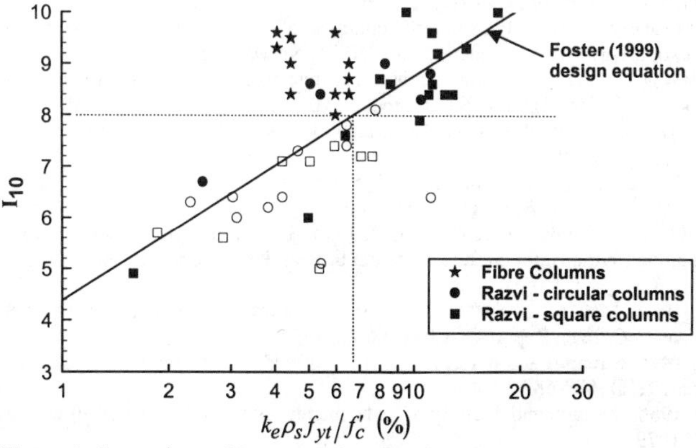

Figure 4. Comparison of I_{10} versus the effective confinement parameter for the fibre columns with HSC columns tested by Razvi and Saatcioglu (1996) and with the design equation of Foster (1999b).

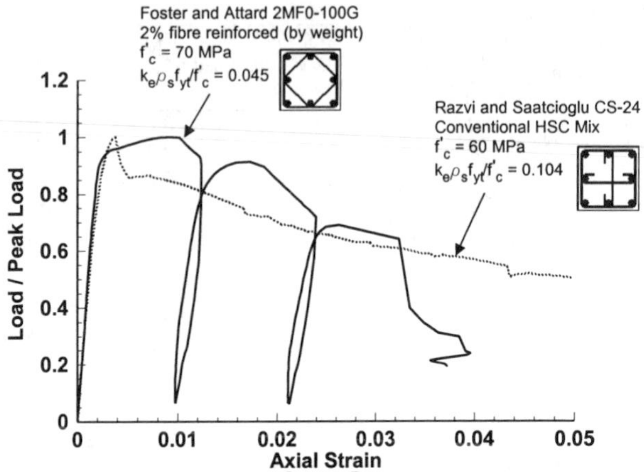

Figure 5. Comparison of load versus axial strain for column 2MF0-100G with that of Razvi and Saatcioglu (1997) specimen CS-24.

4 CONCLUSIONS

In this study 12 HSC columns cast with fibre reinforced concrete were tested in eccentric compression. Results of the tests showed that when at least two percent (by weight) of steel fibres are introduced into the concrete mix the cover does not spall away from the section. The columns showed a superior performance to comparable specimens cast without fibres in the mix design particularly for post failure ductility.

ACKNOWLEDGEMENTS

The research described in this paper was funded via an Australian Research Council Small Grant 1999. The ARC support is gratefully acknowledged.

REFERENCES

Cusson, D. & Paultre, P. 1994. High strength concrete columns confined by rectangular ties. Journal of Structural Engineering, *ASCE*, 120(3): 783-804.

Foster S.J. 1998. Design of HSC columns for strength. International Conference on High Performance High Strength Concrete. *Curtin University of Technology*, Perth, Western Australia: 409-423.

Foster, S.J. 1999a. The falacy of early cover spalling in HSC columns. 5th International Symposium on Utilization of High Strength/High Performance Concrete, Sandefjord, Norway: 282-291.

Foster, S.J. 1999b. Peak and post peak behaviour of high strength concrete columns. *Concrete Institute of Australia,* Proceedings, Concrete 99: Our Concrete Environment: 500-511.

Foster, S.J. & Attard, M.M. 1997. Experimental tests on eccentrically loaded high strength concrete columns, *ACI*, Structural Journal, 94(3): 2295-2303.

Foster, S.J., Liu, J. & Sheikh, S.A. 1998. Cover spalling in HSC columns loaded in concentric compression. Journal of Structural Engineering, *ASCE*, 124(12): 1431-1437.

Paultre, P., Khayat, K.H., Langlois, A., Trudel, A. & Cusson, D. 1996. Structural performance of some special concretes. 4th International Symposium on the Utilization of High Performance Concrete, Paris, France: 787-796.

Razvi, S.R. & Saatcioglu, M. 1996. Tests of high strength concrete columns under concentric loading. Dept. Of Civil Eng., *University of Ottawa*, Report OCEERC 96-03: 147 pp.

Razvi, S.R. & Saatcioglu, M. 1994. Strength and deformability of confined high-strength concrete columns. *ACI*, Structural Journal, 91(6): 678-687.

Sheikh, S.A. & Uzumeri, S.M. 1980. Strength and ductility of tied concrete columns. Journal of Structural Engineering, *ASCE*, 106(5): 1079-1102.

Yong Y.K., Nour, N.G. & Nawy, E. 1988. Behaviour of laterally confined high strength concrete under axial loads. Journal of Structural Division, *ASCE*, 114(2): 332-351.

Steel structures

Mechanics of Structures and Materials, Bradford, Bridge & Foster (eds)
© 1999 Balkema, Rotterdam, ISBN 90 5809 107 4

Finite element modelling of baseplates for steel hollow sections subject to axial tension

Y. Zhuge
School of Engineering, University of South Australia, Adelaide, S.A., Australia

ABSTRACT: There is very little design guidance available on base plate connections subject to axial tension. The current design method is basically adaptation of the models developed for hot-rolled I beam sections. In order to develop a suitable design method for base plate connections joining steel hollow sections under tension force, an investigation project is being carried out. At the first stage of the project, a series of tests have been conducted and it has been found that current design recommendations are non-conservative for the bolt configuration most commonly used in practice. In this paper, a nonlinear finite element model has been developed to further study the failure mechanisms and related design theory. The model has been calibrated by comparison of the results with those obtained from the experiments in stage one.

1 INTRODUCTION

Column baseplates, which are commonly used to transfer the load from a steel column to the concrete foundations, are very important structural elements. The applications of such structures can be found in low-rise conventional construction and in pre-engineered metal building applications. In the past, the attention of many researchers was focused on the problem of end plate connection joining I-section members. Research on baseplate connections joining rectangular or square hollow sections has been limited. Further more, research on such hollow sections subjected to axial tension has been even rare, partly because uplift is not a critical design consideration in many regions of the world. However, in Australia, wind loads are frequently governing the design of those lightly loaded flexible baseplate connections joining steel hollow sections. Therefore, there is a potential need for extensive research into the behaviour of baseplate connections for steel hollow sections under axial tension loads.

The current design method for steel hollow sections is basically an adaptation of the models developed for hot-rolled I beam sections, which rely either on classical plate bending solutions or yield line analysis (Syam & Chapman 1996). In order to develop a suitable design method for baseplate connections joining steel hollow sections under tension force, an investigation project is being carried out at the Department of Civil Engineering, University of South Australia. At the first stage, a series of tests have been conducted to study the failure pattern and the ultimate load of baseplates for square steel hollow sections (SHS) subjected to axial tension with four bolts. It has been found from the testing results that the current design recommendations are non-conservative for the bolt configuration most commonly used in practice (Mills et al. 1997). It is suggested that the yield line loads were significantly lower than those predicted by the existing yield line theory, and the difference was traced to an incompatibility between the failure mechanisms of I-sections and square hollow sections.

At the second stage of the project, a nonlinear finite element model has been developed to further study the failure mechanisms and related design theory for square hollow sections subjected to axial tension loading. In this paper, the development of this nonlinear finite element model, the predicated yield line patterns and the effect of bolt configurations are discussed. The model has been calibrated by comparison of the results with those obtained from the experiments in stage one. Good agreements have been found for both failure patterns of the base plate and the ultimate loads.

2 EXISTING DESIGN THEORY FOR SHS UNDER AXIAL TENSION

There is very little design guidance available on base plate connections subject to axial tension (Syam & Chapman 1996). The current design model is based on yield line analysis method proposed by Murray (1983), which was derived from the testing of hot-rolled I-section base plate connections with a two bolts configuration. The yield line pattern suggested by Murray's analysis is shown in Figure 1, comprising three lines radiating from the centre of the web of the I-section, one line perpendicular to the web and two line at an angle. Based on such yield line pattern, the plate thickness could be calculated by Equations 1or 2 (Murray 1983):

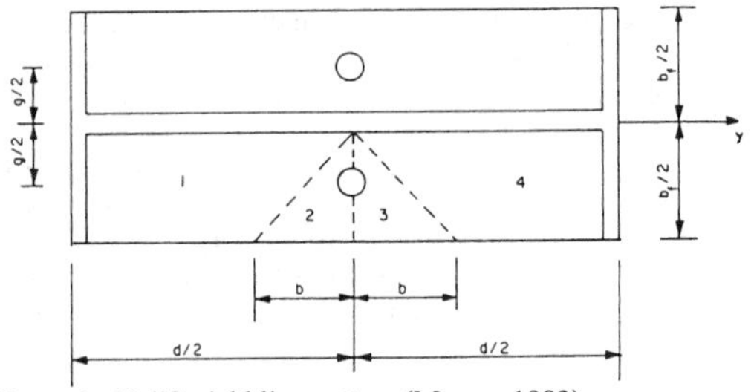

Figure 1. Uplift yield line pattern (Murray 1983)

$$t = \sqrt{\frac{\sqrt{2}P_u g}{4b_f F_y}} \; for \;\; \sqrt{2}b_f \leq d \tag{1}$$

and

$$t = \sqrt{\frac{P_u g d}{F_y (d^2 + 2b_f^2)}} \; for \;\; \sqrt{2}b_f > d \tag{2}$$

where P_u = the total uplift force on the base plate; g = the anchor bolt gage; F_y = yield stress of base plate component; and t = base plate thickness.

Equations 1 and 2 have been slightly modified and extended to four bolts configurations by Hogan & Thomas (1994):

$$\phi N_s = \frac{\phi 4 b_{fo} f_{yp} t_p^2}{\sqrt{2} s_g} \times \frac{n_b}{2} \; for \;\; \sqrt{2}b_{fo} \leq d_c \tag{3}$$

$$\phi N_s = \frac{\phi f_{yp} (d_c^2 + 2b_{fo}^2) t_p^2}{s_g d_c} \times \frac{n_b}{2} \; for \;\; \sqrt{2}b_{fo} > d_c \tag{4}$$

where ϕN_s = design capacity of steel base plate (ϕ = 0.9); $b_{fo} = b_p - b$ (as defined in Fig. 2); f_{yp} = yield stress of component; t_p = base plate thickness; nb = number of anchor bolts; d_c, s_g = as defined in Figure 2.

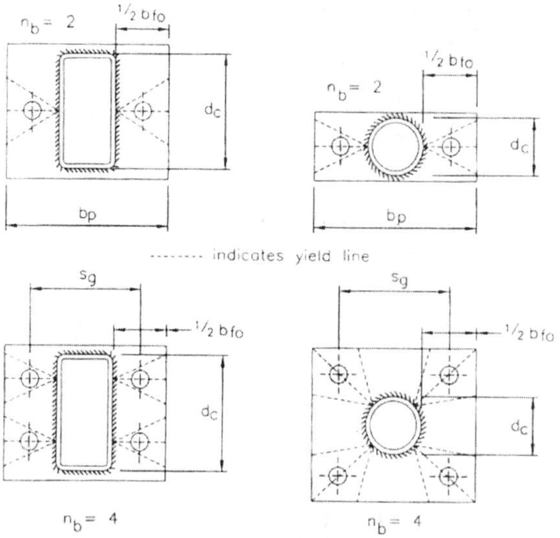

Figure 2. Component design model for axial tension (Syam & Chapman 1996)

It is also suggested that Equations 3 and 4 could be applied to channels, rectangular, square and circular hollow sections based on the assumption that the same failure mechanism might be applicable. However, this assumption needs to be validated by testing (Hogan & Thomas 1994).

3 EXPERIMENTAL RESULTS FOR SQUARE HOLLOW SECTIONS

In order to develop a suitable design model for base plate connections joining cold-formed steel hollow section, a series of tests have been carried out with different bolt configurations. The main objectives of these testing are to determine the yield line patterns and modes of failure. The details of these testing programs have been discussed by Mills et al. (1997).

For all of the testing performed, the column was welded all round to the base plate with an 8 mm fillet weld. The bolt arrangement and observed yield line pattern of the first series of testing is shown in Figure 3. The values of average testing yield loads were compared with the unfactored design recommended loads, in accordance with the current design model (Syam & Chapman 1996). It has been found from such comparison that the design recommendations are conservative. The testing yield loads are about three times higher than the design recommended loads.

The bolt arrangement and observed yield line pattern of the second series of testing is shown in Figure 4. This configuration where bolts located at the corners of the base plate, is found more commonly used in industry practice. Again, the values of average testing yield loads were compared with the unfactored design recommended loads (Syam & Chapman 1996). However, the comparison indicated that the design recommendations are non-conservative. The testing yield loads are only about half of the design recommended values.

It can be concluded from the experimental results that the current design recommendations for SHS base plate connections are questionable. Equations 3 and 4 give either very conservative or unacceptable non-conservative results of design yield load depending on the bolt configurations. The current design model is based on the failure pattern of hot-rolled I section baseplate (Fig.1). However, as indicated in Figure 4, the yield line pattern of bolt configuration 2 for SHS baseplate is quite different to I section baseplate. It is thus evident that new design models are required for hollow section baseplates under axial tension.

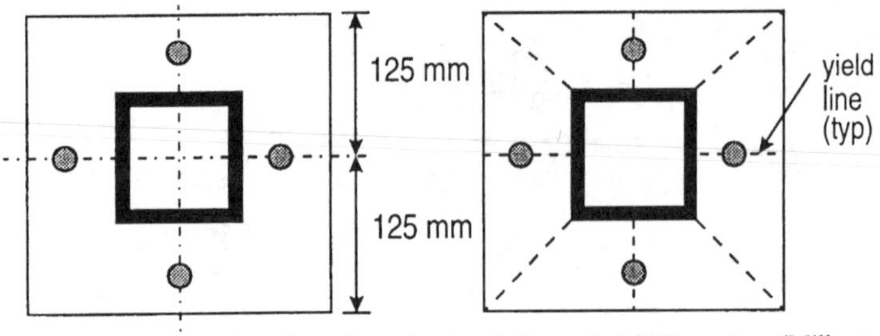

Figure 3. SHS base plate bolt configuration 1 and observed yield line pattern (Mills et al. 1997)

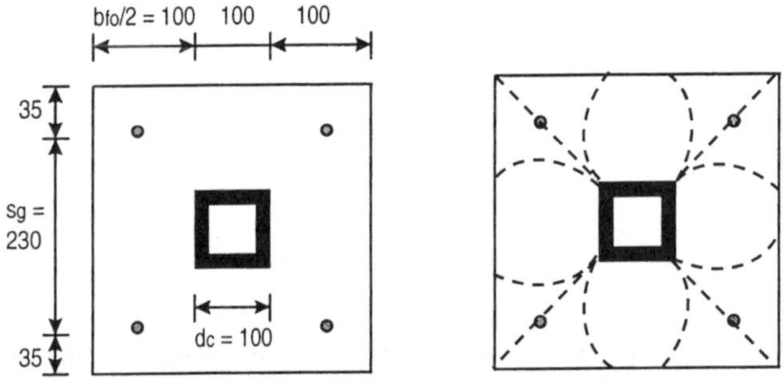

Figure 4. SHS base plate bolt configuration 2 and observed yield line pattern

4 FINITE ELEMENT MODELLING

Preliminary numerical results for the baseplates welded to the square hollow section are contained in this section. The finite element model has been verified by comparing the results with those obtained from existing tests. However, the primary objective of this paper is to assess the performance of the baseplates with two different bolt configurations as discussed earlier in this paper.

In recent years, a number of powerful finite element software packages have become commercially available. They incorporate facilities that enable a wide range of engineering problems to be efficiently modelled. The finite element analysis of baseplate connections discussed in this paper, has been carried out by using one of these packages - Strand 6. The behaviour of the above discussed baseplate connections is three-dimensional. Therefore, 3D solid elements with 16 nodes per element are used for the modelling. On the basis of the geometrical and load conditions of the specimens, the baseplate is doubly symmetric. Therefore, only a quarter of the full plate has been selected for the analysis. The movements of the bolts are ignored in the linear analysis and a perfect connection between the upper surface of the baseplate and the bolt head is assumed.

To simulate the stiffness of the area of the baseplate beneath the square hollow section, the nodes in this region were restrained from any rotation. Symmetrical boundary conditions were also imposed on the finite element model. The experimental results had clearly indicated that as loading occurred the edges of the baseplate heaved to form a smooth curve whose vertex was located at the midpoints between each adjacent bolt. It is therefore suggested that no rotation was occurring in the plane of the curve as well as horizontal displacement. The boundary conditions for bolt configuration 2 are illustrated in Figure 5. It can be applied to bolt configuration 1 as well.

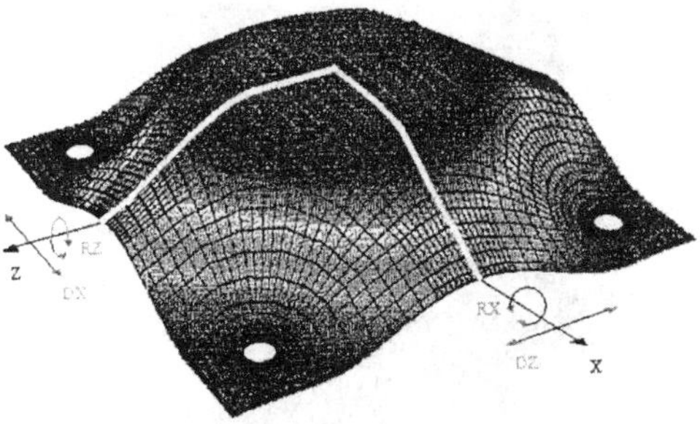

Figure 5. Boundary conditions of the baseplate.

The experimental and analytical load-displacement curves for bolt configuration 2 are shown in Figure 6. There is a good agreement between the experimental and analytical results in the linear elastic range. However, there is some discrepancy in the nonlinear range of the curve. This may be attributed to the simplified model of the bolts. A refined model which considering the bolt effects is currently being developed.

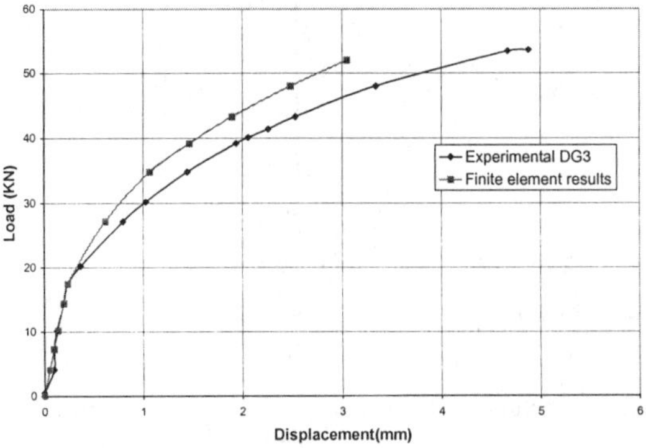

Figure 6. Load-displacement curve of the baseplate

The deformation shapes of the plates with two different bolt configurations have been shown in Figure 7. It can be seen from this figure that the bending deformations are quite different between two bolt configurations. Under the same loading condition, it is indicated from the finite element results that the maximum displacement of bolt configuration 2 is four times greater than bolt configuration 1 which agree with the experimental results. The predicted yield line patterns of the baseplate based on finite element model are presented in Figure 8. These yield line patterns are quite close to those observed from testing (see Fig. 4).

5 CONCLUSIONS

In this paper preliminary experimental and finite element results for the baseplate connections for steel hollow section subject to axial tension have been presented. It has been found from both experimental and finite element results that the current design model, which is based on I beam sections, is not suitable for square hollow sections due to different yield line patterns. The finite element analysis indicated that the bolt configurations play in important role in

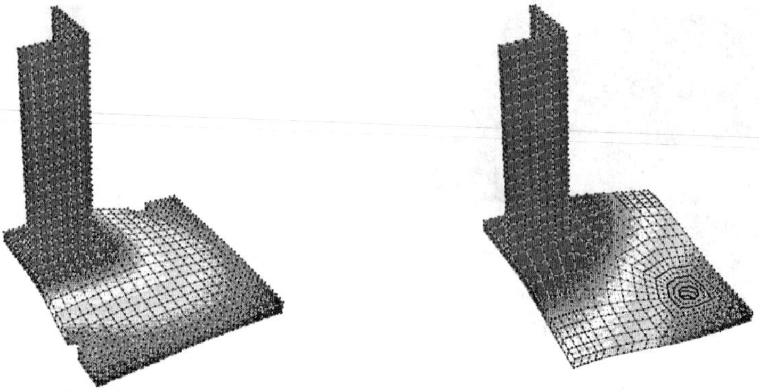

(a) bolt configuration 1 (b) bolt configuration 2

Figure 7 Deformed shape of the finite element model for the baseplate

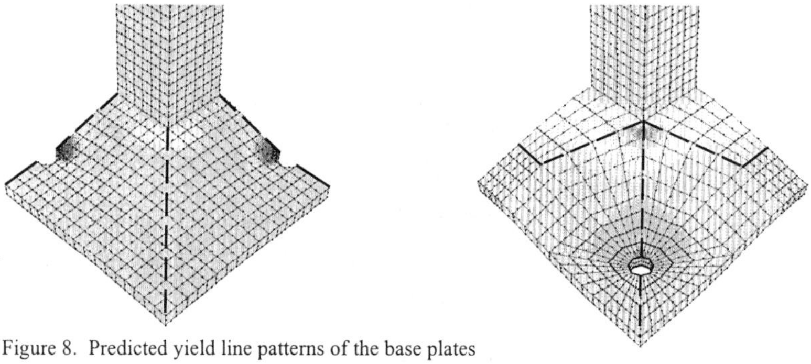

Figure 8. Predicted yield line patterns of the base plates

determining the yield load of baseplate connections. The next step of the research will be the development of a new design model.

ACKNOWLEDGMENTS

The author would like to thank three honour students of the University of South Australia, Mr David Kennedy, Mr Chin Yap and Mr Doan Chau for their assistance with the project.

REFERENCES

Hogan, T.J. & Thomas, I.R. (4th ed.) 1994. Design of structural connections. Sydney: Australian Institute of Steel Construction.

Mills, J.E., Barone, E.T. & Graham, P.J. 1997. Baseplate for structural steel hollow sections subject to axial tension. In R. Grzebieta, R. Al-Mahaidi & J. Wilson (ed.), *Proc. Fifteenth Australiasian conf. mechanics of structures and materials, Melbourne, 8-10 December 1997*. 547-552. Rotterdam: Balkema.

Murray, T.M. 1983. Design of lightly loaded steel column base plates. *Engineering Journal, American Institute of Steel Construction*. 20-4: 143-152.

Syam, A.A. & Chapman, B.G. 1996. Design of structural steel hollow section connections. Sydney: Australian Institute of Steel construction.

Mechanics of Structures and Materials, Bradford, Bridge & Foster (eds)
© 1999 Balkema, Rotterdam, ISBN 90 5809 107 4

Stronger legs for steel towers

M.P. Rajakaruna
School of Engineering, University of South Australia, Adelaide, S.A., Australia

ABSTRACT: Lattice towers made of hot rolled steel angle sections are used to support transmission lines and radio antenna. The legs of these towers are usually strengthened by retrofitting extra parallel bracing members. This paper presents a numerical method for predicting the failure load of single and compound angle sections used in steel towers considering second order and inelasticity effects. Predictions from the method are compared with limited laboratory results.

1 INTRODUCTION

Hot rolled equal angles are widely used in the legs of latticed steel towers, which support transmission lines and radio antenna. In some instances, these towers need to be strengthened to resist increased loads from present and future demands. The most common method of increasing the capacity of a tower is to retrofit extra parallel bracing members of the same cross section to the main compression member (Figure 1). This investigation explores the strength enhancement of compression members by parallel bracing.

2 ANGLES AS COMPRESSION MEMBERS

Although simple to fabricate and erect, angle sections are relatively difficult to analyse. The principal axes of angles do not usually coincide with the direction of imposed loading and the shear centre is not coincident with the centroid causing biaxial bending and twisting of the cross section.

Single or compound angles loaded in compression fail either by flexural buckling about the weaker principal axis; torsional buckling about the shear centre; or by flexural-torsional buckling. Thus, there are three critical values of axial load to be considered of which only the lowest value is of practical interest.

2.1 *Single angles as compression members*

Single equal angles are often used as the main compression members in steel towers. A single equal angle has one axis of symmetry while the centroid and the shear centre do not coincide. An axially loaded single angle member may buckle flexurally about its minor principal axis or by flexural – torsional buckling on a plane perpendicular to the plane of symmetry.

Thus for a single angle, the critical buckling load is the lesser of P_y or P_{cr} obtained from:

$$P_y = \frac{\pi^2 EI_y}{L_y^{\,2}} \tag{1}$$

$$C_w = \frac{b^3 h^2 t}{24}$$

Figure 1 Close up of a tower with parallel braced leg members

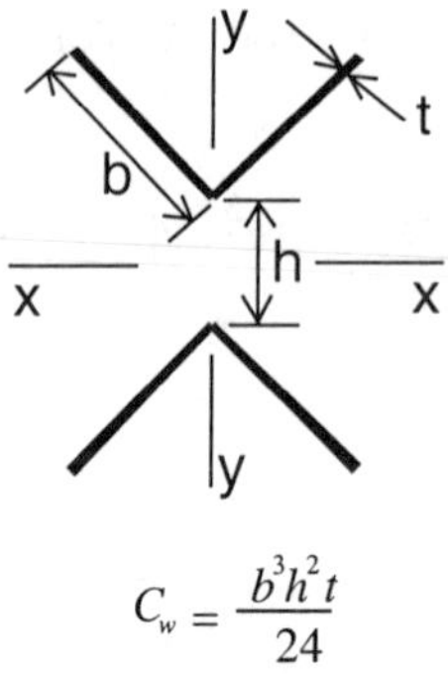

Figure 2 Warping constant of a compound angle

$$\frac{I_{pc}}{I_{ps}} P_{cr}^2 - (P_x + P_t) P_{cr} + P_x P_t = 0 \tag{2}$$

$$P_t = \frac{A}{I_{ps}} \left[GJ + \frac{\pi^2 E C_w}{L_t^2} \right] \tag{3}$$

where P_x and P_y are flexural buckling loads about the major and minor principal axes; P_t and P_{cr} are the critical loads for torsional buckling and flexural torsional buckling; I_{ps} and I_{pc} are polar moments of inertia about the shear centre and centroid respectively. For an equal angle $I_{ps}/I_{pc} = 1.6$ (Timoshenko and Gere 1961).

2.2 Compound angles as compression members

An axially loaded single angle compression member strengthened by retrofitting with a similar angle behaves as an eccentrically loaded doubly symmetric section. The eccentricity of the load (e_y) is about the (major) x-axis. Flexural buckling of the member about the x-axis would occur independently while flexural buckling about the y axis and torsional buckling are coupled.

The buckling load parallel braced compound angle can be estimated as the lesser of P_x and P_{cr} as follows:

$$P_x = \frac{\pi^2 E I_x}{L_x^2} \tag{4}$$

$$P_{cr}^2 \left(1 - \frac{A e_y^2}{I_{ps}} \right) - P_{cr} (P_y + P_t) + P_y P_t = 0 \tag{5}$$

The warping constant (C_w in equation 3) of angles can be found in literature (Madugula and Kennedy 1985); although it is small enough to be neglected for thin walled sections (Trahair 1993). However the warping resistance of a compound angle (Figure 2), is substantial and hence the warping constant is calculated using the method of Timoshenko and Gere (1961).

3 INELASTIC BUCKLING LOAD OF ANGLE MEMBERS

The buckling stress of a member may exceed the proportional limit of the material causing the member to buckle in the inelastic range. The preceding equations derived for elastic buckling can be readily adopted for inelastic buckling by replacing the elastic moduli E and G with the corresponding tangent moduli E_t and G_t (Madugula and Kennedy 1985).

According to the tangent modulus theory, instantaneous values of E_t and G_t are effective when bending and torsional deformations occur during buckling. For inelastic flexural buckling the tangent modulus of elasticity E_t can be expressed by the elastic modulus E and the ratio τ as $E_t = E\tau$ where $\tau = 1$ in the elastic range and $\tau < 1$ when the elastic limit is exceeded.

For torsional and flexural-torsional buckling, there is no consensus on the value of shear modulus to be used in the inelastic range. Bleich (1952) has shown from the assumptions made in the theory of local buckling of plate elements of columns that the tangent modulus G_t and the elastic modulus of rigidity G are related by $G_t = G\sqrt{\tau}$.

3.1 *Newmark's Method*

The buckling loads of columns can be determined by either an eigenvalue approach or a load-deflection (stability) approach. Of the two methods, the latter is appropriate for a geometrically imperfect inelastic system.

Newmark's (1943) method was used in this investigation to obtain the elastic and inelastic buckling load of an imperfect column. It was assumed that failure is always caused by excessive bending in the plane of loading and lateral torsional buckling is prevented. This numerical iterative procedure uses successive approximations to obtain the deflected shape of a member under a given axial load. The member was assumed to have an initial crookedness of 0.001 times the effective length and the initial deflected shape was assumed sinusoidal.

The moment-curvature-thrust (m-ϕ-p) relationship for the section for inelastic bending has to be determined in advance for this method to be used to predict the inelastic buckling load. The elastic plastic behaviour is approximated using the three equations (Figure 3) proposed by Chen (1970). Once the curvatures at all the nodes along the member are known the deflection of the member is determined by numerically integrating the curvature and slope (Chen & Lui 1987). If the calculated deflection is comparable to the assumed deflection, a stable configuration is found. If not, the calculated deflection is used as the assumed deflection and the calculations are repeated until convergence is achieved. When the calculated deflections begin to diverge rather than converge, the column is unstable and the buckling load has been exceeded.

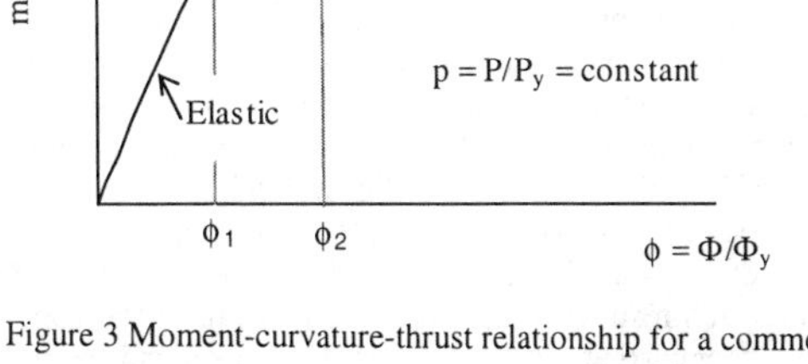

In the elastic regime

$$m = a\varphi$$

In the primary plastic regime

$$m = b - \frac{c}{\varphi^{1/2}}$$

In the secondary plastic regime

$$m = m_{pc} - \frac{f}{\varphi^2}$$

a, b, c, and f are arbitrary constants

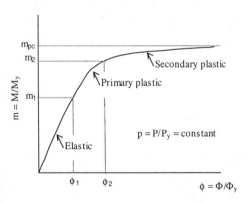

Figure 3 Moment-curvature-thrust relationship for a common structural section

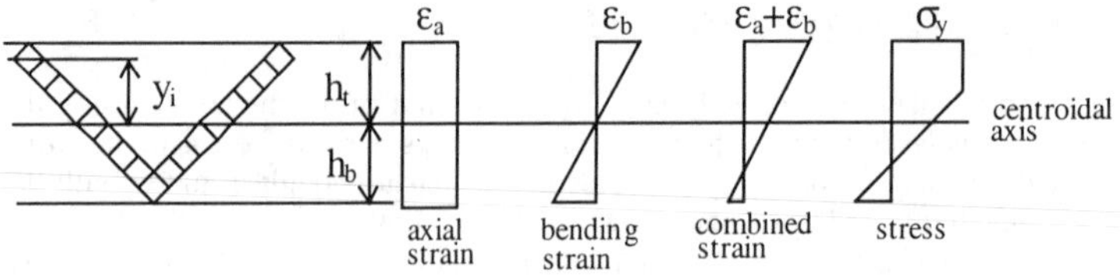

Figure 4 Stress distribution in discretisised cross section

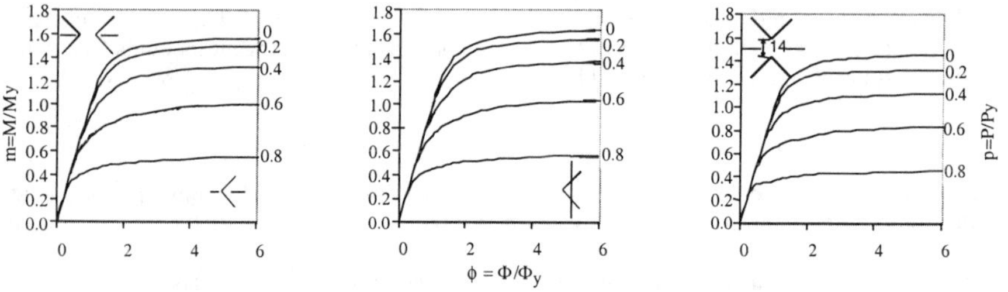

Figure 5 Moment-curvature-thrust relationships for 75x75x5 single and compound angles

3.2 *Moment-curvature-thrust relationship*

Moment-curvature-thrust relationships for a single and compound angle were generated using an approach similar to that adopted by Chen (1970). The angle sections were approximated by rectangles divided into a number of smaller elements as shown in Figure 4.

Once an axial compressive strain (ε_a) and a bending strain at the topmost fibre of the section (ε_b) are specified, the total strain distribution for the entire cross section ($\varepsilon_b\, y_i/ h_t$) can be calculated. Residual stresses are not taken into consideration in this analysis. The stress in any element is the product of the strain in that element and the modulus of elasticity. This value should not exceed the yield stress of the material. The axial force in the section is the sum of the products of stress and area for each element.

The axial force can be adjusted to an arbitrary value (say a proportion of the yield load) by changing the assumed values of axial and bending strains. The bending moment about the centroidal axis is the sum of the products of force and area for each element. The curvature is determined from the assumed strain distribution within the cross section.

The above procedure is readily implemented on a spreadsheet. These relationships expressed in non-dimensional form are shown in Figure 5.

4 LABORATORY TESTS

An experimental testing program was carried out to compare the failure loads with those predicted by the numerical method. A single hot rolled 75x75x5 angle was axially loaded as a pin ended strut over a length of 1.5m and the failure load and the mode of failure was observed. This was compared to the same member strengthened with an identical angle as parallel bracing over the same length and having the same end conditions. The parallel bracing member was attached with flange clamps (Figure 7) at mid height and close to the ends. This method of fixing separated the two members by a distance of 14mm. The strengthened member was loaded at the same position as the original member, which resulted in an eccentric loading of the combined section about its major axis. Five tests were carried out on each configuration and the results are shown in Table 1.

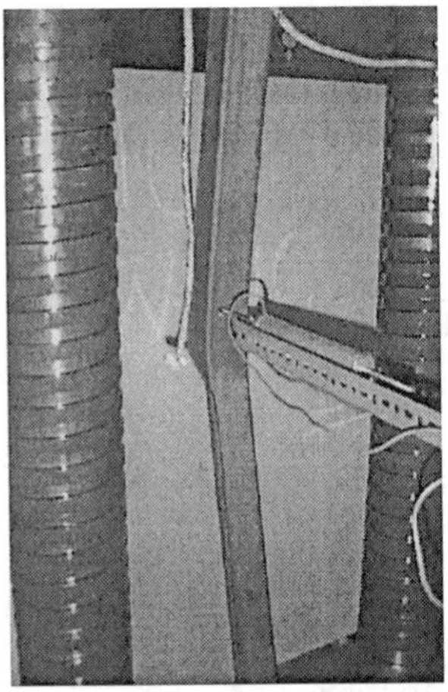

Figure 6 Buckling of a single angle

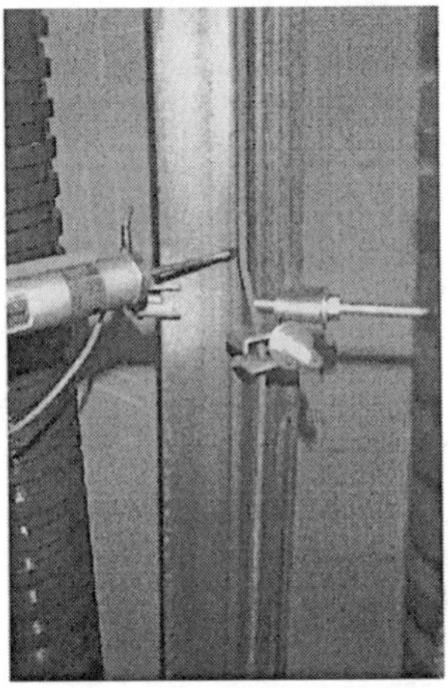

Figure 7 Buckling of a compound angle

Table 1 Comparison between numerical and experimental results

Member	Buckling load (kN) Predicted	Observed (Range)	Buckling mode Predicted	Observed
Single angle 75x75x5	102	114 (108-119)	flexural-torsional	flexural-torsional
Compound angle 2 x (75x75x5)	148	177 (173-197)	flexural-torsional	flexural-torsional

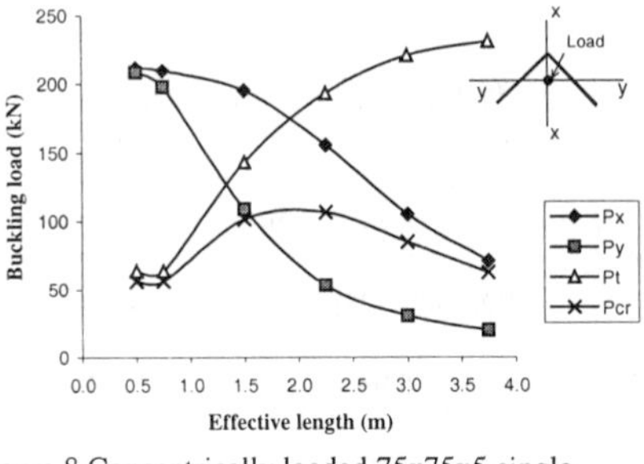

Figure 8 Concentrically loaded 75x75x5 single angle

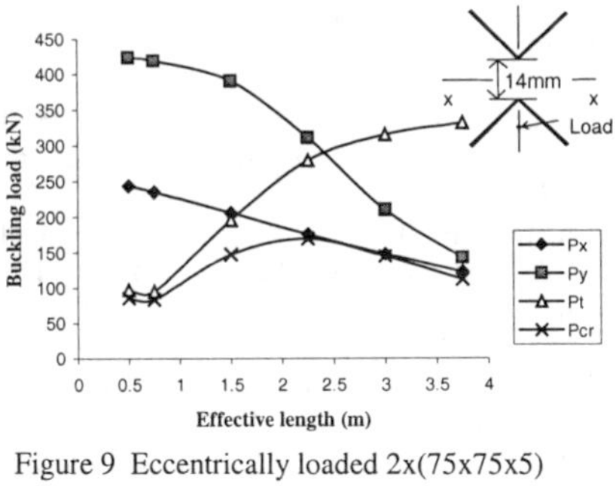

Figure 9 Eccentrically loaded 2x(75x75x5) compound angle

5 RESULTS

The inelastic flexural buckling loads P_x and P_y were obtained by the Newmark's method as described in section 3.1 for both the single and compound angles. The inelastic torsional buckling load P_t was computed using equation (3) with tangent moduli E_t and G_t. The slope of the m vs ϕ curve for the member just before buckling is the ratio $\tau = E_t/E$. This value of τ was used to compute the shear modulus G_t in the inelastic range. For a single angle, flexural buckling about the x axis and torsional buckling are coupled and the ratio τ corresponds to the inelastic flexural buckling load about the x-axis. For a compound angle flexural buckling about the y axis and torsional buckling are coupled and the ratio τ corresponds to the inelastic flexural buckling load about the y-axis.

The flexural torsional buckling loads P_{cr} for single and compound angles were computed with equations (2) and (5) respectively. The buckling load for a single angle is the lesser of P_y and P_{cr} while the buckling load of a compound angle is the lesser of P_x and P_{cr}. A comparison between the numerically determined values and experimental results is made in Table 1.

6 DISCUSSION

Figure 8 shows that an axially loaded 75x75x5 single angle is likely to buckle flexurally about the weaker axis when the effective length is greater than 1.55m. At lengths less than this the failure probably is by flexural torsional buckling.

Figure 9 shows the same angle strengthened by parallel bracing with an identical section would fail by flexural-torsional buckling in most instances except when the effective length is between 2.33 and 2.75m when either flexural buckling or flexural torsional buckling are equally likely.

Results from a limited number of laboratory tests show good agreement with the numerical model. The numerical model accurately predicted flexural torsional buckling for both the single angle and the compound angle at an effective length of 1.5m with the predicted failure loads being marginally lower than the experimental values. Further laboratory testing is needed over a range of effective lengths to confirm the validity of the numerical model for general use.

7 ACKNOWLEDGEMENTS

The assistance of students Paul Haywood and Kieren Chappell in carrying out the tests at the School of Civil Engineering Laboratories of the University of South Australia and Graham Brown of O'Donnell Griffin for his assistance in providing material is gratefully acknowledged.

REFERENCES

Bleich, F., 1952, Buckling Strength of Metal Structures, McGraw Hill Book Company Inc.

Chen, W.F., 1970, General Solution of Inelastic Beam-Column Problem, Proceedings of the American Society of Civil Engineers, August 1970, EM4, pp 421-441.

Chen, W.F., 1971, Further Studies of Inelastic Beam-Column Problem, Proceedings of the American Society of Civil Engineers, February 1971, ST2, pp 529-544.

Chen, W.F., and Lui, E.M.,1987, Structural Stability, Theory and Implementation, P T R Prentice Hall.

Madugula, M.K.S. and Kennedy, J. B., 1985, Single and Compound Angle Members Structural Analysis and Design, Elsevier Applied Science Publishers Ltd.

Newmark, N.M., 1943, Numerical Procedure for Computing Deflections, Moments and Buckling Loads, Transactions, American Society of Civil Engineers, 1943, pp 1161-1188.

Timoshenko, S.P. and Gere, J.M., 1961, Theory of Elastic Stability, McGraw Hill Book Company Inc.

Trahair, N.S., 1993 Flexural Torsional Buckling of Structures, E & FN Spon.

Mechanics of Structures and Materials, Bradford, Bridge & Foster (eds)
© 1999 Balkema, Rotterdam, ISBN 90 5809 107 4

Shear-lag phenomenon in framed tube structures with multiple internal tubes

K.K.Lee, H.Guan & Y.C.Loo
School of Engineering, Griffith University, Southport, Qld, Australia

ABSTRACT: A simple numerical modelling technique is proposed for estimating the shear-lag effects of tube(s)-in-tube structures. The proposed method idealises the framed-tube structures with multiple internal tubes as equivalent multiple tubes (i.e. box beams), each composed of four equivalent orthotropic plate panels. Hence, the tube(s)-in-tube system can be analysed using an analogy approach where each tube is individually modelled by a box beam that can account for the flexural and shear deformations as well as the shear-lag effects. The numerical analysis of shear-lag is based on the minimum potential energy principle in conjunction with the variational approach. In the paper, the additional bending stresses and the shear-lag reversal points are also investigated to estimate the shear-lag behaviour of such a system.

1 INTRODUCTION

Modern highrise buildings of the framed-tube system exhibit a considerable degree of shear-lag with consequential loss of cantilever efficiency. Despite this drawback, the framed-tube structures are accepted as an economical system capable of maximising the structural efficiency for highrise buildings over a wide range of building heights. In particular, the framed-tube structures with multiple internal tubes are widely used due to their high stiffness in resisting lateral loads. In addition, this type of structures shows a reduced shear-lag due to the existence of the internal tubes, the columns of which also participate more effectively in resisting lateral forces.

It has been noted that existing analysis models not only ignore the contribution of the internal tubes to the overall lateral stiffness but also neglect the negative shear-lag effects in the tubes. Thus, these models can cater only for the analysis of the external tube but fail to consider the shear-lag phenomenon of the internal tubes. As a result, the existing models are inadequate in capturing the true behaviour of the tube(s)-in-tube structures. Note also that the existence of the tube-tube interaction coupled with the negative shear-lag in the tubes further complicates the estimation of the structural performance and the accurate analysis of such structures.

The additional bending stresses due to the tube-tube interaction are considered to be a way of explaining the shear-lag phenomenon of the tube(s)-in-tube structures. However, existing simple analytical methods and existing commercial 3-D frame analysis programs do not take into account the additional bending stresses and hence, they cannot be used to interpret the cause of the shear-lag phenomenon existing in the tubes. In view of this, a simple numerical modelling technique is proposed to determine the additional bending stresses and to study the shear-lag reversal points.

Three 40-storey framed-tube structures with single, two and three internal tubes are investigated in this study. The numerical results indicating the additional bending stresses and shear-lag reversal points can then be used to estimate the shear-lag behaviour and its effect on such structures.

2 ANALYSIS OF SHEAR-LAG IN TUBE(S)-IN-TUBE STRUCTURES

The discrete tube(s)-in-tube structure is modelled by an assemblage of equivalent uniform orthotropic plate panels. The framed-tubes can hence be analysed as continuous structures (Lee & Loo 1997). The shape functions adopted herein are the modified Reissner's functions (Lee & Loo 1997, Reissner 1945). Essentially, the modification involves up-grading the displacement functions from parabolic to cubic. This is to account for the independent distribution of vertical displacement in the flange frame panels, thereby encompassing the net shear-lag. The function for estimating the distribution of vertical displacement in the web frame panel is also assumed to be cubic. A pilot study of the modified Reissner's functions indicates that they are adequate to cover the key characteristics of the shear-lag phenomenon in assessing the global behaviour of the tube(s)-in-tube structures as well as the tube-tube interactions (Lee & Loo 1997).

The distributions of vertical displacements, $U_1(z,y)$, for flange frame panel (see Figure 1), and $U_2(z,x)$, for web frame panel (see Figure 2), are assumed respectively as

$$U_1(z,y) = c\left[\frac{dw}{dz} + \left(1 - \left(\frac{y}{b}\right)^3\right)u_1(z)\right] \quad \text{and} \quad U_2(z,x) = \left[\frac{dw}{dz}x + \left(\frac{x}{c} - \left(\frac{x}{c}\right)^3\right)u_2(z)\right] \tag{1}$$

where b and c are the half-widths of the flange and web frame panels respectively; w is the deflection of the structure; $u_1(z)$ and $u_2(z)$ are the functions including shear-lag coefficients due to shear deformation; and x, y and z are the coordinates of axes.

The assumptions are made to simplify the patterns of the vertical displacement distributions in external and internal tubes. Consequently, the complex structural behaviour is reduced to the solution of a single second-order linear differential equation. The numerical analysis is based on the minimum potential energy principle in conjunction with the variational approach. The total potential energy must be minimised using the governing differential equation and the required set of boundary conditions based on the variational approach. The governing differential equations describing the global behaviour of the framed-tube structures with multiple internal tubes (Lee et al. 1998) are

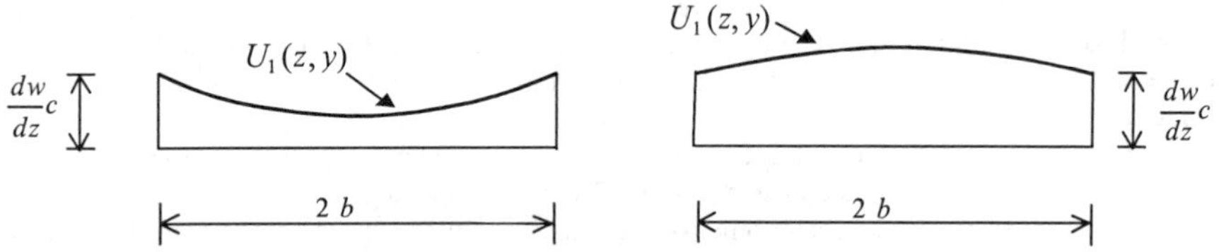

(a) Positive shear lag (b) Negative shear lag

Figure 1. Distribution of vertical displacement in the flange frame panel.

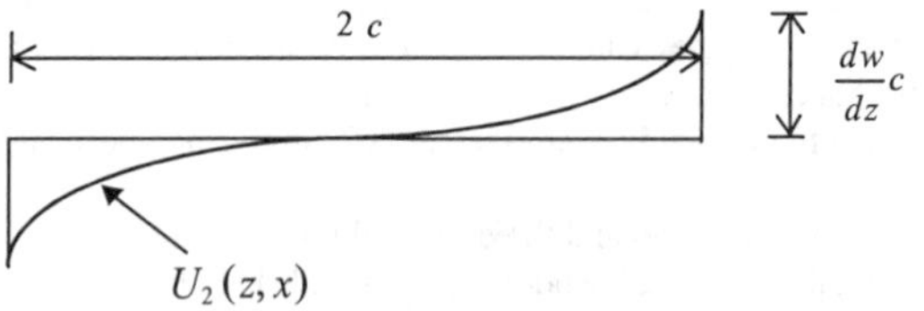

Figure 2. Distribution of vertical displacement in the web frame panel.

$$\frac{d}{dz}\left[u_1'(z)\alpha_2(z)+w''(z)\ \alpha_1(z)\right]-u_1(z)\alpha_3(z)=0\ ;\ -\frac{d}{dz}\left[EI_e w''(z)-u_1'(z)\alpha_1(z)\right]=P_e(z);$$
$$\frac{d}{dz}\left[u_{i1}'(z)\beta_2(z)+w''(z)\beta_1(z)\right]-u_{i1}(z)\beta_3(z)=0;\ -\frac{d}{dz}\left[EI_i w''(z)-u_{i1}'(z)\beta_1(z)\right]=P_i(z) \tag{2}$$

where $u_1(z)$ and $u_{i1}(z)$ are respectively the undetermined functions including shear-lag coefficients of the external and internal tubes; P_e and P_i are the shear forces in the external and internal tubes respectively; α_1, α_2, α_3, β_1, β_2 and β_3 are the constants to be determined; and I_e and I_i are respectively the second moments of area of the external and internal tubes.

Note that the shear-lag phenomenon is due to the distributions of the additional bending stresses, the expressions of which can be derived from those for the axial bending stresses. The axial bending stress distributions including the shear-lag effect are expressed in terms of a series of linear functions by the second moment of area, I, of the entire tube(s)-in-tube system and its corresponding geometric and material properties (Lee et al. 1998). Or,

$$\sigma_{fs}=\left(1-\frac{y^3}{b^3}+\frac{I_N}{I}\right)Ecu_1'(z)+\frac{I_{iN}}{I}Ecu_{i1}'(z) \tag{3}$$

for external flange frame panel of a structure with multiple internal tubes;

$$\sigma_{fis}=\left(1-\frac{y^3}{b_i^3}+\frac{I_{iN}}{I}\right)Ec_iu_{i1}'(z) \tag{4}$$

for internal flange frame panel of a structure with a single internal tube;

$$\sigma_{fis}=\left(1\pm 8\left(\frac{y-n_1}{2b_i}-\frac{1}{2}\right)^3+\frac{I_{iN}}{I_i}\right)Ec_iu_{i1}'(z) \tag{5}$$

for internal flange frame panel of a structure with an even number of internal tubes; and

$$\sigma_{fis}=\left(1\pm 8\left(\frac{y-n_1}{2b_i}-\frac{1}{2}\right)^3+\frac{I_{iN}}{I_i}\right)Ec_iu_{i1}'(z) \tag{6}$$

for internal flange frame panel of a structure with an odd number of internal tubes.

In the preceding four equations, E is the Young's modulus; I_N and I_{iN} are the second moments of area of the flange frame panels in the external and internal tubes, respectively; b_i and c_i are respectively the half-widths of the flange and web frame panels of the internal tube; and

$$u_1'(z)=\frac{qY}{X^2}\left[\frac{\cosh X(H-z)+XH\sinh Xz}{\cosh XH}-1\right]$$

$$u_{i1}'(z)=\frac{qY_1}{X_1^2}\left[\frac{\cosh X_1(H-z)+X_1H\sinh X_1z}{\cosh X_1H}-1\right] \tag{7}$$

and $n_1=\dfrac{(n-1)}{2}a+2\left(\dfrac{n}{2}-1\right)b_i$

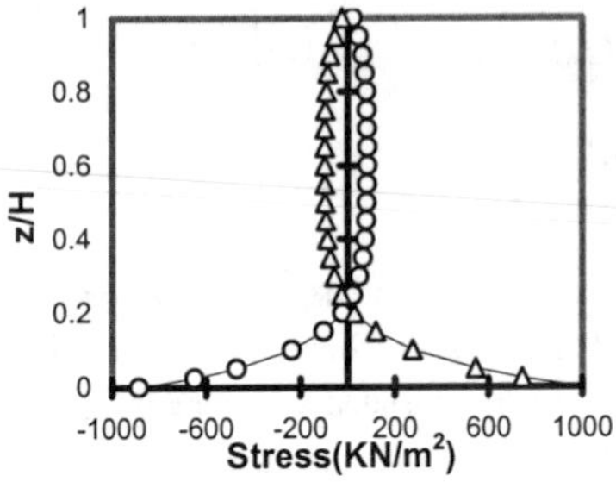

External tube

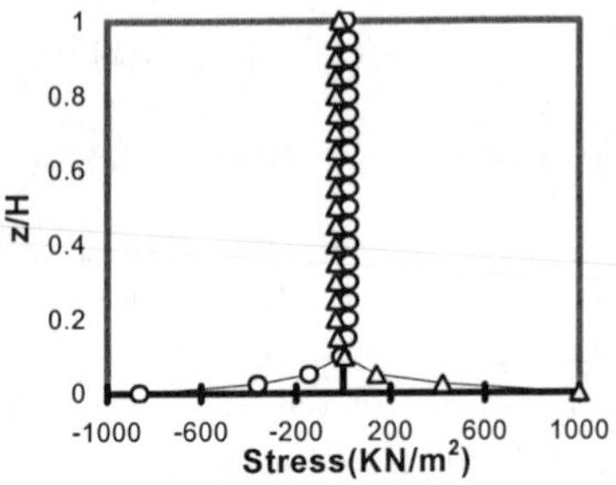

Internal tube

(a) Tube-in-tube structure

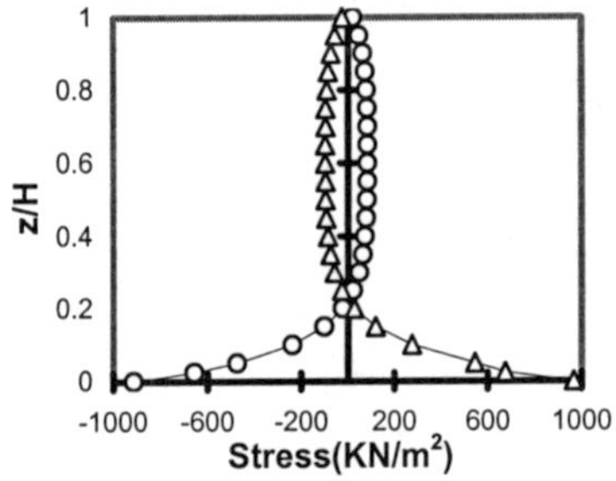

External tube

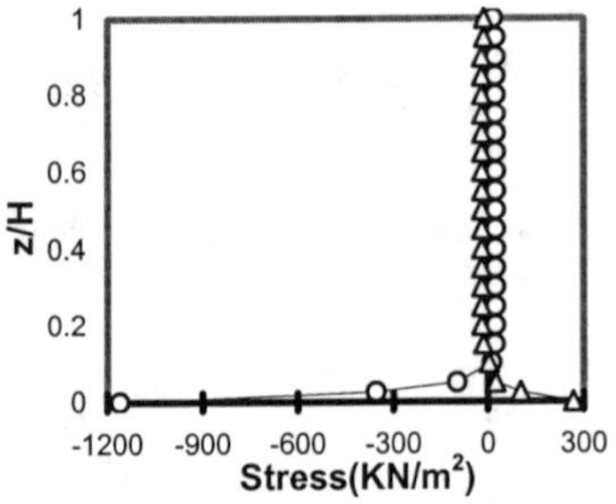

Internal tube

(b) 2 tubes-in-tube structure

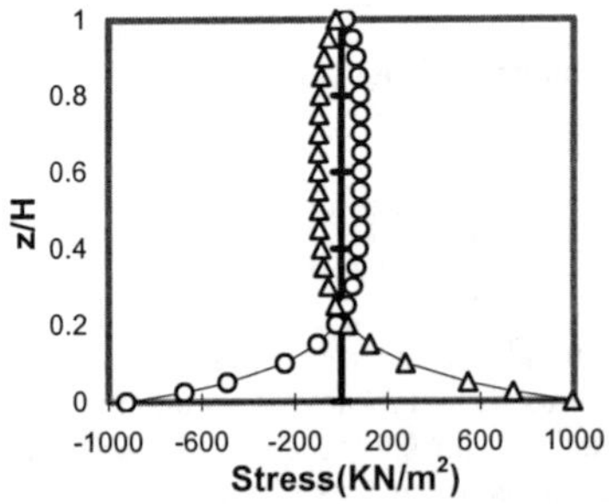

External tube

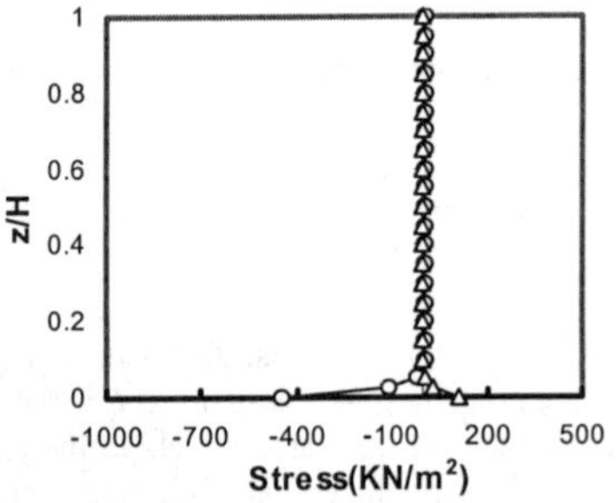

Internal tube

(c) 3 tubes-in-tube structure

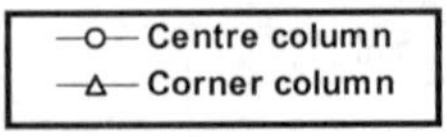

Figure 3. Additional bending stresses in centre and corner columns of the three tube(s)-in-tube structures.

in which H is the total height of the building; q is the uniformly distributed load per unit height; n is a function of N, the total number of internal tubes (Lee et al. 1998); and the variables X, X_1, Y and Y_1 are given in Table 1.

Table 1. Formulas for X, X_1, Y and Y_1.

Variables	External tube	Variables	Internal tubes	
X	$\frac{1}{c}\sqrt{\frac{G}{E}\cdot\frac{1}{\frac{2}{45}\frac{I_N}{I_e}\frac{c}{b}+\frac{4}{21}}}$	X_1	$\frac{1}{c_i}\sqrt{\frac{G}{E}\cdot\frac{1}{\frac{1}{N}(\frac{2}{45}\frac{I_{iN}}{I_i}\frac{c_i}{b_i}+\frac{4}{21})}}$	
Y	$\frac{1}{EI_e}\cdot\frac{1}{\frac{I_N}{I_e}+\frac{30}{7}\frac{b}{c}}$	Y_1	$\frac{1}{EI_i}\cdot\frac{1}{\frac{I_{iN}}{I_i}+\frac{30}{7}\frac{b_i}{c_i}}$ (for single tube)	$\frac{1}{EI_i}\cdot\frac{1}{\frac{3}{4}\frac{I_{iN}}{I_i}+\frac{45}{14}\frac{b_i}{c_i}}$ (for multiple tubes (2,3,...))

Note in Equations (3) to (6) that σ_{fs} and σ_{fis} can be positive or negative depending on the positive or negative shear-lag effect. The reversal point from positive to negative shear-lag can be obtained by setting $u_1'(z)=0$ and $u_{i1}'(z)=0$ in Equations (3) to (6). Or,

$$z_1 = \frac{1}{X}\sinh^{-1}\left[\frac{2XH\cosh XH - \sinh 2XH}{(XH)^2 - 2XH\sinh XH - 1}\right] \quad \text{for external tube} \tag{8}$$

and

$$z_2 = \frac{1}{X_1}\sinh^{-1}\left[\frac{2X_1H\cosh X_1H - \sinh 2X_1H}{(X_1H)^2 - 2X_1H\sinh X_1H - 1}\right] \quad \text{for internal tube} \tag{9}$$

3 ESTIMATION OF SHEAR-LAG BEHAVIOR

To investigate the shear-lag phenomenon of tube(s)-in-tube systems, three 40-storey framed-tube structures with single, two and three internal tubes are analysed. The additional bending stresses and the shear-lag reversal points due to the proposed method are determined to explain the shear-lag phenomenon. The shear-lag reversals obtained using a 3-D frame analysis program (ETABS 1989) are also included for comparison. Note that the 3-D frame analysis program cannot handle the additional bending stresses. The second moment of area of each internal tube is taken to be 90 (m^2), and Young's modulus E and shear modulus G are taken to be 2.06×10^{10} (N/m^2) and 0.2×10^{10} (N/m^2), respectively. To consider the critical case of the structures, a uniformly distributed lateral load of 88.24 (KN/m) is assumed to be applied to the long side frame panel. The geometry of the structures and the applied load are summarised in Table 2.

Table 2. Geometry and loading of the three tube(s)-in-tube structures.

Structures	Tube-in-tube	2 tubes-in-tube	3 tubes-in-tube
Column and beam sizes in external tube ($cm\times cm$)	80 × 80	80 × 80	80 × 80
Column and beam sizes in internal tube ($cm\times cm$)	91 × 91	80 × 80	72 × 72
Size of external tube ($m\times m$)	30 × 15	30 × 15	30 × 15
Size of internal tube ($m\times m$)	15 × 5	2-7.5 × 5	3-5 × 5
Total storey height (m)	120	120	120
Lateral load (kN/m)	88.24	88.24	88.24

Figures 3(a), (b) and (c) show the distributions of the additional bending stresses in each tube of the three tube(s)-in-tube structures. To investigate the shear-lag reversal points and the shear-lag distribution due to the tube-tube interaction, the additional bending stresses in the centre and corner columns of the flange frame panels are plotted along *z/H*. It is found that due to the identical second moments of area of each internal tube, an increasing number of internal tubes results in a gradual reduction in the additional bending stresses between the centre and corner columns. As a result, the shear-lag is also reduced. However, the internal tubes with the same second moment of area have little effect on the additional bending stresses in the external tube. It is further observed that the effect of the positive shear-lag is greater at the bottom of the structures, and the negative shear-lag in the external tube occurs at around 1/4 of the building height. The points of shear-lag reversal for the internal tubes locate at a lower level than those for the external tubes. The storey levels of the shear-lag reversal obtained from the 3-D frame analysis and the proposed formulations (Equations (8) and (9)) are compared in Table 3, where close agreement is observed.

Table 3. Storey levels of shear-lag reversal for the three tube(s)-in-tube structures.

Structures		Reversal level	
		Proposed method	3-D frame analysis (ETABS 1989)
Tube-in-tube	External tube	10	12
	Internal tube	6	7
2 tubes-in-tube	External tube	10	13
	Internal tube	4	4
3 tubes-in-tube	External tube	10	13
	Internal tube	4	4

4 CONCLUSION

A simple numerical method is proposed for the analysis of the framed-tube structures with multiple internal tubes. The proposed method takes into account the additional bending stresses due to the tube-tube interaction in the tubes, which are observed to have significant effect on the shear-lag phenomenon. By quantifying the additional bending stresses and the shear-lag reversals, a better understanding of the shear-lag phenomenon of tube(s)-in-tube structures is provided. The proposed method is considered as suitable for the analysis of the shear-lag phenomenon of tubes-in-tube systems.

REFERENCES

ETABS 1989. *Three dimensional analysis of building system, Computers and Structures Inc.*. California: Berkerky.

Lee, K.K. & Y.C. Loo 1997. Simplified analysis of shear lag in tube-in-tube structures. *Proc. 2nd China-Australia Symposium on Computation and Mechanics, Sydney, Australia, 12-14 February, 1997:* 113-122.

Lee, K.K., Y.C. Loo & H. Guan 1998. Simplified analysis of stress distribution in framed tube structures with multiple internal tubes. *Proc. 5th Int. Conf. Tall Buildings, Hong Kong, 9-11 December, 1998:* 345-350. Hong Kong: University of Hong Kong.

Reisser, E. 1945. Analysis of shear lag in box beam by the principle of minimum potential energy. *Quarterly of Applied Mathematics*, 4(3): 268-278.

Mechanics of Structures and Materials, Bradford, Bridge & Foster (eds)
© 1999 Balkema, Rotterdam, ISBN 90 5809 107 4

The design of pins

R.Q. Bridge
School of Civic Engineering and Environment, University of Western Sydney, Nepean, N.S.W., Australia

ABSTRACT: The rules for the design of pins vary quite considerably from code to code and this has caused some concern in the consulting profession, particularly considering the recent upsurge in the use of pins in visible steel structures. An extensive testing program has been carried out to examine the influence of such parameters such as pin diameter, material properties of the pin, thickness of the loading plates, material properties of the loading plates and the distance of the pin to the edge of the loading plates. The results indicate that the current design rules are in need of modification to accommodate the modes of failure and modified design procedures have been proposed.

1 INTRODUCTION

Pins are similar to bolts except that they have no head, are not threaded (and hence have no nut) and therefore can not sustain any axial forces i.e. pins can only carry shear forces transverse across the pin. Despite this limitation, they are still used in structural applications and have recently been in favour with architects who desire "honesty in structure" for exposed steel structures in applications as canopies, sporting stadiums, convention centres and bridges. In these cases, rotations are relatively small and the pins are essentially subjected to static conditions.

The design procedures for pins can be found in most structural steel codes and standards. However, a review of these codes indicated a relatively wide disparity in the design concepts and design values for the three major conditions: shear of the pin; bearing on the pin; and bearing on the plies (plates) that load the pin. In particular, the Australian Standard AS4100-1998 has an apparently high design value for the strength of a ply (plate) in bearing and yet a low value for the strength of a pin in shear.

The behaviour of pins under load has been examined experimentally to determine the effects of the material and geometrical properties of both the pin and the loading plate on the strength and mode of failure of the pin or plate. The results have been compared with design values from steel design standards and modifications to the design procedures have been proposed.

2 TEST PROGRAMME

The test specimens consisted of a snug-fit single pin loaded in double shear by an interior plate between two exterior cover plates as shown in Figure 1. The lower half of the specimen was bolted and was designed to have a greater capacity than the pin. The ends of the top and bottom interior plates were placed in the grips of a Mohr and Federhaff tensile testing machine (580 kN capacity) and the specimen tested under load control until failure. Deformation of the interior pin plate relative to the exterior cover plates was measured using two dial gauges with a resolution of 0.01mm. The main variables tested were the pin diameter d_f (10, 16 and 27mm), the interior plate thickness t_p (3, 6, 10, 16 and 20mm) and the material properties of the pin. The pins were cut from two types of commercially available steel rod: black rod with a high ductility and a ratio of yield strength f_{yf} to ultimate tensile strength f_{uf} in the range 0.55 to 0.60; and bright rod with a lower ductility and a ratio of yield strength f_{yf} to ultimate tensile strength f_{uf} in the range 0.86 to 0.88. The interior plates were all ductile black plate with a ratio f_{yf}/f_{uf} in the range 0.54 to 0.71. The distance from the pin to the edge of the plate in the direction of loading was kept within the limits of AS 4100-1998 for end plate tear-out to prevent this mode of failure.

The geometrical properties, material properties, test results and modes of failure for the 18 test specimens are shown in Table 1. The primary mode of failure was either shearing of the pin or large bearing deformations of the plate (greater than 70% of the pin diameter). In some cases, large plate bearing deformations were observed prior to shearing of the pin. In other cases, fracture of the plate occurred at the cross-section through the pin. These failures have been labelled as secondary modes of failure. Pin bearing failures were not observed.

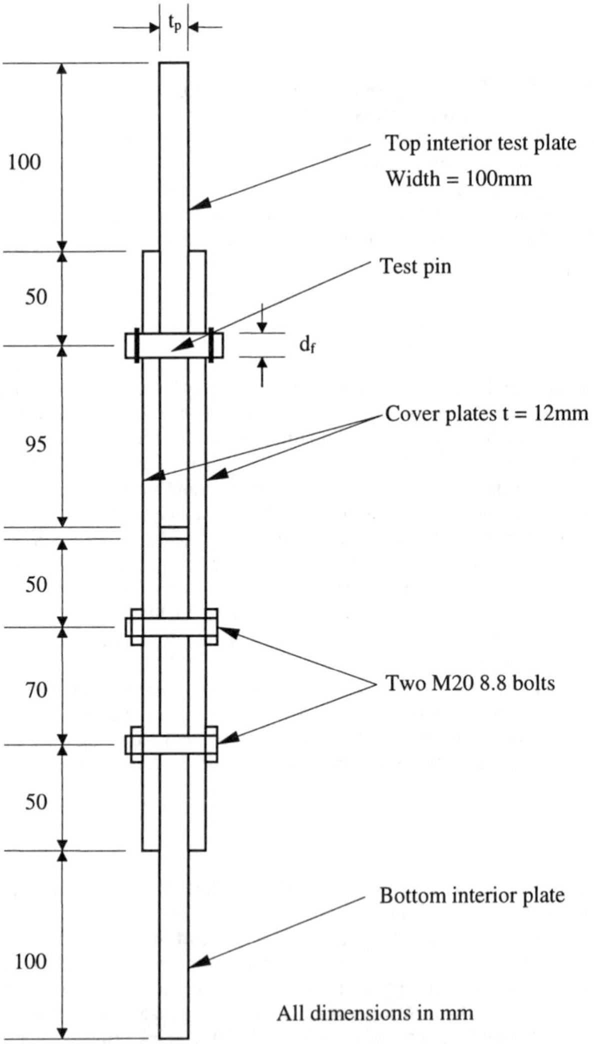

Figure 1. Test specimen for pin in double shear

Table 1. The geometrical and material properties and the test results of the pin test specimens.

Test No.	Pin d_f mm	Pin f_{yf} MPa	Pin f_{uf} MPa	Plate thick mm	Plate f_{yp} MPa	Plate f_{up} MPa	Max. Load kN	Primary Failure	Secondary Failure
1	10.06	250	455	3.12	360	496	53.6	Pin shear	Plate bearing
2	10.04	250	455	5.97	310	469	54.0	Pin shear	
3	10.06	250	455	9.85	260	485	54.3	Pin shear	
4	16.13	300	499	3.23	360	496	97.0	Plate bearing	
5	16.14	300	499	10.05	260	485	150.8	Pin shear	
6	16.13	300	499	15.86	250	460	146.5	Pin shear	
7	26.95	270	485	3.12	360	496	113.0	Plate bearing	Plate fracture
8	26.95	270	485	9.9	260	485	346.0	Pin shear	Plate bearing
9	26.95	270	485	19.93	250	446	344.0	Pin shear	
10	9.97	480	558	3.14	360	496	53.6	Pin shear	Plate bearing
11	10.09	480	558	6.12	310	469	56.8	Pin shear	
12	10.00	480	558	10.11	260	485	56.4	Pin shear	
13	15.97	460	523	3.16	360	496	92.5	Plate bearing	
14	15.97	460	523	9.85	260	485	137.0	Pin shear	
15	15.97	460	523	15.9	250	460	131.0	Pin shear	
16	26.90	450	524	3.12	360	496	110.0	Plate bearing	Plate fracture
17	26.90	450	524	10.17	260	485	352.0	Plate bearing	Plate fracture*
18	26.90	450	524	19.87	250	446	350.0	Pin shear	

*Pin also sheared 25% of diameter

The tests were conducted by van Ommen and Hayward 1992. Typical load-deformation curves are shown in Figure 2 for a 16mm diameter bright pin in two thicknesses of plate. Test specimen 13 with the 3mm plate exhibited a plate bearing failure. With plate bearing failures, hole elongations in excess of 70% of the hole diameter are attained. Test specimen 15 with 16mm exhibited a pin shear failure. With pin shear failures, the shear failure of the pin directly through the diameter of the pin is associated with a shear deformation in the pin itself of about 25% of the pin diameter prior to failure, even for the pins manufactured from bright steel rod with a lower ductility than the black steel rod.

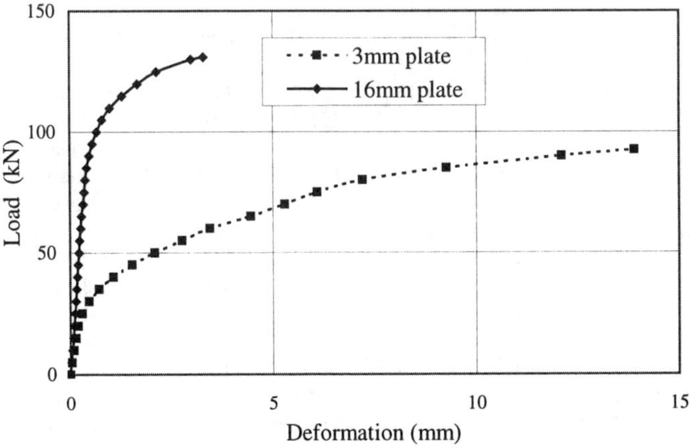

Figure 2. Typical load-deformation behaviour for pin shear and plate bearing failures

Table 2. Comparison of design strengths in steel codes and specifications

Steel code	Pin shear	Pin bearing	Plate bearing
AS4100-1998	$V_f = 0.62 f_{yf} A_f$	$V_b = 1.4 f_{yf} d_f t_p$	$V_b = 3.2 f_{up} d_f t_p$
Eurocode 3-1992	$V_f = 0.60 f_{uf} A_f$	$V_b = 1.5 f_{yf} d_f t_p$	$V_b = 1.5 f_{yp} d_f t_p$
BS5950-1990	$V_f = 0.60 f_{yf} A_f$	$V_b = 1.2 f_{yf} d_f t_p$	$V_b = 1.2 f_{yp} d_f t_p$
AISC-1993	$V_f = 0.60 f_{yf} A_f$	$V_b = 1.4 f_{yf} d_f t_p$	$V_b = 1.4 f_{yp} d_f t_p$

3 COMPARISON WITH DESIGN METHODS

The design strengths for pins according to Australian, European, British and American practice for the three conditions of pin shear, pin bearing and plate bearing are listed below in Table 2. In Table 2, A_f is the cross-sectional area of the pin, d_f is the diameter of the pin, f_{yf} is the yield stress of the steel in the pin, f_{uf} is the ultimate tensile strength of the steel in the pin, t_p is the thickness of the load-bearing plate, f_{yp} is the yield stress of the steel in the plate, and f_{up} is the ultimate tensile strength of the steel in the plate. Most codes are similar with two major exceptions: Eurocode 3 uses the ultimate tensile strength of the pin in calculating the shear strength of the pin (similar to that for bolt strength in most steel codes); and AS4100-1998 uses the ultimate tensile strength of the plate (and a large factor of 3.2) in calculating the bearing strength of the plate. Therefore only AS4100-1998 and Eurocode 3-1992 are used in the following comparisons of the design codes with the test strengths.

The predicted failure modes from Table 3 compare well with actual failure mode in Table 1. However, for test specimens where the actual primary failure was by pin shear, the strength of the pin in shear predicted by AS4100-1998 was markedly lower than the test strengths, particularly for the ductile pins made from black steel rod. For the test specimens 1, 4, 7, 8, 10, 13, 16 and 17 where the primary or secondary failure was by bearing of the plate, the strength predicted by AS4100-1998 was close to the actual test strengths. The predicted bearing strengths for specimens 7, 16, and 17 appear a little high because the full bearing strength of the plate was not attained in the test due to premature fracture of the plate adjacent to the hole.

Table 3. Comparison of test results with design values predicted by AS4100-1998.

Test No.	Max Load kN	V_f Pin kN	Load/V_f Pin	V_b Pin[+] kN	Load/V_b Pin[+]	V_b Plate kN	Load/V_b Plate	Predicted Failure
1	53.6	24.6	2.18	11.0	4.88	49.8	1.08	Pin shear
2	54.0	24.5	2.20	21.0	2.57	90.0	0.60	Pin shear
3	54.3	24.6	2.20	34.7	1.57	153.8	0.35	Pin shear
4	97.0	76.0	1.28	21.9	4.43	82.7	1.17	Pin shear
5	150.8	76.1	1.98	68.1	2.21	251.7	0.60	Pin shear
6	146.5	76.0	1.93	107.4	1.36	376.6	0.39	Pin shear
7	113.0	191.0	0.59	31.8	3.56	133.5	0.85	Plate bearing
8	346.0	191.0	1.81	100.9	3.43	414.1	0.84	Pin shear
9	344.0	191.0	1.80	203.0	1.69	766.6	0.45	Pin shear
10	53.6	46.5	1.15	21.0	2.55	49.7	1.08	Pin shear
11	56.8	47.6	1.19	41.5	1.37	92.7	0.61	Pin shear
12	56.4	46.7	1.21	67.9	0.83	156.9	0.36	Pin shear
13	92.5	114.3	0.81	32.5	2.85	80.1	1.15	Plate bearing
14	137.0	114.3	1.20	101.3	1.35	244.1	0.56	Pin shear
15	131.0	114.3	1.15	163.5	0.80	373.8	0.35	Pin shear
16	110.0	317.1	0.35	52.9	2.08	133.2	0.83	Plate bearing
17	352.0	317.1	1.11	172.4	2.04	424.6	0.83	Pin shear
18	350.0	317.1	1.10	336.7	1.04	762.8	0.46	Pin shear

[+] Pin bearing ignored in predicting failure as none was observed in tests

Table 4. Comparison of test results with design values predicted by Eurocode 3.

Test No.	Max Load kN	V_f Pin kN	Load/V_f Pin	V_b Pin$^+$ kN	Load/V_b Pin$^+$	V_b Plate kN	Load/V_b Plate	Predicted Failure
1	53.6	43.4	1.24	11.8	4.55	16.95	3.16	Plate bearing
2	54.0	43.2	1.25	22.5	2.40	27.87	1.94	Plate bearing
3	54.3	43.4	1.25	37.2	1.46	38.65	1.41	Plate bearing
4	97.0	122.4	0.79	23.4	4.14	28.13	3.45	Plate bearing
5	150.8	122.5	1.23	73.0	2.07	63.26	2.38	Plate bearing
6	146.5	122.4	1.20	115.1	1.27	95.93	1.53	Plate bearing
7	113.0	332.0	0.34	34.1	3.32	45.41	2.49	Plate bearing
8	346.0	332.0	1.04	108.1	3.20	104.05	3.33	Plate bearing
9	344.0	332.0	1.04	217.5	1.58	201.42	1.71	Plate bearing
10	53.6	52.3	1.03	22.5	2.38	16.91	3.17	Plate bearing
11	56.8	53.5	1.06	44.5	1.28	28.71	1.98	Plate bearing
12	56.4	52.6	1.07	72.8	0.77	39.43	1.43	Plate bearing
13	92.5	125.7	0.74	34.8	2.66	27.25	3.39	Plate bearing
14	137.0	125.7	1.09	108.5	1.26	61.35	2.23	Plate bearing
15	131.0	125.7	1.04	175.2	0.75	95.22	1.38	Plate bearing
16	110.0	357.4	0.31	56.7	1.94	45.32	2.43	Plate bearing
17	352.0	357.4	0.98	184.7	1.91	106.69	3.30	Plate bearing
18	350.0	357.4	0.98	360.8	0.97	200.44	1.75	Plate bearing

$^+$Pin bearing ignored in predicting failure as none was observed in tests

It can be seen that failure mode predicted by the code in Table 4 does not compare well with actual test failure modes shown in Table 1. However, for the test specimens 1, 2, 3, 5, 6, 8, 9, 10, 11, 12, 14, 15 and 18 where the primary failure was by pin shear, the strength of the pin in shear predicted by Eurocode 3-1992 was close to the actual test strengths. For the test specimens 1, 4, 7, 8, 10, 13, 16 and 17 where the primary or secondary failure was by bearing of the plate, the strength predicted by Eurocode 3-1992 was significantly less than the actual test strengths.

4 DESIGN RECOMMENDATIONS

From the comparisons with AS4100-1998 and Eurocode 3-1992, it was clear that the AS4100-1998 provided the best model for plate bearing strength based on the ultimate strength of the steel in the plate whereas Eurocode 3-1992 provided the best model for the pin shear strength, again based on the ultimate strength of the steel in the pin.

It is proposed that the strength V_f of a pin in shear should be given by

$$V_f = 0.62 f_{uf} A_f \tag{1}$$

This is similar to the strength of a bolt in shear as given in AS4100-1998. The shear factor of 0.62 on the ultimate tensile strength is used to give the shear strength of the steel in the pin. In the tests, the mean value of this factor for the ductile black steel pins was 0.71 with a cov of 0.08 with factors ranging from 0.74 for the 10mm dia. pins to 0.62 for the larger 27mm pins. For the lower ductility bright steel pins, the mean value of the factor was 0.63 with a cov of 0.03 with factors ranging from 0.65 for the 10mm dia. pins to 0.59 for the larger 27mm pins.

It is proposed that the strength of the plate in bearing should be given by

$$V_b = 3.2 f_{up} d_f t_p \tag{2}$$

This is identical to the current requirements in AS4100-1998 for both pins and bolts. The bearing factor of 3.2 on the ultimate tensile strength is used to give the bearing strength of the steel in the plate. In the two tests that had primary bearing failures without plate fracture, the mean value of the factor was 3.74. In the other three bearing failure tests where premature plate fracture occurred, the mean value of this factor was still 2.67, a value close to 3.2.

Table 5. Comparison of test results with design values predicted by modifications to codes.

Test	Max Load kN	V_f Pin	Load/V_f Pin	Load/V_b Plate	Service Load V_s kN	V_{bs} Plate kN	V_s/V_{bs}	Predicted Failure
1	53.6	44.8	1.20	1.08	20	18.1	1.11	Pin shear*
2	54.0	44.7	1.21					Pin shear
3	54.3	44.8	1.21					Pin shear
4	97.0	126.4		1.17	30	30.0	1.00	Plate bearing
5	150.8	126.6	1.19					Pin shear
6	146.5	126.4	1.16					Pin shear
7	113.0	343.1		0.85	38	48.4	0.78	Plate bearing
8	346.0	343.1	1.01	0.84	125	111.0	1.13	Pin shear*
9	344.0	343.1	1.00					Pin shear
10	53.6	54.0	0.99	1.08	20	18.0	1.11	Plate bearing
11	56.8	55.3	1.03					Pin shear
12	56.4	54.3	1.04					Pin shear
13	92.5	129.9		1.15	29	29.1	1.00	Plate bearing
14	137.0	129.9	1.05					Pin shear
15	131.0	129.9	1.01					Pin shear
16	110.0	369.3		0.83	42	48.3	0.87	Plate bearing
17	352.0	369.3	0.95	0.83	130	113.8	1.14	Pin shear*$^+$
18	350.0	369.3	0.95					Pin shear

*Plate bearing ($^+$pin shear) was a secondary failure mode in the tests.

It is proposed that a new serviceability condition for plate bearing be included in design codes. As shown in Figure 2 for the 3mm plate that failed in bearing, the bearing deformations of the plate at maximum load are very large and typically exceed 70% of the hole diameter. Using a proof load at 2% of the hole diameter as the basis to define the maximum service load V_s that can be sustained prior to the onset of large plate bearing deformations, a mean design value of bearing strength V_{bs} for serviceability conditions has been determined as

$$V_{bs} = 1.6 f_{yp} d_f t_p \quad (3)$$

The value of the factor 1.6 was derived from the eight tests that had primary and secondary bearing failures. It is close to the factors shown in Table 3 for plate bearing strengths based on the yield strength, indicating that this should be a serviceability condition, not a strength condition.

Comparisons of the proposals with the test results are given in Table 5 as ratios of test to design values. Values are shown for both primary and secondary failure modes and indicate reasonable agreement over the test range.

5 CONCLUSIONS

A series of tests have highlighted some deficiencies in the design models in current codes that are used to predict the strength of pins in plated structures. Modifications have been proposed that better model the modes of failure. A new serviceability condition is proposed. Bearing of the pin was not identified as a mode of failure and this aspect needs further examination.

6 REFERENCES

American Institute of Steel Construction AISC-1993. *Load and resistance factor design specification for structural steel buildings – Second edition.* American Institute of Steel Construction, Chicago

Australian Standard AS4100-1998. *Steel structures.* Standards Australia, Sydney

British Standard BS5950-1990. *Structural use of steel in buildings.* British Standards Institution, London

Eurocode 3-1992. ENV *1993-1-1 Design of steel structures – Part1.1: General rules and rules for buildings.* European Committee for Standardization, Brussels.

Van Ommen, M. and Hayward, I.G. 1992. Pins in steel structures. B.E. Thesis, University of Sydney.

Mechanics of Structures and Materials, Bradford, Bridge & Foster (eds)
© 1999 Balkema, Rotterdam, ISBN 90 5809 107 4

Characterising asymmetry of the initial imperfections in steel silos

P.A. Berry & R.Q. Bridge
School of Civic Engineering and Environment, University of Western Sydney, Nepean, N.S.W., Australia

J.M. Rotter
School of Civil and Environmental Engineering, University of Edinburgh, UK

ABSTRACT: Thin cylindrical shells under axial compression are very sensitive to imperfections in the initial geometry. A local axisymmetric depression, which has been shown to be one of the worst imperfection forms, occurs regularly at the circumferentially welded joints in civil engineering shell structures such as steel silos and tanks. Several theoretical studies have examined this imperfection form, but have been limited to the idealised case of perfectly axisymmetric imperfections. Only one study is known to have included the effects of asymmetry, achieved by superimposing single harmonic terms on a sinusoidal axisymmetric imperfection.

This paper presents a parametric study of initial imperfections based on the measured geometry of an experimental specimen. The fabrication procedures were similar to those used in full-scale construction, leading to a more realistic pattern of asymmetry than just a single harmonic term. Several measures of asymmetry are compared and the best characteristic amplitude is shown to be the mean plus one standard deviation.

1 INTRODUCTION

The behaviour of cylindrical shells under axial compression is very sensitive to geometric imperfections. Many imperfection forms have been studied and it is well established that axisymmetric imperfections cause the greatest reductions in strength (Koiter 1963; Yamaki 1984). The form of the initial imperfections depends on the method of fabrication: the circumferential welds used in steel silos lead to regular imperfections that are predominately axisymmetric in form (Bornscheuer and Häfner 1983; Clarke and Rotter 1988; Ding 1992; Coleman *et al.* 1992).

Most previous theoretical studies of circumferential weld shrinkage depressions have been limited to the idealised case of perfectly axisymmetric imperfections. Rotter and Teng (1989) investigated many factors and found the depression amplitude to be the most significant, predicting strength losses of up to 70% for an amplitude of one wall thickness. Teng and Rotter (1992) compared sinusoidal imperfections with inward and outward local depressions, and showed the importance of inward depressions, especially when the cylinder is also pressurised. Rotter (1996) showed that residual stresses are generally beneficial and that the interaction of adjacent weld depressions can lead to a further 15% reduction in strength.

Field measurements show that circumferential weld depressions in full-scale silos also contain significant levels of asymmetry (Ding 1992). The theoretical study by Pederson (1974) considered the effects of adding small asymmetric components, but was restricted to single harmonic terms superimposed on a sinusoidal axisymmetric imperfection.

The present study forms part of a larger investigation into the strength of cylinders with circumferential weld depressions (Berry 1997). Experimental specimens were fabricated using techniques closely matching those of full-scale construction. This paper presents a parametric study based on the initial imperfections of one specimen, and therefore includes a circumferential weld depression with asymmetric components that would be more representative of full-scale construction. The measured imperfections were scaled in different ways to examine the variations in

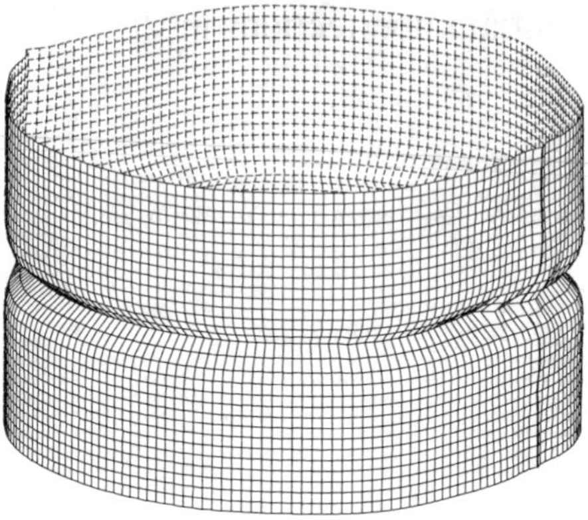

Figure 1: Mid-section of Specimen 801 with deviations from the mean scaled by a factor of 20.

buckling strength due to imperfection amplitude and asymmetry. It is not clear what measure best characterises an imperfection, so several different measures are used for comparison.

2 EXPERIMENTAL PROGRAMME

The experimental equipment, procedures, and results have been described in detail elsewhere (Berry *et al.* 1993, 1996; Berry 1997). The specimens were manufactured from thin steel sheet, which was rolled and welded to form cylinders with a radius of 500 mm and a height of 1500 mm. High welding currents and initial perturbation by a roller were used to ensure a significant circumferential weld depression, so that this imperfection form would dominate the specimen behaviour. Fixed end conditions were achieved by attaching stiffening rings to each end of the specimen prior to removal of the forming mandrel. The specimens were tested under uniform axial compression and the geometry was measured, both initially and at frequent load intervals, using a laser displacement meter driven in the rotational and longitudinal directions by stepper motors. Five specimens were tested covering a range of thicknesses from 0.6 mm to 0.9 mm.

3 EXPERIMENTAL GEOMETRY

The measurement system provided excellent detail of the initial specimen geometry, as demonstrated by Fig. 1, which, for clarity, uses only 5% of the available measurements. The local axisymmetric depression at the circumferential weld is clearly the dominant imperfection, but other imperfection forms are also present: a ridge occurred along each longitudinal weld, and the entire specimen was found to contain a regular asymmetric pattern to its geometry, which can be seen more clearly in the surface development of Fig. 2. These additional imperfection forms are also present in full-scale structures (Ding 1992).

4 FINITE ELEMENT MODELLING OF THE EXPERIMENTS

The ABAQUS finite element analysis package was used to analyse each specimen of the experimental investigation. The nodal geometry was based directly on the measured imperfections, and the material model was an isotropic multilinear approximation to the coupon results.

The specimen ends were fully restrained against both displacement and rotation, which corresponds to C3 in Singer's notation (Yamaki, 1984). A geometrically non-linear static analysis was performed using prescribed displacements at one of these ends to provide axial shortening of the specimen. Due to computing limitations, each specimen was analysed in quarters, justified by

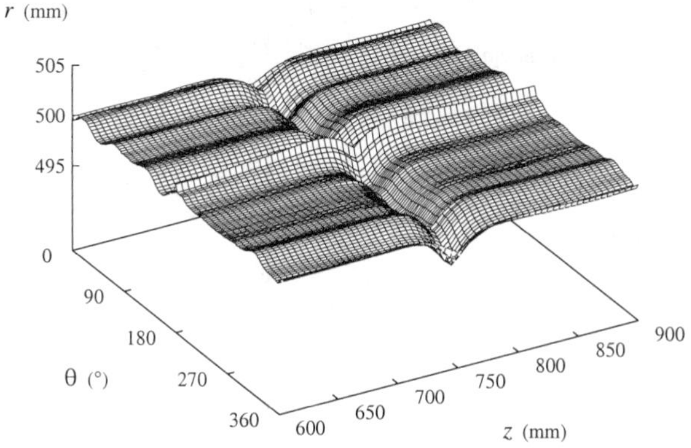

Figure 2: Initial surface development of the mid-section for Specimen 801.

the symmetry observed in the experimental behaviour. The symmetry plane at the circumferential weld was restrained against axial displacement and meridional rotation, while the diametrical symmetry plane through the longitudinal welds was restrained against circumferential rotation only.

The analysis used rectangular 9-noded doubly curved thin shell elements with reduced integration and five degrees of freedom per node (S9R5). The refined mesh had a uniform circumferential nodal spacing of 1° (10 mm) and a meridional spacing near the weld depression of $\lambda_0/10$ (5 mm), where λ_0 is the meridional bending half wavelength. The bending effects at the specimen ends were minor and a meridional spacing of λ_0 (50 mm) was shown to be sufficient for most of the remainder of the specimen.

The finite element analyses modelled the experiments very well. The mean ultimate loads, $\overline{P}_{uf}$, predicted by the finite element analyses for each specimen, were all within ±8% of the experimental values. The best match was achieved for Specimen 801, hence Quarter 4 of this specimen, which was closest to the mean result of all quarters, was chosen for the following parametric study.

5 PARAMETRIC STUDY OF INITIAL IMPERFECTIONS

5.1 *Imperfection amplitude*

The imperfection amplitude was varied by scaling the initial imperfections of Specimen 801, where a scale factor of 100% reproduced the measured geometry. Since the mean weld depression amplitude for this specimen, $\overline{\delta}/t = 3.1$, was considerably higher than would be expected of standard construction, only reduced imperfection levels were studied.

5.2 *Imperfection asymmetry*

For the purposes of this study, the initial geometry of Specimen 801 was divided into two components:

1. An axisymmetric component ($\delta_0/t = 3.1$) with geometry equal to the mean meridian.
2. An asymmetric component, being the deviations at each measurement point from the axisymmetric case.

Different levels of imperfection asymmetry were obtained by scaling this second component, where a scale factor of 100% reproduced the measured geometry. The initial imperfections at the circumferential weld are shown for selected scale factors in Fig. 3.

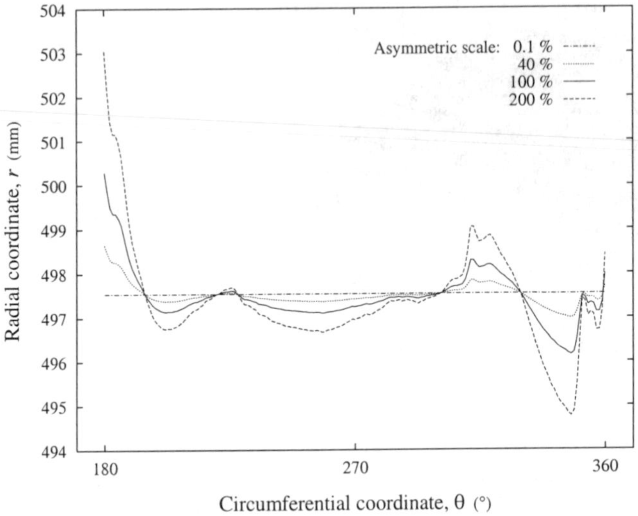

Figure 3: Effect of varying asymmetry on the initial imperfections around the weld depression (where 100% corresponds to Quarter 4 of Specimen 801).

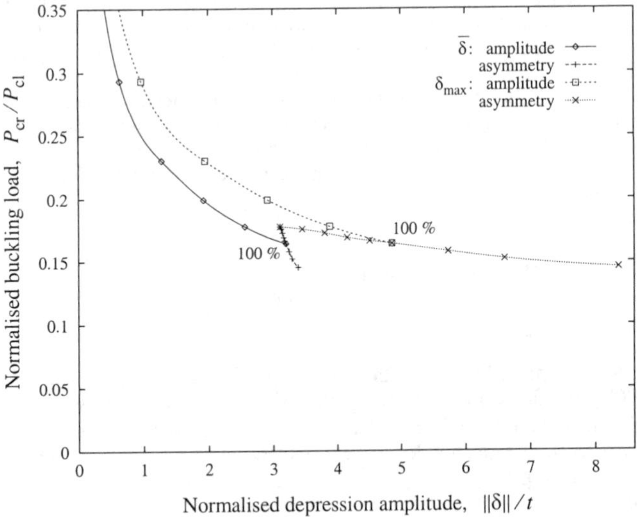

Figure 4: Buckling loads for varying levels of imperfection amplitude and asymmetry using the mean and maximum amplitude as the imperfection measure (where 100% corresponds to Quarter 4 of Specimen 801).

5.3 *Results*

It is useful to compare the results using different measures to characterise the imperfection amplitude, $\|\delta\|$. The buckling loads for both studies are shown in Fig. 4, using the mean depression amplitude and the maximum depression amplitude as two simple measures. Note that if Quarter 4 had been an unbiased sample of Specimen 801, its mean amplitude ($\bar{\delta}/t = 3.2$) would have been

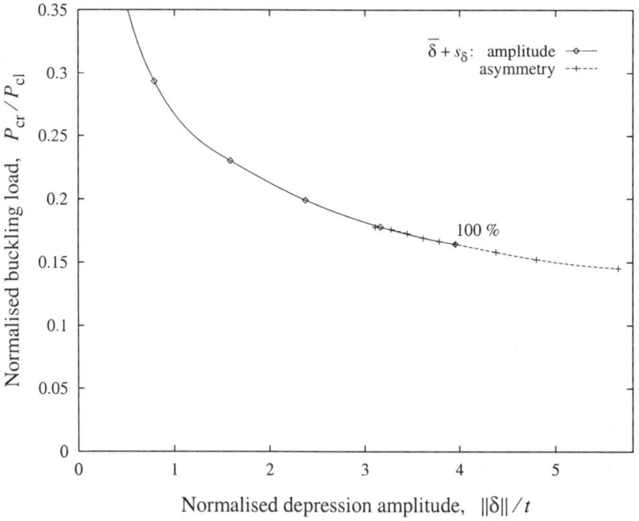

Figure 5: Buckling loads for varying levels of imperfection amplitude and asymmetry using the mean amplitude plus one standard deviation as the imperfection measure (where 100% corresponds to Quarter 4 of Specimen 801).

equal to that for the specimen as a whole ($\delta/t = 3.1$), and the results of the asymmetry study for the mean amplitude measure would have been a vertical line. For the case of 0.1% asymmetry (essentially axisymmetric), the mean and maximum amplitudes are equal, and the results of the asymmetry study for both measures coincide at that point. Both studies used the initial imperfections for Quarter 4 of Specimen 801 as a starting reference (100% amplitude, 100% asymmetry) and hence coincide at that point, regardless of the imperfection measure.

Whilst the imperfection amplitude clearly dominates the buckling behaviour of thin cylinders with weld shrinkage depressions, the imperfection asymmetry also causes further reductions in strength. The decrease in buckling load occurs almost linearly with increasing asymmetry and is of significant magnitude. The measured initial imperfections of Specimen 801, which are more axisymmetric than would be expected of commercial construction, produced an 8% decrease in buckling load compared to the axisymmetric case.

It is apparent that the mean depression amplitude underestimates the severity of an imperfection, while the maximum depression amplitude overestimates it. This implies that there may be an intermediate measure that is more appropriate. The results for both studies are shown again in Fig. 5, using the mean depression amplitude plus one standard deviation, $\overline{\delta}+s_\delta$, as the imperfection measure. The resulting two curves line up almost exactly, showing that in this case the mean amplitude plus one standard deviation is the best imperfection measure.

6 CONCLUSIONS

The imperfection amplitude dominates the buckling behaviour of thin cylinders with weld shrinkage depressions. Imperfection asymmetry leads to further significant reductions in strength and for levels of asymmetry consistent with full-scale construction, the buckling load could decrease by 10% compared to the axisymmetric case.

This study of imperfection measures indicates that the mean amplitude plus one standard deviation is the most appropriate measure for characterisation of circumferential weld shrinkage depressions.

REFERENCES

Berry, P. A. (1997). *Buckling under Axial Compression of Cylindrical Shells with Circumferential Weld Shrinkage Depressions*. PhD thesis, The University of Sydney, Australia.

Berry, P. A., Bridge, R. Q., and Rotter, J. M. (1993). The manufacture and testing of thin-walled cylinders with circumferentially welded joints. In Schmidt, L., editor, *Thirteenth Australasian Conference on the Mechanics of Structures and Materials*, pages 81–88, Wollongong, Australia.

Berry, P. A., Bridge, R. Q., and Rotter, J. M. (1996). Imperfection measurement of cylinders using automated scanning with a laser displacement meter. *Journal of the British Society for Strain Measurement*, 32(1):3–7.

Bornscheuer, F. W. and Häfner, L. (1983). The influence of an imperfect circumferential weld on the buckling strength of axially loaded circular cylindrical shells. In *Third International Colloquium on Stability of Metal Structures*, pages 407–414, Paris.

Clarke, M. J. and Rotter, J. M. (1988). A technique for the measurement of imperfections in prototype silos and tanks. Research Report R565, School of Civil and Mining Engineering, The University of Sydney, Australia.

Coleman, R., Ding, X. L., and Rotter, J. M. (1992). The measurement of imperfections in full-scale steel silos. In *Fourth International Conference on Bulk Materials Storage, Handling and Transportation*, pages 467–472, Wollongong. Institution of Engineers Australia.

Ding, X. L. (1992). *Precise Engineering Surveying — Application to the Measurement of Large Scale Steel Silos*. PhD thesis, The University of Sydney, Australia.

Koiter, W. T. (1963). The effect of axisymmetric imperfections on the buckling of cylindrical shells under axial compression. *Proceedings of Koninklijke Nederlandse Akademie van Wetenschappen, Series B*, 66(5):265–279.

Pederson, P. T. (1974). On the collapse load of cylindrical shells. In Budiansky, B., editor, *Buckling of Structures*, pages 27–39. Springer-Verlag, Berlin.

Rotter, J. M. (1996). Buckling and collapse in internally pressurised axially compressed silo cylinders with measured axisymmetric imperfections: Imperfections, residual stresses and local collapse. In *International Workshop on Imperfections in Metal Silos: Measurement, Characterisation, and Strength Analysis*, pages 119–139, Lyon, France. BRITE/EURAM CA-Silo.

Rotter, J. M. and Teng, J. G. (1989). Elastic stability of cylindrical shells with weld depressions. *Journal of Structural Engineering*, 115(5):1244–1263.

Teng, J. G. and Rotter, J. M. (1992). Buckling of pressurized axisymmetrically imperfect cylinders under axial loads. *Journal of Engineering Mechanics*, 118(2):229–247.

Yamaki, N. (1984). *Elastic Stability of Circular Cylindrical Shells*. North-Holland, Amsterdam.

Mechanics of Structures and Materials, Bradford, Bridge & Foster (eds)
© 1999 Balkema, Rotterdam, ISBN 90 5809 107 4

Strength of eccentrically loaded incomplete penetration butt welds

K.J.R.Rasmussen, J.Ha & K.-Y.Lam
Department of Civil Engineering, University of Sydney, N.S.W., Australia

ABSTRACT: The paper describes the results of a pilot series of tests on incomplete penetration butt welds subjected to combined tension and bending. Nine tension tests were performed using increasing levels of eccentricity. Two distinct failure modes were observed depending on the direction of the eccentricity. In the first, the eccentricity caused the connection to "close-up", inducing compression between the unfused parts of the abutting surfaces. In the second, the eccentricity caused the connection to "open-up", involving a separation of the unfused parts and large bending strains in the weld. A relationship between eccentricity and strength is derived. The results are compared with the design provisions of AS4100:1998.

1 INTRODUCTION

Incomplete penetration butt welds are frequently used in column base connections and splices where lack of access precludes using full penetration butt welds. Designers show aversion from using incomplete penetration butt welds to transfer tensile stresses, partly because penetration cannot be visually inspected at the root and partly because several reported failures have been related to brittle behaviour of incomplete penetration butt welds (Bjorhovde 1987, Bruneau & Mahin 1991, Fisher & Pense 1987, Popov & Stephen 1976).

In the previous Australian steel structures standard AS1250 (1981), incomplete penetration butt welds could only be designed for strength when the weld transferred compressive stresses. While Clause 9.7.2.7(b) of the current standard AS4100 (1998) allows the tensile strength of the weld to be utilised, the nominal strength is to be calculated as that of a fillet weld, so that the strength per unit length of weld (v_w) is

$$v_w = 0.6 f_{uw} t_t \ , \qquad (1)$$

where f_{uw} and t_t are the tensile strength of the weld deposit and design throat thickness respectively. The throat is through the leg for common geometries of incomplete penetration butt welds. However, recognising that it is difficult to achieve full penetration at the root of the weld, the design throat thickness is to be taken as the leg length (l_w) less 3 mm (Clause 9.7.2.3 of AS4100), so that

$$t_t = l_w - 3 \text{ mm} \ . \qquad (2)$$

In the context of fillet welds, the term ($0.6 f_{uw}$) of eqn. (1) represents the shear strength (τ_{uw}) of the weld deposit. In incomplete penetration butt welds, the throat is usually subjected to normal stresses and so the strength should be based on f_{uw} rather than τ_{uw}. Thus the factor (0.6) of eqn. (1) serves a different purpose in the context of incomplete penetration butt welds, where it is seen as a safety factor guarding against brittle fracture and eccentricity of loading.

Eccentric loading is particularly an issue in tension-loaded incomplete penetration butt welds, since separation of the unfused part of the abutting surfaces will necessarily induce bending stresses in the weld. However, AS4100 does not require such bending stresses to be designed for. On the contrary, Section C9.7.3.10 of the Commentary of AS4100 (AS4100-Commentary 1999) states that “The influence of bending moments at the faces of the weld and of normal forces applied longitudinally to the weld cross-section have been shown to have little influence on the weld strength“, giving reference to Clark (1972) and Strating et al. (1971). Noticing that the research referred to in justification of the approach was entirely concerned with fillet welds, and acknowledging that there must be a limit to the level of bending stress that can be tolerated in applying eqn. (1) to incomplete penetration butt welds, a pilot series of tests on eccentrically tension-loaded incomplete penetration butt welds was conducted. The eccentricity was varied to determine the maximum allowable eccentricity as a fraction of the leg length (l_w) that can be accommodated in applying eqn. (1) to determine weld strength.

2 TESTS

2.1 *Test Specimens*

The nominal geometry of the test specimens is shown in Figs 1a and 1b. Two vertical plates were welded to a horizontal plate using a complete penetration butt weld on one side and an incomplete penetration butt weld on the other. This arrangement ensured that failure would occur at the incomplete penetration butt weld. The off-set (*s*) of the vertical plates controlled the eccentricity, or the degree of bending stress, in the incomplete penetration butt weld.

The vertical and horizontal plates were sawn from nominally 20 mm Grade 350 hot-rolled plate with nominal yield stress and tensile strength of 350 MPa and 450 MPa respectively. Circular holes were drilled in the vertical plates for fitting spherical bearings (BSC GE35DO), through which pins could be inserted, and locking rings were tack-welded to the sides of the plates before inserting the bearings to prevent these from pushing out during testing.

The spherical bearings were used to achieve a statically determinate test set-up. It was deemed imperative that the initial eccentricity of loading on the incomplete penetration butt weld be known. However, the bearings allowed unrestrained plate rotations which in actual connections would be partially restrained because of the bending stiffness of connecting plates. Partial restraint ensures a more uniform pull on the weld and hence, a lower bound to the strength of practical welds was achieved by using spherical bearings in the tests.

The nominal dimensions of the incomplete penetration butt weld are shown in Figs 1b and 1c. The nominal values of leg length (l_w) and outside length of the weld (L_w) were 10 mm and 110 mm respectively. The upper vertical plate was bevelled at 45° longitudinally, as shown in Fig. 1c, and transversely at the ends, as shown in Fig 1b, to assist the fusing of the weld deposit. Three weld runs were used, each started and stopped as fillet welds outside the bevelled length (L_w) to achieve a continuous weld along the bevel. The fillet welds and the excess weld deposit of the incomplete penetration butt weld were ground off so that the front face of the weld was

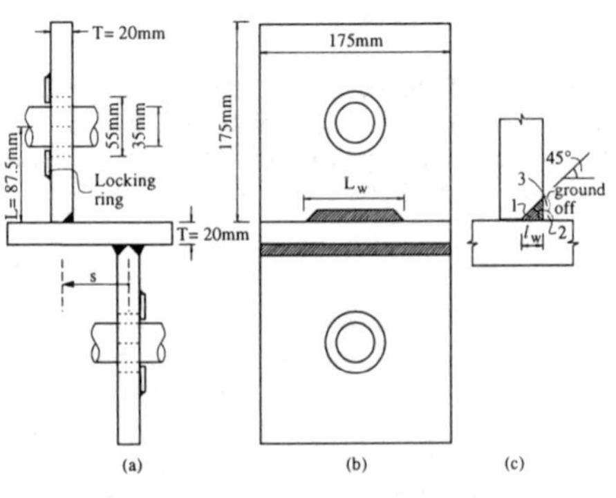

Figure 1: Nomenclature

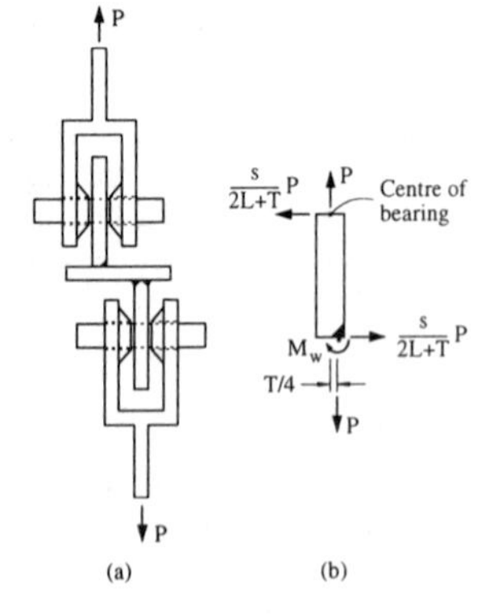

Figure 2: Test set-up

Table 1: Dimensions and ultimate loads of test specimens

Specimen	s_n (mm)	s (mm)	e (mm)	L_w (mm)	l_w (mm)	A_{wf} (mm^2)	P_u (kN)
(1)	(2)	(3)	(4)	(5)	(6)	(7)	(8)
1a	30	30.1	8.5	110	9.6	899	358.5
1b	30	30	8.5	110	9.5	894	385.0
2	20	19.9	3.9	113	10.2	947	404.0
3	10	10	-0.5	111	8.9	828	449.0
4	0	0	-5.0	116	8.6	924	407.0
5	-10	-10.1	-9.5	113	9.3	866	328.2
6a	-20	-20	-14.0	110	9.0	829	188.0
6b	-20	-20	-14.0	115	8.7	815	224.0
7	-30	-29.9	-18.4	112	8.2	829	210.0
Average	-	-	-	112.2	9.1	870	-

flush with the side of the plate, as shown in Fig. 1(c). This procedure defined clearly the nominal leg length of the weld, and hence the design throat thickness.

Table 1 shows the nominal (s_n) and measured (s) values of the off-set (or shift) of the vertical plates, the initial eccentricity of loading (e), defined in the following section, the measured outside length of weld (L_w), measured leg length (l_w), measured fused weld area (A_{wf}) and ultimate load (P_u) of each connection. The length (L_w) was measured on the front face of the incomplete penetration butt weld after fabrication, as shown in Fig. 1b, the leg length (l_w) was obtained as the average width of the fractured weld deposit measured at the ends and at the centre of the weld, and the fused weld area (A_{wf}) was calculated as the full area ((L_w-10mm) l_w) less the area of the unfused regions, which were mainly detected at the root of the weld and between passes. In calculating the full weld area, 10 mm was subtracted from L_w because the length along the centre of the weld was 10 mm shorter than the outside length (L_w).

The welds were laid using Gas Metal Arc Welding (GMAW) procedures. Argoshield 52 gas and a 1.2 mm Autocraft LW1 Cig-Weld wire with a nominal tensile strength of 500 MPa were used for all specimens. The three plates were tack-welded outside the bevelled length before full welding. All weld runs were laid using (closed-circuit) 230 Amps and 19 Volts. The measured speed of the arc was 7.8 mm/s, and hence the average heat input was $Q=AV/v=0.63$kJ/mm. Visual inspection revealed no major defects.

2.2 *Statics*

The test set-up is shown schematically in Fig. 2a. High strength (f_y=690 MPa) steel 35 mm diameter pins were inserted through the spherical bearings and holes in fork-fittings at each end. The end plates of the fork-fittings were gripped by the jaws of the testing machine. The jaws were shifted horizontally to accommodate the off-set of the vertical plates.

Because of the spherical bearings, the line of action of the resultant force applied to the specimen coincided with the line connecting the centres of the bearings. The resultant force consisted of the applied vertical force (P) and the horizontal shear force ($s/(2L+T)P$). Referring to Fig. 2b, the bending moment (M_w) acting initially on the throat of the incomplete penetration butt weld was,

$$M_w = \left(\frac{s\,L}{2L+T} - \left(\frac{T}{2} - \frac{l_w}{2} \right) \right) P, \tag{3}$$

where L=87.5 mm is the vertical distance between the horizontal plate and the centre of the bearing, T=20 mm is the thickness of the plates, l_w=10 mm is the weld leg length, s is the off-set of the plates, and P is the vertical load applied by the testing machine. Defining the eccentricity of the force acting on the incomplete penetration butt weld as,

$$e = \frac{M_w}{P} = \frac{s\,L}{2L+T} - \left(\frac{T}{2} - \frac{l_w}{2} \right), \tag{4}$$

and using that $l_w = T/2$ for the present test series, the off-set providing pure tension stresses on the throat of the incomplete butt weld is obtained as,

$$s = \frac{T}{4}\frac{2L+T}{L} = 11.2 \text{ mm}. \qquad (5)$$

The moment (M_w) and eccentricity (e) defined by eqns (3) and (4) respectively are the initial values. During loading, the plates flexed and large rotations occurred at the incomplete penetration butt weld causing substantial reductions of the initial values. The initial eccentricity (e) is shown in Table 1.

2.3 *Test Procedure*

After gripping the fork-fittings by the upper and lower jaws of the testing machine, a pin was inserted at each end through the fork-fitting and the bearing fitted in the test specimen. A small load was applied to take up slack in the connection before commencing the test. The load was applied by a 2000 kN capacity servo-controlled hydraulic actuator operated in stroke control throughout the test. A very low strain rate was used. The load and the stroke of the ram were recorded at regular intervals.

2.4 *Test Results*

The ultimate loads (P_u) are listed in Table 1. The measured load-extension graphs are shown in Fig. 3. The extension shown on the horizontal axis is the ram position and so includes deformations of the fork-fittings and pins in addition to the elongation of the test specimen, the latter consisting mainly of plastic bearing deformations at the spherical bearings. The jaggedness of the load-extension curves is partly a result of movement of the test specimen along the pins or movement of the spherical bearings in their holes, both resulting momentarily in a drop in load, and partly a result of fracture propagating through the weld. Fracture generally occurred near the ultimate load and/or in the post-ultimate range.

The load-extension curves provide a comparison of the behaviour and ductility of the test specimens. The most ductile tests were those incorporating positive eccentricities (Specimens 1a, 1b and 2). In these tests, the induced moment caused the unfused surfaces of the incomplete penetration butt weld to close-up resulting in compressive stresses between the abutting unfused surfaces. The strongest specimen was no. 3 which incorporated a shift (s) of 10 mm corresponding closely to the value of s=11.2 mm for a concentrically loaded connection, see eqn. (5). The bending moment induced in the tests with non-positive shifts ($s \leq 0$) caused the unfused parts of the incomplete penetration butt weld to open-up as the specimen suffered large bending strains in the weld. The bending strains triggered yielding and an associated plateau in the load-

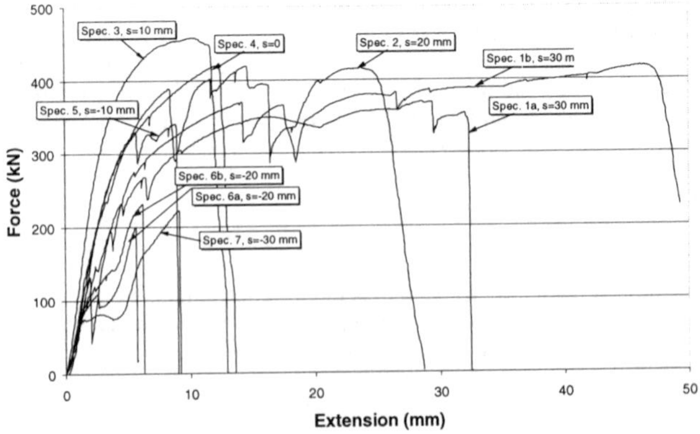

Figure 3: Load-extension graphs

extension graph, as shown for Specimen 7 in Fig. 3. The plateau was less pronounced for the specimens with smaller negative eccentricities. As large bending strains developed in the weld, the specimen deformed such that the incomplete penetration butt weld moved closer to the line of action of the force, thus reducing the bending moment acting on the weld. As the connection straightened, the stiffness increased leading to a steeper load-extension graph, as shown for Specimens 6a, 6b and 7 in Fig. 3. These specimens were generally less ductile than those with smaller negative or positive eccentricities.

The incomplete penetration butt welds were inspected after testing, revealing some lack of penetration at the root of the welds, at surfaces between weld runs and at the 45°-bevelled surface of the vertical plates. Fracture occurred through the welds, rather than through the 20 mm plates, and included areas with lack of penetration. It follows from the average measured leg length (l_w) shown in Table 1 that the unfused areas made up 9% of the nominal value of 10 mm. The fracture surface formed an angle of about 15° with horizontal.

3 COMPARISON WITH AS4100:1998

The test strengths are compared with design strengths in Table 2 and Fig. 4. The third and fourth columns of Table 2 show comparisons of the ultimate loads with strengths based on the measured fused area (A_{wf}) and the nominal area calculated according to AS4100 respectively. The nominal area was calculated according to AS4100 as $A_{AS4100}=(L_w\text{-}10\text{mm})(l_{wn}\text{-}3\text{mm})$, recognising that the length along the centre of the weld was 10 mm shorter than the outside length (L_w) and that the design throat thickness was given by eqn. (2). In both cases, the nominal tensile strength of the weld deposit was used (f_{uw}=500 MPa). The comparisons shown in Table 2 conservatively ignore the shear stresses on the weld throat. When accounting for shear stresses, the ratios shown in columns 3 and 4 of Table 2 are increased by 2 % or less.

It follows from Table 2 that for the nearly concentrically loaded Specimen 3, the test strength exceeded $A_{wf}f_{uw}$ by 8 % suggesting that the tensile strength of the weld deposit was about 8 % higher than the nominal value of 500 MPa. For the eccentrically loaded specimens, the test strengths are lower than $A_{wf}f_{uw}$ in all cases. Thus, the eccentricities reduced the average stress at

Table 2: Comparison of test strengths with design strengths.

Specimen	e/l_{wn}	$P_u/(A_{wf}f_{uw})$	$P_u/(A_{AS4100}f_{uw})$
(1)	(2)	(3)	(4)
1a	0.85	0.80	1.02
1b	0.85	0.86	1.10
2	0.39	0.85	1.12
3	-0.05	1.08	1.27
4	-0.50	0.88	1.10
5	-0.95	0.76	0.91
6a	-1.40	0.45	0.54
6b	-1.40	0.55	0.61
7	-1.84	0.51	0.59

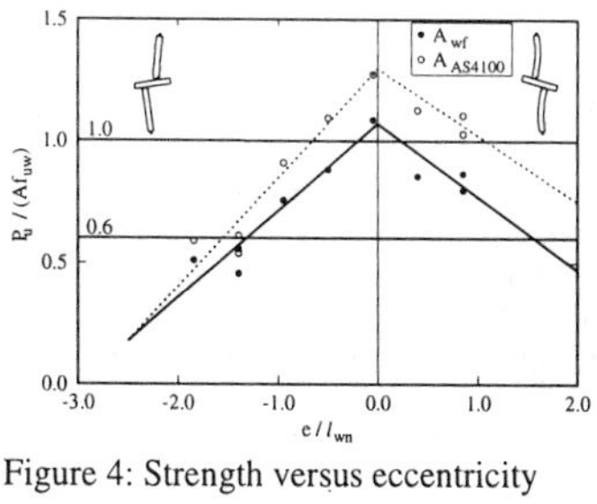

Figure 4: Strength versus eccentricity

failure below the tensile strengths.

When using the nominal area calculated according to AS4100, the test strengths exceeded the strength $A_{AS4100} f_{uw}$ for positive eccentricities $e/l_{wn} \leq 0.85$ and negative eccentricities $e/l_{wn} \geq -0.5$. Thus, for $-0.5 \leq e/l_{wn} \leq 0.85$, the strength can be based on the tensile strength of the weld metal and the nominal area without the need for the 0.6-factor currently specified in AS4100, see eqn. (1). In fact, when the 0.6-factor is used, all the test eccentricities (e/l_{wn}) greater than -1.4 can be accommodated by the current rules of AS4100 for incomplete penetration butt welds, as shown in Table 2. In drawing this conclusion, it is recognised that the area (A_{AS4100}) is significantly reduced for the present test series because 3 mm (30%) is subtracted from the weld leg length according to eqn. (2). This leads to a favourable comparison with test strengths. For larger leg lengths (l_{wn}), the effect of reducing the leg length by 3 mm in determining the design throat thickness becomes less significant and it may not be possible to accommodate as large eccentricities.

Figure 4 shows lines of best fit through the test points in regions of positive and negative eccentricities. The solid and dashed lines are fitted through the strengths nondimensionalised with respect to A_{wf} and A_{AS4100} respectively. According to the intercepts of the dashed lines with the horizontal line $P_u/A f_{uw}=0.6$, the design rules of AS4100 can accommodate eccentricities in the range $-1.5 \leq e/l_{wn} \leq 2.5$. When based on the fused area (A_{wf}), representative of large weld sizes, eccentricities can be accommodated in the range $-1.3 \leq e/l_{wn} \leq 1.5$, as shown by the intercepts of the solid lines with the horizontal line $P_u/A f_{uw}=0.6$.

4 CONCLUSIONS

A pilot series of tests has been described on the strength of incomplete penetration butt welds loaded eccentrically in tension to produce a state of combined tension and bending on the throat of the weld. The current rules of AS4100:1998 specify the strength of incomplete penetration butt welds be calculated using the rules for fillet welds and reducing the leg length by 3 mm in calculating the design throat thickness. It is shown that these rules can accommodate eccentricities of loading in the range $-1.5 \leq e/l_{wn} \leq 2.5$ where negative and positive eccentricities indicate that the abutting surfaces open-up or close-up respectively. For large weld sizes, the range may be reduced to $-1.3 \leq e/l_{wn} \leq 1.5$. Both ranges are likely to cover most eccentricities encountered in practice. Thus, the tests indicate that the design rules of AS4100:1998 incorporate sufficient safety margin to allow for bending stresses in incomplete penetration butt welds.

Using the lines fitted to the test strengths, the results also show that the strength of incomplete penetration butt welds can, in fact, be designed as full penetration butt welds provided the eccentricity lies in the range $-0.6 \leq e/l_{wn} \leq 1$ and the design throat thickness includes a reduction of the leg length of 3 mm. This conclusion applies to leg lengths of the order of 10 mm.

REFERENCES

AS4100. 1998. *Steel structures*, Standards Australia, Sydney.

AS4100-Commentary. 1999. *Steel structures-Commentary*, Standards Australia, Sydney.

AS1250. 1981. *SAA steel structures code*, Standards Australia, Sydney.

Bjorhovde, R. 1987. Welded splices in heavy wide-flange shapes, *Proc. of the National Engineering Conference and Conference on Operating Personnel*, New Orleans. American Institute of Steel Construction.

Bruneau, M. & Mahin, S.A. 1991. Full-scale tests of butt-welded splices in heavy-rolled steel sections subjected to primary tensile stresses. *Engineering Journal*, American Institute of Steel Construction, Vol 28, No. 1, pp 1-17.

Clark, P.J. 1972. Basis of design for fillet welded joints under static loading. *Proc. of Conference on Improved Welding Product Design*, Vol. 2, Welding Institute, pp 85-96.

Fisher, J.W. & Pense, A.W. 1987. Experience with use of heavy W shapes in tension, *Engineering Journal*, American Institute of Steel Construction, Vol 24, No. 2, pp 63-77.

Popov, E.G. & Stephen, R.M. 1976. Tensile capacity of partial penetration butt welds, *EERC Report No 70-3*, Earthquake Engineering Research Center, University of California.

Strating et al. 1971. The influence of a longitudinal stress on the strength of statically loaded fillet welds, *Lasttechniek*, Vol 37, No. 12, pp 236-246.

Mechanics of Structures and Materials, Bradford, Bridge & Foster (eds)
© 1999 Balkema, Rotterdam, ISBN 90 5809 107 4

Comparison of analysis with tests of cold-formed RHS portal frames

T.Wilkinson & G.J.Hancock
Centre for Advanced Structural Engineering, Department of Civil Engineering, University of Sydney, N.S.W., Australia

ABSTRACT: Different types of structural analyses, first order elastic, second order elastic, first order plastic, and plastic zone analysis, were used to simulate the behaviour of 3 portal frames constructed from cold-formed rectangular hollow sections. All analyses underestimated the magnitude of the deflections of the frame mainly due to the non-inclusion of flexible joint behaviour. A second order inelastic analysis, which included gradual yielding, multi-linear stress-strain curves, and member imperfections underestimated the deflections and the hence the second order effects, and therefore slightly overestimated the strength of the frame. A second order inelastic analysis that did not account for member imperfections provided the best estimates of the frame strengths, as it omitted the beneficial effects of the unsympathetic imperfections while underestimating the magnitude of the second order effects.

1 PORTAL FRAME TESTS

As part of a project to investigate the suitability of cold-formed rectangular hollow sections RHS for plastic design (Wilkinson and Hancock 1997, 1998, 1999), three portal frames were tested under simulated gravity and transverse wind load. The general layout of the frames is shown in Figure 1. Each frame spanned 7 metres, with an eaves height of 3 metres, and a total height of 4 metres, constructed from 150 × 50 × 4.0 RHS in either Grade C350 or Grade C450 (*DuraGal*) steel to AS 1163 (Standards Australia 1991). There was a collar tie which joined the midpoint of each rafter. Each frame was pin based. There was a welded internal sleeve connection between the column and rafter as shown in Figure 3. The apex joint is shown in Figure 4. Full details of the connections are in Wilkinson and Hancock (1999). The frame was laterally braced at several critical locations. A downwards vertical load was provided via an MTS actuator connected to a gravity load simulator (which ensured the load remained vertical as the frame swayed). A horizontal point load acted on the north column, via steel cables hung over an auxiliary frame connected to lead weights.

Table 1 briefly summarises the results of the portal frame tests. The value of yield stress presented in Table 1 is from a coupon cut from one web of each specimen. There was variability

Table 1: Summary of Portal Frame Tests

Frame	Made from section	f_y (MPa)	Ratio of vertical to horizontal load V/H	Ultimate load (kN)	
				Vertical	Horizontal
Frame 1	150 × 50 × 4.0 C350	411	40	68.2	1.75
Frame 2	150 × 50 × 4.0 C450	438	40	71.5	1.87
Frame 3	150 × 50 × 4.0 C450	438	3.3	45.7	13.8

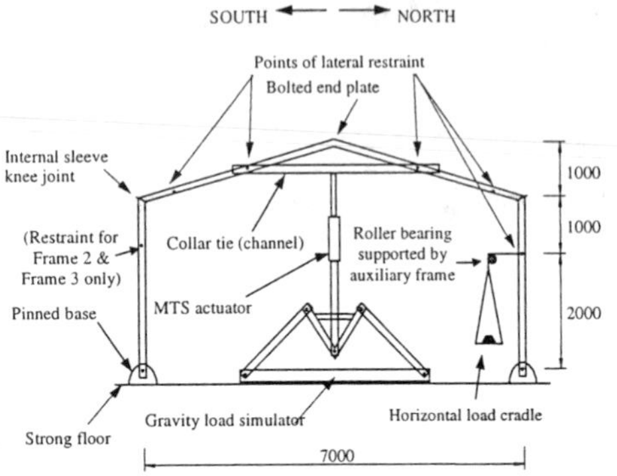

Figure 1: Layout of Portal Frame

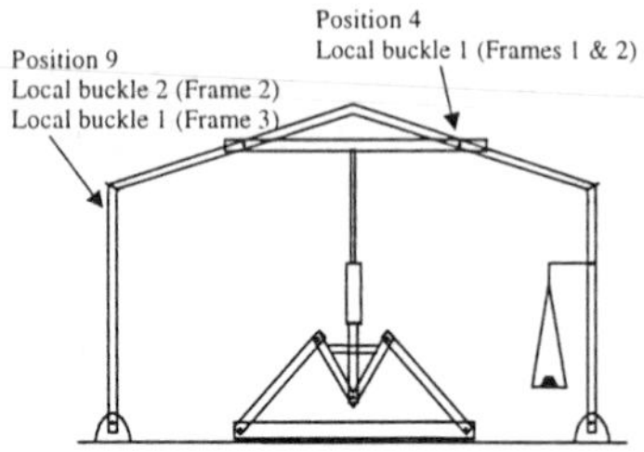

Figure 2: Buckle/Hinge Locations for Portal Frame Tests

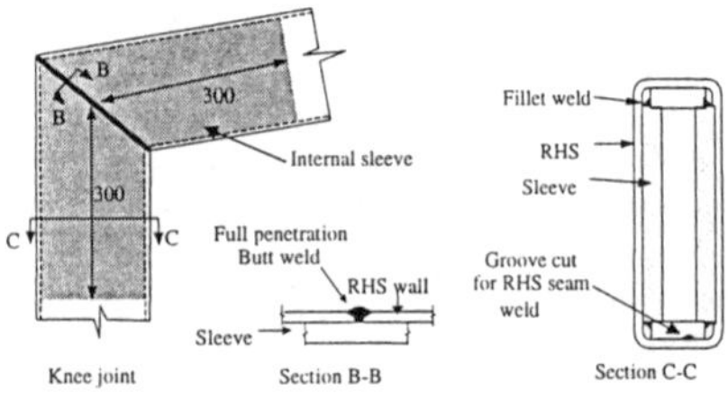

Figure 3: Knee Joint

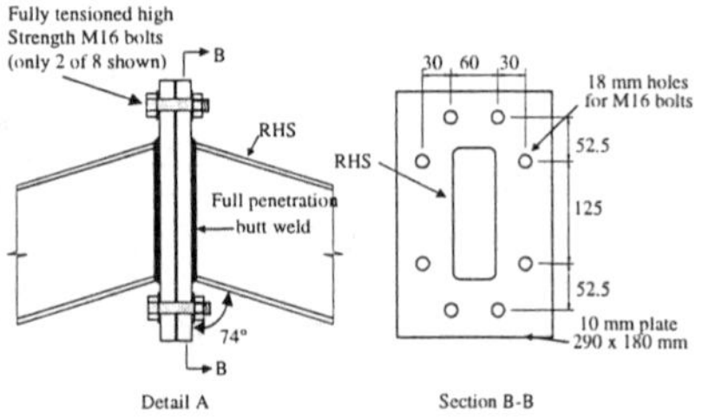

Figure 4: Apex Joint

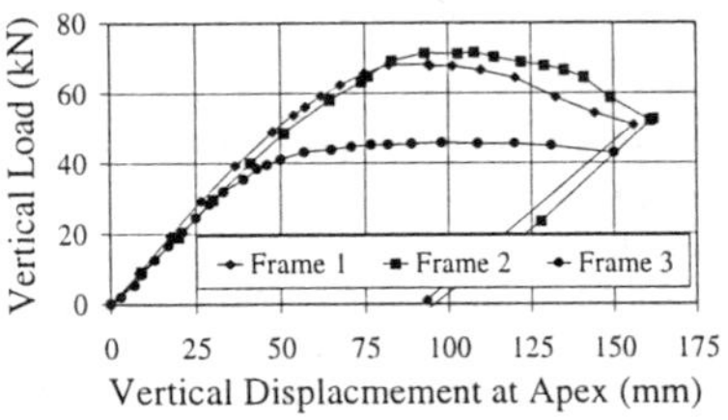

Figure 5: Load - Vertical Displacement

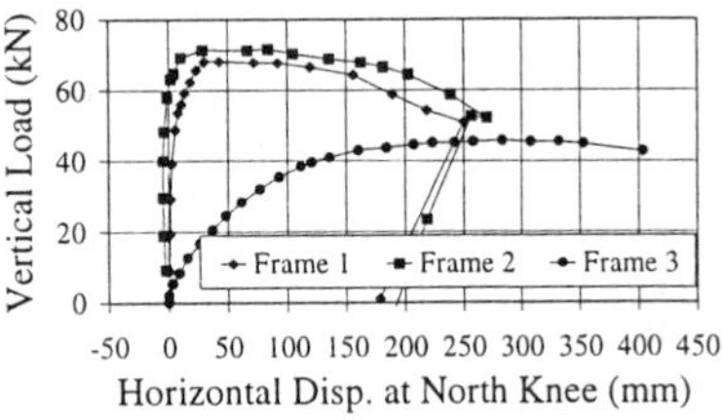

Figure 6: Load - Horizontal Displacement

of yield stress around the section, and considerable strain hardening. Full details on stress-strain curves can be found in Wilkinson and Hancock (1999). Figure 2 indicates where local buckles formed in the frames. Figures 5 and 6 show the load-deflection curves.

2 TYPES OF ANALYSIS

The knee and apex joints were modelled as rigid connections. The extra stiffness caused by the internal sleeve was modelled by defining different member properties in the region of the sleeve. The second moment of inertia (*I*) for the sleeve zone was calculated using the combined thickness of the RHS and the sleeve (14 mm) and assumed prismatic action of the combined RHS/sleeve unit.

The connections at the base and at the end points of the collar tie were pin connections.

First order elastic analysis, using PRFSA (CASE 1997), was performed. The curvature distribution predicted by a first order analysis was compared with the experimental curvature obtained from strain gauges. The general shape of the experimental distribution of curvature matched the curvature distribution predicted by a first order elastic analysis, but the experimental curvature values, and hence the experimental bending moments, were slightly lower than the analysis values at the connections. One reason for lower moments was that the connections were not perfectly rigid, as assumed by the analysis. Full details are in Wilkinson and Hancock (1999).

Simple plastic analysis was carried out using PRFSA (CASE 1997). The analysis assumed no reduction of the plastic moment in the presence of axial force, and simple elastic - plastic material properties were considered. Second order effects were not included.

Second order plastic zone analyses, using the computer program NIFA (Clarke and Zablotskii 1995), were performed. Yielding (or plasticity) is accounted for by a plastic zone approach, rather than assuming infinitely small plastic hinges.

Several different types of analysis were performed using NIFA:
- Second order elastic,
- First order plastic zone using simple elastic-plastic material properties,
- Second order plastic zone using simple elastic-plastic material properties,
- Second order plastic zone using different multi-linear material properties (based on the tensile coupon tests) for the flange, web and corners, and
- Second order plastic zone using different multi-linear material properties, and including the geometric imperfections of the structure.

3 COMPARISON AND DISCUSSION

Figures 7, 8 and 9 plot typical load deflection graphs for two of the frames, and include both the experimental and analytical responses. Other load - deflection curves are in Wilkinson and Hancock (1999). Table 2 summarises the ultimate loads and compares them with the experimental ultimate load.

The simple plastic analysis using the measured yield stress gave a reasonable estimate of the ultimate load, despite all the deficiencies of such an analysis. The absence of second order effects and moment - axial force interaction would tend to give a higher ultimate load, but the lack of strain hardening in the material properties produced a lower load than the experimental values. The two aforementioned effects tended to counteract each other. For Frames 1 and 2, the simple plastic analysis predicted the frame strength to within 2 %. However, in Frame 3, where the second order effects were greatest, the plastic analysis overestimated the strength by nearly 10 %.

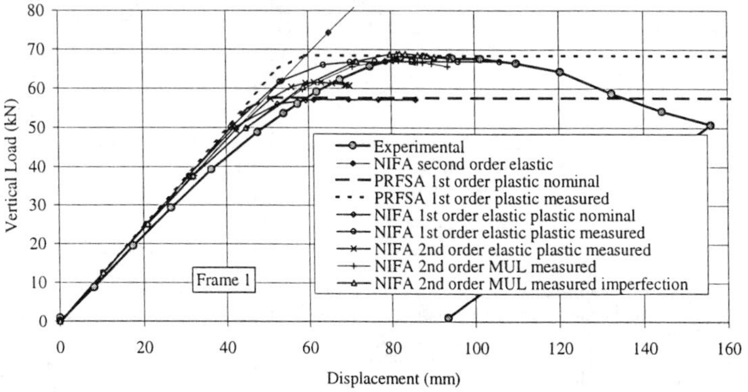

Figure 7: Portal Frame 1: Vertical Deflections at Apex

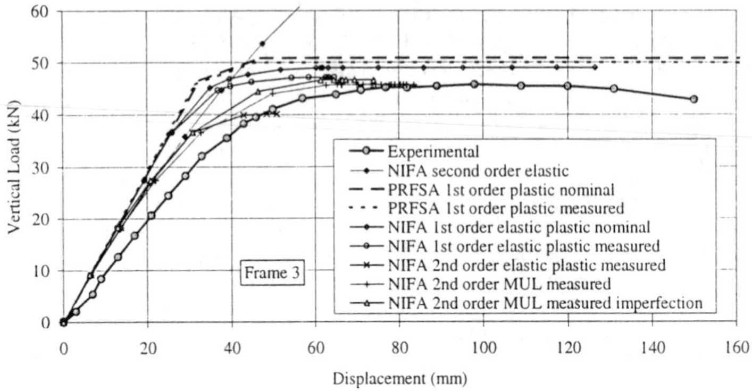

Figure 8: Portal Frame 3: Vertical Deflections at Apex

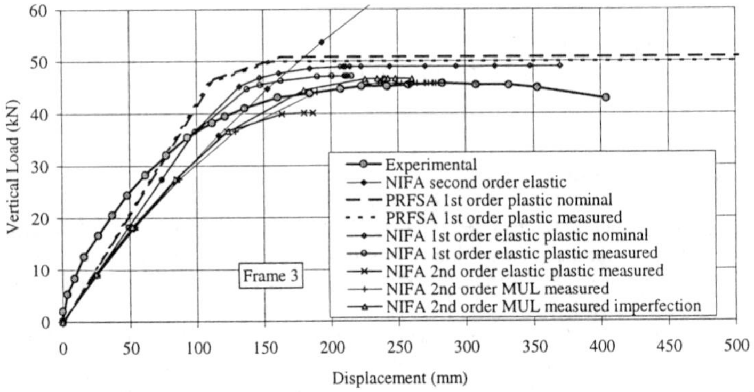

Figure 9: Portal Frame 3: Horizontal Deflections at North Knee

Table 2: Summary of Analysis Results

Program	Type[1]	Props[2]	Mat[3]	Imp[4]	Frame 1		Frame 2		Frame 3	
					Vert[5]	Ratio[6]	Vert[5]	Ratio[6]	Vert[5]	Ratio[6]
	Experimental				68.2		71.5		45.7	
PRFSA	1st plast	nom	e-p	n	57.64	0.85	74.03	1.04	50.82	1.11
PRFSA	1st plast	meas	e-p	n	68.41	1.00	72.76	1.02	49.96	1.09
NIFA	1st pz	nom	e-p	n	57.18	0.84	73.51	1.03	48.94	1.07
NIFA	1st pz	meas	e-p	n	66.89	0.98	70.85	0.99	47.12	1.03
NIFA	2nd pz	meas	e-p	n	61.49	0.90	64.79	0.91	40.12	0.88
NIFA	2nd pz	meas	mul	n	67.16	0.98	73.14	1.02	45.71	1.00
NIFA	2nd pz	meas	mul	y	68.84	1.01	77.05	1.08	46.63	1.02

Notes:
(1) Type of analysis is either 1st order simple plastic, 1st order plastic zone, or 2nd order plastic zone.
(2) Dimensions and yield stress are based on either nominal or measured properties.
(3) Material properties are either elastic-plastic with the properties of the webs used for the entire cross section, or multi-linear approximation to the measured properties with separate properties for the webs, flanges and corners.
(4) The frame imperfections are either included or excluded.
(5) Ultimate vertical load (kN).
(6) Ratio of the analysis load to the experimental load.

The comparison between the first order simple plastic analysis (PRFSA) and the first order plastic zone analysis (NIFA) highlights the effect of interaction of bending moment with axial force. The simple plastic analysis did not account for the reduction of plastic moment in the presence with axial force, while the plastic zone did consider interaction. In addition, the NIFA analysis models the spread of plasticity in the members, while the PRFSA plastic analysis merely considered plastic hinges at distinct points. Since the levels of axial force were relatively low, there was only a small reduction in the bending capacity of the members. Hence an analysis including interaction gave only a slightly lower value. The deflections from the NIFA first order plastic zone analysis asymptotically approached a value just below the simple plastic predictions of PRFSA.

The importance of second order effects was identified by comparing the first and second order NIFA analyses. For Frames 1 and 2, the ultimate load dropped approximately 8 % when second order effects were included. In Frame 3, the horizontal force, and hence the horizontal deflections and second order effects, were greater than in Frames 1 and 2. For Frame 3, the ultimate load was reduced by about 15 % when second order effects were considered.

The shape of the experimental load-deflection curves followed the general trend of the analysis predictions. The experimental vertical deflections in Frame 2 and Frame 3 diverge from the predictions at a very low load - most likely caused by take up of fit in the connections. The maximum load occurs at much larger deflections than those predicted by the various analyses. Each analysis assumes that the joints are perfectly rigid, and any additional non-linearity in the joint behaviour (perhaps caused by slippage, bolts, yielding etc) is not accounted for in the analysis. Wilkinson and Hancock (1998) described tests on the connections used in the portal frame behaviour. The connections exhibited non-linear behaviour, with slight loss of rigidity compared to a plain member (though the observed behaviour would not be described as "semi-rigid"). It was also unknown whether the RHS and sleeve acted prismatically or non-prismatically. (Prismatic action is defined as the two objects acting as a unit with the stiffness of the combined unit, while non-prismatic action occurs when the two elements act independently and slip occurs between the two). The analysis assumed that there was prismatic action, whereas the true behaviour was probably a mixture of prismatic and non-prismatic action. Assuming prismatic action would give smaller deflections.

Generally modern structural analysis software is capable of including flexible joint behaviour, but the lack of information of the moment-curvature behaviour of various connections prevents its widespread inclusion.

When the imperfections were included in the analysis, the ultimate load increased, by 3 % for Frame 1, 6 % for Frame 2, and 2 % for Frame 3. Typically, if an imperfection is in the same direction as the deflections of a structure it will tend to reduce the ultimate load. Such an imperfection is sometimes referred to as a sympathetic imperfection. An unsympathetic imperfection results in an increase in ultimate load. The measured imperfections were found to be unsympathetic (values and shapes of the imperfections are in Wilkinson and Hancock 1999).

In all cases, the 2nd order plastic zone analysis including the imperfections, overestimated the ultimate load, and provided a less accurate prediction of the ultimate load than the same analysis without the imperfections. The major reason for the less accurate prediction was that the analysis underestimated the deflections (due mainly to the non-inclusion of flexible joint behaviour), so that the second order effects in the analysis are less than those in the experimental frames. Hence the predicted results are higher than the experimental loads. Excluding the imperfections gave a more accurate result because the effect of underestimating the deflections counteracted the effect of omitting the imperfections.

If an advanced analysis is to provide a highly accurate prediction of the frame behaviour, the flexible or semi-rigid behaviour of the connections ought to be considered. However, significant information on the semi-rigid joint behaviour needs to be gathered before it can be included in the analysis.

4 CONCLUSION

This paper has compared the results of tests on cold-formed RHS portal frames, with the predictions of various forms of structural analysis. A simple, plastic analysis provided a reasonable estimate of the frame strength since the effects that were omitted from the analysis tended to counteract each other, especially for frames with small horizontal load. A second-order inelastic analysis, which included material non-linearity and member imperfections slightly over predicted the frame strength, since the analysis did not consider the small loss of connection rigidity, which resulted in underestimating the sway deflections and second-order moments within the frame. A second order inelastic analysis which did not account for member imperfections provided the best estimate of the frame strength, but also under-predicted the deflections.

5 REFERENCES

CASE, (1997), "PRFSA: Plane Rigid Frame Structural Analysis", *Users Manual*, Centre for Advanced Structural Engineering, Department of Civil Engineering, University of Sydney, Sydney, Australia.

Clarke, M. J. and Zablotskii S. V., (1995), "NIFA: Non-Linear Inelastic Frame Analysis", *Users Manual*, Centre for Advanced Structural Engineering, Department of Civil Engineering, University of Sydney, Sydney, Australia.

Standards Australia, (1991), Australian Standard *AS 1163 Structural Steel Hollow Sections*, Standards Australia, Sydney.

Wilkinson T. and Hancock G. J., (1997), "Tests for the Compact Web Slenderness of Cold-Formed Rectangular Hollow Sections", *Research Report*, No. R744, Department of Civil Engineering, University of Sydney, Sydney, Australia.

Wilkinson T. and Hancock G. J., (1998), "Tests of Knee Joints on Cold-Formed Rectangular Hollow Sections", *Research Report*, No. R779, Department of Civil Engineering, University of Sydney, Sydney, Australia.

Wilkinson T. and Hancock G. J., (1999), "Tests of Cold-Formed Rectangular Hollow Section Portal Frames", *Research Report*, No. R783, Department of Civil Engineering, University of Sydney, Sydney, Australia.

6 NOTATION

f_y Yield stress
H Horizontal load
I Second moment of inertia
V Vertical load

Mechanics of Structures and Materials, Bradford, Bridge & Foster (eds)
© 1999 Balkema, Rotterdam, ISBN 90 5809 107 4

Study on the residual stress and the elasto-plastic behavior for a circular plate

N. Koike, Y. Kato, T. Nishimura & H. Honma
Department of Mechanical Engineering, College of Science and Technology, Nihon University, Tokyo, Japan

ABSTRACT: The objective of this paper is to investigate how the residual moment caused by the thermal deformation develops over the plate. In our previous study, the thermal stress analysis of a circular plate has been carried out with taking account of the spreading plastic hinge model, and the numerical results agree approximately with the experimental results. However, it is hard to measure accurately the deflection of the plate while the plastic deformation develops on the heating process. The plastic deformation, caused by the statically applied symmetric load instead of the thermal agency, is experimentally investigated in order to identify the behavior of the plastic deformation In this study.

1. INTRODUCTION

The residual moment in a plate member for isolating a thermal agency is discussed in this study. When the plate member, which is subjected to a high temperature, is clamped at the periphery, the thermal stress is induced by the restriction of thermal expansion. If the temperature difference at the back and face of the plate is larger than the mean temperature change over the plate surface, the bending deformation is predominant rather than the membrane deformation. Since the bending deformation may cause the ill condition in a system by the interference to the adjacent element, the bending deformation should be reduced in an actual structure. On the other hand, if the plate member is constrained at the periphery in order to reduce the bending deformation, the bending moment by thermal stresses occasionally exceeds the ultimate bending moment (fully plastic moment), and the plastic strain occurs locally. Thus, the residual moment is produced over the whole area of the plate at the room temperature after the cooling process.

The distribution of the residual moment is closely related with the process of the plastic deformation in the plate. In the former limit analysis of a circular plate [Sawczuk, A. 1989], it is concluded that the circumferential plastic hinge line develops at the outset, when the ratio of the incremental deformation $\dot{\rho}_\theta/\dot{\rho}_r$ approaches to zero for an axial symmetric loading (where, $\dot{\rho}_r$ and $\dot{\rho}_\theta$ denote the incremental curvature of a plate along the radial and circumferential direction, respectively). As the inclination angle at the hinge line develops in accordance with the increasing $\dot{\rho}_r$ the increment of curvature $\dot{\rho}_r$ must be infinitely large, that is, the kinematic mechanism on the plastic deformation is satisfied without the spread of the plastic area. However, the actual plastic deformation does not concentrate at the plastic hinge line because of the thickness of the plate. In this study, it is attempted to resolve how the plastic area spreads under the progressive deformation beyond the elastic limit.

In our former study [Watanab.H. & Kato, Y. 1997], the thermal stress analysis of a circular plate with a circular whole at the center was carried out with taking account of the plastic hinge which spreads over the finite region. However, it is difficult to observe experimentally under a high temperature how the plastic area spreads from the periphery. Then we investigate experimentally the plastic deformation by the statically applied symmetric load instead of a thermal agency.

2. OVERVIEW OF THERMAL DEFORMATION AND EXPERIMENT

In this section, the experimental result of the thermal deformation and the corresponding numerical calculation [Watanabe,H. & Kato, Y. 1997] are stated briefly before discussing the plastic deformation by a external load under the room temperature.

2.1 *Experiment for the thermal deformation*

The schematic view of the experimental system is illustrated in Figure 1. The system installs a rotary blade, a couple of stationary circular plates made of mild steel and a magnetic coil surrounding the rotary blade. In the system, a couple of stationary circular plates isolates a thermal agency caused by the effect of the Eddy Current, which is induced by the rotating blade across the magnetic flux. The rotation of the rotary blade, the intensity of magnetic flux and the flow of the coolant water can adjust adequately the temperature difference at the back and face of the plate. The temperature distribution over the plate is measured by several thermoelectric couples attached on the plate surfaces. The deflections, during the heating process or cooling process, are also measured by the several flection meters.

After completing the heating process, the plate is cooled up to the room temperature and the strain gages are bonded to the deformed plate. The strain, which is induced by a release from the constraint at the outer edge, is measured, and then the residual moment can be resolved by the measured strain.

2.2 *Analysis and numerical result*

Although the temperature distribution obtained by the experiment is not uniform but slightly quadratic over the plate, it is assumed for simplicity that the temperature over the plate surface is constant and the temperature distribution through the thickness of the plate is liner. It is supposed in the analysis that the circular plate is fixed rigidly at the inner periphery and supported simply at the outer edge. Under this boundary condition, the plastic hinge is brought

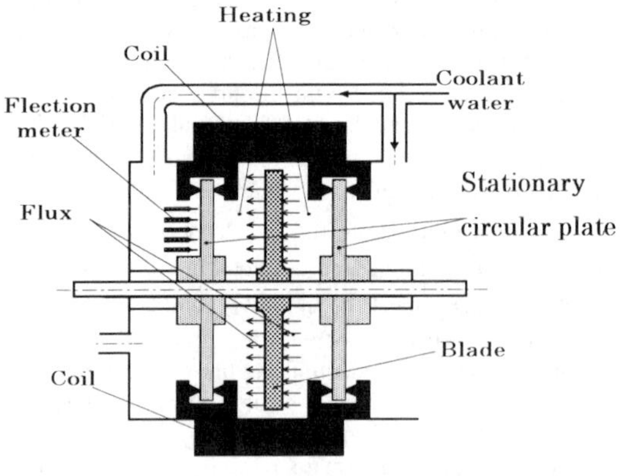

Figure 1. Experimental system

along the inner periphery of the circular plate at the outset. If the temperature difference becomes larger, it is expected that the plastic region tends to spread gradually from the inner periphery to the outer. Hence, the fact of the spreading plastic area is incorporated into the numerical calculation of the thermal deformation in this analysis.

The bending moment M are related with the curvature ρ in the thermal elastic problem by the following equations,

$$M_r = D(\rho_r + \nu\rho_\theta - \rho_{th}), \qquad M_\theta = D(\rho_\theta + \nu\rho_r - \rho_{th}) \tag{1}$$

where, D is the well known flexural rigidity of a plate, and the subscript r and θ denote the radial and circumferential direction. ρ_{th} is the additional curvature caused by the thermal expansion, and depends on the heat convection coefficient and the temperature difference at the face and back surface of a plate.

The generalized bending moment ($\overline{M}_r, \overline{M}_\theta$) and the generalized deformation ($\overline{\eta}_r, \overline{\eta}_\theta$), which are used in the analysis, are defined as follows,

$$\overline{M}_r = M_r / M_p, \qquad \overline{M}_\theta = M_\theta / M_p \tag{2}$$

$$\overline{\eta}_r = (\rho_r + \nu\rho_\theta - \rho_{th}) / \eta_p, \qquad \overline{\eta}_\theta = (\nu\rho_r + \rho_\theta - \rho_{th}) / \eta_p \tag{3}$$

where, M_p is the fully plastic moment of a plate, and $\eta_p = M_p / D$. Hence, the generalized deformations coincide with the corresponding generalized moments within the interaction curve, respectively.

The numerical calculation is carried out up to the maximum temperature deference 185°C, which is observed in the experiment.

The distribution of the residual bending moment corresponding to the maximum temperature difference (185℃) are shown in Figure 2 as an example. The thick solid line in the figures indicates the distribution of the bending moment, derived by the numerical calculation, at the heated state. The calculated residual moment at the room temperature is also shown by the thin solid line. The experimentally obtained residual moment is represented by the symbol "□" in the figure.

However it is hard to measure the actual bending moment distribution at the heated state. Since it is inevitable to know the exact bending moment at the heated state in order to obtain the residual moment exactly, the plastic deformation caused by the statically applied symmetric external load is investigated experimentally instead of the thermal deformation.

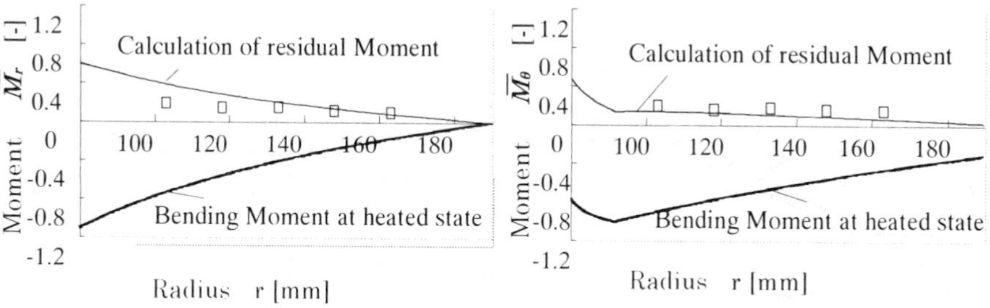

(a) Bending moment $\overline{M}_r$ [-] (b) Bending moment $\overline{M}_\theta$ [-]

Fig.2 Distribution of the residual moment

3 MODELING OF THE PLASTIC HINGE LINE

Figure 3 shows the schematic view of the cross section of the deformed circular plate in the static load experiment. The portion "ab" is the plastic region, and "bc" is the elastic region in the figure. The inner radius and the outer radius are denoted by r_i and r_o, respectively. The radius at the boundary line between the plastic and elastic region is designated by r^*. The width of the spreading plastic area is expressed by dL.

Sawczuk stated that the plastic hinge line develops when

$$\frac{\dot{\rho}_\theta}{\dot{\rho}_r} = \lim_{\Delta r \to 0} \frac{\dfrac{\dot{\theta}}{r_i}}{\dfrac{\Delta \dot{\theta}}{\Delta r}} \to 0 \tag{4}$$

This condition is reasonable at the onset of the yielding, because the inclination angle at the fixed periphery is zero. The plastic deformation occurs only along the plastic hinge line, provided that the bending moment remains on the interaction curve without traveling. However, as mentioned above, the development of curvature is restricted by the geometrical constraint of the actual plate. In other words, the concentrated plastic hinge line does not bring the plate into the kinematic mechanism, and the plate can sustain the increasing load in spite of perfectly plastic body. Consequently, the bending moment in the plastic region tends to travel along the interaction curve, and then Eq.(4) can not be adopted to the progressing plastic deformation. Hence, the progressive plastic deformation can not be resolved without introducing the concept of the spreading plastic area. Therefore, we suggest the spreading plastic hinge model in order to assure the continuity of the bending moment across the boundary line between the elastic and plastic region. Then, an adequate ratio of the incremental deformation is allowable in the plastic area, that is,

$$\alpha = \frac{\dot{\rho}_\theta}{\dot{\rho}_r} \tag{5}$$

where α is not constant but varies with the radial position r and determined by the value of the current bending moment on the interaction curve.

4 EXPERIMENT

The experimental system by the static load is illustrated in Figure 4. The specimen is made of annealed mild steel, which is the same material on the thermal experiment. The dimension, and the physical and mechanical property of the specimen are shown in Table.1. The load is applied at the center of the plate up to 48069 [N]. The distribution of the deformation is measured by the

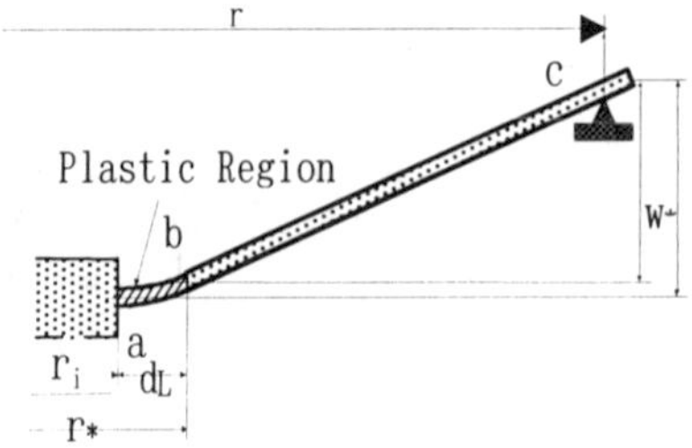

Figure 3. Spreading plastic

Table 1. The dimension and physical Property of the circular plate

Inner radius (r_i)	98[mm]
Outer radius (r_o)	220[mm]
Thickness (h)	8[mm]
poisson's ratio (v)	0.3[-]
young's modulus (E)	204[Gpa]
yield stress	242[Mpa]

three units of strain gages, each of which is composed of five gages arranged serially by 3[mm] apart each other. The first one is attached in the surface of the plate at the radial position r=103[mm]. The second unit is shifted by 7[mm] toward the radial direction from the first and the third one is positioned by 15[mm] apart from the first. The several flection meters are arranged radially in order to measure the deflections. The radial positions of the flection meters are shown in the figure.

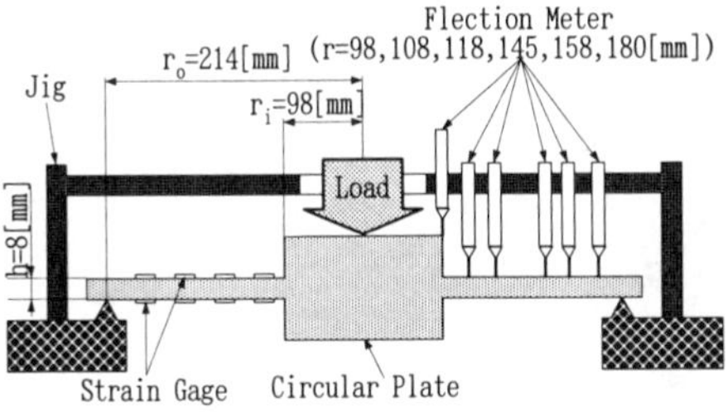

Figure 4. Experimental system

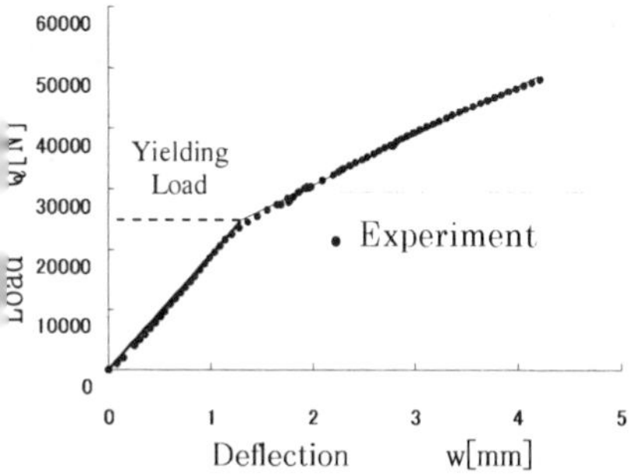

Figure 5. The relation between the load and the deflection

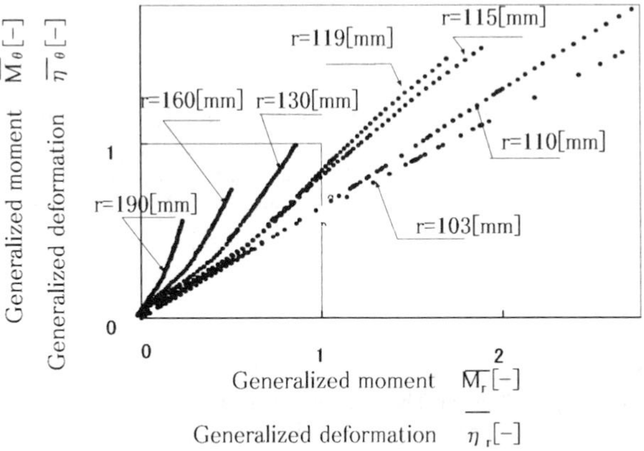

Figure 6. Generalized deformation

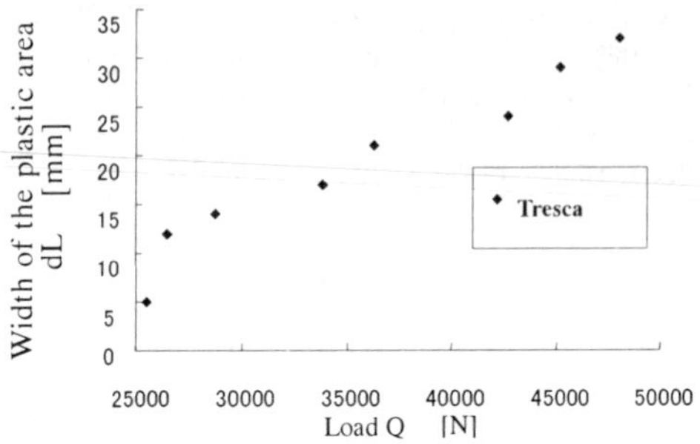

Figure 7. The expanding plastic region

5 EXPERIMENTAL RESULTS

Figure 5 indicates the relation between the deflection at the center of the plate and the applied load. The yielding load (corresponding to the fully plastic moment) can be estimated as 24000 [N]. It should be noted that the applied load still increases beyond the yielding load in accordance with the deflection. It implies that the plastic deformation does not concentrate at the plastic hinge line but the plastic area spreads radially, because the circular plate remains still stable.

Figure 6 shows that the generalized deformations ($\bar{\eta}_r, \bar{\eta}_\theta$) change in accordance with the increasing load up to 48069 [N]. Although the bending deformations are measured at thirteen radial positions by the strain gages, in order to avoid the intricacy of expression the obtained results at the position (r=103,110,115,119,130,160 and 190[mm]) are cited in the figure. It is obvious from the figure that the deformation at the inner periphery is maximum. When the deformation curve intersects Tresca's interaction curve, the yielding occurs and the following all deformational behaviors tend to be bent slightly upward. The fact suggests that the bending moment travels along the interaction curve. Since we can observe the relation between the applied load and the deformation, the spread of the plastic area can be obtained from the figure 6. The obtained results are shown in Figure 7. It is apparent that the plastic region expands gradually with the increase of the load.

6. CONCLUSION

(1) It is confirmed from the experimental results that the plastic hinge line spreads from the inner periphery to the outer.
(2) The distribution of the bending moment in the plastic region can be resolved precisely by introducing the concept of spreading plastic hinge
(3) The residual moment brought by the thermal deformation can be resolved by using the knowledge of the plastic deformation beyond the elastic limit and the moment distribution stated here.

REFERENCE

Watanabe,H. & Kato, Y. 1997. Fundamental Study of Thermal Stress Analysis. (On Case of the Circular Plate with a Circular Hole at the Center). The Second Asian-pacific Conference.
Sawczuk, A. 1989. Mechanics and Plasticity of Structures. Chichester: ELLES HORWOOD LIMITED.

Mechanics of Structures and Materials, Bradford, Bridge & Foster (eds)
© 1999 Balkema, Rotterdam, ISBN 90 5809 107 4

Yield lines in collapse mechanisms of CHS joints

K. Dale, X.-L. Zhao & P. Grundy
Department of Civil Engineering, Monash University, Clayton, Vic., Australia

ABSTRACT: Recent experimental programs at Monash University on CHS KT-joints and CHS YT-Joints have shown incremental collapse occurring at loads up to 18% below the static strength of the joint. Incremental collapse occurs when the joints are subjected to high amplitude, low frequency repeated loads. Little theoretical analysis has been undertaken thus far and current research aims at predicting shakedown loads for these joints. The Upper Bound Theorem of shakedown is used to assess the shakedown limit, and this requires an envelope of elastic response of the structure to all load cases, and a kinematically admissible strain rate. The kinematically admissible strain rate is basically a valid collapse mechanism and this paper examines a possible method of quantifying the strain rate, using yield line analysis.

1 INTRODUCTION

In recent years, test programs on CHS connections at Monash University have resulted in incremental collapse of the connection occurring up to 18% below the ultimate static collapse load of the joint.

The research at Monash on tubular connections under variable repeated loads was begun by Goh (1994). His work involved testing YT-joints and he found that the shakedown limit of the tubular joint was less than the static collapse limit by up to 18%. The work of Milani & Grundy (1996, 1997) followed on, with experimental work on profile cut KT-connections, and innovative KT-connections under variable repeated loads. The findings of this research indicate for the conventional profile cut KT-joints that the shakedown limit was less than the static collapse limit by up to 17%, with a corresponding reduction of 20% in the case of the innovative KT-joints.

The previous work in this area has been primarily experimental, with no attempt to make any comparison with analytically derived shakedown limits.

The aim of the current research is to use the Upper Bound Theorem of Shakedown to assess the shakedown limit. This method requires an elastic stress envelope and a kinematically admissible strain rate. The elastic stress envelope can be obtained using finite element analysis (FEA), and the procedure used is described in another paper by the authors (Dale et al. 1999). The kinematically admissible strain rate is basically a valid collapse mechanism and this paper examines a possible method of quantifying the strain rate, using yield line analysis. The validity of this method was checked using test results obtained from the test database published in 1996 (Makino et al. 1996).

It can be seen that the yield line method can be used successfully to obtain data such as yield line rotations and yield line lengths to quantify the kinematically admissible strain rate. The strain rate data can then be used in conjunction with the elastic stress envelope to obtain the shakedown limit (Dale et al. 2000).

2 SHAKEDOWN AND INCREMENTAL COLLAPSE

The phenomena of shakedown and incremental collapse were first proved theoretically by Melan (1936). Melan had formulated the lower bound or static theorem of shakedown. The static theorem of shakedown is such that shakedown will occur provided that at all times the stresses obtained from an elastic analysis (σ_e) satisfy the following:

$$\sigma_o^- \leq \sigma_e(t) + \sigma_r \leq \sigma_o^+ \tag{1}$$

where σ_r = statically admissible residual stress, σ_o^+, σ_o^- = maximum and minimum yield stresses, and $0 \leq t \leq \mathrm{T}$, where T = design life.

Of more interest in this project is the upper bound or kinematic theorem of shakedown (Koiter 1956). This theorem states that shakedown will not occur if a collapse mechanism exists such that:

$$\int_V \max\{\sigma_e\}.\varepsilon.\mathrm{d}V \geq \int_V \sigma_o.\varepsilon.\mathrm{d}V \tag{2}$$

where ε = strain rate, $\max\{\sigma_e\}$ defines the envelope of stresses from elastic response over the time, T, and the other variables are as defined for Equation 1.

With regard to Equation 2, if we know the strain rate, ($\dot{\varepsilon}$), the yield strength of the material, (σ_o), and, the elastic envelope of stresses, (σ_e), the incremental collapse load factors, or the shakedown limit can be estimated. The incremental collapse load factor (λ), can be found by factoring σ_e by λ and then using the collapse mechanism which minimises λ. Equation 2 is used by performing integration along the yield lines of the product of normal moment by hinge rotation, and normal membrane force by membrane dislocation. The strain rate is localised at hinges, so we require hinge rotations.

3 YIELD LINE ANALYSIS

Yield line analysis was pioneered by Johansen (1952), as a method of analysis of slabs as laterally loaded plates.

Yield line analysis has been used on RHS tubular connections (Morris & Packer 1988, Korol 1982, Packer et al. 1982, Mitri & Korol 1984, Zhao & Hancock 1991), and has been reasonably successful in estimating static yield strengths. Yield line analysis is being used in this research as a method of determining the hinge rotations.

Yield line theory has also been applied to CHS joints (Makino et al. 1989). In particular, it was used to estimate the ultimate strength of CHS X-joints. In that analysis it was assumed that the collapse mechanism is an ellipse in the XY-plane. The yield pattern assumed in this case is similar to that proposed by Makino et al. 1989, as shown in Figure 1. A spreadsheet has been created in which variables such as D (chord diameter), d (brace diameter), β (d/D), T (chord thickness), $\sigma_{y,c}$ (chord yield stress), δ (local chord indentation), and ϕ (in-plane brace rotation), as well as the ellipse parameters a (yield ellipse semi-major axis), and b (yield ellipse semi-minor axis), can be altered to obtain an estimate of the yield load. This analysis uses vector mathematics to calculate plastic hinge rotations by comparing the position of rigid plates before and after the brace has deflected.

4 ANALYSIS METHOD

The method adopted to discretise the yield lines is to identify points on the yield line (positive and negative yield) and create a series of triangular plates between the positive and negative yield lines.

A spreadsheet has been developed that will calculate yield line rotations for T-joints or X-joints in which the brace is loaded axially or rotated in-plane. Variables required for input to the spreadsheet are defined in Section 3.

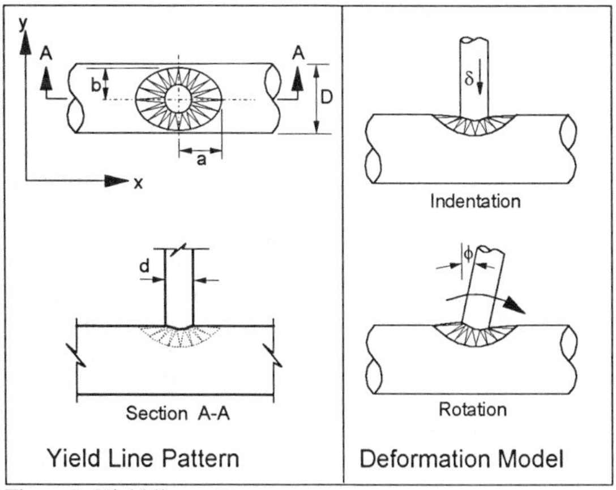

Figure 1: Yield line pattern

The initial function of the spreadsheet is to calculate the coordinates of nodes along the yield lines. Once the nodes have been defined, vector normals can be determined for each of the triangular plates. The brace is then deflected by δ, or rotated by ϕ (Fig. 1), the vector normals are recalculated, and from these, the yield line rotations can be calculated.

Although the yield line method is an accepted analysis method some work was undertaken in an attempt to justify its use in this project.

The work method is used to estimate the yield load on X- and T-joints, so the lengths of the yield lines are required, these are simply calculated as the distances between nodes. Some of the values change slightly after the brace is displaced, so the average of the lengths before and after the deflection is used. The work method equates the internal and external work on the joint. Internal work is provided by the yield lines yielding, and external work is due to the movement of the load applied to the brace, or in the case of in-plane rotation, the rotation of the brace in radians multiplied by the moment to cause the rotation.

The total internal work is the sum of the rotations of the yield lines multiplied by the lengths of the particular yield lines, multiplied by the plastic moment per unit length of the chord material. The plastic moment per unit length is $(\sigma_{y,c}T^2)/4$. Interaction with membrane force is ignored at this stage. This is believed to be of minor significance, owing to the deformability of the circular cross-section.

After calculating the internal work done by the joint, it is divided by the brace displacement, δ, or rotation, ϕ, to give the estimated yield load or in-plane moment applied to the brace.

5 RESULTS

5.1 *Yield load analysis*

Data to be used in the yield line analysis was obtained from the *Database of Test and Numerical Analysis Results for Unstiffened Tubular Joints* (Makino et al. 1996). Data for X- and T-joints in both compression and tension was used, although the data set had to be modified to fit within the limitations of the analysis method. Some of the constraints on the data used are as follows: they must be T- or X- CHS joints (θ=90°), values of N_y and $\sigma_{y,c}$ must be provided, axial loads only, test data only (no numerical modelling), no axial load on the chord, and only values of $0.2<\beta< 1.0$.

With the above constraints in place the data available for use is presented in Table 1.

Table 1: Data available for use

	X-Joints	T-Joints
Compression	21	50
Tension	24	30

The results obtained saw a consistent underestimation in yield load. This is largely a result of problems in defining the yield load.

Values given using the yield line method are "first yield" results, with no account of the effects of membrane action or strain hardening. According to the test database (1996), "Some different definitions of yield strength exist depending on the purposes of research. In this database, when the yield strength are shown by researchers, those values are used as yield strengths. When the load-deformation curves are clearly shown, the yield strength is determined by the method of Kurobane et al (1984)."

Kurobane et al. (1984), define the yield load in compression as the load at which the load-deformation curve intersects the straight line $N=0.779K_n\delta$, where K_n is the initial joint stiffness (Fig. 2). A different method is used in the case where the brace is in axial tension. For X- and T-joints in tension, the load-deformation curve plotted on log-log scales gives two approximately straight lines. The yield strength is defined as the load at the point of intersection of these lines. The two previously mentioned methods will always give values larger than "first yield". In some cases though, researchers will have nominated their own values of yield load, and probably have used a different method of defining the load. In the case of Yamasaki et al. (1979), N_y was defined as the load at which the principal stress at the joint hot spot reaches the yield stress of the chord material.

The above examples demonstrate the difficulty in attempting to use N_y as a base for calibration. In an attempt to reasonably justify this method for use in predicting yield line rotations, load-deformation curves were obtained for some of the tests comprising the data in Table 1. An assumption was made that the value of "first yield" being calculated by our spreadsheet was more likely to be a similar value to the lower circle shown in Figure 2, that is, the point where the load-deformation curve begins to deviate from its initial angle, K_n. In this paper and another by the authors (Dale et al. 1999), one particular test specimen is studied, a T-joint designated "TC-10" in the test database (Makino et al. 1996). TC-10 has a chord diameter of 165.2mm, a brace diameter of 76.3mm, a chord thickness of 4.7mm, and a chord yield stress of 441MPa. The load-deformation curve for this test specimen indicated a "first yield" compressive load of about 40kN applied axially to the brace. When this connection was modelled using the devised spreadsheet, with a local chord indentation of 1mm, and yield ellipse semi-major and -minor axes of 100 and 75mm respectively, a "first yield" value of 45.2kN is obtained. Figure 3 is an isometric view of one quarter of this particular connection.

5.2 *Yield line rotations*

Yield line rotations and lengths are calculated by the devised spreadsheet and it is a simple procedure to obtain rotations and yield line lengths once a yield pattern is chosen. The theory assumes small displacements, and a unit displacement is generally chosen. Calculated yield line lengths and rotations for the connection described in Section 5.1 are displayed in Table 2. The yield lines referred to are numbered as in Figure 3.

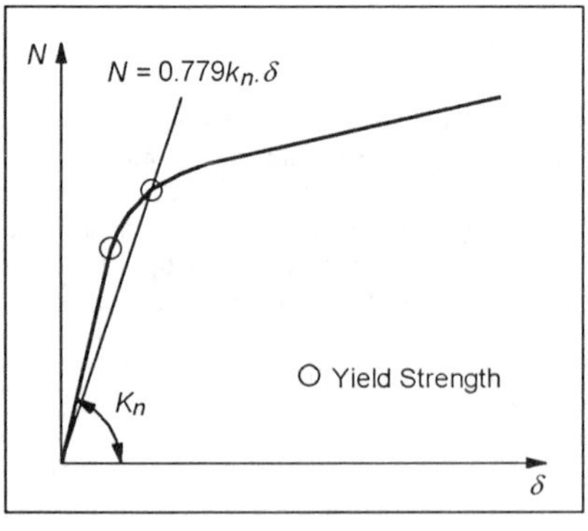

Figure 2: Schematic load-deformation curve

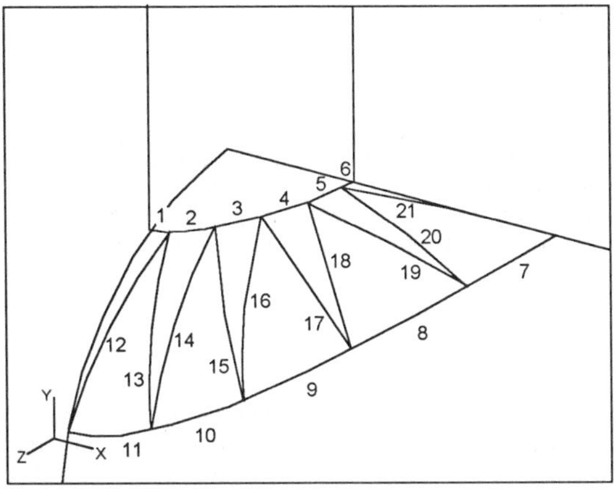

Figure 3: Yield line locations

Table 2: Yield line rotations and lengths

Yield line number	Axial Compression, δ=1mm		In-Plane Bending, ϕ=1°	
	Rotation (rad.)	Length (mm)	Rotation (rad.)	Length (mm)
1	0.013017286	5.967974841	0.012593607	5.967974841
2	0.014189709	12.06746645	0.01097940	12.05286401
3	0.016575375	12.25752047	0.008455638	12.23831585
4	0.017662314	12.23597253	0.009204075	12.2224837
5	0.01677372	12.04543179	0.010447166	12.0411006
6	0.016043832	5.967974841	0.010550042	5.967974841
7	-0.016777686	31.42951915	-0.011030118	31.42951915
8	-0.01651138	32.6627864	-0.009771514	32.6627864
9	-0.015518724	31.0276263	-0.007264371	31.0276263
10	-0.014392822	27.44982005	-0.00430928	27.44982005
11	-0.013615879	24.15587197	-0.001400079	24.15587197
12	0.00259207	53.87727825	0.012425054	54.19623198
13	0.001901199	52.10683025	0.011930408	52.39991693
14	0.002059769	53.28667887	0.01301865	53.521222
15	0.002452255	50.68417161	0.010854159	50.85574581
16	0.000214799	52.11242384	0.012012696	52.25311386
17	0.004121668	53.20234507	0.007514603	53.27033612
18	0.00085658	53.95067589	0.009061744	54.01267885
19	0.005535939	59.19346797	0.002046931	59.20608359
20	0.000774107	59.30401177	0.005831372	59.31969395
21	0.00440052	62.61088493	0.002887549	62.61088473

Negative rotations indicate a "hogging" type of rotation, while positive rotations are "sagging" types of rotations. When the brace is deflected axially there are two planes of symmetry and the yield line rotations and lengths will correspond right around the brace to those shown in Figure 3. In the case of in-plane rotation of the brace, there is one plane of symmetry, and one plane of anti-symmetry. The plane of symmetry is along the centre line of the chord member, with the anti-symmetrical plane being the y-z plane in Figure 1. Yield line rotations across the plane of anti-symmetry will have an opposite sign to their corresponding yield line described in Table 2.

6 DISCUSSION

6.1 *Yield load analysis*

Although it is difficult to validate the proposed model using test data, the use of load-deformation curves allows us to get a better estimate of "first yield" and compare this type of value favourably with the results obtained in this research. This positive result and the wide acceptance and use of the yield line method are taken as justification of the use of the yield line method in this research.

6.2 *Yield line rotations*

Yield line rotations and yield line lengths can be obtained using the method described in this paper. The rotations are the strain rate referred to in Equation 2. Obtaining membrane strain is an issue not discussed in this paper, but it is likely to be estimated by examining the change in length of radial yield lines. This involves second order effects including deformation of the chord out of circular and requires further work.

7 CONCLUSIONS

The yield line method can successfully be used to obtain data such as yield line rotations and yield line lengths to quantify the kinematically admissible strain rate for CHS. The strain rate data can then be used in conjunction with the elastic stress envelope to obtain the shakedown limit.

REFERENCES

Dale, K., Zhao, X.-L., & Grundy, P. 1999. Application of Finite Element Analysis in Shakedown Analysis of CHS Joints. *The Ninth International Offshore and Polar Engineering Conference*. Brest, France.

Dale, K., Zhao, X.-L., & Grundy, P. 2000. Shakedown Analysis of CHS T-Joints, Submitted for *IMPLAST 2000*. Melbourne.

Goh, T.K. 1994. *Shakedown Analysis of Fixed Offshore Structures.* PhD Thesis , Monash University, Melbourne.

Johansen, K.W. 1952. *Brudlinieteorier*. Copenhagen, Denmark, I kommission hos Teknisk Forlag.

Koiter, W.T. 1956. A New General Theorem on Shakedown of Elastic-Plastic Structures. *Koninkl. Nederl. Akad. Wetensh.* 59(1).

Korol, R.M. 1982. Plate reinforced square hollow section T-joints of unequal width. *Canadian Journal of Civil Engineering* 9: 143-148.

Kurobane, Y., Makino, Y. & Ochi, K. 1984. Ultimate Resistance of Unstiffened Tubular Joints. *Journal of Structural Engineering, ASCE* 110(2): 385-400.

Makino, Y., Kurobane, Y. & Tozaki, T. 1989. Ultimate strength analysis of simple CHS joints using the yield line theory. *Third International Symposium on Tubular Structures*, Lappeenranta, Finland.

Makino, Y., Kurobane, Y., Ochi, K., van der Vegte, G.J., & Wilmshurst, S.R. 1996. *Database of Test and Numerical Analysis Results for Unstiffened Tubular Joints.* (IIW Doc. XV-E-96-220), Kumamoto University, Kumamoto, Japan.

Melan, E. 1936. Theorie Statisch unbestimmter Systeme. *Second International Congress of the Bridge and Structural Engineers* , Berlin, Germany.

Milani, N.K. & Grundy, P. 1996. Incremental collapse of KT-joints under variable repeated loading. *Seventh International Symposium on Tubular Structures* , Miskolc, Hungary, A.A.Balkema.

Milani, N.K. & Grundy, P. 1997. Behaviour of Innovative Tubular KT-Joints Under Variable Repeated Loading. *The Seventh International Offshore and Polar Engineering Conference*. Honolulu, USA 4.

Mitri, H.S. & Korol, R.M. 1984. The strength of beam-to-staggered column connections of rectangular hollow sections. *International Journal of Mechanical Engineering Science* 26(6-8): 459-470.

Morris, G.A. & Packer, J.A. 1988. Yield Line Analysis of Cropped-Web Warren Truss Joints. *Journal of Structural Engineering, ASCE* 114(10): 2210-2224.

Packer, J.A., Davies, G. & Coutie, M.G. 1982. Ultimate Strength of Gapped Joints in RHS Trusses. *Journal of the Structural Division, ASCE* 108(ST2): 411-431.

Yamasaki, T., Takizawa, S. & Komatsu, M. 1979. Static and fatigue tests on large size tubular T-joints. *Offshore Technology Conference*, Houston, Texas. 4: 583-591.

Zhao, X.L. & Hancock, G.J. 1991. Plastic mechanism analysis of T-joints in RHS under concentrated force. *Journal of the Singapore Structural Steel Society* 2(1): 31-44.

Mechanics of Structures and Materials, Bradford, Bridge & Foster (eds)

Behavior of a penetrated nuclear steel containment vessel under combined loads

F.S. Fanous & L.F. Greimann
Department of Civil and Construction Engineering, Iowa State University, Ames, Iowa, USA
S.S. Aly
Department of Civil Engineering, Cairo University, Giza, Egypt

ABSTRACT Buckling analysis of a stiffened-penetrated thin steel vessel under internal pressure, dead, thermal and seismic loads was conducted using a 3-D finite element model using the ABAQUS program. Two hundred and seventy-five free vibration modes were extracted to ensure that a sufficient total modal effective masses in the three orthogonal directions. The maximum meridional seismic stress resultants were computed by the response spectrum method based on the structure response spectra for the Safe Shutdown Earthquake (SSE). Two potential buckling regions were identified and analyzed. The applied loads were linearly magnified until the containment became unstable.

1. INTRODUCTION

The steel containment vessel investigated herein is a thin cylindrical shell structure with an approximately 2:1 smooth elliptical head (see Fig. 1). SA537-Class 2 steel plates are used to construct the containment vessel. The cylindrical portion is provided with two T-ring stiffeners as well as an internal stiffener supporting a polar crane that weighs 2824 kN (635 kips). An elliptical head embedded into concrete foundation is also used to enclose the bottom of the vessel.

As the containment vessel is subjected to various loading conditions, regions in the containment would be subjected to compressive membrane forces that may cause the shell to become unstable. In order for the containment vessel to perform its intended safety function and to sustain these loads, a sufficient margin of safety against buckling should exist. This paper summarizes the results of a 3-D finite element analysis that was conducted to investigate the performance of the vessel shown in Fig.1 under a load combination of dead, external pressure, thermal and seismic loads.

2. DESCRIPTION OF THE CONTAINMENT VESSEL

Two equipment hatches and two personnel air locks are located at different elevations within the cylindrical portion of the vessel as shown in Fig. 1. The larger hatch weighs 467 kN (105 kips) and is at Azimuth 67° (measured from the North in the clockwise direction), while, the other equipment hatch weighs 276 kN (62 kips) and is located at Azimuth 126°. Each personnel air lock weighs 311 kN (70 kips) and is located at Azimuth 107°.

3. SOFTWARE SELECTION

The capabilities and the accuracy of two finite element codes ANSYS (1993), and ABAQUS (1994) were reviewed. This was accomplished by investigating the elastic buckling of an axially compressed perfect cylinder. The comparison included the types of shell elements, buckling and vibration analysis procedures, pre- and post-processing capabilities, and software limitations. ABAQUS was selected to perform the three-dimensional buckling analysis of the containment vessel.

To further calibrate the ABAQUS program, an inelastic analysis of a cylindrical shell under uniform axial load was performed. The results were compared to that of the widely recognized BOSOR5 finite difference software (Bushnell, 1995). The inelastic buckling analysis results were also utilized to establish the adequate mesh size for modeling the containment vessel.

4. SEISMIC ANALYSIS

When the containment vessel is subjected to seismic excitation, the generated stress field will be time dependent. Hence, the containment vessel may become unstable under the inertia forces that are induced by vibration motion (Greimann et. al., 1994). To obtain the buckling factor of safety, a time history analysis that incorporates both geometric and material non-linearities is appropriate. However, for practicality, a time history analysis could be replaced by a quasi-static approach in which the seismic excitation is replaced by a set of static forces that reproduce the maximum seismic stress resultants in the structure (Regulatory Guide 1.92,

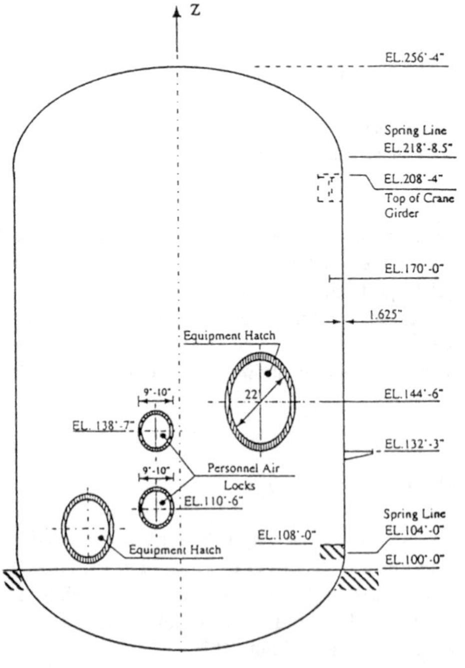

Figure 1. Elevation of the Vessel (Penetrations shown are distorted)

1976). The later was used in this study. In this approach, the Square Root of the Sum Square (SRSS) of meridional and hoop stress resultants, due to the Safe Shutdown Earthquake (SSE) ground excitation was computed. The two horizontal components in the North-South and East-West directions, together with the vertical component of the SSE floor response spectra at Elev. 30.5 m (Elev. 100 ft.) (see Fig. 2) were used in the analysis. Equivalent static forces to regenerate the SRSS meridian stress resultants were developed and utilized in conjunction with other loads to study the stability of the vessel.

Response Spectrum Analysis

Vibration Modes: A 3-D model for the vessel was utilized to compute the mode shapes and natural frequencies of the vessel in Fig. 1. A sinusoidal axisymmetric imperfection with a wave length of 4.0 $\sqrt{rt}$ and a peak-to-peak amplitude equals to one shell thickness was utilized (Greimann et. al., 1995). The subspace iteration method with and without the shift option (ABAQUS, 1994) was utilized to extract two hundreds seventy-five free vibration modes. Figure 3 illustrates some of these vibration modes. The cumulative modal effective masses were 87%, 89% and 74% of the total mass in the X, Y and Z directions, respectively.

Modal Responses: Inertia forces, F_i^{ϕ}, associated with each mode, i, were applied as pseudostatic loads on the containment model. These inertia forces were computed as:

$$\{F_i^{\phi}\} = [M] \ \{\Phi_i\}\omega_i^2 \tag{1}$$

where Φ_i and ω_i is the mode shape and the circular frequency of mode i, respectively, and [M] is the mass matrix. These forces were then applied as static forces to the containment vessel

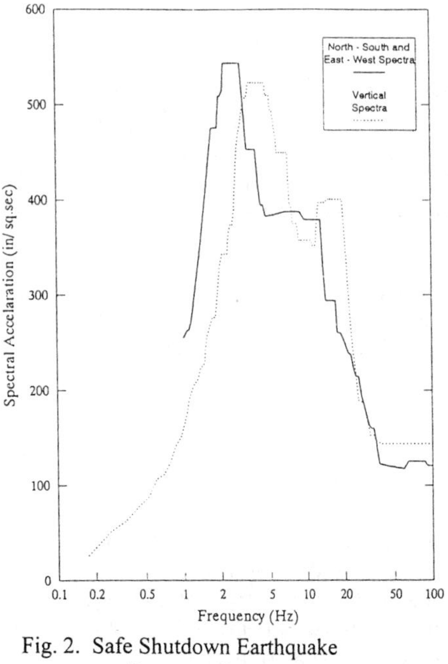

Fig. 2. Safe Shutdown Earthquake Response Spectrum

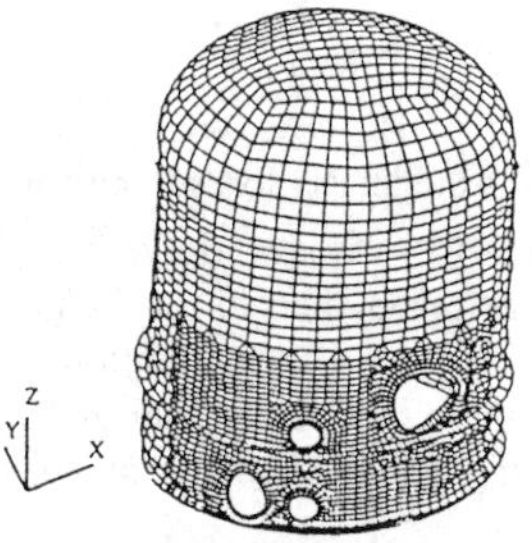

a) First Cantilever Mode f = 5.91 Hz

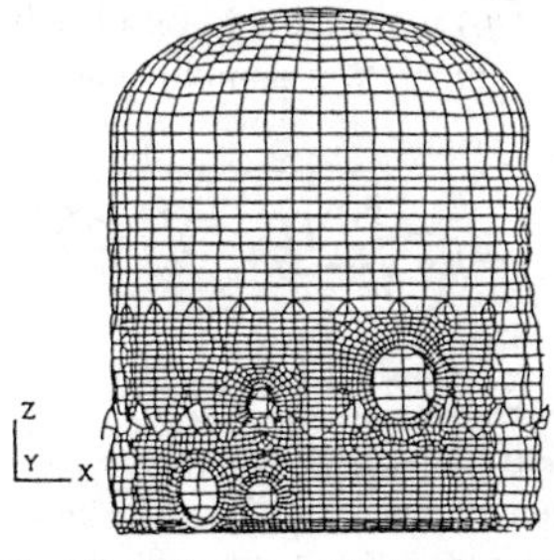

b) First Vertical Mode f = 12.4 Hz

Fig. 3. Selected Modes of Vibration

and the corresponding modal meridional stress resultant, $N1^{\Phi}$, and hoop stress resultant, $N2^{\Phi}$, were calculated. Next, the maximum seismic modal response Nm_{ij} for the ith mode due to the j (j= X, Y or Z and m=1,2) earthquake motion was computed as:

$$Nmij = \frac{Ni^{\phi}\,\mu_{ij}\,S_{ij}}{\omega_i^2} \tag{2}$$

where μ_{ij} is the participation factor of mode i in the direction j, and S_{ij} is the spectral acceleration corresponding to mode i frequency for earthquake component in direction j.

High-Frequency Modes: In order to incorporate the response of high-frequency modes, the procedure outlined in (U.S. Standard Review Plant, 1981) was implemented. This approach assumes that the response is essentially static and the modal inertial forces are calculated from the peak Zero Period Acceleration (ZPA). The fraction of lumped mass per degree of freedom that was not included in the extracted modes was computed as follows:

$$e_k = \sum_{i=1}^{N} (\mu_{ik}\Phi_{ik}) - \delta_{kj} \tag{3}$$

where e_k is the fraction of the remainder mass for degree of freedom (DOF) k, Φ_{ik} is the ith mode shape at DOF k, N is the total number of extracted modes, and δ_{kj} is the Kronecker delta, which is one if DOF k is in the direction of the earthquake, j, and zero if DOF k is a rotation or not in the j direction. The inertial forces associated with the summation of all high-frequency modes for each DOF k was then computed as follows:

$$P_{kj} = ZPA_j \cdot M_k \cdot e_k \tag{4}$$

where P_{kj} is the inertia force to be applied at DOF k in the direction j, M_k is the lumped mass at DOF k. The ZPA_j was conservatively assumed to be equal to the spectral acceleration of the jth spectrum corresponding to the frequency of 20.3 Hz. This was the frequency of the highest extracted mode. The inertia forces in each direction j were applied separately and the containment was statically analyzed in each case to determine stress resultants due to seismic response in the high-frequency modes.

Modal and Directional Combination: The maximum modal responses determined by Eq. 3 and the high-frequency modal responses were combined by the ten percent rule (Regulatory Guide 1.92, 1976) that accounts for closely spaced modes. The maximum combined response for the stress resultants was computed as follows:

$$Nm_{\max j} = \sqrt{\sum_i (Nm_{ij})^2 + 2\sum_i \sum_k |Nm_{ij} Nm_{jk}|} \qquad \text{where } i=k, \text{ and } m=1,2 \tag{5}$$

where N_{maxj}, is the maximum stress resultant response due to ground excitation in the jth direction. Modes of vibration were considered closely spaced if their frequencies differ by less than ten percent. It is to be noted that the modal response of the high-frequency modes constituted 3.49%, 2.69% and 4.46% to the total modal response in the X, Y and Z directions, respective. This means that the total response was not increased by more than 10% of its value when the high-frequency modal response was added. Hence, the requirements of the (U.S.

Standard Review Plant, 1981) were satisfied. The maximum stress resultant N_{max} of the containment, was then computed by combining the responses due to the three earthquake directions by the SRSS method (Regulatory Guide, 1976) as follows:

$$Nm_{max} = \sqrt{\sum_j \left(N_{max\,j}\right)^2} \qquad m=1,2 \text{ and } j=X,Y,Z \tag{6}$$

Equivalent Static Loads: The maximum SRSS meridional stress resultants obtained from the response spectrum analysis can not be directly input into the ABAQUS program. An alternative, was to use equivalent external static loads that produced a stress resultant field that bounded the maximum value of in the region where buckling may occur (Greimann et. al., 1994). The stress resultant field in the remainder of the containment was less than the SRSS stress resultants to ensure that buckling will take place in the regions being investigated.

5. INELASTIC BUCKLING OF THE CONTAINMENT VESSEL

The containment was loaded with the equivalent static forces that regenerate the maximum SRSS meridional stress resultant, change in ambient temperature from 21°C (70° F) to 49° C (120° F), external pressure of 20.7 Pa (3 psig), and gravity loads. Both geometric and material non-linearities were incorporated in the analysis. The constitutive relation utilized for the steel at a temperature 49° C (120° F). The yield stress of 408 MPa (59.2 ksi) and a proportional limit stress was 224 MPa (32.5 ksi). The buckling analysis was conducted using the modified Riks method (ABAQUS, 1994) where all the loads were applied proportionally in increments and the solution was obtained by iterations up to convergence. The load deflection behavior and the Von Mises effective stresses were monitored through out the solution. The analysis was terminated when the containment lost its stability and started to unload.

6. POTENTIAL BUCKLING REGIONS

The maximum seismic stress resultants N_{1max} were peaked at two regions: Region 1 near the base at Azimuth73.125°and Region 2 around all four penetrations. The highest potential for buckling will occur at one of these regions due to the compressive stress resultants. Hence, these two regions were selected to investigate the stability of the vessel.

Buckling in Region 1

The containment lost its stability as the applied loads were magnified by a load factor, λ_L, of 2.54. The deformed shape at buckling is shown in Fig. 4. The radial displacements at a location near the base and in the vicinity of the upper air lock were monitored. These displacements were increasing until instability occurred. In order to identify the location of first buckling, the Von Mises stresses were also examined at each load increment. Examining the results showed that the stresses near the base began to release, while it continued to increase at the other locations. This demonstrated that buckling took place first near the base where the maximum stress resultants were enveloped by the proposed equivalent static forces.

Buckling of Region 2

In this case, the buckling was examined around the penetrations where high stress concentrations were recorded. The containment lost its stability at a load factor equals to 2.62.

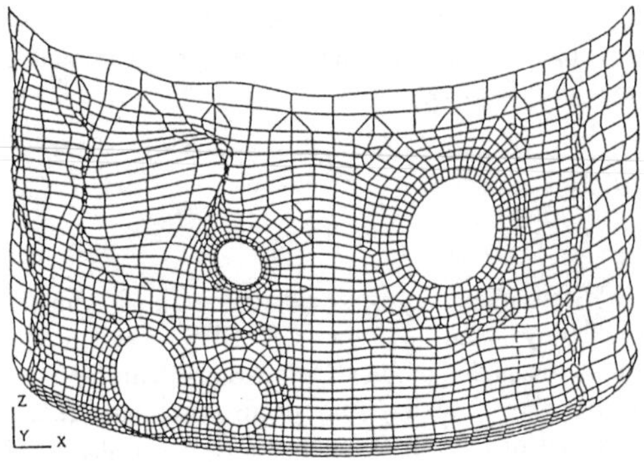

Fig. 4. Deformed Shape for Region 1 at a Load Factor of 2.54

The maximum radial displacement and the peak Von Mises stress took place at location in the shell below the upper equipment hatch-reinforcing collar. This confirms that instability took place at this location away from the upper equipment hatch penetration.

7. SUMMARY

A procedure for the buckling analysis of a stiffened-penetrated thin steel vessel under a load combination was conducted using the ABAQUS program. A 3-D finite element model that includes major penetrations and stiffeners was constructed. Two hundred and seventy-five free vibration modes were extracted to ensure that a sufficient total modal effective masses in the three orthogonal directions. The maximum meridional seismic stress resultants were computed by the response spectrum method based on the structure response spectra for the Safe Shutdown Earthquake (SSE). The effects of high-frequency modes and closely spaced modes were included in the calculations. Two potential buckling regions were identified and analyzed. The applied loads were linearly magnified until the containment became unstable. The Factor of safety against buckling was determined for each region using a quasi-static approach. The analyses revealed that the containment design satisfies the ASME code minimum requirements

8. REFERENCES

Bushnell, D., 1975 "BOSOR 5 Program for Buckling Elastic-Plastic Complex Shells of Revolution Including Large Deflection and Creep," Structural Mechanics Laboratory, Lockheed Missiles and Space, Inc., Palo Alto, CA.

Greimann, L. et. al, 1994, "Buckling Evaluation of System 80+TM Containment," NUREG-CR/6161, NRC, Washington DC.

Greimann, et. al.,1995, "Axisymmetric Buckling Analysis and Performance Beyond Design Basis for the AP600 Standard Plant Containment Vessel," NUREG-CR/6378, NRC, Washington, DC.

Hibbit, Karlson and Sorensen, Inc., 1994, "ABAQUS User's Manual," Newark, CA.

Regulatory Guide 1.92, 1976, "Combining Modal Responses and Spatial Components in Seismic Response Analysis," NRC Washington, DC.

Swanson Analysis Systems, Inc., 1993, "ANSYS User's Manual," Houston, PA.

"U.S. NRC Standard Review Plan (SRP), 1981," Section 3.8.2, 3.8.2.4 and 3.8.3.11.

Mechanics of Structures and Materials, Bradford, Bridge & Foster (eds)
© 1999 Balkema, Rotterdam, ISBN 90 5809 107 4

Benchmark solutions for steel frame structures subject to lateral buckling

M.Mahendran & Y.Alsaket
Physical Infrastructure Centre, School of Civil Engineering, Queensland University of Technology, Brisbane, Qld, Australia

ABSTRACT: Benchmark solutions are needed to verify the accuracy of simplified concentrated plasticity methods suitable for practical advanced analysis of steel frames. This paper contains a set of analytical benchmark solutions for steel frames subject to lateral buckling effects. The benchmark solutions were obtained using a distributed plasticity shell finite element model that explicitly accounts for the effects of gradual cross-sectional yielding, longitudinal spread of plasticity, initial geometric imperfections, residual stresses, and lateral buckling. The distributed plasticity model was validated using experimental results of large scale steel frames.

1 INTRODUCTION

Distributed plasticity methods of analysis are particularly suitable for the analysis of benchmark calibration frames which can be used to verify the accuracy of simplified concentrated plasticity methods of analysis. Various distributed plasticity analytical benchmarks are available for fully laterally restrained steel frames comprising compact sections, ie. not subjected to local and/or lateral buckling effects (Kanchanalai 1977, Vogel 1985). Advanced analysis methods are available for these frames and the current Australian standard AS4100 (SA 1998) permits the use of advanced analysis for these frames. The recent trend towards the use of high strength steel and thin-walled sections has increased the proportion of steel frame structures constructed of non-compact sections subject to local and lateral buckling effects. Recently Avery (1998) has developed adequate benchmark solutions and simplified advanced analysis methods for steel frames comprising non-compact sections subject to local buckling effects. Research was therefore undertaken to produce a series of distributed plasticity analytical benchmarks for steel frames subject to lateral buckling effects and to develop simplified advanced analysis methods. This paper presents a detailed description of the analytical benchmark frames subject to lateral buckling, the distributed plasticity finite element model used to develop these benchmarks and the three large scale experimental frames used to validate the distributed plasticity model.

2 LARGE SCALE EXPERIMENTS

The three test frames used in this investigation were the two-dimensional, single bay, single storey rigid frames (fixed bases and rigid joints), fabricated from Grade 350 steel rectangular hollow sections (RHS). The overall average dimensions of the frames are approximately 4 m between column centrelines, and 3 m from base to beam centreline. The test frames were subject to equal vertical loads applied to the columns, and a horizontal load applied at the base level using a 1500 kN test rig (see Figure 1). Details of the three test frames are as follows:
Test Frame 1: compact 200x100x9 mm RHS (form factor $k_f = 1$ and ratio of effective section modulus to plastic section modulus $Z_{ey}/S_{ey} = 1$), bending about minor axis with full lateral restraint and vertical to horizontal load ratio P/H = 4, Test Frame 2: compact 150x50x5 mm

RHS (k_f = 1 and Z_{ex}/S_{ex} = 1), bending about major axis without full lateral restraint and P/H = 10, Test Frame 3: non-compact 150x50x3 mm RHS (k_f = 0.776 and Z_{ex}/S_{ex} = 1), bending about major axis without full lateral restraint and P/H = 10. For each frame, Table 1 presents the measured yield and ultimate tensile stresses (σ_y, σ_u in MPa), thickness (t) and overall width (b) of web and flange, and in-plane and out-of-plane out-of-plumb imperfections for left and right hand columns. The web was considered the larger side of RHS in Table 1.

The knee connections were stiffened with 12 mm side plates, and base connections had 25 mm base plates welded to test frames and then bolted to floor girder with eight bolts. The base plates were also welded to floor girder. Therefore these connections could be assumed to be rigid. The horizontal load was applied at the base level because it was convenient to fix the position of the vertical jacks, rather than require them to move laterally with the frame as it deforms. The test frame was therefore fixed to a floor girder which was free to move laterally in-plane on greased roller bearings. This means that sway in the frame is induced by displacing the column bases horizontally while the beam and vertical jacks remain in position. The floor

Table 1. Measured section dimensions and column imperfections of test frames

Test Frame	Web				Flange				Out-of-plumb Imperfections (mm)	
	T	b	σ_y	σ_u	t	B	σ_y	σ_u	Left column	Right column
1	8.74	199.8	405	456	8.84	99.9	400	455	2, 10	2, 3
2	4.92	149.7	450	520	4.79	49.8	430	505	5, 9	6, 7
3	2.97	149.8	400	465	2.87	49.9	385	464	11,9	6, 16

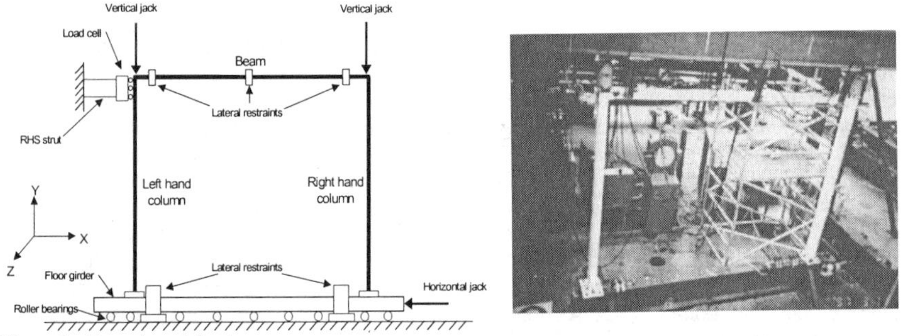

Figure 1. Experimental set-up

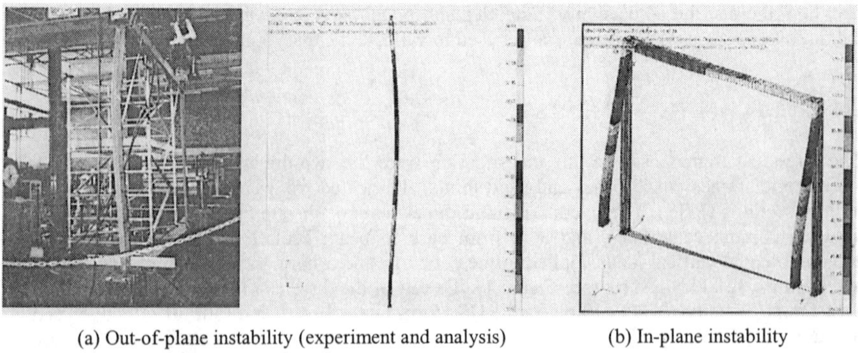

(a) Out-of-plane instability (experiment and analysis) (b) In-plane instability

Figure 2. Failure modes

girder was braced to prevent out-of-plane movement as shown in Figure 1. The horizontal reaction was provided by means of a RHS strut at the top of left hand column. Roller bearings were placed between the vertical jacks and the main girder of the test rig to allow for the small horizontal in-plane deflections at the beam level. The test column imperfections, strains and relative axial and sway deflections were measured. The frame was loaded to failure by displacement controlled hydraulic jacks.

Test Frame 1 failed by inelastic in-plane instability due to reduced stiffness caused by yielding and spread of plasticity whereas Test Frame 2 failed by a combination of out-of-plane buckling and inelastic in-plane instability (see Figure 2). Test Frame 3 comprising noncompact RHS also failed in a similar manner to Test frame 2 before the occurrence of local buckling. The load-deflection curves and ultimate failure loads for each frame are compared with analytical results in the following section.

3 DISTRIBUTED PLASTICITY MODEL

The overall aim of this investigation was to develop a distributed plasticity model that can be used to develop benchmark solutions for steel frame structures subject to lateral buckling effects. Therefore the distributed plasticity shell finite element model was first validated using experimental frame results. The HKS /Abaqus S4R5 element was selected for all analyses. A finer mesh with an aspect ratio of 1.0 was used to accurately model the residual stress distribution, spread of plasticity, and buckling deformations. For the RHS models, number of elements used in the larger and smaller sides of RHS were 8 and 4, respectively. The geometry and finite element mesh of a typical benchmark frame model is illustrated in Figure 3. The bearing plates used to distribute the vertical loads and the horizontal reaction and the beam to column stiffened connections were modelled explicitly as in the experiments. The experimental boundary conditions were modelled using single point constraints (SPC) which prevented specified movements of a particular node or group of nodes. The floor beam was modelled using rigid tie elements that connected all the nodes of the column base to a centre node at the outside flange of right hand column. The horizontal load was applied at this indepenedent node, which was restrained as SPC against out-of-plane displacement, but was free to move in-plane vertically and horizontally. The column base fixed connections were modelled using SPCs eliminating all the degrees of freedom of the nodes located at the base of the columns.

Both local and overall imperfections were included in the analyses. Local imperfections were included in all non-compact sections by modifying the nodal coordinates using a field created by scaling the appropriate buckling eigenvectors obtained from an elastic bifurcation buckling analysis of the model. Since the local flange and web plate imperfections could not be measured accurately, their magnitudes in the non-compact frames were taken as the assumed fabrication tolerances for compression members specified in AS4100 (SA 1998), ie. width of web or flange/150. Similarly, an out-of-straightness imperfection magnitude of length/1000 (SA 1998) was used for frames (Frames 2 and 3) subject to global buckling. However, measured values of in-plane and out-of-plane out-of-plumbness imperfections shown in Table 1 were used.

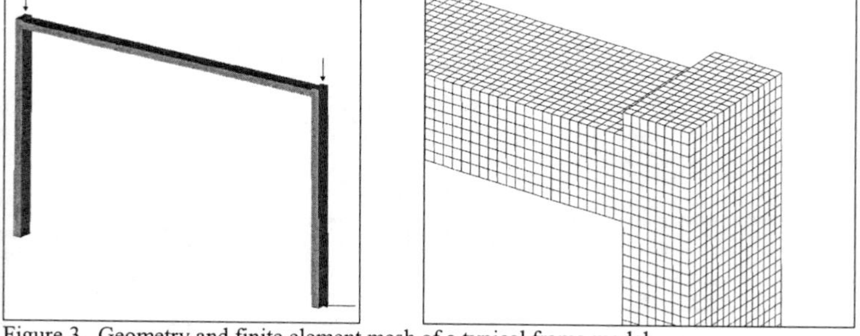

Figure 3. Geometry and finite element mesh of a typical frame model

The assumed residual stress distribution for RHS was recommended by Key and Hancock (1988). The longitudinal bending and membrane residual stresses were the most significant components affecting the frame behaviour and were included in all the analyses. These values were 200 and 30 MPa, respectively. The residual stresses were modelled using the Abaqus *INITIAL CONDITIONS option, with TYPE = STRESS, USER.

In all the analyses, the modulus of elasticity E and Poisson's ratio were taken as 200 GPa and 0.3, respectively, and multi-linear stress-strain curves including strain hardening effects were used based on the results from tensile coupon tests (Alsaket 1999). The rounded corners and associated higher yield stresses and residual stresses were ignored in the analyses.

Three methods of analyses were used: linear static, elastic buckling and nonlinear static. The

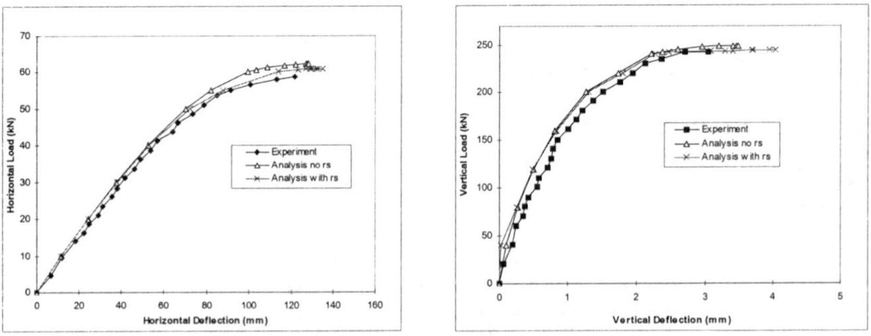

Figure 4. Comparison of experimental and analytical load-deflection curves for Test Frame 1

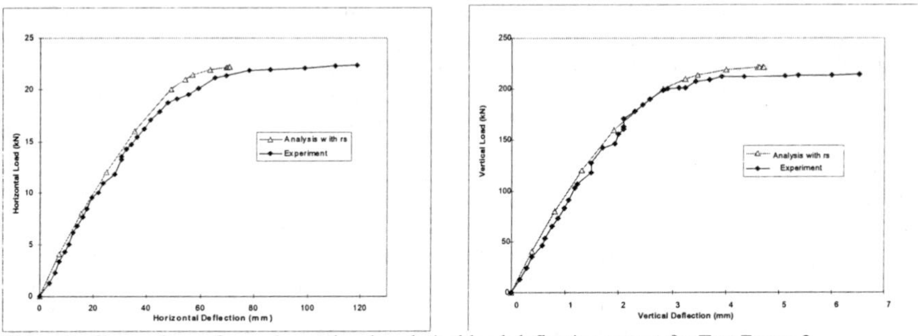

Figure 5. Comparison of experimental and analytical load-deflection curves for Test Frame 2

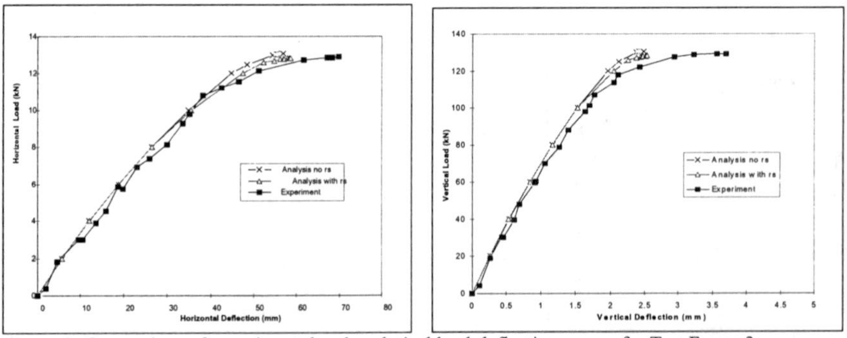

Figure 6. Comparison of experimental and analytical load-deflection curves for Test Frame 3

linear static analyses were used to check the model. The elastic buckling analyses were used to understand the frame buckling behaviour and to obtain the buckling modes that were used to include local and overall imperfections in the nonlinear analyses. The nonlinear static analyses were used to obtain the load-deflection curves and ultimate strength of frames.

The accuracy of this type of distributed plasticity model has been established by Avery (1998) for laterally restrained frames subject to yielding and local buckling effects using experimental and analytical solutions. In this paper it is extended to frames subjected to lateral buckling effects. The results from the distributed plasticity models of the three experimental frames agreed well with the experimental results as shown in Figures 4 to 6 and Table 2.

Figure 2 shows that analysis was able to model the failure mode accurately for each of the three frames. Local buckling was again not observed in the analysis of Test Frame 3 constructed of a noncompact RHS, which failed by out-of-plane buckling. The verification analyses demonstrated that the shell finite element model accurately represents distributed plasticity resulting from the combined effects of applied forces, residual stresses, geometric imperfections, and lateral buckling. Effects of not modelling corners can be considered to be negligible for these frames. As seen in Figure 4, the effect of residual stress on ultimate strength of frames was within 5%.

The AS4100 predicted loads were considerably less than the experimental and analytical loads, however, when measured yield stresses and a capacity reduction factor ϕ of 1 were used, they were increased, but were still less than the experimental and analytical loads. In the case of Test Frame 3, the AS4100 procedure uses a form factor k_f of 0.776 although no local buckling

Table 2. Summary and comparison of experimental and analytical capacities

Method	Test Frame 1		Test Frame 2		Test Frame 3		mean
	P	*H*	*P*	*H*	*P*	*H*	
(1) experiment	242	59	214	22.5	129	12.8	-
(2) analysis	243	61	210	21	130	13.1	-
(3) AS4100	178	45	119	12	78	7.8	-
(4) AS4100**	216	54	148	15	103	10.3	-
(2) / (1)	1.004	1.034	0.981	0.933	1.008	1.023	0.997
(4) / (2)	0.889	0.885	0.705	0.714	0.792	0.786	0.795

** AS4100 predictions based on measured yield stresses and a capacity factor ϕ of 1.

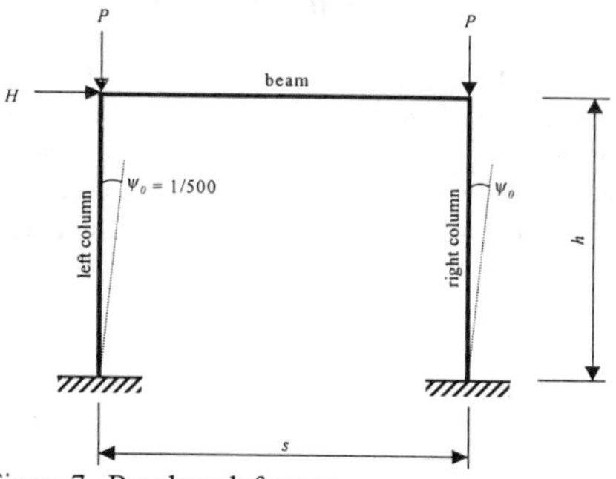

Figure 7. Benchmark frames

Table 3. Summary of analytical benchmark frames and results

Benchmark Frame	Section	Section Properties (mm units)	P_u (kN) P/H = 4	P/H = 10	P/H = 100
Group 1 Minor axis bending	200x100x9	A_g=4800, k_f=1 Z_{ey}=180,000, Z_{ey}/S_y =1	247	465.8	813.4
Group 2 Major axis bending	150x50x5	A_g=1810, k_f=1 Z_{ex}=78900, Z_{ex}/S_x =1	117	194.9	231.6
Group 3 Major axis bending	150x50x3	A_g=4140, k_f=0.776 Z_{ey}=51400, Z_{ex}/S_x =1	65	114.7	175.9

was observed. Therefore in the modified AS4100 calculation, it was taken as 1.0, however, the predicted loads are still less than experimental and analytical loads. All of these observations imply that the AS4100 design procedure is conservative.

4 ANALYTICAL BENCHMARK SOLUTIONS

The validated distributed plasticity model was used to develop three groups of analytical benchmarks comprising cold-formed RHS subject to lateral buckling effects. They are similar to the experimental frames, modified to make them suitable for the verification of simplified methods of analysis. The three groups of benchmark frames are single bay, single storey, fixed base sway frames comprising RHS subject to major or minor axis bending and axial compression (see Figure 7). The frame dimensions are: s = 4 m and h = 3 m. The dimensions and properties of the frames comprising of idealised RHS are given in Table 3.

Although the main aim was to develop benchmark frames subject to lateral buckling effects, one group of frames was also developed for compact and laterally restrained RHS frames for the purposes of comparison. The shell finite element model used to analyse experimental frames was used with the following modifications: Section: nominal section sizes were used with square corners; Material properties: elastic-perfectly-plastic model with identical nominal yield stress of 350 MPa for both web and flange, E = 200 GPa and $\nu = 0.3$; Initial local and overall geometric imperfections were taken as the fabrication tolerances for compression members specified in AS4100, ie. Out-of-straightness = length of member/1000, Out-of-plumbness = frame height/500, Out-of-flatness of web and flange = width/150. Longitudinal membrane and bending residual stress distributions were as for experimental frames. Loads and boundary conditions were essentially as for experimental frames. However, the loads were applied at the intersection of beam and column centrelines and distributed using rigid body elements. These rigid body elements also served to provide rigid connections, and thus there was no need for the beam-column connection to be stiffened. The column bases were modelled as fixed using single point constraints applied to all nodes on the base connections. It must be noted that the horizontal load was applied at the top of left hand column in the benchmark frames as shown in Figure 7.

The failure mode of Group 1 frames was in-plane stability due to cross-section yielding at the column base (similar to that shown in Figure 2 for experimental frames) whereas for both Groups 2 and 3, it was a combination of in-plane and out-of-plane instability (similar to that shown in Figure 2 for experimental frames).

A summary of the ultimate vertical loads (P_u) for benchmark frames is also provided in Table 3. The ultimate horizontal loads can be obtained by dividing the ultimate vertical loads by the appropriate P/H ratios. The vertical and sway load-deflection curves for each benchmark analysis are given in Alsaket (1999). These tabulated load-deflection results are suitable for the calibration and verification of simplified methods of analysis for frames subject to lateral buckling effects. A series of benchmark solutions is also available in Alsaket (1999) for Group 1 frames with multi-linear stress-strain curves.

5 CONCLUSIONS

This paper has described the development of a range of benchmark solutions for steel frames subjected to lateral buckling effects, which can be used to verify the accuracy of simplified concentrated plasticity methods suitable for practical advanced analysis. The benchmark solutions were obtained using a distributed plasticity shell finite element model that has been validated using large scale experimental results.

6 REFERENCES

Alsaket, Y. 1999. *Benchmark Solutions for Advanced Analysis of Steel Frames*, ME thesis, School of Civil Engg, Queensland University of Technology, Brisbane, Australia.
Avery, P. 1998. *Advanced analysis of steel frames comprising non-compact sections*, PhD thesis, School of Civil Engg, Queensland University of Technology, Brisbane, Australia.
Kanchanalai, T. 1977. *The design and behaviour of beam-columns in unbraced steel frames*, AISI Project No. 189, Report No. 2, Civil Engineering/Structures Research Laboratory, University of Texas at Austin, TX, USA, 1977.
Key, P. and Hancock, G.J. 1988. *An experimental investigation of the column behaviour of cold-formed square hollow sections*, Research Report, School of Civil Engineering, the University of Sydney.
Standards Australia (SA) 1998. *AS4100 Steel Structures*, Sydney, Australia.
Vogel, U. 1985. Calibrating frames, *Stahlbau*, Vol. 54, pp. 295-301.

Mechanics of Structures and Materials, Bradford, Bridge & Foster (eds)
© 1999 Balkema, Rotterdam, ISBN 90 5809 107 4

Model for tubular moment end plate connections utilising eight bolts

A.T.Wheeler
Department of Civic Engineering and Environment, University of Western Sydney, N.S.W., Australia
M.J.Clarke & G.J.Hancock
Department of Civil Engineering, University of Sydney, N.S.W., Australia

ABSTRACT: The paper presents a model for the determination of the serviceability and ultimate moment capacities of bolted moment end plate connections utilising rectangular hollow sections joined with eight bolts. The model considers the combined effects of prying action due to flexible end plates, the formation of yield lines in the end plates, and failures due to punching shear and beam section failure. The model is calibrated and validated using experimental data from an extensive test program. The model constitutes a relatively simple method for predicting the serviceability and ultimate moment capacities for the particular type of bolted moment end plate connection described herein.

1 INTRODUCTION

The increase in the use of rectangular hollow sections in mainstream structures coupled with the economics of prefabrication has highlighted the need for simple design methods that produce economical connections for tubular members. In an effort to address this need, the American Institute of Steel Construction has recently published the document *Hollow Structural Sections Connection Manual* (AISC, 1997), and the Australian Institute of Steel Construction has published the document *Design of Structural Steel Hollow Section Connections* (Syam & Chapman, 1996). Both of these publications present design models for commonly used tubular connections, however the eight-bolt moment end plate connection described in this paper is not included in either one since an appropriate design model did not exist at the time of their publication. The eight-bolt connection described in this paper and depicted in Figure 1 represents one of two fundamental bolting arrangements studied by Wheeler (1998). The other bolting arrangement utilises four bolts, with the design model described in Wheeler et al. (1998).

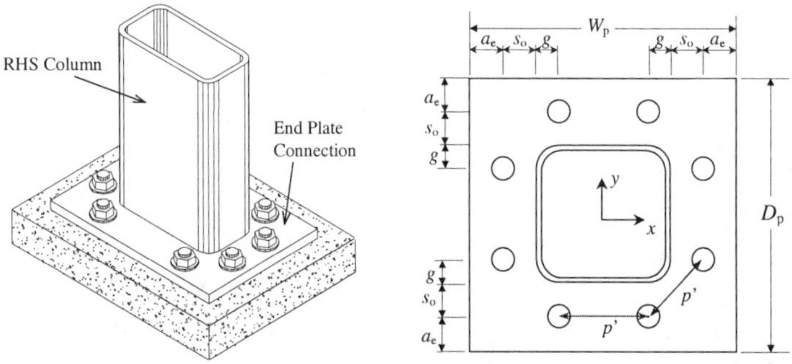

Figure 1. Typical eight-bolt end plate application and layout

In this paper, a theoretical model is presented for the analysis of eight-bolt tubular end plate connections subjected to flexural loading. The model utilises yield line analysis to predict the yield moment of the connection and combines the stub-tee and yield line analysis to predict the ultimate strength of the connection. The model also considers section capacity and punching shear failures, which were observed in the experimental programme. Full details of the model are given in Wheeler et al. (1999). The predictions of the model are compared with the results obtained from an associated experimental program conducted at the University of Sydney (Wheeler et al., 1995).

2 EXPERIMENTAL PROGRAMME

Two basic connection configurations, termed Type A and Type B, were investigated in an experimental programme conducted at the University of Sydney (Wheeler et al., 1995). The Type A connections utilised eight bolts, and the Type B connections employed four bolts. This paper deals only with the eight-bolt connections as depicted in Figure 1.

The parameters varied in the experimental programme include the plate size (W_p, D_p), the plate thickness (t_p), the section shape, and the positions of the bolts with respect to the section flange and web (s_o and g). The dimensions of the end plates and the type of sections (150×150×9 SHS or 200×100×9 RHS) used for the eight bolt specimens are given in Table 1. The end plate material was 350 Grade steel. The bolt and nut assemblies were M20 structural grade 8.8 (Grade 8.8/T), with a measured yield strength and ultimate tensile strength of 195 kN and 230 kN, respectively. The connections were prefabricated using a combination fillet/butt weld joining the section to the end plate, with a nominal fillet leg length of 8 mm.

The connections were loaded in pure flexure using a beam splice connection at mid-span of a four point bending test. As the sections were not susceptible to local buckling, the ultimate load of the specimen was limited to connection failure which occurred when the tensile bolts fractured, when section failure or punching shear failure occurred, or when the end plate deformations were deemed excessive. The experimental ultimate moment (M_{cu}) and the failure mode for each test are listed in Table 1.

3 GENERAL THEORETICAL BASIS OF MODEL

The theoretical model for the moment end-plate connection is based on yield line analysis and the stub-tee analogy to determine the serviceability and ultimate strength capacities. The ultimate strength may also be limited by bending section strength and punching shear. The yield line analysis serves to provide an estimate of the yield moment of the connection (M_{cy}), with prying action not considered. The stub-tee analysis serves to provide an estimate of the ultimate strength (M_{cu}) of the connection through its integration of prying effects with yield line analysis. The particular application of the stub tee analysis described in this paper is termed the "Cumulative Modified Stub-Tee Method" due to the fact it considers prying effects in both orthogonal directions independently.

Table 1. End Plate connection Details and Test Results

Specimen	Section	Plate Dimensions (mm)					Bolt Length (mm)	M_{cu}	Failure
No.	Type	t_p	W_p	D_p	s_o	g	l_B	(kN.m)	Mode*
1	SHS	16	280	280	35	30	75	116.0	Bolt
2	RHS	16	230	330	35	15	75	124.5	Punching
3	SHS	12	280	280	35	30	75	93.9	Bolt
4	SHS	20	280	280	35	30	90	116.0	Bolt
5	RHS	12	230	330	35	15	75	92.7	Punching
6	RHS	20	230	330	35	15	90	136.7	Bolt
7	SHS	16	260	260	25	35	75	106.0	Bolt
8	SHS	16	300	300	45	25	75	97.6	Punching
9	RHS	16	210	310	25	20	75	133.0	Punching
10	RHS	16	250	350	45	10	75	119.3	Punching

* Punching = Failure by section tearing away from plate at toe of weld (punching shear).
Bolt = Failure by bolt fracture.

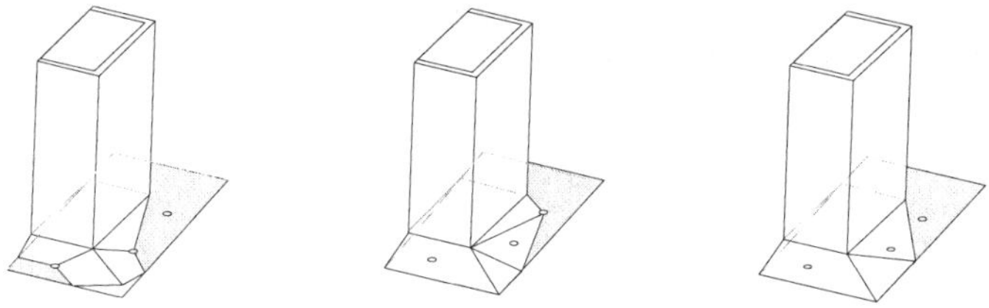

Mechanism 2 *Mechanism 4* *Mechanism 5*

Figure 2. Observed yield line mechanisms

3.1 *Yield Line Analysis*

Numerous plastic mechanisms were considered in the yield line analysis and in many cases the yield line patterns are relatively complicated and make for lengthy expressions defining the collapse moment (M_{yl}). Consequently, the derivations of collapse moments (see Wheeler et al., 1999) are not given in this paper. Three end plate plastic mechanisms were observed in the experimental programme (Figure 2). The plastic collapse of the beam section (referred to as Mechanism 8) also governed the ultimate capacity of a number of tests. The results presented in Table 2 (see Section 4) show the calculated yield moment (M_{yl}), the corresponding collapse mechanism, and the experimentally determined yield moment (M_{cy}).

The results given in Table 2 highlight the effect of the section shape on the mode of failure of the connection. For the SHS connections, where the vertical separation of the bolts is comparatively small compared with the RHS connections (compare Test #1 and Test #2), the dominant failure mode was Mechanism 4. Conversely, the increased vertical separation of the bolts in the RHS connections resulted in the latter collapsing most commonly according to Mechanism 5.

3.2 *Cumulative Modified Stub-Tee Method*

The use of stub-tee analogies to determine the strength of an end plate connection have been used extensively (Nair et al., 1974; Kennedy et al., 1981). The method utilises a simple rigid plastic analysis on an analogous beam that represents the one-dimensional behaviour of the end plate with yield lines parallel to the axis of bending only. In the eight-bolt tubular end plate connections, bending in the end plate occurs about two axes, with the yield lines not necessarily being parallel to either axis of bending. The so-called "Cumulative Modified Stub-Tee Method"

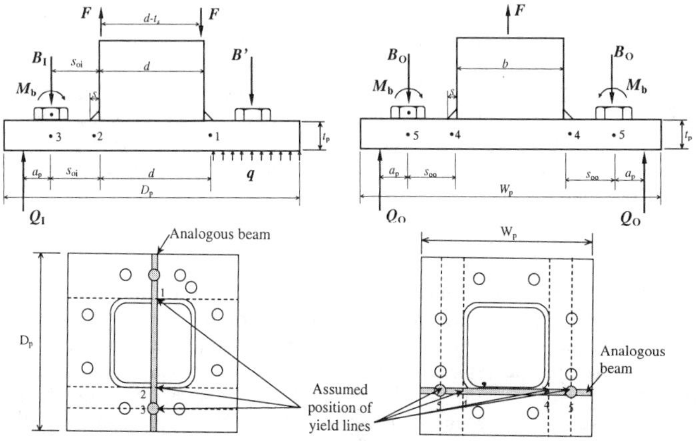

(a) "In-plane Bending" of End Plate *(b) "Out-of-Plane Bending" of End Plate*

Figure 3. Analogous beams for cumulative stub-tee model

utilises analogous beams in both orthogonal directions, together with the principle of superposition, to obtain the resultant connection behaviour.

A simple representation of the two beams used to approximate the connection behaviour is shown in Figure 3. The "In-plane Bending" analysis models the effect of the bolts below the flange of the section, with plastic hinges forming at Points 1, 2 and 3. The second analysis, representing the "Out-of-plane Bending" of the end plate, models the effect of the bolts lying on either side of the section webs. The plastic hinges are assumed to form at Points 4 and 5 on both sides of the hollow section. To simplify the problem, the bolts above the neutral axis (in the compressive zone) are assumed to have a negligible effect on the connection strength.

The behaviour of the connection may be divided into three categories depending on the thickness of the end plate (t_p) and the magnitude of the applied load. These categories are defined as *Thick Plate Behaviour*, *Intermediate Plate Behaviour* and *Thin Plate Behaviour* (Kennedy et. al, 1981), and are identified by the position and number of yield lines. Thick plate behaviour occurs when the connection fails due to bolt fracture prior to yield lines forming at Points 2 or 4. Intermediate plate behaviour occurs when the bolts fracture after the formation of yield lines at Points 1, 2 and 4 (i.e. Plastic Collapse Mechanism 5). Thin plate behaviour corresponds to the formation of yield lines at Points 1, 2, 3, 4 and 5 in the end plate (i.e. Plastic Collapse Mechanism 2), without deformation of the bolts.

As derived by Wheeler et al. (1999), the moment capacity for each mode of behaviour is given by

$$M_{\text{Cthick}} = \left(\frac{M_{1\text{p}} + 2 \cdot B_{\text{I}} \cdot (d + s_{\text{oi}} + h \cdot d)}{d} \right) \cdot (d - t_{\text{s}}) \tag{1}$$

$$M_{\text{Cint}} = \left(\frac{2 \cdot B \cdot a_{\text{p}} + M_{2\text{p}} + 2 \cdot M_{\text{b}}}{(a_{\text{p}} + s_{\text{oi}})} + 2 \cdot \frac{h \cdot B \cdot a_{\text{p}} + M_{4\text{p}} + M_{\text{b}}}{(a_{\text{p}} + s_{\text{oo}})} + \frac{M_{1\text{p}} + M_{2\text{p}}}{d} \right) \cdot (d - t_{\text{s}}) \tag{2}$$

$$M_{\text{Cthin}} = \left(\frac{M_{1\text{p}} + M_{2\text{p}}}{d} + 2 \cdot \frac{M_{5\text{p}} + M_{2\text{p}} + M_{\text{b}}}{s_{\text{oo}}} + \frac{M_{3\text{p}} + M_{2\text{p}} + 2 \cdot M_{\text{b}}}{2 \cdot s_{\text{oi}}} \right) \cdot (d - t_{s}) \tag{3}$$

in which $M_b = \pi d_b f_{yb}/32$ (d_b = bolt diameter, f_{yb} = bolt yield stress) is the moment generated by the bending of the bolts, and M_{ip} is the plastic moment for the ith yield line. Since the yield lines invariably undergo significant rotations prior to the ultimate strength being reached, much of the material is stressed into the strain-hardening range. Consequently, the plastic moment M_{ip} is defined in terms of a "design stress" (f_p) rather than the yield stress (Packer et al., 1989).

$$M_{\text{ip}} = \frac{1}{4} \cdot t_{\text{p}}^2 \cdot f_{\text{p}} \cdot l_i \qquad f_{\text{p}} = \frac{f_{\text{y}} + 2 \cdot f_{\text{u}}}{3} \tag{4}$$

The stub-tee analogy assumes that the yield lines form in a linear fashion, transversely across the end plate. The plastic collapse mechanism analysis, however, indicates that the yield lines rarely occur in this manner for eight-bolt connections. To compensate for this inconsistency, "*equivalent lengths*" (for in-plane and out-of-plane bending) are determined for the yield lines such that the total amount of internal work involved in the mechanism remains unchanged. The equivalent lengths of the yield lines used for the cumulative stub-tee analysis depend on the assumed plastic collapse mechanism. Full details are given in Wheeler et al. (1999).

3.3 *Plastic Section Capacity*

The plastic section capacity of the tubular member may also govern the ultimate moment that the connection can attain. Design specifications generally define the plastic section capacity as the yield stress (f_y) times the plastic section modulus (S). Although appropriate for design, this method of calculating the section plastic capacity does not usually reflect the experimentally measured ultimate moment as the cold working of the section introduces significant strain hardening into the material properties. A more accurate method to predict the experimental plastic section capacity would be to use the design stress (f_p) as defined in Equation 4

$$M_{\text{s}} = S \cdot f_{\text{p}} \tag{5}$$

3.4 *Punching Shear*

Punching shear failure (tearing of the end plate) occurs when the concentrated loads transferred from the section to the end plate exceed the shear capacity of the end plate over a localised region. To model the shear failure a simplified approach is used, assuming that shear failure planes are defined by the geometry of the connection. The connection is considered to have failed in punching shear when the load in the tensile flange and adjacent regions of the section exceed the shear capacity of the corresponding region (termed the "nominal shear length") of the end plate.

The *nominal shear length* is the length around the perimeter of the section that is assumed to fail as a result of the section pulling out from the end plate. This length (l_s) is divided into two regions, corresponding to flange failure (l_{sf}) and web failure (l_{sw}) (i.e. $l_s = l_{sf} + l_{sw}$)

$$l_{sf} = d - 5 \cdot t_s + \frac{\pi}{2} \cdot \left(2.5 \cdot t_s + s - \frac{t_p}{2} \right) \quad l_{sw} = 2 \cdot \left(g - 2.5 \cdot t_s + \frac{d_{bh}}{2} + \frac{\pi}{4} \cdot \left(2.5 \cdot t_s + s - \frac{t_p}{2} \right) \right) \tag{6}$$

In the above equations, s denotes the weld leg length, d_{bh} is the diameter of the bolt head, and it is assumed that the external corner radius of the tubular section is 2.5 times the wall thickness. Utilising the Von Mises yield criterion, the moment capacity of the connection with respect to punching shear failure is expressed as

$$M_{PS} = \frac{f_p}{\sqrt{3}} \cdot t_p \cdot \left(l_{sf} \cdot (d - t_s) + \left(l_{sw} \cdot (d - g) \right) \right) \tag{7}$$

4 GENERALISED CONNECTION MODEL

The cumulative modified stub-tee method identifies two modes of connection failure, which are bolt capacity and end plate capacity. Bolt capacity (which may occur in conjunction with thick or intermediate plate behaviour) occurs when the tensile bolts fracture, while plate capacity (thin plate behaviour) occurs when a plastic mechanism forms in the end plate without deformation of the bolts. The plate capacity is independent of the bolt loads. Other failure modes not considered in the modified stub-tee analysis but which may occur in practice include shear failure of the end plate (punching shear) and plastic section failure. The predicted results for each mode of failure are calculated and the critical values presented in Table 2.

The results shown in Table 2 indicate that although a total of ten eight-bolt tests were performed, four of these were limited in strength by punching shear failure$^{(p)}$ and a further four were governed by plastic section capacity$^{(s)}$. Only two tests were governed by failure of the end plate$^{(t)}$ itself, as computed using the stub tee analysis. While the ultimate failure mode of the specimens was generally punching shear$^{(p)}$ or bolt failure section failure$^{(s)}$ and yielding$^{(y)}$ in the end plate were observed.

The failure criteria and failure loads for the standard RHS tests (Tests #2, #5, #6) are pre-

Table 2 – Theoretical and Observed Results for Yield and Ultimate Connection Moments

Test #	Experimental Yield Moment M_{yl} (kNm)	Calculated Yield Moment & Mode M_{cy} (kNm)	M_{yl}/M_{cy}	Experimental Ult. Moment M_{cu} (kNm)	Calculated Ult. Moment M_{pred} (kNm)	M_{pred}/M_{cu}
1 (SHS)	93.9	102.3④	1.09	116.0 (s)	108.3 (s)	0.93
2 (RHS)	98.6	132.0⑤	1.31	124.5 (p)	116.8 (p)	0.94
3 (SHS)	66.6	68.6②	1.03	93.9 (y)	92.8 (t)	0.99
4 (SHS)	98.0	113.0⑧	1.15	116.0 (s)	108.3 (s)	0.91
5 (RHS)	68.7	106.1⑤	1.54	92.7 (p)	87.6 (p)	0.95
6 (RHS)	102.0	129.0⑧	1.26	136.7 (s)	125.3 (s)	0.92
7 (SHS)	94.2	113.0⑧	1.20	113.2 (y)	108.3 (s)	0.97
8 (SHS)	73.3	92.6④	1.26	97.6 (p)	104.9 (t)	1.07
9 (RHS)	109.0	129.0⑧	1.18	133.0 (p)	123.2 (p)	0.93
10 (RHS)	94.0	119.0⑤	1.27	119.0 (p)	110.0 (p)	0.92

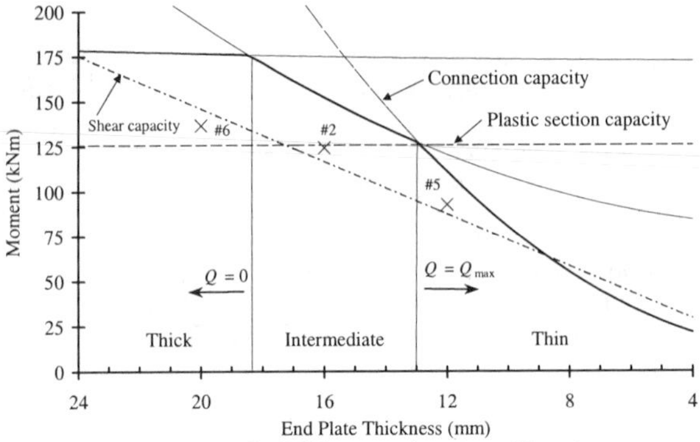

Figure 4. Failure criteria for RHS connections (s_o = 35 mm)

sented in Figure 4. In these figures, the modes of failure corresponding to plastic section capacity, cumulative modified stub-tee capacity, and punching shear capacity are shown.

For rectangular hollow sections, the depth-to-width aspect ratio results in punching shear failure dominating. Connections comprising end plates thicker than 17 mm will attain full plastic section capacity, while end plates thinner than 8 mm will fail as a result of a plastic mechanism forming in the end plate. Connections containing end plates thinner than 17 mm and thicker than 8 mm fail as a result of punching shear. The theoretical results depicted in Figure 4 are consistent with the experimental findings.

5 CONCLUSIONS

In this paper, a design model is presented for prediction of the strength and behaviour of eight-bolt moment end plate connections joining square and rectangular hollow sections. The models consider a variety of failure modes including the formation of a plastic collapse mechanism in the end plate, tensile failure of the bolts, plastic section failure in the connected beam, and punching shear (tear out) failure.

The experimental and analytical results indicate that for the SHS connections, plastic section capacity failure dominates with end plate failure occurring only for the most flexible end plate cases. For the RHS connections, the failure mode is predominantly that of punching shear, with plastic section capacity limiting the strength for the thicker end plates. The transition between the different failure modes, which occurs with changing end plate thickness, is captured very well by the analytical model. The model demonstrates excellent correlation with the test results and is effective in its consideration of all relevant failure modes which can occur.

REFERENCES

AISC (1997), *Hollow Structural Sections Connections Manual*, American Institute of Steel Construction, Inc.

Kennedy, N. A., Vinnakota, S. and Sherbourne A. N., (1981). "The Split-Tee Analogy in Bolted Splices and Beam-Column Connections", Joints in Structural Steelwork, John Wiley &sons, London-Toronto, 1981, pp. 2.138–2.157.

Nair, R. S., Birkemoe, P. C. and Munse, W. H., (1974). "High Strength Bolts Subject to Tension and Prying", *Journal of the Structural Division*, ASCE, **100**(2), pp 351-372.

Packer, J. A., Bruno, L., Birkemoe, P. C. (1989). "Limit Analysis of Bolted RHS Flange Plate Joints" *Journal of Structural Engineering*, ASCE, **115**(9), 2226-2241.

Syam, A. A. & Chapman, B. G., (1996), *Design of Structural Steel Hollow Section Connections. Volume 1: Design Models*, 1st Edition, Australian Institute of Steel Construction, Sydney.

Wheeler, A. T., Clarke, M. J. and Hancock, G. J. (1995) "Tests of Bolted Flange Plate Connections Joining Square and Rectangular Hollow Sections", *Proceedings, Fourth Pacific Structural Steel Conference*, Singapore, 97–104.

Wheeler A. T., Clarke M. J. & Hancock G. J., (1999), "Design Model for Eight Bolt Rectangular Hollow Section Bolted Moment End Plate Connections", *Research Report*, Department of Civil Engineering, The University of Sydney (in preparation).

Wheeler A. T., Clarke M. J., Hancock G. J. & Murray, T. M., (1998), "Design Model for Bolted Moment End Plate Connections Joining Rectangular Hollow Sections", *Journal of Structural Engineering*, ASCE, **124**(2), 164–173.

Wheeler, A. T. (1998), "The Behaviour of Bolted Moment End Plate Connections in Rectangular Hollow Sections Subjected to Flexure", *PhD Thesis*, Department of Civil Engineering, The University of Sydney.

Mechanics of Structures and Materials, Bradford, Bridge & Foster (eds)
© 1999 Balkema, Rotterdam, ISBN 90 5809 107 4

On the analysis and design of submerged steel pipelines

P.Ansourian
Department of Civil Engineering, University of Sydney, N.S.W., Australia

ABSTRACT: Steel pipelines submerged in a fluid, subjected to variable internal pressure and to longitudinal stresses due to 'self-weight' bending are investigated. The pipes are ring-stiffened at regular intervals and are cement lined. The design problem is one of interaction buckling under external pressure and axial stress. The proportions of the ring stiffeners must be such that their local buckling is precluded. An effective length of pipe is assumed to act with the stiffener. The failure mode is triggered by yielding in the inner fibres under combined compression and amplified bending stresses of the initially imperfect ring, assumed deformed in an ovalising mode. The ultimate strength of the shell between stiffeners is evaluated taking account of the reduced buckling strength arising from the finite flexibility of the ring stiffeners and the presence of longitudinal compressive stress from global bending of the shell.

1 INTRODUCTION

In this paper, steel pipelines submerged in a fluid, subjected to variable internal pressure and to longitudinal stresses due to "self-weight" bending are investigated. The pipes are ring-stiffened at regular intervals. Therefore, the design problem for the shell wall is one of interaction buckling effects caused by external pressure combined with axial stress buckling. The spacing and stiffness of the ring stiffeners are crucial parameters that also control the behaviour of the shell. Steel pipelines are internally protected with a cement lining that can contribute to the buckling strength. The effectiveness of the lining must be assessed when evaluating the properties of the steel/concrete composite, most likely separately for membrane or bending actions. A conservative allowance for the contribution of the cement lining to the stiffness the pipe is made by assuming full effectiveness in membrane action only. A review is made of the buckling response of the composite shell subjected to uniform and non-uniform external pressure, taking into account both the spacing and the stiffness of the ring stiffeners, and the level of imperfection. The geometric proportions of the ring stiffeners must be such that their local buckling is precluded; they should be designed as 'heavy' stiffeners, so that they would buckle in an ovalising mode.

Stresses in the stiffeners under ultimate conditions consist of combined membrane and bending stresses. For determining this stress state, an effective length of pipe is assumed to act with the stiffener. While the elastic buckling stress of a circular ring under uniform compression is well known, the failure mode is not one of elastic buckling, but is triggered by the attainment of yield in the inner fibres under combined compression and amplified bending stresses of the initially imperfect ring. The extreme fibre stress is evaluated in a second-order analysis of the ring, in which the ring is assumed initially deformed in an ovalising mode. Finally, the ultimate strength of the shell between stiffeners is evaluated taking account of the reduced buckling strength arising from the finite flexibility of the ring stiffeners and the presence of longitudinal compressive stress from global bending of the shell.

2 DESIGN PHILOSOPHY FOR THE SHELL

2.1 *Shell buckling*

The pipeline sustains a head of water which varies linearly from the crown to the invert. The skew or pressure factor ρ (Figure 1) depends on the depth of immersion and the internal pressure. The cement lining is conservatively assumed to be effective in membrane action only. Thus the effective wall thickness is t for bending action and $t + t_c/mr$ for membrane action, where mr is the modular ration of the concrete.

A long unrestrained pipe under uniform radial pressure has buckling pressure:

$$p_{cr} = \frac{E}{4(1-\nu^2)}\left(\frac{t}{R}\right)^3 \qquad (1)$$

independent of length, assuming the pressure remains perpendicular to the shell surface during buckling. The resistance of an unrestrained pipe to external pressure is very low when compared with the resistance of the same tube stiffened with rings at regular intervals. These rings must however have sufficient stiffness to remain approximately circular during buckling of the pipe. With fully effective rings, the elastic buckling pressure of the stiffened pipe is inversely proportional to the distance L between stiffeners; an approximate expression (Trahair et al 1983) is:

$$p_{crit} = \frac{184{,}000}{\left(\frac{L}{R}\right)\left(\frac{R}{t}\right)^{2.5}} \quad \text{MPa} \qquad (2)$$

The allowable pressure on the pipe is calculated by multiplying this critical pressure by a reduction factor not greater than 0.7 to allow for imperfections, and by a load factor.

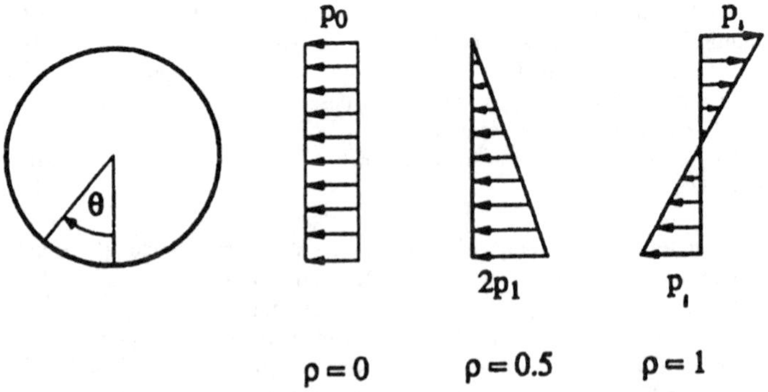

Figure 1. Fluid pressure distribution and skew factor ρ

Many theoretical analyses of the buckling of circular cylinders under external pressure have been performed within the framework of the Donnell linear stability theory with simple supports. It is however more accurate to apply Flügge's equations (Vodenitcharova and Ansourian 1996, 1998) in coupled form to the buckling problem for both uniform and variable pressure. In this more refined theory, the in-plane displacements are not neglected when compare with transverse deflections of the surface. Differences in the two solutions are exacerbated as the length of the cylinder increases. A further advantage of the method is that solutions can be obtained for non-classical edge conditions. For a thin, elastic circular cylindrical shell of length L, radius R and wall thickness t, compressed by a non-uniform lateral pressure $p(\theta)$ with fluid pressure linearly distributed with height, the skew coefficient or pressure factor ρ is defined as:

$$\rho = \frac{p_1}{p_0 + p_1} \tag{3}$$

where p_0 is the pressure at mid-height of the cylinder. The pressure is linearly distributed with height and is given by:

$$p = p_0 + \frac{\rho}{1-\rho} p_0 \cos\theta \tag{4}$$

$\rho = 0$ corresponds to uniform pressure; $\rho = 0.5$ to zero pressure at the top, and $\rho = 1$ to $p = -p_1$ at the top and $+p_1$ at the bottom. The hydrostatic pressure can also be expressed as:

$$p = \sum_{r=0}^{r=1} a_r \cos(r\theta) \quad \text{where} \quad a_0 = p_0, \quad a_1 = \frac{\rho}{1-\rho} p_0 \tag{5}$$

Values of $(p_{st}/p_{cr}-1)$ in the case of hydrostatic pressure are plotted against R/h in Figure 2 for $L/R = \pi$ and ρ = 0.1, 0.3 and 0.55; more complete results are given in Table 1 which also contains buckling pressures for the case of uniform pressure.

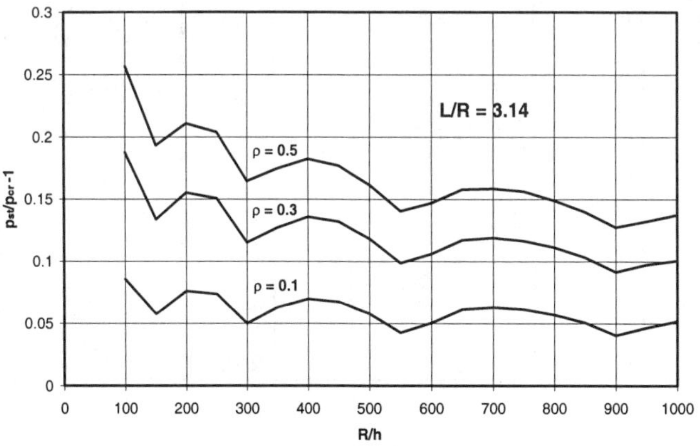

Figure 2. Variation of hydrostatic stagnation pressure $(p_{st}/p_{cr} - 1)$ versus R/t $(L/R = \pi)$

TABLE 1. Uniform pressure (p_{cr}) and hydrostatic pressure at buckling (p_{st}) (non-dimensionalised by young's modulus).

L/R	R/t	Uniform		Hydrostatic					
		$p_{cr}/E \times 10^{-9}$	n_{cr}	ρ	$p_{st}/E \times 10^{-9}$	ρ	$p_{st}/E \times 10^{-9}$	ρ	$p_{st}/E \times 10^{-9}$
	100	13,960	9		14,540		15,070		15,440
	200	2,333	11		2,425		2,502		2,557
	300	832.0	13		857.4		882.5		900.3
	400	399.7	13		411.2		422.6		430.5
π/4	500	225.7	14	0.1	232.9	0.3	239.0	0.5	243.3
	600	141.8	15		146.5		150.2		152.8
	700	96.29	16		99.01		101.4		103.1
	800	68.47	16		70.56		72.23		73.39
	900	50.93	17		52.34		53.54		54.39
	1,000	38.94	17		40.09		40.97		41.60
	100	3,001	4		3,258		3,565		3,772
	200	528.2	4		568.4		610.1		639.6
	300	195.2	5		205.0		217.7		227.3
	400	92.4	5		98.84		105.0		109.2
π	500	53.31	5	0.1	56.4	0.3	59.61	0.5	61.91
	600	33.97	5		35.69		37.57		38.96
	700	22.73	6		24.16		25.43		26.33
	800	16.33	6		17.26		18.14		18.76
	900	12.34	6		12.84		13.47		13.91
	1,000	9.37	6		9.86		10.31		10.66
	100	1,577.0	4		1,695		1,854		1,995
	200	256.4	4		279.6		312.4		336.5
	300	96.71	5		103.9		112.6		119.5
	400	45.27	5		49.21		54.07		57.36
2π	500	26.11	5	0.1	28.20	0.3	30.67	0.5	32.44
	600	17.20	5		18.06		19.34		20.39
	700	11.43	6		12.23		13.10		13.77
	800	8.04	6		8.67		9.34		9.80
	900	5.99	6		6.43		6.93		7.26
	1,000	4.61	6		4.94		5.30		5.56
	100	803.4	4		881.5		1,011		1,114
	200	122.7	4		135.3		159.7		180.5
	300	47.39	5		51.69		58.20		63.71
	400	24.17	5		26.07		28.60		30.79
4π	500	13.10	5	0.1	14.33	0.3	16.14	0.5	17.43
	600	8.10	5		8.87		10.07		10.90
	700	5.49	6		6.00		6.77		7.33
	800	3.97	6		4.33		4.83		5.21
	900	3.03	6		3.27		3.60		3.86
	1,000	2.40	6		2.54		2.76		2.94

A study of Table 1 (Vodenitcharova and Ansourian 1998) shows that for the case $\rho = 0.5$ corresponding to the maximum skew in the table, the greater percentage rise in buckling pressure calculated relative to the uniform buckling pressure occurs in the longer and thicker cylinders. For the case $L/R = 2\pi$, the rise is 26.5% at $R/h = 100$, and 20.6% at

R/h = 1000. The rise is only 6.8% at L/R = π/4 , R/h = 1000 (short cylinders), and naturally decreases with reducing skew.

2.2 *Siffeners for shell buckling resistance*

The minimum moment of inertia I of a stiffener at an end of a length of shell L required to effectively satisfy radial restraint requirement for shell buckling is evaluated as (Showkati and Ansourian 1995):

$$I_{\min} = 1.1\left\{0.15\left(\frac{R}{t}\right)^{0.516}\left(\frac{R}{L}\right)^{0.97} - 0.091\right\}t^3 L \qquad (6)$$

Stiffener inertia less than given by equation 6 will result in significant loss of buckling strength.

2.3 *Interaction*

In addition to circumferential compression due to external pressure, the shell normally sustains longitudinal stress from the global bending of the pipeline and hydraulic conditions. As the two effects are combined in the shell and are crucially governed by buckling failure, an interaction expression must be used. A relatively simple expression (e.g. ECCS 1988) has been used in design to account for this highly complex interaction phenomenon:

$$\frac{p_d}{p_u} + \frac{\sigma_d - \dfrac{Rp_u}{2t}}{\sigma_u - \dfrac{Rp_u}{2t}} \leq 1.0 \qquad (7)$$

where p_d is the design external pressure, p_u the ultimate external pressure, σ_d the design axial stress and σ_u the ultimate axial load. The latter may be evaluated by (AS4100 1990) for a section slenderness $\lambda_s = \dfrac{d_o f_y}{250t}$ and yield limit λ_{ey}, giving an effective section modulus for the pipeline in bending, corresponding to an effective ultimate stress σ_u as a fraction of the yield stress.

3 DESIGN OF STIFFENERS

The above method relates to the shell between ring stiffeners, which themselves require a degree of compactness for the development of yield before local buckling. Compact ring stiffeners then buckle in an ovalising mode (n = 2). Stresses in the ring assumed integral with a short effective length of shell, consist of combined membrane and bending stresses under factored loads. The theoretical global elastic buckling pressure of the compact stiffener then is:

$$p_n = \frac{3EI}{R^3 L} \qquad (8)$$

To ensure survival of the stiffener, the pressure to be used in its design must be greater than the design strength of the imperfect shell between stiffeners. The combined extreme fibre stress in the stiffener under these ultimate conditions should be limited to the yield stress of the stiffener; The peak stress, calculated in a second order analysis assuming a stiffener imperfection of oval form 0.005Rsin(2θ) or greater, is given by:

$$\sigma_{membrane} + \sigma_{bending} \leq F_y \tag{9}$$

where the amplified moment in the stiffener is $\dfrac{0.005pR^2}{1-\dfrac{p}{p_n}}$.

4 CONCLUSION

In this paper, cement lined steel pipelines submerged in a fluid have been investigated. They are subjected to variable external pressure and meridional stress due to global bending. The design problem is one of interaction buckling under external pressure and axial stress. The proportions of the ring stiffeners must be such that their local buckling is precluded. An effective length of pipe is assumed to act with the stiffener. The failure mode is triggered by yielding of the inner fibres under combined compression and amplified bending stresses of the initially imperfect ring, assumed deformed in an ovalising mode. The ultimate strength of the shell between stiffeners is evaluated taking account of the reduced buckling strength arising from imperfections and flexibility of the ring stiffeners.

5 REFERENCES

AS 4100 1990. 'Steel structures', Standards Australia.

American Petroleum Institute 1987. 'Bulletin on Stability Design of Cylindrical Shells', Bulletin U2.

ECCS (European Convention for Constructional Steelwork) 1988. 'Buckling of Steel Shells European Recommendations', Brussels.

Showkati, H. and Ansourian, P. 1995. 'Practical Stiffening of Cylindrical Shells against Buckling Collapse', Proc. 5th Intl. Conf. on Bulk Materials Storage, Handling and Transportation, IEAust. Pub. 95/04, July, 439-444.

Trahair, N.S., Abel, A., Ansourian, P., Irvine H.M. and Rotter, J.M. 1983. 'Structural Design of Steel Bins for Bulk Solids', AISC 1983.

Vodenitcharova, T. and Ansourian, P. 1994. 'Influence of boundary conditions on he bifurcation static instability of circular cylindrical shells subject to uniform lateral pressure', Research Report R701, October, The University of Sydney, School of Civil and Mining Engineering, Centre for Advanced Structural Engineering, Sydney, Australia, 1-136.

Vodenitcharova, T. and Ansourian, P. 1996 'Buckling of Circular Cylindrical Shells Subject to Uniform Lateral Pressure'. Engineering Structures 18, No. 8, 604 - 614.

Vodenitcharova, T. and Ansourian, P. 1998. 'Hydrostatic, Wind and Non-Uniform Lateral Pressure Solutions for Containment Vessels', Thin-walled Structures, Vol. 31, No. 1-3, May-July, 221-236.

Mechanics of Structures and Materials, Bradford, Bridge & Foster (eds)
© 1999 Balkema, Rotterdam, ISBN 90 5809 107 4

Nonlinear analysis of I-section beams curved in plan

Y-L. Pi & M.A. Bradford
School of Civil and Environmental Engineering, The University of New South Wales, Sydney, N.S.W., Australia

ABSTRACT: The design of I-section beams curved in plan, as may be used in highway bridges, usually assumes that the behaviour is linear. However, when such a beam is subjected to transverse (vertical) loading, it experiences primary bending and nonuniform torsion actions. As a result, the vertical deflections couple with the twist rotations, and the primary bending and torsion actions (with their associated deformations) then couple to produce second order bending actions in and out of the plane of the curved beam. The interaction between these actions can grow rapidly, and produce early nonlinear behaviour and even yielding. This paper describes a curved-beam finite element that is able to account for geometric and material nonlinearity in the analysis of I-section beams curved in plan. The focus is on the behaviour of beams for which the in-plan curvature may be small or large, and it is shown that this effect is of high significance in the structural behaviour of beams curved in plan.

1 INTRODUCTION

The accurate analysis of I-section girders curved in plan requires consideration of the coupling between primary bending and torsion actions. The Structural Stability Research Council Task Group 14 (SSRC 1991) identified research on the nonlinear inelastic behaviour of horizontally-curved beams and girders as an important field of investigation, since the 'linear' analysis of these beams (that ignores the coupling, or treats the behaviour as that of flexural-torsional buckling) can lead to very erroneous results. In response to this identified area of research, Pi, Bradford & Trahair (1999) have developed a 3-D curved beam element model for the geometric and material nonlinear analysis of I-section beams and girders. This paper describes briefly the development of the 3-D element, which is subsequently used to illustrate the differences in the structural response of curved beams whose in-plan curvature varies from very small to large.

2 THEORY AND VALIDATION

The formulation of the nonlinear finite element model is based on the following assumptions and considerations:

- The Euler-Bernoulli theory of bending and Vlasov's theory of torsion;
- A statement of the nonlinear strain-displacement relationships for large displacements, twists and rotations;
- The influence of height of application of the on nonlinear behaviour;
- An elastic-plastic incremental stress-strain relationship derived from the von Mises yield criterion, with the associated flow rule and an isotropic strain hardening parameter;
- Sinusiodal initial crookedness along the beam axis; and
- Longitudinal residual stresses.

The longitudinal strain at P (Figure 1) is expressed as (Pi, Bradford & Trahair 1999)

$$\varepsilon_P = \tilde{w}' + \frac{1}{2}\tilde{u}'^2 + \frac{1}{2}\tilde{w}'^2 - x\left[\tilde{u}''C + \tilde{v}''S + (1+\tilde{w}')\kappa_o C - \kappa_o\right] + y\left[u''S - v''C + (1+\tilde{w}')\kappa_o S\right]$$
$$+\frac{1}{2}\left(x^2+y^2\right)\left(\phi' - v'\kappa_o\right)^2 - \omega\left[\phi'' - v''\kappa_o + \frac{1}{2}\left(\tilde{u}''v' - \tilde{u}'v''\right)\right] \quad (1)$$

where κ_o = $1/R$ is the initial curvature of the beam (R = the initial radius of the beam), $\tilde{u}' = u' + w\kappa_o$, $\tilde{w}' = w - u\kappa_o$, $C = \cos\phi$, $S = \sin\phi$, u,v,w = the displacements of the centroid in the x,y,s directions (Figure 1), ϕ = twist rotation of cross-section about the s axis, and ω = the warping function (Trahair & Bradford 1998). Furthermore, the uniform shear strain may be expressed as

$$\gamma_P = -2t_P\left[\phi' - v'\kappa_o + \frac{1}{2}\left(\tilde{u}''v' - \tilde{u}'v''\right)\right] \quad (2)$$

where t_P = the distance from the mid-line of the wall thickness.

The elastic-plastic incremental stress-strain law can be derived from the von Mises yield criterion, together with the associated flow rule, and the isotropic strain hardening rule as

$$\{\Delta\sigma\} = [E]^{ep}\{\Delta\varepsilon\} \quad (3)$$

where $[E]^{ep}$ is the elastic-plastic incremental stress-strain matrix (Pi, Bradford & Trahair 1999), and $\{\Delta\varepsilon\}$ and $\{\Delta\sigma\}$ are the increments in the strain and stress vectors respectively. In a Lagrangian framework, the nonlinear incremental-iterative equilibrium equations can be obtained by invoking the principle of virtual work as

$$[k_T]_i\{\Delta r\}_i^j = \{\Delta p_e\}_j^i + \{\Delta p_r\}_i^{j-1} \quad (4)$$

where i denotes the load step, j the iterative cycle, $[k_T]_i$ = the tangent stiffness matrix obtained from the last (i-1 th) converged load step and used at the current load step i, $\{\Delta r\}_i^j$ = the increments of the nodal vector in the current iteration j, $\{\Delta p_e\}_i^j$ = the increments of the external load vector $\{p_e\}$ at the current iteration j, and $\{\Delta p_r\}_i^j$ = the out-of-balance force vector of the previous (j - 1 th) iteration.

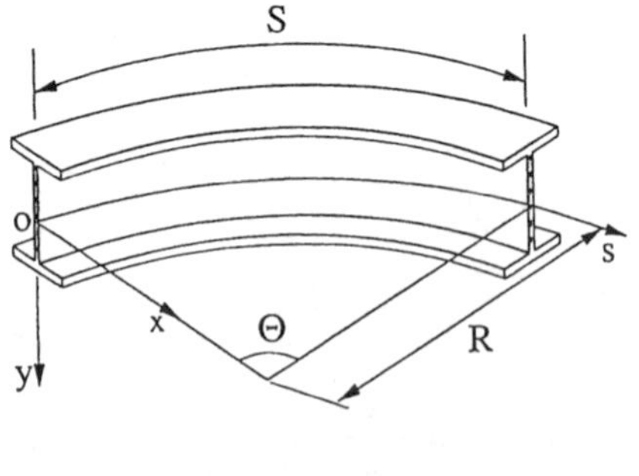

Figure 1. Axis system of I-section beam curved in plan

Eqs. 4 are solved using a Modified Newton-Raphson method, in conjunction with an arc-length control iterative strategy (Crisfield 1981) and an automatic incrementation strategy. A tangent predictor and radial return technique is used to ensure that the increments in the stresses are deduced correctly from the increments in the strains. The trial stress vector is given by

$$\{\sigma_t\}_i^j = \{\sigma\}_i^{j-1} + [E]^e \xi \{\Delta\varepsilon\}_i^j + [E]^{ep}(1-\xi)\{\Delta\varepsilon\}_i^j \tag{5}$$

where $[E]^e$ = the elastic property matrix, and ξ is a measure of yielding (Pi, Bradford & Trahair 1999) such that the strain increments are divided into elastic portions $\xi\{\Delta\varepsilon\}$ and plastic portions $(1-\xi)\{\Delta\varepsilon\}$ $(0 \le \xi \le 1)$. The radial return is performed using

$$\{\sigma\}_i^j = \{\sigma_t\}_i^j + \{\delta\sigma\}_i^j \tag{6}$$

where

$$\{\delta\sigma\}_i^j = -\frac{\sigma_e g}{\sigma_P^2 + 9\tau_P^2}\begin{Bmatrix}\sigma_P \\ \tau_P\end{Bmatrix}_i^j \tag{7}$$

$$g = \sigma_e - f_y \tag{8}$$

$$\sigma_y = f_y + \int H' \mathrm{d}\varepsilon_e^p \tag{9}$$

in which σ_e = the effective stress, f_y = the uniaxial yield stress, H' = the hardening parameter, and the equivalent plastic strain increment $\mathrm{d}\varepsilon_e^p$ is given by

$$\mathrm{d}\varepsilon_e^p = \frac{\sigma_e\left(E\sigma_P\Delta\varepsilon_P + 3G\tau_P\Delta\gamma_P\right)}{H'\sigma_e^2 + E\sigma_P^2 + 9G\gamma_P^2} \tag{10}$$

where E and G are the elastic and shear moduli respectively.

The radial return can be performed iteratively until a predetermined accuracy is obtained, and the enforcement of consistent conditions is performed at every iteration rather than at the end of a load step. In order to avoid spurious elastic unloading in the numerical scheme, Nyssen's (1981) assumption of incremental reversibility is used. Full details of the model, and of the method of solution of the resulting nonlinear equations (Eqs. 4) is given in Pi, Bradford & Trahair (1999).

Experimental verification of the numerical model has been reported in Pi, Bradford & Trahair (1999). The first comparison reported in this study was that of Fukumoto & Nishida (1981), who tested six horizontally-curved I-beams made from structural steel with an elastic-fully plastic-strain hardening constitutive response. Residual stresses appropriate to the cutting of the flanges and bending of the web were used. Figure 2 shows the variation of the absolute value of the twist rotation $|\phi_c|$ with the load Q from the tests, the present theory and the results of the ABAQUS shell element used by Tan *et al.* (1992), which required 140 shell elements to obtain a converged solution. The results of the present curved beam element, which only needs 8 elements to obtain a sufficiently-convergent solution, are in good agreement with the tests; more so than the ABAQUS results.

Shanmugam *et al.* (1995) also tested hot-rolled I-beams that were cold-bent, as well as some welded beams fabricated from mild steel plate. The variations of the vertical displacement v_q with the vertical load Q for two of the beams are shown in Figure 3, together with the numerical solution of the present study and the results of the ABAQUS shell element used by Shanmugam *et al.* (1995). The very favourable accuracy of the curved-beam finite element model is evident from this figure.

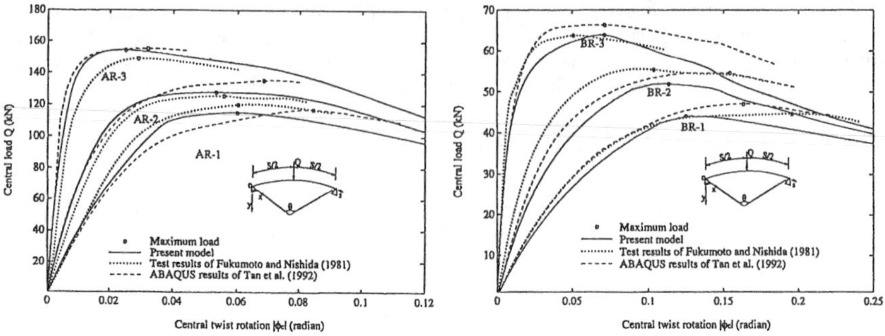

Figure 2. 'AR' and 'BR' beam results (Fukumoto & Nishida 1981)

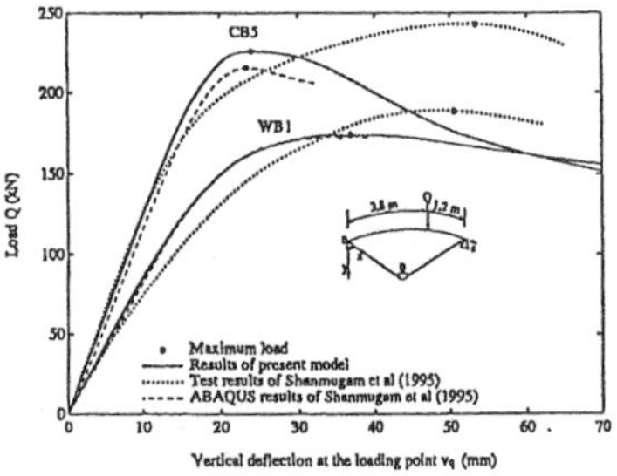

Figure 3. Shanmugam's beams

3 NONLINEAR INELASTIC BEHAVIOUR OF CURVED I-BEAMS

The nonlinear inelastic flexural-torsional response of a group of curved beams with an included angle varying from 0.2° to 90° has been investigated using the present finite element method. The geometry and material properties used was the same as those of Fukumoto & Nishida (1981) with a developed length of 2.0 m, so that the initial curvature κ_o varied from 1/573 m^{-1} to 1/1.273 m^{-1}. These beams were subjected to equal and opposite moments M_x which produce a bending moment and torque at a cross-section θ as

$$M = M_x \frac{\cos(\Theta/2-\theta)}{\cos\Theta/2} \qquad M_s = M_x \frac{\sin(\Theta/2-\theta)}{\cos\Theta/2} \tag{11}$$

where Θ is the inclined angle of the curved beam. The maximum bending moment occurs at midspan, while the maximum torque occurs at the beam ends.

The variations of the dimensionless lateral displacements u_c/B, the central twist rotations ϕ_c and the dimensionless vertical displacements v_c/D with the dimensionless end moments M_x/M_p are shown in Figures 4 to 6 respectively, where B is the flange width, D is the beam depth and M_p is the full plastic moment. The curved beam with a small included angle (0.2°) is almost straight, and so the primary coupling between the vertical deflection and twist rotations is small and bending is the dominant action. Because of this, its inelastic flexural-torsional behaviour is very similar to the inelastic flexural-torsional buckling and postbuckling response of a straight beam ($\Theta = 0$). The lateral displacements u_c (Figure 4) and twist rotations ϕ_c (Figure

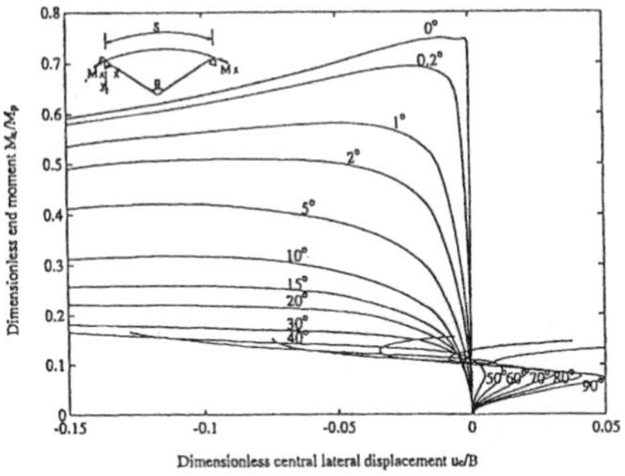

Figure 4. Variation of central lateral deflection

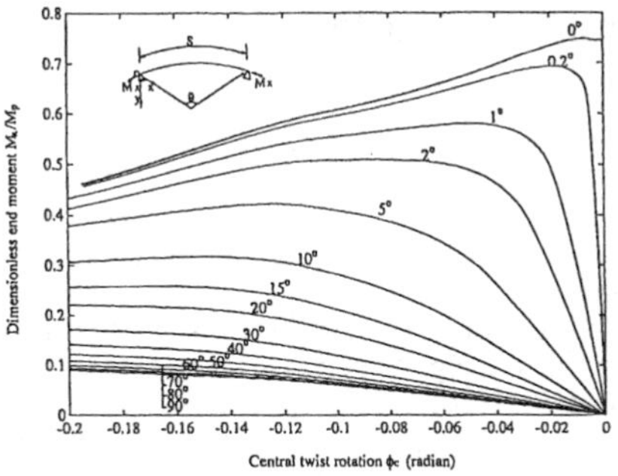

Figure 5. Variation of central twist rotation

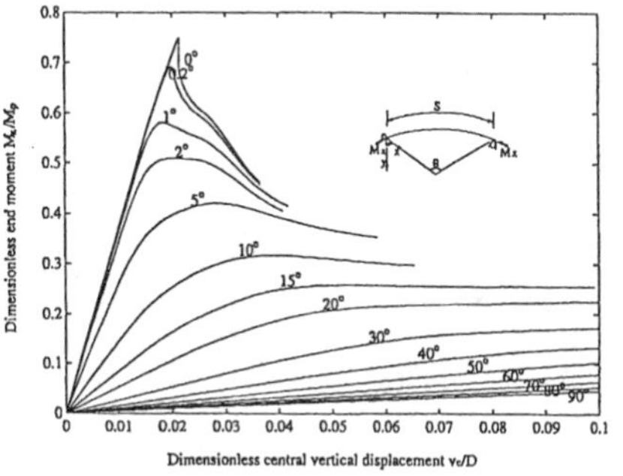

Figure 6. Variation of central vertical deflection

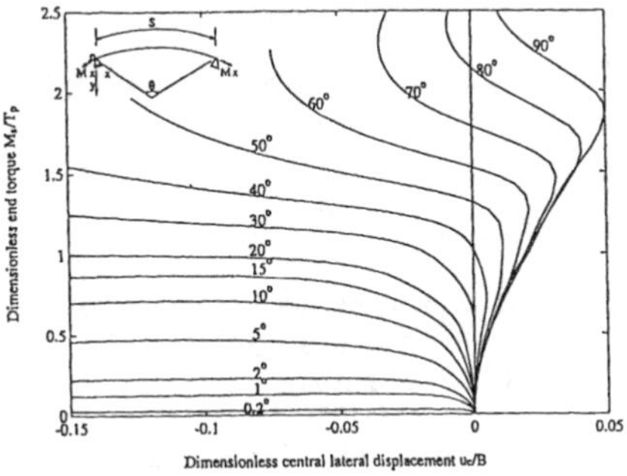

Figure 7. Variations of central lateral deflections with end torques

5) are very small until the end moments approach the inelastic "buckling" moment, after which u_c and ϕ_c increase rapidly while the end moments M_x decrease. This "near flexural-torsional buckling" behaviour only occurs for almost straight beams for which $\Theta \leq 1^\circ$. For curved beams with an included angle $1^\circ < \Theta \leq 20^\circ$, the lateral displacements and twist rotations (Figures 4 and 5 respectively) become substantial before the ultimate moment is reached. After the ultimate moment is attained, the lateral displacements and twist rotations increase rapidly while the end moments M_x decrease slightly. For curved beams with an included angle $\Theta > 20^\circ$, torsion becomes the major action, as can be seen from Figure 7 where u_c/B is plotted as a function of the dimensionless end torque M_s/T_p, where T_p is the upper bound of the full plastic torque (the sum of the full plastic uniform torque and the full plastic warping torque) given by Trahair & Bradford (1998). For these highly curved beams, the twist rotations and vertical deflections are substantial even at small end moments M_x, and no maximum moments are observed. When the end moments reach certain values, the lateral displacements (Figure 5), twist rotations (Figure 6) and vertical displacements (Figure 7) increase rapidly while the end moments have little increase. These values of the end moments can be considered sensibly as the ultimate moment-carrying capacities of these beams. As the end moments M_x increase further, the deflections and twist rotations stiffen owing to the increase of the torsional stiffness associated with large twist rotations and Wagner effects (Trahair & Bradford 1998). The Wagner effects are related to the longitudinal normal strain term $(x^2 + y^2)(\phi' - v'\kappa_o)^2/2$ in Eq. 1 which increases rapidly for curved beams with large included angles.

4 CONCLUDING REMARKS

This paper has summarised the theoretical development of a curved finite element that can be used to investigate the geometric and material nonlinear behaviour of I-section beams curved in plan. The element allows for coupling of bending and twist actions and deformations, and the nonlinear stiffness equations are solved by recourse to the Modified Newton-Raphson Method in conjunction with an arc-length control iterative strategy and an automatic incrementation strategy.

The finite element solutions have been verified against independent experiments, and used to study one particular test beam parametrically. It was found that for beams with an included angle less than 1°, the flexural-torsional response resembled that of inelastic flexural-torsional buckling. For an included angle varying from 1° to 20°, the lateral and twist deformations are large before the ultimate moment is reached, while for beams with an included angle greater than 20°, torsion dominates and it is harder to define an ultimate moment-carrying capacity.

5 ACKNOWLEDGEMENT

The work reported in this paper was supported by the Australian Research Council under the ARC Large Grants Scheme.

REFERENCES

Crisfield, M.A. 1981. A fast incremental/iterative solution procedure that handles snap-through. *Comps. & Structs.* 13:55-62.

Fukumoto, Y. & Nishida, S. 1981. Ultimate load behavior of curved I-beams. *J. Eng. Mechs. Div., ASCE.* 107(EM2):367-385.

Nyssen, C. 1981. An efficient and accurate iterative method, allowing large incremental steps, to solve elasto-plastic problems. *Comps. & Structs.* 13:63-71.

Pi, Y-L, Bradford, M.A. & Trahair, N.S. 1999. Nonlinear inelastic analysis and behaviour of steel I-beams curved in plan. *UNICIV Report* R-377, School of Civil & Environmental Engineering, The University of New South Wales, Sydney.

Shanmugam, N.E., Thevendran, V., Liew, J.Y.R. & Tan, L.O. 1995. Analysis and design of I-beams curved in plan. *J. Struct. Engg., ASCE.* 121(2):249-259.

SSRC 1991. Horizontally curved girders - a look into the future. SSRC Task Group 14, Annual Report. Chicago: SSRC.

Tan, L.O., Thevandran, V., Liew, J.Y.R. & Shanmugam, N.E. 1992. Analysis and design of I-beams curved in plan. *J. Singapore Struct. Steel Soc.* 3(1):39-45.

Trahair, N.S. & Bradford, M.A. 1998. *The Behaviour and Design of Steel Structures to AS4100.* 3rd edn., London: E&FN Spon.

Mechanics of Structures and Materials, Bradford, Bridge & Foster (eds)
© 1999 Balkema, Rotterdam, ISBN 90 5809 107 4

Steel frame design using advanced analysis

Z.Yuan, M.Mahendran & P.Avery
Physical Infrastructure Centre, School of Civil Engineering, Queensland University of Technology, Brisbane, Qld, Australia

ABSTRACT: The use of advanced analysis methods permits comprehensive assessment of the actual failure modes and ultimate strengths of frames in practical design situations. One of the advanced analysis methods, the pseudo plastic zone method, has been developed recently for steel frame structures subject to local buckling. It uses accurate analysis parameters obtained from finite element analysis of stub beam-column models to form the element force-displacement relationships. This paper presents the results of FEA models of a range of universal beam (UB) sections, which are subject to local buckling. Using these results, the pseudo plastic zone method can be applied to steel frame structures comprising any of the current range of Australian hot-rolled UB sections.

1 INTRODUCTION

The refined plastic hinge method is one of the advanced analysis techniques for steel frame design that combines analysis and design tasks into one step. The method has significant advantages over the FEA method, that is, it does not require excessive computer resources for the program execution. It is a reasonably accurate technique for the advanced analysis of steel frame structures comprising non-compact sections, and significantly superior to the conventional design procedure based on elastic analysis (Avery 1998). However, comparison with benchmark solutions identified a number of limitations and sources of error in the model due to simplifying approximations (initial yield, imperfections, softening, flexural stiffness reduction function, tangent modulus, and elastic stability functions) and the use of model parameters based on empirical specifications equations (tangent modulus, section capacity, plastic strength, and effective section properties). Therefore Avery and Mahendran (1998) proposed a rational and more accurate method for advanced analysis of steel frame structures comprising non-compact sections. Analytically "exact" model parameters were derived from distributed plasticity analyses of a stub beam-column model, which were then used for the formulation of frame element force-displacement relationships. The new method, referred to as pseudo plastic zone analysis, is verified by comparison with the analytical benchmark solutions provided by Avery (1998). However, it was based on the results from only three non-compact I-sections. Therefore it was extended to include most of the Australian UB sections (BHP, 1998). This paper presents the details of the formulation of the new method including the stub beam-column model analysis and the analytical results for a range of hot-rolled UB sections.

2 DISTRIBUTED PLASTICITY ANALYSIS OF STUB BEAM COLUMN MODEL

Distributed plasticity analysis can be used to calibrate accurate section capacity and stiffness reduction functions for use in concentrated plasticity models. This approach was adopted in the past for a typical compact section (King and Chen, 1994). Finite element stub beam-column can model the gradual formation of a plastic hinge in a member of a structural system and was used in this study (see Figure 1). In order to explicitly model local buckling deformations and spread of plasticity effects, 3-D shell elements were used (ABAQUS S4R5). This element is a thin isoparametric quadrilateral shell with four nodes and five degrees of freedom per node. Due to the lack of experimental data involving strain hardening, elastic perfect plasticity material properties were used for the stub beam-column models. In non-linear FEA of noncompact sections, local imperfections and residual stresses must be included. It is most critical when local imperfection shape coincides with the critical eigenvector obtained from an elastic buckling analysis. The magnitudes of local web and flange imperfections were taken as $1/150^{th}$ of the plate width based on the fabrication tolerances for compression members specified in AS4100 (SA 1998). The residual stress distribution included in the stub beam-column model was based on the recommendations of ECCS Technical Committee 8 (1984).

In order to distribute the axial force and bending moment to the stub beam column member, a loading plate was used to apply loads to the beam-column model. The ABAQUS R3D4 element was adopted (see Figure 1). The rigid plate is attached to one end of the beam-column and a concentrated nodal force P and a bending moment M is applied to the centroid of the plate. The effects of stress concentrations caused by the rigid elements were eliminated by including an elastic strip. A quarter of one local buckling wavelength was taken as the length of the stub beam-column model. At the end of the model representing the plane of symmetry, the boundary condition used was "345" for all nodes on the cross-section. At the other end adjacent to the loading plate a single point constraint of "126" was applied to the centroidal node, representing an ideal pinned support. The degrees of freedom notation – 123 correspond to translation in x, y and z axes whereas 345 relate to rotation about x, y and z axes (Figure 1).

There were 2 stages of FEA: the preliminary and systematic. The 310UB32.0 was chosen for the preliminary analysis, because it is the most slender UB section (ie. larger local buckling effects). From the preliminary analyses, the following important conclusions were drawn.

- The stub beam-column length equal to half of the web depth is suitable for all analyses.
- Local imperfections obtained from pure compression elastic buckling analysis are appropriate for all load cases, if the imperfection magnitudes of the web and flange are $d_w/150$ and $b_f/150$, respectively.

Of the hot-rolled Universal Beam sections produced in Australia by BHP Steel (1998), there are 5 sections which are non-compact for pure bending and pure compression, and 15 sections which are non-compact for pure compression only (ie. $k_f < 1$). All of these sections are therefore subject to the effects of local buckling. In this study, all five sections from the first group and another five sections from the second group were included. In the second stage of analysis, four types of finite element non-linear static analysis were performed under 13 different load combinations of P and M for each section to obtain the required analysis parameters.

1. Second-order inelastic analysis with no local imperfections. The plastic strength interaction curve was obtained by multiplying the ultimate load factors for each load case by the corresponding applied loads.
2. Second-order inelastic analysis with local imperfections. The section capacity curve was obtained by multiplying the ultimate load factors for each load case by the corresponding

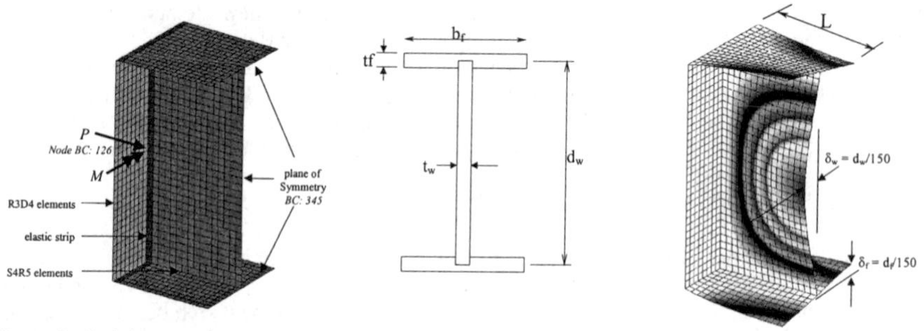

Figure 1. Stub beam-column model

applied loads. The axial displacements (u_i), major axis rotations (θ_i), and load increments were also obtained.
3. Second-order elastic analysis with local imperfections. The axial displacements (u_e), major axis rotations (θ_e), and load increments were obtained The section tangent moduli were then calculated for each load increment, using the equation following:

$$e_{tf} = \frac{\dot{\theta}_i}{\dot{\theta}_e} \qquad e_{ta} = \frac{\dot{u}_i}{\dot{u}_e} \tag{1}$$

4. Second-order inelastic analysis with local imperfections using the RIKS arc-length method. The results of these analyses were used to plot the softening regions of the load-deflection and moment-rotation curves, and to determine the softening moduli.

3 RESULTS AND EVALUATION

The combined effects of material and geometric non-linearity can be represented by the following force-displacement relationship for a pseudo plastic zone (PPZ) frame element:

$$\begin{Bmatrix} \dot{M}_A \\ \dot{M}_B \\ \dot{P} \end{Bmatrix} = \frac{\zeta EI}{L} \begin{bmatrix} \phi_A\left[s_1' - \dfrac{s_2'^2}{s_1'}(1-\phi_B)\right] & \phi_A\phi_B s_2' & 0 \\ \phi_A\phi_B s_2' & \phi_B\left[s_1' - \dfrac{s_2'^2}{s_1'}(1-\phi_A)\right] & 0 \\ 0 & 0 & \dfrac{e_{ta}A}{\zeta I} \end{bmatrix} \begin{Bmatrix} \dot{\theta}_A \\ \dot{\theta}_B \\ u \end{Bmatrix} + \dot{\mathbf{f}}_{lp} \tag{2}$$

Avery and Mahendran (1998) illustrate how the inelastic stability functions (s_1' and s_2'), and flexural stiffness reduction factors (ϕ_A, ϕ_B) are obtained from the important parameters such as plastic strength (p_{ps}), section capacity (p_{sc}), initial yield (p_{iy}), tangent moduli (e_{tf}), and hinge softening modulus (e_s) that are derived from the stub beam-column FEA. Avery and Mahendran (1998) also present the equation for the imperfection reduction factor ζ and other details of the method whereas this paper presents the results from the stub beam column analyses. The applied axial loads and moments and the corresponding capacities are non-dimensionalised with respect to axial yield capacity and plastic moment capacity, respectively. Similarly, tangent and softening moduli are non-dimensionalised with respect to the modulus of elasticity.

3.1 Plastic strength, section capacity, and initial yield curves

Using stub beam column model with non-linear static analysis, the effective section ratios (k_f and Z_e/S) can be derived. An attempt was made to express the effective section ratios as a function of the plate element slenderness ratios, as shown in Equation 3.

$$\begin{cases} k_f = -0.1798\left\{\dfrac{d_1/t_w}{45} + \dfrac{(b_f - t_w)/2t_f}{16}\right\} + 1.1948 & ; \quad k_f \le 1 \\[2ex] \dfrac{Z_e}{S} = -0.0581\left\{\dfrac{d_1/t_w}{82} + \dfrac{(b_f - t_w)/2t_f}{9}\right\} + 1.0711 & ; \quad Z_e/S \le 1 \end{cases} \tag{3}$$

For convenience of computer implementation, the normalised plastic strength can be defined as a function of the non-dimensional axial load to moment *(p/m)* ratio using two series of cubic equations in the following form:

$$p_{ps} = \begin{cases} 0.9978t - 0.0724t^2 - 0.9649t^3 \; ; & \dfrac{p}{m} < 0.2 \\[2ex] 0.0047 + 1.0379t - 0.6268t^2 + 0.2373t^3 \; ; & \dfrac{p}{m} > 0.2 \end{cases} \tag{4}$$

$$m_{ps} = \frac{p_{ps}}{(p/m)} \qquad t = \tan^{-1}(p/m)$$

The constants of the cubic equations were derived from a least squares regression analysis of the stub beam-column model results for each section. The variable t represents the angle between the horizontal m axis and the load path (Figure 3). The variable t varying from 0 to $\pi/2$ is preferable for curve fitting purposes while simulating the p/m variation from 0 to ∞. The plastic strength corresponding to any applied load can be evaluated from Equation 4 without solving a polynomial equation as is required for alternative functions such as those proposed by Attalla et al (1994). The normalised section capacity p_{sc} of member subjected to combined actions can be derived in a similar manner (Equn. 5). A summary of the section capacity constants for 3 UB sections is provided in Table 1. Other results are given in Yuan et al. (1999).

$$p_{sc} = b_0 + b_1 t + b_2 t^2 + b_3 t^3 \quad ; \quad m_{sc} = p_{sc}/(p/m) \quad ; \quad t = \tan^{-1}(p/m)$$
$$p_{sc} = k_f \quad (m = 0) \quad ; \quad m_{sc} = Z_e/S \quad (p = 0) \qquad (5)$$

Table 1. Constants of Section Capacity Cubic Equation (selected results)

Section	p/m < 0.2				p/m > 0.2			
	b_0	b_1	b_2	b_3	b_0	b_1	b_2	b_3
610UB125	0.0	1.0064	-0.1857	-1.0528	0.0323	0.8406	-0.3691	0.1303
360UB44.7	0.0	0.9782	-0.1221	-1.7529	0.0266	0.8263	-0.386	0.139
200UB18.2	0.0	0.9888	-0.1028	-1.5829	0.0223	0.8788	-0.4481	0.1694

The constants above give near-perfect fitting for individual section capacity curves (Figure 2(a)). A generalised section capacity equation was derived by introducing the section slenderness parameters (ie. k_f and Z_e/S) as variables and curve fitting the coefficients obtained for each of the 10 analysed sections. In order to minimise the error, it was necessary to divide the curve into five segments (Equation 6):

$$p_{sc} = (-1.554\, t^3 - 0.206\, t^2 + 0.976\, t)\, \{k_f\, t/0.889 + Z_e\, (1-t)/S/0.976\} \quad ; p/m \le 0.2$$
$$p_{sc} = (0.392\, t^3 - 0.791\, t^2 + 1.1\, t + 0.001)\, \{k_f\, t/0.889 + Z_e\, (1-t)/S/0.976\} \quad ; 0.2 < p/m \le 0.5$$
$$p_{sc} = (0.392\, t^3 - 0.791\, t^2 + 1.1\, t + 0.001)\, \{k_f^2/0.889 + (Z_e/S)^2/0.976\}/(k_f + Z_e/S) \quad ; 0.5 < p/m \le 1 \qquad (6)$$
$$p_{sc} = (-0.013\, t^3 + 0.142\, t^2 + 0.231\, t + 0.234)\, \{k_f^2/0.889 + (Z_e/S)^2/0.976\}/(k_f + Z_e/S) \quad ; 1 < p/m \le 2.5$$
$$p_{sc} = (-0.013\, t^3 + 0.142\, t^2 + 0.231\, t + 0.234)\, \{k_f\, (1-1/t)/0.889 + Z_e/S\, /(0.976\, t)\} \quad ; p/m > 2.5$$

Avery (1998) proposed the initial yield curve as Equation 7 (see Figure 2(b)):

$$p_{iy} = \frac{k_f(1-\sigma_r/\sigma_y)}{1 + \frac{k_f}{(Z/S)} \frac{1}{p/m}} ; m_{iy} = \frac{p_{iy}}{p/m} \qquad (7)$$

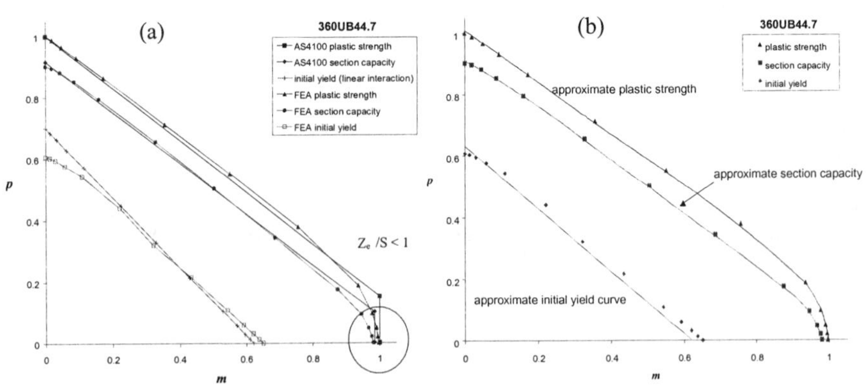

Figure 2. (a) Plastic strength, Section capacity, and Initial yield curve (k_f<1, Z_e/S<1)
(b) Comparison of approximate equations and FEA results

3.2 Section tangent moduli

Flexural and axial tangent modulus curves (Equation 1) derived from the stub beam-column clearly indicate that: The tangent modulus is a function of the p/m ratio, and the flexural stiffness reduction differs significantly from the rate of axial stiffness reduction The tangent modulus is equal to 1.0 before initial yielding occurs. Subsequent to the commencement of yielding, the tangent modulus can be defined by Equation 8 (Avery, 1998):

$$e_t = 1 - c_1 \alpha'^{c_2} - (1 - c_1)\alpha'^{c_3} \tag{8}$$

An effective plastic load state parameter (α') is used to define this function. This parameter varies from zero at the point of initial yield to one when the section capacity is reached. The m-p interaction diagram (Figure 3) illustrates the relationship between the load state parameter (α) and the effective load state parameter (α'). The parameter (α') is defined by Equation 9.

$$\alpha' = \frac{(\alpha - \alpha_{iy})}{(\alpha_{sc} - \alpha_{iy})} = \frac{(m - m_{iy})}{(m_{sc} - m_{iy})} \qquad \text{For } \alpha > \alpha_{iy}$$

$$\alpha' = 0; \qquad \text{For } \alpha <= \alpha_{iy} \tag{9}$$

The constants, c_1, c_2, and c_3 were determined by a least-squares regression analysis for each load combination. Table 2 presents these constants for some load combinations. Other results are given in Yuan et al (1999). Tangent moduli for intermediate p/m ratios can be evaluated using linear interpolation, using $t = \tan^{-1}(p/m)$ as the interpolation variable (shown in Figure 4).

A comparison of different sections with the same p/m ratio reveals some interesting aspects of the relationship between the tangent modulus and the section slenderness. 1) For non-compact sections, increasing section slenderness is associated with a more abrupt stiffness reduction (eg. 310UB32 section). 2) For sections with similar slenderness, the tangent modulus functions are similar (eg. 200UB18.2, 200UB25.4 and 250UB25.7 sections).

Table 2. Constants of the Tangent Modulus Equation (200UB25 section)

Load combination p/m	flexural e_t coefficients			axial e_t coefficients		
	c_1	c_2	c_3	c_1	c_2	c_3
0	0.500	1.003	2.538			
0.02	0.456	1.003	2.722	0.654	0.313	0.000
5	0.865	0.904	16.380	0.585	2.086	56.029
20	1.139	0.515	1.704	0.659	1.998	57.000
∞ (ie. m=0)				0.753	2.059	57.398

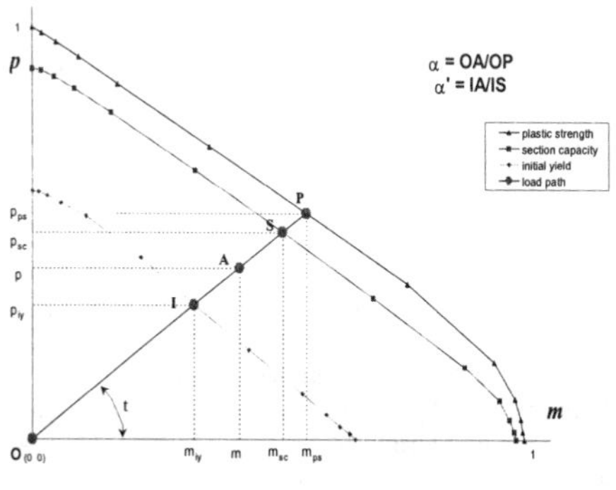

Figure 3. m-p interaction diagram

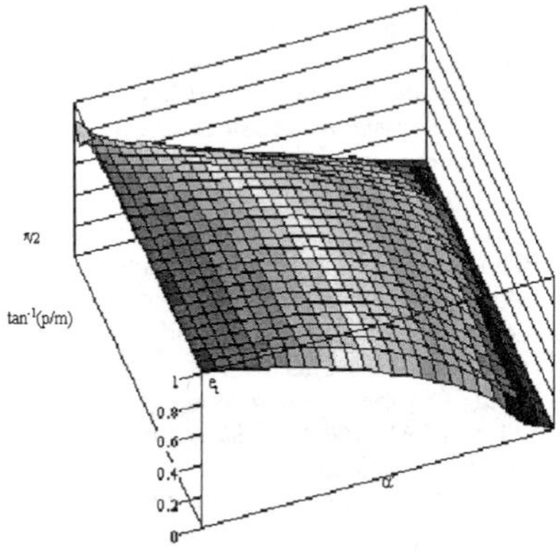

Figure 4. Tangent modulus as a function of α' and $\tan^{-1}(p/m)$

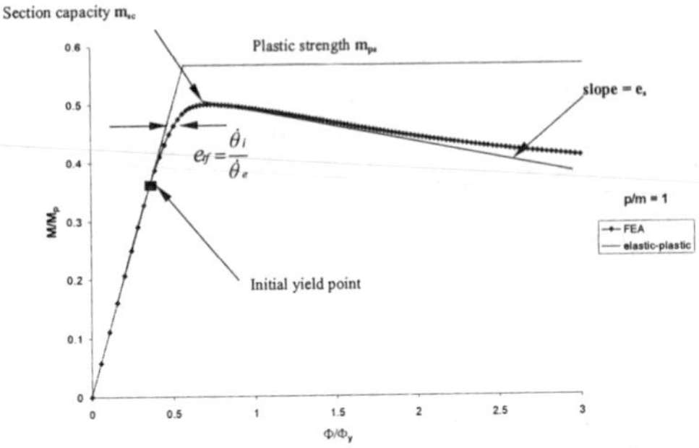

Figure 5. Comparison of analytical and approximate flexural moment-rotation curves

3.3 Hinge softening

After the formation of a plastic hinge, the reduction in the section capacity and flexural stiffness due to hinge softening can be approximately modelled by replacing the tangent modulus with a negative softening modulus. The softening modulus (e_s) can be determined from the analytical moment-curvature curves for each section and load combination. A generalised equation was again derived based on the FEA results for all sections and p/m ratios. This equation (Equation 10) can be used to represent the approximate hinge softening modulus for all sections studied.

$$e_s = -0.2635\sin(t)^3 + 0.4303\sin(t)^2 - 0.2258\sin(t) - 0.0486$$
$$t = \tan^{-1}(p/m) \tag{10}$$

4 CONCLUSION

Concentrated plasticity analysis is a computationally efficient method that provides similar accuracy to a non-linear shell finite element analysis. The effects of local buckling for a range of Australian hot-rolled UB sections can be included in the pseudo plastic zone method by application of the parameters and equations presented in this paper: Effective section ratios (Equation 3), Initial yield function (Equation 7), Plastic strength interaction curves (Equation 4), Section capacity interaction curves (Equation 6), Tangent modulus functions (Equations 8,9), Softening modulus function (Equation 10).

5 REFERENCES

Avery, P. 1998. *Advanced analysis of steel frames comprising non-compact sections*, Ph.D. thesis, School of Civil Engineering, QUT, Brisbane, Australia.
Avery, P. and Mahendran, M. 1998. *Refined plastic hinge analysis of steel frame structures comprising non-compact sections*, Physical Infrastructure Centre, Research Monograph 98-5, Queensland University of Technology, Brisbane, Australia
Attalla, M. R., Deierlein, G. G., and McGuire, W., Spread of plasticity: quasi-plastic-hinge approach, *J. of Struct. Engineering, ASCE*, Vol. 120, No. 8, 1994, pp. 2451-2473.
BHP (1998) Hot Rolled and Structural Steel products, Melbourne
ECCS 1984. *Ultimate limit state calculation of sway frames with rigid joints*, Tech. Committee 8 - Structural Stability Technical Working Group 8.2 - System, Publication No. 33.
King, W. S., and Chen, W. F. 1994. Practical second-order inelastic analysis of semi-rigid frames, *Journal of Structural Engineering, ASCE*, Vol. 120, No. 7, 1994, pp. 2156-2175.
SA (1998), "AS4100-1998 Steel Structures", Standards Australia, Sydney, Australia.
Yuan, Z., Mahendran, M. and Avery, P. 1999. *A Parametric Study of Non-compact I-Sections for the Pseudo Plastic Zone Analysis Method*, Physical Infrastructure Centre, Research Monograph 99-4, Queensland University of Technology, Brisbane, Australia

Composite structures

Mechanics of Structures and Materials, Bradford, Bridge & Foster (eds)
© 1999 Balkema, Rotterdam, ISBN 90 5809 107 4

Influence of partial shear connection on the behaviour of composite joints

A.A.Diedricks, B.Uy & M.A.Bradford
School of Civil and Environmental Engineering, The University of New South Wales, Sydney, N.S.W., Australia

D.J.Oehlers
Department of Civil and Environmental Engineering, University of Adelaide, S.A., Australia

ABSTRACT: This paper investigates the effect of partial shear connection on the behaviour of composite joints. By varying the degree of shear connection it is possible to alter certain behavioural properties such as stiffness, strength and ductility of the joint. Two full-scale web side plate composite joints were fabricated and tested, using the standard cruciform testing procedure, which represented full and partial shear connection respectively. Investigation into the influence of partial shear connection on the composite behaviour in the location of the joint region has been minimal. Comparisons are made between the theoretical and experimental results obtained concerning serviceability, strength and rotational requirements. Furthermore, predictions of stiffness, strength and ductility of the joint are also made.

1 INTRODUCTION

This paper outlines tests carried out on composite joints for semi-continuous composite beam behaviour. The two tests adopted a cruciform arrangement used to simulate the negative moment region of a semi-continuous beam. A number of factors were investigated with these tests, to determine the influence of partial shear connection on composite joint behaviour. Currently international codes of practice specify that it is necessary for the designer to provide full shear connection at the location of the negative support (British Standards Institution 1994). However, through the adoption of partial shear connection stiffness, strength and ductility characteristics of composite beams can be varied.

This paper also investigates the force transfer mechanism of the web side plate composite joint at the beam-column location. The advantages associated with composite joints have been widely researched and discussed by notable researchers around the world (Nethercot et al. 1996, Leon 1998). However, of the experimental tests carried out, the quantity that incorporated the web side plate connection has been limited. There is thus a need to investigate how this type of joint is influenced at the negative support region since this type of connection is considered to be pinned and to have minimal moment carrying capacity in steel design. Thus producing design solutions that are overly conservative.

The difference between full and semi continuous composite beams is characterised by the amount of moment capacity that can be transferred to the support location. Fully continuous beams are generally associated with welded type joints such as full and partial depth end plate connections. Conversely semi-continuous or partial strength joints are generally bolted type joints such as the cleat and web side plate connections. The benefits of adopting fully continuous joint designs are that it is possible to take advantage of redistribution effects, which reduce the mid-span moment and thus allow smaller sections to be adopted. However, these types of joints are costly to fabricate and erect. Therefore, by adopting joints, which are less fabrication intensive, it is possible to reduce costs and produce more economical design solutions.

2 EXPERIMENTS

Two full-scale cruciform composite joint tests were conducted. Both test specimens were designed for gravity loading as part of a braced frame adopting a typical 8 metre by 8 metre grid system, with a series of secondary beams and a one way composite slab. Office loading according to the Australian Standard AS1170.1-1989 (Standards Australia 1989) was assumed. Furthermore, flexural strength was determined according to AS2327.1-1996 (Standards Australia 1996) with minor modifications for semi-rigidity and partial strength.

2.1 *Description of test set-up*

The two tests were conducted on typical composite joint arrangements as shown in Figure 1(a). Both experiments adopted similar web side plate connection arrangements. Figure 1(b) schematically represents the steel connection arrangement with pertinent dimensions and reinforcing quantities.

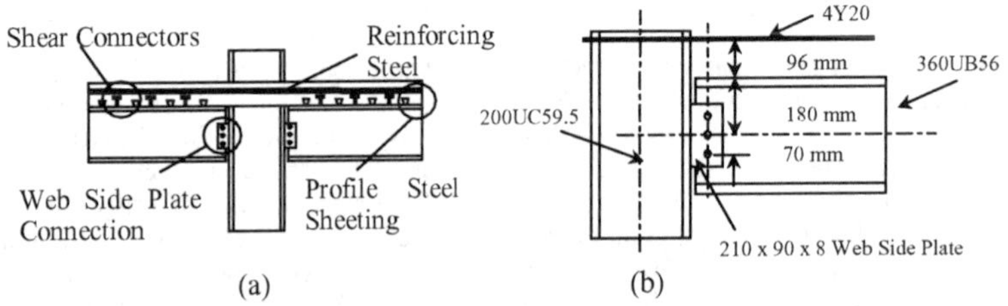

Figure 1. (a) Typical cruciform composite joint set-up, (b) Schematic of the composite joint

Prior to the cantilever beams being loaded in a symmetrical manner, the column was loaded axially. Ahmed and Nethercot (1998) in a series of tests considered the effect of axial loads on the behaviour of the composite joint. The conclusion reached was that there is a change in the behaviour of the joint when the column is loaded and it is particularly prevalent when the joint is subjected to a non-symmetric loading pattern.

2.2 *Material properties*

Before comparisons between experimental results and theoretical calculations can be made it is necessary to ascertain the true material properties. Tensile tests were carried out on steel beam coupons, steel reinforcing bars and profiled steel sheeting coupons. Compression testing of the concrete at various stages throughout the curing process was also conducted, as well as an indirect tensile test to observe the tensile capacity of the concrete. Push tests were also conducted to determine the load-slip characteristics and ultimate capacity of the shear connectors.

2.2.1 *Tensile tests – structural steel*

All tensile testing was carried out in an Instron Tensile Testing Machine. The first series of tests were for the determination of the tensile capacity of the steel beam coupons. Standard coupons were obtained from four locations on the steel beam. Figure 2 identifies the locations at which the coupons were taken. Table 1 identifies the cross-sectional area (A), yield stress (f_y) and ultimate stress (f_u) associated with each coupon specimen.

Table 1. Tensile coupon results

Coupon Location	A (mm^2)	f_y (MPa)	f_u (MPa)
A – Web centre	280	296	358
B – Web offset	280	279	361
C – Flange edge	455	292	369
D – Flange offset	455	286	369
Mean		288	364

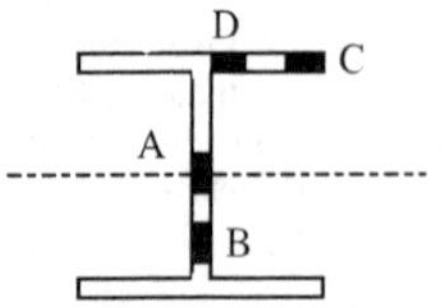

Figure 2. Location of tensile coupons

2.2.2 *Tensile tests – reinforcing steel*
A series of eight tensile tests were conducted so that the mean yield and ultimate capacities could be determined. The results are outlined in Table 2 and identify a mean value for yield stress of 448 MPa which is greater than the nominal yield value of 400 MPa.

2.2.3 *Tensile tests – profiled steel sheeting*
The profiled steel sheeting was tested to ascertain it's tensile capacity. This was deemed necessary, because if the neutral axis was found to be lower than the composite slab then it will be necessary to include in considering the contributing tensile force of the profiled steel sheeting. Two tensile specimens were tested, with the mean yield and ultimate stresses attained being 402 and 480 MPa respectively.

2.2.4 *Compression tests - concrete*
Compressive strength tests were conducted at intermediate times during the curing process and on the day of each connection test. The mean compressive capacities associated with each stage are outlined in Table 3.

Table 2. Steel reinforcement results

Specimen No.	f_y (MPa)	f_u (MPa)
1	450	530
2	453	524
3	452	524
4	442	525
5	445	527
6	444	528
7	444	530
8	450	530
Mean	448	527

Table 3. Results obtained from compressive tests

Time, t (days)	$f'_{c(t)}$ (MPa)
7	14.2
14	21.1
21	23.9
28	27.6
Day of Test 1, t = 30	29.6
Day of Test 2, t = 33	32.3

2.2.5 *Tensile tests - concrete*
A series of standard indirect tensile tests were conducted. The mean tensile capacity obtained from the three specimens tested was 2.96 MPa which could be used to evaluate the cracking moment.

2.2.6 *Push tests – stud shear connectors*
Push tests were designed and constructed according to Eurocode 4 (British Standards Institution 1994), with the testing procedure outlined by Oehlers and Bradford (1995) being adopted. The general arrangement is shown in Figure 3. The push tests were tested to failure and it was found that the ultimate load attained by the three specimens was 450, 483 and 475 kN respectively. This indicated that the mean capacity per shear connector was 117 kN which is 40% greater than the nominal capacity derived using the Australian Standard AS2327.1-1996, (Standards Australia 1996).

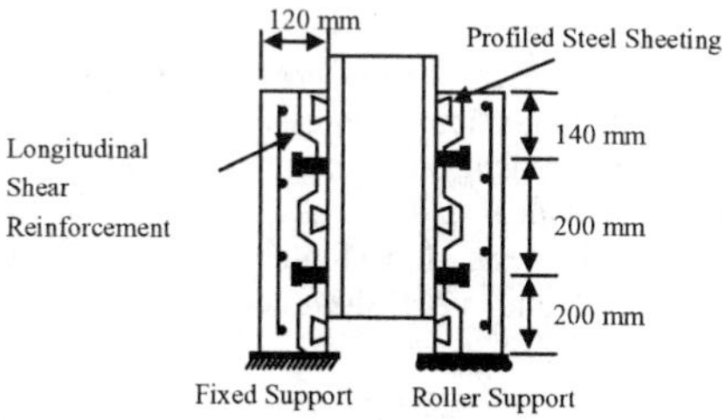

Figure 3. Testing arrangement for push tests

3 RESULTS

The most accurate representation of the behaviour of joints is their moment-rotation curves, since these curves identify the stiffness, yield and ultimate moment capacities of a joint. Figure 4(a) represent the moment-rotation curves for both the full shear connection (SS1) and partial shear connection (SS2) tests. Figure 4(b) presents these in non-dimensionalised form were M_{fsc} and M_{psc} represent the theoretical full and partial shear connection moment capacities respectively. When considering the non-dimensionalised curves $\overline{M}$ and $\overline{\phi}$ represents the non-dimensionalised moment and rotation respectively as outlined in Uy et al (1998). From these graphs it is possible to see that the joints do exhibit semi-rigid or partial strength behaviour, when the curves are compared to Eurocode 4 (British Standards Institution 1994) definitions. It is also possible to see that the yield and ultimate moment before contact are all greater than the theoretical values that were calculated using the rigid plastic model, which is explained henceforth.

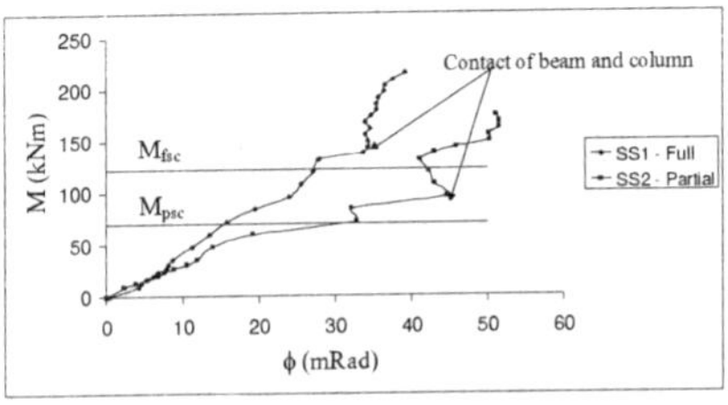

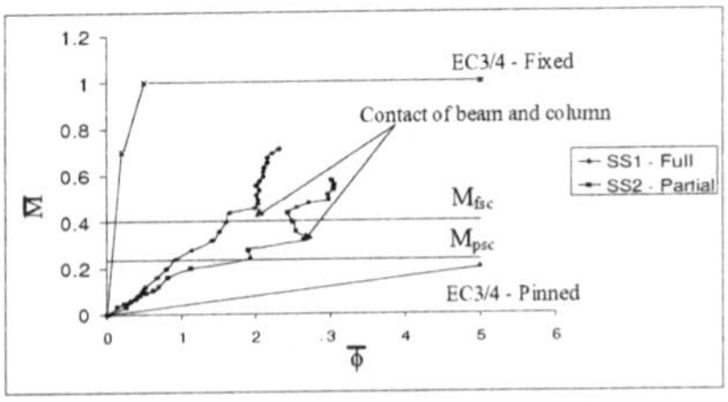

Figure 4. (a) Moment rotation curve (b) Non-dimensionalised moment rotation curve

4 COMPARISON WITH BASIC THEORY

In producing design guidelines for practising engineers there are three characteristics which need to be assessed for composite joints in semi-continuous beam systems. These three characteristics are the stiffness, strength and ductility requirements.

4.1 *Stiffness comparison*

In the assessment of deflection and vibration behaviour, service-loading conditions need to be considered. A model developed by Crisinel and Carretero (1996) gives a method in which one can predict the moment-rotation characteristics by determining the stiffness of the beam-to-column composite joints during the service load regime. The rotational stiffness S_j, is calculated as

$$S_j = \frac{z_s^2}{\frac{1}{k_s} + \frac{1}{k_v} + \frac{1}{k_c}} \tag{1}$$

This equation describes the influence of the factors that effect the stiffness of a composite joint. The stiffness of the slab, shear connectors and steel connection compression zone is given by k_s, k_v and k_c respectively. The distance between the tensile and compressive forces is represented by the parameter z_s. Table 4 outlines the individual and overall joint stiffness, as well as the stiffness obtained from the experimental joint tests. It is found that the model provides a conservative estimate of stiffness for both the full and partial shear connection tests.

Table 4. Stiffness comparisons

	S_J (10^6 Nm)
Theoretical	1.78
Full shear connection	3.24
Partial shear connection	2.47

4.2 *Strength comparison*

In evaluating the ultimate strength capacity of the joint a rigid plastic model was adopted for the full shear connection test. For the partial shear connection test it was assumed that the shear studs would fracture prior to concrete crushing. Figure 5 identifies the force distribution adopted for the rigid plastic model.

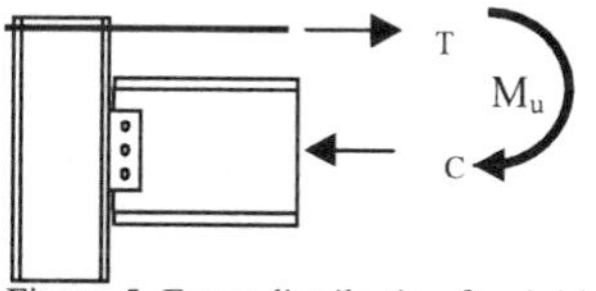

Figure 5. Force distribution for rigid plastic model

The results in Table 5 indicate that the rigid plastic model provides a conservative estimate of the ultimate capacity of these types of joints. Table 5 also outlines the differences, which were encountered between the theoretical and experimental values for the capacity of the shear connectors. This variation in capacity influenced the resultant degree of shear connection for each joint, which is outlined in the table. In comparing the test and theoretical results for the partial shear connection test it was still concluded that partial shear connection was exhibited.

Table 5. Comparison of theoretical and experimental values for full and partial shear connection tests

	Theoretical Values		Experimental Values	
	SS1 - Full	SS2 - Partial	SS1 - Full	SS2 - Partial
Connector Capacity (kN)	84	84	117	117
β (%)	100	57	145	83
Moment Capacity (kNm)	122	70	138	96

4.3 *Ductility comparison*

In order to adopt a plastic design philosophy for semi-continuous composite beams it is required that the joint exhibit sufficient rotational capacity. Nethercot et al. (1995) developed a model to evaluate the rotation required to promote a plastic hinge to form at the centre of the composite beam. The required rotation can be determined using the relationship in Equation 2.

$$\phi_{required} = \left[0.344 - 0.212\frac{M_{d,c}}{M_{d,b}} + 0.561\left(\frac{M_{d,b} - M_{y,b}}{M_{p,b} - M_{y,b}}\right)^2 \frac{1}{\sqrt{1+\frac{M_{d,c}}{M_{d,b}}}} \right] \frac{M_{d,b}L}{EI} \qquad (2)$$

Where $M_{d,c}$ and $M_{d,b}$ are the joint and span design moments respectively, while $M_{y,b}$ and $M_{p,b}$ represents the beams' yield and plastic moment capacities respectively. The associated span length is given as L and EI is the flexural rigidity. From Equation 2, it was evaluated that 17 mRad will be required to promote a plastic hinge to form at the mid-span of the beam. Table 6 outlines the experimental rotations attained at both yield and failure for both the full and partial shear connection tests. From the values tabulated it is possible to conclude that both tests reached the required rotation prior to failure. However, it can be seen from Table 6 that for the partial shear connection test the yield rotation is substantially greater than the calculated required rotation. The consequences of this are that the mid-span plastic hinge would be formed prior to the connection attaining its full capacity. Thus one must check that the rotation capacity of the mid-span region is sufficient to produce the full moment capacity at the joint region.

Table 6. Rotational comparisons

	ϕ_{yield} (mRad)	$\phi_{failure}$ (mRad)	$\phi_{required}$ (mRad)
Full Shear Connection	18	34	17
Partial Shear Connection	32	45	17

5 CONCLUSIONS AND FURTHER RESEARCH

The first aim of the testing program was to investigate the influence of partial shear connection on the behaviour of the joint. It was found that there is a reduction in the stiffness and strength of the joint and an increase in the rotational capacity of the joint when partial shear connection was adopted. The second aim was to identify the load transfer mechanism, which is set up in the web side plate composite joint and it was found that the rigid plastic model gives a good conservative prediction of the joint moment capacity.

These tests were the second series of tests in this program and it is envisaged that the testing program will be extended to include full-scale two span beam systems. This is required so that the effect of partial shear connection, which is a function of the span length, can be investigated more rigorously to allow for this behaviour. Furthermore, a mathematical model is also being developed to describe this behaviour and will be reported at a later date.

6 ACKNOWLEDGEMENTS

The authors would like to thank the Australian Research Council Large Grant Scheme which provided the funds for the experiments and a scholarship for the first author. Thanks are also extended to Messrs I. Bridge and G. Matthews for their assistance throughout the testing program.

7 REFERENCES

Ahmed, B. & D.A. Nethercot 1998. Effect of Column Axial Load on Composite Connection Behaviour, *Engineering Structures*, 20,(1-2): 113-128.

British Standards Institution. 1994 Eurocode 4 : Design of composite steel and concrete structures, Part 1.1 General rules and rules for buildings, BS DDENV 1994.1.1.

Crisinel, M. & A. Carretero 1996, Simple Prediction Method for Moment-Rotation Properties of Composite Beam-to-Column Joints. *Proceedings of Composite Construction in Steel and Concrete Conference III, Irsee, Germany, June 9-14*: 823-835, New York: ASCE.

Leon, R.T. 1998. Composite Connections. *Progress in Structural Engineering and Materials*. 1(2): 159-169.

Nethercot, D.A. Ahmed, B. & Y. Xiao 1996. Design Procedures for Composite Connections. *Proceedings of Composite Construction in Steel and Concrete Conference III, Irsee, Germany, June 9-14*: 794-808, New York: ASCE.

Nethercot, D. A. Li, T. Q. & B. S. Choo 1995. Required Rotations and Moment Redistribution for Composite Frames and Continuous Beams, *Journal of Constructional Steel Research*, 35: 121-163.

Oehlers, D. J. & M. A. Bradford 1995. *Composite Steel and Concrete Members - Fundamental Behaviour*. Oxford :Pergamon.

Standards Australia 1989. *Minimum design loads on structures – Part 1: Dead and live loads and load combinations, AS1170.1-1989*. Sydney.

Standards Australia 1996. *Composite structures – Simply supported beams*, AS2327.1-1996. Sydney.

Uy, B. Diedricks, A. A. Bradford, M. A. & D. J. Oehlers 1998. Behaviour of semi-rigid composite connections for continuous composite beams in braced frames. *Proceedings of the Australasian Structural Engineering Conference. Auckland, October*: 221-228.

Mechanics of Structures and Materials, Bradford, Bridge & Foster (eds)
© 1999 Balkema, Rotterdam, ISBN 90 5809 107 4

Moment resisting connections for composite frames

A.P.Gardner & H.M.Goldsworthy
School of Civil and Environmental Engineering, University of Melbourne, Vic., Australia

ABSTRACT: The analysis and design of composite sway frames for seismically induced loads, depends on the strength and ductility of the connections, and of the members adjacent to the joint. However, one of the difficulties in using this type of construction is the problem of how to connect concrete-filled circular steel columns to composite beams without site welding. A proposed connection is presented. This connection represents details used to transfer the tension or compression component of the moment in the beam, to the column. The end result will be an economically designed joint that may include some welding at the fabrication stage, but is essentially a bolted connection.

1 INTRODUCTION

Millions of dollars are spent each year in Australia on the construction of low-rise (less than ten storeys) non-residential and commercial buildings. Most of these buildings use nominally pinned joints and add bracing or structural walls to resist lateral loads such as those arising from earthquakes or wind. However, a moment-resisting frame can resist lateral loads while the requisite continuous beams can allow the use of longer spans and help minimise floor vibrations.

2 SEISMIC DESIGN ISSUES

In areas of high seismicity the preferred failure mechanism is one where hinges form at the ends of beams, except for the lowest storey where hinges are required to form at the base of the columns. Some hinging may also develop in columns at other locations because of the effects of higher modes and inelastic moment distribution on the actual seismic response of the building (Tomii, Sakino and Xiao (1987)). This beam sidesway mechanism is shown in Figure 1(a). Designers aim for columns to have a higher flexural strength than beams to achieve the so-called 'strong column-weak beam' hierarchy.

In regions of low seismicity such as Australia, design of low to medium rise buildings is dominated by gravity type loading such as dead and live loads, which often leads to flexurally strong beams and relatively weak columns. In recognition of this behaviour of gravity dominated frames, designers are encouraged to provide a building that can achieve a mixed-mode mechanism, as shown in Figure 1(b). In this type of failure mechanism, hinges form at the ends of beams at exterior joints, while for interior joints and the lowest storey, hinges are assumed to form in the columns.

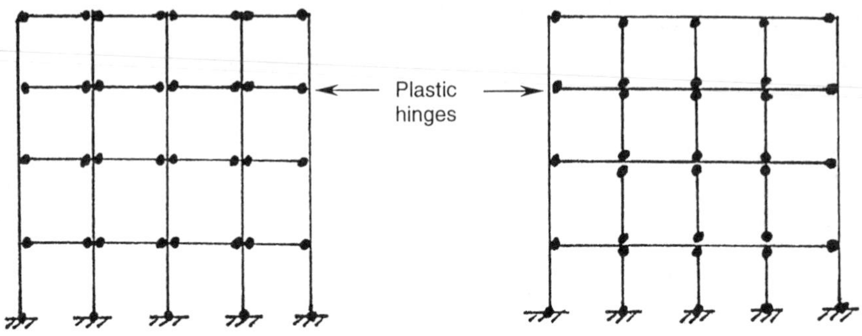

Figure 1: (a) Beam sidesway mechanism (b) Mixed mode mechanism

3 CIRCULAR CONCRETE-FILLED STEEL TUBE COLUMNS

In the event of a plastic hinge forming at the end of a column, it is important that the column maintains its strength while deforming plastically. The concrete-filled circular steel tube has an advantage over comparable steel and reinforced concrete columns in this regard in that it is inherently ductile. Many researchers have validated the ductility of composite columns. Ricles and Paboojian (1996) found the displacement ductility of their composite columns varied from 4.3 to 6.4, Akbas, Theeravat and Shen (1996) measured it at 5, while Elnashai and Broderick (1996) found that their specimens reached displacement and rotational ductilities as high as 6 before limits of their experimental equipment were exceeded. Kawano and Matsui (1996), O'Shea and Bridge (1997), Kawaguchi, Morino, Atsumi and Yamamoto (1991), Usami and Ge (1994), and Matsui and Tsuda (1987) found that filling tubes with concrete makes them more ductile than their bare steel equivalents. Tomii, Sakino and Xiao (1987) and Itani (1996) found that circular concrete-filled tubes are more ductile than an equivalent reinforced concrete column using spiral hoops as lateral reinforcement. Itani recorded curvature ductility ϕ_u/ϕ_y of between 35 and 40 for his concrete-filled circular steel tubes. Chai (1996) found an increase in curvature ductility from 10 for a normally reinforced column to 79 for a steel-jacketed column. Kim and Lu (1992) and Elnashai and Broderick (1996) found that moment –resisting frames constructed of steel beams with a composite slab and composite column show post-yield strength and ductility greater than that of their bare steel counterparts.

Ductility comparisons between square and circular steel tubes have found that the circular tube is more ductile than the square tube. This is because the circular shape is more efficient at confining the concrete inside the tube (Shakir-Khalil (1993), Tomii, Sakino and Xiao (1987), Bergmann (1994), Schneider (1998)).

Other advantages of this type of composite column include:

- High stiffness
- High strength
- Steelwork provides permanent formwork for concrete
- Tube prohibits concrete spalling
- Smaller cross-section than reinforced concrete means more lettable space in the building
- Construction time quicker than traditional reinforced concrete construction.

So using circular concrete-filled steel tubes as columns in a moment-resisting frame would appear to be an efficient structural system. The question remains as to how we can join these circular columns to a typical composite floor beam (universal beam section topped by steel decking and a reinforced concrete slab), so that we have a reliable and economical connection.

4 WELDED CONNECTIONS

A common detail in China and Japan is to provide external and/or internal diaphragms welded to the tube and the beam (Matsutani et al. (1991), Uchida and Tohki (1997), Ruoyu and Chao (1994), Storozhenko (1994), Shanzhang and Zhijun (1994), Fukumoto and Sawamoto (1997), Shanzhang and Zhijun (1994), Choi et al. (1995), Kato et al. (1992), and Morio et al. (1992)). However, it is difficult to install internal diaphragms into columns less than 750mm in diameter, and external diaphragms are essentially only loading the steel tube, which can cause excessive distortion of the tube wall.

Schneider (1997) tested six different connections of circular concrete-filled tubes to universal beam sections to investigate the inelastic flexural behaviour under cyclic deformation. He reported that inelastic connection behaviour improved significantly when more of the beam forces were transferred to the concrete core. His Type I connection, shown in Figure 2(a), could not develop the plastic bending strength of the beam. The beam was directly welded to the steel tube, which developed high local distortions near this weld. Fracture of the connection stub flange occurred at a rotation of 1.5%. The crack progressed from the outside edge of the plate towards the beam web, and initiated degradation of the flexural strength. This fracture placed a high demand on the beam web plate, which led to fracture of the weld between the beam web and the tube. Thus the beam lost all shear capacity shortly after flange failure. At the other flange the steel tube pulled away from the concrete core. Schneider recommended against using this type of connection in areas of moderate to high seismic risk. However, at an internal joint in an area of low seismicity, the beam is not intended to reach its plastic moment capacity, so this sort of detail may suffice.

Schneider's Type IV connection, shown in Figure 2(b) consisted of the Type I connection with the addition of deformed bars embedded in the concrete core through holes drilled in the steel tube. Local flange buckling in the beam was observed at 3.5% rotation suggesting that the connection was strong enough to initiate yield in the beam flange. Failure of this connection was recorded at 5% rotation when the deformed bars fractured and Schneider concludes that this detail could be used in areas of moderate to high seismic risk. In areas of low seismic activity, of course, it could be used at exterior connections where the mixed-mode mechanism requires a hinge in the beam.

5 PROPOSED BOLTED CONNECTION

These connections rely heavily on welding, much of which, especially in the U.S., is site welding. During the recent Northridge and Kobe earthquakes steel moment-resisting frames experienced significant damage to welded joints connecting beams to columns, suggesting that there are still

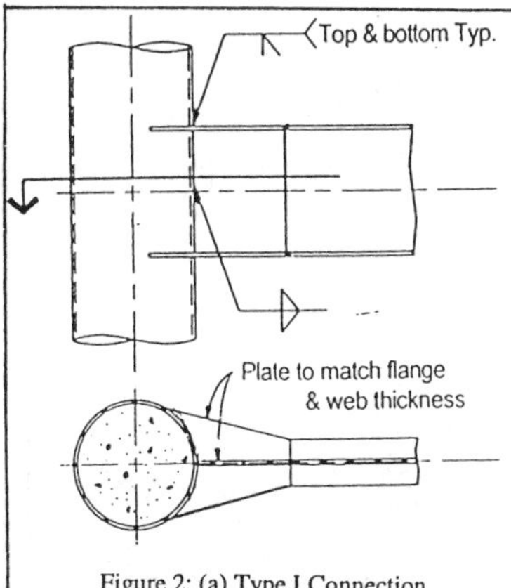

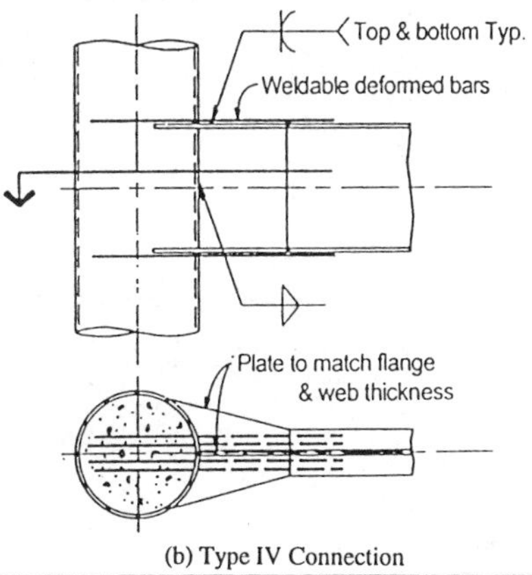

Figure 2: (a) Type I Connection (b) Type IV Connection

some issues to be resolved regarding the ability of these site welded connections to behave in a ductile manner. Site welding is also very expensive due to the stringent material testing requirements, high quality workmanship, and strict inspection procedures.

In developing the proposed connection, the authors have placed the emphasis on shop welding and site bolting to make the connection more economical in the Austalian context. A tentative joint design has been developed by the authors in collaboration with Dr. Saman Fernando from the Ajax Technology Centre and Dr. Mark Patrick from BHP Research Laboratories. This joint is shown in Figure 3.

The joint consists of two T-stub-like details connecting the steel beam flanges to the column. The end plates of these T-stub details are bolted to the column using the Ajax blind bolts which may include extensions into the concrete of the column depending on whether the connection is an interior or exterior joint in a region of high or low seismic activity. The development lengths of these extensions will be less than is quoted in AS3600 for normal reinforced concrete because the concrete in the columns is subject to some confinement. The flange plates of these T-stub details may be shop welded to the flanges of the steel beams. A web cleat is included to provide a residual load path for gravity loads (G + 0.4Q) in the event of significant deterioration of the flange connection. Reinforcement is provided in the concrete slab which forms part of the top flange of the composite beam. This reinforcement is continuous through the connection but is not relied on under earthquake loading.

For the design of the bolts and the end plate, the flanges are assumed to transmit all the bending moment, the compression force being resisted by bearing and the tension force being resisted by the top bolts. These bolts in the tension flange are subject to a 'prying action' effect which increases the force on these bolts. The magnitude of the prying force is influenced by the stiffness of the end plate, the stiffness of the bolt, the number of lines of bolts and the stiffness of the support. For a thick end plate, only small deformations of the end plate occur and so only low prying forces develop. In a more flexible flange the tensile load initially reduces the contact pressure between the end plate and the column until separation occurs at the bolt line and bending at the ends of the end plate develops prying forces. The AISC manual notes that precise evaluation of the effect of prying is very complicated and, at present, a purely analytical approach is out of the question; it goes on to recommend an allowance of 20 – 30% for prying.

Proposed details for an internal and an external connection will be tested with the column in the subassemblage subjected to constant axial load and cyclic lateral loading. The experiments will test detailing appropriate for an area of low seismic risk, but the test programme will then

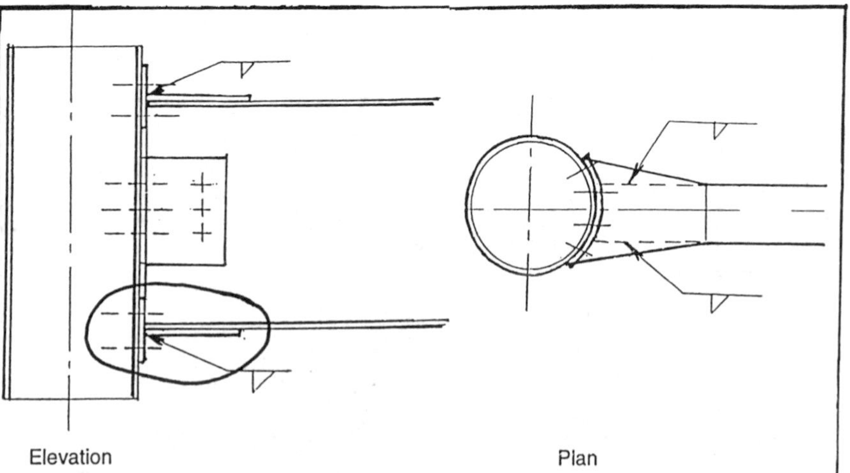

Figure 3: Proposed Bolted Connection. Note- length of bolts may vary. Concrete slab omitted for clarity.

encompass detailing appropriate for high seismic risk areas. Considerable focus will be placed on the failure hierarchy of the connection to evaluate the option of the failure mechanism occurring in the end plate of the joint to minimise cost and time involved in retrofitting a building after a significant earthquake.

6 INITIAL TEST PROGRAMME

In an attempt to resolve some of the design issues associated with this connection and to validate the failure hierarchy the authors will be testing the flange plate detail highlighted in Figure 3. Important parameters varied in these tests include the axial load on the column and the embedded length of the bolt extensions.

7 CONCLUSION

A moment-resisting frame consisting of circular concrete-filled steel tube columns and composite steel/concrete beams is an efficient structural system for resisting lateral loads, due in part to the inherent ductility of this type of composite column. Another important factor in the behaviour of these frames is the strength and ductility of the connections. This paper presented a shop welded, site bolted joint to connect the beams and columns. An experimental programme at the University of Melbourne will determine the response of this connection to cyclic loading. Design parameters can then be recommended for this joint applicable to areas of either low or high seismic activity.

REFERENCES

Akbas B., Theeravat K. and Shen J., 1996. Energy aspects of composite/hybrid frames. *Proc. 11th World Conf. on Earthquake Eng., Mexico, June 1996: paper no. 211.* Elsevier Science.

Bergmann R., 1994. Load introduction in composite columns filled with high strength concrete. Grundy, Holgate and Wong (eds). *Tubular Structures VI – Proc. 6th Inter. Symposium on Tubular Structures, 1994: 373 – 380.* Rotterdam:Balkema.

Chai Y.H., 1996. An analysis of the seismic characteristics of steel-jacketed circular bridge columns *Earthquake Eng. and Structural Dynamics.* 25:149-161.

Choi S., Shin I., Eom C., Kim D., and Kim D., 1995. Elasto-plastic behaviour of the beam to concrete filled circular steel column connections with external stiffener rings. Y.C.Loo (ed) *Building for the 21sst Century – Proc. 5th East Asia-Pacific Conf. on Structural Eng. and Construction, Gold Coast, 25-27 July, 1995:451 – 456.* Australia: EASEC –5 Secretariat

Elnashai A.S. and Broderick B.M., 1996. Seismic response of composite frames-II. Calculation of behaviour factors. *Engineering Structures. 18(9):707-723.* Elsevier Science.

Fukumoto T. and Sawamoto Y., 1997. CFT beam-column connection with high strength materials. *Composite Construction – Conventional and Innovative Proc. Conference, Innsbruck, 16-18 September, 1997:463-468.*

Hogan T.J. and Thomas I.R. 1994. *Design of Structural Connections.* Sydney: Australian Institute of Steel Construction.

Imamura T., Hatato T., Taga A., Watanabe T., Yoshino S., Teranishi K., Miyaki S.,and Iwaoka S., 1994. Concrete-filled Steel Tubular Columns: Study and Construction Practice. *Steel Concrete Composite Structures – Proc. 4th ASCCS International Conference, Slovakia, 1994: 84-87.*

Itani A.M., 1996. Future use of composite steel-concrete columns in highway bridges. *Engineering Journal Third Quarter, 1996:110:115.*

Kato B., Kimura M., Ohta H. and Mizutani N., 1992. Connection of beam flange to concrete-filled tubular column. . W. S. Easterling and W. M. K. Roddis (eds). *Composite Construction in Steel and Concrete II – Proc. of an Engineering Foundation Conf., Missouri, 1992:528-537.* ASCE.

Kawaguchi J., Morino S., Atsumi H. and Yamamoto S., 1991. Strength deterioration behaviour of concrete-filled steel tubular beam-columns under repeated horizontal loading. . *Steel Concrete Composite Structures - Proc. 3rd ASCCS International Conference, Fukuoka, 26-29 September, 1991: 119-124.*

Kawano A. and Matsui C., 1996. Buckling behaviour and aseismic property of concrete-filled tubular members under cyclic axial loading. *Composite Construction III – An Engineering Foundation Conf., Irsee, 9-14 June, 1996:15-28.* New York:ASCE.

Kim W. and Lu L.W., 1992. Cyclic lateral load analysis of composite frames. W. S. Easterling and W. M. K. Roddis (eds). *Composite Construction in Steel and Concrete II – Proc. of an Engineering Foundation Conf., Missouri, 1992.* ASCE.

Matsui C. and Tsuda K. , 1987. Strength and behaviour of concrete-filled steel square tubular columns with large wideth-thickness ratio. *Proc. of Pacific Conf. On Earthquake Eng., New Zealand, August, 1987:1-9.*

Matsutani T., Ishida J., Katagihara K., Kuroki Y., Kamisawa H., and Mori H., 1991. A study on super-high-rise building with concrete-filled tubular columns. *Steel Concrete Composite Structures - Proc. 3rd ASCCS International Conference, Fukuoka, 26-29 September, 1991:171-176.*

Morino S. Kawaguchi J., Yasuzaki C. and Kanazawa S., 1992. Behaviour of concrete-filled steel tubular three-dimensional subassemblages. . W. S. Easterling and W. M. K. Roddis (eds). *Composite Construction in Steel and Concrete II – Proc. of an Engineering Foundation Conf., Missouri, 1992:726-741.* ASCE.

O'Shea M.D., and Bridge R.Q., 1997. The influence of varying concrete strengths on the performance of concreete filled steel tubes. *Proc. Concrete 97, Australia, 1997:167-174.* Concrete Institute of Australia.

Park J.W., Shin B.C., Kim K.S. and Kim D.J., 1994. An experimental study on the elasto-plastic behaviour of concrete filled steel tube columns under axial force and bending moment. *Steel Concrete Composite Structures – Proc. 4th ASCCS International Conference, Slovakia, 1994: additional papers 8-13.*

Ricles J.M. and Paboojian S. A., 1996. Behaviour of composite columns under seismic conditions. *Earthquake Eng. – Proc. 10th World Conf. on Earthquake Eng., Maddrid, July,* 1992. Rotterdam:Balkema.

Ruoyu M. and Chao X., 1994. The transfer behaviour and working mechanism of rigid joint for concrete filled steel tubular framed structures. *Steel Concrete Composite Structures – Proc. 4th ASCCS International Conference, Slovakia, 1994:202-205.*

Schneider S. P., 1997. Detailing requirements for concrete-filled steel tubes connections. *Composite Construction – Conventional and Innovative Proc. Conference, Innsbruck, 16-18 September, 1997:609-614.*

Schneider S.P., 1998. Axially loaded concrete-filled steel tubes. *Journal of Structural Eng. 124(10):1125-1138.* ASCE.

Shakir-Khalil H. ,1993. Pushout strength of concrete-filled steel hollow sections. *The Structural Engineer 71(13):230-233.*

Shanzhang Y. and Zhijun L., 1994. The experimental research on rigid aseismatic beam-column joints of concrete filled steel tubular frame. *Steel Concrete Composite Structures – Proc. 4th ASCCS International Conference, Slovakia, 1994:535-538.*

Storozhenko L., 1994. Design Experience and construction of concrete filled steel tubular structures. *Steel Concrete Composite Structures – Proc. 4th ASCCS International Conference, Slovakia, 1994:80-83.*

Tomii M., Sakino K. and Xiao Y., 1987. Ultimate moment of reinforced concrete short columns confined in steel tube. *Proc. of Pacific Conf. On Earthquake Eng., New Zealand, August, 1987:11-22.*

Uchida N. and Tohki H. , 1997. Design of high-rise building using round tubular steel composite columns. *Composite Construction – Conventional and Innovative Proc. Conference, Innsbruck, 16-18 September, 1997:373-378.*

Usami T. and Ge H., 1994. Ductility of concrete-filled steel box columns under cyclic loading. *Journal of Structural Eng. 120(7):2021-2040.* ASCE.

Mechanics of Structures and Materials, Bradford, Bridge & Foster (eds)
© 1999 Balkema, Rotterdam, ISBN 90 5809 107 4

Steel-concrete composite beams under repeated loading

G.Taplin & P.Grundy
Department of Civil Engineering, Monash University, Melbourne, Vic., Australia

ABSTRACT: Composite steel and concrete beams rely for their strength and stiffness on the stud shear connection between the steel and the concrete. Tests have been undertaken on composite beams in order to investigate the behaviour of the shear connection when it is subjected to repeated loading. In particular, the effect of repeated loading on slip in the shear connection has been studied. The tests, reported in this paper, show that repeated loading of the beams causes slip damage to accumulate in the shear connection with a consequent loss of stiffness, and an increase in the deflection of the composite beam. An analytical model is presented which enables this effect to be quantified, and comparisons are made between the test results and the analytical model.

1 INTRODUCTION

The calculation of displacements caused by static loads on structures which have linear elastic behaviour is straight forward and reliable. Repeated loading can lead to increases in displacement, and even when repeated loading is anticipated, the calculation of the increased displacement is complex and subject to large uncertainty. The increase in displacement due to repeated loading is commonly referred to as *incremental collapse.* Incremental collapse occurs because of local inelastic material behaviour within a structure which is otherwise elastic. The process of structural design does not, in general, focus on *local* material behaviour, being concerned with *member* actions and capacities. This approach is successful, because ductility of the materials used in structural engineering ensures that if local inelastic behaviour occurs, a redistribution of the stresses will follow, such that at a member level elastic behaviour is observed. However when a load that causes local inelastic material behaviour is repeated, one of two phenomenon will occur;

- the local material behaviour will be elastic because of the stress redistribution that occurred as a result of the previous load cycles – this is *shakedown*
- the inelastic behaviour will recur and a small additional displacement will result due to the local plastic deformation – this leads to *incremental collapse*

Many repetitions of the load under the latter circumstances will see the displacements continue to increase, and serviceability problems may result.

This phenomenon has been recognised for many years as shakedown behaviour, whereby for indeterminate structures, there is a load level, in between first yield and plastic collapse, beyond which repetition of the load will lead to incremental collapse. This is termed the shakedown limit load. If the shakedown limit load is exceeded, permanent displacement will increase with every load cycle, and eventually the beam will fail to meet the serviceability requirements. If the peak loads in the loading spectrum are less than the shakedown limit load, the increments of displacement will reduce until a stable situation is reached, and no further displacement will occur. This situation, known as shakedown, is desirable.

An experimental investigation to determine the shakedown limit load of a composite beam subject to variable repeated loading found that the experimentally determined shakedown limit load was much less than that predicted by the classical theory. Measured shakedown limit loads were in the range of 53% to 69% of the static collapse load, whereas classical shakedown theory predicted values in the range 94% to 98% (Thirugnanasundralingham (1991)). This finding implied that incremental collapse might occur for loads that are within the design service load range of composite beams subject to variable repeated loading (of which bridge beams are the most common example). There was no explanation of why the shakedown loads were significantly less than the values predicted by classical shakedown theory, although it was hypothesised that slip in the stud shear connection might be the cause of the discrepancy. Similar results were reported in tests by Leon & Flemming (1997).

To understand why composite beams demonstrate incremental collapse at such comparatively low load levels, it is necessary to recognise the additional sources of inelastic behaviour that exist in a composite beam as compared to a steel or a reinforced concrete beam. Whereas incremental collapse of a steel beam is associated with yielding of the outer fibres, and the formation of a plastic hinge at a cross section, and incremental collapse of a reinforced concrete beam is associated with yielding of the steel reinforcing (for the normal case of an under reinforced beam), composite beams have an additional source of inelastic behaviour in the yielding of the stud shear connection, and local crushing of the concrete at the base of the stud. Under a single loading event this local inelastic behaviour is not significant at a member level, because redistribution of the stresses occurs and the observed behaviour is linear elastic. Under repeated loading however the inelastic behaviour causes slip at the stud shear connection to accumulate, and deflection of the beam increases - incremental collapse occurs.

Various researchers have investigated the effect of repeated loading on slip in the shear connection. See for example Hallam (1978), Roderick & Ansourian (1976), Oehlers & Foley (1985), Oehlers & Coughlan (1986) and Gattesco & Giuriani (1990). Taplin & Grundy (1995, 1997) reported on testing of the stud shear connection under repeated loading, in a series of 'push-out' tests. These tests demonstrated that incremental slip occurs under repeated loading. The tests reported in this paper establish that the incremental collapse of composite beams occurs due to this incremental slip in the stud shear connection under repeated loading, and furthermore that the loading does not have to be variable repeated loading (which is a prerequisite for classical shakedown to occur). The incremental collapse is due to an accumulation of 'damage' in the stud shear connection under repeated loading. The incremental slip in the stud shear connection, and consequent incremental collapse of the composite beam, is primarily a serviceability issue, as the observable effect of this phenomenon will be an increase in the deflection of the composite beam. It is appropriate therefore that incremental collapse is investigated at service loads, not at ultimate loads. That approach has been adopted in the tests reported here.

2 RESULTS OF PUSH-OUT TESTING

Tests on stud shear connectors under repeated loading have been reported elsewhere (Taplin & Grundy, 1995, 1997), and only the results of those tests are summarised here. The tests were conducted on 12.5 mm welded shear studs, under both symmetric cyclic and unidirectional cyclic loading. The rate of slip growth that occurred in those tests is summarised in Figures 1 and 2.

The rate of slip growth for the two cases of symmetric and unidirectional cyclic loading, using the stud and concrete of these tests, is given in equations (1) and (2).

$$\text{symmetric cyclic loading, slip growth per cycle} = 10^{\left(3.60 \times \frac{P}{P_u} - 4.22\right)} \text{ mm/cycle} \tag{1}$$

$$\text{unidirectional cyclic loading, slip growth per cycle} = 10^{\left(3.91 \times \frac{P}{P_u} - 4.71\right)} \text{ mm/cycle} \tag{2}$$

Looking in more detail at the load versus slip response for a symmetric cyclic push-out test, it has a characteristic pinched hysteresis behaviour. The load-slip relationship can be idealised as a region of zero stiffness (a damage region) combined with regions of constant stiffness (Figure 3). Under repeated loading the damage region grows, while the constant stiffness region has the same stiffness as the linear elastic stiffness seen in a monotonic push-out test (90 kN/mm).

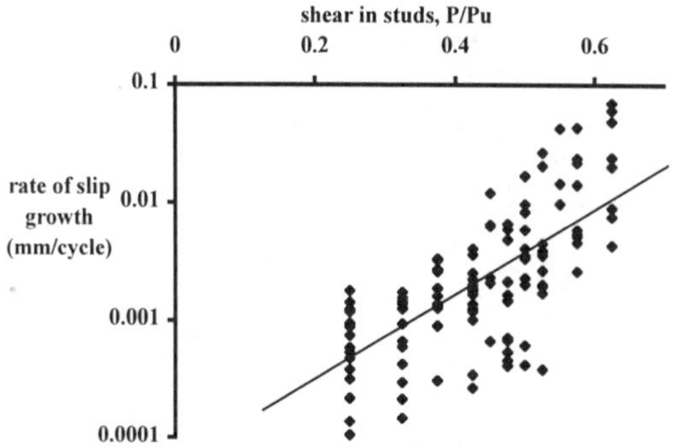

Figure 1. Shear in studs versus rate of slip growth – symmetric cyclic tests

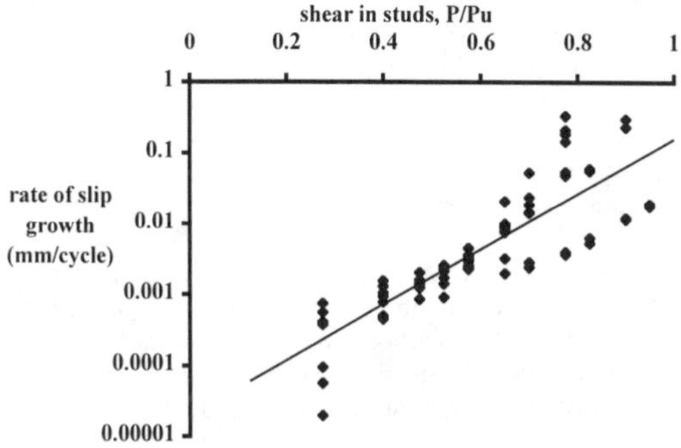

Figure 2. Shear in studs versus rate of slip growth – unidirectional cyclic tests

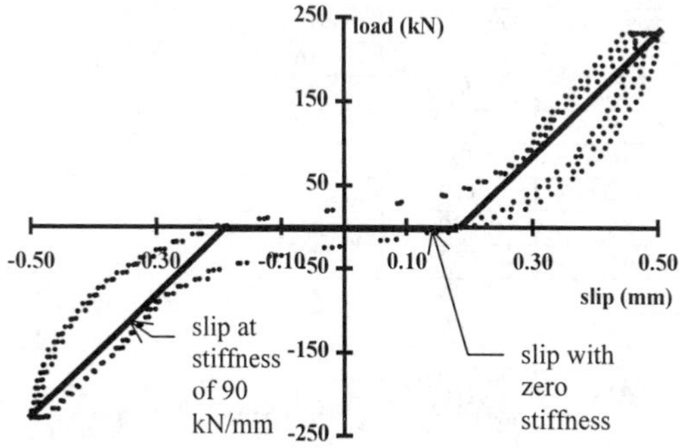

Figure 3. Load-slip behaviour – push-out test

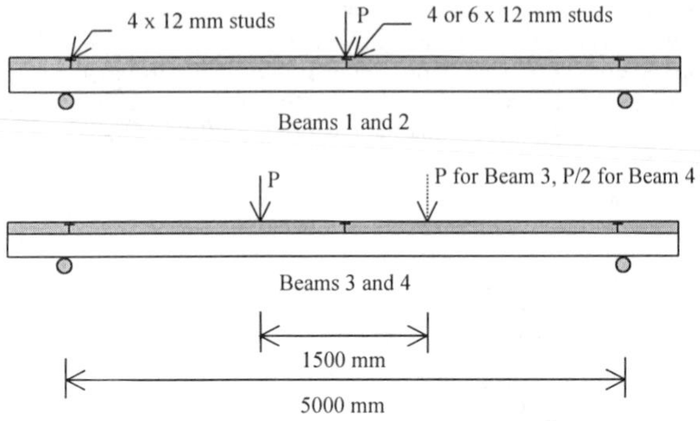

Figure 4. Summary details of the four beam tests

3 COMPOSITE BEAM TESTS

Four composite beams were tested to investigate the behaviour under repeated loading. The loading arrangement is described in Figure 4.

Beam 1 was tested under monotonic loading, with a central point load. Beam 2 was tested under repeated loading, with a central point load. Beam 3 was tested under repeated loading, with the alternate application of two point loads located symmetrically about the midspan of the composite beam, the two point loads being of equal magnitude. Beam 4 was tested under repeated loading, with the alternate application of two point loads located symmetrically about the midspan of the composite beam, one point load being half of the magnitude of the other.

The studs were 12.5 mm welded studs, as used in the push-out tests previously described. The concrete slab was 500 mm x 100 mm and was reinforced to prevent splitting. The concrete strength was 60 MPa at the time of test. The steel beam was a 200UC60 section from BHP Steel with a flange yield strength of 305 MPa. As shown in Figure 4, the studs were grouped at the centre and ends of the beam in order to ensure that the force in the studs was readily measurable during the testing. Beam 1 was tested under monotonic loading and failed by stud fracture at a peak load of 219 kN. The peak loads for beams 2, 3 and 4 (tested under repeated loading) were 209, 219, and 216 kN respectively. Table 1 summarises the applied loads.

Table 1. Peak loads for Beams 2, 3 and 4

load set	peak load (kN)				
	Beam 2	**Beam 3**		**Beam 4**	
		east jack	west jack	east jack	west jack
a	82	38	38	37	16
b	99	58	58	57	27
c	116	79	79	78	37
d	133	100	100	98	47
e	150	121	120	119	57
f	167	141	141	139	67
g	184	162	162	159	77
h	201	180	180	180	88
i	209	200	200	199	99
j		210	210	206	101
k		219	218	216	106

4 THEORETICAL MODEL

A theoretical model to describe the behaviour of composite beams subjected to repeated loading has been previously developed by Oehlers & Singleton (1986) and Oehlers & Carroll (1987). The model was based upon the slip versus cycles relationship proposed by Oehlers & Coughlan (1986). Gattesco (1997) developed a model based upon the slip versus cycles model proposed by Gattesco & Giuriani (1990). Using an idealisation of stud behaviour derived from their push-out tests, Taplin & Grundy developed a new model based upon a non-linear interface element to model the behaviour of the stud shear connection under repeated loading. The key assumptions of the modelling were:

1. The behaviour of both the steel and the concrete materials, other than at the studs, is linear elastic.

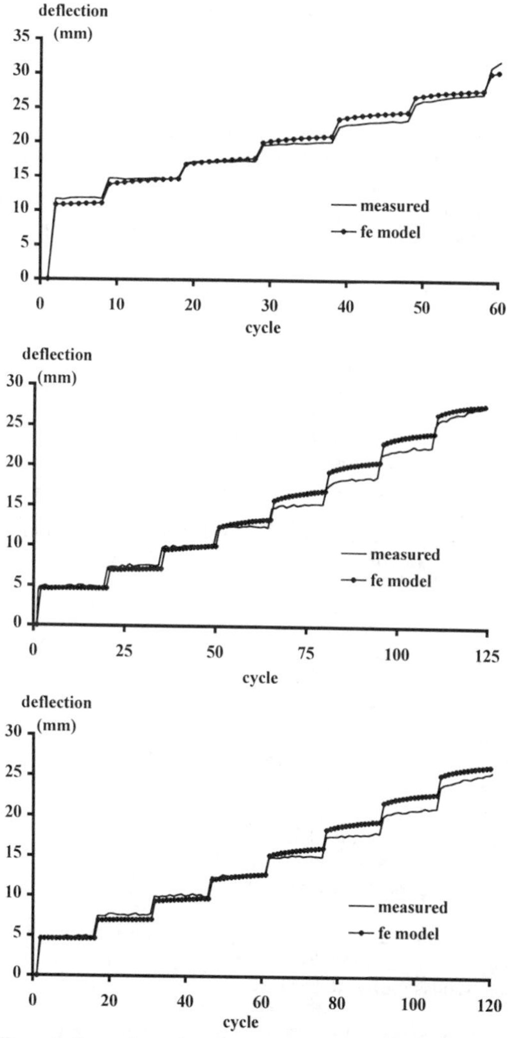

Figure 5. Comparison of measured and theoretical cycles-deflection plots for Beams 2, 3 and 4.

2. The stiffness of the shear connection is non linear, being zero when the magnitude of the slip is between the damage limits, and equal to the linear elastic stiffness of the studs when the slip magnitude is outside the damage limits (this is consistent with the behaviour illustrated in Figure 3).
3. The damage limits update after every load increment, so that the stiffness of the stud must update after every load increment.

The code written to describe the stud shear connection was compiled into the finite element program 'Diana' (TNO 1996).

5 COMPARISON BETWEEN MEASURED RESULTS AND THEORETICAL MODEL

The theoretical model gave excellent predictions of the behaviour of the composite beam under repeated loading. Plots of deflection versus cycle for Beams 2, 3 and 4 are given in Figure 5.

6 CONCLUSIONS

It is confirmed that the incremental collapse of composite beams that had previously been observed in tests can be explained by incremental slip in the shear connection, and furthermore that the magnitude of the deflections that will result from a given loading history can be quantified using the analytical model outlined herein. The inputs required for the model are the linear stiffness of a stud shear connector, and the rate of slip growth under repeated loading, both of which can be obtained from push-out testing.

REFERENCES

Gattesco, N. 1997. Fatigue in stud shear connectors. *Proceedings of the IABSE International Conference Composite Construction – Conventional and Innovative*, Innsbruck, pp. 139-144.

Gattesco, N. & Giuriani, E. 1990. Analysis of steel and concrete composite beams under repeated loads. *Technical Report, University of Udine*, Udine.

Hallam, M.W. 1978. The behaviour of shear studs under repeated loading. *Civil Engineering Transactions IEAust*, vol. CE 20, no. 1, pp. 28-36.

Leon, R.T. & Flemming, D.J. 1997. Shakedown performance of composite beams with partial interaction. *Proceedings of the IABSE International Conference Composite Construction – Conventional and Innovative*, Innsbruck, pp. 241-246.

Oehlers, D.J. & Foley, L. 1985. The fatigue strength of stud shear connections in composite beams. *Proceedings ICE*, Pt. 2 79, June, pp. 349-364.

Oehlers, D.J. & Coughlan, C.G. 1986. The shear stiffness of stud shear connections in composite beams. *Journal of Constructional Steel Research*, vol. 6, pp. 273-284.

Oehlers, D.J. & Singleton, W.M. 1986. The simulation of simply supported composite beams tested under fatigue loads. *Proceedings ICE*, Pt 2 81, December, pp. 647-657.

Oehlers, D.J. & Carroll, M.A. 1987. Simulation of composite beams subjected to traffic loads. *Composite construction in steel and concrete*, C.D. Buckner & I.M. Viest ed, pp. 450-459.

Roderick, J.W. & Ansourian, P. 1976. Repeated loading of composite beams. *University of Sydney School of Civil Engineering Research Report*, R 280, February.

Taplin, G. & Grundy, P. 1995. The incremental slip behaviour of stud shear connectors. *Proceedings of the Fourteenth Australasian Conference on the Mechanics of Structures and Materials*, Hobart

Taplin, G. & Grundy, P. 1995. Incremental slip of stud shear connectors under repeated loading. *Proceedings of the International Conference on Composite Construction*, Innsbruck, Austria

Thirugnanasundralingham, K. 1991. Continuous composite beams under moving loads. *PhD Thesis*, Monash University

TNO Building and Construction Research, Amsterdam 1996. Diana Finite Element Analysis Users Manual release 6.1

Mechanics of Structures and Materials, Bradford, Bridge & Foster (eds)
© 1999 Balkema, Rotterdam, ISBN 90 5809 107 4

The effects of local buckling and confinement in concrete filled circular steel tubes

M.D.O'Shea
Hyder Consulting (Australia) Pty Limited, St.Leonards, N.S.W., Australia
R.Q.Bridge
School of Civic Engineering and Environment, University of Western Sydney, Nepean, N.S.W., Australia

ABSTRACT: For economy in high-rise structures, concrete filled steel tubes tend to be thin-walled and filled with high strength concrete. This raises questions about the effects of local buckling of the thin-walled steel tube and the effects of confinement on the concrete, particularly with the use of high strength concrete. An extensive series of tests has now been completed which has enabled design procedures to be developed. However, the behaviour on which the rules were developed was often complex, particularly that concerning local buckling, the effects of confinement, and the influence of bond, or lack of it. This paper examines each of the concepts and explains the phenomena that were observed in the tests, thereby providing the fundamental basis from which the design rules were derived.

1 INTRODUCTION

Concrete filled steel tubes, when used as columns in high rise structures where the load effects are predominantly axial, provide an economical solution, particularly with the use of thin-walled steel tubes and high strength concrete infill. An extensive series of tests on short axially-loaded concrete-filled circular steel tubes has been completed (O'Shea & Bridge 1997a,b,c) which examined parameters such as the diameter to wall thickness ratio D/t of the steel tube, the yield stress of the steel f_y, the strength of the concrete f_c, and the influence of bond between the steel and the concrete. The normal loading condition assumed in design is for the steel and concrete to be loaded simultaneously. This is shown in Figure 1 where a concrete filled tube is loaded by an axial force P through stiff end platens.

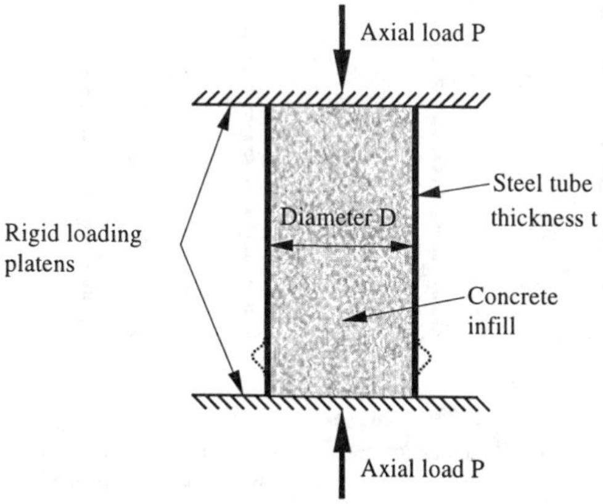

Figure 1. Axially loaded tube with concrete and steel loaded simultaneously

To examine the effect of the concrete infill in restraining the local buckling of the steel tubes and the influence of the steel tube in providing confinement to the concrete thereby enhancing the strength of the concrete filled steel tubes, additional tests were carried out for a variety of end loading and bond conditions as shown at (a), (b) and (c) in Figure 2.
A careful examination of the test results revealed a number of interesting phenomena. For tubes with only the steel loaded, the concrete infill did not increase the local buckling strength of the steel tube. The increase in concrete strength due to confinement of the concrete is greatest when only the concrete is loaded and the steel tube is not loaded axially. For tubes with steel and concrete loaded simultaneously, the bond between the steel and concrete is generally sufficient to prevent local buckling of the thin-walled steel tubes. For tubes with steel and concrete loaded simultaneously where the bond did not prevent local buckling and slip between the steel and concrete occurred, the strength of the composite tube was greater than similar tubes where local buckling was prevented by bond. Simple design rules ignoring local buckling of the steel tube can be used to give conservative estimates of axial strength of composite tubes.

2 STEEL LOADED TUBES

The failure mode of thin-walled bare steel tubes subjected to axial load is by an outwards buckle of short wavelength L_b around the circumference as indicated at (a) in Figure 2. This is generally known as an "elephant's foot" buckle as it usually occurs near an end (top or bottom) although some buckles near mid-height were observed in the tests.

The load-axial shortening behaviour of the tube depends on the length of the tube. This is indicated at (a) in Figure 3 for different ratios of tube length L to buckle length L_b. If the tube is short, the local buckle (even if elastic initially) will form a plastic mechanism with large deformations and the post-ultimate unloading response will be stable and gradual as indicated for $L/L_b = 1$. For long tubes, while the region around the local buckle is plastic in the unloading region and will shorten, the remainder of the tube is elastic and will elastically unload with a resulting elongation. This elongation can be greater than the shortening of the local buckled region resulting in an unstable response and even the phenomena known as "snap-back" as shown at (a) in Figure 3 for $L/L_b = 50$.
For tubes with only the steel loaded, the strength of bare steel tubes (loaded as at (a) in Figure 2) is compared Figure 3 (b) with the strength of tubes with unbonded concrete infill (loaded as at

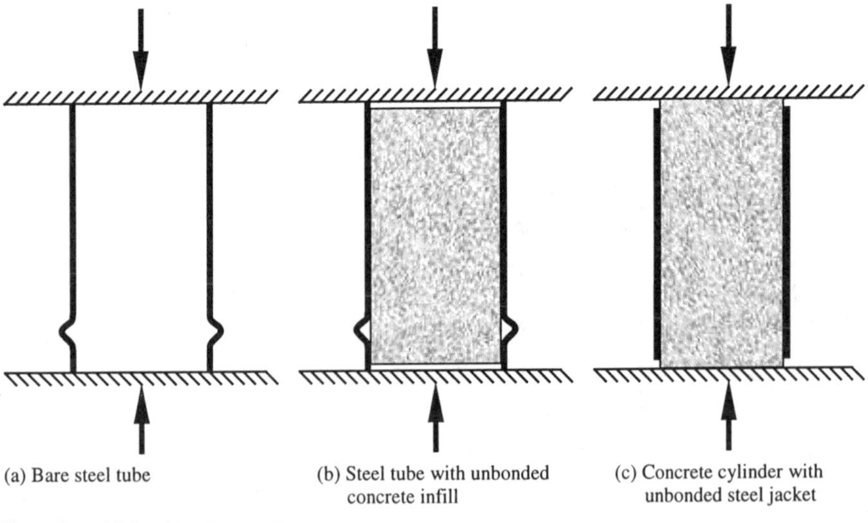

Figure 2. Additional loading conditions used in the test program

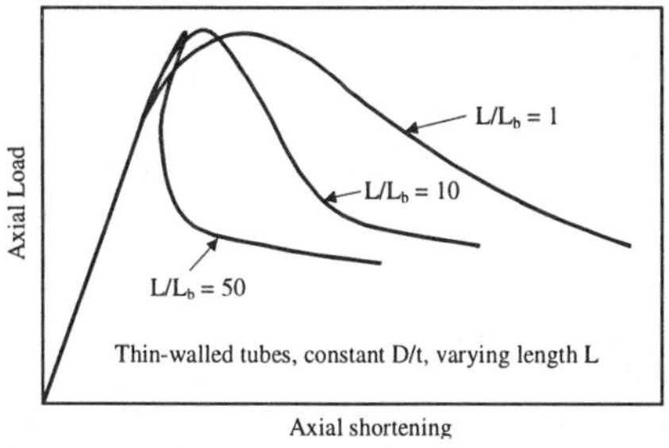

(a) Load-axial shortening response for steel tubes

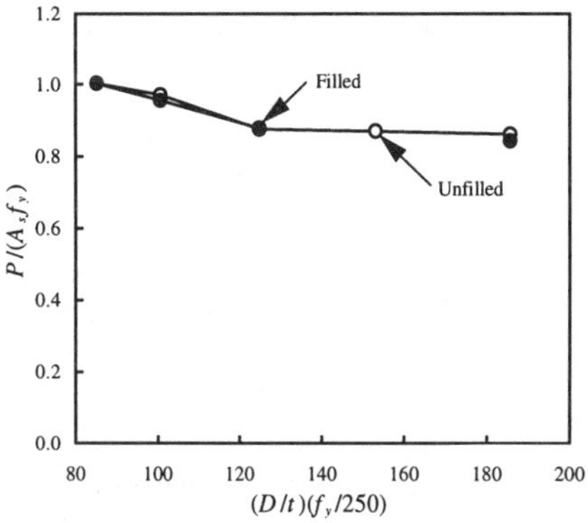

(b) Comparison of strengths for steel-loaded tubes

Figure 3. Behaviour of steel loaded tubes

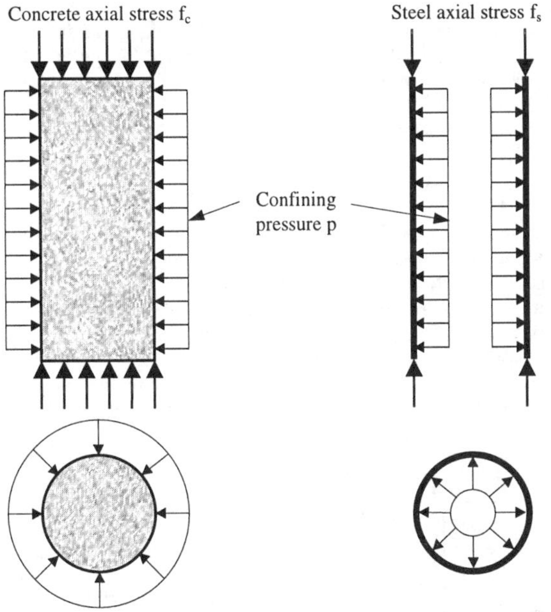

Figure 4. Confining pressures on the internal concrete cylinder and the external steel tube

(b) in Figure 2). It can be seen that the concrete infill had no influence on the strength of steel tubes as the local buckle is predominantly outwards and is not restrained by the concrete.

O'Shea and Bridge 1997c found that a good estimate of the local buckling strength for all values of the plate slenderness parameter $(D/t)(f_y/250)$ was given by American Standard AISI-LRFD-1991 while Australian Standard AS4100-1998 was quite conservative, particularly for plate slenderness values greater than 110.

3 CONCRETE LOADED CYLINDERS AND TUBES

When a concrete cylinder surrounded by a steel tube is loaded by an axial stress f_c as shown in Figure 4, the dilation of the cylinder is restrained by the steel jacket resulting in a confining pressure p along the length and around the circumference of the cylinder. The steel tube is subjected to an equilibrating internal pressure p resulting in circumferential hoop tension in the walls of the steel tube. If only the concrete is loaded and there is no bond between the steel and concrete as shown at (c) in Figure 2, there will be no axial stress f_s in the steel tube. If there is no steel tube, there will be no confining pressure p and the concrete cylinder is unconfined.
Stress-strain curves for unconfined concrete cylinders of varying strengths are show at (a) in Figure 5. For very high strength concrete (110 MPa or greater), the unloading response is very steep (brittle) and can even exhibit "snap-back" while the unloading response for lower strength concrete can be relatively more ductile.

Shown at (b) in Figure 5 are the stress-strain curves for 190mm diameter concrete cylinders with an unconfined cylinder strength of 38 MPa restrained by unbonded steel tubes of thickness 0.86mm (S10, f_y = 210 MPa), 1.13mm (S12, f_y = 186 MPa), 1.94mm (S20, f_y = 256 MPa) and 2.82mm (S30, f_y = 363 MPa). It can be seen that the thicker the tube, the more confinement and the higher the maximum confined concrete strength with ratios f_{cc}/f_c of confined to unconfined concrete strength ranging from 1.15 for the thinnest tube to 2.15 for the thickest tube. A model for the full confined stress-strain curves has been developed as a modification to existing models (O'Shea and Bridge 1997d) based on the test results. This requires a knowledge of the confining pressure p at maximum strength f_{cc}. The confining pressure p can be determined directly from the diameter D and the thickness t of the steel tube, the yield stress f_y of the steel and the unconfined strength f_c of the concrete.

4 TUBES WITH CONCRETE AND STEEL LOADED SIMULTANEOUSLY

When the concrete infill and the steel tube are loaded simultaneously by an axial force P acting through rigid ends (Figure 1), the overall axial shortening of the steel and concrete will be identical resulting in an axial stress f_c in the concrete and an axial stress f_s in the steel tube as shown in Figure 4. Depending on the relative dilation of both the concrete infill and steel tube, a confining pressure p can be induced between the steel tube and the concrete infill.

Shown at (a) in Figure 6 are the stress-strain curves for 190mm diameter concrete infill with an unconfined cylinder strength of 38 MPa restrained by the loaded steel tubes of thickness 0.86mm (S10, f_y = 210 MPa), 1.13mm (S12, f_y = 186 MPa), 1.52mm (S16, f_y = 256 MPa), 1.94mm (S20, f_y = 256 MPa) and 2.82mm (S30, f_y = 363 MPa). These concrete curves were

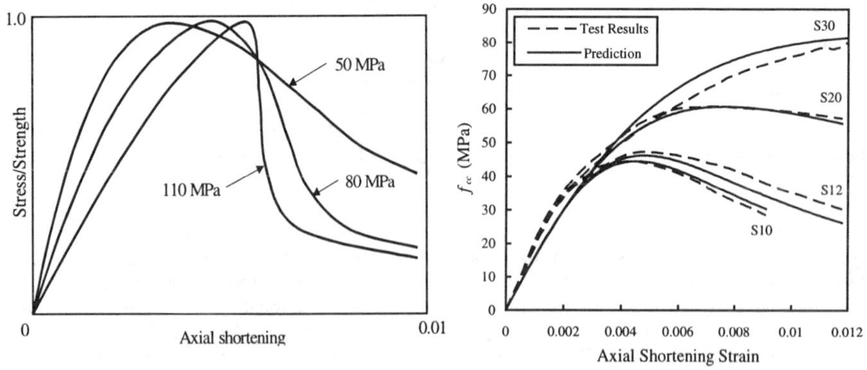

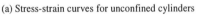

(a) Stress-strain curves for unconfined cylinders

(b) Stress-strain curves for 38 MPa cylinders confined by steel tubes

Figure 5. Stress-strain curves for concrete cylinders with only the concrete loaded obtained by subtracting the steel axial load component from the applied load using an incremental analysis based on measured strains combined with a yield failure surface for the steel tube (O'Shea & Bridge 1997a,b). No local buckling was observed in these tests indicating the bond was sufficient to prevent local buckling. Again, it can be seen that the thicker the tube, the more confinement with ratios f_{cc}/f_c of confined to unconfined concrete strength ranging from 1.10 for the thinnest tube to about 1.23 for the thickest tube. This is much lower than the strength enhancement that was achieved for the concrete loaded specimens where the unbonded steel tube provided confinement to the concrete but carried no axial load. In fact, the concrete loaded specimens carried a higher maximum axial load P than similar specimens where both the concrete and steel were loaded simultaneously. The extra concrete enhancement provided by the unbonded steel tube more than compensates for the total loss in axial load capacity of the steel tube.

The lower enhancement in concrete strength in the specimens with concrete and steel loaded simultaneously is due to two factors. The steel has a similar Poisson's ratio to concrete and dilates at a similar rate in the elastic range until micro-cracking occurs in the concrete which effectively causes it to dilate at higher rate than the steel resulting in contact pressure between the concrete and steel. The steel in the loaded tubes is subject to combined axial compression and circumferential tension and which lowers the effective circumferential yield stress of the steel and hence reduces the maximum confining pressure that the steel can exert on the concrete.

Shown at (b) in Figure 6 are the stress-strain curves for 190mm diameter concrete infill with an unconfined cylinder strength of about 80 MPa restrained by the loaded steel tubes of thickness 0.86mm (S10, f_y = 210 MPa), 1.13mm (S12, f_y = 186 MPa), 1.52mm (S16, f_y = 256 MPa), 1.94mm (S20, f_y = 256 MPa) and 2.82mm (S30, f_y = 363 MPa). These concrete curves were also obtained by subtracting the steel axial load component from the applied load using an incremental analysis based on measured strains (O'Shea & Bridge 1997a,b). It can be seen generally that the thicker the tube, the more confinement with ratios f_{cc}/f_c of confined to unconfined concrete strength ranging from about 1.00 for the thin S12 tube to about 1.18 for the thickest S30 tube.

However, there was an apparent anomaly in these tests with 80 MPa concrete. The ultimate load of the thinner S10 specimen was significantly higher than the ultimate load of the equivalent thicker S12 and S16 specimens and was similar to that for the thick S20 and S30 specimens. A local buckle was observed in this S10 specimen which did not occur in the other four specimens. This local buckle is indicated as a dotted line in Figure 1 and has a small wavelength, similar to that for an identical bare steel tube or a steel loaded tube with unbonded concrete infill. For this local buckle to form, slip between the steel and concrete must have occurred. This was confirmed by strain measured on the steel. The subsequent large plastic deformations in the local buckled region are associated with a reduction in axial load carried by

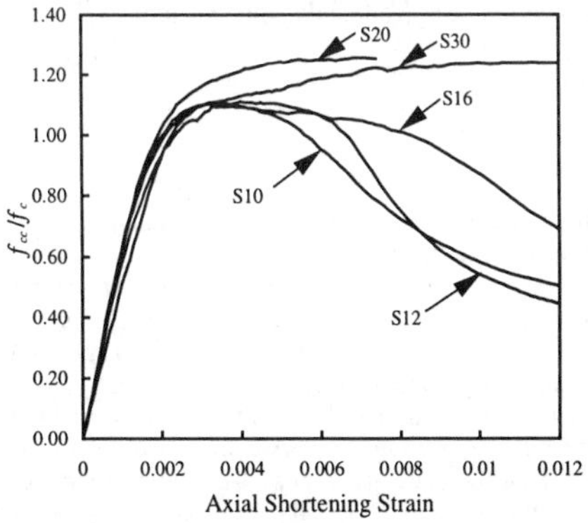

(a) Tubes with 38 MPa concrete infill

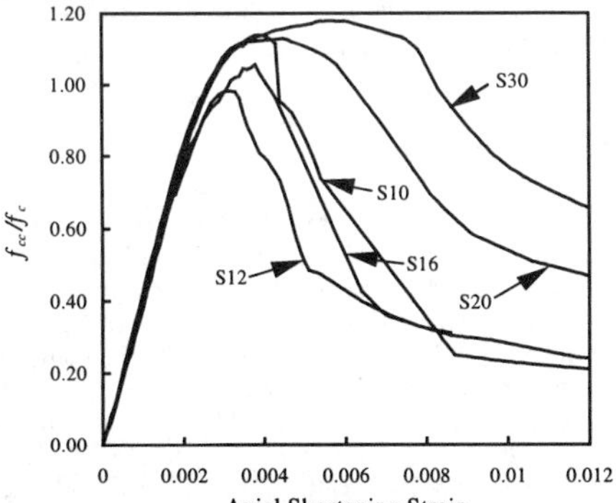

(b) Tubes with 80 MPa concrete infill

Figure 6. Stress-strain curves for tubes with the steel and concrete loaded simultaneously the steel tube, with the tube outside the buckled region elastically unloading. This reduction in axial stress in the steel tube in contact with the concrete allows greater circumferential tensile stress and hence internal pressure to be resisted by the steel tube thereby providing additional confinement to the internal concrete. As a result, this test was similar to the concrete loaded tests with unbonded concrete infill (Section 3 above) with a corresponding higher load capacity.

For the 190mm diameter steel tubes with very high strength concrete infill having an unconfined cylinder strength of 108 MPa, little confinement was provided by the steel tubes with the ratio f_{cc}/f_c of confined to unconfined concrete strength being close to unity for the thinnest S10 tube to the thickest S30 tube. This is because the high strength concrete has close to a linear elastic stress-strain response up to maximum strength with little micro-cracking and hence dilation required to mobilise the restraint offered by the steel tube.

5 DESIGN RECOMMENDATIONS

Eurocode 4-1992, which accounts for concrete confinement by the steel tube, has been found (O'Shea & Bridge 1997d) to give conservative estimates of the axial load capacity of circular tubes with both the concrete and steel loaded for concrete strengths up to 80 MPa provided no account is taken of local buckling (ignoring local buckling for thin-walled tubes). For concrete strengths beyond 80 MPa, Eurocode 4-1992 has been found to give conservative estimates of the axial load capacity of provided no account is taken of both local buckling and concrete confinement (ignoring both local buckling for thin-walled tubes and concrete confinement).

6 CONCLUSIONS

The axial load behaviour of circular concrete filled steel tubes has been examined for a range of different geometrical and material properties, bond conditions and end loading arrangements. The effects on the axial load capacity of the local buckling of thin walled tubes and the confinement provided by the steel have been investigated in detail. The observed phenomena have been discussed and despite the apparent complexity of the behaviour, relatively simple rules in Eurocode 4-1992 can be used for design provided local buckling is ignored and provisions for concrete confinement are also ignored for concrete strengths above 80 MPa.

7 ACKNOWLEDGMENTS

The work was funded by an ARC Collaborative Grant with BHP Steel and Connell Wagner, and additional support was provided by Boral Research Laboratory and Palmer Tubemills. The testing was carried out in the JW Roderick Laboratory for Materials and Structures in the School of Civil Engineering at the University of Sydney.

8 REFERENCES

AISI-LRFD 1991. *Cold formed steel design manual.* American Iron and Steel Institute, Washington.

Australian Standard AS4100-1998. *Steel structures*, Standards Australia, Sydney.

Eurocode 4-1992. *PrENV 1994-1-1 Eurocode 4, Design of composite steel and concrete structures Part1.1 - General rules and rules for buildings*, European Committee for Standardisation, Brussels.

O'Shea, M.D & Bridge, R.Q. 1997a. Tests on circular thin-walled steel tubes filled with medium and high strength concrete. *Research Report*, R755, School of Civil Engineering, University of Sydney, Sydney

O'Shea, M.D & Bridge, R.Q. 1997b. Tests on circular thin-walled steel tubes filled with very high strength concrete. *Research Report*, R754, School of Civil Engineering, University of Sydney, Sydney

O'Shea, M.D & Bridge, R.Q. 1997c. Local buckling of thin-walled circular steel sections with or without internal restraint. *Research Report*, R740, School of Civil Engineering, University of Sydney, Sydney

O'Shea, M.D & Bridge, R.Q. 1997d. Design of thin-walled concrete-filled steel tubes. *Research Report*, R758, School of Civil Engineering, University of Sydney, Sydney

Mechanics of Structures and Materials, Bradford, Bridge & Foster (eds)
© 1999 Balkema, Rotterdam, ISBN 90 5809 107 4

Elastic buckling modes in unpropped continuous composite tee-beams

Z.Vrcelj, M.A.Bradford & B.Uy
School of Civil and Environmental Engineering, The University of New South Wales, Sydney, N.S.W., Australia

ABSTRACT: One of the most difficult beam buckling problems to analyse accurately is the buckling of a steel joist in a continuous composite beam. These difficulties arise because the buckling mode is lateral-distortional rather than lateral-torsional, because the joist is subjected to combined bending moments and axial forces that vary along the length of the span, and because there is no evidence that the conversion of the elastic buckling load factor to a strength load factor that incorporates yielding is the same for lateral-distortional buckling and the familiar lateral-torsional formulation used in national steel design standards. This paper presents the results of a numerical buckling analysis of a two-span composite tee-beam that is cast unpropped, as would normally be the case for highway overpass bridges. The elastic solutions presented indicate the effect of restraint conditions over the interior support and of bracing of the bottom flange of the composite beam. Although significant research has been devoted to this topic, it is concluded that this is still a grey area of research for which the development of even moderately accurate design rules for practising structural engineers has not been achieved.

1 INTRODUCTION

Continuity in composite tee-beams can result in economic designs, with such continuity existing in propped beam-to-column connections in buildings, or in unpropped bridge beams with an internal support. Continuity is associated with hogging bending, where the desirable attributes of a tee-beam are reversed, with the concrete being subjected to tension and liable to cracking, and the joist subjected to compression and liable to buckling. Significant economies can be achieved in bridge beams with compact joists that are designed using rigid-plastic principles (Oehlers & Bradford 1995, 1999), and while cross-sectional proportioning to achieve the necessary moment redistribution in the hogging region has been quantified fairly accurately (Bradford & Kemp 1999), the problem of overall member buckling has still to be addressed properly. Overall member buckling in a composite beam must necessarily be lateral-distortional rather than lateral-torsional, since the slab restrains the top flange of the steel joist at its attachment through the shear connectors, and the bottom compressive flange is restrained in the hogging region only by the stiffness of the web.

While elastic distortional buckling of members subjected to pure bending has received a good deal of attention (Bradford 1992), the joist in a continuous composite beam is subjected not only to bending moments, but also to varying axial actions that can be compressive at the internal support and tensile near the simply supported end support. The effects of combined actions on the distortional buckling of isolated members has received very little attention (Bradford 1990), and even studies of lateral-torsional buckling under combined actions when the axial force varies along the member appear to be rare (Trahair 1993). Although a number of models, which are essentially a modification of the 'Inverted U-Frame Approach' (Oehlers & Bradford 1995, 1999) that appear in some design codes for treating distortional buckling in composite beams have been proposed, their accuracy is questionable, and recourse must be made to rational elastic distortional buckling software (Bradford & Ronagh 1997) for accurate results. Whereas this software can determine the accurate *elastic* buckling load factor for distortional buckling with combined actions, its conversion to a *strength* load factor has not been sufficiently researched, and a proposal that makes use of the α_c and modified α_s formulae in the AS4100 (1998) has been suggested by Oehlers & Bradford (1999) in the absence of more accurate information.

The present paper does not embrace the issue of the buckling *strength* of a continuous composite beam. Rather, it uses an in-plane analysis of two-span continuous composite beams (Bradford *et al.* 1999) and a rational model for out-of-plane distortional buckling (Bradford & Ronagh 1997) to investigate the elastic distortional buckling of composite beams cast unpropped over one internal support. The method allows for the combined actions of bending and varying axial force, and the effect of restraint conditions at the interior support are investigated, as are the effects of bracing in the negative moment region.

2 THEORY

2.1 *General*

Conventionally, design against the limit state of lateral-torsional buckling is usually based on the results of an in-plane elastic (second order) analysis, or a plastic analysis. The rationale of an elastic analysis is conservative, and applicable to bare steel members for which elastic analysis is appropriate. The extension of an 'elastic' analysis to composite tee-beams could be open to debate, since in the negative region of a composite beam the slab is invariably cracked and so an elastic analysis (using transformed area principles) is invalid.

Nevertheless, in this paper an 'elastic-cracked' analysis of a two-span continuous composite beam is undertaken first in order to determine the stress resultants that act in the steel joist. A rational beam-type finite element analysis is then invoked using these stress resultants as input to perform an elastic distortional buckling analysis, so as to determine the load factor against distortional buckling. Since the in-plane and out-of-plane analyses are well-documented, they are described very briefly in the following two sub-sections.

2.2 *In-plane analysis*

A flexibility method of analysis developed by Bradford *et al.* (1999) has been used in this study to determine the short-term moments and axial actions in a two-span continuous composite beam, whose spans may have different lengths and with concentrated loads placed at specified positions within each span. This method allows for propped or unpropped construction, with the latter being considered in this study. The method is in essence 'linear elastic', but accounts for cracking of the slab in the negative moment region where the tensile strength of the concrete is specified. Basically, the method uses transformed area techniques, but as the composite beam has different rigidities in the positive and negative moment regions, it resembles a close to nonuniform stepped beam. The position of the step where the rigidities changes is at the point of contraflexure (if for simplicity in the argument the tensile strength of the concrete is ignored), but this position is not known *a priori* and so an iterative scheme must be invoked to converge on its position and hence on the final bending moment in the composite beam. While the entire composite cross-section is not subjected to axial actions, these are present in the slab and in the joist, with equilibrium being maintained by between these two components by means of the shear connection. Thus under a given geometry and loading, the in-plane analysis is able to generate the bending moment diagram and shear force diagram for the steel joist.

2.3 *Out-of-plane analysis*

The beam or line-type finite element method for elastic distortional buckling analysis of I-sections developed by Bradford & Ronagh (1997) is used for the out-of-plane analysis. Each end of the line element has eight buckling degrees of freedom, corresponding to the lateral displacements of the top and bottom flanges, and their respective rates of change with respect to the beam longitudinal axis. The web is allowed to distort as a cubic curve during buckling, with its flexural displacements being related to the flange buckling freedoms by imposing displacement and slope compatibility at the top and bottom of the web. All freedoms relating to buckling deformations in the top flange of the joist are suppressed in the present analysis, owing to the rigid restraint assumed to be provided by the slab and the shear connection.

Because of the linearity of the in-plane analysis, the geometric stiffness matrix [S] is assembled from the moments and axial forces in the joist due to a set of initially applied loads. These loads are then scaled by a buckling load factor λ, which is the eigenvalue of the well-known buckling problem

$$([K]-\lambda[S])\cdot\{Q\}=\{0\} \quad (1)$$

in which [K] and {Q} are the elastic stiffness matrix and vector of buckling displacements respectively.

3 RESULTS

3.1 *General*

Distortional buckling loads are dependent on a multiplicity of geometric and material properties, which when coupled with the in-plane analysis prohibit general solutions. Hence an illustrative beam has been chosen for this study, that being a 250UB37.3 joist supporting a 1500 mm×130 mm slab cast unpropped with 0.17% reinforcement throughout positioned 25 mm from the top of the slab. The short-term elastic modulus of the concrete was taken as 25,250 MPa with a tensile strength of 3.0 MPa.

3.2 *In-plane behaviour*

The comparison between bending and axial actions has been determined by plotting the ratio $\sigma_{axial}/\sigma_{bending}$ in the bottom flange along the beam length, where $\sigma_{axial} = N_{steel}/A_{steel}$ and $\sigma_{bending} = M_{steel}/Z_{steel}$, with N_{steel} and M_{steel} being the axial force and moment respectively in the joist at a particular section obtained from the in-plane analysis, and with A_{steel} and Z_{steel} being its respective area and elastic section modulus. For the plot, compressive stresses are taken as positive, and a beam with two equal 5 m spans with a central support was considered.

Figure 1a shows the stress ratio when self-weight is ignored, while Figure 1b plots this ratio when self weight is included. The sign of the ratio is the same as the sign of the total stress ($\sigma_{bending} + \sigma_{axial}$) to indicate which portions of the beam are in sagging and hogging bending. The influence of self weight is important in unpropped construction, as it generates bending stress only in the joist and no axial stress. In the absence of self weight, the beam acts as if propped, and in Figure 1a it can be seen that the axial stress is twice the bending stress over most of the beam. On the other hand, the ratio in Figure 1b only 0.4, as the inclusion of self weight increases the bending stress but not the axial stress as composite action is not achieved under self-weight.

Although many buckling models (including that used here) do not include the effects of self weight, this clearly is important for composite beams if the buckling load factor is low. It is worth noting that in both cases the 'stabilising' influence of the sagging region of the span against buckling (where the bending stresses in the bottom flange are tensile) is enhanced somewhat by tensile stresses that equilibrate with compression in the slab. On the other hand, the compressive actions in the hogging region act with the hogging moment to destabilise that region against buckling.

3.3 *Elastic buckling analysis*

The stress resultants obtained from the in-plane analysis have been used to assemble the stability matrix [S] in Eq. 1 so that the eigenvalue or buckling load factor λ may be found. The

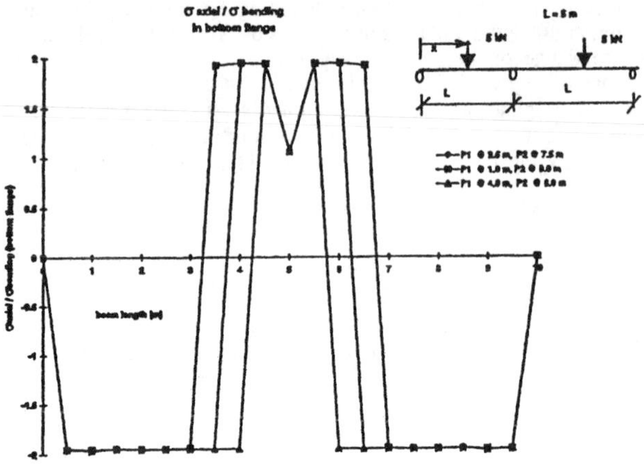

Figure 1a. In-plane stress ratio excluding SW

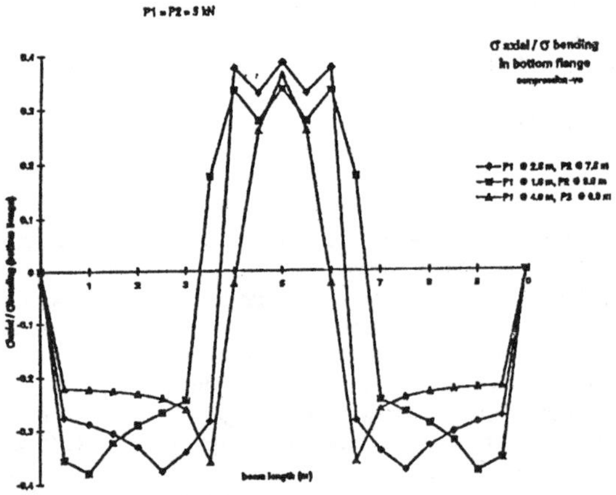

Figure 1b. In-plane stress ratio including SW

eigenvector {Q} in Eq. 1 represents the buckled shape, and this has also been calculated. The solution of Eq. 1 is extremely rapid on a contemporary personal computer.

In modelling the buckling restraints for a continuous beam, it is assumed that the top flange is completely restrained by the slab, and at the simple end supports that bottom flange displacement and twist (but not their rates of change) are also fixed. At the interior support, the bearing support would restrain the lateral displacement and twist of the bottom flange, but there may be some elastic restraint provided against the in-plane lateral rotation (u'_b) of the bottom flange. Figure 2 shows the normalised buckling mode shapes for two identical 5 m spans with central concentrated loads when u'_b is completely fixed or restrained. It can be seen that the bottom flange buckling deformations are almost identical for the two cases, except close to the internal support region.

Bracing of the bottom flange can have ramifications on the buckling loads and modes. Figure 2 shows the bottom flange translation buckling mode for a symmetric two-span beam with a central concentrated load in each span, and with a bottom flange brace positioned $\alpha_b L$ from the end support of a span of length L. The braces provide full restraint against lateral

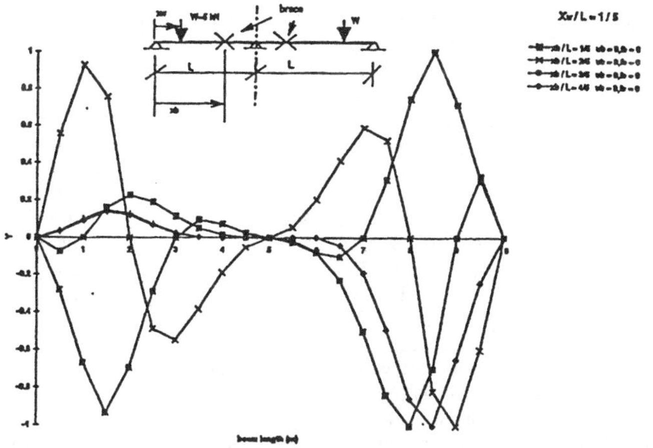

Figure 2. Buckling modes in symmetric braced continuous composite beam

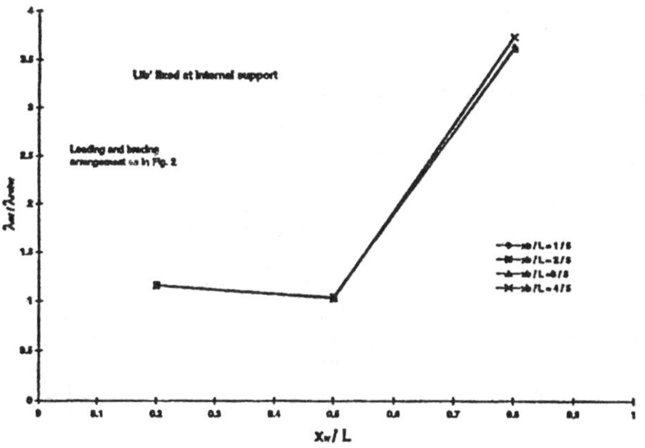

Figure 3. Effect of brace and load positions on elastic buckling load factor

displacement and twist, and braces in the sagging region are also included as these would often be cross-braces used to prevent lateral-torsional buckling of the unpropped beam when it supports unset concrete. In Figure 2, the internal supports provides complete restraint against buckling deformations.

The influence of the brace position and load position in a symmetric two-span beam has also been investigated. For this, the brace is positioned $\alpha_b L$ from the free end and the concentrated load $\alpha_W L$ from the free end, with the brace providing complete restraint against lateral deflection and twist. Such bracing may be typical of conventional cross-bracing. A reference buckling load factor, λ_{ref}, has been taken when $\alpha_W = 0.5$ in the braced beam.

Figure 3 shows the effect of single brace positions as the load moves along the beam ($0 \leq \alpha_W \leq 1$). When normalised with respect to the load position $\alpha_W = 0.5$, the figure indicates that the position of the brace has little effect on the scaled buckling load factor $\lambda_{cr}/\lambda_{ref}$, where λ_{cr} is the elastic buckling load factor for a given combination set of (α_W, α_b) values. Although the effect of the brace position is somewhat disguised by the normalisation used, it is also a reflection of the presence of axial actions in the joist. When the loads are placed in the positive region up to midspan, the destabilising hogging moments and the associated hogging region result in an elastic buckling load factor that increases only up to about 20% (at $\alpha_W = 0.2$) compared with that at $\alpha_W = 0.5$, and the effect of various bracing positions is indistinguishable. On the other hand, as α_W increases towards unity and the hogging moment increases at the internal support (which of course is restrained from buckling), the extent of the hogging region decreases, and the larger sagging moment region (whose bottom flange tensile bending stress is enhanced by the axial tension in equilibrium with the slab compression) is very significant in restraining the beam against elastic buckling. Consequently when $\alpha_W = 0.8$, the buckling load factor for the braced continuous beam has increased fourfold over that when $\alpha_W = 0.5$ for which the hogging region is quite extensive. When $\alpha_W = 0.8$, the hogging region is only small, so providing a brace in this region will increase the buckling load factor, as can be seen in Figure 3 for the brace positioned at $\alpha_b = 0.8$. Braces further away from the internal support than this are in the sagging region, and have negligible effect on increasing the buckling load factor.

It is also worth noting that if the internal bearing is unable to provide lateral rotational restraint, the counterpart to Figure 3 is almost identical to the latter figure.

Finally, although the out-of-plane analysis has been for an unpropped continuous beam that did not include self-weight in the finite element modelling, the elastic buckling loads were very high compared with the self-weight. As a consequence, the omission of self-weight in the buckling analysis was justifiable for the beams studied.

4 CONCLUDING REMARKS

The overall buckling of the steel joists of continuous composite beams is of practical importance, yet research into its prediction is far from complete. The present paper has shown how a rational in-plane analysis of a two-span continuous beam which may have different span lengths with arbitrary positions of the loads may be married with a rational out-of-plane beam-type finite element procedure to determine the elastic buckling load factors. Converting the elastic buckling loads into design *strengths* is another problem which requires recourse to inelastic distortional buckling solutions that are calibrated against test results, and has not been considered herein.

The illustration of the prowess of the method presented here has been very limited, and indeed used only for symmetrically-loaded equal two-span beams which may be modelled at propped cantilevers. Fuller use is being made of the method to investigate a multiplicity of geometrical, loading and bracing configurations, as well as propped and unpropped construction, with the final goal being a design proposal that is simple and of much higher accuracy than those currently used by design engineers.

5 ACKNOWLEDGEMENT

The first author is the recipient of the Faculty of Engineering's 'Women in Engineering' postgraduate award, as well a Dean's Scholarship provided by The University of New South Wales. This financial support is gratefully acknowledged.

6 REFERENCES

Bradford, M.A. 1990. Stability of monosymmetric beam-columns with thin webs. *Thin-Walled Structs.* 6:287-304.

Bradford, M.A. 1992. Lateral-distortional buckling of steel I-section members. *J. Const. Steel Res.* 23:97-116.

Bradford, M.A. & Kemp, A.R. 1999. Buckling of composite beams. *Progress in Struct. Eng. & Mats.* (to appear).

Bradford, M.A. & Ronagh, H.R. 1997. Generalised elastic buckling of restrained I-beams by the FEM. *J. Struct. Engg.*, ASCE, 123(12):1361-1367.

Bradford, M.A., Vu Manh, H. & Gilbert, R.I. 1999. Continuous composite beams under service loads. Part 1: Numerical analysis. (submitted for publication).
Oehlers, D.J. & Bradford, M.A. 1995. *Composite Steel and Concrete Structural Members: Fundamental Behaviour.* Oxford: Pergamon.
Oehlers, D.J. & Bradford, M.A. 1999. *Elementary Behaviour of Composite Steel and Concrete Structural Members.* Oxford: Butterworth-Heinemann.
Standards Australia 1998. *AS4100 Steel Structures.* Sydney: SA.
Trahair, N.S. 1993. *Flexural-Torsional Buckling of Structures.* London: E&FN Spon.

Mechanics of Structures and Materials, Bradford, Bridge & Foster (eds)
© 1999 Balkema, Rotterdam, ISBN 90 5809 107 4

Finite element modelling of cold-formed steel wall frames lined with plasterboard

Y.Telue & M.Mahendran
Physical Infrastructure Centre, School of Civil Engineering, Queensland University of Technology, Brisbane, Qld, Australia

ABSTRACT: Gypsum plasterboard is a common lining material for steel wall frame systems comprising cold-formed steel studs (unlipped or lipped C-sections). These steel wall frame systems are commonly used as load bearing walls in the residential and commercial building construction. However the design of these wall frames ignores the strengthening effects of plasterboard in carrying axial loads. Therefore a detailed investigation was conducted using finite element analyses and full scale experiments. The finite element model was validated by comparing its results with experimental results. This paper presents the details of the finite element model, the results and comparison with experimental results.

1 INTRODUCTION

The cold-formed steel wall frame systems made of unlipped or lipped C-sections and Gypsum plasterboard lining are commonly used as load bearing walls in the residential and commercial building construction. However, the design of these wall frames does not utilise the full strengthening effects of plasterboard in carrying axial loads. The Australian standard for steel structures AS/NZS 4600 (SA 1996) considers the lining material to provide only lateral and rotational support to the steel studs in the plane of the wall. Therefore a detailed investigation was conducted using finite element analyses (FEA) of lined wall frames. The finite element model was validated by comparing its results with full-scale experimental results. The results were compared with predictions from the Australian standard AS/NZS 4600 and the American specification AISI 1996. This paper presents the details of the finite element model, the results and comparison with experimental results.

2 FINITE ELEMENT MODEL

It is important that the FEA models are validated before their use in detailed parametric studies. Therefore, the wall frames used in the experimental study (Telue and Mahendran 1998) were first used in the FEA. These wall frames were made of three unlipped C-section studs and two tracks as shown in Figure 1. The test set-up is also shown in Figure 1. The finite element analyses were carried out using MSC/Abaqus & Patran programs. The finite element model was simplified by modelling the middle stud and half of the top and bottom tracks on either side of the tracks as shown in Figure 2. Three groups of frames were considered. They were the unlined frames, and frames with plasterboard lining on one side and both sides.

For the unlined frames, the track and the steel studs were modelled as shell elements using ABAQUS S4R5 elements. These elements have 4 nodes and 5 degrees of freedom per node, are shear flexible and use reduced integration (with 5 integration points). This element is only suitable for thin elements with small strains, however, large displacement and/or rotation are allowed. The S4R5 elements are significantly less expensive since they use the reduced

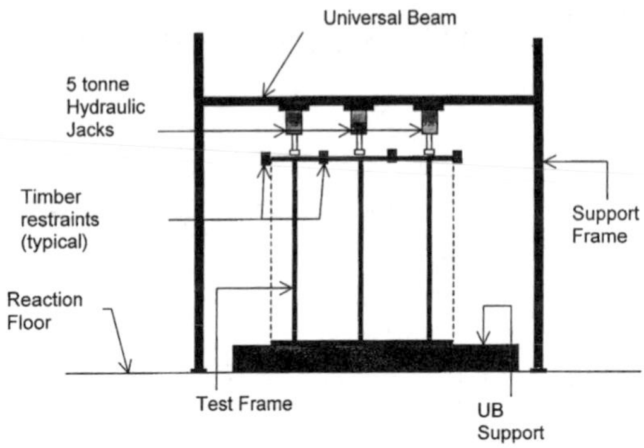

Figure 1. Experimental wall frame

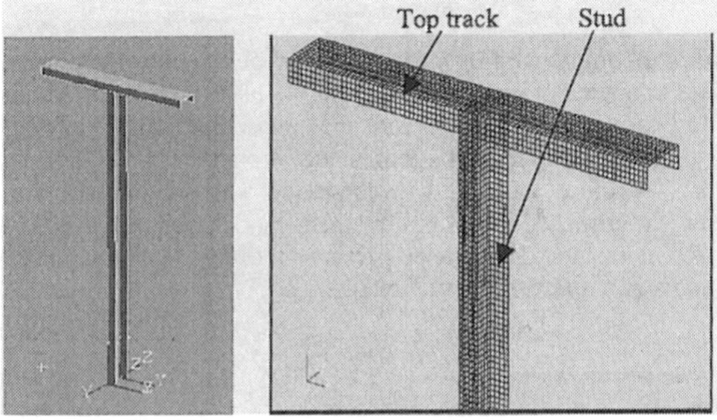

Figure 2. Finite Element Mesh of Unlined Frames

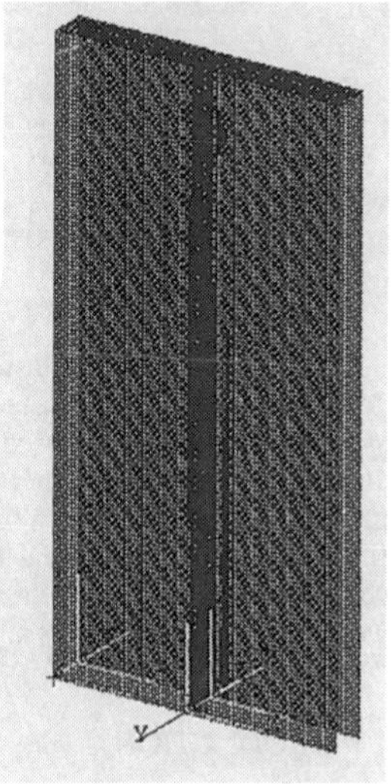
Figure 3. Mesh of Lined Frame

integration rule. The aspect ratio of the mesh was kept closer to 1.0 throughout. The studs and tracks were also modelled using S4R elements, which are similar to S4R5 elements but allow large strains. The ultimate strength results obtained were within 1%, and therefore S4R5 elements were used throughout the FEA.

At the top track to stud connection, the screws were modelled as beam elements in space (ABAQUS B31 element) with 2 nodes and 6 active degrees of freedom per node. The B31 elements allow the transverse shear deformation. A rigid body ABAQUS R3D4 element with 4 nodes was used to model the steel sheets that were used to pack the gaps between the track and the studs. In this model the reference node adopted throughout the study was the node at the top of the track in which the load was applied. The reference node has 6 "master" degrees of freedom and governs the motion of the rigid body. These elements were required to transfer the axial load to the entire stud area without any rotational restraint. Local coordinate system was specified for all the stud elements to enable the residual stresses to be applied in this coordinate system. The local x-axes of the web and flanges are along the longitudinal axis of the stud (ie. parallel to the global z-axis of the stud).

Two additional elements were required to complete the model for the lined frames. These were the plasterboard and the screws connecting plasterboard to studs. The screws were modelled as ABAQUS B31 beam elements, which are similar to those used in the stud to track connection. These beam elements were introduced at the screw locations along the stud. The plasterboard was also modelled as S4R5 elements as for the studs. A thickness of 10 mm was specified for the plasterboard. Local coordinate system was specified for the plasterboard. The local x-axis was specified parallel to the machine direction of the plasterboard with the local y-axis perpendicular to the machine direction. Figure 3 shows the plasterboard (S4R5) shell elements attached to the studs to form the model for studs lined with plasterboard.

The material properties used in this FEA for the steel studs and tracks were based on tests reported in Telue and Mahendran (1998): modulus of elasticity E = 200,000, 203,000 MPa, yield stress σ_y = 179, 572 MPa for G2 and G500 grades of steel. The actual tensile and shear strengths of the screws were over 800 MPa and 450 MPa, respectively. In the FEA the following properties of the screws were assumed, E = 200,000 MPa, σ_y = 450 MPa. These stresses were not exceeded in the FEA. An elastic perfectly plastic model was assumed for steel. The modulus of elasticity (E_p) values of the plasterboard were obtained from tests: 200 and 140 MPa in the directions parallel to and perpendicular to the machine direction, respectively. The shear modulus was 180 MPa in both directions. A compressive strength of 3.2 MPa was adopted as the stress in the machine direction and 2.3 MPa was adopted as the stress perpendicular to the machine direction.

The load was applied at a point (node) on the tracks that coincided with the geometric centroid of the stud. Boundary conditions were applied at points of symmetry on the tracks restraining displacement in the x and y directions and allowing displacement in the z direction. The track was free to rotate about the x, y and z-axis. At the mid-height of the studs, the displacement in the z direction and the rotations about the x and y-axis were restrained. In the lined frames additional boundary conditions were applied at the locations where the plasterboard was cut. Since the top end of the stud was not rigidly connected to the track web, the web and flange of the tracks in contact with the corresponding web and flange elements of the stud were modelled as contact pairs with zero friction. This allowed for any interface movement of the two surfaces when they came into contact during loading. The same procedure was carried out for the plasterboard that was in contact with the track and stud elements in the case of lined frames.

The geometric imperfections in the studs were applied by modifying the nodal coordinates using a field created by scaling the appropriate buckling eigenvectors obtained from an elastic buckling analysis of the model. This method was successfully used by Avery and Mahendran (1998) in the finite element analysis of steel frame structures with non-compact sections. The maximum web and flange imperfections d_1 and d_2 (for the local and distortional buckling respectively) were estimated using Equations (1) and (2) (Schafer and Pekoz 1996).

$$d_1 = 6\,t\,e^{-2t} \tag{1}$$

$$d_2 = t\,(0.014\,w/t + 0.5) \tag{2}$$

where t = thickness and w = element width

The member out of straightness (for global buckling) was in the order of L/700 to L/1000 (where L = Length of the stud). A value of (at least) L/700 was recommended by AISI (1996) about the weak axis and L/350 about the strong axis. The imperfection due to the rotation about the longitudinal axis of the stud was set to 0.008 radians. This value was selected based on values measured and reported by Young and Rasmussen (1995). This value is well above the AISI (1996) recommended value of at least L/(d*10,000). For lined frames, the geometric imperfections for studs were applied in two ways. The first method was to adopt geometric imperfections from those for the unlined frames. This includes the local and distortional buckling of web and flanges, the global buckling about the weaker axis and twisting. The second method was to conduct a separate buckling analysis for the studs with lining attached and scale the appropriate eigenvectors corresponding to the lowest buckling mode. The results obtained suggest that there is very little difference in the ultimate loads predicted using both methods to estimate the imperfection.

In the FEA, the residual stresses in bending of 8 and 17% of σ_y were applied to the flat regions in the flange and web respectively, while a higher value of 33% of σ_y were applied to the elements in the corner regions. These values were also based on Schafer and Pekoz (1996). The residual stresses were applied using the ABAQUS command; *INITIAL CONDITIONS option, with TYPE=STRESS, USER. The user defined initial stresses were created using the SIGINI FORTRAN user subroutine, which defines the local components of the initial stress as a function of the global coordinates.

3 VALIDATION OF FINITE ELEMENT MODEL

Two methods of analysis were used; elastic buckling and non-linear static. The elastic buckling analyses were used to obtain the eigenvectors for the geometric imperfections and to obtain the buckling loads. The non-linear static analyses including the material and geometric non-linear effects and residual stresses were used to obtain the ultimate load capacity and load-deflection curves of the stud.

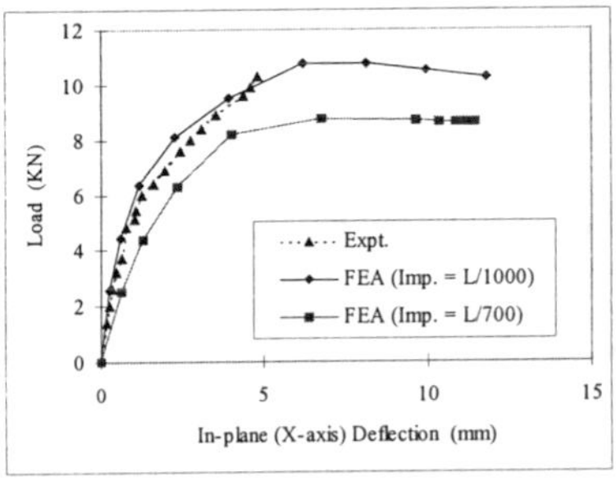

(a) Unlined Frame

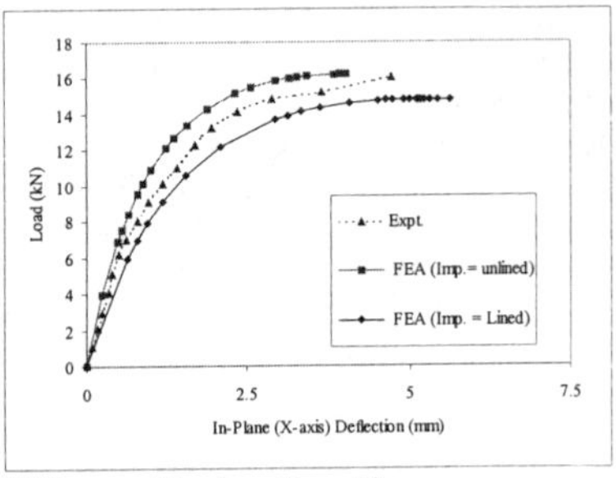

(b) Frames Lined on One side

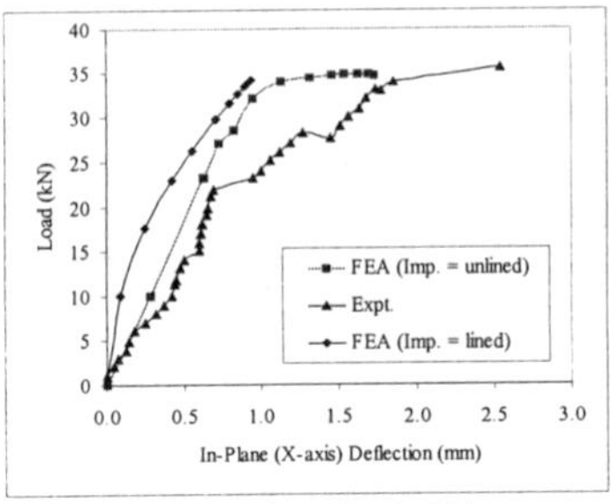

(c) Frames Lined on Both Sides

Figure 4: Typical Load versus In-plane (X-axis) Deflection Curves

Table 1. Comparison of Ultimate Loads from FEA and Tests for Unlined Frames

Frame	Studs	Stud Size (mm)			Steel	Ultimate Load (kN)		FEA/
		Web	Flange	Thickness	Grade	FEA	EXPT.	EXPT
1	1	75	30	1.15	G2	7.3	-	-
2	1,2	75	30	1.20	G500	8.5	7.8, 7.2	1.136
3	1,2	200	35	1.15	G2	10.6	10.7, 8.3	1.134
4	1,2,3	200	35	1.20	G500	10.8	10.8, 10.8,	1.013
				MEAN = 1.08 Coefficient of Variation = 0.10				

Table 2. Comparison of Ultimate Loads from FEA and Tests for Frames lined on one side

Frame	Studs	Stud Size (mm)			Steel	Ultimate Load (kN)		FEA/
		Web	Flange	Thickness	Grade	FEA	EXPT.	EXPT
5	1,2,3	75	30	1.15	G2	16.2	17.0,16.5,16.7	0.968
6	1,2,3	75	30	1.20	G500	27.9	28.4,28.2,28.4	0.984
7	1,2,3	200	35	1.15	G2	14.9	14.6,17.1,13.4	0.999
8	1,2,3	200	35	1.20	G500	18.2	18.3,18.0,18.3	1.000
				MEAN = 0.988 Coefficient of Variation = 0.054				

Table 3. Comparison of Ultimate Loads from FEA and Tests for both sides lined Frames

Frame	Studs	Stud Size (mm)			Steel	Ultimate Load (kN)		FEA/
		Web	Flange	Thickness	Grade	FEA	EXPT.	EXPT
1	1,2,3	75	30	1.15	G2	17.5	21.2,22.2,20.6	0.821
2	1,2,3	75	30	1.20	G500	34.9	35.3,35.3,36.5	0.978
3	1	200	35	1.15	G2	23.9	22	1.086
4	1,2,3	200	35	1.20	G500	34.5	41.5,41.5,42.2	0.826
MEAN = 0.896 Coefficient of Variation = 0.11								

Figures 4 (a) to (c) show the load versus in plane (x-axis) deflection curves whereas Tables 1 to 3 present the ultimate strength results and compare them with those from tests. They show that the test and the FEA results are in good agreement. Figures 5 to 7 show typical deformed shapes at failure for the studs in the three groups of frames. The deformed shapes of the studs from the tests are also shown next to those from the FEA. In all cases, it can be seen that the deformed shapes are also in good agreement.

In Figure 5 the unlined frames failed by global buckling, whereas Figure 6 shows the flexural torsional failure of the stud in the one-side lined frame. Figure 7 shows the strain distribution at the ultimate load for the frames lined on both sides. It can be seen that the maximum strains occur at the top fastener location. This is consistent with the tests where the failure mode of the stud is between the fasteners near the top with the plasterboard exceeding the ultimate strain of 0.007. During the tests it was observed that there was pull through of the screws at failure. Both the FEA and tests exhibit similar failure modes.

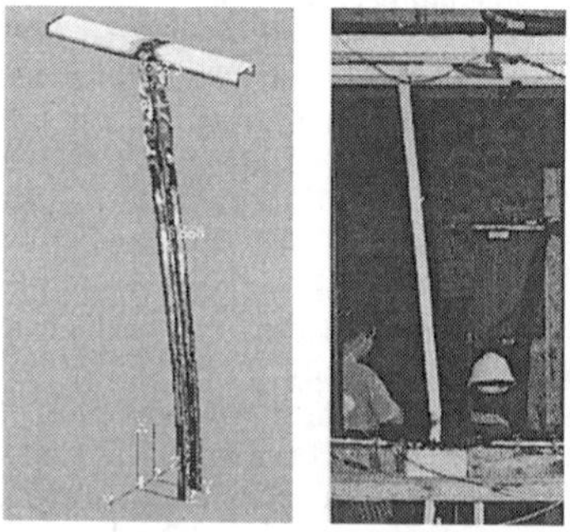

FEA Test

Figure 5. Deformed Shape at Failure for Unlined Frames

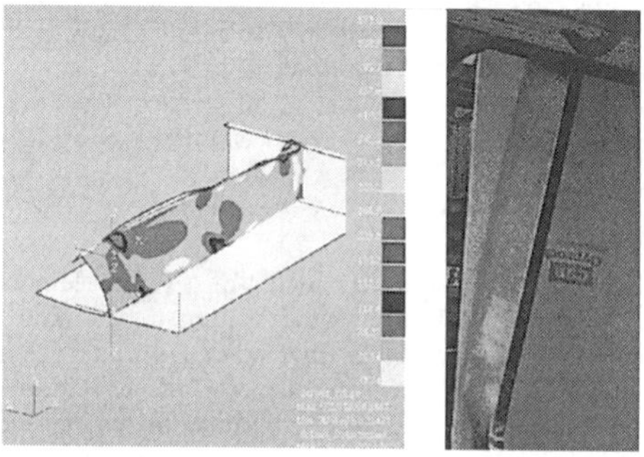

FEA Test

Figure 6: Stress Distribution and Deformed Shape at Failure for Frames Lined one side

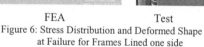

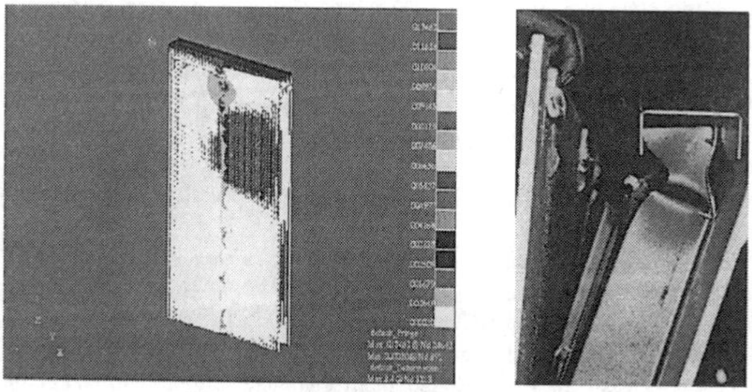

FEA Test

Figure 7: Strain Distribution and Failure Modes in Both Sides Lined Frames

4 PARAMETRIC STUDIES AND DESIGN RULES

Experimental results were used by Telue and Mahendran (1998) to develop approximate design rules based on appropriate effective length factors for unlined and lined frames. They recommended a value of 0.75 for the effective length factors (K_x, K_y, K_t) of unlined frames whereas the following values were recommended for both sides lined frames: $K_x = 0.75$ $K_y = K_t = 0.1$. Their results showed that the American design method based on a shear diaphragm model was inadequate. In the case of one side lined frames, predicted failure loads were improved with the use of effective length factors $K_x = 0.75$ $K_y = 0.1$ $K_t = 0.2$, however, the 200 mm stud failure loads could not be predicted well. Therefore a series of parametric studies is currently being undertaken using the finite element model described in the previous sections in order to improve the understanding of the behaviour of both unlined and lined studs under axial compression and to develop design rules based on this improved understanding. The parametric studies are also being used to investigate the effect of following parameters: (1) number and size of screws connecting stud to track, track section and material properties, and stud section and material properties for all three groups of frames, (2) spacing of screws connecting plasterboard to studs, the location of first screw, thickness and material properties of plasterboard for lined frames.

For unlined frames, Young and Rasmussen (1998) reported that the studs can be assumed fixed (effective length factor of 0.5) if the rotational restraint at the stud ends is greater than 3EI/L. However, the track to stud connections are neither fixed nor pinned. The results from parametric studies have led to the development of appropriate effective length factors as a function of the ratio of flexural rigidity of track to stud. For the studs undergoing local and/or global buckling, this factor is found to be independent of the flexural rigidity ratio when it exceeds 7.5. The parametric studies have also produced useful results for the effects of various parameters mentioned above. In particular, they have shown that ultimate strength of stud is not improved by reducing the plasterboard screw spacing below 200 mm or by placing the first plasterboard screw at the stud end or by increasing the plasterboard thickness (>10 mm) or some of its properties such as Poisson's ratio. Attempts to develop appropriate design models for lined frames are currently under way. These results will be presented at the conference.

5 CONCLUSIONS

This paper has presented details of a finite element analysis of unlipped C-section studs in a cold-formed steel wall frame system under axial compression. The study forms part of an ongoing investigation to develop appropriate design rules for both unlined wall frames and frames lined with Gypsum plasterboard. The finite element models of unlined and lined studs have been validated using experimental results and are being used in a series of parametric studies with the overall aim of developing appropriate design rules. Details of model validation and some results are presented in the paper.

6 REFERENCES

American Iron and Steel Institute. 1996. Specification for the Design of Cold-formed Steel Structural Members, Washington, USA

Avery, P. and Mahendran, M. 1998. Advanced Analysis of Steel Frame Structures Comprising Non-Compact Sections. *Proc. of the Australasian Struct. Engineering Conf.*, Auckland, Vol.2, pp. 883-890.

Standards Australia (SA) 1996. *AS/NZS 4600, Cold-formed Steel Structures*, Sydney

Schaefer, B. and Pekoz, T. 1996. *Geometric Imperfections and Residual Stresses for use in the Analytical Modelling of Cold-formed Steel Members*, Proceedings of 13the International Specialty Conference on Cold-formed Steel Structures, St. Louis, Mo, 649-664.

Telue, Y.K. and Mahendran, M. 1998. *Behaviour of Cold-formed Steel Stud Wall Frames*, Proceedings of 2nd International Conference on Thin-Walled Structures, Singapore, 111-119.

Young, B. and Rasmussen, K.J.R. 1995, *Compression Tests of Fixed-ended and Pin-ended Cold-formed Plain Channels*, Res. Rep. No. R714, School of Civil Engineering, The University of Sydney, Sydney.

Young, B. and Rasmussen K.J.R. 1998. Tests of Fixed-Ended plain channel Columns. *Journal of Structural Engineering* ASCE 124:2, 131-139.

Mechanics of Structures and Materials, Bradford, Bridge & Foster (eds)
© 1999 Balkema, Rotterdam, ISBN 90 5809 107 4

The numerical simulation of shear connection

B. Kim, H. D. Wright & R. Cairns
Department of Civil Engineering, University of Strathclyde, Glasgow, UK
M. A. Bradford
School of Civil and Environmental Engineering, The University of New South Wales, Sydney, N.S.W., Australia

ABSTRACT: This paper describes a numerical analysis of the behaviour of shear connection between steel beams and composite slabs which use through-deck welded shear connectors. An experimental study involving three push-out test specimens was carried out and the test results are discussed. The movement of the concrete slab relative to the steel beam (slip), the stress in the concrete slab and the concrete crack pattern have been studied and compared with the test results.

1 INTRODUCTION

In composite beams shear connection between the steel beam and concrete slab is of great importance as it resists the separation and transmits longitudinal shear between the two. Mechanical shear connectors are the most common form of connection for composite beams. Headed stud shear connectors have been widely used due to their rapid and easy construction.

The shear strength and stiffness of stud shear connectors can be obtained from push-out tests. The connection resistance of a stud, Q_u can be determined from an empirical code formula.

$$Q_u = K r \phi_s A_s (f_c E_c)^{0.5} \leq \phi_s A_s f_u \qquad (1)$$

where ϕ_s is the performance factor for the stud, (taken as 0.8), A_s is the cross sectional area of the stud, f_c is the cylinder strength of the concrete, E_c is the modulus of elasticity of the concrete, and f_u is the specified ultimate tensile strength of the stud. K and r are the empirical factor of connection resistance and the reduction factor for slabs incorporating profiled steel sheeting respectively. A value of 0.46 in Eurocode 4 (BSI, 1992) and a value of 0.5 in the American code (AISC, 1986) are employed for K whereas the reduction factor, r depends on the configuration of the stud and the orientation and geometry of the profiled steel sheeting. The stud strength from the AISC is therefore approximately 10% higher than that from Eurocode 4. Being empirical there is some question as to the accuracy of these formulae and experimental results are preferable.

However, push-out tests require test facilities and are costly to be carried out. To predict the behaviour of shear connectors numerical analysis has therefore been proposed. Up to date, a great deal of research has been conducted on push-out tests or composite beam tests. There has been relatively little study of the numerical analysis of composite beams. Moreover the research available generally deals with solid slabs. No study of the numerical analysis of push-out specimens or composite beams incorporating profiled steel sheeting has been found by the authors.

2 CONCRETE PULL-OUT FAILURE

In some cases Equation (1) does not predict actual strength from tests. This is especially in the case of the push-out tests with deck ribs perpendicular to the steel beam, which concrete pull-out failure governs the connection strength. The concrete pull-out strength, Q_p was first proposed by Hawkins and Mitchell (1984).

$$Q_p = K_p \lambda A_p f_c^{0.5} \tag{2}$$

where K_p is the empirical factor of the concrete pull-out strength, λ is the factor dependent upon the type of concrete used (1 for normal-weight concrete and 0.85 for lightweight concrete) and A_p is the concrete pull-out failure surface area dependent upon the geometry of the sheeting (see the reference). They conducted on 13 push-out test specimens with 38mm and 76mm deep decks and proposed a pyramid cone failure around the stud and a value of 0.45 for K_p. The cone started from the top of the stud making an angle of 45° and bounded by the underside of the stud head. The pull-out failure surface area and the pull-out strength are functions of the geometries of the stud and the sheeting, such as, overall stud height, the deck rib height, the average rib height and the stud spacing.

Equation (2) was re-evaluated by Jayas and Hosain (1988). Separate coefficients were proposed for K_p: a value of 0.61 for 38mm deep decks and a value of 0.35 for 76mm deep decks, instead of 0.45.

The study on concrete pull-out failure was extended by Lloyd and Wright (1990). They observed a wedge cone failure from their 42 push-out tests with 50mm deep decks and normal-weight concrete. They derived a value of 0.36 for K_p and simplified the formula of connection resistance (see Equation (3)). Their pull-out surface area was 1.49-2.23 times as large as Hawkins and Mitchell's pyramid cone area and was again very sensitive to the change of geometry of the profile sheeting and the stud height.

$$Q_p = A_p^{0.34} f_{cu}^{0.17} \tag{3}$$

where f_{cu} is the cube strength of the concrete.

3 THREE PUSH-OUT TESTS

A set of push-out tests was carried out to determine the behaviour of 13mm stud shear connectors welded through deck in a composite slab. These tests were used to establish the strength and stiffness of the connection for two composite beam tests (Kim *et al.*, 1999). Three push-out specimens were fabricated and tested in the Heavy Structures Laboratory in the University of Strathclyde in Glasgow. The objectives of the push-out tests were to determine the load-slip curve, the ultimate shear capacity and the slip capacity for the shear connection of the composite beams.

An arrangement for the push-out tests is shown in Figure 1. 178x102 UB 19 was used for the steel beam. Headed studs of 13mm diameter and 65mm long were welded to both sides of the steel beam through profiled steel sheeting. A stud was welded in each trough of the sheeting. The geometry of the profiled sheeting is shown in Figure 2. All the concrete slabs were 450mm wide and 425mm high. Reinforcement of A142 (6mm bars, 200 mm each way) was placed on the top of the profiled sheeting. A concrete depth of 75mm allowed for 15mm concrete cover from the stud head after welding.

The concrete was made using ordinary Portland cement and 10mm normal-density aggregate and cast in a horizontal position. The water/cement (W/C) ratio and slump were 0.55 and 105mm respectively. The specimens were cured and tested after 14 days.

Each Push-out specimen was set up on two pieces of fibreboard in order to restrain the movement of the bottom of the specimen. The fibreboard may have a similar effect to gypsum

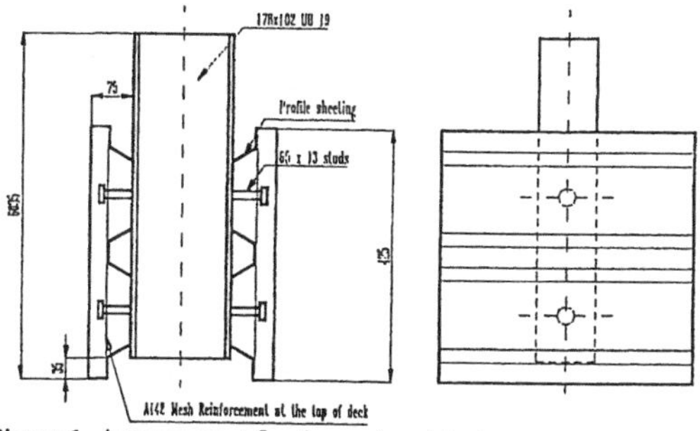

Figure 1. Arrangement for the push-out tests.

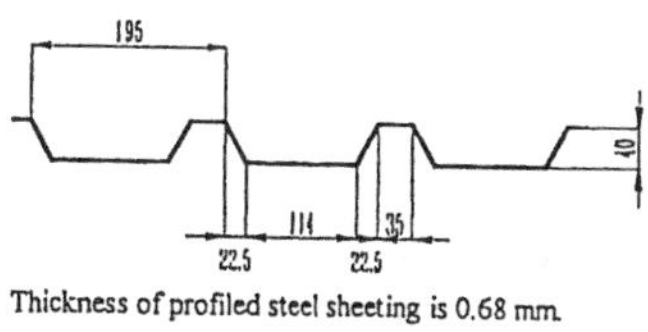

Figure 2. Geometry of the sheeting.

or mortar. Load was applied on the upper part of the steel beam. The movement of the concrete slab relative to the steel beam was measured by six dial gauges, which were attached to either the profiled steel sheeting near the studs or to the concrete top surface.

Table 1 shows the results of the push-out tests including their failure modes. The connection resistance of the stud from the tests is compared with the existing codes in Table 2. The resistance of the stud in Eurocode 4 and AISC was evaluated taking into account a 75% reduction of concrete strength for the 100mm cubes in order to convert these to cylinder strengths. The value from the test is lower than that in BS 5950 (1990) and AISC, but close to the value in Eurocode 4. It is therefore seen that BS 5950 and AISC overestimates the connection resistance. This may be due to the eccentricity of shear forces caused by the shallowness of the concrete slab.

Table 1. Test results.

ID	f_{cu} N/mm^2	f_{ct} N/mm^2	E_c kN/mm^2	E_s kN/mm^2	E_{ss} kN/mm^2	Maximum Load/stud (kN)	Failure modes SLAB1	SLAB2
P1	34.5	2.42	22.05	189	184	41.50	CP	CC
P2	34.5	2.42	22.05	189	184	38.75	SS and CP	CC
P3	34.5	2.42	22.05	189	184	37.25	CP	CC
Mean	34.5	2.42	22.05	189	184	39.17		

CP stands for concrete pull-out failure, SS stud shear failure and CC local concrete crushing around the foot of the stud.

Table 2 Comparisons of connection resistance.

	Test	BS5950	Eurocode 4	AISC
Connection strength (kN)	39.17	48.5	37.0	50.1
Comparison (Code/Test)	1.00	1.24	0.94	1.28

4 NUMERICAL ANALYSIS

The push-out specimen shown in Figure 1 was modelled with 2-dimensional finite elements using LUSAS FE programme. The steel beam, the concrete slab and the studs were modelled using plane stress elements and the profiled steel sheeting using bar elements. The stud was assumed to be of a rectangular cross-section. The properties of the circular stud cross-section were therefore transferred to an equivalent rectangular cross-section.

A half specimen was modelled due to the geometrical symmetry against the steel web of the push-out specimen. Figure 3 shows the finite element mesh and the deformed mesh. Two restraints were given to the model as shown by the arrows in Figure 3. One was the web of the steel beam in horizontal direction. Only vertical displacement was therefore allowed at the steel web. The other was at the slab base in the vertical direction since the base resists the compression load as reactions. This is also because the base was restrained by the

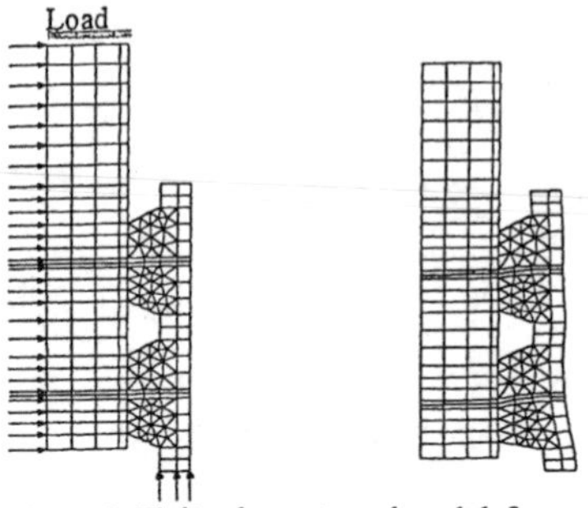

Figure 3. Finite element mesh and deformed mesh.

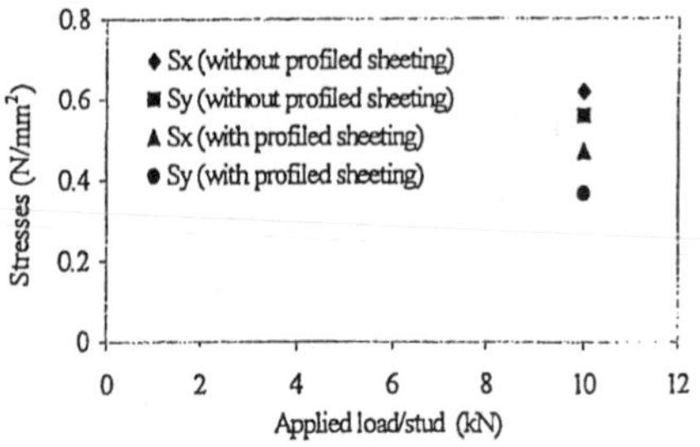

Figure 4. Stresses in the concrete.

fibreboard in the push-out tests. A load was applied to the upper steel beam and distributed along the web and flange of the steel beam

The material properties obtained from the tests were given to the numerical model. Half stiffness was given to the bottom elements of the stud to take account of the stud yielding before failure as used by Johnson and Oehlers (1981). It was observed in the tests that the stud had separated from the concrete on the upper surface of the stud. The separation was first pointed out in the work of Johnson and Oehlers (1981) and simulated by giving zero stiffness to the concrete elements behind the stud. A similar method was used by Kalfas *et al.* (1997). They used a double grid of nodes to simulate the separation. This method was also used here.

It was also observed in the push-out tests that the profiled sheeting separated from the concrete at a certain load and had little effect on the concrete once the separation occurred. It was however difficult to show the separation in the FE model since the same nodes were used for the profiled sheeting and the concrete. To take account of the separation a reduced stiffness, one hundredth of the stiffness of the profiled sheeting was therefore applied in the non-linear analysis. However for the linear analysis the full stiffness of the sheeting was used in order to study the effect of the inclusion of profiled sheeting at the elastic stage.

Initially a linear analysis was carried out and then extended to a non-linear analysis. The LUSAS non-linear concrete model was used to simulate the behaviour of the concrete and the Von Misses criteria to model the stud and the profiled sheeting. Linear material properties were however given to the steel beam since it did not yield even at the failure load in the tests. The Newton-Raphson method was used for the iteration procedure. The effect of the inclusion of profiled steel sheeting in the linear analysis, the effect of the width of the concrete slab, concrete cracking and load-slip relationship were examined.

5 RESULTS AND COMPARISONS

5.1 *The effect of the inclusion of profiled steel sheeting in the linear analysis*

The stresses in the concrete decreased, as shown in Figure 4, when the profiled sheeting was included. The decrease was as much as 24% and 34% in the X and Y-directions respectively. This is because the profiled sheeting has eight times higher stiffness than the concrete and is very significant since it results in less concrete cracking.

5.2 *The effect of the width of the concrete slab*

Figure 5 shows the effect of the width of the concrete slab on the failure load. The values in the Y-axis are the ratios of the failure loads from the FE analysis to the actual failure load from the

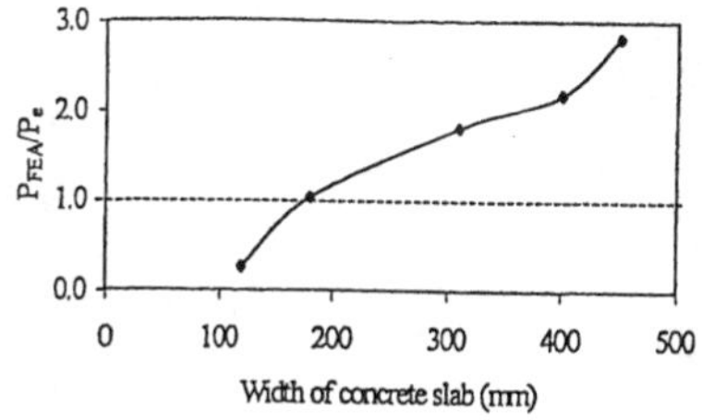

Figure 5. Effective width of concrete.

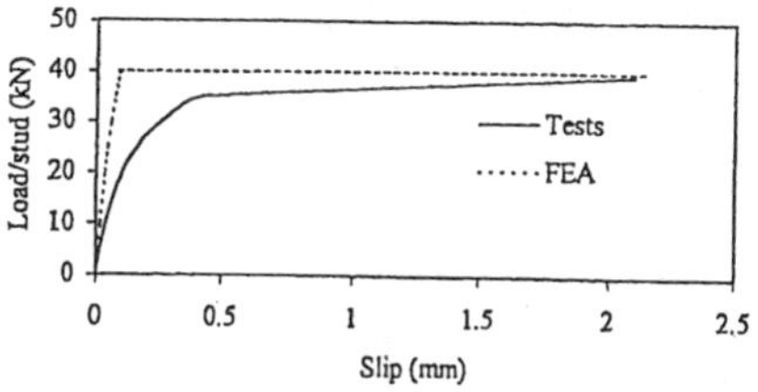

Figure 6. Load-slip curve.

experiment. In the FE analysis the failure load was evaluated as the tensile strain limit of the concrete. The thicker the concrete slab, the higher a failure load. For the effective concrete width a value of 180mm was chosen for which the failure load was within 3% deviation of the actual failure load. In fact the failure load at the FE model with 450mm concrete width was 2.8 times higher than the actual one. So it is clearly seen that in 2-dimensional analysis a direct simulation, which has the same condition as in the experiment, results in higher stiffness.

5.3 *Concrete crack pattern*

When the tensile stress in the concrete reaches the limit of tensile stress, the concrete produces cracks. Once the concrete cracks, the cracked concrete no longer has tensile stiffness unless a shear retention factor is set up. A value of 0.4 was use for the shear retention factor. 40% of shear stresses will therefore be transmitted through the cracked concrete. Figure 7 shows the concrete crack pattern according to the increase of the applied load. The FE model failed at the applied load of 40kN and its concrete crack is shown in Figure 7 (g). Most concrete cracking occurred in the bottom trough. This may probably be because of the relative stiffness of the concrete near the stud. The simulated concrete crack pattern shows a similar shape to the concrete pull-out failure surface obtained from the test and shown in Figure 8. The measured concrete pull-out area in the tests was 30% higher than that suggested by Hawkins and Mitchell (1984) and 41% less than that suggested by Lloyd and Wright (1990). Their pull-out strengths are 1.46 to 1.91 times higher than the test and FEA strength (see Table 3.).

Table 3. Comparisons of the concrete pull-out area and strength.

	Hawkins and Mitchell	Lloyd and Wright	Current Tests	FEA
Area (mm^2)	25095	54886	32522	-
Strength (kN)	57.4	75.0	39.2	40

5.4 *Load-slip relationship*

The movement of the concrete relative to the steel beam (slip) was plotted against the applied load and compared with the test results in Figure 6. The test results are three times larger than the FE results. This is probably because the 2-D FE model was employed. Further, the slip was little influenced by either the inclusion of the profiled sheeting or the effective concrete width.

6 CONCLUSIONS

In order to explore the behaviour of the shear connection of composite beams a numerical analysis has been studied. The push-out test specimen was simulated using

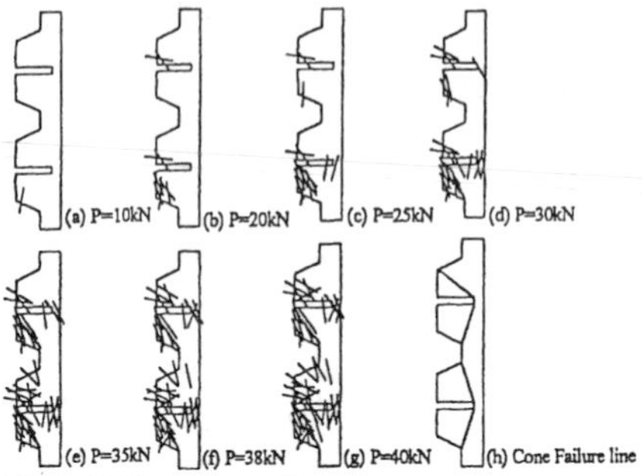

Figure 7. Concrete crack.

Figure 8. Concrete pull-out failure.

2-dimensional finite elements. From the FE analysis, it can be concluded that:

The inclusion of profiled steel sheeting is very significant in the linear analysis since it resulted in less concrete cracking due to the contribution of the tensile strength of the sheeting.

To model the concrete cracking behaviour in the 2D analysis it is recommended that an effective slab width is used. (For the specific tests in this paper a width of 180mm was found appropriate.)

The concrete crack pattern from the FE analysis gave rise to a good agreement with the concrete pull-out failure in the tests.

To model a load-slip relationship it was found that the 2D analysis gave high shear stiffness.

The main failure mode in the FE analysis was the concrete pull-out failure. The analysis did not show the yield of the studs although non-linear material properties were given to the studs.

7 ACKNOWLEDGEMENT

The authors wish to thank the Bone Steel Limited. for providing steel material and also acknowledge the help and assistance given by the stud welders at the Studweld Pro.UK Limited in Nottingham and the technical staff in the University of Strathclyde.

8 REFERENCES

American Institute of Steel Construction 1986. *Manual of Steel Construction: Load and Resistance Factor Design. Part 6.* AISC, Chicago.

British Standards Institution. 1990. *BS 5950: Part3: Section 3.1. Code of practice for design of simple and continuous composite beams.* BSI, London.

British Standards Institution. 1992. DD ENV 1994-1-1, *Eurocode 4, Design of composite steel and concrete structures. Part 1.1, General rules and rules for building.* BSI London.

Hawkins, N. M. and Mitchell, D. 1984. Seismic response of Composite Shear Connections. *Journal of the Structural Engineering Division, ASCE*, Vol. 110, No. 9, pp. 2120-2136.

Jayas, B. S. and Hosain, M. U. 1988. Behaviour of headed studs in composite beams: push-out tests. *Canadian Journal of Civil Engineering*, Vol. 15, 1988, pp. 240-253.

Johnson, R. P. and Oehlers, D. J.1981. Analysis and design for longitudinal shear in composite T-beams. *Proc. Instn. Civil Engrs.*, Part 2, pp. 989-1021.

Kalfas, C. and Pavlidis, P. 1997. Load-slip Curve of Shear Connectors Evaluated by FEM Analysis. *International Conference, Composite Construction-Conventional and Innovative*, Innsbruck, Austria, 16-18 Sep., pp.151-156.

Kim, B., Wright, H. D. and Cairns, R. 1999. A Study of a New Continuous Stem Girder System, *The First International Conference on Advances in Structural Engineering and Mechanics*, Seoul, Korea.

Lloyd, R. M. and Wright, H. D. 1990. Shear Connection between Composite Slabs and Steel Beams. *J. Construct. Steel Research*, Vol. 15, pp. 225-285.

Mechanics of Structures and Materials, Bradford, Bridge & Foster (eds)
© 1999 Balkema, Rotterdam, ISBN 90 5809 107 4

Reverse-cycle testing of stud shear connectors

R. Seracino, D.J. Oehlers & M.F. Yeo
Department of Civil and Environmental Engineering, University of Adelaide, S.A., Australia

ABSTRACT: The vast majority of fatigue tests carried out in the past were uni-directional and very little research has been carried out to investigate the behaviour of shear connectors when subjected to reverse-cycle fatigue loading as occurs in practice. As a result, a new type of push test for reverse-cycle fatigue loads has been developed, the results of which confirm the following two points about the behaviour of stud shear connectors: the strength and stiffness of the connectors reduce throughout the fatigue life; and that for a given range of load, connectors subjected to reverse-cycle loading last longer than connectors subjected to uni-directional cyclic loads. The tests also showed that the increase in slip per cycle is constant over approximately 3/4 of the design life after which the slip increases rapidly, providing adequate warning of failure.

1 INTRODUCTION

Stud shear connectors are used to provide the composite action in steel-concrete bridge beams which are subjected to repeated, fatigue type loading, resulting from the traversal of vehicles. In the design of new composite bridge beams, it must be ensured that the strength of the shear connection at the end of the anticipated life of the bridge is adequate to resist the maximum static design load. As a result, an extensive amount of research by numerous researchers world wide have investigated the fatigue behaviour of stud shear connectors experimentally (Slutter and Fisher 1966, Mainstone and Menzies 1967, Oehlers and Foley 1985, Oehlers 1990a, Oehlers 1990b, Gattesco and Giuriani 1996). The vast majority of the classical push-out tests, however, subjected the specimens to either monotonically increasing static loads or uni-directional fatigue loading.

A few researchers have carried out a limited number of tests with reverse-cycle fatigue loading (Slutter and Fisher 1966, Mainstone and Menzies 1967), and more recently, a series of low-cycle fatigue tests were also carried out (Gattesco and Giuriani 1996). It has been shown (Oehlers and Bradford 1995, Seracino 1999) that the shear connection in simply supported beams is subjected to a reversal of the shear load, even when partial-interaction analyses are performed. This is illustrated in Fig. 1, which plots the maximum positive and negative longitudinal shear force along the steel-concrete interface of a simply supported beam of length L resulting from the traversal a concentrated load. The results of three different analysis procedures are shown in Fig. 1, and the range of shear force resisted by the connectors R, is given by the difference between the maximum and minimum longitudinal shear force at a design point. With a full-interaction analysis, the entire length of the beam is subjected to reverse-cycle loading and, even when a more realistic linear partial-interaction analysis is performed, approximately half of the span experiences reverse-cycle loading. Furthermore, when incremental set (Oehlers and Foley 1985) is included in a partial-interaction analysis and

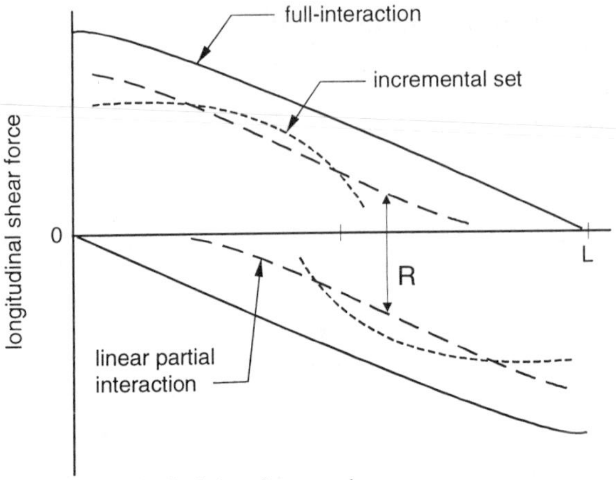

Figure 1. Longitudinal shear force envelopes.

redistribution of the longitudinal shear force occurs, the central portion of the span continues to be subjected to reverse-cyclic loads.

It has also been shown (Gattesco, Giuriani, and Gubana 1997) that for relatively large values of slip, the shear connectors may be loaded beyond the elastic range while the rest of the structure remains elastic. This implies that a shear force in the opposite direction occurs as the loads are removed and the structure returns to its original configuration due to the inelastic unloading deformations of the connectors.

As a result, an experimental programme was developed to further investigate the behaviour of stud shear connectors subjected to reverse-cycle fatigue loading. The results obtained from the tests carried out by Gattesco and Giuriani (1996) are significant. However, by accurately modelling the load conditions present along the interface of a composite beam, their specimens were complex and extensively instrumented which is perhaps why so few tests were carried out. In addition, fatigue failure occurred after a very small number of cycles ranging from about 10 - 600 as a result of the large load range applied. In the current investigation, a simple specimen was desired so the manufacturing cost could be kept low and, hence, twenty specimens were cast and tested so that a wide range of load conditions could be applied. It was also decided to subject the specimen to a large number of cycles and, hence, low load ranges, as the shear connection in composite bridges are typically subjected to millions of cycles over the life of the structure.

The tests on stud shear connectors performed by Mainstone and Menzies (1967) involved a large number of cycles ranging from about 6 thousand to 2 million, however, they formed only a part of an extensive series of tests which also included channel and bar connectors. The specimens were tested under reverse-cycle loading by using a set of four springs that were tensioned prior to testing to induce the load in the reverse direction.

1.1 *Specimen details*

A total of 15 cyclic tests were performed, 4 of which were uni-directional. Each specimen contained three stud shear connectors, 75 x 12.7 mm diameter, which were supplied and welded onto a steel plate (flange) at 300 mm spacing by a local contractor and embedded in the concrete. A Linear Variable Displacement Transducer (LVDT) was glued to the concrete near the end of the flange so that the relative movement between the steel and concrete, or slip, could be measured as the cyclic or static loads were applied. The loads were applied through

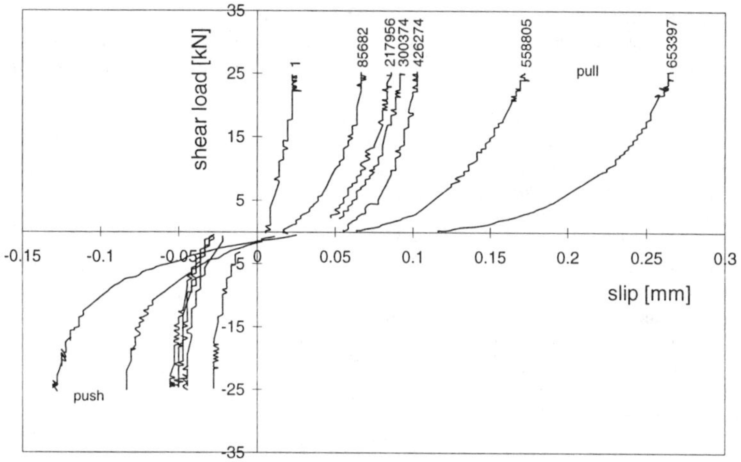

Figure 2. Typical reverse-cycle load-slip curves.

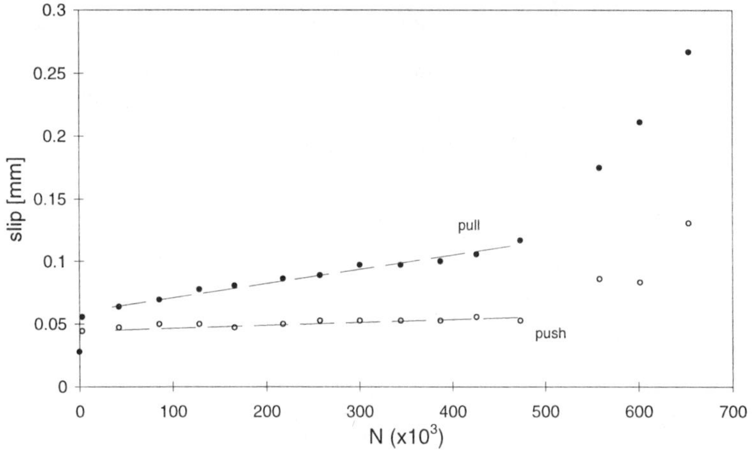

Figure 3. Typical reverse-cycle slip-N curves.

the steel flange and the concrete component was restrained from moving by corner supports fastened to the strong floor.

2 RESULTS

A typical load-slip curve of a reverse-cycle test is shown in Fig. 2 where a positive shear load represents the pull branch of the cycle meaning that the load was orientated such that the actuator was pulling on the flange of the specimen. The specimen failed in fatigue after 661.5 x 10^3 cycles and was subjected to a total range R of 50 kN and a maximum positive load P_{max} of 25 kN, so that the ratio P_{max}/R was equal to 0.5. The slip shown is that of the unloading branch of the cycle, and the numbers above each branch indicate the number of cycles that have elapsed when the reading was taken. Figure 2 clearly shows that the slip increases as the number of cycles increase, which is referred to as the incremental set (Oehlers and Foley 1985). Incremental set is a loss of energy resulting from the damage sustained by the shear connection upon cyclic loading

Another way of representing the data presented in Fig. 2 is by plotting the maximum positive and negative slips against the number of cycles elapsed N, as shown in Fig. 3. Figure 3 illustrates that the rate of increase in slip is constant over most of the fatigue life of the shear connection, as indicated by the broken lines. There is a rapid increase in the slip at the start of the fatigue life where the initial softening of the connection takes place prior to stabilising. The damage that occurs involves local crushing or powdering of the concrete in the vicinity of the connector and the initiation and propagation of fatigue cracks within the connector or the steel flange near the weld collar. More significantly, it is evident that the maximum slip, hence, the incremental set and the fatigue damage increase very rapidly near the end of the fatigue life of the connection, providing a reasonable amount of warning of eminent failure. Although the slips recorded during the pull branch of the cycle were consistently greater than those of the push branch, it is interesting to note that the slip begins to increase rapidly in both of the branches at approximately the same time.

3 ANALYSIS OF RESULTS

As the number of cycles increase, the strength of the shear connection reduces. Failure will occur when either P_{max} is equal to the residual strength after N cycles, or when an overload occurs at $N^* < N$ cycles and the overload exceeds the residual strength.

It is therefore evident that the experimentally determined number of cycles to cause failure, for a given range, is inherently related to the magnitude of the peak load. To eliminate this effect, one can define the asymptotic endurance (Oehlers and Bradford 1995) E_a given by the following expression, which was derived from a statistical analysis of fatigue data

$$E_a = 10^{\left(3.12 - \frac{0.70}{\sqrt{n}}\right)} \left(\frac{R}{D_{max}}\right)^{-5.1} \tag{1}$$

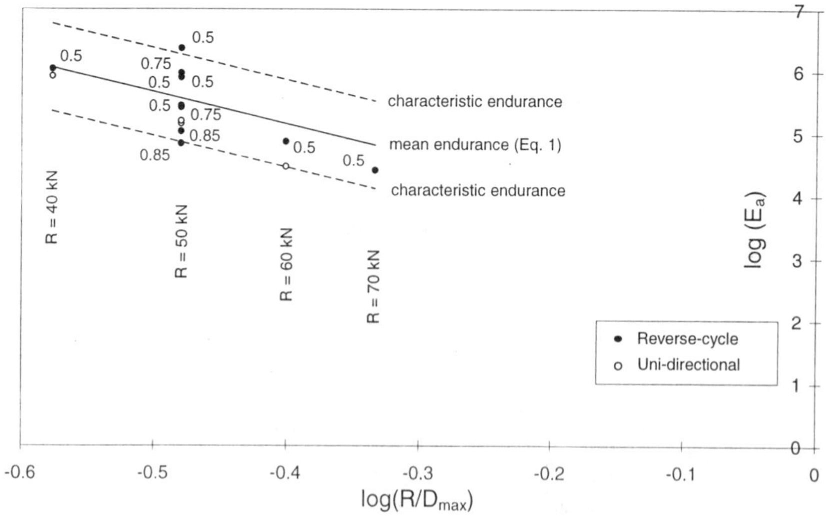

Figure 4. Log(E_a) vs log(R/D_{max}) for the current investigation.

where n is the number of connectors to fail as a group and D_{max} is the initial static strength of the shear connection.

It follows then that the asymptotic endurance can be plotted against the non-dimensional parameter R/D_{max}. Figure 4 plots the $\log(E_a)$ vs $\log(R/D_{max})$ data for the current investigation along with the asymptotic endurances predicted by Eq. 1. The mean endurance line is determined using Eq. 1 when n approaches ∞, which represents the case when all of the connectors fail as a group. The two characteristic endurance bounds represents the endurance of a single connector, hence, n is equal to 1. The coefficients shown beside the reverse-cycle data points are the P_{max}/R values.

The results plotted in Fig. 4 all fall within the bounds defined by the characteristic endurance, with the exception of the test where P_{max}/R was equal to 0.5, which was loaded monotonically to failure after 616.8 x 10^3 cycles. It can also be seen that the reverse-cycle tests have higher endurances than the corresponding uni-directional test with the exception of the two tests where P_{max}/R was equal to 0.85, however, it was subsequently observed that all three of the studs in each specimen were not adequately welded.

4 SUMMARY AND CONCLUSIONS

It is clear that the strength and stiffness of the shear connection is constantly reducing throughout the fatigue life from initial application of cyclic loads. This is evident by the constant increase in slip measured as the number of cycles increased. This information can be useful when trying to establish the remaining fatigue life of existing composite bridges as the rate of increase in slip remains constant until approximately 50% of the asymptotic endurance remains which would provide adequate warning of failure.

The second significant conclusion that can be made is that the fatigue life of a shear connection subjected to reverse-cycle loading is longer than one subjected to uni-directional cyclic loading of the same total range. As the scatter is quite large, the increase in the endurance can not be quantified, however, the increase is certainly substantial ranging from about 1.5 to 15 times the uni-directional asymptotic endurance.

The results reinforce the fact that current design procedures are conservative as the fatigue endurance of connectors subjected to reverse-cycle loading are predicted using relationships obtained from uni-directional tests. More testing of reverse-cycle push-out specimens would be required to reduce the scatter so that the increase in the endurance could be quantified.

Additional testing is also required to quantify the slip-N curve more accurately which could be used in computer simulations to help predict the remaining strength and endurance of existing structures.

5. REFERENCES

Gattesco N. & Giuriani E. 1996. Experimental study on stud shear connectors subjected to cyclic loading, *Journal of Constructional Steel Research*, Vol. 38 No. 1, pp 1-21.

Gattesco N., Giuriani E. & Gubana A. 1997. Low-cycle fatigue test on stud shear connectors, *Journal of Structural Engineering*, ASCE, Vol. 123 No. 2, pp 145-150.

Mainstone R.J. & Menzies J.B. 1967. Shear connectors in steel-concrete composite beams for bridges. Part 1: Static and fatigue tests on push-out specimens, *Concrete*, London, Vol. 1, pp 291-302.

Oehlers D.J. 1990a. Deterioration in strength of stud connectors in composite bridge beams, *Journal of Structural Engineering*, ASCE, Vol. 116 No. 12, pp 3417-3431.

Oehlers D.J. 1990b. Methods of estimating the fatigue endurances of stud shear connections, *IABSE Proceedings*, P-145/90, pp 65-84.

Oehlers D.J. & Bradford M.A. 1995. *Composite steel and concrete structural members: Fundamental behaviour*. Oxford, Pergamon Press.

Oehlers D.J. & Foley L. 1985. The fatigue strength of stud shear connections in composite beams, *Proceedings of the Institution of Civil Engineers*, Part 2, 79, June, pp 349-364.

Seracino, R. 1999. *Partial-interaction behaviour of composite steel-concrete bridge beams subjected to fatigue loading*. Ph.D. thesis to be submitted to the Department of Civil and Environmental Engineering, The University of Adelaide, Australia.

Slutter R.G. and Fisher J.W. 1966. Fatigue strength of shear connectors. *Highway Research Record 104*, HRB, pp 65-88.

Building and construction products

Mechanics of Structures and Materials, Bradford, Bridge & Foster (eds)
© 1999 Balkema, Rotterdam, ISBN 90 5809 107 4

The use of recycled crushed concrete as road aggregates

A. Nataatmadja & Y. L. Tan
School of Engineering, Griffith University, Gold Coast, Qld, Australia
M. Deacon
Geoff Maiden and Partner, Gold Coast, Qld, Australia

ABSTRACT: This paper describes the status of the Recycled Crushed Concrete (RCC) in Australia and presents the results of a recent investigation on the performance of a typical RCC aggregate. The material, obtained by crushing normal strength concrete, was reconstituted from three different size fractions to satisfy the grading requirements for a sub-base material. Triaxial specimens of 200 mm diameter were tested under repeated loads using an electro-pneumatic equipment with on-specimen displacement transducers. From the results of the resilient modelling, it was found that the resilient response of the RCC was not dissimilar to that of commonly used fresh road aggregates. Furthermore, a particle size distribution analysis carried out subsequent to the Los Angeles Abrasion test and repeated loading test suggested that material degradation was not significant. The results hence indicated that RCC may be utilised as a sub-base or base course material if it can be produced to consistently meet product the quality standards.

1 INTRODUCTION

For many years now, since the end of the Second World War, the recycling of concrete has been practised on a large scale in Europe and Japan (Paul & Warwick 1996). The practice was initiated as a means for using the waste generated from extensive bombing of buildings. It was then continued as a means of utilising the vast quantities of concrete tied up in disused airstrips spread across the continent.

Today, many developed and developing countries are finding it necessary to find alternative sources of aggregates as their reserves of natural aggregate slowly diminish. With the emphasis of this decade being on recycling in order to preserve the environment, it is a logical progression to include recycling of demolition waste. The recycling of demolition waste is, however, dependent on there being an economically and environmentally superior product made than those already available.

In Australia, due to the increasing shortage of suitable landfill sites and the spiraling cost of tipping fees, considerable effort is being put into utilising waste products such as the Recycled Crushed Concrete (RCC) or Recycled Concrete Aggregate (RCA) for road pavements. The difficulty with the non-standard materials of this kind is the limited information available regarding their performance, particularly under the application of wheel loading. Moreover, the variability of the materials may affect the reliability of the prediction of their future in-service performance.

2 USE OF RCC IN ROADS

Several municipalities in Melbourne and Sydney began recycling concrete at their landfill sites in the late 1980's (Paul & Warwick 1996). Since then, a number of demolition companies have started commercial production of RCC for various purposes. At present, it is estimated that approximately 400,000 tonnes of concrete is recycled annually in Sydney and approximately 350,000 tonnes in Melbourne.

Until now, RCC has been used in several road construction projects around the Melbourne metropolitan area. While the RCC was initially used as drainage backfill material and bedding layer for footpaths, it soon found application as a sub-base material. In 1992, VICROADS developed standard specification clauses for the production of both upper and lower sub-base materials (Paul & Warwick 1996). Table 1 shows the grading and test requirements for a 20 mm RCC.

Table 1. Grading and test requirements for RCC.

Sieve Size (mm)	Upper Sub-base		Lower Sub-base
	Target Grading	Grading Limits	Grading Limits
26.5	100	100	100
19.0	100	95-100	
13.2	85	75-95	
9.5	75	60-90	
4.75	59	42-76	42-76
2.36	44	28-60	
0.425	19	10-28	10-28
0.075	6	2-10	2-14

Test	Upper Sub-base	Lower Sub-base
Liquid Limit % (max)	35	40
Plasticity Index (max)	10	20
CBR % (min)	30	15
Los Angeles Abrasion (max)	35	40

While current data are still insufficient to prove that RCC can be produced to consistently meet the above requirements, research is continuing in many road organisations and universities to further define the performance of RCC, particularly with regards to its behaviour under traffic loading.

In line with the current practice in mechanistic pavement design procedure, and in agreement with Nunes et al. (1996), the present authors have used the repeated load triaxial test (RLTT) to evaluate resilient and permanent deformation characteristics and static triaxial test to determine ultimate strength of various waste materials. In the following sections, the results of a laboratory investigation on a typical RCC material are presented.

3 EXPERIMENTAL WORK

The RCC material was supplied by in three fractions namely, coarse (4.75-19.0 mm), medium (0.15-13.2 mm) and fine aggregates (< 9.5 mm)[1]. The Los Angeles Abrasion values were 28% and 24% for B grading and K grading, respectively. All aggregate fractions were subsequently prepared for particle size analyses and mixed through a trial and error procedure to produce a combined particle size distribution according to the Talbot's equation:

$$p = 100\,(d/D)^n \qquad (1)$$

where p = the resulting percent passing, d = sieve size, D = maximum particle size, and n = grading exponent.

A mix consisting of 15% coarse fraction, 25% medium fraction and 60% fine fraction was subsequently selected to produce a gradation that satisfies the requirement (Table 2). Compac-

[1] Supplied by Alex Fraser Pty Ltd. from its Brisbane depot.

tion test (modified energy) was performed on the reconstituted RCC, which resulted in a maximum dry density of 2.0 t/m^3 and an optimum moisture content of 9.5%. Based on the result, triaxial specimens were prepared at a moisture content of about 8.5% and compacted to 98% maximum dry density.

Table 2. Particle size distribution of the reconstituted RCC.

Sieve Size (mm)	% passing
19.0	100
13.2	85
9.5	71
4.75	52
2.36	40
0.425	14
0.075	2

Static triaxial tests were carried out on 100 mm specimens (as compacted) to establish the static shear strength of the material. Based on the shear strength parameters (Table 3), the material can be classified as a class 1 base material. Note that for the RLTT, a 'reduced shear strength' was established by multiplying the static deviator stress at failure by a factor of 0.75. The repeated deviator stress in the RLTT was kept below the maximum stress associated with the reduced shear envelope to avoid a premature failure (Nataatmadja *et al* 1996).

Repeated load triaxial testing was carried out using a large diameter triaxial cell developed by the first author at Monash University (Nataatmadja & Parkin 1988). The School of Engineering, Griffith University, now owns the triaxial cell and the pneumatic loading system. The specimens, 200 mm diameter and 400 mm high, were compacted using a Kango hammer to the maximum dry density. An on-specimen deformation measuring system was used to capture the variation of resilient deformation during testing.

While a test with cyclic confining pressure can simulate the field conditions more closely, the use of constant confining pressure is recommended by Standards Australia. To further simplify the test procedure, air pressure was used to apply confinement. The use of air pressure may be dangerous for high-pressure application; however, in this particular case the confining pressure values were kept below 175 kPa. The repeated load triaxial testing was conducted with the drainage open. Under the application of a constant confining pressure, the resilient modulus (M_r) is simply the ratio of the repeated axial load and the recoverable (resilient) axial strain.

Table 3. Static shear strength of the RCC.

Parameters	Actual shear strength from static triaxial	Reduced shear strength for repeated triaxial
Cohesion intercept (kPa)	153	114
Friction angle, ϕ (deg)	53	37

4 REPEATED LOAD TRIAXIAL TEST RESULTS

4.1 *Resilient response*

It is known that the resilient modulus is a stress dependent parameter. Until recently, the resilient modulus variation has been frequently modelled using the well-known K-Theta model (Hicks, 1970):

$$M_r = K1\ (\theta)^{K2} \tag{2}$$

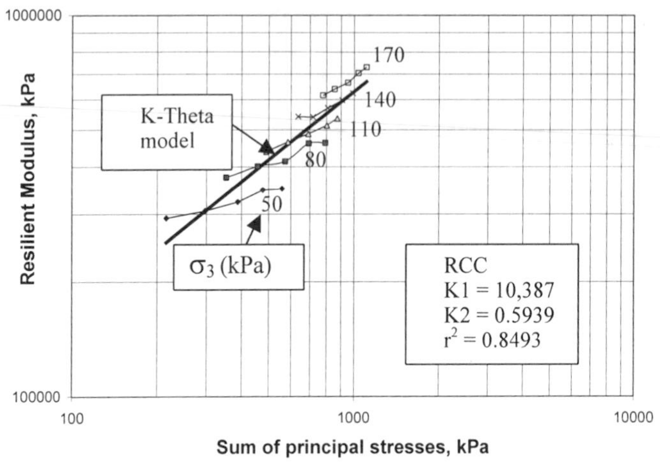

Figure 1. Resilient modulus variation of the RCC using the K-Theta model.

where M_r = resilient modulus, θ = sum of the principal stresses, and K1 and K2 = experimental coefficients. Figure 1 shows a typical resilient modulus variation of the RCC with the sum of the principal stresses.

Despite its simplicity, the K-Theta model is not an accurate or correct model. This is because the resilient modulus is not only dependent on the sum of the principal stresses but also significantly affected by the repeated deviator stress. An equally simple but more accurate empirical Two-Parameter model has been proposed by the first Author based on the results of an extensive testing of crushed rocks (Nataatmadja 1992):

$$M_r = (\theta/q_r)\,(A + B\,q_r) \tag{3}$$

where M_r = resilient modulus, θ = sum of the principal stresses, q_r = repeated deviator stress, and A and B = experimental coefficients. Figure 2 presents a typical variation of the resilient modulus of the RCC using the Two-Parameter model.

Resilient Poisson's ratio is also a stress dependent parameter. Figure 3 presents the variation of resilient Poisson's ratio with the ratio of the deviator stress to the confining pressure. It is seen that, similar to that of fresh aggregates (Hicks 1970), the Poisson's ratio of the RCC varies between 0.2 for a low stress application to 0.45 under a high stress application.

4.2 *Comparison with typical road aggregates*

To evaluate the performance of the RCC, it is necessary to compare the above results with those obtained from some typical road aggregates. Table 4 presents the tabulated values of coefficients K1 and K2 and the corresponding coefficient of determination (r^2). It is seen that because of the inaccuracy of the model, it is difficult to comment on the performance of the RCC. On the other hand, when the coefficients A and B of the Two-Parameter model for each of the above material are compared, it is apparent that the RCC is as good as if not better than the fresh aggregates even when moisture content was high (Table 5).

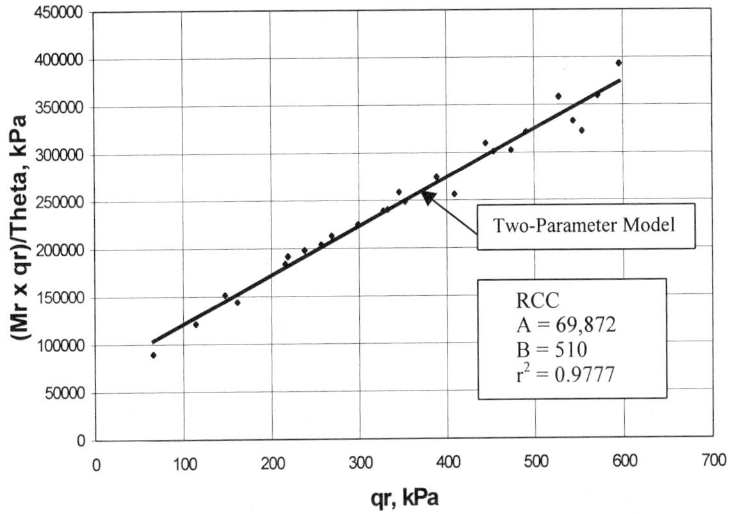

Figure 2. Resilient modulus variation of the RCC using the Two-Parameter model.

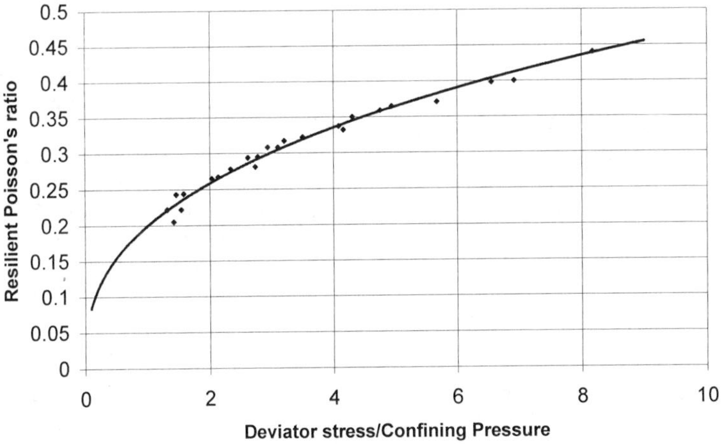

Figure 3. Poisson's ratio variation of the RCC.

Table 4. Comparison of performance using the K-Theta model.

Material Coefficients	RCC	Hicks' Base-course (Hicks 1970)	Dry Rhyolite (Nataatmadja & Parkin 1989)	Uzan's Dense Graded Aggregate (Nataatmadja & Parkin 1989)
K1 (kPa)	10,387	3982	5,104	40,681
K2	0.5939	0.6951	0.67	0.3528
r^2	0.8493	0.9466	0.9137	0.5619

Table 5. Comparison of performance using the Two-Parameter model.

Material Coefficients	RCC	Flooded RCC*	Dry Rhyolite (Nataat-madja & Parkin 1989)	Sandstone (Nataat-madja 1994)	Hick's Base-course (Nataat-madja & Parkin 1989)	Uzan's Dense Graded Aggregate (Nataatmadja & Parkin 1989)
A (kPa)	69,8726	59,434	24,200	44,300	20,4200	38,310
B	510	517	560	350	412	445
r^2	0.9777	0.9758	0.9801	0.9593	0.9691	0.9500

* The specimen was flooded with water and allowed to drain for 24 hrs before testing to simulate the worst condition *in-situ*.

4.3 *Degradation under repeated loading*

It is known that tests such as the Los Angeles Abrasion, Aggregate Crushing Value, etc. cannot give a good representation of service under traffic (Lay, 1990). Therefore, in this investigation, a specimen was tested under 50,000 cycles of 500 kPa. It was found that the resilient modulus reached a stable level after about 1,000 cycles. A particle size analysis carried out after the test indicated that the degradation of the RCC was not significant.

5. CONCLUSIONS

Based on the test results, it can be concluded that the performance of the RCC may be comparable to that of fresh road aggregates. This also supported by data reported by Paul & Warwick in 1996. The well-graded RCC may even produce a higher resilient modulus under low deviator stresses as compared with other materials, possibly due to high pore suction values. The effect of moisture in the RCC was found to be less significant than that of the fresh aggregates. This may be due to higher particle surface roughness, which reduces the potential slipping.

Considering the limited information available on the long-term performance of the RCC under the application of wheel loading, future research will be focused on the effects of concrete strength and number of cycles on the resilient modulus and permanent deformation of the RCC.

REFERENCES

Hicks, R.G. 1970. Factors influencing the resilient properties of granular materials. *Ph.D Thesis*, University of California, Berkeley.

Lay, M.G. 1990. *Handbook of road technology*. Second Edition. New York: Gordon and Breach.

Nataatmadja, A. 1992. Resilient modulus of granular materials under repeated loading. *Proceedings 7th International Conference on Asphalt Pavements*, Nottingham (1): 172-185.

Nataatmadja, A. 1994. On the response of granular unbound pavement materials under repeated loading. *Proceedings 17th ARRB Conference*, Gold Coast (2): 99-114.

Nataatmadja, A., Clarke, P. & Singh, A. 1996. Some aspects of dynamic testing of soils using MATTA equipment. *Proceeding Roads 96 (18th ARRB) Conference*, Christchurch (2): 211-226.

Nataatmadja, A. and Parkin, A.K. 1988. a large cell for repeated load triaxial testing of base course materials. *Proceedings of the 14th ARRB Conference*, Canberra (7): 85-93.

Nataatmadja, A. & Parkin, A.K. 1989. Characterisation of granular materials for pavements. *Canadian Geotechnical Journal* (26): 725-730.

Nunes, M.C.M., Bridges, M.G. & Dawson, A.R. 1996. Assessment of secondary materials for pavement construction: technical and environmental aspects. *Waste Management* 16(1-3): 87-96.

Paul, R. & Warwick, R. 1996. Use of recycled crushed concrete for road pavement sub-base. *Proceedings Roads 96 (18th ARRB) Conference*, Christchurch (2): 93-106.

Mechanics of Structures and Materials, Bradford, Bridge & Foster (eds)

A lightweight cement-bonded building material

P.G. Lowe
Department of Civil and Resource Engineering, University of Auckland, New Zealand

ABSTRACT: There are many challenges ahead in the new millennium for engineers to design and build ductile, strong and durable structures and products from materials which make lesser demands on resources, produce less waste during the building process and use a proportion at least of recycled material. The studies reported here use waste wood fibre as the 'aggregate' in cement-bonded building materials. Essentially the material behaves like reconstituted wood, has about the same density and strength, and can be cast into the required shapes.

1 INTRODUCTION

1.1 *The aims*

The building process makes heavy demands on a whole range of resources. The engineering profession is intimately involved in all aspects of the decision–making in the building process from the initial conception to the finished product.

There is an urgent need to review the extent of these demands and to reassess the priorities, in order to lessen the demands and to make available to the public a greater range of building materials. In particular, two aspects need to be addressed. First, there is a need to perfect further, new, viable, lighter weight cement-bonded materials. Secondly, there is a need to ensure that these and other bulk building materials make use of at least some recycled materials in the new processes of manufacture, as a contribution towards an environmentally sustainable industry.

1.2 *The choice made in this study*

Here we report on studies to use waste newsprint as the 'aggregate' in a cement-bonded bulk building material. Several factors contributed to the choice of newsprint as the waste material to be experimented with.

The first factor was availability. In Auckland, a city of modest size by world standards, there is just one widely circulated daily newspaper. This dominates the scene. There are many much smaller circulation suburban papers. From these sources roughly one thousand tonnes of newsprint are used in the city every week. The bigger the city the larger the amount used. Some of this newsprint is recycled but much is not. As will be described later, the non-recycled newsprint could potentially provide a substantial slice of the building materials market, if suitably processed.

A second factor in the choice was to experiment with lightweight cement-bonded materials that do not require special man-made aggregates. Newsprint derived waste paper appealed as a possible contender. Another related choice, here only a second choice, is to recycle waste treated timber. Due to the treatment it is difficult to re-use such timber in any other constructive way.

2 THE MATERIAL

2.1 *Initial studies*

These studies began some years ago. At first they were very simplistic attempts to substitute recycled newsprint for regular aggregates in a concrete mix. Both cylinders and hollow blocks were made. There was much experimentation with the proportions for the ingredients, and the mixing and placing capability of the paper pieces.

Almost immediately it became evident that new mixing techniques were needed. The normal concrete mixing equipment is designed to handle the heavy, rounded materials normally used. In the case of newsprint as the 'aggregate', this material is light and in sheet form and these characteristics mean it is difficult to work with when using conventional mixers. Another practical difficulty is the quite different character of dry as compared with wet paper.

Despite the difficulties encountered in mixing the material, acceptable products can be made which have some desirable characteristics. The hollow blocks were 400x200x140 mm in overall size. Though not strong in the sense of a regular hollow concrete block, these paper-cement blocks have regularly shown capacities of many tonnes in compression. The main focus has been on the block as the first choice end-use.

2.2 *Later studies*

More recently, blends of wood particles derived from waste treated timber with newspaper have been used. The wood particles, as might be expected, behave more like regular aggregates, and can be relatively easily mixed with conventional mixing equipment. The material made from all-wood 'aggregate', and no paper, is much nearer in properties to conventional concrete, including remaining a brittle material. The all-paper material is essentially ductile. Indeed, this ductility property is a most welcome feature of the paper-cement composites. This is discussed and illustrated later.

Thus far there has been no scope to have a purpose-made mixer constructed. It seems clear however that a specially adapted mixer is really essential for further progress, and when available, will provide scope to greatly improve the quality of the paper/cement-bonded material.

3 TYPICAL EXPERIMENTAL DATA

3.1 *The two most important parameters*

The two most important parameters for the paper-cement composite are first the density and secondly the strength of the material. For many applications, normal weight concretes are not only excessively heavy but they are also unnecessarily strong. In particular we are supposing that paper–cement composites could be suitable for interior, non-structural applications, such as partitioning, non-load bearing walls and as a filling for various applications.

3.2 *Typical densities*

It is quite easy to make the paper-cement material in a range densities from around 600 kg/m^3 for the lightest, to two or three times this figure. To achieve the higher densities more cement and/ blending with mineral aggregates is probably needed. Questions of cost and returning brittleness with such blends suggest that the ideal density range is from about 600 to about 1200 kg/m^3, that is, about one quarter to a half the density of normal weight concretes. This density range puts the material into a lighter range than is achievable when using most man-made mineral aggregate-concretes, though it is within the range of concretes possible using plastics as aggregates. The most comfortable density range for paper based products seems to be 1000 - 1200 kg/m^3, although densities down to 600 kg/m^3 have been experimented with.

3.3 *Typical strengths*

Strength is a more difficult parameter to deal with. The amount of cement used will largely determine how much strength the composite has. Generally this study has restricted the amount of cement used to be about the same as in a similar sized member or component made from regular concrete. This is a quite severe restriction. Even with this constraint, the strength of the material made varies quite markedly depending upon the method and quality of the mixing and the amount of water used.

Typical strengths are a few MPa in compression, with an upper limit of perhaps 8 – 10 MPa. This may seem too little for many applications. However when standard sized 400x200x140 mm hollow blocks are made with paper-cement the wall thickness can be increased to 50 mm or more compared to about 30 mm for the regular concrete block. This means that the load capacity of a typical paper-cement block with a wall thickness of 50 mm can be made equivalent to a 20 – 25 MPa conventional concrete block with a 30 mm wall thickness. Current block dimensions, and especially the wall thickness, are determined by block weight. A block with 50 mm wall thickness, made with a conventional concrete, would be too heavy to pick up, single-handed, for laying. The paper-cement block with a 50 mm wall is however lighter than the (approx.)10 kg of the conventional block. The typical paper-cement blocks weigh in the range 5 – 6 kg.

Recently a typical block, with a density of 1066 km/m^3 and made about a year ago, was put back into the test machine and loaded once again to the maximum load, as indicated by the near constant load as the block deformed. This load and the deformation of the block were essentially the same as when first loaded six months earlier. The block remains undamaged. This is some measure of the 'ductility' of the material. The maximum load of 67 kN equates to about 1.5 MPa stress in compression. From the deformation of about 2 mm over the linear portion on a 149 mm gauge length, the elastic modulus was about 110 MPa. This is two orders of magnitude lower than for a typical, regular weight concrete. Hence the material is neither strong nor stiff compared with regular concretes, but it is probably strong enough, stiff enough, and light enough for a range of applications.

3.4 *General points*

The water requirements to make the paper-cement material are critical for several reasons. First is the need to preserve as much cement strength as possible. Too much water will erode this. Also important is the slow drying of the relatively large thicknesses of material in the specimens being experimented with. Excess water must be removed by some means. Quite large amounts are needed though precisely how much will not be discussed here. The drying characteristics of the material are also quite different to conventional concrete. Further study is needed.

Another aspect of the concentration on block making with this paper-cement material in these studies has been the desire to employ the finished block in an assembly that essentially does away with the mortar joint in the finished, assembled wall or panel. The usual mortar joint is the site for shrinkage cracks and sliding failures: If it can be avoided, so much the better.

If block-work technology can cope with a near zero thickness of joint between blocks, which was the norm in the best of stone masonry for hundreds of years, and if the individual blocks are not brittle, then there is scope to achieve a homogeneous panel or wall which will not be subject to shrinkage of the bedding mortar or preferential sliding along the block interfaces, and will have a very large deformation capacity. The sliding tendency can also be arrested or eliminated if suitable keys are added to the assembly in the hollow cavities, spanning from block to block. These have been some of the aspects experimented with in the present studies.

A no-mortar joint can be made with glues if required or desired. Very thin cement mortar joints can also be used. But no jointing scheme will be satisfactory if adequate dimensional tolerance of the blocks is not achieved. Experience has been that quite strict dimensional tolerances, to better than a mm, can be achieved with the cement-bonded paper material.

3.5 *Ductile or brittle?*

A feature of these paper/cement materials is that they have substantial capacity to deform and are not brittle. This is unlike most conventional cement-based materials that are frequently very brittle. With the conventional material the aggregates are coated by the cement mortar but the mortar does not penetrate the aggregate. The mortar-aggregate interface then becomes the site for separation and fracture to occur and this is what endows these materials with their brittleness. In the paper-based materials, the cement does penetrate the paper and hence there is less tendency for local fracture and as a result, far less tendency for brittle behaviour. In Appendix A there is further discussion of brittle behaviour.

3.6 *Other features*

Most conventional concretes can only be cut with power saws using special and often expensive blades. The paper-cement material can be cut with a conventional wood saw, either hand or power, though no doubt there is merit in experimenting with blade types to achieve best results and good saw life. Another positive feature is that fixings of all types, including nails and screws, can be driven or fixed without the use of inserts or special requirements.

Blocks and block assemblies of various sorts have been constructed and tested using this material. The material can also be used as the filling in tubular constructions such as Externally Reinforced Concrete, which has been described at earlier ACMSM conferences, Lowe and Choong 1990 and Lowe 1993. In the New Zealand context lighter weight cement-bonded 'concretes' have the potential to reduce the dead weight in our structures and this would lessen our exposure to harmful effects in earthquakes. This single aim is a sufficient stimulus to continue the search for lighter weight materials.

3.7 *Production considerations*

We have noted earlier that there are massive amounts of waste newsprint and other paper-like products available. Frequently these materials become a burden on disposal regimes. It does not seem too much to ask for the profession and industry to be more creative, and devise uses for this material within the general sphere of building products.

If half of the discarded newspapers from the one major newspaper publisher in Auckland was used in a product such as the paper-cement block discussed above, then approximately 100,000 of these blocks could be made every week. This is far more than the current manufacturers' of concrete blocks own capacity, and there is no ready market for this amount of product. It, however, illustrates the scope to mass-produce. In environmental terms this activity would not be making demands on quarries to supply the aggregates but instead would be largely a recycling operation with consequent environmental benefits.

There are other community-related factors that could also be considered. Many changes have been visited on engineering-based industries in recent years. In New Zealand much engineering infrastructure, both in personnel and plant, has been discarded, or relocated away from the smaller towns, or has gone off-shore. There are probably also parallels in the Australian socioeconomic scene.

It is suggested that some of our building practices contribute to these socio-economic trends. For example, sheet-type building materials are likely to be manufactured in the large factory environment whereas block and brick-like building products could be manufactured in smaller units. These smaller units could be replicated in many places, the smaller towns for example, so producing new employment opportunities in the building materials sector. The social ills of unemployment are not usually discussed in papers at technical conferences such as the present one. But much of what we discuss at a technical level is only one step removed from socio-economic issues. Those of us who are engaged in the educational sector have more opportunities to influence the thinking of the young engineers than colleagues in other sectors. But we may not agree about the sorts of influences that should be promoted.

A combination of circumstances, including the new millennium, sustainable use of resources, resource demands, recycling, new materials, new uses for old materials, and many other considerations, will impact on the engineering profession and associated employment increasingly in the future. A personal view is that the profession can do much, and has much to offer the community, by considering other alternatives, such as promoting the search for new building materials which incorporate a greater recycled proportion of what are at present bulk waste materials.

4 CONCLUSIONS

The lightweight material described in his paper is in the earliest stages of investigation, and for that reason may be an inappropriate choice for discussion at a conference such as the present one. Here the point of view adopted is that there is an onus on, and an urgency for, the engineering profession to look more closely than has been the case in recent times at other, perhaps quite different, choices for some engineering materials, if we are to make an adequate response to the use of recycled products in the building materials of the future. There are many potential choices and all will require much effort to understand and develop.

In terms of the time spent on the present investigations thus far, the paper–cement choice has been a satisfactory, perhaps even a good choice. None of the usual funding sources were successfully applied to and no 'public good' monies have been applied in the programme. There is much scope for coming up with novel ideas and improvements to combine the materials into a final composite in some sort of optimum way. Student projects are becoming a more integral part of the teaching scene. Inexpensive experimental topics such as the present one, are an alternative to computer modelling based projects.

5 ACKNOWLEDGEMENTS

Several students have undertaken projects based on the general concept of this paper. Their collective skills and perseverance have produced results that earlier we thought could not be achieved. The laboratory staff helped in a variety of ways. It is a pleasure to acknowledge the contributions of both of these groups.

6 REFERENCES

Lowe, P.G. 1986. Strain-based failure criteria and reinforced concrete structure response. *The tenth Australasian conference on the mechanics of structures and materials* 347–352. Adelaide.

Lowe, P.G. and Choong, K.C 1990. Externally Reinforced Concrete. *The twelfth Australasian conference on the mechanics of structures and materials* 177-182.Brisbane.

Lowe, P.G. 1993. Externally reinforced Concrete–II. *The thirteenth Australasian conference on the mechanics of structures and materials* 541-547. Wollongong.

Todhunter, I. and Pearson, K. 1886 (Vol. 1), 1893 (Vol 2, Parts 1 and 2) *A history of the theory of elasticity and the strength of materials,* Cambridge: University Press, Cambridge.

APPENDIX

A SOME FURTHER NOTES ON BRITTLE MATERIALS

A.1 *An earlier paper.*

In a paper to the 10th ACMSM in Adelaide, Lowe 1986, proposals were made to describe brittle materials, such as concrete, through a tensile strain limitation criterion. It is now appreciated that similar proposals were made very much earlier.

A.2 *Earlier precedents*

In the nineteenth century, when many parts of our subject were being developed, one of the most widely used materials was cast iron, and this was a brittle material. Only later did concrete appear on the scene, or at least did it take on an important role as another brittle material in general use.

B. de Saint Venant (1797 – 1888), a leading engineer, theoretician and educator of the day, made many studies in which he investigated the failure conditions in engineering materials. In particular he studied the conditions to cause failure in brittle materials such as cast iron. He was also well informed about earlier studies. Todhunter and Pearson, 1886 and 1893, made an extensive study of all the literature of the time and earlier. From the account given in these volumes it is clear that the tensile strain limit criterion was discussed by St. Venant, or possibly by his countryman E. Mariotte (1610 – 1684), a near contemporary of Robert Hooke of the elastic law. Hence they are the originators of the strain (rather than stress) based failure criteria. As ductile materials, such as steel, made their way into the industry and into general use, the preoccupation with brittleness receded and so too, we conclude, did study and knowledge of the related literature.

A.3 *Another derivation for the 1986 proposals*

The core concept in the paper (Lowe 1986) was the proposal that a brittle material such as concrete is governed at the limit (cracking) by a principal tensile strain (hence non-dimensional) parameter, denoted by ε_t. It was also proposed in 1986 to obtain the uniaxial tensile stress , f_t, at this limiting strain from the relationship $f_t = n.\varepsilon_t$ where n is a material parameter, a tensile modulus.

The quantity of most interest is usually the unconfined compressive cylinder strength, f_c'. In the unconfined compression cylinder there is always circumferential tensile strain present, but not tensile stress. When this tensile strain reaches the tensile limit, ε_t, the (longitudinal) cracking of the compression cylinder occurs. The subsequent behaviour is governed by the onset of instability of the slender, axial columns of cracked concrete. As with other instability phenomena, we expect that the compression elastic modulus, E_c, will be present in the relationship. Thus the expression $f_c' = k. E_c. \varepsilon_t$ was proposed in 1986, though after a different line of argument. Here k is a second non-dimensional parameter, to be determined experimentally.

To complete the description we prefer now to expect a relationship of the form $E_c = \alpha. f_t$ rather than the alternative expression given in the 1986 paper. This third non-dimensional parameter, α, is a numerical coefficient with a value of about 7. 10^3. Typically, n could be expected to have a value of around 20 GPa, k a value of about 7 and ε_t to be around 200 microstrain. The values of all four parameters must be found experimentally, and would probably change as the composition and quality of the material changes.

When these relations are analysed, it is found that E_c (GPa)= Sqrt (n. f_c'), with n in GPa. Now it is clearer that the compression elastic modulus ,E_c, with which we are familiar, is, in this development at least, an explicit function of the tensile elastic modulus, n. If there is merit in elaborating the present approach any further, then it seems likely that an anisotropic elastic material description will be needed.

Mechanics of Structures and Materials, Bradford, Bridge & Foster (eds)
© 1999 Balkema, Rotterdam, ISBN 90 5809 107 4

Development of a new structural core material for composites in civil engineering

S.R. Ayers & G.M. Van Erp
Faculty of Engineering and Surveying, University of Southern Queensland, Toowoomba, Qld, Australia

ABSTRACT: This paper presents preliminary findings of an ongoing research program into the development of a new structural core material for fibre composites in civil engineering applications. The material under development is a particulate filled resin based on micro-spherical glass and ceramic fillers in a polymer matrix. Static flexural and compressive test results are presented for a range of different fillers and resin types. Different behavioural trends are identified and discussed.

1 INTRODUCTION

Standard beam and plate theory shows that for materials to be used efficiently, they must be located as far as possible from the neutral axis of the section. In traditional structural materials such as steel and aluminium this has resulted in extensive use of I sections and box beams. In these members the flanges carry the bending moment while the webs are used to space the flanges and provide the necessary shear strength. These principles are essentially material independent and should also be applied in the development of efficient fibre composite members. As a result, structural items in composites have often taken the form of accepted structural shapes. Pultruded sections using this approach have been available now for a number of years. However, this approach has failed to provide the expected high performance (Maji et al 1997). Due to their low stiffness, pultruded beams exhibit structural instability such as local buckling of the flange and web long before ultimate failure of the constituent materials (Gilchrist et al. 1996). Introduction of carbon fibre into pultruded beams can significantly improve the stiffness but these hybrid structures are still far too expensive. Usage of pultruded beams in primary applications is also hindered by their limited capability to deal with concentrated loads, bolt holes, and accidental damage.

Developments under way at USQ are based on a different approach. High strength unidirectional laminates are used for the top and bottom flanges of the members and these flanges are held apart by a continuous, damage-tolerant core material. The advantages of this 'sandwich' approach lie in improved section stability and reduction in cost through reduction of structural demands on the core material by spreading the load across a large volume of material.

Most core materials available today have been designed with weight reduction in mind. However, weight is normally not the critical issue that drives civil engineering designs, cost and load carrying capacity are far more important. Other important issues are damage tolerance and ability to carry high concentrated loads. Most aerospace and boating core materials fail to meet these requirements. Researchers at USQ are presently experimenting with a particulate filled resin as a core material. The results obtained to date look extremely promising. The properties and cost of this filled resin system depend heavily on the type and amount of filler and base resin used. In order to obtain a better understanding of the fundamental issues that determine the behaviour of particulate filled resin systems an experimental research program was initiated.

2 EXPERIMENTAL RESEARCH PROGRAM

The aim of this ongoing research program is to develop a comprehensive understanding of mechanisms influencing the mechanical performance of particulate filled resin systems. The initial subject of investigation is the behaviour of such materials under static loading. The material properties that have been considered so far are Modulus of Elasticity, Flexural modulus, compressive strength, tensile strength and strain to failure.

2.1 *Investigated materials*

2.1.1 *Filler materials*

Five grades of glass and four grades of ceramic microsphere fillers were investigated in this study. Glass microspheres were supplied by 3M Specialty Additives (Australia) from their Scotchlite Glass Bubble Range. Four fillers (K1, K15, S38, S60) were from the General Purpose Series. H50 filler is from the Floated Series and uses an epoxy-compatible silane surface treatment. Table 1 outlines the different filler grades used in testing and gives particle properties of density, size and strength.

Table 1. Scotchlite glass bubble fillers used in experimentation (Source: 3M Specialty Additives).

Product Grade	Nominal Particle Size Range (microns)	Median Particle Size (microns)	Particle Density (g/cc)	Isostatic Crush Strength (MPa)
K1	15 – 125	70	0.125	1.72
K15	15 – 125	70	0.15	2.07
S38	8 – 88	45	0.38	27.58
S60 / 10000	6 – 88	30	0.60	68.95
H50 / 10000	6 – 125	30	0.50	68.95

Ceramic microspheres used in this study were from the E-Sphere SL Series and were supplied by Envirospheres Pty Ltd. The four different grades investigated differ primarily in terms of the particle size range. Table 2 outlines the different grades investigated and their respective particle size distributions. Average densities for each grade are also listed.

Table 2. E-Spheres Ceramic Fillers Used in Experimentation (Source: Envirospheres Pty Ltd).

Product Grade	Nominal Particle Size Range (microns)	Approximate Mean Particle Size (microns)	Particle Density (g/cc)	Filler Cost ($/kg)
SLG	20 – 300	130	0.696	0.90
SL150	20 – 150	100	0.701	1.02
SL125	12 – 125	80	0.706	1.50
SL75	12 – 75	45	0.648	4.20

2.1.2 *Polymer Resins*

Two different epoxy vinylester resins were used in this study. Hetron 922 was supplied by Huntsman Chemical Company Pty Ltd. Hetron 922 is a general purpose vinylester resin. The resin was supplied pre-promoted with Cobalt Naphthenate / Dimethylaniline (CoNap/DMA) and contained a thixotropic additive. Derakane 8084 was supplied by DOW Chemical (Australia). Derakane 8084 is an elastomer modified epoxy vinylester resin and was selected primarily due to its increased strain to failure (10 – 12%) over standard vinylester resins (Hetron 922 – 6.5%). The Derakane 8084 resin was promoted prior to fabrication with 0.3% CoNap and 0.025%DMA. Both vinylester resins were cured using the initiator Norox MEKP-9 (The Norac Company, Inc.). MEKP-9 is a solution of methyl ethyl ketone peroxide (MEKP) in dimethyl phthalate. An initiator level of 1% was used and found to yield a gel time of approximately 60 minutes.

A number of epoxy systems were also investigated. ADR246TX, a formulated laminating system, was supplied by ATL Composites Pty Ltd. ADR246TX is a DGEBA based epoxy blend with the addition of a BisF epoxide and a diluent for viscosity modification. The resin also includes a thixotropic additive. The resin was reacted with ADH160 hardener also from ATL Composites. ADH160 is a proprietary amine blend. DER331 was supplied by DOW Chemical (Australia). DER331 is a pure DEGBA epoxy. It was reacted with a triethylenetetramine (HY951) hardener supplied by Ciba Specialty Chemicals.

2.2 *Specimen Preparation.*

A pre-calculated mass of the specified resin was firstly weighed into an appropriate container. The initiator/hardener was then added and thoroughly mixed. Vinylester mixtures were then allowed to stand until the cessation of the MEKP's initial foaming action. The calculated mass of filler for a specified filler loading was then slowly added to the resin under continuous hand stirring. Once all the filler was added to the resin and wet, the mixture was mechanically blended to ensure a consistent distribution of filler.

Compression specimens were cast into individual cylindrical moulds (diameter 30mm, height 48mm). Flexure specimens were cast in a single 180 x 180 x 16 mm block in a vertical steel mould. This technique was found to offer good control of exotherm temperatures through the large exposed surfaces available for heat dissipation. In some instances air cooling of the moulds was used to further assist heat dissipation. The castings were allowed to cure at room temperature for a period of 48 hours. After this time the castings were removed from their respective moulds and post-cured at 80°C for a period of four hours. Castings were then cut and finished to final size.

2.3 *Mechanical Testing*

Flexural behaviour was assessed using a 3-point bending test in accordance with ISO 178:1993. Specimens dimensions were; l = 180mm, b = 15 mm and h = 9 mm. The supported span was set at L = 144 mm. A total of seven specimens were tested for each material combination. Applied load and corresponding mid-span deflection were recorded continuously during testing.

Five cylindrical compression specimens of each material combination were tested. Testing was conducted using a constant displacement rate of 2 mm/min. Applied load and corresponding deflection were recorded continuously throughout the test.

3 EXPERIMENTAL RESULTS

3.1 *Flexural Modulus*

Modulus of elasticity is a critical parameter in engineering design as it determines member deflection under load. The majority of composite structural elements under development at USQ function in bending, thus initial experimental work has focused on examination of the modulus of elasticity in flexure (flexural modulus). Flexural modulus varies slightly from pure tensile modulus as it also incorporates compressive behaviour and possibly some shear deformation effects.

The effect of filler and resin type, particle size, particle strength and filler loading level on flexural modulus has been examined. Experimental results indicate that flexural modulus behaviour is significantly influenced by the type of particle used. Results from specimens using ceramic SL150 microspheres indicate an increase in modulus with increasing filler levels (see Figure 1). Data from mixtures with lower strength glass particles (K15) indicates an opposite trend with modulus decreasing with increased filler levels. (see Figure 1) The increase of flexural modulus with increasing particle strength is further demonstrated in testing of glass fillers with varying strengths. Results from these tests indicate that flexural modulus is increased through use of a higher compressive strength particle (see Figure 2).

Assessment of the influence of resin type on flexural modulus was undertaken with K15 glass microsphere fillers at varying loadings (see Figure 3). Results indicate that while differing resin properties affect modulus at low filler levels, this effect diminishes with increasing filler content. At filler level in excess of 40% it appears that flexural modulus is affected more by filler properties than those of the resin.

The effect of particle size variation within fillers was also investigated using E-Spheres fillers in vinylester resin. Test data indicates that at the filler levels considered, there is no significant influence of particle size on flexural modulus (see Figure 4).

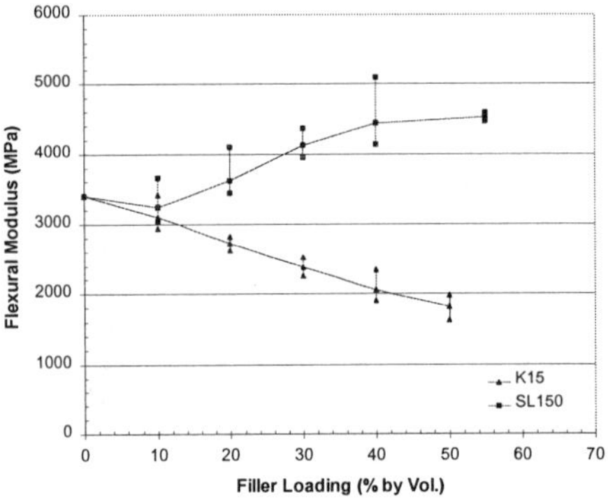

Figure 1. Variation of Flexural Modulus with Filler Loading (Hetron 922 Resin)

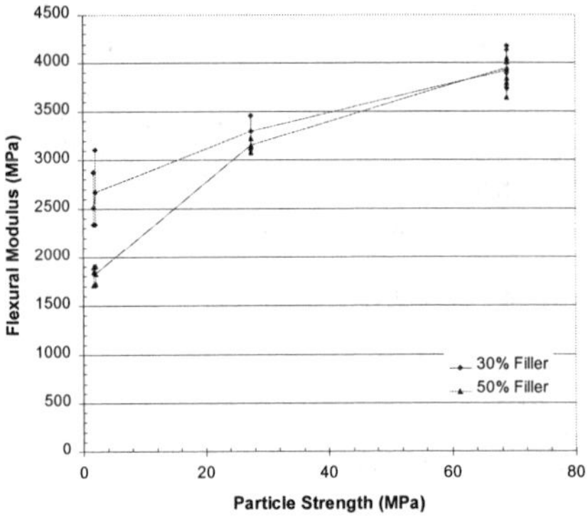

Figure 2. Variation of Flexural Modulus with Particle Strength (Glass Microspheres)

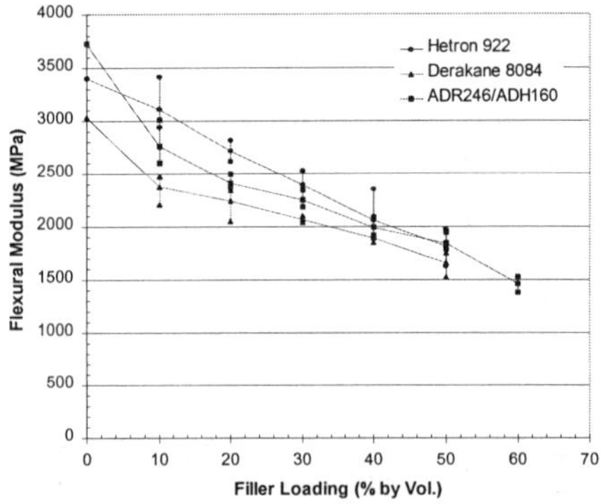

Figure 3. Flexural Modulus Variation with Filler Loading (K15 Microspheres)

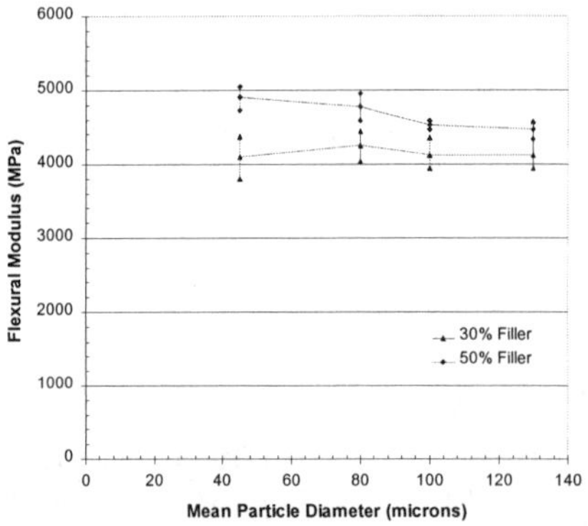

Figure 4. Flexural Modulus Variation with Particle Size (E-Spheres/Hetron 922)

3.2 *Flexural Strain to Failure*

A sound understanding of strain to failure properties is also important in engineering design. In structural development work at USQ, new members are designed on the assumption that the ultimate strain level of the core exceeds that of the reinforcing skins. This allows utilisation of the cores structural properties through the complete service range of the member.

The effect of filler and resin type, particle size, particle strength and filler loading level on flexural strain has been examined. Test data indicates that resin properties exhibit significant influence on ultimate strain behaviour at low filler loadings (see Figure 5.). However, these differences decrease with increasing filler content. Strain levels are seen to decrease with increasing filler contents. This decrease in strain level is significantly more pronounced in mixtures using the higher strength ceramic (SL150) particles than in K15 mixtures (see Figure 6). Further examination of the relationship between strain levels and particle strength confirms a decrease in strain levels (see Figure 7) as well as an increased rate of property change with increasing strength.

The effect of particle size on ultimate strain levels was also investigated. Experimental data indicates little variation between ultimate strain levels at the two filler loadings examined (see Figure 8)

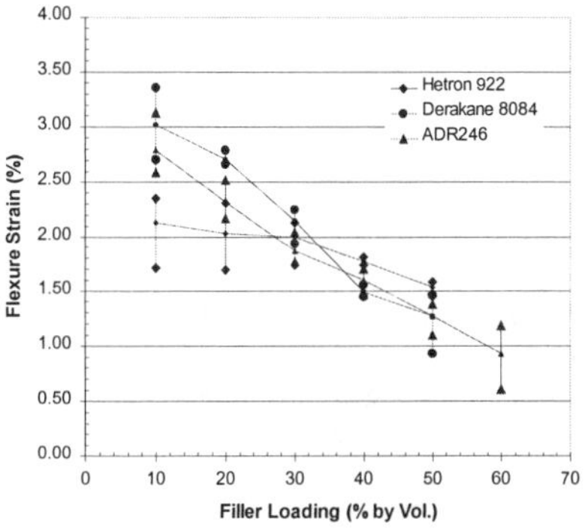

Figure 5. Variation of Flexure Strain with Filler Level (K15 Glass Microspheres)

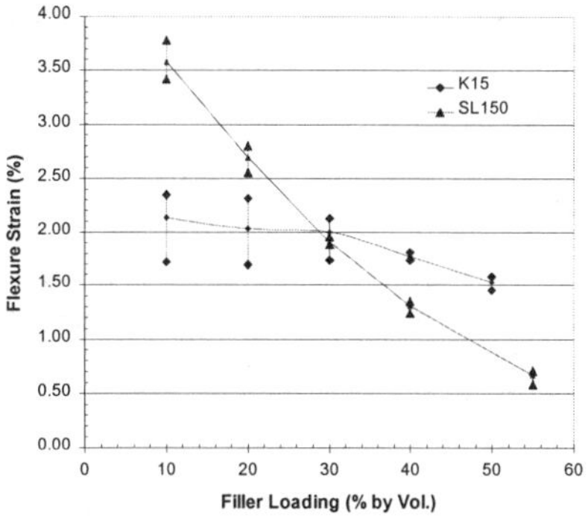

Figure 6. Variation of Flexure Strain with Filler Type (Hetron 922 Resin).

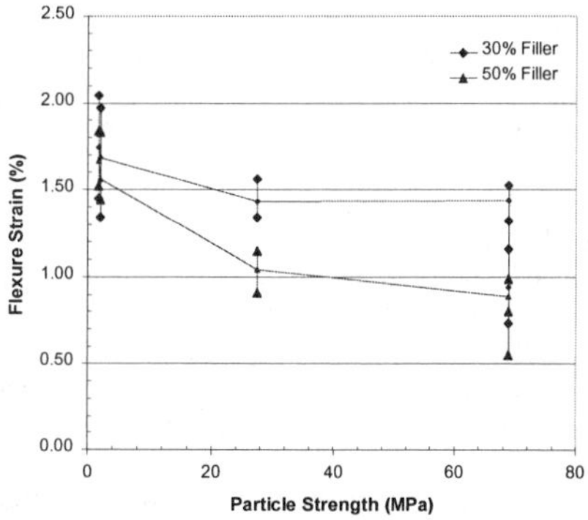

Figure 7. Variation of Flexure Strain with Particle Strength (Scotchlite Glass Bubbles/Hetron 922)

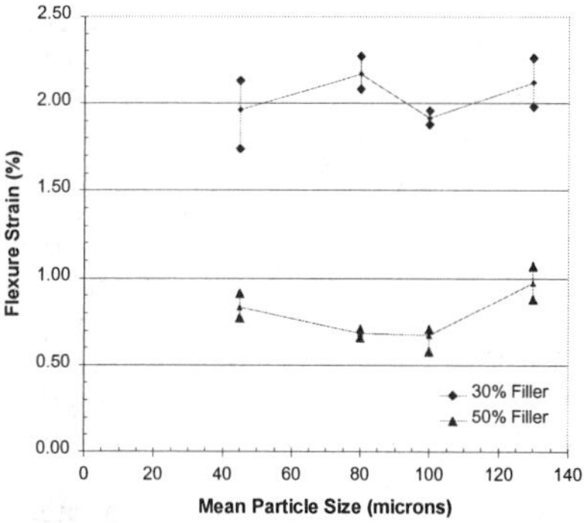

Figure 8. Variation of Flexure Strain with Mean Particle Size (E-Spheres/Hetron 922)

3.3 *Flexural Strength*

Flexural strength has also been studied in this experimental work. The effect of resin and filler type, particle size and particle strength have been examined though space restrictions do not permit graphical representation of the results.

Experimental data indicates that the type of resin used has little effect, with samples displaying similar strengths across the 10 to 50% filler range. Strength values are seen to decrease with increasing filler loading. The use of higher strength particles has been shown to increase the material's flexure strength. Examination of the relationship between filler size and flexural strength has once again shown that at the filler levels in question (30 and 50%) particle size has a minimal effect on material properties.

3.4 *Compressive Properties*

Compressive strength is a key property in the ability of the core/spacer material to successfully accommodate concentrated loads in a structural section. Poor compressive strength will result in localised crushing. At the time of writing, only initial results from experimental study of compressive strength relationships with filler loading levels were available.

This data indicates that compressive modulus is relatively unaffected by filler levels in SL150 mixtures but experiences significant decrease in modulus with an increasing percentage of K15 filler. Compressive strength properties are seen to decrease almost linearly with increasing K15 content, though in SL150 mixtures strength remains relatively unchanged until around 30% filler loading. Above 30% filler loading SL150 mixtures are seen to rapidly loose compressive strength. Compressive strain properties are seen to decrease almost linearly with increasing filler content in both K15 and SL150 mixtures.

Experimental investigation of the effect of resin and filler type, particle size and particle strength on compressive properties is continuing.

4 DISCUSSION AND CONCLUSIONS

A new particulate filled resin system has been developed at USQ for use as a core/spacer material in fibre composite structural members. An experimental study into the static behaviour of this material has been initiated. Initial results on flexural and compressive behaviour have been presented in this paper. From these results it can be seen that resin and filler type, particle strength, particle size and filler loading level all influence the material's flexural and compressive properties.

Experimental data indicates that resin type only exerts a significant influence on flexural properties in situation of low filler loading. At filler levels over 40% test data indicates a convergence of properties for different resins. It may then be concluded that there is little benefit to be gained from using more expensive high-performance resins in this filler range.

The use of a general-purpose vinylester resin would result in significant economic benefits. However, despite testing indicating the strong flexural performance of these materials, there are several issues which must be addressed for successful exploitation of these resins. These issues include shrinkage and exotherm during cure and the compatibility of these materials with main laminates in structural sections.

Experimental data has shown E-Spheres based mixtures generally offering higher levels of structural performance than the glass microsphere blends. Data has also indicated that there is little difference between mixtures based on different E-Spheres grades. This is significant as E-Spheres offers important benefits in terms of both cost and environmental issues. E-Spheres are extracted from fly-ash, a power station by-product, and are thus relatively inexpensive compared with glass microspheres which must to be manufactured. The lack of variation between filler grades yields cost savings due to the price differences between grades (SLG = \$0.90/kg , SL75 = \$4.20/kg).

While several notable trends have been indicated by this experimental study, further research is required to improve understanding of the mechanisms involved in these behaviours. Further investigation into flexural, compressive and tensile behaviour is currently under way.

REFERENCES

Gilchrist M D, 1996, Mechanical Performance of Carbon Fibre and Glass Fibre Reinforced Epoxy I-Beams, *Composites Science and Technology*, Vol 56, No 1 : p 37 – 53.

ISO 178:1993 Plastics – Determination of Flexural Properties, International Organisation for Standardisation, Switzerland.

Maji A K et al, 1997, Evaluation of Pultruded FRP Composites for Structural Applications, *Journal of Materials in Civil Engineering*, Vol 9, No. 3 : p 154 – 158.

Mechanics of Structures and Materials, Bradford, Bridge & Foster (eds)
© 1999 Balkema, Rotterdam, ISBN 90 5809 107 4

Enhancing the performance of laminated veneer lumber beams with carbon fibre reinforcement

A.G.Greenland
University of Technology, Sydney, N.S.W., Australia
S.L.Bakoss
University of Technology, Sydney, Australia and Centre for Built Infrastructure Research, Sydney, N.S.W., Australia
K.I.Crews
Timber Engineering Studies, University of Technology, Sydney, N.S.W., Australia

ABSTRACT: Laminated Veneer Lumber (LVL) beams are being increasingly used in engineered timber structures and as part of flooring systems to replace unserviceable bridge decks and trafficable floors of warehouses, wharves and industrial structures. The serviceability and strength limit states of 'T'-beams manufactured using LVL webs can be significantly enhanced by the provision of carbon fibre tension reinforcement. This paper presents the results of a program of component and prototype testing which indicates that the application of carbon fibre composite (AFC) tensile reinforcement to LVL beams and 'T'-beams with laminated timber flanges and LVL webs can produce effective gains in ductility, stiffness, and strength. Predicted strength limit states and deflections are shown to agree with measured values. The influence of AFC reinforcement on the dynamic response of the beams tested is also presented. The systems described in the paper provides flexibility for designers for the rehabilitation and retrofitting of heavy duty flooring systems and bridge decks as well as for new construction. The system is readily adaptable to pre-fabrication or to site construction and can have considerable weight advantage over reinforced concrete or prestressed concrete alternatives.

1 INTRODUCTION

Laminated Veneer Lumber (LVL) beams are being increasingly used in engineered timber structures and as part of deck and flooring systems in the replacement of unserviceable bridge decks and trafficable floors of warehouses, wharves and industrial structures. Manufactured from renewable plantation timbers in a wide range of depths and lengths, LVL beams have considerably less variability in their engineering characteristics than is the case for solid timber beams.

LVL has been used effectively in Australia and elsewhere to longitudinally stiffen stress laminated timber bridge decks (see Figure 1). The resulting 'built-up' structural cross-section is commonly referred to as a 'T'-beam or 'stress laminated T-beam deck'. In this type of deck, deep LVL beams form webs separated by (smaller) flange laminates, such as solid radiata pine elements.

Used appropriately, 'stress laminated T-beam' decks and floors can offer several advantages over reinforced and prestressed concrete structures, particularly for the rehabilitation of existing timber structures, most notably cost and considerable weight savings.

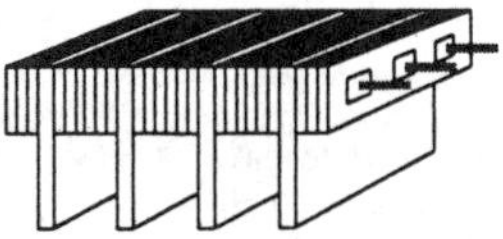

Figure 1: Conceptual sketch of Stress Laminated 'T'-beam Deck

The applications for Advanced Fibre Composites (AFC's) such as Carbon Fibre – Epoxy Composites in Civil Engineering applications are also rapidly expanding. While structural materials in their own right, many benefits from the use of AFC's to enhance the properties of "traditional" construction materials such as concrete, masonry, and timber can be realised.

The main applications of AFC's for the enhancement of 'traditional' materials to date has been in the rehabilitation of deteriorating reinforced concrete and masonry structures, and the retrofitting of such structures to improve their behaviour under seismic loads. Recently, renewed research and development efforts in Europe and the United States of America have investigated the reinforcement of timber with AFC's.

AFC reinforcements have the potential to increase the strength, stiffness, ductility, durability, and cost-effectiveness of engineered timber structures. Likely applications include use in the repair and rehabilitation of existing structures, use to enable smaller sections or longer spanning members, and use to enable lower grade or quality timbers such as plantation grown softwoods in applications that would otherwise require higher quality/grade timbers (University of Maine Advanced Engineered Wood Composites Center 1998).

The use of AFC reinforcements for structures in Australia is very new and their potential for rehabilitation and repair of timber structures is highly significant. For example, many of the estimated 10,000 timber bridges still in service in Australia are far older than their original life expectations and require major rehabilitation or replacement now or in the near future. Similarly, many other aging timber structures such as wharves, warehouses, and buildings are also in poor condition.

AFC reinforcements have been used in research and commercial applications to reinforce sawn and laminated rectangular beams. Research by Meier et al. 1996, Tingley et al. 1996, Chajes et al. 1996, University of Maine Advanced Engineered Wood Composites Center 1998, and work by the Authors 1998 has demonstrated that attaching AFC's to the tensile face (or tensile and compressive faces) can significantly increase the bending strength and stiffness of timber beams.

There is also considerable scope for AFC's to compliment stress laminated timber (SLT) technology, which is now well established in Australia and has been used successfully for a number of prototype timber bridge rehabilitations (Greenland 1996, Crews et al 1996). AFC's have been used in place of steel prestressing bars and strands in SLT decks, and current research indicates that externally bonded AFC reinforcements have considerable potential to improve the tensile capacity of SLT T-beam webs.

Both SLT techniques and the application of AFC reinforcements to timber structures are readily adaptable to prefabrication or site construction.

2 UTS TESTING PROGRAM

Since late 1996, the University of Technology, Sydney has been undertaking a research program to characterise the response of AFC tensile reinforced timber when subjected to bending. The research program has included determination of the tensile properties of the AFC's and flexural properties of the timbers used in experiments, and characterisation of the response of sawn rectangular beams and prototype size LVL beams subject to bending when reinforced with various proportions of AFC reinforcements. Tests examining the short term effects of moisture on reinforced and unreinforced beams were also conducted. The materials used in research at UTS to date have been limited to Australian Radiata Pine products and Carbon Fibre – Epoxy Composites.

This paper reports the short term behaviour of full size LVL and full size stress laminated timber T-beams constructed of LVL webs and solid radiata pine flanges with and without AFC web reinforcement.

2.1 *AFC Properties*

The reinforcements for the LVL and stress laminated T-beams tested to date have utilised a unidirectional carbon fibre-epoxy composite material manufactured by Mitsubishi Chemical Co. Mitsubishi Replark is a proprietary, carbon fibre-epoxy prepreg system, that is applied in conjunction with their proprietary epoxy resin to achieve cure without the application of pressure and at ambient temperatures. Each ply of Replark used contained 200g/m^2 of carbon fibres. The tensile properties of Replark, as stated by the manufacturer and as determined in accordance with ASTM D3039-95a, are presented in Table 1.

2.2 *LVL Beams*

Laminated Veneer Lumber (LVL) specimens, 400mm deep by 45mm thick, with spans of approximately 6 metres, were tested for strength and stiffness. Beams reinforced with a 0.15% proportion of carbon fibre area to timber area were compared against beams tested without reinforcement. The short term effects of moisture on reinforced and unreinforced beams were also investigated.

The reinforcement configuration of these beams is illustrated in Figure 2.

For the purposes of this study, 'dry' behaviour is defined as the behaviour of a beam when its moisture content (MC) is at or near its equilibrium moisture content (EMC). 'Wet' behaviour is defined as the behaviour of a beam when its moisture content approaches fibre saturation.

2.3 *T Beam Specimens*

Six timber 'T'-beams, constructed with LVL webs and flanges built up from solid radiata pine beams have been constructed and tested, three with reinforcement and three without reinforcement. Composite action between the web and the flanges was maintained by transversely prestressing the beams at 800mm centres. The general arrangement of the beams is shown in Figure 3. Not all prestressing bars are shown in this figure. Similarly to the LVL beams, the 'T'-beams were reinforced on the sides within the tensile zone rather than on the bottom face.

Each beam comprised a 400mm x 65mm LVL web in between six 140mm x 45mm sawn radiata pines laminates forming the flange. Each beam spanned 6.1 metres and was loaded at third points.

Three 'T'-beams were progressively reinforced with the Mitsubishi composite so the effect of various proportions of AFC to timber on stiffness could be investigated. Proportions of reinforcements tested include (approximately) 0.13%, 0.2%, and 0.3% respectively.

Table 1. Mitsubishi Replark Type 20

Design Properties		Properties as Tested	
Design Thickness	0.11mm	Design Thickness	0.11mm
Tensile Strength	>3,000 MPa	Tensile Strength	
Tensile Modulus	235 GPa	5th percentile	3,400 MPa
		CoV	9.0 %
		Mean	3,800 MPa
		Tensile Modulus	
		5th percentile	223,000 MPa
		CoV	8.0 %
		Mean	255,000 MPa

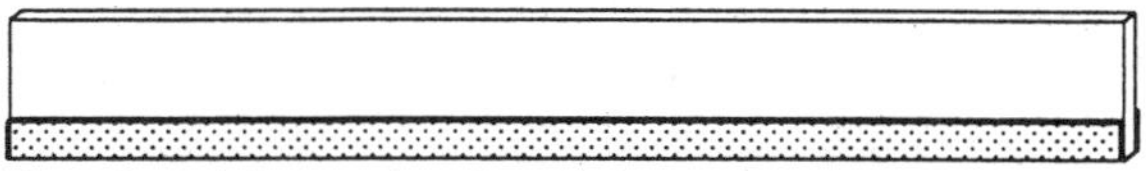
Figure 2: Reinforcement configuration for LVL beams

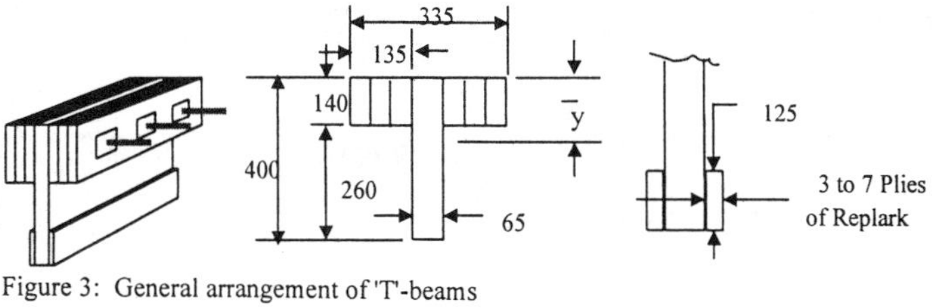

Figure 3: General arrangement of 'T'-beams

3 MODELLING BEAM BEHAVIOUR

Within the service load range, the load-deflection response of both LVL and reinforced LVL beams, and 'T'-beams and reinforced 'T'-beams is linear-elastic. This type of behaviour can be modelled using a transformed sections analysis and well established, linear relationships.

However, beyond the service load range, the behaviour of reinforced beams can be non-linear.

3.1 *Reinforced LVL Beams*

There is scope for non-linear behaviour in reinforced, rectangular beams due to 'compressive wrinkling' or 'compressive yielding' and 'progressive tensile failure'.

'Compressive wrinkling' or 'compressive yielding' occurs when compressive strains in the beam exceed the 'compressive yield strain' of the timber. Bodig et al. 1982, Tingley et al. 1996, Hernadez et al. 1997, and the authors 1999 have previously described non-linear load-deflection behaviour due to this effect. It is assumed that once the 'compressive yield strain' of the timber is exceeded, the stress in that part of the cross-section remains constant (where the stress-strain relationship for the timber can be reasonably approximated by a bi-linear relationship).

'Progressive tensile failure' may occur if the tensile capacity of the AFC allows the moment capacity of the reinforced section to increase beyond the moment which would cause the timber to crack or fail in tension. As the tensile strain at failure of the timber is typically significantly less than the tensile strain at failure of an AFC, the section may continue to carry load after the timber has started to fail in tension, leading to more ductile failures.

In a discretised model, used by the authors to examine beam behaviour beyond the service-load range, 'compressive yielding' and progressive tensile failure' can be illustrated by Figure 4.

This model calculates the stress in each discrete "strip" of a section, based on a linear strain distribution and known stress-strain relationships. The model assumes timber has a bi-linear stress-strain relationship in compression, and a linear-elastic stress-strain relationship in tension (Bodig et al. 1982). The stress-strain relationship for a unidirectional AFC is linear-elastic to failure. For each strip of timber, the model allows compressive stresses to remain constant if the strain in that strip exceeds a "compressive yielding strain", and tensile stresses to be zero if the strain in that strip exceeds the "ultimate tensile strain" of the timber. The process is iterative – the position of the neutral axis determines the strain distribution, the strain distribution determines the stress distribution, and the stress resultants must satisfy internal equilibrium requirements.

3.2 *Reinforced 'T'-Beams*

Some scope exists for non-linear behaviour of 'T'-beams where the AFC can allow progressive tensile failure of the timber web. However, the 'compressive yielding' observed in tests on rectangular beams was not observed in the 'T'-beams due to the large area of the flange and consequent position of the neutral axis in a T-beam. This results in lower strains in the flange relative to those in the web, generally resulting in the tensile strength of the web being reached well before "wrinkling" of the flange can commence.

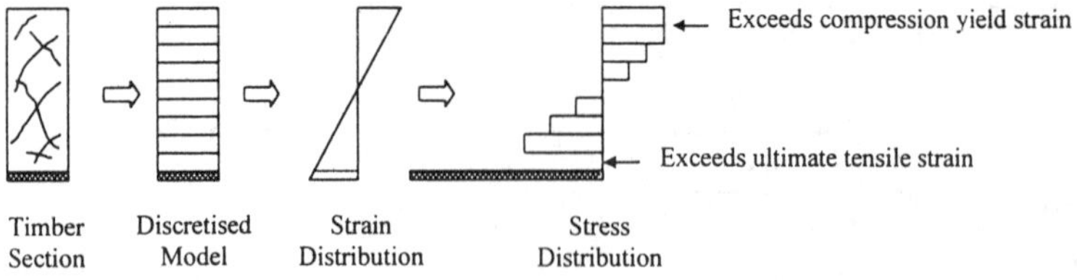

Figure 4: Discretised model of a reinforced rectangular beam

Some non-linear load-deformation behaviour may also be observed in T-beam decks due to load sharing between webs, and inter-laminate slip between flange elements or between flanges and webs.

The discretised model used to predict beam behaviour beyond the service-load range is shown in Figure 5.

Failure of AFC reinforced SLT T-beams is expected to occur with tensile failure of either the AFC, the timber web, or a combination of both. In addition to having some scope to delay the onset of failure by shifting the position of the neutral axis, the AFC has the potential to redistribute stresses as tensile cracking of the timber web progresses, increasing the capacity and ductility of the beam.

4 EXPERIMENTAL PROCEDURE

Bending tests were conducted using a four point loading configuration based on AS/NZ 4063:1992 "Timber – Stress Graded – Ingrade Strength and Stiffness Evaluation" (Standards Australia /Standards New Zealand 1992).

4.1 *LVL and Reinforced LVL beams*

Each LVL and Reinforced LVL beam was tested over a span of six metres, and loaded at third points. The flexural strength and stiffness of the reinforced beams was compared against the flexural strength and stiffness of (non-reinforced) LVL beams. The test setup is illustrated in Figure 6. Lateral supports have been omitted for clarity.

4.2 *'T'-beams and Reinforced 'T'-beams*

Each 'T' beam and Reinforced 'T'-beam spanned 6.1 metres and was loaded at third points.

The flexural strength and stiffness for reinforced 'T'-beams was compared against:

- the stiffness of the 'T'-beams prior to being reinforced
- the strength of the 'control' or unreinforced 'T'-beams.

Due to the small sample sizes, results for each beam are presented in Tables 4-6. The calculated mean and coefficient of variation of results are also presented.

Limited dynamic response tests were also conducted to quantify changes in natural frequency and damping characteristics brought about by the addition of the composite reinforcement.

The dynamic tests were conducted by measuring the vertical accelerations at mid span of the beams with an accelerometer, following the sudden release of a displacement. The beams were simply supported on knife-edge supports.

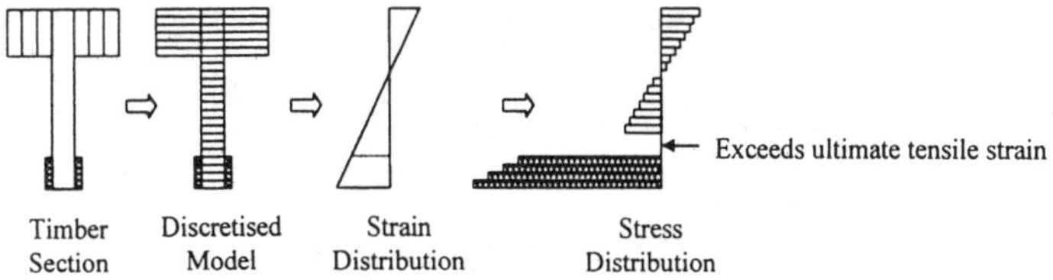

Figure 5: Discretised model of reinforced 'T'-beams

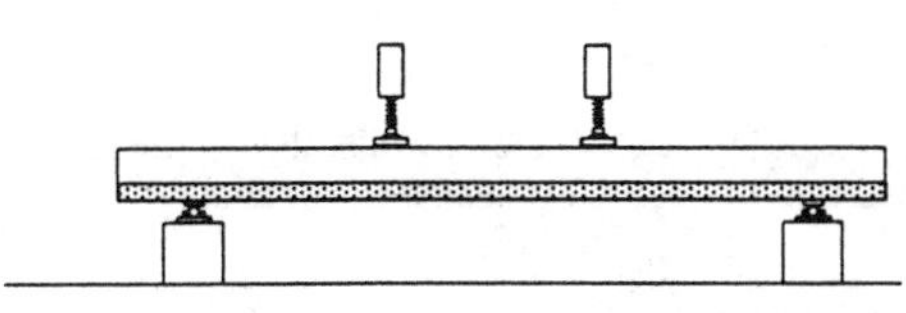

Figure 6: Test setup for LVL and Reinforced LVL Beams

Table 2: Results of testing LVL and Reinforced LVL

LVL 400mm x 45mm x 6,000mm			
Specimen Description	Mean	CoV %	5th %tile
Unreinforced Bending Modulus (MPa)	12,900	7	11,800
Unreinforced Ultimate Moment (kNm)	63	9	48
Reinforced Bending Modulus (MPa)	13,200	5	12,400
Reinforced Ultimate Moment (kNm)	74	6	62

Table 3: Progressive stiffness increase in reinforced 'T'-beams

Beam	0.13% (6 plies) Re-plark	0.2% (10 plies) Replark	0.3% (14 plies) Replark
Beam 3	12%	16%	28%
Beam 4	10%	16%	*na*
Beam 6	14%	22%	31%
Average	12%	18%	30%

Table 4: Moment Capacity of 'T'-beams and reinforced 'T'-beams

Unreinforced Beams		Reinforced Beams	
Beam	Ultimate Moment (kNm)	Beam	Ultimate Moment (kNm)
Beam 1	129.2	Beam 3	192.7
Beam 2	150.4	Beam 4	*na*
Beam 5	104.4	Beam 6	197.6
Average	128.0	Average	195.2
CoV	18%	CoV	2%

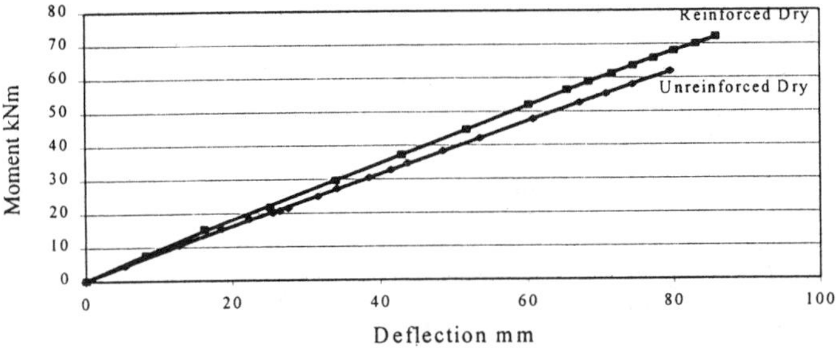

Figure 7: Typical load-deflection behaviour of LVL and reinforced LVL beams

5 EXPERIMENTAL RESULTS

5.1 *LVL and Reinforced LVL*

Summary results of the testing of LVL beams are presented in Table 2, whilst the typical load deformation response of both the unreinforced and reinforced LVL beams tested under 4 point bending, is shown in Figure 7.

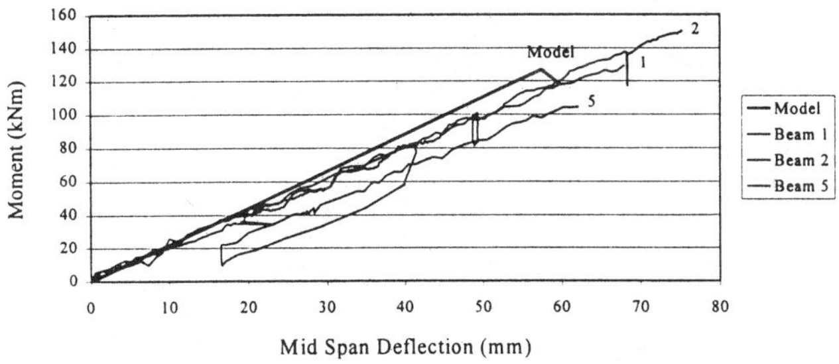

Figure 8: Load-deflection response of (unreinforced) 'T'-beams

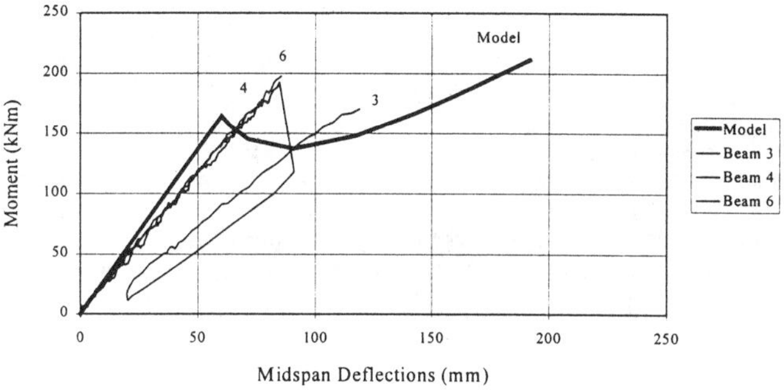

Figure 9: Load-deflection response of reinforced 'T'-beams

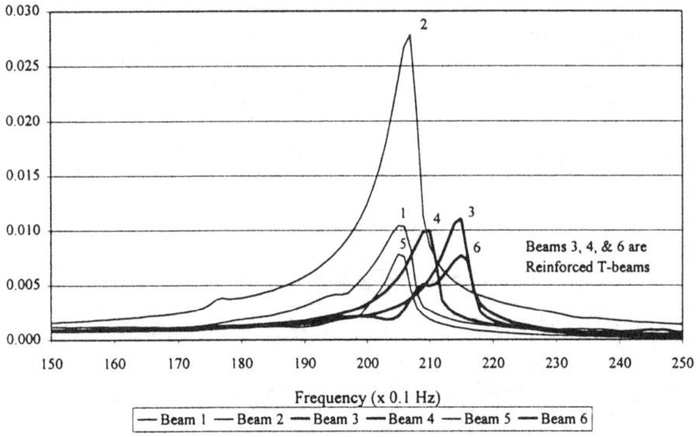

Figure 10: Dynamic Response of 'T'-beams and reinforced 'T'-beams

Table 5: Natural Frequencies of 'T'-beams and reinforced 'T'-beams tested

Unreinforced Beams		Reinforced Beams	
Beam	Natural Frequency (Hz)	Beam	Natural Frequency (Hz)
Beam 1	20.6	Beam 3	21.5
Beam 2	20.7	Beam 4	21.0
Beam 5	20.5	Beam 6	21.5
Average	20.6	Average	21.3
CoV	0.5%	CoV	1.4%

5.2 *T Beams and AFC Reinforced T Beams*

Stiffness increases for the three reinforced T-beams as they were progressively reinforced from unreinforced to 0.3% reinforcement are presented in Table 3. The moment capacity of reinforced beams are compared to the moment capacity of unreinforced beams in Table 4.

Moment-Mid span deflection for the T-beams and the reinforced T-beams are shown in Figure 8 and Figure 9 respectively.

5.3 *Dynamic Response of T Beams and AFC Reinforced T Beams*

Changes in the fundamental natural frequency are summarised in Table 5 and illustrated in Figure 10.

6 DISCUSSION AND REVIEW OF RESULTS

6.1 *LVL and Reinforced LVL*

Reinforcing LVL with a ratio of 0.15% carbon fibres resulted in an indicative improvement in the bending stiffness or effective MOE of the LVL beams of approximately 5% (Table 2). The reduction in variability (COV) was of the order of 30%.

Bending strength was improved by up to 30 percent at the 5th percentile level. Variability (COV) of bending strength was reduced by one third. The mode of failure was predominantly tensile.

While strength and stiffness was improved by such small proportions of AFC, little observable effect on ductility was apparent (Figure 7). Variability improvements of approximately 40% were discerned from these results.

6.2 *T Beams and Reinforced T-Beams*

Results of stiffness tests of T-beams indicates significant increases in beam stiffness can be brought about with small proportions of AFC reinforcements (Table 3). Average stiffness increases of approximately 12%, 18%, and 30% are indicated with reinforcement proportions of 0.13%, 0.2% and 0.3% respectively.

The reinforcement of stress laminated timber T-beams with a 0.3% reinforcement proportion produced an average increase in beam bending strength of 52% (Table 4). Beam 4 failed prematurely through loss of composite action between the web and the flanges at a bending moment of 178.4 kNm and so was excluded from the average.

Composite action between the flange elements and the webs was maintained in beams 3 and 6 through to failure.

Behaviour of beam 6 was principally linear elastic through to failure. Three events appeared to occur simultaneously on failure of beam 6. (1) The timber web cracked in tension, (2) the fibre composite delaminated from one end of the beam, and (3) the fibre composite ruptured near the area that the timber cracked.

Beam 3 behaved linearly elastically through until the timber web cracked in tension. Some localised debonding of the composite either side of the crack occurred when the timber cracked (at a moment of 192.7kNm). Due to the nature of the hydraulic loading arrangement, hydraulic pressure, and consequently applied load, dropped at this point. Failure of the composite followed later as the load was brought up almost to the previous level (bending moment of 170.5kNm). On failure of the composite, delamination occurred from one end of the beam, followed by fracture of the composite near the area that the timber cracked. The final mid span deflection of the beam was 119mm, a large increase on the 85mm at the initial failure.

More reinforced T-beams are currently being tested to confirm the failure sequence and provide more data.

6.3 *Dynamic Response of T-Beams and Reinforced T-Beams*

The results in Table 5 indicate that the reinforcement of T-beams with 0.3% carbon fibre increases the fundamental frequency of the beams by an average of 0.7Hz, or approximately 3.5%.

7 CONCLUSIONS

The results of testing to date indicate that the provision of small proportions of AFC reinforcements can significantly enhance the serviceability and strength limit states of stress laminated 'T beam' bridge decks and floors.

Non-linear numerical modelling also suggests that the application of these materials within the tensile region of the web can significantly improve the ductility of these beams. Current tests at UTS aim to verify these ductility improvements.

While the behaviour of AFC reinforced SLT 'T-beams' under long-term loading needs to be investigated, preliminary investigation by overseas researchers suggest that the long term performance of the these beams will be satisfactory (University of Maine Advanced Engineered Wood Composites Center 1998, Tingley et al. 1996, Plevris et al. 1995).

Overseas experience also indicates that AFC reinforcements can be cost-effective for beams used in both new construction and rehabilitation applications. As demand for AFC's increase, particularly for civil infrastructure applications, the cost-effectiveness of these reinforcements should also increase dramatically (Tingley et al. 1996).

REFERENCES

Bodig J, Jayne BA 1982, "*Mechanics of Wood and Wood Composites*", Van Nostrand Reinhold, New York; London. 1982.

Chajes M, Kaliakin V, Meyer Jr. A 1996, "*Behaviour of Engineered Wood-CFRP Beams*", International Conference on Composites in Infrastructure, 1996.

Crews KI, Walter G 1996, "*Five Years of Stress Laminated Timber Bridges in Australia – A Review of Development and Application*", International Conference on Wood Engineering – October 1996, New Orleans, LA, USA.

Greenland AG 1996, "*Cost Effective Rehabilitation of a Reinforced Concrete Bridge with Timber*", Institute of Municipal Engineers Australia New South Wales Division Annual Conference, Sydney, Australia, March 1996.

Greenland A, Crews K, Bakoss S 1997, "*Enhancing Timber Structures with Advanced Fibre Reinforced Composite Reinforcements*", 5th World Conference on Timber Engineering, Lausanne-Montreux, Switzerland, August 17-20, 1997.

Greenland A, Bakoss S, Crews K 1999, "*Enhancing the Flexural Behaviour of Stress Laminated T-Beam Bridge Decks and Industrial Floors Using Advanced Fibre Composite Reinforcements* ", Submitted for publication, Faculty of Engineering, Univesity of Technology, Sydney. March 1999.

Greenland A, Crews K, Bakoss S 1999, "*Application of Advanced Fibre Composite Reinforcements to Structural Timber* ", Pacific Timber Engineering Conference, Rotorua, New Zealand, March 15-19, 1999.

Hernadez R, Davalos JF, Sonti SS, Kim Y, Moody RC 1997, "*Strength and Stiffness of Reinforced Yellow-Poplar Glued-Laminated Beams*" United States Department of Agriculture Forest Products Laboratory, Research Paper FPL-RP-554, July 1997.

Meier U, Deuring M, Meier H, Schwegler G 1996, "*Strengthening of Structures with Advanced Composites*", EMPA Swiss Federal Laboratories for Material Testing and Research, Duebendorf, Switzerland.

Plevris and Triantafillou 1995, "*Creep Behaviour of FRP-Reinforced Wood Members*", Journal of Structural Engineering, American Society of Civil Engineers, February 1995.

Standards Australia / Standards New Zealand 1992, *AS/NZ 4063:1992 "Timber - Stress Graded – In-grade Strength and Stiffness Evaluation"*, Standards Australia, Homebush NSW, Australia.

Tingley DA, Cegelka S 1996, "*High-Strength-Fiber-Reinforced-Plastic Reinforced Wood*", International Wood Engineering Conference, New Orleans, October 1996.

University of Maine, Advanced Engineered Wood Composites Center 1998, http://www.aewc.um.maine.edu/awec/ (March 1998).

US Department of Agriculture Forest Service, Forest Products Laboratory 1998, "*Wood Transportation Structures Research*", http://www.fpl.fs.fed.us/wit/ (March 1998)

Mechanics of Structures and Materials, Bradford, Bridge & Foster (eds)

Rapidwall – A new wall technology

G.W.Wyett
Rapidwall, Sydney, N.S.W., Australia
D.R.Mahaffey
Mahaffey Associates, Consulting Engineers, Sydney, N.S.W., Australia

ABSTRACT: RAPIDWALL combines the benefits of precast, plasterboard linings and hollow blockwork in a unique wall panel system developed in Australia. RAPIDWALL is a hollow panel manufactured 12 m x 2.85 m x 120 mm thick from GYPCRETE. Gypcrete is a mixture of gypsum plaster, glass rovings and waterproofing additives to create a panel which is fire rated, load bearing, sound rated, termite resistant, water resistant and easily placed using small cranes or forklifts. 46,000 sq metres of RAPIDWALL has been placed in an eight storey Sydney residential development. The use of prefabricated RAPIDWALL panels together with a precast floor system has resulted in fast erection times and early access to following trades. Prototype testing in accordance with A3600 has been carried out to determine RAPIDWALL's load bearing capacity. The test method and results form part of the paper.

1. INTRODUCTION

A series of full scale tests were conducted on 120mm thick x 1.0 metre wide RAPIDWALL panels in accordance with the prototype testing requirements of AS3600 'Australian Concrete Code'. Concentric and eccentric loads were applied to varying height unfilled or concrete filled panels. Typical 1/3 point loads were applied to horizontal panels to determine the panel's flexural strength. The test data was reviewed and a series of load capacity design equations developed to allow for variation in:

(a) panel height
(b) concrete strength
(c) eccentricity

The test results complement earlier work on concrete filled 100mm thick panels and have allowed the RAPIDWALL to be used as a load bearing wall in buildings to 9 storeys.

2. BACKGROUND

RAPIDWALL was originally developed as a low cost, plaster panel primarily for the non-load bearing partition market. Subsequent improvements produced a wall panel suitable as a load bearing single skin external/internal wall capable of two storey residential construction.

Strong interest was generated overseas in the area of low cost housing due to the relatively low level of capital required to establish a factory capable of producing 500,000 sp metres/year in a two shift operation. The ability to sell the technology overseas was hampered by the lack of major high profile projects particularly on the eastern seaboard.

In early 1997 the specific markets of cinema acoustic division walls and load bearing party walls for medium-rise residential construction were earmarked as growth areas. Additional sound, load and fire testing was commissioned to specifically produce two systems:

(a) A load bearing wall capable of meeting the BCA requirement for residential unit party walls and load bearing to 5 storeys.

(b) A cinema division was which was light, fast to erect and had a minimum 65STC sound rating and a 3 hr fire rating.

3. RAPIDWALL DESCRIPTION

RAPIDWALL is a precast, hollow wall, manufactured in a patented process from GYPCRETE. GYPCRETE is a mixture of gypsum plaster, glass rovings, waterproofing chemicals and various minor additives. The process produces a panel which is 12.0 m long x 2.85 high x 120 m thick and weighs 38 kg/sq metre.

The panels consist of two skins 13 mm thick formed together with a 20 mm rib at 250 mm centres. The panel is cast on a flat table with the voids being formed with collapsible plugs which are removed once the plaster has reached initial set. Thirty minutes after the first mixing of plaster powder and water the casting table is tilted and the 12.0 m panel is picked up by a forklift and placed in a rack for drying. Panels are subsequently cut to length and height using a computer controlled saw.

4. LOAD BEARING WALL SYSTEM

A RAPIDWALL panel filled with 25 to 40 MPa concrete provides a pre-finished, load bearing panel wall which meets the BCA requirement of a 45STC sound rating and a 90/90/90 FRL fire rating. At a wall spacing of 5.5m the load bearing capacity is 8 storeys. At lower levels of load the fire rating is 4 hours.

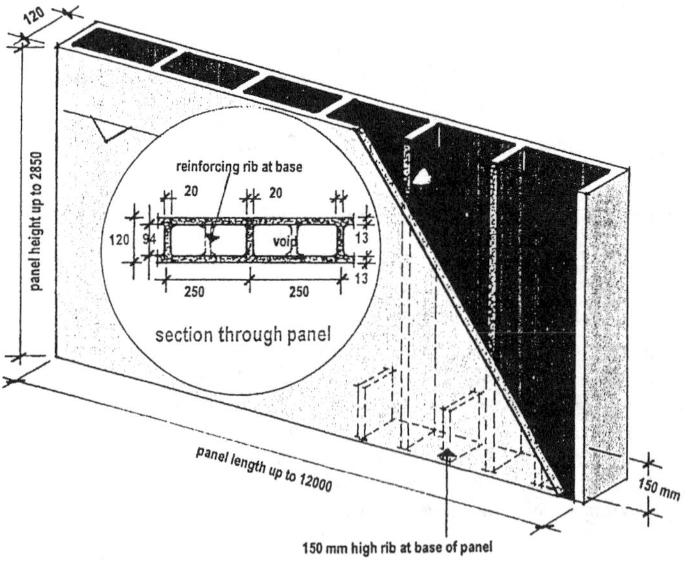

Figure 1. Typical panel.

Following the production of detailed shop drawings similar to precast concrete the wall layout is marked out on the concrete slab using chalk lines and survey pins. A stillage (frame carrying 8 panels) is loaded onto the deck at strategic locations.

The same crane or a smaller floor crane is then used to lift the panel into position. Since typically 30 panels/day can be lifted into position the system is most suited to long straight walls. Generally other non-load bearing walls less than 1.0 m would be done in traditional materials. The panels are braced by cross walls or raking props at 4.0 m centres.

RAPIDWALL can be used with traditional formwork or floor systems such as Ultrafloor, Bondek or Smartfloor. These lost formwork systems are typically faster and more economical than traditional formwork.

The completed formwork provides access to concrete fill the walls. The wall is first filled to 500 mm and after initial set of this layer filling is completed on the same day. Since the unfilled wall is load bearing, systems such as Ultrafloor can be achieved without propping. Following reinforcement placement to the floor and concrete pouring the cycle begins again. The only reinforcement required for the RAPIDWALL is typically a dowel, top and bottom at 1.0 centres.

5. CINEMA DIVISION WALL

A 65 STC division wall is achieved by installing a double wall of 6.0 m long x 2.85 m high x 120 mm thick RAPIDWALL panels on a light frame of steel columns and girts. The voids of the RAPIDWALL are prefilled with recycled cellulose fibre insulation with fire retardant. A 100 mm blanket is fixed to the cavity face of one wall assembly.

A 73 STC sound rating can be achieved with the addition of plasterboard to one side only. The steel frame and panels are erected using forklifts and scissor lifts with the elimination of scaffold.

6. SYDNEY PARK VILLAGE – CASE STUDY

The project consisted of 370 units to 8 storeys high and used 46,000 square metres of concrete filled RAPIDWALL.

The project was split into two approximately equal stages. The walls to Stage 1 commenced in late July 1997 and were completed by December 1997. The walls to Stage 2 commenced in December of 1997 and were completed prior to Easter of 1998. A number of problems were encountered in Stage 1 which were resolved in Stage 2. The use of RAPIDWALL and Ultrafloor saved 3 months on the Head Contract program for both Stages 1 and 2. The key issues learnt from the project were:

1. Accurate Architectural drawings with walls dimensioned on centreline and related to grid lines are essential.
2. Panel erection quantities are almost entirely dependent on continuous crane availability and duration.
3. Floors must be cast to the relevant tolerances in the Australian Standard.
4. A stronger panel capable of load bearing to plus 8 storeys and capable of being filled in a single lift was desirable.
5. Internal walls should be delivered with both sides finished to minimize field setting of the 'B-side'.
6. The system is best suited to wall lengths greater than 2.0 m.

7. TESTING

In principal, the testing was carried out in accordance with clause 21.2 of AS3600, "Prototype Testing". Whilst there is no specific provision for the testing of walls in this section of the concrete structures code, guidance can be provided in the general principles that are set out therein.

The purpose of this testing was to determine the load carrying capacity of the RAPIDWALL panels. Various tests were conducted which determined the respective longitudinal and transverse strengths of filled and unfilled RAPIDWALL panels. The longitudinal tests were conducted with various load eccentricities.

The steps in the procedure were as follows.

1. Setting Up the Panels
2. Filling with Concrete
3. Establishing Concrete Properties
4. Determining the Test Ages
5. Carrying Out the Load Testing At the Appropriate Test Ages

The details of each phase of the testing were as follows.

7.1 Setting up the panels

The panels were supplied and erected by Australasian Concrete Services (ACS). Testing was carried out at the laboratory of Mahaffey Associates (MA). The panels were erected so that they were vertical and level, to ensure that they were filled in even layers, and that the concrete did not slope on any of the cast surfaces. The test panels were nominally 2.85 m high or 2.50 m and 1.03 m wide, and they were erected with the cores vertical. For those panels which were filled, the concrete was placed into them from the top.

7.2 Filling the panels

The panels were filled by an experienced concreting subcontractor. The concrete for the filling was a nominal 32 MPa concrete with a sufficient flow to ensure that the concrete could be well compacted in the panels.

The concrete was placed into the panels in three layers. This is because it was thought that the hydrostatic pressure of concrete that is placed the full height of the panels in one lift could be enough to break the walls of the cores at the base of the panels.

7.3 Determining the test ages

Panel tests were carried out when the air cured cylinders reached a strength of 25 MPa.

7.4 Carrying out load testing at appropriate ages

On the day that the required strength was achieved, load testing was commenced.
Longitudinal testing was carried out in a purpose built load frame that was designed with the following parameters in mind:

The frame was able to apply a load to the panels such that a vertical load of 1600 kN could be applied.

The load was to be applied to the test panels such that it was uniformly distributed along both the bottom and top edges. To this end, the loading and bearing members were sufficiently stiff such that the deflection during the test was effectively zero.

The frame was constructed such that the load could be applied with various eccentricities. The load frame was designed by Taylor Lauder Bersten (TLB) to achieve the strength and deflection requirements of the test program. Details of the frame as constructed are shown in Figure 2. The members used were selected from available materials such that they provided equivalent strength and stiffness to the members designed by TLB.

The frame was designed for the panels to sit vertically, and when required, the eccentricity was achieved by placing spacers between the back stops of the load frame and the panels. The loads were therefore applied through the axis of the load frame, but eccentric to the panels.

The load was applied to the panels using two 60 tonne, or two 100 tonne hydraulic rams placed at third points along the loading beam. These were actuated by a hand operated two stage pump. The force was measured using pressure gauges that measure the oil pressure in the system, and this pressure was converted to force using calibration curves. These curves were developed by placing the rams into an "A" grade laboratory testing machine and registering the forces that coincide with particular pressures.

The upper and lower faces of the test panels were uneven due to the cutting of the panels and the finishing of the concrete. To ensure that even bearing was provided to these surfaces throughout the testing, strips of pyneboard were placed between the ends of the panel under test, and the 125 mm by 19 mm loading bars of the load frame.

Transverse testing was carried out with the panels lying horizontally. A purpose built test rig was used for this testing. A span of 2700 mm was used to support the panel, whilst the load was applied over two bars located below the panel. These bars were spaced 1000 mm apart, and pushed upward on the lower face of the panel.

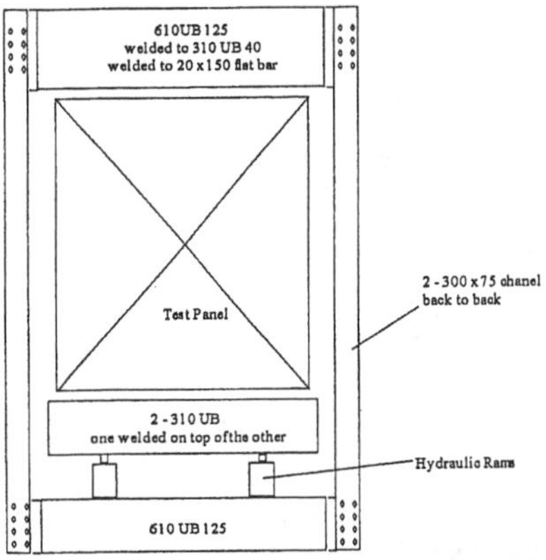

Figure 2. Load Frame Used for Testing Wall Panels

For both transverse and longitudinal tests, the load was applied to the panels at a steady uniform rate until either the panel failed and could not sustain any further load, or the capacity of the loading equipment was reached. Full records of all test data were kept, including –

- Date of Test
- Age of concrete at test
- Strength of concrete at test
- Identification of hydraulic loading system components
- Load achieved at failure (or statement that maximum capacity of the test equipment was applied without physical evidence of failure)
- If failure occurred, the form of the failure.

8. LOAD TEST RESULTS

The results of the load testing are shown in Table 1.

Table 1. Load Test Results

Test No.	Max Load (kN)	Broke or Held	Concrete Strength (MPa)	Panel Height (mm)	Filled or Unfilled	Type Of Test	Eccen-tricity (mm)	Mode of Failure
1	1650	held	37.0	2850	filled	Longitud.	0	1
2	1650	held	37.0	2850	filled	Longitud.	0	1
3	1650	held	35.5	2850	filled	Longitud.	0	1
4	770	broke	37.5	2850	filled	Longitud.	20	2
5	820	broke	38.5	2850	filled	Longitud.	20	2
6	1010	broke	37.5	2850	filled	Longitud.	20	2
7	1110	held	-	2850	filled	Longitud.	20	1
8	830	broke	-	2850	filled	Longitud.	20	3
9	800	held	-	2500	filled	Longitud.	0	4
10	920	broke	35.0	2500	filled	Longitud.	20	2
11	890	broke	28.0	2500	filled	Longitud.	20	2
12	766	broke	29.0	2500	filled	Longitud.	20	2
13	86	broke	na	2850	unfilled	Longitud.	10	5
14	145	broke	na	2500	unfilled	Longitud.	0	5
15	151	broke	na	2850	unfilled	Longitud.	0	6
16	28.6	broke	39.0	2850	filled	Transverse	Na	2
17	28.6	broke	42.5	2850	filled	Transverse	Na	2
18	18.1	broke	41.0	2850	filled	Transverse	20	2

Legend: 1. na 2. Broke approximately mid span 3. Diagonal shear of top 100 mm 4. Concrete permanently deformed 3-4mm under top loading bar. 5. Local crushing top and bottom. 6. Failed in compression 300 mm from bottom.

9. FUTURE DIRECTIONS

The RAPIDWALL system is being continually researched and tested to find new uses and improve panel qualities. Some of the matters under research include:

(a) Production of a double A-sided panels within the forming process.
(b) Increasing surface hardness of the panel.
(c) Alternative materials to Gypsum plaster for panel production

Future products include:

(a) A 150mm thick panel for + 10 storey construction.
(b) A 90 mm thick panel for the domestic market.

10. CONCLUSION

Pre-fabricated structural systems which minimise on-site labour are popular in regions with high labour costs or shortage of skilled tradesman. The lightweight properties of RAPIDWALL allow complete walls to be erected within minimum labour and modest craneage. The prefinished panel combined with the speed of construction allows significant savings in construction time over traditional methods of construction.

Design codes

Mechanics of Structures and Materials, Bradford, Bridge & Foster (eds)
© 1999 Balkema, Rotterdam, ISBN 90 5809 107 4

Aspects of code development for the assessment of existing structures

D.V.Val
School of Engineering, James Cook University, Townsville, Qld, Australia
M.G.Stewart
Department of Civil, Surveying and Environmental Engineering, University of Newcastle, N.S.W., Australia

ABSTRACT: As civil infrastructure is ageing, the assessment of existing structures is becoming increasingly important. Current building codes have been developed for new design and are not efficient for assessment because of significant differences between these two situations. One of the main differences is the possibility to update information about an existing structure. The paper proposes how updating can be taken into account in the context of current code safety formats such as the LRFD and partial factor formats. Different sources of information for updating including on-site inspection, proof load testing and past performance of a structure are considered.

1 INTRODUCTION

Assessment of existing structures is usually required if there is a change of ownership or structural use, evident signs of deterioration, or as part of regular monitoring program. As civil infrastructure is ageing, the assessment of existing structures is becoming increasingly important. At the same time current building codes have been explicitly developed for new design and so are not appropriate for assessment. There are significant differences between these two situations. First, there are different uncertainties associated with design and assessment. In design, uncertainties arise from the prediction *a priori* of load and resistance parameters of a new structure. In assessment an existing structure can be inspected/tested, so that these parameters can be measured on-site. However, other uncertainties then arise from on-site inspection/testing. Second, conservative design does not result in a significant increase in structural cost, while a conservative assessment may result in unnecessary and costly repairs or replacement.

Therefore, there is a clear need for technical rules to be developed specifically for the assessment of existing structures. Currently, deterministically based guidelines are available for structural assessment (e.g., ISE 1980), and work is now in progress to develop reliability- based (or probabilistic) code type documents (e.g., Allen 1991, Vrouwenvelder 1997, ISO 1998a). The present paper considers aspects of the code development related to the treatment of additional information on structural resistance, which can be obtained from inspection, testing or past performance of a structure.

2 CODE FORMAT

It is important that codes for structural assessment be compatible with codes for new design, i.e., be based on limit state analysis and safety factor format. In this context, the criterion for assessment can be generally formulated as (ISO 1998b)

$$R_d \geq S_d \tag{1}$$

where R_d is the design resistance and S_d the design action effect. The resistance side of Eq. (1) can be represented in one of the two formats:

$$\text{LRFD Format: } R_d = \phi R_n(f_y, f_c', \ldots) \qquad \text{Partial Factor Format: } R_d = R(\phi_s f_y, \phi_c f_c', \ldots) \tag{2}$$

where R_n is the nominal (characteristic) resistance, ϕ an overall capacity (strength) reduction factor, f_y and f_c' characteristic values of steel and concrete strengths, and ϕ_s and ϕ_c partial reduction (safety) factors for these strengths. The load and resistance factor design (LRFD) format is adopted in Australia and in the USA, while the partial factor format is used in Europe (it is more conventional to define the partial factors as $\gamma_s = 1/\phi_s$ and $\gamma_c = 1/\phi_c$). Both the formats are similar, although the overall resistance factor cannot account for the relative importance of different constituents of a composite material as efficiently as the partial factors.

The characteristic resistance R_n (usually represents 0.05 fractile) and the design value R_d can be determined as (ISO 1998b)

$$R_n = F_R^{-1}(0.05) \qquad R_d = F_R^{-1}[\Phi(-\alpha_R \beta)] \tag{3}$$

where F_R is a distribution function of resistance, Φ the standard normal distribution function, α_R the sensitivity factor and β the target reliability index. Generally, the values of α_R should be found from detailed reliability analysis (e.g., Thoft-Christensen & Baker 1982). However, this is out of the scope of this study and a standardised value $\alpha_R = 0.8$ (ISO 1998b) is used further in the paper. The target reliability index for existing structures may differ from that used in design (e.g., Allen 1991). However, this problem is not considered herein and it is assumed that $\beta = 3.8$ (ISO 1998b). The reduction factor ϕ can then be estimated as

$$\phi = \frac{R_d}{R_n} = \frac{F_R^{-1}[\Phi(-\alpha_R \beta)]}{F_R^{-1}(0.05)} \tag{4}$$

If the partial factor format given by is adopted, then the partial factor ϕ_x for an individual material is

$$\phi_x = \frac{x_d}{x_k} = \frac{F_X^{-1}[\Phi(-\alpha_R \beta)]}{F_X^{-1}(0.05)} \tag{5}$$

where x denotes strength of the material (e.g., $x \equiv f_s$ or $x \equiv f_c'$), F_X its distribution function, and x_k and x_d characteristic and design values of the strength, respectively.

3 UPDATING INFORMATION

Updating information on structural properties is an essential part of an assessment procedure. Two different approaches to updating can be distinguished:

(i) Collecting data about individual structural properties by conducting on-site inspection;

(ii) Checking performance of the whole structure (or its components) by load testing or by using information about past performance of the structure. It should be noted that no rational method has been yet developed for incorporating information about past performance into assessment criteria (Allen 1991).

4 ON-SITE INSPECTION

The aim of an on-site inspection is to collect new data on individual parameters affecting performance of a structure. Based on these new data, distribution functions of the parameters can then be updated and substituted directly into Eq. (5) to estimate new partial factors. A more complicated reliability analysis is required to find a new overall reduction factor. Thus, the partial factor format is more convenient for updating based on-site inspection data than the LRFD format and is used herein. In the following, updated distribution functions will be derived using a Bayesian statistical approach.

Non-destructive and if necessary partially destructive techniques are usually employed in on-site inspections. In most cases, when these techniques are used, a parameter of interest (x) cannot be measured directly and a surrogate parameter (y) is measured instead. Thus, before taking measurements on-site it is necessary to establish the relationship between x and y. For example, a correction factor is required for converting core strengths (y) to equivalent on-site compressive strength of concrete (x). In many cases a simple linear relationship between x and y can be assumed

$$x = ay \tag{6}$$

where the coefficient a is determined from calibration. The uncertainty associated with the calibration includes: (i) instrument error, i.e., specimens being tested under the same conditions and having the same y, but the measurements taken by the instrument vary; and (ii) calibration (or "model") uncertainty, since Eq. (6) assumed between x and y is simply an approximation of a real relationship. The later is represented by a, which is treated as a random variable. Since calibration is carried out in a laboratory, in the following it is assumed that the means and coefficients of variation of a have been obtained from a large number of prior "calibration studies" so that statistical uncertainty can be ignored.

The uncertainty associated with the measurements taken on site includes: (i) instrument error (the same as for calibration); (ii) inherent variability of the measured parameter y; and (iii) statistical uncertainty due to a limited number of measurements.

As it can be noticed, uncertainty associated with instrument error affects measurements both in the laboratory and on site, i.e., it affects evaluation of statistical parameters of a as well as on-site measurements of y. In the present study it is assumed that the uncertainty associated with instrument error is small compared with "model" uncertainty and the inherent variability of y, so that a and y can be treated as independent random variables.

For convenience, rewrite Eq. (6) in terms of logarithms as $X = A + Y$, where $X = \log x$, $A = \log a$ and $Y = \log y$. It is then assumed that x, a and y are lognormally distributed and so this means that X, A and Y are normal random variables.

In the Bayesian approach distribution parameters of X - its mean, μ_X, and standard deviation, σ_X, are considered as random variables, so that a distribution of X is conditional on μ_X and σ_X, i.e., $f(X \mid \mu_X, \sigma_X)$. Thus, to derive an unconditional distribution of X (also referred to as a "predictive" distribution) a prior distribution of μ_X and σ_X based on prior information and their posterior distribution based on on-site inspection data need to be defined. For a normal random variable such as X a conjugate normal-gamma distribution is usually used as the prior distribution, $f'(\mu_X, \sigma_X)$, of its mean and standard deviation (e.g., Raiffa & Shlaifer 1961).

Let n on-site measurements of y be taken and a vector $\mathbf{Y} = (Y_1, \ldots, Y_n)$ represents the logarithm of the measured values. Y is a normal random variable with the mean, $\mu_Y = \mu_X - \mu_A$, and variance, $\sigma_Y^2 = \sigma_X^2 - \sigma_A^2$, where μ_A and σ_A^2 are mean and variance of the logarithm of the calibration factor a, respectively. The likelihood function, $L(\mu_X, \sigma_X \mid \mathbf{Y})$, which represents the knowledge gained from the on-site test data **Y,** is then proportional to

$$L(\mu_X,\sigma_X \mid \mathbf{Y}) \propto \left(\sigma_X^2-\sigma_A^2\right)^{-n/2} \exp\left\{-\frac{1}{2\left(\sigma_X^2-\sigma_A^2\right)}\left[\nu s^2+n\left(\mu_X-\mu_A-m\right)^2\right]\right\} \quad (7)$$

where m is the mean and s the standard deviation of $\mathbf{Y}$, respectively. According to Bayes' theorem, posterior distribution, $f''(\mu_X,\sigma_X)$, of μ_X and σ_X given $\mathbf{Y}$ is

$$f''(\mu_X,\sigma_X) = cL(\mu_X,\sigma_X \mid \mathbf{Y})f'(\mu_X,\sigma_X) \quad (8)$$

where c is a normalising factor. The predictive distribution, $f(X)$, of X can then be obtained as

$$f(X) = \int\int f(X \mid \mu_X,\sigma_X)f''(\mu_X,\sigma_X)d\mu_X d\sigma_X \quad (9)$$

Since the variance σ_Y^2 cannot be negative (i.e., $\sigma_X^2-\sigma_A^2 \geq 0$), the lower bound of integration over σ_X in Eq. (9) in this case is equal to σ_A. This limits the use of prior information, since the prior distribution is now truncated at σ_A. Based on the predictive distribution of X, the updated distribution function of x can be found and then substituted into Eq. (5) to estimate a new partial factor.

To illustrate the approach described above and to examine the influence of uncertainties associated with on-site inspection on partial factors a sensitivity analysis has been carried out. Input parameters varied in the analysis include: a coefficient of variation of test data, $V_y \approx s$ (represents inherent variability of the measured parameter y), a coefficient of variation of the calibration factor a, $V_a \approx \sigma_A$ (represents "model" uncertainty), a number of measurements n

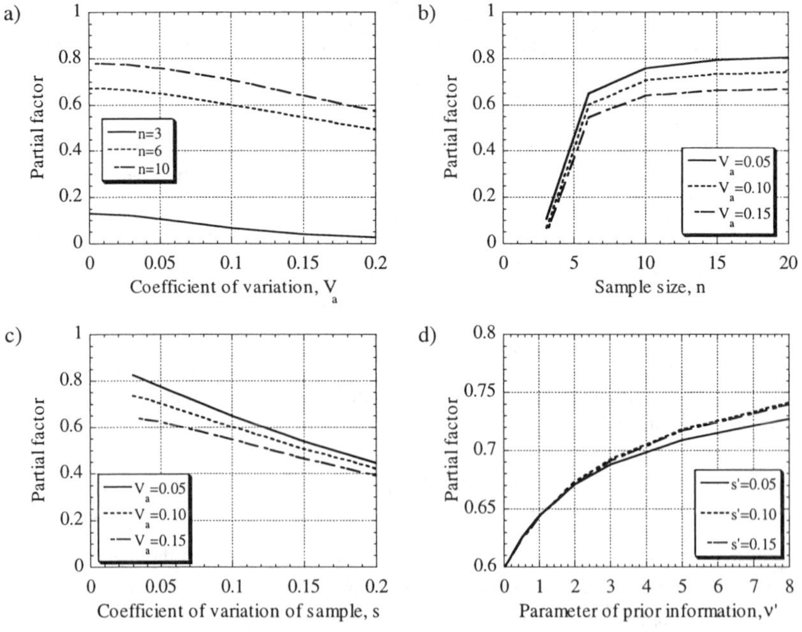

Figure 1. Partial factors ϕ_X: a) $s=0.10$; b) $s=0.10$; c) $n=6$; d) $s=0.10$, $V_a=0.10$ and $n=6$.

(represents statistical uncertainty), and the parameter ν' in the prior distribution $f'(\mu_X, \sigma_X)$ (represents the effect of prior information). The partial factors ϕ_x, which account for uncertainties in structural assessment, are outcomes of the analysis. Results of the analysis are presented in Figure 1 (prior information is taken into account only in Figure 1d).

5 PROOF LOAD TESTING

Proof load testing is used to verify resistance of an existing structure (or its components). The observation that a structure has survived a proof load test indicates only that the minimum resistance of the structure is greater than an applied load effect – it does not reveal the actual resistance of the structure, nor does it provide a meaningful measure of the structure safety. Since proof load testing provides information about overall resistance of a structure (or its components), the LRFD format is more suitable in this case.

If a structure has survived a known proof load then the distribution function, F'_R, of structural resistance is simply truncated at this known load effect, Q_{PL}, so that the updated distribution function, F''_R, is given by

$$F''_R(r) = \frac{F'_R(r) - F'_R(Q_{PL})}{1 - F'_R(Q_{PL})} \qquad r \geq Q_{PL} \tag{10}$$

The updated characteristic resistance, R''_n, and reduction factor, ϕ'', can then be estimated by substituting F''_R into Eqs. (3) and (4), respectively.

The relative values of the characteristic resistance $\rho_{Rn} = R''_n / R'_n$ and reduction factor $\rho_\phi = \phi''/\phi'$ (where R'_n and ϕ' are the characteristic resistance and reduction factor used in design) have been calculated as functions of load level (assuming the structure survives the proof load test). The resistance is treated as a lognormal random variable with three different coefficients of variation $V_R = 0.10$, 0.15 and 0.20. It is assumed that dead load in an existing structure can be estimated with sufficient accuracy, so that it can be treated as a deterministic variable. As dead load and proof load are both deterministic, their load effect can be presented by a single variable Q_{PL}. The load level is determined as a fraction of the characteristic resistance used in design. For example, the load level equal to 0.7 means that $Q_{PL} = 0.7R'_n$. Results of the analysis are shown in Figure 2a. The probabilities of failure of a structure associated with the proof load test (also referred to as "test risk") see Figure 2b.

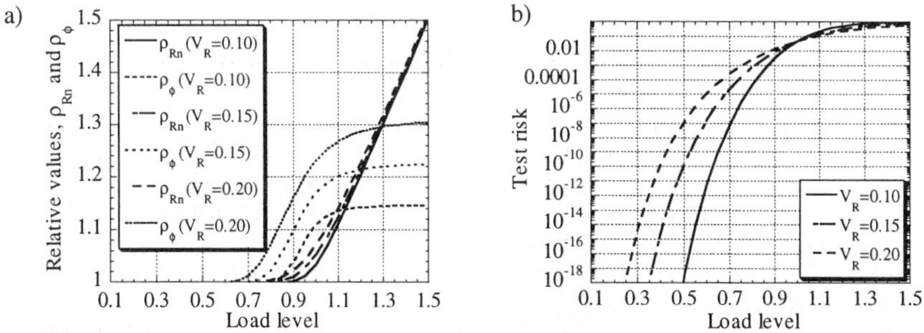

Figure 2. a) Relative values of characteristic resistance and reduction factor; b) test risk.

6 PAST PERFORMANCE

Satisfactory performance of a structure during T years in service means that the structural resistance is greater then the maximum load effect over this period of time. The updated distribution function of structural resistance at time T is then given by (Hall 1988)

$$F_R''(r) = \frac{\int_0^r F_Q^T(r) f_R'(r) dr}{\int_0^\infty F_Q^T(r) f_R'(r) dr} \tag{11}$$

where F_Q^T is the distribution function of the load effect experienced up to time T and f_R' the density function of resistance prior to loading. As was the case of proof load testing, the updated characteristic resistance, R_n'', and reduction factor, ϕ'', can then be estimated by substituting F_R'' into Eqs. (3) and (4), respectively.

The load effect, Q, due to live load is represented by a Gumbel distribution. If μ_Q and σ_Q are the mean and standard deviation of this distribution for annual load, the distribution of the live load effect within a period of T years is also a Gumbel distribution with the mean, $\mu_Q^T = \mu_Q + \left(\sqrt{6}/\pi\right)\sigma_Q \ln(T)$, and the standard deviation, $\sigma_Q^T = \sigma_Q$ (e.g., Thoft-Christensen & Baker 1982). It is assumed that the coefficient of variation of annual live load is $V_Q = 0.35$. Dead load is treated as a deterministic variable and is assumed equal to the mean value of dead load used in design. The mean values of dead load and live load are normalised with respect to the characteristic resistance, R_n', used in design. They are determined from the Australian design condition $\phi' R_n' = 1.25 G_k + 1.5 Q_k$ ($\phi' = 0.8$) depending on the ratio $\rho = Q_k / G_k$, where G_k is the mean dead load and Q_k is the characteristic value of live load. Q_k is taken as the 98-th percentile of the distribution function of live load during intended design life, T_d, of a structure (it is assumed that $T_d = 100$ years). Structural resistance is treated as a lognormal random variable with a coefficient of variation $V_R = 0.15$.

The updated characteristic resistance and reduction factors have been estimated for different values of ρ and for different structural ages. According to results of the analysis, for the scenarios considered herein, dead loads and live loads within the range expected in design do not have any noticeable effect on updating the characteristic resistance and reduction factor.

REFERENCES

Allen, D.E. 1991. Limit states criteria for structural evaluation of existing buildings. *Can. J. Civ. Eng.* 18:995-1004.

Hall, W.B. 1988. Reliability of service-proven structures. *J. Struct. Eng., ASCE* 114(3):608-624.

ISE 1980. Appraisal of existing structures. The Institution of Structural Engineers, London.

ISO 1998a. Assessment of existing structures. ISO/TC98/SC2/WG6, Doc. N24, 8-th Draft.

ISO 1998b. ISO 2394: General principles of reliability for structures. International Organization for Standardization, Geneva, Switzerland.

Raiffa, H. & R.. Schlaifer 1961. *Applied statistical decision theory*. Cambridge, MA: Harvard University Press.

Thoft-Christensen, P. & M.J. Baker 1982. *Structural reliability theory and its applications*. Berlin: Springer-Verlag.

Vrouwenvelder, T. 1997. The JCSS probabilistic model code. *Struct. Safety* 19(3): 245-251.

Mechanics of Structures and Materials, Bradford, Bridge & Foster (eds)
© 1999 Balkema, Rotterdam, ISBN 90 5809 107 4

The proposed concrete model code for Asia and the Australian concrete structures standard – A comparison

S. Fragomeni & Y.C. Loo
School of Engineering, Griffith University, Southport, Qld, Australia

ABSTRACT: The second draft of the Concrete Model Code for Asia has been published recently; it is intended for use in Asian countries. This paper serves to introduce the Model code by providing a brief history of its origins. The contents are discussed and evaluated. Comparisons are made with the well-established Australian Concrete Standard-AS3600 (1994) on philosophical and general grounds.

1 INTRODUCTION

Since its inception in 1992, the Concrete Model Code for Asia has undergone 7 years of preparation and development work, culminating in the publication of the second draft in 1999 (International Committee on Concrete Model Code, 1999).

The code aims to:

(i) provide a guidance for concrete engineers and academics writing their national codes;
(ii) help construct more efficient infrastructures in harmony with the natural and social environment; and
(iii) foster increased cooperation in technological exchanges and research within Asia.

In preparation of the draft an international committee, drawn from various Asian countries including Australia, has met at various international workshops in recent years for constructive discussions on aspects of the code. A survey of state of the art practice on design, construction and maintenance of concrete structures in various Asian regions was carried out and, based on the results, a framework of the model code was drafted.

The benefits of one unified code of practice in the Asian region has been highlighted extensively (Loo, 1994). In a time where national trade boundaries are breaking down and many large projects have international involvement in design, construction and maintenance, it does make sense to have a unified document.

Ideally, the aim is to provide a code, which is all encompassing and is as detailed as possible but it must have the basic ingredient of being user-friendly. In this regard it would be advantageous to incorporate well-accepted simplified design methods from various recognised codes which practising engineers are already familiar with.

In this paper a brief history of the new Asian Model Code development is given, followed by a general overview of the proposed contents and layout. The content is also evaluated by comparing it to the well-established Australian Concrete Structures Standard, AS3600 (1994), one of the most widely used codes in Australasia. The comparison is made in a philosophical and general way, although in some instances more direct comparisons are made. In addition to focusing on design aspects of both codes, reference is also made to material and construction requirements.

2 CONCRETE MODEL CODE HISTORY

The development of codes and standards, relating to concrete structures, in different Asian countries have taken divergent paths, frequently with strong ties to their respective colonial past. The result is many Asian countries adopted internationally accepted documents such as ACI Codes and the British Standards. These codes do not necessarily reflect local conditions of countries in the Asian region (Uomoto, 1999).

In order to achieve a more applicable concrete code in the Asian region a Research Committee on the Concrete Model Code for Asia was set up by the Japan Concrete Institute (JCI) in May 1992. The committee revealed that academics and engineers in the region recognised the need for and supported the idea of having a model code with an Asian emphasis. It should encompass the differences in economic, climatic and cultural environments of the region (Yamazaki, et. 1993). In 1994, the committee evolved into an international committee independent of JCI, even though JCI has continued to provide financial support throughout the past years.

Since its beginnings the committee's activities have also been supported by the following organisations (Uomoto, 1999):

- Association of Structural Engineers of the Philippines (ASEP)
- China Civil Engineering Society (CCES)
- Engineering Institute of Thailand (EIT)
- Indian Concrete Institute (ICI)
- Indonesian Society of Civil & Structural Engineers (HAKI)
- Japan Concrete Institute (JCI)
- Korea Concrete Institute (KCI)
- Sri Lankan Standards Institute (SSI)

Current membership of the committee consists of representatives from Australia, Bangladesh, Cambodia, China, India, Indonesia, Japan, Korea, Pakistan, Papua New Guinea, Philippines, Singapore, Taiwan and Thailand.

In April 1994, a Code framework was discussed and accepted in principle by the Research Committee in its meeting held in Tokyo. Subsequent meetings were conducted in Bangkok in December 1994 (where the Research Committee was reorganised and renamed the International Committee on Concrete Model Code); Tokyo, again, in March 1995; Gold Coast in July 1995; Jakarta in March 1996; Dalian, China in October 1996; Hyderabad, India in March 1997; Jakarta, again, in August 1997.

These international discussions culminated in the publication of the "First Draft" of the Code at the Taipei meeting in January 1998. This was followed by a meeting in Singapore in August 1998 where discussions centred on the plan to update and enhance the "First Draft" (Loo, 1999). A subsequent "Second Draft" was released in March 1999, coinciding with the IABSE Colloquium on the Concrete Model Code for Asia in Phuket, Thailand.

3 CONCRETE MODEL CODE FORMAT AND CONTENTS

The Concrete Model Code for Asia comprises of three major parts. Volume I contains Design, Volume II has specifics on Materials and Construction, and Volume III gives guidance on Maintenance and Repair. The contents have been developed by three specialised task groups from the various countries in Asia.

Each of the three volumes is subdivided into three levels. Level 1 and 2 documents generally cover the basic knowledge and fundamental provisions. The site-dependent recommendations and case studies are presented as Level 3 works. Whereas the first 2 levels of contents are applicable in all countries, the Level 3 components may be country-specific or of a problem solving nature. The documents will be continually upgraded and refined, with new developments and practical experiences gained published from time to time as Level 3 documents to enhance the universality of the code. The document levels and the intended readership may be summarised in Table 1 (Loo, 1999).

Table 1. Level of presentation and intended readership

Level 1	Level 2	Level 3
Engineering managers; Technicians; university and technical college Students; Architects; Code writers	Technical college teachers; University lecturers and Students; Designers; Construction engineers; Architects; Code writers	Researchers; designers; University lecturers and Students; Code writers; Construction engineers

Table 2 gives an overview of the contents in the Concrete Model Code. The major headings are as shown, each of which has further sub-headings.

Table 2. Table of Contents for Asian Concrete Model Code

Section	General Contents
Level 1 Vol. 1 - Design	General/Scope, General Principles, Requirements, Materials, Actions, Analysis, Examination of Performance, Evaluation of Performance
Vol. 2 -Materials & Construction	General Scope, Essential Requirements. Basic Requirements for Formwork , Reinforcement, Concrete and Prestressed Concrete (with regard to materials, workmanship, quality control and assurance, and records).
Vol. 3 - Maintenance & Repair	General, Basis of Maintenance, Inspection, Deterioration Mechanism and its prediction, Evaluation and decision making, remedial action, records
Level 2 Vol. 1 - Design	Design for Actions in Normal Use, Design for Wind Actions, Design for Seismic Actions, Design for Environmental Actions
Vol. 2 -Materials & Construction	General Scope, Essential Requirements. Basic Requirements for Formwork, Reinforcement, Concrete and Prestressed Concrete (with regard to materials, workmanship, quality control and assurance, and records).
Vol. 3 - Maintenance & Repair	General, Basis of Maintenance, Inspection, Deterioration Mechanism and its prediction, Evaluation and decision making, remedial action, records
Level 3 Vol. 1 - Design	Provides guidelines for load-bearing design, Seismic design & its performance evaluation and durability design & its performance evaluation. Acts like a Design Manual and is prepared for a particular type of structure or particular region or country
Vol 2 & Vol 3	T.B.A.

A close look at the design part (Volume 1) of the Concrete Model Code for Asia reveals that the performance-based concept is used. This concept is similar to limit state design except that limit state design is based on engineering-specific limitations, whereas the performance concept describes non-engineering limitations which are readily understood by the users.

For example crack limit state is associated with allowable crack width, and non-engineers may not understand this concept. In performance based design this comes under serviceability and more particularly aesthetics of a structure. Hence limits for crack width and length are used to indicate whether a structure looks good or not in performance based design (Ueda, 1999).

As such a performance is quantified by some index, so that the required level is explicitly expressed. Therefore, the general formula to be satisfied is:

$$PI_P > PI_R \qquad (1)$$

That is, the performance index possessed by the structure or structural element, PI_P, must be greater than the performance index required, PI_R. A typical example of PI_P and PI_R is the bending strength of an element and the maximum bending moment acting on the element respectively.

Not only the examination of the Performance Index is undertaken through Eq. (1), but the evaluation of the performance possessed by the structure can also be conducted. The evaluation can tell us how good the performance actually is using the ratio of PI_P / PI_R (Ueda, 1999).

The required performance, under the effects of considered actions, is investigated under three categories:

i) serviceability - to provide adequate functionality and not to cause an unpleasant environment;

ii) restorability - for the ability of a structure or element to be repaired physically, economically and safely;

iii) safety - to ensure no harm comes to the users and others in the vicinity of the structure.

4 AUSTRALIAN STANDARD–AS3600

The Australian Concrete Standard, AS3600 (1994) applies to reinforced members with, or without some degree of prestressing, while separate requirements are given for unreinforced (plain) concrete. The current version of the code, as with previous versions, has adopted the limit state design approach. The appropriate functional states, such as strength and serviceability, and the corresponding performance limits are presented as a function of the design action effects (including bending moments and shear forces) and the corresponding design resistances. The design action effect at ultimate condition such as moment, shear, torsion, etc., is collectively known as S^* and the ultimate design resistance of the reinforced concrete section under that action is ϕR_u. For design purposes the general formula to be satisfied is:

$$\phi R_u \geq S^* \tag{2}$$

In short, loads and load combinations are given in Section 3 of AS3600 with further reference to the loading code, AS1170 (1989). Design action effects are determined from analysis in accordance with Section 7. Design resistances are then determined for the various elements from Sections 8, 9,10 and 11 as appropriate. The various sections are listed for completeness in Table 3.

A tiered approach to member design rules is included allowing for more flexibility in choice of design methods to suit a particular project. Simplified rules, for common applications within certain limits are presented first, followed by more complex rules having wider applications.

The current version has introduced separate chapters on Durability and Fire Resistance. Previously they were included in various design rules or given only as recommendations in the appendices. Additional rules and requirements for concrete structures to be designed for earthquake loads, previously given in AS2121 Earthquake Code, are now provided in Appendix A of the Standard.

The Concrete Standard refers to various other Australian standards including AS 1170 (1989) Minimum design loads on structures; AS 1379 (1991) The specification and manufacture of concrete; and AS 3735 (1991) Concrete Structures for retaining liquids. AS1170 gives provisions for dead, live, snow, wind and earthquake loading. AS 1379 covers all aspects of the manufacture of concrete from specification of materials, through batching and mixing, to discharge of plastic concrete on site. The standard also takes into account the recent advances in materials technology, continuing research into structural behaviour and the availability of computer-aided methods of analysis.

Some of the provisions relating to good construction practices and workmanship have not been included in AS3600 because of the likely variety of project situations and alternative practices that could be adopted. It is considered that specific construction practices and workmanship can be dealt with more effectively in project specifications and included in handbooks and guides on good practice.

Table 3. AS3600 Table of Contents

Section	Contents
1	Scope and General
2	Design Requirements and Procedures (for Stability, Strength, Serviceability, etc)
3	Loads and Load Combinations for Stability , Strength and Serviceability
4, 5, 6	Design for Durability, Fire Resistance and Properties of Materials respectively.
7	Methods of Structural Analysis
8	Design of Beams for Strength and Serviceability
9	Design of Slabs for Strength and Serviceability
10	Design of Columns for Strength and Serviceability
11	Design of Walls
12	Design of Non-Flexural Members, End Zones and Bearing Surfaces
13	Stress Development and Splicing of Reinforcement and Tendons
14	Joints, Embedded Items, Fixings and Connections
15	Plain Concrete Members
16	Concrete Pavements, Floors and Residential Footings
17, 18	Liquid Retaining Structures, Marine Structures
19	Material and Construction Requirements
20	Testing & Assessment for Compliance of Concrete Specified by Comp. Strength
21	Testing of Members and Structures
App. A	Additional Requirements for Structures subject to Earthquake Actions
App. B	Referenced Documents

5 COMPARISONS AND COMMENTS

The Concrete Model Code for Asia seems to be a more detailed document than AS3600. This would be expected considering that it is a model code intended for adoption in various countries, some of which may have their own codes of practice. Users of the Model Code would expect to find the relevant data in one document rather than being referred to a number of sources. As mentioned earlier, the AS3600 code refers to many other Australian standards, which is convenient for Australian designers who would have easy access to such documents.

Also, AS3600 is contained in one volume and does not have different levels whereas the draft Model Code is divided into three volumes (i.e. design; materials and construction; maintenance and repair) and is presented in three levels. AS3600 focuses more on design and detailing topics and gives only general guidance on aspects such as construction, maintenance and repair. It is considered in Australia that because of the likely variety of project situations and alternative practices that could be adopted, specific construction practices and workmanship can be dealt with more effectively in project specifications and publications other than AS3600.

In terms of design, it is highlighted in sections 3 and 4 that performance-based design is used in the Model Code as opposed to the limit state method adopted in Australia. The idea is to make the Model Code more accessible to a wider group of people as identified in Table 1. The Australian Concrete Standard, on the other hand, is a more specific document where its intended readership is limited to practising designers and researchers.

Both codes provide adequate guidance on the strength and serviceability design of standard concrete elements such as beams, columns slabs and walls. Areas which are scrutinised herein are seismic design, durability and fire resistance. The Model Code gives a thorough coverage of seismic design with detailed guidance given on performance indices and analysis methods. AS3600 gives limited guidance on earthquake design in its Appendix A. The particular emphasis in seismic design in the Model Code is a direct consequence of the seismic activity which is prevalent in many parts of the Asian region.

An interesting recent development is that both codes have sections devoted to durability design. The Model Code has this aspect detailed in its level 2 document under, "Design for Environmental Actions". With the increasing use of high performance concretes the committees of both codes have realised the importance of designing a durable structure. In previous years, durability was not such a big problem as experience in the use of normal strength concrete led to appropriate concrete cover to reinforcement and mix designs. The range of concrete available today has made researchers and practitioners more concerned about durability considerations.

One aspect that does not seem evident in the Model Code content is the design for fire resistance. In Australia, companies such as BHP have undertaken a significant amount of research on designing concrete buildings for fire resistance in the past decade. The result is that the current version of AS3600 has a new section on fire resistance. The Standard provides deemed-to-satisfy clauses for beam, slab, column and wall elements. In short the elements must satisfy certain concrete cover to reinforcement requirements for various fire resistance periods and in some cases minimum thicknesses for fire insulation purposes.

In order to simplify specifications and requirements given in Levels 1 and 2, the Model Code presents in the Level 3 document practical design examples from various countries in the Asian region. This will be of great benefit to designers who can use the Code not only as a specification but also as a design manual. That is, the Code will eventually become all encompassing. The current AS3600 code does not provide design examples but its main strength is in the provision of simplified design methods. This document along with accompanying design manuals, published by organisations such as the Cement and Concrete Association and Concrete Institute of Australia, are reasonably easy to use.

6 CONCLUSION

This paper serves to introduce the Concrete Model Code for Asia. An overview and general evaluation of its contents are presented on general and philosophical grounds. A comparison is also made with the Australian Concrete Standard. This is to highlight some differences and similarities. It is envisaged that most countries in the Asian region will, in due time, adopt the Concrete Model Code.

7 REFERENCES

AS3600 (1994). *Concrete Structures*, North Sydney: Standards Association of Australia.

International Committee on Concrete Model Code. 1999. *Asian Concrete Model Code, 2nd Draft,* Tokyo, Japan Concrete Institute.

Loo, Y.C. 1999. Concrete Model Code for Asia – editorial philosophy and promotional strategy, *IABSE Colloquium Phuket 1999: Concrete Model Code for Asia*, Phuket, Thailand, March: pp. 29-34.

Ueda, T. 1999. Concrete Model Code for Asia: Design, *IABSE Colloquium Phuket 1999: Concrete Model Code for Asia*, Phuket, Thailand, March: pp. 35-46.

Uomoto, T. 1999. Concrete Model Code for Asia – the Needs, Development and Details, *IABSE Colloquium Phuket 1999: Concrete Model Code for Asia*, Phuket, Thailand, March: pp. 1-7.

Yamazaki, J., Noguchi, H. Kabeyasawa, T. and Ueda, T. 1993. Towards Model Code, Proceedings of the Fourth East Asia- Pacific Conference on Structural Engineering and Construction, Seoul, Sept: pp. 481-486.

Mechanics of Structures and Materials, Bradford, Bridge & Foster (eds)
© 1999 Balkema, Rotterdam, ISBN 90 5809 107 4

Suggested revisions to Australian Standard (AS3600) requirements on design for longitudinal shear in concrete beams

A. K. Patnaik
School of Civil Engineering, Curtin University of Technology, Perth, W.A., Australia

ABSTRACT: Requirements of designing and detailing for transfer of longitudinal shear forces, across interface shear planes through webs and flanges of composite concrete beams and across shear planes through flanges cast monolithically, are specified in Clause 8.4 of Australian Standard AS3600. Detailing of the interface in a composite concrete beam to resist longitudinal shear is crucial to develop monolithic action in such a beam. Recent research established that the provisions of Clause 8.4 in AS3600 are conservative. This paper critically reviews the existing design provisions of the standard for normal strength concrete (below 50 MPa). These design provisions are evaluated with respect to the insight developed from the latest research at Curtin University of Technology and elsewhere for both normal and high strength concrete. Suitable changes to the clauses of the standard are suggested for normal strength and higher strength concrete for consideration during the forthcoming revision to the standard.

1 INTRODUCTION

Requirements of designing and detailing for transfer of longitudinal shear forces, across interface shear planes through webs and flanges of composite concrete beams and across shear planes through flanges cast monolithically, are specified in Clause 8.4 of Australian Standard AS3600 (AS3600 1994). Recent research has established that these provisions are conservative, and do not represent the true behaviour of such beams. Longitudinal shear strength of composite concrete beams mainly depends on the *roughness* of the top surface of the precast girder over which the in-situ concrete is placed, the amount of tie reinforcement crossing the interface, and the concrete strength of the two elements. Good understanding of the interface behaviour was developed by Patnaik (1998, 1999 & 1992) and Loov and Patnaik (1994a and 1994b) for composite concrete beams with a *rough* interface. However, test data reported on composite beams with a *smooth* interface has so far been limited.

2 PREVIOUS RESEARCH WITH COMPOSITE BEAMS

Of the many beams tested in the earlier investigations, most test beams failed in diagonal shear, flexure, in bond of reinforcing bars, or due to poor detailing. All these modes prevented any possibility of failure of the test beams in longitudinal shear. The test beams had potential to resist larger longitudinal shears if the other modes of failure could be prevented. It was considered in these studies that the longitudinal shear *strength* of test beams is the longitudinal shear *stress at failure load* even though the beams failed in other modes. In many earlier studies, the longitudinal shear failure in beams was considered to have occurred when the slip at the level of the interface between the flange and the web exceeded 0.13 mm (0.005″). This arbitrary limit on slip, which refers to working stress philosophy of design, introduced a severe restriction on the utilisation of longitudinal shear strength. Substantially greater longitudinal shear strength is developed if no limit is placed on the slip at the interface (Patnaik 1992).

3 DESIGN LONGITUDINAL SHEAR FORCE

The evaluation of design longitudinal shear force according to AS3600 is as follows:

(a) For a shear plane through a flange, equal to $V^* A_1/A_2$, where, V^* is the design shear force at a section in Newton.

   (i) for a flange in compression - A_1/A_2 = the ratio of the area of flange outstanding beyond the shear plane to the total area of flange;
   (ii) for a flange in tension - A_1/A_2 = the ratio of the area of longitudinal reinforcement in the flange outstanding beyond the shear plane to the total area of longitudinal tensile reinforcement.

(b) For a shear plane through the web, equal to V^*.

4 LONGITUDINAL SHEAR STRENGTH

AS3600 recommends the following design equation for longitudinal shear strength:

$$V_{uf} = \beta_4 A_{sf} f_{sy}\, d/s + \beta_5\, b_f\, d\, f'_{ct} \leq 0.2 f'_c\, b_f\, d \tag{1}$$

where

V_{uf} = ultimate longitudinal shear strength at an interface, Newton
β_4, β_5 = shear plane surface coefficients (see Table 1 given below)
A_{sf} = cross-sectional area of tie reinforcement anchored each side of the shear plane, mm^2
f_{sy} = yield strength of the tie reinforcement crossing the shear plane, MPa
d = effective depth of the composite beam, mm
s = spacing of reinforcement crossing the shear plane, mm
b_f = width of the shear interface, mm
f'_{ct} = characteristic principal tensile strength of the concrete = $0.4\sqrt{f'_c}$, MPa
f'_c = characteristic compressive cylinder strength of concrete at 28 days, MPa

5 SHEAR PLANE SURFACE COEFFICIENTS AND SURFACE CONDITIONS

The shear plane surface coefficients and the definition of different interface surface classification as per AS3600 are given in Table 1.

Table 1 Shear plane surface coefficients

Surface Condition of the shear plane	Coefficients β_4	Coefficients β_5
A smooth surface, as obtained by casting against a form, or finished to a similar standard	0.6	0.1
A surface trowelled or tamped, so that the fines have been brought to the top, but where some small ridges, indentations or undulations have been left; slip formed or vibro-beam screeded; or produced by some form of extrusion technique	0.6	0.2
A surface deliberately roughened – (a) by texturing the concrete to give a pronounced profile; (b) by compacting but leaving a rough surface with coarse aggregate protruding but firmly fixed in the matrix; (c) by spraying when wet, to expose the coarse aggregate without disturbing it; or (d) by providing mechanical shear keys	0.9	0.4
Monolithic construction	0.9	0.5

6 DISCUSSION OF CURRENT REQUIREMENTS AND SUGGESTED CHANGES

Clause 8.4.1 regarding the application of Clause 8.4 on longitudinal shear in beams is misleading as this clause gives an impression that transfer of longitudinal shear forces must be considered for all types of beams including those cast monolithically. Beam design for monolithic construction may not be governed by this requirement for most normal situations.

6.1Design shear force

The method of the standard for determination of design longitudinal shear force is quite different to the other conventional methods and the methods given in other international design codes and standards. In this clause, longitudinal shear is based on vertical shear force and is therefore flawed. Figure 1 shows an example of a free-body diagram that must be considered in determination of longitudinal shear. Design longitudinal shear force (or stress, v_l) must be calculated based on equilibrium condition by *computing the actual change in compressive or tensile force in any segment, and provisions made to transfer that force as longitudinal shear to the supporting element.* This relationship can be expressed as:

$$v_l = \frac{C}{b_f l_v} \tag{2}$$

where C is the total compression (or tension) in the flange in Newton and l_v is the length of the shear interface (see Figure 1).

6.2 Design shear strength

The shear strength equation of AS3600 (see Eq 1) results in conservative designs and unrealistic strengths as the equation refers to the area of the vertical cross-section ($b_f d$) rather than the area of the horizontal shear interface ($b_f l_v$). The equation also relates the longitudinal shear strength to the ratio *d/s*. There is no published theoretical or experimental basis to establish the effect of *d/s* on longitudinal shear strength. For example, if *d/s* is about 3 or 4 (quite normal in practice), the first part of strength equation is increased by a factor of 3 or 4. This is a serious flaw as there is not enough evidence to prove this order of increase in strength. This can be unsafe if a beam was to fail in a longitudinal shear mode. This aspect has not been a serious one so far as the equation itself leads to conservative designs. The upper limit of the equation is related to ($b_f d$) which is similarly flawed. The upper limit is applicable to all the types of interface preparations and therefore unrealistic when applied to design of *smooth* interfaces.

6.2.1 *Smooth* Interface

Figure 2 shows the equations of ACI code (ACI Code 1995) and Canadian Standard (CSA A23.3 1995) for a *smooth* interface. There are several push-off tests reported by others, but none have conducted a systematic study of the effect of different factors on shear transfer

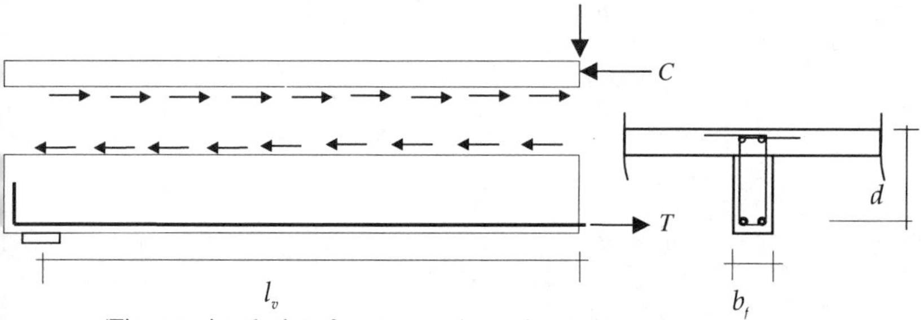

(Ties crossing the interface are not shown for clarity)

Figure 1 Equilibrium method for calculating longitudinal shear stress strength of a *smooth* interface. Figure 2 shows that both the equations are conservative. AS3600 equation has not been plotted as it is flawed for reasons mentioned earlier. The maximum clamping stress of the test beams is about 3.3 MPa, but the equations do not specify a realistic upper limit on strength. A detailed experimental and theoretical investigation on behaviour of composite beams with a *smooth* interface is currently being carried out at Curtin University of Technology. The results of the study will be available by the end of the year. The dependence of shear strength on d/s is questionable and the upper limit of $0.2 f'_c b_f d$ of AS3600 must be changed to refer to the shear interface rather than the area of the beam cross-section. From the few test data plotted in Figure 2, it is difficult to establish a trend. However, the simplest representation of test data may be by a linear equation with a Y-axis intercept of about 0.4 MPa and slope of about 1.0 as a lower bound. However, until the results of further tests are available, it is proposed to use CSA Standard formula which is as follows:

$$v_{uf} = 0.25 + 0.6\, \rho_v f_{sy} \tag{3}$$

where

v_{uf} = ultimate longitudinal shear stress strength at an interface, MPa

$\rho_v f_{sy}$ = clamping stress, MPa

ρ_v = steel ratio = $A_{sf}/(b_f s)$

6.2.2 *Rough* Interface

The following general equation was suggested recently (Patnaik 1998 & 1999) for design of composite concrete beams with a *rough* interface and ties crossing perpendicular to the plane of the interface:

$$v_{uf} = 0.55\sqrt{(0.25 + \rho_v f_{sy}) f'_c} \le 0.25 f'_c \text{ and 9 MPa} \tag{4}$$

The maximum limits of 9 MPa or $0.25 f'_c$ are recommended as longitudinal shear strength greater than these values was not achieved in any of the tests conducted so far. Furthermore, beams required to resist longitudinal shear stresses greater than 9 MPa are quite unlikely in routine designs.

A factor 0.5 is suggested in Eq. 4 so as to use a slightly lower shear strength to allow for the possibility of *smoother* interfaces. The strength equation then takes the following form:

$$v_{uf} = 0.5\sqrt{(0.25 + \rho_v f_{sy}) f'_c} \le 0.25 f'_c \text{ and 8 MPa} \tag{5}$$

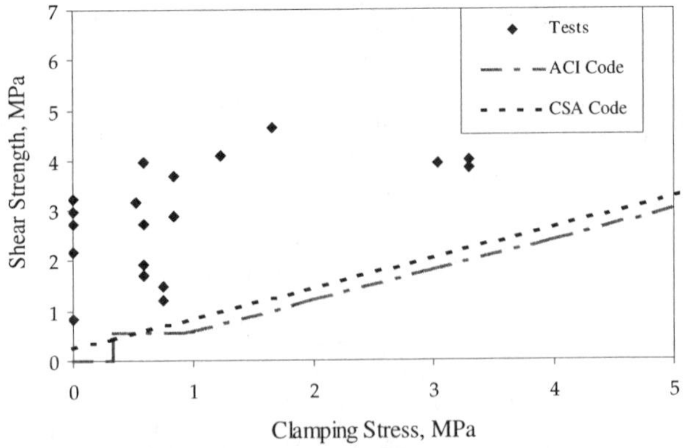

Figure 2 Comparison of all the test results with ACI code and CSA standard equations for a *smooth* interface

All the test results of recent studies and the results of previous successful tests for beams with ties and without ties are plotted in Figures 3 and 4. The concrete strength of the test beams in Figure 4 ranges from approximately 17 MPa to 62 MPa. The test beams have clamping stresses which cover a wide range of practical values. As Figure 4 is plotted with the axes expressed in dimensionless terms, the suggested equation becomes a straight line with a slope of 0.55 for Eq. 4 and 0.5 for Eq. 5. The line plotted with Eq. 5 represents the test data as a lower bound.

6.2.3 *Monolithic* shear interface

Eq. 4 was developed as a practical representation of the test results developed from systematic studies (Patnaik 1992, 1998 & 1999) for composite beams with a *rough* interface. A *monolithic* beam (which is considered to be capable of transferring greater shear at a *monolithic* shear interface) will achieve the shear transfer strength defined by Eq. 4. Therefore, shear transfer for a *monolithic* surface condition can be represented by Eq. 4.

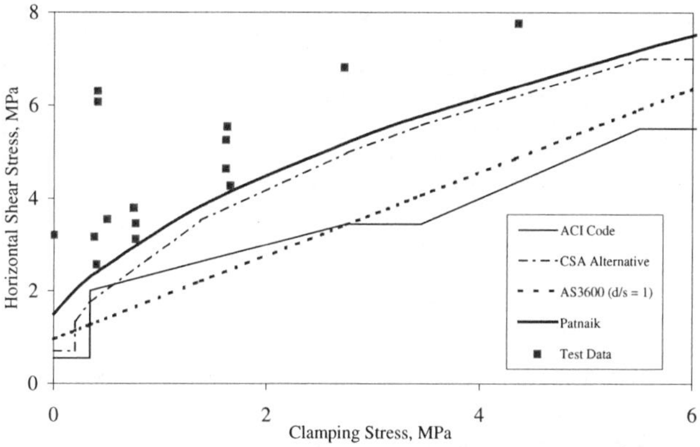

Figure 3 Comparison of results of successful tests with various design equations for $f'_c \approx 35$ MPa and a *rough* interface

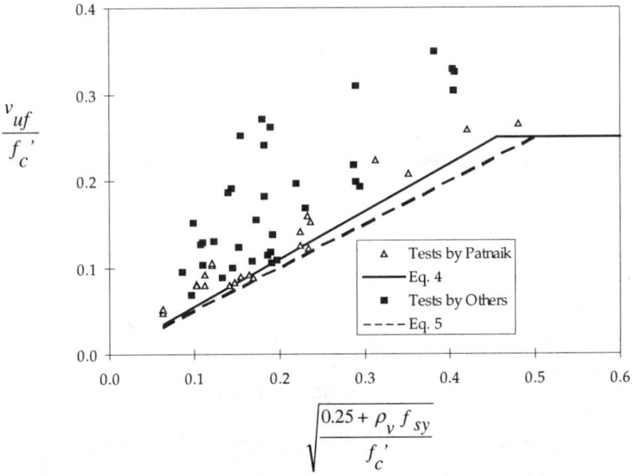

Figure 4 Comparison of results of all successful tests with the suggested design equations for a *rough* interface

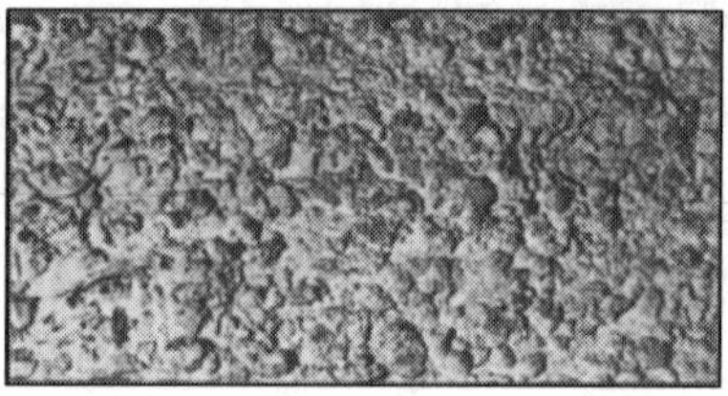

Figure 5 Typical *roughness* recommended for *rough* classification of a surface

6.3 Other clauses

AS3600 specifies two different *smooth* surfaces with almost insignificant difference between the two corresponding shear transfer strengths. There is no published experimental or theoretical basis for this difference. It is suggested that the two surface conditions be classified as just one *smooth* surface. It is important to provide guidance on achieving a *rough* surface (such as that shown in Figure 5) in practice in order to develop strength predicted by the use of Eq. 5. The minimum average thickness of structural components subjected to interface shear is required by the standard to be greater than 50mm with local thickness of 30 mm being admissible. This may mislead the users of the standard into considering 50mm thick floor toppings as structural components (integral part) of concrete beams. This clause needs further clarification.

7 CONCLUSIONS

This paper identifies a number of unrealistic requirements of Clause 8.4 of AS3600 dealing with design for longitudinal shear and suggests changes to the relevant clauses for better utilisation and representation of strength and behaviour. The two important recommendations of the suggested changes are as follows:

1. Reference to vertical cross-sectional area in the determination of longitudinal shear must be changed to horizontal shear interface area.
2. The longitudinal shear strength equation of AS3600 is conservative and unrealistic. Design equations are suggested in this paper for *smooth*, *rough* and *monolithic* surface conditions of shear planes.

REFERENCES

ACI code 1995. *Building Code Requirements for Reinforced Concrete*, ACI Standard: 318-M95, 1995, American Concrete Institute, Detroit, Michigan, USA.

AS3600 1994. *Concrete Structures*, Standards Association of Australia, 1994, Standards House, 80 Arthur Street, North Sydney, NSW.

CSA Standard A23.3 1994. *Design of Concrete Structures for Buildings*, Canadian Standards Association, 1994, Rexdale, Ontario, Canada.

Loov, R.E. & Patnaik, A.K. 1994a. Horizontal Shear Strength of Composite Concrete Beams with a Rough Interface, *PCI Journal*, Jan.-Feb. 1994: 48-69.

Loov, R.E. & Patnaik, A.K. 1994b. Authors' Closure to Comments on Horizontal Shear Strength of Composite Concrete Beams with a Rough Interface, *PCI Journal*, Sep-Oct: 106-109.

Patnaik, A.K. 1992. *Horizontal Shear Strength of Composite Concrete Beams with a Rough Interface*, Ph.D. Thesis, Department of Civil Eng., The University of Calgary, Dec. 1992.

Patnaik, A.K. 1998. Horizontal Shear Strength of High Strength Composite Concrete Beams with a Rough Interface, in *High Performance High Strength Concrete*, Proceedings of the International Conference, Perth, Aug 1998, (Eds: Rangan, B.V. and Patnaik, A.K.): 565-579.

Patnaik, A.K. 1999. Longitudinal Shear Strength of Composite Concrete Beams with a Rough Interface and no Ties, *Australian Journal of Structural Engineering*, IEAust, 1999:1(3), 157-166.

Mechanics of Structures and Materials, Bradford, Bridge & Foster (eds)
© 1999 Balkema, Rotterdam, ISBN 90 5809 107 4

AS / NZS4600-1996 – Does it cover design cold-formed roof panels?

A.J.Castle
Rickard and Partners Pty Limited, Sydney, N.S.W., Australia

B.Samali
Centre for Built Infrastructure Research, Faculty of Engineering, University of Technology, Sydney, N.S.W., Australia

ABSTRACT: Design tables were required for a cold-formed roof panel. Currently, authorities require proof testing to be performed on roofing panels in order to approve their use. Proof load tests were therefore performed on the said cold-formed roof panel and useful data obtained. The new cold-formed structures design code: AS/NZ4600-1996 has provisions by which design loads for particular profiles can be established. This paper investigates the correlation between the test data and the new code provisions with a view to determine the suitability of using the code for roof profiles which are not covered by the code.

1 GENERAL BACKGROUND

1.1 *Requirement for testing*

Northern parts of Australia are prone to severe cyclone activities. Therefore, roof sheeting intended for use in such regions, including north Queensland, is required to satisfy code provisions for cyclic loading at the nominated design wind pressures. Local authorities accept design tables for the roof sheeting only on the basis of proof testing. Hunter Douglas Limited have prepared design tables for one of their products supported by a significant amount of testing performed on their Shademaster Roof Panel.

1.2 *Comparison with ASNZ4600-1996*

The new cold-formed structures design code: AS/NZ4600-1996 has provisions for predicting the ultimate design pressure of the roof panels for certain profiles. Since load test data was available on Shademaster Roof Panels, it was worthwhile to investigate how successfully the new code, AS/NZ4600–1996, can predict the ultimate design pressure of this particular roof panel. The analysis was performed using relevant data provided by the code. Comparisons between the results obtained using code provisions and those obtained from testing have then been made.

2 TESTING OF ROOF PANELS

2.1 *General testing information*

The benchmarks for all comparisons in this study are the results gained during the destructive testing of the roof panels. Test results were obtained from two sources. The first source was the testing undertaken by CSIRO in Sydney during June 1997, and the second was the testing undertaken by the Cyclone Structural Testing Station of James Cook University during July 1998.

2.2 *General testing methodology*

Both cyclic and non-cyclic tests were performed on the Hunter Douglas Roof Panel. During cyclic testing, a design pressure (pd*) is selected. This is the pressure at which failure is desired. Once this pressure is determined, then the sheeting is cycled at various percentages of this design pressure as indicated in Table 1.

Table 1. Test regime according to AS1170.2-1989

Range	Number of cycles
0 to 0.40 pd*	8,000
0 to 0.50 pd*	2,000
0 to 0.65 pd*	200
0 to 1.00 pd*	1
0 to 1.30 pd*	1

The objective of cyclic testing is to fail the roof sheet during the last load increment. This testing regime is recommended in AS4040.3-1993 and AS1170.2-1989 as being suitable for fatigue testing of structures in cyclonic wind conditions. During non-cyclic testing, load is increased until failure of the sheeting occurs.

2.3 *Testing material*

The roof panels tested were the Hunter Douglas 0.42mm steel Shademaster Roof Panels. These panels are rolled out of BHP cold-formed steel sheeting, grade 550, into a 47mm trapezoidal profile. The panels are approximately 300mm wide, and the edges of adjoining panels interlock with one another.

The roof panel is supported by two different means. On one end of the roof panel, a typical beam used in the normal construction of the Hunter Douglas panels supports the sheeting. This beam consists of two 125mm x 50mm cold-formed beams clipped together to form a rectangular section. The other end was supported by a channel attached to the testing rig. The channel was 50mm high, with a top flange measuring 80mm and a bottom flange measuring 50mm, and was constructed out of 1.0mm steel. The profile of the roof sheeting is shown below in figure 1.

2.4 *Testing methods*

The method used by CSIRO involved using a vacuum, cyclic chamber. With this apparatus, the roof panel is fixed into place using the Hunter Douglas fixings and support beams. The roof sheeting is then placed above a vacuum chamber and sealed over with plastic sheeting. The pressure on the sheeting is then applied by removing the air in the vacuum chamber.

The Cyclone Structural Testing Station of James Cook University used a passive, air bag system. In this system, roof sheeting is placed in a steel-testing rig. The sheeting is attached by means of Hunter Douglas fixings and support beams. Passive air bags are then placed over the roof sheeting. Pressure is applied to the roof sheeting through the air bag. The air bag is loaded via a hydraulic ram loading a timber plate, which is placed over the air bag.

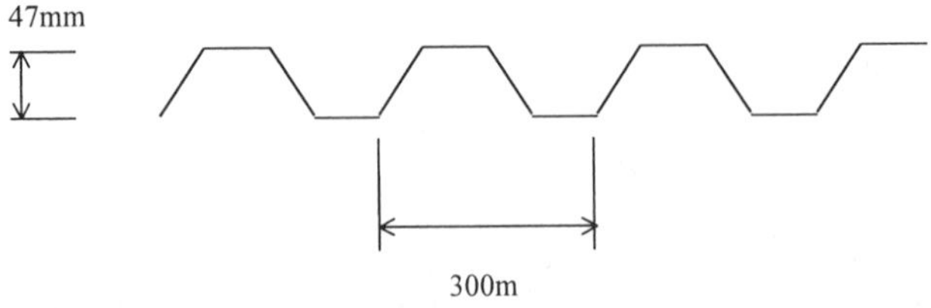

Figure 1. The profile of roof sheeting

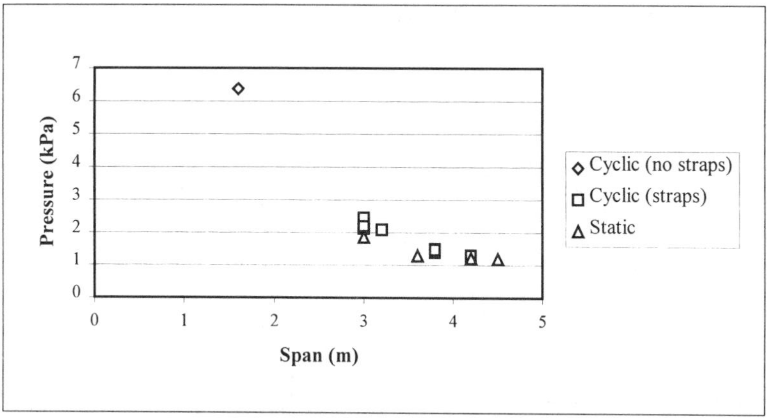

Figure 2. Comparisons of test results

2.5 *Testing results*

The maximum loads obtained from both sets of testing have been plotted in Figure 2. This graph indicates two features. Firstly, there is a good correlation between the results obtained from the two testing laboratories, and secondly, there is very little difference between the results obtained from cyclic and static testing. The results were in fact the opposite of what would be expected if there was a large amount of fatigue in the testing.

2.6 *Observations made during testing*

The mode of failure for all simply supported panels was mid-span buckling. There was no apparent fatiguing near the fasteners of the roof panels. During loading of the roof panel, "oil panning", or rippling of the webs, occurred.

3 CAPACITY OF HUNTER DOUGLAS SHADEMASTER ROOF PANEL ACCORDING TO AS/NZS4600-1996

3.1 *Background to AS/NZS4600-1996*

AS/NZS4600-1996 is the only design code available which is capable of determining the capacities of cold-formed steel sections. The code has been developed by investigating 'Z', 'C' and 'box' sections. Therefore the code naturally categorizes sections according to these types and leaves very little scope for the analysis of sections that do not fit into these categories.

The Hunter Douglas Shademaster Roof Panel does not fit into any of the above categories. This necessitates a reasonable level of interpretation of the code in order to apply the principles contained therein.

3.2 *Effective Section of the Roof Panel*

For the computation of the effective web, the code seems to be rather vague. The approach undertaken in this study is to consider the width of the element being considered as the width of the web in compression only. The reason behind this is that effective widths usually relate to local buckling of that part of the section. Since local buckling only occurs in the compression zones, the section in tension should not be reduced.

Section 2 of AS/NZS4600-1996 sets out the method for calculating the effective section of cold-formed sections.

3.3 *Nominal member capacity*

Member moment capacity is considered for all simply supported cases. However, in continuous panels, maximum moment occurs over the support and this support is assumed to contribute to restraint at this location.

Clause 3.3.3.2 of AS/NZS4600-1996 sets out the method for determining member moment capacity (M_b). According to this section, M_b is determined by the non-dimensional ratio (κ_b) as follows:

$$\kappa_b = \sqrt{\frac{M_y}{M_o}} \qquad (1)$$

where M_y = moment to cause yield using the full section; and M_o = Elastic Buckling Moment.

To determine how sensitive the code provision is, different widths of the panel were considered and the capacities calculated. The widths considered are one rib width (150mm), one panel width (300mm) and the width tested (900mm).

3.4 *AS/NZS4600-1996 results*

As can be seen from Figure 3, the failure pressures calculated by AS/NZS4600-1996 are lower than those tested, however they generally follow the same trend. The results for all three widths are reasonably close, indicating that the code is not very sensitive to the width of the panel considered.

4 ULTIMATE DESIGN PRESSURES

4.1 *Design test pressure*

In order to use the test pressure to develop a set of design charts, the failure loads determined during the testing have to be appropriately factored down. This is done as follows:

Design Pressure = Test Pressure x Material Factor / Number of Tests Factor (2)

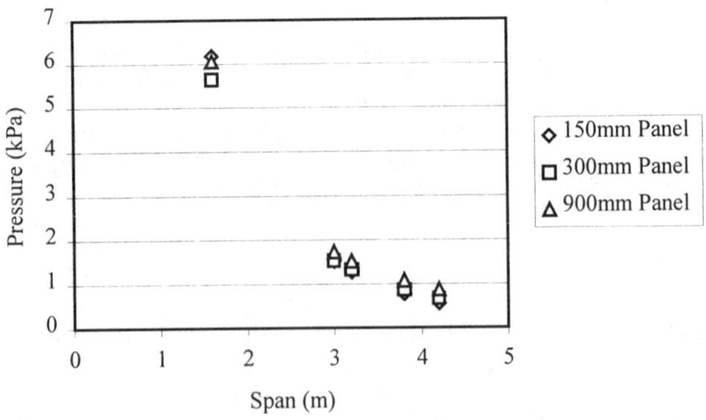

Figure 3. AS/NZ4600 – 1996 Ultimate pressure

where material factor = ϖ = 0.9; number of tests factor = a factor set out by AS1170.2 to account for whether 1, 2 or 3 tests have been conducted. This factor is 1.3 if only one test is performed, 1.2 if two tests are performed, or 1.0 if three or more tests are performed.

Table 2 shows the tested pressure, the number of tests, the appropriate factors and the ultimate design pressures as derived from the testing.

Table 2. Ultimate design pressures as derived from testing

Span M	Number of tests	Test pressure KPa	ϖ	Number of tests factor	Ultimate design pressure kPa
1.6	1	6.37	0.9	1.3	4.41
3.0	3	2.10	0.9	1.0	1.89
3.2	1	2.08	0.9	1.3	1.44
3.8	3	1.41	0.9	1.0	1.27
4.2	1	1.22	0.9	1.3	0.84
1.6 double	1	7.35	0.9	1.3	5.08
3.3 double	1	1.95	0.9	1.3	1.35

4.2 *AS/NZ4600–1996 ultimate design pressures*

AS/NZS4600-1996 has been written in accordance to limit state design philosophy, therefore the ultimate design pressure can be determined. This is accomplished by multiplying the failure pressure by ϖ. The ultimate design pressures according to AS/NZS4600-1996 are shown in Table 3.

Table 3. Ultimate design pressures as derived from AS/NZS4600

Span M	Failure pressure Kpa	Ultimate design pressure Kpa
1.6	6.06	5.45
3.0	1.72	1.54
3.2	1.51	1.35
3.8	1.07	0.96
4.2	0.87	0.78
1.6 double	8.08	7.27
3.3 double	1.90	1.71

While the ultimate design pressures obtained through the code and testing seem to compare well, it must be noted that many of the tests had a high uncertainty factor (number of test factors = 1.3) due to the number of tests undertaken. The results obtained during testing appeared to be reasonably repeatable. Therefore it is not unlikely that if more tests had been undertaken then the uncertainty factor would be decreased. Table 4 compares the Ultimate design pressure as determined from AS/NZS4600 and the test results with the number of test factor set at 1.

Table 4. Ultimate design pressures as derived from AS/NZS4600

Span m	AS/NZS4600 ultimate pressure Kpa	Testing ultimate pressure Kpa
1.6	5.45	5.73
3.0	1.54	1.89
3.2	1.35	1.87
3.8	0.96	1.27
4.2	0.78	1.09
1.6 double	7.27	6.60
3.3 double	1.71	1.75

The above table shows a reasonable correlation between the ultimate pressure derived from both the testing and AS/NZS4600. Close comparison shows that the code is only slightly conservative in comparison with the testing results. This provides the designer confidence that by designing the structures in accordance with the code, their design will not be overly conservative or under-designed.

5 CONCLUSION

AS4600–1996 can be used to calculate the ultimate pressure on a cold-formed roof sheet. It was shown that different widths of roof panel could be used without greatly affecting the results. The results obtained from the code are expectedly more conservative in relation to test results on the Hunter Douglas Roof Panel.

Even though the code was originally written considering only 'Z', 'C' and box sections, the results obtained in this study have shown that the code can be effectively used to predict the failure behaviour of cold formed roof sheet sections.

It can be said that the code provides an adequate platform for designing cold formed roof panels and that it does not overly design such sections.

6 REFERENCES

Australian Standard 1989. 1170.2 Minimum design loads on structures (known as the SAA Loading Code) Part 2: Wind Loads. Sydney: The Standards Association of Australia.

Australian Standard 1993. 4040.3 Methods of testing sheet roof and wall cladding - Method 3: Resistance to wind pressures for cyclone regions. Sydney: The Standards Association of Australia.

Australian/New Zealand Standard 1996. Cold-formed steel structures. Sydney: The Standards Association of Australia.

Reardon, G.F. 1998. Shademaster roofing panels - Fatigue loading for Hunter Douglas. Townsville: Cyclone Structural Testing Station.

Sankaran, R. & Tappouras, J. 1997. CSIRO Report Number DTA474: Performance testing on the Hunter Douglas Limited deep ridge steel roofing panel. Sydney: CSIRO.

Mechanics of Structures and Materials, Bradford, Bridge & Foster (eds)
© 1999 Balkema, Rotterdam, ISBN 90 5809 107 4

Limit states purlin design to AS/NZS 4600:1996

M.J.Clarke & G.J.Hancock
Department of Civil Engineering, University of Sydney, N.S.W., Australia

ABSTRACT: The limit states Australian/New Zealand Standard *AS/NZS 4600:1996 Cold-formed steel structures* was published in late 1996 and supersedes the corresponding permissible stress Standard AS 1538–1988. AS/NZS 4600:1996 is based mainly on the 1996 edition of the American Iron and Steel Institute (AISI) Specification (AISI, 1996). In some cases, design rules from AS 1538–1988 have been carried through to the new standard, allowing for the appropriate conversions to limit states format.

One of the main applications of cold-formed steel is purlins and girts in metal roof and wall systems. The design rules for these structural members have been refined over the years and procedures are now available which allow the beneficial effects of the sheeting restraint to be incorporated.

This paper outlines three approaches to purlin design to AS/NZS 4600:1996 and describes the relative merits and drawbacks of each. The ultimate load capacities computed using the various design models are compared with test results obtained from vacuum rig testing at the University of Sydney over a period of more than 10 years. A simple modification to one of the design procedures, which results in improved design capacities, is proposed as an amendment to AS/NZS 4600:1996.

1 INTRODUCTION

Roof systems comprising high strength steel profiled sheeting fastened to high strength steel cold-formed purlins of lipped channel or Z-section are common in Australia and throughout the world. In Australia, the design of such systems, which is usually governed by wind uplift, is performed according to the provisions of the limit states Australian/New Zealand Standard *AS/NZS 4600:1996 Cold-formed steel structures* (SA/SNZ 1996). The design procedures in the latter standard have been developed and verified using an extensive database of test data obtained from more than 10 years of testing in the vacuum testing rig at the University of Sydney. These tests have covered a wide range of parameters including: single, double and triple spans; inwards and outwards loading; zero, one and two rows or bridging per span; screw fastened and concealed fixed sheeting systems; and cleat and flange bolting of purlins to rafters (Hancock et al., 1990; 1992, 1994, 1996).

With the publication in late 1996 of the limit states standard for cold-formed steel design, AS/NZS 4600:1996, it is appropriate to revisit the design procedures for purlin–sheeting systems since there are several new or modified design rules in the latter standard which were not in its permissible stress predecessor AS 1538–1988 (SA, 1988). With specific reference to purlin systems, these new provisions are summarised as follows:

- High strength galvanised cold-reduced steels of strength grades G450, G500 and G550 (not less than 0.9 mm thick) to Australian Standard AS 1397 are included in the list of material specifications in Clause 1.5.1.1.

- Calculation of the effective area of stiffened elements subjected to a stress gradient is covered in Clause 2.2.3.
- Calculation of the effective area of unstiffened elements subjected to a stress gradient is covered in Appendix F.
- Edge and intermediate stiffeners may now be regarded as partially effective as detailed in Clause 2.4, rather than completely inadequate or fully adequate as in AS 1538–1988.
- Distortional buckling is included for members subjected to bending in Clause 3.3.3.3, and concentrically loaded compression members in Clause 3.4.6.
- An R-factor design approach is included in Clause 3.3.3.4 for beams (purlins) having one flange through fastened to deck or sheeting.
- Lateral bracing requirements for channel and Z-sections with one flange connected to sheeting and subjected to wind uplift are included as Clause 4.3.3.2, and are based on Australian test data.

The purpose of this paper is to describe the basis of purlin design to AS/NZS 4600:1996, and to evaluate the effectiveness of the design provisions by comparing predicted ultimate load capacities with tests performed at the University of Sydney since the late 1980s. Three different approaches for lateral buckling design may be applied to purlin systems and the relative merits and drawbacks of these are outlined. Based on the comparisons with test data, a simple modification to one of the lateral buckling design procedures is proposed which will produce improvements in capacities from zero to several percent.

2 SUMMARY OF TEST DATA ON PURLIN–SHEETING SYSTEMS

In 1988, a large vacuum test rig was commisioned in the Centre for Advanced Structural Engineering at the University of Sydney using funds provided by the Metal Building Products Manufacturers Association (MBPMA) for the purpose of providing test data on metal roofing systems. The test rig uses a conventional vacuum box to simulate wind uplift or inwards load. While the early series of tests were "generic" by virtue of their funding through the MBPMA, later test programs have been performed specifically for individual companies who have nevertheless made their results available in the public domain. The test programs which have been conducted are summarised in Table 1. More detailed information on specific tests is listed in Table 2 presented later.

Table 1. Purlin–Sheeting Test Programs Performed at the University of Sydney

Series	Loading	Spans*	Bridging†	Sheeting Type	Rafter Fixing
S1	Outwards	3-span lapped	0, 1, 2	Screw fastened	Cleats
S2	Outwards	2-span lapped	0, 1, 2	Screw fastened	Cleats
S3	Outwards	Simply supported	0, 1, 2	Screw fastened	Cleats
S4	Inwards	3-span lapped	0, 1	Screw fastened	Cleats
S5	Outwards	Simply supported	0, 1, 2	Concealed fixed	Cleats
S6	Outwards	3-span lapped	1	Concealed fixed	Cleats
S7	Outwards	Simply supported	0, 1, 2	Screw fastened	Cleats
S8	Outwards	Simply supported 3-span lapped	1, 2	Screw fastened	Cleats
CP	Outwards	3-span lapped	0, 1	Screw fastened	Flange

* 3 × 7.0 m spans with 900 mm laps between bolt centres for 3-span lapped configuration.
2 × 10.5 m spans with 1500 mm laps between bolt centres for 2-span lapped configuration.
1 × 7.0 m span for simply supported configuration.

† 0: Zero rows of bridging in each span
1: One row of bridging in each span
2: Single and double spans: Two rows of bridging in each span
Triple spans: Two rows of bridging in the end spans, one row in the central span.

3 LATERAL BUCKLING DESIGN CRITERIA FOR PURLIN SYSTEMS

For conventional purlin systems which are fastened to rafters by bolting through the web to cleat plates, the relevant design checks to AS/NZS 4600:1996 include: section capacity in combined bending and shear (AS/NZS 4600 Clause 3.3.5); member capacity governed by distortional buckling (Clause 3.3.3.3) (Hancock, 1998); member capacity governed by lateral buckling (Clause 3.3.3.2); bolts in shear at the cleat connection (Clause 5.3.5.1); and ply in bearing at the cleat connection (Clause 5.3.4).

Of all the design criteria mentioned in the preceding paragraph, the one which is most influenced by the restraint provided by sheeting is that of lateral buckling strength. Design provisions for beam lateral buckling are prescribed in Section 3.3.3 of AS/NZS 4600:1996 and may be classified into three distinct approaches which are herein referred to as the "C-factor" approach; the "FELB" (Finite Element Lateral Buckling) approach; and the "R-factor" approach.

3.1 *C-factor approach*

The so-called C-factor approach is outlined in Clause 3.3.3.2(a) of AS/NZS 4600:1996. For each segment of the purlin system between brace positions, a C_b factor based on the shape of the bending moment diagram is determined. This C_b factor is then used in conjunction with beam effective lengths to determine the elastic buckling moment (M_o) of the segment. The corresponding beam strength curve is shown labelled as Clause 3.3.3.2(a) in Figure 1. The advantage of the C-factor approach is that it is universally applicable to all purlin systems, including all section shapes, loading types, sheeting types and bridging layouts. It does not, however, consider the effect of load height or the lateral and torsional restraint provided by sheeting on the buckling moment. This latter aspect is a source of considerable conservatism when it is applied to purlin systems.

3.2 *FELB approach*

In the so-called FELB approach, a finite element lateral buckling analysis (CASE, 1997) of the purlin system is performed to obtained the elastic buckling moment (M_o). The advantages of the FELB approach are that the load height and sheeting restraint effects can be accounted for in the buckling analysis. A disadvantage is that the corresponding beam strength curve, labelled as Clause 3.3.3.2(b) in Figure 1, is generally lower than the Clause 3.3.3.2(a) curve used with the C-factor aproach. The reason for the use of two beam strength curves in AS/NZS 4600:1996 is more historical than rationally based, with Clause 3.3.3.2(a) reproduced from the AISI (1996) Specification and Clause 3.3.3.2(b) reproduced from AS 1538–1988.

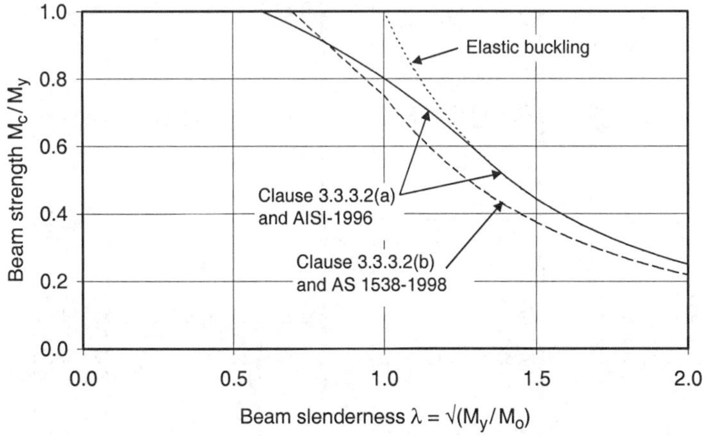

Figure 1. Beam strength curves in AS/NZS 4600:1996

Table 2a. Purlin Test Results and Comparison with AS/NZS 4600:1996 Design Models — Single and Double Span Tests

Test	Section	Bridging	Sheeting	Test Result		FELB – 3.3.3.2(b)		C-Factor		R-Factor		FELB – 3.3.3.2(a)	
				f_y (MPa)	q_T (kN/m)	q_{FB} (kN/m)	q_T/q_{FB}	q_C (kN/m)	q_T/q_C	q_R (kN/m)	q_T/q_R	q_{FA} (kN/m)	q_T/q_{FA}
S3T1R	Z-20024	0	TD	529	3.28	1.03	3.18	0.63	5.21	3.02	1.09	1.17	2.80
S3T4	C-20024	0	MC	518	3.63	1.02	3.56	0.64	5.67	3.01	1.21	1.16	3.13
S5L1	Z-20025	0	KL	525	2.57	1.05	2.45	0.62	4.15	3.09	0.83	1.19	2.16
S5S1	Z-20019	0	SD	517	2.17	0.72	3.01	0.43	5.05	2.17	1.00	0.81	2.68
S7T1	Z-20015	0	SH	527	1.85	0.50	3.70	0.32	5.78	1.46	1.27	0.57	3.25
S7T2	C-20015	0	SH	548	1.70	0.49	3.47	0.30	5.67	1.41	1.21	0.55	3.09
						Mean	3.23	Mean	5.25	Mean	1.10	Mean	2.85
						SD	0.46	SD	0.62	SD	0.16	SD	0.40
S3T2	Z-20024	1	TD	529	3.69	2.36	1.56	2.41	1.53	3.42	1.08	2.69	1.37
S3T5	C-20024	1	MC	518	3.63	2.39	1.52	2.44	1.49	3.41	1.06	2.84	1.28
S5L2	Z-20025	1	KL	525	4.19	2.40	1.75	2.42	1.73	3.50	1.20	2.82	1.49
S5S2	Z-20019	1	SD	517	2.28	1.60	1.43	1.61	1.42	2.46	0.93	1.91	1.19
S7T3	C-20015	1	SH	512	1.77	1.09	1.62	1.10	1.61	1.54	1.15	1.28	1.38
S8T2	C-20015	1	TD	480	1.71	1.09	1.57	1.10	1.55	1.48	1.16	1.28	1.34
S8T3	C-15012	1	TD	582	0.83	0.42	1.98	0.42	1.98	0.85	0.98	0.49	1.69
						Mean	1.63	Mean	1.62	Mean	1.08	Mean	1.39
						SD	0.18	SD	0.19	SD	0.10	SD	0.16
S3T3	Z-20024	2	TD	529	4.76	3.63	1.31	3.58	1.33	3.72	1.28	3.72	1.28
S3T6	C-20024	2	MC	518	4.71	3.59	1.31	3.52	1.34	3.82	1.23	3.67	1.28
S5L3	Z-20025	2	KL	525	4.90	3.60	1.36	3.54	1.38	3.89	1.26	3.69	1.33
S5S3	Z-20019	2	SD	517	2.74	2.54	1.08	2.54	1.08	2.54	1.08	2.54	1.08
S7T5	C-20015	2	SH	510	1.95	1.65	1.18	1.65	1.18	1.65	1.18	1.65	1.18
S8T1	C-20015	2	TD	500	1.98	1.63	1.21	1.63	1.21	1.63	1.21	1.63	1.21
S8T4	C-15012	2	TD	578	0.93	0.76	1.22	0.78	1.19	0.90	1.03	0.84	1.11
						Mean	1.24	Mean	1.25	Mean	1.18	Mean	1.21
						SD	0.10	SD	0.11	SD	0.09	SD	0.09
S2T1	Z-30025	0-0	MC	485	4.33	2.51	1.73	1.23	3.52	2.59	1.67	2.83	1.53
S2T2	Z-30025	1-1	MC	485	4.93	3.80	1.30	3.20	1.54	3.02	1.63	4.01	1.23
S2T3	Z-30025	2-2	TD	485	5.77	4.01	1.44	4.01	1.44	3.45	1.67	4.01	1.44
						Mean	1.14	Mean	1.65	Mean	1.27	Mean	1.07
						SD	0.72	SD	1.41	SD	0.78	SD	0.66

Table 2b. Purlin Test Results and Comparison with AS/NZS 4600:1996 Design Models — Triple Span Tests

Test	Section	Bridging	Sheeting	Test Result		FELB – 3.3.3.2(b)		C-Factor		R-Factor		FELB – 3.3.3.2(a)	
				f_y (MPa)	q_T (kN/m)	q_{FB} (kN/m)	q_T/q_{FB}	q_C (kN/m)	q_T/q_C	q_R (kN/m)	q_T/q_R	q_{FA} (kN/m)	q_T/q_{FA}
S1T1	Z-15019	0-0-0	MC	487	2.31	1.01	2.29	0.48	4.81	2.08	1.11	1.17	1.97
S1T4A	Z-20015	0-0-0	TD	520	2.58	1.54	1.68	0.65	3.97	2.31	1.12	1.72	1.50
S1T7	Z-20019	0-0-0	TD	495	3.51	2.00	1.76	0.86	4.08	3.29	1.07	2.31	1.52
						Mean	1.91	Mean	4.29	Mean	1.10	Mean	1.66
						SD	0.33	SD	0.46	SD	0.03	SD	0.27
S1T2	Z-15019	1-1-1B	MC	487	2.63	1.77	1.49	1.67	1.57	2.36	1.11	2.03	1.30
S1T5	Z-20015	1-1-1A	TD	520	2.94	2.56	1.15	2.36	1.25	2.62	1.12	2.75	1.07
S1T8	Z-20019	1-1-1A	TD	495	4.28	3.28	1.30	3.05	1.40	3.73	1.15	3.69	1.16
S6L1	Z-15019	1-1-1A	KL	615	2.56	1.87	1.37	1.84	1.39	2.88	0.89	2.13	1.20
S6L2	Z-20019	1-1-1A	KL	517	3.81	3.59	1.06	3.60	1.06	3.97	0.96	3.94	0.97
S6S1	Z-20015	1-1-1A	SD	529	2.64	2.38	1.11	2.40	1.10	2.51	1.05	2.58	1.02
S6S2	Z-15019	1-1-1A	SD	527	2.71	1.90	1.43	1.85	1.46	2.56	1.06	2.16	1.25
						Mean	1.27	Mean	1.32	Mean	1.05	Mean	1.14
						SD	0.17	SD	0.19	SD	0.09	SD	0.12
S1T3	Z-15019	2-1-2B	TD	487	2.98	2.47	1.21	2.11	1.41	2.58	1.16	2.54	1.17
S1T9	Z-20019	2-1-2B	TD	495	4.55	3.96	1.15	3.77	1.21	3.96	1.15	3.96	1.15
S8T5	Z-20015	2-1-2A	TD	529	2.93	2.65	1.11	2.65	1.11	2.65	1.11	2.65	1.11
S8T6	Z-15019	2-1-2B	SR	546	3.37	2.47	1.36	2.25	1.50	2.85	1.18	2.58	1.31
						Mean	1.21	Mean	1.31	Mean	1.15	Mean	1.18
						SD	0.11	SD	0.18	SD	0.03	SD	0.09
S4T3	Z-20015	0-0-0	TD/MC	480	2.90	2.90	1.00	2.90	1.00	2.67	1.09	2.90	1.00
S4T4	Z-20015	0-0-0	MC	480	2.94	2.90	1.01	2.90	1.01	2.67	1.10	2.90	1.01
S4T5	Z-15019	0-0-0	MC	480	2.92	2.28	1.28	2.66	1.10	2.34	1.25	2.38	1.23
						Mean	1.10	Mean	1.04	Mean	1.15	Mean	1.08
						SD	0.16	SD	0.05	SD	0.09	SD	0.13
S4T1	Z-20019	1-1-1A	TD	480	3.97	4.22	0.94	4.22	0.94	3.89	1.02	4.22	0.94
S4T2	Z-20019	1-1-1A	MC	480	4.42	4.22	1.05	4.22	1.05	3.89	1.14	4.22	1.05
S4T6	Z-15019	1-1-1A	MC	480	2.69	2.66	1.01	2.66	1.01	2.34	1.15	2.66	1.01
						Mean	1.00	Mean	1.00	Mean	1.10	Mean	1.00
						SD	0.05	SD	0.05	SD	0.07	SD	0.05

Table 3. Explanation of Sheeting and Bridging Symbols used in Table 2

Sheeting Symbol in Table 2	Sheeting Name	Bridging Symbol in Table 2	Bridging Locations
KL	Klip-Lok	1-1-1A	2800 mm from end supports
MC	Monoclad	1-1-1B	2890 mm from end supports
TD	Trimdeck	2-1-2A	2370 mm and 4195 mm from end supports
SD	Speed Deck	2-1-2B	2410 mm and 4250 mm from end supports
SH	Spandek Hi-Ten	Note: All bridging is located centrally in the centre span	
SR	Spanrib		

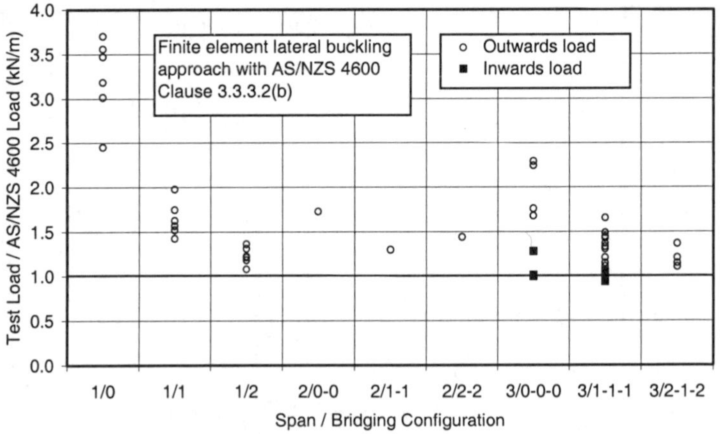

Figure 2. Comparison of FELB design model (Clause 3.3.3.2(b)) with test data

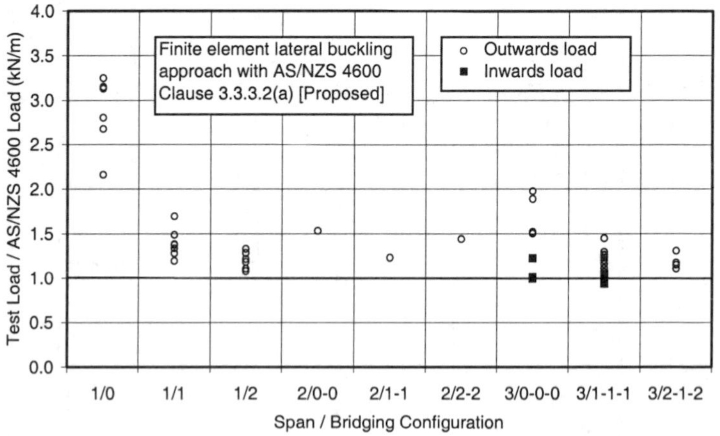

Figure 3. Comparison of modified FELB design model (Clause 3.3.3.2(a)) with test data

3.3 *R-factor approach*

The so-called R-factor approach (Clause 3.3.3.4) is an empiricaly based procedure whereby a single reduction (R) factor is used to account for the lateral and distortional behaviour of purlin systems with screw fastened sheeting. No lateral buckling calculations are required as the R factor is applied directly to the bending section capacity. The advantage of the R-factor approach is that it is calibrated directly from test results and therefore generally provides higher design capacities than the C-factor and FELB approaches. The major disadvantage of the R-factor approach is that it apples only to specific purlin system configurations (concealed fixed sheeting is excluded for example), and consequently is quite restrictive in its use.

4 COMPARISON OF LATERAL BUCKLING APPROACHES AND A PROPOSED MODIFICATION

All cleat fastened purlin tests performed at the University of Sydney have been analysed using the three design approaches outlined in Section 3. The results of the studies are summarised in Table 2, in which the symbol q refers to the uniformly distributed load capacity of the purlin system. The measured yield stress (f_y) and cross-section dimensions, together with capacity factors (ϕ) of unity, were used in all comparisons. The symbols used in Table 2 to define sheeting and bridging configurations are defined in Table 3. The columns of Table 2 headed "FELB – 3.3.3.2(b)", "C-factor" and "R-factor" refer to the direct application of the methods as described in Section 3. Ratios of test load (q_T) to predicted load are provided for all methods. In the finite element lateral buckling analyses (FELB approach), the laps over internal supports and the load height were modelled, and a minor axis rotational restraint (k_{ry}) of 1000 kN (representing the elastic restraint provided by the sheeting to the purlins) was employed.

As foreshadowed in Section 3, it can be seen in Table 2 that the R-factor approach provides the best comparisons with the test results, and the lowest variability. The strength predictions based on the C-factor approach are generally the most inferior and have the highest variability. The FELB approach (with the Clause 3.3.3.2(b) beam curve) may therefore be viewed as the most practically useful design methodology, since it is applicable to all purlin system configurations and incorporates the beneficial effects of sheeting restraint on the elastic buckling moment.

It can be observed in Table 2 that the predictions of the C-factor and FELB approaches improve as the number of rows of bridging increases. This conclusion is also clearly apparent in Figure 2 which has been plotted specifically for the FELB results in conjunction with the Clause 3.3.3.2(a) beam curve. This trend is a reflection of the fact that as the purlin segments between braces become less slender, the mode of failure shifts from lateral buckling to distortional buckling or even combined bending and shear. The latter two modes of failure are evidently captured quite well through the provisions of AS/NZS 4600:1996. However, the conservatism of the strength predictions when failure is clearly premised on lateral buckling is due to the fact that there is apparently more lateral and torsional restraint being provided by the sheeting to the purlins than is included in the models (no torsional restraint from sheeting is included in the FELB predictions, for example).

In view of the conservatism of the strength predictions based on the FELB approach with the *lower* beam curve specified in Clause 3.3.3.2(b), it is appropriate to consider a simple modification to AS/NZS 4600:1996 whereby the FELB approach is permitted to be used in conjunction with the *higher* beam curve specified in Clause 3.3.3.2(a). This should have the effect of improving the computed capacities of those tests which are governed by lateral buckling (which are currently quite conservative), while having little or no effect on the more heavily braced purlin systems for which the strength is governed by distortional buckling or combined bending and shear (which are currently acceptably accurate). The effectiveness of this proposal is encapsulated in the last two columns of Table 2, and in Figure 3. Use of the higher beam curve in conjunction with a FELB lateral buckling analysis evidently provides improvements in computed capacities which range from zero to several percent without introducing any unconservatism.

5 CONCLUSIONS

This paper has outlined three current approaches to the design of purlin systems to AS/NZS 4600:1996 which are referred to herein as the C-factor, R-factor and FELB approaches. As currently stated in AS/NZS 4600:1996, the use of a finite element lateral buckling (FELB) analysis to determine the lateral buckling moment (M_o) is coupled with the beam strength curve specified in Clause 3.3.3.2(b). This beam curve is lower than the one adopted in Clause 3.3.3.2(a), which is used in conjunction with the C-factor approach.

This paper has proposed the use of the higher beam curve (Clause 3.3.3.2(a)) in conjunction with a rational elastic buckling (FELB) analysis of the purlin system. The rationality and effectiveness of the proposal has been confirmed by comparing the predicted strengths with the results of all cleated purlin tests performed at the University of Sydney since the late 1980s.

REFERENCES

AISI 1996. *Specification for the Design of Cold-Formed Steel Structural Members*, 1996 Edition (Printed 1 June 1997), Cold-Formed Steel Design Manual—Part V, American Iron and Steel Institute, Washington DC.

CASE 1997. *PRFELB—A Computer Program for Finite Element Flexural-Torsional Buckling Analysis of Plane Frame*, Version 3.0, Centre for Advanced Structural Engineering, Department of Civil Engineering, The University of Sydney, April 1997.

Hancock, G. J., Celeban, M., Healy, C., Georgiou, P. N. and Ings, N. 1990. "Tests of Purlins with Screw Fastened Sheeting under WInd Uplift", *Proceedings, Tenth International Specialty Conference on Cold-formed Steel Structures*, St. Louis, Missouri, U.S.A., pp. 393–419.

Hancock, G. J., Celeban M., and Healy, C. 1992. "Tests of Continuous Purlins Under Downwards Loading", *Proceedings, Eleventh International Specialty Conference on Cold-formed Steel Structures*, St. Louis, Missouri, U.S.A., pp. 155–179.

Hancock, G. J., Celeban, M. and Healy, C. 1994. "Tests of Purlins with Concealed Fixed Sheeting", *Proceedings, Twelfth International Specialty Conference on Cold-formed Steel Structures*, St. Louis, Missouri, U.S.A., pp. 489–511.

Hancock, G. J., Celeban, M. and Popovic, D. 1996. "Comparison of Tests of Purlins with and without Cleats", *Proceedings, Thirteenth International Specialty Conference on Cold-formed Steel Structures*, St. Louis, Missouri, U.S.A., pp. 155–175.

Hancock, G. J. 1998. *Design of Cold-Formed Steel Structures*, 3rd Edition, Australian Institute of Steel Construction, Sydney.

SA 1988. *AS 1538–1988 Cold-Formed Steel Structures*, Standards Australia, Sydney.

SA/SNZ 1996. *AS/NZS 4600:1996, Cold-Formed Steel Structures*, Standards Australia/ Standards New Zealand, 1996.

Mechanics of Structures and Materials, Bradford, Bridge & Foster (eds)
© 1999 Balkema, Rotterdam, ISBN 90 5809 107 4

ColdSteel/4600 – Software for the design of cold-formed steel structures

M.J.Clarke
Department of Civil Engineering, University of Sydney, N.S.W., Australia

ABSTRACT: The limit states Australian/New Zealand Standard *AS/NZS 4600:1996 Cold-formed steel structures* was published in late 1996 and supersedes the corresponding permissible stress Standard AS 1538–1988. AS/NZS 4600:1996 is based mainly on the 1997 edition of the American Iron and Steel Institute (AISI) Specification. In some cases, design rules from AS 1538–1988 have been carried through to the new standard, allowing for the appropriate conversions to limit states format. Reflecting the research output in the intervening years, there are also some new design provisions in AS/NZS 4600:1996, such as rules for distortional buckling, which do not appear in either AS 1538–1988 or the AISI Specification.

At the time AS/NZS 4600:1996 was in draft form for public review, the Centre for Advanced Structural Engineering at the University of Sydney embarked on a project to develop a user-friendly computer program, **ColdSteel/4600**, to assist manufacturers and practising engineers apply the new standard. Concomitant objectives were the education of students and engineers in the principles of cold-formed design, and promotion of the latter in Australia and New Zealand.

This paper describes the scope and features of the program **ColdSteel/4600** and provides a brief example of its use.

1 INTRODUCTION

This paper describes the development, scope and features of the user-friendly program **ColdSteel/4600** for the design of cold-formed steel structures to the current Australian/New Zealand Standard AS/NZS 4600:1996 (SA/SNZ 1996, 1998). The present standard is in limit states format and therefore constitutes a major revision of the preceding (permissible stress) standard AS 1538–1988. Although AS/NZS 4600:1996 is based on the current edition of the American Iron and Steel Institute (AISI) Specification, there are also several new design provisions not contained in the latter document.

Apart from the change from permissible stress to limit states format, another significant philosophical change in AS/NZS 4600:1996 is the adoption of the so-called "unified concept" for the calculation of effective cross-section properties. In essence, the unified concept requires the calculation of effective sections for all types of member and section capacities based on the assumed stress distribution at the ultimate strength limit state. In the case of a compression member, for example, the effective cross-section used in the calculation of section capacity is determined assuming the whole section is subjected to a uniform stress equal to the yield stress (f_y). The effective cross-section used in the calculation of member capacity, on the other hand, is determined assuming the cross-section is subjected to a uniform stress equal to the inelastic buckling stress (f_n). As f_n reduces, the effective cross-sectional area increases until the member is sufficiently slender that the cross-section is fully effective. This continual calculation of effective sections, which may involve an iterative procedure for sections in bending, is a major source of complexity in AS/NZS 4600:1996.

It is clear that to facilitate the design of cold-formed steel structures, a user-friendly computer program is required. **ColdSteel/4600** is such a program. **ColdSteel/4600** is intended to be used as a cold-formed steel design "calculator" that facilitates the semi-automated design of cold-formed steel structural members by freeing the engineer from the complex and error prone details of effective section, distortional buckling stress and other detailed design computations.

The following sections of this paper describe the scope, capabilities and method of operation of the program, and give brief examples of its application.

2 SCOPE

2.1 *Range of cross-sections*

Unlike hot-rolled sections which are available in only a few standard shapes (I-sections, channel sections, angle sections, T-sections, etc), cold-formed structural members may be manufactured by cold-forming thin sheet steel into an infinite variety of cross-sectional shapes. The procedures used to determine the effective cross-sections depend on the nature of the various constituent plate elements, which may be classified as fully stiffened, partially stiffened by edge or intermediate stiffeners, or unstiffened. For partially stiffened elements, the effective width calculations are influenced by the number, shape and size of the stiffening elements.

In view of the preceding remarks, it is apparent that it is virtually impossible to develop a cold-formed steel design program for an arbitrary "generic" cross-section for which the geometry is input by the user as a series of "nodes" connected by "elements". **ColdSteel/4600** is therefore necessarily restricted in operation to the range of sections for which it has been specifically programmed. The current version of **ColdSteel/4600** (Version 1.1) incorporates the range of sections shown in Figure 1. Although only the basic shapes shown in Figure 1 are available, the relevant characteristic dimensions can be customised by the user provided the assumed structural behaviour of the section remains valid.

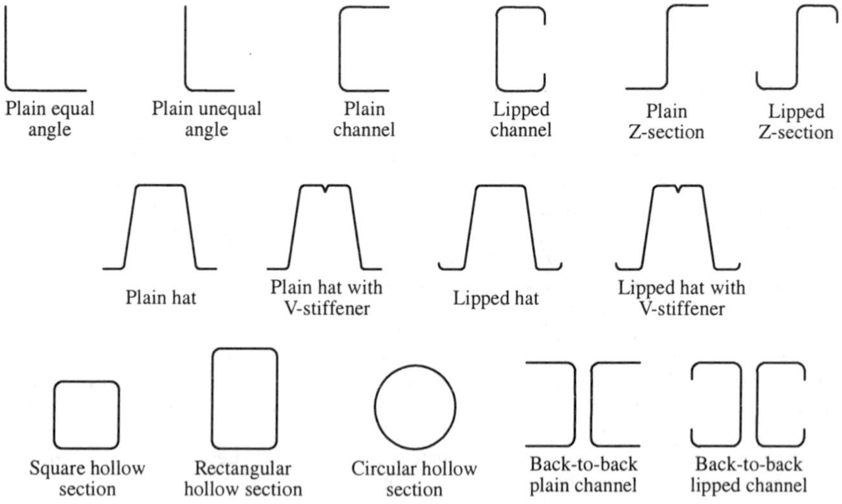

Figure 1. Cross-sections incorporated into **ColdSteel/4600**

2.2 *Design rules incorporated*

ColdSteel/4600 is based on the member strength design rules specified in AS/NZS 4600:1996. The rules incorporated include the principal clauses of Section 2 (Elements) and Section 3 (Members). No connector or connection design rules are incorporated in **ColdSteel/4600**. The member design rules included relate to design against:

- yielding and fracture in pure tension;
- cross-section yielding, and local, distortional and lateral buckling of beams in bending;
- yielding and buckling of webs in shear, and combined bending and shear;
- yielding and buckling of webs in bearing, and combined bending and bearing;
- cross-section yielding, and local, distortional and flexural-torsional buckling of compression members; and
- failure of members under combined bending and tensile or compressive axial force.

At the present time, **ColdSteel/4600** is not linked to any structural analysis program. However, the software has been developed in such a manner that it would be a relatively simple task to fully integrate it with any analysis program.

3 PROGRAM DESCRIPTION

3.1 *Main form*

After the initial splash screen, the main form of **ColdSteel/4600** is displayed as shown in Figure 2. The majority of the data that is required to perform a member strength check or design is displayed on the Main form. However, since some of the design actions may be zero, it may not be necessary to enter data for every input parameter. At all times, the relevant data items are clearly delineated, and the unnecessary items are shaded the same colour as the Main form.

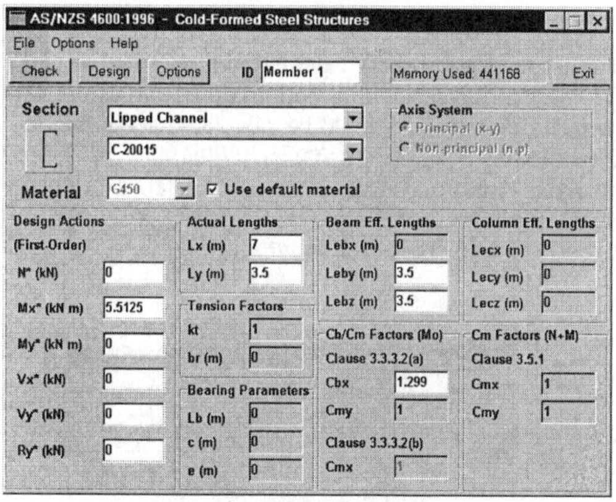

Figure 2. Main form of **ColdSteel/4600**

3.2 *Check and design modes*

ColdSteel/4600 may be operated in either check or design modes. Clicking on the Check button on the Main form will instruct **ColdSteel/4600** to perform a strength check using the currently chosen cross-section, material, design actions and other parameters. All relevant strength limit states specified in AS/NZS 4600:1996 are examined. The result of the strength check is displayed in summarised form in a window which indicates the cross-section class, designation and material, together with the governing "load factor" (λ) and the governing mode of failure. A load factor of at least unity indicates that the member has satisfactory strength.

Clicking on the Design button on the Main form instructs **ColdSteel/4600** to perform a design using the currently chosen cross-section type, design actions and other parameters. The cross-section designations for the chosen cross-section type (e.g. lipped channel) are ranked in order of mass per unit length and a strength check proceeds from the lightest to heaviest sections until a satisfactory design ($\lambda \geq 1.0$) is found. The load factor is displayed, and an indication of the criterion governing the capacity is also given.

3.3 *The ColdSteel/4600 database*

The **ColdSteel/4600** database file **ColdSteel.ini** is a text file containing the material and dimensional information of all the sections as input by the user of the program. Other section specific data, such as distortional buckling stresses (f_{od}) in compression and/or bending, can also be defined in the **ColdSteel/4600** database. Each of the cross-section types has a defined data format as shown in Figure 3 for the case of a lipped channel section. The user is not required to enter any cross-sectional properties (area, second moments of area, warping constant, etc.), into the **ColdSteel/4600** database as all the relevant properties are calculated "on-the-fly" from the basic cross-sectional dimensions.

3.4 *Program options*

The Options form, invoked by clicking on the Options button on the Main form, enables the user to set several fundamental parameters which control the program operation and facilitate access to the more unusual or advanced features. The particular options which are available are grouped in the following categories:

- General options
- Compression options
- Bending options
- Distortional buckling options.

As indicated in Figure 4, the General options form enables the user to specify preferences relating to units, the nature of design actions, section property calculation and effective width determination in unstiffened elements. The Distortional buckling options form allows the user to select whether the relevant distortional buckling stresses should be calculated according to a simplified model such as the one described in Appendix D of AS/NZS 4600:1996, or whether the results of a rational elastic buckling analysis (e.g. finite strip) of the whole plate assemblage should be employed. All options are context sensitive in the sense that a particular option can only be selected if it is relevant to the currently chosen cross-section type.

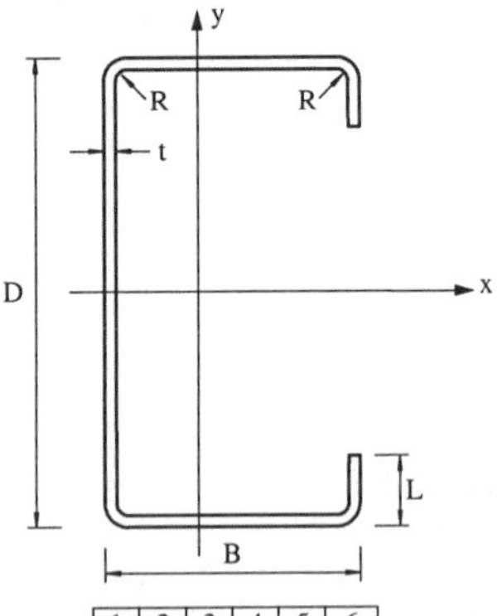

1	2	3	4	5	6
t	D	B	L	R	f_{odx}

Figure 3. [LippedChannel] definition

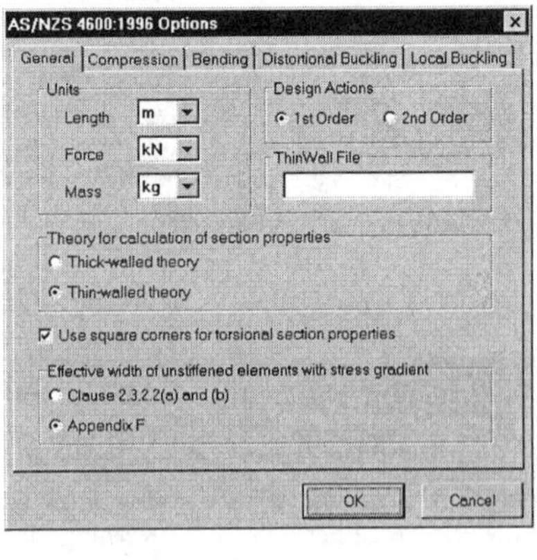

Figure 4. General options form

3.5 *Visualisation of full and effective sections*

A major educational feature of **ColdSteel/4600** is that it enables the user to visualise the full section geometry, and the effective sections under a range of loading conditions including pure compression and pure bending about both cross-section axes in both a positive and negative sense. For each type of loading condition, the effective section corresponding to both section capacity (where the maximum stress is equal to the yield stress) and member capacity (where the extreme compressive fibre stress is typically less than yield) can be drawn. Examples of the effective sections for a C-20015 channel section are shown in Figure 5.

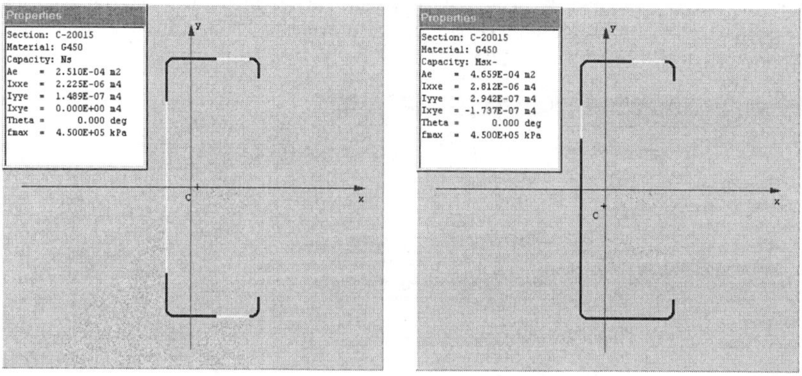

Figure 5. Effective sections for C-20015 channel section in compression and bending

4 EXAMPLES

4.1 *Simply supported lipped channel section purlin*

This example is adapted from the one described in Section 5.8.1 of Hancock (1998), in which full calculation details may be found.

Consider a C-20015 purlin simply supported over a 7 m span with one brace at the centre. The member is subjected to a uniformly distributed upwards load of 0.9 kN/m as shown in Figure 6. The following describes how **ColdSteel/4600** can be used to determine whether the member is satisfactorily designed to AS/NZS 4600:1996 with respect to all relevant strength limit states.

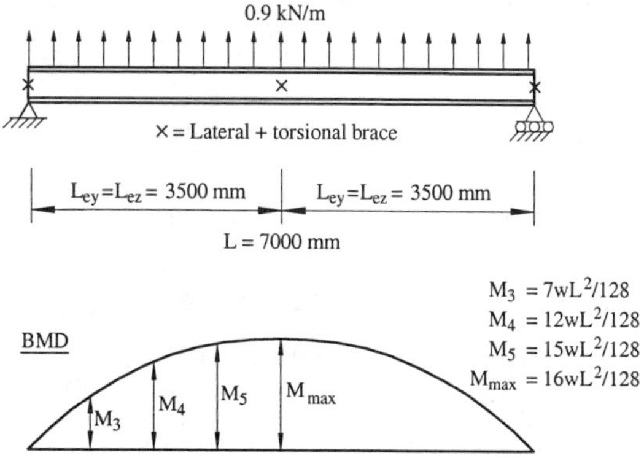

Figure 6. Simply supported purlin with central brace

The relevant C-20015 lipped channel section is a specific instance of the [LippedChannel] section class defined in the **ColdSteel/4600** database. The maximum design moment is $qL^2/8 = 5.5125$ kNm, the beam effective lengths are $L_{ey} = L_{ez} = 3.5$ m, and the moment modification factor (C_{bx}) is 1.299 for the segment from the end support to the central brace. The options within **ColdSteel/4600** are set to prescribe the use of thin-walled theory for section property calculations, square corners for torsional section properties, and Appendix D of AS/NZS 4600:1996 for distortional buckling stress calculations.

The completed Main form for this example is shown in Figure 2, and the result of performing the strength check is indicated in Figure 7. With a computed load factor of 1.10, the member is evidently safely designed, and the most critical design condition is that of lateral buckling (Clause 3.3.2b). Clicking on the Full Details button provides a listing of all relevant input parameters, section properties and design capacities.

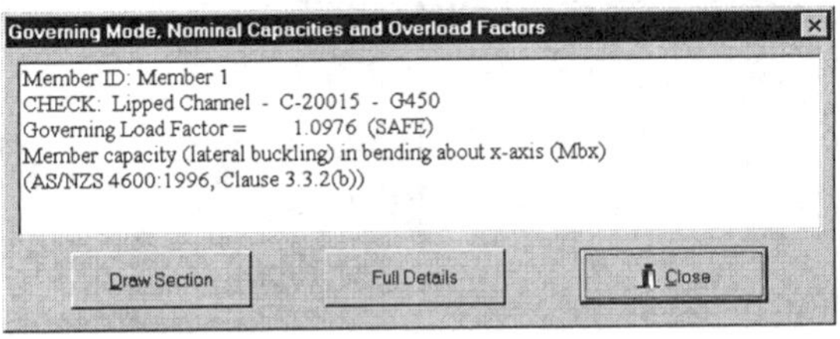

Figure 7. Design check for simply supported purlin example

5 CONCLUSIONS

The recently published limit states Australian/New Zealand standard AS/NZS 4600:1996 *Cold-formed steel structures* is considerably more complex than its permissible stress predecessor AS 1538–1988. Designers and manufacturers who wish to take advantage of the new standard will therefore benefit from the development of a computerised design aid. Such a design aid will also serve an important role in the education of students undertaking advanced-level courses in structural steel design.

This paper describes the user-friendly program **ColdSteel/4600** which has been developed by the Centre for Advanced Structural Engineering at the University of Sydney for the design of cold-formed steel structures to AS/NZS 4600:1996. The program incorporates a wide range of commonly used cold-formed sections and performs strength design checks for all relevant section and member strength limit states. As part of this process, all the relevant full and effective section properties are calculated. Users can input their own cross-section dimensions.

Although the current version of **ColdSteel/4600** runs in a stand-alone mode, future versions will be able to be linked to structural analysis programs to provide fully integrated analysis and design of cold-formed steel structures.

REFERENCES

AISI 1996. *Specification for the Design of Cold-Formed Steel Structural Members*, 1996 Edition (Printed 1 June 1997), Cold-Formed Steel Design Manual—Part V, American Iron and Steel Institute, Washington DC.

Hancock, G. J. 1998. *Design of Cold-Formed Steel Structures*, 3rd Edition, Australian Institute of Steel Construction, Sydney.

SA 1988. *AS 1538–1988 Cold-Formed Steel Structures*, Standards Australia, Sydney.

SA/SNZ 1996. *AS/NZS 4600:1996, Cold-Formed Steel Structures*, Standards Australia/ Standards New Zealand, 1996.

SA/SNZ 1998. AS/NZS 4600 Supplement 1:1998, Cold-Formed Steel Structures—Commentary (Supplement 1 to AS/NZS 4600:1996), Standards Australia/ Standards New Zealand, 1998.

Environmental loadings

Mechanics of Structures and Materials, Bradford, Bridge & Foster (eds)
© 1999 Balkema, Rotterdam, ISBN 90 5809 107 4

A simplified approach for calculating fire resistance of steel frames under fire conditions

K.H.Tan, W.S.Toh & Z.Yuan
School of Civil and Structural Engineering, Nanyang Technological University, Singapore

ABSTRACT: The Merchant-Rankine formula is a simple approach for calculating the failure load of a frame subjected to increasing loads and constant ambient temperature. It takes account of the interaction of stability and strength and the calculated failure load is close to that obtained using second order elasto-plastic finite element analysis. This paper presents a novel approach to predict the fire resistance of steel frames using the Merchant-Rankine formula. The predictions for a series of six frames are compared with the finite element analyses and actual test results. Apart from one test result, the comparison shows that the application of this formula to frames subjected to fire conditions yields consistently good agreement with finite element method and actual test results.

1 INTRODUCTION

At ambient temperature, the failure load factor of a steel frame can be approximated using the simplified Merchant-Rankine approach (Horne & Merchant 1965) that makes use of two simple programs, viz. rigid-plastic analysis and elastic buckling analysis. This is because the Merchant-Rankine load is based on the rigid-plastic collapse load factor λ_p and the lowest buckling load λ_c. The Merchant-Rankine formula is simply stated as:

$$\frac{1}{\lambda_f} = \frac{1}{\lambda_p} + \frac{1}{\lambda_c} \tag{1}$$

where λ_f denotes the actual failure load factor; λ_p denotes the rigid plastic collapse load factor; λ_c denotes the critical buckling load factor.

Figure 1 shows the various types of analysis available for the computation of load-deflection behavior of a structure. Although both the rigid-plastic analysis and the elastic buckling analysis give upper bound values for λ_p and λ_c, respectively, the Merchant-Rankine load gives a lower bound estimate of actual failure load factor λ_f close to that obtained using second-order elastic-plastic analysis. These have been proved to be valid for a wide range of structures (Horne 1951).

In the Merchant-Rankine approximation of λ_f (Horne 1963) the following assumptions are involved:

(a) The deflected form for the frame at collapse is assumed to be similar in the elastic buckling analysis as well as the rigid-plastic analysis.
(b) Local and lateral torsional buckling of members are not considered.
(c) Finite deformations have no effects on the equilibrium state.
(b) The frame is uniformly heated and there is no thermal gradient across a section.
(c) The effect of axial load on the plastic moment of resistance at a section can be ignored.
(d) The axial loads in the members of a structure are proportional to the load factor, irrespective of the state of structure, be it in the elastic or elastic-plastic state.

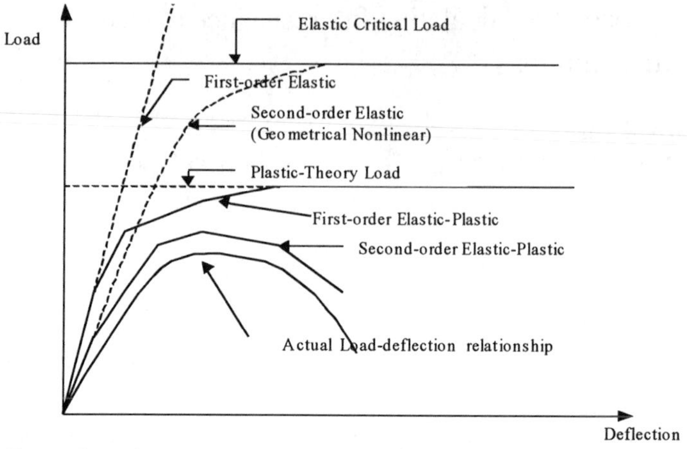

Figure 1 The load-deflection of the structure

Table 1. Reduction Factors For Stress-Strain Relationship

T (°C)	$k_y(T) = f_y(T)/f_y$	$k_E(T) = E(T)/E$
20	1.00	1.00
100	1.00	1.00
200	1.00	0.90
300	1.00	0.80
400	1.00	0.70
500	0.78	0.60
600	0.47	0.31
700	0.23	0.13
800	0.11	0.09
900	0.06	0.0675
1000	0.04	0.0450
1100	0.02	0.0225
1200	0.00	0.00

For frames subjected to fire conditions, the applied loads remain constant but now the temperature increases beyond the ambient temperature. Consequently, both the elastic modulus E and the material yield strength f_y deteriorate with continuous increase in temperature T, that is, they are now functions of temperature T. The reductions of E and f_y are adopted from the Eurocode 3 (1995) and are included in Table 1. This means that both the buckling load and the rigid-plastic collapse load reduce with T, while the working loads remain constant. To apply the Merchant-Rankine formula to frames under fire conditions, modifications to the load factors in (1) are required.

$$\frac{1}{\lambda_f(T)} = \frac{1}{\lambda_p(T)} + \frac{1}{\lambda_c(T)} \tag{2}$$

Here, $\lambda_f(T)$ denotes the actual failure load factor at critical temperature T; $\lambda_p(T)$ denotes the rigid-plastic collapse load at critical temperature T; $\lambda_c(T)$ denotes the first elastic critical buckling load at temperature T. Thus, instead of determining the actual failure load factor λ_f, we determine the critical temperature T_c at which a structure fails under constant loads.

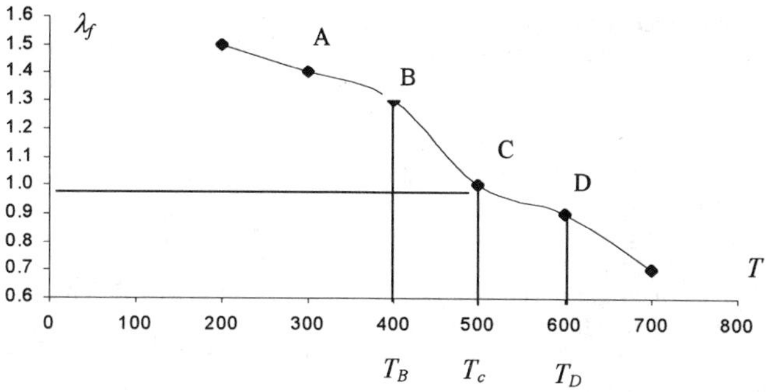

Figure 2 Finding the critical temperature T_c

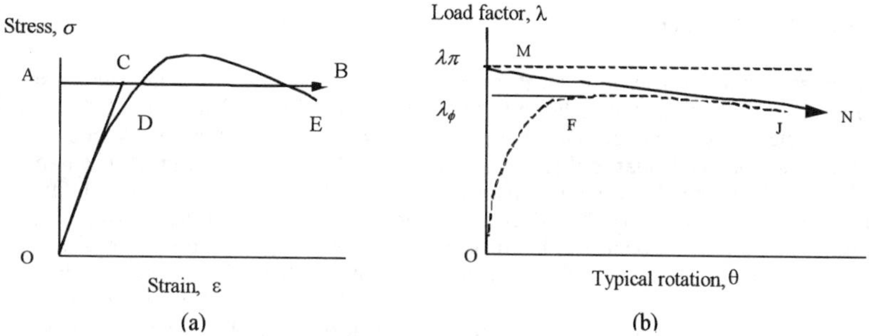

Figure 3 Stress-strain relationship and λ_p vs rotation θ

2 GENERAL DESCRIPTION

At a particular temperature T_A, the frame has a set of material properties $E(T_A)$ and $f_y(T_A)$. The reductions of E and f_y follow the recommendation by the Eurocode 3 (1995) (Table 1). The approach analyses both the rigid plastic and elastic buckling load factors at temperature T_A using the reduced material properties $E(T_A)$ and $f_y(T_A)$. If the failure load factor $\lambda_f(T)$ corresponding to that temperature T_A exceeds unity, that means the elevated temperature T_A has exceeded critical temperature T_c. Conversely, if $\lambda_f(T)$ is less than unity, the structure will remain stable upon further increase in temperature T.

Figure 2 shows six computations of $\lambda_f(T)$ performed at six different temperatures T_A, T_B, T_C, etc. The theoretical critical temperature for the structure fail corresponds to $\lambda_f(T) = 1.0$. Suppose the load factor $\lambda_f(T)$ has a value above unity at temperature T_B and below unity at another temperature T_D. Clearly, the failure temperature T_c corresponding to $\lambda_f(T) = 1.0$ lies in between T_B and T_D. Thus, T_C can be calculated either by linear interpolation or, for more accurate prediction, by using more trial temperatures within the region T_B and T_D. In this manner, the modified Merchant-Rankine formula takes account of the deteriorating material properties with elevated temperatures.

3 RIGID-PLASTIC ANALYSIS

The rigid-plastic load of a structure is derived from the rigid plastic theory, which is based on the rigid-plastic stress-strain relationship (Horne 1951). This is represented by the curve OAB in Figure 3(a). It is noteworthy that the rigid-plastic theory assumes zero deformation under increasing load factor λ, until λ reaches the rigid plastic collapse load factor λ_p. At this stress level, plastic hinge of constant moment of resistance is assumed to form at the section with the maximum moment. In reality, actual material stress-strain relationship is closer to the non-linear curve ODE (Fig. 3(a)). Thus, the idealised curve OCB represents an approximation of purely elastic-plastic relationship.

When the rigid-plastic behaviour is pursued through finite deformations, the load-deflection curve for a portal frame subjected to top and side loads is depicted by the drooping curve OMN in Figure 3(b). And when elastic deformations are taken into account, the load-deflection relationship follows the curve OFJ (Figure 3(b)). Thus, in structures with drooping rigid-plastic load deflection curves, λ_p is an overestimate of the actual failure load factor λ_f.

It is noteworthy that in applying the rigid-plastic analysis for frames under fire conditions, the effect of induced thermal forces is akin to that of initial imperfections or residual stresses for the conventional plastic analysis at constant ambient temperature. That is to say, initial imperfections or residual stresses may affect the sequence of plastic hinge locations or even the collapse mechanism, but not the ultimate collapse load (Fung et al. 1999).

4. ELASTIC CRITICAL LOAD

When an ideal structural element or an ideal frame is subjected to concentric axial compressive load, theoretically, it will remain straight up to the limit of critical load (Tan et al. 1999). Beyond which, the element or structure can either remain straight (unstable equilibrium), or simply buckle in a direction perpendicular to that of load application (neutral equilibrium). The transition point from a straight configuration to a buckled shape is referred to as the point of bifurcation of equilibrium (Chen & Lui 1987). Thus, by setting the determinant of the global stiffness matrix to zero, the bifurcation point can be identified. This approach is also known as the eigenvalue approach, which is used for finding eigenvalues of a matrix. The critical conditions are represented by the eigenvalues of the global stiffness matrix, and the displaced configurations by the eigenvectors. The lowest eigenvalue is the critical load of the system. It is noteworthy that Wong & Patterson (1996) has used a similar method to obtain the buckling of resistance of frames under elevated temperatures. This paper considers the combined effects of stability and strength. Two programs are used in the Merchant-Rankine approach, viz. rigid plastic analysis which assumes small deflection theory and undeformed geometry, and elastic critical buckling analysis which assumes small defection theory and infinite elastic deformations (Smith & Griffiths 1988).

5. RESULTS AND DISCUSSION

The Merchant-Rankine approach is verified by finite element analyses (Toh 1998) and actual test results based on a series of six EHR frames (Rubert and Schaumann 1986). These two-member braced frames were all fully and uniformly heated. Details of the frames are included in Table 2. All members in the frame are restrained against torsional displacement and out-of-plane deformation by providing adequately spaced stiffener elements indicated in Figure 4. The finite element method consists of an assemblage of co-rotational beam elements in two-

Table 2 The dimensions and loadings for all EHR frames (Rubert & Schaumann 1986)

No	F1 (kN)	F2 (kN)	H (m)	L (m)	E (kN/mm^2)	A (mm^2)	I (cm^4)	EA (kN)	EI (kNm^2)	M_p (kNm)
1	56	14	1.17	1.19	210	764	80.14	160440	168.29	9.172
2	84	21	1.17	1.24	210	764	80.14	160440	168.29	9.172
3	112	28	1.17	1.24	210	764	80.14	160440	168.29	8.870
4	20	5	1.50	1.25	210	764	8.49	160440	17.83	2.264
5	24	6	1.50	1.25	210	764	8.49	160440	17.83	2.264
6	27	6.7	1.50	1.25	210	764	8.49	160440	17.83	2.264

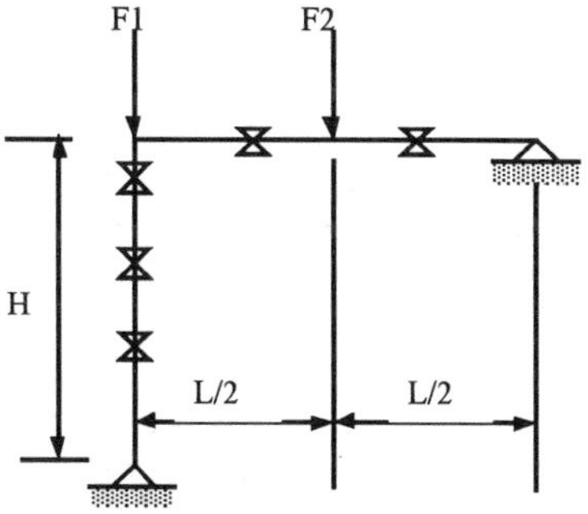

Figure 4 Layout of EHR frame (Rubert & Schaumann 1986)

Table 3 Comparison of predicted critical temperature with measured test results

Frame ID	$T_{c(TEST)}$ (°C)	$T_{c(MR)}$ (°C)	$T_{c(FE)}$ (°C)	$\frac{T_{c(MR)}}{T_{c(TEST)}}$	$\frac{T_{c(FE)}}{T_{c(TEST)}}$
EHR 1	600	630	620	1.050	1.034
EHR 2	530	544	545	1.026	1.028
EHR 3	475	451	458	0.949	0.964
EHR 4	562	522	531	0.929	0.944
EHR 5	460	458	479	0.996	1.040
EHR 6	523	403	425	0.771	0.813

Table 4: Comparison of calculated T_c and measured T_c for frames EHR 1 to EHR 5 (Rubert & Schaumann 1986)

	$T_{c(MR)}/T_{c(TEST)}$	$T_{c(FE)}/T_{c(TEST)}$
Mean	0.990	1.002
Std deviation	0.051	0.045
Coef. of variation	0.051	0.045

dimensional plane. Geometric nonlinearities are modelled with a close-form geometric stiffness matrix and an updated reference geometry, while material nonlinearities are modelled with a set of nonlinear stress-strain relationships for steel with respect to temperature. The material-induced and temperature-dependent nonlinearities on the strain state are taken into account by means of section discretisation at the element nodes in the form of a fibre model.

The predictions of collapse temperature T_c from (i) Merchant-Rankine approach and (ii) finite element analyses and actual test results are summarised in Table 3. The material modelling for $E(T)$ and $f_y(T)$ are adopted from the Eurocode 3 (1995) (Table 1). Using actual test results as a yardstick, the theoretical predictions by the Merchant-Rankine approach compares favourably with those obtained from the second-order-elasto-plastic-large-displacement finite element analysis. A comparison of the ratio of $T_{c(MR)}/T_{c(TEST)}$ and $T_{c(FE)}/T_{c(TEST)}$ is shown in Table 4. Of the six tests, only the last predicted value lies outside of 10% error range. The finite element prediction for this case (Toh 1998) also shows that the last predicted value is outside of 10% error range. The deviation may be largely due to experiment error. Based on the comparison study for the first five frames (EHR 1-5), the mean value of the ratios $T_{c(MR)}/T_{c(TEST)} = 0.990$, indicating that the proposed approach yields very accurate critical temperatures for the tested frames under elevated temperatures.

In the finite element analysis, the stress-strain relationship of mild steel changes continuously with elevated temperature. Plasticity spreads from the outer fibre towards the neutral axis under elevated temperatures. Thus, a more realistic material modelling is employed in the finite element analysis. Nevertheless, apart from the EHR 6 frame, the errors for both methods are well within 5% and thus, they are deemed satisfactory for engineering application.

6. CONCLUSION

The Merchant-Rankine formula is a simple approach for calculating the collapse temperatures of steel frames subjected to elevated temperatures. It takes account of the interaction of stability and strength. The predictions for a series of six frames are compared with the finite element analyses and actual test results. The comparison study shows that the application of this formula to frames subjected to fire conditions yields consistently good agreement with finite element method and actual test results.

ACKNOWLEDGMENTS

The Nanyang Technological University, Singapore provided the funding for this research through the Applied Research Project RG39/96.

REFERENCES

CEN (1995). Eurocode 3: Design of steel structures. Part 1.2: General Rules - Structural Fire Design. *ENV 1993 – 1-2.* European Committee for Standardisation.

Chen, W.F. & Lui, E.M. (1987). Structural Stability:theory and implementation. Prentice Hall.

Horne, M.R. (1951). Fundamental proportions in the plastic theory of structures. *J. Instn Civ. Engrs.* Vol. 34, 174.

Horne, M.R. (1963). "Elastic-plastic failure loads of plane frames", *Proc. Roy. Soc. A.*, 274, 343-364.

Horne, M.R., and Merchant, W. (1965). *The stability of frames*, Pergamon Press, Oxford.

Fung, T.C., Tan, K.H. & Toh W.S. (1999). Rigid plastic analysis of frames under fire conditions. Submitted to *J. Structural Engineering & Mechanics.*

Rubert, A. & Schaumann, P. (1986). Structural steel and plane frame assemblies under fire action. *Fire Safety Journal.* Vol. 10, pp. 173-184.

Smith, I.M. & Griffiths, D.V. (1988). Programming the finite element method. Second edition, *Wiley,* New York.

Tan, K.H., Fung, T.C. & Toh, W.S. (1999). A simple approach to strength and stability of steel columns in fire. Conference Proceedings of EASEC-7, Japan.

Toh W.S. (1998). Stability and strength of steel structures under thermal effects. Interim research report submitted to Nanyang Technological University.

Wong, M.B. & Patterson, N.L. (1996). Unit load factor method for limiting temperature analysis of steel frames with elastic buckling failure mode. *Fire Safety Journal.* Vol. 27, 113-122.

Mechanics of Structures and Materials, Bradford, Bridge & Foster (eds)
© 1999 Balkema, Rotterdam, ISBN 90 5809 107 4

Fire testing of high strength concrete filled steel columns

N.L.Patterson, X.L.Zhao, M.B.Wong, J.Ghojel & P.Grundy
Department of Civil Engineering, Monash University, Melbourne, Vic., Australia

ABSTRACT: Over the past few decades, composite columns have become a popular component of the structural skeleton. One such composite column is manufactured by filling steel hollow section tubes with high strength concrete. The use of steel and concrete together in this configuration has many advantages. However the current technique of providing adequate fire resistance to this type of composite column is by external fire protection which significantly adds to the cost of construction. Monash University has conducted research on the feasibility of using high strength concrete filled steel columns without any external fire protection in which part of the column perimeter is directly exposed to the elevated temperature atmosphere. This is the first experimental study with this temperature exposure condition for these types of columns. Various composite column sections were used in the furnace tests to study their behaviour and performance. This paper reports some results obtained from this experimental study.

1 INTRODUCTION

The introduction and general acceptance of high strength concrete by the construction industry has led to further advantages in composite construction. However with new types of structural elements being introduced into the building skeleton much research and testing must be conducted to investigate the performance of these elements under various loading scenarios.

One type of composite column, which is receiving increasing attention, is manufactured by filling steel hollow section tubes with concrete. The use of steel and concrete together in this configuration has many advantages. The concrete infill eliminates the possibility of inward buckling of the steel tube, which adds to the compressive strength of concrete by confining it. The positioning of the steel shell at the extreme tension side of the column is most effective in resisting bending. The steel tube also acts as permanent formwork, reducing the amount of labour required and increasing the speed of construction.

The dimensions of the columns can also be smaller when high strength concrete is used as infill than for a reinforced concrete column of the same strength, thereby increasing the available floor space. However the current technique of providing adequate fire resistance to this type of composite column is by external fire protection which significantly adds to the cost of construction and reduces the available floor space. The reason for adopting this conservative approach is due to the lack of experimental data on the performance of these members in fire conditions.

Monash University has recently conducted elevated temperature research on the feasibility of using high strength concrete as an infill for the steel tubes without any external fire protection. This is believed to be the first experimental study on this type of composite column. The columns were subjected to the ISO834 fire curve while eccentrically loaded in compression. The columns were non-uniformly exposed to the elevated temperatures (on three sides only). This is believed to more realistically represent a fire scenario and has been considered as being more critical for these types of columns. Variations in the steel tube column sections include width (or diameter), load eccentricity and thickness.

2 PREVIOUS RESEARCH

Better performance in fire resistance is a major benefit in the use of structural hollow sections when they are infilled with concrete. As the member is heated up a higher proportion of the load will be transferred from the steel shell to the somewhat cooler concrete core.

Many elevated temperature tests have been conducted on reinforced concrete columns, plain concrete filled steel columns and steel fibre reinforced steel tubes. Sakumoto et al (1994) conducted fire performance tests on eccentrically loaded rectangular hollow section columns filled with 38 MPa concrete. They found that these types of columns are very susceptible to failure when the load is applied eccentrically. Some of their experimental tests involved externally protected columns in which the time to failure was significantly prolonged.

In the past decade the use of high strength concrete as an infill in these columns has slowly been accepted by engineers and room temperature testing has been conducted (Rangan 1991 and Rangan & Joyce 1993). Since very little research has been conducted on their elevated temperature behaviour high strength concrete filled columns, which are fitted with external fire protection, are more than adequate for fire resistance purposes.

To the authors' knowledge no experimental work had been conducted on high strength concrete filled steel tubular sections, which were exposed to a non-uniform temperature distribution.

3 TEST SPECIMENS - PREPARATION & INITIAL MEASUREMENTS

A total of ten specimens were tested. Of these two were conducted at room temperature and the remaining were tested under the standard fire time-temperature conditions according to ISO834. The specimens consisted of both rectangular and circular sections with maximum cross-sectional dimensions of between 100 mm and 219.1 mm. The thicknesses of these tubes varied between 3.6 mm and 6.5 mm. All specimens were 3600 mm in length and were provided by Tubemakers, Australia.

The initial imperfections (length, thickness and external dimensions) of the tubes were measured along with the mechanical properties of both the steel and concrete. The initial maximum out-of-straightness was also measured prior to testing. These are reported in Table 1 below. Yield strength values for the square hollow sections were taken from both the flats and the corners of the tubes.

To commence column preparation one of the end plates were welded onto the steel tube so that the upper and lower faces could be marked for the placement of thermocouples. The end plate was aligned so the longitudinal weld of the steel tube was at 90 degrees to the upper face.

Table 1. Test Specimen Details

		Strength (MPa)			Initial Imperfections (mm)			
ID No.	Dimensions (mm)	Yield Steel Proof, UTS	Compressive Concrete 28 Day	DOT	Average Thickness	Avg Length	Diam/ Width	Mid height Bow
SR1	100.0x4.0	410,489	77.27	86.34	+0.150	-3	-0.2	1.75
CR1	114.3x3.6	446,480	77.27	86.34	+0.133	-1	-0.05	1.45
SF1	100.0x4.0	482,551 509,547 cnr	87.89	93.64	-0.055	-	+1.0	1.250
SF2	100.0x6.0	505,561 537,560 cnr	88.77	116.61	+0.046	-	+1.0	3.825
SF3	150.0x6.0	407,484 587,635 cnr	89.64	96.22	-0.127	-7	-0.3	2.172
SF4	200.0x6.5	490,556 566,612 cnr	89.64	93.17	-0.189	-2	-0.2	3.380
CF1	114.3x4.8	398,460	85.58	112.83	+0.07	+2	+0.7	1.863
CF2	114.3x6.0	396,445	91.76	109.94	+0.015	+1	+0.7	3.220
CF3	168.3x6.4	505,605	89.64	95.102	0.204	-4	+0.1	2.559
CF4	219.1x6.4	488,629	89.64	97.670	0.114	-2	+0.15	2.704

DOT - Day of column test

Thermocouples were placed at various positions throughout one half of the cross-section to obtain an estimate of the temperature profile through the column with time. A diagram showing the approximate position of the thermocouple positions is shown in Figure 1. The thermocouples were placed at a distance of no more than 300 mm from the mid-span of the steel tube. Holes were drilled at various positions around the tube cross-section and the thermocouples were strung together and inserted into the holes so that one end protruded out of the column side. The ends of the thermocouple groups were then fixed to the column so that they were not dislodged during concrete placement.

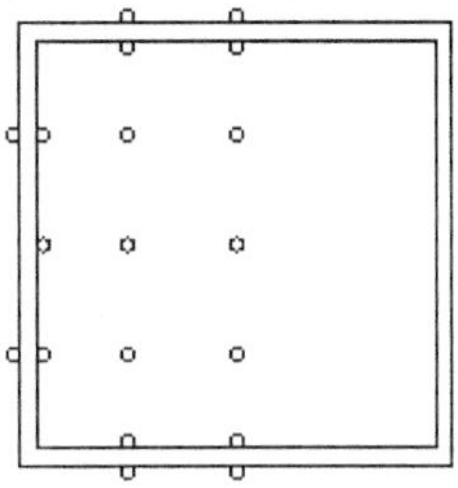

Figure 1. Example of thermocouple placement

Once the thermocouples were placed in the cross-section the tubes were positioned on a slight inclination so that they could be filled with concrete. The four smaller sections were filled using concrete that was batched up at the Monash Clayton laboratories. The larger four columns were filled with concrete that had been batched at a CSR mixing plant. All columns were filled at the Civil Engineering laboratories at Monash University. Tight quality control was exercised on the journey from the mixing plant site to Monash to ensure that no extra materials were added to the mix without knowledge. The design mix for the concrete is shown in Table 2. The target concrete strength was 100 MPa with very little slump.

Table 2. Concrete Design Mix Proportions

Material	Density (kg/m^3)
Cement	550 kg
Water	165 kg
Sand	520 kg
Coarse Aggregate	1340 kg
Super Plasticiser	8.7 litres

Before concrete placement each tube was fitted with a 300-mm tube cut from a 76.1x3.6 CHS, approximately 500 mm from the base. During placement of the concrete a poker vibrator was inserted in this tube to facilitate concrete compaction. The feasibility of using this method for compaction was verified by cutting open a trial tube to observe the compaction of the concrete. When the concrete level reached halfway up the height of the tube another poker vibrator was inserted in the column to aid in compaction for the upper half of the column. The other endplate was welded onto the tube approximately 7 days after concreting.

Prior to testing the concrete filled tubes in the furnace, small holes were drilled along the steel tube to aid in the egress of moisture and to avoid an enormous build up of steam pressure in the steel tube. The number of holes drilled in the tubes varied depending on the external dimensions of the tubes. All holes were 5 mm in diameter and were positioned on the upper and lower faces in four positions along the length of the column. The number of holes in each position varied from three to five.

4 TEST METHOD AND RESULTS

A diagram of the test rig is given in Figure 2. In this diagram the column located at the top is a high strength concrete filled 125x125x5 SHS which is loaded concentrically in compression. The lower column is the test specimen, which is attached to the test rig through steel pins and loaded eccentrically according to Table 3. The middle member was a high tensile CT-Stress bar, which had a diameter of 40 mm and a nominal tensile strength of 1030 MPa. A hydraulic jack was placed on one end of this bar and a tensile force applied. This force created compressive forces in both the upper and test columns. One of the rig posts was allowed to move freely by use of Teflon strips placed under the post simulating a roller support. The movement of the rig post was incurred due to the expansion and contraction of the test column from the induced thermal loads.

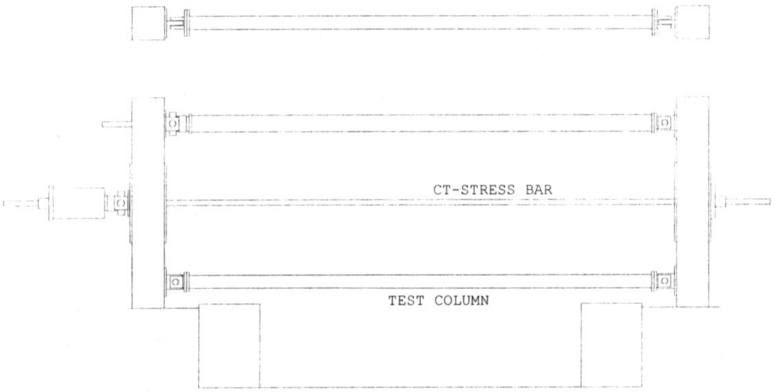

Figure 2. Testing Rig

To ensure that the rig would perform adequately in the furnace tests, two room temperature tests were conducted on the laboratory floor. The failure load and eccentricities of these tests and the fire tests are reported in Table 3.

The fire tests were conducted in the furnace at BHP Melbourne Research Laboratories. This furnace measures 5.1m long by 2.5m wide. The testing rig was placed above the furnace and the test column positioned such that it was exposed to the temperature on one half of its cross-section only. The furnace was sealed by positioning concrete slabs on either side of the specimen and sealing the remaining gaps and the upper surface of the column with caowool.

4.1 *Loading*

The specimens were subjected to eccentric compressive loading under increasing temperature, which in turn induced thermal loads in the test column. The initial loads applied were between 2% and 7% of the specimens room temperature squash load and between 5.5% and 41% of the room temperature buckling load calculated according to Rangan and Joyce (1992). The details of the loading and eccentricity for each test are specified in Table 3.

In the furnace tests the load was applied in 30 kN increments up to the test load before the furnace program was commenced. The load was kept at the test load for a minimum of 3 minutes before the furnace was turned on.

It was noticed that the load in the test column significantly increased once the furnace was turned on. This is due the thermal stresses which are induced in the column caused by its thermal expansion.

The column was seen to move longitudinally with the expansion in the column. This movement was measured and the maximum expansion is recorded in Table 3. As mentioned previously one of the test rig bases were placed on Teflon strips to provide a friction-less surface. The deformation with time in the longitudinal direction was measured for the four larger columns. For the smaller four columns the lateral deformation was only observed and the results were seen to be similar to the final four tests.

Table 3. Test conditions

	Section Type	Eccentricity (mm)	Test Load (kN)	Test time (mins)	Longitud'l Expansion	Residual Deflection
SR 1	SHS	20	247.52	-	-	-
CR 1	CHS	20	255.45	-	-	-
SF 1	SHS	20	110	17.833	-	89 mm
SF 2	SHS	20	96	23.250	-	101 mm
SF 3	SHS	40	100	34.416	-7/14.2	135 mm
SF 4	SHS	50	104.5	48.000	-15.9/12.8	176 mm
CF1	CHS	20	97	*31.000	-	103 mm
CF2	CHS	20	96	20.420	-	88 mm
CF3	CHS	40	104	37.583	-14.3/8.2	108 mm
CF 4	CHS	65	105	42.830	-13.2/16	145 mm

4.2 *Temperatures*

The furnace was programmed to generate a time-temperature curve that followed the ISO834 Standard Fire Curve. An example of the fire curves generated from the test is shown in Figure 3. The heavy line in Figure 3 is the ISO834 fire curve and the line just beneath is the temperature curve for the furnace which was recorded during the test. The next line down shows the temperature of the steel directly exposed to the furnace atmosphere and the lower line shows the temperature in the centre of the concrete core with time.

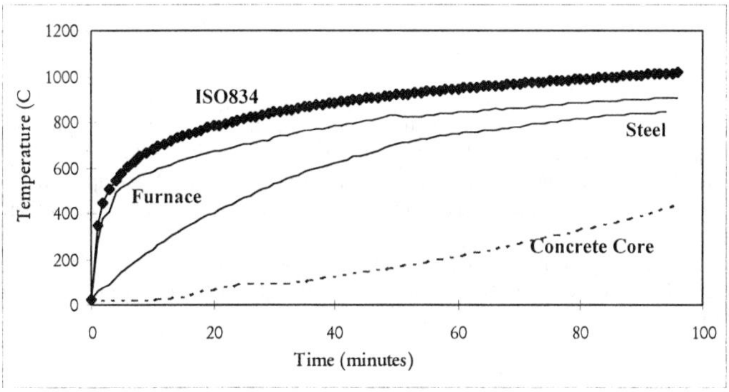

Figure 3. Fire Curves for SF4 Test

It was assumed that the temperature profile would be symmetrical in the cross-section of the specimens so only half of the cross-section was fitted with thermocouples.

The internal temperatures were measured with Type K thermocouples which had a fibreglass sheathing whereas the external temperatures were measured with MIMS thermocouples. This was done because the fibreglass-sheathed thermocouples could not sustain temperatures in excess of 200°C. To ensure that the sheathing would not disintegrate when in contact with the high temperatures, (especially when the deflection was significant), the fibreglass sheathed thermocouples were wrapped in caowool. This method was successful for the majority of the thermocouples although some burnt out just prior to the column failing when the rate of deflection increased rapidly.

Steam was seen to egress from the holes in the tubes at a very fast rate once the furnace had been turned on. Prior to all tests small gaps were created in the positioning of the caowool just above the holes to assist in the escape of the steam.

4.3 *Deflections*

The tubes started to deflect as soon as the furnace had been started. The rate of deflection remained at a constant level and then rapidly increased as the column began to fail. The variation in the axial movement of the test column was observed at various stages throughout the test program. This deflection was measured via a string-pot located on the roller end of the test rig. In general it was noticed that soon after the ISO834 fire curve was applied that the column expanded longitudinally with the induced thermal stresses. Once the column started to fail, the column started to retract as the lateral deflection increased rapidly.

The residual deformation of the column was measured once the column had cooled down. This result is also recorded in Table 3.

5 CONCLUSIONS

The use of high strength concrete filled steel tubes is gaining increasing acceptance in today's construction industry. At present these columns are conservatively fitted with fire protection measures in order to conform to fire resistance standards. These fire protection measures can significantly add to the cost of construction.

Although experimental work for fire resistance purposes is very costly the authors believe that more experimental work needs to be conducted on the structural elements of buildings. This will help in both the design of the elements and the understanding of the behaviour of these elements at elevated temperatures. The tests, which were conducted at Monash University, provided an insight into the behaviour of these columns in elevated temperature conditions. However, further studies need to be conducted to obtain a greater understanding on the fire resistance of these columns.

6 ACKNOWLEDGMENTS

The authors would like to acknowledge Tubemakers for the provision of the steel tubes, CSR for the provision of the concrete mixes, and Melsteel Pty Ltd for their help in the fabrication of the test rig. Thanks also to staff at the Monash Clayton Civil Engineering laboratories for all their help with the preparation of the test rig and test specimens and to BHP Melbourne Research Laboratories for the use of their furnace test facilities.

The authors are especially grateful to technician Jeff Dodrell at Monash University for all his help at the Monash laboratories and BHP furnace site and to Mike Culton at BHP-MRL for all his help and advice in conducting the furnace tests.

7 REFERENCES

ISO834, *Fire-resistance tests – Elements of Building Construction*, Part I, General Requirements for Fire Resistance Testing, Underwriters' Laboratories of Canada, Scarborough, Ontario, 1989.

Rangan, B. 1991. Design of slender hollow steel columns filled with concrete, *International Conference on Steel & Aluminium Structures. 22-24 May 1991, Singapore*, Composite Steel Structures, 104-112.

Rangan, B. & M. Joyce 1992. Strength of eccentrically loaded slender columns filled with high strength concrete, *ACI Structural Journal*, 89(6): 676-681.

Sakumoto, Y., T. Okada, M. Yoshida & S. Tasaka 1994. Fire Resistance of concrete-filled steel-tube columns, *J. Materials in Civil Engineering*, 6(2): 169-184.

Mechanics of Structures and Materials, Bradford, Bridge & Foster (eds)
© 1999 Balkema, Rotterdam, ISBN 90 5809 107 4

Buckling of unilaterally constrained mild steel plates at elevated temperatures

S.T. Smith
Department of Civil and Structural Engineering, Hong Kong Polytechnic University, People's Republic of China

M.A. Bradford
School of Civil and Environmental Engineering, The University of New South Wales, Sydney, N.S.W., Australia

D.J. Oehlers
Department of Civil and Environmental Engineering, University of Adelaide, S.A., Australia

ABSTRACT: An inelastic Rayleigh-Ritz method, developed previously by the authors, is used to analyse the local buckling response of unilaterally constrained mild steel plates at elevated temperatures. Unilateral restraint arises when steel plates are bolted to the sides of reinforced concrete beams in order to both strengthen and stiffen them, and although utilised in other situations, the deployment of bolted steel plates in the upgrade of reinforced concrete structures is the motivation for the study presented in this paper. The benefits that accrue to side-bolting of steel plates may be lost in conditions of fire when the plates may buckle, with a consequent reduction in strength of the composite steel-concrete beam. The paper aims to present an understanding of this phenomenon theoretically, as the cost of fire testing is usually excessive. Both the decrease in the yield strength and elastic modulus of mild steel plates at elevated temperatures are represented by a series of linear stress-strain constitutive relationships. The change in material properties at elevated temperatures is shown to influence the unilateral local buckling of mild steel plates in this unique form of side-plated composite behaviour.

1 INTRODUCTION

Steel plates can be bolted to the sides of reinforced concrete beams to both strengthen and stiffen them (Oehlers *et al.* 1997), as well as in the retrofit of existing concrete structures for seismic loading. Side plating using bolts is preferable to soffit plating, since the ductility of the resulting composite beam is enhanced with steel in both compressive as well as tensile zones (Smith & Bradford 1995), so that the beam does not become 'over-reinforced'. Moreover, the congestion of conventional flexural reinforcement within a reinforced beam near its soffit presents obvious difficulties if bolt holes are to be drilled into the concrete beam at this position. Side plating also enhances the shear resistance of reinforced concrete beams.

Generally, the plates in side-plated reinforced concrete beams experience compressive and shear actions, and are therefore susceptible to buckling. The buckle must necessarily be of a unilateral type, in which the juxtaposition of the plate with the concrete core prevents buckling from occurring into the concrete, and the edge restraints provided by bolting (as well as a gradient of longitudinal stress through the depth of the plate) results in a local rather than flexural (or Euler-type) buckle. This unilateral buckling falls into a class of contact problems, and its modelling with combined bending, axial and shear actions is difficult.

In the side-plating technique considered for this study, it is envisaged that mild steel plates would be glued initially to the side faces of concrete beams, and then bolted. In the advent of breakdown of the glue, due to adverse environmental conditions such as elevated temperatures that would be experienced in a fire, the bolts would solely provide the shear connection. The efficiency of such a 'belt and braces' approach therefore relies on the plate to remain unbuckled when the bolt shear connection is effected. It is well-known that steel plates can buckle prematurely at elevated temperatures, and apart from the study of Uy & Bradford (1995) that considered unilateral buckling in cold-formed steel at elevated temperatures subjected to bending and compression only, there appears to be little work reported in the published literature related to inelastic unilateral local buckling at elevated temperatures.

This paper thus considers the bifurcative unilateral local buckling of steel side plates when subjected to elevated temperatures caused by fire. The pb-2 Rayleigh-Ritz method developed by Liew & Wang (1993), and modified subsequently for inelastic unilateral lateral buckling by Smith *et al.* (1999a) forms the basis of the study. In order to incorporate the effects of elevated temperatures, the constitutive relationship for the plates is obtained by modifying the well-known Ramberg-Osgood formulation.

2 STABILITY ANALYSIS

2.1 *General*

The theoretical and experimental elastic buckling behaviour of steel plates bolted to the sides of reinforced concrete beams has been studied by the authors (Smith *et al.* 1999b,c). This was the modified to include inelasticity by Smith *et al.* (1999a). This modelling uses the Rayleigh-Ritz method, with a polynomial-based flexural displacement function in conjunction with discrete tensionless springs placed at grid points over the plate to enforce the unilateral restraint. The displacement function used was a two-dimensional Ritz function, developed initially by Liew & Wang (1993) in their pb-2 Rayleigh-Ritz analysis of plates. This idealisation has been shown to behave well numerically for elastic analysis of unilateral buckling problems (Smith *et al.* 1999d). The unilateral restraint condition is enforced in the authors' method by means of a penalty function, which introduces an additional stiffness matrix but does not increase the number of degrees of freedom. The analytical tool allows the flexibility of defining various boundary conditions, and produces rapid solutions on contemporary personal computers.

In formulating the analysis for buckling at elevated temperatures, the inelastic buckling model developed by Smith *et al.* (1999a) is used herein, but with an appropriate adjustment of the constitutive relationships for the mild steel plates that change with increasing temperature. Residual stresses are neglected, and the von Mises yield criterion is used to relate the applied in-plane stresses to the yield strength of the plate (Smith *et al.* 1999a), so that plate slenderness limits can be determined for initially flat mild steel plates.

2.2 *Material properties*

The mechanical properties of steel change with temperature, and increasing temperature causes both the yield stress and elastic modulus to decrease. These are given in many national steel structures standards by presenting simplified mathematical formulations for the stress-strain response of mild steel at elevated temperatures, and that presented in the Australian AS4100 (SA 1998) will be used in this study. The relationship between the yield stress $f_y(T)$ at a temperature T^o C and $f_y(20)$ is given as

$$\frac{f_y(T)}{f_y(20)} = \begin{cases} 1.0 & 0 \le T \le 215^o\,C \\ (905 - T)/690 & 215^o\,C < T \le 905^o\,C \end{cases} \qquad (1)$$

$f_y(20)$ can be taken as the ambient yield strength.

The modulus of elasticity of the steel also chances with T, and is given in AS4100 as

$$\frac{E(T)}{E(20)} = \begin{cases} 1.0 + \dfrac{T}{2000\, ln(T/1100)} & 0 \le T \le 600^\circ C \\ \dfrac{690(1 - T/1000)}{T - 53.5} & 600^\circ < T \le 1000^\circ \end{cases} \tag{2}$$

Again, $E(20)$ can be taken as the ambient elastic modulus.

2.3 *Constitutive matrix*

In the Rayleigh-Ritz method developed for inelastic buckling, the in-plane stress and strain are related by the following property matrix $[D]$.

$$\begin{Bmatrix} \sigma_x \\ \sigma_y \\ \sigma_s \end{Bmatrix} = \begin{bmatrix} D_x & D_1 & 0 \\ D_1 & D_y & 0 \\ 0 & 0 & D_{xy} \end{bmatrix} \begin{Bmatrix} \varepsilon_x \\ \varepsilon_y \\ \gamma_{xy} \end{Bmatrix} \tag{3}$$

where σ_x and σ_y are the in-plane stress in the x and y directions respectively, σ_s is the shear stress, and ε_x, ε_y and γ_{xy} are the corresponding associated strains. In the elastic range of structural response, the mild steel plate is isotropic but in the inelastic range the plate properties in Eq. 3 become orthotropic.

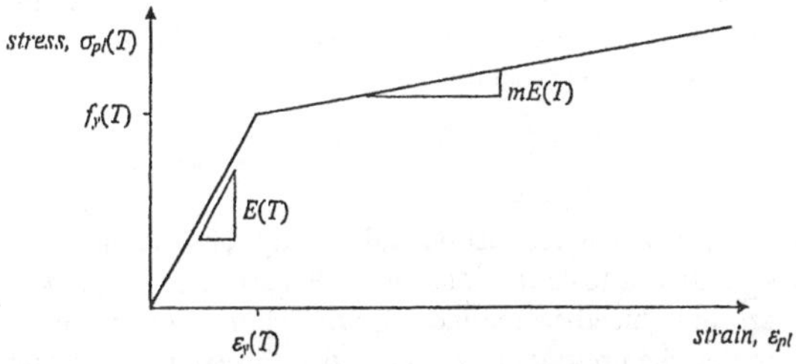

Figure 1. Idealisation of stress-strain behaviour of steel at elevated temperatures.

The linear elastic-strain hardening stress-strain relationship shown in Figure 1 is utilised in this study where m is the slope of the strain hardening region. Traditionally, a Ramberg-Osgood formulation (Olawale & Plank 1988) is used in this modelling, however the bilinear representation in Figure 1 has been adopted as this makes the execution of the inelastic computer program (Smith *et al.* 1999a) more efficient. At ambient temperatures, a linear elastic-perfectly plastic stress-strain curve is assumed, and the 'plastic' region increases to a 'strain hardened' region in accordance with the strain hardening modular ratio m as the temperature is increased. This is depicted in Table 1, where AS4100 (SA 1998) provides reduced yield stress and elastic modulus limits as described in Eqs. 1 and 2. The second, third and fourth columns of Table 1 provide a summary of these properties with varying temperature for Grade 250 mild steel plate. The strain hardening modular ratios were obtained from the study of Olawale & Plank (1988) that has been approximated in the British standard for the post-yielded region. In summary, the constitutive model used in this study is thus a pseudo Ramberg-Osgood approximation, and the properties obtained from this are used in the property matrix $[D]$ in Eq. 3, whose terms are stated explicitly in Smith *et al.* (1999a).

Table 1. Stress-strain relationship for mild steel at elevated temperatures

T °C	$f_y(T)$ MPa	$E(T)$ MPa	$\varepsilon_y(T)$ $\mu\varepsilon$	m	$mE(T)$ MPa
20	250	200,000	1250	0	0
100	250	195,830	1277	0	0
300	250	188,268	1328	0	0
300	219	176,910	1239	0.19	33,613
400	183	160,458	1140	0.22	35,301
500	147	136,585	1074	0.25	34,146
600	111	101,012	1094	0.11	11,111

3 NUMERICAL STUDIES

3.1 *General*

A parametric study has been undertaken to investigate the local buckling behaviour of unilaterally constrained mild steel plates at elevated temperatures. The curves generated are for normalised critical strains and moments at various temperatures as a function of the plate slenderness (b/t). This allows limits defining compact and non-compact sections to be chosen.

The parametric study was performed at temperatures ranging from ambient ($T = 20$°C) to a temperature of $T = 600$°C. It was assumed that the temperature throughout the plate remained constant. Under a fire condition, it is also assumed that the concrete and bolt shear connectors remain suitably intact so as to effect buckling of the side plate. In reality, the concrete beam would spall at high temperatures, and also the connectors would lose their strength. This secondary scenario was not considered in the parametric study.

3.2 *Accuracy of numerical model*

The pb-2 Rayleigh-Ritz approach with a tensionless foundation and penalty function has been calibrated in the elastic range with the results of tests (Smith *et al.* 1999b,c). The analysis is nonlinear, and further nonlinearities are introduced in the inelastic modelling. Although no benchmark data are available for calibrating the present numerical model, the extensive testing programme undertaken by Smith *et al.* (1999c) has validated the approach for elastic unilateral buckling. Further studies using the so-called local buckling push test developed by the authors (Smith *et al.* 1999e) has validated the inelastic model, albeit at ambient temperatures. In the absence of expensive test results at elevated temperatures, these experimental validations of the above numerical studies are cited as a basis for the accuracy of the model used herein.

3.3 *Buckling curves*

Plates bolted to the sides of concrete beams buckle with a node at the discrete bolt position. However, practical bolt spacings required to achieve the desired degree of shear connection (Oehlers *et al.* 1997) have indicated that the lines between bolts can be treated as nodal lines during buckling, and this has been confirmed experimentally (Smith *et al.* 1999c). Because of this, a plate with clamped (C) loaded edges and free (F) unloaded edges will be studied. Moreover, the influence of bending and axial actions in obtaining the required degree of shear connection tends to be more significant than that of shear, so for illustrative purposes and for brevity, only the cases of pure bending and bending plus axial actions will be considered for plates whose boundaries are modelled as C-F-C-F and with an aspect ratio of unity. The results of other studies can be found in Smith (1998).

Figures 2 and 3 show the results for a plate subjected to uniform bending. The curves in Figure 2 represent the critical strain normalised with respect to the yield strain at 20°C, while Figure 3 shows the critical moment in the plate $M_{pl}(T)$ normalised with respect to the yield moment $M_y = (b^2t/6)f_y(20)$ at 20°C. It can be seen that for a given value of b/t, the buckling strain and moment decrease with increasing temperature. Figure 2 allows one to make an assessment of the ductility of a plate at increasing temperatures, while Figure 3 permits an estimate of the section classification of the plate. For example, at $b/t = 20$, the critical moment at ambient temperatures is 1.5 times the yield moment, and this indicates that the plastic moment can be reached (and the plate is compact) since the shape factor is 1.5. However, for the same value of b/t the plastic moment is not attained at temperatures above 200°C, and indeed the yield moment is not attained prior to buckling for temperatures above about 430°C, indicating at temperatures higher than this that the section classification of the plate has deteriorated from compact at ambient temperatures through non-compact to slender.

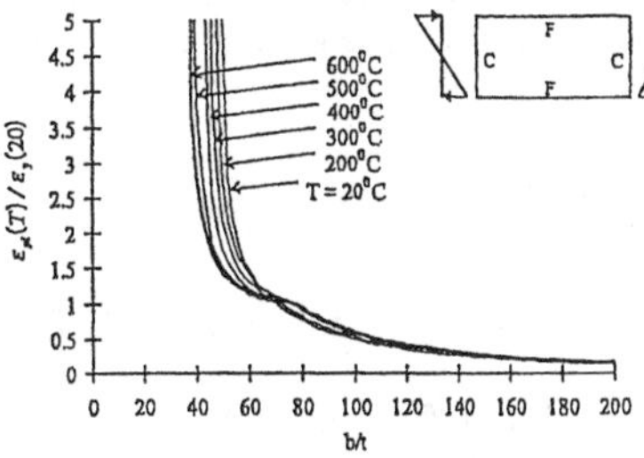

Figure 2. Critical strains under pure bending

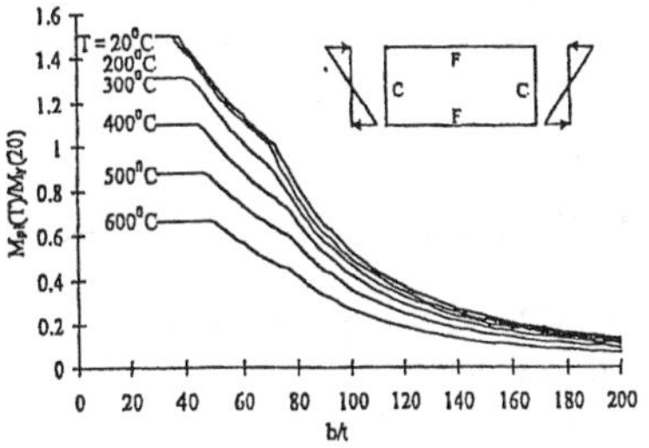

Figure 3. Critical moments under pure bending

A plate with as aspect ratio of unity with C-F-C-F edge conditions subjected to combined bending and tensile actions, as are often experienced in the partial interaction-partial shear connection analysis of bolted side plates (Oehlers *et al.* 1997) was also studied, with the tensile strain being 20% of the applied bending strain at the edge of the plate. Figures 4 and 5 are the respective counterparts of Figures 2 and 3 for this loading condition. Again, it an be observed that the inelastic buckling load is decreased for a given plate slenderness as the temperature increases. It is interesting to compare Figures 3 and 5 which plot the maximum moment. The plateaux achieved in both figures for stocky plates indicate that the plastic moment has been reached prior to buckling, and as expected there is a slight reduction in this plastic moment in the presence of tension (Trahair & Bradford 1998). On the other hand, the buckling loads of much more slender plates is increased above that for pure bending due to the benign effects of tension, so that the critical moment at a given temperature for large values of b/t is in higher for combined tension and bending than for pure bending alone.

4 CONCLUDING REMARKS

The use of rational numerical methods to ascertain the behaviour of composite structural elements at elevated temperatures is a valuable complement to expensive and limited testing. This paper has attempted to present such a numerical model for studying the inelastic unilateral buckling of mild steel plates at elevated temperatures, with the motivation being the loss of strength of the steel plates in the upgrade of existing reinforced concrete beams, that are side-bolted to the side of the beam, in the advent of fire. The results of the pb-2 Rayleigh-Ritz method coupled with a penalty function indicate that the buckling moment is reduced substantially at high temperatures, both for plates in pure bending and in combined bending and tension that is encountered in a partial interaction-partial shear connection analysis of this form

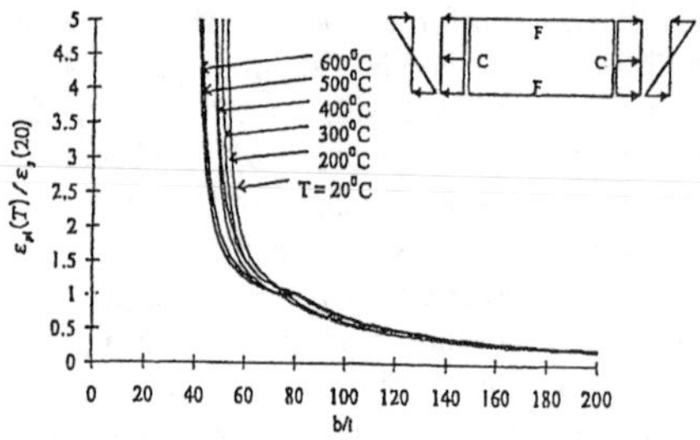

Figure 4. Critical strains under bending and tension

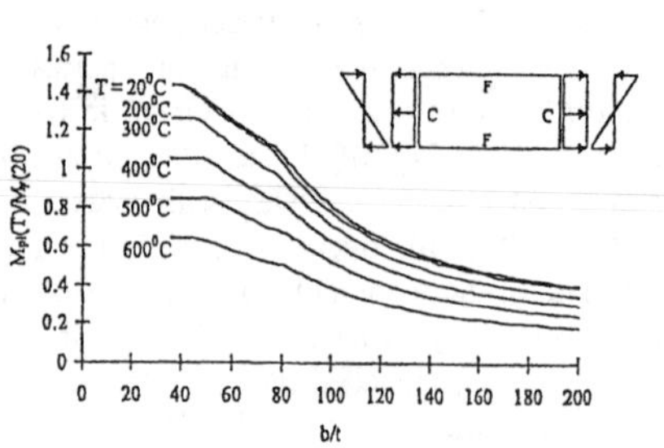

Figure 5. Critical moments under bending and tension

of composite construction. The illustrative curves presented enable the specification of slenderness limits defining compact, non-compact and slender plates, and these decrease with increasing temperature.

5 REFERENCES

Liew, K.M. & Wang, C.M. 1993. Pb-2 Rayleigh-Ritz method for general plate analysis. *Engg. Structs.* 15:55-60.

Oehlers, D.J., Nguyen, N.T., Ahmed, M. & Bradford, M.A. 1997. Lateral and longitudinal partial interaction in composite bolted side-plated reinforced concrete beams. *Struct. Engg. & Mechs.* 5(5):553-564.

Olawale, A.O. & Plank, R.J. 1988. The collapse analysis of steel columns in fire using a finite strip method. *Int. J. Num. Meth. Engg.* 26:2755-2764.

Smith, S.T. 1998. *Local buckling of steel side plates in the retrofit of reinforced concrete beams*. PhD Thesis, The University of New South Wales, Sydney.

Smith, S.T. & Bradford, M.A. 1995. Ductility of reinforced concrete beams with side and soffit plates. In A. Beazley et al. (eds), *Proceedings of ACMSM14*, Hobart, Dec. 1995:428-432 Hobart: University of Tasmania.

Smith, S.T., Bradford, M.A. & Oehlers, D.J. 1999a. Inelastic unilateral local buckling of rectangular plates. Submitted for publication.

Smith, S.T., Bradford, M.A. & Oehlers, D.J. 1999b. Local buckling of side-plated reinforced concrete beams. I: Theoretical study. *J .Struct. Engg.*, ASCE 125(6) (in press).

Smith, S.T., Bradford, M.A. & Oehlers, D.J. 1999c. Local buckling of side-plated reinforced concrete beams. II: Experimental study. *J. Struct Engg.*, ASCE 125(6) (in press).

Smith, S.T., Bradford, M.A & Oehlers, D.J. 1999d. Numerical convergence of simple and orthogonal polynomials for the unilateral plate buckling problem using the Rayleigh-Ritz method. *Int. J. Num. Meth. Engg.* 44:1685-1707.

Smith, S.T. Bradford, M.A. & Oehlers, D.J. 1999e. Local buckling push-tests for bolted steel plates in composite construction. Submitted for publication.

Standards Australia 1998. *AS4100 Steel structures*. Sydney:SA.

Trahair, N.S. & Bradford, M.A. 1998. *The behaviour and design of steel structures to AS4100.* 3rd edn., London:E&FN Spon.

Uy, B. & Bradford, M.A. 1995. Local buckling of cold formed steel in composite structural elements at elevated temperatures. *J. Const. Steel Res.* 34:53-73.

Mechanics of Structures and Materials, Bradford, Bridge & Foster (eds)
© 1999 Balkema, Rotterdam, ISBN 90 5809 107 4

Can the influence of mechanical impedance improve structural crashworthiness?

E.W. Brell, G.M. Van Erp & C.P.S. Snook
Faculty of Engineering and Surveying, University of Southern Queensland, Toowoomba, Qld, Australia

ABSTRACT: A novel device that may enhance structural crashworthiness is presented. This device relies on removing overall crash energy from a collision. The device, called the impedance column, improves energy removal efficacy proportionally to closing velocity. Using elementary one-dimensional wave theory, the device is performance compared to an on-board battering ram, then cost compared according to a societal cost model based on head injury criteria (HIC). The potential fields of application are train carriages or bus coaches.

1 INTRODUCTION

The effects of mechanical impedance are pervasive throughout our physical world from the short wavelength rain droplet (Johnson 1972 P83) that stings a motorcycle rider's face to the bursting of a long pipeline caused by a sudden valve closure (Swaffield & Boldy 1993), at the other end of the wavelength spectrum.

The effects of mechanical impedance manifest a localization of stress. The theory, having been well established in the 19th century, did not find applicability to crashworthiness until 1964 when Macaulay & Redwood demonstrated this localization on a model railway coach (Johnson 1972).

In this paper the response to a crash event of a collision partner to an impedance device is presented promising an improvement in passenger passive safety, the focus of structural crashworthiness. The device is called the impedance column. The work presented forms part of the Masters of Engineering study of the first author.

2 THE IMPEDANCE COLUMN.

2.1 *General*

The principle of the impedance column involves the interspersal of an impedant media between the impact interface (the piston head) and the impedance carriage reaction point (the cap) to create a delay. For the duration of this delay, called the defensive phase, the impedance carriage is unaffected while the collision partner's momentum is changed on account of the impulse imposed. A schematic of a carriage fitted with an impedance column is depicted in Figure 1. The velocity of the collision partner is lowered during the defensive phase causing a lowering of the full-face approach velocity. Half of this benefit flows to the impedance carriage. The process relies on the compressible nature of the impedance media and the characteristically rapid rise in force in response to velocity of impact. The rapid rise was shown by Al-Mousawi *et al* (1987) and Kida *et al* (1992) who reported rise times in the order of 80 microseconds.

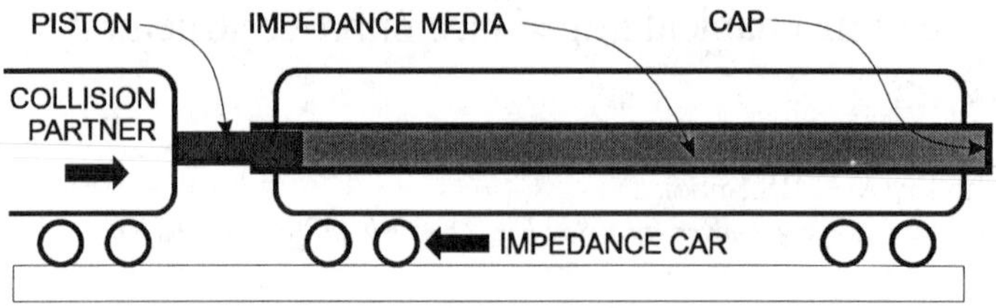

Figure 1. Schematic sketch of a simple impedance column

2.2 *Impact interface velocity*

A collinear crash of identical carriages will produce a stationary impact interface velocity. However, the impact interface will have velocity, if the impedance of the collision partners is not identical. This velocity will be in the direction of the side with the lower impedance media. The forces on both sides of the impact interface will of course be equal numerically and opposite vectorially. Since the steel impedance is very much larger than the media impedance, the collision partner can be taken as perfectly rigid with little error for the simplified analysis that follows.

From classical impact theory (Timoshenko & Goodier 1970), stress (S) is proportional to the velocity (V) of impact, the constant of proportionality being mechanical impedance (Z), the product of density (ρ) and celerity (c) (celerity is often called acoustic velocity). These relationships, expressed in equation format, become:

$$S = Z \, x \, V \tag{1}$$

where

$$Z = \rho \, x \, c. \tag{2}$$

The equation of motion of the collision partner and also the impact interface on account of the collision partner's rigidity then, can be represented by:

$$M \frac{dV_i}{dt} = -A\left(ZV_i + ZV_m + P\right). \tag{3}$$

Where subscripts i and m refer to interface and media, M is the mass of the collision partner, A the area of the impedance column and P is a charging pressure for the column.

Integration of equation 3 with respect to time t yields

$$-\frac{M}{A}\int_{\sim V_0}^{\sim V_i} \frac{1}{ZV + ZV_m + P} dV = \int_{\sim 0}^{\sim t} 1 dt. \tag{4}$$

Carrying out the integration and re-arranging finally results in

$$V_i = \left(V_0 + V_m + \frac{P}{Z}\right)\exp\left(-\frac{tAZ}{M}\right) - V_m - \frac{P}{Z}. \tag{5}$$

The reader will recognize in equation (3) the impulse-momentum exchange arguments of the three forces decelerating the mass M of the collision partner, namely (a) the impedance media reaction to the collision partner's velocity, (b) the interface reaction to the impedance media velocity and (c) the charging pressure.

2.3 *Defensive Phase Duration*

The impedance carriage is not entirely devoid of influence from the impedance column. Since, this paper relies only on the momentum change of the collision partner to show the column's performance, the response of the impedance carriage is not an essential element. Notwithstanding, the impedance carriage is retarded by frictional interaction of the media with the pipe wall and the sudden release of the columns' longitudinal stresses (caused by the charging pressure) at the shear of the seal welds including the actual shearing of the seal welds. Ignoring these has the effect of understating the columns' performance since their effects reduce the closing velocity at the point of major impact. The duration of the defensive phase is to some extent apparatus dependent. Accordingly, a preferred embodiment is defined below:

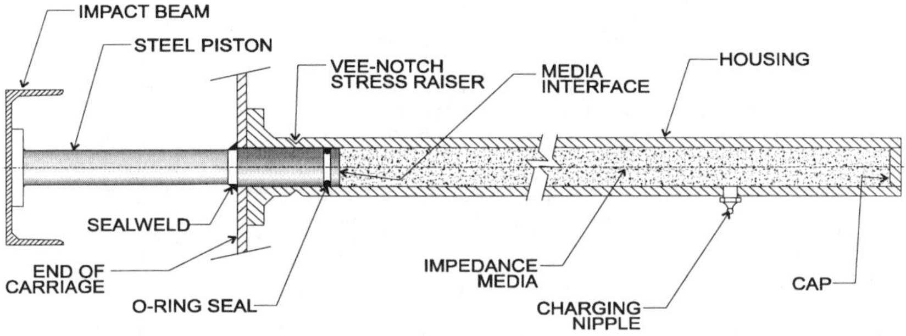

Figure 2. Sketch of preferred embodiment of impedance column

The body of the housing is connected to the carriage at one end only (except for sliding support along the length). The pulse must travel the full length through the media after which a tensile pulse returns from the cap along the pipe housing to snap the vee-notch stress raiser near the end of the carriage so marked in Figure 2. An unloading tensile wave in the housing returns to the cap and initiates an unloading pulse in the media to annul the reflected compressive wave at the intersection of the dotted line at approximately L / Lo = 0.25. This is more easily followed with the aid of the characteristic diagram shown in Figure 3.

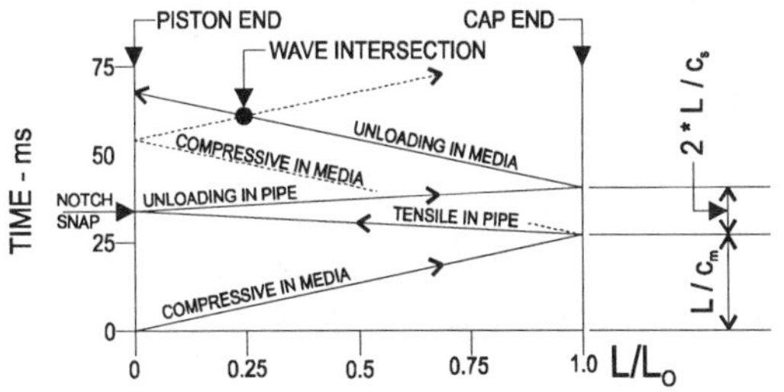

Figure 3. Typical Characteristic Diagram for Wave Progression & Reflection

Although the piston head up to this point is not exposed owing to the depth of incursion of the piston into the housing, the media and housing are now not connected to the carriage and can dissipate any residual stored energy with no effect on carriage momenta. When the piston stroke limit has been reached, the two carriages are set to engage in the major phase of the collision. As shown in Figure 3 the duration of the defensive phase is made up of slightly more than two media traverses.

2.4 *Performance prediction*

In order to get a feel for the performance of the impedance column, two hypothetical carriages are impacted head-on in a collinear trajectory. The carriages notionally weigh 32,000 kg. One carriage, the impedance carriage, carries two impedance columns of 35 m length and 100 mm diameter. The velocity of the carriages is 8 m/s. The resulting closing velocity of the collision partners of 16 m/s corresponds with Scholes (1994) for a "heavy collision". A nominal passenger density of 50 passengers per carriage is assumed. Space limitations prevent presentation of the detailed solution however, it is calculated for different media and tabled below:

Table 1. Defensive Phase Duration.

Media	c - Celerity	ρ - Density	Impedance	Duration -ms	Ref (c & ρ)
Steel	5150 m/s	7750 kg/m^3	39,912,500	-	Johnson 1972
Aluminium	5100 m/s	2660 kg/m^3	13,566,000	20	
Water/Sand	1582 m/s	2024 kg/m^3	3,201,968	51	Charlie 1989
Water	1438 m/s	996 kg/m^3	1,432,248	55	Johnson 1972
Bitumen	1300 m/s	1000 kg/m^3	1,300,000	61	Whiteoak 1990
Kerosene	1290 m/s	814 kg/m^3	1,050,060	61	Swaffield 1993

It is shown in Table 1 that the longest defensive phase duration was created by the kerosene media and the shortest, by the aluminium media (steel excepted).

2.5 *Collision partner's momentum change*

The most effective media is the one which reduces the collision partner's momentum; ie. velocity, the most. Equation (5) is plotted against time in Figure 4 to compare the velocity decay of the collision partner from the initial 8 m/s.

The miscellaneous group in Table 1 comprises water, kerosene and bitumen. Since it is not convenient on Figure 4 to compare the different media at each of its own phase duration, the velocity drop of the collision partner is re-calculated using Equation (5) at the phase duration appropriate for the media and displayed in Table 2. In addition, a column is added to Table 2 to reflect the influence of 70 MPa prior pressurization:

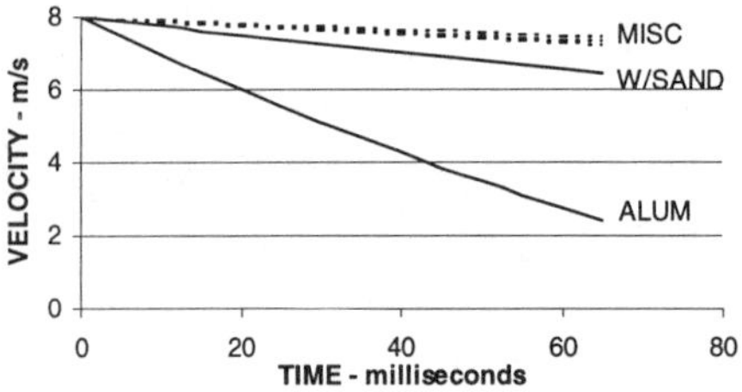

Figure 4. Collision partner's velocity decay from 8 m/s initial velocity

Table 2 – Velocity drop of collision partner from 8 m/s initial velocity for defensive phase duration

Media	Celerity m/s	Density kg/m^3	Impedance	Duration of Phase - ms	ΔV Media Only	ρ/ΔV (Media)	ΔV (Incl. Pressure)
Aluminium	5100	2660	13,566,000	20	2.00 m/s	1330	2.64 m/s
Water/Sand	1582	2024	3,201,968	51	1.23 m/s	1645	2.92 m/s
Water	1438	996	1,432,248	55	0.61 m/s	1630	2.46 m/s
Bitumen	1300	1000	1,300,000	61	0.61 m/s	1640	2.67 m/s
Kerosene	1290	814	1,050,060	61	0.50 m/s	1630	2.56 m/s

The best performance is achieved from the water/sand medium at 2.92 m/s. The ratio $\rho/\Delta V$ correlates well with fluid media density suggesting that density may be the dominant factor.

2.6 *Battering ram comparison*

The carriage of impedance columns adds a mass burden to the tare of the carriages. To consider the battering ram effect of this mass, the kinetic energy of this mass at 8 m/s is calculated and compared with the energy removal imposed on the collision partner. Table 3 compares these energies for the now familiar impedance media:

Table 3. Extra energy removal from collision partner over battering ram energy at 8 m/s

Media	Mass Kg	Housing Mass	Total Burden	Burden Energy	ΔV (Incl. Pressure)	Energy Removal	Extra Energy Removed
Aluminium	1490	1200	2690 kg	86 kJ	2.64 m/s	122 kJ	26 kJ
Water/Sand	1133	1200	2333 kg	75 kJ	2.92 m/s	149 kJ	74 kJ
Water	556	1200	1756 kg	56 kJ	2.46 m/s	106 kJ	50 kJ
Bitumen	560	1200	1760 kg	56 kJ	2.67 m/s	125 kJ	69 kJ
Kerosene	456	1200	1656 kg	53 kJ	2.56 m/s	115 kJ	62 kJ

Table 3 shows that the impedance column removes more energy than its own burden kinetic energy for all media. Accordingly, the impedance column is more than a battering ram.

3 INJURY MITIGATION

An improvement has been shown in carriage velocity change and overall energy reduction. The effect on passengers will now be considered. Due to space limitation, detailed calculations can not be included.

A seated unrestrained passenger will go into 'free-flight' until contact is made with the interior of the carriage (Galganski 1993 & Simons *et al* 1998), the so-called secondary impact. Typically, the severity of impact is mitigated by a compliant seat back ahead. The passengers velocity differential then, is taken as the difference between the initial carriage velocity and carriage velocity at the secondary impact point determined here to be the common velocity of the carriages after impact. Rebound is not considered. Passenger velocity differential is calculated in Table 4 below with the first line showing derivation:

Table 4. Passenger velocity differential from 8 m/s initial velocity

Media	ΔV	Collision Partner Velocity After Impedance Column Effect	Velocity Differential At Passenger/Seat Impact
Aluminium	2.64 m/s	8 – 2.64 = 5.36	8 – 2.62/2 = 6.68
Water/Sand	2.92 m/s	5.08 m/s	6.54 m/s
Water	2.46 m/s	5.54 m/s	6.77 m/s
Bitumen	2.67 m/s	5.33 m/s	6.67 m/s
Kerosene	2.56 m/s	5.44 m/s	6.72 m/s

Simons *et al* (1998) provides a convenient graph for direct reading of head injury criteria (HIC) from velocity differential with a seat-back. The reference HIC is 860 for a secondary impact velocity differential of 8 m/s. A Societal Cost Model developed as part of the first author's Master of Engineering study was used to determine cost of death and injury on a societal scale for the 50 passengers involved at the HIC exposure level determined from passenger velocity differential designated as ΔV_P. The table below shows a risk-adjusted saving for a carriage with impedance columns fitted as against one not so fitted. The size of saving is more significant than the individual differences. The cost of fitment is estimated to be less than a fifth of the savings predicted.

Table 5. Cost of death and injury comparison

Media	ΔV_P	HIC	Societal Cost $1000	Risk (Bing 1993)	Risk-Adjusted * Indifference Cost	Risk-Adjusted Saving**
Reference	8 m/s	860	3,600	0.125	$450,000	Reference
Aluminium	6.68 m/s	677	2,300	0.125	$288,000	$162,000
Water/Sand	6.54 m/s	659	2,200	0.125	$275,000	$175,000
Water	6.77 m/s	689	2,400	0.125	$300,000	$150,000
Bitumen	6.67 m/s	676	2,300	0.125	$288,000	$162,000
Kerosene	6.72 m/s	683	2,300	0.125	$288,000	$162,000

*Product of Societal Cost and Risk. **Reference Indifference Cost minus Indifference Cost for media.

4 CONCLUSIONS

The general principles of the impedance column have been demonstrated. It was shown that the performance of the impedance column is much greater than the equivalent mass of an on-board battering ram. Modest (as compared to crush zones) energy dissipation has also been demonstrated. It was shown that impedance media density may be a more dominant performance factor than its celerity. In addition, substantial risk-adjusted savings were demonstrated. Research is continuing to generate experimental verification of impedance column behaviour.

REFERENCES

Al-Mousawi, M.M. & H.R. Harrison 1987. Experimental investigation of transient flexural waves in beams with discontinuities of cross-section. Exp. Mech. 27/4:404-413.

Bing, A.J. 1993. *Collision avoidance and accident survivability – collision threat.* US Dept. of Trans. DOT-VNTSC-FRA-93-2.I.

Charlie, W.A., G.E. Veyera & D.O. Doehring 1989. *Dilatational-wave-induced pore-water Pressure in Soil.* Exp. Mech. 29/3:437-442.

Galganski, R.A. 1993. *Collision avoidance and accident survivability.* US Dept. of Trans. DOT-VNTSC-FRA-93-2.III.

Johnson, W. 1972. *Impact strength of materials.* London: Edward Arnold.

Kida, S. & H. Kakishima 1992. *Impact wave and stress.* Exp. Techn. Mar/April 92:32-35.

Kirk, N.E., A.F. Kalton, E.G. Candy, G.C. Newell, C.E. Nicholson & C.Wilson 1998. Modifications to existing rolling stock to improve crashworthiness. *IJCRASH'98 Int'l Crashworthiness Conf Proc* Cambridge: Woodhead Publishing.

Lewis, J.H. 1998. Rail vehicle crashworthiness – role of analysis and testing in the design process. *IJCRASH'98 Int'l Crashworthiness Conf Proc.* Cambridge: Woodhead Publishing.

Scholes, A., J.H. Lewis & W.G. Rasaiah 1995. *Crashworthiness Collision Tests – Presentation to the Railway Industry* Derby: British Rail Research.

Scholes, A. & J.H. Lewis 1993. *Development of crashworthiness for railway vehicle structures.* Proc. Inst. Mech Engs Vol 207.

Scholes, A. 1994. *Structural Design Loading of Railway Vehicle Bodies.* Railtech Seminar 1 London: IMechE

Shukla, A. & C. Damania 1987. *Experimental Investigation of Wave Velocity and Dynamic Contact Stresses in an Assembly of Disks.* Exp. Mech. 27/3:268-281.

Simons, J.W. & S.W. Kirkpatrick 1998. High-speed train crashworthiness and occupant survivability. *IJCRASH'98 Int'l Crashworthiness Conf Proc.* Cambridge: Woodhead Publishing.

Swaffield, J.A. & A.P. Boldy 1993. *Pressure surge in pipe and duct systems.* Aldershot: Avebury Tech.

Timoshenko, S. & J.N. Goodier 1970. *Theory of elasticity.* New York: McGraw Hill.

Tyrrell, D., K. Severson & B. Marquis 1251998. *Crashworthiness of passenger trains.* US Dept. of Trans. DOT-VNTSC-97-4.

Whiteoak, D. 1990 *Shell Bitumen Handbook.* UK: Shell Bitumen.

Mechanics of Structures and Materials, Bradford, Bridge & Foster (eds)
© 1999 Balkema, Rotterdam, ISBN 90 5809 107 4

Finite element modeling of moisture transport through the brick-mortar interface

Y.Z.Totoev & H.O.Sugo
Department of Civil, Surveying and Environmental Engineering, University of Newcastle, N.S.W., Australia

ABSTRACT: The Initial Rate of Absorption (IRA) of bricks and water retention properties of the fresh mortar are two major parameters affecting the extraction of moisture from fresh mortar and subsequently the strength of the brick-mortar bond. The scientific fundamentals of this process have been studied at the University of Newcastle and at Delft Technical University in the Netherlands. In this paper an attempt is made to develop a finite element model for this phase of the moisture transport process. It is intended as a tool for optimization of the water absorption/retention properties of masonry constituents and hence achieve stronger bond.

1 INTRODUCTION AND BACKGROUND

Masonry is a composite material consisting of masonry units ("bricks or blocks") and mortar joints. The achievement of an effective bond between mortar and brick is the most important aspect of masonry construction. If adequate bond is not achieved, the mortar joints will act as planes of weakness, and cracking along these planes will occur due to wind or minor earthquake tremor. Good bond strength is also required for the adequate performance of masonry under normal service loads. Significant stresses can be introduced in masonry by foundation soil movements, mine subsidence, and differential movements caused by temperature effects or moisture. The potential for damage from these effects is high. For example, a recent survey by the Public Works Department of NSW has shown that 50% of serviceability failures are due to cracking of masonry.

The fundamental mechanism of bond creation between fresh mortar and brick has been studied both in Australia and overseas (Lawrence et al. 1987, Grandet 1973, Sugo et al. 1997). All these studies agree that bond is mainly physical in nature and is achieved by mechanical interlock between bricks and hardened mortar. Some chemical bonding may also occur. It is generally agreed that this interlock is created by brick suction acting on the mortar paste, which carries small particles from the fresh mortar into the pores of the brick. This process also creates a cement-rich layer at the brick-mortar interface.

The effects of moisture on bond and moisture transport across the brick-mortar interface have been studied experimentally and also combined with attempts to model this phenomenon using Unsaturated Flow Theory (UFT) (Groot 1991, Pel et al. 1997). However, a reliable model to simulate moisture transport between the fresh mortar and brick does not yet exist. The major obstacle to its development is the lack of knowledge of the pressure changes in the mortar as a combined result of two processes: physical (moisture extraction accompanied by transport of fine particles to the interface); and chemical (hydration of cement).

Current masonry construction practices are mostly empirical and based on a superficial understanding of the phenomenon of moisture transport. Fundamental understanding of the process will enable hydraulic compatibility to be obtained between bricks and mortar, thus maximising the bond strength. The development of analytical and computer model of moisture transport in masonry is one of the essential elements in this research.

2 MOISTURE TRANSPORT IN MASONRY: CONCEPT

The initial extraction of moisture from the mortar is a physical process driven by the brick suction. This extraction is very rapid at the beginning, during the first one or two minutes, due to the large difference in suction between the (unsaturated) brick and saturated mortar paste. The moisture content of brick increases at the interface as well as permeability while the capillary suction reduces.

Changes of hydraulic parameters, which occur in the mortar, are more complex. It is known that the moisture content in mortar joints reduces, especially at the interface, while the suction increases. It is evident that liquid, which flows through pores between the mortar aggregate, carries small cement particles towards the brick-mortar interface forming a thin layer of the cement rich paste on the brick surface. The porosity and the permeability of this layer are reduced compared to the rest of the joint. This aspect of the flow process is believed to be beneficial to the development of good bond.

On the other hand, the flow is not uniform through the depth of the joint. The initial moisture extraction from mortar is more intense at the interface. It is reasonable to assume that the transport of cement particles, which accompanies the moisture flow, is also more intense at the interface. It largely occurs from the layer of mortar adjoining to the interface. There is some evidence, that adjacent layer becomes depleted in paste and thus has higher porosity and reduced strength. The reduced strength of the paste depleted mortar layer may become a plane of weakness in the masonry joint. Therefore this aspect of the flow process has an important effect on the bond strength.

The initial stage of the moisture extraction from the fresh mortar continues until a balance is reached between the capillary suction of the brick and the water retention forces in the mortar. This stage determines the amount of moisture that is left in mortar and is available for curing.

Australian Standard for Masonry Structures (AS3700 1998) requires that "masonry shall be constructed from compatible combinations of units and mortar" in order to achieve the necessary strength and durability. Mortar is considered compatible with the brick type if it has good workability during construction and the masonry meets the required bond strength. To improve compatibility of mortar with high suction bricks addition of water retention admixtures to mortar or prewetting bricks to reduce suction is recommended. If this is not done and dry bricks are used, the intensive and massive moisture extraction affects workability of mortar.

Prewetting is done in Australia by sprinkling or hosing bricks for an extended period of time followed by a short period of drying. This technique reduces the suction gradient between brick and mortar and, therefore, it significantly reduces the overall moisture extraction from mortar.

The typical European prewetting technique is different. The bricks are submerged into water for a short time just before laying. This technique reduces the suction gradient at the interface and the rate of the initial extraction with minimal effect on the overall flow.

3 UNSATURATED FLOW THEORY (UFT)

UFT, originally developed in soil water hydrology, will be the main analytical method used in this project. Previous work (Hall 1977) has verified that UFT is also applicable to porous building materials such as fired clay brick.

UFT is usually described by the Philip and de Vries model (1957), which is an extension of Darcy's law for the simultaneous transport of liquid water, water vapour, and heat, governed by gradients of moisture content and temperature. The total mass balance equation for the isothermal moisture transport is

$$\frac{\partial \theta}{\partial t} = \nabla(D\nabla\theta) \tag{1}$$

where: θ is the total moisture content, D is the moisture diffusivity, and t denotes time. This model cannot be applied to describe the moisture transport across the interface of two porous materials because the governing potential is discontinuous at the interface. To overcome this problem capillary pressure equilibrium can be considered as the governing potential. To do so, the microscopic Young-Laplace equation, which relates the capillary pressure to the diameter of the pores, is averaged over the volume. Moisture transport is then attributed to the capillary pressure forces only (Richards 1931) and Equation 1 can be rewritten as

$$\frac{\partial \theta}{\partial t} = \nabla(P \nabla \varphi) \qquad (2)$$

where: P is the permeability of the material and φ is the macroscopic capillary pressure. All parameters in Equation 2 can be measured experimentally.

4 FINITE ELEMENT APPROXIMATION

The finite element space discretization of the governing differential equation for moisture movement in unsaturated porous material (2) leads to the following matrix system of non-linear ordinary differential equation (Smith 1982).

$$\mathbf{K}(u) \cdot u + \mathbf{B}(u) \cdot \dot{u} = \mathbf{Q}(u,t); \qquad u(0) = u_0, \qquad (3)$$

where: $\mathbf{K}(u)$ and $\mathbf{B}(u)$ are porosity and permeability matrices respectively, the coefficients are generally dependent on u; $u = \log(\varphi/g)$ is the vector of nodal values of the logarithmic suction; g is the gravitational acceleration; u_0 is the initial suction profile; $\mathbf{Q}(u, t)$ is the vector of sources and sinks which may be a function of time and u; and the overdot denotes time derivative.

In the case of the initial moisture extraction, it has been assumed that brick-mortar system is closed and the above equation becomes homogeneous. A linear relationship has been assumed between u and $\log \theta$, which implies a non-linear relationship between the capillary pressure and moisture content. The effective porosity and permeability has been assumed constant for both brick and mortar. These assumptions reflect current lack of information about hydraulic properties of masonry constituents. They reduce (3) to

$$\mathbf{K} \cdot u + \mathbf{B} \cdot \dot{u} = \mathbf{0}; \qquad u(0) = u_0. \qquad (4)$$

5 COMPUTER SIMULATIONS

Three examples presented are computer simulations of moisture extraction from fresh mortar for three typical construction practices: (i) by dry bricks, (ii) by bricks prewetted for 3 hours with subsequent 10 minutes drying, and (iii) by bricks dipped into water for 1 minute just before laying.

A 1D finite element model of the masonry couplet shown in Figure 1 was used. The masonry unit was a fired clay brick and a masonry cement mortar with addition of lime was used as the mortar. Hydraulic properties of materials were taken to match, where possible, materials experimentally studied by Brocken (1998). Namely: brick, specified as FCB_1, and mortar, specified as $MC_{45}M$. The relative dry density of 1630 kg/m^3 was assumed for brick and 2100 kg/m^3 – for mortar. The saturated hydraulic conductivity of 27.7x10^{-10} m/s was assumed for brick and 16.7x10^{-9} m/s for mortar. The time step was 1 min.

Results of the first simulation converted from the suction to moisture are presented in Figure 15% at 90s). The water front propagation in brick was 7.5mm over the first 3min (experiment ~7mm over 3min). Moisture in the middle of mortar joint has dropped from the initial 14.1% to

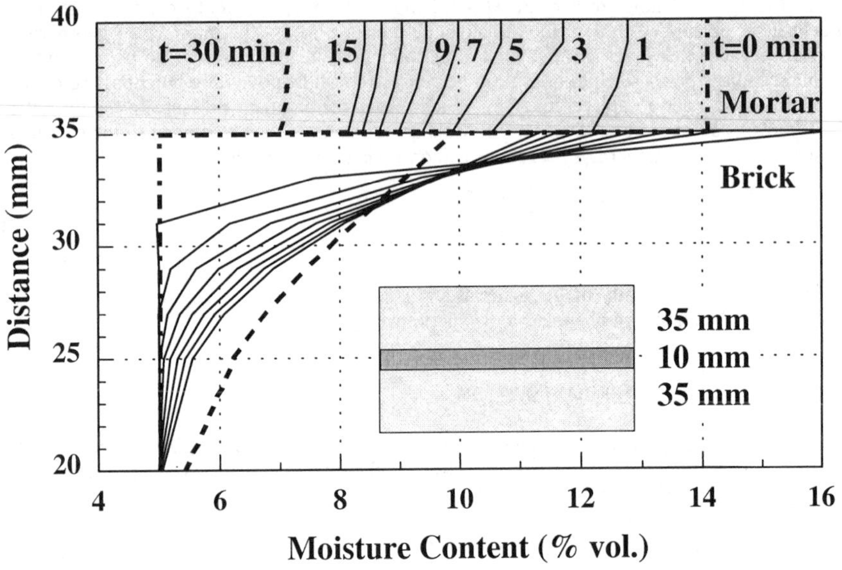

Figure 1. Moisture extraction by dry brick. Simulated moisture profiles.

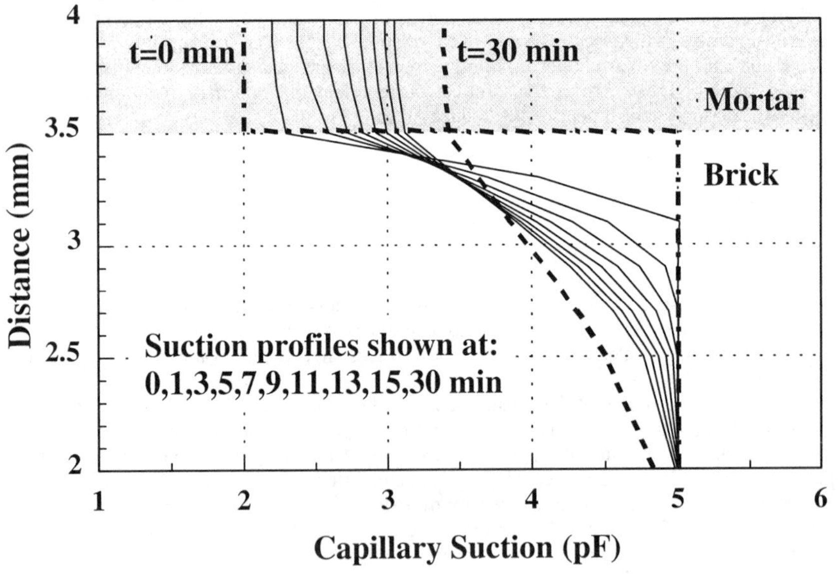

Figure 2. Moisture extraction by dry brick. Simulated suction profiles.

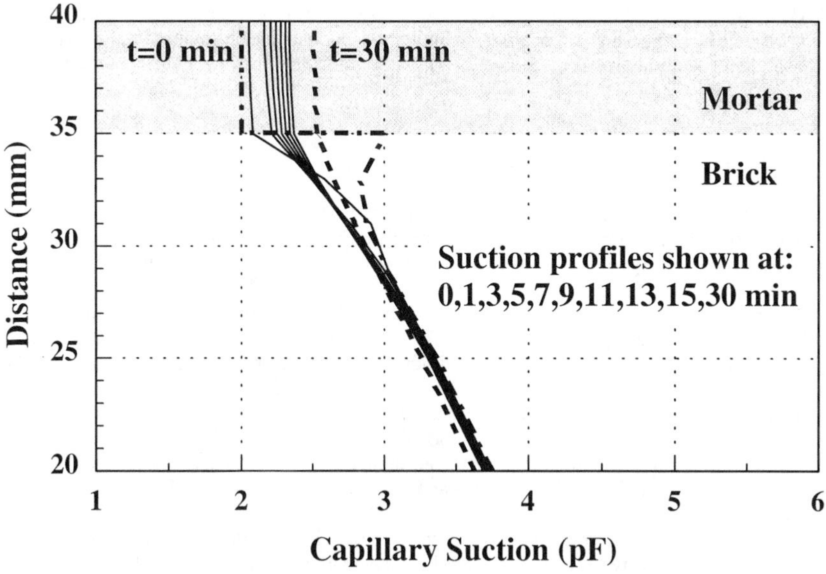

Figure 3. Moisture extraction by brick prewetted for 3 hours with subsequent drying for 10 minutes. Simulated suction profiles.

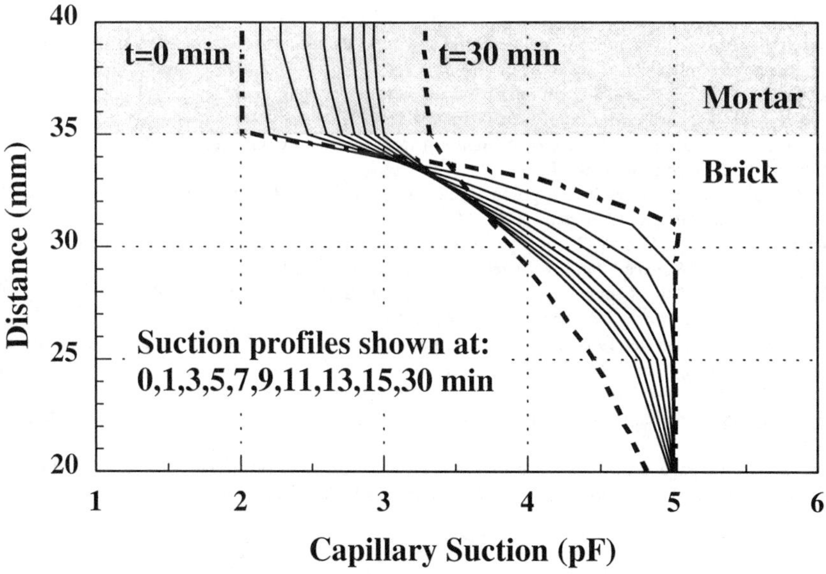

Figure 4. Moisture extraction by brick dipped into water for 1 minute just before laying. Simulated suction profiles.

~12.3% at 90s, and to 7.1% at 30min (experiment: ~15%, ~7%, ~6%). Simulated profiles have similar shape to the experimental profiles, including the distinct discontinuity at the interface.

The simulated suction profiles (Figures 2 - 4), unlike moisture profiles, have no discontinuity at the interface.

6 SUMMARY AND CONCLUSIONS

The attempt was made to develop a finite element model for the initial extraction of moisture from the fresh mortar by brick. Moisture extraction for three typical construction practices was simulated. From the experience gained in the modelling process and the results of simulation the following conclusions can be drawn:

- The capillary suction as the governing potential for the FE model of moisture flow across the material interface appears to be better then the moisture content, mainly because it does not discontinue at the interface. In addition, the permeability is more suitable as the transport parameter than the diffusivity, because it is easier to measure.
- The model based on the UFT gives reasonable results for moisture transport in brick, but it is oversimplified for the mortar, which undergoes a phase change from a saturated paste to a solid during moisture extraction. Moreover, the UFT is unable to account for the transport of fine cement particles to the interface and the chemical reaction of cement hydration in mortar.
- There is very little data available in literature on hydraulic properties of bricks and mortars.
- Future work in this area should be focused on establishing the suction and the permeability curves for various types of brick and studying hydraulic parameters of mortar during the processes of hardening and curing.

REFERENCES

AS3700 1998. *Masonry Structures.* Standards Australia, Sydney.

Brocken, H.J.P. 1998. *Moisture transport in brick masonry: the gray area between bricks.* Ph.D. thesis, the Eindhoven University of Technology.

Grandet, J. 1973. Physikalisch-chemische mechanismen der haftung zwischen ziegel und zement. In *Proc. 3rd Int. Brick Masonry Conf.* Essen, Germany, 217-221.

Groot, C.J.W.P. 1991. First minutes water transport from mortar to brick. In *Proc. 9th Int. Brick/Block Masonry Conf.* Berlin, Germany, 71-78.

Hall, C. 1977. Water movement in porous building materials – I. Unsaturated flow theory and its applications. *Building and Environment* 12: 117-125.

Lawrence, S.J. & H.T. Cao 1987. An experimental study of the interface between brick and mortar. In *Proc. 4th N. Am. Masonry Conf.* Los Angeles, California, paper 48, 1-14.

Pel L., H. Brocken & K. Kopinga 1997. Moisture and salt transport in brick: A NMR study. *Proc. 11th Int. Brick/Block Masonry Conf.* Shanghai, China, 50-58.

Philip, J.R. & D.A. de Vries 1957. Moisture movements in porous materials under temperature gradients. *Trans. Am. Geophys. Un.* 38: 222-232.

Richards, L.A. 1931. Capillary conduction of liquids through porous mediums. *Physics* 1: 318-333.

Smith, I.M. 1982. *Programming the finite element method with application to geomechanics.* Chichester: Wiley.

Sugo, H.O., A.W. Page & S.J. Lawrence 1997. Characterisation and bond strengths of mortars with clay masonry units. In *Proc. 11th Int. Brick/Block Masonry Conf.* Shanghai, China, 59-68.

Mechanics of Structures and Materials, Bradford, Bridge & Foster (eds)
© 1999 Balkema, Rotterdam, ISBN 90 5809 107 4

Behaviour of multi-storey steel frames in fires

P.J. Moss
Deparment of Civil Engineering, University of Canterbury, Christchurch, New Zealand
G.C. Clifton
Heavy Engineering Research Association, Manukau City, New Zealand

ABSTRACT: In order to investigate the effect of fire on a multi-storey steel frame building, a particular 17 storey building was studied to determine the behaviour as a fire spread throughout one floor of the building. Resistance to wind or seismic lateral forces acting along the major axis of the building is provided by moment resisting frames on two opposite sides while at right angles the lateral resistance is provided by eccentrically braced frames. A gravity load carrying frame runs along the major axis of the building. The columns and braces were fully fire protected, while the beams (acting compositively with the concrete floor slab) were unprotected. For the analyses carried out, the fire occurred on only the second level and spread from one quarter of the building.

1 INTRODUCTION

Under the old Building Control System (prior to 1992), design of multi-storey steel framed buildings for fire resistance was undertaken on the incorrect premise that the building will fail to meet the required levels of fire safety unless the beams and columns are insulated from temperature rise under fully developed fire conditions. The advent, in 1992, of a performance based regulatory system and set of fire safety requirements which foster the use of rational fire engineering design, has led to the now routine use of unprotected steel beams, and sometimes columns, in multi-storey buildings where the structural fire severity is low (such as in car parking buildings) or where it is moderate and the steel members are shielded from immediate exposure to the fire by non-fire-rated linings (such as in hotels and apartment buildings). This paper describes advanced analyses being undertaken as part of an international research programme aimed at determining the extent to which unprotected steel beams can be used when the structural fire severity is high.

2 BUILDING BEING MODELLED

The building is a 17 storey office building located in the Auckland Central Business District. It was built in 1988 and incorporates a perimeter based seismic-resisting system and an internal gravity load-carrying system.

Fig. 1 shows a view of the building looking towards the north-west corner. The west side is to the right and the north side to the left. Fig. 2 shows the reflected floor plan and the third floor level, which forms the top surface of the fire floor. The access for services and stairs is enclosed in a separate firecell in the south-east corner; its shape and extent is shown in Fig. 3.

The lateral load-resisting system comprises perimeter moment-resisting frames (MRFs) along the north and south sides, with eccentrically braced frames along the east and west sides, as shown in Fig. 2.

The floor system comprises a 120mm thick concrete slab, with $f_c' = 25$MPa, cast onto Hibond decking and made composite with 410UB54 Grade 250 supporting secondary beams.

Fig. 1 View of building looking towards the north-west corner

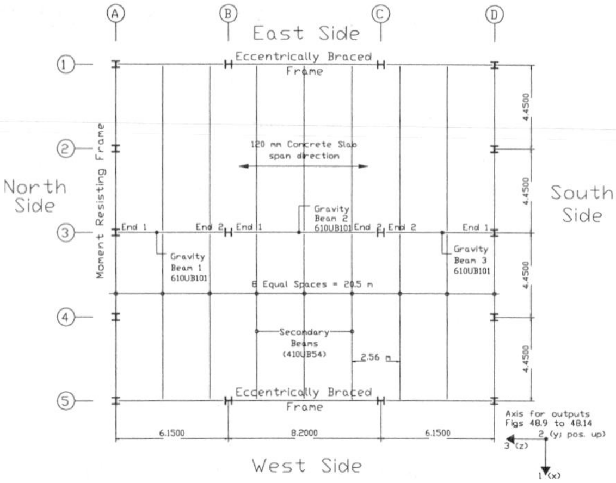

Fig. 2 Reflected floor plan of building over the fire floor (second level) showing gravity load-carrying and lateral load-resisting systems

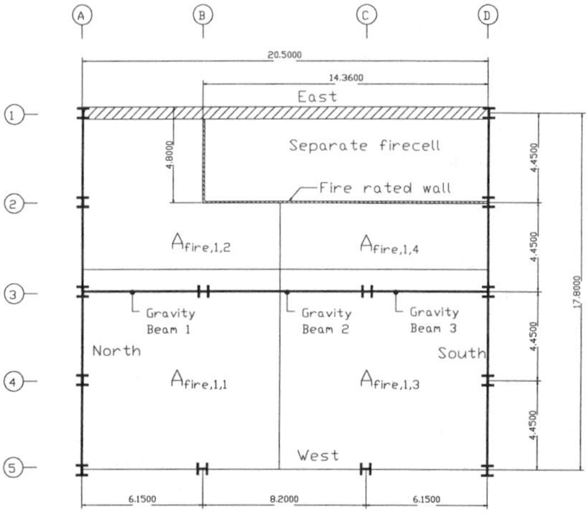

Fig. 3 Plan of fire floor (second level) showing extent of firecell and the four areas of fire

These span from a central primary beam along gridline 3 out to the east and west sides, as shown in Fig. 2.

Because the floors will function as effective fire separations, experience from actual fires shows that a fully developed fire spreads to adjacent floors via the windows in external walls, and in the absence of fire service intervention, will take a minimum of 30 minutes. For this reason, the fires were modelled as occurring in one storey only, as the structural effects will occur principally on a floor by floor basis.

Storey 2 was chosen, for three reasons. First, it is the lowest typical floor. The ground floor (storey 1) has greater ventilation and lower fire load, being intended for access and reception, and hence will generate a significantly lower structural fire severity. The second storey is therefore the lowest level on which high structural fire severity could realistically be generated.

Putting the fire on the lowest level also means it affects the most heavily loaded columns and has the potentially greatest destabilising effect on the floors above.

3 ANALYTICAL MODELS USED

3.1 *Fire model*

The fire model used was the large firecell fire model from HERA Report R4-83 (Clifton 1996).

For the building response modelling required in this project, a fire model is required which accounts for the fact that, in a large firecell, not all the firecell will be subjected to fully developed fire conditions at any one time. Instead, if conditions are favourable for fire growth, the fire will reach full development, at its point of origin, then migrate throughout the remainder of the firecell. Thus, at any one time after full development is reached at the point of fire origin, there will be areas of the firecell not yet subjected to fully developed fire, other areas under full development and other areas which have been burned out.

3.2 *Heat transfer model*

The programme TASEF-2 (Temperature Analyses of Structures Exposed to Fire-Two Dimensional Version) was used to determine the temperature rise in the steel members and the concrete slab (the concrete slab applied to the gravity beams on gridline 3, which have been modelled as fully composite members). Once steel and, where appropriate, concrete temperatures were determined from TASEF-2, they could be used as input for the elements of the cross sections in the structural model.

3.3 *Structural model*

The structural model and analyses used the general finite element analysis program ABAQUS (Hibbit et al 1997). This program allowed the following aspects to be taken into account:

(1) The effect of elevated temperatures on the mechanical properties of the steel and concrete. The input values for concrete were taken from Liu (1998). Those for steel, in the first 16 runs, were taken from Stevenson (1993); however, in the later 11 runs, the much more accurate stress-strain-time model of Poh (1996) was used.
(2) The effect of lateral restraint and continuity provided by the floor slab to the columns. The floor slab was modelled as a membrane element.
(3) The effect of composite beam action on the central primary beams along gridline 3. These three beams were modelled as fully composite members.
(4) Column members, which are steel I-sections, were modelled using standard I-beam elements which incorporated temperature dependant material properties.
(5) Connections between perimeter gravity beams A1-B1, C1-D1, A5-B5 and C5-D5 and their supporting columns (see Fig. 2) were modelled as continuous.
(6) Connections between primary beams A3-B3, B3-C3 and C3-D3 and their supporting columns were modelled as realistically as practicable as temperature-dependant rotational and axial springs.

4 SCOPE OF ANALYSES UNDERTAKEN

Twenty-seven different scenarios were analysed, determining the influences of the following parameters:

(1) Two different levels of fire load
   - e_f = 800 MJ/m^2 floor area (office design fire load)
   - e_f = 1200 MJ/m^2 floor area (office maximum credible fire load or office library or office storage design fire load)
(2) Three different ventilation conditions, obtained by applying different window heights around the three sides with curtain walling.

(3) The extent of fire spread on building behaviour. This involved runs keeping all loading and ventilation parameters constant and applying the fire over all four areas of fire, then over $A_{fire,1,1}$ and $A_{fire,1,2}$ only, then over $A_{fire,1,1}$ only (see Fig. 3).
(4) The presence of a non-fire-rated suspended ceiling forming a radiation barrier between the unprotected beams above and the fire below.
(5) Applied lateral loading (simulated wind loading) acting concurrently with fire; two levels of loading were considered.
(6) Variation in the elevated temperature mechanical properties (stress-strain-time) of the steel, as mentioned in 3.3(1) above.
(7) Variation in the connection strengths between the primary beams on gridline 3 and the supporting columns.

5 KEY FEATURES OF THE BUILDING'S BEHAVIOUR

5.1 *General*

This section gives a brief overview of key features of the building's behaviour and shows six figures (Figs. 4 - 9) from the range of outputs described above.

These figures are taken from the second set of 11 runs, which represent the most realistic input conditions. This is particularly the case with regard to the elevated temperature mechanical properties of the steel, with the model of Poh (1996) being used. This allows the stress-strain relationship to be derived for a given temperature and length of time over which that temperature is applied. In the case of the runs shown in Figs. 4 - 9, the length of time used for each temperature was derived from the fire time-temperature curve for $A_{fire,1,1}$ corresponding to the level of fire load and ventilation conditions

5.2 *Behaviour of the gravity beams*

Fig. 4 shows the midspan deflection of the 8.2m span gravity beam 2 (beam B3-C3) over the two hour analysis time. The beam midspan deflection reaches a maximum of 340mm, recovering to 255mm after the fire. These values are for the fire load of e_f = 800 MJ/m^2 floor area; for e_f = 1200 MJ/m^2 floor area, the pattern of behaviour is the same but the peak midspan deflection is 390mm, recovering to 280mm after the fire.

Fig. 5 shows the deflected shape before the fire, at the time of maximum deflection and after the fire. The small midspan deflection of 8mm before the fire comes from the gravity ($G + Q_u$) loading. The deflected shape during and after the fire is not quite symmetrical because gravity beam 2 straddles two areas of fire (see Fig. 3) and so the heating condition is not uniform along the beam in the analyses; nor would it be in practice.

During the heating phase, the beams pushed against the end supports as they underwent a nett expansion. The compression force induced by this reduced during the cooling phase and then became tensile, with a final post-fire residual tensile force remaining through the connection into the supporting columns.

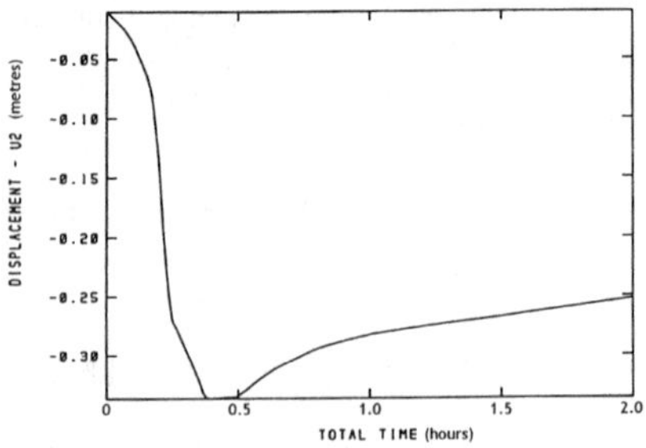

Fig. 4 Midspan deflection of gravity beam 2 versus time (h = 1.8m, e_f = 800 MJ/m^2)

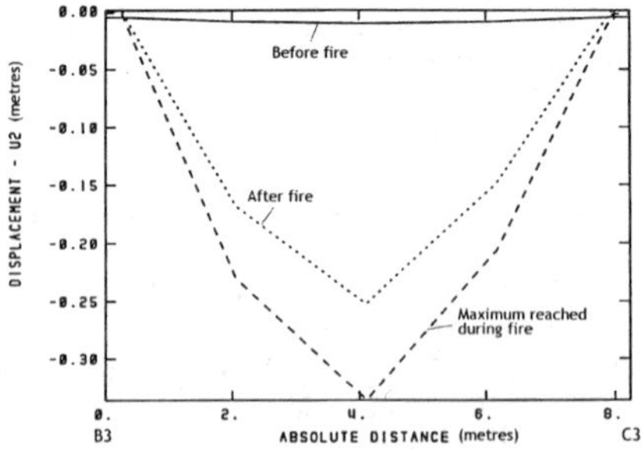

Fig. 5 Deflected shape of gravity beam 2 (h = 1.8m, e_f = 800 MJ/m^2)

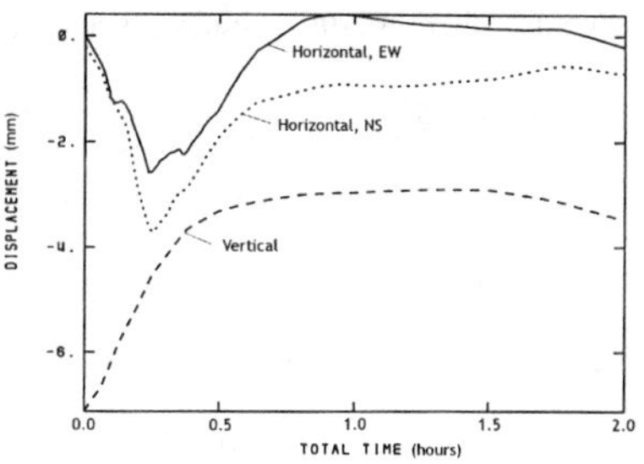

Fig. 6 Deflection versus time at top of building, column A3 (h = 1.8m, e_f = 800 MJ/m^2, no lateral load)
NS = North-South direction,
EW = East-West direction

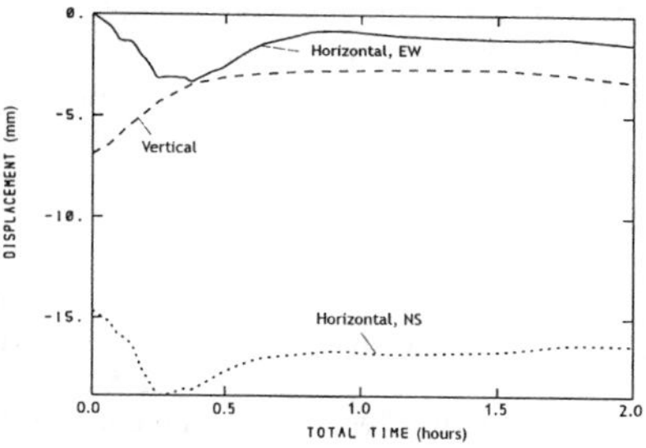

Fig. 7 Deflection versus time at top of building, column A3 (h = 1.8m, e_f = 800 MJ/m^2, 0.5 kPa lateral load on North face)

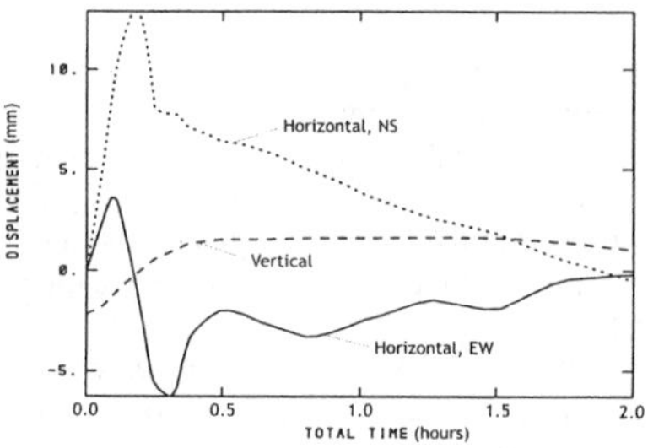

Fig. 8 Deflection versus time fire floor ceiling level, column A3 (h = 1.8m, e_f = 800 MJ/m^2, no lateral load)

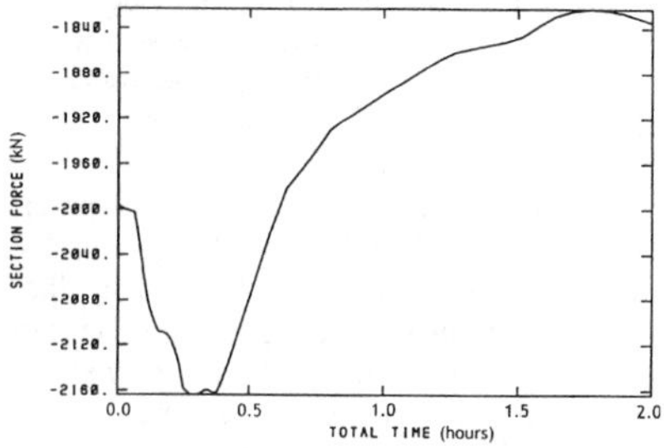

Fig. 9 Axial force in column A3 at fire floor versus time (h = 1.8m, e_f = 800 MJ/m^2)

5.3 *Behaviour of the supporting columns*

The maximum temperature reached from TASEF-2 in any of the supporting columns, all of which are protected, was 130°C. This is consistent with the UK test results (Kirby 1996). Because of this, the columns remained elastic. Axial compression force in column B3, on the fire floor, for the run involving e_f = 800 MJ/m^2 floor area, increased from 4,980 kN prior to the fire to 5,250 kN at time of maximum column temperature rise. This corresponded to a growth in column length of 6 mm. A similar situation occurred with column A3, on the fire floor, as shown in Fig. 9. In this case, the growth in column length was 4 mm and the axial force changed by ± 8% from its initial value. The moment input into any of the supporting columns from fire-induced actions was also much less than the section moment capacity of the column. The supporting columns therefore remained elastic during each of the analyses.

5.4 *Lateral stability of the overall building*

As previously mentioned, one of the key outputs desired from this research was an indication of the extent to which the building above the fire floor would be destabilised by the fire. The building's high aspect ratio in both plan directions and the exposure of the lateral load-resisting systems on 3 of the 4 sides to the fire meant that the fire would be expected to have more of a destabilising effect on this building than it would on buildings with lower aspect ratios or buildings with a central, lateral load-resisting core (which invariably houses services and access routes and so forms part of a separate firecell).

Fig. 6 shows the deflection versus time in each direction at the top of the building, location column A3, with no applied lateral load. The vertical deflection (upwards) is due to the lengthening of column A3 on the fire floor. The magnitude of lengthening (3.3mm) reflects the fact that this column is more restrained against lengthening than gravity column B3 and also remains cooler during the fire.

There is a very small horizontal movement in each direction, of less than 4mm, at the peak of the fire, returning to the pre-fire location on cooling. Overall, with no applied lateral loading, the fire at level 2 appears to have no noticeable influence on lateral movement of the overall building above the fire floor.

Fig. 7 shows the same structural model and design fire, this time with a lateral load (pressure) of 0.5 kPa applied against the north face. This magnitude of pressure is approximately that associated with serviceability wind loading on the building. The effect on the vertical movement is negligible. The effect is just noticeable on the east-west horizontal movement. As expected, the "wind" causes the top of the building to move laterally 15mm southward and this influence remains constant during the analysis. This can be seen by the fact that the horizontal north-south deflection with time shown in Fig. 7 is simply that from Fig. 6 plus a constant value of minus 15mm due to the lateral load. (The minus value is due to the wind-induced deflection being towards the south; i.e. in the negative z-axis direction, see Fig. 2).

If the magnitude of lateral load is increased to 1.0 kPa (i.e. slightly above the serviceability level), the pattern of north-south deflection remains that of Fig. 6, but now with a constant offset of minus 28mm due to the lateral load.

Fig. 8 shows the deflection versus time in each direction at the top of column A3 at the fire floor level. Note that the north-south deflection is greater than at the top of the building (Fig. 6) but still well within the limits for which the column remains stable and elastic. The maximum value of north-south outward movement of the column at the fire floor ceiling level corresponds to 1/300 of the storey height. This illustrates that the lateral stability of the overall building is not adversely affected by the fire with or without the presence of a concurrent lateral load corresponding to the serviceability wind loading level. It has been conservatively assumed in these analyses that the nature of fire is not affected by the wind. In practice, its effect on the structure would be lessened because of the reduced structural fire severity resulting from the increased ventilation and cross-flow driven by the wind.

6 CONCLUSIONS

The research (Clifton et al 1999) has shown that this particular multi-storey steel frame building, if constructed using unprotected beams, would possess an adequate level of fire safety. It is expected that this conclusion would apply to a much wider range of multi-storey steel framed buildings with protected columns and unprotected beams.

7 ACKNOWLEDGEMENTS

The contribution of all people/organizations involved in HERA's fire research is acknowledged. The principal past and ongoing funding for this project was provided by the Foundation for Research, Science and Technology

8 REFERENCES

Clifton G.C. 1996. *Fire Models for Large Firecells*; HERA, Manukau City, HERA Report R4-83.

Clifton, G.C. et. al.; 1999. *Behaviour of a Multi-Storey Steel Building in Fully Developed Natural Fires*; HERA, Manukau City, HERA Report R4-95.

Hibbit, Karlsson and Sorensen, Inc; 1997. *ABAQUS Finite Element Computer Program*, Version 5.7.

Kirby, B.R. 1998 *The Behaviour of a Multi-Storey Steel Framed Building Subject to Fire. Attack-Experimental Data*, British Steel Swinden Technology Centre, United Kingdom.

Liu; 1998.Three Dimensional Modelling of Steel/Concrete Composite Connection Behaviour in Fire; Second World Conference on Steel in Construction; *Journal of Constructional Steel Research*, 46:1-3.

Poh, K.W. 1996. *Modelling Elevated Temperature Properties of Structural Steel*; BHP Research, Melbourne, Australia, Report BHPR/SM/R/055.

Stevenson, P.L. 1993. *Computer Modelling of Structural Steel Frames in Fire*, Master of Engineering report, University of Canterbury, Christchurch, 1993.

Mechanics of Structures and Materials, Bradford, Bridge & Foster (eds)
© 1999 Balkema, Rotterdam, ISBN 90 5809 107 4

Local buckling of steel plates at elevated temperatures

M.B.Wong
Department of Civil Engineering, Monash University, Melbourne, Vic., Australia
K.H.Tan
School of Civil and Structural Engineering, Nanyang Technological University, Singapore

ABSTRACT: An investigation into the local buckling characteristics of flat plate elements at elevated temperatures is carried out. An equation to calculate the yield limit of flat plate at elevated temperatures is derived and its variations due to the use of different design codes are compared. Using the concept of effective width for plate design, formulation for the calculation of effective width of plate elements at elevated temperatures is presented and its variations using different codes are also compared.

1 INTRODUCTION

The behaviour of steel members in structures at elevated temperatures has been studied in many ways, mostly based on the fundamental properties, including the stress-strain relationship of steel at high temperatures. In these studies, emphasis is usually placed on the global behaviour of the members which are assumed to maintain strength according to the variation of the yield strength with temperature. These studies lead to the "load ratio method" (BSI 1990) which is commonly used by engineers to calculate the limiting temperature at which the member is assumed to fail.

When using the load ratio method, the member's ultimate capacity, usually in bending, is calculated and assumed to vary with temperature in accordance with the yield strength–temperature curve (SA 1990). However, this assumption is questionable because the local buckling behaviour of the steel member at elevated temperatures may or may not behave in a way similar to that at room temperature. A compact steel section which exhibits no local buckling effect at room temperature may transform into a section which has different local buckling characteristics due to the effects of rising temperature.

This paper describes a preliminary theoretical investigation into the effects of elevated temperatures on the local buckling characteristics of steel plate elements. The theoretical development is based on the classical plate buckling theory and only simple elastic cases are examined. It is shown that the yield limit which determines the initiation of local buckling of a plate element under uniform temperature can be expressed as a function of both the modulus of elasticity and yield stress at elevated temperatures. For plates under uniform compression, the effective width concept applied to plates can also be established as a function of temperature. Formulas for variation of modulus of elasticity and yield strength with temperature adopted by different codes are used in this investigation and results compared. In the context of this paper, the effect of the residual stresses in plates has been ignored and only the theoretical local buckling analysis will be discussed.

2 VARIATION OF YIELD LIMIT WITH TEMPERATURE FOR PLATES

The critical buckling stress f_{ol} of a plate element is derived from plate bending theory (for example, Bulson 1970) as

$$f_{ol} = \frac{\pi^2 Ek}{12(1-\nu^2)\left(b/t\right)^2} \tag{1}$$

where E = Young's modulus, k = buckling coefficient, ν = Poisson's ratio, b = width of plate, and t = thickness of plate.

It is common to establish a yield limit λ_{ey} at which the transition between buckling and yielding can be monitored and checked for design. Hence, for buckling to occur before yielding at temperature T, the following inequality is established:

$$\sqrt{\frac{f_{yT}}{f_{ol}}} = \frac{b}{t}\sqrt{\frac{f_{yT}}{E_T}\frac{12(1-\nu^2)}{\pi^2 k}} \geq 1.0 \tag{2}$$

where

$$\frac{E_T}{E_{20}} = \frac{\text{Young's modulus at T}}{\text{Young's modulus at room temperature}} = \phi_{ET}\ , \tag{3}$$

$$\frac{f_{yT}}{f_{y20}} = \frac{\text{yield stress at T}}{\text{yield stress at room temperature}} = \phi_{yT}\ . \tag{4}$$

It can be shown from Equation 2 that, for E_{20} = 200,000 MPa and ν = 0.3,

$$\lambda_e \geq \lambda_{ey} \tag{5}$$

where $\lambda_e = \frac{b}{t}\sqrt{\frac{f_{y20}}{250}}$ and $\lambda_{ey} = 26.89\sqrt{k}\sqrt{\frac{\phi_{ET}}{\phi_{yT}}}$.

Equation 5 gives a means to determine whether the yield strength of a plate across a broad temperature range can be fully developed before buckling sets in. The buckling coefficient k is a function of geometry and boundary conditions of the plate. Thus, the yield limit λ_{ey} of a plate at any temperature can be obtained if ϕ_{ET} and ϕ_{yT} are known.

Although the functions ϕ_{ET} and ϕ_{yT} vary from one country to another when stipulated in design codes, the difference is usually very little. In Figure 1(a), the variation of the term $\sqrt{(\phi_{ET}/\phi_{yT})}$ with temperature in Equation 5 is plotted using the formulas for ϕ_{ET} and ϕ_{yT} stipulated in AS4100 (SA 1990). The same curve is compared with those adopted by ASCE (Lie, 1992) and BS5950 in Figure 1(b) which shows that the value of the term $\sqrt{(\phi_{ET}/\phi_{yT})}$ is in general increasing with temperature.

A best-fit curve is also plotted showing the trend of this variation for functions stipulated in AS4100. The best-fit curve is found to be

$$\sqrt{(\phi_{ET}/\phi_{yT})} = 0.972 + 0.00014T \tag{6}$$

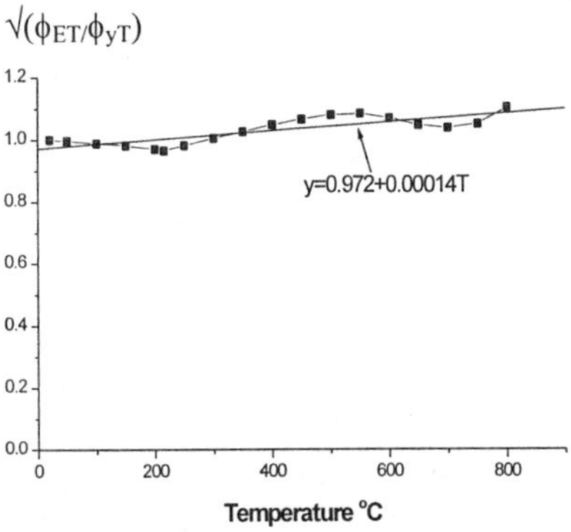

(a)

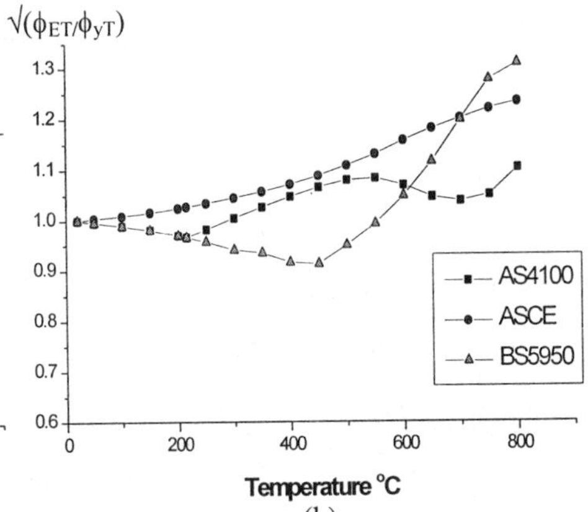

(b)

Figure 1. Variation of $\sqrt{(\phi_{ET}/\phi_{yT})}$ with temperature

Substituting Equation 6 into 5, the yield limit can be obtained as a linear function of temperature:

$$\lambda_{ey} = (26.14 + 0.00376T)\sqrt{k} \geq 26.89\sqrt{k} \quad (7)$$

Equation 7 provides a convenient means to calculate the yield limit at elevated temperatures for plates with various boundary conditions provided that the value of the buckling coefficient k is known.

3 EFFECTIVE WIDTH FOR FLAT PLATES AT ELEVATED TEMPERATURES

A common practice to estimate the maximum strength of plates is by the use of effective cross-section through the concept of "effective width". For plates under uniform compression with the edges simply supported, the ratio of the effective width b_e to the original width b of a plate can be expressed as a function of f_{ol}/f_{yT} by various semi-empirical methods (refer to Tan, et. al., 1993 for equation details). Three of these expressions are chosen here for comparison purposes:

$$\frac{b_e}{b} = \sqrt{\frac{f_{ol}}{f_{yT}}} \quad \text{(von Karman)} \quad (8)$$

$$\frac{b_e}{b} = \sqrt{\frac{f_{ol}}{f_{yT}}}\left(1 - 0.22\sqrt{\frac{f_{ol}}{f_{yT}}}\right) \quad \text{(AISI 1968)} \quad (9)$$

$$\frac{b_e}{b} = 0.15 + 0.85\frac{f_{ol}}{f_{yT}} \quad \text{(TNO Holland)} \quad (10)$$

Using Equation 2 for the ratio (f_{ol}/f_{yT}), Equations 8-10 can be related to temperature T in terms of (ϕ_{ET}/ϕ_{yT}).

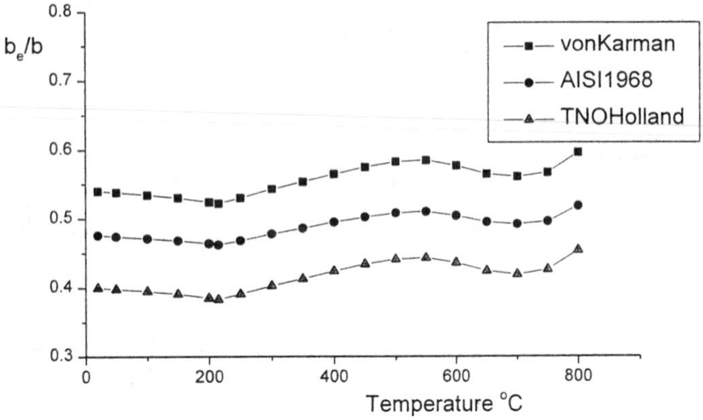

Figure 2a. Comparison of effective width formulas using AS4100

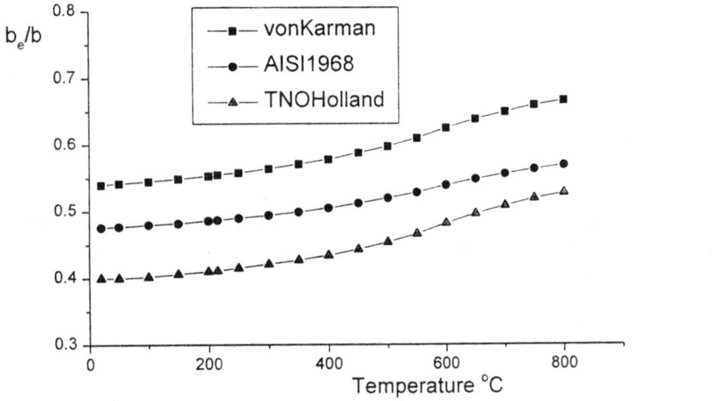

Figure 2b. Comparison of effective width formulas using ASCE standard

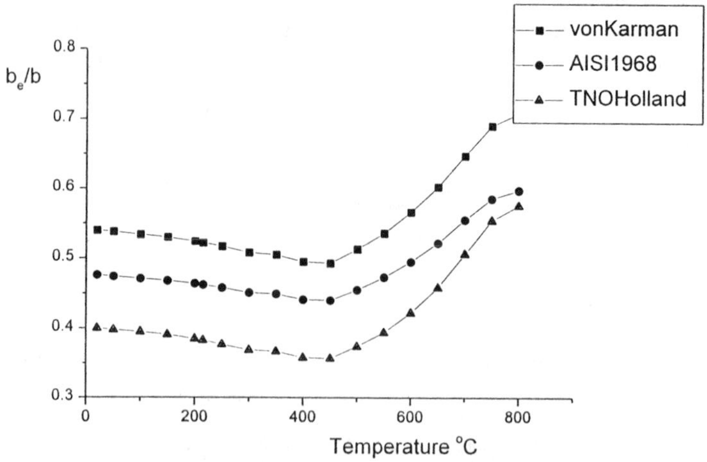

Figure 2c. Comparison of effective width formulas using BS5950

4 EXAMPLE

As an example, a flat plate with b/t = 100 is used for its effective width assessment within a range of temperatures. Through the use of Equation 2 with k = 4, E_{20} = 200,000 MPa and ν = 0.3, Equations 8 to 10 become

$$\frac{b_e}{b} = \sqrt{\frac{f_{ol}}{f_{yT}}} = \frac{54}{b/t}\sqrt{\frac{\phi_{ET}}{\phi_{yT}}} = 0.54\sqrt{\frac{\phi_{ET}}{\phi_{yT}}} \qquad \text{(von Karman)} \qquad (11)$$

$$\frac{b_e}{b} = 0.54\sqrt{\frac{\phi_{ET}}{\phi_{yT}}}\left[1 - 0.22(0.54\sqrt{\frac{\phi_{ET}}{\phi_{yT}}})\right] \qquad \text{(AISI 1968)} \qquad (12)$$

$$\frac{b_e}{b} = 0.15 + 0.25(\frac{\phi_{ET}}{\phi_{yT}}) \qquad \text{(TNO Holland)} \qquad (13)$$

To examine the variation of the effective width ratio at high temperatures using various design codes, Equations 11 to 13 are plotted for each design code using Figure 1b. The results are shown in Figure 2.

It can be seen that there is a significant difference in calculating the effective width of a plate element when different effective width formulas, as given in Equations 11 to 13, are used. However, the curves in Figure 2 indicate a general trend of increasing effective width when different design codes are used.

5 CONCLUSIONS

Formulations for local buckling characteristics of flat plate elements are presented. It is shown that the yield limits of plates at elevated temperatures can be conveniently expressed in terms of temperature rise. The variations of yield limits with temperature are compared when different design codes are used. It has been found that the yield limits tend to increase slightly with temperature.

Similarly, the effective width of plates can also be expressed in terms of temperature rise and its variations are given when using different design codes. Again, the effective width of plates tends to increase with temperature when different design codes are used.

Therefore, the assumption that the local buckling characteristics applied to plate at room temperature can also be adopted for plates at elevated temperatures will give safe but conservative results in design.

6 REFERENCES

British Standards Institution (BSI) 1990. *BS 5950: The structural use of steelwork in buildings. Part 8: Code of practice for fire resistant Design.*

Standards Australia (SA) 1990. *AS4100 - 1990, Steel Structures, SA.*

Bulson, P.S. 1970. *The stability of flat plates.* Chatto & Windus.

Lie, T.T. 1992. *Structural Fire Protection*, American Society of Civil Engineers.

Tan K.H., Lok, T.S. and Chiew, S.P. 1993. A simplified method for ultimate load prediction of all-steel sandwich panels. *Thin-Walled Structures.* 17:27-44.

Composites engineering and new materials

Mechanics of Structures and Materials, Bradford, Bridge & Foster (eds)

 ISBN 90 5809 107 4

On the mathematical modeling of magneto-piezoelastic anisotropic materials

A. Basu
Department of Mechanical Engineering, Wollongong University, N.S.W., Australia
A. L. Kalamkarov
Department of Mechanical Engineering, Daltech, Dalhousie University, Halifax, N.S., Canada

ABSTRACT: This paper is concerned with the basic mathematical aspects of the coupled magneto-piezoelastic anistropic materials. It is shown that Green's function method is applicable to time harmonic magneto-elastic-piezoelectricity problems using the boundary integral technique. Solutions are obtained through generalisation of Betti's reciprocal theorem and boundary value problem. As an application, a two dimensional static plane-strain problem has been dealt with to find the effect of magnetic field on piezoelectric material.

1 INTRODUCTION

Combined action of piezoelectricity, continuum mechanics and magnetism is open for criticism although mathematical development for possible applications is feasible for many engineering problems. As far as the theoretical development of this method is concerned, not much work has been reported in this area. The object of this study is to propose an idealized mathematical model to describe the behaviour and interrelations of physical phenomenon combining all these three fields. Every model is inherently approximate but its adequacy or degree of accuracy depends on physical reality in a limited range as complete detail is never known in a broad sense. In the direct piezoelectric effect, the application of an external mechanical loading induces an electrical response in the material. In the converse effect, an applied electrical field makes the material strained. It is known that an applied electromagnetic field induces currents in a body, which in turn give rise to Lorentz body force $\mathbf{J} \times \mathbf{B}$ which comes as an external body force in the magneto-piezoelastic equation of motion. Electromagnetic and elastic fields in a piezoelectric medium are fully described by the equations of motion of a continuous medium

$$\sigma_{ij,j} + F_i = \rho \ddot{u}_i \tag{1}$$

combined with Maxwell's equations

$$\begin{aligned} \text{curl} \quad \mathbf{E} &= -\frac{\partial \mathrm{B}}{\partial t}, \quad \text{div} \quad \mathbf{D} = \mathbf{0} \\ \text{curl} \quad \mathbf{H} &= -\frac{\partial \mathrm{D}}{\partial t}, \quad \text{div} \quad \mathbf{B} = \mathbf{0} \end{aligned} \tag{2}$$

for the situation when no free charges and no external currents are present. The stresses σ_{ij}, electrical displacement D and magnetic flux B are related to the strains ε_{ij} and the fields E and H through constitutive equations.

2 MATHEMATICAL PRELIMINARIES

We consider a homogeneous magneto-elastic-piezelectric anisotropic body B with boundary Γ is subject to a uniform magnetic field $\mathrm{H}(\mathrm{B} = \mu \mathrm{H})$. The equations of motion and Gauss' law are given by (see e.g., Basu, 1994)

$$\mathrm{div}\ \sigma + (\mathbf{J} \times \mathbf{B}) + \mathbf{f} = \rho \ddot{\mathbf{u}} \ , \ \mathrm{div} \quad \mathbf{D} = \mathbf{q} \tag{3}$$

where $\sigma = \{\sigma_{ij}\}, \mathbf{u} = \{u_i\}, \mathbf{D} = \{D_i\}, \rho, \mathbf{f} = \{f_i\}, (i = 1,2,3), \mathrm{q}, \mathbf{J}, \mathbf{B}$

denote stress, elastic displacement, electric displacement, mass density, body force per unit volume, electric charge density, induced current and magnetic flux density respectively. Let e and E be the strain and electric fields respectively, given by

$$\varepsilon = 1/2(\nabla_u + \nabla_u^{\ T})E = -\nabla\phi\sigma_{zz} \tag{4}$$

where ϕ is the electric potential. Also μ is known as the magnetic permeability. The constitutive relations for linear piezoelectricity are (see e.g.,Kalamkarov, 1992)

$$\sigma_{ij} = C_{ijkl}\varepsilon_{kl} - e_{kij}E_k \qquad D_i = e_{ikl}\varepsilon_{ki} + \varepsilon_{ik}E_k \tag{5}$$

where C_{ijkl}, e_{ijk}, and ϵ_{ij} are the elastic, piezoelectric and dielectric material constants respectively, satisfying the following symmetry relations:

$$c_{jikl} = c_{ijlk} = c_{jikl} = c_{klij} \qquad e_{kij} = e_{kji} \qquad \epsilon_{ik} = \epsilon_{ki} \tag{6}$$

The combination of equations (3) to (6) results in a system of four partial differential equations coupling the displacement components and electric potential, namely

$$\begin{aligned} c_{ijkl}u_{k,lj} + e_{kij}\phi_{,kj} + (\mathrm{J} \times \mathrm{B})_i + f_i &= \rho\ddot{u}_i \\ e_{ikl}u_{k,li} - \epsilon_{ik}\phi_{,ki} &= q \end{aligned} \tag{7}$$

3 TWO-DIMENSIONAL MAGNETO-PIEZOELASTIC PROBLEM

The constitutive equations for the plane-strain case when $\epsilon_{yy}, \epsilon_{xy}, \epsilon_{zy} = 0$ is

$$\begin{aligned} \sigma_{xx} &= c_{11}\epsilon_{xx} + c_{13}\epsilon_{zz} - e_{31}E_z, \quad \sigma_{zz} = c_{13}\epsilon_{xx} + c_{33}\epsilon_{zz} - e_{33}E_z \\ \sigma_{xz} &= 2c_{44}\epsilon_{xz} - e_{15}E_x, \quad D_x = 2e_{15}\epsilon_{xz} + \varepsilon_{11}E_x \\ D_z &= e_{31}\epsilon_{xx} + e_{33}\epsilon_{zz} + \varepsilon_{33}E_z \end{aligned} \tag{8}$$

$$\begin{aligned} \frac{\partial \sigma_{xx}}{\partial x} + \frac{\partial \sigma_{xz}}{\partial z} + f_x + (\mathrm{J} \times \mathrm{B})_x &= \rho\frac{\partial^2 u_x}{\partial t^2} \\ \frac{\partial \sigma_{xz}}{\partial x} + \frac{\partial \sigma_{zz}}{\partial z} + f_z + (\mathrm{J} \times \mathrm{B})_z &= \rho\frac{\partial^2 u_z}{\partial t^2} \\ \frac{\partial D_x}{\partial x} + \frac{\partial D_z}{\partial z} &= q \end{aligned} \tag{9}$$

Also

$$
\begin{aligned}
E_x &= -\frac{\partial \phi}{\partial x} \qquad E_z = -\frac{\partial \phi}{\partial z} \qquad \epsilon_{xx} = \frac{\partial u_x}{\partial x} \qquad \epsilon_{zz} = \frac{\partial u_z}{\partial z} \\
\epsilon_{xz} &= \frac{1}{2}\left(\frac{\partial u_x}{\partial z} + \frac{\partial u_z}{\partial x}\right)
\end{aligned} \tag{10}
$$

Assume H = (H1, 0, 0), substitute (10 into (9)), then

$$
\begin{aligned}
c_{11}\frac{\partial^2 u_x}{\partial x^2} &+ c_{44}\frac{\partial^2 u_x}{\partial z^2} + (c_{13}+c_{44})\frac{\partial^2 u_z}{\partial x \partial z} + (e_{31}+e_{15})\frac{\partial^2 \phi}{\partial x \partial z} \\
&+ f_x = \rho\frac{\partial^2 u_x}{\partial t^2} \\
c_{44}\frac{\partial^2 u_z}{\partial x^2} &+ c_{33}\frac{\partial^2 u_z}{\partial z^2} + (c_{13}+c_{44})\frac{\partial^2 u_x}{\partial x \partial z} + e_{15}\frac{\partial^2 \phi}{\partial x^2} \\
&+ e_{33}\frac{\partial^2 \phi}{\partial z^2} + \mu_e H_1 \nabla^2 u_z + f_z = \rho\frac{\partial^2 u_z}{\partial t^2} \\
e_{15}\frac{\partial^2 u_z}{\partial x^2} &+ e_{33}\frac{\partial^2 u_z}{\partial z^2} + (e_{31}+e_{15})\frac{\partial^2 u_x}{\partial x \partial z} \\
&- \varepsilon_{11}\frac{\partial^2 \phi}{\partial x^2} - \varepsilon_{33}\frac{\partial^2 \phi}{\partial z^2} = q
\end{aligned} \tag{11}
$$

Assume the general solution of the homogeneous equation with

$$
\begin{aligned}
f_x &= f_z = q = 0, \quad \frac{\partial^2 u_x}{\partial t^2} = \frac{\partial^2 u_z}{\partial t^2} = 0 \\
u_x &= \frac{1}{2\pi}\int_{-\infty}^{\infty} i\xi\left(A_1 e^{-\alpha|\xi|z} - A_2 e^{\alpha|\xi|z}\right)e^{-i\xi x}d\xi \\
u_z &= \frac{1}{2\pi}\int_{-\infty}^{\infty} |\xi|\left(B_1 e^{-\alpha|\xi|z} + B_2 e^{\alpha|\xi|z}\right)e^{-i\xi x}d\xi \\
\phi &= \frac{1}{2\pi}\int_{-\infty}^{\infty} |\xi|\left(C_1 e^{-\alpha|\xi|z} + C_2 e^{\alpha|\xi|z}\right)e^{-i\xi x}d\xi
\end{aligned} \tag{12}
$$

The substitution of equation (12) into (11) gives α as the root of the characteristic equation

$$
\det[P] = 0, \quad \text{or} \quad \alpha^6 + d_1\alpha^4 + d_2\alpha^2 + d_3 = 0 \tag{13}
$$

4 SOLUTIONS FOR LOADS APPLIED TO THE BOUNDARY

Assume a magneto-piezoelastic medium subjected to vertical line load and electric charge at the surface, see Figure 1. The consideration of regularity conditions of field variables as $z \to \infty$ implies that $H_j = 0$. In the first case of a verical load of magnitude P_0 per unit length applied to the surface the boundary conditions are

$$
\sigma_{xz}(x,o) = 0; \; \sigma_{zz}(x,o) = -P_o\delta(x) \quad D_z(x,o) = o \tag{14}
$$

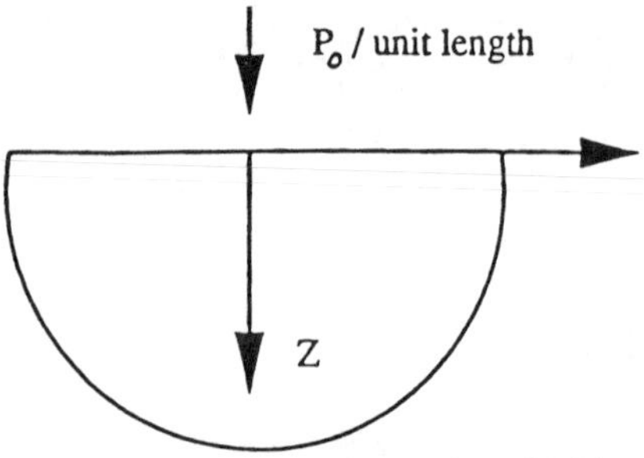

Figure. 1 concentrated line load applicable to magneto-piezoelastic medium with free foundary

The graphical solutions are given in Figures 2 & 3 for three materials.

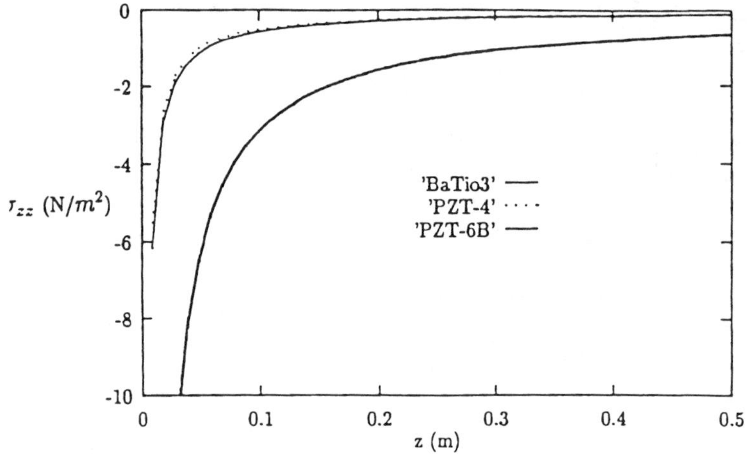

Figure. 2 Vertical stress in different magneto-piezoelastic solids due to a concentrated vertical line load of intensity 1.0 Nm^{-1}

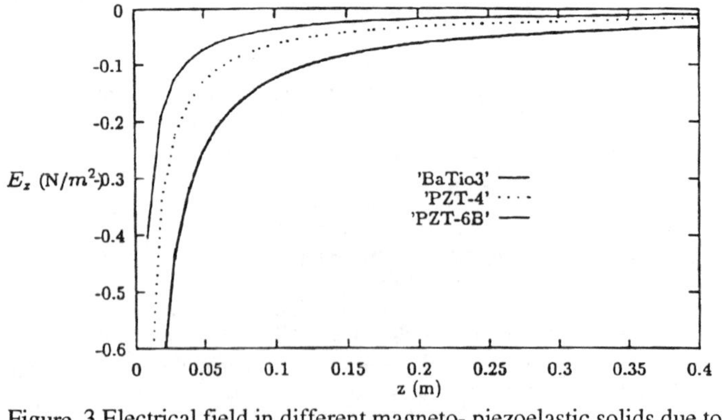

Figure,3 Electrical field in different magneto- piezoelastic solids due to a vertical line load of intensity 1.0 Nm-1

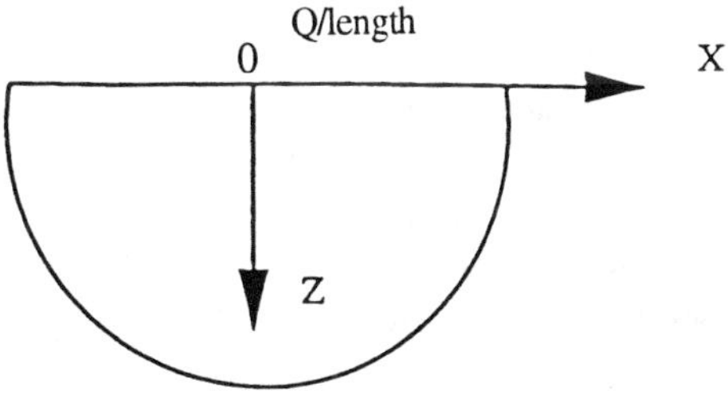

Figure 4 Concentrated electric charge applicable to magneto - piezoelastic medium with free boundary.

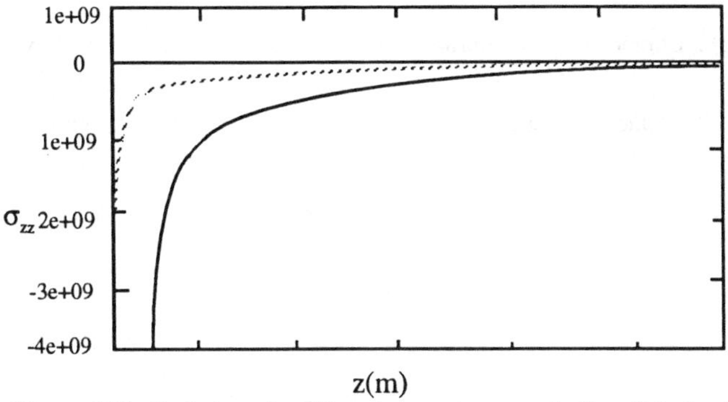

Figure 5 Vertical stress in different magneto-piezoelastic solids due to aline electric charge of intensity 1.0 C m^{-1} applied to the surface.

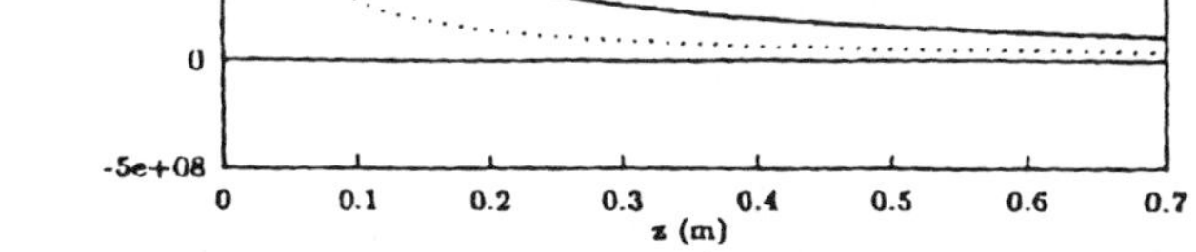

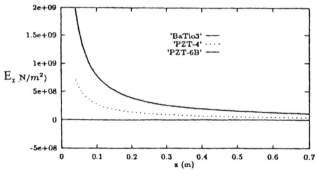

Figure 6 Electric fild in different magneto-piczoelastic solids due to a line electric charge of intensity 1.0Cm^{-1} applied to the surface

5 CONCENTRATED ELECTRIC CHARGE APPLIED WITH FREE BOUNDARY

The boundary conditions are (see Figure 4)

$$\sigma_{zz}(x,0) = 0 \quad \sigma_{zx}(x,0) = 0 \quad D_z(x,0) = Q_0\delta(x) \tag{15}$$

Solutions are given in Figures 5 and 6.

6 CONCLUSIONS

Figures 2,3 5 and 6 show the variation of σ_{zz} and E_z along the vertical axis of a magneto-piezoelastic solid which is subjected to a vertical line load of intensity 1.0 Nm^{-1} and an electric charge of intensity 1.0 Cm^{-1} at the top surface as shown in Figures 1 and 4. The vertical stress and the vertical electric field along the z-axis are nearly indentical for the three materials. Both σ_{zz} and E_z decay rapidly with the vertical distance. Thise results are very new in the composite area.

REFERENCES

(Basu, A.,1994) Boundary Value problems in magneto-thermo-elasticity and magneto-hydrodynamics, *ASTAM Publ., Wollongong, Australia,*

(Kalamkarov, A.L 1992)., Composite and reinforced elements of construction," *Wiley: Chichester, New York,.*

(Rajapakse, R.K.N.D 1997)., Plane strain/stress solutions for piezoelectric solids, *Composite Engineering: Part B, 28(B), , pp. 385-396.*

Mechanics of Structures and Materials, Bradford, Bridge & Foster (eds)
© 1999 Balkema, Rotterdam, ISBN 90 5809 107 4

Introduction to durable concrete in the arid and hot climates of Iran

M. Damghani & L. Niayesh
Department of Civil Engineering, Tehran-Boston Consulting Engineers, Iran

ABSTRACT: There are vast hot and arid regions in Iran, where the durability of concrete structures is seriously reduced due to the severe climatic conditions. In recent years, the Iranian concrete technologists have been researching on how to design durable concrete mixes resistant to the harsh environment in those regions. Tehran – Boston Engineers Ltd. was selected by Iran's Bureau of Plan and Budget to investigate the problem and provide solutions. This paper reviews methods of designing, constructing and maintaining concrete structures the environs of Iran and gives recommendations for concrete mix design for some special concretes.

1 INTRODUCTION

Tests on the soil and salty underground water of the arid regions in Iran have shown that they heavily contain harmful elements, and that high temperature and humidity accelerate the process of deterioration. Chloride attack by chemical agents within both concrete and the surrounding environment is to blame for most concrete deteriorations in the southern part of Iran's central desert near the Persian Gulf. The important factors which the acceleration, retardation and prevention of deterioration in concrete depend on are: oxygen, alkalinity of concrete, chloride ions, porosity and permeability, electrical conductivity, humidity, temperature, cover to reinforcement, cement type, water/cement ratio and curing.

2 CLIMATIC CONDITIONS

The hot regions of Iran are either arid or humid. The humid regions, mainly on the coastal parts of the Persian Gulf and the Oman sea have a climate much harsher to concrete structures. Temperature, humidity, sun radiation, wind and soil are the main factors affecting concrete works in those regions.

2.1 *Temperature*

Temperature and its variation directly affect the quantity of water required, workability, crack formation and chemical reactions in concrete.

2.2 *Humidity*

It reduces the rate of evaporation of water and the shrinkage of fresh concrete. But, it accelerates the sulfate, chloride and alkali reactions in concrete. In general, the structures in the hot and

Table 1. Causes of deterioration of concrete

Cause	Environment	Effect on concrete
External causes	High temperature and successive variation in humidity	Workability is reduced. Shrinkage and drying cracks are formed. Corrosive chloride and sulfate ions and carbonic acid permeate concrete rapidly. Corrosion starts rapidly and its rate increases.
	Water existing in the Surrounding; underground water; sea water; soft water	Corrosion starts by chloride ions existing in underground and sea water. Surface deteriorates due to deposition of expanding salts. Concrete surface is deeply washed off by soft water in hot and humid regions.
	Biochemical effect of creatures on marine structures	Corrosion starts more rapidly by chloride ions due to reduction of concrete cover.
	Sand storm	Erosion Cracks
Internal causes and unsuitable materials	Sulfate resisting portland cement	Little resistance to chloride ions.
	Porous coarse aggregate and sand	Workability is reduced. More penetration of corrosive ions. Increase in shrinkage due to drying and creep of concrete. Less apparent durability due to rapid Deterioration of concrete surface caused by sedimentation of salts.
	Aggregates contaminated with clay, dust	More water is required in the mix causing an increase in drying shrinkage, creep and permeability and reducing compressive strength.
Practical mistakes	Coarse aggregates and sand containing sulfate and chloride	Chloride corrosion starts. Concrete surface deteriorate due to Sedimentation of expanding salts.
	Sharp edged coarse and fine aggregates with uniform grading	Loss of cement paste from concrete. Drying shrinkage. Increase in the quantity of water required increasing the concrete porosity.
	Alkali reactive aggregates	Cracks and sapling of concrete reducing cover to reinforcement.
	Improper aggregates and reinforcing bars	Aggregates or rebars containing sulfates or chlorides before mixing or placement cause the corrosion of reinforcement, and the salts formed reach the surface of concrete and cause more deterioration.
	Improper aggregates and reinforcing bars	Aggregates or rebars containing sulfates or chlorides before mixing or placement cause the corrosion of reinforcement, and the salts formed reach the surface of concrete and cause more deterioration.
	Improper compaction of concrete	Voids are formed inside concrete causing more rapid penetration of the deteriorating agents.
	Inadequate curing	Concrete permeability increases. Using chloride containing water for curing concrete cause rapid penetration of chloride ions.

humid regions are less resistant to chloride attacks than in the arid areas for the corrosion of steel develops if the surrounding humidity is about 70%-85%.

2.3 *Sun radiation*

In the hot and arid regions of Iran, cracks are formed on the surface of concrete structures due to the high average daily heat resulted from the radiation of Sun.

2.4 *Wind*

Most deserts of Iran are windy. Increasing the rate of evaporation of water from the surface of fresh concrete, heavy wind damages concrete by making it lose moisture so rapidly. Wind also causes further damage to concrete by carrying dust to its surface.

2.5 *Soil*

The bottom and external surfaces of concrete are in contact with soil, which normally contains chlorides and sulfates.

3 CAUSES OF DETERIORATION

Generally, the high temperature and wind are harmful for both concrete placing and curing in the hot regions. The combined effect of high temperature of both air and concrete and high wind velocity which causes the concrete surface water to evaporate at a rate of more than 0.5 kg/h per square meter, makes it necessary to find ways of preventing the early drying of concrete.

As for the chemically caused deteriorations of concrete, the main types of deterioration in the order of importance are chloride corrosion, sulfate attack, alkali reactions of aggregates and carbonation of concrete. The general causes of deterioration are described in the Table 1.

4 MATERIALS AND METHODS

The properties of the materials in a concrete mix are essential in affecting the durability of concrete in the hot regions. There is abundant calcareous and siliceous aggregate in the hot regions of Iran, which are completely controlled for their chemical contamination.

This section of the main report includes comprehensive data and descriptions of concrete components, i.e. aggregates, water, various types of cement, pozzolans, additives and steel reinforcement. There are also tables showing the acceptable limits in the quantities of harmful substances in aggregates and water and their physical properties.

Because of the high temperature in hot regions, generally the quantity of water in concrete mix is increased in order to maintain its workability. The fresh concrete temperature must also be reduced. The temperature of fresh concrete can be determined from

$$T = \frac{0.22(T_a W_a + T_c W_c) + T_w W_w + T_{wa} W_{wa}}{0.22(W_a + W_c) + W_w + W_{wa}} \tag{1}$$

where T = temperature of fresh concrete (deg. C), T_a = aggregate temperature (deg. C), T_c = cement temperature (deg. C), T_w = water temperature (deg. C), Twa = temperature of water absorbed by aggregate (deg. C), W_a = mass of aggregate per cubic meter of mix, W_c = mass of cement per cubic meter of mix, W_w = mass of water per cubic meter of mix, W_{wa} = mass of water absorbed by aggregate per cubic meter of mix.

If the temperature is too high, ice can be used as part of the concrete mix water so that the fresh concrete temperature doesn't reach 30 degrees Celsius. In that case, taking account of the mass of ice and its latent heat, the above formula can be modified by 80 calories per gram.

Provided that mixing and curing are carried out properly, the use of certain additives is recommended for their effect in reducing the permeability of concrete. These additives include: natural pozzolans, which are compacted volcanic ash and abundant in Iran, microsilica, electric arc furnace by-products, blast-furnace slag, rice husk and fly ash.

For its high cost, galvanised steel is used for reinforcement in special cases, and epoxy coated reinforcement is used for normal cases. Formworks should be made of plywood; and should be water tight and free from sharp corners. Controlled permeable formworks (CPF) are recommended in some places. The temperature should be controlled during placing and curing concrete. Water and curing blanket and membrane can also be used for curing concrete.

5 UTILIZATION AND MAINTENANCE OF CONCRETE IN HOT CLIMATES

As the concrete structures in these regions are prone to severe chemical (mainly chloride and sulfate) attacks from the environment, regular inspections, tests, and if required repairs, which the efficient service life time of a structure depends on, are necessary. The factors affecting the durability of the repair works are illustrated in Figure 1.

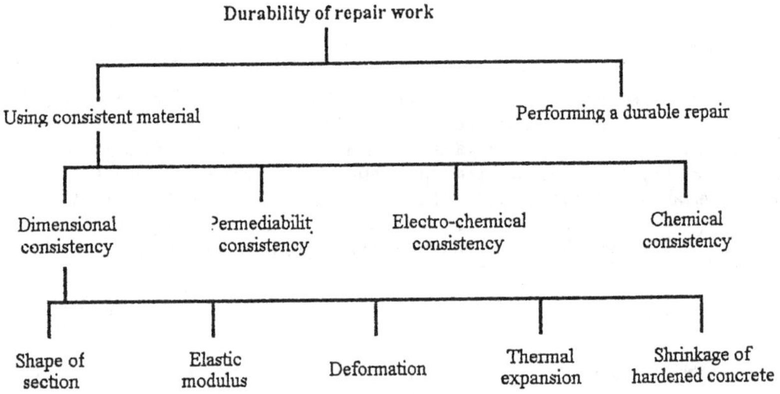

Figure 1. Factors affecting durability in concrete repair work.

6 OBSERVATIONS AND CASE STUDIES

The conclusions of the studies carried out on three large concrete structures located in the hot and environmentally aggressive regions of Iran show that the durability of concrete in such environments depends on:

i. studying and fully understanding the environmental effective factors;
ii. suitable concrete mix resistant to the chloride and sulfate attacks. This was achieved by using pozzolanic materials and microsilica to reduce the permeability and superplasticizers in order to keep a W/C ratio of about 0.45.
iii. suitable structural design and geometry, reinforcing bars, cover to reinforcement, thermal reinforcement, relaxation, creep.

7 HIGH PERFORMANCE CONCRETE

High performance concrete (HPC) is increasingly used in the hot and humid regions of Iran. The following are four types of HPC generally recommended for the structures in these regions.

7.1 *Concrete with high fly ash content*

It contains fly ash and a low quantity of lime of about 60% by weight of cement. The ratio of water to total cementitious materials is between 0.28 and 0.32. The use of superplasticizers is necessary. This type of concrete gains a compressive strength of 7 to 9 MPa in one day, and its compressive strength reaches 70 MPa in one year. It is highly durable and resistant to thawing and freezing and most chemical attacks for its low permeability.

The permeability test (ASTM C1202) on this concrete has shown an electric flux of less than 500 q. This type of concrete requires thorough damp curing for at least 7 days and an overall higher quality control than for ordinary concrete. Because of its high compressive strength and cost, it is suitable for high rise buildings in the hot regions of Iran.

7.2 *Concrete with high content of blast-furnace slag*

The permeability of this type of concrete is low. It is resistant to chloride ions and has low electrical conductivity. It is highly resistant to oxygen diffusion, and its expansion due to chemically reactive aggregates is low. It is also resistant to getting washed off by sea water and marine biological attacks. Because its ultimate compressive strength is medium and reached very slowly, this type of concrete is suitable in structures for which very high compressive strength is not required.

7.3 *Concrete with microsilica content*

Compact silica fume containing more than 85% silica (SiO2) is classified as microsilica. For its minute particles of about 0.1 micron in size, microsilica fills the voids, reacts with the excess Ca(OH)2 in concrete and reduces its permeability. The electric flux passing through this concrete is between 100 and 1000 q, which is an indication of a very low permeability.

This is the most suitable type of concrete used in the hot regions of Iran for its high compressive strength, resistance to chemical attacks, and low cost. Comparisons of the effect of microsilica on permeability and on compressive strength are given in Figure 2.

7.4 *Polymer modified concrete*

Using 'Latex' which is a polymer, as an additive in cement reduces the permeability of concrete and increases its electrical resistivity. The recommended Latex polymers are Styrene Bu-

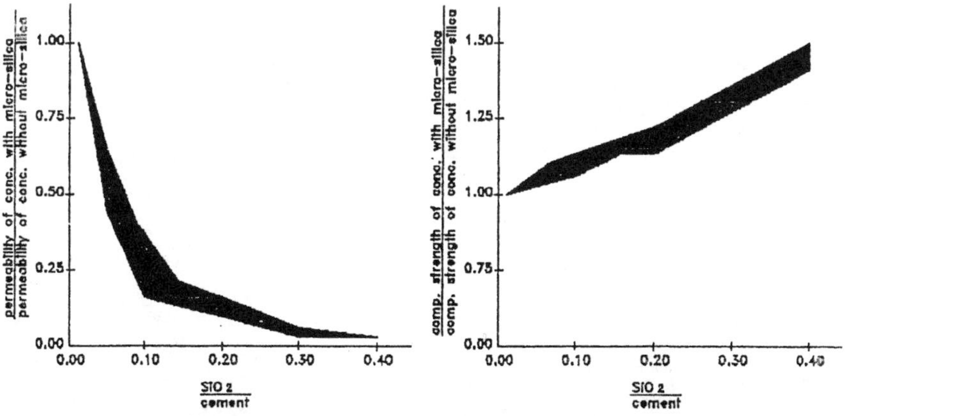

(a) **(b)**

Figure 2. The effect of microsilica on a) concrete permeability and, b) compressive strength.

tadiene Rubber (SBR), Acrylic, and Poly Vinyl Stat. The advantages of this concrete are its easy mixing, high workability with low water/cement ratio, quick curing and less care needed after placing, good adhesion, and high resistance to permeation of water and gas. The modulus of elasticity is about 15% less than that of ordinary concrete. Because of its formability and low cost, it is suitable for concrete repair works in the hot regions of Iran.

8 TESTS AND STANDARDS

The quality control and assurance of the concrete components and mixing, placing and curing concrete are very important for increasing its durability. Various tests at different stages of concreting are, therefore, necessary to ensure that the target quality is achieved. The Iranian standards and the relevant tests carried out in this country are generally based on the ASTM; hence, are not mentioned in this paper.

REFERENCES

Malhotra V.M. 1993. Flyash, slag, silica fume and rice Huskash in concrete: a review. *ACI,* Concrete international, April.

Hover K.C. 1994. Evaporation of surface moisture. Proceedings of the third international RILEM conference.

Ozyildirim C. 1994. Laboratory invasion of low-permeability concretes containing slag and silica fume. *ACI,* Materials Journal.

Mechanics of Structures and Materials, Bradford, Bridge & Foster (eds)
© 1999 Balkema, Rotterdam, ISBN 90 5809 107 4

A new fibre composite beam – An experimental study of the structural behaviour

G.M.Van Erp
Faculty of Engineering and Surveying, University of Southern Queensland, Toowoomba, Qld, Australia

ABSTRACT: This paper presents preliminary findings of an experimental study into the structural behaviour of a new type of fibre composite beam recently developed by the author. The innovative design eliminates most problems associated with standard composite beams and is capable of carrying very high loads at a cost similar to steel and concrete beams. The principles of the new design are discussed together with the test results for three different beams.

1 INTRODUCTION

Fibre composites have been around now for some fifty years and in this time sufficient work has been done to alleviate any doubts that they are structurally capable materials (Goldsworthy 1995). Although originally developed for the aerospace industry, fibre composites have found their way into a much wider range of applications, such as transportation (automotive, railway, boats), sports (tennis, golf, skiing, bicycles), and more recently civil engineering (bridges, offshore structures, frames).

1.1 *The civil engineering environment*

Advantages such as high strength, light weight and excellent durability should make fibre composites a serious contender for many civil and structural engineering projects. However, penetration into this large market has been very slow to date. There are many reasons for this, cost is certainly an important one. It is simply not enough to offer civil engineers a reduction in weight and the promise of reduced maintenance costs if the initial outlay is greater than other proven alternatives. Fibre composites have to be competitive in initial cost and subsequently superior when assessed over the life of the structure to provide the necessary incentive for civil engineers to try a new material. Furthermore, standard work practices on a building site and limited inspections during the life of the structure require significantly more damage tolerance than in the aerospace industry. Most fibre composite products currently on the market do not satisfy these criteria.

Lack of formal design standards and unfamiliarity of civil and structural engineers with fibre composites are other major problems. As a result the current tendency is to try to find justification for rejecting them, rather than exploring the possibility of an advantageous new approach. However, history shows that development and appropriate use of new materials have always been critical areas that underpin, and often enable, progress in engineering and technology. Civil and structural engineers now have a wider range of increasingly sophisticated materials at their disposals. In order not to waste this opportunity, new ways are required to incorporate these advanced materials into civil and structural designs.

1.2 *Pultruded composite beams*

The beam as a fundamental element is widely used in civil and structural engineering. Therefore, development of strong and economical fibre composite beams could spearhead the introduction of fibre composites into this field.

Glass fibre reinforced composite beams have been used for many years in industrial applications involving corrosive environments or weight driven designs. To date, most of these beams have taken the form of the accepted structural shapes in steel and aluminium such as I-beams, channels, angles, tubes and so on. In most countries these types of composite beams are readily available as standard commercial items from pultrusion companies. However, cost and low stiffness have been key obstacles to the usage of these pultruded beams in primary civil engineering applications. Due to their low stiffness, pultruded beams are susceptible to structural instability such as local buckling, resulting in structural failure long before the ultimate load carrying capacity of the reinforcement is attained (Barbero et al 1991, Gilchrist et al 1996). Introduction of carbon fibre into pultruded beams can significantly improve their stiffness but these hybrid structures are still far too expensive.

Usage of pultruded beams in primary civil applications is also hindered by their limited capability to deal with concentrated loads, bolt holes, and accidental damage. However, most beams in civil engineering structures will have to support purlins, ceiling panels, light fittings, audio equipment etc and therefore should be able to deal with concentrated load along their total length. Within limits, it should also be possible to screw, drill and cut beams without creating major problems. Standard pultruded beams do not satisfy these requirements. Yet, if fibre composites are to penetrate into the construction industry it is important that these requirements are taken into account.

The following section describes a new type of fibre composite beam recently developed by the author. This damage tolerant beam does not suffer from the problems mentioned above and is capable of carrying very high loads at a cost similar to standard steel and concrete beams.

2 NEW BEAM DESIGN

Standard beam theory shows that for beam materials to be used efficiently they have to be located as far as possible from the neutral axis. In steel and aluminium this has resulted in the extensive use of I-beams and box beams. In these beams the flanges carry the bending moment and the webs are used to keep the flanges apart and provide the necessary shear strength. The same principles should be applied to fibre composite beams. They should have unidirectional flanges at the extreme distance from the neutral axis and these flanges should be held apart by a spacer/core material that provides the necessary shear strength and torsional rigidity.

2.1 *Adopted approach*

Figure 1a shows the cross section of the new beam and 1b shows a schematic of the reinforcement. The beam's flanges are located as far as possible from the neutral axis and are fully encased in a particulate filled resin – a combination of epoxy and ceramic microspheres filler. This filled resin plays an important role in the beam's overall performance, as it:

- forms the web of the beam,
- protects the reinforcement from environmental and accidental damage,
- prevents buckling and edge delamination of the reinforcement,
- carries all concentrated loads, including support reactions,
- significantly contributes to the beam's overall load carrying capacity, and
- is capable of supporting screws and bolts.

All filled resin (except for a 5mm thick protective layer on the outside) is confined by double bias reinforcement resulting in extra strength and stiffness. Circular voids are included to save material and weight, also providing a compressed air duct for controlling exotherm during the curing process. The filled resin is damage tolerant and weighs about 800kg/m^3, it has a tensile strength of approximately 25MPa and a compressive strength of 70 MPa.

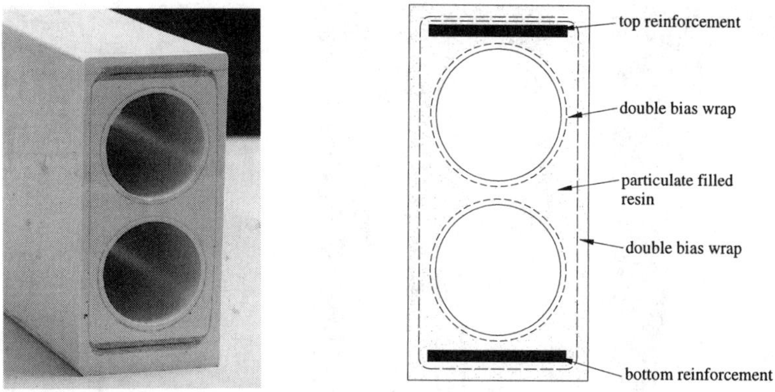

Figure 1. (a) Cross section of the new beam (b) Schematic of beam reinforcement

2.2 *Reinforcement*

It is of paramount importance that fibre composite beams are designed such that they give reasonable and adequate warning of failure prior to reaching their ultimate load carrying capacity. Unlike metals, fibre composites do not exhibit gross yielding and therefore the design should ensure that excessive deflection occurs before failure. Other types of warning such as local buckling can also be used but often result in less economical beams.

Even though fibre composites do not yield, they are also not classic brittle materials. Under static load many laminates show non-linear characteristics attributed to sequential ply failure. For example, a glass/carbon hybrid laminate exhibits non-linear behaviour and significant displacements before failure when loaded in tension. This pseudo ductile behaviour is the result of the different failure strains of glass and carbon. The failure strain of glass (2%) is about twice that of carbon (1%), therefore when the carbon fails, the glass has reached only half of its failure stress. The amount of carbon and glass can be tailored such that the laminate will continue to carry load after failure of the carbon. This residual load carrying capacity is associated with a large increase in displacement. Figure 2 shows the load-strain behaviour of a typical glass/carbon $[0_G/0_G/0_G/0_C]_S$ laminate (Van Erp 1999). This type of hybrid laminate has been selected for the top and bottom flange of the current beam.

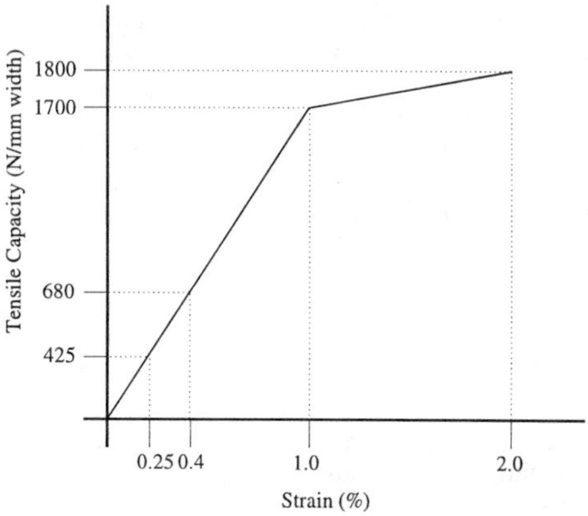

Figure 2. Pseudo ductile tensile behaviour of a $[0_G/0_G/0_G/0_C]_S$ hybrid laminate.

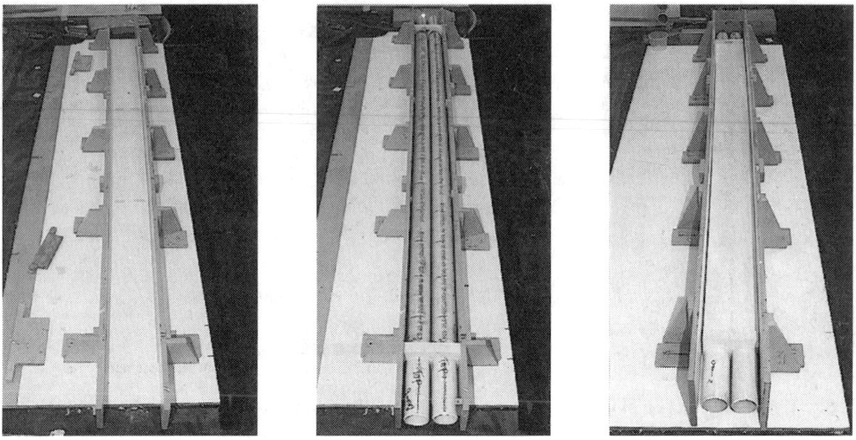

3a 3b 3c

Figure 3. The production process

3 THE PRODUCTION PROCESS

The beams are produced on their side on a flat table using two vertical timber boards (see Figure 3a). The boards are supported by adjustable brackets which allow for the introduction of a camber in the beam (the present 2m long test beams were all produced with a 6mm camber). PVC tubes are used as sacrificial formwork to produce the circular voids in the beam. The different production steps are as follows:

1. Cut PVC tubes to length and wrap the tubes with double bias reinforcement.
2. Wet-out the double bias wrap, place the wet tubes in the formwork and close off the formwork with two end dams (Figure 3b).
3. Fill the formwork with the particulate filled resin (Figure 3c). For the present set of test beams the distance between the pipes and formwork is 5mm.
4. Let the filled resin cure and demould.
5. Lightly sand the beam surface and apply top and bottom caps. Apply peel ply on the caps.
6. Once the caps have cured, remove the peel ply and apply double bias wrap to the beam.
7. Wet-out the wrap and replace the beam with the wet double bias wrap in the form work.
8. Pour another 5mm of resin around the beam, let it cure, demould and post cure.

The final beam has a good quality finish and requires no further work.

4 DESIGN OF TEST BEAMS

Three test beams were designed to investigate the load carrying behaviour of the new beam. The following materials were used:

- $448g/m^2$ T300 Carbon Heatset Uni from Hexcel,
- $450g/m^2$ Heatset E-glass Uni from Colan,
- Epoxy resin ADR246TX and ADH160 hardener from ATL Composites, and
- SL150 Envirospheres.

Coupon testing was used to determine the tensile strength of the carbon/epoxy and glass/epoxy laminae. The load carrying capacities per mm width were as follows:

- Carbon/epoxy lamina 480N/mm width
- Glass/epoxy lamina 340N/mm width.

The failure strain of the carbon lamina was approximately 1% and that of the glass 2%.

4.1 *Test beam 1*

During the design of the first test beam it was assumed (based on limited experience with the new core material) that the particulate filled resin core would act as a 'standard' core material ie. it would carry the shear load and keep the top and bottom flange apart but its contribution to the bending moment capacity would be minimal. Hence, the bending moment capacity of the cross section was determined using the capacity of the top and bottom reinforcement only.

The first test beam was 100mm wide and 180mm deep (outside dimensions). The width of the top and bottom flange was 65 mm and the lever arm between the two flanges 170mm. The same laminate was used for the top and bottom flange, 2x 0° glass/0° carbon/0° glass, resulting in an overall symmetric beam. At failure of the carbon, the glass has reached only half of its load carrying capacity, namely 340/2=170 N/mm width. Hence, the estimated bending moment at failure of the carbon equals:
480x65x170 + 3x170x65x170 = 10.94 kNm (the associated deflection is 40mm).
The ultimate bending moment of the cross section (moment at failure of the glass) equals:
3x340x65x170 = 11.27 kNm (the associated maximum deflection is 80mm).

4.2 *Test beam 2*

Test beam 2 was designed after beam 1 had been tested. As will be discussed later, beam 1 failed in a brittle manner at a relatively low strain level. In order to get a more ductile behaviour the amount of glass in the top and bottom flange of beam 2 was increased to twelve layers, the amount of carbon was kept the same. The estimated bending moment capacity of beam 2 at failure of the carbon equals:
480x65x170 + 12x170x65x170 = 27.8 kNm (the associated deflection is 40mm).
The ultimate bending moment (the moment at failure of the glass) equals:
12x340x65x170 = 45.08 kNm (the associated maximum deflection is 80mm).

4.3 *Test beam 3*

Beam 3 was made similar to beam 2 except for the reinforcement in the top flange. In order to investigate the capability of the filled resin to provide the necessary compression force on its own, the top reinforcement was reduced to two layers of glass only. These layers were applied to stop possible spalling of the filled resin. No estimates were generated for this beam.

5 DISCUSSION OF TEST RESULTS

All beams were tested in four point bending using a simply supported geometry and a 1800mm span. Load was applied at 1/3 and 2/3 of the span. Displacements were measured at midspan. No constraints were utilised to prevent local buckling or lateral torsional buckling. All beams failed through rupture of the bottom reinforcement in the middle third of the beam. A typical failure mode is shown in Figure 4.

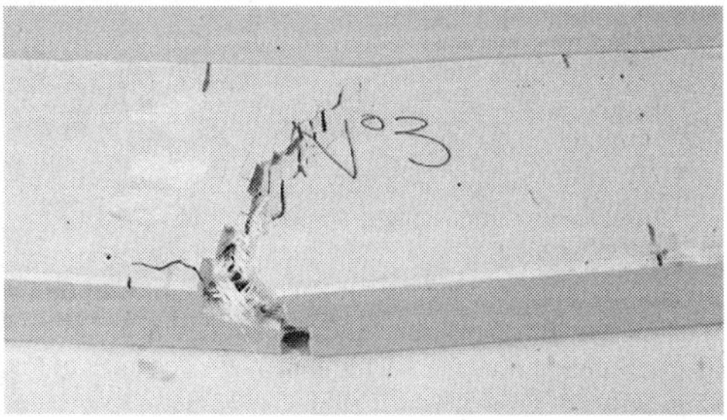

Figure 4. Typical failure mode

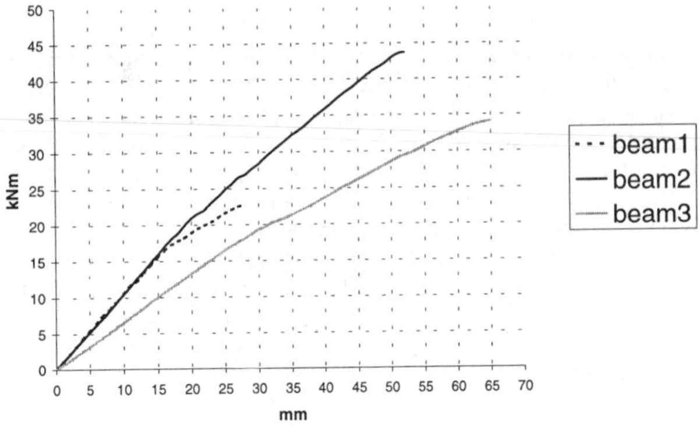

Figure 5. Moment-deflection behaviour

Beam 1 failed at 22.92 kNm which is twice the moment predicted. It is obvious from this result that the filled resin contributes significantly to the load carrying capacity of the beam. The deflection at failure was 28mm which is significantly lower than predicted. The strain at the bottom of the beam at 28mm deflection is approximately 0.7%. Testing has shown that this is approximately the failure strain of the filled resin. At 22.92 kNm the beam has significantly exceeded the load that can be carried by the glass/carbon reinforcement alone. Therefore when the filled resin fails, the whole beam fails in a brittle manner.

Beam 2 failed at 43.69kNm, which is very close to the predicted failure moment of 45.08kNm. The first noise was heard at 20mm deflection and continued until the beam failed at 52mm deflection. This corresponds to a strain of 1.4%, which is less than the predicted 2%. An explanation for this discrepancy has not yet been found.

Beam 3 also failed through rupture of the tensile reinforcement, which indicates that the filled resin is capable of carrying very high compressive loads. Because the neutral axis has shifted downwards for this beam, the 62 mm deflection again represents a failure strain in the reinforcement of 1.4-1.5%.

6 CONCLUSIONS

The test results clearly show the potential of this new type of beam. All beams failed through ultimate failure of the tensile reinforcement, no local buckling problems were experienced. The load carrying capacity of the particulate filled resin has exceeded all expectations. Further research is required to determine how the filled resin is affected by creep and temperature.

ACKNOWLEDGEMENT

The author gratefully acknowledges the financial assistance and professional guidance provided by Australian Consulting Engineers Connell Wagner Pty Ltd.

REFERENCES

Barbero, E. J. Fu, S. H.and Raftoyiannis, I. 1991, Ultimate Bending Strength of Composite Beams, *Journal of Materials in Civil Engineering*, v3, n4, 292-306.

Gilchrist, M.D. Kinloch, A.J. Matthews F.L. Osiyemi, S. O. 1996, Mechanical Performance of Carbon Fibre and Glass Fibre Reinforced Epoxy I-beams, *Composite Science and Technology*, v56, n1, 37-53.

Goldsworthy, W. B. 1995, Composites, Just Another Building Material – Only Better, *Proceedings of the 40th International SAMPE Symposium, Anaheim California*, pp 504-512.

Van Erp, G.M. 1999. Design and Analysis of fibre Composite Beams for Civil Engineering Applications, *Proc. 1st Int ACUN Composites Conf*, Sydney, pp 229-238.

Mechanics of Structures and Materials, Bradford, Bridge & Foster (eds)
© 1999 Balkema, Rotterdam, ISBN 90 5809 107 4

Influence of fine aggregates on the elastic modulus of mortar

E.S. Bernard & M. Logan
School of Civic Engineering and Environment, University of Western Sydney, Nepean, N.S.W., Australia

ABSTRACT: The elastic modulus of concrete is known to be influenced by the mechanical characteristics of the coarse aggregate and hydrated cement paste fractions. However, little attention has been paid to the influence of the fine aggregate, even though this may comprise a significant volume fraction of a mix. The present investigation has addressed this issue by examining the influence of four different types of fine aggregate on the elastic modulus of a mortar produced with constant mix proportions.

1 INTRODUCTION

Although the strength of concrete is an important material property, most reinforced concrete structures are designed to function at stresses significantly below that required to cause failure. Of all the constitutive properties of concrete, the elastic modulus, E_c, is the most important in determining the short-term behaviour of a reinforced concrete structure under in-service stresses (Gilbert and Mickleborough 1991). Despite this, the elastic modulus of concrete is seldom measured experimentally, but instead is calculated on the basis of an approximate correlation with compressive strength (AS3600 1994). Whilst strength has been found to correlate well with elastic modulus, this is not the only parameter that influences behaviour. Some codes of practice (BS8110 1985) have attempted to account for the influence of coarse aggregate type on the elastic modulus of concrete, in recognition of tests conducted on coarse aggregates of varying mineralogy and mechanical behaviour (Alexander and Milne 1995).

The importance of the elastic modulus is based on the influence it has on deformation and stability in structures. An increase in elastic modulus will result in decreased short-term deflections under a given set of loads. Although not immediately obvious in current codes of practice for reinforced concrete structures (AS3600 1994), an increase in elastic modulus will also reduce the susceptibility of a slender member to buckling (Ghali and Favre 1994). High modulus concrete is therefore desirable for the construction of slender columns (Bridge 1988), walls (Swartz and Rosenbraugh 1974), and membranes (Schlaich 1996).

The magnitude of the elastic modulus is not the only issue of structural interest in concrete: its linearity throughout the stress-strain envelope is also important. Structures made of materials with highly non-linear constitutive properties are difficult to model analytically. To simplify analysis, especially when instability is possible, materials should ideally exhibit a linear stress-strain relationship. Neville (1997) noted that the disparity in modulus between the paste and coarse aggregate fractions of most concretes may play a significant role in determining the degree of non-linearity in E_c apparent for a mix. High strength concretes typically display more elastic behaviour (with a relatively constant modulus up to approximately 80 per cent of ultimate strength) than normal strength concrete because the modulus of the paste is closer in magnitude to the modulus of the aggregate. In normal strength concrete the modulus of the paste is often substantially lower than that of the aggregate. Lightweight structural concrete has also been observed to behave in a more linear-elastic manner because the modulus of manufactured lightweight aggregates is lower than that of natural aggregates and therefore closer to that of the paste (Zhang and Gjørv 1991).

Based on the advantages described above, it is desirable to identify a method of enhancing the modulus of mortar and concrete. Apart from studies of the influence of water/cementitious ratio (Alexander and Milne 1995), investigations of other methods of achieving a high modulus, such as the use of high modulus fine aggregate, have been limited to date. However, the influence of fine aggregate on other properties of concrete has received attention. Goodhew and Sullivan (1995) examined the influence of fine aggregate on workability and strength, with particular attention to the use of crusher fines of various compositions. The motivation for using crusher fines was twofold: traditional sources of sand are becoming exhausted in many parts of Australia, and large quantities of crusher fines left over from the crushing of coarse aggregates are proving to be a disposal problem in quarries. The replacement of sand with crusher fines was therefore proposed as an obvious solution.

Crusher fines are available in a wide range of particle sizes depending on the composition and grain size of the source rock. They are similar to natural sands in many respects, and are typically available as ungraded residues from the crushing process, but may occasionally be separated by grade. Considering the demonstrated influence of coarse aggregates on the elastic modulus of concrete, an examination was instigated into whether fine aggregates based on crusher fines have a significant influence on the elastic modulus of mortar. Concrete was not used because it was believed that variations in the properties of the coarse aggregate may mask the effect of the fines. Instead, a series of experiments involving mortar made with natural sand and crusher fines was carried out.

2 EXPERIMENTAL PROGRAM

A natural sand and three types of crusher fines were selected for use in the investigation. The source and type of parent rock was chosen to represent a wide range of rock compositions whilst also being restricted to economically important aggregate sources near Sydney. A summary of the sources is included in Table 1.

Table 1. Sources of crusher fines, and their characteristics.

Type	Moisture absorption (%)	Density (kg/m^3)	E (GPa)	f'_c (MPa)
Quartz sand	1.3	2550	-	-
Limestone	0.7	2687	65	94
Basalt	1.2	2968	102	216
Ignambrite	1.9	2574	70	194

The following petrographic description of the parent rocks from which the crusher fines were derived was provided by Smith (1998).

Natural Sand

The natural sand was a river sand obtained from the Nepean quarry, Emu Plains, NSW. It consisted primarily of quartz grains, with some feldspar. It could be described as coarse and angular. The mechanical characteristics of this aggregate were not determined, nor could the parent rock be traced.

Basalt

This rock was obtained from the Peat's Ridge Quarry, operated by Boral Concrete and Quarries Ltd., north of Sydney, NSW. It consisted of coarse crystals of plagioclase and olivine (up to 2 mm) set in a matrix of very fine plagioclase, pyroxene, iron-titanium oxides, and volcanic glass. The olivine was partially altered to expandable clay along cracks, and many of the small crystals were strongly aligned, possibly causing a plane of weakness within the rock.

Limestone
This rock was obtained from the Boral Marulan Quarry, NSW. It consisted of the mineral calcite in the form of coarse crystals. The coarse grain size may have exacerbated the potential for grains of calcite to fail in tension or shear on their internal cleavage surfaces.

Ignambrite
This rock, also known as a welded Tuff, was obtained from the Boral Seaham Quarry, near Newcastle, NSW. It consisted of the minerals quartz and alkali feldspar, and volcanic rock fragments, set in a fine aligned matrix of crystals and glass in the form of flattened and welded grains of pumice. The strong alignment of the matrix may have produced a slight strength anisotropy.

2.1 Sample Preparation

The grading curves for the aggregates as received differed, because the characteristics of crusher fines are generally not controlled. To eliminate the possibility that variations in grading characteristics might influence rheology in the wet mix, and consequently the modulus, the grading of each aggregate was adjusted so that they matched each other. The grading curve for the sand, after the -0.075 mm and +2.36 mm fractions had been removed, was taken as the base curve. The grading for the other aggregates was adjusted to match this by sieving a sufficient quantity of each to separate the size fractions, and then re-combining them to develop the required curve.

One mortar was made using the natural sand, as well as three sets of five mortars using varying proportions of each of two crusher fines. In the first mix for each set of five mortars, the proportions of the two fines was selected as 0% of the first and 100% of the second. For the second set, the proportions were 25%:75%, then 50%:50%, and so on. The natural sand was used only once because the modulus of this material was not known. It was therefore used primarily to represent a 'common' mortar.

All the mortar mixes had equal proportions of fine aggregate and cement, and a water/cement ratio of 0.35. In an early trial, it was found that high workability was required to avoid the entrapment of air bubbles which could significantly influence apparent modulus. A superplasticiser was therefore used at a dosage rate of 1% by weight of cement to enhance workability. To produce a sufficient number of cylindrical specimens without having to grade large quantities of aggregate, the mortar was cast into ∅50×100 mm cylinder moulds. For each mix, a quantity of materials sufficient to produce 8 cylinder specimens, plus 100 per cent wastage, was weighed and placed in a mixing tray. A conventional mortar mixing bowl was not used because this was believed likely to entrap air. The ingredients consisted of 3.20 kg of fine aggregate, 3.20 kg of cement, 1.12 Litres of water, and 32 grams of superplasticiser. The materials were combined, mixed with a hoe, and placed on a vibrating table to eliminate entrapped air. The mortar was then scooped progressively into the cylinder moulds, leaving time to consolidate each lift on the vibrating table. It is believed that this was very effective in preventing the inclusion of air bubbles.

The filled moulds were capped with lids and put in a sealed box with 10 mm of water covering the base to cure for two days before stripping. The stripped samples were marked for identification and immersed in lime water at 23°C until tested. Immediately prior to testing, the specimens were prepared by first grinding and then lapping each end to ensure they were flat and parallel. Modulus measurements were undertaken at the relatively late ages of 80 and 180 days because of previous findings that modulus development in concrete is a relatively slow process compared to strength development (Shideler 1957).

2.2 Measurement of Elastic Modulus

Hardened mortar cylinders were tested in an Instron 6027-R5500 electro-mechanical Universal Test Machine at an effective load rate of 20 MPa/min. The maximum stress applied to each specimen was only 12.7 MPa, which was low compared to the estimated 60 MPa compressive strength of the mortar. Strain was measured using a pair of Instron 2630-100 electronic axial extensometers placed diametrally at mid-height on each specimen A spherical seat was used to eliminate possible stress concentrations due to non-parallel ends. Load and strain were logged at 1 kN increments of load and the results were used to calculate the secant modulus between 1 and 4 MPa on the stress-strain curve.

3 RESULTS

A total of 16 mixes were produced and tested. The results consisted of average modulus estimates for each set of 8 specimens, plotted against aggregate proportions, as presented in Figs. 1 and 2 (representing results at 80 and 180 days). The use of eight samples to determine each estimate of modulus permitted standard deviations to be determined. These are plotted as error bars (at one standard deviation above and below each point) on the same figures. Each set of data indicate the variation in modulus as the proportion of aggregate was varied from 0% through to 100% of the first listed aggregate (ie. 100% through to 0% of the second). The 80 day results for the limestone/basalt series, for example, varied from 30784 MPa for 100% basalt, to 30915 MPa for 100% limestone, indicated by the square markers. The three sets of results for the crusher fines have been staggered slightly to show the error bars more clearly.

The results for the mortars made with crusher fines appear to indicate that fine aggregate composition has little influence on modulus, regardless of age. Although the average within-batch Coefficient of Variation for the 32 sets of modulus measurements was a relatively modest 3.2%, this was large enough to mask any significant difference between the mortars made with the different crusher fines. However, the modulus for the mortar made with quartz sand had a mean value of 37720 MPa at 80 days and 38626 MPa at 180 days. This was significantly greater than the modulus of any mortar made with crusher fines.

A plot of density versus elastic modulus at 80 days for each specimen is indicated in Fig. 3. Whilst the modulus appeared to be independent of density for the mixes made with crusher fines, the natural sand-based mortar specimens showed a significantly higher modulus for a given density than the rest. The crusher fines made using ignambrite resulted in the lowest density mortar, whilst that based on basalt resulted in the highest density mortar.

The density of each mortar made with 100% of just one source of fines is plotted against the density of each source aggregate in Fig. 4. Although the mortars made with crusher fines showed an approximately linear correlation with source aggregate density, the mortar made with natural sand had a high density despite the relatively low specific gravity of quartz. The fact that the sand-based mortar possessed such high density suggests that the sand was incorporated into the paste in a more effective manner. Assuming no errors have been made in procedure, this may be indicative of a fundamental difference in the way quartz sand interacts with cement paste compared with other aggregates.

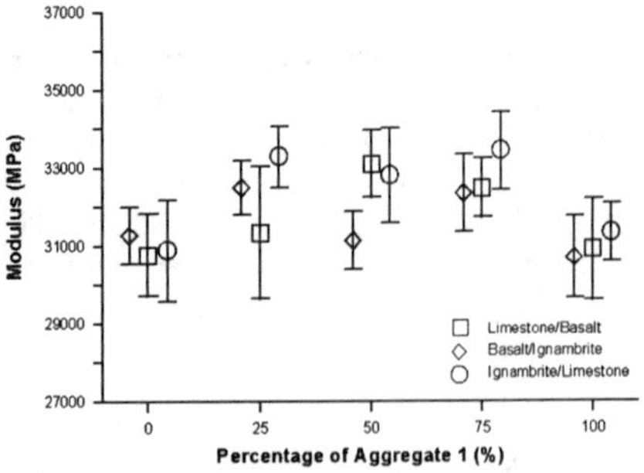

Figure 1. Elastic modulus versus aggregate composition at age 80 days

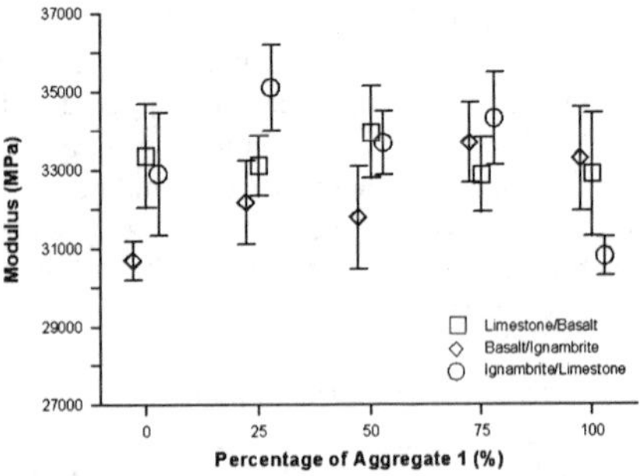

Figure 2. Elastic modulus versus aggregate composition at age 180 days.

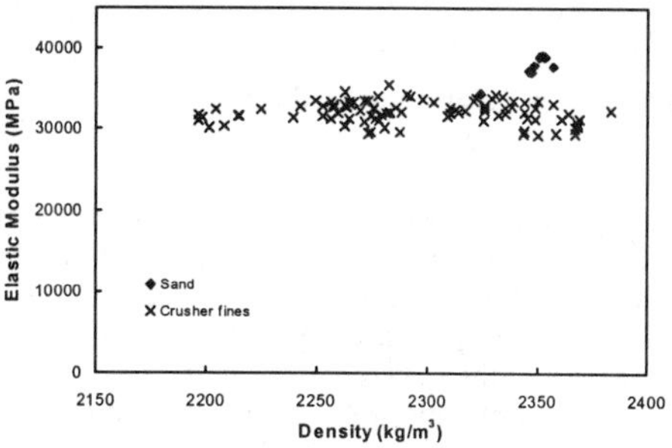

Figure 3. Elastic modulus vs. density for mortars.

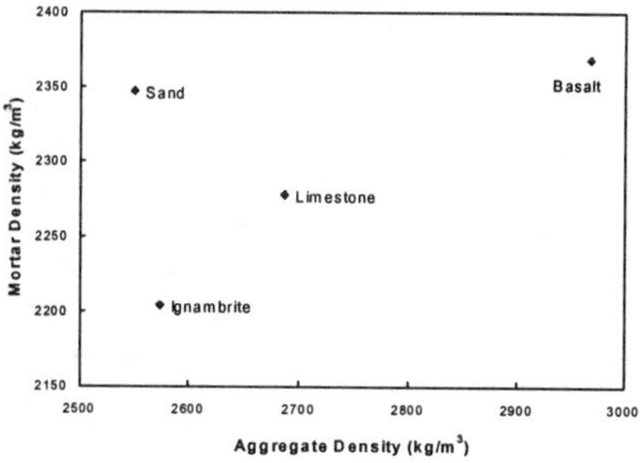

Figure 4. Density of aggregates and mortars.

4 DISCUSSION

This investigation appeared to demonstrate that the elastic modulus of mortar is independent of the composition or modulus of the constituent fine aggregate, at least for the crusher fines presently used. However, a possible explanation for this result is that the mix proportions were based on the oven-dried mass of each fine aggregate. Since the dry densities of each fine aggregate varied, the mix incorporating the high density basalt had a lower volume proportion of fine aggregate than any of the other mixes, and the mortar made with low density natural sand had the highest. This may have resulted in the low density aggregates such as ignambrite exerting a greater influence on the elastic modulus of the mortar than the more dense aggregates, as all of the source rocks had a modulus that was greater than commonly observed for mortars. However, the mortar made with sand behaved quite differently to the other mixes. Although sand has a density similar to ignambrite, and therefore was added at a similar volume proportion, it produced a mortar with substantially higher modulus.

Effective differences in w/c ratio for the mortars were unlikely, despite the dry state of the ingredients prior to mixing. This is because the moisture absorption values for each aggregate were all low and relatively constant (see Table 1). The maximum possible variation in absorption was therefore small compared to the amount of water present in each mix.

Whilst no microscopic investigations were presently undertaken of the interface between the hydrated cement paste and fine aggregate particles, the findings of previous studies may provide clues to the basis for the similarity in results for the crusher fines. Yuji (1988) undertook micro-hardness studies on a number of mortars and found that the transition zone between aggregate and bulk paste can be up to 50 μm thick. This thickness increased slightly with particle size, but was approximately uniform across the range of fine aggregate particle sizes. The hardness of the transition zone was significantly lower than that of the bulk paste, hence it can be assumed that it is also more compliant. The transition zone surrounds each aggregate particle and acts as an interface between aggregate and paste, so the modulus of the mortar may be governed by the modulus of this zone. The characteristics of the fine aggregate may therefore not be significant to the modulus of the mortar. The effect of the transition zone is less significant for coarse aggregate particles because the volume of this zone is less significant for large particles.

Although this model may explain the behaviour of the mortars made with crusher fines, it does not support the evidence of the increased modulus associated with the quartz sand. Further work will be required to establish whether sand produces a mortar with a consistently higher modulus than crusher fines.

5 CONCLUSION

An experimental investigation of the influence of fine aggregate on the elastic modulus of mortar was undertaken. The aggregates included a natural quartz sand and three types of crusher fines derived from basalt, limestone, and ignambrite.

The influence of fine aggregate type on modulus was studied by varying the proportions of each of two types of aggregate in a mix with constant cement factor and w/c ratio. However, the mix design was based on mass, so the volume fraction of aggregate compared to cement varied with the density of each aggregate. This appeared to mask any possible relationship between the modulus of the aggregate and the modulus of the corresponding mortar because the aggregate with the highest modulus also had the highest density. It therefore appeared that the fine aggregates derived from crusher fines had little influence on modulus.

In contrast, the natural sand produced a mortar with a modulus significantly higher than that made with any of the other aggregate sources. This suggests a fundamental difference in the way sand behaves in mortar compared to crusher fines

6 ACKNOWLEDGEMENTS

The authors gratefully acknowledge the support of Boral Concrete and Quarries Ltd. in providing aggregate for this investigation. Dr. John Smith of Southern Cross University, Lismore, NSW, is also thanked for his petrographic analyses of the parent rocks from which the crusher fines were derived.

7 REFERENCES

Australian Standard AS3600 1994. *Concrete Structures*, Standards Australia, Sydney.

Alexander, M.G. and Milne, T.I., 1995. "Influence of cement blend and aggregate type on stress-strain behaviour and modulus of elasticity of concrete", *ACI Materials Journal*, Vol. 92, No. 3: 227-235.

Bridge, R.Q., 1988. "Design of Columns to Australian Standard AS 3600", *Designing Columns and Walls Seminar*, Concrete Institute of Australia, IEAust, Sydney, pp8.1-8.36.

British Standard BS8110 1985. *Structural Use of Concrete: Code of Practice for Design and Construction*, BSI, London.

Ghali, A. and Favre, R., 1994. *Concrete Structures: Stresses and Deformations*, 2nd Ed., E&FN Spon, London.

Gilbert, R.I. and Mickleborough, N.C., 1991. *Design of Prestressed Concrete*, Unwin Hyman, London.

Goodhew, A.J. and Sullivan, B.W.P, 1995. "Basalt crusher fines as a replacement for sand in concrete", *Australian Civil Engineering Transactions*, IEAust, Vol. CE37, No. 2: 183-188.

Neville, A.M. 1997. "Aggregate bond and modulus of elasticity of concrete", *ACI Materials Journal*, Vol. 94, No. 1: 71-74.

Schlaich, J. 1996. *The Solar Chimney: Electricity from the Sun*, Edition Axel Menges, Stuttgart.

Shideler, J.J. 1957. "Lightweight aggregate concrete for structural use", *Journal of ACI*, Vol. 54: 299-328.

Smith, J.V. 1998. "Petrographic Investigation of Rock Aggregates Used in Experimental Concrete Mixtures, with Special Reference to Aggregate-Cement Bond Strength", Southern Cross University, Australia.

Swartz, S.E. and Rosebraugh, V.H. 1974."Buckling of reinforced concrete walls", *Journal of the Structural Division*, ASCE, Vol. 100, ST1: 195-208.

Yuji, W. 1988. "The effect of bond characteristic between steel slag fine aggregates and cement paste on mechanical properties of concrete and mortar", *Proceedings of the Materials Research Society Symposium*, ed. S. Mindess and S. Shah, pp 49-54.

Zhang, M-H. and Gjørv, O.E. 1991. "Mechanical Properties of High Strength Lightweight Concrete", ACI Materials Journal, Vol. 88, No. 3: 240-247.

SYNOPSIS

The elastic modulus of concrete has been shown to be influenced by the characteristics of the coarse aggregate and hydrated cement paste fractions of a mix. However, little attention has been paid to the influence of the fine aggregate fraction, even though this may comprise a significant proportion of the volume of a mix. The present investigation has addressed this by examining the influence of four different types of fine aggregate on the elastic modulus of a mortar with constant mix proportions. Three of these fine aggregates were based on crusher fines, which are the fine residue left over after the crushing of coarse aggregates. The results indicate that the elastic modulus of the fine aggregate has little effect on the modulus of mortar when the mix design is based on the mass of each ingredient.

KEYWORDS

Modulus, aggregate, mortar, concrete, measurement, testing, basalt, limestone, sand

Mechanics of Structures and Materials, Bradford, Bridge & Foster (eds)
© 1999 Balkema, Rotterdam, ISBN 90 5809 107 4

Structural behaviour of monocoque fibre composite trusses

M.F.Humphreys & G.M.Van Erp
Faculty of Engineering and Surveying, University of Southern Queensland, Toowoomba, Qld, Australia

C.H.Tranberg
Connell Wagner Consulting Engineers, Brisbane, Qld, Australia

ABSTRACT: This paper presents early experimental results of an ongoing research project into the structural behaviour of monocoque fibre composite trusses. The testing of three different types of truss joints will be discussed, along with the structural behaviour of the tension and compression elements. Descriptions of the fabrication, test procedures, results and design/analysis of the different jointing methods are also presented.

1 INTRODUCTION

Trusses are amongst the most efficient structural forms available in terms of material and weight. History has shown that the popularity of trusses reached a maximum during the Great Railway era of the 19th century and subsequently died with the advent of suspension bridges and post-tensioned concrete (Gregory 1971). One factor that contributed to the waning popularity of truss structures was the increase in labour costs during manufacture. Truss joints are very labour intensive and require a large proportion of the manufacturing time. By the mid 20th century, labour costs had increased to such a level that material cost was no longer the governing component in the choice of the structure.

The technology of trusses was initially developed by people such as Ithiel Town, William Howe and Thomas Pratt, however truss use has been noted as early as the Roman times. The first scientifically designed truss was a modification of the Howe truss by Thomas Pratt. The parallel top and bottom chords of Pratt's truss were made of timber as were the vertical web members. The diagonals were made of wrought iron.

As fibre composite technology continues its steady transfer into the field of civil engineering, opportunities arise for a new and innovative approach to conventional truss design. In particular, the suitability of fibre composites to monocoque construction creates many new possibilities such as the elimination of expensive truss joints which should result in structures that are more efficient.

2 NEW TRUSS DESIGN

To date very few fibre composite trusses have been built. Those that have are generally based on existing bolted steel or timber designs or use monolithic joint inserts to transfer the load from member to member (Agematzu 1998). This approach fails to recognise the special characteristics of fibre composites. As emphasised in the literature no other material has the potential of fibre composites for improving the performance of structures (Gibson 1994, Tsai 1992). However, to achieve this potential fibre composite structures should not imitate steel equivalents, but rather exploit the considerable advantages of the new material (King 1994, Head 1998). In line with this philosophy a new and innovative planar monocoque fibre composite truss (see Figure 1) was recently developed by the second author. The new truss consists of two

Figure 1. Monocoque fibre composite truss

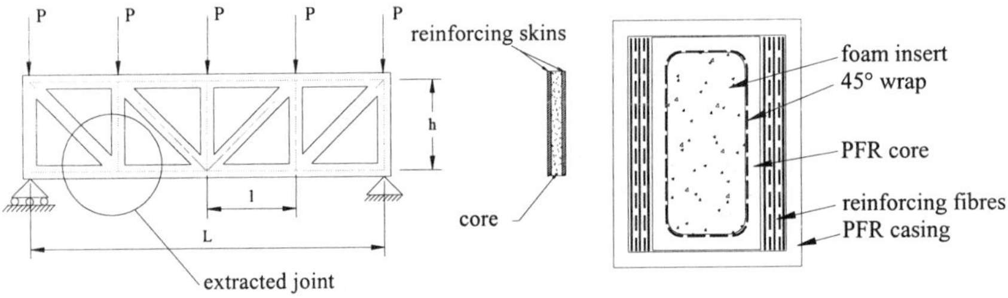

Figure 2. Typical monocoque fibre composite truss, cross section and element section

external skins of glass and/or carbon fibre over a hollow particulate filled resin core, refer Figure 2. The truss is constructed from a cast core material molded in the form of the truss and the reinforcement is applied to the core as individual strips that overlap at the joints. An important part of the truss design is the force transfer between the web members and the top and bottom chords. A number of innovative connections have been developed to ensure failure of the truss will occur in the members rather than in the joints.

2.1 *Truss details*

The panel aspect (h/l) of the present truss is chosen as 1:1 to produce a diagonal tension member at 45° to the top and bottom chords. The reinforcement layers in the top and bottom chords of the truss are continuous along the length of the truss, making pullout of the diagonal tension members the critical failure mechanism. The present truss is simply supported at each end, with the top (compression) chord restrained laterally at the panel points and this type of configuration is often called a "Pony" truss.

2.2 *Preliminary truss testing*

Preliminary testing was carried out on two trusses, an end grain balsa cored truss and a foam cored truss. Both trusses use epoxy resin and 50mm wide uni-directional E-glass tape manufactured by ATL Composites. Four point bending tests have shown that the shear modulus of the core material has a significant effect on the buckling capacity of the truss. The dimensions of the trusses are shown in Figure 3.

Both trusses were tested to destruction. The balsa cored truss achieved a maximum ram load of 91kN and failed by rupture of the outer panel diagonal tension element, while the foam cored truss buckled at a ram load of 75kN. Figure 3 shows the balsa truss in the purpose built test rig.

The initial truss tests clearly demonstrated the potential of this type of truss. However, they also showed some deficiencies in both core materials. The balsa and foam tended to crush easily and required hard points in the vicinity of loads and supports. To overcome these problems the core was replaced by a hollow, particulate filled resin (PFR) core. Tests have been carried out on a series of joints using a PFR core to determine the load capacity of this new apporach.

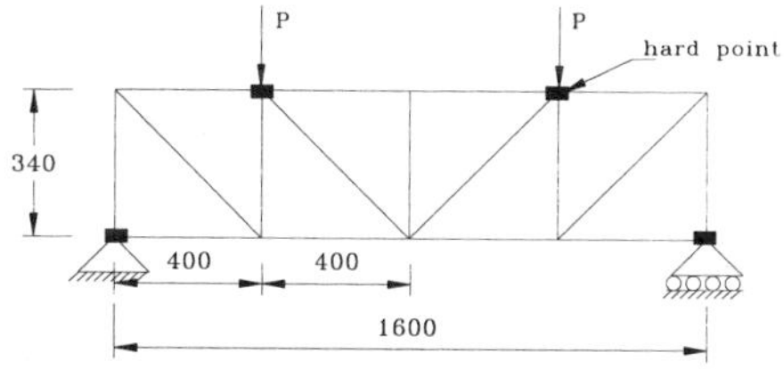

Figure 3. Global truss testing

(All dimensions in mm)

2.3 *Joint configuration*

The monocoque truss joints use a combination of fill layers and strength layers to ensure load is transferred from member to member whilst minimising the generation of out of plane forces. Three types of joints were investigated, differing mainly in the termination of the diagonal tensile reinforcement. The fibre orientation for each of the joint types is presented below. Note that layer 1 is always the first layer to be applied to the core.

Table 1. Joint fibre architecture

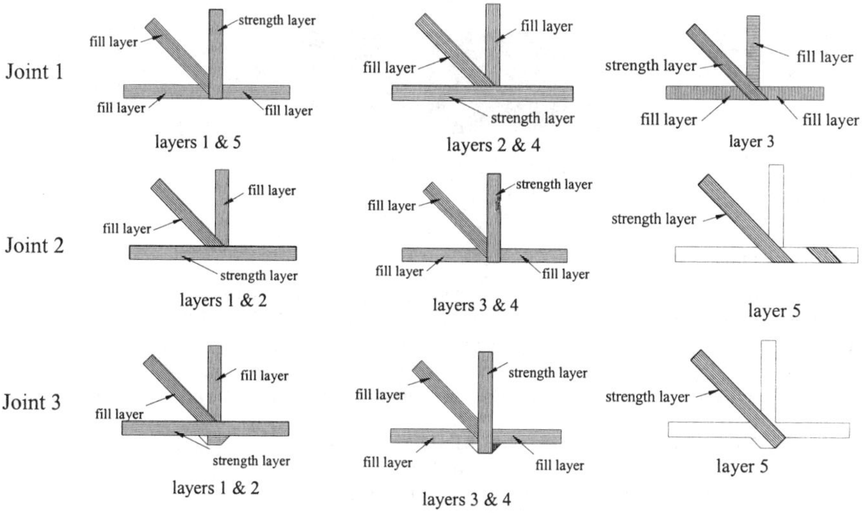

Joint 1 comprises five layers of reinforcing fibres. The strength fibres of the diagonal tension element reside at the mid-layer of the laminate (layer three). At the intersection point, either side of the strength layer, the bottom chord reinforcement is located (layers two and four), which are continuous along the length of the bottom chord. The reinforcing fibres of the vertical web member (layers one and five) are located on the outside of the bottom chord reinforcement. The force in the diagonal member is transmitted to the adjacent members through inter-laminar shear. Joints 2 and 3 were developed to reduce the reliance on interlaminar shear for the anchorage of the tensile reinforcement as is the case with joint 1.

Joint 2 uses a spiral wrap to terminate the main tensile reinforcement around the bottom chord. The stacking sequence differs from that of joint 1 as the two bottom chord reinforcing layers are applied first, followed by the vertical compression web member layers and finally the strength diagonal. Consequently, the amount of fill material is reduced.

Joint 3 has a stacking sequence similar to that of joint 2. The main diagonal reinforcement terminates by looping the fibres beneath the bottom chord and compression member to lap back on the diagonal tension member. This approach also uses less fill material than joint 1.

2.4 *Element configuration*

A schematic cross section of a typical member is presented in Figure 2. The members comprise a foam void former that has been wrapped with uni-directional E-glass tape at 45° to the longitudinal axis of the member. The intention of the wrap is to reinforce the inside of the layer of particulate filled resin (PFR) in which the foam insert is encased. The reinforcing skins that are bonded to the external faces of the core are hand laid and encased in a protective coating of PFR. The outer protective casing has been excluded in the present series of joint tests to allow for observation of the reinforcing fibres.

3 TEST JOINT FABRICATION

The core of each joint was cast in a flat open mould shown in Figure 4. The wrapped foam inserts were cured at room temperature, cut to length and located in the pre-waxed and assembled mould. The foam was prevented from floating and the mould was filled with PFR. After the joints were cured at room temperature for 24 hrs, they were removed from the mould, cleaned and reduced to the correct thickness in preparation for application of the reinforcing skins.

The uni-directional E-glass skins were hand laid and wet-out by manual rolling. The skins were applied to each side of the joint on consecutive days and were allowed to cure at room temperature for 24 hrs after which time they were post-cured at 60 degrees centigrade for 8 hrs.

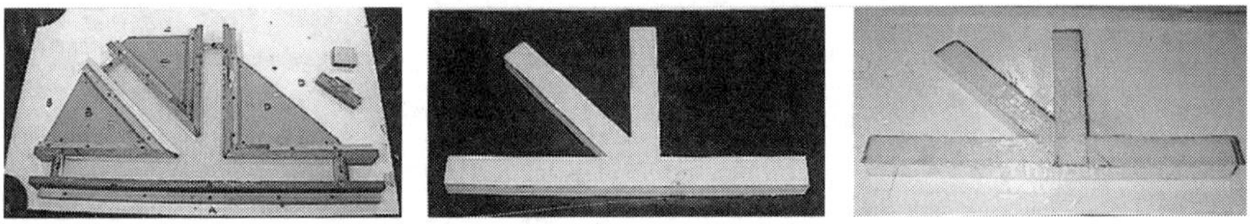

Figure 4. Casting mould, bare core, finished joint

3.1 *Test joint materials*

The material used for the production of the test joints was 50mm wide, 450g/m^2 heat-set uni-directional E-glass tape from Colan Products Pty Ltd. and ADR246TX epoxy resin with ADH160 hardener, both from ATL Composites, were used to bind the fibres. Coupon testing was used to determine the tensile strength of the uni-directional glass laminate. The load carrying capacity per layer per millimetre width was found to be 320N/mm. The strain at failure is approximately 2%.

ADR246TX epoxy resin was used with ADH160 and fly-ash microspheres provided by Envirosphere to produce the PFR. The finished density of the filled resin is approximately 800 kg/m^3. Research into alternative filled resin formulations is continuing at USQ. All joints were post-cured at 60 degrees centigrade for 8 hrs.

4 STRENGTH ESTIMATES

Generally, some difference is noted in the modulus of elasticity between compression and tension tests. For reasons of simplicity, the material properties in this investigation are assumed to be identical in tension and compression. The resulting error is estimated to be of the order 5%. By assuming the skins and the core undergo the same strain during loading, Hookes law can be used to determine the ratio of force carried by the skins and core of the truss element:

$$F_{skins} = F_{core} \cdot (A_{skins} E_{skins} / A_{core} E_{core}) \quad (1)$$

For the element under investigation:
$A_{skins} = 250mm^2$; $E_{skins} = 30E3MPa$
$A_{core} = 885mm^2$; $E_{core} = 3E3MPa$
$F_{skins} = 2.83 . F_{core}$

Therefore:
F_{core} = 26.1% of applied force
F_{skins} = 73.9% of applied force
$F_{total} = F_{core} + F_{skins}$

With an ultimate tensile failure strength of the epoxy/glass material of 320N/mm width and one 50mm wide layer of diagonal tensile reinforcement on each side of the truss, the failure load of the reinforcing skins is:

F_{skins} = 320 x 50 x 2 = 32.0kN
F_{core} = 26.1/73.9 x 32.0 = 11.3kN
F_{total} = 32.0 + 11.3 = 43.3kN

5 TESTING OF TRUSS JOINTS

The aim of the testing program is to closely examine the failure mechanisms of different types of joints and to provide data for the validation of numerical models. At the time of preparation of this paper, nine joints had been tested. These joints were loaded in static tension using an Avery universal testing machine and were tested to destruction.

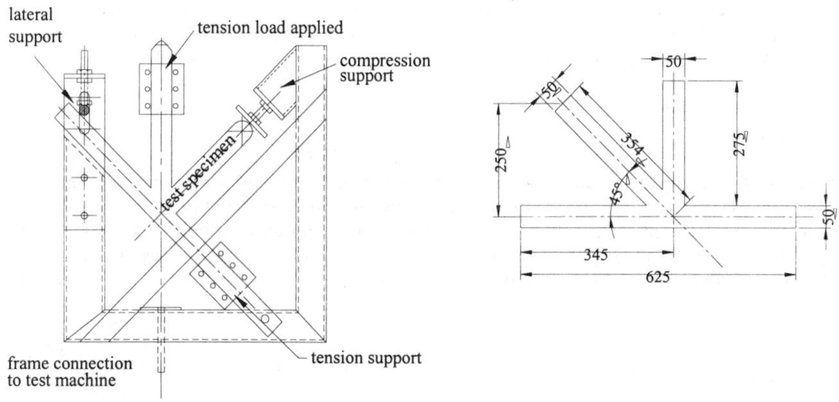

Figure 5. Joint test rig

5.1 *Joint testing procedure*

In the four panel Pratt truss shown in Figure 2, the critical tension members are the outer-most diagonals. Under load the diagonals tend to pull out of the top and bottom chord and a purpose built test rig was developed to model this type of behaviour. The test rig is shown in Figure 5.

5.2 *Results*

The table below shows the ultimate failure loads of the nine tests carried out on the joints. A predicted load is given for each joint as well as the achieved load from the test. The predicted load was obtained using the approach discussed in section 4 resulting in the same load for all joints.

Table 2 – Predicted and observed load capacity

Joint	1-1	1-2	1-3	2-1	2-2	2-3	3-1	3-2	3-3
Predicted (kN)	43.3	43.3	43.3	43.3	43.3	43.3	43.3	43.3	43.3
Observed (kN)	64.7	60.6	62.2	55.2	70.7	53.5	59.4	56.8	60.7

Figure 6 - Joint failure modes

In all cases the observed failure load was much higher than the predicted failure load. The contribution of the core to the overall strength of the joint is significant. Cracks noted in the core during loading are evidence that strain relaxation has occurred and this is not taken into account in the strength prediction.

Joint 1 failed through tensile rupture of the diagonal strength layer at the face of the joint. The overlap of fibres onto the adjacent members appears to disrupt the failure mode. The line of failure was not distinct and included some pull-out of the fill layers.

Joint 2 showed evidence of a new critical section resulting from the omission of two fill layers in the bottom chord. Failure was initiated in this member by a combination of tension and bending moment. During the test the joint opened up and cracking of the core occurred at the junction of the web compression member with the tension diagonal and bottom chord.

Joint 3 failed at the junction of the bottom chord and the compression member in a mode similar to that of joint 2. Failure was due to tensile failure of the bottom chord reinforcement possibly because of the influence of secondary stresses.

6 CONCLUSIONS

The test results confirm the potential of the present truss design and joint configuration. In all cases failure was initiated by fibre failure in the member rather than joint pull-out.

The experimental ultimate load carrying capacity of the joints was compared with the predicted load carrying capacity of each joint and found to be higher than the predicted capacity inall cases. Further research is required to obtain a more fundamental understanding of the joint behaviour.

7 ACKNOWLEDGEMENTS

The author gratefully acknowledges the technical and financial support by Connell Wagner Pty. Ltd., Consulting Engineers, Australia, which continues to make this research possible.

8 REFERENCES

Gibson, Ronald 1994. Principles of Composite Material Mechanics. New York: McGraw Hill.

Gregory, M 1971. History and Development of Engineering. Great Britain: Longman.

Head, Peter 1998. New Construction Techniques and New Forms of Structure Using Advanced Materials. London: Maunsell Engineering Group.

Tsai, Stephen 1992. Theory of Composites Design. USA: Think Composites.

King, Richard 1994. Designers Beware, New Thinking Needed. *Journal of the Institution of Structural Engineers Vol. 72 No. 9, UK, SETO.*

Agematzu. 1998. The First Carbon Fibre Truss App. *SAMPE Journal, Vol. 34, No.3, USA.*

Mechanics of Structures and Materials, Bradford, Bridge & Foster (eds)
© 1999 Balkema, Rotterdam, ISBN 90 5809 107 4

Estimation of the Bauschinger effect under the multi-axial loading

T. Nishimura, H. Hoshi & Y. Furukawa
Department of Mechanical Engineering, College of Science and Technology, Nihon University, Tokyo, Japan

ABSTRACT: The Bauschinger effect is well known as a stress hysterisis phenomenon, which is the nonlinear behavior on a reloading path after a plastic deformation is given preliminarily. Although so many studies of the Bauschinger effect under the uni-axial loading have been carried out, there are few studies under the multi-axial loading. In this paper, the concept of the Bauschinger ratio is suggested in order to explain the influence of Bauschinger effect and the experimental study under the multiaxial loading is mentioned

1 INTRODUCTION

The Bauschinger effect is well known as a phenomenon of stress hysteresis. The Baushinger effect is nonlinear deformational behavior, which is occasionally observed on reloading process. Although the Bauschinger effect had been studied under the uni-axial loading, the study under the multi-axial loading has hardly been seen.

It is already published that the Bauschinger effect depends on the magnitude of pre-deformation, that is, lager pre-deformation causes the predominant Bauschinger effect. It is expected that the Bauschinger effect on multi-axial loading is also affected especially by the reloading direction, that is, the deviation from the pre-loading direction. Therefore, in this study, the Bauschinger effect is investigated experimentally concerning the reloading direction that deviates from a pre-loading.

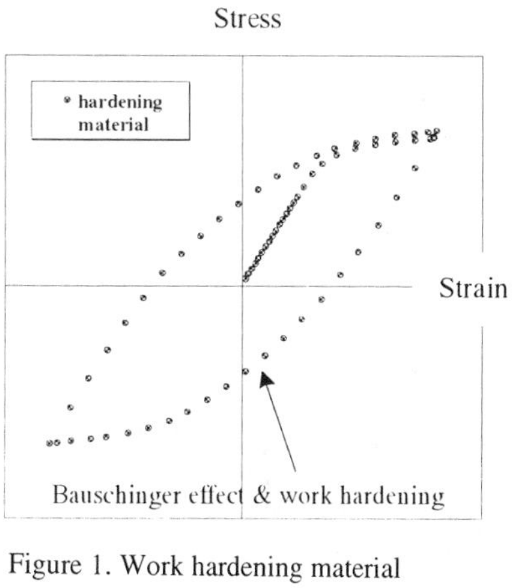

Figure 1. Work hardening material

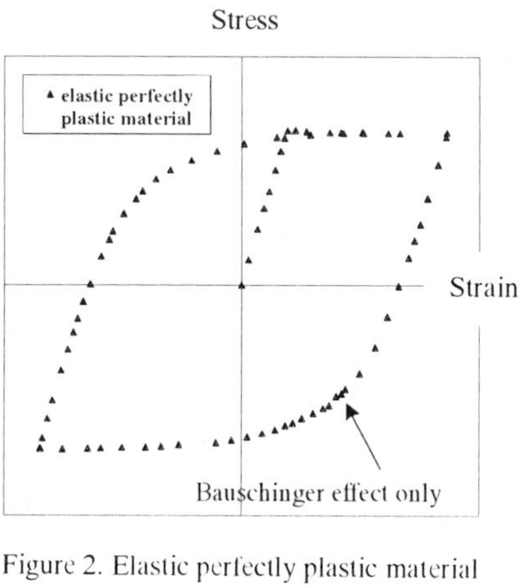

Figure 2. Elastic perfectly plastic material

It is difficult to isolate alone the influence of Bauschinger effect by the experiment on a hardening material, because the non-linear behavior of work hardening, as shown in Figure 1, resembles to that of the Bauschinger effect shown in Figure 2. Consequently, the only Bauschinger effect can not be discriminated from the non-linear behavior on work hardening material. Therefore an elastic and perfectly plastic material is preferably useful for the study of the Bauschinger effect. It is known that annealed mild steel has a yield plateau, which extends over about eight times length of the elastic limit strain. Therefore, since mild steel is regarded as an elastic perfectly plastic material within the yield plateau, annealed mild steel is adopted in this experimental study of the Bauschinger effect.

The influence of the Bauschinger effect is investigated experimentally under the different reloading direction from the pre-loading direction. Moreover the concept of Bauschinger ratio is introduced in this paper, in order to clear the evaluation of Bauschinger effect.

2 EXPERIMENT

It is confirmed from our previous research (Nishimura et al. 1997 & Furukawa et al. 1998) that annealed mild steel yields according with the Tresca's yield criterion under the combination of a normal stress and a shearing stress. It is well known that the Tresca's yield curve is represented by an elliptic formula shown below,

$$(\sigma_x)^2 + (2\tau_{xy})^2 = \sigma_y^{\ 2}$$

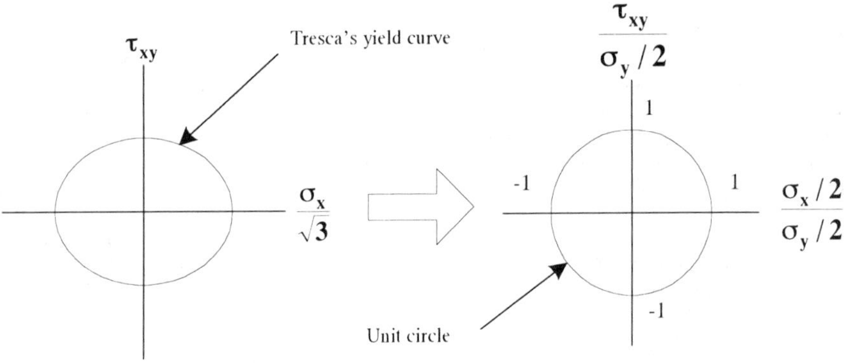

Figure 3. Transformation of Tresca's yield curve

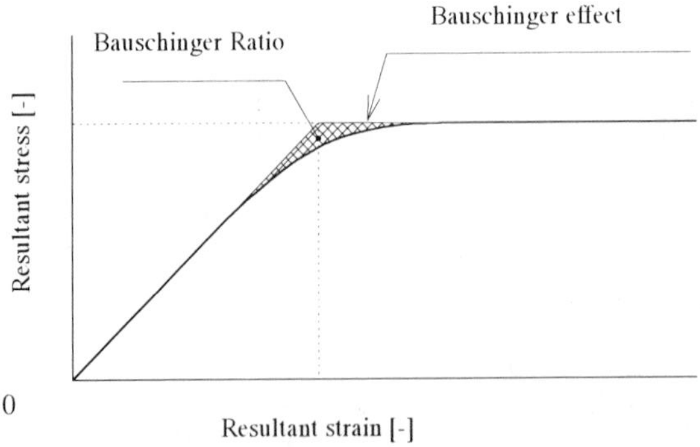

Figure 4. Definition of Bauschinger Ratio

The magnitude of the yielding stress vector is different on each radial loading direction. Accordingly, it is inconvenient to compare the deformational behavior on the different radial loading. Here the Tresca's yield ellipsoid is transformed into a circle, which is depicted in the stress space where the shearing axis is expanded by twice. In this warped stress space, all radial deformations can be explained by a single deformational behavior regardless of loading directions. Moreover, the circular yield curve is transformed into the unit circle, where each stress is divided by the yield stress of individual specimen in order to avoid the deviation of yield stress on each specimen, since the value of yield stress is slightly scattered on actual specimen. The resultant stress and the resultant strain are defined by the magnitude of the stress vector and the strain vector in the suggested normalized stress space, respectively.

The concept of Bauschinger Ratio is introduced in order to evaluate the Bauschinger effect exactly. The area of the shaded portion in Figure 4 defines the Bauschinger Ratio, which explains the intensity of Baushinger effect.

A thin-walled circular tube, shown in (Fig.5), is subjected to tensile force and twisting moment simultaneously on the experimental equipment (Fig.6) for investigating the Bauschinger effect under multi-axial loading. The specimen is annealed under the temperature 900 Kelvin for 3 hours. The geometrical configuration and the mechanical property of the specimen are shown in Figure 5.

Since a plastic deformation dose not occur uniformly over the specimen within the yield plateau, but spreads locally and irregularly over the specimen, it is hard to measure a strain on the yield plateau by a strain gauge. The displacement sensors shown in Figure 6, which are assured experimentally to have enough accuracy, are attached directly to the protrusions formed on the specimen in order to measure precisely the deformation without the interference of the warping on apparatus. The following three types of loading path are investigated:

(a) Repetitive radial loading beyond the elastic limit strain along the radial direction (Fig.7-1and Fig.7-2); The experimented loading-paths are 0 [deg] and 30 [deg] from the tensile direction.

(b) Reloading toward deviated directions from the pre-loading (Fig.7-3);
After giving preliminary deformation up to a certain value beyond elastic limit, a load is removed and reloaded toward the radial direction inclined by the angle θ from the pre-loading direction.

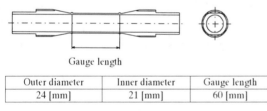

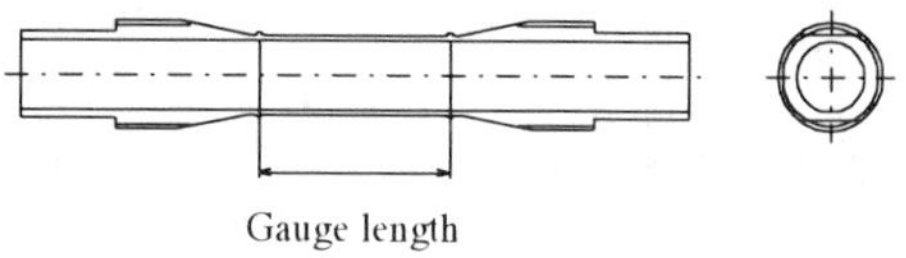

Outer diameter	Inner diameter	Gauge length
24 [mm]	21 [mm]	60 [mm]

E=193[MPa]

G=79[GPa]

ν=0.3[-]

σ_y =260[MPa]

Figure 5. Specification of Specimen

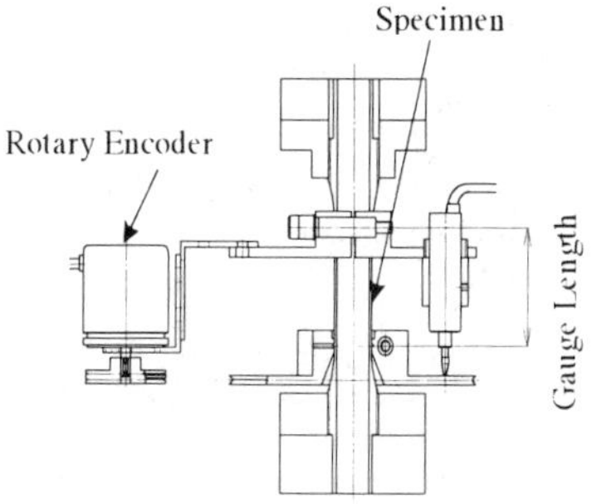

Figure 6 Experimental System

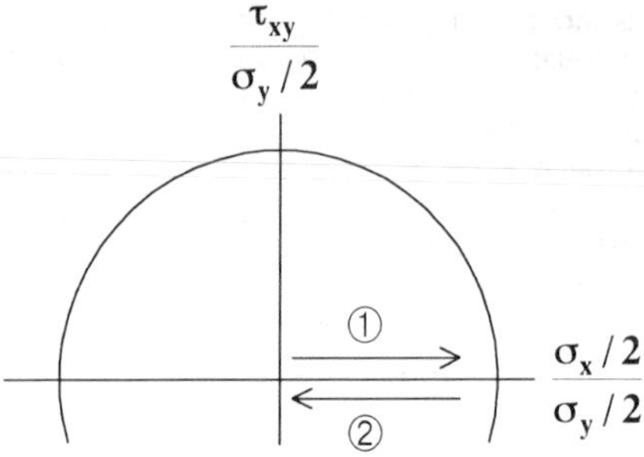

Figure 7-1

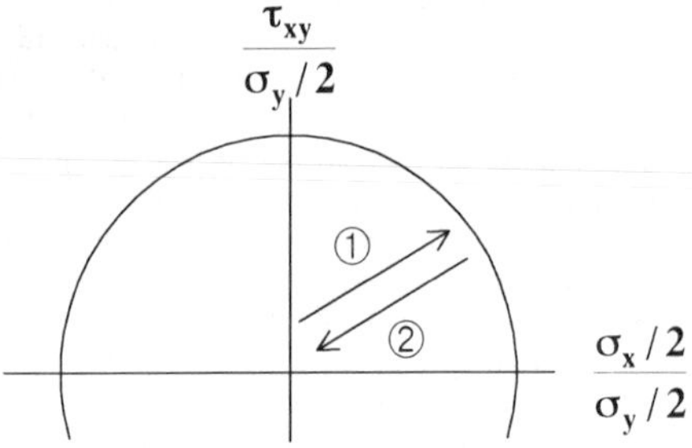

Figure 7-2

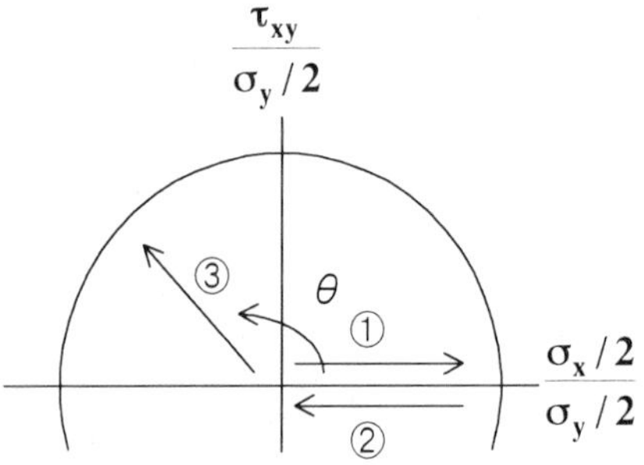

Figure 7-3

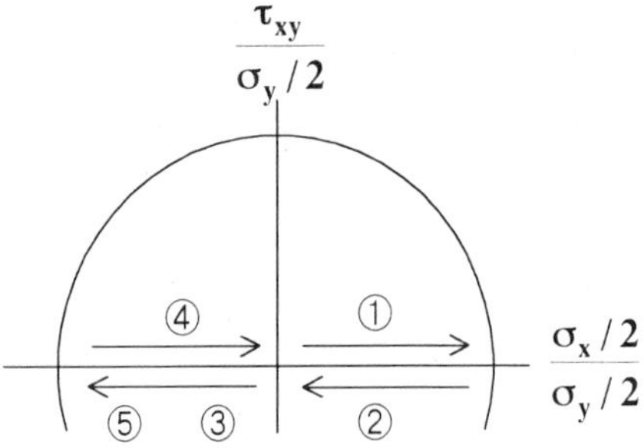

Figure 7-4

Figure 7. Loading path

(c) Unloading and reloading again on the same direction after once recognizing the Bauschinger effect (Fig.7-4).

The Bauschinger effect is studied in the experiment on the three cases of pre-strain 1.5, 2 and 3 as the resultant strain. The semi-circle in the figures represents the Tresca's yield curve, and the solid arrow and its associated number explain the loading direction and the order of loading, Respectively.

3 EXPERIMENTAL RESULTS

Figure 8 and 9 show the results of the experiment (a). The abscissa in the figurers denotes a resultant strain and the ordinate denotes a resultant stress. It is clear that the unloading and reloading behavior coincide with each other within the elastic region, and no Bauschinger effect is observed. Consequently, if the reloading direction is the same to the unloading direction, the Bauschinger effect dose not occur regardless of loading direction.

The results on the loading path (b) are shown in Figure 10, which represents the deformational behavior on the given resultant pre-strain 3. The observed deformational behaviors on each reloading direction are superposed in the figurers. It is apparent from the experimental results that non-linear behavior on deformation increases with growth of the deviation angle θ. Figure 12 shows the relation between the deviation angle θ and the Bauschinger Ratio, which is calculated from the results of experiment (b). It can be seen from the figure that the Bauschinger Ratio increases with growth of the deviation and becomes higher also for larger magnitude of given pre-strain. The no-linear deformational behavior by the Bauschinger effect depends on not only the magnitude of pre-strain but also on the deviation angle of reloading direction from the pre-deformation.

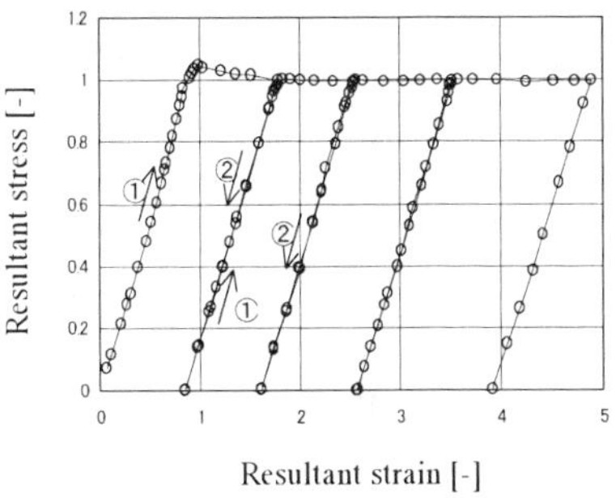

Figure 8. Result of experiment (a)

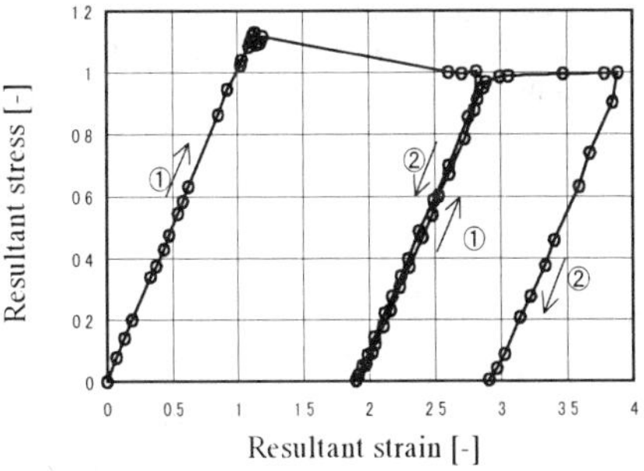

Figure 9. Result of Experiment (a) 30[deg]

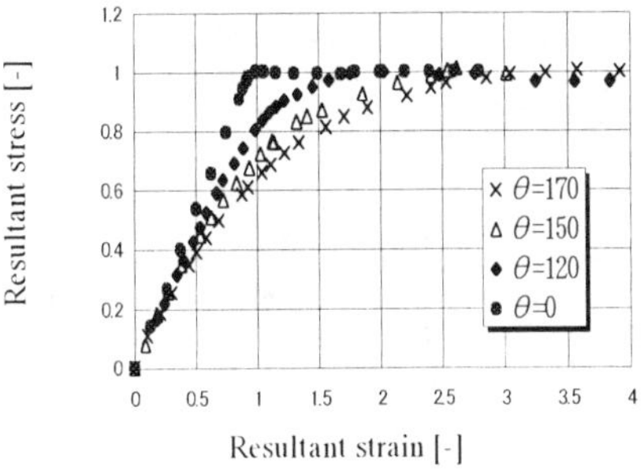

Magnitude of pre-strain [3.0]

Figure 10. Result of experiment (b)

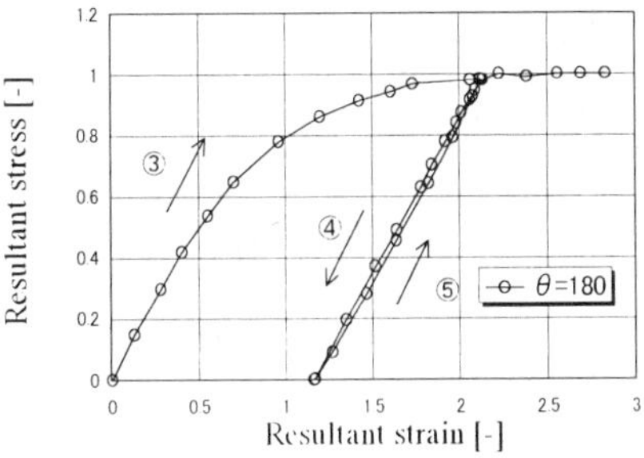

Figure 11. Result of experiment (c)

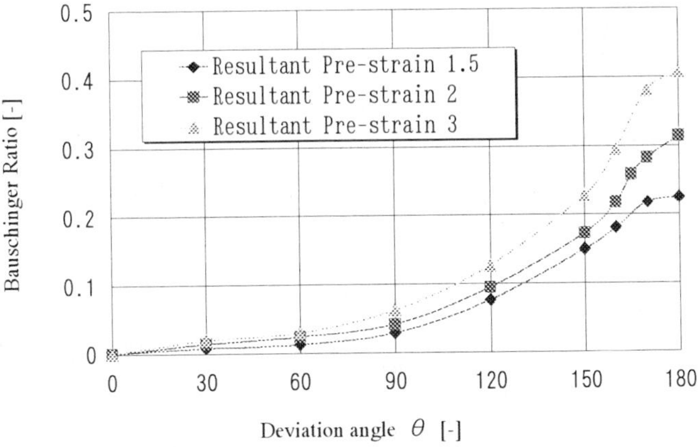

Figure 12. Bauschinger Ratio

The loading cycle in the experiment (c) is depicted in Figure 7-4. Tensile load is added beyond the elastic limit (①), and the reversal deformation (compression) is given until the stress attains to the yield stress (② and ③) as the first step. It is no matter to say that the predominant Bauschinger effect is observed. Moreover, after removing the load (④), the compression load is added again (⑤). The deformational behavior on the experiment (c) is shown in Figure 11. It should be noted that the Bauschinger effect is no more observed on the final loading process, although the no-linear deformation occurs through the first step.

The similar phenomenon is also observed on the different reloading direction (θ). If the reloading direction is the same to the latest reloading where the non-linear deformation is induced, the Bauschinger effect is cancelled on the current reloading. The fact seems to suggest that the Bauschinger effect is strongly affected by the latest loading direction.

4 CONCLUSION

When the reloading direction is the same to the pre-loading, the Bauschinger effect is not observed on a repetitive loading as shown in Figure 8 and 9.

The suggested Bauschinger Ratio represents the Bauschinger effect concerning the influence of deviation angles θ and the magnitude of pre-strain under the multi-axial loading in Figure 12. It is expected that the deviation angle of principal axis of residual strain, which is caused by pre-loading, against the reloading principal direction is important factor for the Bauschinger effect. The large deviation gives the intensive non-linear deformation.

It is imagined from the experimental result (c) that the Bauschinger effect is strongly affected by the latest loading direction (Fig.11), because the latest loading erases the effect of the former residual strain and produces the new residual strain.

5 REFERENCE

Nishimura, T., Sakai, Y.& Saito, T. 1997. On a proceeding direction of plastic strain. *The Second Asian-Pacific Conference on Aerospace Technology and Science* (APCAT'97): 36-41.

Furukawa,Y & Nishimura,T. 1998. Study of Bauschinger effect under multi-axial loading. *2nd Pacific-Asia Conference on Mechanical Engineering*: 579-584.

Mechanics of Structures and Materials, Bradford, Bridge & Foster (eds)

The behavior of expanded polystyrene (EPS) under repeated loading

Y. Zou, C.J. Leo & E.S. Bernard
School of Civil Engineering and Environment, University of Western Sydney, Nepean, N.S.W., Australia

ABSTRACT: EPS geofoam is an ultra-lightweight filling material frequently used in building road embankments and as a load-bearing subgrade layer on very soft soils. In these and other geotechnical applications, the EPS geofoam is subjected to repeated loading arising from the motion of vehicles. The response of EPS geofoam to cyclic loading has been studied in a series of unconfined cyclic loading tests at various stress levels. Useful relationships describing the number of loading cycles to cause failure for a given peak stress, and the permanent strain accumulation with number of load repetitions, have been established.

1 INTRODUCTION

EPS (Expanded Polystyrene) is a rigid closed-cell cellular plastic commonly made from expandable polystyrene bead. It is very lightweight and compressible, and exhibits good insulation properties. As a ultra-lightweight material, EPS geofoam is an excellent fill for road embankments and retaining structures located on soft soils. The use of EPS geofoam as a replacement material for poor quality subgrade in road construction is gaining acceptance.

Current engineering design methods for geotechnical application of EPS geofoam remains based on empiricism and experience, even though the material has been used successfully for nearly 30 years in Europe, North America and Japan. EPS materials exhibit a visco-plastic behavior and therefore suffer from permanent deformation under repeated loading, even at small stress levels. The long-term performance and durability of EPS remain as technical barriers to its wider acceptance and use in geotechnical applications.

In this study, a series of unconfined cyclic loading tests at various peak stress levels have been carried out and the response of EPS geofoam to cyclic loading is studied. The effects of stress level and number of load repetitions on EPS material are investigated. Two relationships have been established: (1) permanent strain accumulation with the number of load repetitions; and (2) the number of cycles to cause failure, N, for various cyclic stress level, S (the classical S-N relationship of Wohler (in Timenshenko, 1953)).

2 LABORATORY TESTS ON THE BEHAVIOUR OF EPS

A number of studies have recently investigated the mechanical behavior of EPS, including triaxial compression tests, unconfined/confined compression tests, hydrostatic compression testing, shear tests, creep tests and cyclic unconfined compression tests (Hamada and Yamanouchi, 1987, Eriksson and Trank, 1991, Preber et al., 1994, Horvath, 1995, Zou and Leo, 1998). Field monitoring of existing EPS sites (Refsdal, 1985, Aaboe, 1987) have generally indicated that

there are no mechanical problems with EPS fill after several years of traffic loading. However, these results are confined to materials subject to low magnitude stresses and strains. At larger stresses and strains, the long-term effects of repeated loading appear less certain.
When EPS material was subjected to the repeated loading condition, both the maximum stress level and number of repetitions affect the plastic deformation of the material. The effects of number of load repetitions can be investigated in cyclic loading tests where the maximum cyclic stress was kept constant. The effects of stress levels can be studied by conducting of cyclic loading tests at different peak stresses.

2.1 *Definitions of key parameters*

In unconfined cyclic loading tests, the key parameters of interest for the stress-strain curve are defined as follows, and are illustrated in Figure 1.

peak/maximum cycle stress, σ_n: the maximum stress imposed during loading cycle number
compressive strength, σ_c: compressive stress at 10% of axial strain on stress-strain curve of monotonic compression test
peak strain, ε_a: the axial strain exhibited in response to the maximum cycle stress
residual strain, ε_r: the plastic (permanent) strain after unloading of each cycle
residual strain, ε_1: the plastic (permanent) strain after unloading of the first cycle
recoverable strain, ε_0: the peak strain minus plastic strain, $\varepsilon_0 = \varepsilon_a - \varepsilon_r$
initial tangent Young's modulus, E_{ti}: the slope of the initial linear-elastic portion of the stress-strain curve
resilient modulus, E_{rs}: is defined as the ratio of the maximum cycle stress, σ_n, to recoverable (resilient) strain, ε_0: $E_{rs} = \sigma_n/\varepsilon_0$
stress ratio, $S = \sigma_n/\sigma_c$

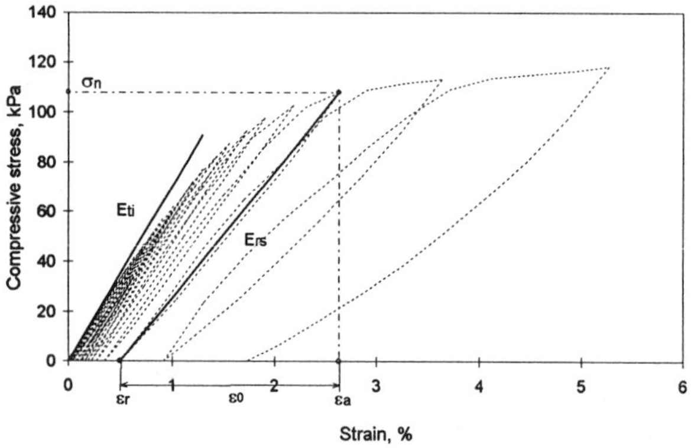

Figure 1 Parameters describing stress-strain relation in cyclic test

2.2 *Test conditions*

Unconfined cyclic compression tests were performed on cylindrical specimens of dimensions Ø50× 50 mm with a density of 21kg/m^3. Test conditions were those of a laboratory environment (approximately 20-25°C and 50% relative humidity). The tests were displacement controlled to produce a rate of loading of 10% of the thickness of the specimen per minute (5 mm/min) in accordance with Australian Standard AS2498.3 (1993) and ASTM Standard D 1621-73/79.
The cyclic tests were carried out using an INSTRON 6027-R5500 Electro-mechanical Universal Testing Machine equipped with high speed digital control and data acquisition. Axial loading and unloading was effected by moving the crosshead on which the base of specimen was seated

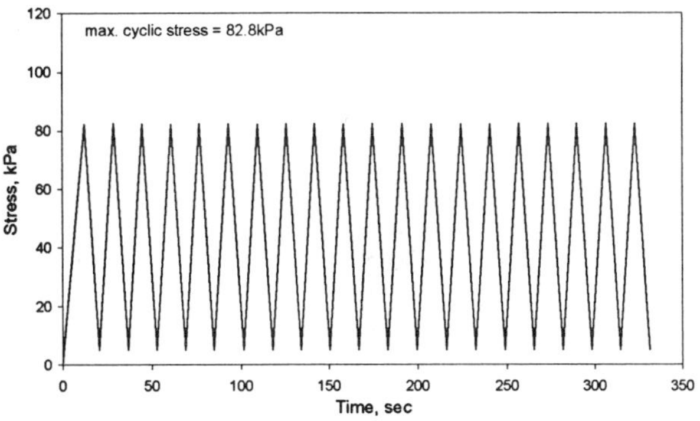

Figure 2 Typical loading waveform of cyclic test

Table 1 Loading Condition and Results for Constant Stress Cyclic Loading Tests

Max. load, N	Max cyclic stress, kPa	Stress ratio*, %	ε_1	No. of cycles to failure
240	123.3	112.1	10.1	1 ($\varepsilon_r > 10\%$)
220	112.6	102.4	3.09	31 ($\varepsilon_r > 10\%$)
200	102.7	93.4	1.247	240 ($\varepsilon_r > 10\%$)
180	92.4	84.0	0.225	8400 ($\varepsilon_r > 10\%$)
160	82.8	74.7	0.135	31500 ($\varepsilon_r > 10\%$)
100	51.1	46.7	0.02	3.1×10^8 (projected)

* Stress ratio = Max cyclic stress/compressive strength, where the compressive strength of 110kPa (at 10% axial strain) was obtained from the monotonic compression test.

up and down. The magnitude of loading was measured by a calibrated load cell and the test system was monitored using a proprietary program called *Merlin*. The saw-tooth loading waveform used for this study is shown in Figure 2.

2.3 *Test results*

A total of six samples were used in constant stress cyclic loading tests where the maximum cyclic stress was keep constant. Failure was achieved in all tests except for the case of when the maximum cyclic stress was 51.1 kPa. In this test, the number of cycles to failure was based on a projection of existing data as the test would otherwise have required a very long time to reach failure.

The details of the test conditions and test results are listed in Table 1. Figure 3 shows some typical stress-strain curves developed in the constant stress cyclic loading tests. As expected, it was found that the higher the stress level, the higher the initial strain and residual strain rate in term of the number of cycles. The strains induced at all levels of loading of EPS are not fully recoverable upon unloading. This is true even for small levels of peak stress, σ_n.

The resilient moduli, developed from the data shown in Figure 3, are shown in Figure 4. This indicates that in the post-yield range the resilient modulus, E_{rs}, does not change much with the number of loading cycles. It suggests that the recoverable behavior of EPS will not be affected by repeated loading.

The residual strain accumulation was measured as a function of the number of load cycles for 5 specimens. The results are shown in Figure 5. This curve shows an approximately linear relationship between $\log(\varepsilon_r)$ and log(N). The nature of this plot suggests that the relationship between ε_r and N can be described by:

$$\log\varepsilon_r = m \times \log N + \log\varepsilon_1 \qquad \text{or} \qquad \varepsilon_r/\varepsilon_1 = N^m \tag{1}$$

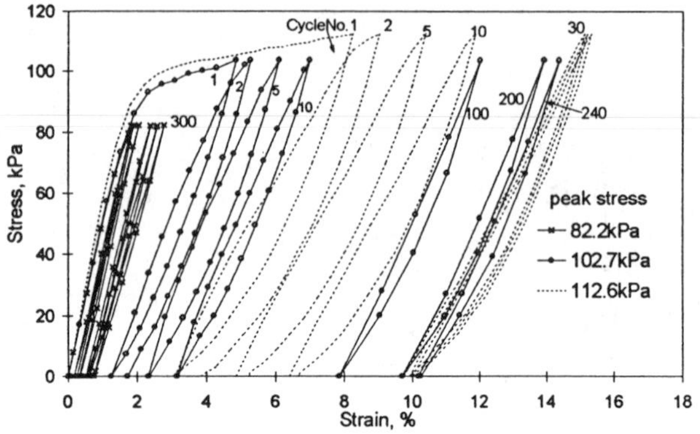

Figure 3 Typical stress-strain curve in constant stress cyclic test

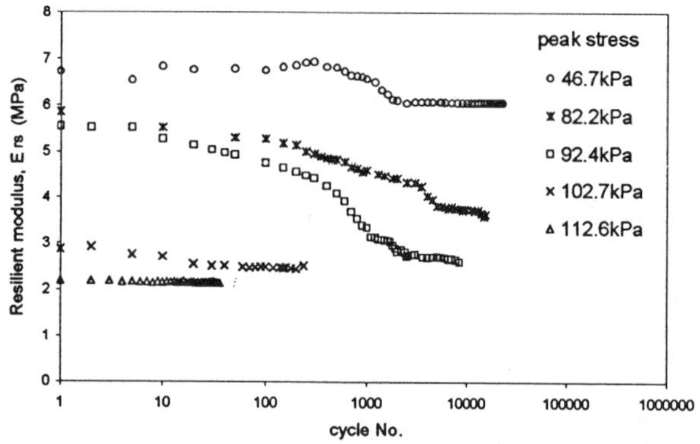

Figure 4 Resilient modulus in constant stress cyclic test

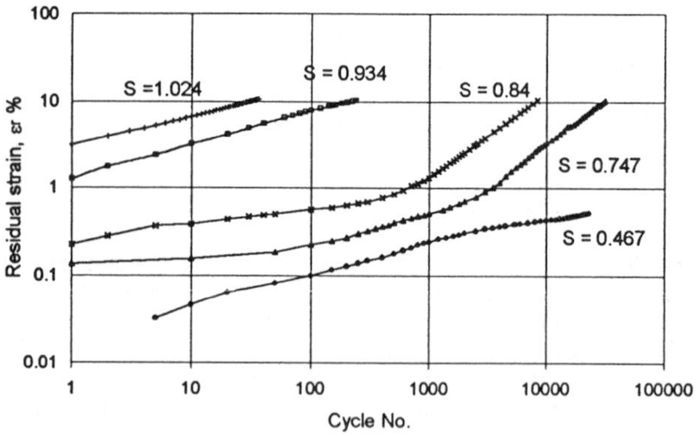

Figure 5 Residual strain in log diagram for various stress ratio ($S=\sigma_n/\sigma_c$)

where ε_1 is residual strain in the first cycle (N=1), and m is an experimentally derived parameter. The normalized residual strain, $\log(\varepsilon_r/\varepsilon_1)$, is plotted against the logarithm of the number of cycles, logN, on a log-log curve in Figure 6. The parameter m is the slope of the linear regres-

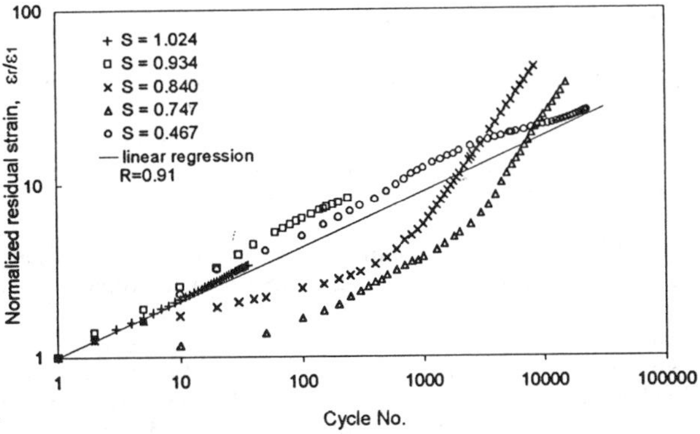

Figure 6 Normalized residual strain in log diagram

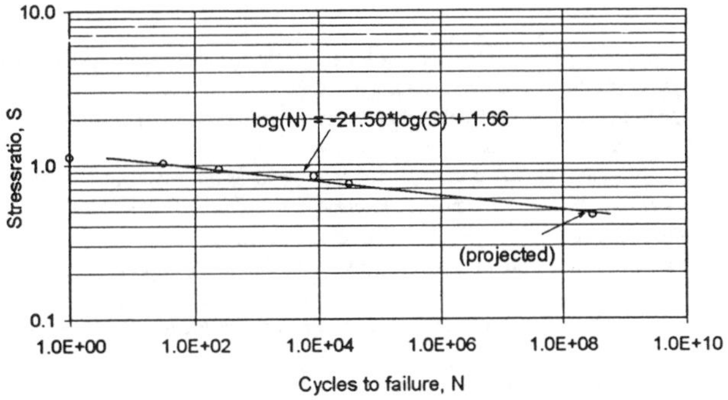

Figure 7 Cyclic stress ratio versus number of cycles to cause failure

sion for all the test data. Based on the result of the current cyclic tests, an *m* value of 0.318 was obtained for all stress levels.

The cyclic test results show that a decrease in the maximum stress level requires a significant increase in the number of cycles to cause failure. At lower stress levels, the post-yielding stage may be absent. For stress levels lower than 50%, it may take several years to complete a single cyclic test to failure. Equation (1) may be used to forward extrapolate the approximate number of load cycles to yield a strain at failure of 10%. Thus, the predicted number of cycles to failure for a stress level of 51.1kPa (stress ratio, S=0.467) in accordance with Equation (1) is approximately 3.1×10^8.

Figure 7 shows the relation between the number of cycles to failure, N, and the stress ratio, S, for data obtained from the five unconfined constant stress cyclic tests conducted in this investigation. Failure means that either the prescribed maximum stress under cyclic loading cannot be achieved or the accumulated residual strain exceeds a threshold value (e.g. 10%). An approximately linear relationship between log(N) and log(S) was obtained as:

$$\log(N) = -21.50\log(S) + 1.66 \quad \text{or} \quad N = (1.195/S)^{21.5} \tag{2}$$

Equation (2) can be used to predict the cycles of loading to cause failure for EPS material under a given stress level.

3. CONCLUSION

From unconfined cyclic compression test results, the influence of different stress levels and number of loading cycles on EPS behavior have been investigated. The conclusions were:

1. The permanent deformation of EPS is affected by both the applied stress level and the number of loading and unloading cycles. The cumulative plastic (residual) strain ε_r, is increased as the number of cycles increases. The rate of increase of residual strain is also proportional to the peak stress level.
2. The permanent deformation of EPS accumulates with the number of loading and unloading cycles, even in the so-called elastic range of the stress-strain curve obtained from a monotonic compression test.
3. An approximate relationship between normalized strain and number of load cycles can be developed to predict the approximate residual strain for a given number of load repetitions.
4. An approximate relationship between stress level and number of cycles to failure can be developed to predict the approximate numbers of cycles to failure (10% of residual strain).

REFERENCES

Aaboe, R. 1987. 13 years of experience with expanded polystyrene as a lightweight fill material in road embankments. *Internal Report, Norwegian Road Research Laboratory.* Oslo, Norway.

Eriksson, L. And Trank, R. 1991. Properties of expanded polystyrene, laboratory experiments. *Internal publication of the Swedish Geotechnical Institute.* Linkoping, Sweden.

Hamada, E. And Yamaouchi, T. 1987. Mechanical properties of expanded polystyrene as a lightweight fill material. *9th Southeast Asian Geotechnical Conference.* pp35-47. Bangkok, Thailand.

Horvath, J.S. 1995. *Geofoam Geosynthetic.* Horvath Engineering, P.C. Scarsdalc, N.Y. 217p.

Preber, T., Bang, S., Chung, Y. & Cho, Y. 1994. Behavior of expanded polystyrene blocks. *Transportation Research Record,* n1462, pp36-46.

Refsdal, G. 1985. Plastic foam in road embankments: future trends for EPS use. *Internal Report, Norwegian Road Research Laboratory.* Oslo, Norway.

Timoshenko, L.P. 1953. *History of strength of material.* pp162-170. McGraw Hill

Zou, Y. & Leo, C.J. 1998. Laboratory studies on the engineering properties of expanded polystyrene (EPS) material for geotechnical applications. *Proceedings of the 2nd International Conference on Ground Improvement Techniques.* Singapore.

Mechanics of Structures and Materials, Bradford, Bridge & Foster (eds)
© 1999 Balkema, Rotterdam, ISBN 90 5809 107 4

Mechanical properties of recycled railway ballast

D. Ionescu
School of Management, Technology and Environment, La Trobe University, Bendigo, Vic., Australia

ABSTRACT: Three model fractions of recycled railway ballast have been subjected to a series of isotropically consolidated, triaxial compression tests conducted at low confining pressures and their findings are reported here. The test results have been used for the prediction of the shear strength of ballast as a function of the particle maximum size and grains shape and distribution. The effect of maximum principal stress ratio on the deformation and degradation of recycled material is also studied. It was observed that, non-linear behaviour characterizes the shear strength, dilation rate and degree of particle crushing for different principal stress ratio, and specially at low level of confining pressure (characteristic to railway tracks). In order to determine the advantages and disadvantages of recycled ballast as a construction material for railway tracks, the results of this study have been compared with the behaviour of new railway ballast.

1 INTRODUCTION

Rail transport in Australia is under increasing pressure to be more reliable and to be competitive with road transportation. This has encouraged the introduction of higher loads, longer trains, greater speeds and the need to examine ways of reducing maintenance costs. To date these goals have been incompatible as the increase in axle loads and speed has produced a rapidly increasing maintenance bill.

The management and possible reduction of the maintenance costs can only be achieved when the performance of the various track components is understood and when their life cycle becomes predictable. The ballast layer is one of the key components contributing to the life and performance of the track. The performance of the ballast body will depend not only on its physical parameters such as: the density or initial compaction, particle size distribution of the ballast, the type of parent rock, but also on track geometric parameters: the spacing of sleepers, the thickness of the ballast bed, and the width of shoulders that provide confinement to the ballast bed.

The search for cost effective solutions and in line with the increasing pressure from the environmentally conscious community to prevent further quarrying have considered the use of waste products for railway track construction. During its life in the track, ballast is subjected to several cycles of maintenance. Some of the maintenance processes (cleaning and renewal of ballast) produce large quantities of waste material, which in the past was used as infill material. Recent regulations of the Environmental Protection Authority reduced the options of disposal, causing accumulation of considerable quantities of waste material in Rail Services Australia (RSA-NSW) stockpiles.

The present study is part of a major research project initiated by the University of Wollongong in collaboration with the Rail Services Australia. The study of degradation and settlement characteristics of the ballast is the main concern of this project. The present paper presents the results of a large number of CID tests performed on recycled ballast. The effect of type of ballast, the shape of the grains, particle gradation, and the level of confining pressure on the stress-strain behaviour and particle crushing characteristics were evaluated. The finding of the tests are analyzed and compared with results from a similar study carried on fresh ballast in order to determine the suitability of recycled ballast for reconstruction of railway tracks.

2 MODELLED MATERIALS

2.1 *Ballast type*

The recycled ballast sample was supplied from a stockpile at St Mary's, NSW. The worn material was obtained by screening particles larger than 73 mm and smaller than 20 mm, from the spoil resulting from ballast cleaning/undercutting process. Three model gradations have been used for the purpose of this study, where gradations A and B are parallel to the upper and lower limit of RSA specifications, respectively and gradation C is representative of the recycled material in the stockpile. The particle size distribution for the three modelled materials (before and after tests) is presented in Table 1, together with the upper and lower boundary of the specification for ballast warranted by AS 2758.7 - 1996. A careful examination of hand specimens permitted the identification of the source of grains for each particle fraction, and the composition of test sample so determined is presented in Table 2. On average, the sample comprised of about 61% of angular crushed rock fragments, the remaining consisting of partly rounded crushed river gravel reported to be from a gravel quarry at Emu Plains NSW. Table 3 summarises the characteristics of the model samples before testing. The C_u value of 1.5 - 1.8 corresponds to a uniform ballast.

Table 1. Particles size distribution of modeled ballast

Grains fraction	Gradation A		Gradation B		Gradation C		Specifications
(mm)	Before (%)	After (%)	Before (%)	After (%)	Before (%)	After (%)	AS 2758.7-96 (%)
73					100	100	
63					98	98.9	100
53	100	100	100	100	82.9	83.9	85 – 100
37.5	44	45.1	85	85.5	36.8	42.9	20 – 65
26.5	8.8	13.1	30	37	12.1	17.2	0 – 20
19	2.4	6.1	5	11	2.3	7.68	0 – 5
13.2	0	2.4	1	4	0.2	3.98	0
9.5		1.2	0	2	0	2.71	
4.75		0.5		1		1.44	
2.36		0.3		0.6		1.02	

Table 2. Composition of test specimens based on grains source for each particle fraction

Grains fraction	63	53	37.5	26.5	19	13.2	9.5
Particles source – grains nature	(mm)	(mm)	(mm)	(mm)	(mm)	(mm)	(mm)
Emu Plains – crushed river gravel	22.7	21.2	39.9	46.1	42	41	39.2
NSW Quarries – mixed crushed rock	77.3	78.8	60.1	53.9	58	59	60.8

Table 3. Grain size characteristics of tested materials

Gradation	Particle shape	d_{max} (mm)	d_{10} (mm)	d_{30} (mm)	d_{50} (mm)	d_{60} (mm)	C_u	C_c	Size ratio
A	Highly angular (fresh ballast)	53	27.1	32.6	38.9	41.3	1.5	0.9	5.7
B	Angular-Partly	53	20.7	26.7	30.3	32.8	1.6	1.0	5.7
C	rounded (recycled ballast)	73	25.2	34.6	41.5	44.6	1.8	1.1	4.1

The ballast was compacted to simulate the realistic densities measured in the field (frequently used tracks) where the mean bulk unit weight of the compacted specimens was determined to be 16.5 kN/m^3. The achieved relative densities were 70 – 99%, which corresponded to a porosity of 38% to 42%. Prior to testing, each specimen was soaked for 24 hours in order to determine the effect of ponding water on railway track sections having poor drainage.

2.2 *Ballast characteristic tests*

In this study, a worn material has been tested as potential railway ballast. Table 4 summarises the physical properties of this material, as warranted by standard ballast tests (AS 2758.7, 1996). It is not surprising that the degradation results obtained for worn ballast were slightly lower that those obtained from fresh ballast. This is due to the loss of sharp edges (quarried material) while performing its function in track, apart from the fact that river gravel is harder due to natural selection, and exhibits lower degradation even in fresh state. Based on the above properties and in relation to the standard recommendations, one would expect less degradation from recycled ballast than from fresh aggregates when used as a construction material for railway tracks.

Table 4. Ballast characteristic tests

Characteristic test \ Source of ballast	Recycled material (St Mary's stockpile)	Fresh ballast (Bombo quarry latite)	Standard specifications (AS 2758.7 – 1996)
Durability:			
1. Aggregate Crushing Value - ACV (%)	11	12	<25
2. Los Angeles Abrasion - LAA (%)	12	15	<25
3. Wet Attrition Value - Deval (%)	5	8	<6
Shape:			
4. Flakiness (%)	22	25	<30
5. Misshapen Particle (%)	18	20	<30
6. Fractured Particle (%)	85	100	>95

3 TEST EQUIPMENT

Previous research showed that in the case of larger granular materials, the use of the conventional triaxial equipment leads frequently to inaccurate or misleading deformation behaviour and failure modes due to the disparity between the particle sizes in the field and in the triaxial specimens. This is due to the inevitable size-dependent dilation and different mechanisms of particle crushing. Therefore, in order to obtain more realistic stress-deformation and degradation characteristics, a large-scale equipment was employed to carry out the test program. The large-scale triaxial testing apparatus accommodates specimens of 300 mm diameter x 600 mm high. Detailed description of the equipment is presented elsewhere (Indraratna, 1996).

Isotropically consolidated, drained triaxial compression tests (CID) were conducted, with the effective confining pressure varying from 1 to 240 kPa. This pressure range is adequate to simulate the typical confining pressures generated within the ballast bed (particles interlocking as well as loading from other sleepers). The maximum deviator stress was recorded for each test specimen, and the post-peak stress-strain behaviour was monitored up to about 20% axial strain. The response at peak deviator stress $(\sigma_1'-\sigma_3')_p$ was considered to be the 'failure', because no distinct failure plane was observed.

4 ANALYSIS OF TEST DATA

4.1 *Shear strength*

From current test data, the strength envelopes have been plotted (Fig. 1) using a non-linear failure criterion proposed by Indraratna et al. (1993) for granular media (rockfill). The criterion is independent of the initial test conditions and is unaffected by the system of units:

$$\frac{\tau'_f}{\sigma_c} = a\left(\frac{\sigma'_n}{\sigma_c}\right)^b \quad (1)$$

where a and b = dimensionless parameters, and σ_c = uniaxial compressive strength of intact parent rock. The test data fall within a narrow band of which upper and lower boundaries are defined in Fig. 1. The recycled ballast test results are distributed near the lower boundary. The differences in shear strength between fresh and recycled ballast are more marked in the lower range of confining pressure, which is critical for railway tracks (eg. low confining pressure on the ballasted foundation, higher risk of loss of stability of track).

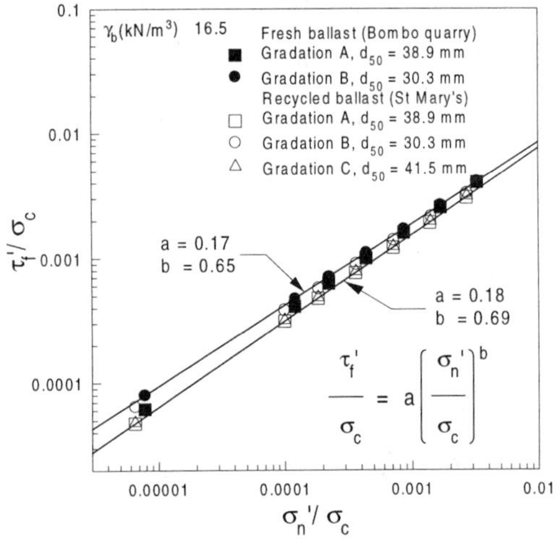

Figure 1. Normalised shear strength - normal stress relationship for railway ballast

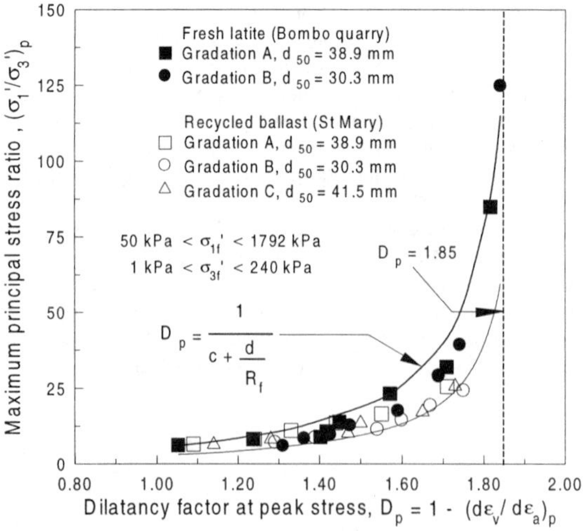

Figure 2. The strength – dilatancy relationship for railway ballast

4.2 *Effect of confining pressure on sample deformation*

At lower levels of confining pressure, the volume of particulate media can expand upon loading, and this behaviour is more pronounced for dense specimens. Figure 2 illustrates the effect of $(\sigma_1'/\sigma_3')_p$ on the rate of dilation, where the dilatancy factor (D_p) is defined by the expression $1-(d\varepsilon_v/d\varepsilon_a)_p$, as discussed by Rowe (1962). It is verified that the relationship between $(\sigma_1'/\sigma_3')_p$ and D_p is highly non-linear for the range of applied confining pressure, before becoming asymptotic to $D_p = 1.85$ for higher values of maximum principal stress ratio. By using a hyperbolic fit, the following equation is obtained to describe the relationship between the ballast dilatancy factor and the maximum principal stress ratio:

$$D_p = \frac{1}{c + \dfrac{d}{R_p}} \tag{2}$$

where, D_p = the dilatancy factor at the maximum principal stress ratio, R_p = the ratio of principal stress at peak and *c, d* = test constants. As shown in Figure 2 at the same level of principal stress ratio, the recycled ballast exhibits larger dilatancy behaviour in comparison to the fresh ballast, and this behaviour is more obvious at a lower range of confining pressure ($\sigma_3' < 100$ kPa).

4.3 *Effect of confining pressure on particle degradation*

The degree of particle crushing affects the deformation and the ultimate strength characteristics of any ballast material (Ionescu et al. 1996), which in turn influences the track performance. In the first stage of loading, local crushing at interparticle contacts is initiated, followed by complete fracture of weaker particles on further increase in load. This grain breakage contributes to differential track settlement and lateral deformation. In addition, the long-term accumulation of fines and decreasing porosity of ballast in certain depths can cause undrained failures during and after heavy rainfall (Indraratna et al. 1998).

After each test, the specimens were sieved and the changes in grading were recorded. For granular materials, Marsal (1973) introduced a measure for grain crushing, breakage index (B_g), which was equal to the sum of the positive values of ΔW_k (ΔW_k = the difference between the percentage by weight of each grain size fraction before and after the test). This index was adopted here to estimate the extent of ballast grains degradation during shear test. Figure 3 illustrates the variation of the particle breakage index with the change in $(\sigma_1'/\sigma_3')_p$. As the term $(\sigma_1'/\sigma_3')_p$ decreases (hence increase of σ_3'), the magnitude of B_g varies from 0.23% to 10.58%.

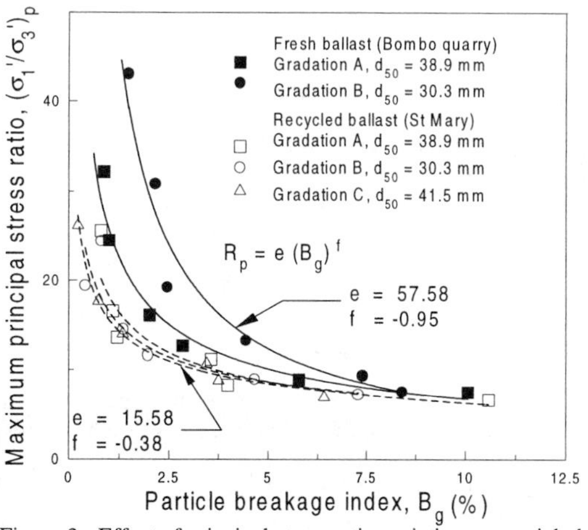

Figure 3. Effect of principal stress ratio variation on particle breakage for Bombo latite and for recycled ballast from St Mary's stockpile

As indicated in Figure 3, a greater extent of degradation was observed for fresh ballast due to sharp edges and corners. The largest degree of degradation was exhibited by gradation B, attributed to the greater initial density upon compaction. The difference in magnitude of B_g for fresh and recycled ballast is marked at a lower range of σ_3'. At higher values of σ_3', the change in B_g is insignificant; hence, the grains shape is not a determining factor for ballast behaviour. A non-linear relationship between the maximum principal stress ratio and the particle breakage index is given by the following equation:

$$R_p = e\,(B_g)^f \tag{3}$$

where R_p = the maximum principal stress ratio, and e and f = empirical coefficients.

5 CONCLUSIONS

The study described in this paper elucidated some of the mechanical properties of railway ballast and the effect of gradation and type of ballast on engineering behaviour. It was shown that the deformation and shear behaviour of railway ballast at low levels of confining pressure (< 100 kPa) depart significantly from the behaviour at larger confining pressure. The particle size distribution, angularity of grains, and the degree of packing mainly influence the shear strength and the crushing of particles. It is shown that ballast with a broader gradation having highly angular grains undergoes a greater degree of degradation upon loading. It was also confirmed, that there are slight differences between the behaviour fresh ballast and recycled ballast. These differences are more marked at lower range of confining pressure (< 100 kPa) characteristic to railway track, especially in crib and shoulder ballast. However, the effects of dilation and degradation seem to be of secondary importance in contrast to the effect of confining stress on the shear strength, which is the most influential feature for both fresh and recycled ballast.

The current test results show that there are some advantages in using the recycled ballast over the fresh ballast. These include the reduced maintenance cost, the reduced particle degradation during loading, and finally, the obvious environmental implications of preserving natural resources. The principal disadvantage of using the recycled ballast is its larger volumetric deformation during shear that has a significant bearing on the measured strength, especially at low levels of confining pressure that is expected in the shoulders and crib ballast. For this reason, the use of this material should be limited to the lower sections of ballast bed, where enough confinement is provided, hence, the expansive behaviour is minimised.

REFERENCES

Australian Standards, 1996. *Aggregates and rock for engineering purposes: Railway ballast.* AS 2758.7 - 96. Sydney, Standards Australia.

Indraratna, B., Wijewardena, L. S. S. and Balasubramaniam, A. S., 1993. Large-scale triaxial testing of greywacke rockfill, *Geotechnique*, 43 (1): 37-51.

Indraratna, B., 1996. Large-scale triaxial facility for testing non-homogeneous materials including rockfill and railway ballast. *Australian Geomech.*, 30 (12): 125-126.

Indraratna, B., Ionescu, D. & Christie, D., 1998. Shear behaviour of railway ballast based on large-scale triaxial tests. *J Soil Mech. and Found. Engg.*, 124(5): 439-449.

Ionescu, D., Indraratna, B. and Christie, H. D., 1996. Laboratory evaluation of the behaviour of railway ballast under static and repeated loads, *Proc. 7th ANZG Conference*, Adelaide, 1-7 July 1996: 86-91. Adelaide: IE Aust.

Ionescu D., Indraratna B. and Christie H. D., 1998. Mechanical Properties of Railway Ballast From Laboratory Tests, *Proc. 2nd Int. Symp. Hard Soils - Soft Rocks*, Naples, 12 -14 October 1998, (1): 551-560. Rotterdam: Balkema.

Marsal, R. J., 1973. *Mechanical properties of rockfill.* New York: Wiley.

Rowe, P. W., 1962. The stress-dilatancy relation for static equilibrium of an assembly of particles in contact, *Proc. Roy. Soc.* A 269: 500-527.

Mechanics of Structures and Materials, Bradford, Bridge & Foster (eds)
© 1999 Balkema, Rotterdam, ISBN 90 5809 107 4

Durability of high strength steel fibre reinforced concrete

R. Dhanasekar & C. Hudson
Faculty of Engineering and Physical Systems, Central Queensland University, Rockhampton, Qld, Australia

ABSTRACT: Concrete matrix reinforced with steel fibres, known as steel fibre reinforced concrete(SFRC) has several desirable engineering properties. In spite of the several positive aspects of the SFRC, there is some uncertainty on its durability. This paper aims at studying the effect of seawater attack on SFRC specimens. The strength characteristics and cracking pattern at failure were studied and included in the paper. A chemical composition analysis under Scanning Electron Microscope (SEM) was also carried out and the results explained. Both plain and SFRC specimens exhibited significant increase in the weight percentage of sodium and chlorine after a month of seawater attack. Within the study period of four months all the ninety specimens exhibited significant reduction in the weight percentage of calcium that has the potential to make the concrete and brittle.

1 INTRODUCTION

Concrete matrix reinforced with steel fibres, known as steel fibre reinforced concrete (SFRC), has several desirable engineering properties. In the SFRC steel fibre is distributed more uniformly throughout the concrete mix thus minimising the crack width due to either loading or shrinkage. The crack arresting properties of SFRC has encouraged the researchers to use it as a grout material in masonry construction (Dhanasekar & Steedman 1998). The strength and properties of SFRC as reported in the literature.(Romualdi & Batson 1963, Shah & Rangan 1971, Edginton & Hannant 1972) shows the benefit of the addition of steel fibre (as coarse aggregate replacement) that results in the increase of the compressive and tensile strengths. Volume fraction of steel fibre has substantial influence on the properties of SFRC (Shah 1991). This paper describes an ongoing research at the Central Queensland University (CQU) where 3% steel fibre is used as an optimal value from the strength and workability points of view (Bako 1997). As the cover to steel fibres in SFRC is not well defined, there is no clear understanding of its durability. Durability of SFRC is only sparsely discussed in the literature. The available literature is again scattered due to the interdisciplinary nature of the durability research. This range from environmental conditions such as industrial atmosphere, sea water, and sewage water contact to pure acidic attack such as sulphur and chlorine. This paper focuses on the sea water attack on SFRC. Similar work is also reported by other researchers (Mangat & Gurusamy 1987a, 1987b, 1987c, 1987d, Mangat & Gurusamy 1988) using laboratory and beach situations.

2 EXPERIMENTAL INVESTIGATION

A total of 90 cylindrical specimens (45 plain concrete and 45 SFRC) were constructed and tested at the materials engineering laboratory of CQU. The steel fibre used in this experimental study was FS186EE in which 'FS' and 'EE' stands for fibre steel and enlarged end respectively.

The nominal dimensions of the fibres are 18 mm length, 0.6 mm width and 0.3 mm thickness with an average tensile strength of 850MPa (BHP).

2.1 *Preparation of Specimens*

The mix used for the manufacture of steel fibre reinforced and plain concrete specimens is given in Table 1. The amount of fibre added in the fibre reinforced specimens was 3% of the total weight of aggregate. The W/C ratio and the A/C ratio have remained constant for all the 90 specimens.

Table 1. Mix Proportion .

Constituent Material	Plain Concrete Mix (kg/m^3)	SFRC Mix (kg/m^3)
Cement	310.0	310.0
Fly ash	165.0	165.0
Water	172.0	172.0
Fine sand	35.0	33.9
Coarse sand	515.0	499.5
10 mm gravel	310.0	300.7
20 mm gravel	895.0	868.1
Steel fibre	0.0	52.9

Needle compaction method was used in the manufacture of the specimens. The specimens were compacted carefully to ensure, that all the steel fibres were covered with concrete. The slump of the mix was in the range of 50 mm - 100 mm. The specimens were cured in a water tank for 28 days. After 28 days of curing the specimens were taken out of the tank and their density surface-dry condition was monitored. Strength tests on five plain concrete and five SFRC specimens were carried out. The remaining 40 plain and 40 SFRC specimens were placed in two separate polypropylene tanks filled with seawater collected from Yeppoon, near Rockhampton. Testing of plain concrete and steel fibre reinforced concrete (SFRC) specimens immersed in seawater were carried out at regular interval of two weeks for a period of 4 months. The seawater was replaced at a regular interval of 7 days.

2.2 *Compressive Strength Testing*

Compressive strength tests were carried out using an Avery Testing Machine of 1800kN capacity. The result is presented in Figure 1.
Figure 1 shows an increase of 21% of compressive strength for both the plain and the SFRC specimens subjected to 3 month sea water exposure (after 28 days of fresh water curing). After 3 months of seawater exposure, however, the compressive strength of plain concrete specimens

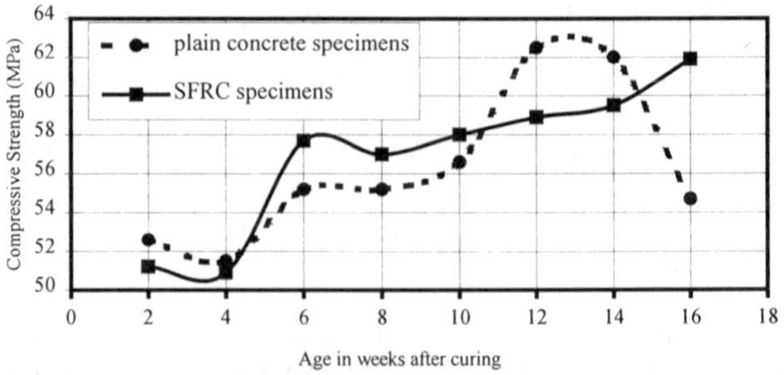

Figure 1 Effect of Seawater Exposure on Plain and SFRC Specimens

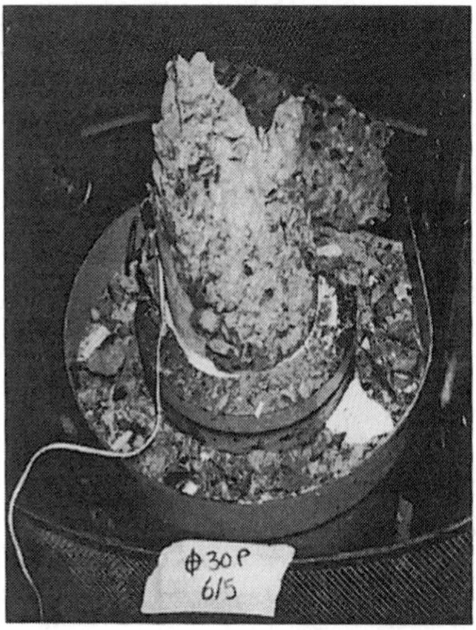

Figure 2 Failure of Plain Concrete

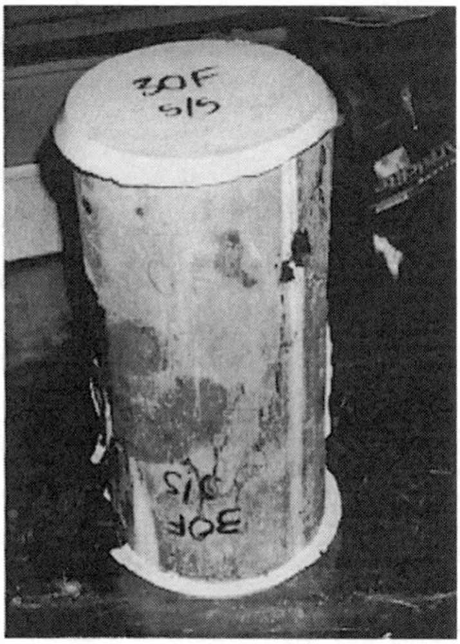

Figure 3 Failure of SFRC

has reduced more rapidly than the SFRC specimens. Similar finding was also reported by Mangat 1988 and is in good agreement with the current result.

2.3 *Failure Mode*

After 6 weeks of seawater exposure most of the plain concrete specimens failed in an explosive manner. SFRC specimens remained intact after failure. This may be due to the enlarged head of the steel fibres that resists lateral tension of the specimen to greater extent. Typical modes of failure of specimens at 10weeks of exposure to seawater is shown in Figures 2 and 3 for plain and SFRC specimens respectively.

2.4 *Weight Loss/ Gain*

The weight of the specimens was monitored for the full four months of testing. As only surface-dry weight was considered, there was no significant loss or gain in the weight noticed. As each specimen was not cured in independent tanks, there was no possiblity of collecting the loss of solid particles. The measurement of surface-dry weights, therefore, did not give meaningful result. The maximum loss or gain in weight was found to be 2%.

3 MORPHOLOGICAL AND CHEMICAL COMPOSITION ANALYSIS

Scanning electron microscope (SEM), JEOL 53001V, equipped with an energy dispersive X-ray analyser was used to study the morphology and the chemical composition of samples. For this study Back-scattered Electron Image (BEI) Detector was used. The BEI provides X-ray Dot Maps, in which elements with atomic weight higher than the atomic weight of sodium (Na) can only be viewed. In the X-ray dot maps the area that has heavier elements appear bright and the areas of the sample constituted of low atomic number elements appear dark. This technique has assisted in the determination of the distribution of particular elements across the sample surface.

The dimension of the specimens examined under SEM was approximately 25 mm in diameter with a thickness of about 5 mm. They were sampled from thin slices of concrete slabs of thickness about 5 mm that were cut from the plain and SFRC samples at a minimum distance of

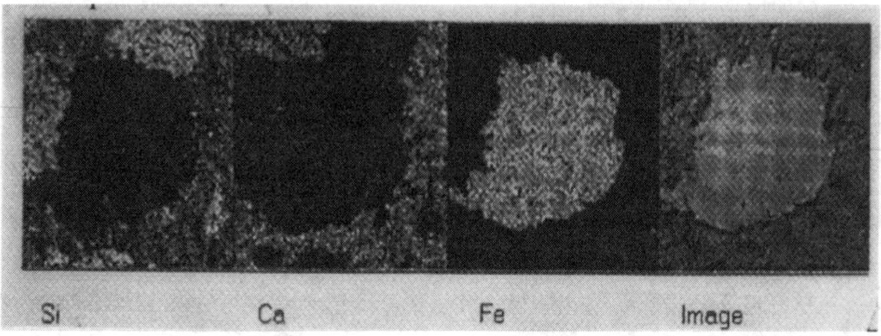

Figure 4 Electron Image and the X-ray Dot Maps of SFRC Specimen

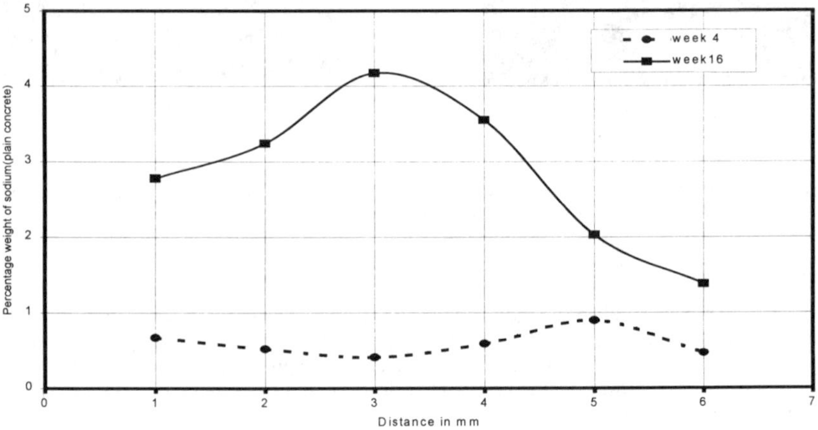

Figure 5 Variation of Sodium (Na) Content within SFRC Specimen

10 mm from the top surface across the cross section of the specimens.

Examination of the specimens under SEM was carried for each square mm for a distance of six mm starting from the edge towards the centre of the specimens. The electron image and the X-ray dot maps of a fibre reinforced specimen exposed to seawater for a period of four weeks after curing is shown in Figure 4. The magnification factor of the electron image was 500.

The percentage weight of potassium, calcium, chlorine, magnesium, silicon, aluminium, sulphur, iron and sodium present in each square mm of the specimens were observed. The observed values were plotted on a graph with distance in mm along the X-axis and the percentage of element present along the Y-axis. Typical graphs that show the presence of sodium and chlorine in the SFRC specimen is shown in Figures 5 and 6 respectively.

The graph in Figs. 5 and 6 shows the dispersion of sodium and chlorine within the body of the SFRC after four and sixteen weeks of exposure to seawater. From the graphs it can be observed that during a period of 4 weeks exposure there is only a marginal variation (almost uniform spread) of sodium along the distance examined. The uniform value means that the specimen is not affected by seawater (predominantly sodium chloride) and also it proves that the concrete mix has been properly prepared with the elements spread uniformly throughout the body of the specimen. After 16 weeks of exposure, remarkable increase in the percentage of sodium element has occurred. This shows that the chemical properties has changed due to the sea water diffusion. Both the plain and the fibre samples showed the similar trend.

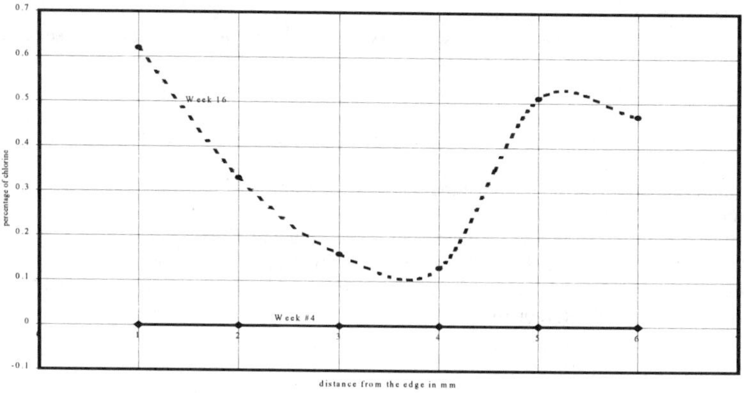

Figure 6 Variation Chlorine (Cl) within SFRC Specimen

Table 2. The percentage of elements within 6mm from the edge of the specimen

Elements	Percentage weight of Elements 4 Weeks of Seawater Attack		Percentage weight of Elements 16 Weeks of Seawater Attack	
	Plain Concrete	SFRC	Plain Concrete	SFRC
Sodium (Na)	0.59	0.74	2.9	3.4
Iron (Fe)	2.9	4	2.4	3
Aluminium (Al)	5.3	5.7	6.3	6.5
Silicon (Si)	27.8	25.6	34.4	35.3
Chlorine (Cl)	0	0	0.35	0.33
Calcium (Ca)	26.0	24.9	9.3	11.7
Potassium (K)	0.9	6	1.2	1.4

Even though the line for week 16 in Figs. 5 and 6 shows significant fluctuation (along the length measured), no firm conclusion on the trend could (and should) be derived because each point on the line is just a single measurement and is not the average of several measurements. The best we could conclude is the average increase in chlorine and sodium levels compared to 4 and 16 weeks of seawater attack on SFRC.

Energy dispersive spectra (EDS) provided the percentage of weight of different elements above the atomic weight of sodium. The elements aluminium, potassium, silicon and magnesium showed similar trend, while iron and calcium showed the opposite trend of reduction in percentage of weight with the increase in duration. The average weight in percentage of some of the elements within the body of the specimen is given in Table 2

From Table 2 it could be seen that the major elements, silicon and calcium, has undergone significant changes in their composition within the period of 12 weeks (ie, 4 weeks to 16 weeks of sea water attack). Plain concrete lost 64% of calcium whilst SFRC lost 54%. As the measurements are not averaged, no definitive conclusion on the superiority of SFRC over plain – concrete or vice-versa could be made. However loss of significant amount of calcium will make concrete brittle and would affect its structural response.

4 SUMMARY AND CONCLUSIONS

A pilot study on the effect of seawater attack on SFRC was carried out at CQU. Ninety specimens were constructed and tested over a period of four months. Strength and Scanning Electron Microscope (SEM) examinations were carried out.

From the pilot study it is concluded that

(1) the plain concrete specimens have shown an explosive failure pattern after six weeks of sea water attack.

(2) the SFRC specimens have remained intact after failure even after sixteen weeks exposure to sea water

(3) both the plain and SFRC specimens have not been affected by the sea water attack for a period of one month exposure.

(4) the Sodium and Chlorine levels have increased significantly within the body of concrete for the duration of the testing period.

(5) the silicon content has shown some increase in both plain and SFRC concretes

(6) the calcium level has recorded significant reduction in both the plain and SFRC concretes due to four month sea water attack. The reduction in calcium is expected to affect the structural properties of concrete by making it brittle.

REFERENCES

Bako, D. R. 1997. An experimental investigation on the effects of compaction methods on the properties of steel fibre reinforced concrete. *Final year Bachelor of Engineering thesis.* Central Queensland University, Australia.

BHP Reinforcing products, *Fibre steel,* NSW, Australia.

Dhanasekar, R. & M. Steedman 1998. Performance of concrete masonry prisms filled with steel fibre rein forced grout. *5th Australasian masonry conference proceedings*: 105-114. Queensland, Australia.

Edginton, J. & D. J. Hannant 1972. Steel fibre reinforced concrete, the effects on fibre orientation of compaction by vibration. *Materials and structures, research and testing (RILEM)*, Vol 5: 41-44.

Mangat, S. & K.Gurusamy 1987a. Long term properties of steel fibre reinforced marine concrete. *Jour nal of materials and structures/materials at constructions*, Vol 20: 273-282.

Mangat, P. S. & K. Gurusamy 1987b. Chlorides diffusion in steel fibre reinforced marine concrete. *Ce ment and concrete research*, Vol 17: 385-396

Mangat, P. S. & K. Gurusam 1987c. Chloride diffusion in steel fibre reinforced concrete containing PFA. *Cement and concrete research*, Vol 17: 640-650

Mangat P. S. & K. Gurusamy 1987d. Pore fluid composition under marine exposure of steel fibre rein forced concrete. *Cement and concrete research*, Vol 17: 734-742.

Magnet P. S. & K. Gurusamy 1988. Long term properties under marine exposure of steel fibre reinforced concrete containing PFA. *Journal of materials and structures/materials at constructions*, Vol 2: 352-358

Ramualdi, J. P. & G. Batson 1963. Mechanics of crack arrest in concrete. *ASCE journal*, V89: 147-168.

Shah, S. P. 1991. Do fibres increase the tensile strength of cement based matrixes?. *ACI Materials Jour nal*, Vol 88: 595-602.

Shah, S. P. & V. B. Rangan 1971. Fibre reinforced concrete properties. *ACI journal proceedings*. Vol 68: 126-135.

Mechanics of Structures and Materials, Bradford, Bridge & Foster (eds)
© 1999 Balkema, Rotterdam, ISBN 90 5809 107 4

Composite materials? The 3000MPa question

P.E.Simpson & G.M.Van Erp
Faculty of Engineering and Surveying, University of Southern Queensland, Toowoomba, Qld, Australia

ABSTRACT: Fibre reinforced composite materials are very different in comparison to conventional structural materials. The materials are brittle, have extremely high tensile strengths and relatively low moduli, which is in complete contrast to structural steel. Consequently, it is important for engineers to "think composites" when contemplating their use in structural engineering. This paper discusses many pertinent issues that have surfaced when considering composite materials for structures such as bridges within Australia.

1 INTRODUCTION

When compared to structural steel, carbon fibre reinforced polymers (CFRP) have disparate properties that seem to be conflicting. The ultimate tensile strength (UTS) of CFRP is in excess of 1500MPa while the modulus is relatively low at around 150GPa. The properties of the material are such that very efficient highly optimised structures must be utilised in order to use the material properly ie. the structures must be designed to utilise the very high tensile capacity while still providing the necessary serviceability deflection through section properties.

Table 1. A comparison of material properties.

Material	Cost $/m^3$	UTS MPa	E GPa	E_n	Depth	Max beam Stress	Increase in Volume	Cost $AUD
Steel	15700	250(y)	200	1	1 x	1	1x	15 700
CFRP	100000	1500	150	0.75	1.1 x	0.825	1.1x	110 000
GFRP	25000	800	30	0.15	1.9 x	0.285	1.9x	47 500

Note: Costs vary depending upon country, quantity etc and are used to demonstrate relative values only. Carbon refers to "high strength" (common variety). GFRP stands for Glass Fibre Reinforced Polymer.

For a deflection-based design, which is often the limiting case in civil structures, FRP materials present some interesting challenges. Choosing to do a direct material replacement of CFRP over steel would only result in an uneconomical structure. This can be easily demonstrated by considering a beam with a solid rectangular section, which is bound primarily by deflection limits. The different moduli can by normalised against that of steel since it has the highest value. This is presented in the E_n column of Table 1. To meet the same deflection criterion the depths of the CFRP and GFRP beams have to increase relative to the steel section (the width is held constant) see Table 1. The stresses in the extreme fibres can be determined using Hooke's law. It is interesting to note that the stress in the steel beam is higher than in the corresponding carbon beam! The volume increases by the same amount as the depth and the relative cost of the beams (last columns in Table 1). These costs make it clear why different structural materials are used in different forms. A $110 000 solid carbon fibre beam is difficult to justify when equivalent performance can be achieved with a $15700 steel beam. Similar conclusions can be reached for I-beams and box beams (Van Erp 1999a).

2 THE CORE ISSUE IN STRUCTURAL COMPOSITES

In composite aerospace applications, sandwich structures are by far the most common form of construction (Gibson 1994). In sandwich construction, very thin (relative to the core) laminates are bonded to lightweight core materials, usually in the form of a honeycomb or solid foam. Thin skins are designed to provide the bending and longitudinal stiffness usually consisting of CFRP or GFRP laminates. The core contributes very little in the way of global bending stiffness due to its low Modulus of Elasticity and consequently is often neglected during calculations. The core does however provide the shear stiffness and local structural performance to the sandwich structure. Such structures can be considered the composite equivalent of steel I-beam sections where the flanges and web are the laminates and core respectively.

In civil engineering applications, however, typical aerospace sandwich structures are often difficult to utilise when trying to meet necessary objectives. Some typical problems encountered include:

- Inadequate shear stiffness of available core materials necessary for high loads and short spans.
- Inability to provide sufficient bearing strength at loading and support points.
- Relatively poor local performance under the application of localised loads (wheels for eg.).
- Difficult to provide anchorage points for fixing.
- Expensive, with the focus placed on saving weight (ie. aerospace).

The main goal amongst the plethora of researchers using composites for primary applications such as bridges is how to effectively utilise the performance potential of fibre composite reinforcement. It has been demonstrated that the material can undergo significantly higher strain than structural steel and that solid carbon fibre structures are highly inefficient.

The core of the structure needs to be of significantly lower cost than the composite flanges to reduce the overall cost of the structure to that of a steel equivalent. A suitable core material should, in addition, be able to overcome the problems associated with aerospace cores. Currently, research at the University of Southern Queensland into alternative core materials centres around the use of a particulate filled resin (PFR). For a complete description, the reader is directed to papers by Van Erp and Ayers et al. in the seminar proceedings.

The PFR core is very attractive in that it is "castable" which allows for the production of both limitless cross-sectional shapes and variations along the length of a structural element. The PFR core is used to not only separate the carbon fibre flanges, but also to fully encapsulate them, thus providing full buckling restraint and protection from vandalism and construction damage (Van Erp 1999b).

3 THE MYTH OF 3000MPA

For structural steel and many other homogeneous materials, stress is an excellent parameter or descriptor to use when assessing a material or designing a structure. Fibre reinforced composite materials provide some difficulty when attempting to produce data in this form. First, the material is inhomogeneous, being composed of a matrix and fibres, which are in the form of "mats" and "fabrics". These are often wavy in profile, making thickness measurements difficult. This is compounded by the variation of fibre volume fraction often within the same structure or between manufacturing processes when using the same amounts of fibre reinforcement. Table 2 presents the variation in tensile stress obtained between a variety of specimens manufactured by different processes.

From initial inspection, it may be inferred that to get the best strength from a laminate the autoclave consolidation process should be employed in the manufacturing technique. However, in the autoclaved laminate all excess resin has been "squeezed out" in an attempt to reduce the weight, resulting in a significantly reduced thickness. Consequently, for the same volume of fibre reinforcement the autoclaved laminate is considerably thinner than the RTM processed equivalent, resulting in different failure stresses.

Table 2. Comparison of physical properties of glass fibre epoxy composites produced using autoclave consolidation and resin transfer moulding.

Property	RTM	Autoclave	Difference (%)
I.L.S.S. (Mpa)	17.73	10.45	-41
Tensile Stress (Mpa)	274	332.9	21.5
Tensile Modulus (Mpa)	13910	16360	17.6
Fibre Fraction (%)	50.4	63.9	26.8
Thickness (mm)	1.775	1.394	-21.4
Void Content (%)	1.507	1.566	3.9

(Adapted from Abraham et al, 1998.)

To provide a better understanding of this issue the stress and Modulus of Elasticity should not be compared directly, but rather load carrying capacity (stress x thickness) and tensile stiffness (E x t). These values are presented in Table 3.

Table 3. Alternative presentation of composite materials data.

Property	RTM	Autoclave	Difference (%)
Tensile Capacity per unit width (N/mm) (= tensile strength x thickness)	486.4	464.1	-4.6
Tensile Stiffness per unit width (E x t)	24690	22805	-7.6

It is apparent (see table 3) that there is very little difference between the "high tech" autoclaved laminate and the mass-produced RTM product. In fact, the RTM product displays superior load capacity probably as a result of the slight reduction in void content.

The use of stress values is often misleading and confusing and results in failure stress' of 3000MPa and higher to be quoted. In reality, these values can only be obtained for very high fibre volume fractions, which are in most cases impractical.

Conversely, strain is of paramount significance when evaluating fibre reinforced composite materials. As mentioned previously, FRP materials have very high failure strains (CFRP 0.9% and GFRP 2.5%) compared to structural steel (0.13% yield). The fact that fibre reinforced composite materials only attain their "trade mark" high strengths for high strains has major implications for the style of structure that should be designed.

4 SMART STRUCTURAL FORMS

The previous discussion highlighted the fact that composite materials should not be used in forms that mimic conventional structural elements. This is no more apparent than in some structural bridge elements in the United States that employ carbon fibre reaching a maximum serviceability stress of 80 MPa, providing a safety factor of roughly 35 (Hazen 1998).

Many composite elements, despite having large apparent reserves of tensile capacity, fail well before ultimate reinforcement rupture due to localised effect such as delamination, crushing or local buckling. The large gap that exists between serviceability capacity and ultimate capacity is a direct result of inappropriate and inefficient use of low modulus fibre reinforcement.

In an effort to obtain high strain levels in a structure while limiting the overall depth, special "composite" construction techniques should be used. For example, a T-section with the neutral axis close to the horizontal flange has similar strain levels at the bottom of the web as a rectangular section with twice the depth. For bridge structures, this implies that the decking and girders should operate as one structural component through some form of shear connection between them. When trying to produce sufficient structural depth in bridge girders, for a ten metre span bridge subjected to AUSTROADS loading consisting of the T44 truck plus dynamic effects, the girders, when made composite with the deck, can be very slender (see figure 1b). This may prompt the use of a composite truss as an alternative (see Humphreys et al. in seminar proceedings), or a hybridised beam-arch as shown in Figure 1a.

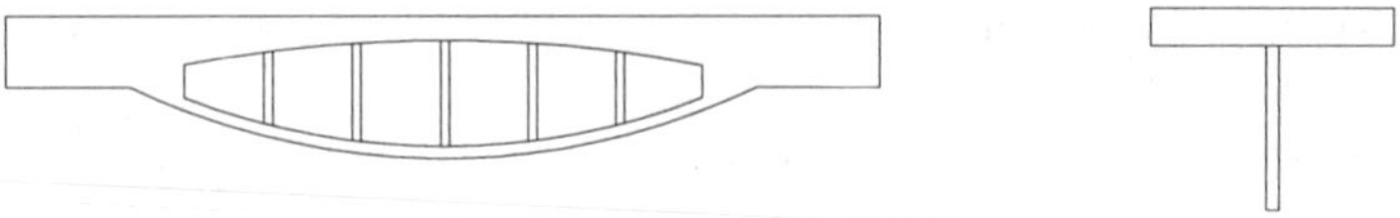

a) b)

Figure 1. a) Hybrid Beam-Arch. b) Slenderness of composite bridge girders.

The hybridised beam-arch provides the required structural depth at the point of maximum bending moment while allowing for a far-reduced depth at the abutments. The use of a "castable" core such as PFR allows for the production of a hybrid beam-arch whose cross section varies in both depth and width over its length. The tensile reinforcement, following a parabolic path, provides exceptional stiffness under the action of distributed loads while shear reinforcement is used extensively at the end regions. Such a structural element has the ability to be integrated into existing structures (provided there is sufficient clearance) since little or no modification is required at the abutments. Tests on this type of beam are currently in progress at USQ.

5 LOAD & DEFLECTION

Most reference books on fibre composite structures use laminated plate theory to predict the mechanical properties of composite beams (Gibson 1994, Powell 1994). This approach is rather complicated and normally requires the use of special finite element software. The use of PFR as a structural core and encasing material for the reinforcement removes most of the premature failure modes associated with local crushing, local buckling or general instability. Consequently, most beams fail only due to tensile failure of the reinforcement and simple beam theory seems to predict this very well. In order to verify this, a PFR beam test was conducted. The beam was loaded in four point bending and the strains in the top and bottom reinforcement were measured using strain gauges (at midspan and centre, see fig. 3). Predictions for the strain were determined using the simple method discussed in the next paragraph and summarised in Figure 2.

The moment of resistance is derived by summing the contributions offered by the composite reinforcement and the core respectively. The individual contribution offered by the fibre reinforcement may be determined theoretically by estimating the thickness of the laminate and the modulus given the strain, which is essentially a stress approach since area is required. Alternatively, load capacity per layer of the reinforcement may be used which removes the uncertainties that may result through manufacturing.

Area: $A = b.t$

Modulus of Elasticity: E

Moment of Resistance:

$M_{Rtotal} = \Sigma\, \varepsilon EA.d$

Alternatively,

$M_{Rtotal} = \Sigma\, F.d$

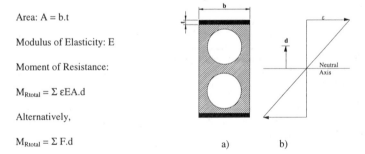

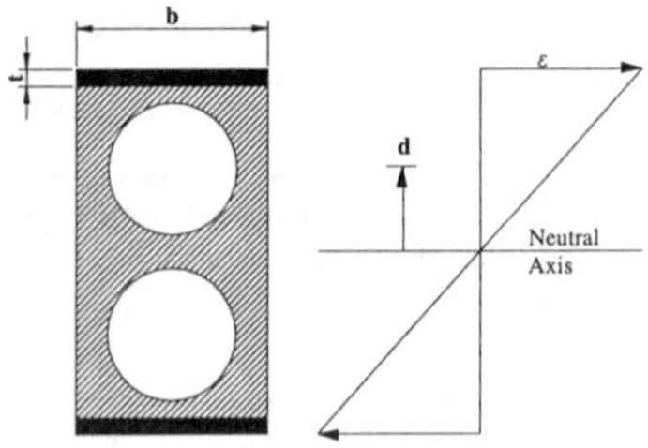

a) b)

Figure 2. a) Experimental beam for design verification. b) Assumed strain profile through section.

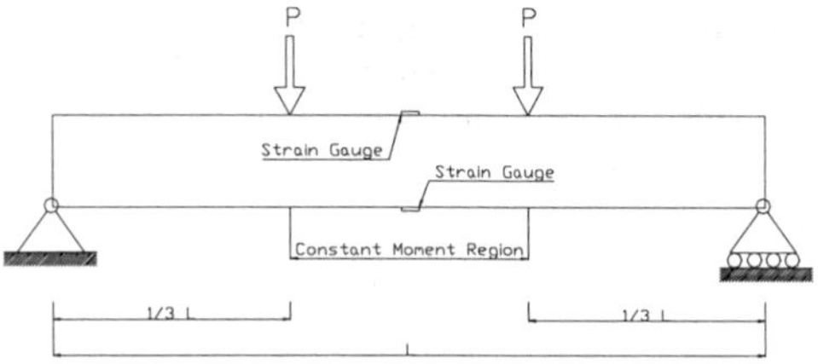

Figure 3. Four point loading used in experiment.

6 EXPERIMENTAL INVESTIGATION

The test beam and the location of strain gauges are shown in Figure 3.
For the case of four point bending and ignoring the effects of shear deformation, the following relationship between deflection and strain can be derived using simple beam theory.

$$\varepsilon = 4.683\frac{h}{L^2}\delta \qquad (1)$$

Using a simple spreadsheet, the moment deflection relationship can be calculated for any cross section. Since all beam dimensions and material properties are defined in terms of parameters, comparison between designs is quick and simple.
The experimental results from the test were promising in that it produced good correlation with the predicted results. Differences between experimental results and theoretical predictions can be mainly attributed to the sensitivity of stain gauges and extensometer under low load conditions. At this initial loading there was some apparent "settling" of the experimental apparatus. This fundamental test suggests that a simple elastic bending approach could aid in the design of PFR (bridge) girders. Shear contributes little to the overall beam deformation, due to the comparatively high shear modulus of the core, and can potentially be neglected in calculations. The PFR shear modulus is approximately 50 times higher than a typical foam core.

Table 4 presents some experimental results with the addition of tolerances placed upon

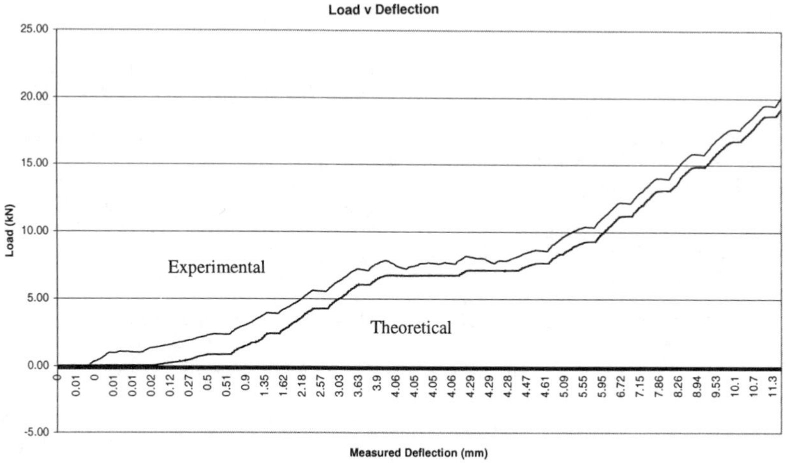

Figure 4. Experimental & theoretical load v displacement curves.

physical dimensions. The tolerances on physical depth are included due to the variation that may exist in the real structure, and are used to demonstrate its effect on the predicted results.

The variation between measured strain at the top and bottom of the test beam (see table 4) is a result of the slight shifting of the beams neutral axis due to the differing compression and tension moduli of the material.

Table 4. Strain at 11.5mm deflection and 20kN.

Depth	Predicted	Measured			Error
	Top & Bottom	Top	Bottom	Average	
+1mm	2.815E+03 με				4.8 %
Nominal	2.80E+03 με	-2.63E+03 με	2.73E+03 με	2.68E+03 με	4.2 %
-1mm	2.785E+03 με				3.78 %

7 DISCUSSION AND CONCLUSIONS

Current applications of composite materials often under utilise the potential of the fibre reinforcement. It has been demonstrated that composite structures, which mimic those of conventional materials, often result in uneconomical solutions regardless of other issues such as corrosion. The quoted tensile strengths of 3000MPa for carbon fibre laminates are unrealistic since it relies on other variables such as fibre volume fraction. Composite laminate data would be more useful when presented as load capacity per layer of reinforcement since this uncouples the design and manufacturing processes, which are critical in composite design and construction.

A particulate filled resin (PFR) core has been developed at the University of Southern Queensland and is being utilised extensively in the development of civil engineering structural elements. The resulting structures are free from premature failure mechanisms such as delamination, local crushing and general instability through buckling.

A fundamental approach has been presented providing good correlation with experimental results for the PFR composite beam. The method is robust in that variations in geometric tolerances can be incorporated without dramatic increases in error. This is essential for civil engineering structures.

REFERENCES

Abraham D. Matthews S. McIlhagger R. 1998. *A Comparison of Physical Properties of Glass Fibre Epoxy Composites Produced by Wet Lay Up with Autoclave Consolidation and Resin Transfer Moulding.* Composites Part A. 29A 795-801. Elsevier Science.

Ayers S.R. Van Erp G.M. 1999. *Development of a new Structural Core Material for Composites in Civil Engineering.* Proc. ACMSM 16. UNSW Sydney Australia.

Gibson R.F. 1994. *Principles of Composite Material Mechanics.* McGraw-Hill. New York.

Hazen J.R. 1998. *Composites for Infrastructure - A guide for Civil Engineers.* Ray Publishing. pp38.

Humphreys M.F. Van Erp G.M. Tranberg C.H. 1999. *Structural Behaviour of Monocoque Fibre Composite Trusses.* Proc. ACMSM 16. UNSW Sydney Australia.

Powell P.C. 1994. *Engineering With Fibre-Polymer Laminates.* Chapman & Hall. London.

Van Erp G.M. 1999a. *Design and Analysis of Fibre Composite Beams for Civil Engineering Applications.* Proc. The 1st ACUN International Composites Meeting. UNSW Sydney Australia. pp229-238

Van Erp G.M. 1999b. *A New Type of Carbon Fibre Composite Beam for Civil Engineering Applications.* Proc. The 1st ACUN International Composites Meeting. UNSW Sydney Australia. pp23-26

Mechanics of Structures and Materials, Bradford, Bridge & Foster (eds)
© 1999 Balkema, Rotterdam, ISBN 90 5809 107 4

Development of a small scale fibre composite semi-trailer for transport applications

R.A.Coker & G.M.Van Erp
University of Southern Queensland, Toowoomba, Qld, Australia

ABSTRACT: This paper introduces work that forms a part of ongoing PhD research into the development of a small scale fibre composite semi-trailer. This trailer utilises a new beam design, complemented by a construction method free from mechanical fasteners. The trailer was designed with weight reduction as its primary goal, with attention also given to the reduction of assembly complications. This paper discusses the general design approach that has been followed and the design, analysis and testing of the flat bed trailer which was developed as part of this research.

1 INTRODUCTION

Increasingly over the last decade, the transport industry in Australia has become extremely competitive. Many companies push their vehicles and drivers to the limits in order to increase productivity. Some drivers drive in excess of 14 hours per day, and it is not uncommon for trailers to carry loads well in excess of the legal limits. Some trailers have been weighed in at over 110 tonnes, which is over twice the legal gross vehicle mass (GVM) in this country.

Various attempts have been made to legally increase the payload capacity of semi-trailers, including the use of aluminium and a revision of 'good trailer design'. These have succeeded to varying degrees, but none of these methods have increased the payload by any significant percentage.

The University of Southern Queensland, in conjunction with Wagners Transport Pty Ltd, is investigating the use of advanced fibre composites for the construction of semi trailers. Because of their high strength and low weight, fibre composites have the potential to reduce the tare weight significantly. This paper introduces the approach adopted in this project to date, and discusses the design, analysis and testing of a small scale, fibre composite, flat bed trailer which was developed as part of this project.

2 DESIGN PHILOSOPHY

2.1 *Challenging Current Thinking*

While the basic design of a typical trailer has not been challenged for some decades, there has been a period of refinement through trial and error. However the original concept does not appear to have been questioned to any great extent. The introduction of composites to the transport industry presents an opportunity to take a fresh and unbiased look at this situation. The expert literature (Tsai 1992, Gibson 1994, Schwartz 1995) emphasises that no other material has the potential of fibre composites for improving the performance of structures. Fibre composite structures should not replicate steel equivalents, but rather utilize the considerable advantages of the new material. A new design philosophy must be developed from first principles. Due regard should be given to available methods of manufacture, since many properties can be determined to some extent by the manufacturing process.

2.2 *Applying Composites*

It has long been recognised that fibre composites possess tremendous strength, but they lack the stiffness of steel. While high modulus carbon possesses comparable stiffness properties, its high cost and brittleness remove it at present from consideration as a feasible alternative. E-glass has a stiffness of around 30 GPa, almost one seventh that of steel. So in order to achieve similar stiffness values to those of a steel structure, the structure must possess a second moment of area seven times that of a steel structure. This will never be achieved economically if composites are merely used as a direct substitute for steel in the form of pultruded I-beams and the like. A more sophisticated approach is required. The ability to form composites into irregular and customised shapes increases the number of alternatives to traditional trailer designs. Reinforcement can be placed in the desired areas, resulting in structures that are efficient and lightweight, possessing strength and stiffness properties comparable to steel structures.

2.3 *Industry Design Standards*

The greatest obstacle to the development of a new trailer lies in the lack of appropriate design standards. The design and manufacture of trailers has always been accomplished through use of little more than prior experience. The experience used to manufacture steel trailers is of little use when dealing with composites. So the question arises as to the origin of the design criterion. To wait until the appropriate standards are produced is not feasible, hence an alternative method must be found to generate acceptable parameters.

These parameters can be identified by two methods. The first involves the use of strain gauges to record the loads imposed on an existing steel trailer. These loads could then be used as guidelines in the development of the composite trailer. This method has been attempted by Henry (1995) and Willing (1995). The results presented by Willing in a paper on the development of a composite tanker, included readings from strain gauges placed on the prototype. Henry took a large amount of readings from strain gauges placed on a livestock trailer. He lists a large number of difficulties associated with this method. In both cases, readings taken from the strain gauges were influenced heavily by white noise, and the results taken from these test were disregarded. Because of these experiences, it was decided not to attempt this particular method.

Table 1. Structural Characteristics of Various Steel Flat-Bed Trailers

Trailer Make	Year of Manufacture	Chassis Rail Min. I_{xx} (mm^4)	Chassis Rail Max. I_{xx} (mm^4)	Cross Member I_{xx} (mm^4)
Haulmark	1982	6.79E+07	1.87E+08	1.46E+06
Fruehauf	1985	4.77E+07	1.81E+08	1.43E+06
Loughlin	1989	9.57E+07	4.98E+08	3.27E+06
Haulmark	1992	7.51E+07	2.75E+08	1.46E+06
Lusty Allison	1995	6.62E+07	3.42E+08	1.50E+06
Average Properties		7.05E+07	2.97E+08	1.82E+06

The second approach for parameter identification uses the properties of existing steel trailers. Although the lay out of a typical flat bed trailer has changed very little throughout the last two decades, small improvements have been made in a gradual manner. Therefore it is reasonable to assume that the current designs have reached an optimum level in terms of structural efficiency. It would be advantageous then, to examine some existing trailers to determine their structural characteristics. Since they can be considered to be structurally efficient, these characteristics will give a good indication of required structural performance. Table 1 lists a number of trailers that were measured for their structural characteristics.

As can be seen from this table, there is little variation in trailer properties, with the exception of the Loughlin trailer. There does not seem to be any definite trend in these properties with respect to age, probably due to the fact that there are a number of different manufacturers.

The property variation does highlight the need for standards in the industry. While major dimensions such as length and width have been standardised, the table shows that each manufacturer uses their own knowledge in its trailer design.

Despite their spread, these properties can be used in the development of the trailer. In order to obtain a good representation of the properties, their average was determined. Because the Fruehauf and Loughlin trailers do not show the level of refinement displayed by the other models, they were discarded from the calculation of the average properties.

3 SMALL SCALE TRAILER DEVELOPMENT

As a first step in the investigation of composites and their potential in transportation, the design, manufacture and preliminary testing of a small scale, flat bed trailer was undertaken. In order to gain maximum benefit and insight from this prototype, the trailer was to replicate the basic layout of a flat bed trailer, including the chassis configuration, suspension and hitch details.

A new beam design was produced for use in this trailer. Each beam consisted of a sandwich panel with end caps of reinforcement. The end caps were manufactured by draping the reinforcement over the ends of the panel, as can be seen in Figure 1. A 50 mm thick core with generous radii (10 mm) was chosen. The top and bottom caps were designed to carry the bending moment, and the material in the midsection was designed to carry the shear stress. The design was optimised using finite element analysis with laminate plate elements.

One of the disadvantages of composites is their sensitivity to stress concentrations. The use of mechanical fasteners is therefore a problem. To overcome this disadvantage, a modular approach was adopted. The absence of bolted joints results in less stress concentrations and therefore the trailer was designed in such a way so as to minimise the number of mechanical fasteners required in the assembly of the trailer. This was achieved by removing sections of the

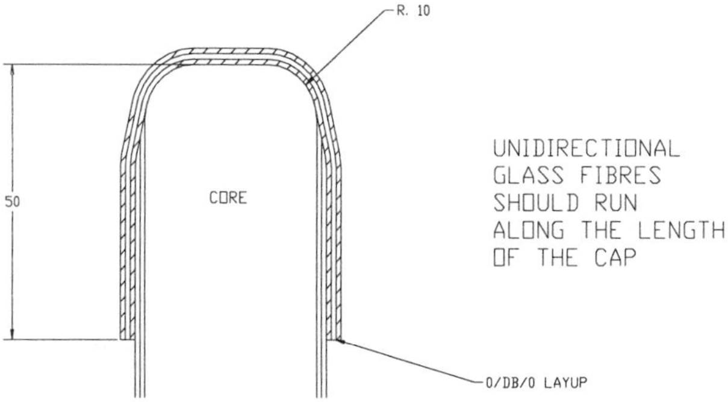

Figure 1. End cap lay-up

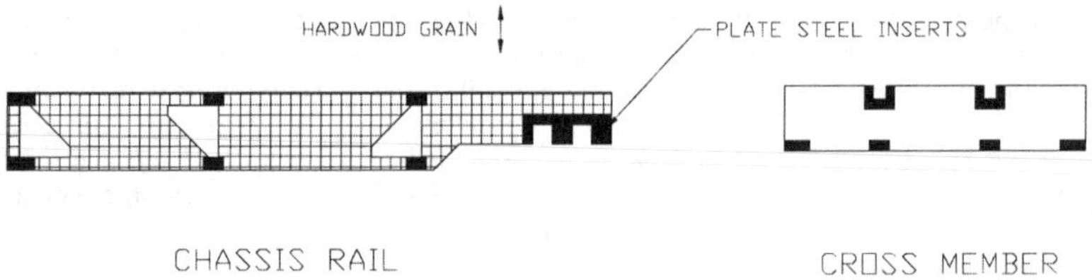

Figure 2. Hard point placement

web from the chassis rail to accommodate the cross members. Notches were cut into the cross members to allow them to be located and partially fixed to the chassis rails. To assemble the frame, the cross members were slid through the gap in the chassis rails in a diagonal manner. The member was then rotated to an upright position and locked into place. The remaining gaps in the rails were then refilled, fixing the cross members in place.

Hardwood hard points were inserted into the core at all of the bearing points. Stainless steel hard points were introduced at the front end of the trailer around the hitch connection point. This was to aid in the transfer of the load from the chassis to the hitch.

Typically on a semi-trailer, the suspension system utilises springs or airbags that are mounted to the chassis rails. Unfortunately, due to the small size of the trailer, no suitable suspension of such a scale was available. Therefore leaf spring suspension was chosen, as is typically found on box trailers of this size. This suspension could not be mounted to the chassis rails, as they are positioned too close together, resulting in instability. Hence, the suspension was mounted at the outer most ends of the cross members using purpose built spring hangers.

A special hitch and dolly were developed to provide towing characteristics similar to that of a semi-trailer. 'Fifth wheel' hitch assemblies were not considered practical since the trailer was to be towed by a passenger vehicle. The hitch and dolly were designed to allow rotational freedoms similar to those provided by a 'fifth wheel'.

In order to provide a deck, sandwich panels (10 mm thick) were adhesively bonded to the chassis. Similar panels were added to form side panels on the trailer, and provided additional stiffness to the frame. Further panels were added to the frame to prevent racking. These sandwich panels, (35 mm thick), were laminated into place as shown in Figure 3.

After assembling of the trailer, the frame was weighed and found to have a mass of 60 kg. Preliminary testing was then undertaken. This included loading the trailer up to 125% of its design load of 800 kg. The trailer was pulled over various difficult terrains and obstacles such as speed humps. This testing proved extremely successful, as the trailer withstood all testing without any signs of excessive strain. However, the maximum load was limited by the suspension system. In order to overcome this problem, the trailer was statically loaded under labora-

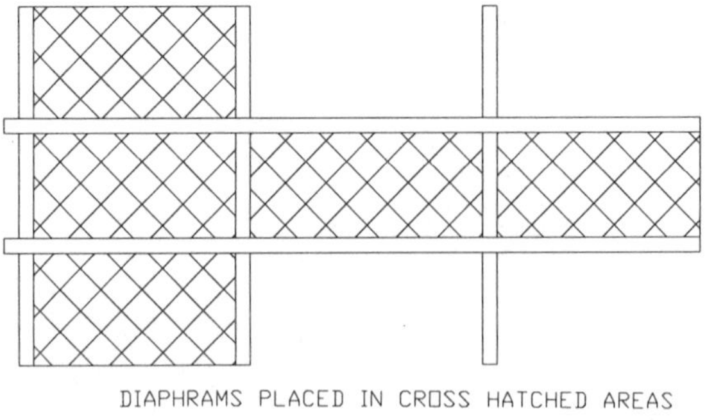

Figure 3. Diaphrams used to prevent racking

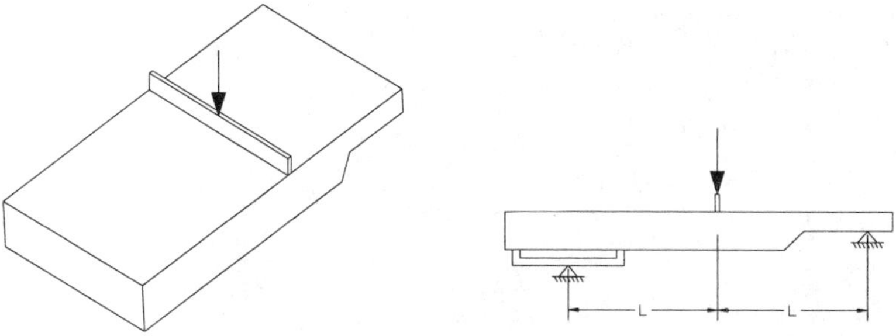

Figure 4. Laboratory loading of trailer

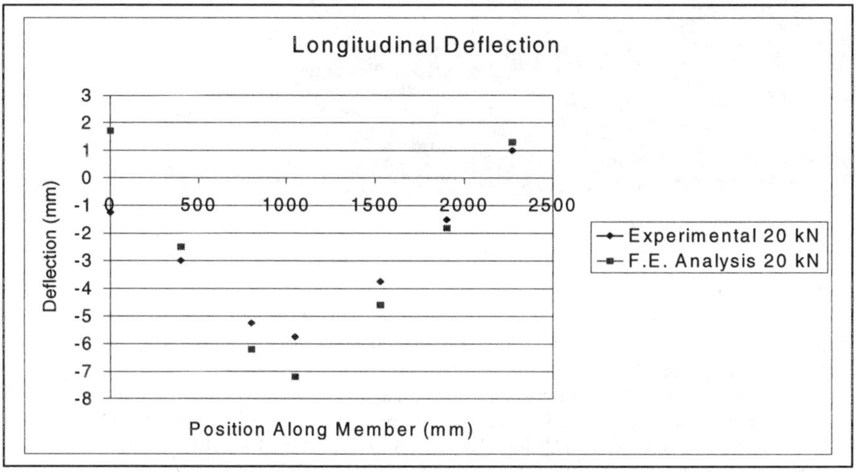

Figure 5. Deflection of chassis rail under 20 kN load

tory conditions to determine the load-deflection characteristics of the structure. The load was applied through a horizontally positioned beam placed midspan, as shown in Figure 4. A comparison of the experimental results and the finite element analysis can be seen in Figure 5. The lower deflection measurements yielded by experimentation is reasonable due to the stiffening effect of the deck and shear panels which were not included in the finite element analysis.

4 DISCUSSION AND CONCLUSIONS

The transportation industry is constantly searching for ways to increase efficiency and productivity. Composites provide an alternative that possesses a high strength-to-weight ratio, a property that is highly sought after. The paper has presented the development of a small scale, fibre composite trailer, which is an important first step in the evolution of fibre composite semi-trailers. The trailer, with a mass of 60 kg and a failure load capacity in excess of 2000 kg, demonstrates the potential of this trailer for greater applications.

A new beam configuration was introduced which allows a modular construction of the frame. This modular approach was successfully used to eliminate the stress concentrations associated with mechanical fasteners. Field and laboratory testing were performed, and the results of the finite element analysis showed good correlation with the experimental results.

Figure 6. Completed Trailer

The scale of the trailer prevented the use of suspension equivalent to a full size trailer, but leaf spring suspension was used with success. A deck and two side panels were adhesively bonded to the frame, however the flexible nature of the adhesive determined the addition of shear panels to prevent racking.

A number of issues are still outstanding. Damage tolerance and creep resistance are areas which have not yet been addressed. The fatigue performance must also be investigated, as it plays a large part in the life of a trailer. These areas will all form part of further research which will see the development of a full scale prototype in the next few years.

"Automotive design engineers seem to agree with market analysts, 89% of whom consider metal- and fiber-reinforced plastic composites the materials most likely to see increased use in automotive manufacturing." (Schwartz 1994)

REFERENCES

Gibson, R. F. 1994. *Principles of Composite Material Mechanics.* Mcgraw-Hill Series in Mechanical Engineering.

Henry, B. 1995. Design and analysis of the main structural flanges of a low tare weight livestock trailer. *A Report Submitted Towards the Award of Associate Diploma in Mechanical Engineering.* University of Southern Queensland

Schwartz, M. M. 1994. *Joining of composite matrix materials.* Materials Park, Ohio: ASM International.

Tsai, S. W. 1992. Theory of Composites Design. *Think Composites.* Dayton, Ohio

Willing, J. 1995. Development of an aramid fiber reinforced road tanker with increased loading capacity. SAE paper 960173. *SAE Transactions.* Society of Automotive Engineers.

Dynamic analyses of structures

Mechanics of Structures and Materials, Bradford, Bridge & Foster (eds)
© 1999 Balkema, Rotterdam, ISBN 90 5809 107 4

Seismic response analysis of three dimensional asymmetric buildings

Y. Arfiadi & M.N.S. Hadi
Department of Civil, Mining and Environmental Engineering, University of Wollongong, N.S.W., Australia

ABSTRACT: This paper discusses seismic response analysis of asymmetric structures with centre of mass uncertainty. In order to account for an accidental torsion, the eccentricity of the centre of mass is varied and dynamic analysis is then conducted. In this case it is not necessary to calculate the centre of rigidity which is usually done in static analysis. In the analysis, the building is modelled as a structure composing of three-dimensional elemental elements connected with a rigid floor as a diaphragm. The responses of earthquake resisting elements at certain locations that represent elements nearby and far from the centre of mass location are then observed. The results may be used for the estimation of the increase in the internal forces due to the additional torsion.

1 INTRODUCTION

In practical design usually a building is analysed by extracting the structure into several plane frame structures then a static analysis is performed to such individual frames. The resulting seismic lateral force is then corrected by accounting for the additional torsion due to the eccentricity between the centre of mass (CM) and centre of rigidity (CR). Another approach is by performing three-dimensional static analysis with the accidental torsion due to the additional eccentricity between CM and CR is also taken into account. The approaches above require the computation of CM and CR of the buildings. Nevertheless, it has been recognised that the centre of rigidity or centre of stiffness of the building is not clearly defined and such a centre is not uniquely present in the building or it is present in the case of a special class of buildings (Hejal & Chopra 1987). In that case the analysis of additional torsional moment in the building has been devoted to the complexity of finding the centre of rigidity of the buildings.

In this paper, instead of computing the additional eccentricity between CM and CR, the dynamic analysis is performed by varying the location of CM to cover the additional eccentricity between those two references. In the dynamic analysis conducted in this paper, it is assumed that the building posses rigid floors such that each floor has three degrees of freedom: two displacements in the orthogonal axes and rotation with respect to the vertical axis. The correlation between the variation of CM and the internal force of the earthquake resisting elements is then observed to estimate the effect of the additional torsion on the internal forces.

2 BUILDING MODEL AND ANALYSIS

The building is modelled as a structure containing members connected with a rigid floor. Each floor has three degrees of freedom consisting of two displacements in two orthogonal axes and rotation about a vertical axis. The mass and rotational mass of the building are lumped at the floor levels. In the building model, all necessary members are included in the analysis as three-

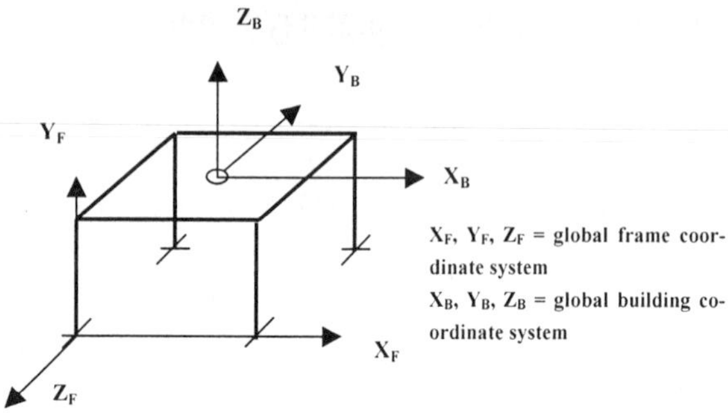

Figure 1. Three dimensional model, with global frame coordinate and global building coordinate systems

dimensional space frame elements. The computation of the stiffness matrix and the equations of motion are discussed in Arfiadi & Hadi 1999 and is adopted here.

The stiffness matrix is formed by first transforming elemental stiffness matrix into a global frame coordinate system where this reference can be located anywhere in the system as shown in Figure 1. The element stiffness matrices are then summed up for all members. Note that in assembling the stiffness matrix, the order of degrees of freedom is taken such that degrees of freedom that do not relate to the rigid floor displacement are grouped in the upper part of the vector. After assembly, the stiffness matrix of the structure (in terms of global frame coordinate systems) can be written as

$$\begin{bmatrix} \mathbf{K}_{CC} & \mathbf{K}_{CR} \\ \mathbf{K}_{RC} & \mathbf{K}_{RR} \end{bmatrix} \begin{Bmatrix} \Delta_{FC} \\ \Delta_{FR} \end{Bmatrix} = \begin{Bmatrix} \mathbf{R}_{FC} \\ \mathbf{R}_{FR} \end{Bmatrix} \tag{1}$$

in which Δ_{FC} contains joint displacements that do not relate to the rigid floor displacements, $\Delta_{\mathbf{FR}}$ contains joint displacements that relate to the rigid floor displacements. $\mathbf{K}_{CC}$, $\mathbf{K}_{CR}$, $\mathbf{K}_{RC}$ and $\mathbf{K}_{RR}$ are submatrices as elements of the stiffness matrix in terms of global frame coordinate systems, respecyively; $\mathbf{R}_{FC}$ and $\mathbf{R}_{FR}$ are load vector corresponding to Δ_{FC} and Δ_{FR}, respectively. The joint displacements that do not relate to the rigid floor displacements are then condensed out according to the static condensation technique such that we found the system equation

$$\mathbf{K}^*_{RR}\Delta_{FR} = \mathbf{R}^*_{FR} \tag{2}$$

where

$$\mathbf{K}^*_{RR} = \mathbf{K}_{RR} - \mathbf{K}_{RC}\mathbf{K}^{-1}_{CC}\mathbf{K}_{CR}\,,\ \mathbf{R}^*_{FR} = \mathbf{R}_{FR} - \mathbf{K}_{RC}\mathbf{K}^{-1}_{CC}\mathbf{R}_{FC} \tag{3-4}$$

in which the first sub-matrix equation of equation (1) of the form

$$\Delta_{FC} = \mathbf{K}^{-1}_{CC}\left(\mathbf{R}_{FC} - \mathbf{K}_{CR}\Delta_{FR}\right) \tag{5}$$

has been used to obtain equation (2).

Joint displacement Δ_{FR} in equation (2) can be transformed into floor degrees of freedom $\mathbf{X}_{3D}$ in which floor degrees of freedom are measured at the global building coordinate systems. Since dynamic force due to earthquake loading is applied at the centre of mass, the global building coordinate system may be located at the centre of mass of the building. The transformation of joint displacements to the floor degrees of freedom then can be performed as

$$\Delta_{FR} = \mathbf{a}\ \ \mathbf{X}_{3D} \tag{6}$$

in which

$$\Delta_{FR} = \begin{Bmatrix} \Delta_{FR,1} \\ \Delta_{FR,2} \\ \vdots \\ \Delta_{FR,NF} \end{Bmatrix}_{3NTF\times 1}, \quad \mathbf{a} = \begin{bmatrix} \mathbf{a}_1 & \mathbf{0} & \cdots & \mathbf{0} \\ \mathbf{0} & \mathbf{a}_2 & \cdots & \mathbf{0} \\ \vdots & \vdots & \ddots & \vdots \\ \mathbf{0} & \mathbf{0} & \cdots & \mathbf{a}_{NF} \end{bmatrix}_{3NTF\times 3NF}, \quad \mathbf{X}_{3D} = \begin{Bmatrix} \mathbf{X}_{3D,1} \\ \mathbf{X}_{3D,2} \\ \vdots \\ \mathbf{X}_{3D,NF} \end{Bmatrix}_{3NF\times 1} \tag{7a-c}$$

where $\Delta_{FR,i}$ = joint displacements at ith floor, $\mathbf{a}_i$ = transformation at floor i, $\mathbf{X}_{3D,i}$ = floor degrees of freedom at level i, NF = total number of floors and NTF = total number of joints. The sub-matrices in eqs. (7) of particular floor-n have the form

$$\Delta_{FR,n} = \begin{Bmatrix} \Delta^1_{FR,n} \\ \Delta^2_{FR,n} \\ \vdots \\ \Delta^{NJn}_{FR,n} \end{Bmatrix}_{3NJn\times 1}, \quad \mathbf{a}_n = \begin{bmatrix} \mathbf{a}^1_n \\ \mathbf{a}^2_n \\ \vdots \\ \mathbf{a}^{NJn}_n \end{bmatrix}_{3NJn\times 3}, \quad \mathbf{X}_{3D,n} = \begin{Bmatrix} r_{x,n} \\ r_{y,n} \\ r_{\theta,n} \end{Bmatrix}_{3\times 1} \tag{8a-c}$$

in which $\Delta^j_{FR,n}$ = jth joint displacement at floor-n, $\mathbf{a}^j_n$ = matrix transformation for joint-j at floor-n, $r_{x,n}$ = nth floor displacement at x direction, $r_{y,n}$ = nth floor displacement at y direction, $r_{\theta,n}$ = nth floor rotation and NJn = total number of joints at floor-n. Joint displacement $\Delta^j_{FR,n}$ has the form $\Delta^j_{FR,n} = \begin{bmatrix} \Delta^j_{FR1,n} & \Delta^j_{FR2,n} & \Delta^j_{FR3,n} \end{bmatrix}^T$ where it contains displacements in the x and y direction of the global building coordinates and rotation with respect to the vertical axis. With the assumption that the X-global frame coordinate and the X-global building coordinate system axes are in parallel, the relation between joint displacements at a particular joint j after condensation and floor degrees of freedom at a particular level n has the form

$$\begin{Bmatrix} \Delta^j_{FR1,n} \\ \Delta^j_{FR2,n} \\ \Delta^j_{FR3,n} \end{Bmatrix} = \begin{bmatrix} 1 & 0 & d_x \\ 0 & -1 & d_z \\ 0 & 0 & 1 \end{bmatrix} \begin{Bmatrix} r_{x,n} \\ r_{y,n} \\ r_{\theta,n} \end{Bmatrix} \tag{9}$$

where d_x = distance from the origin of the global building coordinates to the line of action of x-displacement of the joint (Δ_{FR1}) and d_z = distance from the global building coordinates to the line of action of z-displacement of the joint (Δ_{FR2}). The values of d_x and d_z are taken positive if the building rotation causes positive joint displacements. In case the above axes are not in parallel the transformation matrix presented in Arfiadi & Hadi 1999 can be employed.

Utilising equation (9) the static equilibrium can be found as

$$\mathbf{K}_{3D}\,\mathbf{X}_{3D} = \mathbf{R}_{3D} \tag{10}$$

where

$$\mathbf{K}_{3D} = \mathbf{a}^T\mathbf{K}^*_{RR}\,\mathbf{a}, \quad \mathbf{R}_{3D} = \mathbf{a}^T\mathbf{R}^*_{FR} \tag{11-12}$$

Note that the size of $\mathbf{K}_{3D}$ in equation (10) is $3NF$ x $3NF$, where NF = number of floors, and $\mathbf{X}_{3D}$ is the degree of freedom associated with the floor displacements.

The equations of motion can be written as

$$\mathbf{M}_{3D}\ddot{\mathbf{X}}_{3D} + \mathbf{C}_{3D}\dot{\mathbf{X}}_{3D} + \mathbf{K}_{3D}\mathbf{X}_{3D} = -\mathbf{M}_{3D}\mathbf{e}_o\ddot{x}_g \tag{13}$$

in which $\mathbf{M}_{3D}$ = mass matrix, $\mathbf{C}_{3D}$ = damping matrix, x_g = ground displacement, dot (.) is the derivative with respect to time and vector $\mathbf{e}_o$ has a form

$$\mathbf{e}_o = \begin{bmatrix} \mathbf{e}^T_{o,1} & \mathbf{e}^T_{o,2} & \cdots & \mathbf{e}^T_{o,n} \end{bmatrix}^T \quad \mathbf{e}_{o,i} = \begin{bmatrix} \cos\theta & \sin\theta & 0 \end{bmatrix}^T, \; i = 1, 2, \cdots, n \tag{14a-b}$$

in which θ = the direction angle of earthquake measured from the X-global building coordinate systems.

3 UNCERTAINTY OF THE CENTRE OF MASS

In most building codes, the static analysis requires that the lateral earthquake force be applied at a certain distance from the centre of rigidity. This distance is known as the design eccentricity, which should be taken as

$$e_{dj} = \alpha e_{sj} + \beta b_j, \quad e_{dj} = \delta e_{sj} - \beta b_j \tag{15a-b}$$

where e_{sj} = distance between the floor centre of mass (CM) and centre of rigidity (CR), b_j = floor-plan dimension of the building perpendicular to the direction of earthquake, α, δ and β = specified coefficients. Though the formula is simple the implementation of those formulas in practice is rather cumbersome. In addition, the location of the centre of rigidity is usually not unique and dependent on the external force (Hejal & Chopra 1987). Therefore, only a special class of buildings actually may utilise the design formula to account for such a design eccentricity. Although it is also possible to carry out the analysis that takes into account the design eccentricity without knowing the location of the centre of rigidity (Goel & Chopra 1993), the approach might be restricted to the special class of building in which the static analysis can be performed. For general buildings, the dynamic analysis offers alternative analysis and therefore is adopted here.

The building under consideration is a single storey building having asymmetrical plan as shown in Figure 2. The structural properties are as follows: $A_b = 0.204$ m^2, $I_{yb} = 5.074 \times 10^{-3}$ m^4, $I_{zb} = 4.623 \times 10^{-3}$ m^4, $J_{xb} = 2.99 \times 10^{-3}$ m^4, $A_c = 0.09$ m^2, $I_{yc} = 6.75 \times 10^{-4}$ m^4, $I_{zc} = 6.75 \times 10^{-4}$ m^4 and $J_{xc} = 1.141 \times 10^{-3}$ m^4, in which subscript b and c denote beam and column, respectively; A, I_y, I_z and J_x stand for area, second moment of area about the y-member axis, second moment of area about the z-member axis and torsional rigidity, respectively. The height of the column is 4 m.

The natural frequencies of the structure are 17.334, 17.372 and 25.063 rad/s. The damping matrix is assumed to be proportional to the stiffness matrix with the first modal damping ratio of 1.5 %. The CM of the building has the coordinate (x_{Fm}, z_{Fm}) = (4.17 m, 4.17 m) with respect to the global frame coordinate systems. To see the effect of the variation of the centre of mass to the responses of the structures the computed CM is assumed to have a variation such that the value of x_m varies with the variation of 0.7, 0.8, 0.9, 1.1, 1.2 and 1.3 of the original position. The structure is then subjected to uni- and bi-directional El Centro 1940 NS excitation. The responses of the structure are then computed and the results are normalised to the results of the original structure without having variation of the CM. In figures 3-5 M_{lx} and M_{ly} denote moment about member local axis-l due to earthquake in the x and y directions, respectively and M_{lbi} stands for moment about the local axis-l due to the bi-directional earthquake.

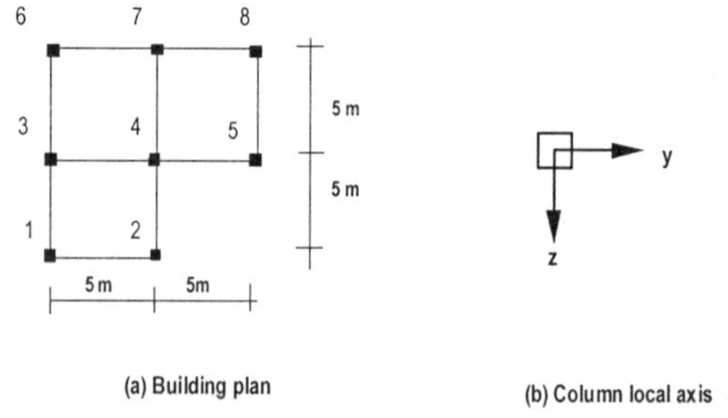

Figure 2. Plan of the building.

Because we vary the location of x_m ranging from 0.7 to 1.3 of the originally computed x_m, the earthquake in the y-direction is the main concern. In this case the column moment M_{yy} is the significant value to be observed. From the results in figures 3-5, it can be seen that the variation in peak moment M_{yy} for the column near to the CM (column 4) is smaller than the variation of the peak moment M_{yy} for the column farther from CM (columns 1 and 8). From figure 3, the maximum amplification for column 4 is about 1.11 whereas from figures 4 and 5, maximum amplifications for columns 1 and 8 are 1.29 and 1.52, respectively. Note that the variation of x_m has also increased the amplification of the column moment in the orthogonal axis (M_{zy}) where the amplification can be as high as 7.41. Nevertheless, the nominal value of moment M_{zy} due to y-earthquake is usually small. As expected, the y-direction excitation has a little effect to the amplification of M_{zx} due to the variation of x_m. But, it has a significant effect to the amplification of M_{yz}. As can be seen from figures 3-5, the effect of the bi-directional earthquake reflects the influence of each uni-directional earthquake.

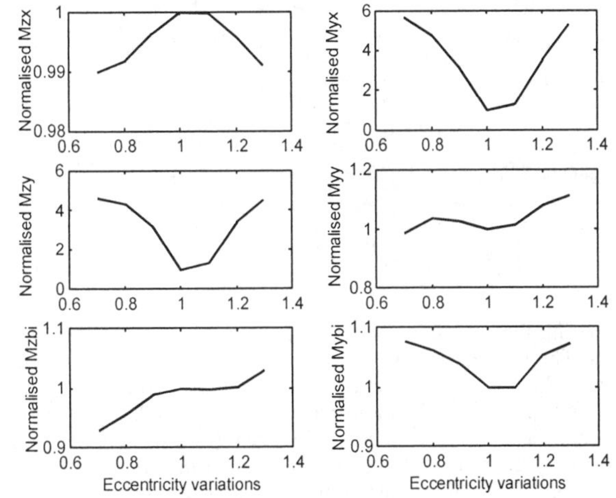

Figure 3. Normalised peak moment, column 4 due to CM uncertainty

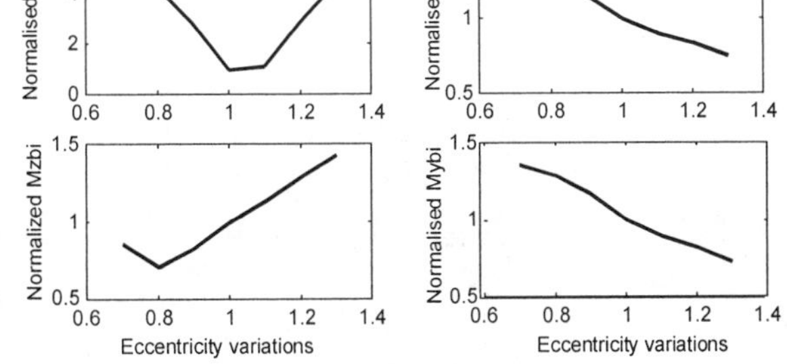

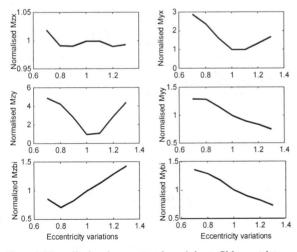

Figure 4. Normalised peak moment, column 1 due to CM uncertainty

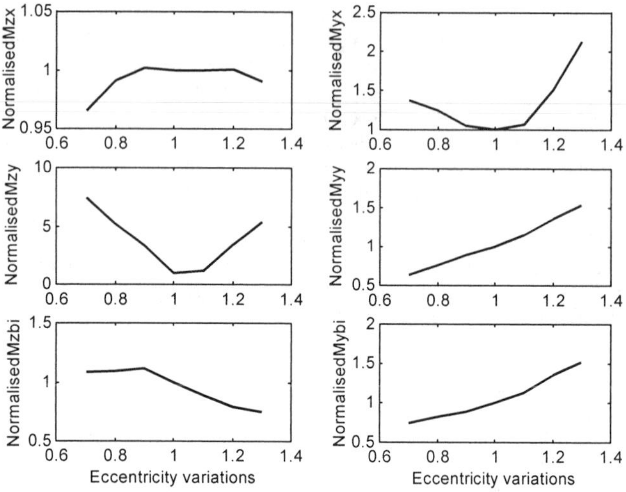

Figure 5. Normalised peak moment, column 8 due to CM uncertainty

4 CONCLUSSION

Seismic response analysis considering the uncertainty in the location of CM has been conducted utilising dynamic analysis. The moment magnification of the earthquake resisting elements has been observed as regard to those CM variations. From the numerical results to the building subject to El Centro 1940 NS excitation with the variation of CM from 0.7 – 1.3 of the original x_m position, the moment amplification of the associated moment (in which the direction and the variation of CM are governed) is about 1.52. The resulting amplification for the moment in another axis can be as high as 7.41. If this particular moment is large in magnitude, this has to be considered in the design process.

REFERENCES

Arfiadi, Y. & Hadi, M. N. S. 1999. Passive and active control of three dimensional buildings. *Earthquake Engrg. And Struct. Dyn.* (submitted).

Goel, R. K. & Chopra, A. K. 1993. Seismic code analysis of buildings without locating centers of rigidity. *J. of Struct. Engrg.*, ASCE, 119(10), 3039-3055.

Hejal, R. & Chopra, A. K. 1987. *Earthquake response of torsionally coupled buildings. Report no. UCB/EERC-87/20.* Berkeley, Calif :Earthquake Engrg. Res. Cr., Univ. of California.

Mechanics of Structures and Materials, Bradford, Bridge & Foster (eds)
© 1999 Balkema, Rotterdam, ISBN 90 5809 107 4

Seismic response of steel frames containing hierarchical friction-dissipating joints

J.W. Butterworth
Department of Civil and Resource Engineering, University of Auckland, New Zealand

ABSTRACT: Friction dissipating joints with slotted bolt holes have been used in concentrically braced frames (linear sliding) and more recently, in moment resisting frames (rotational sliding). Such joints have the ability to provide many cycles of ductile energy dissipation with little or no primary structural damage and permit the decoupling of the strength and stiffness of connected members. Suggestions are offered on the use of linear sliding joints in K-braced frames where they could lead to cheaper, stiffer structures with high levels of ductility. Rotating sliding bolted joints extend the benefits of damage-free energy dissipation to moment-resisting frames. The decoupling of beam stiffness and end moment strength avoids over-sizing columns to deal with beam over-strength moments. The performance of a rotational slotted joint having a hierarchy of two distinct moment levels at which limited rotational slip can occur is discussed. The basic characteristics of the joint are described and some observations made on the seismic response of some sample frames to seismic ground motion.

1 INTRODUCTION

The design of structures for earthquake resistance typically relies on the provision of adequate ductility and a secure path for gravity loads within a proven, stable structural form. Conventional construction provides ductility by means of hysteretic energy dissipation resulting from inelastic action within members and joints. Such inelastic action causes damage and if continued long enough the cumulative effect may exhaust the ductile capacity leading to fracture. Although this approach has been very successful at preventing collapse, the cost of damage repair can be very high.

Currently, the most widely used system for providing earthquake resistance without significant accompanying damage is that of seismic isolation (base isolation). However, not all structures are suited to this form of protection, for example slender or relatively flexible buildings, and there is an associated cost. Another alternative that is beginning to receive more attention involves the use of passive energy dissipating elements within the structure. There are a number of different types of these devices, utilising viscous, visco-elastic or coulomb friction effects (Aiken et al, 1993). The viscous and visco-elastic devices have no activation threshold and are able to generate damping at very low displacements, but tend to be relatively complex and expensive. The devices considered in more detail in this paper are of the friction-dissipating type, in particular those that employ components bolted together with slotted bolt holes allowing relative sliding movement.

2 SLIDING BOLTED JOINTS

A sliding bolted joint (SBJ) that is to provide satisfactory performance as an energy-dissipating element in a seismic resistant structure must be capable of repeated cycles of displacement without loss of strength, stability or energy dissipation ability. A suggested testing regime should include at least 50 complete cycles to the displacement corresponding to the design structure displacement ductility factor and 10 cycles to the device's maximum displacement. Important factors that influence the satisfactory performance of a sliding bolted joint include:

1. Maintenance of contact pressure between the sliding surfaces,
2. Maintenance of a more or less constant coefficient of friction between the sliding surfaces,
3. Avoidance of brittle failure when the joint reaches the limit of its sliding range.
4. The joint should also be simple, cheap and easy to construct and maintain.

A variety of sliding bolted joints have been tried, some using special surface facing materials such as brake lining and others using direct steel to steel contact. Direct steel to steel contact tends to produce a joint with an undesirably variable friction coefficient, with a high static break-away level requiring associated components to be sized for the friction over-strength. Brake lining has been found to work satisfactorily (Pall & Marsh, 1982) although the friction coefficient may be a little low. A joint using steel/brass contact surfaces has been found to perform well (Popov, Grigorian & Yang, 1995), and has subsequently been tested in New Zealand in a slightly different form (Clifton, Butterworth & Weber, 1998).

2.1 *Linear sliding joint using a steel/brass interface*

The basic joint described in (Popov, Gregorian & Yang, 1995) was intended to provide stable energy dissipation during linear displacement cycles. The details are shown in Fig. 1 below.

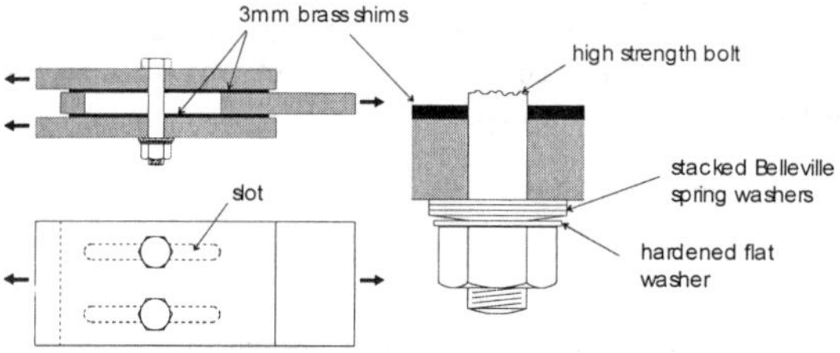

Figure 1. Basic sliding joint details

Note the use of two clamping plates providing symmetric loading when the joint is subjected to axial loading. Having two friction interfaces also doubles the slip force, N_{slip}, to $N_{slip} = 2nN_b\mu$, where n is the number of bolts, N_b is the tension force in one bolt and μ is the coefficient of friction. The shims were of half hard cartridge brass (UNS-260). The tension in the bolts was maintained during joint deformation by means of the stack of Belleville washers placed under each nut. Bolt tension was initially set by means of load-indicating washers under the bolt heads. During test the joint was subjected to displacement cycles of increasing amplitude at constant velocity, and simulated earthquake motions. In all cases the hysteresis loops were essentially rectangular with a kinetic friction coefficient in the range 0.27 to 0.31. There appeared to be no upper bound on dissipation of hysteretic energy – i.e. there was no evidence of a cumulative damage effect. Displacement cycles that exceeded the slot length were not reported.

Clifton et al (1998) reported on a beam-column joint in which brass shims were placed between beam flange and column cleat and between beam web and web cleat, with bolts passing through normal clearance holes. Bolts were tightened by the part-turn method against stacked Belleville washers. Stable sliding behaviour was observed, despite the asymmetric loading on the bolt interface. Loading beyond the limited slip range provided by the 2mm clearance holes caused inelastic deformation and finally cracking of the cleat plates. In another study it was noted that the localised yielding around the bolt holes in web side plate connections provided a very ductile and stable energy dissipating mechanism provided the bolts were sized to avoid shear failure prior to plate yielding (Murray & Butterworth, 1990). In relation to sliding bolted connections this plate yielding mechanism can be expected to furnish extra energy absorption under extreme conditions when the joint displacement exceeds the slot length, provided the bolt diameter to plate thickness ratio is controlled appropriately.

2.2 *Use in braced frames*

Linear SBJs have most commonly been used in braced frames with the SBJ placed at one end of each diagonal brace and designed to slip prior to yield or buckling of the brace. Single diagonal braces have been used, but the need for a compression capacity in excess of the SBJ slip load results in very heavy members. A K-braced configuration, on the other hand still requires compression-capable braces, but they would be shorter and lighter than a full diagonal brace. Both forms have performed well under test (Yang & Popov, 1995). An ingenious application of the linear slider is embodied in the Pall device (Pall & Marsh, 1982). The rectangular mechanism is placed at the intersection of a pair of tension cross-braces and maintains tension in both diagonal braces, confining compression to the short diagonals within the rectangle, with frictional sliding occurring at the centre. A patented version of this device has been used in both new and retrofitted structures. There appears to be scope for investigating other alternatives for incorporating linear sliding joints into K-braced frames. Some configurations that are currently under investigation are shown in Figs. 2 and 3.

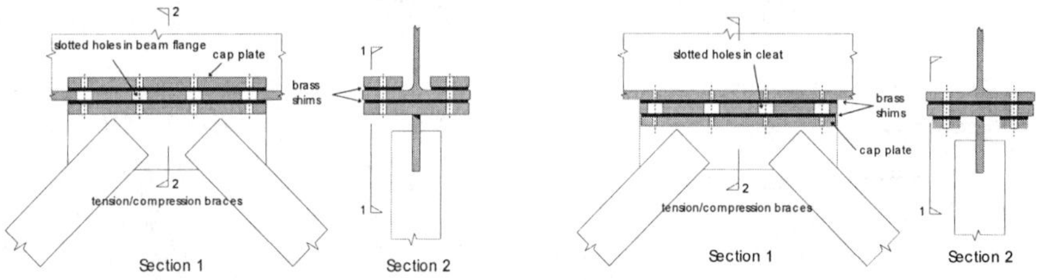

(a) Detail requiring slotted holes in beam flange (b) Detail avoiding slotted holes in beam flange

Figure 2. Sliding connection configurations at junction of K-brace and beam soffit

The connection shown in Fig. 2(a) requires slotted holes in the beam flange and has a sliding action that is not symmetric in that the top plate is activated by the bolts from below and there is also likely to be a small moment due to brace eccentricity. Tests on similar asymmetric sliding junctions in beam to column connections have shown promise and further testing is in progress.(Clifton et al, 1998). Gravity load on the beam would maintain a small clamping force across the friction interface in the event that the bolt tension was lost. The braces could be tension-only without prejudice to the reversibility of the sliding, but would then introduce a significant shear force into the beam at mid-span. Propping action to the beam could be exploited if desired, leading to a smaller beam size. The connection in Fig. 2(b) is similar to the preceding case but avoids the need for slotted holes in the beam flange.

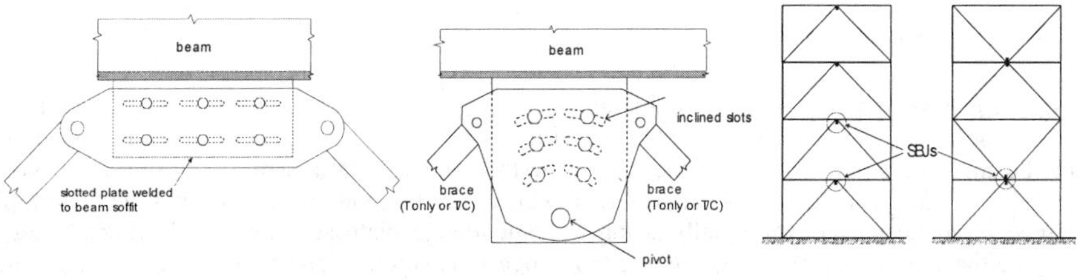

(a) Linear slider (b) Rotational slider (c) K and X-braced styles

Figure 3. Vertically oriented sliding interfaces for braced frame dissipators

The version shown in Fig. 3(a) moves the sliding plates into the vertical plane and restores some symmetry to the loading on the bolts and sliding plates. However, the bolts bearing to some extent on the sides of their slots might require sleeves around the bolts to facilitate sliding. Fig. 3(b) shows a prototype rotational dissipator exploiting curved or inclined slots. Either type could be used with tension-only braces, with the double storey X-brace arrangement being preferred to the K-brace in this case to avoid the large mid-span beam shear that would otherwise occur when the compression braces shed load.

Significant benefits accrue from the use of braced frames, especially the high stiffness, simpler beam-column joints, faster construction and lower cost. It also follows that if the slider capacities can be matched reasonably closely to the seismic storey shears then the column shears can be kept relatively small.

3 ROTATING SLIDING JOINT

Although braced frames appear to offer the best opportunities for exploiting friction dissipators, moment-resisting frames (MRFs) are a popular form of construction relying on flexural action for lateral resistance. The appropriate form of friction dissipator for such structures is one with a rotational action. Fig. 4(a) shows the details of such a joint developed at UC Berkeley (Yang & Popov, 1995). Tests were carried out on the configuration shown in Fig. 4(b) using cycles of steadily increasing displacement resulting in the near perfect hysteresis loops shown in Fig. 4(c).

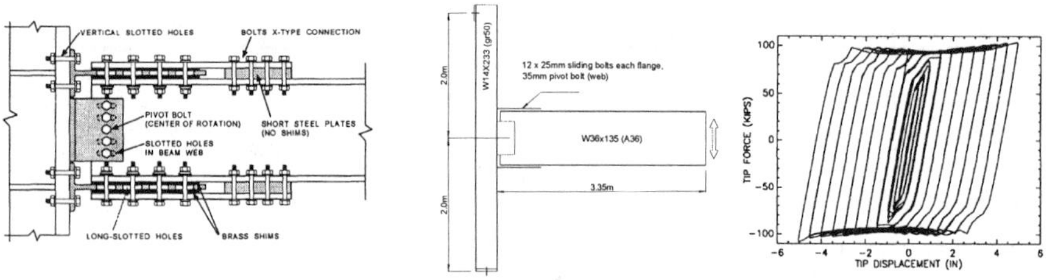

(a) Joint detail (b) Test configuration (c) Hysteresis loops

Figure 4. Rotational slotted bolted connection (RSBC) (Yang & Popov, 1995)

Under the action of a beam moment the joint behaves elastically like a rigid connection until the beam flange forces reach the slip level of the sliding connections. The beam then rotates at near constant moment about the larger central (pivot) bolt in the web (which has only a normal clearance hole) until the bolts reach the end of their slots. The threshold moment M_{slip} is given quite closely by $M_{slip} = 2nN_b\mu D$, where n is the number of bolts in one flange, N_b is the clamping force of one bolt, μ is the friction coefficient and D is the beam depth. The factor of 2 results from the number of friction interfaces.

4 DUAL LEVEL JOINT

Fig. 5(a) shows details of a proposed sliding rotating joint in which the beam flanges contain slotted holes and the cover plates are attached to the extra wide flange plates on either side of the beam. A bolted T is used for the top flange to facilitate joint assembly. Clifton et al proposed a similar joint (Clifton et al, 1998) in which only the bottom flange can slide. The top flange is connected conventionally using a single flange plate and normal clearance holes, forcing the centre of rotation close to the top flange and avoiding damage to the reinforced concrete floor slab that is usually present. The simplified version in Fig. 5(b) which exploits a slip interface driven from one side rather than from the centre is currently the subject of detailed modelling with physical testing of some prototype specimens scheduled for the latter half of 1999.

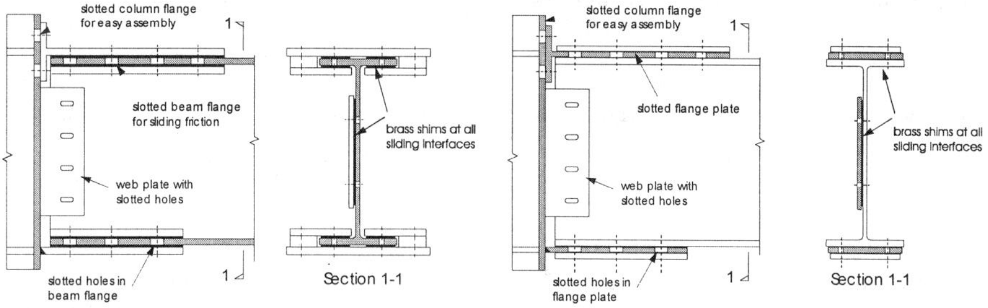

(a) Slip driven from centre plate

(b) Slip driven from non-central plate

Figure 5. Dual slip level joint details

The dual slip level capability of the joint is achieved by making the top flange slip force higher than that of the bottom flange by means of a higher clamping force. Under increasing beam moment the joint responds elastically until the bottom flange connection starts to slip at the threshold moment M_b causing 'plastic' rotation about the top flange. This continues until the bolts reach the ends of the slots in the bottom flange, after which the moment increases to M_t causing the top flange to start slipping. From this point 'plastic' rotation occurs about the bottom flange until the top flange bolts reach the ends of their slots. Further increase in moment will eventually lead to inelastic deformation of the joint components, concentrated at the slot ends.

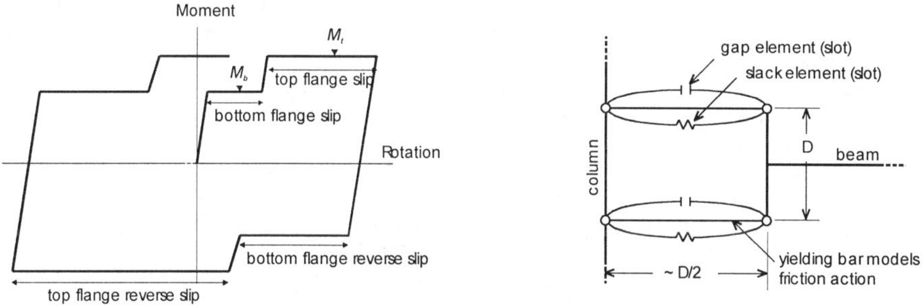

(a) Moment-rotation response

(b) Components of finite element model

Figure 6. Dual level joint - idealised moment-rotation response

The lower slip threshold is intended to meet the strength and ductility demands imposed by the design level earthquake with the upper level providing a reserve of non-damaging inelastic deformation for extreme events, accompanied by an alteration in structure characteristics that may help avoid resonant response. Provided the upper level is not activated, all the slip will occur at the bottom flange, minimising floor slab damage.

Slot length L, is related to the beam depth D, and angle of inelastic rotation θ, by $L=D\theta+d$, where d is the bolt diameter. For example, a 450mm deep beam with a maximum expected inelastic rotation of ±0.03rad and 24mm bolts would require slotted holes 51mm long.

4.1 *Time-history analysis*

Some preliminary studies of frame response have been made using the nonlinear dynamic analysis program DRAIN-2DX (Allahabadi & Powell, 1988). Inelastic time-history analyses of perimeter frame models of either 2 or 6 storeys were conducted under a range of scaled earthquake records using the friction dissipator joint model shown in Fig. 6(b). Details of the models are reported elsewhere (Butterworth, 1999). The main parameters varied were the slot lengths, threshold moments and number of levels with dissipators. Response values monitored included post-earthquake permanent deformation, ductility demand, activation of friction sliders, energy dissipation, member actions and base shear. Although it is not possible to present details in this paper, some general observations based on the results of the nonlinear time-history analyses are included in the conclusions.

5 CONCLUSIONS

The present investigation on hierarchical friction-dissipating joints is at an early stage and few firm conclusions have been able to be reached. However there are strong indications of trends, which together with evidence from earlier investigations leads to the following observations.

Friction-dissipating steel joints utilising slotted holes, brass shims and tension-maintaining bolts show considerable promise as a means of virtually damage-free energy dissipation under seismic loading. Large numbers of virtually identical loading cycles can be achieved. In practice, a residual offset left in a sliding joint after structural loading would be readily repairable by slackening the bolts, re-aligning and re-tightening (possibly replacing) the bolts.

The use of linear sliding joints in K or X-braced frames could lead to cheaper, stiffer structures, possibly using tension-only bracing, and with sufficiently high levels of ductility to make them suitable for seismic zones.

Rotating sliding bolted joints extend the benefits of damage-free energy dissipation to the popular moment-resisting frame. The decoupling of beam stiffness and end moment strength avoids the possible need for over-sized columns to deal with beam over-strength moments.

Rotating sliding joints can be readily made to 'yield' at two distinct threshold moments by varying the slip levels of the top and bottom flange connections. In addition, by varying the lengths of the slotted holes the amount of 'plastic' rotation available at each moment level can be controlled.

Limiting slot length was helpful in controlling permanent offset, but making the lower slots too short will cause upper flange slip and possible slab damage.

The time-history case studies all showed a first mode dominated response, with dissipators in the upper storeys contributing little to the overall energy dissipation.

Removing the dissipators from the upper storeys tended to reduce permanent offset of the structure but with an accompanying increase in member actions in the upper storeys.

Issues such as corrosion resistance, long term retention of bolt tension and the performance of unsymmetric two-plate connections have yet to be addressed.

6 REFERENCES

Aiken, I.D., Nims, K., Whittaker, A.S. and Kelly, J.M. 1993. Testing of passive energy dissipation systems. *Earthquake Spectra*, Vol. 9, No. 3,.

Allahabadi, R. and Powell, G.H. 1988. DRAIN-2DX User Guide. *Report No. UCB/EERC-88/06, University of California, Berkeley, 1988.*

Butterworth, J.W. 1999. Seismic response of moment-resisting steel frames containing dual-level friction dissipating joints. *Technical Conference, NZ Society for Earthquake Engineering, March, 1999, pp. 59-65.*

Clifton, G.C., Butterworth, J.W. and Weber, J. 1998. Moment-resisting steel framed seismic-resisting systems with semi-rigid connections. *SESOC Journal*, Vol. 11, No. 2.

Murray, C.M. and Butterworth, J.W. 1990. Inelastic rotation capacity of ductile web side plate connections. *Report, Heavy Engineering Research Association 88/R6.*

Pall, A.S. and Marsh, C. 1982. Response of friction damped braced frames. *J. Struct. Div., ASCE*, Vol. 108, No. ST6.

Popov, E.G., Grigorian, C.E. and Yang, T-S. 1995. Developments in seismic structural analysis and design. *Engineering Structures*, Vol. 17, No. 3.

Yang, T-S. and Popov, E.P. 1995. Experimental and analytical studies of steel connections and energy dissipators. *Report No. UCB/EERC-95/13, University of California, Berkeley.*

Mechanics of Structures and Materials, Bradford, Bridge & Foster (eds)
© 1999 Balkema, Rotterdam, ISBN 90 5809 107 4

The bifurcation behaviour of vertically irregular multi-storey buildings subject to earthquake excitations

C.Chen, N.Lam & P.Mendis
Department of Civil and Environmental Engineering, University of Melbourne, Vic., Australia

ABSTRACT: Recent investigations by the authors have shown that existing displacement predictive procedures could grossly underestimate the inelastic storey-drifts of a multi-storey building under ultimate conditions if the collapse mechanism bifurcates from one principal elastic displacement shape (i.e. the fundamental mode shape) into two principal displacement shapes following the formation of certain plastic hinges. Such bifurcation behaviour is demonstrated in this paper with the analysis of an eight-storey building, and illustrated with analyses of a number of two-degree-of-freedom models. It has been further identified that excessive storey-drifts associated with bifurcation behaviour pertain to irregular multi-storey buildings where a soft storey is located at the upper level in the building (eg. secondary structures supporting penthouses in high-rise apartment buildings).

1 INTRODUCTION

Vertical irregularities in a multi-storey building such as the presence of a "soft-storey" or sudden change in geometry and stiffness (Fig.1) have been recognised widely as undesirable features of a building from an earthquake resistant viewpoint (Paulay & Priestley, 1992). However, such irregularities are commonly found in buildings in regions of low and moderate seismicity where earthquake resistant considerations are seldom considered in the schematic design of the building. Earthquake loading standard (eg. AS1170.4 1993) generally prefers dynamic analyses, including elastic modal analyses, to static analyses in the analyses of vertically irregular structures. However, the conventional elastic modal analysis procedure has the shortcoming that it cannot model the dynamic behaviour of an inelastically responding structure.

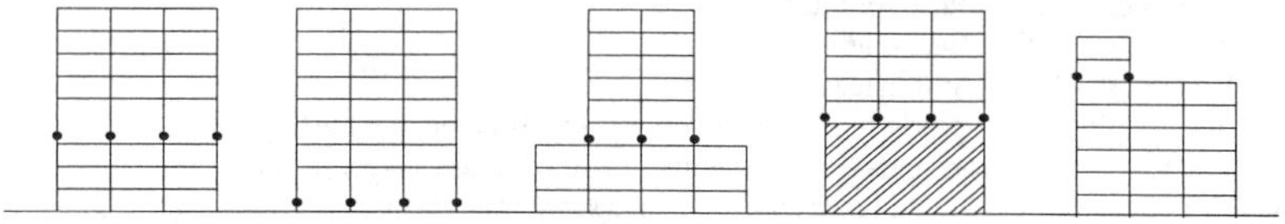

Figure 1. Common vertically irregular buildings and the locations of plastic hinges.

Inelastic storey-drift is an important parameter in the seismic performance evaluation of a building at the ultimate limit state since the parameter is directly related to (i) maximum strains developed in the plastic hinges (ii) damages to attached non-structural components and (iii) "P-Δ effects". Recently, displacement based design procedures have been proposed to predict inelastic storey-drifts of both regular and irregular building structures (eg. Priestley, 1995; Calvi & Pavese, 1995). These procedures draw the analogy between the displacement of a single-degree-of-freedom (SDOF) system and that of a multi-degree-of-freedom (MDOF) system based on an assumed displacement shape. It is considered that such procedures would take into account the post-elastic behaviour of the building so long as the assumed displacement shape of the building has been realistically represented. Push-over analyses have been used to determine inelastic displacement shapes which are affected significantly by plastic hinge formation (Chen, Lam & Mendis, 1998).

Significantly, it has been found from recent investigations by the authors that such procedures could grossly under-predict storey-drifts if the collapse mechanism of the building bifurcates from one principal elastic displacement shape into two principal displacement shapes following the formation of certain plastic hinges (Fig. 2). Consequently, existing displacement predictive procedures based on the SDOF analogy has major limitations when applied to building structures which have not been designed to develop a well defined collapse mechanism under ultimate conditions.

This paper reviews existing procedures in the evaluation of inelastic storey-drifts in irregular building structures, and demonstrates, with examples, the effects of bifurcation behaviour. The paper is outlined as follow:

1. Review of existing procedures in storey-drifts predictions.
2. Bifurcation behaviour demonstrated by an example.
3. Further illustrations by two-degree-of-freedom models.

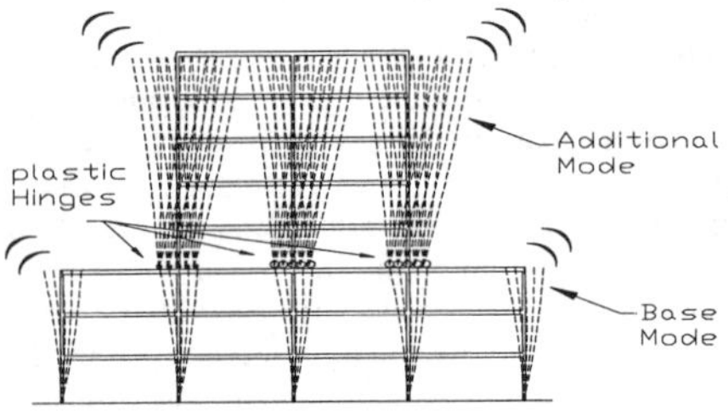

Figure 2. Illustrated bifurcation behaviours.

2 REVIEW OF EXISTING PROCEDURES IN STOREY-DRIFTS PREDICTIONS

A number of methods have been developed to predict inelastic storey-drifts as summarised in the followings:

(a) Australian Earthquake Loading Standard (AS1170.4, 1993) procedure :
In this procedure, storey displacement δ_x (at level x) and storey-drift (Δ) is obtained using the expression:

$$\delta_x = K_d \delta_{xe} \quad (1a)$$
$$\Delta = \delta_x - \delta_{x-1} \quad (1b)$$

where:

δ_x = inelastic displacement at level x.
Δ = storey-drift between level "x" and "x-1".
K_d= displacement amplification factor.
δ_{xe}= displacement predicted by elastic analyses of the design earthquake forces.

Note, the effect of inelastic behaviour is deemed to have been taken into account by the K_d factor which is dependent on the classification of the structural form of the building. However, there is no explicit allowance in the Standard for the effect of structural irregularity and the associated bifurcation behaviour.

(b) Effective Height Method (Priestley, 1995) :
According to the Displacement-Based procedure recommended by Priestley, storey drifts (δ_s) in a building can be predicted relatively easily using the "effective height" principle. The effective height (H_e) depends on the assumed displacement shape of the building under ultimate conditions. For example:

$H_e = 0.64H$ (for "beam-sway" mechanism) (2a)
$H_e = H_s$(for "column-sway" mechanism developed at a bottom soft-storey) (2b)

The earthquake induced "effective displacement" of the building at the effective height (H_e) is taken to be equal to that of a suitably modelled substitute structure which is a SDOF system. Consequently, the effective displacement can be obtained directly from an elastic displacement response spectrum. The effective displacement can be used to obtain displacements at other floor levels and storey-drifts in accordance with the assumed displacement shape. Significantly, this method is only applicable to buildings developing one of the classical collapse mechanisms. Thus, there is no allowance for bifurcation behaviour in the procedure.

(c) Generalised SDOF System Method

A more generalised procedure for predicting storey drifts is by "push-over analysis" whereby the displacement shape of the building at collapse is determined by push-over analysis in which the building is subjected to a monotonic slowly increasing lateral load. Importantly, the shape of the applied lateral load must be consistent with the associated displacement shape. Thus, iterations may be required. By employing similar principles used in elastic modal analyses, the force-displacement relationship of the MDOF system is related to that of an equivalent SDOF system. The storey drift demand is then obtained from the time-history analysis of a SDOF system with non-linear hysteretic characteristics which has been modelled in accordance with results from the push-over analysis (Chen, Lam & Mendis, 1998).

This method gives more realistic predictions of the inelastic storey-drift than the elastic modal analysis method since the effect of the plastic hinges on the displacement shape of the building is captured by the push-over analysis. However, this method is incapable of capturing the dynamic bifurcated response behaviour of the building. Hence, the storey-drifts predicted by this method can be grossly unconservative in certain situations as demonstrated by examples presented in this paper.

(d) Time-History Analysis Method

In principle, the time-history analysis (THA) method is capable of modelling inelastic storey drift and bifurcation behaviour, provided that the finite element model of the structure represents realistically the cyclic inelastic deformation behaviour of the structure. THA is not commonly used in practice due to its high running costs and the requirements for a sufficient number of representative accelerograms. In this study, THA has been used to illustrate and quantify the effects of bifurcation behaviour. Accelerograms which have been used in the analyses are listed in Table 1.

Table 1. Earthquake accelerograms used in this paper.

Location of rupture	Date of rupture	Magnitude	Distance (km)	PGA(g)
Nahanni	23/12/85	6.4	7.5	1.34
San Fernando	09/02/71	unknown	unknown	1.24

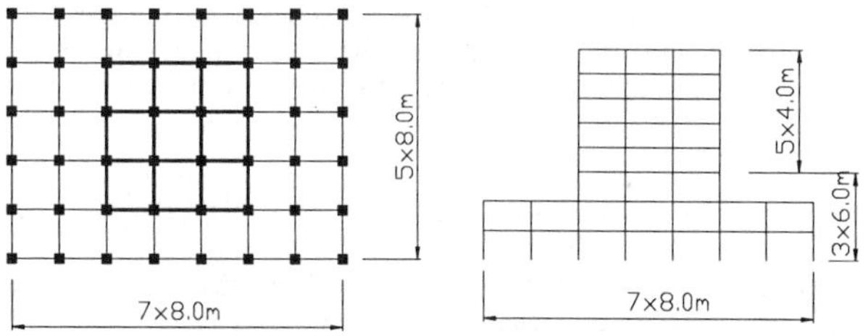

Figure 3. General Arrangement Details of the Eight-Storey Building.

3 BIFURCATION BEHAVIOUR DEMONSTRATED BY AN EXAMPLE

In this section, an eight-storey building designed in accordance with the Australian Standard for Concrete Structures (AS3600, 1994) and the Australian Standard for Loadings in Buildings

(AS1170) is analysed by both the generalised SDOF system method and the time-history analysis method to demonstrate the bifurcation behaviour.

The eight-storey building considered in the example is supported by a three-bay moment resisting frame with a two-storey podium structure (Fig.3). All bays are 8.0 m wide and 6.0 m high for the podium structure and 4.0 m high for the tower structure. In addition to the self-weight of the structure, the assumed gravity loads include live loads of 1.0 kPa for the roof and 3.0 kPa for the typical floors. The analyses are based on a 2-dimensional frame model that carries a tributary weight of about 3250 kN at each level below the podium, 1250 kN at the roof level and 1400 kN at each of the other floor levels.

The lateral load resistance of the structure has been designed in accordance with the Australian Earthquake Loading Standard (AS1170.4, 1993) based on a site factor of 1.0, an acceleration coefficient of 0.08g, a structural response factor (R_f) of 4.0 and a deflection amplification factor (K_d) of 2.0. Capacity design principles have not been applied in reflection of the common practices in low seismicity regions, such as Australia. Static analyses were carried out using Space-Gass (a general-purpose structural analysis program) to determine the required strength of the structural members.

Program Ruaumoko (1998) was then used for the push-over analysis and time-history analyses with the following assumptions:

- Bilinear hysteretic behaviour with 10% strain hardening at the plastic hinges.
- 5% critical damping.
- Inverted triangular lateral force pattern for the push-over analysis.

The generalised SDOF system method, involving push-over analyses, was first used to predict the storey drift demands in the building based on the accelerograms described in Section 2. The procedure, which is described in detail by (Chen, Lam & Mendis, 1998), is briefly summarised in the following:

Step 1: Use push-over analysis to determine the displacement shape of the building in the inelastic range.

Step 2: Develop an equivalent SDOF model based on the load-displacement behaviour observed in Step 1.

Step 3: Perform time-history analyses on the equivalent SDOF system using selected earthquake accelerograms to determine the “effective displacement”.

Step 4: Identify the relationships between the “effective displacement” and the storey-drifts of the MDOF system based on the displacement shape obtained in Step 1.

Step 5: Obtain inelastic storey-drifts based on the results of Steps 3 & 4.

Non-linear time history (THA) analyses were then applied to the MDOF (finite element) model based on the same accelerograms used in Step 3 above. The inelastic storey-drifts predicted by the two analysis methods were then compared (Fig.4). Discrepancies in the predictions by the two methods indicated are evidences of bifurcation behaviour. Excessive storey-drifts associated with the bifurcation behaviour were consistently shown in the comparative analyses.

4 FURTHER ILLUSTRATIONS BY TWO-DEGREE-OF-FREEDOM MODELS

It is shown in Section 3 that for buildings which develop bifurcation behaviour, the additional displacement shape developed above the plastic hinges at the podium level can only be captured by non-linear time-history analyses, and not by push-over analyses. Further studies on the bifurcation behaviour were subsequently carried out on a number of two-degree-of-freedom (2DOF) models to illustrate the principal trends (Fig.5 & 6).

In each of the 2DOF models (of a 2-storey building), the lumped mass (m_1) at the lower floor weighed approximately 30 tonnes whilst the lumped mass (m_2) at the upper floor weighed 9 tonnes ($0.3m_1$), 30 tonnes ($1.0m_1$) or 60 tonnes ($2.0m_1$) depending on the model.

Storey-drifts predicted from the generalised SDOF system analyses (involving push-over analyses) and from the non-linear time-history analyses were compared. Both Nahanni and San Fernando accelerograms were employed again. Significant discrepancies between the two pre-

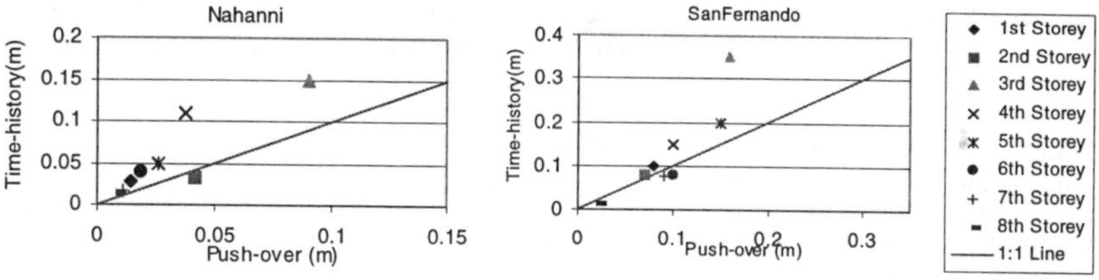

Figure 4. Comparisons of storey drifts.

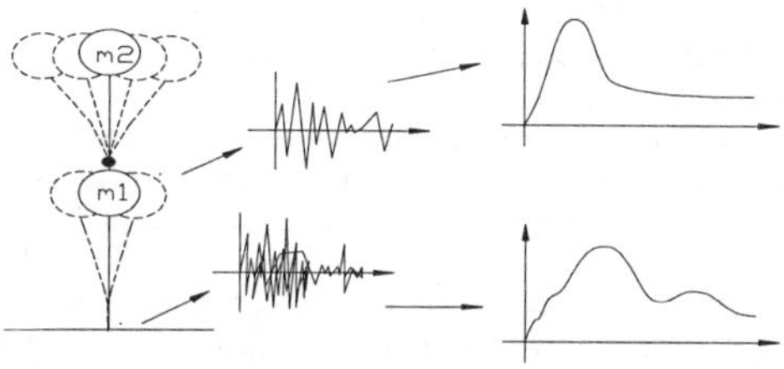

Figure 5. 2-DOF model and response spectra at different levels.

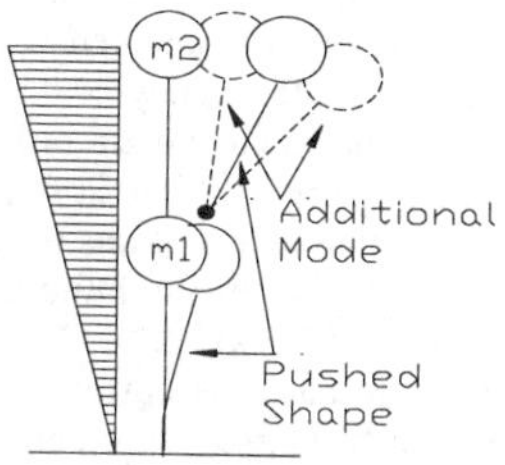

Figure 6. Bifurcated displacement shapes

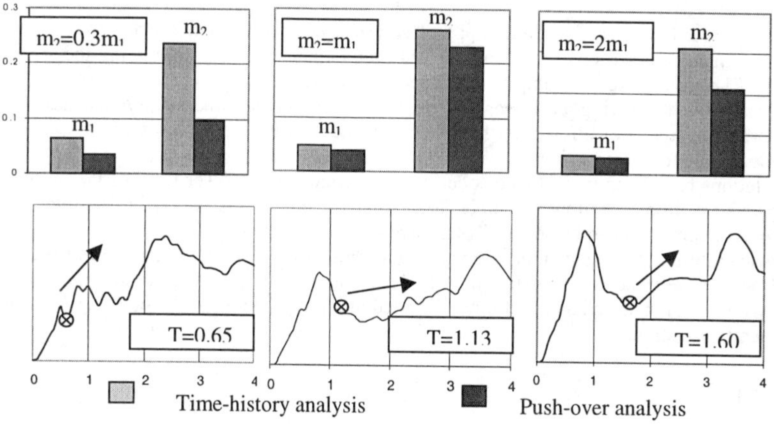

Figure 7. Comparisons of storey drifts from Time-history and Push-over analysis (in meters), and the effects of B-level response spectrum. (Results from Nahanni earthquake only)

dictions indicate bifurcation behaviour. Due to the limitation of space, and very similar results obtained from both earthquake excitations, the results from Nahanni earthquake are presented here only in Fig. 7. Clearly, bifurcation behaviour pertains to models in which m_2 was very much less than $m_1(m_2 = 0.30m_1)$.

To explain this phenomenon, it is proposed that the dynamic response behaviour of the upper floor of a building is largely dependent on the computed motion at the lower floor of the same building. In other words, motions of the lower floor are treated as "ground motions" which excite the upper floor of the same building. To identify the frequency contents of such motions, the computed response time-histories of the lower floor were analysed to obtain their elastic displacement response spectra. It is observed in Fig.7 that the dominant period of the displacement response spectra (T_1) were insensitive to variations between the models. To predict storey-drifts, the natural period (T_2) of the upper floor of the building (considered as an isolated structure) has been identified on each of the displacement spectra.

An interesting analogy can be drawn between the natural periods described above (T_1 and T_2) with the natural period of a SDOF system model (T_s) and the dominant natural period of a soft soil site (T_g) respectively. Comparisons of T_s with T_g have been used to describe the trends in the ductility demand behaviour of inelastic responding SDOF structures (Lam, Wilson & Hutchinson, 1998). It has been shown that excessive ductility demand tend to occur when $T_s < T_g$. Interestingly, building models identified with significant bifurcation behaviour pertains to situations where $T_1 < T_2$ eg. ($m_2 = 0.30m_1$). Thus, the excessive ductility demand of low period buildings situated on soft soil sites and the excessive storey-drifts associated with bifurcation behaviour could be attributed to similar phenomenon.

It is postulated that when plastic hinges form at the lower floor level, the natural period (T_2) of the structure supporting the upper floor increases as a result of inelastic behaviour. This period shift is often associated with changes in the displacement demand of the upper floor depending on the frequency content of the excitations at the lower floor level. The arrows shown on the displacement spectra in Fig. 7 indicate significant increases in the displacement demand as result of period shift if $T_1 < T_2$ (or $m_2 < m_1$). Thus, excessive storey-drift associated with bifurcation behaviour pertains to irregular multi-storey buildings where the soft-storey is located at the upper level of the building (eg. secondary structures supporting penthouses in high-rise apartment buildings).

Further research is currently being undertaken to derive simple procedures to predict and quantify the effects of bifurcation behaviour in multi-storey buildings which have not been designed to fail with a well defined ductile collapse mechanism.

5 CONCLUSIONS

1) Existing procedures in the prediction of inelastic storey-drifts have been reviewed.
2) The significance of bifurcation behaviour has been demonstrated by comparing the storey-drifts of an eight storey building as predicted by the generalised single-degree-of-freedom system method (involving push-over analyses) and by non-linear time-history analyses.
3) The principal trends of bifurcation have been illustrated with the analyses of a number of two-degree-of-freedom models.
4) Postulations have been made by considering the motions of the lower floor of the building as "ground motions" exciting the upper floor of the same building. Interesting analogy with SDOF structures situated on soft soil sites have also been drawn.
5) Significant bifurcation behaviour pertained to models in which the lumped mass of the upper floor was very much less than that of the lower floor. This is possible in irregular multi-storey buildings in which a soft-storey is located at an upper level of the building (eg. certain penthouse structures).

6 REFERENCES

AS1170.4 1993. *Minimum design loads on structures - earthquake loads.* Standards Australia.

AS3600 1994. *Concrete structures code.* Standards Australia.

Calvi, G.M, Pavese, A. (1995), *Displacement Based Design of Building Structures,* European Seismic Design Practice, 1995.

Carr, A.J., (1998), *Ruaumoko*, Computer Program Library, Department of Civil Engineering, University of Canterbury, New Zealand.

Chen, C., Lam, N, Mendis, P, *Prediction of Storey Drifts Induced by Earthquake and Pulse Excitation*, Proceedings, Australasian Structural Engineering Conference, 1998, Auckland, NZ. pp.697-704.

Lam, N., Wilson, J.L. and Hutchinson, G.L., 1998*: The ductility reduction factor in the seismic design of buildings*, Earthquake Engineering and Structural Dynamics, Vol.27, pp749-769.

Lawson, R.S., Vance, V., Krawinkler, H., *Nonlinear Static Push-over Analysis-Why, When and How,* Proceeding, 5th U.S. National Conference on Earthquake Engineering, Vol.1, pp283-292.

Paulay, T., Priestley, M. J. N. *Seismic Design of Reinforced Concrete and Masonry Buildings*, 1992, John Wiley & Sons.

Priestley, M.J.N., *Displacement-Based Seismic Assessment of Existing Reinforced Concrete Buildings,* Proceeding, Pacific Conference on Earthquake Engineering, Australia, November 1995, Vol. 2, pp. 225-244.

Mechanics of Structures and Materials, Bradford, Bridge & Foster (eds)
 ISBN 90 5809 107 4

Structural system identification of MRWA Bridge 617

N. Haritos
Department of Civil and Environmental Engineering, University of Melbourne, Parkville, Vic., Australia

I. Chandler
School of Civil Engineering, Curtin University of Technology, Perth, W.A., Australia

ABSTRACT: This paper describes the performance of structural system identification studies on MRWA Bridge 617 located on the Great Eastern Highway near Meckering, Western Australia. Three test state conditions of this flat slab bridge were investigated: (i) with the original (suspect) 'spraycrete' transverse bandbeams introduced over each pier, in 1995, in an attempt to improve the stiffness and shear capacity of the bridge (ii) with these original bandbeams removed to bring the bridge to its original (pre-retrofit) condition and (iii) after remedial retrofit of a new set of transverse bandbeams over each pier location, in order to gauge the efficacy of this type of remedial strategy on flat slab bridges. Simplified Experimental Modal Analysis (SEMA) studies, using the dynamic response measurements of the bridge from natural traffic, greatly assisted with the tuning of structural parameters in the Finite Element Analysis (FEA) modelling of this bridge and in providing this assessment.

1 INTRODUCTION

1.1 *Background*

The Department of Civil & Environmental Engineering at The University of Melbourne was commissioned by Curtin University of Technology, School of Civil Engineering, to perform structural system identification studies from dynamic testing, using traffic excitation, of Main Roads Western Australia (MRWA) Bridge 617, located on the Great Eastern Highway near Meckering. This bridge, of Reinforced Concrete (RC) slab on pier construction, consists of four continuous spans, (successively 7.1m, 7.8m, 7.5m and 7.2m in length), cast on piers, themselves each consisting of 4 prestressed concrete columns approx. 360mm square and 2.6 m apart. The bridge, constructed in the early 1970's, is simply supported on bearings at its abutment ends. Slab thickness is near uniform at 300mm. The distance between kerb beams is approx. 8.51m.

An attempt was made, over 3 years ago, to use 'spraycrete' to form transverse bandbeams, approx. 2200mm wide and between 250 and 300mm deep, across the bridge at each pier location, for the purpose of improving its stiffness and shear strength at these support locations, (see Figure 1). The spraycrete had since been found to be fissured and to contain voids to the extent that MRWA were concerned over its integrity and effectiveness for this purpose. A new contract was therefore let to remove the existing spraycrete bandbeams, replacing them with new work that met the acceptability requirements of the MRWA.

The testing proposed for this bridge was designed to gauge the relative influence on bridge stiffness (and strength), separately posed by the original spraycrete bandbeams and the more recent remedial work.

The three test scenarios that were identified for this purpose, are described as:

- Test 1 - Original bandbeams in place (testing performed in March, 1998)
- Test 2 - Original bandbeams removed (testing performed in April, 1998)
- Test 3 - Replacement bandbeams introduced (testing performed in December, 1998)

Figure 1. Photograph of MRWA Bridge 617 under Test 1 conditions (original spraycrete bandbeams)

1.2 *Structural system identification strategy*

The strategy behind the approach for the structural system identification of Bridge 617 was one in which the dynamic modes, predicted from a Finite Element Analysis (FEA) model of the bridge, were to be 'matched' to estimates of these modes obtained from its dynamic testing in each of its three test scenarios.

FEA model parameters that were to be 'tuned' in this form of calibration procedure were chosen from:

- E_1, the Young's modulus of Concrete (based upon an uncracked section) of the deck
- E_2, the Young's modulus of Concrete of the combined section of bandbeam plus deck
- E_3, the Young's modulus of the elastomeric bearings supporting the deck at the abutment ends of the bridge

The 'tuned' versions of the FEA models that would result from this calibration would be considered to be reliable representations of the bridge and its properties in each of these test states. Differences observed in E_2 between Test 1 and Test 3 conditions would provide a 'measure' of the efficacy of the replacement set of spraycrete bandbeams over the original set.

2 DYNAMIC TESTING TECHNIQUE

2.1 *Basic description of the Simplified Experimental Modal Analysis technique*

The University of Melbourne has accumulated a great deal of experience with the dynamic testing of a number of bridge superstructures, (Haritos 1996) in order to determine their dynamic characteristics (modal frequencies, mode shapes and damping ratios). Wherever economically feasible, performance of this testing has been in the form of an Experimental Modal Analysis (EMA) of the bridge, (Ewins 1985), using controlled forced excitation from a linear hydraulic shaker or an impact device. More recently, a Simplified form of EMA (identified as 'SEMA') has been developed by the lead author, which is based upon using natural traffic as the source of dynamic excitation of the bridge, (Haritos & Aberle 1995).

The SEMA technique requires that measurements of the dynamic response of a bridge be made over a pre-determined grid of points when the bridge is excited by natural traffic excursions over the bridge. This is normally achieved using a set of accelerometers that are re-located to different positions on this grid whilst retaining at least one accelerometer (the ‘reference’) in its original location, throughout the testing. The ‘roving’ accelerometers need be re-located a number of times (‘setup’ sequences) as the number of points on the measurement grid normally far exceeds the complement of accelerometers available/ number of data channels on the Data Acquisition System (DAS) used to perform digital logging of the response measurements. In addition, usually a large number of repeat test records (typically a minimum of 32) would be required to improve on the statistical accuracy of the modes being identified using this technique, (Haritos & Dao 1995, Haritos 1997, Haritos & Abu-Aisheh 1998).

2.2 *Implementation of the Simplified Experimental Modal Analysis technique*

The SEMA technique is conceptually straightforward in its implementation. It can be shown that the Relative Response Function, (RRF), at point ‘q’ relative to a reference point ‘o’, and given by, $R_{qo}(\omega)$, is dominated by the modal contribution of mode ‘i’ when the circular frequency ω corresponds to one of the natural frequencies ϖ_i, (Haritos & Aberle 1995), hence:

$$R_{qo}(\varpi_i) = \frac{X_q(\varpi_i)}{X_o(\varpi_i)} \approx \frac{\Phi_{qi}}{\Phi_{oi}} \tag{1}$$

in which $X_q(\varpi_i)$ and $X_o(\varpi_i)$ represent the Fourier Transforms, at circular frequency ϖ_i, of time traces of response at points ‘q’ and ‘o’, respectively, and Φ_{qi} and Φ_{oi} the mode shape amplitudes of mode ‘i’ at these points.

The approximation in Equation 1 will improve in accuracy when ensemble averaged versions of $R_{qo}(\omega)$ from multiple realisations of response time traces are used in the description. In addition, the more separated in frequency that mode ‘i’ happens to be from its adjacent modes, the closer this approximation becomes to the modal amplitude ratio of mode ‘i’ at these points. However, the converse is also true, i.e. the closer in frequency that mode ‘i’ happens to be to its adjacent modes, the further this approximation deviates from the modal amplitude ratio of mode ‘i’ at these points. (The resultant modal ratio, under these latter circumstances, becomes a hybrid combination of these contributing adjacent modes).

Specialist software for performing data acquisition on the DAS used by The University of Melbourne is capable of also performing ensemble averaging of the RRFs of the time traces of acceleration response measurements captured during the field testing of a bridge in preparation for the performance of SEMA.

3 PERFORMANCE OF ACCELERATION RESPONSE MEASUREMENTS

3.1 *Selection of Accelerometer Measurement Grid*

The basic 120 point grid chosen for performing acceleration response measurements of Bridge 617 for the purpose of conducting a SEMA is depicted in Figure 2 as it applied to the Test 3 configuration. (The grid versions adopted for the Test 1 and Test 2 configurations were essentially similar except for the measurement of response of the deck close to the abutments).

All measurements were performed from the underside of the bridge deck. Steel plates were fixed to the grid point locations to allow the accelerometers to be magnetically attached thereon.

A letter label, ‘A’ to ‘O’ in alphabetical sequence, was adopted to identify each of the fifteen Dytran Model 3951 accelerometers used in the testing. The last three accelerometers in this sequence (‘M’, ‘N’ and ‘O’) were positioned at the same fixed reference locations adopted for the Test 1 and 2 series, (ie in spans#1, #2 and #4, respectively). The remaining 12 accelerometers were relocated to new positions, identified using a ‘setup’ sequence identifier using a numeral/letter following the letter ID symbol of each accelerometer.

Setup conditions 1 to 9 represented coverage of the measurement grid as originally chosen for Tests 1 and 2. Setup ‘A’ and ‘C’ were used to obtain repeat measurements for accelerometer locations that either exhibited faulty records during earlier recording sequences or were located on the new grid points in each of the Test series (e.g. points close to the abutments).

The Setup ‘B’ configuration was chosen on the basis that it provided an effective reduced grid for performing a simplified ‘first cut’ SEMA of the bridge.

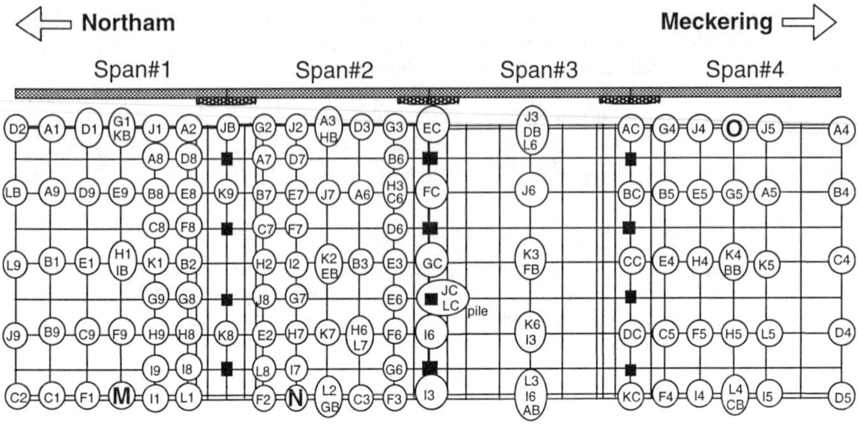

Figure 2. Accelerometer measurement grid adopted for Test 3

3.2 *Recording Procedure for Accelerometer Response Measurements*

A sampling rate of 128 Hz over all 16 possible data channels of the DAS and a 16 second period of recording was adopted for all data sample sets resulting in 2048 data points per channel per record. Manual triggering for data capture of vehicular (both 'car' and 'truck') excursions over the bridge was performed using the TSPECTRA data acquisition software resident on the DAS.

Reviewing of data traces for each accelerometer in each record capture permitted 'on the spot' rejection of those record sets considered not to be suitable for subsequent SEMA analysis and allowed identification of faulty records in which accelerometers were 'misbehaving' for one reason or another. Those accelerometer locations with significant numbers of such instances were chosen as candidates for repeat test measurements in Setups 'A' and 'C'.

By using this approach, it was possible to ensure that most files that were actually stored on hard disk contained data records that conformed to SEMA analysis requirements, at the stage of data capture. A minimum of 64 records of vehicular excursions over the bridge (separate 'truck' and 'car' records and combinations of both forms of traffic) were recorded in each setup sequence for all 15 accelerometers.

Figure 3 depicts a 'car' crossing over the bridge with a 'truck' approaching the bridge from the opposite direction in the far background. Figure 4 depicts typical time traces for reference accelerometer 'O' in the case of a single vehicular excursion of a 'car' and of a 'truck', respectively. It is clearly observed in this figure that the levels of acceleration response are much higher from a 'truck" excursion than they are from a 'car' excursion over the bridge.

3.3 *Identification of modal frequencies*

Modal frequencies in each Test series configuration of the bridge were identified from the frequencies corresponding to peaks in response energy of the bridge. The bridge response energy, as a function of frequency, was described by separately averaging the Auto-spectral density functions of all accelerometer records in each of the Test series, for the 'car' and 'truck' records. Figure 5 presents the resultant variations in the case of the 'car' records for the Test 1 and Test 2 data and for the Test 3 and Test 2 data separately. This figure set can be used to gauge the relative influence of the introduction of the 'remedial' spraycrete bandbeams on the bridge against that resultant from the introduction of the 'original' bandbeam set.

It is clear from the differences in the level of frequency shift in corresponding peaks in going from the Test 2 to Test 1 and Test 3 configurations, that the 'remedial' spraycrete bandbeams were much more effective in stiffening Bridge 617 than the 'original' versions of these bandbeams. For the conditions associated with this bridge, the level of frequency shift is largely attributable to modal stiffness increase in each mode, with the frequency at peak energy locations corresponding to the corresponding modal frequency itself.

Figure 3. Single vehicular excursion over bridge ('Car' followed by a 'Truck' in the opposite direction)

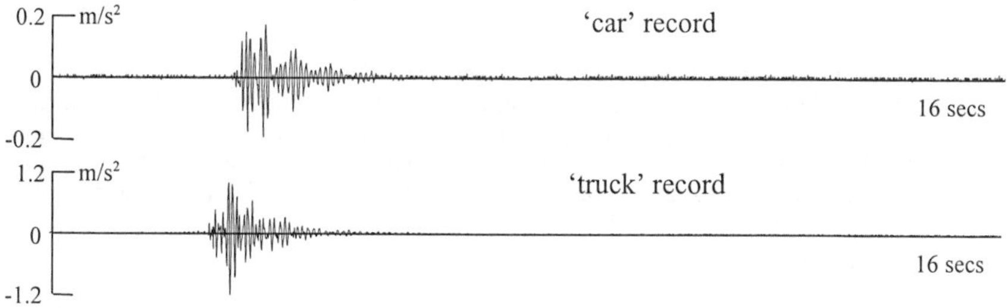

Figure 4. Sample response traces for a 'car' & 'truck' excursion over Bridge 617 for accelerometer 'O'

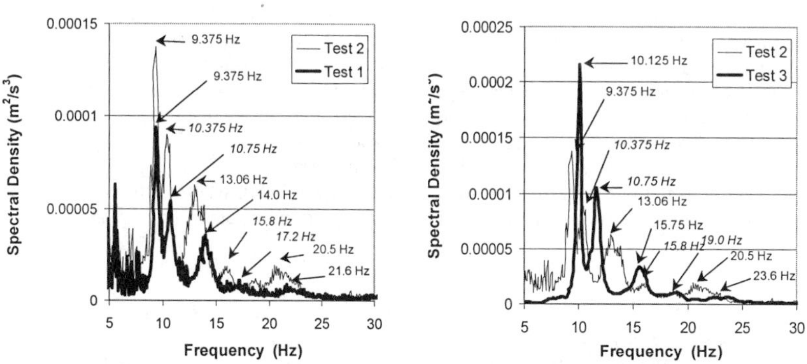

Figure 5. Comparisons of Auto-spectral density functions - Test 1 with Test 2 and Test 3 with Test 2

4 PERFORMANCE OF SEMA INVESTIGATION FOR MODE SHAPE IDENTIFICATION

Figure 6 depicts a sample of mode shapes identified from the SEMA investigation using the entire 'car' record data set, comparing them with their 'tuned' FEA model prediction counterparts. The match between the observed and predicted forms, in this figure, is considered to be quite good, both for mode shapes and frequencies.

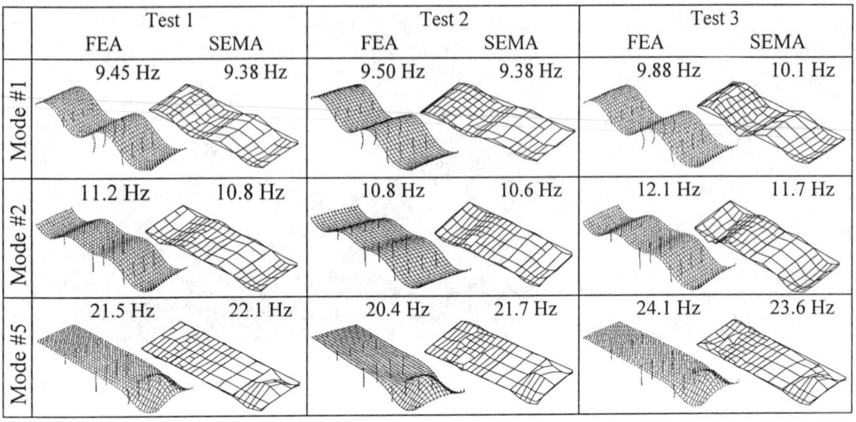

Test 1
FEA
SEMA
Test 2
FEA
SEMA
Test 3
FEA
SEMA
Mode #1
9.45 Hz
9.38 Hz
9.50 Hz
9.38 Hz
9.88 Hz
10.1 Hz
Mode #2
11.2 Hz
10.8 Hz
10.8 Hz
10.6 Hz
12.1 Hz
11.7 Hz
Mode #5
21.5 Hz
22.1 Hz
20.4 Hz
21.7 Hz
24.1 Hz
23.6 Hz

Figure 6. Comparison of selected modes in all three Test configurations – 'tuned' FEA with SEMA

In addition, it is observed that the spraycrete bandbeams have only a minor influence on the mode shapes, but have a more significant influence upon the corresponding modal frequencies.

Values required for the major 'tuning' parameters in the FEA models of the bridge in all three Test states, were: E_1=19 GPa, E_3=20 MPa, (all Test states), and E_2=8 GPa, 19 GPa and 24 GPa, in the sequence Test 1 to Test 3. The results for E_2 imply that the mechanical properties of the original spraycrete transverse bandbeams were inferior to those of the deck whereas the converse is implied for the remedial spraycrete transverse bandbeams.

5 CONCLUDING REMARKS

Dynamic testing of MRWA Bridge 617 was performed to establish calibrated Finite Element Analysis models that represented the behaviour of this bridge in each of its three test scenarios: Test 1 - original spraycrete bandbeams at each pier, Test 2 - original spraycrete bandbeams removed from each pier and Test 3 - after introduction of a remedial set of spraycrete bandbeams at each pier. The Simplified Experimental Modal Analysis technique was used to identify the bridge's modal characteristics, over a number of modes of response, using a grid of acceleration response measurements from vehicular excursions over it.

Results indicated that the remedial work was instrumental in substantially increasing the effective stiffness of the bridge (estimated to be generally over 30%), whereas the original work only had a marginal stiffening effect on this bridge, (estimated to be generally less than 10%).

REFERENCES

Ewins, D. J. 1985. *Modal Testing: Theory and Practice*, New York: John Wiley.

Haritos, N. 1996. Application of EMA testing to the identification of in-service condition of bridge superstructures, *Proceedings of Asia-Pacific Symposium on Bridge Loading and Fatigue*, Dec., Monash University, 91-100.

Haritos, N. 1997. Ambient Traffic Excitation of Bridge Superstructures, *Proc. AUSTROADS 1997 Bridge Conference,* Dec., Sydney University, Vol 1, 259-274.

Haritos, N. & Aberle, M. 1995. Using traffic excitation to establish modal properties of bridges, Proc. MODSIM '95 Conf., Nov., Newcastle, Aust., 1:243-248.

Haritos, N. & Abu-Aisheh, E. 1998. Dynamic testing techniques for structural identification of bridges, *Proc. Australasian Structural Engineering Conf.*, Sept./Oct., Auckland, New Zealand, 1:117-124.

Haritos, N., & Dao, M. N., 1997. Ambient Vibration Testing of Bridge Superstructures, In R.H. Grzebieta et al (eds) *Mechanics of Structures and Materials*, *Proc. Conf.*, 8-10 Dec., Melbourne, 265-270 Rotterdam: Balkema.

Mechanics of Structures and Materials, Bradford, Bridge & Foster (eds)
© 1999 Balkema, Rotterdam, ISBN 90 5809 107 4

Study of dynamic response in an elasto-plastic beam

N. Murakami, T. Nishimura, M. Kimura, K. Matushima & K. Miura
Department of Mechanical Engineering, College of Science and Technology, Nihon University, Tokyo, Japan

ABSTRACT: The static plastic analysis of structures has been fully investigated, but there are few studies of the dynamic plastic analysis except for impact problems. Since the small plastic deformation may be allowable for some kinds of structures, it is effective to investigate the dynamic response beyond the elastic limit in such cases. In our previous study, the dynamic behavior of a simple spring-mass-damped system was analyzed numerically. In this paper, the dynamic behavior of a cantilever to a forced external agency is investigated to ascertain whether the deformational response is stable or unstable. As a result, limits in pure elastic response, and in the stable elastic response to steady state motion are identified.

1 INTRODUCTION

It is necessary to study the deformational behavior beyond the elastic limit, since relatively small plastic deformation is allowable for some kinds of structures. The static plastic analysis of structures has been fully investigated, but there are few studies concerned with dynamic plastic analysis except for impact problems. The authors have an interest in the dynamic elasto-plastic response of structures to an oscillatory external agency.

In our previous study (Nishimura et al. 1998), the dynamic behavior of a simple spring-mass-damped system, which is subjected to an oscillatory external agency, is resolved numerically. It becomes clear from the calculated results that the oscillatory motion of the system is characterized by four kinds of behavior, which are described in the next section in detail. The same study performed on a beam is more effective for actual application to structures.

If a beam is forced to vibrate intensively, plastic deformation may be induced in the beam, that is, a plastic hinge is produced. Since a plastic hinge on an elastic and perfectly plastic beam attains a fully plastic bending moment, the beam tends to rotate around the plastic hinge for the proceeding deformation. Hence the resultant normal mode and natural frequency differs from the realizations for a pure elastic beam. This fact complicates plastic analysis, as compared with the widely known elastic analysis for beam vibration.

In this paper, we discuss the dynamic response of a cantilever to a forced external agency. The stable and unstable motion of an elastic and perfectly plastic cantilever is investigated with respect to the frequency and amplitude of an external oscillatory displacement.

2 ANALYTICAL RESULT ON A SPRING-MASS-DUMPED SYSTEM

The dynamic response of a spring-mass-damped system, where the spring is supposed to be elastic and perfectly plastic, is treated in this section. Figure 1 represents the numerically calculated result of the dynamic response to an oscillatory external force. In the figure, the abscissa denotes the non-dimensional amplitude of a given external agency, which is divided by the elastic limit force of the spring, and the ordinate represents the non-dimensional frequency of the external agency, which is expressed by a ratio to the natural frequency. The calculated result introduces the four typical characteristic deformational behaviors depending on the magnitude and the frequency of the external agency. They are 1) the pure elastic, 2) the elastic and plastic, 3) the re-elastic and 4) the divergent region. The elastic and plastic deformation is observed alternately in the region 2), and the oscillatory motion is stable. In the re-elastic region 3), the steady state motion is in elastic and the motion is also stable, even if several plastic deformations are induced in the transient phase. In the divergent region, large plastic deformation is produced abruptly and the deformational motion is unstable. The solid lines, which are obtained analytically, are the division lines of each region.

The calculated result suggests that the allowable condition of a structural design be expanded, if small plastic deformation is acceptable.

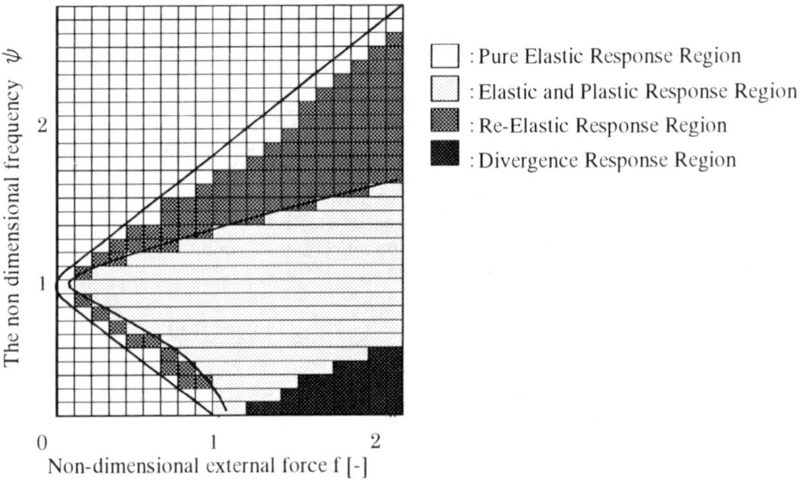

Figure 1. Typical Region of Response

3 ANALYSIS OF BEAM VIBRATION

The cantilever discussed in this paper has a concentrated mass **m** at the free end, and is subjected to an external oscillatory force at the free end as shown in Figure 2. It is assumed that the cantilever is homogeneous, and has a uniform cross section. Therefore, the density ρ, the cross sectional area **A**, and the flexural rigidity **EI** of the beam are constant. The beam is supposed to be elastic and perfectly plastic. The bending moment on an elastic and perfectly plastic beam converges asymptotically to the fully plastic moment beyond the elastic limit. However it is supposed for simplicity that the bending moment grows linearly up to the fully plastic moment through the elastic limit and remains constant at the fully plastic moment for the proceeding curvature after attaining the fully plastic moment. The well-known equation of beam vibration is given as follow,

$$\frac{\partial^4 w}{\partial x^4} + \frac{\rho A}{EI}\frac{\partial^2 w}{\partial t^2} = 0 \qquad (1)$$

where **w** denotes the lateral deflection of the beam. Although the plastic hinge may be produced at any cross section depending on the given frequency, the plastic hinge is induced usually at the built-in end for relatively lower frequency such as the first natural frequency. Therefore, in this paper, we discuss the case where the plastic hinge is brought at the built-in end. Eq.(1) can be solved under the following boundary conditions,

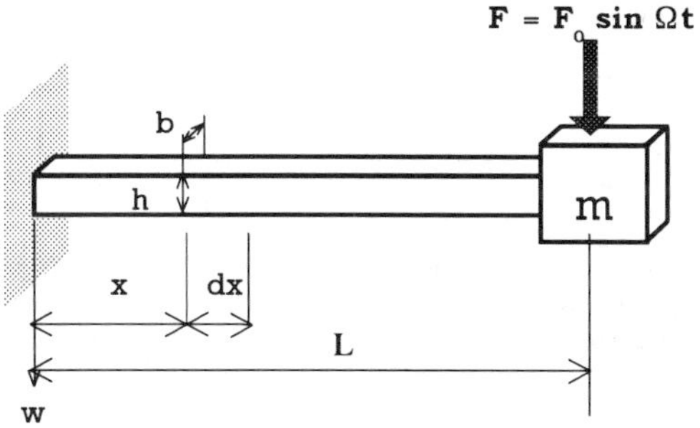

Figure 2. Analyzed cantilever

Boundary conditions:

a) Pure Elastic

$$w=0,\ \frac{\partial w}{\partial x}=0:\quad \text{at } x=0,\quad \text{and}\quad \frac{\partial^2 w}{\partial x^2}=0,\ \frac{\partial^3 w}{\partial x^3}=\frac{m}{EI}\frac{\partial^2 w}{\partial t^2}:\quad \text{at } x=L \tag{2}$$

b) Plastic Hinge at the Build-In End ; (Plastic Phase)

$$w=0,\ \frac{\partial^2 w}{\partial x^2}=0:\quad \text{at } x=0,\quad \text{and}\quad \frac{\partial^2 w}{\partial x^2}=0,\ \frac{\partial^3 w}{\partial x^3}=\frac{m}{EI}\frac{\partial^2 w}{\partial t^2}:\quad \text{at } x=L \tag{3}$$

In the plastic phase, the plastic hinge at the built-in end rotates under the fully plastic bending moment while the curvature proceeds in motion. Under the above boundary conditions, the natural frequency and the normal mode can be derived by solving the following equations,

a) Pure Elastic Phase

$$1-\frac{m}{\rho A}\sqrt{\lambda}\sin(\sqrt{\lambda}L)\cosh(\sqrt{\lambda}L)+\frac{m}{\rho A}\sqrt{\lambda}\cos(\sqrt{\lambda}L)\sinh(\sqrt{\lambda}L)$$
$$+\cos(\sqrt{\lambda}L)\cosh(\sqrt{\lambda}L)=0 \tag{4}$$

b) Plastic Phase

$$\sin(\sqrt{\lambda}L)\cosh(\sqrt{\lambda}L)-\cos(\sqrt{\lambda}L)\sinh(\sqrt{\lambda}L)+\frac{2m}{\rho A}\sqrt{\lambda}\sin(\sqrt{\lambda}L)\sinh(\sqrt{\lambda}L)=0 \tag{5}$$

Here, λ satisfying Eq.(4) and (5) is related to the natural frequency ω, by:

$$\omega=\sqrt{\frac{EI}{\rho A}}\lambda \tag{6}$$

It is known that the general solution to Eq.(1) can be represented by an appropriate summation of homogenous solutions and a particular solution. The homogenous solutions w_h can be written as,

$$w_h=\sum_{j=1}^{\infty} w_{hj} \tag{7}$$

where w_{hj} corresponds to each normal mode given by:

$$w_{hj}=\{C_{1j}\cos(\sqrt{\lambda_j}x)+C_{2j}\sin(\sqrt{\lambda_j}x)+C_{3j}\cosh(\sqrt{\lambda_j}x)+C_{4j}\sinh(\sqrt{\lambda_j}x)\}$$
$$\times(B_{1j}\sin\omega_j t+B_{2j}\sin\omega_j t) \tag{8}$$

The coefficient C_{ij} can be resolved from Eq.4 and 5 and B_{ij} is determined from the initial conditions on each phase. After once experiencing the plastic hinge, the built-in end starts to rotate in the plastic phase, although the inclination does not vary in the elastic phase. Hence additional terms shown by Eq.(9) are required in the plastic phase.

$$w_i = -\frac{M_p \rho AL}{40L^3 EI(3m+\rho AL)}x^5 + \frac{M_p(2m+\rho AL)}{4LEI(3m+\rho AL)}x^3 - \frac{M_p}{2EL}x^2 + \left(\frac{3M_p \rho AL}{2L^3 \rho A(3m+\rho AL)}t^2 + \phi\right)x \qquad (9)$$

where M_p is the fully plastic moment and ϕ designates the inclination of the built-in end at the beginning of the plastic phase. The first three terms in Eq (9) represent the flexural deformation of a cantilever caused by the inertia force distributing itself along the beam. The remaining terms denote a rigid body rotation of a cantilever about the plastic hinge. Finally, the general solution is given by,

$$w_g = \sum_{j=1}^{\infty} w_{hj} + w_p + w_i \qquad (10)$$

The response of a cantilever can be calculated numerically by Eq.(10) for a given frequency and amplitude of external force. The concentrated mass m is assumed to be the total beam mass times six is the example treated here. The calculated results are shown in Figure 3. In the figure, the abscissa denotes the non-dimensional bending moment of the forced external agency, and the ordinate represents the non-dimensional given frequency. The pure elastic motion is realized in the region □, the elastic and the plastic deformations are repeated alternately in the region ▩, and in the region designated by ■, the accumulated plastic deformation in the plastic hinge grows over 8 times that of the elastic limit deformation and the motion is unstable.

However, the steady state motion of actual structures, which exhibit a non-zero structural

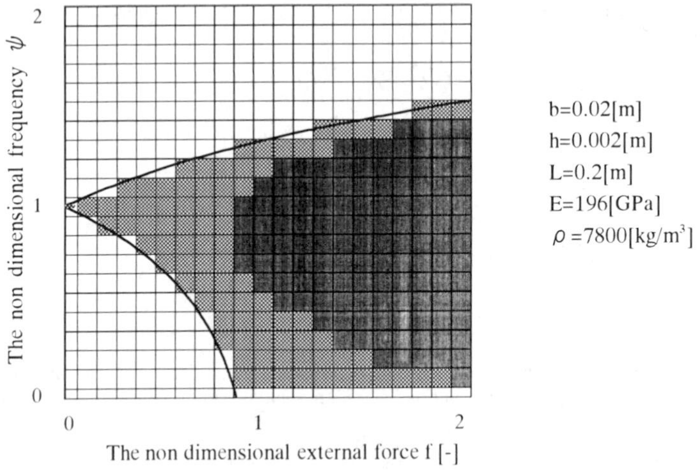

Figure 3. Calculated result

damping factor in nature, can not be explained by Eq.(1). Here a discrete model of the beam is introduced in order to resolve the beam motion with a damping factor. The second natural frequency is highly separated the first for the beam discussed here. Since we are studing the dynamic response of the beam to a relatively low frequency, it is sufficient to use only two concentrated masses in our discrete model for the beam. Accordingly, the cantilever is replaced by a structure composed of massless beams and two concentrated masses as shown in Figure 4. The following equations of motion are derived by referring to the depiction of the equilibrium model in the figure, where the external oscillatory displacement is given at the built-in end.

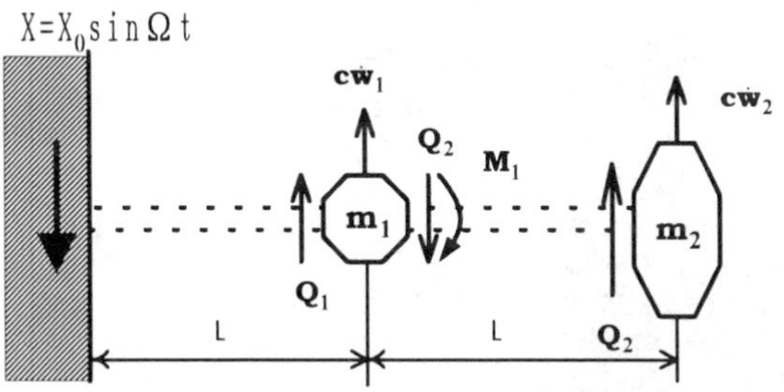

Figure 4. The discrete model

$$
\begin{aligned}
m_1\ddot{w}_1 &= Q_2 - Q_1 - c(\dot{w}_1 - \dot{X}) \\
m_2\ddot{w}_2 &= -Q_2 - c(\dot{w}_2 - \dot{w}_1)
\end{aligned}
\tag{11}
$$

The deflections of each mass are given by the following equation after consideration of elementary beam theory.

$$
\begin{bmatrix} w_1 - X \\ w_2 - X \end{bmatrix} = \frac{L^3}{6EI}\begin{bmatrix} 2 & 3 \\ 5 & 11 \end{bmatrix}\cdot\begin{bmatrix} Q_1 \\ Q_2 \end{bmatrix} + L\begin{bmatrix} 1 \\ 2 \end{bmatrix}\cdot i_0 \tag{12}
$$

Where i_0 is the inclination at the built-in end. The inclination is constant in the elastic phase, but varies with time and is related with the fully plastic moment M_P in the plastic phase.

$$
i_0 = \frac{6}{4L}w_1 - \frac{1}{4L}w_2 - \frac{5}{4L}X - \frac{7LM_p}{24EI} \tag{13}
$$

Consequently, the following equations are derived from Eq.(11), (12) and (13) to give the natural frequency for each phase.

a) Pure Elastic

$$
m_1m_2\lambda^4 + c(m_1+m_2)\lambda^3 + \left(c^2 + \frac{EI}{L^3}\left(\frac{12}{7}m_1 + \frac{96}{7}m_2\right)\right)\lambda^2 + \frac{78EI}{7L^3}c\lambda + \frac{252E^2I^2}{49L^6} = 0 \tag{14}
$$

b) Plastic Phase

$$
m_1m_2\lambda^4 + c(m_1+m_2)\lambda^3 + \left(c^2 + \frac{EI}{L^3}\left(\frac{3}{2}m_1 + 6m_2\right)\right)\lambda^2 + \frac{9EI}{2L^3}c\lambda = 0 \tag{15}
$$

It should be noted that the natural frequency and the normal mode are different to each other in both phases, and the first natural frequency is zero in the plastic phase. Finally, the general solution is given by

$$
\begin{bmatrix} w_1 \\ w_2 \end{bmatrix} = \sum_{i=1}^{4} C_i\begin{bmatrix} 1 \\ A_i \end{bmatrix}\exp(\lambda_i t) + \left\{\begin{bmatrix} B_1 \\ B_2 \end{bmatrix}X_0\cos(\Omega t) + \begin{bmatrix} B_3 \\ B_4 \end{bmatrix}X_0\sin(\Omega t)\right\} + D\begin{bmatrix} 1 \\ r+2 \end{bmatrix} \tag{16}
$$

where $\mathbf{A}_i$ in the first term of the right side equation explains each normal mode, and the vector **B** in the second term can be resolved directly from consideration of the external oscillatory displacement subjected to the built-in end. The factor **r** in the last term takes on the value of zero in the elastic phase but $r=L^2M_p/9EI$ in the plastic phase. The five unknown coefficients C_i and D can be determined by the initial velocity and deflection of the two masses and the initial inclination at the built-in end, which can be given by Eq.(11), (12) and (13), where the term "initial" implies the onset of the elastic or plastic phase. Therefore these are given by the final values on the latest phase.

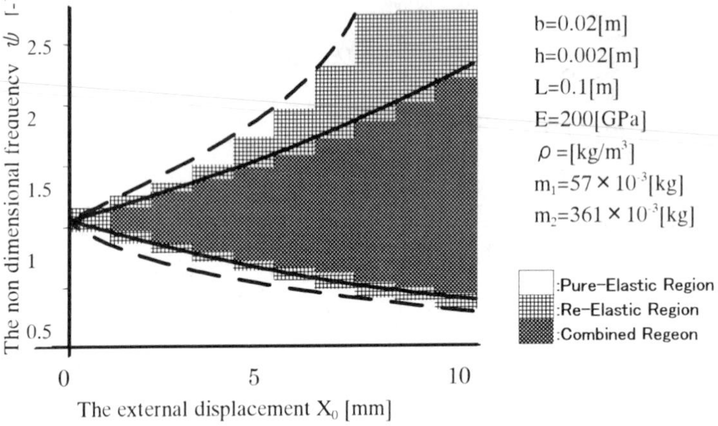

Figure.5 typical region of response

The response of the cantilever can be calculated numerically by Eq.(16) on a given frequency and amplitude of forced external displacement. The calculated result is shown in Figure 5. The calculated dynamic responses are divided into three typical characteristic regions, that is, the pure-elastic region, the elastic or re-elastic region and combined elastic-plastic region. The solid line in the figure denotes the boundary of the elastic limit obtained by the steady state solution of the discrete model as shown in Figure 3. In the combined elastic-plastic region, the plastic deformation is induced alternately. A broken line in the figure shows the division line of the pure elastic limit. The solution of the continuous model, which dose not include a damping factor, gives the transient solution. Therefore in the figure, the pure-elastic limit is obtained by the solution of the continuous beam, which is subjected to the dynamic external displacement at the built-in end.

4 EXPERIMENT

We have also investigated the dynamic response of a cantilever in the vicinity of the first natural frequency in an experiment. The cantilever was vertically fixed to the vibrator in order to avoid the vertical initial deflection induced in this experiment by the gravity force. The motion of a cantilever was observed by a high-speed video camera (1/200-sec). The deflection of the cantilever was measured from digital image data on a computer display.

5 CONCLUSION

The result of numerical analyses shows that the deformational behavior of the cantilever subjected to an oscillatory external agency can be divided into typically three characteristic regions, that is, the pure elastic region, the re-elastic region in the steady state and the combined elastic and plastic region in which the oscillatory motion induces the plastic hinge alternately. Hence, the allowable condition of structural design can be expanded by taking account the frequency and amplitude of a given external agency, provided small plastic deformation is acceptable.

REFERENCE

T. Nishimura & K. Matsushima 1998. Dynamic response of an elastic and perfectly plastic structure by forced vibration: Non-destructive pasting and experimental stress analysis of concrete structures: 357-362. *In Slovakia, Proceedings of the RILEM-IMEKO International Conference 1998*: EXPERT CENTRUM

Mechanics of Structures and Materials, Bradford, Bridge & Foster (eds)
© 1999 Balkema, Rotterdam, ISBN 90 5809 107 4

Robustness of fuzzy control algorithms for the control of civil engineering structures under earthquake loading

B. Samali
Centre for Built Infrastructure Research, Faculty of Engineering, University of Technology, Sydney, N.S.W., Australia

S.C. Khoo & K.C.S. Kwok
Department of Civil Engineering, University of Sydney, N.S.W., Australia

ABSTRACT: The robustness of the control effort generated by the fuzzy control algorithm is investigated and compared with that of the instantaneous optimal control algorithm. Numerical analysis is conducted on a three dimensional five storey building with varying masses to demonstrate that the fuzzy control algorithm is more robust than the optimal control algorithm. An earthquake loading is the source of excitation. Classical control algorithms (eg instantaneous optimal control) are determined from a finite element model of the structure that is to be controlled. Hence, variations in the mass or stiffness of the structure could compromise the stability as well as the efficiency of the control effort. The robustness of the fuzzy control algorithm arises from the fact that the finite element model of the structure is not used in the derivation of the control algorithm. Instead, the parameters of the fuzzy control algorithm are derived from the structural response of the building.

1 INTRODUCTION

Among the many existing control algorithms used to determine the appropriate control force required, given a particular set of state conditions, there is a branch of control theory which is collectively classified as soft computing. Soft computing refers to the application of fuzzy logic and neural networks to control theory. These new control methods are now being used in the active control of civil engineering structures. These approaches are believed to be less rigid and more robust than conventional control methods such as optimal control.

Optimal control is certainly a very popular control approach. The optimal control algorithm determines the control force according to a cost function which minimizes the control force and maximizes the controlled structural response. A number of researchers have made significant advances in the field. Yang et al (1987) introduced the instantaneous optimal control algorithm which allowed the optimal control law to be implemented to control processes where the complete time history of the loading is not known a priori. Researchers have also investigated ways of solving the non-linear ricatti differential equation in matrix form which arises in the solution of the optimal control problem as well as the possibility of enhancing the performance and capability of the optimal controller by incorporating fuzzy logic into the control law (Abe, 1996).

The fundamental concepts in fuzzy logic were first developed over thirty years ago by Zadeh (1965). Mamdani and Assilian (1975) were among the first researchers to apply fuzzy logic by using this strategy to control a steam-engine plant. Fuzzy logic is often used in electronic equipment from camcorders to rice cookers. New applications for fuzzy logic continue to be discovered for example, in pattern recognition exercises such as voice and face recognition. The use of fuzzy controllers in the field of structural control is relatively new. Casciati and Giorgi (1996) examined the way in which fuzzy controllers have been implemented in structural control systems. Abe (1996) used fuzzy rule based algorithms in conjunction with equations derived from optimal control theory using a perturbation method to actively control a tuned mass damper. Subramaniam et al (1996) and Casciati (1997) used fuzzy controllers to control base isolated structural systems while Faravelli and Yao (1996) investigated ANFIS based fuzzy controllers. Fuzzy controllers are also incorporated in part in structural control systems. Hora et al (1996) used fuzzy logic to implement a gain-schedule control law in the active control mode of their hybrid mass damper system as a means of regulating the degree of control gain.

Jang et al (1997) have been researching in the soft computing aspect of the field and have developed the widely used software ANFIS (1993), "embedding the fuzzy inference system into the framework of adaptive networks". ANFIS (Adaptive-Network-based Fuzzy Inference System) uses a neural network to improve the "if-then" rules used in fuzzy reasoning through training.

2 FINITE ELEMENT MODEL

The degree of control that is achieved with optimal control and fuzzy control algorithms for the control of a civil engineering structure under earthquake excitation is analyzed in this paper. A model of a five storey building was selected as an appropriate control example. The model is to be used as an experimental benchmark frame for structural control studies.

The five storey frame has a rectangular cross section with a width of 1m and a depth of 1.5m. It has a height of 3m with uniform individual floor heights of 0.6m. The frame is constructed from beams with a cross sectional area of 1.536×10^{-3} m^2 and a cross section moment of inertia of 2.3634×10^{-6} m^4, and solid square columns with a cross sectional area of 6.25536×10^{-4} m^2 and a cross section moment of inertia of 3.2552×10^{-8} m^4. The Young's Modulus, E, for steel of 2×10^5 MPa was used. The joints between columns and beams as well as the connection between the frame and the base are fully fixed. Figure 1 shows the five storey frame model. The structural damping ratio ζ of the five storey building is assumed to be 0.01.
Following a finite element analysis of the model, its first three natural frequencies of the structure were obtained as : $\omega_1 = 20.5$ rad/sec $\quad \omega_2 = 59.9$ rad/sec $\quad \omega_3 = 94.3$ rad/sec

The input load vector U(n) is formed from the control force vector $F_C(n)$ and the vector of earthquake forces $F_\ell(n)$. The optimal and fuzzy control algorithms have been designed to control the motion of the five storey frame due to the Northridge earthquake loading shown in Figure 2.

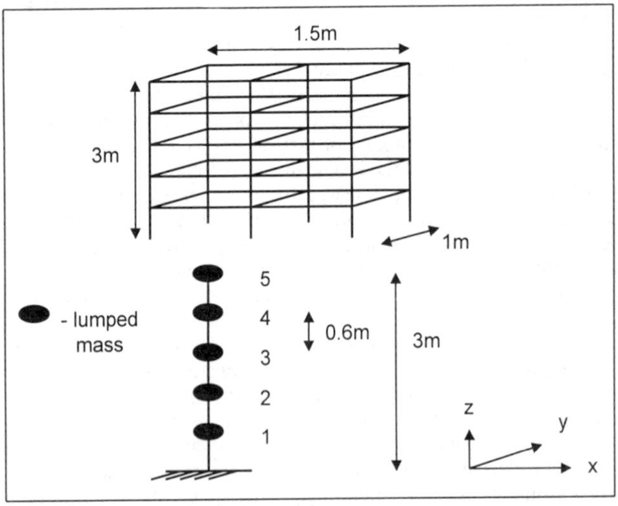

Figure 1. The five storey frame model

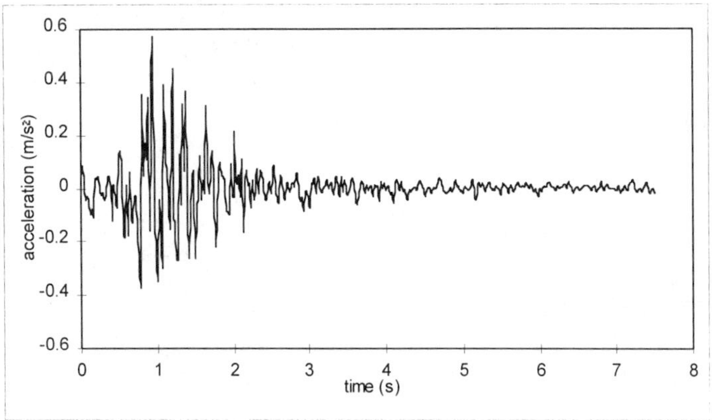

Figure 2. Northridge Earthquake Time History

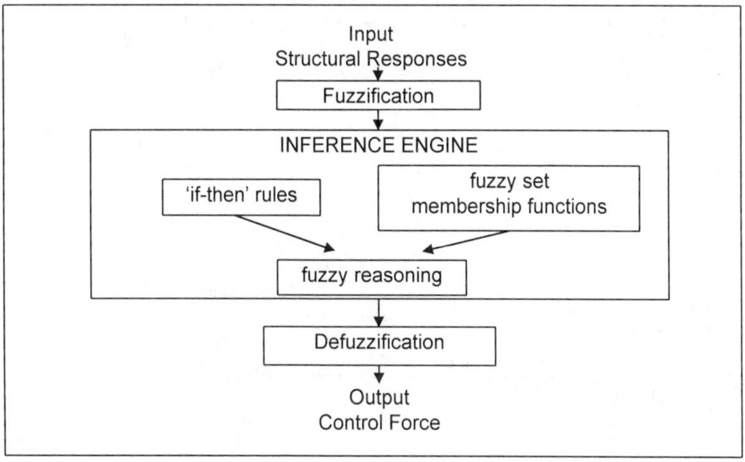

Figure 3. Fuzzy controller

The earthquake excitation forces are applied to the five nodes of the five storey frame and are obtained by multiplying the ground accelerations with the lumped masses located at the nodes. The last term in the input vector is the combined earthquake excitation and control force that is applied to the node located at the top of the five storey frame structure.

3 FUZZY CONTROL

The fuzzy controller in a structural control system applies a control force to counteract the dynamic loading on the structure in order to reduce its vibration. The fuzzy control algorithm determines the necessary magnitude and direction of control force from the displacement and velocity of the structure. The fuzzy controller essentially consists of three main components, a fuzzification unit, a knowledge-based fuzzy inference engine, and a defuzzification unit (see Fig.3).

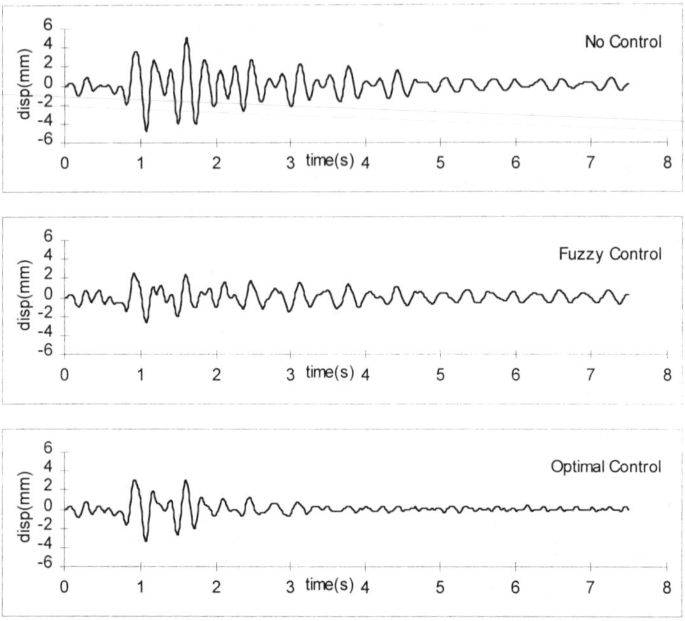

Figure 4. Displacement response of the 5 storey frame without any reduction in mass

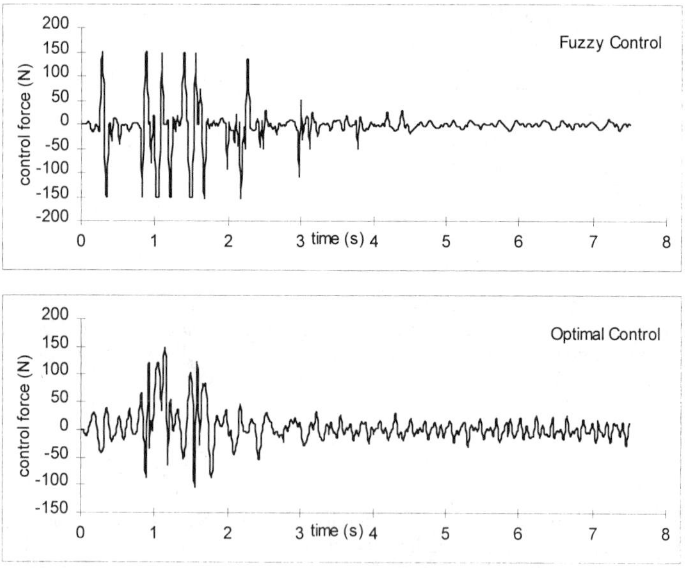

Figure 5. Control forces applied to the 5 storey frame without any reduction in mass

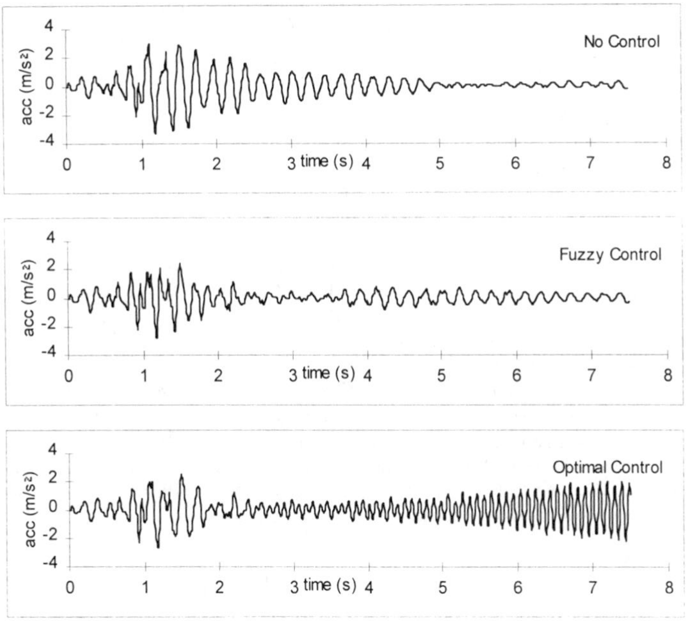

Figure 6. Acceleration response of the 5 storey frame with 20% reduction in mass

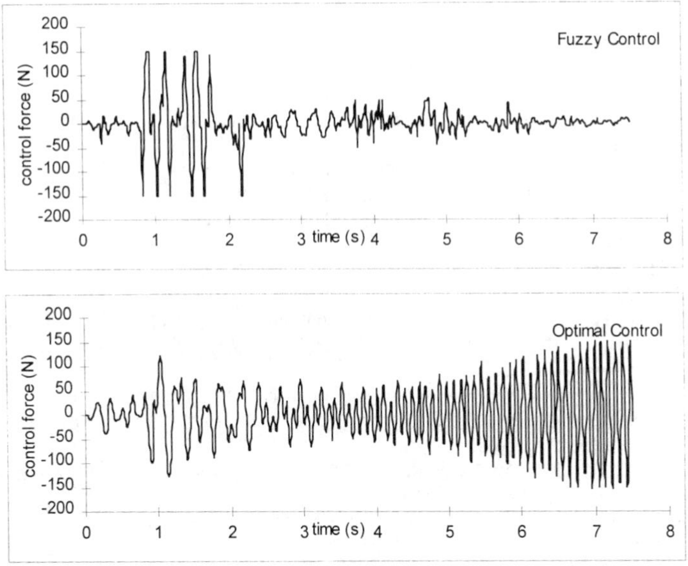

Figure 7. Control forces applied to the 5 storey frame with 20% reduction in mass

4 RESULTS

The displacement responses of the five storey frame using the optimal and fuzzy control algorithms are shown in Figure 4, when the mass of the structural system was not altered. The optimal and fuzzy control forces that were applied to the structure to counter the earthquake excitation forces are shown in Figure 5. The maximum control force that is applied by the fuzzy controller is equal to the maximum control force applied by the optimal controller. The results show that the controlled response achieved using the fuzzy control algorithm is better than that of the optimal control algorithm. The peak response reductions are greater for the fuzzy control algorithm.

The robustness of the control algorithms were also studied for the control of the five storey frame using the Northridge Earthquake loading. The controlled response of the structure was simulated for a total structural mass reduction of 6, 12 and 20 percent. The acceleration response of the structure with a 20 percent reduction in mass is shown in Figure 6. The corresponding control forces are presented in Fig. 7.

The control algorithms were designed to control the five storey frame without any reduction in the mass of the structure. The results show that not only does the fuzzy control algorithm achieve a greater peak response reduction but it is also able to maintain the stability of the structure when the properties of the structural system are altered. The controlled response of the optimal controller becomes increasingly unstable as the mass of the structural system varies from the specified structural mass that the optimal controller was designed on.

5 CONCLUSION

The performance of a fuzzy controller was compared to that of an optimal controller to control the response of a five storey frame subjected to Northridge earthquake excitation. The fuzzy controller was not only able to control the peak response better with the same level of applied control force, it was also more robust to variations in structural properties and hence more suitable for applications where structural properties are likely to vary.

6 REFERENCES

Abe, M. 1996. Rule-Based Control Algorithm for Active Tuned Mass Dampers. *Journal of Engineering Mechanics*, 122(8):705-713.

Casciati, F. & Giorgi F. 1996. Fuzzy controller implementation, *Proceedings of the 2nd International Workshop on Structural Control*,119-125

Casciati, F. 1997. Checking the stability of Fuzzy Controller for Nonlinear Structures. *Microcomputers in Civil Engineering*, 12:205-215.

Faravelli, L. & Yao, T. 1996. Use of Adaptive Networks in Fuzzy Control of Civil Structures, *Microcomputers in Civil Engineering*, 11: 67-76.

Hora, H., Miyano, H., Omni, T., Hitomi, Y. & Fujita, T. 1996. Application of Hybrid Mass Damper With Convertible Active and Passive Modes to High-Rise Buildings, *Proceedings of the 2nd International Workshop on Structural Control*, 233-240.

Jang, J.-S.R., Sun, C.-T. & Mizutani, E., 1997. *Neuro-Fuzzy and Soft Computing*, Prentice-Hall Inc.

Jang, J.-S.R. 1993. ANFIS: Adaptive-Network-Based Fuzzy Inference System, *IEEE Transactions on Systems, Man and Cybernetics*, 23: 665-685.

Mamdani, E.H. & Assilian, S., 1975. An Experiment in Linguistic Synthesis with a Fuzzy Logic Controller, *International Journal of Man-Machine Studies*, 7 : 1-13.

Subramaniam, R.S., Reinhorn, A.M., Riley, M.A. & Nagarajaiah, S. 1996., Hybrid Control of Structures using Fuzzy Logic, *Microcomputers in Civil Engineering*, 11: 1-17

Sugeno, M., 1995. *Industrial Applications of Fuzzy Control*, Elsevier Science Publishing Company

Yang, J.N., Akbarpour, A. & Ghaemmaghami, P. 1987., New Optimal Control Algorithms for Structural Control, *Journal of Engineering Mechanics, ASCE*, 113(9) : 1369-1386.

Zadeh, L.A., 1965. Fuzzy Sets, *Information and Control*, 8:338-353.

Mechanics of Structures and Materials, Bradford, Bridge & Foster (eds)
© 1999 Balkema, Rotterdam, ISBN 90 5809 107 4

Seismic response behaviour of unreinforced masonry walls by shaking table testings and time-history analyses

B. Rodolico, N. Lam & J. Wilson
Department of Civil and Environmental Engineering, University of Melbourne, Vic., Australia
K. Doherty & M. Griffith
Department of Civil and Environmental Engineering, University of Adelaide, S.A., Australia

ABSTRACT: The 1989 Newcastle Earthquake clearly demonstrated the seismic vulnerability of Unreinforced Masonry (URM) Structures in Australia. The extensive use of masonry in construction requires the competent understanding of the response of masonry to seismic loading. In order to facilitate a better comprehension of the structural dynamic processes that a URM structure experiences, a joint study by the University of Adelaide and the University of Melbourne has been in progress for the past two years. A major consideration in the structural response of an URM wall, is the out-of-plane behaviour. This paper will review the current methods used in determining the out-of-plane response behaviour of URM walls, and describe the experimental and analytical processes being utilised by the authors to further develop these methods.

1 INTRODUCTION

The Magnitude 5.5 earthquake which struck Newcastle, New South Wales, Australia in 1989 caused 11 deaths and widespread damages to unreinforced masonry (URM) infrastructures. The seismic vulnerability of URM walls, which are widely used in the construction of both residential and commercial buildings in Australia, was clearly demonstrated by the earthquake. Investigations have been undertaken jointly by the University of Adelaide and the University of Melbourne since 1997 to address potential problems associated with the seismic performance of URM walls in Australia. The objective of the investigation is to understand the physical process governing the collapse behaviour of URM walls and to establish effective analysis and design procedures for practical applications. It has been recognised that failure by out-of-plane bending is the major consideration requiring research input.

The aim of this paper is to (i) review critically a number of existing approaches which has been used in modelling the out-of-plane bending behaviour of URM walls (Section 2 & 3) and (ii) outline the experimental and analytical research strategy adopted by the investigators to develop better modelling approaches (Section 4).

The section headings for this paper are as follows:
Section 2: Static Analysis Procedures
Section 3: Dynamic Analysis Procedures
Section 4: Shaking Table Testing and Time-History Analyses

2 STATIC ANALYSIS PROCEDURES

The response of an URM wall to earthquake induced base excitations is a complex dynamic process. However, the analysis has often been simplified by considering the maximum acceleration occurring instantaneously at the critical snapshot of the response. The instantaneous acceleration is then represented by the associated inertia force in a quasi-static analysis. Earthquake resistant codes and standards typically specify analysis methods, which are based on this quasi-static philosophy. The two common methods to be briefly reviewed in this section are, (a) The linear elastic analysis method, and (b) The rigid body equilibrium analysis method.

The linear elastic analysis method, which is adopted by the Masonry Code of Australia (Standards Australia 1998), calculates the tensile and compressive stresses developed across the critical section of the wall based on the applied bending moment and axial forces. Thus, the earthquake resistant capacity of the wall is proportional to the design tensile strength of the mortar and the axial pre-compression at the cross-section of the wall.

The assumption of an uncracked wall and elastic stress-strain behaviour is a major limitation of the linear elastic analysis method. Significantly, the formation of flexural tensile cracks does not necessarily result in failure of the wall as implied by the linear elastic analysis method. Instead, the neutral axis would shift away from the centre-line of the cross-section in the direction of the compression zone. This neutral-axis shift associated with crack propagation can lead to a "tip-toe" situation in which the wall is supported on its edge [refer Figure 1].

Thus, the post-cracked seismic performance of an URM wall is more realistically analysed by the rigid body equilibrium analysis method, which compares the overturning moment with the restoring moment taken about the rocking edge of the wall. The overturning moment is obtained in the same manner as in the linear elastic analysis method, whereas the restoring moment is obtained in accordance with the self weight of the wall together with the imposed loads from the upper floors as shown in Figure 1.

It must be recognised that neither the linear elastic analysis method nor the rigid body equilibrium analysis method takes into account the time-dependent nature of the response of the wall since only the instantaneous acceleration occurring at the critical snapshot has been considered. Thus, in reality, the wall will not necessarily overturn even if the quasi-static force exceeds the

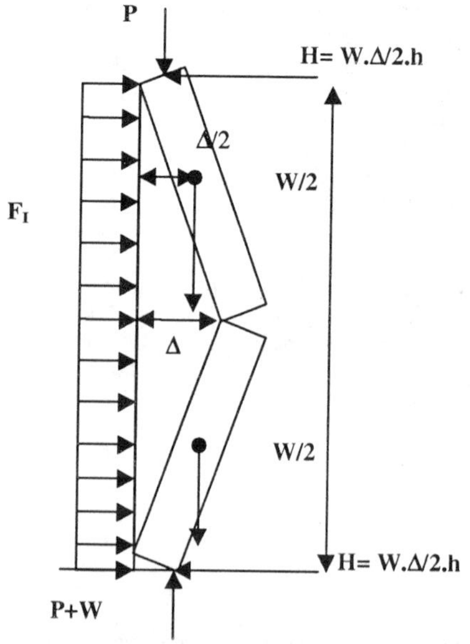

Figure 1. Static Equilibrium Analysis (Pinned -End Supports)

resistant capacity of the wall, as obtained by either of the methods just described. More realistic dynamic modelling approaches, which account for the transient nature of the earthquake-induced response, are reviewed in the next section of the paper.

3 DYNAMIC ANALYSIS PROCEDURES

Non-linear time-history analysis (THA) based on time-step integration is the most direct and reliable method to account for the time-dependent nature of the wall response and the applied excitations, provided that the force-displacement and damping properties have been accurately represented in the analytical model. Special computer softwares have been developed by the authors to operate the non-linear THA's which are being verified by comparison with results obtained from shaking table experiments. Refer Section 4 for further details.

Whilst THA is used in parametric studies for research purposes, simplified analytical procedures are generally required for design applications. Dynamic analyses can be simplified in different ways. For example, the quasi-static analysis procedures described in Section 2 are based on representing the maximum instantaneous acceleration by equivalent static forces.

Alternatively, dynamic analyses can be simplified in accordance with the equal-energy principle. The potential energy absorptive capacity of a wall undergoing large displacement is first quantified by determining the area under the force-displacement (f-d) curve as the wall is pushed from the vertical non-displaced position to the maximum displaced position without overturning. A procedure to predict such "f-d" curve of a load-bearing URM wall for the purpose of applying the equal-energy procedure has been presented by Priestley (Priestley 1985). The maximum kinetic energy (KE) demand is then derived from the maximum spectral velocity as obtained from a elastic response spectrum to compare with the PE capacity. Thus, it is assumed that the wall would remain uncracked and behave linearly elastic up to the time when the maximum KE is reached. The wall is then assumed to crack and rock as the PE is increased to absorb the KE. Importantly, ground excitations are assumed to have stopped once rocking commences.

The abrupt transition from the linear elastic response (in the uncracked state) to the non-linear rocking response (in the cracked state) is a major assumption in the equal energy procedure described above. In reality, the dynamic behaviour of the wall prior to the commencement of rocking is not always linearly elastic. It has been demonstrated by dynamic testings that the stiffness and natural period of an URM wall can be very amplitude dependent even in the uncracked state. Further, it is possible that the wall cracks a few cycles before the critical condition is reached, or the wall has already cracked before the earthquake. Thus, the "elastic" natural period of vibration of the wall, which governs its spectral velocity (and hence its maximum KE), cannot be predicted with certainty. Another major shortcoming of the equal energy procedure is that the accumulated effects of multiple pulses on the wall during rocking have not been accounted for.

An additional alternative to quasi-static analyses is based on the displacement criterion. The displacement capacity of a cracked URM wall is ascertained in accordance with the permissible displacement of the centre of gravity (COG) of the wall. For example, a parapet wall will overturn if the displacement of the COG exceeds half the wall thickness. The displacement demand of the wall can be assumed to be the maximum elastic response spectral displacement at a representative level of damping. The assumption of equal displacement of the COG of the wall and the ground motion has been shown by previous tests performed on parapet walls at the University of Melbourne and on Simply Supported (SS) walls at the University of Adelaide under sinusoidal excitation to be well correlated (Lam 1995).

Significantly, the Equal Displacement observations have been based on periodic excitation, with a 180-degree phase lag between the response of the wall and the excitation. The effect of the difference in the phase angle of the pulses and the wall's response requires additional investigation.

4 SHAKING TABLE TESTING AND TIME-HISTORY ANALYSES

One of the objectives of the shaking table is to the anomalies that are current in the 'equal energy' and the "equal displacement' principles.

Currently under development by the authors is a time-history-based design program, which utilises the iterative "Constant Average Acceleration" numerical method to derive the rocking behaviour of the wall, and relates the capacity of an URM Wall to the ability of the wall to undergo a maximum displacement.

In order to address the questions in the two models, and allow the accurate definition of the critical parameters to be utilised in the developing Time-History Program, the following testing regime is currently in progress at the two Universities

(1) Low Intensity Pulse Tests

The dynamic response behaviour are tested are tested by loading the wall with a low intensity pulse, by utilising a rubber mallet.

(2) Static Push-Over Tests

The load-displacement behaviour of cracked URM Walls, which are indicative of the amplitude dependency of the of the wall's natural period and damping characteristics are determined by the static pushover tests. Dynamic Tests

(3) Dynamic Tests

(a) Single Pulse Dynamic Tests

The response of URM walls to sinusoidal excitation has been well documented, with good agreement of equal displacement of the COG of the wall and the base excitation in test results. To further investigate this, non-periodic dynamic excitation will be applied to the wall in the form of single pulse excitations. Gaussian Pulses of varying intensities and pulse widths will be used.

(b) Dual Pulse Dynamic Tests

The application of dual pulses is to investigate the effect of phase lag between the individual pulses. These sets of tests utilise two single Gaussian pulses.

(c) Earthquake Records Dynamic Tests

Both the single and dual pulse tests will provide valuable data on distinct aspects of the URM walls. Ultimately we are interested in the response of the walls to earthquakes ground motion that will denote the accumulation of displacements and pulse arrival times on the wall's response. The El Centro, Taft and Nahanni earthquake accelerograms will be used as the applied base excitations.

A total of eight specimens will be tested at both Universities, four with a H/t ratio of 15, and four with a H/t ratio of 30. The height of the walls undergoing testing at the University of Melbourne will be for 1.5 metres and 3.0 meters respectively. (Figure 2) The specimens are of single wythe construction, using typical masonry units supplied by Boral Bricks, with a mortar composition of 1:1/2:6

Instrumentation consisted of Solartron Transducers (Inductive Type) and Sundstrand Accelerometers. The measurements were taken relative to the reference frame, which has been installed for safety. The static push-over test measurements were recorded and logged utilising the data acquisition system "Orion", while the dynamic data was logged and stored by the data acquisition system "T-Spectra" (a 16 bit 16-channel system). In conjunction with the data acquisition systems, additional data was recorded using digital video and photos, this was also enhanced by personal notes relating the observations of the authors.

Figure 2. URM Wall Construction University of Melbourne

For the dynamic excitation of the testing program, the MTS System Controlled Shaker Table system (Ghad 1997) design and built at the University of Melbourne was employed, with the testing regime at the University of Adelaide undertaking a similar testing program. The 2m x 2m Table has a load capacity of approximately 3 tonnes and a stroke capacity of 100mm.

5 CONCLUSIONS

1. The current quasi-static analyses methodology employed in the design of unreinforced masonry walls are conservative design procedures and do not realistically represent the potential seismic performance behaviour of URM walls.
2. Simplified procedures that are based on dynamic analysis principles, which provide realistic and reliable results, have been reviewed and will be developed further.
3. In order to be able to utilise the simplified forms of dynamic analysis, the accurate definition of the critical parameters in the dynamic response of URM walls is essential. The development of the Time-History Program by the authors is a crucial means of predicting the response behaviour of an URM Wall.

REFERENCES

Gad, E F. & Siah, M. 1997. Structural Laboratory Facilities. *Research Report University of Melbourne.* RR/Struct/01/97: 11-12.

Lam, N. & Wilson, J. & Hutchinson, G. 1995. Modelling of an unreinforced masonry parapet wall for seismic performance evaluation based on dynamic testing. *Departmental Report*. RR/Struct/03/95

Priestley, M.J.N. 1985. Seismic Behaviour of Unreinforced Masonry Walls. *Bulletin of the New Zealand National Society for Earthquake Engineering*, Vol 18: No.2: 191-205

Standards Australia. 1998 Masonry Structures.

Mechanics of Structures and Materials, Bradford, Bridge & Foster (eds)
© 1999 Balkema, Rotterdam, ISBN 90 5809 107 4

Determination of earthquake response spectra in low and moderate seismicity regions using new methodologies

N. Lam, J. Wilson, R. Koo & G. Hutchinson
Department of Civil and Environmental Engineering, University of Melbourne, Vic., Australia

ABSTRACT: Intraplate earthquakes occur away from tectonic plate boundaries in areas of low and moderate seimicity, such as Australia. The occurrences of these intraplate events appear randomly in both space and time. The conventional empirical approach of obtaining response spectra for these regions is often problematic due to the typical lack of representative strong motion data. The alternative seismological model, which was originally developed in the United States, addresses the physical processes in the generation of earthquake ground motions and hence does not rely on indigenous strong motion data in the prediction of earthquake excitations. The model has been further developed by the authors to obtain earthquake response spectra for rock sites and bedrock. A separate procedure has been developed to predict response spectra which take into account the effects of resonance in the soil. Such recently developed procedures are summarised in this paper.

1 INTRODUCTION

The conventional empirical approach of obtaining response spectra for regions of low and moderate seismicity, such as Australia, is often problematic due to the typical lack of representative strong motion data. The alternative seismological model was originally developed in the United States within a theoretical framework and hence it addresses the underlying physical processes in the generation and transmission of seismic shear waves (Atkinson & Boore, 1998; Boore & Joyner, 1997). The model has been developed by the authors to determine response spectra for rock sites and bedrock in areas of low and moderate seismicity (e.g. Lam, Wilson & Hutchinson, 1999a, b). In the procedure proposed herein, the Velocity, Displacement and Acceleration Response Spectra are constructed separately in accordance with the tri-linear velocity spectrum model. The response spectra are defined by the following parameters: (i) Effective Peak Ground Velocity (EPGV), (ii) Effective Peak Ground Displacement (EPGD), and (iii) Effective Peak Ground Acceleration (EPGA). Having determined the bedrock response spectra, the Frame Analogy Soil Amplification (FASA) model is applied to construct response spectra which take into account the effects of resonance in the soil (Lam, Wilson, Edwards & Hutchinson, 1997).

These recently developed response spectrum modelling methodologies are introduced in this paper which is outlined as follows:
(i) The tri-linear velocity response spectrum for rock sites and bedrock (Section 2)
(ii) Predictions of the ground motion parameters for rock sites and bedrock (Section 3)
(iii) Seismicity Modelling (Section 4)
(iv) The Frame Analogy Soil Amplification Model (Section 5)

2 THE TRI-LINEAR VELOCITY RESPONSE SPECTRUM FOR ROCK SITES AND BEDROCK

In this section, the tri-linear velocity response spectra (Fig.1b), associated displacement (Fig.1a) and acceleration response spectra (Fig.1c) are defined, and the relationships of these response spectra with the relevant ground motion parameters (EPGV, EPGD and EPGA) are described.

The tri-linear velocity spectrum (in the logarithmic tripartite form) is defined by the two corner periods T_1 and T_2, as shown in Figure 1b. The peak response spectral velocity, RSV_{peak}, defining the flat (constant) part of the spectrum is, by definitions, related to EPGV based on the expression (Lam, Wilson, Chandler, & Hutchinson, 1999):

$$RSV_{peak} = 2.0 \text{ EPGV} \tag{1}$$

The corresponding displacement response spectrum (RSD) in the bi-linear form is defined by the corner period T_2 (Fig.1a). The peak RSD in the flat part of the spectrum is taken to be equal to EPGD, whilst the sloping part is defined by the expression:

$$RSD = (RSV_{peak} / 2\pi)\, T \quad \text{or} \quad RSD = (2.0EPGV// 2\pi)\, T \tag{2}$$

where EPGD is defined as the response spectral displacement of a single-degree-of-freedom oscillator possessing a natural period of 5 seconds and 5% damping.
The corner period, T_2, is defined by the expression:

$$T_2 = (2\pi/2.0)\, EPGD/EPGV = 3.15\, EPGD/EPGV \tag{3}$$

The corresponding acceleration response spectrum (RSA) is represented in the usual flat-hyperbolic form (Fig.1c). The hyperbolic part of the spectrum between the two corner periods T_1

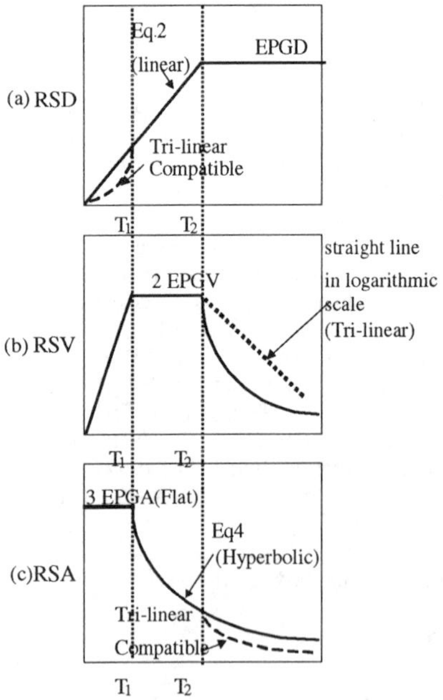

Figure 1a-1c Idealised Displacement, Velocity & Acceleration Spectra and T_2, is defined by the expression:

$$RSA = 2\pi\, RSV_{peak} / T \quad \text{or} \quad RSA = 2\pi\,(2.0\, EPGV / T) \tag{4}$$

RSA_{peak}, which defines the flat part of the spectrum ($T<T_1$), is related to EPGA by the relationship (Lam, Wilson, Chandler, & Hutchinson, 1999):

$$RSA_{peak} = 3.0\, EPGA \tag{5}$$

The corner period, T_1, is defined by the expression:

$$T_1 = 0.42/(EPGA/EPGV) \tag{6}$$

where EPGA/EPGV is in the common units of g/(m/sec)

3 PREDICTION OF THE GROUND MOTION PARAMETERS FOR ROCK SITES AND BEDROCK

In this section, expressions to determine the ground motion parameters introduced in Section 2 are presented. These expressions have been found to be consistent with a number of existing empirical models (Lam, Wilson, Chandler, & Hutchinson, 1999; Lam, Wilson, Edwards & Hutchinson, 1998; Lam, Wilson, & Hutchinson, 1999b). Significantly, recent seismological investigations indicate that the regional dependent frequency characteristics of earthquake ground motions are largely attributed to regional variations in the crustal properties. In comparison, regional variations in the average source properties of the generated seismic shear waves are much more moderate (Atkinson & Boore, 1998). Brief qualitative descriptions of the two generic crustal conditions are given below, whilst more precise definitions can be found in Boore & Joyner (1997) and Atkinson & Boore (1998).

3.1 *The Generic Crustal Conditions.*

Generic Hard Rock refers to the very old and hard sedimentary and metamorphic rocks in the stable "shield" regions (e.g. Eastern North America and Western Australia). This class of rocks has very high shear wave velocities and excellent wave transmission. Shear wave velocity is well in excess of 1000m/sec, even for exposed rocks. Crystalline volcanic rock (e.g. granite) possesses similar qualities. The velocity and displacement components of the seismic shear waves transmitted to the ground surface, through Generic Hard Rock, are practically identical to those generated at the source of the earthquake except for allowances associated with the geometrical spreading of energy. The acceleration components of the seismic shear waves are only moderately attenuated by energy absorption along the wave travel path. Thus, high frequency spikes are commonly observed in earthquakes occurring in regions covered by Generic Hard Rock.

In contrast, Generic Rock refers to young sedimentary rocks in geologically active and heavily folded continental margins (e.g. Western North America and Eastern Australia). The shear wave velocity is generally much lower, with much steeper gradient. Shear wave velocity in exposed Generic Rock is only of the order of 600m/sec. The wave transmission qualities of Generic Rock vary between regions, and are generally much poorer than Generic Hard Rock, particularly within the upper 3 - 4 km of the earth crust. Due to crustal amplification associated with the steep velocity gradient in Generic Rock (Boore & Joyner, 1997), both the velocity and displacement components of the seismic shear waves are significantly amplified. Suitable crustal factors have been derived accordingly to allow for the amplifications (Lam, Wilson, & Hutchinson, 1999b). In contrast, the acceleration components of the earthquake ground motions in Generic Rock are attenuated much more than in Generic Hard Rock. Thus, earthquake ground motions in Generic Rock is typified by a low frequency content.

3.2 *EPGV*

EPGVs (in mm/sec) can be predicted for the Generic Hard Rock model (Boore & Joyner, 1997):

$$EPGV = 50\,\alpha_V(M)\,\beta(R) \tag{7a}$$

Where $\alpha_V(M) = 0.35 + 0.65\,(M-5)^{1.8}$ (7b)

$$\beta = 30/R \tag{7c}$$

M and R refer to the Moment Magnitude and the site-source distance. To allow for the crustal effects of the Generic Rock conditions, EPGV is modified by the following crustal factor:

$$\gamma_V(M,R) = 1.6 + (30-R)/100 - (6-M)/10 \tag{8}$$

3.3 *EPGV/EPGD (V/D) ratios*

The EPGV/EPGD (V/D) ratio depends on the frequency content of the generated seismic shear waves, which is a function of M, R and the crustal properties. However, parametric studies show that the V/D ratio is insensitive to R and regional variations in the crustal properties (Table 1). Clearly, the larger the earthquake magnitude, the lower the V/D ratio.

Table 1: V/D ratios ($secs^{-1}$)

Moment Magnitude	V/D ($secs^{-1}$)
5	6.0*
5.5	5.0*
6	3.5
6.5	2.5
7	2.0

* The V/D ratio can be very sensitive to rounding off errors associated with both the modelling of the EPGD and EPGV, particularly for small magnitude events.

3.4 *EPGA/EPGV (A/V) ratios*

The high frequency properties of the seismic shear waves are very sensitive to regional anelastic attenuation properties of the earth crust. Seismological simulations by the authors for Generic Hard Rock (assuming a Quality factor (Q) of 680 $f^{0.36}$) have derived the following expression for the EPGA/EPGV (A/V) ratio (Atkinson & Boore, 1998):

$$A/V\ [g/(m/sec)] = 6 + (30-R)\,\{3+0.15(M-5)\}/90 + 1.2(6-M)$$ (

The results of magnitude and distance dependent A/V ratios are shown in Table 2.

The A/V ratio is shown to increase from 3 to 8 depending on the earthquake magnitude and distance, which is generally very high due to the excellent wave transmission quality of the Generic Hard Rock. It is noted that the Chicultimi Nord earthquake records of the 1988 Saguenay earthquake event (in Quebec, Canada) shows an A/V ratio in the order of 5 at a distance of approximately 43km from the source (of about magnitude 6), which is in general agreement with Table 2.

Table 2. A/V Ratios for Generic Hard Rock [units: g/(m/sec)]

M	R=10km	R=30km	R=50km	R=85km
5	8.0	7.0	6.5	5.5
5.5	7.5	6.5	6.0	4.5
6	6.5	6.0	5.5	4.0
6.5	6.0	5.5	4.5	3.5
7	5.5	5.0	4.0	3.0

Similar simulations for Generic Rock showed A/V ratios in the order of 0.4 times the values shown in Table 2. A Quality factor of 204 $f^{0.56}$ and a magnitude dependent upper crust attenuation parameter Kappa (k) of between 0.035 and 0.050 were assumed in the simulations (Atkinson & Boore, 1998).

4 SEISMICITY MODELLING

A procedure to calculate the response spectral parameters in terms of M and R has been presented in Sections 3. In high seismicity regions, spectral attenuation functions are typically combined with an assumed source zone model, using Cornell-McGuire Integration, to determine the site seismic hazard. However, low and moderate seismicity regions are characterised by large diffused area source zones with uniform seismicity. In these situations, Cornell-McGuire Integration may not be necessary and the site seismic hazard can be determined in accordance with a list of M-R combinations, which are related to the level of seismicity through a probabilistic function. The derivation of the probabilistic function is as follows:
First, the seismicity level is defined by the Richter-Gutenberg magnitude recurrence relationship:

$$\log_{10} N(M) = a - b\,M \qquad (10a)$$

or

$$\log_{10} N(M) = a_5 - b\,(M-5) \qquad (10b)$$

where N(M) may be defined as the expected number of earthquakes of Magnitude, M, or greater which occur within an area of 100,000km^2 over a 100 year period. a_5 is the logarithm of the total number of earthquakes with magnitude 5 or greater.

Consider that the number of earthquakes, N^*, generated within a circular area, S' (with a radius R_S), within a source zone surrounding a given site, is proportional to the size of that area (πR_S^2) and the average return period, T_{RP} (years). Hence, N(M) can be defined by the following relationship (based on proportionality):

$$N^* = N(M)\,(\pi R_S^2 T_{RP})/(100 \text{ years x } 100{,}000 \text{ km}^2) \qquad (11)$$

A specific source area $S' = \pi R_S^2$ (km^2) is needed to produce one event, that is $N^* = 1$, of magnitude M or larger, in a period of T_{RP} (years). Hence the design earthquake magnitude, M, for given values of R_S, a_5 and b can be determined by substituting equation (11), assuming $N^* = 1$, into equation (10b), and rearranging the terms as follows:

$$M = 5 + \{\log_{10}(\pi\, R_S^2 T_{RP}) - 7 + a_5\}/b \qquad (12)$$

We now ask at what average distance R would occur in the epicentre of an event of magnitude M or larger, from a point site that floats in the open-ended region characterized by seismicity parameters a_5 and b. Expressing the total area S' in terms of a median-probability (50-percentile) distance R (Jacob, 1997), one obtains $\pi R^2 = \pi\, R_S^2/2 = S'/2$. Hence

$$R = R_S / \sqrt{2} \qquad (13)$$

Thus, equation (12) can be rewritten as follows:

$$M = 5 + \{\log_{10}(2\pi R^2 T_{RP}) - 7 + a_5\}/b \qquad (14)$$

It can be shown that if $a_5 = 1.6$, $b = 0.9$ and TRP = 500 years are substituted into eqn (14), the M-R Combinations shown in Table 3 are obtained.

Interestingly, the listed M-R Combinations, when substituted into equations 7a-7c, would predict a EPGV value in the order of 60 mm/sec which corresponds approximately to an acceleration coefficient of 0.08g as specified by AS 1170.4 (1993) for most Australian Capital Cities.

Table 3. M-R Combinations associated with $a_5 = 1.6$, $b = 0.9$ and TRP = 500 years

M	R (km)
5	10
5.5	20
6	30
6.5	50
7	70

The procedure to determine the response spectrum for rock sites and bedrock is now summarised as follows:

(i) Determine M-R combinations based on the given Return Period and seismicity level (Sect. 4).
(ii) Determine the response spectral parameters (EPGV, EPGD & EPGA) based on the given M-R combinations and regional crustal classification (Sect. 3).
(iii) Construct response spectra based on the tri-linear velocity spectrum model and the parameter values obtained in Step (ii) (Sect. 2).

5 THE FRAME ANALOGY SOIL AMPLIFICATION MODEL FOR SOIL SITES

Having determined the bedrock response spectrum, the Frame Analogy Soil Amplification (FASA) model can be applied to construct response spectra, which take into account the effects of resonance in the soil. The FASA model is based on drawing an analogy between the dynamic response of a soil column and a moment resisting frame developing "beam-sway mechanism (Lam, Wilson & Hutchinson, 1997).

A key component of the procedure is the determination of the maximum response spectral acceleration (Sa_{max}) at the natural period of the site (T_g). Refer to Fig. 2.
Sa_{max} can be obtained by the following relationships:

$$Sa_{max} = 3\ a_{max} \tag{15}$$

$$a_{max} = 1.2\ \beta Sa_i \tag{16}$$

where a_{max} = max. ground acceleration of site
= 3, conventional spectral amplification factor
= 1.2, participation factor based on a parabolic displacement profile
β = correction factor for damping other than 5%
Sa = bedrock spectral acc. at site natural period for 5% damping.

A correction factor for damping other than 5% can be established using the following expression:

$$\beta = (7\% / (2 + \xi\%))^{1/2} \tag{17}$$

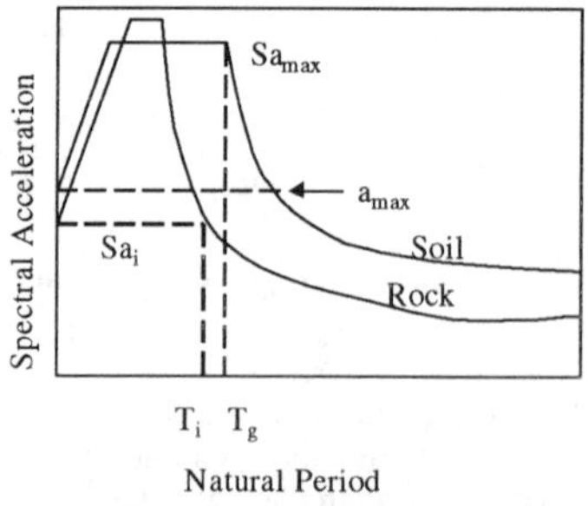

Figure 2 Soil and Rock Spectrum Diagram

Equations 15, 16 and 17 can be combined into the following expression:

$$Sa_{max} = 3(1.2)Sa(7\% / (2 + \xi\%))^{1/2} \quad (18)$$

If significant shifts are expected to occur, equation (18) is rewritten as:

$$Sa_{max} = 3(1.2)Sa_i(T_i / T_g)(7\% / (2 + \xi\%))^{1/2} \quad (19)$$

where Sa_i = spectral acceleration at T_i, T_i = initial site period and T_g = final site period
For soil columns significantly exceeding 15m, an additional amplification factor λ should be used, where $\lambda = 1 + 0.03(H - 15)$ and H is the depth of the soil column.
Refer to Fig. 2 for diagrammatic illustration of the response spectrum construction.

6 CONCLUSIONS

1. The velocity, displacement and acceleration response spectra based on the tri-linear velocity spectrum model have been introduced.
2. The relationships between the response spectra and the spectral parameters (EPGV, EPGD and EPGA) have been defined.
3. Spectral attenuation functions of M and R in the determination of the spectral parameters have been derived from seismological model simulations.
4. The relationship to derive suitable M-R combinations from the given Return Period and seismicity level has been introduced.
5. The procedure to construct response spectra for rock sites and bedrock has been summarised.
6. The frame analogy soil amplification model has been introduced to construct response spectra, which take into account the effects of resonance in the soil.

7 ACKNOWLEDGEMENTS

The procedure described in this paper has been developed as part of a project funded by the Australian Research Council (large grant), entitled: " Earthquake Design Parameters and Design Methods for Australian Conditions"(AB89701689). This support is gratefully acknowledged.

8 REFERENCES

Atkinson, G.M. & Boore, D.M. (1998), "Evaluation of Models for Earthquake Source Spectra in Eastern North America", Bulletin of the Seismological Society of America, Vol.88 (4), pp917-934.

Boore, D.M. & Joyner, W.B. (1997), "Site amplifications for generic rock sites", *Bulletin of the Seismological Society of America*, Vol.87 (2), pp327-341.

Jacob, K.H. (1997), "Scenario earthquakes for urban areas along the Atlantic seaboard of the United States", NCEER-SP-0001, National Centre for Earthquake Engineering Research, Buffalo, New York.

Lam, N.T.K., Wilson, J.L., Chandler, A.M. & Hutchinson, G.L. (1999), "Response Spectrum Modelling in Low and Moderate Seismicity Regions Combining Velocity, Displacement and Acceleration Predictions", Earthquake Engineering and Structural Dynamics (under review).

Lam, N. T. K., Wilson, J. L. & Hutchinson, G. L. (1997), "Introduction to a New Procedure to Construct Site Response Spectrum", Proceedings of ACMSM Conf., Melbourne, 8-10 December, pp345-350.

Lam, N. T. K., Wilson, J. L., Edwards, M. R. & Hutchinson, G. L. (1998), "A Displacement Based Prediction of Seismic Hazard for Australia", Australian Earthquake Engineering Society Seminar, Perth.

Lam, N.T.K., Wilson, J.L. & Hutchinson, G.L. (1999a), "Generation of Synthetic Earthquake Accelerograms based on the Seismological Model ", Journal of Earthquake Engineering (in press).

Lam, N.T.K., Wilson, J.L. & Hutchinson, G.L. (1999b), "A Generic Displacement Spectrum Model for Rock Sites in Low Seismicity Regions", Journal of Earthquake Engineering (in press).

Mechanics of Structures and Materials, Bradford, Bridge & Foster (eds)
© 1999 Balkema, Rotterdam, ISBN 90 5809 107 4

Vibrations of laminated rectangular plates with intermediate line-supports

Y.K.Cheung
Department of Civil Engineering, University of Hong Kong, People's Republic of China
D.Zhou
School of Mechanical Engineering, Nanjing University of Science and Technology, People's Republic of China

ABSTRACT: In this paper, a computationally efficient and highly accurate numerical method is proposed to analyze the free vibrations of symmetrically laminated rectangular composite plates with intermediate line supports. A set of admissible functions are developed from the static solutions of a beam with intermediate point-supports under a series of sinusoidal loads, in which the beam may be considered to be a strip of unit width taken from the plate in a direction parallel to the edges of the plate. In addition to satisfying both the geometric boundary conditions of the plate and the zero deformation conditions at the line-supports, this set of static beam functions, being different from the existing admissible functions, can also properly describe the discontinuity of the shear forces at the line-supports, so that more accurate results can be expected for the dynamic analysis of laminated rectangular plate with intermediate line supports. The governing eigenfrequency equation is derived by using the Rayleigh-Ritz approach. Several test problems are solved to demonstrate the accuracy and the convergency of the proposed method. Some data known for the first time are given and can serve as the benchmark for further research.

1 INTRODUCTION

Composite materials, especially laminated composite plates, have been widely used in various kind of engineering such as aeronautic, astronautic and marine structures and so on. In addition to their high strength and light weight, another advantage of laminated composite plates is the controllability of the structural properties through changing the fibre angle and the number of plies and/or selecting proper composite materials, over a wide range.

It is obvious that in the governing differential equation [1] of symmetrically laminated rectangular composite plates, the presence of odd derivatives in the terms containing D_{16} and D_{26} (flexural stiffnesses causing bending-twisting coupling) prevents the equation from exact solution, even for the case when all edges of the plate are simply supported. Therefore, numerical methods must be used for solution. A brief review for the vibration study of such problems has been given by Leissa and Narita [2]. Bert [3] gave an approximate formula of the fundamental frequency for orthotropic plates. Han and Petyt [4] used the finite element method, Whitney [5] used Fourier series expanding method and Chow et al. [6] used Rayleigh-Ritz method to study the vibration behavior of symmetrically laminated rectangular composite plates. Moreover, some investigators [7-10] studied the free vibration of isotropic rectangular plates with intermediate line-supports. A review of literature shows that no information is available for the vibration of symmetrically laminated rectangular composite plates with intermediate line-supports except that Kong and Cheung [11] used the finite layer method to study the shear-deformable rectangular plates with intermediate line supports, but, only for cross-ply laminated thick plates with $D_{16} = D_{26} = 0$.

In this paper, a set of admissible functions, which is composed of static beam functions [10,12], is developed for the vibration analysis of symmetrically laminated rectangular composite plates. Unlike the conventional admissible functions, in which only the geometric boundary conditions on the edges and the zero displacement conditions at the line-supports can be satisfied, the discontinuity of the shear forces at the line-supports may also be properly described by using this set of proposed functions. The Rayleigh-Ritz method is employed to determine the vibration behavior of the composite plates. Comparison and Convergency studies show that the present approach has rather high accuracy and efficiency. Because there is a lack of vibration solutions for symmetrically laminated rectangular composite plates with intermediate line-supports, some new data, which can serve as the benchmark for further research, are presented.

2 A SET OF STATIC BEAM FUNCTIONS

Consider a uniform beam with J intermediate point-supports under an arbitrary load $q(x)$. The length of the beam is l and the point-supports are, respectively, at $x_j, j=1,2,...,J$. The deflection $z(x)$ of the beam in the z direction should satisfy the differential equation

$$EId^4z/dx^4 = \sum_{j=1}^{J} p_j\delta(x-x_j) + q(x), \quad 0<x<1 \tag{1}$$

where EI is the flexural rigidity of the beam, p_j is the reaction of the jth point-support and $\delta(x-x_j)$ is the Dirac delta function. Letting

$$\xi = x/l,\ \xi_j = x_j/l,\ P_j = p_j l^3/(EI),\ Q(\xi) = q(l\xi)l^4/(EI) \tag{2}$$

one has

$$d^4z/d\xi^4 = \sum_{j=1}^{J} P_j\,\delta(\xi-\xi_j) + Q(\xi), \quad 0<\xi<1 \tag{3}$$

For an arbitrary load $Q(\xi)$, it can be expanded into a Fourier sine series as follows

$$Q(\xi) = \sum_{i=1}^{\infty} Q_i(i\pi)^4 \sin i\pi\xi, \quad 0<\xi<1 \tag{4}$$

where Q_i are the unknown coefficients which may be decided uniquely if $Q(\xi)$ is given.
It is clear that for the linear problem considered here, the solution $z(\xi)$ of equation (3) can take the form of

$$z(\xi) = \sum_{i=1}^{\infty} Q_i z_i(\xi) \tag{5}$$

Substituting equations (4) and (5) into equation (3), one obtains the general solution of $z_i(\xi)$ as follows

$$z_i(\xi) = \sum_{k=0}^{3} C_k^i \xi^k + \sum_{j=1}^{J} P_j^i(\xi-\xi_j)^3 U(\xi-\xi_j)/6 + \sin i\pi\xi \tag{6}$$

where $C_k^i\ (k=0,1,2,3)$ and $P_j^i\,(j=1,2,...,J)$ are unknown coefficients, and $U(\xi-\xi_j)$ are the Heaviside functions. It should be noted that the order of the polynomials in the above equation is never higher than cubic and is independent of both the number of the intermediate point-supports and the number of terms of the Fourier sinusoidal series.
For convenience, two capital letters are used to indicate the boundary conditions at the two ends of the beam by letting the letters *C, S* and *F* denote, respectively, clamped, simply supported and free ends. For the *CC, CS, CF, SS* beams with or without intermediate point-supports, the *SF* beam with no less than one intermediate point-support and the *FF* beam with no less than two intermediate point-supports, the coefficients $C_k^i(k=0,1,2,3)$ and $P_j^i(j=1,2,...,J)$ in equation (6) can be uniquely decided by the boundary conditions of the beam and the zero deflection conditions at the internal point-supports, which may be written in matrix form of

$$\begin{bmatrix} A & D \\ T & G \end{bmatrix} \begin{bmatrix} C^i \\ P^i \end{bmatrix} = \begin{bmatrix} R^i \\ S^i \end{bmatrix} \tag{7}$$

where A is a $J\times 4$ matrix, D is a $J\times J$ matrix and R^i is a $J\times 1$ matrix, and they refer, respectively, to the values of the first series, the second series and the third term on the right side of equation (6) at $x_j, j=1,2,\cdots,J$. T is a 4×4 matrix, G is a $4\times J$ matrix and S^i is a 4×1 matrix, and they refer, respectively, to the values of the first series, the second series and the third term on the right side of equation (6) for the boundary conditions of the beam. C^i and P^i are the unknown coefficient matrices as follows

$$C^i=[C_0^iC_1^iC_2^iC_3^i]^T,\quad P^i=[P_1^iP_2^i...P_J^i]^T \tag{8}$$

Without losing generality It may be assumed that $\xi_j<\xi_k$ if $j<k$, and the matrices D, A and R^i may be, respectively, written as

$$D=\begin{bmatrix} 0 & 0 & 0 & \cdots & 0 \\ (\xi_2-\xi_1)^3/6 & 0 & 0 & \cdots & 0 \\ (\xi_3-\xi_1)^3/6 & (\xi_3-\xi_2)^3/6 & 0 & \cdots & 0 \\ \vdots & \vdots & \ddots & \ddots & \vdots \\ (\xi_J-\xi_1)^3/6 & (\xi_J-\xi_2)^3/6 & \cdots & (\xi_J-\xi_{J-1})^3/6 & 0 \end{bmatrix},$$

$$A=\begin{bmatrix} 1 & \xi_1 & \xi_1^2 & \xi_1^3 \\ 1 & \xi_2 & \xi_2^2 & \xi_2^3 \\ \vdots & \vdots & \vdots & \vdots \\ 1 & \xi_J & \xi_J^2 & \xi_J^3 \end{bmatrix},\quad R^i=\begin{bmatrix} -\sin i\pi\xi_1 \\ -\sin i\pi\xi_2 \\ \vdots \\ -\sin i\pi\xi_J \end{bmatrix} \tag{9}$$

For convenience, the symbols $t_{kl}(k=1,2,3,4, l=1,2,3,4)$, $g_{kj}(k=1,2,3,4, j=1,2,...,J)$ and $s_k^i(k=1,2,3,4)$ are used to represent the elements in matrix T, G and S^i, respectively. According to the boundary conditions of the beam, there are $t_{11}=t_{22}=1$, $s_2^i=-i\pi$ for the beam with a C left end, $t_{11}=1$, $t_{23}=2$ for the beam with a S left end and $t_{13}=2$, $t_{24}=6$, $s_2^i=(i\pi)^3$ for the beam with a F left end, and $t_{31}=t_{32}=t_{33}=t_{34}=1$, $t_{42}=1$, $t_{43}=2$, $t_{44}=3$, $g_{3j}=(1-\xi_j)^3/6$, $g_{4j}=(1-\xi_j)^2/2$, $s_4^i=-(-1)^i i\pi$ for the beam with a C right end, $t_{31}=t_{32}=t_{33}=t_{34}=1$, $t_{43}=2$, $t_{44}=6$, $g_{3j}=(1-\xi_j)^3/6$, $g_{4j}=1-\xi_j$ for the beam with a S right end and $t_{33}=2$, $t_{34}=6$, $t_{44}=6$, $g_{3j}=1-\xi_j$, $g_{4j}=1$, $s_4^i=(-1)^i(i\pi)^3$ for the beam with a F right end. The other elements, which are not listed here, are all equal to zero. Solving the linear equation group (7) gives all the unknown coefficients.

For a FF beam with an intermediate point-support, the rigid rotation around the point-support exists. In this case, the coefficient matrices C^i and P^i cannot be directly determined from equation (7). However, one may add the rigid rotation mode to the static beam functions, for example, assuming that $z(\xi)=\sum_{i=1}^{\infty}Q_i\bar{z}_i(\xi)$ then taking $\bar{z}_1(\xi)=\xi-\xi_1$ and $\bar{z}_i(\xi)=z_{i-1}(\xi)\ (i\geq2)$ which are the static beam functions of a SF beam with the corresponding point-support. For a SF or FF beam without intermediate point-supports, similar handling

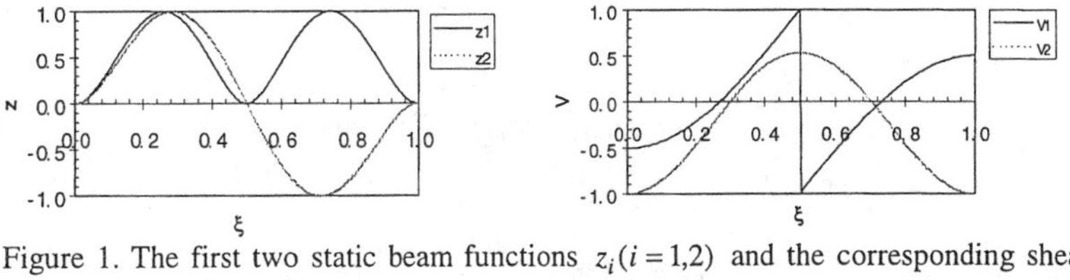

Figure 1. The first two static beam functions $z_i(i=1,2)$ and the corresponding shear-force distributions $V_i(i=1,2)$ for a CC beam with a mid-point support.

method can be used, which has been described in reference [12] in detail. It is important to note that the matrix on the left side of equation (7), which is uniquely determined by the boundary conditions of the beam and the locations of the intermediate point-supports, is independent of the series variable i. So only one inverse calculation to the coefficient matrix of equation (7) is needed to determine the coefficients $[C_i\ P_i]^T$ for all i, which will result in very small computational cost. It can be seen that this set of static beam functions can describe the discontinuity of the shear forces at the point-supports exactly. As an example, the first two static beam functions (divided by the maximum deflection) and the corresponding shear-force distributions (divided by the maximum shear force) for a *CC* beam with a mid-point support are given in Figure 1.

3. EIGENFREQUENCY EQUATION

Consider a thin, symmetrically laminated rectangular composite plate with P and L line-supports in x and y directions respectively. The plate, with thickness h in the z direction, consists of n layers of orthotropic plies. The fibre orientation within a layer with respect to the x axis is indicated by the angle β. The moduli of elasticity for layers parallel to the fibres is E_1 and for those perpendicular to the fibres is E_2.
The maximum strain and kinetic energies for free vibration of the plate can be expressed as

$$U_{\max} = \frac{1}{2}\int_0^a\int_0^b \{D_{11}[\frac{\partial^2 W}{\partial x^2}]^2 + 2D_{12}[\frac{\partial^2 W}{\partial x^2}\frac{\partial^2 W}{\partial y^2}] + D_{22}[\frac{\partial^2 W}{\partial y^2}]^2 + 4D_{16}[\frac{\partial^2 W}{\partial x^2}\frac{\partial^2 W}{\partial x\partial y}] +$$

$$4D_{26}\frac{\partial^2 W}{\partial y^2}\frac{\partial^2 W}{\partial x\partial y}] + 4D_{66}[\frac{\partial^2 W}{\partial x\partial y}]^2\}dydx\,,\quad T_{\max} = \frac{1}{2}\rho h\omega^2\int_0^a\int_0^b W^2(x,y)dydx \tag{10}$$

where ρ is the material density, ω is the angular eigenfrequency, W is the dynamic displacement function of the plate and D_{ij} (i,j=1,2,6) are the bending stiffnesses of the plate, as defined in reference [1].
Letting $\xi = x/a$, $\eta = y/b$ and applying the method of separation of variables, the displacement function $W(\xi,\eta)$ may be expressed as

$$W(\xi,\eta) = \sum_{i=1}^{M}\sum_{j=1}^{N} C_{ij}X_i(\xi)Y_j(\eta) \tag{11}$$

where C_{ij} is the unknown coefficient, M and N are the truncated orders of the admissible functions in ξ and η directions, respectively. Here, $X_i(\xi)\,(i=1,2,...,M)$ are the static beam functions satisfying both the geometric boundary conditions of the plate and the zero deflections at the intermediate line-supports in the ξ direction , as described in the last section. Similarly, $Y_j(\eta)(j=1,2,...,N)$ are those in the η direction. It is obvious that this set of admissible functions can properly represent the discontinuity of the shear forces at the intermediate line-supports, as shown in Figure 1.
Substituting equation (11) into equation (10) and applying the Rayleigh-Ritz procedure

$$\partial(U_{\max} - T_{\max})/\partial C_{ij} = 0\,,\qquad i=1,2,\ldots,M \text{ and } j=1,2,\ldots,N \tag{12a}$$

leads to the following governing eigenfrequency equation

$$\sum_{i=1}^{M}\sum_{j=1}^{N}[K_{imjn} - \lambda^2 F_{im}^{00}G_{jn}^{00}]C_{ij} = 0\,,\qquad m=1,2,\ldots,M \text{ and } n=1,2,\ldots,N \tag{12b}$$

where

$$K_{ijmn} = \{D_{11}F_{im}^{22}G_{jn}^{00} + \gamma^4 D_{22}F_{im}^{00}G_{jn}^{22} + \gamma^2 D_{12}[F_{im}^{02}G_{jn}^{20} + F_{im}^{20}G_{jn}^{02}] + 2\gamma D_{16}[$$

$$F_{im}^{21}G_{jn}^{01} + F_{im}^{12}G_{jn}^{10}] + 2\gamma^3 D_{26}[F_{im}^{01}G_{jn}^{21} + F_{im}^{10}G_{jn}^{12}] + 4\gamma^2 D_{66}F_{im}^{11}G_{jn}^{11}\}/D_0 \tag{12c}$$

$$F_{im}^{rs} = \int_0^1\int_0^1 \frac{d^rX_i(\xi)}{d\xi^r}\frac{d^sX_m(\xi)}{d\xi^s}d\xi\,,\quad G_{jn}^{rs} = \int_0^1\int_0^1 \frac{d^rY_j(\eta)}{d\eta^r}\frac{d^sY_n(\eta)}{d\eta^s}d\eta\,,\quad r,s=0,1,2 \tag{12d}$$

$\gamma = a/b$, $\lambda^2 = \rho h \omega^2 a^4 / D_0$, $D_0 = E_1 h^3 / [12(1 - \nu_{12}\nu_{21})]$ (12e)

The eigenfrequencies and coefficients for the mode shapes described by equation (11) can be obtained by solving eigenvalue equation (12b). Moreover, because the order of the polynomials in this set of static beam functions is always lower than 4, the numerical instability of high order polynomials in the numerical computation is therefore avoided. Therefore, the present method is especially suitable for the dynamic analysis of plates with a large number of the intermediate line supports and/or when higher vibrating modes need to be calculated.

4. NUMERICAL RESULTS

In this section, some numerical results are presented for symmetrically laminated square plates with a mid-line support in each direction, which are made up of three-layer composite with the fibre orientations $[-\beta, \beta, -\beta]$. Each layer is a Graphite/Epoxy material with the mechanical properties: major elastic modulus $E_1 = 138(Gpa)$; minor elastic modulus $E_2 = 8.96(Gpa)$; shear modulus $G_{12} = 7.1(Gpa)$; major Poisson's ratio $\nu_{12} = 0.3$.

4.1. ACCURACY AND CONVERGENCY STUDY

To demonstrate the validity, accuracy and convergence of the proposed method, a comparison study was first carried out on isotropic plates with one mid-line support in each direction. Three different boundary conditions are considered in the analysis: *SSSS*, *CCCC* and *FFFF*. In the numerical computations, four terms of the admissible functions in each direction are used for the *SSSS* and *CCCC* plates and five terms are used for the *FFFF* plate. Table 1 shows the computational results which are compared with values available from references [8,9]. Very close agreements are observed. Table 2 shows the convergence patterns of the frequency parameters of symmetrically laminated square plates with a fibre orientation angle $\beta = 30^o$, with respect to the number of terms of the admissible functions used. The plates are simply supported or clamped at all edges. It is shown that the convergency is rather rapid and the eigenfrequencies monotonically decrease with the increase of the using terms of the admissible functions. A small number of terms of the admissible functions may give sufficiently accurate results.

Table 1. The first six eigenfrequencies of the isotropic square plates with a mid-line support in each direction

B. C.	λ_1	λ_2	λ_3	λ_4	λ_5	λ_6
SSSS	78.957	94.590	94.590	108.240	197.392	197.392
Ref. [7]	78.96	94.68	94.72	108.44	197.40	198.96
Ref. [8]	78.958	94.826	94.826	108.41	197.50	197.50
Exact [9]	78.957	94.585	94.585	108.22	197.39	197.39
CCCC	108.299	127.417	127.417	144.109	242.818	243.778
FFFF	13.563	21.566	21.571	27.919	70.847	77.837

Table 2. The convergency of the first six eigenfrequencies of symmetrically laminated square plates (three layers and $\beta = 30^o$) with a mid-line support in each direction

Terms	λ_1	λ_2	λ_3	λ_4	λ_5	λ_6
			SSSS			
4×4	53.144	56.182	63.439	63.896	101.864	106.103
5×5	52.973	55.965	62.473	63.753	95.198	103.034
6×6	52.368	55.282	62.188	62.350	94.841	100.509
7×7	52.234	55.168	61.996	62.023	93.973	99.707
8×8	52.179	54.986	61.733	61.815	93.245	99.200
9×9	52.128	54.880	61.521	61.711	92.768	98.912
			CCCC			
4×4	70.476	73.412	82.577	83.755	123.716	127.917
5×5	69.776	72.569	81.260	83.046	115.803	122.434
6×6	69.289	72.025	81.232	81.591	115.218	120.464
7×7	69.180	71.996	81.133	81.440	114.894	119.932
8×8	69.150	71.856	80.896	81.317	114.146	119.398
9×9	69.134	71.801	80.830	81.286	113.958	119.267

Table 3. The first six eigenfrequencies of symmetrically laminated square plates (three layers) with a mid-line support in each direction

β (degree)	λ_1	λ_2	λ_3	λ_4	λ_5	λ_6
			SSSS			
15	47.370	49.512	63.795	63.967	80.622	85.930
30	52.234	55.168	61.996	62.023	93.973	99.707
45	54.639	58.862	61.218	61.860	96.385	107.861
			CCCC			
15	67.782	69.827	87.999	88.575	102.684	107.875
30	69.180	71.996	81.133	81.440	114.894	119.932
45	70.060	74.686	77.564	78.117	116.776	128.700
			FFFF			
15	6.584	7.927	14.816	15.132	27.651	28.728
30	9.093	9.732	14.501	15.679	30.522	35.110
45	10.259	10.740	14.641	16.245	31.345	37.869

4.2 NUMERICAL EXAMPLES

Some first-time known numerical results are presented in Table 3 for the symmetrically laminated square plates with one mid-line support. The fibre orientation angle β varies from 0° to 45° in 15° increment. Seven terms of the admissible functions in each direction are used to obtain the first six eigenfrequencies for all cases. It can be seen that the fibre orientation angle β has an important effect on the vibration characteristics of the composite plates.

5. CONCLUDING REMARKS

In this paper, a set of static beam functions are used as the admissible functions to study the vibratory characteristics of symmetrically laminated rectangular composite plates. This set of static beam functions is made up of the static deflections of a beam with intermediate point-supports under a series of sinusoidal loads distributed along the length of the beam. Unlike conventional admissible functions, apart from the fact that the geometric boundary conditions of the plates and the zero deflections at the intermediate line-supports can be satisfied directly, this set of admissible functions may also properly describe the discontinuity of the shear forces at the line-supports, so that more accurate results can be expected. As an application example, some first-time known data for free vibration of symmetrically laminated square composite plates (three layers) with a mid-line support in each direction are presented.

REFERENCES

1. Kim, D.-H., Composite Structures for Civil and Architectural Engineering, E & FN Spon, London, 1995.
2. Leissa, A. W. and Narita, Y., Vibration studies for simply supported symmetrically laminated rectangular plates, J. Compos. Struct. 12, 1989, 113-132.
3. Bert, C. W., Fundamental frequencies of orthotropic plates with various planforms and edge conditions, Shock Vibr. Bull. 47, 1977, 89-94.
4. Han, W. and Petyt, M., Linear vibration analysis of laminated rectangular plates using the hierarchical finite element method — I. Free vibration analysis, Comput. Struct. 61, 1996, 705-712.
5. Whitney, J. M., Free vibration of anisotropic rectangular plates, J. Acoust. Soc. Am. 52, 1971, 448-449.
6. Chow, S. T., Liew, K. M. and Lam, K. Y., Transverse vibration of symmetrically laminated rectangular composite plates, Compos. Struct. 20, 1992, 213-226.
7. Wu, C. I. and Cheung, Y. K., Frequency analysis of rectangular plates continuous in one or two directions, Earthq. Eng. Struct. Dyn. 3, 1974, 3-14.
8. Kim, C. S. and . Dickinson, S. M., The flexural vibration of line supported rectangular plate systems, J. Sound Vibr. 114, 1987, 129-142.
9. Leissa, A. W., The free vibration of rectangular plates, J. Sound Vibr. 31, 1973, 257-293.
10. Zhou, D. and Cheung, Y. K., Free vibration of line supported rectangular plates using a set of static beam functions, J. Sound Vibr., 1999 (in press).
11. Kong, J. and Cheung, Y. K., Vibration of shear-deformable plates with intermediate line supports: a finite layer approach, J. Sound Vibr. 184, 1995, 639-649.
12. Zhou, D., Natural frequencies of rectangular plates using a set of static beam functions in Rayleigh-Ritz method, J. Sound Vibr. 189, 1996, 81-87.

Mechanics of Structures and Materials, Bradford, Bridge & Foster (eds)
© 1999 Balkema, Rotterdam, ISBN 90 5809 107 4

Analysis of rail track curved in plan

Y.Q. Sun
Centre for Railway Engineering, Central Queensland University, Rockhampton, Qld, Australia
M. Dhanasekar
School of Advanced Technologies and Processes, Central Queensland University, Rockhampton, Qld, Australia

ABSTRACT: Significant rail wear occurs at curves of rail tracks. As the forces and moments in rails contribute to rail wear, in this paper we have attempted to study the effect of rail curvature on its internal actions. A simplified computer model using beam and spring elements to represent the stiffness of various track components was established for this purpose. The output of the tangential track model compared well with the analytical results of the conventional beams on elastic foundation formulation. The model was therefore extended for various radii of curves in plan. It is shown that the increase in rail curvature increases the internal actions of rails, in particular the vertical and horizontal shear. Sensitivity of the model parameters has then be studied and reported in this paper. Increase in fastener stiffness has the potential to reduce rail forces whilst reduction (including voiding) of ballast stiffness has only a marginal effect on rail forces.

1 INTRODUCTION

With a view to increasing productivity, the management and engineering bodies of public and private railways in Australia now allow higher axle load and higher operation speed for the trains. To match the design of such railway vehicles appropriate rails with sufficient ultimate strength must be selected that would have a significant influence upon the service life of track.

Historically the Beams on Elastic Foundation (BOEF) method that conceptualises the rail of a specified flexural stiffness as a beam resting on a continuous linear elastic foundation of a specified track modulus has been used in the rail selection process. The rail bending moment caused by the design wheel loads are calculated using analytical expressions and the stresses in rails calculated using simple beam theory. Higher bending stresses are, therefore, evaluated at fibres far away from the neutral axis such as railhead and rail foot (Profillidis 1995 and Tew et al 1991).

Although the finite element method (FEM) is successfully used in evaluating the stress distribution in the rail body, the method has not become popular in the field due to complexities and indeterminate nature of the loading and support conditions and the general perception of the FEM as an academic tool in the infrastructure departments of the railways (Hagaman and Dhanasekar 1998 & Tew 1986). Rail track analysis is still popularly carried out on the basis of the BOEF method. However the BOEF method has some limitations as follows: (1) it can only analyse tangential tracks. (2) it considers all components other than the rail as a "track foundation" and lumps all stiffness together.

The BOEF method therefore could neither be used for the analysis of tracks curved in plan nor be used for the evaluation of the sensitivity of stiffness of the track components on rail forces and moments. The continuity or coupling of the rails and fastening system, the fastening system and sleeper, the sleeper and ballast, and ballast and subgrade layers that make up the track components could also be not studied using the BOEF method. The magnitudes of the coupling depend upon the type of rail, fastening system, the sleeper size, the sleeper spacing, the ballast depth, and the subgrade properties (Tew 1986).

As the static analysis plays an important role in track design and the BOEF method is technically sound for the evaluation of rail forces and moments, the BOEF model is extended in this paper using the computer modelling approach. The track system is modelled using beam and spring elements and the details of various track components included. The model allows for varying curve radius and track component assembly and stiffness. The sensitivity of these parameters are evaluated and presented.

2 TRACK COMPONENTS

For the purpose of analysis a typical "S" class track structure of the Queensland Rail (QR) is used (Queensland Rail, 1995). The "S" class lines are generally used for the transport of coal and minerals. The axle load of the trains range from about 18 tons to 25 tons. A typical coal train consists of 6 locomotives and 148 wagons with the total mass in excess of 10,000 tons. The operational speed is usually 80 km/h. The main components of the majority of this kind of track include 60 kg/m continuously welded rails (CWR); "Fist" or "Pandrol" resilient fasteners; post tensioned prestressed concrete sleepers (150 mm deep × 230 mm wide × 2150 mm long), ballast and other subgrade.

2.1 *Rail*

The rail used by the state railway systems in Australia are manufactured to Australian Standard A. S. 1085 Part 1 and designated in size by their mass per linear metre. The current rail sizes used in Queensland are 31, 41, 50 and 60 kg/m. The 60 kg/m rail is used in the heavy haul coal and mineral lines. Rail is now supplied in 27.4m lengths and the majority of these are flash butt welded into 100m length for main lines. The rail is modelled using beam elements in this paper.

2.2 *Fasteners*

All modern designs of concrete sleepers utilise resilient fasteners. The elastic fastenings provide an excellent opportunity to achieve and maintain good track geometry and stability. The fundamental to all concrete sleeper application is the requirement for an anchorage to secure the fastening and for a rail pad to provide the impact attenuation, the electrical insulation and the means of modifying the stiffness of the track structure. The rail fasteners are modelled as spring elements that connect rails with sleepers in this paper. Three spring stiffness values – 8 kN/mm (low), 19 kN/mm (normal) and 30 kN/mm (high) are chosen in consistent with the type of fastener used in QR.

2.3 *Sleepers*

All major track routes of QR are now laid with concrete sleepers. Concrete sleepers manufactured in Queensland are monoblock, prestressed concrete units of 280 kg mass with 18 mm X 5.08 mm diameter tendons tensioned to 24.5 kN each. The sleepers are made from concrete of minimum compressive strength 50 MPa. The concrete sleepers are modelled using beam elements in this paper.

2.4 *Ballast and Subgrade*

Ballast consists of clean, durable, and angular crushed rock. The ballast bed under the sleepers provides an elastic structure. The elasticity of the ballast bed is an essential feature of the railway track structure. The ballast is compacted adequately to provide a stable foundation for the track structure. The track modulus is defined as the force per unit deflection per unit track length. A large amount of experimental results have been acquired under the conditions of different tracks that defines the spring constant of ballast to vary from 50 kN/mm to 300 kN/mm.

3 TRACK MODELLING

Modelling track is a complex task because of the requirement of deciding on an appropriate length of track based on the length of the wagon or locomotive. Inclusions of all components that make the track also pose challenge particularly in relation to identification of appropriate property sets. The length of the track structure has been decided from the length of QR's 100 tonne coal train. A one and a half wagon length has been chosen to investigate the effect of the influence of both the rear wheel of the leading wagon and the front wheel of the trailing wagon on the track forces. This has resulted in thirty-six sleeper spacings (each of 685 mm) in the model. The gauge length has been kept as 1067 mm consistent with the QR gauge system.

Beam elements have been used to represent the rail and sleeper whilst spring elements have been used to represent the pad, fastener, ballast and subgrade. Two beam elements each have been used to model the spacing of rail between sleepers and a total of 148 beam elements have been used for the two rails in the full length of the model. Each sleeper on the other hand has been modelled using five beam elements constituting 180 beam elements for all the sleepers. The two rails and the sleepers are connected by vertical and horizontal spring elements representing the vertical and horizontal stiffness of the pad and fasteners respectively. All nodes on the sleeper are linked to the ground by vertical and horizontal spring elements representing the vertical and lateral stiffness of ballast and subgrade respectively.

The track model has been used to represent both the tangential and curved tracks. The curves represent arc of circle with radii varying from 200 m to 600 m in steps of 100 m. A model for the radius of 200 m radius of curvature track is shown in Fig. 1.

All longitudinal displacement degrees of freedom at one end of the model have been suppressed to reflect symmetry and at the other end all degrees of freedom of two nodes on the two rails have been fixed to prevent the rigid body motion.

3.1 *Loading*

The vertical dynamic load P has been determined empirically from the static wheel load, P_s.

$$P = \Phi \cdot P_s \tag{1}$$

Where Φ = dimensionless impact factor (>1) calculated as follows:

$$\Phi = 1 + d \times n \times t \tag{2}$$

in which d = track roughness factor, n = speed factor, t = a number corresponding to a specified upper confidence limit. In this paper Φ was calculated as 1.684 and the vertical dynamic wheel load was determined as 206.3 kN. The vertical loads applied at both the high and low rails for the different radii are listed in Table 1

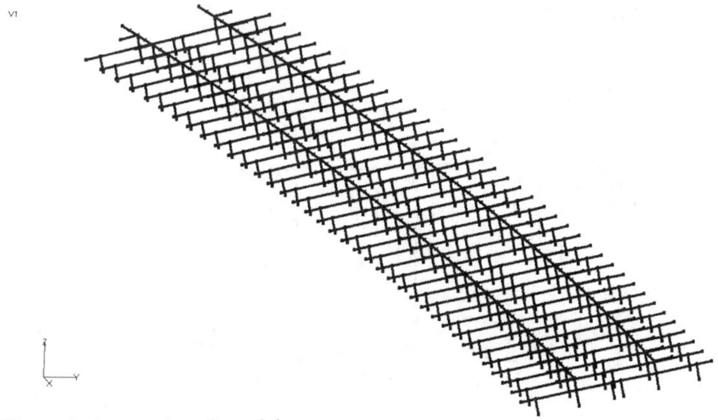

Figure 1. A curved track model

Table 1. Vertical Load.

Radius	200	300	400	500	600
Load on High Rail (kN)	270.3	249.0	238.2	231.7	227.4
Load on Low Rail (kN)	154.6	169.4	177.8	183.2	186.9

3.2 *Results*

The track with the largest curvature (smallest radius of 200 m) was first analysed and the deformation of rail examined for the effect of the influence of the load sequence. The deflection profile presented in Fig. 2 shows that the deformation of rail is not adversely affected due to load sequence. No evidence of the influence of deflection due to one load on the other was seen both in the shape of the deflection profile and in the magnitude of resultant deflection close to the single and double bogies. For example, the resultant maximum deflection at the two bogie section was 8.04 mm and the same at the single bogie section was 7.90 mm (representing reduction of only 1.7%). As significant number of elements exhibited nil deformation, it was decided to further simplify the model by reducing the length of the model and considering only a single bogie (four wheel loads).

3.3 *Simplified Model*

The model with reduced length is shown in Fig. 3.

This represents one-third the length of the 100 tonne wagon. The deformation of the model is also shown in the figure. The maximum deflection of this model is 8.02 mm which is extremely close to the value of 8.04 mm for the section under two bogie loading of the large model (8.04mm).

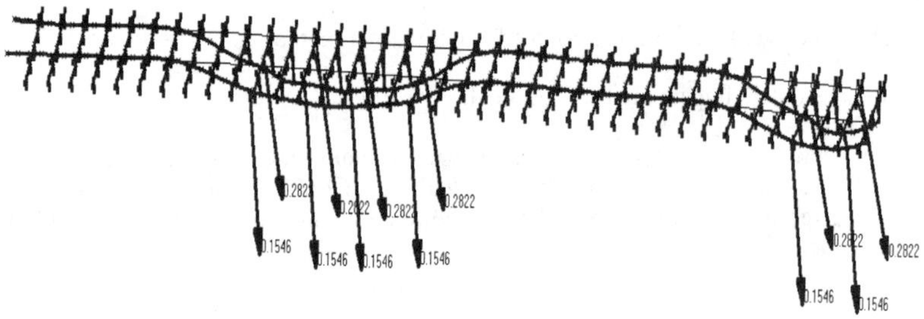

Figure 2 Deformation of rail track model

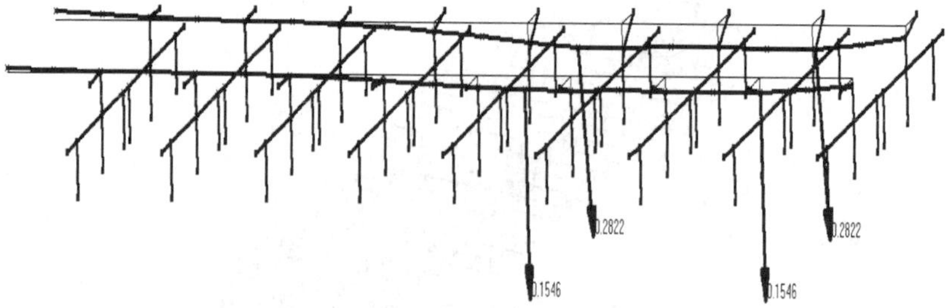

Figure 3 Simplified model illustrating total translation

3.4 *Model verification*

The deflection of the tangent track model was verified using the theory of BOEF. As experimental or field measurement of these parameters for the curved track are not available either from our own work or in the literature, we could not calibrate the results of the curved track analysis. However, with the validation of the tangent track results, it is believed that the curved track results must be representative as the parameters were not changed between the models.

The solution for the rail deflection (Y_x), at any position x from the load point is:

$$Y_x = \frac{P\beta e^{-\beta x}}{2K}(\cos \beta x + \sin \beta x) \qquad (3)$$

where P = concentrated wheel load (206.3 kN), β = relative stiffness of rail to foundation modulus. In this problem, $\beta = 1*10^{-3}$ 1/mm. The maximum deflection calculated under two vertical loads $Y_{xtotal} = 4.126 + 0.756 = 4.882$ mm whilst the tangent track vertical deflection is 5.03 mm which gives an error –2.9%. The computer model was, therefore, considered to be sufficiently accurate.

To further calibrate the model, the stiffness of the horizontal springs were varied from 20% of the vertical stiffness to 100% of the vertical stiffness. The vertical (and resultant) deflection of rail was found to be unaffected by the variation of the stiffness.

A second calibration was performed for the assumed lateral load to vertical load ratio of 0.3. By varying the ratio from 0 to 0.3 in steps of 0.1 (by keeping the vertical load constant and increasing the horizontal load from 0 to 30% in steps of 10% increment) it has been found that the lateral moment and lateral shear force increase linearly with the ratio whilst the vertical moment and vertical shear force almost have remained unchanged. It has therefore been concluded that the results reported in the paper (for 30% lateral load) could be used for any other lateral load value by linear interpolation or extrapolation.

4 ANALYSIS OF RESULTS

The rail forces determined from the tracks of varying radius of curvature is plotted in Fig. 4. From the Fig. 4 it can be seen that the vertical and lateral shear forces vary with the curve radii significantly whilst the moments in both vertical and lateral directions change only marginally. Whilst the lateral force has increased from -4.2 kN in tangential line to 39.7 kN on the track of

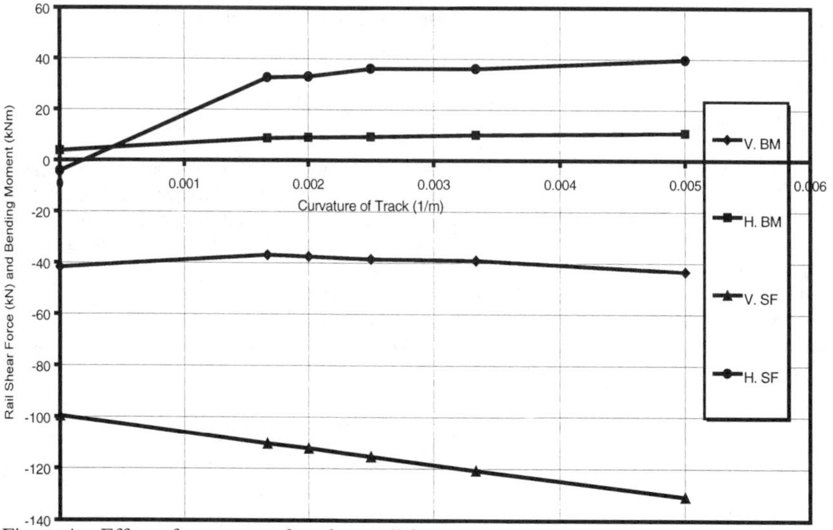

Figure 4 Effect of curvature of track on rail forces

radius 200 m, the absolute value of vertical shear force has increased from 99.6 kN to 130.8 kN. This shows that the shear forces make a great contribution to the rail stresses. In a refined analysis of rails for stresses, we have used solid modelling – the results of which are not included here.

4.1 *Sensitivity analyses*

The influence of fastener and ballast spring stiffness on rail moment and shear force on curve of 400 m was investigated. It was found that the lowest fastener spring stiffness of 8 kN/mm has produced the bending moment of -41.5 kNm and the shear force of -120.4 kN whilst the highest stiffness of 30 kN/mm lead to the bending moment of -37 kNm and shear force of-113.3 kN respectively. Based on these results it could be concluded that the fastener with higher stiffness will reduce rail forces and hence would increase rail life.

A similar analysis of "hard" or "soft" ballast springs, indicated no significant influence of ballast stiffness on rail forces. In order to investigate the effect of ballast-foul on rail forces, the original model on the curve of 200 m radius was used. At the location where the wheel loads act, the eight ballast spring elements under the sleeper were eliminated in a sequential order to simulate the ballast-foul percentages. It was assumed that 25% foul was formed when two ballast spring elements below the low rail were removed. It has been found that when the foul is lass than 75% no significant increase in rail force has occurred. However when the foul was 100%, the high rail vertical moment increased by 19.4%.

5 CONCLUSIONS

The BOEF model for rail track has been extended in this paper to allow for the variations in the stiffness of pad, fasteners and ballast. The extended model has given good agreement with the analytical BOEF formulation. The extended model has predicted increases in horizontal and vertical shear forces with the increase in rail curvature. Stiffer fasteners have found to reduce rail moment and shear whilst the ballast stiffness and partial ballast-foul are found to not influence rail forces significantly.

6 ACKNOWLEDGEMENT

The work presented in this paper was supported financially by the Centre for Railway Engineering internal grant NR XY U 100 to the second author.

REFERENCES

Hagaman, B. and Dhanasekar, M. (1998), "New Generation Rails", Proc. CORE98 Conf., 521-528

Jeffs, T & Tew, G.P (1991), *A review of track design procedures - Volume 2 : Sleepers and Ballast,*

Profillidis, V. A. (1995), *Railway Engineering*, Avebury Technical, Ashgate Publishing Limited.

Tew, G. P., (1986), *Prediction of rail behaviour by finite element modelling*, Proc. Sixth Inter. Rail Track Conf., Rail Track Assoc. Australia, Melbourne..

Tew, G. P., Marich, S. & Mutton, P. J., (1991), *A review of Track Design Procedures - Volume 1 : Rails*, BHP Research - Melbourne Laboratories, Railways of Australia.

Queensland Rail, (1995), *Railway Civil Engineering Course 1995 - Track Design Volume 3.* Brisbane, Australia.

Mechanics of Structures and Materials, Bradford, Bridge & Foster (eds)
© 1999 Balkema, Rotterdam, ISBN 90 5809 107 4

Evaluation of cyclic moment-curvature relationship of reinforced masonry

M. Dhanasekar
School of Advanced Technologies and Processes, Central Queensland University, Rockhampton, Qld, Australia

ABSTRACT: Moment - Curvature relation is an important information that defines the deformation capacity of structural members and the corresponding ductility of the cross section. A layer method of analysis of lightly reinforced masonry cross section is presented as part of an ongoing research. In this method, the reinforced masonry cross section is divided into a number of layers each representing a masonry shell, or shell and grout (confined or unconfined) or shell, grout and steel. The response of each lamina to induced curvature increment depends on the stress-strain characteristics of the material of the lamina. For the analysis of sections subjected to cyclic curvature increments, cyclic stress-strain relations have been obtained from experiments on masonry prisms. A computer program was developed and its output on the hysteretic moment-curvature relations of lightly reinforced masonry cross sections have been validated against an experimental result. A refined stress-strain curve and more validation are required.

1 INTRODUCTION

Reinforced masonry (RM), as used in Australia, is lightly reinforced and lightly loaded in the vertical direction. The vertical cores in RM are filled with cement grout and reinforced with a single steel bar (occasionally two bars are also used). Under combined axial and lateral loading these cores are subjected to lateral shear, bending and axial force. Only empirical methods based on experiments are reported in the literature for the prediction of the failure and deformation of the RM. This paper reports a theoretically sound method based on the fundamental layer approach to predict the moment-curvature relations of the reinforced cores. Combined bending and axial forces are considered. The result of the method is calibrated against an experimental result reported elsewhere (Dhanasekar and Shrive, 1998).

2 A LAYERED MODEL FOR MOMENT-CURVATURE RELATION

This section describes the moment-curvature relations of the grouted reinforced masonry subjected to vertical and lateral loading. The vertical load simulates the dead and live load corresponding to single or double storey construction and is kept constant throughout the loading. The horizontal load is cycled until failure occurs. Failure of the cores occur at the mortar joint subjected to the maximum bending moment. At ultimate stage, the failure plane extends to the full width of the core and the vertical steel acts as the dowel that connects the two rigid bodies.

A theoretical method commonly known as the laminae or layer method (Paulay and Priestley, 1992) is used. An appropriate uniaxial stress-strain relation of the various constituent materials that makes up the RM core is required for this method. Through a series of experimental investigation (Dhanasekar et al, 1997) on concrete masonry prisms (with and without grout filled), the stress strain relations of the concrete masonry shell, grout, grouted masonry and grout have been evaluated. Others have reported similar work for concrete (Watanabe, 1971 & Watson et al, 1994). These equations when implemented in the theoretical model have predicted the monotonic moment-curvature relations of the RM accurately. This paper describes prediction of cyclic moment-curvature relations.

2.1 *The model*

The theoretical model is developed based on the following assumptions:
1. Plane sections remain plane during bending
2. The masonry is capable of resisting compression and very limited tension
3. Steel resists tension and compression
4. There exists perfect bond between the grout and the steel reinforcement
5. There exists perfect bond between the grout and the concrete shell

Even though assumptions 4 and 5 are questionable, this method predicts appropriate moment-curvature relations of the RM that is reproducible in the experiments.

The cross section of the grouted RM is shown in Fig. 1. The cross section is composed of the masonry shell, grout and vertical steel. In the theoretical method, the cross-section is subdivided into a number of laminae. Each lamina runs along the horizontal direction because the moment-curvature relation in the weaker direction is evaluated. The outer most laminae contained masonry shells only. Other laminae are composed of shell, and grout. The lamina at the middle of the cross section is composed of shell, grout and steel. From the stress-strain equations derived for the components (shell, and grout) and from the well-known stress-strain relations of the steel, moment-curvature relation of the cross section subjected to axial load and curvature increment was evaluated.

In Fig. 1 the layer is schematically shown in light line. The longitudinal reinforcing steel is shown at the centre of the core for illustration only (the method allows the positioning of the steel at any other location). The cross section shown in Fig. 1 is a typical reinforced masonry section as used in the northern Australia.

It is assumed that the cross section is subjected to the axial load (N) and the curvature (ϕ). The moment (M) and the axial strain (ε) are evaluated from the stiffness of the cross section which in turn is derived from the stress-strain relations of the constituent materials. As the problem is non-linear, all quantities involved are defined in the incremental form as shown in Eqn. 1.

$$\begin{vmatrix} \Delta N \\ \Delta M \end{vmatrix} = \begin{vmatrix} p & q \\ q & r \end{vmatrix} \begin{vmatrix} \Delta \varepsilon_o \\ \Delta \phi \end{vmatrix} \qquad (1)$$

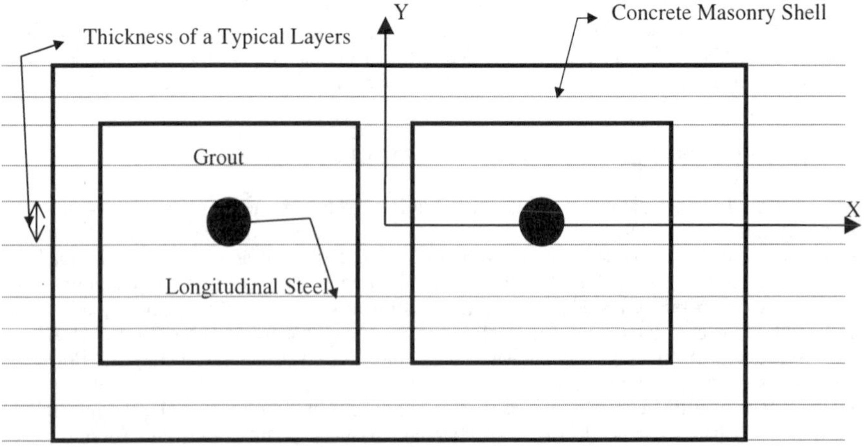

Figure 1. Laminae method of cross section analysis

in which

$$p = \left[\sum E_{\sec i} A_i\right] \quad (1a)$$

$$q = \left[\sum E_{\sec i} A_i y_i\right] \quad (1b)$$

$$r = \left[\sum E_{\sec i} A_i y_i^2 + \sum E_{\tan i} I_i\right] \quad (1c)$$

where

A_i = Cross sectional area of the i - th layer

I_i = Moment of inertia of the i - th layer

$E_{\tan i}$ = Tangent modulus of the i - th layer

The section is subjected to curvature increments while the axial force N is kept constant. For a prescribed N, the M-Φ relationship is calculated using the procedure given below:

(a) N is divided into a number of small increments ΔN .

(b) p, q and r are calculated using the E_{sec} and E_{tan} values that have been obtained from the previous increment. For the first increment the input value has been used.

(c) Incremental and cumulative strain are calculated from Eqn. 2.

$$\Delta \varepsilon_o = \frac{\Delta Ne}{p} \quad (2a)$$

$$\Delta \varepsilon_i = \Delta \varepsilon_o + y_i \Delta \phi \quad (2b)$$

$$\varepsilon_i = \varepsilon_{i-1} + \Delta \varepsilon_i \quad (2c)$$

(d) From the stress-strain curves of steel, grout and masonry the elastic modulii are calculated.

(e) p, q and r values are re-evaluated using the elastic modulii obtained at step (d)

(f) $$\Delta \varepsilon_o = \frac{\Delta Ne}{p} \quad (3)$$

(g) Incremental strain obtained in step (f) is compared to that obtained at step (b). If the difference is within an acceptable tolerance, the solution is considered converged and the solution process is diverted to step(h). Otherwise steps (c) to (g) are repeated.

(h) Load is further incremented and steps (a) to (g) are repeated until the value of N is achieved. At this stage the section is in equilibrium under the axial load N.

(i) Curvature is then incremented and the strain is calculated from Eqn. 4.

$$\Delta \varepsilon_o = \frac{-q \Delta \phi}{p} \quad (4a)$$

$$\Delta \varepsilon_i = \Delta \varepsilon_o + y_i \Delta \phi \quad (4b)$$

$$\varepsilon_i = \varepsilon_{i-1} + \Delta \varepsilon_i \quad (4c)$$

(j)From the stress-strain relationship of steel, grout and masonry, the tangent and secant modulii are calculated.

(k) p , q and r values are re-evaluated using the elastic modulii obtained at step (j)

(l) The incremental strain is calculated as in Equation 5.

$$\Delta \varepsilon_o = \frac{-q \Delta \phi}{p} \quad (5)$$

(m) The incremental strain obtained at step (l) is compared with that obtained at step (i). If the difference is within an acceptable tolerance, the solution is considered converged and the solution process has been diverted to step (n). Otherwise steps (k) to (m) are repeated.

(n) ΔM and M are calculated using Eqn. 6.

$$\Delta M = q \Delta \varepsilon_o + r \Delta \phi \tag{6a}$$

$$M_i = M_{i-1} + \Delta M \tag{6b}$$

(o) Curvature is incremented and steps (k) to (n) are repeated until the desired φ value is achieved. At this stage the section is in equilibrium under the applied axial load and the curvature increment.

(p) The M-Φ relation for the given level of axial load N is the output.

Using this algorithm the moment-curvature relations of reinforced masonry is evaluated. A computer program was developed based on the algorithm. The input for the program was set up in three files - one each for geometric properties, masonry and grout properties and steel properties. In addition to the moment-curvature relations, the program was capable of reporting stress and strain levels at selected fibres. The stress-strain information reported by the program was useful in monitoring the actual behaviour of the constituent materials (layers). All the output data were graphically viewed using the standard spreadsheet program, EXCEL.

2.2 *Stress-strain relations*

The stress - strain relation of the grout filled concrete masonry is very important for the accurate determination of the moment-curvature relations. Steel was assumed to be elastic - perfectly plastic material. As there was no information reported in the literature on the confined grouted concrete masonry, stress-strain relations have been evaluated from experiments on masonry prisms (Dhanasekar et al, 1997 & Dhanasekar and Shrive, 1998). The cyclic stress-strain relations have been broken into four components, namely, the envelope, common-point, unloading and reloading curves. Eqn. 7 represents the envelope and common-point curves.

$$y = y_{max} \left(\frac{(1 + u_0 (1 + u_1)) x}{u_0 (1 + u_1 x) + x^{(u_0 + 1)}} \right) \tag{7}$$

In which "y" is normalised stress and "x" is normalised strain. To normalise the stresses and strains, the average of the maximum stress and strain values from the group of experimental data on masonry prisms has been used. Typical value of y_{max} for the envelope curve was 1.0 whilst that of the common-point curve is 0.85.

The unloading and reloading curves have been formulated using the approach proposed by Watanabe (1971) for reinforced concrete as first approximation. The unloading curve, in this model, is composed of a straight line from the point of unloading on the envelope curve to the point where it intersects the common-point curve and a parabolic from that point to the X axis where complete unloading of stress takes place. The reloading curve is simplified as a simple straight-line connecting the point on the X axis where unloading terminated and the point of intersection between the unloading and common-point curve.

The algorithm presented in this section was programmed in FORTRAN under WIN-NT.

3 RESULTS AND VERIFICATION

The computer program was used in the analysis of RM sections. As the main objective of the analysis was to validate the algorithm, an experiment carried out by the author at the University of Calgary, Canada (Dhanasekar and Shrive, 1998) was used as an example. Care was taken to exactly represent the loading path (curvature history) that was applied in the experiment to obtain better comparison of the experimental and simulated moment-curvature curves.

3.1 *The test specimen*

The test specimen had the same shape of the RM cross section shown in Fig. 1. The gross dimension of the cross section was 390 mm wide X 190 mm deep. Each grouted core was approximately of size 140 mm wide X 110 mm deep. Each core was centrally reinforced with a Y12 bar. The compressive strengths of masonry shell, grout and steel were 12.4 MPa, 22.8 MPa, and 410 MPa respectively

3.2 *Loading*

The specimen was loaded to a monotonically increasing vertical load (corresponding vertical stress 1.2 MPa). The curvature cycle was applied laterally (in the weaker direction) whilst the vertical load was kept at its constant level. As curvature was not explicitly monitored due to limitations in the loading device, a symmetric curvature history could not be applied. Lateral load was monitored to decide and manually control the curvature reversal points on the loading path. The resulting path was unsymmetrical as shown in Fig. 2.

3.3 *Discussions*

The predicted and experimentally obtained moment-curvature relations of the reinforced masonry is shown in Fig. 3.

The prediction is reasonable, however, at large curvature levels experimental curve is much fatter than the predicted curve (not shown). This shows that the prediction does not allow for

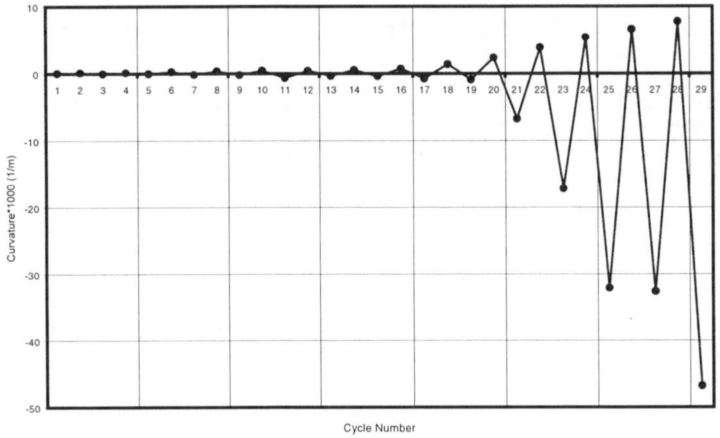

Figure 2 Cyclic curvature history

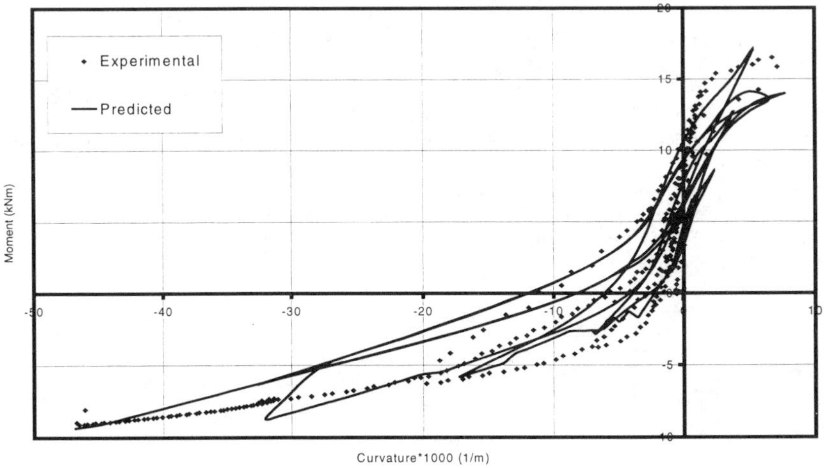

Figure 3 Moment – curvature relations of a reinforced masonry section

much energy dissipation as the actual experimental result and hence may be described as conservative. The real reason for the difference is largely attributed to the unloading and reloading curves. Work is under progress to fit better curves for unloading and reloading components. Pilot studies indicate that the new curves tend to make the moment-curvature prediction fatter.

In spite of the deviation between the results (which is largely a function of the assumed stress-strain curves, particularly the unloading and reloading components), the algorithm appears to work well and predicts the cyclic moment-curvature relations of the RM cross-section. Considering the cross section being composed of several materials, the prediction appears to be good.

At large curvature stage appreciable bond failure (between steel reinforcement and grout) occurred that invalidated some of the assumptions made in the development of the algorithm. The curvature increment was, therefore, stopped without any further increase to the history shown in Fig. 1.

4 CONCLUSIONS

An algorithm based on classical layer approach is presented for the analysis of reinforced masonry cross sections subjected to axial load and curvature increment. Envelope and common-point components of the cyclic compressive stress-strain curve of grouted masonry were experimentally derived and used in the algorithm. Unloading and reloading components of the curve was according to Watanabe et al (1971). The predicted reversed cyclic moment-curvature relations was in reasonable agreement with the experimental curve. More comparison and better representation of unloading and reloading curves are necessary for full validation of the algorithm.

5 ACKNOWLEDGEMENTS

The author acknowledges ARC (small) 96/04 grant administered by CQU, Professor Nigel Shrive and Professor Robert Loov of the University of Calgary, Canada and Dr. Benny Assa of Sam Ratulangi University, Manado, Indonesia.

REFERENCES

Dhanasekar, M., Loov, R.E., McCullough, D., & Shrive, N.G., (1997). "Stress-Strain Relations for Hol low Masonry under Cyclic Compression" *Proc. 11th IB^{2}MaC*, Shanghai, 1269-1278.

Dhanasekar, M., Loov, R.E., and Shrive, N. G., (1997), "Stress-Strain Relations for Grouted Masonry under Cyclic Compression", *Proc. 16ACMSM, Balkema, Rotterdam*, 663-668.

Dhanasekar, M., and Shrive, N. G., (1998), "Failure of the Reinforced Cores of the Partially Reinforced Masonry", *Proc. 5AMC, CQU Press*, Rockhampton, 95-104..

Paulay, T. and Priestley, M.J.N., (1992),"Seismic Design of RC and Masonry Buildings", *John Wiley & Sons*, , 744pp.

Watanabe, F. (1971), "Complete Stress-Strain Curve for Concrete in Concentric Compression", Proc. Int. Conf. On Mechanical Behaviour of Materials, Vol. 4, Kyoto, Japan, 153-161.

Watson, S., Zahn, F.A., & Park, R. (1994),"Confining Reinforcement for Concrete Columns", *J. of Structural Engineering, ASCE*, V120, 6, 1798-1823.

Mechanics of Structures and Materials, Bradford, Bridge & Foster (eds)
© 1999 Balkema, Rotterdam, ISBN 90 5809 107 4

Dynamic response of semi-continuous composite beams

B.Uy
School of Civil and Environmental Engineerng, The University of New South Wales, Sydney, N.S.W., Australia

P.F.G. Belcour
Société d'Etudes en Infrastructures, Nouméa, New Caledonia

ABSTRACT: This paper is concerned with the dynamic response of semi-continuous composite beams. A brief review of acceptable natural frequencies of floor systems to provide occupancy comfort is conducted. A method for modelling the dynamic response of semi-continuous composite beams using a beam finite element program is then presented and this is augmented with the appropriate spring stiffness associated with the semi-rigid joints existent at the beam-column locations. The stiffness of the continuous composite beams is modelled using a transformed section analysis procedure coupled with the Eurocode 4 model for the required regions of cracked and uncracked sections. A parametric study is conducted to determine beam natural frequencies for various span configurations and these are compared with acceptable values with favourable outcomes. The paper concludes by outlining further research, including experimental calibration and consideration of the effects of partial shear interaction and connection on the dynamic response of semi-continuous composite beams.

1 INTRODUCTION

Semi-continuous composite beams consist of steel-concrete composite beams, which are connected to vertical steel columns by semi-rigid joints. These semi-rigid joints are generally formed by a web-side plate connection and coupled with tensile reinforcement over the support. This form of connection is termed semi-rigid for serviceability as it produces structural behaviour of beams which is not classified as pinned or fixed with regard to end conditions. A typical semi-continuous beam system is illustrated in Figure 1, which highlights the presence of the steel reinforcement over the support.

Semi-continuous composite beams have been consistently researched throughout the world in the last decade, with a significant amount of work being produced related to the strength aspects of these beams, (Nethercot 1997 and Leon 1998). In many structural systems, serviceability conditions often govern and thus it is necessary to provide a proper treatment of this behaviour. One aspect of serviceability behaviour is the dynamic response or vibration performance of beam

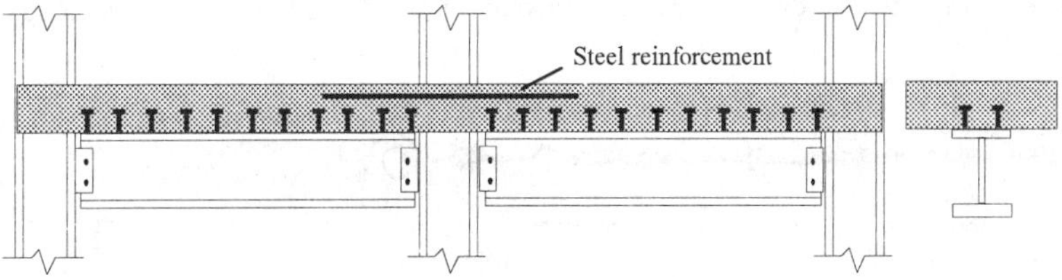

Figure 1. Semi-continuous composite beam in a steel framed building

systems in building floor systems to provide occupancy comfort and in bridges to limit possible fatigue damage. This paper deals with the dynamic response of semi-continuous composite beams in typical building floors where occupancy comfort is of primary importance.

This paper firstly reviews acceptable natural frequencies to ensure the comfort of occupants in building floor systems. A model is then proposed which employs the use of a beam finite element program and a lumped mass model to determine fundamental natural frequencies. The program is initially calibrated for simply supported beams with a closed form solution. The model is then augmented to allow for the inclusion of semi-rigid joints. A detailed parametric analysis is then carried out to consider the effects of the type of modelling, percentages of reinforcement and span length on the dynamic response. The detailed study is then compared with acceptable natural frequencies and recommendations are then made on the suitability of current designs. Further research is then identified which includes the investigation of issues, which relate to the determination of the dynamic response of beams with partial interaction and partial shear connection between the steel beam and concrete slab.

2 ACCEPTABLE NATURAL FREQUENCIES

Prior to undertaking a structural analysis of semi-continuous composite beam systems it is important to ascertain the spectrum of natural frequencies which are generally acceptable for floor systems. The type of loading, which is present, usually determines this. In office buildings the most prevalent form of loading is live loading exhibited from persons walking, which can promote excitation of floor systems. The Australian Standard AS 2327.1-1980 (Standards Association of Australia 1980) provides guidance on the acceptable ranges of natural frequencies, which are outlined herein.

The problem caused by these vibrations is the annoyance resulting from the promotion of nausea and sound transmission arising from the oscillation of floors. People walking alone or together can create period forces in the frequency range of 1 to 4 Hz. However, for very repetitive activities such as dancing, it is possible to excite frequencies of around 10 Hz.

In general, natural frequencies less than 5 Hz should be avoided otherwise walking resonance may occur. However, it is recommended that for very repetitive activities such as dancing, the natural frequency of floors be set at 10 Hz or more. It is also suggested that for spans greater than 6 metres, especially for those without partitions and with natural frequencies smaller than 15 Hz, that a more rigorous analysis is undertaken.

3 STRUCTURAL MODEL

The structural model described herein employs a beam finite element approach which incorporates the varying beam stiffness and is augmented with semi-rigid joints by invoking linear springs at the beam column locations. The general structural model is illustrated in Figure 2.

3.1 *Description of lumped mass model*

For the lumped mass model, mass, stiffness and span length are considered when computing the natural frequencies of the composite beams. Here, the vibration analysis is carried out using the lumped mass model concept where the mass due to dead load of the beam systems only was considered. In this model, discrete masses, referred to as lumped masses, representing

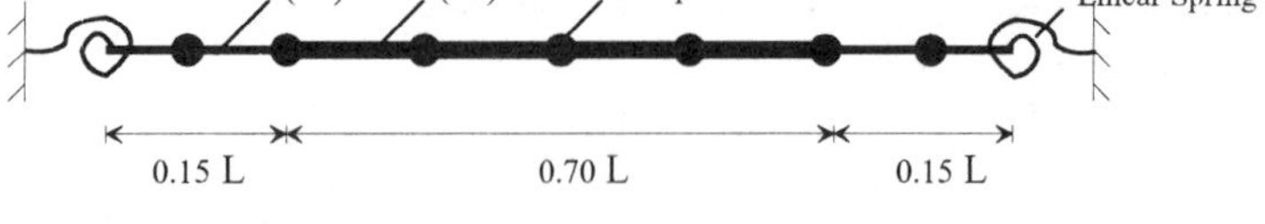

Figure 2. Generalised structural model for dynamic analysis of semi-continuous composite beam

appropriate fractions of the dead load of the composite beams are assigned to the nodes. These nodes are placed at regular intervals along the length of the beam as illustrated in Figure 2.

3.2 *Description of beam stiffness*

In the analysis the short-term beam stiffness is utilised which considers the beam as either cracked or uncracked. The positive moment region, where the concrete slab is primarily in compression is taken to be uncracked and equal to $(EI)^+$. The negative moment region, where the concrete slab is in tension is taken to be cracked and equal to $(EI)^-$. The beam stiffness throughout both regions is calculated using the transformed section approach, where the beam is converted to an equivalent area of structural steel. The reinforcing steel in the negative moment region is also included in this calculation. The regions over which the positive and negative flexural rigidities are applied are based on the recommendations of serviceability provisions prescribed in Eurocode 4 (British Standards Institution 1994) and these are illustrated in Figure 2.

3.3 *Description of semi-rigid joint stiffness*

In the modelling of the semi-continuous composite beam, a linear spring is used to represent the stiffness of the end of the beam to the column. This spring is assumed to have a linear stiffness for the serviceability behaviour of the beam. This linear stiffness is determined using the approach of Crisinel and Carretero (1996) where the stiffness of the joint, S_j is calculated using Equation 1

$$S_j = \frac{z_s^{\,2}}{\frac{1}{k_s} + \frac{1}{k_v} + \frac{1}{k_c}} \qquad (1)$$

where z_s = the lever arm between the compressive and tensile forces; k_s = the stiffness of the reinforced concrete slab; k_v = the stiffness of the shear connection; and k_c = the stiffness of the compression zone.

3.4 *Calibration of model*

In order to calibrate the lumped mass model for natural frequency, the number of lumped masses was varied on a typical simply supported composite beam. A simply supported composite beam with a span length of 6 m was chosen. This beam was assumed to be part of a typical 6 metre x 6 metre floor grid. The properties of the beam were I=3.25x10^6 mm^4; A=23,523 mm^2; G_s=7,850 kg/m^3. Of these properties I represents the transformed second moment of area, A represents the transformed cross-sectional area and G_s represents the density of the structural steel.

The Australian Standard AS2327.1-1980 (Standards Association of Australia 1980) also provides guidance for calculating the fundamental natural frequency, f of simply supported composite beams using the relationship in Equation 2

$$f = K\sqrt{\frac{E_s I'_{cs}}{wL^4}} \qquad (2)$$

where K = 155 for simply supported beams; E_s = the modulus of elasticity of the steel joist (N/mm^2); I'_{cs} = the second moment of area of the composite section (mm^4); w = the dead load on the beam (kN/m); and L = the beam span length (mm).

The number of nodes and lumped masses was increased until the natural frequency converged. A plot of the results is shown in Figure 3 and this illustrates that convergence occurs when about 7 lumped masses are used. This was thus invoked for all further analyses conducted in this paper.

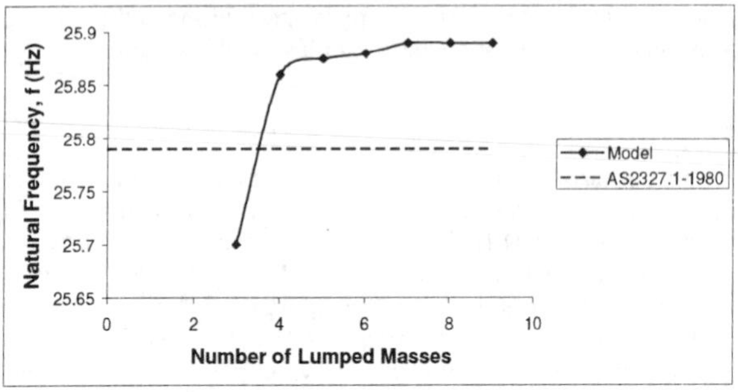

Figure 3. Calibration of model with closed form solution

4 PARAMETRIC STUDY

The model has been successfully calibrated for simply supported composite beams in the previous section. To consider the dynamic behaviour of semi-continuous composite beams a range of parameters must be considered which reflect real design situations in practice. A parametric study is undertaken here to consider the effects of cracking, reinforcement percentage and span length on the dynamic response of semi-continuous composite beams. The parametric study has been conducted on beams which have been designed for typical office floor loading according to Australian Standard AS1170.1-1989 (Standards Australia 1989) in a gravity loaded multi-storey braced frame. The beams thus satisfy both the serviceability and strength limit states of the Australian Standard AS2327.1-1996 (Standards Australia 1996) with appropriate modifications to account for semi-rigid and partial strength joints.

4.1 *Effect of cracking*

A comparison of the natural frequencies for cracked and uncracked beam sections was undertaken. The results are presented in Table 1 and are determined using Equation 3, where ϕ=the fundamental natural frequency ratio; $f_{cracked}$= the fundamental natural frequency which includes cracking in the negative support region; and $f_{uncracked}$= the fundamental natural frequency which ignores cracking in the negative support region.

$$\phi = \frac{f_{cracked}}{f_{uncracked}} \tag{3}$$

The results in Table 1 illustrate that the inclusion of cracking in the negative support region does have an effect on the fundamental natural frequency. The effects of cracking influence the

Table 1. Effects of cracking on beam natural frequency

	ϕ=Fundamental Natural Frequency Ratio				
	Span Length, L				
Reinforcement, p (%)	6 m	8 m	10 m	12 m	14 m
0.35	0.99	0.97	0.98	0.99	0.99
0.45	0.98	0.97	0.98	0.98	0.99
0.7	0.97	0.96	0.97	0.97	0.98
1.0	0.96	0.96	0.96	0.97	0.97
1.5	0.94	0.94	0.95	0.96	0.96
2.0	0.93	0.93	0.94	0.95	0.95

fundamental natural frequency by at most 10 % for the cases considered. The effects of cracking were thus included for all of the remaining analyses considered in this paper.

4.2 *Effect of percentage of reinforcement*

The effect of percentage of reinforcement area as a function of the slab dimensions was considered for percentages, p varying from 0.0% to 2.0% for span lengths varying from 6 metres to 14 metres. Figure 4 illustrates the effect of reinforcement, which can increase the natural frequencies by up to about 10 % when using about 2.0% of reinforcing steel in the slab. Thus by including the effect of the semi-rigid joint for serviceability conditions, one can provide a more accurate determination of the dynamic response of semi-continuous composite beams which results in improved behaviour.

4.3 *Effect of span length*

The most important parameter to influence the natural frequencies of semi-continuous composite beams is the beam span length. The effect of span length on fundamental natural frequency for the fundamental mode is illustrated in Figure 5. This shows that the fundamental natural frequency decreases as the span length increases. Span lengths, which are commonly, used in composite construction range from 10 to 12 metres. It is therefore important to note that for such span lengths, the natural frequencies of the beams are in the range of 18.5 to 20 Hz, which is above the 15 Hz limit suggested by the Australian Standard AS2327.1-1980 (Standards Association of Australia 1980).

The fundamental frequencies for the longest span of 14 metres were found to be greater than 16 Hz. Therefore, vibration sensitivity is not considered to be a critical issue for the design of semi-continuous beams. However, an extrapolation of the curves in Figure 5, would suggest that the natural frequencies for beams exceeding 14 metres would drop below 15 Hz.

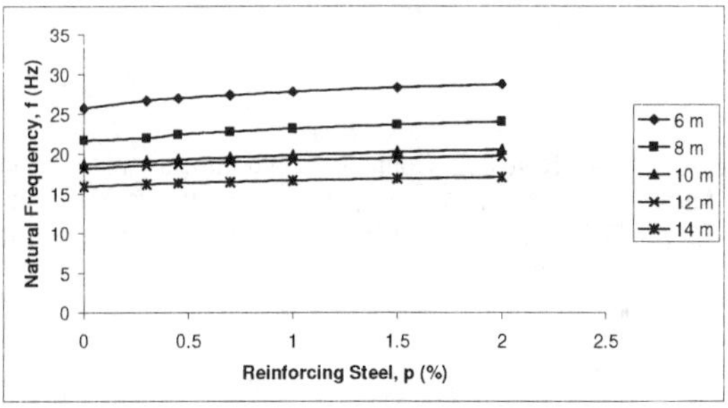

Figure 4. Effect of reinforcement percentage on beam natural frequency

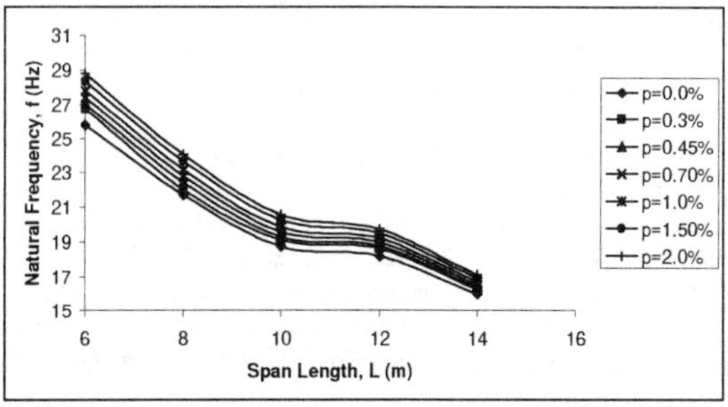

Figure 5. Effect of span length on beam natural frequency

5 COMPARISONS

As discussed previously, all the analyses undertaken for typical spans and loading in buildings were found to provide natural frequencies in the acceptable range of structural response. All natural frequencies were found to be greater than the maximum accepted limit of 15 Hz. By extrapolation, it can be clearly seen that for spans greater than 14 metres, natural frequencies could fall below 15 Hz and thus resonance effects would need to be checked more closely according to the imposed loading conditions. This may include considering the effects of damping provided by partitions and other parts of the structure.

6 CONCLUSIONS AND FURTHER RESEARCH

This paper has considered the effect of semi-rigid connections on the vibration performance of continuous composite beams. The inclusion of semi-rigid composite connections was found to improve the performance of these beams and could potentially allow increased spans for structural systems, which have been previously limited. The effects of including cracking were considered and found to be quite important in modelling the dynamic response of these beams. The effects of reinforcement area were also considered and typical increases of about 10% in the fundamental natural frequency were achieved by adopting modest amounts of reinforcing steel. The most important parameter to affect the dynamic response of semi-continuous composite beams was span length and it was concluded that the use of these beams for spans up to 16 metres would provide acceptable vibration performance in building floors.

Further research is required into the serviceability behaviour of semi-continuous composite beam systems, so designers can confidently predict the advantageous effects of the semi-rigid connection on the dynamic behaviour. In particular, full-scale tests on the dynamic behaviour of semi-continuous composite beams would allow calibration of the proposed model. In addition to this, the effects of partial interaction and partial shear connection need to be investigated to assess the effects of these aspects on the dynamic response on semi-continuous beams.

7 ACKNOWLEDGEMENTS

The authors would like to thank the Australian Research Council, which funded this project through the Large Grants Scheme. Furthermore, the authors would like to thank Mr. John Morgan from Engineering Systems for providing the software program Microstran that was utilised in undertaking the dynamic analysis in this paper.

8 REFERENCES

British Standards Institution 1994. Eurocode 4, ENV 1994-1-1 1994. Design of composite steel and concrete structures, Part 1.1, General Rules and Rules for Buildings.

Crisinel, M. & A. Carretero 1996. Simple prediction method for moment-rotation properties of composite beam-to-column joints, *Proceedings of the Engineering Foundation Conference on Composite Construction:Composite Construction III, Irsee, 9-14 June 1996:* 823-835. New York: ASCE.

Leon, R.T. 1998. Composite connections, *Progress in Structural Engineering and Materials, Construction Research Communications,* 1: 159-169.

Nethercot, D.A. 1997. Behaviour and design of composite connections, *Composite Construction, Conventional and Innovative, Innsbruck, 16-18 September 1997:* 657-662. Zurich: IABSE.

Standards Association of Australia 1980. *AS2327.1-1980, SAA Composite Construction Code, Part 1: Simply supported beams.*

Standards Association of Australia 1989. *AS1170.1-1989, SAA Loading Code, Part 1: Dead and live loads and load combinations.*

Standards Association of Australia 1996. *AS2327.1-1996, Australian Standard, Composite Structures, Part : Simply supported beams.*

Mechanics of Structures and Materials, Bradford, Bridge & Foster (eds)
© 1999 Balkema, Rotterdam, ISBN 90 5809 107 4

The seismic performance of reinforced concrete band beam frames in Australia

J.S.Stehle, K.Abdouka, H.Goldsworthy & P.Mendis
Department of Civil and Environmental Engineering, University of Melbourne, Vic., Australia

ABSTRACT: Reinforced concrete wide band beams are used widely in Australia even though their performance under seismic loading has previously not been researched. A series of tests and analytical work have been undertaken at the University of Melbourne to assess the performance of such construction for the level of seismicity that can be expected in Australia. The structural system appears to be generally adequate, however some minor detailing changes are recommended to improve performance at very little extra cost.

1. EXPERIMENTAL WORK

1.1 *Design according to Australian codes*

An experimental program was conducted at the University of Melbourne which included the testing of two interior and two exterior half scale joint subassemblages. The design of the members was based on a 4 storey, 6 bay frame. The first series of specimens were designed in accordance with AS3600 (SAA 1994) and AS1170.4 (SAA 1993). The details of the specimens are shown in Figure 1 and Figure 2. Such a design is not recommended for regions of high seismicity according to pre-existing codes and research (Abdouka and Goldsworthy 1997).

All specimens were tested taking into account loading on the beams and axial load in the column due to dead load and 0.4 live load (Figure 3). A quasi-static cyclic lateral displacement was subsequently applied to the bottom column of the specimen.

The hysteretic response of the specimens are shown in Figure 5 and Figure 6. Important to note here is the strength reduction that occurs in the exterior specimen when the beam is subjected to positive moment and the bottom bars lose their anchorage. This loss of anchorage also occurs for the interior specimen, however due to the static redundancy of the specimen, the drop in strength of the overall subassemblage is not as pronounced.

Torsional cracking was a major concern and it is shown to be more dramatic in the exterior specimen than the interior specimen (see Figure 4). In both specimens, the large crack widths that are associated with torsional cracking are unacceptable since high local strains in the reinforcement occur which can lead to steel fracture.

1.2. *Improved detailing*

No structural damage occurs for an inter-storey drift ratio of less than 1% with the existing detailing. Time history analyses have shown that for Australian levels of seismicity, drift ratios of more than 1% are unlikely (Stehle et al, 1997, 2000) so it can be concluded that existing

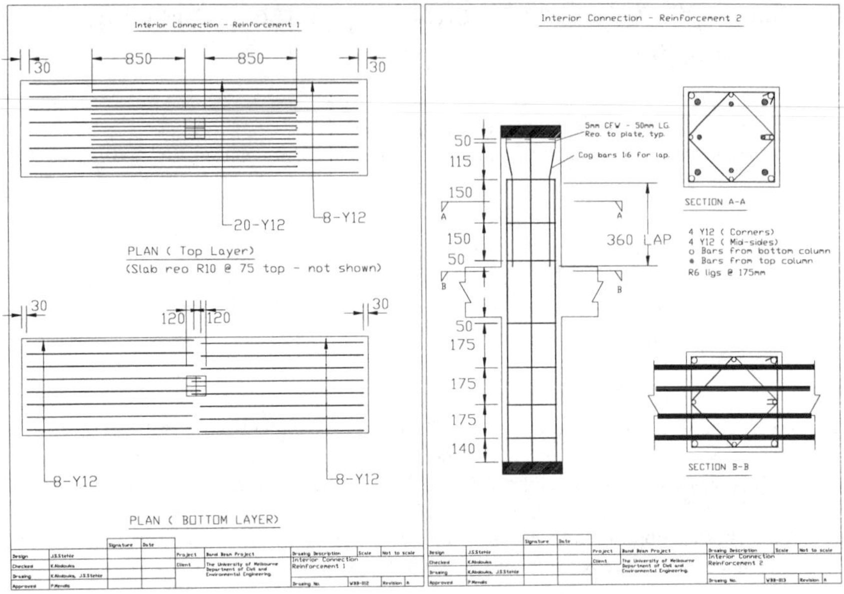

Figure 1: Details of first interior specimen

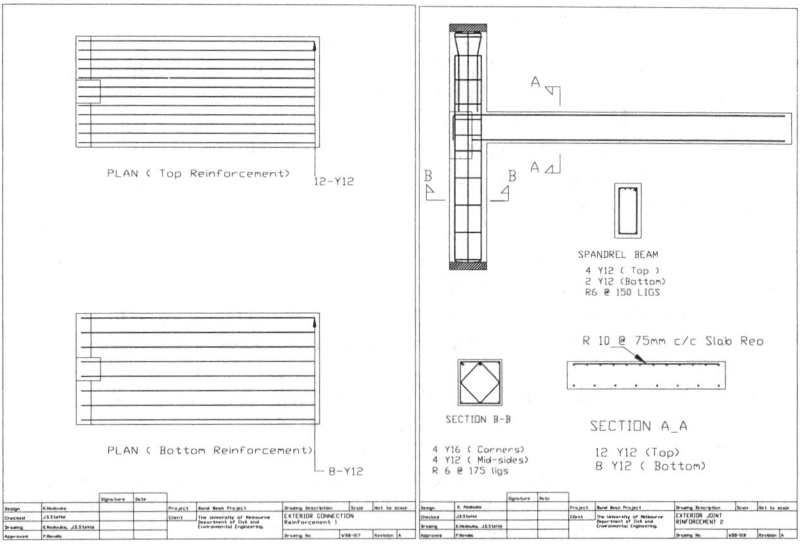

Figure 2: Details of first exterior specimen

design practice is adequate. Since there is a lot of uncertainty associated with estimating seismic risk, it is recommended that some detailing changes are made to improve the seismic performance. These changes are fairly inexpensive.

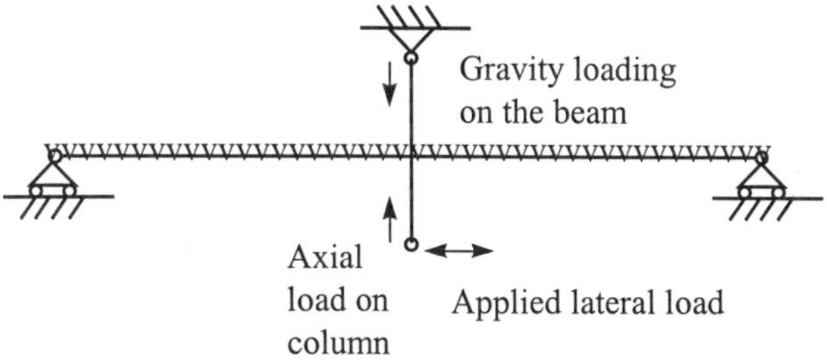

Figure 3: Diagrammatic test setup

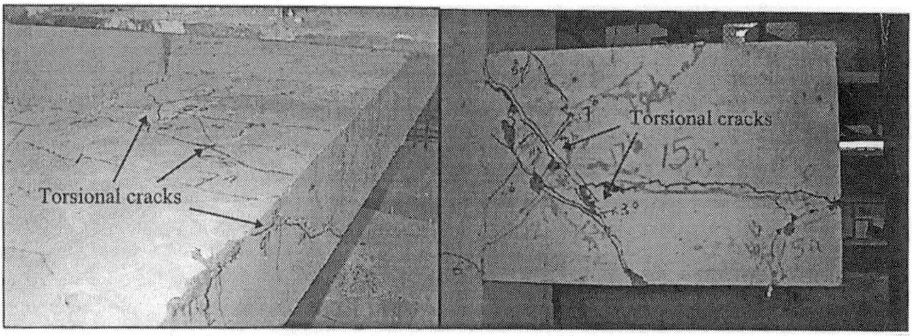

Figure 4: Cracking in first series of tests (interior - left, exterior - right)

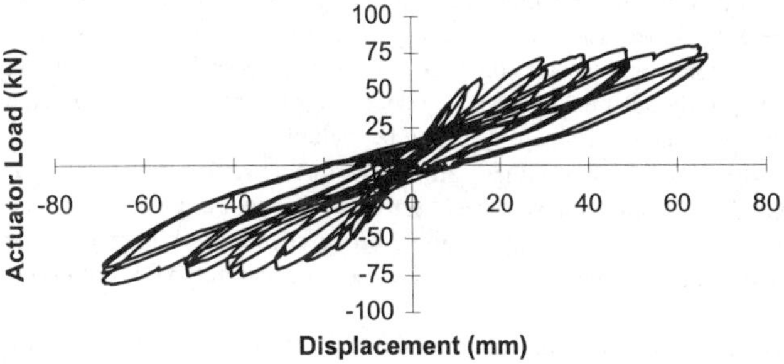

Figure 5: Hysteretic response of first interior specimen

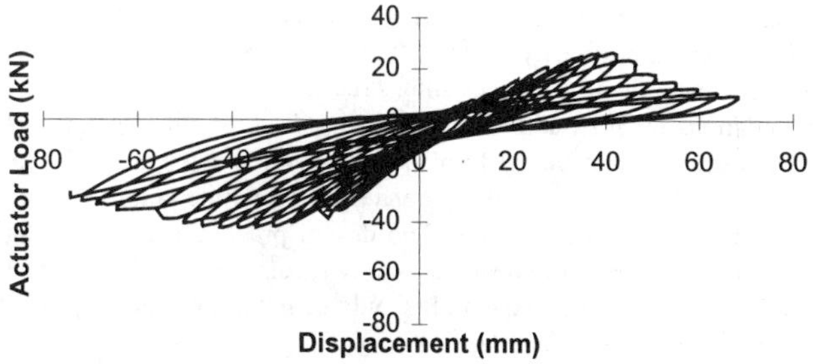

Figure 6: Hysteretic response of first exterior specimen

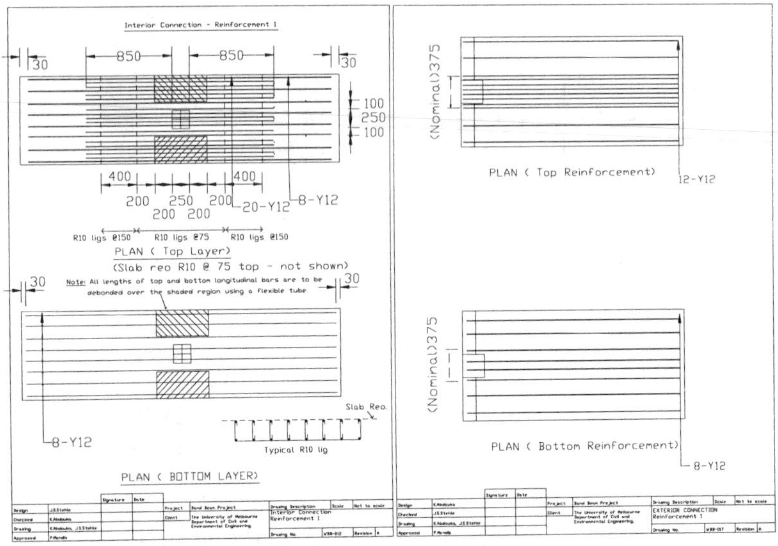

Figure 7: Details of second series of tests (interior - left, exterior - right)

For the second series of tests, a number of inexpensive detailing modifications were made. For the exterior specimen, the bottom bars were fully anchored so as to avoid bar pullout, and the beam bars were concentrated nearer the column to reduce the torsional demand on the sides of the joint (see Figure 7). For the interior specimen, the bottom bars were fully anchored also, and the outer beam bars were debonded using plastic tubing. This was so that the bars slipped through the concrete in this region and no torsion was developed. These bars still remain effective under gravity loading (Abdouka et al. 1999). By debonding the bars, a reduced beam shear width occurred, and hence some beam ligatures had to be included.

The hysteretic response of the improved specimens are shown in Figure 9 and Figure 10. It can be seen that, for the exterior specimen, the specimen's strength does not degrade due to bar pullout. Also, since there is better bar anchorage in the interior specimen, higher ultimate strengths were attained. There is only a minor increase in the interior specimen's flexibility due to the debonding of bars.

Figure 8 shows that torsional cracking still did occur in the exterior specimen, however, the crack widths were much smaller. Hence, concentrating the beam bars closer to the column was effective. For the interior specimen, it can be seen in Figure 8 that no torsional cracking occurred. Debonding the bars was successful. For both specimens, large displacement capacities were evident which shows that there is a possibility of the use of wide band beams in regions of high seismicity, even if only part of a secondary framing system which has sufficient displacement capacity to deflect with a primary force resisting structure.

In terms of design recommendations for Australia, it appears that providing full anchorage of bottom beam bars is easily achieved and that the performance improvement is quite marked. Concentrating longitudinal beam reinforcement at exterior connections near the column is also a cheap improvement, and hence it is advised for Australian design practice also. Avoiding torsional cracking by debonding bars at interior connections is only recommended in regions of higher seismicity where seismic events are more frequent. It should be noted that severe torsion occurs on an exterior joint under normal gravity loading.

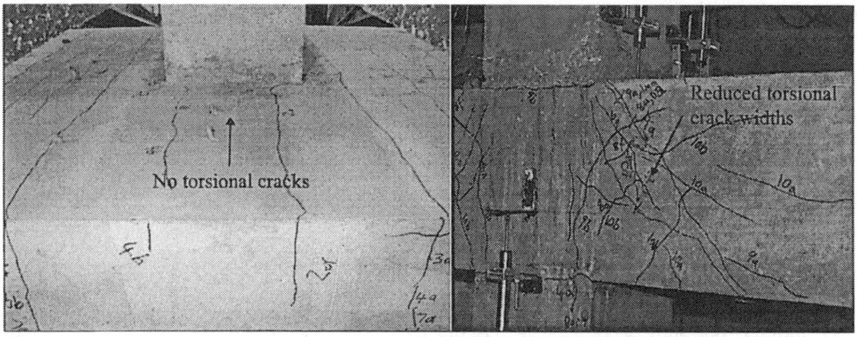

Figure 8: Cracking in second series of tests (interior - left, exterior -right)

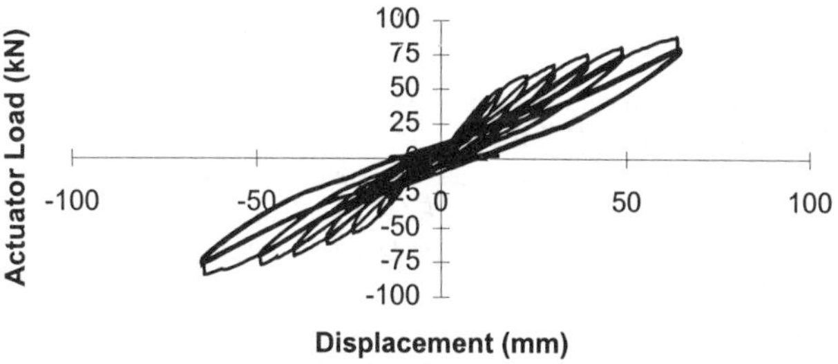

Figure 9: Hysteretic response of 2nd interior specimen

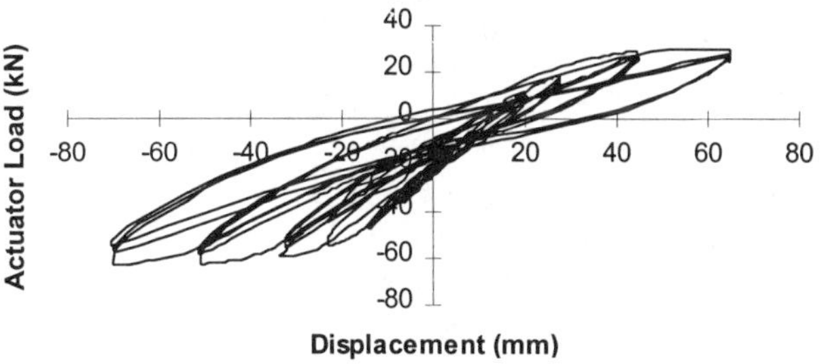

Figure 10: Hysteretic response of 2nd exterior specimen

It is recommended also, for Australian design, that extra shear reinforcement in columns should be provided in case of a soft-storey collapse mechanism occurring which would prevent a brittle failure. This is required since, with such large, strong band beams relative to column strength, hinging is possible in the column.

2. CONCLUSIONS

(1) Band beams constructed in Australia, using current design practices, may perform adequately under an Australian design level earthquake.

(2) A number of minor detailing changes are recommended for Australian design. They are the following:
   (i) full anchorage of bottom beam bars at supports.
   (ii) shear reinforcement in the columns so that in case of column hinging, a brittle shear failure will not occur.
   (iii) for the case of an exterior joint, the concentration of beam longitudinal bars near the column
(3) Wider beams than that currently allowed may be used in regions of high seismicity.

ACKNOWLEDGMENTS

The authors gratefully acknowledge the financial assistance received from the Australian Research Council under large grant no. 896018878.

REFERENCES

Abdouka K., & Goldsworthy H. 1997. Seismic Effects on Wide Band Beam Construction in Australia: Review. *15th ACMSM*. Melbourne: 311-316.
Abdouka, K., Stehle J. S., Goldsworthy H. & Mendis P. 1999. Seismic Performance of Reinforced Concrete Frames with Wide Band Beams. *NZSEE 99*. Auckland: 171-178.
Standards Association of Australia 1993. *Minimum Design Loads on Structures, Part 4: Earthquake Loads AS 1170.4*. Sydney: SAA
Standards Association of Australia 1994. *Concrete Structures Code AS3600*. Sydney: SAA
Stehle J. S., Abdouka, K., Mendis P., & Goldsworthy H. 1999. Seismic Performance of Reinforced Concrete Frames with Wide Band Beams. *WCEE February 2000*. Auckland: accepted for publication.
Stehle J. S., Mendis P., Goldsworthy H. & Wilson J. 1997. Seismic Effects on Wide Band Beam Construction in Australia- A Preliminary Analysis. *15th ACMSM*. Melbourne: 351-356.

Structural analysis and stability

Mechanics of Structures and Materials, Bradford, Bridge & Foster (eds)

 ISBN 90 5809 107 4

The influence of weld-induced localised imperfections on the buckling of cylindrical thin-walled shells

M. Pircher & R.Q. Bridge
School of Civic Engineering and Environment, University of Western Sydney, Nepean, N.S.W., Australia

ABSTRACT: Axisymmetric imperfections in cylindrical thin-walled shell structures under axial load have been known to result in particularly severe reductions in buckling strength. Imperfections in the vicinity of circumferential welds in steel silos and tanks fall into this category and therefore deserve special attention. Finite element models were used to analyse imperfect cylindrical shells and special care was taken to model the weld-induced imperfection. The geometry was calibrated against data gained from measuring such imperfections on existing silos, and residual stresses were taken into account. A number of parameters were evaluated in respect to their influence on the buckling behaviour of the modelled cylinders. While the shape of the weld depression and the height of the individual strakes contributed to variations in the buckling behaviour, residual stresses were found to have only a small influence on the buckling load.

1 INTRODUCTION

Axisymetric imperfections in circular cylindrical steel shell structures such as silos or tanks, occur during construction when rolled steel plates are formed into a series of individual strakes and joined together by circumferential welds (Figure 1). Deformations at each circumferential joint are partly caused by the rolling process and from weld-shrinkage of the heated zone. A close look at the area in the immediate vicinity of this axisymmetric imperfection shows that the geometric imperfection is accompanied by various other interacting influences (Figure 2). Residual stresses result mainly from weld shrinkage and reach yield in the tension zone close to the weld. For equilibrium reasons, these tension stresses have to be accompanied by compression stresses further away from the weld. The properties of the filler material of the weld and the material in the heat affected zone may be different from the material properties of the actual shell structure. Overall deviations from the cylindrical shape also play a role in reducing the buckling strength of the structure even further. Moreover, each individual strake is typically not high enough to isolate the effects of one weld imperfection from its neighbours.

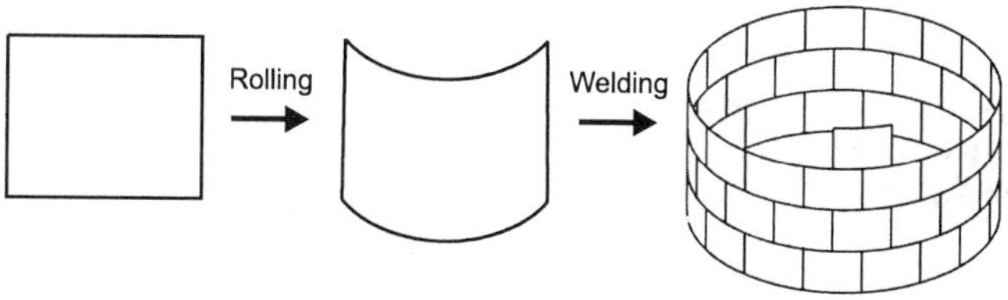

Figure 1. Erection of a Circular Silo or Tank

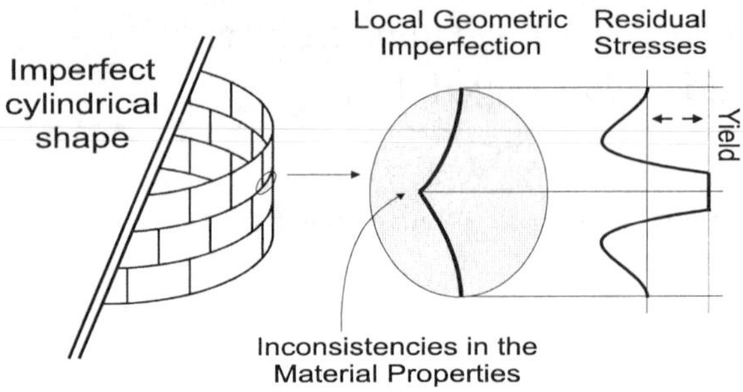

Figure 2. Circumferential weld imperfection

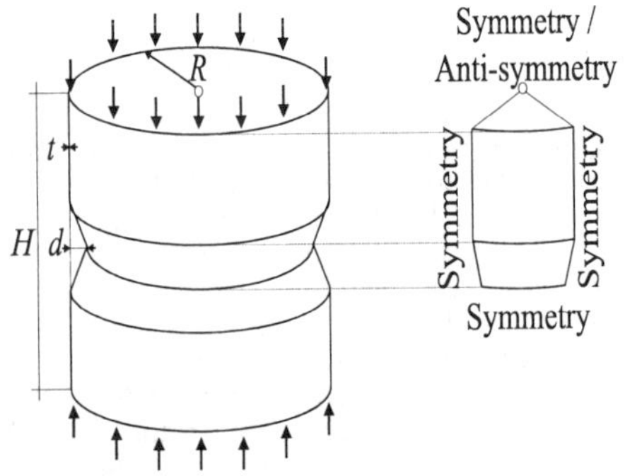

Figure 3. Analysed system

Two papers are known to have looked at circumferential welds in circular cylindrical shell structures including the effects of residual stresses (Bornscheuer et al (1983), Rotter (1996)). These two papers reach different conclusions. Bornscheuer et al (1983) report a decrease of the buckling strength of their model by up to 10% while Rotter (1996) concludes that "Circumferential residual stresses in the welded joint, developed by shrinkage of the weld, appear to increase the buckling strength...". The motivation behind this paper was to clarify the role of circumferential weld induced depressions, in combination with the effects of the accompanying residual stresses, in the buckling of axially compressed cylinders.

2 MODELLING THE IMPERFECT CYLINDER

2.1 *Cylinder geometry*

The case study performed in this paper was based on the geometric properties of a silo in Port Kembla (Australia) on which weld imperfections have been carefully measured (Clarke, Rotter (1988)). Therefore, the steel silo for the study was represented by a cylinder of a radius R of 12m. The wall thickness t was taken to be 12mm. A single strake height H of 3m was modelled and axial load was taken to be constant along this limited partition of the structure. This approach led to a computer model of one circumferential joint with half a strake height on each side of the joint (Figure 3).

2.2 *Depression shape*

The shape of these depressions has been the subject of some research. In the following, names in brackets with quotes signify the name of the authors who have proposed particular shape functions. Rotter & Teng (1989) proposed two different shape functions based on theoretical arguments ('Rotter a', 'Rotter b') and used them in some of their work (Rotter & Teng (1989), Teng & Rotter (1992)). These shape functions were revised later ('Rotter final', 'Rotter closed') on the basis of data gathered at the reference silo in Port Kembla using empirical techniques by Rotter (1996). More suggestions on how to discribe the shape of such a weld-induced depression can be found in Bornscheuer et al. (1983) ('Häfner'), White & Dwight (1977) ('W&D'), Fritschi (1995) ('Fritschi') and Steinhardt & Schulz ('Schulz'). In addition to these shapes one more shape function ('Shrink') was added which is based on a detailed FE-analysis of the shrinkage process of the weld (Pircher & Bridge, 1997).

2.3 *Adjacent joints*

The strake heights typically used in silos or tanks are small enough to ensure interaction between adjacent weld imperfections. Rotter (1996) suggested three patterns of interaction between neighbouring buckling strakes (Figure 4). He showed that assuming symmetry at mid-strake height ('SS') yields the highest buckling values and assuming anti-symmetry on both sides ('AA') the lowest buckling values.

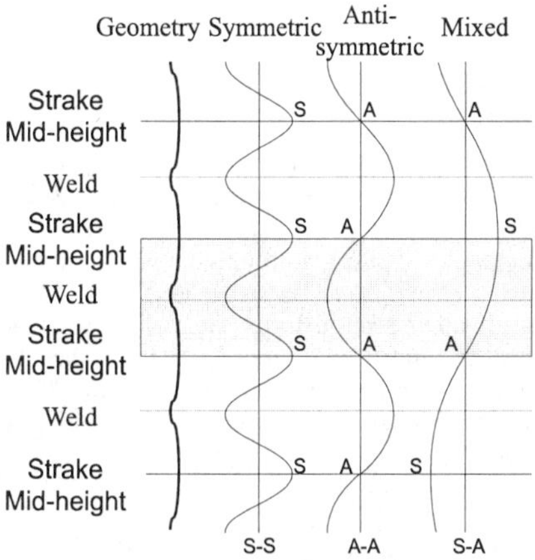

Figure 4. Strake midheight boundary conditions and meridional buckling modes

2.4 *Residual stresses*

All analyses in this paper were based on a bi-linear material law with ultimate stress reached at a plastic strain of 0.2%. Steel with a yield stress of 200 MPa was chosen. A preliminary analysis looked at different ways of modelling the shrinkage of the weld which is the primary source of the residual stresses found in the vicinity of the depression. In a first step, material properties of the cylinder were modelled in a detailed way, including temperature dependencies. The cooling process of the weld was then analysed. This was compared to an alternate method of modelling the shrinkage problem which follows a procedure described by Rotter (1996). A length of the shell of 50mm above and below the weld was subjected to a shrinkage strain equal to the yield strain (0.00125). These strains led to stresses which the system attempted to relieve by deforming. Therefore the geometry of the weld depression was changed by the application of residual strains. As a result of this, the depth of the geometric imperfection was scaled to an amplitude of 9.48mm after which the residual strains were applied, resulting in the desired imperfection of a magnitude of one wall thickness. The two methods of achieving residual stress patterns and geometric imperfections gave similar results and compared well to measured results. The second method was used for all ensuing analyses.

2.5 *FE-Model*

Boundary conditions of the computer model were set as shown in Figure 3 so that one single buckling wave was obtained. Several different buckling modes needed to be investigated to determine the mode with the smallest load that leads to failure and to account for the fact that many possible modes lay within a very small difference in load.

Boundary conditions of the computer model were set as shown in Figure 3 so that one single buckling wave was obtained. Several different buckling modes needed to be investigated to determine the mode with the smallest load that leads to failure and to account for the fact that many possible modes lay within a very small difference in load.

To determine the effects that residual stresses have on the buckling behaviour, two sets of models were investigated:

(1) The first set included stresses and deformations induced by weld shrinkage followed by the application of axial load.

(2) The second set used the same induced weld deformations (ignoring the residual stresses) followed by the application of axial load.

Finite Element code ABAQUS was used for all computer investigations for this paper.

3 RESULTS

3.1 *Weld shrinkage*

On the basis of empirical studies, Faulkner (1977) has suggested using a tension zone of a magnitude of 3 to 6 wall thicknesses on each side of the weld with tensile stresses close to yield. A value of 0.2 to 0.4 is suggested for the ratio of tension stress at the weld to compression stress away from the weld. Both methods of generating residual stresses around the weld depression used in this paper resulted in stress fields that fell well within these suggested limits. In addition to these circumferential membrane forces, all other section forces contribute to achieve equilibrium. Starting the analysis with a local circumferential imperfection of the shape 'closed' with a depth of 9.48mm resulted in a groove with an amplitude of 12.11mm. The geometry of this groove was recorded and used as an initial stress-free imperfection for the second set of cylinders modelled for this investigation.

3.2 *Buckling modes*

A significant aspect of the buckling behaviour of cylindrical thin walled systems under axial load is the close proximity of bifurcation points along the load deflection curve. Each bifurcation point corresponds to a different buckling mode. Obviously, the bifurcation point which requires the smallest load is the one of most interest. The buckling strength of an imperfect cylinder is commonly related to the classical elastic critical stress σ_{cl} which is given by

$$\sigma_{cl} = \frac{E}{\sqrt{3\left(1-\upsilon^2\right)}}\frac{t}{R} \approx 0.605\frac{Et}{R} \qquad (1)$$

in which E stands for the elastic modulus and ν for Poisson's ratio. In a first analysis several buckling modes were investigated for the stress-free model as well as for the model which included residual stresses. The results compared well with results gained by Rotter (1996) and contradict the results of Bornscheuer et al. (1983). When residual stresses are considered in the buckling analysis, the critical load is slightly higher than in the stress-free systems of the same geometry. Also, a different buckling mode applies for the two systems. This has been explained by the partially alleviating stress patterns of the two load conditions. The stress-free system is under meridional and circumferential compression at the point of buckling whereas the residual circumferential tension stresses in the other system have a stabilising effect until the bifurcation point is reached. The parts of the shell a small distance away from the weld are under circumferential compression stresses due to weld shrinkage which are gradually relieved as the axial load increases. A more detailled explanation can be found in Pircher & Bridge (1998).

3.3 *Boundary conditions*

The influence of neighbouring strakes is dependent on the length of the individual strakes and the boundary conditions assumed at strake mid-height as shown in Figure 4. Function 'Rotter closed' was used to investigate the influence of the boundary conditions on the buckling load as it is the most conservative (see Figure 5) apart from shape function 'Fritschi' which has been shown to be rather unrealistic (Ummenhofer & Knödel, 1996). It was found that boundary condition 'S-S' provides a much stronger model than boundary condition 'A-A'. Weld-induced residual stresses had a strengthening effect with both sets of boundary conditions.

Table 1: Buckling strength σ_b/σ_{cl} of silo strakes with boundary conditions A-A and S-S.

	Boundary condition A-A	Boundary condition S-S
Initially stress free	0.245	0.452
Incl. residual stresses	0.273	0.467

3.4 *Shape of the weld imperfection*

Several different shape functions were used to model the imperfection, using boundary condition 'A-A', and the influence of weld-induced residual stresses on the buckling strength was determined for these geometries. As can be seen in Figure 5, the buckling load varies strongly depending on the shape function used. Buckling loads have been found to increase consistently when weld-induced residual stresses were taken into account. The amount of this strengthening effect of residual stresses varied, depending on the shape function. The number of circumferential buckling waves m varied between m = 19 and m = 22.

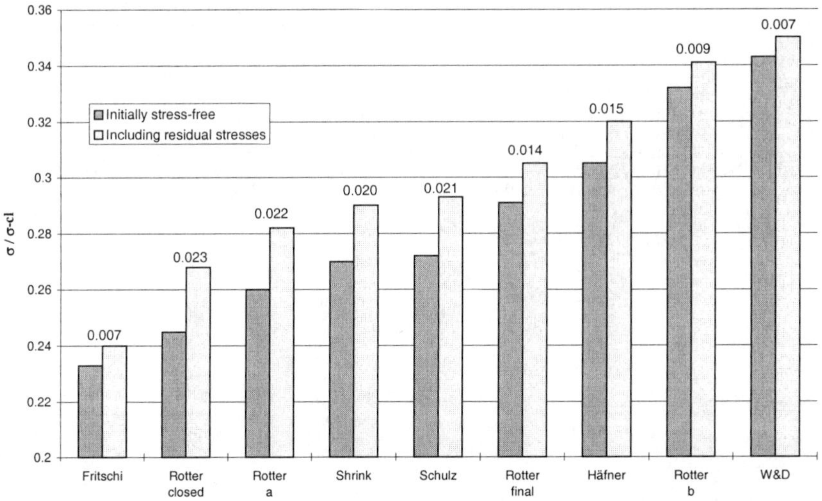

Figure 5. Buckling strength and influence of residual stresses using different imperfection shapes

4 CONCLUSION

The paper studied the influence of weld induced residual stresses in the vicinity of circumferential joints on the buckling behaviour of cylindrical thin-walled structures under axial load. A case study was undertaken that used the geometric properties of an existing silo in Port Kembla (Australia). It was found that on an axisymmetric model for local weld imperfection, residual stresses increased the buckling strength of the studied cylinder by a small amount. Several different shape functions were used to model the axisymmetric imperfections. The buckling strength of the cylinder varied strongly with on the shape of the weld imperfection. Interaction between neighbouring strakes was taken into account and the interaction pattern that leads to the lowest buckling strength was determined.

REFERENCES

Bornscheuer F.W., Häfner L., Ramm E. 1983. Zur Stabilität eines Kreiszylinders mit einer Rundschweissnaht unter Axialbelastung. *Der Stahlbau* v52, n10, pp 313-318

Clarke, M.J. & Rotter, J.M. 1988. A Technique for the Measurement of Imperfections in Prototype Silos and Tanks. *Research Report R565*, School of Civil and Mining Engineering, University of Sydney, Australia

Faulkner, D. 1977. Effects of residual stresses on the ductile strength of plane welded grillages and of ring stiffened cylinders. *J. Strain Analysis for Engng Design*, v12, pp 130-139

Fritschi H. 1995. Shape Deviations in the Vicinity of Circumferential Welds at Standing Cylindrical Steel Shell Structures. *Research Report*, Versuchsanstalt für Stahl, Holz und Steine, University of Karlsruhe, Karlsruhe

Karman T. v., Dunn L. G., Tsien H. S. 1941. The Buckling of Thin Cylindrical Shells under Axial Compression. *J. Aeronaut. Sci*, v8, pp 303-312

Koiter W. T. 1945. The Stability of Elastic Equilibrium, *Dissertation at the Technische Hooge School, Delft.* English translation by E. Riks: *Technical Report AFFDL-TR-70-25*, Air Force Flight Dynamics Laboratory, Air Force Systems Command, Wright-Patterson Air Force Base, Ohio, 1970

Koiter W. T. 1963. The Effect of Axisymmetric Imperfections on the Buckling of Cylindrical Shells under Axial Compression. *Proc. Koninklijke Nederlandse Akademie van Wetenschappen*, pp 265-279

Pircher M., Bridge R. 1997. Modelling the Effects of a Weld-induced Circumferential Imperfection on the Buckling of Cylindrical Thin-walled Shells. Proc. Eleventh Australasian Compumod Conference, Kuranda, pp 180-189

Pircher M., Bridge R. 1998. The Influence of Weld-Induced Residual Stresses on the Buckling of Cylindrical Thin-Walled Shells. *Proc. Thin Walled Structures Conf.*, Singapore, pp 671-678

Steinhardt O., Schulz U 1970. Zur Beulstabilität von Kreiszylinderschalen. *Bericht der Versuchsanstalt fuer Stahl, Holz, Steine*, Universität Karlsruhe, Karlsruhe

Rotter, J.M. 1996. Buckling and Collapse in Internally Pressurised Axially Compressed Silo Cylinders with Measured Axisymmetric Imperfections: Imperfections, Residual Streasses and Local Collapse. *Imperfections in Metal Silos Workshop*, Lyon, France, pp 119-139

Rotter J.M. & Teng J.G. 1989. Elastic Stability of Cylindrical Shells with Weld Depressions. *Journal of Structural Engineering, ASCE*, v 116, n 8, pp 1244-1263

Teng, J.G. & Rotter J.M. 1992. Buckling of Pressurized Axisymmetrically Imperfect Cylinders under Axial Loads. *Jnl. of Engineering Mechanics*, v118, n 2, pp 229-247

Ummenhofer T., Knödel 1996. Typical Imperfections of Steel Silo Shells in Civil Engineering. *Imperfections in Metal Silos Workshop*, Lyon, France, pp 103-118

White J.D., Dwight J.B. 1977. Weld Shrinkage in Large Stiffened Tubulars. *Proc. Conf on Residual Stresses in Welded Constructions*, The Welding Institute, London, pp 337-348

Mechanics of Structures and Materials, Bradford, Bridge & Foster (eds)
© 1999 Balkema, Rotterdam, ISBN 90 5809 107 4

Doubly, singly and point symmetric fixed-ended columns undergoing local buckling

K.J.R. Rasmussen
Department of Civil Engineering, University of Sydney, N.S.W., Australia

B. Young
School of Civil and Structural Engineering, Nanyang Technological University, Singapore

ABSTRACT: The paper sets out the governing equations for the overall buckling of locally buckled columns compressed between fixed ends. The governing equations incorporate the stiffnesses of the locally buckled section against overall minor and major axis flexure as well as warping torsion. Doubly, singly and point symmetric cross-sections are considered. Buckling curves are obtained for an I-section, U-section (or channel section) and Z-section, representing doubly, singly and point symmetric sections respectively. The buckling curves show that local buckling can change the critical overall mode from a flexural-torsional mode to a purely flexural mode in the case of U-sections. For Z-sections, the flexural modes are uncoupled for a non-locally buckled section but become coupled when the section locally buckle.

1 INTRODUCTION

Thin-walled columns are prone to local buckling in their ultimate limit states. A rational design of thin-walled columns requires an understanding of how local buckling influences the overall behaviour.

Bifurcation analyses are useful for categorising the buckling behaviour of columns. Local buckling can be taken into account by using the stiffnesses of the locally buckled section in the overall bifurcation analysis, as described in Rasmussen (1997). Figure 1 illustrates the overall bifurcation of a locally buckled U-section into a flexural-torsional mode.

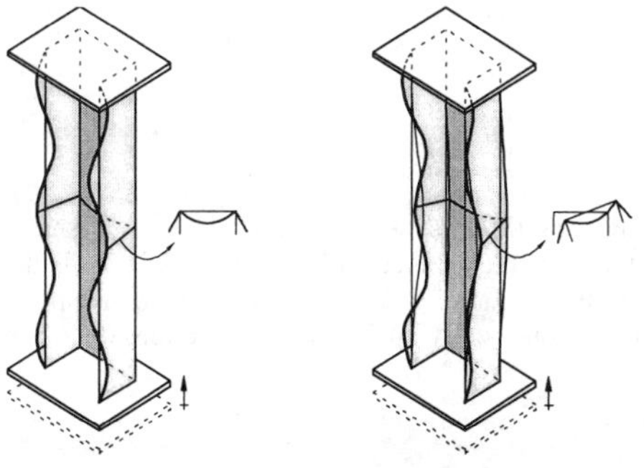

(a) Nonlinear locally buckled fundamental state

(b) Overall bifurcated state (flexural-torsional buckling mode)

Figure 1: Overall flexural-torsional buckling of a U-section

Determining the stiffness of the locally buckled section involves solving the governing nonlinear plate equations for a short length of column, as demonstrated in Young and Rasmussen (1997), where an elastic nonlinear finite strip analysis was employed to obtain the stiffnesses of U-sections against overall flexure and torsion. The same type of analysis was used in Rasmussen and Hasham (1997) to study the influence of local buckling on the interaction curves for I-section beam-columns.

The purpose of this paper is to present an overview of the overall buckling of fixed-ended thin-walled columns. It is shown how double, single and point symmetry affect the terms of the overall bifurcation equation leading to different couplings between minor axis flexure, major axis flexure and torsion. Expressions for the overall buckling loads are established, and buckling curves are presented for an I-section, U-section and Z-section to show the reductions in overall buckling capacity and changes in critical buckling mode resulting from local buckling.

2 ANALYSIS

2.1 *Overall buckling analysis*

As shown in Rasmussen (1997), the overall bifurcation equation for locally buckled columns takes the form,

$$\int_L \begin{Bmatrix} w_b' \\ -u_b'' \\ -v_b'' \\ -\varphi_b'' \\ -\varphi_b' \end{Bmatrix}^T \begin{bmatrix} (EA)_t & (ES_y)_t & (ES_x)_t & (ES_\omega)_t & 0 \\ (ES_y)_t & (EI_y)_t & (EI_{xy})_t & (EI_{y\omega})_t & 0 \\ (ES_x)_t & (EI_{xy})_t & (EI_x)_t & (EI_{x\omega})_t & 0 \\ (ES_\omega)_t & (EI_{y\omega})_t & (EI_{x\omega})_t & (EI_\omega)_t & 0 \\ 0 & 0 & 0 & 0 & (GJ)_t \end{bmatrix} \begin{Bmatrix} \delta w_b' \\ -\delta u_b'' \\ -\delta v_b'' \\ -\delta\varphi_b'' \\ -\delta\varphi_b' \end{Bmatrix} \mathrm{d}z + \int_L \begin{Bmatrix} u_b' \\ u_b'' \\ v_b' \\ v_b'' \\ \varphi_b \\ \varphi_b' \end{Bmatrix}^T \begin{bmatrix} N_0 & 0 & 0 & 0 & 0 & y_S N_0 \\ 0 & 0 & 0 & 0 & 0 & 0 \\ 0 & 0 & N_0 & 0 & 0 & -x_S N_0 \\ 0 & 0 & 0 & 0 & 0 & 0 \\ 0 & 0 & 0 & 0 & 0 & 0 \\ y_S N_0 & 0 & -x_S N_0 & 0 & 0 & (x_S^2 + y_S^2) N_0 + W_0 \end{bmatrix} \begin{Bmatrix} \delta u_b' \\ \delta u_b'' \\ \delta v_b' \\ \delta v_b'' \\ \delta\varphi_b \\ \delta\varphi_b' \end{Bmatrix} \mathrm{d}z = 0 \quad (1)$$

where w, u, v, φ are the displacements in the longitudinal z-direction, principal x-axis direction, principal y-axis direction and twist rotation respectively, $(EA)_t$ … $(EI_\omega)_t$ are the tangent rigidities at the current load level, (x_S, y_S) are the shear centre coordinates, and (N_0, W_0) are the fundamental state (pre-buckling) stress resultants, defined as

$$N_0 = \int_A \sigma_0 \, \mathrm{d}A \quad (2)$$

$$W_0 = \int_A \sigma_0 (x^2 + y^2) \mathrm{d}A \quad (3)$$

In eqn. (1), the prescript δ denotes a virtual quantity and the subscripts $()_b$ and $()_0$ denote fundamental state and bifurcated (buckled) state variables respectively. In eqns (2,3), σ_0 is the fundamental state normal stress which varies nonlinearly on the cross-section in the post-local buckling range. In the case of an undistorted (non-locally buckled) cross-section, σ_0 is constant and so,

$$W_0 = N_0 r_0^2 \quad \text{(non-locally buckled sections)} \quad (4)$$

where

$$r_0^2 = \frac{I_x + I_x}{A} \quad (5)$$

2.2 *Tangent rigidities*

The tangent rigidities $(EA)_t$ ... $(EI_\omega)_t$ can be determined by a nonlinear post-local buckling analysis of a length of member equal to the local buckle half-wavelength. The analysis produces average rigidities for this length of section. In this paper, the rigidities are determined using the elastic finite strip analysis described by Hancock (1985).

In the post-local buckling analysis, the section is uniformly compressed incrementally, and at each increment, small additional axial compression, curvature about the *x*-axis, curvature about the *y*-axis and warping is superimposed to determine the resulting axial force, moment about the *x*-axis, moment about the *y*-axis and bimoment, allowing the tangent rigidities to be determined as the ratios of resulting actions to applied strains. For instance, the tangent rigidity $(EI_{x\omega})_t$ is determined as,

$$(EI_{x\omega})_t = \frac{\Delta B}{-\Delta v''} = \frac{\Delta M_x}{-\Delta \varphi''} \tag{6}$$

where $-\Delta v''$, $-\Delta\varphi''$, ΔB and ΔM_x are applied curvature about the *x*-axis, applied twist, resulting bimoment and resulting moment about the *x*-axis respectively. The equality of $\Delta B/-\Delta v''$ and $\Delta M_x/-\Delta\varphi''$ follows from the reciprocal theorem. Further details can be found in Rasmussen (1997).

The tangent torsional rigidity $(GJ)_t$ is assumed equal to the full torsional rigidity (GJ).

3 GOVERNING EQUATIONS

3.1 *Doubly symmetric cross-sections*

In the case of double symmetry, the following rigidities vanish (Rasmussen 1997),

$$(ES_x)_t = (ES_y)_t = (ES_\omega)_t = (EI_{xy})_t = (EI_{x\omega})_t = (EI_{y\omega})_t = 0 \tag{7}$$

Substituting eqn. (7) into eqn. (1), using $x_S=y_S=0$ and integrating by parts, the governing equations are derived as,

$$\left((EA)_t w_b'\right)' = 0 \tag{8}$$

$$\left((EI_y)_t u_b''\right)'' + \lambda_c\left(\overline{N} u_b'\right)' = 0 \tag{9}$$

$$\left((EI_x)_t v_b''\right)'' + \lambda_c\left(\overline{N} v_b'\right)' = 0 \tag{10}$$

$$\left((EI_\omega)_t \varphi_b''\right)'' - \left((GJ)_t \varphi_b'\right)' + \lambda_c\left(\overline{W} \varphi_b'\right)' = 0 \tag{11}$$

where N_0 and W_0 have been replaced by $-\lambda_c\overline{N}$ and $-\lambda_c\overline{W}$ respectively, $\overline{N}$ and $\overline{W}$ being reference values positive in compression and λ_c a critical value of the load factor (λ).

3.2 *Singly symmetric cross-sections*

Choosing the *x*-axis as the symmetry axis, the following rigidities vanish (Young and Rasmussen 1997),

$$(ES_x)_t = (ES_\omega)_t = (EI_{xy})_t = (EI_{y\omega})_t = 0 \tag{12}$$

Substituting eqn. (12) into eqn. (1), using $y_S=0$ and integrating by parts, the governing equations are derived as,

$$\left((EA)_t w_b'\right)' - \left((ES_y)_t u_b''\right)' = 0 \tag{13}$$

$$-\left((ES_y)_t w_b'\right)'' + \left((EI_y)_t u_b''\right)'' + \lambda_c\left(\overline{N} u_b'\right)' = 0 \tag{14}$$

$$\left((EI_x)_t v_b''\right)'' + \left((EI_{x\omega})_t \varphi_b''\right)'' + \lambda_c\left(\overline{N} v_b'\right)' - \lambda_c\left(\overline{N} x_S \varphi_b'\right)' = 0 \tag{15}$$

$$\left((EI_{x\omega})_t v_b''\right)'' + \left((EI_\omega)_t \varphi_b''\right)'' - \left((GJ)_t \varphi_b'\right)' - \lambda_c\left(\overline{N} x_S v_b'\right)' + \left(\left(\lambda_c \overline{N} x_S^2 + \lambda_c \overline{W}\right)\varphi_b'\right)' = 0 \tag{16}$$

3.3 *Point symmetric cross-sections*

Because of the point symmetry, the following rigidities vanish,

$$(ES_x)_t = (ES_y)_t = (EI_{x\omega})_t = (EI_{y\omega})_t = 0 \tag{17}$$

Substituting eqn. (17) into eqn. (1), using $x_S = y_S = 0$ and integrating by parts, the governing equations are derived as,

$$\left((EA)_t w_b'\right)' - \left((ES_\omega)_t \varphi_b''\right)' = 0 \tag{18}$$

$$\left((EI_y)_t u_b''\right)'' + \left((EI_{xy})_t v_b''\right)'' + \lambda_c\left(\overline{N} u_b'\right)' = 0 \tag{19}$$

$$\left((EI_{xy})_t u_b''\right)'' + \left((EI_x)_t v_b''\right)'' + \lambda_c\left(\overline{N} v_b'\right)' = 0 \tag{20}$$

$$-\left((ES_\omega)_t w_b'\right)'' + \left((EI_\omega)_t \varphi_b''\right)'' - \left((GJ)_t \varphi_b'\right)' + \left(\lambda_c \overline{W} \varphi_b'\right)' = 0 \tag{21}$$

4 BUCKLING LOADS FOR FIXED-ENDED COLUMNS

4.1 *Displacement functions*

The boundary conditions for fixed-ended columns are that all displacements, rotations and warping are restrained at both ends, except for the longitudinal displacement which is free at one end and restrained at the other. These conditions and the governing equations set out in Section 3 can be satisfied by the following displacement field,

$$\frac{u_b}{C_u} = \frac{v_b}{C_v} = \frac{\varphi_b}{C_\varphi} = 1 - \cos\left(\frac{2\pi z}{L}\right) \tag{22}$$

where C_u, C_v and C_φ are arbitrary constants. In the case of doubly symmetric cross-sections, the longitudinal buckling displacement (w_b) is zero. In the case of singly and point symmetric cross-sections, it is given by,

$$\frac{w_b}{C_w} = \sin\left(\frac{2\pi z}{L}\right) \quad \text{(singly and point symmetric sections)} \tag{23}$$

where C_w is an arbitrary constant. Substituting eqns (22-23) into the governing equations yields a determinant equation for each type of cross-section, from which the critical load factors can be determined.

4.2 *Buckling loads for doubly symmetric cross-sections*

The buckling loads for doubly symmetric cross-sections are given by (Rasmussen 1997),

$$\lambda_{c_u}\overline{N} = \frac{4\pi^2 (EI_y)_t}{L^2} \tag{24}$$

$$\lambda_{c_v}\overline{N} = \frac{4\pi^2 (EI_x)_t}{L^2} \tag{25}$$

$$\lambda_{c_\varphi}\overline{N} = \left(\frac{4\pi^2 (EI_\omega)_t}{L^2} + (GJ)_t\right)\underline{\overline{N}/\overline{W}} \tag{26}$$

According to eqn. (4), the underlined term ($\overline{N}/\overline{W}$) in eqn. (26) reduces to $1/r_0^2$ in the case of a non-locally buckled section. The buckling loads ($\lambda_{cu}\overline{N}$, $\lambda_{cv}\overline{N}$, $\lambda_{c\varphi}\overline{N}$) correspond to flexural buckling about the *y*-axis, flexural buckling about the *x*-axis and torsional buckling about the longitudinal *z*-axis respectively. There is no coupling between the buckling displacements in this case.

4.3 *Buckling loads for singly symmetric cross-sections*

The buckling loads for singly symmetric cross-sections are given by (Young and Rasmussen 1997),

$$\lambda_{c_u}\overline{N} = \frac{4\pi^2}{L^2}\left((EI_y)_t - \underline{\frac{(ES_y)_t^2}{(EA)_t}}\right) \tag{27}$$

$$\lambda_{c_{v\varphi}}\overline{N} = \frac{-B \pm \sqrt{B^2 - 4AC}}{2A} \tag{28}$$

The underlined term in eqn. (27) and the underlined terms in equations to follow vanish in case of an undistorted cross-section. The buckling loads ($\lambda_{cu}\overline{N}$, $\lambda_{cv\varphi}\overline{N}$) correspond to flexural buckling about the *y*-axis and flexural-torsional buckling involving flexure about the *x*-axis and torsion respectively. There is no coupling between flexural buckling about the *y*-axis and flexural-torsional buckling. The constants (*A*,*B*,*C*) in the expressions for $\lambda_{cv\varphi}\overline{N}$ are given by,

$$A = 1 - \left(\frac{x_S}{r}\right)^2 - \underline{\frac{r_0^2 - \overline{W}/\overline{N}}{r^2}} \tag{29}$$

$$B = -(P_x + P_z) + \underline{P_x \frac{r_0^2 - \overline{W}/\overline{N}}{r^2} - \frac{2x_S}{r^2}\left(\frac{2\pi}{L}\right)^2 (EI_{x\omega})_t} \tag{30}$$

$$C = P_x P_z - \underline{\left(\frac{2\pi}{L}\right)^4 \frac{(EI_{x\omega})_t^2}{r^2}} \tag{31}$$

where

$$P_x = \frac{4\pi^2 (EI_x)_t}{L^2} \tag{32}$$

$$P_z = \left(\frac{4\pi^2 (EI_\omega)_t}{L^2} + (GJ)_t\right)/r^2 \tag{33}$$

$$r^2 = r_0^2 + x_S^2 \tag{34}$$

4.4 *Buckling loads for point symmetric cross-sections*

The buckling loads for point symmetric cross-sections are given by,

$$\lambda_{c_\varphi}\overline{N} = \left(\frac{4\pi^2(EI_\omega)_t}{L^2} + (GJ)_t - \frac{4\pi^2}{L^2}\frac{(ES_\omega)_t^2}{(EA)_t}\right)\overline{N}/\overline{W} \tag{35}$$

$$\lambda_{c_{uv}}\overline{N} = \frac{-B \pm \sqrt{B^2 - 4AC}}{2A} \tag{36}$$

The buckling loads ($\lambda_{c\varphi}\overline{N}$, $\lambda_{cuv}\overline{N}$) correspond to torsional buckling about the z-axis and coupled flexural buckling about the x- and y-axes respectively. There is no coupling between torsional buckling and flexural buckling. The constants (A,B,C) in the expressions for λ_{cuv} are given by,

$$A = 1 \tag{37}$$

$$B = -(P_x + P_y) \tag{38}$$

$$C = P_x P_y - \left(\frac{2\pi}{L}\right)^4 (EI_{xy})_t^2 \tag{39}$$

where

$$P_x = \frac{4\pi^2(EI_x)_t}{L^2} \tag{40}$$

$$P_y = \frac{4\pi^2(EI_y)_t}{L^2} \tag{41}$$

5 ELASTIC BUCKLING CURVES

5.1 *General*

An I-section, U-section and Z-section are chosen for the present study as representative of doubly, singly and point symmetric cross-sections respectively. For each cross-section, the material is assumed to be elastic with values of Young's modulus and Poisson's ratio equal to $E=2\times10^5$ MPa and $\nu=0.3$ respectively. The dimensions as well as the local buckle half-wavelength (l) and elastic local buckling stress (σ_l) of each cross-section are shown in Fig. 2.

In the post-local buckling analysis, the sections include a geometric imperfection in the shape of the local buckling mode, as determined from a buckling analysis. The magnitude of the imperfection is taken as 2 % of the plate thickness.

5.2 *I-sections*

The buckling loads ($N_{cr}=\lambda_c\overline{N}$) are plotted against the column length (L) in Fig. 3. The buckling loads are nondimensionalised with respect to the elastic local buckling load ($N_l=\sigma_l A$). Figure 3 shows the buckling loads for the distorted (locally buckled) and undistorted (non-locally buckled) cross-sections. The latter curves are shown dashed. It follows from the figure that flexural buckling (u) about the minor axis is critical for both the distorted and undistorted cross-

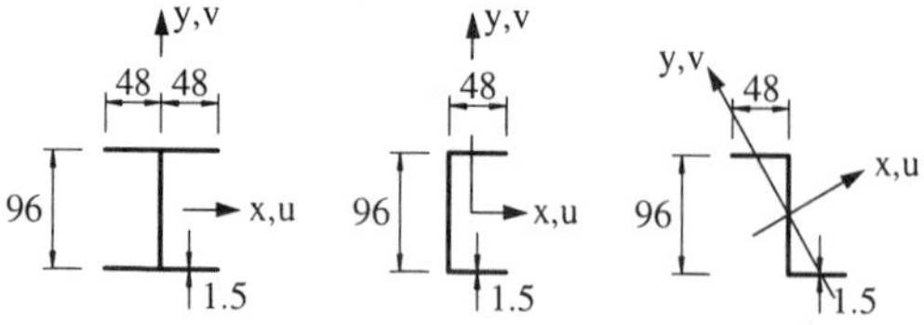

σ_l = 116 MPa, l = 145 mm — (a) I-section
σ_l = 128 MPa, l = 125 mm — (b) U-section
σ_l = 128 MPa, l = 125 mm — (c) Z-section

Figure 2: Cross-section dimensions in mm and local buckling details

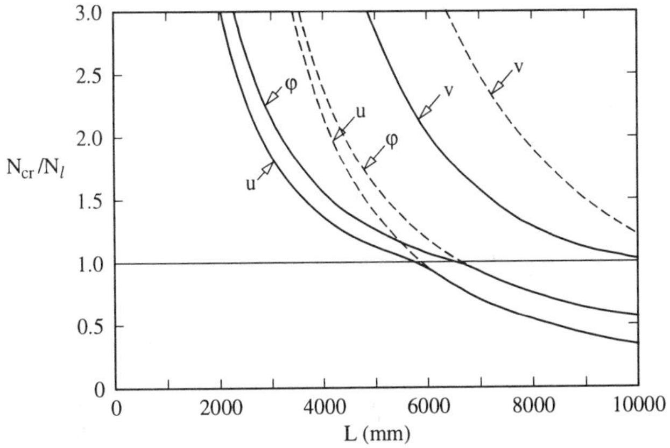

Figure 3: Buckling curves for I-section

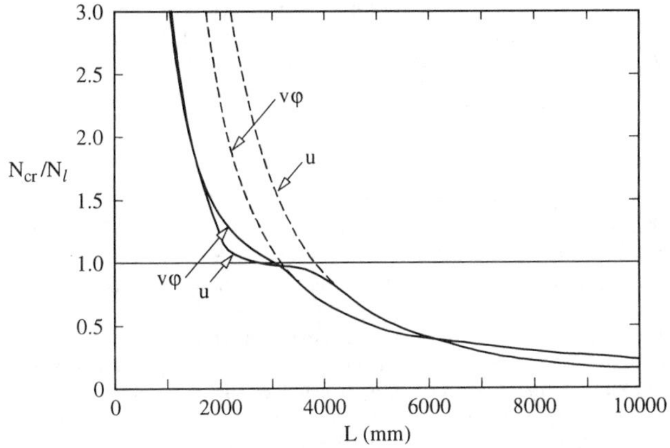

Figure 4: Buckling curves for U-section

sections. Local buckling causes comparable reductions in overall buckling resistance for minor and major axis flexural buckling and torsional buckling (φ).

5.3 *U-sections*

The buckling loads ($N_{cr}=\lambda_c\overline{N}$) are plotted in Fig. 4. Only the lower value of the flexural-torsional buckling loads given by eqn. (28) is shown. For the undistorted cross-section, flexural-torsional buckling is critical ($v\varphi$) for lengths less than about 6200 mm. However, flexural buckling (u) is critical for the distorted cross-section in the range [1500 mm - 3200 mm], as shown in Fig. 4. As discussed in Young and Rasmussen (1997), the change of critical mode is the result of a greater reduction in minor axis flexural rigidity (EI_y) resulting from local buckling, compared to the reductions in major axis flexural rigidity (EI_x) and warping rigidity (EI_ω).

5.4 *Z-sections*

The buckling loads ($N_{cr}=\lambda_c\overline{N}$) are plotted in Fig. 5. Only the lower value of the coupled flexural buckling loads given by eqn. (36) are shown. Flexural buckling (u,uv) is critical for both the

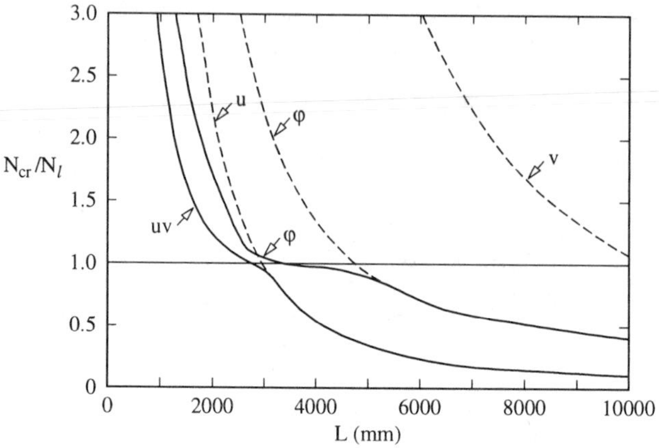

Figure 5: Buckling curves for Z-section

distorted and undistorted cross-sections. However, local buckling reduces the warping rigidity (EI_ω) more significantly than the minor and major axis flexural rigidities ((EI_y), (EI_x)) and the critical loads are not significantly different at lengths of about 2800 mm.

6 CONCLUSIONS

The governing equations and overall buckling loads for locally buckled fixed-ended columns have been summarised. The equations involve the tangent rigidities of the locally buckled cross-section against flexure and torsion. The buckling equations have been applied to an I-section, U-section and Z-section, representing doubly, singly and point symmetric cross-sections respectively. The following conclusions have been drawn:

- For locally buckled doubly symmetric cross-sections, the minor and major axis flexural buckling modes and torsional buckling mode are uncoupled as in the case of undistorted cross-sections. Local buckling reduces the buckling resistances against minor and major axis flexural buckling and torsional buckling by comparable amounts.
- For locally buckled singly symmetric cross-sections, the minor axis flexural mode is uncoupled from the flexural-torsional modes, as in the case of undistorted cross-sections. However, local buckling may produce greater reductions in minor axis flexural rigidity compared to major axis flexural rigidity and warping rigidity, causing the critical buckling mode to change from a flexural-torsional mode to a minor axis flexural mode.
- For locally buckled point-symmetric cross-sections, the torsional mode is uncoupled from the flexural modes, as in the case of undistorted cross-sections. However, the minor and major axis flexural buckling modes become coupled. The reduction in torsional buckling resistance is more significant than the reduction in coupled flexural buckling resistance.

REFERENCES

Hancock, GJ. 1985. Non-linear analysis of thin-walled I-sections in bending. In *Aspects of Analysis of Plate Structures*, ed. DJ Dawe, RW Horsington, AG Kamtekar & GH Little, 251-268.

Rasmussen, KJR. 1997. Bifurcation of locally buckled columns, *Thin-walled Structures*, 28(2), 117-154.

Rasmussen, KJR, & Hasham, AS. 1997. Flexural and flexural-torsional bifurcation of locally buckled beam-columns, *Thin-walled Structures*, 29(1-4), 203-233.

Young, B, & Rasmussen, KJR. 1997. Bifurcation of singly symmetric columns, *Thin-walled Structures*, 28(2), 155-177.

Mechanics of Structures and Materials, Bradford, Bridge & Foster (eds)

Buckling of sectorial Mindlin plates

Y.Xiang
School of Civic Engineering and Environment, University of Western Sydney, Nepean, N.S.W., Australia

C.M.Wang
Department of Civil Engineering, National University of Singapore, Kent Ridge, Singapore

ABSTRACT: This paper presents first-known buckling solutions for sectorial Mindlin plates subjected to an isotropic inplane load. The buckling loads were determined using the Ritz method. The method was made automated using Ritz functions consisting of the product of mathematically complete 2-D polynomial functions and plate boundary equations raised to appropriate powers. The latter equations ensure the satisfaction of the plate geometric boundary conditions. Buckling results are presented for sectorial plates of various boundary conditions, included angles and thickness-to-radius ratios.

1 INTRODUCTION

The use of classical thin plate theory for buckling analysis of plates leads to an over-estimation of buckling loads when the plates are moderately thick. The error is due to the neglect of the effect of transverse shear deformation that becomes significant in thick plates. For a more accurate prediction of the buckling loads, one may adopt the Mindlin plate theory that allows for the aforementioned effect. Using this plate theory and the Ritz method, the authors (Wang *et al.* 1994, Xiang *et al.* 1994) have been successful in obtaining and documenting extensive buckling results of Mindlin plates of various shapes including polygons, parallelograms, triangles, ellipses and semi-circles.

An important and common plate shape, but not been well investigated for its buckling behaviour, is the sectorial plate. In the literature, we find many papers on the static and dynamics analyses of sectorial plates (e.g. Cheung and Chan 1981, Mansfield 1989, Kim and Dickinson 1989, Wang and Lim 2000), but only a few papers (e.g. Rubin 1978, Harik 1985) have been written on buckling of such sectorial and annular sectorial plates. Moreover, the theoretical formulations found in most of these papers are based on the classical thin plate theory. This study fills this lacuna by presenting the Ritz formulation for the buckling of sectorial Mindlin plates. The Ritz functions are formed from the product of mathematically complete 2-D polynomial functions and boundary equations raised to appropriate powers (Liew *et al.* 1998). The latter equations ensure the satisfaction of the geometric boundary conditions which is necessary for convergence of the upper bound results to the correct solutions. By adjusting the powers for the boundary equations, the Ritz method becomes automated for handling sectorial plates of any combination of boundary conditions. Buckling factors are presented for sectorial Mindlin plates of different combinations of edge support conditions with various included angles and thickness-to-radius ratios. These buckling results should be useful to both researchers and engineers working on such a plate shape.

2 GOVERNING EIGENVALUE EQUATION

2.1 *Total potential energy functional*

Consider a sectorial plate of uniform thickness *h*, radius *R*, included angle α, modulus of elasticity *E*, Poisson's ratio ν and shear modulus $G = E/[2(1+\nu)]$. The plate is moderately thick and is subjected to an isotropic inplane compressive load N as shown in Figure 1. The edges may be simply supported, clamped or free. The problem is to determine the buckling loads N_{cr} of such plates.

The strain energy of a sectorial Mindlin plate due to bending is given by (Liew *et al.* 1998)

$$U = \frac{1}{2}\int_0^\alpha\int_0^R \left\{ D\left[\left(\frac{\partial\psi_r}{\partial r}\right)^2 + 2\frac{\nu}{r}\frac{\partial\psi_r}{\partial r}\left(\frac{\partial\psi_\theta}{\partial\theta}+\psi_r\right) + \frac{1}{r^2}\left(\frac{\partial\psi_\theta}{\partial\theta}+\psi_r\right)^2 + \frac{1-\nu}{2r^2}\left(\psi_\theta - r\frac{\partial\psi_\theta}{\partial r} - \frac{\partial\psi_r}{\partial\theta}\right)^2\right] + \kappa^2 Gh\left[\left(\frac{\partial w}{\partial r}+\psi_r\right)^2 + \frac{1}{r^2}\left(\frac{\partial w}{\partial\theta}+r\psi_\theta\right)^2\right]\right\} r\,dr\,d\theta \tag{1}$$

in which $w = w(r,\theta)$ is the transverse deflection at the midsurface of the plate; $\psi_r = \psi_r(r,\theta)$ the bending slope in the radial plane; $\psi_\theta = \psi_\theta(r,\theta)$ the bending slope in the circumferential plane; (r,θ) are the radial coordinates; and $D = Eh^3/[12(1-\nu^2)]$ is the flexible rigidity of the plate.

The work done by the isotropic inplane load *N* is given by

$$W = \frac{1}{2}\int_0^\alpha\int_0^R N\left[\left(\frac{\partial w}{\partial r}\right)^2 + \left(\frac{1}{r}\frac{\partial w}{\partial\theta}\right)^2\right] r\,dr\,d\theta \tag{2}$$

The total potential energy functional of the sectorial plate system is given by

$$\Pi = U - W \tag{3}$$

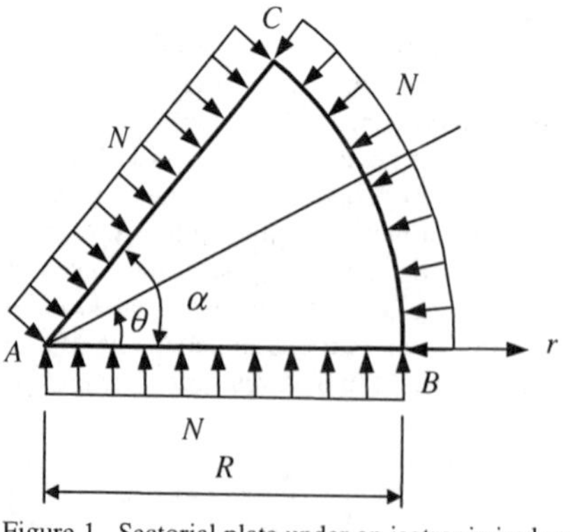

Figure 1. Sectorial plate under an isotropic inplane load

2.2 *Governing eigenvalue equation*

Using the Ritz method (Liew *et al.* 1998), the transverse displacement $w(r,\theta)$ and the bending slopes $\psi_r(r,\theta)$ and $\psi_\theta(r,\theta)$ may be parameterised by

$$w(r,\theta) = \sum_{q=0}^{p_1}\sum_{i=0}^{q} c_m \left[r^i \theta^{q-i} \phi_1^w(r,\theta)\right] \tag{4}$$

$$\psi_r(r,\theta) = \sum_{q=0}^{p_2}\sum_{i=0}^{q} d_m \left[r^i \theta^{q-i} \phi_1^r(r,\theta)\right] \tag{5}$$

$$\psi_\theta(r,\theta) = \sum_{q=0}^{p_3}\sum_{i=0}^{q} e_m \left[r^i \theta^{q-i} \phi_1^\theta(r,\theta)\right] \tag{6}$$

in which $p_s, s = 1,2,3$ is the degree set of the complete polynomial space; c_m, d_m, e_m are the unknown Ritz coefficients; and *m*, the subscript, is defined by

$$m = \frac{(q+1)(q+2)}{2} - i \tag{7}$$

and the total numbers of c_m, d_m, e_m can be determined by Eq. (7) with q replaced by p_1, p_2 and p_3, respectively.

The basic functions $\phi_1^w, \phi_1^r, \phi_1^\theta$ in Eqs. (4-6) are defined by the product of the plate edge equations raised to appropriate powers (Liew *et al.* 1998), i.e.

$$\phi_1^w(r,\theta) = (r-R)^{\Omega_1}(\theta+0)^{\Omega_2}(\theta-\alpha)^{\Omega_3} \tag{8}$$

$$\phi_1^r(r,\theta) = (r-R)^{\Omega_4}(\theta+0)^{\Omega_5}(\theta-\alpha)^{\Omega_6} \tag{9}$$

$$\phi_1^\theta(r,\theta) = (r-R)^{\Omega_7}(\theta+0)^{\Omega_8}(\theta-\alpha)^{\Omega_9} \tag{10}$$

where the values of Ω_i (i = 1, 2, ..., 9), depending on the supporting edge conditions, are given in Table 1. The basic functions in Eqs. (8-10) satisfy the geometric support conditions of the plate at the outset.

The minimization of the total potential energy functional (Eq. 3) with respect to the unknown Ritz coefficients

$$\left\langle \frac{\partial \Pi}{\partial c_m}, \frac{\partial \Pi}{\partial d_m}, \frac{\partial \Pi}{\partial e_m} \right\rangle = \langle 0,0,0 \rangle \tag{11}$$

leads to the governing eigenvalue equation

$$\left(\begin{bmatrix} \mathbf{K}^{cc} & \mathbf{K}^{cd} & \mathbf{K}^{ce} \\ & \mathbf{K}^{dd} & \mathbf{K}^{de} \\ & & \mathbf{K}^{ee} \end{bmatrix} - N \begin{bmatrix} \mathbf{G}^{cc} & \mathbf{G}^{cd} & \mathbf{G}^{ce} \\ & \mathbf{G}^{dd} & \mathbf{G}^{de} \\ & & \mathbf{G}^{ee} \end{bmatrix} \right) \begin{Bmatrix} \mathbf{c} \\ \mathbf{d} \\ \mathbf{e} \end{Bmatrix} = \begin{Bmatrix} \mathbf{0} \\ \mathbf{0} \\ \mathbf{0} \end{Bmatrix} \tag{12}$$

in which **c**, **d** and **e** are column matrices containing the unknown Ritz coefficients, **K** is the stiffness matrix and **G** is the geometric matrix of the plate.

Table 1. The values of Ω_i (i = 1, 2, ..., 9) in the basic functions.

Edge Condition	Edge BC	Edge AB	Edge AC
Free	$\Omega_1 = 0, \Omega_4 = 0, \Omega_7 = 0$	$\Omega_2 = 0, \Omega_5 = 0, \Omega_8 = 0$	$\Omega_3 = 0, \Omega_6 = 0, \Omega_9 = 0$
Simply Supported	$\Omega_1 = 1, \Omega_4 = 0, \Omega_7 = 1$	$\Omega_2 = 1, \Omega_5 = 1, \Omega_8 = 0$	$\Omega_3 = 1, \Omega_6 = 1, \Omega_9 = 0$
Clamped	$\Omega_1 = 1, \Omega_4 = 1, \Omega_7 = 1$	$\Omega_2 = 1, \Omega_5 = 1, \Omega_8 = 1$	$\Omega_3 = 1, \Omega_6 = 1, \Omega_9 = 1$

The buckling load N_{cr} can be obtained by solving the generalised eigenvalue equation defined in Eq. (12).

3 NUMERICAL RESULTS AND DISCUSSIONS

We present buckling results for sectorial Mindlin plates with Poisson's ratio $\nu = 0.3$, shear correction factor $\kappa^2 = 5/6$, various included angles α (= 30° to 180°) and thickness-to-radius ratios *h/R* (= 0.001, 0.1 and 0.2) and different combinations of edge conditions. Note that results associated with *h/R* = 0.001 may be taken to correspond to classical thin plate results. In order to avoid ill-condition in the calculations, the integration limits along the r direction in Eqs. (1-2) have been taken as $0.00001R$ to R. A set of symbols has been employed to represent plate edge support conditions. For instance, a symbol *SCF* denotes that edge *AB* is simply supported, edge *BC* is clamped and edge *CA* is free (see Figure 1). The buckling load is expressed in the form of a buckling load factor defined as $\lambda = N_{cr}R^2/(\pi^2 D)$.

A convergence study has been carried out to establish the degree set of polynomial p_s ($s = 1,2,3$) for converged buckling loads. Table 2 presents the variations of buckling load factors with respect to the degree sets of polynomial for a *CCC* sectorial Mindlin plate. The buckling load factor decreases monotonically as the degree set of polynomial increases. The buckling load factor converges more rapidly for plates with larger included angles and for thicker plates. It is found that $p_s = 14$ is necessary to ensure converged buckling load factors.

Table 2. Convergence study for *CCC* sectorial Mindlin plate

p_s	$\alpha = 30^\circ$			$\alpha = 90^\circ$			$\alpha = 180^\circ$		
	h/R			*h/R*			*h/R*		
	0.001	0.1	0.2	0.001	0.1	0.2	0.001	0.1	0.2
4	19899.2	176.898	68.521	3396.4	65.892	39.831	502.24	42.332	28.120
5	1264.57	159.772	66.116	115.74	60.868	38.840	46.046	37.901	29.189
6	500.587	155.185	65.418	78.378	58.792	38.459	44.973	37.788	28.045
7	335.373	152.510	65.083	74.981	58.679	38.423	42.781	37.651	28.009
8	298.753	151.121	64.937	72.050	58.627	38.413	42.629	37.632	28.004
9	294.654	150.861	64.902	71.785	58.619	38.412	42.609	37.627	28.002
10	283.174	150.739	64.892	71.753	58.617	38.411	42.597	37.626	28.002
11	281.831	150.726	64.890	71.752	58.616	38.411	42.592	37.625	28.002
12	281.264	150.719	64.890	71.744	58.615	38.411	42.591	37.625	28.002
13	280.943	150.717	64.890	71.741	58.615	38.411	42.591	37.625	28.002
14	280.904	150.717	64.890	71.741	58.615	38.411	42.591	37.625	28.002

Table 3. Buckling load factors λ for *SSS* and *CCC* sectorial Mindlin plates

α	*SSS* Plate			*CCC* Plate		
	h/R			h/R		
(degree)	0.001	0.1	0.2	0.001	0.1	0.2
30	96.916	75.887	45.968	280.904	150.717	64.890
45	55.748	48.077	34.031	159.483	106.623	54.842
60	38.854	34.961	26.883	111.274	82.577	47.563
75	29.930	27.563	22.279	86.476	68.074	42.295
90	24.497	22.886	19.117	71.741	58.615	38.411
105	20.874	19.691	16.832	62.156	52.084	35.483
120	18.318	17.385	15.116	55.534	47.383	33.231
135	16.456	15.648	13.784	50.767	43.896	31.466
150	15.009	14.258	12.693	47.242	41.252	30.062
165	13.697	13.065	11.735	44.593	39.214	28.928
180	12.693	12.231	11.059	42.591	37.625	28.002

Table 4. Buckling load factors λ for *SFS* and *CFC* sectorial Mindlin plates

α	*SFS* Plate			*CFC* Plate		
	h/R			h/R		
(degree)	0.001	0.1	0.2	0.001	0.1	0.2
30	35.767	30.128	22.614	103.057	65.452	36.326
45	14.943	13.534	11.523	45.525	35.448	23.908
60	7.733	7.223	6.509	25.467	21.711	16.534
75	4.459	4.234	3.932	16.241	14.525	11.942
90	2.724	2.612	2.469	11.272	10.373	8.966
105	1.709	1.649	1.576	8.312	7.793	6.968
120	1.077	1.040	1.001	6.424	6.100	5.587
135	0.672	0.636	0.615	5.159	4.945	4.608
150	0.398	0.348	0.336	4.285	4.136	3.903
165	0.158	0.114	0.105	3.667	3.558	3.387
180	-	-	-	3.226	3.142	3.006

Therefore, $p_s = 14$ will be used for generating all the buckling results in this study.

Table 3 gives the buckling load factors λ for *SSS* and *CCC* sectorial Mindlin plates. It can be seen that the buckling load factor decreases significantly with increasing included angles α for both plates with a given thickness-to-radius ratio *h/R*. This is because the constraints contributed by the edge supports are reduced as the included angle α increases. The buckling load factor also decreases as the thickness-to-radius ratio *h/R* varies from 0.001 to 0.2 for plates with a given included angle α. The decrease is due to the increasing effect of transverse shear deformation as the plate gets thicker.

Table 4 shows the buckling results for *SFS* and *CFC* sectorial Mindlin plates. Similar trends are observed for the variations of the buckling load factors against the included angles α and thickness-to-radius ratios *h/R*. Note that the special case of *SFS* plate with the included angle $\alpha = 180°$, in which the plate becomes unstable as the two simply supported straight edges are on the same line.

4 CONCLUSIONS

The buckling loads of sectorial Mindlin plates subjected to an isotropic inplane load have been determined using the Ritz method. These first-known buckling load factors are obtained for sectorial Mindlin plates with different combinations of edge conditions, various included angles and thickness-to-radius ratios. The variations of the buckling load factors against the included angles and thickness-to-radius ratios are also discussed.

5 REFERENCES

Cheung, M.S. and Chan, M.Y.T. 1981. Static and dynamic analysis of thin and thick sectorial plates by the finite strip method. *Computers and Structures* 14(1-2): 79-88.

Harik, I.E. 1985. Stability of annular sector plates with clamped radial edges. *Journal of Applied Mechanics* 52: 971-972.

Kim, C.S. and Dickinson, S.M. 1989. On the free, transverse vibration of annular, circular, thin, sectorial plates subject to certain complicating effects. *Journal of Sound and Vibration* 134(3): 407-421.

Liew, K.M., Wang, C.M., Xiang, Y., and Kitipornchai, S. 1998. *Vibration of Mindlin Plates-Programming the p-Version Ritz Method.* Elsevier Science Ltd, Amsterdam, The Netherlands.

Mansfield, E.H. (1989). *The Bending and Stretching of Plates.* 2nd Edition, Cambridge University Press, Cambridge.

Rubin, C. 1978. Stability of polar orthotropic sector plates. *Journal of Applied Mechanics,* 45: 448-450.

Wang, C.M., Xiang, Y., Kitipornchai, S. and Liew, K.M. 1994. Buckling solutions for Mindlin plates of various shapes. *Engineering Structures* 16(2): 119-127.

Wang, C.M. and Lim, G.T. 2000. Bending solutions of sectorial Mindlin plates from Kirchhoff plates. *Journal of Engineering Mechanics, ASCE,* submitted.

Xiang, Y., Wang, C.M., Kitipornchai, S. and Liew, K.M. 1994. Buckling of triangular Mindlin Plates under isotropic in-plane compression. *Acta Mechanica* 102: 123-135.

Mechanics of Structures and Materials, Bradford, Bridge & Foster (eds)
© 1999 Balkema, Rotterdam, ISBN 90 5809 107 4

Inelastic lateral-distortional buckling of continuously restrained beam-columns

D.-S. Lee & M.A. Bradford
School of Civil and Environmental Engineering, The University of New South Wales, Sydney, N.S.W., Australia

ABSTRACT: Overall buckling of steel I-section members is usually in a lateral-torsional mode, in which the Vlasov assumption that the cross-section does not change shape during buckling is assumed. While in many cases this assumption is vindicated, members whose compression flange is restrained only by the stiffness of the web may buckle in a lateral-distortional mode, in which the mode of buckling is usually associated with distortion of the web in the plane of the member cross-section. Lateral-distortional buckling is prevalent in the hogging region of continuous composite beams, in beams with partial end restraint, and in columns and rafters in industrial portal frame buildings where the cladding combined with purlins or girts forms a diaphragm restraint to the tension flange. This paper presents a study of the inelastic distortional buckling of hot-rolled I-section beam-columns with continuous full translational but elastic torsional restraint applied to the tension flange. The method allows for the inclusion of the well-used 'simplified' distribution of residual stresses in the section. Whereas previous studies have indicated that elastic distortional buckling is important in these restraint cases, the extension in this paper to inelastic distortional buckling also indicates the importance of cross-sectional distortion on predicting the strength of the member to the overall buckling limit state.

1 INTRODUCTION

The majority of the wealth of theoretical studies of lateral-torsional buckling have been based on the use of Vlasov's thin-walled section theory (Trahair 1993). One underlying assumption in this theory is that the cross-section remains undistorted during buckling. If an I-section member was loaded in uniform bending with its tension flange completely restrained against buckling, then the Vlasov assumption (in the absence of local buckling) would predict the elastic overall buckling moment to be infinite. This has been shown not to be the case (Bradford 1997), and indeed if the analysis was inelastic rather than elastic, the Vlasov assumption would predict that the strength of the member would be its plastic moment of resistance if the cross-section was compact. Again, theoretical studies (Bradford 1999) have shown this not to be the case, and the buckling of a member with a fully-restrained tension flange must be accompanied by distortion of the cross-section. Generally speaking, it is the slender web rather than the stocky flange that distorts out of the plane of the cross-section (Bradford 1992).

For the majority of I-section members with restrained ends and whose spans are unrestrained, the destabilising effects of cross-sectional distortion are minimal (Bradford 1992), and indeed may be ignored for hot-rolled universal sections. This is generally not the case for members with partial end restraint and/or continuous elastic or full restraint of the tension flange. Here the buckling mode is characterised by cross-sectional distortion. AS4100 (SA 1998) allows for 'design by buckling analysis' (Trahair & Bradford 1998), where the plastic moment M_p and the elastic buckling moment are converted to a design bending strength by use of the slenderness reduction factor α_s, or where the squash load N_s and elastic buckling load are converted a design compressive strength by use of the slenderness reduction factor α_c. In cases where buckling is accompanied by severe cross-sectional distortion, there is little evidence that the formulae for α_s and α_c given in AS4100 should be applicable to distortional buckling. The results of an inelastic distortional buckling analysis are necessary to investigate this hypothesis, and to provide alternative formulae (such as those of Bradford & Johnson 1987 and Bradford 1989 for composite beams) if the AS4100 approach is found to be incorrect.

Although research into elastic distortional buckling has been quite extensive, its extension to the inelastic range of structural response has been reported in only a few studies. The first rational analysis appears to be that of Bradford (1986), which although using an efficient finite element formulation, was still rather restrictive in its application owing to the limited prowess of contemporary accessible computers at the time. Little work has appeared in the open literature since then (Bradford & Kemp 1999), except energy-based methods applied to welded beams. Recently, the authors (Lee & Bradford 1999) have developed an energy-based approach for studying the inelastic distortional buckling of I-section beam-columns that incorporate either the 'polynomial' or 'simplified' residual stress models, the P-δ effect in non-sway members, and elastic restraints against translation, lateral rotation, twist rotation and 'warping' applied at either the tension or compression flange. This paper summarises briefly the method of the energy-based approach, validates its accuracy with independent studies, and considers the effect of restraining the tension flange fully against translation and lateral rotation, but elastically against twist rotation, which models the diaphragm restraint provided by cladding combined with purlins or girts in industrial portal frame buildings.

2 THEORY

2.1 *General*

The inelastic distortional buckling analysis of I-section beam-columns with equal and opposite end moments has been described by the authors (Lee & Bradford 1999). The method requires firstly an in-plane analysis that incorporates residual stresses appropriate to a hot-rolled section, and then an out-of-plane analysis based on the energy method that requires stiffness [k], elastic restraint [r] and stability [s] matrices to be developed. The P-δ effect of combined 'uniform' bending and axial compression is included. The solution strategy is incremental and iterative, and is rapid on a conventional personal computer. Because the method has been documented fully elsewhere (Lee & Bradford 1999), only a brief summary of the theory is presented below.

2.2 *Residual stresses*

Residual stresses play an important part in inelastic overall buckling (Trahair & Kitipornchai 1972). The models used for hot-rolled universal sections have evolved from tests (Lee *et al.* 1967) that have supplemented theoretical expressions chosen essentially to satisfy equilibrium equations. The so-called 'polynomial' pattern has been found to be more applicable to Australian and European sections (Bradford & Trahair 1985), with the 'simplified' pattern being more applicable to North American sections. In this study the simplified model is used, as shown in Figure 1.

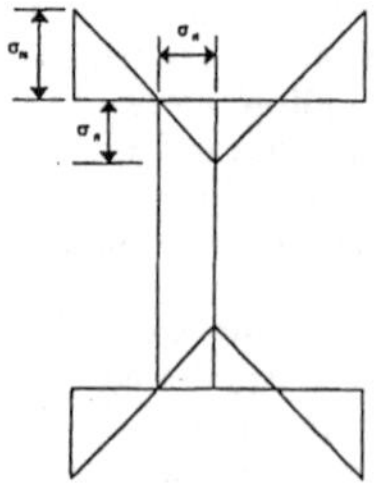

Figure 1. 'Simplified' residual stress model

In Figure 1, the flange tip residual compressive stress is assumed to be $\sigma_{rc} = 0.3\sigma_y$, and the tensile residual stress throughout the web (and at the flange/web junctions) is

$$\sigma_{rt} = \left[\frac{BT}{BT + t_w(D-T)} \right] \sigma_y \qquad (1)$$

In Eq. 1, B is the flange width, T is its thickness, D is the overall depth of the section, t_w is the web thickness and σ_y is the yield stress. Equation 1 satisfies the equilibrium condition of an unloaded member that the axial force and moments about both principal axes vanish, but it does not enforce the torsional equilibrium condition

$$\int_A (x^2 + y^2) \sigma_r \mathrm{d}A = 0 \qquad (2)$$

Satisfaction of Eq. 2 is taken into account in the out-of-plane analysis.

2.3 *In-plane analysis*

The beam-column of length L is subjected to equal and opposite end moments M and an axial compressive force N, and is assumed to be elastic-plastic-strain hardening. The elastic region of modulus E = 200 GPa extends to the yield strain ε_y, is then fully plastic to a strain ε_{st}, after which the steel enters the strain hardening range with modulus $E_{st} = E/h'$. The applied strain is then

$$\varepsilon_a = \varepsilon_o + (y + \bar{y})\rho + \varepsilon_r(x, y) \qquad (3)$$

from which the stress distribution is

$$\sigma(x, y) = \int_{\varepsilon_r}^{\varepsilon_a} E_t \mathrm{d}A + E\varepsilon_r(x, y) \qquad (4)$$

where ε_o is the strain due to the axial force, $\bar{y}$ is the coordinate of the neutral axis, ρ is the curvature, ε_r is the residual strain σ_r/E, and E_t is the tangent modulus. For a given compressive force N and an assumed curvature ρ, Eqs. 3 and 4 are solved by iteration for the position of the neutral axis by satisfying the equilibrium condition

$$N = \int_A \sigma(x, y) \mathrm{d}A \qquad (5)$$

The moment M_{max} at midspan corresponding to the assumed curvature using the value of $\bar{y}$ determined in Eq. 5 is then

$$M_{max} = \int_A \sigma(x, y) y \mathrm{d}A \qquad (6)$$

Finally, the end moments M are obtained from the midspan moment M_{max} iteratively by assuming firstly that the member is subjected to a constant curvature $\rho = M_{max}/(EI)_s$, where $(EI)_s$ is the secant flexural rigidity. The secant rigidity is taken initially as the elastic rigidity, and variation of deflections v are found (Trahair & Bradford 1998) at four equally-spaced stations, and the new curvatures at each of these stations are taken as $\rho^{(1)} = (M + Nv^{(1)})/(EI)_s$ where M is initially guessed. Integrating the curvatures at the four stations leads to an estimate of the (known) midspan deflection. If the iterations do not converge to the midspan deflection known *a priori*, a new value of M is chosen and the process repeated until the correct deflection v_{max} at midspan is obtained. The end moment is then $M = M_{max} - Nv_{max}$, so that this geometric and material nonlinear process accounts for the P-δ effect.

2.4 *Buckling analysis*

The energy-based buckling analysis assumes that the top and bottom flange/web junctions deflect and twist as sine curves with n harmonics as shown in Figure 2. The flanges are

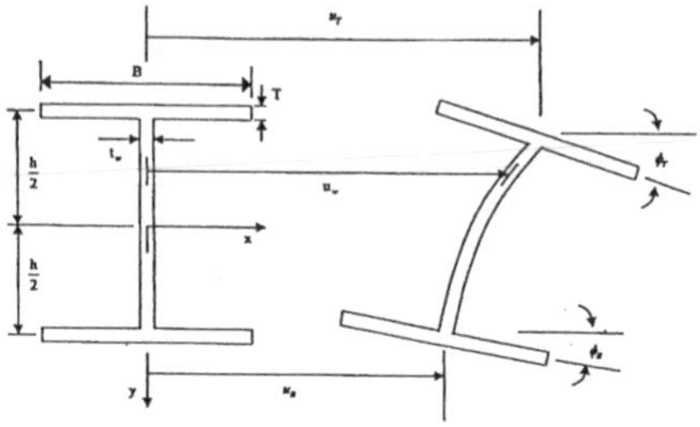

Figure 2. Distortional buckling deformations of the cross-section

assumed to be rigid bars whose minor axis flexural and torsional rigidities are based on tangent modulus theory (Bradford & Trahair 1985). The web on the other hand is modelled as a plate whose orthotropic property matrix is assembled from Haaijer's inelastic model (1957). The tangent torsional rigidity $(GJ)_t$ for the whole cross-section is changed to

$$(GJ)_t - \int_A \sigma_r \left(x^2 + y^2\right) dA$$

so as to enforce the 'torsional equilibrium' requirement of the residual stresses.

The method also allows for an elastic restraint matrix [r] to be included. This matrix is constant, but the stability matrix [s] as well as the stiffness matrix [k] developed as above from tangent modulus and isotropic plate theory are functions of the applied curvature ρ. As the curvature is increased in small steps, the familiar eigenproblem in Eq. 7 must be solved:

$$\left([k(\rho)] + [r] - [s(\rho)]\right) \cdot \{q\} = \{0\} \tag{7}$$

where {q} is the vector of buckling degrees of freedom. The incremental and iterative solution of Eq. 7 for the critical curvature ρ_{cr} is that used by Uy & Bradford (1994), from which the midspan buckling moment M_{max} may be obtained from Eq. 6, and hence the buckling end moments M_{cr} from the iterative process described in Section 2.3.

3. VERIFICATION

Abdel-Sayed & Aglan (1973) reported the inelastic lateral-torsional buckling solution for a simply supported unrestrained North American 8WF31 beam-column assuming the simplified residual stress pattern. Details of the material properties used can be found in their paper. Although the present method has been developed for lateral-distortional buckling, it may be modified easily for lateral-torsional buckling by scaling the strain energy stored during flexure of the web in the plane of its cross-section by a large number (say 10^8) (Lee & Bradford 1999), thereby suppressing distortion. Figure 3 compares the results of this study with that of Abdel-Sayed & Aglan, where the end buckling moment is normalised with respect to the yield moment M_y and the axial force is constant at $0.6N_s$, where N_s is the squash load. The slenderness ratio for which the inelastic buckling moments are obtained is the major axis slenderness ratio L/r_x.

Abdel-Sayed & Aglan used tangent modulus theory, with the appropriate E_t in the yielded and strain hardened regions in their buckling model. The present method has adopted this assumption, where the agreement with the independent solution is generally very good. Also shown in Figure 3 is the solution using the present method with $E_t = E_{st}$ in the plastic and strain hardened regions, where the onset of strain hardening buckling occurs at a higher slenderness ratio than that using the more conservative assumption of $E_t = 0$ in yielded regions. Abdel-Sayed & Aglan's solution is somewhat unconservative compared to the present model for higher values of L/r_x. This is probably due to the neglect of the P-δ effect in the former model. Indeed, using the approximate magnifier of $1/(1-N/N_{ox})$ (Trahair & Bradford 1998) at $L/r_x = 50$, where N_{ox} is the elastic major axis buckling load, indicates that independent solution is about 20% unconservative, and incorporation of this effect renders the independent solution very close to that of the present method.

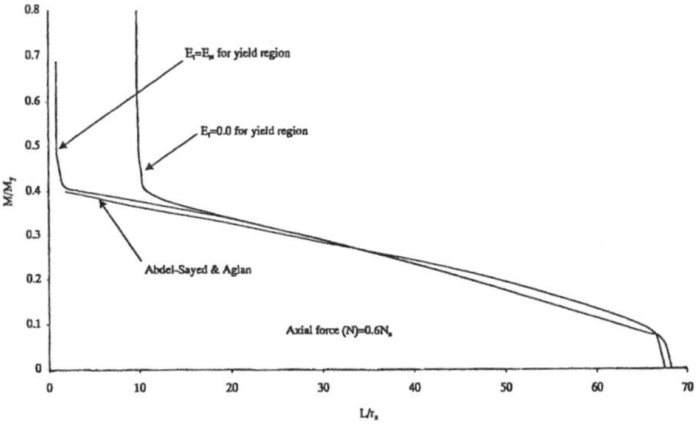

Figure 3. Inelastic lateral-torsional buckling comparison with Abdel-Sayed & Aglan (1973)

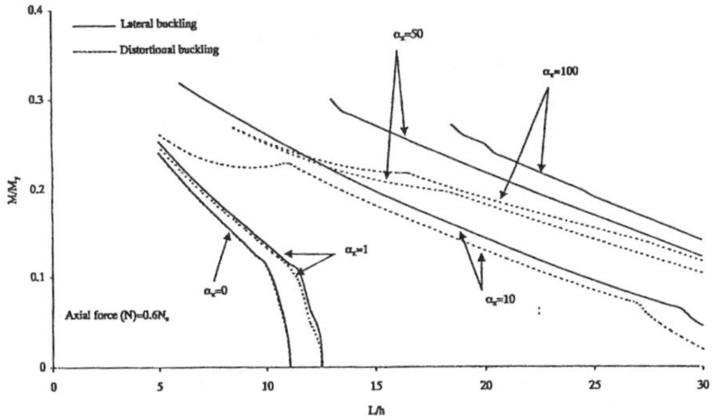

Figure 4. Inelastic buckling of a 610UB101 with elastic twist rotation restraint

4 ELASTIC TORSIONAL RESTRAINT

The inelastic buckling of a 610UB101 has been investigated, where the tension flange is fully restrained against translation and lateral rotation, but where the restraint against twist rotation during buckling is elastic with a stiffness value of k_z. This twist restraint has been normalised here as $\alpha_z = k_z/(\pi^2 GJ/L^2)$. The relevant material properties were $\sigma_y = 250$ MPa, $\varepsilon_{st} = 10\ \varepsilon_y$, h' = 33 and an elastic Poisson's ratio of 0.3.

Figure 4 shows the normalised buckling moments when the beam-column is subjected to a constant axial force of $0.6N_s$. Both inelastic lateral-torsional and lateral-distortional results are plotted, and as expected the lateral-torsional buckling moments overestimate the true (distortional) buckling moments as the degree of torsional restraint α_z increases. In design, the member strengths determined in accordance with AS4100 would be based on the inelastic lateral-distortional buckling curves (which are modified empirically, Trahair & Bradford 1998), but clearly these strengths are unconservative.

For $\alpha_z = 100$, there are four harmonics represented in the solution (ie. the analysis must be performed for increasing numbers of harmonics (n) and the minimum solution adopted, ie. n = 4 in this case). Compared with the elastic solution (Bradford 1997) which is again dependent on the number of harmonics, the transition in the solution between adjoining harmonics is much smoother for inelastic buckling than elastic buckling, where the curves are of the characteristic local buckling 'garland' form.

Finally, it is worth noting that increasing the degree of torsional restraint increases the strength of the beam-column subjected to a constant axial compression. In Figure 4, it is clear that the beam-column has no reserve of bending capacity for L/h > 11 in the absence of torsional restraint, but for large values of α_z the beam-column has a considerable reserve of bending capacity for large member lengths, when it buckles into a number of harmonics.

5 CONCLUSIONS

Although elastic distortional buckling has been studied quite extensively, the application of the solutions to determine member strengths, as are represented in national limit states codes, is reliant upon the results of an inelastic distortional buckling analysis. The latter analysis is a grey area of research, which hitherto has received limited attention. This is even more true for beam-columns, where little work has been reported in the published literature.

The present paper has reported on a method of analysis developed by the authors for the inelastic distortional buckling of universal I-section beam-columns with the so-called 'simplified' pattern of residual stresses. When distortion is suppressed, the solutions agree well with independent studies. A particular case in which distortional buckling is important is in diaphragm-type restraint of the tension flange in a beam-column, where the flexural rigidity of the diaphragm provides torsional restraint to the beam-column, but in which the shear and translational stiffness of the diaphragm prevent translation and lateral rotation of the tension flange. This is a simplistic model of a member in an industrial portal frame building. One sample member has been analysed, and it has been shown that neglect of the effects of cross-sectional distortion for high degrees of torsional restraint can lead to unconservative predictions of the buckling strength if the latter is based on 'rational' lateral-torsional buckling assumptions. This deficiency, believed to be inherent in all limit states standards, requires further research to be fully quantified for design purposes.

6 REFERENCES

Abdel-Sayed, G. & Aglan, A.A. 1973. Inelastic lateral-torsional buckling of beam-columns. *Publications*, IABSE, 33-II:1-16.

Bradford, M.A. 1986. Inelastic distortional buckling of I-beams. *Comps. & Structs.* 14(6):923-933.

Bradford, M.A. 1989. Buckling strength of partially restrained I-beams. *J. Struct. Engg., ASCE* 115(5): 1272-1276.

Bradford, M.A. 1992. Lateral-distortional buckling of steel I-section members. *J. Constr. Steel Res.* 23:97-116.

Bradford, M.A. 1997. Lateral-distortional buckling of continuously restrained columns. *J. Constr. Steel Res.* 42(2):121-139.

Bradford, M.A. 1999. Strength of compact steel beams with partial restraint. Submitted for publication.

Bradford, M.A. & Johnson, R.P. 1987. Inelastic buckling of composite bridge girders near internal supports. *Proc. ICE, London*, Part 2, 83:143-159.

Bradford, M.A. & Kemp, A.R. 1999. Buckling of composite beams. *Progress in Struct. Eng & Mats.*, in press.

Bradford, M.A. & Trahair, N.S. 1985. Inelastic buckling of beam-columns with unequal end moments. *J. Constr. Steel Res.* 5:195-212.

Haajer, G. 1957. Plate buckling in the strain hardening range. *J. Engg. Mech. Div., ASCE* 83(EM2):1-47.

Lee, D-S & Bradford, M.A. 1999. Inelastic distortional buckling of hot-rolled I-section beam-columns. Submitted for publication.

Lee, G.C., Fine, D.S. & Hastreiter, W.R. 1967. Inelastic torsional buckling of H-columns. *J. Struct. Div., ASCE* 93(ST5):295-307.

Standards Australia 1998. *AS4100 Steel structures*. Sydney:SA.

Trahair, N.S. 1993. *Flexural-torsional buckling of structures*. London:E&FN Spon.

Trahair, N.S. & Bradford, M.A. 1998. *The behaviour and design of structures to AS4100.* London:E&FN Spon.

Trahair, N.S. & Kitipornchai, S. 1972. Buckling of inelastic I-beams under uniform moment. *J. Struct. Div., ASCE* 98(ST11):2551-2566.

Uy, B. & Bradford, M.A. 1994. Inelastic local buckling behaviour of thin steel plates in profiled composite beams. *The Struct. Engr.* 72(16):259-267.

Mechanics of Structures and Materials, Bradford, Bridge & Foster (eds)
© 1999 Balkema, Rotterdam, ISBN 90 5809 107 4

Elastic buckling of rectangular plates with mixed boundary conditions

G.H.Su & Y.Xiang
School of Civic Engineering and Environment, University of Western Sydney, Nepean, N.S.W., Australia

ABSTRACT: A numerical model is proposed for buckling analysis of rectangular plates with mixed edge support conditions. In order to accommodate the mixed boundary conditions, a plate is decomposed into two subdomains and the continuity conditions for transverse deflections and slopes along the interface of the two subdomains are imposed. The automated Ritz method is employed to derive the governing eigenvalue equation for the plate system. The validity and accuracy of the method is verified by convergence and comparison studies. Buckling solutions are presented for several rectangular plates with mixed edge support conditions.

1 INTRODUCTION

The load carrying capacity against buckling is one of the most important aspects in designing plate structures in civil, mechanical and marine engineering. Numerous studies have been reported in open literature on buckling of plates subjected to various loading and support conditions (for example: Bulson, 1970, Column Research Committee of Japan, 1971, Reddy and Phan, 1985, Xiang *et al.*, 1995).

Buckling of plates with mixed boundary conditions has also been investigated by many researchers. Bartlett (1963) conducted a research on the vibration and buckling of a simply supported circular plate with partially clamped edge. Hamada *et al.* (1967) studied buckling of simply supported rectangular plates with partially clamped edges. Keer and Stahl (1972) used Fredholm integral equations of the second kind to study buckling and vibration of rectangular plates having mixed support conditions. Sakiyama and Matsuda (1987) applied the numerical integral method for buckling of rectangular Mindlin plates with mixed edge supports. Karamanlidis and Prakash (1989) developed a modified Ritz approach for analysing the buckling and vibration of thick orthotropic plates subjected to mixed boundary conditions. Using the spline element method, Mizusawa and Leonard (1990) studied buckling and vibration of rectangular and skew plates having mixed edge supports.

This paper presents a domain decomposition approach to treat rectangular plates with mixed boundary conditions. A plate is first decomposed into two subdomains along a straight line. Employing the automated Ritz method (Xiang *et al.*, 1996, Liew *et al.*, 1998), the Ritz function for the transverse deflection of a subdomain consists of the product of a basic function and a 2-D complete polynomial. The basic function is made of the product of the boundary equations of the subdomain raised to appropriate powers and ensures automatic satisfaction of the geometric boundary conditions of the subdomain. The continuity conditions at the interface of the two subdomains for the transverse deflections and slopes are imposed. Governing eigenvalue equation is derived by minimizing the total potential energy functional of a plate with respect to the unknown Ritz coefficients. Convergence and comparison studies are carried out to verify the validity and accuracy of the method. Buckling solutions for several rectangular plates with mixed edge support conditions are presented.

2 MATHEMATICAL MODELLING

2.1 *Total potential energy functional*

Consider an isotropic thin rectangular plate subjected to a uniaxial uniform inplane load N (see Figure 1). The plate may be partitioned into two subdomains at location $x = x_0$ and may be supported with mixed edge conditions. The aim of the study is to determine the buckling load of a rectangular plate with mixed boundary conditions.

The bending strain energy U and the potential energy of the inplane load V of the plate can be derived as:

$$U = \frac{D}{2}\left[\iint_{A^{(1)}}\left\{\left(\frac{\partial^2 w^{(1)}}{\partial x^2}+\frac{\partial^2 w^{(1)}}{\partial y^2}\right)^2 - 2(1-\upsilon)\left[\frac{\partial^2 w^{(1)}}{\partial x^2}\frac{\partial^2 w^{(1)}}{\partial y^2} - \left(\frac{\partial^2 w^{(1)}}{\partial x \partial y}\right)^2\right]\right\} dA^{(1)} \right.$$
$$\left. + \iint_{A^{(2)}}\left\{\left(\frac{\partial^2 w^{(2)}}{\partial x^2}+\frac{\partial^2 w^{(2)}}{\partial y^2}\right)^2 - 2(1-\upsilon)\left[\frac{\partial^2 w^{(2)}}{\partial x^2}\frac{\partial^2 w^{(2)}}{\partial y^2} - \left(\frac{\partial^2 w^{(2)}}{\partial x \partial y}\right)^2\right]\right\} dA^{(2)}\right] \tag{1}$$

$$V = -\frac{N}{2}\left[\iint_{A^{(1)}}\left(\frac{\partial w^{(1)}}{\partial x}\right)^2 dA^{(2)} + \iint_{A^{(2)}}\left(\frac{\partial w^{(2)}}{\partial x}\right)^2 dA^{(2)}\right] \tag{2}$$

where $w^{(1)}(x, y)$ and $w^{(2)}(x, y)$ are the transverse deflections at the midsurface of the plate in subdomains 1 and 2; $D(= Et^3/[12(1-\upsilon^2)])$ is the flexural rigidity of the plate; t is the plate thickness; $A^{(1)}$ and $A^{(2)}$ are the areas of subdomains 1 and 2; E is the Young's modulus; and υ is the Poisson ratio. The total potential energy functional of the plate is given by

$$\Pi = U + V \tag{3}$$

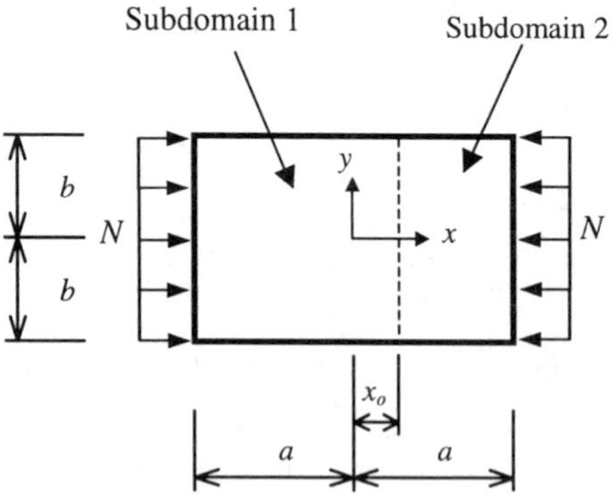

Figure 1: Geometry and coordinate system of a rectangular plate with two subdomains

2.2 *Treatment of mixed boundary conditions*

Using the automated Ritz method (Xiang *et al.*, 1996), the transverse deflections for subdomains 1 and 2 can be parameterized by,

$$\{\Lambda(x,y)\}^T_{1\times(m+n-2)} = \lfloor B^{(1)}(x)B^{(1)}(y)\{\Phi(x,y)\}^T_{1\times m} \quad \{0\}^T_{1\times(n-2)} \rfloor \tag{15}$$

$$\{\Gamma(x,y,x_0)\}^T_{1\times(m+n-2)} = \lfloor\{\Gamma_a(x,y,x_0)\}^T_{1\times m} \quad \{\Gamma_b(x,y,x_0\}^T_{1\times(n-2)}\rfloor \tag{16}$$

$$\{C\}_{(m+n-2)\times 1} = \begin{bmatrix} \{a\}_{m\times 1} \\ \{b_{p2}\}_{(n-2)\times 1} \end{bmatrix} \tag{17}$$

2.3 *Governing eigenvalue equation*

Substituting Eqs. (13) and (14) into Eq. (3), the total potential energy functional for the plate can be expressed as

$$\Pi = \frac{1}{2}\{C\}^T_{1\times(m+n-2)}\left([K]_{(m+n-2)\times(m+n-2)} - N[G]_{(m+n-2)\times(m+n-2)}\right)\{C\}_{(m+n-2)\times 1} \tag{18}$$

in which $[K]$ and $[G]$ are the stiffness matrix and geometric matrix of the plate.

Minimizing the total potential energy functional Π in Eq. (18) with respect to the unknown Ritz coefficients leads to the governing eigenvalue equation

$$\left([K]_{(m+n-2)\times(m+n-2)} - N[G]_{(m+n-2)\times(m+n-2)}\right)\{C\}_{(m+n-2)\times 1} = \{0\}_{(m+n-2)\times 1} \tag{19}$$

The buckling load N can be obtained by solving the generalised eigenvalue problem defined by Eq. (19).

3 NUMERICAL RESULTS AND DISCUSSIONS

The proposed numerical model has been employed to study rectangular plates of Poisson's ratio $\upsilon = 0.3$, aspect ratios a/b = 0.5, 1 and 2 and different combination of mixed edge support conditions. The buckling load N is expressed in terms of a non-dimensional buckling factor $\lambda = Nb^2/(\pi^2 D)$. Symbols F, S and C are used in Tables 1-3 to denote free, simply supported and clamped edges, respectively.

3.1 *Convergence and comparison studies*

Convergence and comparison studies have been carried out for rectangular plates with two sub-domains to check the validity and accuracy of the proposed method.

Table 1 presents the buckling factors λ versus the number of polynomial terms m and n employed in the Ritz functions (Eqs. (4) and (5)) for three square plates (a/b = 1) of different combination of edge conditions. The number of polynomial terms m and n increases with increasing degree of 2-D complete polynomial p ($m = n = (p + 1)(p + 2)/2$). The locations of the lines partitioning the plate are at x_0/a = -1/4, 0 and 1/4. The buckling factors in the last row in the table are results obtained by Mizusawa and Leonard (1990).

It is observed that the buckling factors decrease monotonically as the number of polynomial terms increases for all cases. The buckling factors are convergent as m and n are equal to or greater than 15 (or $p \geq 4$) for simply supported plates. For plates with mixed edge support conditions, the buckling factors converge to an acceptable level as m and n are equal to 91 (or p = 12). Therefore, all results presented in the paper are calculated using $m = n = 91$.

The validity of the proposed model is verified as the present buckling results for a simply supported square plate and a square plate with mixed boundary conditions are in excellent agreement with the ones reported by Mizusawa and Leonard (1990).

$$w^{(1)}(x,y) = B^{(1)}(x)B^{(1)}(y)\sum_{i=1}^{m} a_i \Phi_i(x,y) = B^{(1)}(x)B^{(1)}(y)\{\Phi(x,y)\}_{1\times m}^T \{a\}_{m\times 1} \tag{4}$$

$$w^{(2)}(x,y) = B^{(2)}(x)B^{(2)}(y)\sum_{i=1}^{n} b_i \Phi_i(x,y) = B^{(2)}(x)B^{(2)}(y)\{\Phi(x,y)\}_{1\times n}^T \{b\}_{n\times 1} \tag{5}$$

where $\{\Phi(x,y)\}_{m\times 1}$ and $\{\Phi(x,y)\}_{n\times 1}$ are column matrices containing 2-D complete polynomials of m and n terms; $\{a\}_{m\times 1}$ and $\{b\}_{n\times 1}$ are column matrices containing the unknown Ritz coefficients; and $B^{(i)}(x)B^{(i)}(y)$, i = 1 and 2, are the basic functions that consist of the edge equations of the subdomains and satisfy the geometric boundary conditions of the subdomains at outset (Xiang *et al.*, 1996).

The compatibility conditions for the transverse deflections and slopes in the x direction are imposed at the interface of the two subdomains, where $x = x_0$:

$$w^{(1)}(x_0, y) = w^{(2)}(x_0, y), \quad \frac{\partial w^{(1)}(x_0, y)}{\partial x} = \frac{\partial w^{(2)}(x_0, y)}{\partial x} \tag{6, 7}$$

By making use of the compatibility conditions in Eqs. (6) and (7), the Ritz function for subdomain 2 can be re-arranged as:

$$w^{(2)}(x,y) = \{\Gamma_a(x,y,x_0)\}_{1\times m}^T \{a\}_{m\times 1} + \{\Gamma_b(x,y,x_0)\}_{1\times(n-2)}^T \{b_{p2}\}_{(n-2)\times 1} \tag{8}$$

where

$$\{\Gamma_a(x,y,x_0)\}_{1\times m}^T \{a\}_{m\times 1} = \sum_{i=1}^{m} \Bigg[F_1(x,y,x_0)\Phi_i(x_0,y)$$

$$+ F_2(x,y,x_0)(x-x_0)\Phi_i(x_0,y) + F_1(x,y,x_0)(x-x_0)\left(\frac{\partial \Phi_i(x,y)}{\partial x}\right)\Bigg|_{x=x_0} \Bigg] a_i \tag{9}$$

$$\{\Gamma_b(x,y,x_0)\}_{1\times(n-2)}^T \{b_{p2}\}_{(n-2)\times 1} = \sum_{i=3}^{n} \Bigg[B^{(2)}(x)B^{(2)}(y)(\Phi_i(x,y) - \Phi_i(x_0,y))$$

$$- B^{(2)}(x)B^{(2)}(y)(x-x_0)\left(\frac{\partial \Phi_i(x,y)}{\partial x}\right)\Bigg|_{x=x_0} \Bigg] b_i \tag{10}$$

$$F_1(x,y,x_0) = \frac{B^{(1)}(x_0)B^{(1)}(y)B^{(2)}(x)}{B^{(2)}(x_0)} \tag{11}$$

$$F_2(x,y,x_0) = B^{(1)}(y)B^{(2)}(x)\left[\frac{\partial}{\partial x}\left(\frac{B^{(1)}(x)}{B^{(2)}(x)}\right)\right]\Bigg|_{x=x_0} \tag{12}$$

In view of Eqs. (4)-(5) and (8)-(12), the transverse deflections of the two subdomains can be expressed as

$$w^{(1)}(x,y) = \{\Lambda(x,y)\}_{1\times(m+n-2)}^T \{C\}_{(m+n-2)\times 1} \tag{13}$$

$$w^{(2)}(x,y) = \{\Gamma(x,y,x_0)\}_{1\times(m+n-2)}^T \{C\}_{(m+n-2)\times 1} \tag{14}$$

in which

Table 1. Convergence and comparison studies for square plates with mixed boundary conditions.

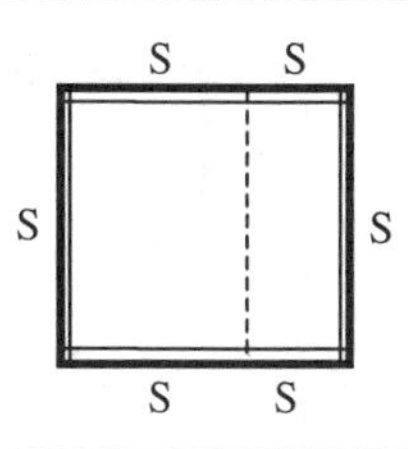

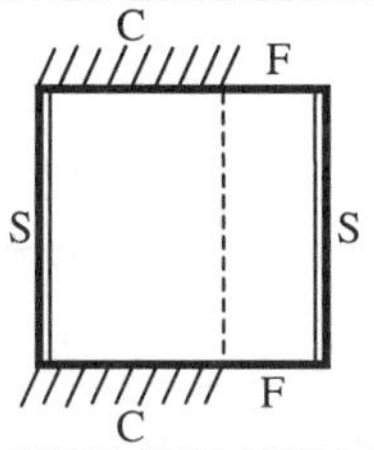

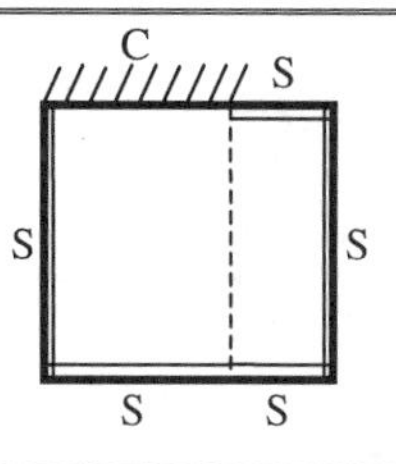

	x_0/a			x_0/a			x_0/a		
$m = n$	-1/4	0	1/4	-1/4	0	1/4	-1/4	0	1/4
3	4.400	4.458	4.400	8.706	8.572	8.706	5.881	6.063	6.667
6	4.042	4.028	4.104	4.841	7.969	8.611	5.704	5.824	5.790
10	4.002	4.002	4.002	3.409	6.257	7.949	4.889	5.631	5.744
15	4.000	4.000	4.000	3.230	6.009	7.715	4.705	5.472	5.735
21	4.000	4.000	4.000	3.201	5.976	7.700	4.617	5.386	5.729
28	4.000	4.000	4.000	3.078	5.746	7.651	4.551	5.325	5.722
36	4.000	4.000	4.000	3.006	5.670	7.639	4.505	5.280	5.716
45	4.000	4.000	4.000	2.955	5.557	7.599	4.472	5.247	5.712
55	4.000	4.000	4.000	2.898	5.428	7.568	4.448	5.222	5.708
66	4.000	4.000	4.000	2.845	5.339	7.550	4.430	5.202	5.705
78	4.000	4.000	4.000	2.806	5.273	7.530	4.416	5.186	5.703
91	4.000	4.000	4.000	2.775	5.215	7.514	4.404	5.174	5.701
Other		*4.000*						*5.198*	

3.2 *Buckling factors for rectangular plates with mixed boundary conditions*

Tables 2 and 3 present buckling factors for six rectangular plates of aspect ratios a/b = 0.5, 1.0 and 2.0 and different combination of mixed edge conditions. The locations of the partition lines are at x_0 / a = 1/4, 0 and 1/4. The edge supports of the two parallel edges on subdomain 1 are arranged to be stronger than the corresponding edges on subdomain 2.

It is seen that the buckling factors increase as the location of domain partition lines x_0 / a moves from –1/4 to 1/4 for all cases in Tables 2 and 3. When x_0 / a = 1/4, the buckling factors for plates with a/b = 0.5 and 1 in Cases 1-5 are approaching the ones with uniform boundary conditions on the two parallel edges. For instance, the buckling factors are approaching 6.25 and 4.0 in Case 1 (see Table 2).

Table 2. Buckling factors for rectangular plates with mixed boundary conditions

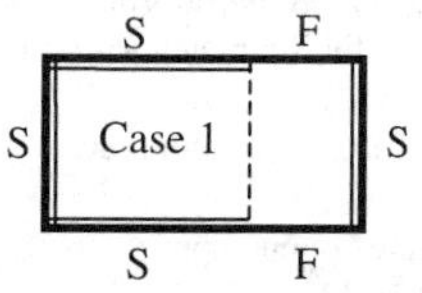

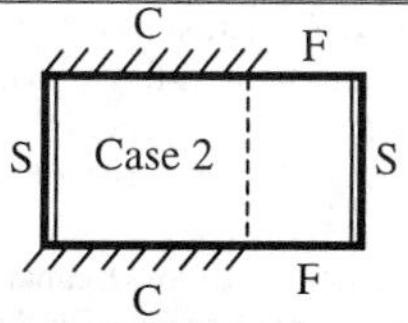

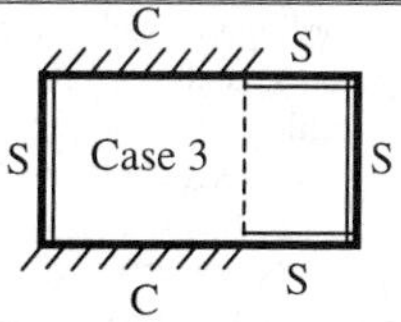

	x_0/a			x_0/a			x_0/a		
a/b	-1/4	0	1/4	-1/4	0	1/4	-1/4	0	1/4
0.5	5.698	6.187	6.249	6.456	7.369	7.670	6.714	7.376	7.671
1	2.281	3.647	3.995	2.775	5.215	7.514	4.750	6.078	7.506
2	0.722	1.501	3.744	0.784	1.677	1.734	4.381	4.479	5.809

Table 3. Buckling factors for rectangular plates with mixed boundary conditions

	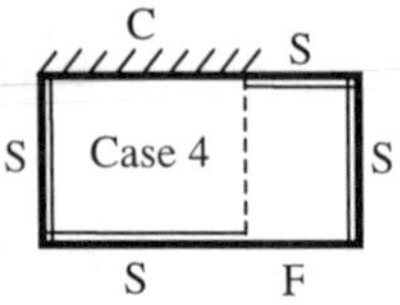			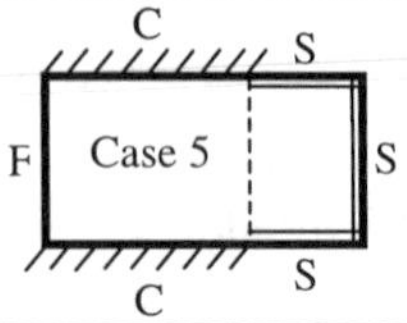			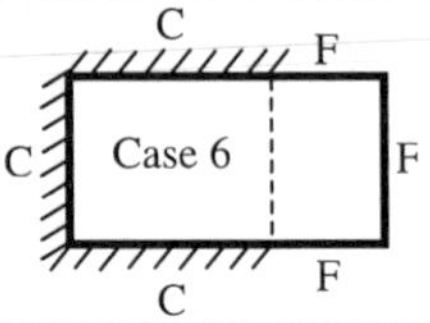		
	x_0/a			x_0/a			x_0/a		
a/b	-1/4	0	1/4	-1/4	0	1/4	-1/4	0	1/4
0.5	6.115	6.700	6.846	4.192	4.219	4.220	1.185	1.904	3.376
1	2.892	4.689	5.697	3.350	3.714	3.850	0.337	0.673	1.824
2	1.204	2.025	4.654	3.730	3.863	3.882	0.094	0.198	0.665

4 CONCLUSIONS

A numerical model based on the automated Ritz method is presented for buckling analysis of rectangular plates with mixed edge support conditions. The plate may be decomposed into two subdomains and different boundary conditions may be imposed on each subdomain. Convergence and comparison studies are carried out to verify the validity and accuracy of the method. Buckling results are presented for several rectangular plates having different combinations of mixed edge support conditions. Further work is being carried out for plates with multiple partitions.

5 REFERENCES

Bartlett, C.C. 1963. The vibration and buckling of a circular plate clamped on part of its boundary and simply supported on the remainder. *Journal of Applied Mechanics* 16:431-440.

Bulson, P.S. 1970. *The Stability of Flat Plates*. Chatto and Windus, London, U.K.

Column Research Committee of Japan. 1971. *Handbook of Structural Stability*. Corona, Tokyo, Japan.

Hamada, H., Inoue, Y. and Hashimoto, H. 1967. Buckling of simply supported but partially clamped rectangular plates uniformly compressed in one direction. *Bulletin of Japan Society of Mechanical Engineers* 10:35-40.

Karamanlidis, D. and Prakash, V. 1989. Buckling and Vibrations of Shear-Flexible Orthotropic Plates Subjected to Mixed Boundary Conditions. *Thin-walled structures* 8(4):273-293.

Keer, L.M. and Stahl, B. 1972. Eigenvalue problems of rectangular plates with mixed edge conditions. *Journal of Applied Mechanics* 39:513-520.

Liew, K.M., Wang, C.M., Xiang, Y., and Kitipornchai, S. 1998. *Vibration of Mindlin Plates-Programming the p-Version Ritz Method.* Elsevier Science Ltd, Amsterdam, The Netherlands.

Mizusawa, T. and Leonard, J.W. 1990. Vibration and buckling of plates with mixed boundary conditions. *Engineering Structures* 12(4):285-290.

Reddy, J.N. and Phan, N.D. 1985. Stability and vibration of isotropic, orthotropic and laminated plates according to a higher-order shear deformation theory. *Journal of Sound and Vibration* 98:157-170.

Sakiyama, T. and Matsuda, H. 1987. Elastic buckling of rectangular Mindlin plates with mixed boundary conditions. *Computers and Structures* 25:801-808.

Xiang, Y., Wang, C.M. and Kitipornchai, S. 1995. Shear buckling of skew Mindlin plates. *AIAA Journal* 33(2):377-378.

Xiang, Y., Wang, C.M. and Kitipornchai, S. 1996. Optimal location of point supports in plates for maximum fundamental frequency. *Structural Optimization* 11(3-4):171-177.

Mechanics of Structures and Materials, Bradford, Bridge & Foster (eds)
© 1999 Balkema, Rotterdam, ISBN 90 5809 107 4

Levels of nonlinearity in frame analysis

J. Petrolito & K.A. Legge
Division of Physical Sciences and Engineering, La Trobe University, Bendigo, Vic., Australia

ABSTRACT: This paper discusses the development and accuracy of various models for nonlinear frame analysis. Starting with a general extensible elastica model, we derive a family of simplified models by truncation of the nonlinear terms. We consider various examples to demonstrate the range of validity and accuracy of the resulting models.

1 INTRODUCTION

Until recently, linear analysis had been the dominant method used for the analysis of structural frames. Nonlinear effects, if included, have usually been considered at the member level rather than at the complete structure level. However, recent changes to codes of practice, such as the Australian Steel Structures Code (Standards Australia 1998), have led to a burgeoning interest in nonlinear frame analysis, to the extent that most commercial packages now include some nonlinear options.

While the intent of the new codes may have been to produce requirements that lead to a more accurate analysis, there has been little guidance given to the type and implementation of such an analysis. Inherent to many nonlinear techniques are assumptions that lead to significant approximations. It has been shown previously (Petrolito & Legge 1994) that the level of approximation used can introduce further complications into the analysis that lead one to question the validity of the nonlinear solution.

In traditional methods of nonlinear frame analysis, two levels of approximations are used. Firstly, various assumptions are made to ignore terms that are considered "small". Secondly, an iterative numerical technique is used to solve the problem, and this often involves the use of an updated geometry strategy that leads to further approximations. It is difficult to determine what the combined effects of these approximations are, and the underlying theoretical model for such approaches is uncertain. Moreover, an inappropriate deletion of terms may have a significant effect on the results. For example, it is interesting to note that McKenna (1999) has recently suggested that an inappropriate trigonometric approximation has led to persistent confusion regarding the cause of collapse of the Tacoma Narrows bridge.

In a previous paper (Petrolito & Legge 1996), the authors developed a unified analysis procedure to enable nonlinear effects in frame analysis to be rationally incorporated. While this procedure can treat any level of nonlinearity consistently, it may be too general in cases where the nonlinear effects are mild. In these cases, it is more appropriate to use a simpler nonlinear model to reduce the computational cost. In this paper, we investigate the use of such simplified models of nonlinearity, and examine their accuracy and range of validity.

2 GOVERNING EQUATIONS

We begin by developing the extensible elastica theory as the most general nonlinear model for frame analysis. Systematic truncation of the nonlinear terms in the governing equations will result in a family of simplified theories, some of which are closely related to established theories.

Consider a straight beam whose centroidal axis is initially along the X axis. A point on the beam's axis is specified by length S with $0 \leq S \leq L$ where L is the length of the beam. After deformation, the beam axis deforms into a smooth curve s. A point P with initial coordinates $(X, 0)$ is mapped to a point P^* with coordinates (x, y). The functions $x(S)$ and $y(S)$ define the geometry of the deformed axis of the beam. The angle between s and the X axis at P^* is denoted by $\phi(S)$. The beam displacements are $u(S) = x(S) - X(S)$ and $v(S) = y(S)$.

An element $dS = dX$ on the original axis is deformed into an element ds. From the geometry of the deformation, we have

$$\frac{dx}{ds} = \cos\phi, \qquad \frac{dy}{ds} = \sin\phi \tag{1}$$

The strain measures for the deformation of the beam are the extensional strain, e, and the bending strain, μ, which are defined as

$$e = \frac{ds}{dS} - 1 = \sqrt{(1+u_S)^2 + v_S^2} - 1, \qquad \mu = \frac{d\phi}{dS} = \frac{(1+u_S)v_{SS} - v_S u_{SS}}{(1+u_S)^2 + v_S^2} \tag{2}$$

where a subscript S denotes differentiation with respect to S.

The force resultants on the beam are the horizontal and vertical force components H and V and the bending moment M, which is taken as positive if it acts anticlockwise. The equilibrium equations for the beam are

$$H_S + p_X = 0, \qquad V_S + p_Y = 0, \qquad M_S - Hv_S + V(1+u_S) = 0 \tag{3}$$

where p_X and p_Y are the distributed loads acting on the beam in the X and Y directions.

Assuming that the material behaviour is linear-elastic, the force-strain relationships are taken as

$$N = EAe, \qquad M = EI\mu \tag{4}$$

where N is the axial force in the beam, E is Young's modulus, A is the cross-section area and I is the moment of inertia.

The above theory results in a nonlinear system of ordinary differential equations of order six. Exact solutions of the governing equations are only possible in simple cases (Antman 1968, Libai & Simmonds 1988), and numerical methods are usually required to solve practical problems. It is the inherent difficulty of working with the general theory that has led to simplifying assumptions being made to achieve a more tractable formulation.

3 SIMPLIFIED THEORIES

We can develop a family of simplified theories by expanding the trigonometric terms in the governing equations as Taylor series, and truncating the series as required. It should be noted that we do not make any assumptions about the size of the strains. The expansions of $\sin\phi$ and $\cos\phi$ are

$$\sin\phi = \phi - \frac{\phi^3}{3!} + \frac{\phi^5}{5!} - \cdots, \qquad \cos\phi = 1 - \frac{\phi^2}{2!} + \frac{\phi^4}{4!} - \frac{\phi^6}{6!} + \cdots \tag{5}$$

The use of these expansions in the governing equations produces an nth order theory when terms of ϕ^{n+1} and higher are dropped from the expansions. The first and second-order theories thus obtained are closely related to the beam-column and the beam-column with bowing theories respectively (Petrolito & Legge 1994). However, these theories make further assumptions regarding the size of various terms in the equations.

4 EXAMPLES

We now present a comparison of the simplified theories for several structures that display varying levels of nonlinearity in their response. The problems have been solved using the numerical procedure previously developed by the authors (Petrolito & Legge 1996), and the results produced by the procedure are exact within the context of the underlying theory. For simplicity, all members in the structures are assumed to have the same properties. The relative influence of the axial and bending deformations is governed by the slenderness ratio $\lambda = L/r$ where L is a characteristic length of the structure and r is the radius of gyration of the cross-section. A constant value $\lambda = 100$ is used in the problems. The results are quoted in terms of non-dimensional displacement and force quantities that are defined as

$$\overline{u} = \frac{u}{L}, \quad \overline{v} = \frac{v}{L}, \quad \overline{P} = \frac{PL^2}{EI} \tag{6}$$

where P is an applied concentrated load. All displacements are quoted at reference point a in the structure.

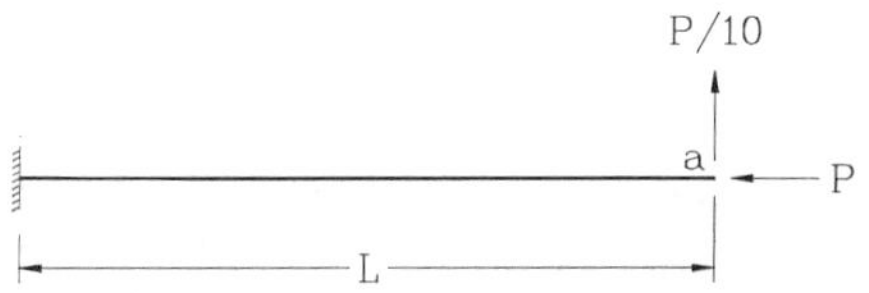

Figure 1. Cantilever under combined loads.

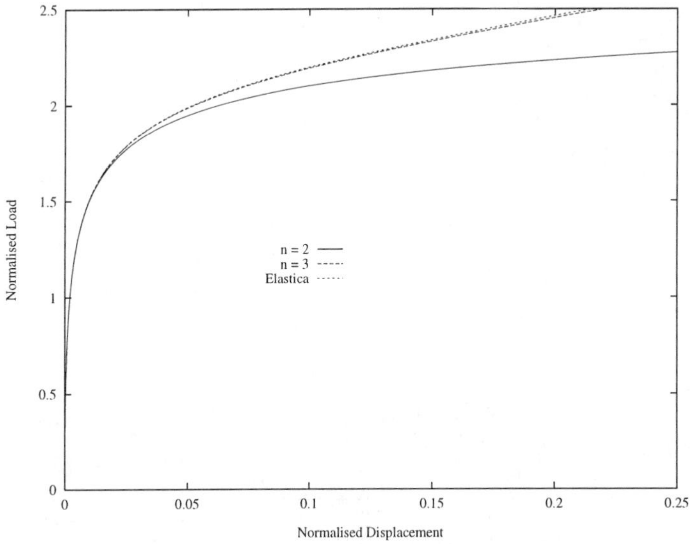

Figure 2. Results for cantilever.

4.1 Cantilever under point loads

As a first example, we consider a cantilever under the combined effects of an end compressive load and shear as shown in Figure 1. The variation of the horizontal displacement is shown in Figure 2. The first-order theory, which is not shown in the figure, erroneously predicts a linear relationship between the displacement and load. Higher-order theories are needed to correctly predict the bowing effect on the horizontal displacement, and in this case the third-order theory is sufficiently accurate for the range shown.

4.2 Toggle under point load

Figure 3 shows a shallow toggle frame subjected to a point load in the centre. The frame exhibits a snap-through behaviour under increasing load. The load-deflection curves are shown in Figure 4. The first-order theory is obviously inadequate as it fails to correctly predict the response of the structure, and significantly over-estimates the limit load. However, the second-order theory is sufficient, and shows little difference from the elastica theory.

4.3 L frame under point load

Figure 5 shows a two-member frame under a point load. The response of this structure is highly nonlinear and low order approximations can only adequately predict the response at low load levels (see Figure 6). In particular, the first-order theory incorrectly predicts

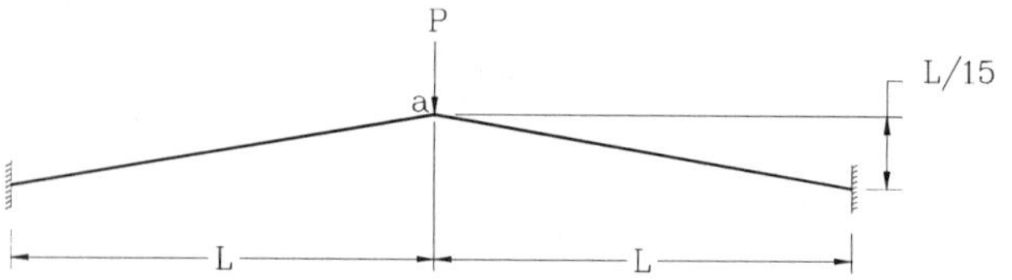

Figure 3. Toggle frame.

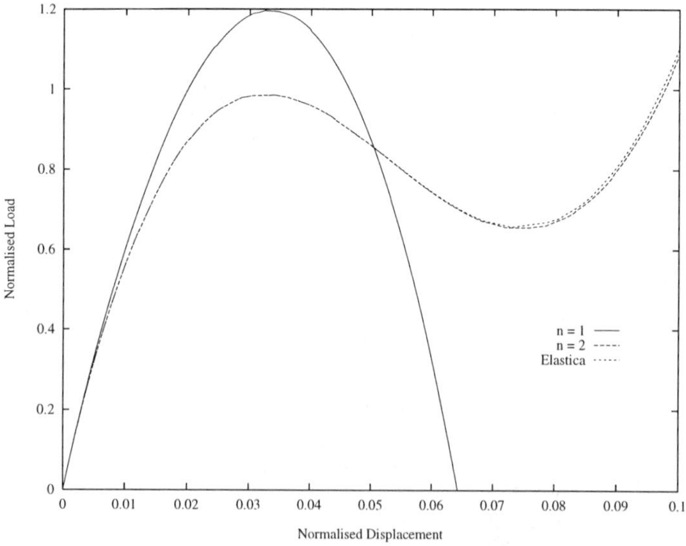

Figure. 4 Results for toggle frame.

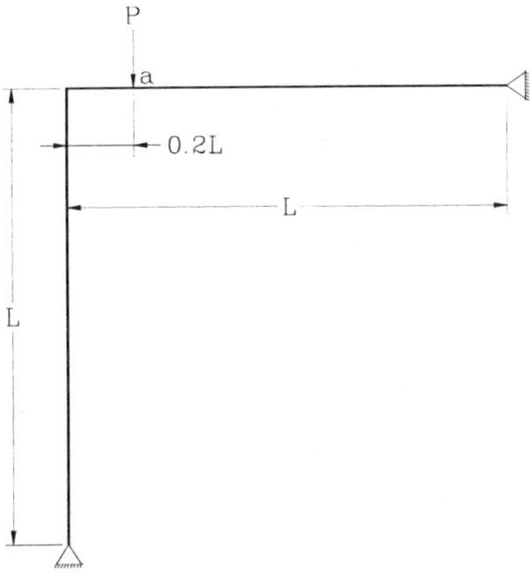

Figure 5. L frame.

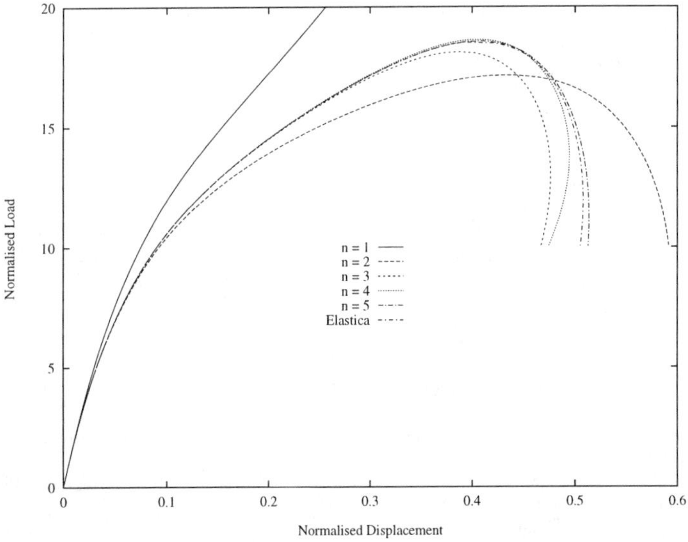

Figure 6. Results for L frame.

a significant post-buckling strength of the structure. The inclusion of higher-order terms avoids this erroneous prediction, with a fourth-order theory being required to obtain an accurate solution for the limit load.

This example also dramatically illustrates how the accuracy of the approximations decreases as the range of the response curve required increases. Figure 7 shows the response of the structure up to the dynamic snap point. It can be seen that severe inaccuracies are present until a sixth-order theory is used. In particular, the results for the fourth-order theory show how sensitive the structure is to modelling assumptions.

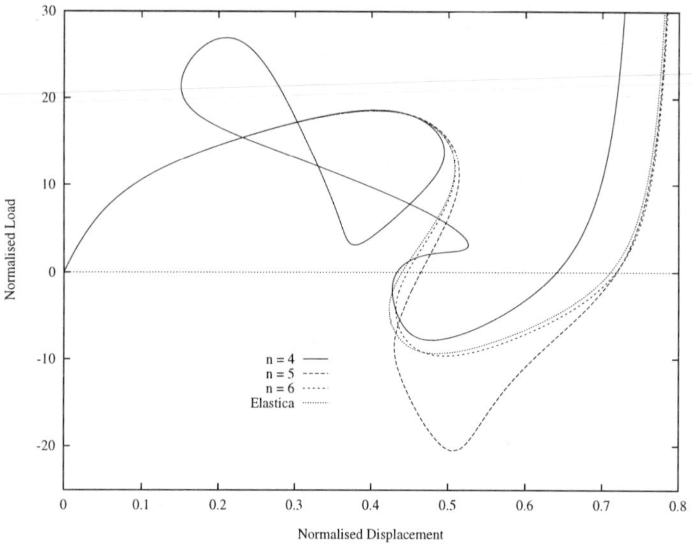

Figure 7. Expanded results for L frame.

5 CONCLUSIONS

Nonlinear analysis is included in codes of practice to improve the accuracy of structural analysis. Generally, this analysis has been done by a combination of low order truncation of nonlinear terms combined with a numerical scheme. However, this paper has shown that indiscriminate truncations have a propensity to limit the accuracy of the analysis, and potentially lead to erroneous conclusions about the structural behaviour.

REFERENCES

Antman, S.S. 1968. General solutions of plane extensible elasticae having nonlinear stress-strain laws. *Quarterly of Applied Mathematics*, 26:33–47.

Libai, A. & Simmonds, J.G. 1988. *The Nonlinear Theory of Elastic Shells: One Spatial Dimension*, Boston: Academic Press.

McKenna, P.J. 1999. Large torsional oscillations in suspension bridges revisited: fixing an old approximation, *American Mathematical Monthly*, 106:1–18.

Petrolito, J. & Legge, K.A. 1994. Balanced approximations in nonlinear analysis, *Proc. Computational Techniques and Applications Conference: CTAC93*: 420–427.

Petrolito, J. & Legge, K.A. 1996. Unified nonlinear elastic frame analysis. *Computers and Structures*, 60:21–30.

Standards Australia 1998. *AS4100: Steel Structures*. Sydney: Standards Australia.

Mechanics of Structures and Materials, Bradford, Bridge & Foster (eds)
© 1999 Balkema, Rotterdam, ISBN 90 5809 107 4

Analysis, design and construction of the Olympic sculptures, AMP Tower, Sydney

R.J. Facioni & E.T. Smith
Hyder Consulting, Sydney, N.S.W., Australia

ABSTRACT: As part of AMP's sponsorship of the Sydney Olympics three Olympic sculptures were erected onto the turret of AMP Tower, Australia's tallest building. The sculptures had to be designed to be strong enough to withstand ultimate wind speeds of 230km/h, yet light enough to be lifted into place by helicopter. Unique "conical" shaped connectors were developed to allow a simple and fast connection to the Tower once each sculpture was lowered into place during the helicopter lifting process. Through the use of sophisticated three dimensional modelling, ongoing design refinement with the sculptor and full load testing prior to erection, Hyder were able to develop a light yet robust design.

1 INTRODUCTION

Watched on by the client and artist, with champagne glasses in hand, the lifting of the sculptures took place Sunday morning, 26 July 1998, to the cheers of hundreds of onlookers. In keeping with AMP Tower being the tallest building in Australia, the sculptures are also the largest of their kind in the world. The three sculptures portray an Olympic gymnast, sprinter and a Paralympic basketballer. Each human figure stands an impressive 14m tall, over seven times life size. To put this in perspective, just the basketballer's ball is taller than a normal adult. In addition to the sculptures, a 12m x 6m Olympic display board on the west face of the tower along with smaller AMP logos on the north and south faces were erected.

Structural engineering input was provided by Hyder Consulting. The figures and signs were designed to resist 50 year return period winds. At the top of the tower, and allowing for airflow acceleration around the turret, the design wind velocity is 230km/hr. Very early in the design process it was decided that solid figures would generate too much drag. Wind tunnel tests were carried out by the University of Sydney, Department of Civil Engineering, to determine the reduced drag potential of wire frame models.

One of the main engineering hurdles faced by Hyder was to make each sculpture both strong enough and light enough. Light weight was essential as each sculpture had to be lifted by helicopter into its position on the turret. The helicopter is a Russian Kamov double rotor design, imported for the project, with a maximum safe lifting capacity of 4.5 tonne.

In all there are over 400 structural members in each sculpture. Each one was analysed during design using a Strand finite element package. Each one was shop detailed using Autocad 3-D drafting. Fabrication of this jigsaw was carried out in Melbourne by Haulmaster and Applied Contract Engineering, companies with ship building backgrounds.

Each sculpture, before leaving the fabricators, was load tested to 100% of its design capacity. Five semi trailers were required to transport the sculptures to Sydney. At this stage heads, arms, legs and torsos were not connected so that the loads would not exceed road-haul width limits. Final assembly took place in the Domain, open parkland within 5 minutes flying time of the tower (Figure 1).

Figure 1. Sculptures In The Domain

Large support structures were also designed by Hyder Consulting, to form the connections between the sculptures and the Tower turret roof. Each fixing point had to be simple enough so that there would be no difficulty docking each sculpture in its correct position while it was suspended on a 65m tether below the helicopter.

2 PROJECT BEGININGS

Artist Dominique Sutton, who won an art competition for the right to design AMP's Olympic gift to the city of Sydney, was involved from the early stages of the project. The project began with the artist and fabricator developing a prototype of the sprinter's leg. The initial objective of the prototype leg was for the artist, client and engineers to assess the quality of what the fabricator was proposing to build. The artist and client had to be satisfied with the appearance whilst the engineers has to be satisfied with the structural integrity. The prototype leg was subsequently mounted on top of AMP Centre and used to test paint finishes and for lighting displays.

The artist then proceeded to create 1.5m wire frame models of each sculpture. The artist was guided at all times by the need for structural integrity in the placement of the wire frame elements. The basic structural form aimed at stringers running longitudinally along the legs, arms and torso, and hoops running perpendicular to the longitudinals forming body cross-sections. Hatching was added later to highlight various muscle groups but was of limited structural benefit. The models were then used as templates for the design and construction of full-scale sculptures.

3 DESIGN DEVELOPMENT

The complex nature of the sculptures meant an appropriate analysis investigation program had to be adopted that would give an understanding of their true behaviour. Specifically, linear, non-linear, buckling and natural frequency analysis were used to determine the overall stability of the sculptures and to assess their behaviour. A true representation of each sculpture had to be modelled in the analysis to ensure accurate results. An accurate representation was obtained for all members by digitising the wire frame models. The digitised information provided 3-dimensional coordinates which were then imported into Strand finite element software for analysis and design. From the analysis, the sculptures were found to behave essentially in a linear manner with small second order effects. The sculptures were globally stable with local member buckling the limiting factor in an overstress situation. The natural frequency of each sculpture was found to be outside the range susceptible to resonant excitation by the wind.

Due to uncertainties in the construction of the structures it was impossible to ensure that the sculptures would be built exactly as designed. To provide comfort in any deviations to the structural analysis and design a non-linear, elasto-plasto push-over analysis was performed. In this model analysis the load was increased until the structure collapsed. As the load increased so did member stresses until a plastic hinge formed. Forces then redistributed and more plastic hinges formed until final collapse. It was found the load could be approximately double that designed for, before the structure became unstable. This information gave confidence that the sculptures were highly redundant with a global factor of safety of two.

From the various analysis investigated and checking implemented it was determined that the sculptures were redundant structures which in an overstressed situation would experience large deformations resulting in a shift to a lower loading. This knowledge was used in developing an appropriate design philosophy for the sculptures and their support frames.

The sculptures' design life on top of AMP Tower is a maximum of three years. In keeping with the requirements of AS 1170.2 structures only need be designed for a probability of failure of 5% in the life of the structure. Hence, a reduced 50 year return period wind loading was used for the ultimate design of the sculptures. This was deemed to be acceptable by all parties, as the nature of failure would be buckling. All members and welded connections within sculptures were designed in accordance with the provisions of AS4100 to resist the 50 year return period winds. All other connections and the sculpture support frames were designed for three times the ultimate wind load and in accordance with AS4100. This was to ensure the failure load of the connections would be significantly higher than the sculptures.

The design philosophy adopted ensured that the sculptures remained light and economical, two of the main requests by the client.

4 LOGISTICAL CONSTRAINTS AFFECTING DESIGN

Several logistical constraints had a large impact on the structural design of the sculptures. The only feasible method to lift the sculptures was with the use of a helicopter. This had limitations in terms of weight restrictions on the sculptures and a need for simplistic connections between the sculptures and their support frames to allow rapid helicopter release.

The helicopter, a Russian Kamov double rotor design, had a maximum safe lifting capacity of 4.5 tonne. This imposed severe weight restrictions on the sculptures and limited the size of members that could be used. What appears as a wire frame structure, in fact consists of structural steel elements ranging in cross section from 75 x 6 strip, to Tees or I-beams welded together from 75 x 6 strip, to 16mm thick plate hoops. All steel was Grade 400 and all members were welded together with 6mm fillet welds.

The connections between the sculptures and the Tower were designed to be simple pin type connections thus requiring a minimum number of bolts. This was essential to minimise the hover time for the helicopter during the lifting operations. The connections were designed with sloped faces in a conical configuration, which allowed them to be self-guiding. The connections were designed with a large positional tolerance to assist the helicopter pilot who was trying to manoeuvre the sculptures into place as they swung on tether cables 65 metres below.

Another hurdle was how to transport the sculptures from the fabricator in Melbourne to Sydney. The solution was to transport the sculptures along the highway with low loaders. Strict guidelines had to be adhered to in transporting the sculptures. Width, height and weight restrictions meant the sculptures had to be spliced in several locations to allow stacking on trucks. Within each of the sculptures, areas of low stress were identified for locating the spliced connections. The splices had to be simple bolted connections as welding was not an option after delivery to Sydney. The spliced connections were designed with a factor of safety of 3.

The support frames of the sculpture had to be in place prior to the lift operation. To maintain a low public profile, helicopters could not be used to lift the support frames. All the support frames, therefore, had to be lifted in the Tower's goods lifts. The lift was small in size, which meant no member could be more than 1.8m in length. All the members were then spliced together with full moment capacity bolted connections.

5 WIND TUNNEL TESTING FOR DESIGN PARAMETERS

Wind design velocities are based on statistical information and vary in intensity depending on location, direction, surrounding environment and height above ground. Most of this information can be extracted from AS1170.2, however there is project specific data that requires wind tunnel testing to get design parameters.

As wind blows around the Tower's turret, it accelerates and the velocity varies depending on position from angle of attack. Wind tunnel tests were carried out at the Boundary Layer Wind Tunnel at Sydney University to review the turret multiplier to be used in design. A multiplier of 1.4 was determined for a position 75° from the angle of attack. The wind tunnel test of the turret also verified that after separating at the leading edge, wind flow closely followed the shape of the turret and thus the wind loading on the sculptures would predominantly be normal to the side of the sculptures.

Wind tunnel tests were also conducted to assess the drag coefficient on each of the sculptures. The actual wire models were placed in the Sydney University wind tunnel. The models were tested with wind normal to the side of the sculpture for two situations, unwrapped and wrapped. The two values were recorded so that the true drag (which is solidity dependent) of the full-scale sculptures could be interpolated from the results.

In the design of each sculpture, wind velocities from all 360° around the Tower had the turret multipliers applied which, when coupled with the positioning of the sculpture with respect to the wind angle of attack, provided a wind velocity (Hyder 1998(a)). The wind velocity assessed for each sculpture was applied to the drag coefficient and a design force was established.

6 CONSTRUCTION OF SCULPTURES

Because of the complex arrangement of members, the sculptures were to be assembled like building a ship. The keelsun first then the hoops would be placed on the keelsun, this would continue until all components had been placed into each figure. All components were clearly labelled and labelled templates were provided to the fabricator. Utilising digitised information from the wire models, accurate CAD files were produced for all the sculptures components. The fabricator had profile cutting equipment available which was able to decode the CAD files. The CAD files were e-mailed to the fabricator and then down loaded into the profile cutter and the shapes were cut out of varying plate thickness depending on where they were on the body of the athlete.

7 MEMBER DUCTILITY TESTS AND FULL SCALE TESTING OF SCULPTURES

One of the assumptions made during design of the sculptures was that the failure mechanism would be a ductile one. The nature of the fabrication meant there was a high degree of welding from stitch welding of T sections to butt welding of spliced members. To give designers and client confidence that the members were still ductile a series of destructive tests were performed at the fabricator's plant (Hyder 1998(b)). A number of T sections were fabricated similar to those in the sculptures. These were then loaded from various directions until failure. In each case failure was ductile with members buckling.

At the completion of fabrication of each of the sculptures it was deemed prudent to full scale test each of them to their full design load (Hyder 1998(c)). The full-scale tests were conducted at the fabricator's factory by attaching the sculptures to a strengthened section of building. Loads were applied by attaching cables at various locations on the sculptures to closely simulate the true wind loading. A crane with load cell was then used to apply the load to the cables. Between 3 and 5 tonnes deadweight was applied through pulleys and multiple bridles to achieve 100% design load.

Overall the full-scale tests verified the global stability of each sculpture under the design wind load. Displacements at various locations through the sculptures were measured and cross-referenced to the results of computer analysis. For all three sculptures there were good agreement between analysed and test recorded displacements.

8 INSTALATION OF SCULPTURES

Safety of the public and all associated with the lifting operation was a high priority for the client. Winton Associates coordinated the lifting operation to ensure the safety of human life and that of the client's artwork. To ensure success, HeviLift, the most highly qualified professionals in the industry with over 10,000hrs of flight time were chosen for the helicopter lifting operation. Pre-lift briefings were held to identify the location of all personnel involved in the operation and to establish the sequence of events for the day. A sterile zone was coordinated with the pilot, council and police to provide a safe flight path for the helicopter. The sterile zone was restricted to personnel involved in the project and had to be secured to ensure no public breaches during the operation.

A final briefing was held at 5:00 am the morning of the lift to notify all parties of weather conditions and establish whether the lift would proceed. At the completion of the final briefing all personnel took up their assigned positions, while security and police proceeded with closing roads and securing the sterile zone. From that point on Hevi Lift's load master had total authority over the lift and would direct the helicopter from on top of the Tower. Riggers were on the Tower to receive the load and fix it into position. All had harnesses attached and safety lines attached to tools. Hyder engineers were positioned on the Tower to ensure connections were sufficiently secured and provide approval for the safe release of payload.

Two helicopters were used for the operation, the Kamov for the sculptures and a smaller Bell for the AMP signage. The lifting of the sculptures took place from the Domain between the hours of 6:30 am and 9:30 am on July 26 (Figures 2 and 3). The AMP signs were lifted into place one week later. Each letter was an individual lift. With near-military precision, the sculptures and signage each slotted onto their mounts. Each of the operations were completed inside one and a half hours.

Figure 2. Kamov Helicopter Lifting the Gymnast

Figure 3. Bell helicopter Lifting the AMP Signage

9 SUMMARY AND CONCLUSIONS

This paper has discussed the design and testing program Hyder Consulting adopted to allow three of the most highly irregular structures to be erected in Australia. New techniques were adopted to obtain electronic representations of the structures which were then incorporated into sophisticated three dimensional computer models. Testing was shown to be an integral part of the project from the assessment of design parameters to the full scale testing of the sculptures. Some of the challenges faced in the design were explored and it was shown that with a little lateral thinking design can overcome almost any constraint. Hyder Consulting's analysis and design allowed the fabrication, transportation and erection of the sculptures to proceed with no problems.

10 REFERENCE

Hyder Consulting 1998(a). Centrepoint Sculptures Wind Design Philosophy, *Report No. 6436/9802142.*

Hyder Consulting 1998(b). Centrepoint Sculptures Ductility Testing, *Report No. 6436/9805021.*

Hyder Consulting 1998(c). Test Load Sculptures, *Report No. 6436/9807108.*

Strand, Finite Element Analysis System – G & D Computing.

AS1170.2 – 1989, Australian Standard – *Part 2: Wind Loads.*

AS4100 – 1998, Australian Standard – *Steel Structures.*

Mechanics of Structures and Materials, Bradford, Bridge & Foster (eds)
© *1999 Balkema, Rotterdam, ISBN 90 5809 107 4*

A theoretical study of plastic buckling of circular cylindrical shells under combined axial compression and torsion

G. Lu
School of Engineering and Science, Swinburne University of Technology, Vic., Australia

R. Mao
Department of Engineering Mechanics, Shanghai Jiao Tong University, Shanghai, People's Republic of China

ABSTRACT: This paper presents a theoretical study of plastic buckling of a circular cylindrical shell under axial and torsional loads. The transverse shear is taken into account by a first-order theory with a correction factor. In equilibrium equations, the contributions of both *v* (circumferential displacement) and *w* (normal displacement) are included and the final eigenvalue problem is in *v* and *w*. Both J_2 deformation theory and J_2 flow theory of plasticity are used. With the existence of torsional load, the plastic in-plane stress-strain relations are anisotropic and the solution in trigonometric functions is no longer applicable. A five-point, instead of three-point, finite difference scheme is used. The numerical results of an example show the interactive role of the axial load and torsion in the plastic buckling.

1 INTRODUCTION

The problem of plastic buckling of a circular cylindrical shell has long been widely studied. Among other recent works, studies by Andrews et al (1983), Ore & Durban (1992) and Lin & Yeh (1994) are worth mentioning. They are all about the buckling under axial compression and are typical of the experimental, analytical and numerical methods, respectively. However, few works can be found on plastic buckling under combined loads especially when torsion is involved. This is partly because of the fact that due to the existence of torsion the prebuckling stress state includes shear stress and in the plastic range the stress-strain relations become anisotropic. In this case, the equation of buckled equilibrium can not be solved in simple trigonometric or exponential functions, but must be treated with numerical methods. Teng and Rotter (1989) give a detailed account of the finite element method for the bifurcation analysis of elastic-plastic axisymmetric shells under combined loads including torsion. Lee and Ades (1957) presented a series of experimental results accompanied by a theoretical analysis with the energy method after some simplifications for the stress-strain relation.

In the present paper, the plastic buckling of a circular cylindrical shell under combined axial and torsional loads is studied "semi-analytically". First order shear deformation theory is used, which allows for the transverse shear of the shell. In the equations of buckled equilibrium, the contribution of the circumferential displacement *v* is included. This may improve the accuracy in predicting the critical load especially in the case of column-type buckling. Numerical results are given with special focus on the mutual effects of the loads as stabilising or destabilising factors.

2 DERIVATIONS

Using a right-handed axis system for a circular cylindrical shell of radius *R* and length *L*, let *x*, *y* and *z* be the coordinates for any point *P* in the shell wall, where *x* is the axial distance of the point to one end of the shell, *y* is the circumferential coordinate in length and z is the distance of the point to the middle surface, which is positive when the point is outside the middle surface.

The displacement field of the first-order shear-deformation theory assumes that

$$u_x(x,y,z)=u(x,y)+z\psi(x,y)$$
$$u_y(x,y,z)=v(x,y)+z\varphi(x,y) \quad (1)$$
$$u_z(x,y,z)=w(x,y)$$

where $\{u_x, u_y, u_z\}$ is the displacement vector in the (x, y, z) coordinate system at the point $P(x,y,z)$. Thus the strain field can be expressed as

$$\varepsilon_x^*(x,y,z)=\varepsilon_x(x,y)+z\frac{\partial\psi(x,y)}{\partial x}$$
$$\varepsilon_y^*(x,y,z)=\varepsilon_y(x,y)+z\frac{\partial\varphi(x,y)}{\partial x} \quad (2)$$
$$\gamma_{xy}^*(x,y,z)=\gamma_{xy}(x,y)+z(\frac{\partial\psi(x,y)}{\partial y}+\frac{\partial\varphi(x,y)}{\partial x})$$

where

$$\varepsilon_x=\frac{\partial u}{\partial x},\varepsilon_y=\frac{\partial v}{\partial y}+\frac{w}{R},\gamma_{xy}=\frac{\partial u}{\partial y}+\frac{\partial v}{\partial x} \quad (3)$$

and they represent the strains at the middle surface. By neglecting the nonlinear terms, the equations of buckled equilibrium can be obtained as (Mao & Williams 1998)

$$\frac{\partial T_x}{\partial x}+\frac{\partial T_{xy}}{\partial y}=0 \quad (4a)$$

$$\frac{\partial T_{xy}}{\partial x}+\frac{\partial T_{xy}}{\partial y}+\frac{Q_y}{R}+T_x^0\frac{\partial^2 v}{\partial x^2}=0 \quad (4b)$$

$$\frac{\partial Q_x}{\partial x}+\frac{\partial Q_y}{\partial y}-\frac{T_y}{R}+T_x^0\frac{\partial^2 w}{\partial x^2}+2T_{xy}^0\frac{\partial^2 w}{\partial x\partial y}+T_y^0\frac{\partial^2 w}{\partial y^2}=0 \quad (4c)$$

$$\frac{\partial M_x}{\partial x}+\frac{\partial M_{xy}}{\partial y}-Q_x=0 \quad (4d)$$

$$\frac{\partial M_{xy}}{\partial x}+\frac{\partial M_y}{\partial y}-Q_y=0 \quad (4e)$$

where T_x^0, T_y^0 and T_{xy}^0 are prebuckling constant membrane forces per unit length. The definitions of the stress resultants are as usual. As in conventional formulation of a buckling problem, the stress resultants and displacements in Equations 4a-e are the increments of these quantities due to buckling (from the prebuckling state to a neighbouring buckled state). Therefore, for the study of initial buckling these increments can be infinitesimal.

For plane stress state and an anisotropic stress-strain relation the following general equations can be written

$$\begin{Bmatrix} T_x \\ T_y \\ T_{xy} \end{Bmatrix}=\begin{bmatrix} A_{11} & A_{12} & A_{16} \\ A_{21} & A_{22} & A_{26} \\ A_{61} & A_{62} & A_{66} \end{bmatrix}\begin{Bmatrix} \varepsilon_x \\ \varepsilon_y \\ \gamma_{xy} \end{Bmatrix} \quad (5)$$

$$\begin{Bmatrix} M_x \\ M_y \\ M_{xy} \end{Bmatrix}=\begin{bmatrix} D_{11} & D_{12} & D_{16} \\ D_{21} & D_{22} & D_{26} \\ D_{61} & D_{62} & D_{66} \end{bmatrix}\begin{Bmatrix} \chi_x \\ \chi_y \\ \chi_{xy} \end{Bmatrix} \quad (6)$$

$$\begin{Bmatrix} Q_y \\ Q_x \end{Bmatrix} = \begin{bmatrix} A_{44} & A_{45} \\ A_{54} & A_{55} \end{bmatrix} \begin{Bmatrix} \gamma_{yz} \\ \gamma_{xz} \end{Bmatrix} \tag{7}$$

where

$$\chi_x = \frac{\partial \psi}{\partial x}, \chi_y = \frac{\partial \varphi}{\partial y}, \chi_{xy} = \frac{\partial \varphi}{\partial x} + \frac{\partial \psi}{\partial y} \tag{8}$$

Equation (4a) can be satisfied by introducing a stress function F such that

$$T_x = \frac{\partial^2 F}{\partial y^2}, T_{xy} = -\frac{\partial^2 F}{\partial x \partial y} \tag{9}$$

Let F, v, w, ψ and φ be the five basic unknowns of the buckling problem considered. Then through Equations 5, 9 and 3, T_y can be expressed by these unknowns:

$$T_y = \frac{1}{\overline{A}_{22}}(\varepsilon_y - \overline{A}_{12} T_x - \overline{A}_{26} T_{xy}) \tag{10}$$

where $[\overline{A}_{ij}] \equiv [A_{ij}]^{-1}$. Hence the five basic equations, ie. the equations of compatibility and the equations of equilibrium (4b-e) cam be expressed in terms of the five basic unknowns.

First, the equation of compatibility

$$\frac{\partial \varepsilon_x}{\partial y} = \frac{\partial}{\partial x}(\gamma_{xy} - \frac{\partial v}{\partial x}) \tag{11}$$

can be expressed as

$$(\overline{A_{11} A_{22}} - \overline{A_{12}}^2) \frac{\partial^3 F}{\partial y^3} - 2(\overline{A_{16} A_{22}} - \overline{A_{12} A_{26}}) \frac{\partial^3 F}{\partial x \partial y^2} + (\overline{A_{66} A_{22}} - \overline{A_{26}}^2) \frac{\partial^3 F}{\partial x^2 \partial y}$$

$$= \overline{A_{26}} \frac{\partial}{\partial x}(\frac{\partial v}{\partial y} + \frac{w}{R}) - \overline{A_{12}} \frac{\partial}{\partial y}(\frac{\partial v}{\partial y} + \frac{w}{R}) - \overline{A_{22}} \frac{\partial^2 v}{\partial x^2} \tag{12}$$

Similarly, Equations 4b-e are transformed into

$$\frac{1}{\overline{A_{22}}}(\frac{\partial^2 v}{\partial y^2} + \frac{1}{R}\frac{\partial w}{\partial y} - \overline{A_{12}} \frac{\partial^3 F}{\partial y^3} + \overline{A_{26}} \frac{\partial^3 F}{\partial x \partial y^2}) - \frac{\partial^3 F}{\partial x^2 \partial y}$$

$$+ \frac{1}{R}[A_{44}(\frac{\partial w}{\partial y} + \varphi - \frac{v}{R}) + A_{45}(\frac{\partial w}{\partial x} + \psi)] + T_x^0 \frac{\partial^2 v}{\partial x^2} = 0 \tag{13}$$

$$A_{45}(\frac{\partial^2 w}{\partial x \partial y} + \frac{\partial \varphi}{\partial x} - \frac{1}{R}\frac{\partial v}{\partial x}) + A_{55}(\frac{\partial^2 w}{\partial x^2} + \frac{\partial \psi}{\partial x}) + A_{44}(\frac{\partial^2 w}{\partial y^2} + \frac{\partial \varphi}{\partial y} - \frac{1}{R}\frac{\partial v}{\partial y})$$

$$+ A_{45}(\frac{\partial^2 w}{\partial x \partial y} + \frac{\partial \psi}{\partial y}) - \frac{1}{R\overline{A_{22}}}(\frac{\partial v}{\partial y} + \frac{w}{R} - \overline{A_{12}} \frac{\partial^2 F}{\partial y^2} + \overline{A_{26}} \frac{\partial^2 F}{\partial x \partial y})$$

$$+ T_x^0 \frac{\partial^2 w}{\partial x^2} + 2S \frac{\partial^2 w}{\partial x \partial y} + T_y^0 \frac{\partial^2 w}{\partial y^2} = 0 \tag{14}$$

$$D_{11}\frac{\partial^2\psi}{\partial x^2}+(D_{12}+D_{66})\frac{\partial^2\varphi}{\partial x\partial y}+D_{16}(2\frac{\partial^2\psi}{\partial x\partial y}+\frac{\partial^2\varphi}{\partial x^2})+D_{26}\frac{\partial^2\varphi}{\partial y^2}+D_{66}\frac{\partial^2\psi}{\partial y^2}$$
$$-A_{45}(\frac{\partial w}{\partial y}+\varphi-\frac{v}{R})-A_{55}(\frac{\partial w}{\partial x}+\psi)=0 \tag{15}$$

$$D_{22}\frac{\partial^2\varphi}{\partial y^2}+(D_{12}+D_{66})\frac{\partial^2\psi}{\partial x\partial y}+D_{26}(2\frac{\partial^2\varphi}{\partial x\partial y}+\frac{\partial^2\psi}{\partial y^2})+D_{16}\frac{\partial^2\psi}{\partial y^2}+D_{66}\frac{\partial^2\varphi}{\partial x^2}$$
$$-A_{44}(\frac{\partial w}{\partial y}+\varphi-\frac{v}{R})-A_{45}(\frac{\partial w}{\partial x}+\psi)=0 \tag{16}$$

Equations 12 -16 are five basic equations in five basic unknowns for the buckling problem considered.

3 CONSTITUTIVE EQUATIONS OF PLASTICITY

The most commonly used constitutive equations of plasticity for buckling problems, J_2 flow theory and J_2 deformation theory, are employed, although there is a paradox about their theoretical correctness and practical effectiveness. The J_2 flow theory is

$$d\varepsilon_{ij}^{p}=\frac{3}{4J_2}(\frac{1}{E_t}-\frac{1}{E})S_{ij}S_{kl}d\sigma_{kl} \tag{17}$$

where, and in Equations 18 and 19, the subscripts range from 1 to 3 and the repeated subscript means summation over it from 1 to 3. The tensor S_{ij} is stress deviator and $J_2=S_{mn}S_{mn}/2$. E_t is the tangent modulus. For the case of plane stress as in shell buckling problems, the constitutive equations are greatly reduced by the condition $\sigma_{23}=\sigma_{31}=\sigma_{33}=0$.

The J_2 deformation theory usually takes the form

$$\varepsilon_{ij}=\frac{3}{2E_s}\sigma_{ij}-(\frac{1}{2E_s}-\frac{1-2v}{3E})\sigma_{kk}\delta_{ij} \tag{18}$$

where E_s is the secant modulus. Since buckling problems need stress-strain relation in incremental form, Equation 18 should be differentiated to give

$$d\varepsilon_{ij}=\frac{3}{4J_2}(\frac{1}{E_t}-\frac{1}{E_s})S_{ij}S_{kl}d\sigma_{kl}+\frac{3}{2E_s}(d\sigma_{ij}-\frac{1}{3}d\sigma_{kk}\delta_{ij})+\frac{1-2v}{3E}d\sigma_{kk}\delta_{ij} \tag{19}$$

As for the J_2 flow theory, in the case of plane stress Equation 19 reduces to a set of three equations.

4 SOLUTION OF THE BASIC EQUATIONS

The solution is assumed to be of the following form (e.g. for F and w):

$$\begin{aligned}F(x,y)&=e_c(x)\cos n\theta+e_s(x)\sin n\theta\\ w(x,y)&=w_c(x)\cos n\theta+w_s(x)\sin n\theta\\ &\vdots\end{aligned} \tag{20}$$

where n is an integer. Substituting Equation 20 into Equation 12-16 and then letting the coefficients of $\cos n\theta$ and $\sin n\theta$ be zero for each equation yields ten ordinary differential equations. These ten lengthy equations can be written in a compact matrix form. All the coefficient matrices are 2 x 6 constant matrices. Their expressions are not presented in this paper for the sake of conciseness. $\lambda \geq 0$ is a load parameter related to the prebuckling membrane forces in the following way:

$$T_x^0 = -P\lambda, T_y^0 = -Q\lambda, T_{xy}^0 = S\lambda \tag{21}$$

P, Q are the normal stresses and S is the shear stress. A five point finite difference scheme is used to solve the above eigenvalue problem. For any given circumferential wave number n in Equation 20, the solution of the eigenvalue problem yields the eigenvalues, the smallest of which, λ_n^*, is the buckling load parameter corresponding to that n. The minimum of all these λ_n^* is the critical load parameter, λ_{cr} of the shell. As the values of P, Q and S are given for a specific problem, the buckling loads can be found from Equation 21.

5 A NUMERICAL EXAMPLE

For the computation of numerical examples, two typical sets of boundary conditions are considered. One is fully clamped and is expressed as

$$T_x = 0, v = 0, w = 0, \psi = 0 \text{ and } \varphi = 0 \text{ at } x = 0, L \tag{22}$$

The other is called simply supported and is expressed as

$$T_x = 0, v = 0, w = 0, M_x = 0 \text{ and } \varphi = 0 \text{ at } x = 0, L \tag{23}$$

The conditions (22) are all in terms of the five basic unknowns and can be directly used in the solution procedure described in the previous section. In the conditions (23), however, M_x is not an unknown variable and the condition M_x=0 should be transformed into one expressed by the five basic unknowns.

The stress-strain relation of the aluminium alloy used for the cylinder in this example is expressed by the Ramberg-Osgood equation

$$\varepsilon = \frac{\sigma}{E}[1 + \frac{3}{7}(\frac{\sigma}{\sigma_y})^{N-1}] \tag{24}$$

where the three parameters are Young's modulus E, yield stress σ_y , and the hardening parameter N. Here E=70 GPa, σ_y = 503.66 MPa and N = 10. The Poisson's ratio is ν=0.32. Calculations of buckling stress under either axial compression or torsion alone compare very well with the experiments by Lee (1962) and Lee & Ades (1957). Here only the results under combined loadings are shown in Figure 1. The ratio of radius to thickness is fixed at R/h =15.

Figure 1 demonstrates the variation of the buckling stresses for varying $P/(P+S)$. The value of $P/(P+S)$ determines the proportion of the prebuckling axial compressive stress (P) and torsional shear stress (S). For instance, $P/(P+S)=0$ and $P/(P+S)=1$ are the two extremities, corresponding to pure torsion and pure axial compressive, respectively. Each curve in the figure has two sections corresponding to the wave number n =1 and n =2, respectively. For buckling axial stress, there is a clear sudden turning between the two sections on each curve. For buckling shear stress, there is no such sudden turning, so a star is used to separate the neighboring sections.

The curves in Figure 1 show that when the prebuckling shear stress is not large enough compared with the prebuckling axial stress [approximately $P/(P+S) \geq 0.8$], its influence on the compressive buckling stress is negligible. Similarly, when the prebuckling axial stress is not

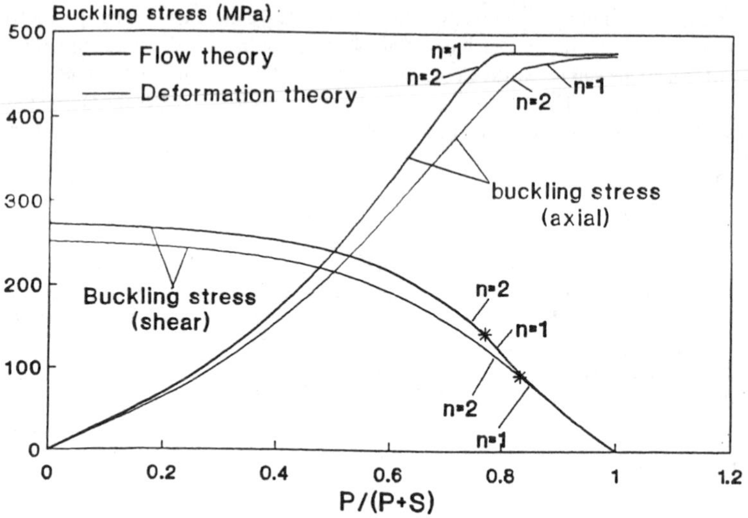

Figure 1. Buckling stresses at combined compressive and torsional buckling.

large enough compared with the prebuckling shear stress [approximately $P/(P+S) \leq 0.5$], its influence on torsional buckling stress is negligible.

6 CONCLUSION

A theoretical study is performed on the plastic buckling of cylindrical shells under combined axial load and torsion. First order shear deformation is successfully incorporated. A five point finite difference scheme is employed for solutions. A numerical example demonstrates interesting features on the interactive roles of both torsion and axial compression.

ACKNOWLEDGEMENT The authors wish to thank the Australian Research Council for the financial support for this project.

REFERENCES

Andrews, K.R.F., England, G.L. & Ghani, E. 1983. Classification of the axial collapse of cylindrical tubes under quasi-static loading. *Int. J. Mech. Sci.* 25: 687-696.
Lee, L.H.N. 1962. Inelastic buckling of initially imperfect cylindrical shells subject to axial compression. *J. Aero. Sci.* 29: 87-95.
Lee, L.H.N. and Ades, C.S. 1957. Plastic torsional buckling strength of cylinders including the effects of imperfections. *J. Aero Sci.* 24: 241-248.
Lin, M.C. and Yeh, M.K. 1994. Buckling of elastoplastic circular cylindrical shells under axial compression. *AIAA Journal* 32: 2309-2315.
Mao, R. and Williams, F.W. 1998. Nonlinear analysis of cross-ply thick cylindrical shells under axial compression. *Int. J. Solids Strut.* 35: 2151-2171.
Ore, E. and Durban, D. 1992. Elastoplastic buckling of axially compressed circular cylindrical shells. *Int. J. Mech. Sci.* 34: 727-742.
Teng, J.G. and Rotter, J.M. 1989. Non-symmetric bifurcation of geometrically nonlinear elastic-plastic axisymmetric shells under combined loads including torsion. *Computers & Structures* 32: 453-475.

Mechanics of Structures and Materials, Bradford, Bridge & Foster (eds)
© 1999 Balkema, Rotterdam, ISBN 90 5809 107 4

Shaping, stiffness and ultimate load capacity of domical space trusses

L.C.Schmidt & G.Dehdashti
Department of Civil, Mining and Environmental Engineering, University of Wollongong, N.S.W., Australia

ABSTRACT: The paper discusses laboratory tests and the comparative ultimate loads and stiffnesses resulting from the shaping of two different layouts of steel chord members from an initally flat condition to a gently curved domical shape (ie., with positive Gaussian curvature). Two trusses have out-of-plane web members in all panels, and two trusses have out-of-plane web members located only at the periphery of the trusses. The shaping operation in each case is carried out by a post-tensioning operation. Ultimate load tests on models of about 3m span are described, and the overall responses of the four tests are discussed and compared. It is shown that the use of out-plane-web members enhances considerably the ultimate load capacity and stiffness of the two shaped trusses when compared with the same layouts without internal web members. In each case the shaping operation is able to be carried out readily.

1 INTRODUCTION

The process of shaping a doubly-curved surface from an intersecting set of chords that are effectively connected together at their points of intersection requires that in-plane distortion be possible. Previous work by Schmidt (1989), Schmidt and Dehdashti (1993), Schmidt and Li (1995) has shown that this double curving process is possible.

In order to achieve substantial load carrying capacities however, it has been necessary to use out-of-plane web members, as indicated in Fig. 1, to improve performance over a system without these web members.

It is the prime purpose of this paper to compare the structural performances of shaped domical systems with and without out-of-plane web systems.

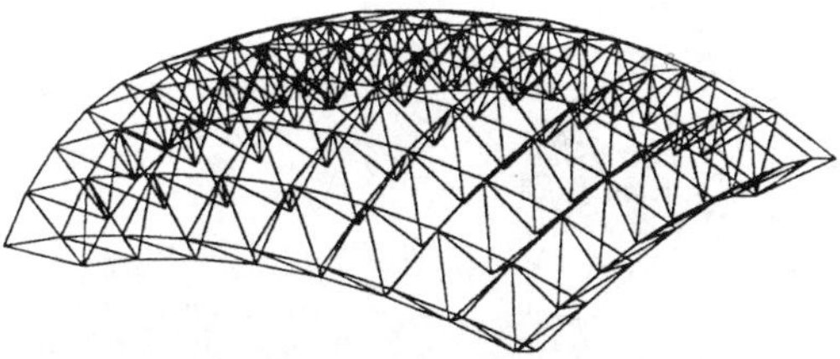

Figure 1. Dome with out-of-plane web members.

2 SHAPING PROCESS

The space trusses with out-of-plane web systems can be be shaped by a process of using post-tensioning cables that pass through certain members that are too short, thereby closing the gaps between members. These members are at the level of the lower points of intersection of the web members, and are usually placed around the periphery of the truss. No other lower chord members are used.

The in-plane distortion that is necessary to form the doubly-curved surface requires that the original layout of the chords cannot be triangulated, but higher order polygons are satisfactory. Although the joints could be pinned, in reality the joints will be at least semi-rigid, and the shaping process will involve some flexure of the chords.

In the other cases considered herein, where no web members are used in the interior region of the truss, the chords need to be connected by rigid or semi-rigid joints so that the shaping process can occur by flexure of the chords. If perfect pinned joints were used then the system would be unstable, as this system is essentially a flexural one.

3 MODEL TESTS

The layouts of the two sets of laboratory test models are shown in Figs. 2 and 3. In Fig. 2a the square plan dome is referred to as a single-chorded space-truss dome (SCSTD), while the comparative structure in Fig. 2b, without interior web members, is referred to as a single layer dome (SLD). These two domes are called Set 1. Likewise the circular plan domes in Figs. 3a and b are referenced with the same notation, and are called Set 2.

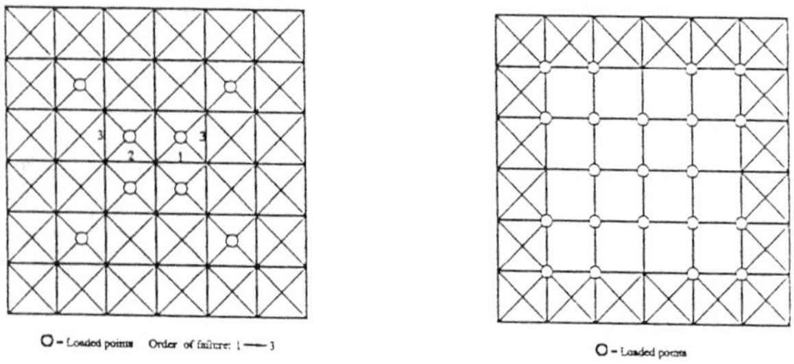

(a) SLSTD (b) SLD

Figure 2. Square plan domes (Set 1)

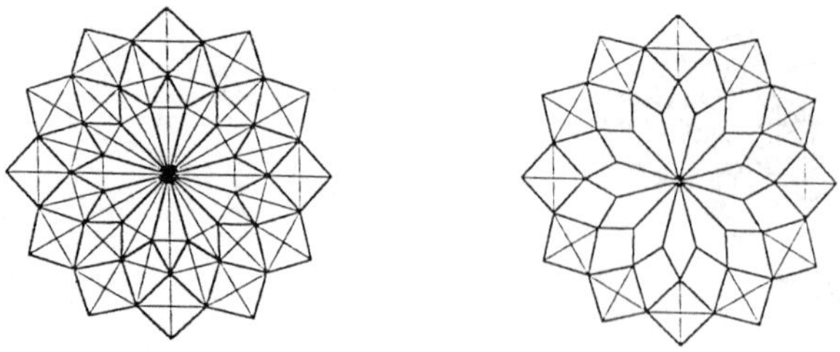

(a) SLSTD (b) SLD

Figure 3. Circular plan domes (Set 2)

In order that the shaping procedure for the SCSTD can proceed it is necessary that a mechanism or near-mechanism condition exists in the planar layout. Because of the use of semi-rigid joints the term 'near-mechanism' is used herein. Even then, in this condition, the post-tensioning force necessary for the shaping process is relatively small.

In order to assess this situation it is necessary to consider the following condition (Calladine 1978), assuming the joints to be pinned.

$$R - S + M = 0 \qquad (1)$$

where R is the degree of redundancy, S is the number of independent prestress states, and M is the number of independent, infinitesimal mechanisms that could exist.

Table 1 gives the results of the application of Eqn. (1) to the SCSTD for Sets 1 and 2, when the number of restraints is that used in the load-test condition.

Table 1 Number of mechanisms for each SCSTD

Set number	Number members	Nodes	Restraints	R	S	M
1	248	85	7	0	0	0
2	240	85	15	0	0	0

Before the post-tensioning operation (i.e., before the gaps in the lower peripheral chords were closed) in the SCSTD, a near-mechanism condition existed, so that the shaping could readily occur for each case. The SLD for Sets 1 and 2 also required small post-tensioning forces, as only flexure of the grid was needed in each case.

3.1 *Model dimensions, shaping, supports and loading*

3.1.1 *Set 1*

The member sizes of the steel used were 13x13x1.8 SHS for the top chords, 13.5x2.3 CHS for the webs and 13.5x2.3 CHS for the peripheral bottom chords. The length of the top chords was 3120mm, and the depth of the trusses was 360mm.

The support for the frames was at the four corners only. The loading was applied to lower nodes as shown in Figs. 2a and b, through a whiffletree system. The loading patterns were not identical, but they were close enough to allow the assessment of the overall comparative performance of the two types of system.

3.1.2 *Set 2*

The member sizes for the circular plan domes were the same as those used for Set 1. The outer diameter of the dome in the planar condition was 2760mm, and the depth was 250mm.

The curvature of the domes was considered to be gentle, and was determined by the pre-defined gaps in the lower chords. The formation radius of curvature of each dome was approximately 2.5 m.

Supports were provided at all lower chord peripheral nodes. In each case a central concentrated load was applied.

Fig. 4 shows the shaped form of the SCSTD of Set 2 before it was placed in the loading rig. Fig. 5 shows the SLD of Set 2 in the loading rig under the central concentrated load.

3.2 *Load - test results*

The monotonic loading of the trusses furnished the maximum loads and stiffnesses as indicated in Table 2. The load versus central deflection curves are shown in Figs. 6 and 7 for Sets 1 and 2, respectively.

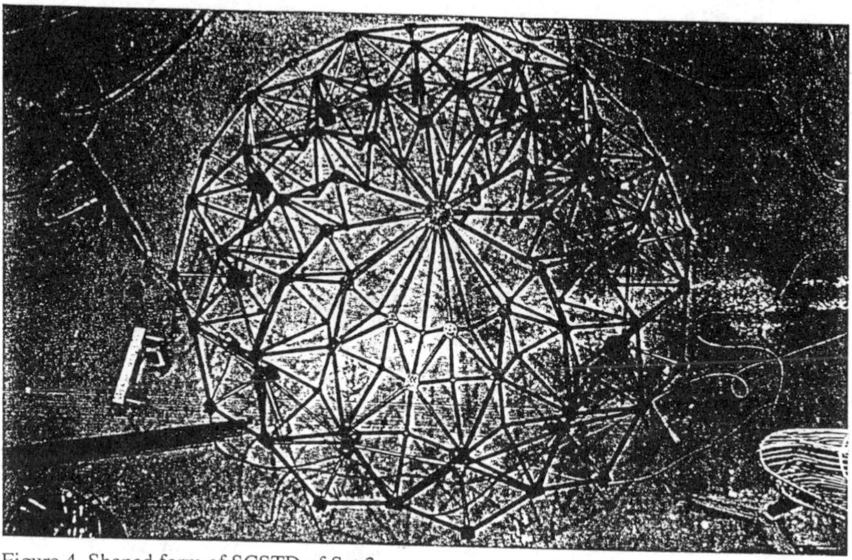

Figure 4. Shaped form of SCSTD of Set 2.

Figure 5. Loading of SLD of Set 2.

Table 2 Load-Test results

Set number	Truss type	Ultimate load (kN)	Stiffness (kN/mm)
1	SCSTD	62	1028
	SLD	15	42
2	SCSTD	62	3604
	SLD	5	37

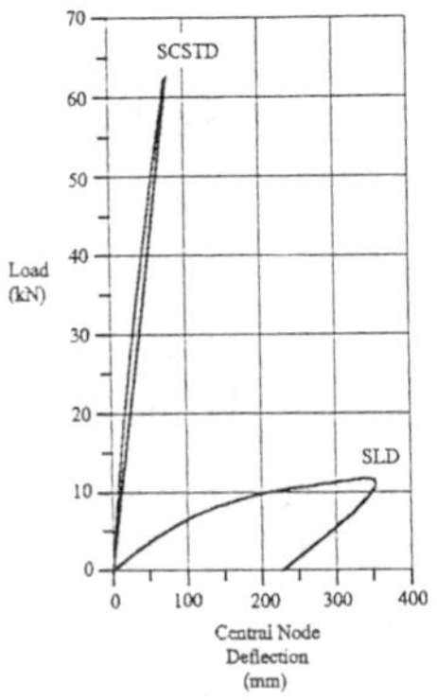

Figure 6. Set 1 load - deflection curves

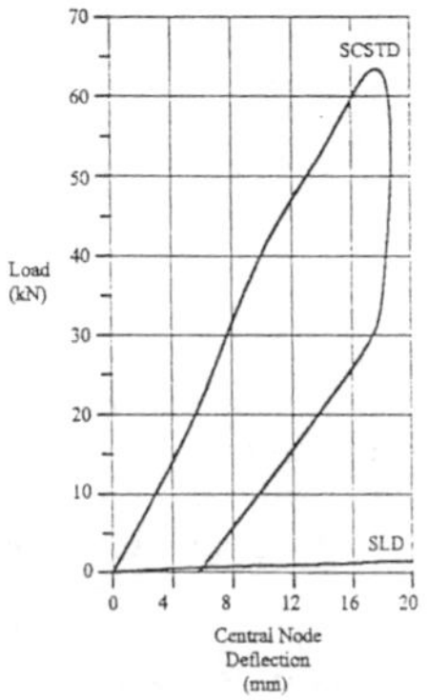

Figure 7. Set 2 load - deflection curves

The maximum load for the SCSTD of Set 1 was 61.7 kN and 14.7 kN for the SLD, being a ratio of 4 to 1. The secant stiffnesses up to the respective maximum loads were 1030 N/mm and 42 N/mm, respectively, being a ratio of 24 to 1.

The maximum load for the SCSTD of Set 2 was 62 kN and 5 kN for the SLD, being a ratio of 12 to 1. The secant stiffnesses up to the respective maximum loads were 1030 N/mm and 42 N/mm, respectively, being a ratio of 97 to 1.

Although the SLDs of Sets 1 and 2 were of different forms, it can be expected that a significant reason for the SLD of Set 1 having a greater load capacity than that of the SLD of Set 2 is that a distributed load was used in the former case while a concentrated load was used in the latter case.

3.3 *Failure modes*

The failure mode of the Set 1 - SCSTD involved the axial buckling of a chord member between panel points. As the truss was just statically determinate, the maximum load was reached as this member buckled. The truss acted a space truss under primary axial member forces. Fig. 2a indicates the order of chord member failure after the first failure. The failure mechanism of the Set 1 - SLD was in the form of flexure of members in the central region of the grid, axial action increasing in importance as the deflection became larger due to dishing or catenary-like behaviour being developed in the central region of the grid.

The failure mode of the Set 2 - SCSTD was a shear failure, with one web member buckling at a support, which determined the maximum load. The failure mode of the Set 2 - SLD was the same as that of the Set 1 - SLD.

4 CONCLUSIONS

Prior work has shown that it is possible to shape planar layouts of intersecting chord members into doubly curved shapes that have load carrying capacity.

The principal points to note from the current investigation, involving two sets of results from two different layouts of chord members and two different configurations of web members (SCST and SLD), are as follows.

Use of an outer ring of out-of -plane web members allows post-tensioning forces to be applied conveniently to shape an initially planar layout of chords from either a SCST or a SLD. The layout of the chords needs to form quadrilaterals or higher order polygons.

The structural performance of the finally shaped SCST far surpasses that of the SLD in both strength and stiffness, highlighting the advantages in using the out-of-plane web system throughout the truss. Such a system allows easy shaping. while providing significant strength.

ACKNOWLEDGMENTS

The writers are grateful to the Australian Research Council for financial assistance with this research project. S.Selby fabricated the trusses used in this work.

REFERENCES

Calladine, C.R.. 1978. Buckminster Fuller's "Tensegrity" structures and Clerk Maxwell's rule for the construction of stiff frames. *Int. J. of Solids and Structures*, 14 (3): 161-172.

Schmidt, L.C. 1989. Erection of space trusses of different forms by tensioning of near mechanisms. *Proc. IASS Congress*, Cedex-Laboratorio Central De Estructuras Y Materiales, Madrid, Spain, 4, 12p.

Schmidt, L.C. and Dehdashti, G. 1993. Shape Creation and Erection of Metal Space Structuresby means of Post-Tensioning. *Space Structures 4*, Eds. G.A.R. Parke and C.M. Howard, ThomasTelford, London, 1: 69-77.

Schmidt, L.C., and Li, H. 1995. Shape Formation of Deployable Metal Domes. *International Journal of Space Structures*, 10 (4): 189-194.

Foundation engineering

Mechanics of Structures and Materials, Bradford, Bridge & Foster (eds)
© 1999 Balkema, Rotterdam, ISBN 90 5809 107 4

Modelling reactive soil movements for the study of cracking in masonry structures

M.J. Masia, Y.Z. Totoev & P.W. Kleeman
Department of Civil, Surveying and Environmental Engineering, University of Newcastle, N.S.W., Australia

ABSTRACT: Shrink - swell movements in reactive soils have long been a cause of serviceability problems in masonry structures, with the cracking of masonry walls supported by footings being the most obvious. Research at The University of Newcastle aims to produce rational serviceability design criteria for masonry structures. This paper describes a numerical model for simulating soil moisture movements in three dimensions.

1 INTRODUCTION

The shrink - swell action of reactive soils is one source of structure foundation movement which has historically caused widespread problems with respect to the serviceability performance of masonry structures. The most obvious problem is that of cracking of masonry walls supported by footings.

A research project at The University of Newcastle aims to produce rational design criteria for serviceability design of masonry structures using reliability based structural design (Masia et al. 1998, Melchers 1996, Page & Kleeman 1993). This work has resulted in the development of numerical models for soil movement and structural response.

The numerical model for soil movements is capable of generating 3D soil suction profiles beneath a structure using real climatic records and representative soil properties. These soil suction profiles are then used to model ground movements in reactive soils and a combined soil/structure model is used to study the structural response to these movements. By incorporating the simulated behaviour into a suitable structural reliability framework, various probabilistic aspects of cracking in structures can be studied.

This paper details the way in which the soil model is used to estimate the continuous record of ground movements that a structure is likely to experience during its design life from time of construction. The way this information is included in the probabilistic study for cracking is also outlined.

2 THEORETICAL BASIS AND ASSUMPTIONS FOR SOIL SUCTION MODEL

The governing equation for moisture movement in unsaturated soils on which the model is based as well as the way in which the model incorporates the processes of interception, infiltration and evapotranspiration are detailed elsewhere (Totoev and Kleeman 1998, Muniruzzaman and Totoev 1998).

It will assist in understanding the use of the soil suction model however, if the form of the model inputs and outputs are briefly discussed. The inputs to the model consist of:

- Record of daily precipitation depths and durations for a number of years.
- Monthly averages of pan evaporation, relative air humidity and air temperature.
- A set of soil parameters, the most influential of which is soil permeability. In its current form the model adopts the simplifying assumption that permeability remains constant with soil suction.

• An initial soil suction profile through depth. The 'At depth' value of soil suction (at the lower end of the profile) is held constant for the entire run at a value approximating the middle of the surface suction range.

The model steps through time to solve for daily changes in soil suction at the ground surface which are then used as boundary conditions for the finite element computation of the soil suction profile through depth at each day.

The output therefore consists of 3D distributions of soil suction horizontally in two directions and through depth. It is assumed that soil volume changes are proportional to changes in soil suction. If the change in soil suction from start of run to any given time is integrated over depth at some point in plan then a measure of the ground movement relative to time zero is obtained at the point.

For a 1D simulation a continuous measure of open ground movement versus time can be plotted over the complete duration of the available rainfall record.

For a complete 3D analysis with simulated climatic shielding by a structure, measures of differential ground movements under a footing can be computed against time. This allows the dates to be identified on which peak annual doming or dishing movements occurred beneath the footing. The associated profiles of changes in soil suction over depth and along the footing can therefore be estimated.

3 MODELLING ONE DIMENSIONAL SOIL COLUMN - CALIBRATION

The movement of soil moisture through depth can be studied in 1D using a column of finite elements.

The estimate for the initial soil suction profile at time zero will inevitably differ from the actual profile which would have resulted from the real rainfall record leading up to time zero. The result is that plots of ground movement versus time display an initial shift and establish a mean position away from zero ground movement. The simulation time taken for this initial shift to stabilise depends on soil permeability and is typically in the range 3-5 years. The magnitude of the observed

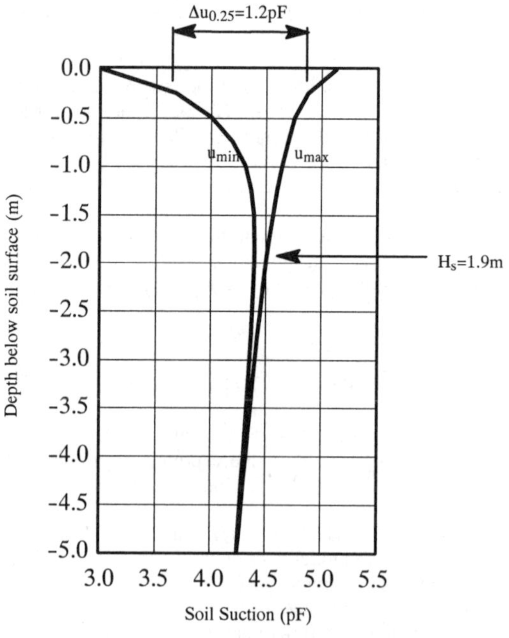

Figure 1. Range in soil suction with depth for simulation period

shift can be significantly reduced by re-running the model for the same rainfall record but using an initial profile from the same date, but in a later year. This profile will be a better estimate for the real profile at the time of year of run commencement. In any case, results for changes in soil suction and resulting measures of ground movement must be computed relative to a date chosen after initial effects have stabilised.

If the 1D soil suction profiles over depth are obtained, say monthly, for all years after the chosen stabilised date, plots of the minimum (u_{min}) and maximum (u_{max}) soil suction against depth for this simulation period can be obtained (Figure 1). Directly from these follow estimates of surface suction range $\Delta u_{0.25}$ and active depth of moisture variation H_s. The surface suction range is taken at depth Z=0.25m to ensure better representation of the state of the soil and the active depth is taken equal to depth at which $u_{max} - u_{min} = 0.1 \times \Delta u_{0.25}$. The model can then be calibrated by adjusting the soil parameters until the $\Delta u_{0.25}$ and H_s values are consistent with those observed experimentally for the region of interest.

Using calibrated soil properties and climatic data appropriate to Melbourne for a 25 year period starting 1/1/1973 the measure of ground movement versus time is shown in Figure 2. Note that the vertical axis displays values of integrated soil suction profile through depth. The actual ground movement will depend on soil reactivity. For a highly reactive soil, a multiplier of around 80 would yield typical values in millimetres.

The simulated plot in Figure 2 clearly displays a seasonal variation, random in nature. Field measurements for surface movement versus time also typically display a seasonal variation as shown in Figure 3 for Marylands near Newcastle. However, one noticeable feature of the field data is that a state of maximum swell is reached during wet periods. The 'upper limit' on ground movement is due to saturation of the soil profile. The absence of this upper ceiling on swell movements for the Melbourne simulation is thought to be due to the simulated profile wetting up too slowly in response to rain events. It is suspected that this results from the low value of soil permeability, assumed constant, required to achieve the correct calibrated H_s value. In practice soil permeability has been found to increase with increased moisture content (Wray 1998).

Current attempts to achieve more rapid wetting of the soil profile include the injection of moisture at upper level subsurface nodes to represent the infiltration of surface runoff via shrinkage cracks or alternatively, the use of varying soil permeability with depth to represent the more porous nature of 'near surface' soil due to the presence of shrinkage cracks and plant root paths.

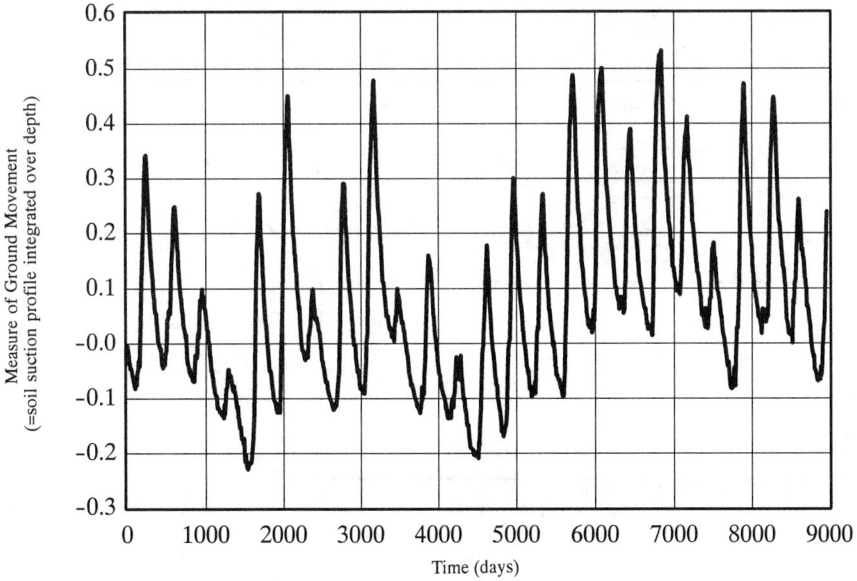

Figure 2. Measure of ground movement versus time for Melbourne for period commencing 1/1/1973.

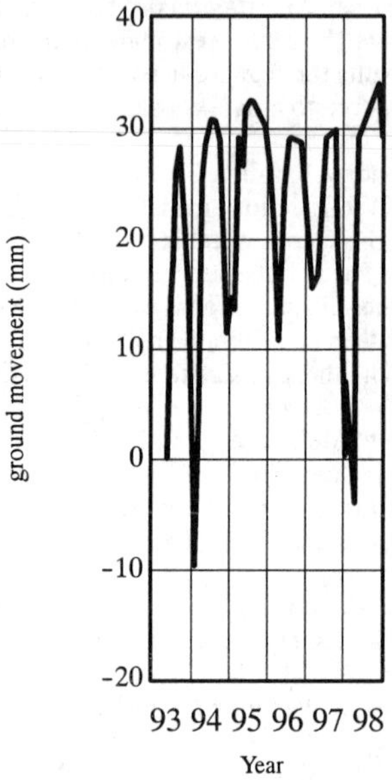

Figure 3. Measured ground movement versus time for Marylands (Allman et al. 1998)

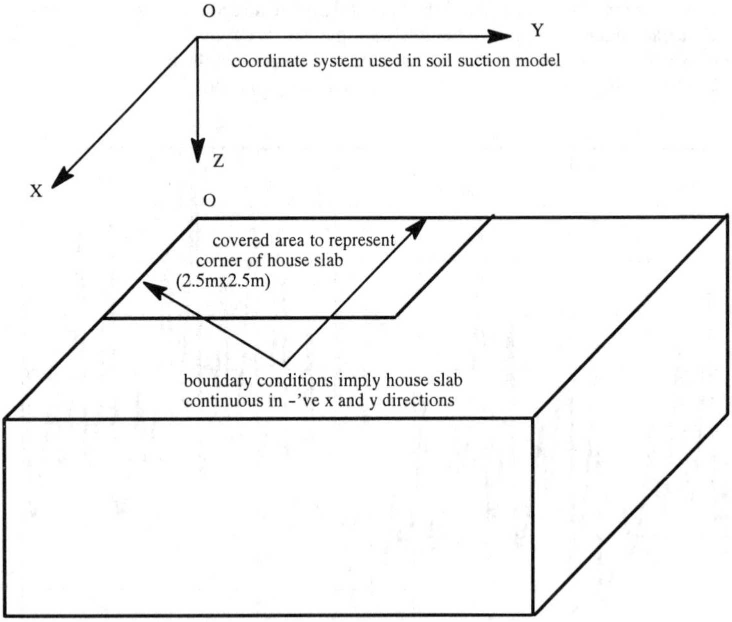

Figure 4. Block of soil simulated using 3D finite element mesh.

4 3D MOUND SHAPE MODELLING

The moisture movements and resulting soil suction changes with time in three directions beneath a structure can be modelled using a 3D finite element mesh. The rainfall record is specified at all ground surface points except for a 2.5m x 2.5m square region at the corner of the mesh in plan (Figure 4). This simulates the climatic shielding provided by the corner of a structure (this being where critical differential footing movements occur). The distance of 2.5m has been found to be sufficient for all corner and edge effects to have dissipated.

A systematic approach for the 3D analyses has been developed, which allows statistical information regarding ground movements to be obtained. The steps are as follows:

1. The soil properties are calibrated using a 1D soil model as described previously.
2. The 1D model is run with a real rainfall record for sufficient time for initial effects to stabilise.
3. Continuing to run the 1D model past the stabilisation date, the soil suction profile is recorded at the start of each month for the next two years. This provides 24 different soil suction profiles which are used as construction date profiles for the 3D runs.
4. The 3D model is run from each of the 24 construction dates to the end of the rainfall record using the appropriate construction date profile as the initial soil suction profile over depth at all points in plan across the 3D mesh. The construction date represents a point in time when the impermeable ground cover provided by a structure is placed and it is assumed that the constructed footing is straight and level at this time. During each run, differential movements relative to the construction date under the footing are computed at weekly intervals. The annual maximum doming and dishing movements and the associated profiles of change in soil suction from construction date are recorded.

The need to consider a range of construction dates arises from the observation that the annual peak ground movements relative to the levels at the construction dates, although occurring at similar times, will vary in magnitude for different construction dates due to the differing initial suction profiles. For example, a very dry initial profile will result in more subsequent swell than shrink movements since the soil profile will be wetter on average than the initial condition.

The effect of introducing a ground cover at the construction date is a transient one. There is a period of 3-5 years required to establish the long term dynamic equilibrium beneath the structure. After this period of moisture redistribution the 3D profiles of soil suction will be the same for all 24 runs but the profiles of suction change will differ due to different initial profiles.

The date at which all profiles become approximately equal can be taken as a reference date. An alternative approach for the 3D analyses (which will allow a reduction in computational effort) is to run each of the 24 3D runs up to this reference date and then continue only one of the runs past the reference date to the end of the rainfall record.

5 RELIABILITY CONCEPTS

A separate numerical model which simulates soil/structure interaction has already been developed (Masia & Kleeman 1998). This takes as input from the soil suction model the profiles of change in soil suction beneath the footing relative to the construction date. The changes in soil suction are converted to equivalent soil strains to simulate ground movements and the associated structural response, including wall cracking locations and widths. At this stage soil strains are assumed proportional to suction changes.

From the 3D analyses using the soil suction model profiles of changes in soil suction can be obtained beneath a footing at regular intervals (monthly, for example) for each construction date. The structural response model can then be used to determine firstly, if masonry cracking is likely to occur, and if so, what the maximum crack width is, anywhere in the wall, for each monthly profile. The profiles associated with peak annual doming and dishing movements can be used to compute critical crack widths between the regular monthly values.

The result are records of wall crack width versus time for each construction date. These can be used to determine the largest crack width likely during the duration of available rainfall records. The outcrossing rate for a given crack size can also be obtained as well as an estimate of the cumulative time which the structure spends in an unserviceable state.

6 CONCLUSIONS

The soil suction model can be calibrated to any site provided appropriate meteorological data can be assumed. The structural response model has the flexibility to study a variety of wall geometries and material properties.

The approach presented is therefore a useful tool for evaluating the serviceability performance of a range of structural systems for domestic masonry construction.

7 REFERENCES

Allman, M.A., M.D. Delany & D.W. Smith 1998. A field study of seasonal ground movements in expansive soils. *Proceedings, Second International Conference on Unsaturated Soils, Beijing, China, 27-30 August, 1998.* Vol 1, 309-314.

Masia, M.J. & P.W. Kleeman 1998. A numerical model for cracking in masonry for use in reliability modelling. *Proceedings, 8th Canadian Masonry Symposium, Jasper, Alberta, 31 May - 3 June, 1998.* 84-95.

Masia, M.J., P.W. Kleeman & A.W. Page 1998. Serviceability design for masonry structures subjected to foundation movements. *Proceedings, 5th Australasian Masonry Conference, Gladstone, Queensland, Australia, 1-3 July, 1998.* 255-264.

Melchers, R.E. 1996. Probabilistic crack prediction for masonry structures. *Congress Report, 15th Congress, International Association for Bridge & Structural Engineering IABSE, Copenhagen, 16-20 June, 1996.* 621-626.

Muniruzzaman, A.R.M. & Y.Z. Totoev 1998. Generation of 3D mound shapes for the analytical study of masonry on reactive soil. *Australian Civil / Structural Engineering Transactions, I.E. Aust.* (to appear).

Page, A.W. & P.W. Kleeman 1993. Development of serviceability criteria for masonry structures. *Proceedings, 13th Australasian Conference on the Mechanics of Structures and Materials, University of Wollongong, Australia, 1993.* 657-664.

Totoev, Y.Z. & P.W. Kleeman 1998. An infiltration model to predict suction changes in the soil profile. *Water Resources Research,* Vol. 34, No. 7, July 1998. 1617-1622.

Wray, W.K. 1998. Mass transfer in unsaturated soils: a review of theory and practices. *Proceedings, 2nd International Conference on Unsaturated Soil, Beijing, China,* (to appear in Volume 2).

Mechanics of Structures and Materials, Bradford, Bridge & Foster (eds)
© 1999 Balkema, Rotterdam, ISBN 90 5809 107 4

Design of shallow foundations in sands: State-of-the-art

N. Sivakugan & A. E. Burke
Civil and Environmental Engineering, James Cook University, Townsville, Qld, Australia

ABSTRACT: Designs of shallow foundations in granular soils are generally governed by settlement considerations than the bearing capacity ones. The current methods for settlement predictions significantly overestimate the settlements, thus underestimating the allowable design pressures. A new method, using artificial neural networks, appears to give better predictions than the traditional methods. A very versatile software *Footset*, written in Visual Basic, simplifies the design process by incorporating five different settlement prediction methods.

1 DESIGN CRITERIA

Shallow foundations are defined as having breadth greater than depth. These include pad, strip footings and raft foundations. In weak ground conditions and when the column loads are high, shallow foundations may become inadequate other options such as deep foundations will be considered.

Shallow foundations in sands are designed to satisfy bearing capacity and settlement criteria. While ensuring adequate safety against possible bearing capacity failure within the surrounding soil, it is necessary to limit the future settlement. Safety with respect to bearing capacity failure is achieved by providing a safety factor (F) of about three. Limiting the total settlements limits the differential settlements, which in turn limits the structural or architectural damage to the building. The current practice is to limit the total settlements to 25 mm.

A graphical representation of the design criteria, is shown in Fig. 1. Here, the straight line segment OX is developed using bearing capacity considerations (i.e., F = 3) and the XY segment is developed using settlement considerations (i.e., settlement = 25 mm). In granular soils, the point X often lies within the foundation width of 1.5 m. Therefore, settlement considerations govern the design of shallow foundations in granular soils.

In the early days, the design charts proposed by Peck et al. (1974) were quite popular among the practicing engineers. It is considered that the charts, based on bearing capacity and settlement considerations, are very conservative (Burke and Sivakugan 1999).

2 SETTLEMENT PREDICTIONS

Settlement of a shallow foundation depends on the pressure applied to the ground, foundation dimensions, and the soil stiffness. The soil stiffness is indirectly measured by the penetration resistance from in situ tests such as standard penetration test or cone penetration test.

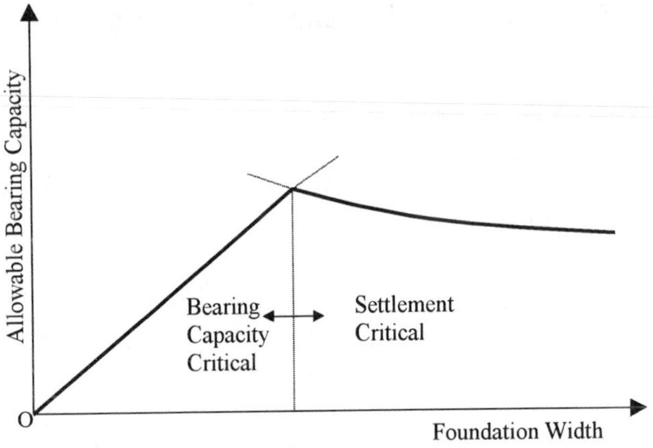

Figure 1. Bearing capacity and settlement criteria governing the design

2.1 *Settlement prediction methods*

More than forty different methods have proposed to predict settlements of shallow foundations in sands (Douglas 1986). Several researchers have compared the predicted and the observed settlements and found the current state-of-the-art to be far from adequate (Jeyapalan & Boehm 1986, Tan & Duncan 1991, Sivakugan et al. 1998).

Terzaghi & Peck (1967) proposed the following empirical relationship relating the settlement of a square footing to that of a 300 mm square plate.

$$settlement_{footing} = settlement_{plate}\left(\frac{2B}{B+0.3}\right)^2\left(1-\frac{1}{4}\frac{D}{B}\right)$$

Here, B and D are the width and depth of the foundation respectively. The settlement of the plate, corresponding to a blow count (N) and applied pressure can be obtained from their pressure-settlement curves. Sivakugan et al. (1998) reported that this method overestimates the settlements by more than 200%. Although the method is backed by the plate load test data, it lacks theoretical basis.

Schmertmann (1970) proposed a more rational method based on theory of elasticity, and suggested the following expression for the settlement.

$$settlement = C_1C_2\ q\sum\frac{I_z dz}{E_z}$$

Here, q is the net applied pressure, E_z is the Young's modulus of the sand at depth z below the footing, and I_z is the influence factor proposed by Schmertmann. C_1 and C_2 are correction factors to account for the stress relief and creep. Sivakugan et al. (1998) reported that this method overestimates the settlements by more than 300%.

Burland & Burbidge (1985) carried out a statistical analysis of over 200 settlement records, and presented an empirical method, incorporating correction factors for the foundation shape, presence of stiff strata at shallow depth, and creep. Application of this method requires the knowledge of whether the sand is normally consolidated or overconsolidated. In practice, it is very difficult to determine the preconsolidation pressure of a sand deposit, and to check whether it is overconsolidated.

In all the above methods, the soil stiffness is lumped into the penetration resistance (blow count or cone resistance) from the in situ tests. In reality, soils being elasto-plastic materials, the soil stiffness depends on the strain level, which is not considered in the above methods. Berardi & Lancellotta (1991) developed an iterative procedure to estimate the settlement, taking the strain level into account. This method will not be discussed within this paper.

2.2 *Accuracy and reliability of the methods*

Accuracy and reliability are the two desired features of any settlement prediction method. Tan & Duncan (1991) defined accuracy as the ratio of predicted to actual settlement, and reliability as the probability that the predicted settlement is greater than the actual settlement. In other words, accuracy is a measure of how close the prediction is to the actual settlement; and reliability is the degree of conservatism. The most desired value for accuracy is 1, and the most desired value for the reliability is 100%.

The accuracy and reliability values for several settlement prediction methods are summarised in Fig.2. The methods having high reliability have very poor accuracy, and the methods having good accuracy have very low reliability. The geotechnical engineer must strike the right balance in selecting a method for settlement computations.

2.3 *Settlement prediction session – Settlement '94*

The American Society of Civil Engineers, Federal Highway Administration (US) and A&M University at Texas jointly organized a prediction symposium to assess the state-of-the-art for predicting settlements of shallow foundations in sands (Briaud & Gibbens 1994). An extensive site investigation, involving nine different in situ and laboratory tests, was carried out in a 12 m x 28 m granular site. Five isolated square pad footings, widths varying between one and three meters were cast at a depth of 0.75 m.

Thirty one participants from different countries were presented with all the above data and

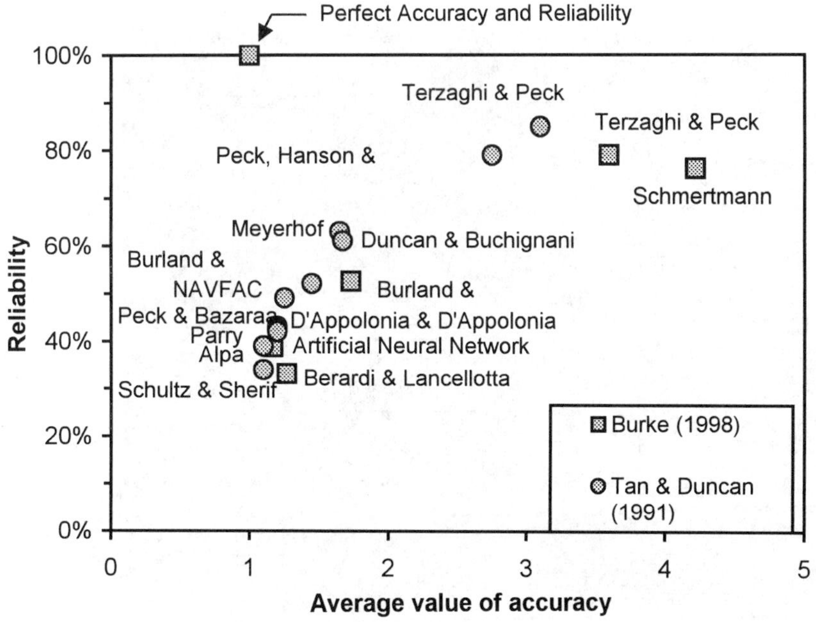

Figure 2. Accuracy and reliability of settlement prediction methods

were requested to make "class A" predictions of the loads that would produce 25 mm settlement. 22 different methods were used, Schmertmann's (1970) and Burland & Burbidge's (1985) methods being the most popular. In spite of having access to such an extensive site investigation data, the predictions were overall poor. The loads were significantly underestimated, as the consequence of most methods overestimating the settlements.

It is widely believed that in granular soils, the foundation designs are governed by the settlement considerations than the bearing capacity ones (Berardi & Lancellotta 1994, Jeyapalan & Boehm 1986, Tan & Duncan 1991). However, for each of the five footings loaded to failure, the allowable load based on bearing capacity considerations (i.e., failure load/3) was less than the load producing 25 mm settlement. In other words, bearing capacity governs the design than the settlements, for all the above footings. Therefore, the point X in Fig. 1 can lie at foundation width greater than 3 m. This implies an upward shift in the XY segment in Fig. 1, thus pointing to the underestimation of settlements by the current techniques.

3 NEW TECHNOLOGIES

The authors and co-researchers at JCU have used artificial neural networks (ANN) as a potential tool in settlement predictions and the results are promising. The predictions appear to be far better than those from the traditional methods (Arnold & Sivakugan 1997, Sivakugan et al. 1998).

3.1 *Artificial neural network applications*

Artificial neural networks had been used quite successfully in several disciplines, including geotechnical engineering. In problems where the analytical models do not work well, ANN appears to be a very good alternative, provided there is enough data to train the model.

ANN developments require special software, The authors to develop the settlement

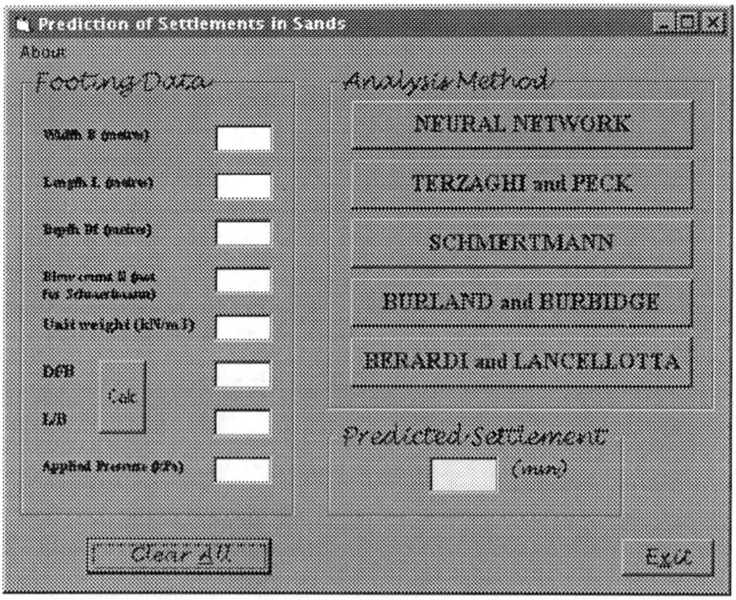

Figure 3. Main program screen of *Footset*

prediction software used *Predict* (Neuralware Inc., USA) and *NeuroShell 2* (Ward Systems, USA). Settlement records of more than 100 prototype footings were used to "train" the neural network, the input values for each pattern being B, B/L, D/B, N and q. The output was the settlement. Once trained with a large database, the model worked very well on unseen data.

3.2 *Visual Basic software Footset*

Footset is a software program developed at James Cook University to simplify the design process for shallow foundations in granular soils. Written in Visual Basic the program allows the foundation engineer to predict the settlement of a foundation under specific loading conditions. Requiring several easily attainable inputs, the settlement can be easily calculated using any one of the settlement prediction methods discussed in this paper. The main program screen of *Footset* is seen in Fig. 3.

4 SUMMARY AND CONCLUSION

With the design of shallow foundations in sands governed by settlement considerations, and the current state of the art in settlement prediction methods lacking in accuracy, a new settlement prediction method was developed using Artificial Neural Network technology. The new method returns accurate predictions and has been incorporated into the software Footset, allowing the foundation designer a more rapid calculation of settlements using several different prediction methods.

5 ACKNOWLEDGEMENTS

The Authors acknowledge the financial support provided by Australian Research Council, DEETYA and James Cook University.

6 REFERENCES

Arnold, M. and Sivakugan, N. (1997). "Artificial neural networks applied to the prediction of settlements in sands," Proceedings of the Second International Symposium on Structures and Foundations in Civil Engineering, Hong Kong, 118-123.

Berardi, R. & Lancellotta, R. (1991). "Stiffness of granular soils from field performance," Geotechnique, 41(1), 149-157.

Berardi, R. & Lancellotta, R. (1994). "Prediction of settlements of footings on sands: accuracy and reliability," Vertical and Horizontal Deformations of Foundations and Embankments, ASCE, 1, 640-651.

Briaud, J-L. & Gibbens,R.M. (1994). Predicted and measured behaviour of five spread footings on sand, Geotechnical Special Publication 41, ASCE.

Burke, A. & Sivakugan, N. (1999). "Design charts for shallow foundations in sands", Proc. 16th ACMSM, Sydney.

Burland, J.B. & Burbidge, M.C. (1985). "Settlement of foundations on sand and gravel," Proc. Institution of Civil Engineers, 1(78), 1325-1381.

Douglas, D. 1986. State-of-the-art, *Ground Engineering*, 9(2), 2-6.

Jeyapalan, J. K. & Boehm, R. (1986). "Procedures for predicting settlements in sands," *Settlements of shallow foundations on cohesionless soils: design and performance*, ASCE, 1-22.

Schmertmann, J.H. (1970). "Static cone to compute static settlement over sand," *Journal of Soil mechanics and Foundation Division*, ASCE, 96(3), 1011-1043.

Schmertmann, J.H., Hartmann, J.P. & Brown, P.R. (1978). "Improved strain influence factor diagrams," *Journal of the Geotechnical Engineering Division*, ASCE, 104(8), 1131-1135.

Sivakugan, N., Eckersley, J.D. & Li, H. (1998). "Settlement predictions using neural networks," Australian Civil Engineering Transactions, CE40.

Tan, C.K. & Duncan, J.M. (1991). "Settlement of footing on sands – accuracy and reliability," *Geotechnical Engineering Congress 1991*, Eds. F.G.McLean, D.A. Campbell & D.W. Harris, ASCE, 446-455.

Mechanics of Structures and Materials, Bradford, Bridge & Foster (eds)
© 1999 Balkema, Rotterdam, ISBN 90 5809 107 4

Design charts for shallow foundations in sands

A.E.Burke & N.Sivakugan
Civil and Environmental Engineering, James Cook University, Townsville, Qld, Australia

ABSTRACT: Design charts are a valuable tool in the design process of a shallow foundation in sands. These charts incorporate bearing capacity and settlement considerations, the two essential components in the performance of a shallow foundation. The first, and still most widely used, design charts were developed by Peck et al. (1974) using bearing capacity equations and settlement prediction methods developed by Terzaghi. Studies have been performed on the accuracy of the prediction method developed by Terzaghi showing the method has a strong tendency to over-predict the settlement and hence under-predict the allowable load a foundation can accommodate. With this in mind the design charts have been updated using a settlement prediction method based on Artificial Neural Network technology, developed by Sivakugan & Arnold (1998). The settlement prediction used has been shown to have high accuracy in prediction of settlements. Results indicate an increase of approximately 350% in the allowable bearing capacity by substituting the ANN based settlement prediction method.

1 INTRODUCTION

Today's common practice in the design of shallow foundations in granular soil is to use a design chart (eg. Peck et al.) relating the SPT blow count of the soil, the foundation size and the allowable soil pressure. This is a tried and tested method used extensively through out

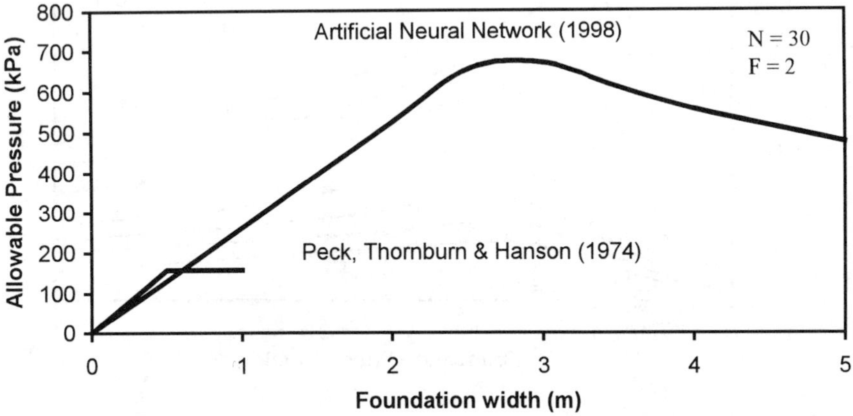

Figure 1. Comparison of ANN based design chart to Peck et al.'s

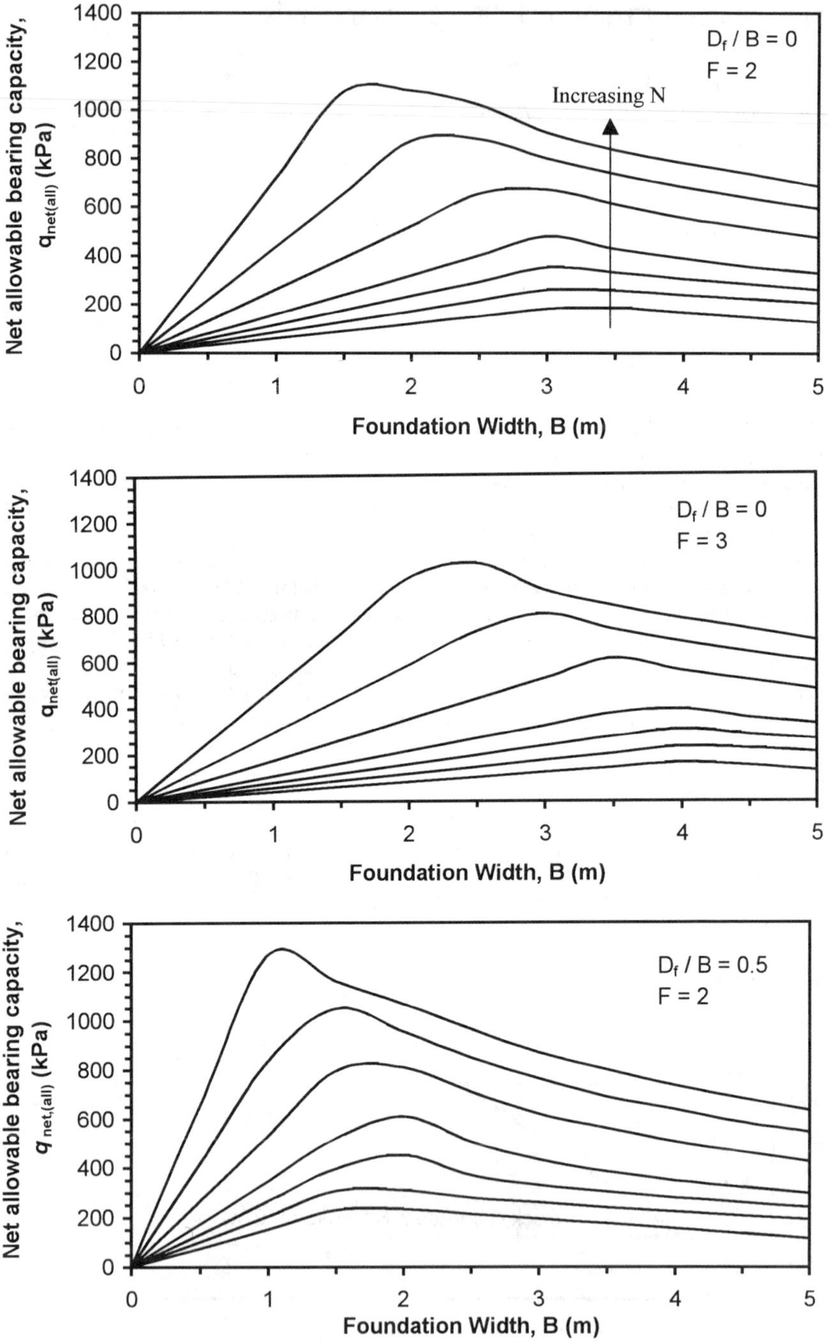

Figure 2. Design charts developed for different D_f/B ratios and Factors of safety (F)

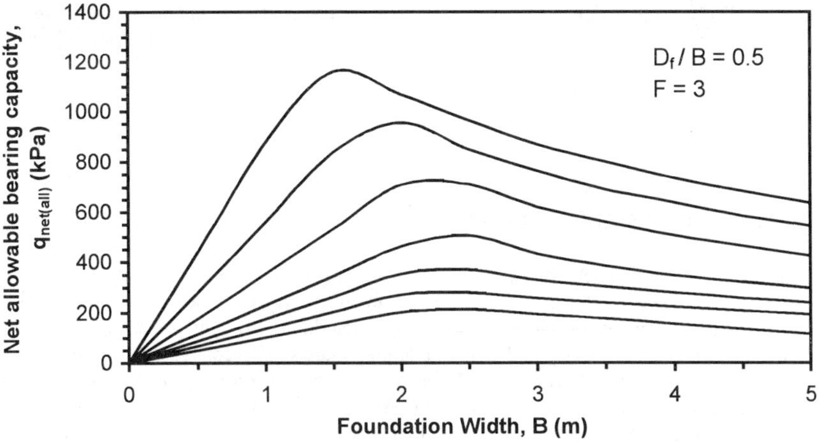
D_f / B = 0.5
F = 3
Net allowable bearing capacity, $q_{net(all)}$ (kPa)
Foundation Width, B (m)
0
200
400
600
800
1000
1200
1400
0
1
2
3
4
5

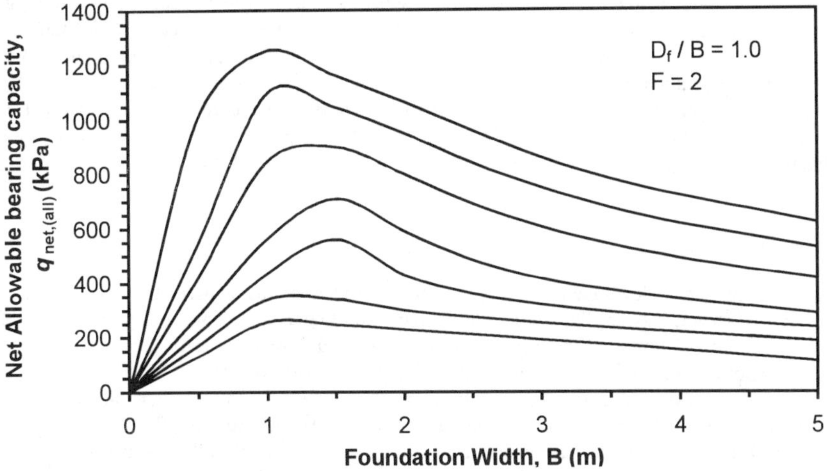
D_f / B = 1.0
F = 2
Net Allowable bearing capacity, $q_{net,(all)}$ (kPa)
Foundation Width, B (m)
0
200
400
600
800
1000
1200
1400
0
1
2
3
4
5

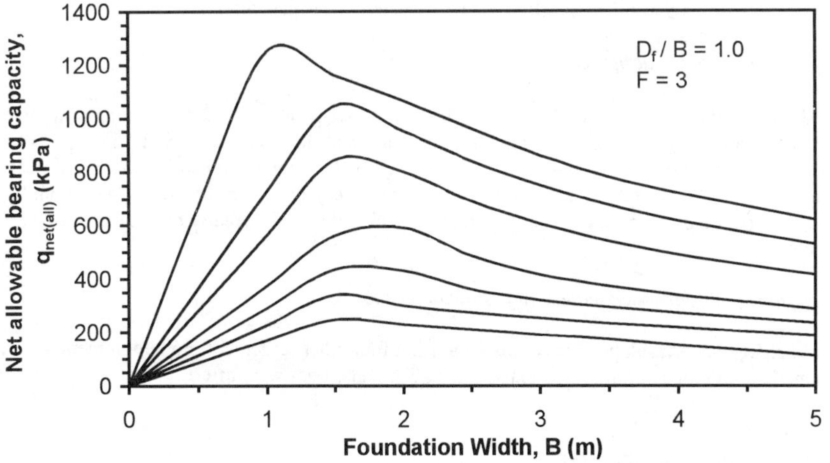
D_f / B = 1.0
F = 3
Net allowable bearing capacity, $q_{net(all)}$ (kPa)
Foundation Width, B (m)
0
200
400
600
800
1000
1200
1400
0
1
2
3
4
5

engineering. With the prevalence of the use of these design charts and taking into account the studies that have been performed on the accuracy of settlement prediction methods (Tan & Duncan 1991, Burke 1998) an investigation into the currently used design charts was undertaken. Based on the results of this examination, new design charts were developed using a more accurate settlement prediction method (Sivakugan & Arnold 1998) and Meyerhof's bearing capacity equation.

2 ACCURACY OF DESIGN CHARTS (Peck et al. 1967)

The settlement prediction method used in the generation of the design charts by Peck et al. (developed by Terzaghi & Peck) has been shown to over-predict the settlement of a loaded shallow foundation on sand by an average of 360% (Burke 1998). This leads to the under-estimation of the allowable load and hence a larger, more costly foundation is required to support a given load. These chart err on the conservative, which is an acceptable solution to a problem that is still not fully understood. With the Artificial Neural Network prediction method (Sivakugan & Arnold 1998) demonstrating a high level of accuracy in recent studies, a new chart was developed using this method to compare the allowable bearing capacities of a footing, founded in the same soil using the two different methods. The results can be seen in Figure 1. Further, Peck et al chart was developed only for footings smaller than 1.5m. In the charts developed for this paper include a more realistic and wider range of footing widths.

There is a clear increase in allowable pressure for a foundation so as to limit its settlement to less than 25mm. This leads to a reduction in the foundation size to transmit a specific load, thus reducing the cost of the foundation necessary.

3 DESIGN CHARTS

Design charts were developed using the ANN based settlement prediction method and Meyerhof's bearing capacity equation. These charts were developed using different factors of safety (F) in the calculation of the bearing capacity and for different values of the depth to width ratio (D_f/B) of the foundation. The results can be seen in the following figure2.

In figure 2, each envelope represents the maximum value of bearing pressure that can be applied to a soil of a given SPT blow count, increasing form 5 to 10, 15, 20, 30, 40 then to 50 in an upward direction, to limit the settlement to 25mm and to give a safety factor against bearing capacity failure of at least 3. This 25mm limit on the maximum allowable settlement of the foundation, ensures that the total differential settlement to less than 20mm.

3.1 *Verification of the design charts*

The design charts were verified by plotting load test data points obtained from literature onto the appropriate design chart, only the data points that had the exact same SPT blow count as those used in deriving the charts were used. Some 48 data points were plotted on the chart having $D_f/B = 0$ and $F = 2$. Visual inspection of these plots indicated only 3 of the 48 points lay on the incorrect side of the envelope, above the line if settlement is greater than 25mm or below if less.

3.2 *Design charts for other settlement prediction methods*

Design charts were developed for comparison using four other settlement prediction methods, in conjunction with the two aforementioned charts. These methods were after:

- Terzaghi & Peck (1948)
- Schmertmann et al (1978)

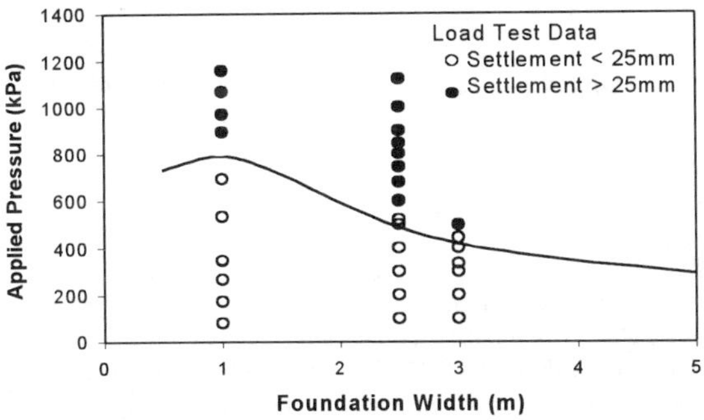

Figure 3. Verification of Design Charts

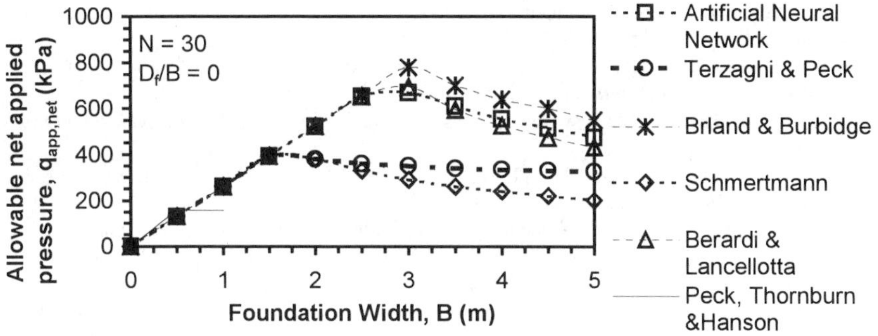

Figure 4. Comparison of design charts using different methods

- Burland & Burbidge (1986)
- Berardi & Lancellotta (1991)

Figure 3 was generated using these methods for N=30 and $D_f/B = 0$.
The above figure demonstrates the variability of the settlement prediction methods used in current practice. With the ANN method being shown to be the statistically most accurate prediction method (Burke 1998) it was chosen to generate the design charts to give the best approximation of the actual settlement under loading.

4 SUMMARY AND CONCLUSIONS

Design charts assist greatly in the process of design. With the sometimes complex and lengthy calculations need to predict the settlement of a shallow foundation in sand, these charts lend themselves well to the task. Charts expedite the design process giving quick and reasonably accurate predictions of the settlement expected under the specified loading conditions. A set of design charts were developed using the ANN settlement prediction method to replace those developed by Peck et al (1967). These new charts allow for more accurate estimations of settlement, foundation width or allowable bearing capacity for a foundation on a sand with a known SPT blow count.

From the new charts it can be inferred that the settlement consideration governs the design of shallow foundations when the foundation width exceeds 1.5 m and not approximately 0.5 m as previously proposed by Peck et al.

5 ACKNOWLEDGEMENTS

The Authors acknowledge the financial support provided by ARC, DEETYA and James Cook University.

6 REFERENCES

Arnold M.A. (1998) "*Settlements of shallow foundations in granular soils*" Masters of Engineering Science Thesis, James Cook University, Submitted for publication.

Berardi R, Jamiolkowski, M and Lancellotta, R (1991) "Settlement of shallow foundations in sand, selection of stiffness on the basis of penetration resistance" Geotechnical Engineering Congress 1991, *Geotechnical special publication No. 27*, ASCE, New York, pp 185 – 200

Burke A (1998) "*Settlement of Shallow Foundations on granular soils*" Bachelors of Engineering Honours Thesis, James Cook University

Burke A, Johnson K and Sivakugan N (1999) " A probabilistic approach to settlement predictions"

Burland, J.B. and Burbidge, M.C. (1985) "Settlement of Foundations on Sand and Gravel" *Proceedings of the Institution of Civil Engineering Part 1* (78) pp 1325-1381.

Meyerhof, G.G. (1965) "Shallow foundations" *Journal of soil mechanics and foundations division, ASCE*, New York

Peck, R.B., Hanson, W.E., and Thornburn, T.H. (1974), "*Foundation engineering*", Second edition, John Wiley & Sons.

Schmertmann, J.H. (1970) "Static Cone to Compute Static Settlement over Sand" *Journal of the Soil Mechanics and Foundations Division*, ASCE, Vol. 96, No. SM3., pp 1011-1043

Schmertmann, J.H., Hartman, J.P. and Brown, P.R. (1978) "Improved Strain Influence Factor Diagrams" *Journal of the Geotechnical Engineering Division,* ASCE, Vol. 104, No. 8, pp 1131-1135

Sivakugan, N. and Arnold, M. (1997) "Recent Advances in Settlement Predictions of Shallow Foundations in Sands" *Proceedings of the Second International Symposium on Structures and Foundations in Civil Engineering* Hong Kong University of Science and Technology Hong Kong

Tan, C.K. and Duncan, J.M. (1991) "Settlement of Footings on Sands – Accuracy and Reliability" *Geotechnical Engineering Congress 1991* ASCE Geotechnical Special Publications No. 27 New York USA

Terzaghi, K., Peck, R. and Mesri, G. (1996) "*Soil Mechanics in Engineering Practice*" John Wiley and Sons Inc. New York USA

Mechanics of Structures and Materials, Bradford, Bridge & Foster (eds)
© 1999 Balkema, Rotterdam, ISBN 90 5809 107 4

Prediction of deflections of Burrinjuck Dam using finite element analysis

A.K.Patnaik & A.Khan
School of Civil Engineering, Curtin University of Technology, Perth., W.A., Australia

ABSTRACT: Burrinjuck Dam is a concrete gravity dam located in a narrow gorge on the Murrumbidgee River in NSW. The maximum height of the dam is 93 m above the lowest point of the foundation, and the maximum base width is 61 m. The crest length of the non-overflow section between the sector gates is 152 m and the total length of the dam is 233 m. Plummets and targets were used in the dam to monitor the vertical and horizontal movement of the dam at a number of locations during and after extensive upgrading works. The data collection was commenced before the upgrading started and almost 8 years of data were available. This paper outlines the deflections measured in the dam due to temperature variations and comparison of the measured thermal deflections with those predicted from a finite element analysis of the dam.

1. INTRODUCTION

Burrinjuck Dam is a concrete gravity dam located in a narrow gorge on the Murrumbidgee river in NSW, downstream of Canberra and upstream of Gundagai. It is owned by the NSW Department of Land and Water Conservation, and is operated, monitored and maintained by State Water, the Department's bulk water delivery business. Upgrading work on the dam, to allow it to pass a much larger design flood, was commenced in February 1990 and was completed in August 1994. The upgrading works included raising of the existing dam crest by 12.2 m using reinforced concrete, installation of 128 post-tensioned cables each of 63 – 15.2 mm diameter strands from the top of the dam through the height of the wall into the foundation of the dam, raising and stabilising by post-tensioning of the upstream sections of the existing spillway training walls, and construction and/or modification of other related structures. Plummets and targets were used in the dam to monitor the vertical and horizontal movements of the dam during and after the entire upgrading works. The deflections of the dam were recorded at a number of stations. The data collection was commenced in the 1990's before the upgrading started and has continued ever since. This paper outlines the instrumentation used and discusses the data recorded for the dam. The measured deflections are compared with those predicted from a finite element analysis of the dam.

2. HISTORY

The complete details of the history and the background of the dam structure, and construction can be found from the references cited in this paper. The main features of the dam are as follows.

2.1 Original Dam

Construction of the original dam commenced in 1907 and supply of irrigation water commenced in 1912. After delays due to World War I, and changes in spillway design (required in the light of floods experienced during the construction period, notably in 1925), the construction of the original dam was completed in 1928.

The original dam wall was built in blocks of cruciform shape, each a maximum of 9 m long and 11 m wide. On completion, the original dam was 75.3 m high with the crest at 365.33 m A.H.D up to the top of the parapet wall. The wall up to 329.76 m A.H.D consists of cyclopean concrete containing large granite "plums". Each plum weighed 4 to 5 tonnes. The plums occupied approximately 28% of the volume in the dam concrete (Report 1963). The original Burrinjuck Dam featured two side channel spillways on the abutments, central low level outlet works near the valley floor and a hydro-power station some 600 m downstream.

Further strengthening and enlargement of the original dam were done in the early 1950s and in the 1990s. These modifications included:

- raising of the fixed crest of the spillway by about 1.5 m to A.H.D 361.66 m.
- removing the ends of the main wall and fitting 2 x 15.2 m (northern end) and 1 x 24.4 m (southern end) sector gates (each gate is 4.6 m high).
- providing a curtain of cement grout in the dam and foundations about 3 m from the upstream face of the wall to reduce the flow of seepage water.
- drilling of 75 mm diameter vertical drainage holes to the newly constructed galleries; these holes were downstream of the grout curtain, 5 m from the upstream face and spaced at about 6 m centres.
- thickening of the effective cross section of the wall by constructing concrete buttresses on the downstream face.
- increasing the height of the crest of the main wall by 16.2 m to A.H.D 382.7 m.
- installation of one hundred and twenty eight post-tensioned cables.

2.2 Main wall

The maximum height of the main wall is 93 m above the lowest section of the foundation, and the maximum base thickness is 61 m. The crest length of the non-overflow section (between sector gates) is 152 m; the total length of the dam including the spillways is 233 m. The dam wall is curved in plan with a radius of curvature of about 366 m.

2.3 Foundation

The dam is located within a large mass of granite known as the Burrinjuck Granite Complex. The foundation comprises predominantly massive coarse grained granite with medium to fine grained granite occurring locally under the dam wall and at the downstream ends of the spillway channels on each side of the main wall. The foundation rock is very strong with unconfined compressive strength ranging from 136 to 292 MPa.

A grout curtain was provided as part of the 1950's strengthening work, and comprises a three row curtain to a depth of 15 m. The average cement take-in in the downstream row was 0.18 L/cm, and 0.04 L/cm in the other rows.

3. INSTRUMENTATION

3.1 Seepage

Seepage is monitored in the two upper galleries, in the lower gallery, and at the lower gallery shaft sump. Seepage is not monitored in the sector gate galleries at the spillway drains or the dam face.

3.2 Uplift pressure

The uplift pressure monitoring system comprises 22 hydraulic tips, mostly located at the dam-foundation interface. One is located just below and between the high level penstocks.

3.3 *Deflections (Plummets)*

Until 1986 there was no deflection or deformation survey/monitoring of the dam. In 1986 and later there were various attempts to drill straight enough holes to hang plummets; most of these attempts were unsuccessful. In October 1986, after successfully drilling a vertical hole, reading of one of the plummets (P2) commenced. The plummet gave deflections at the left hand side of the South Gallery. Readings were taken on a plummet terminating at the right hand side of the lower gallery (P21) beginning in August 1989. Five inverted plummets were installed in July 1993, immediately prior to post-tensioning, measuring relative deflections between the anchor slab and the lower gallery.

4. MONITORING AND BEHAVIOUR

4.1 Concrete main wall deflection

At plummet P2, which is 50m long, seasonal deflections of 10mm were recorded. At plummet P21, which is 80m long, seasonal deflections of 15mm were recorded prior to post-tensioning.

4.2 Concrete main wall deflection after post-tensioning

The five new plummets were installed in holes specifically drilled for the purpose as a part of the upgrading contract, just prior to the beginning of the post-tensioning. It was estimated from the plummet readings that the post-tensioning operation caused the dam, at the anchor slab level, to move upstream by the following amounts:
Centre of dam = 17mm±4mm, and at the plummets closest to the abutment = 8 mm ±3mm

4.3 Deformation survey

The horizontal movements were monitored by targets installed at the anchor slab level on the upstream side of the dam, using a Wild T300 + D12002 levelling instrument. The initial survey was done in September 1992. The measured horizontal movements from the survey were very similar to those indicated by the plummets. The vertical settlement of the dam was monitored by precise levelling of 5 reference points that were installed at the crest level. The crest (midway) settled 6 mm during the post-tension stressing period. The total crest settlement up to September 1994 was 8mm. Most of this settlement was probably due to the stressing operation.

4.4 Movement of the dam with air temperature variation

The movements of the dam parellel and perpendicular to the longitudinal axis of the dam were measured with the help of plummets. A brief summary of the average measured movements is given in Table 1 (U/S stands for up-stream) for an estimated dam temperature.

Table 1 Average horizontal thermal movement of Burrinjuck dam

	Estimated temperature	Measured deflection, mm		Deflection from analysis, mm	
		Mid-way	Crest end	Mid-way	Crest end
Summer time	30°	21mm (U/S)	9mm (U/S)	19.3mm (U/S)	13.5mm (U/S)
Winter time	15°	16mm (U/S)	9mm (U/S)	12.2mm (U/S)	8.0mm (U/S)

5. FINITE ELEMENT ANALYSIS

5.1 Scope

The data of the measured deflections from the plummets is very valuable in understanding the dam behaviour. The main cause of deflections of the dam as given in Table 1 was isolated to be

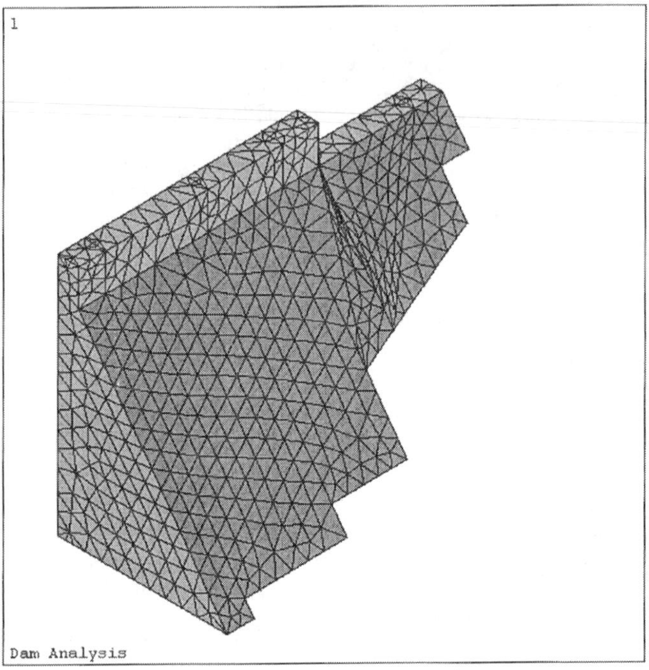

Figure 1 Finite element model of the dam used in the analysis

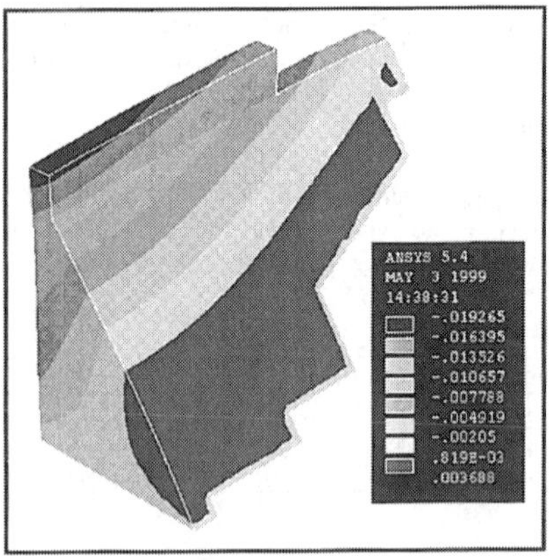

Figure 2 Contours of deflections for summer temperature

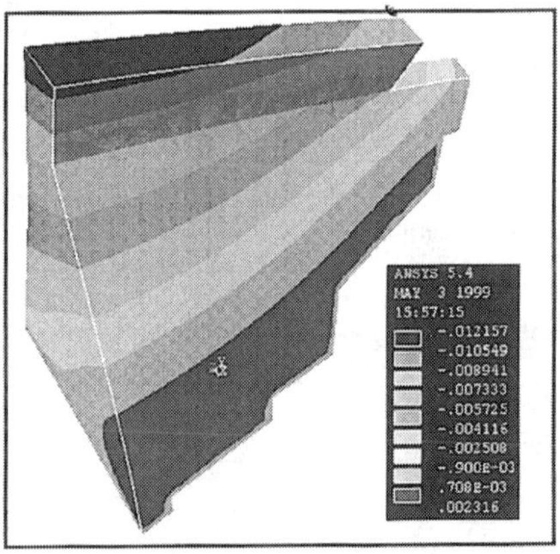

Figure 3 Contours of deflections for winter temperature

the temperature variations in the vicinity of the dam. Theoretical prediction of dam deflections by conventional theories of mechanics due to temperature variation is almost impossible. Therefore, a finite element model of Burrinjuck Dam was developed with the help of a commercial finite element program known as ANSYS (ANSYS5.4 1998). This analysis mainly concentrated on the effects of air temperatures on dam deflections (movements) in the direction lateral to the axis of the dam (see section 4.4). The analysis outlined in this paper is only preliminary in nature. More rigorous analysis including most of the practical and physical features of the dam is being conducted to develop a proper comparison of the measured deflections for other stages of construction, modification, and service detailed earlier with the predicted deflection by finite element analysis.

The scope of this study was to develop the deflections corresponding to the conditions given in Table 1. Two seasonal conditions were studied, one for typical summer temperature and one for typical winter temperature. For the summer condition, the foundation temperature was considered to be 15° at the deep end and 20° close to the surface. The surface of the dam in contact with water was considered to be at a temperature of 20°. The rest of the exposed surface of the dam was considered to be at 30°. A thermal analysis was performed with these surface temperatures. The temperatures at different nodes developed from the thermal analysis were transferred to the structural model and stress analysis performed for predicting the deflections. A summary of deflections developed from the finite element analysis is also shown in Table 1.

Analysis was repeated for typical winter temperature with the foundation temperature considered to be 20° at the deep end and 15° close to the surface. The dam surface in contact with water was considered to be at 8° and the rest of the surface of the dam was considered to be at 15°. A summary of deflections developed from analysis for this condition is shown in Table 1.

The behaviour of the dam depends on several factors. The correct data on these factors is unavailable at the moment. Therefore, the dam analysis presented in this paper was conducted using some simplifications.

5.2 Finite element model

The finite element model follows very closely the geometry of the actual foundation and surface profile of the dam. The finite element model of the dam is shown in Figure 1. Some simplifications of the geometry were made for this preliminary analysis. The dam was considered symmetric at approximately mid way along the length of the dam. The curvature of dam was not included in this finite element model. The buttresses provided on the downstream face of the dam were not included in the model. The openings for galleries and vertical shafts were disregarded. The prestressing of the dam also was not included for the present model as it is unlikely to have significant a effect on thermal movements of the dam. This assumption regarding the effect of prestressing will be confirmed with the help of the detailed models that are being developed. Detailed models will also include other features of the dam not included in this preliminary model.

The 3D finite element model was developed with the help of the 10-node tetrahedral thermal solid element (Solid 87 of ANSYS) for the steady state thermal analysis of the dam model. The results of the thermal analysis were transferred to the structural model that was developed from the 10-node tetrahedral structural solid (Solid 92 of ANSYS) which is compatible with the thermal solid. Automatic mesh generation of ANSYS was used to mesh the model.

5.3 Material properties of dam concrete

The following material properties of the concrete (dam structure) were used in the analysis:
Modulus of elasticity of concrete in the dam = 18 GPa
Poisson's ratio = 0.15
Coefficient of thermal expansion of concrete = 10^{-6} per °C
Mass density of concrete = 2400 kg/m^3
Specific heat = 1000 J/(kg °C)
Thermal conductivity = 1.75 W/(m °C)
Emissivity = 0.9

5.4 Thermal analysis

Steady state thermal analysis of the dam was conducted to develop the temperature distribution in the body of the dam. The two seasonal conditions described earlier were considered.

5.5 Structural analysis

The temperature distribution from the thermal analyses was read into the structural models for stress analysis. The deflection contours developed from the stress analysis are shown in Figures 2 and 3 for the two seasonal conditions.

5.6 Comparison of deflections and discussion of results

Considering that the finite element model used in this study is a preliminary model with a number of simplifications, the contours shown in Figures 2 and 3 and the values given in Table 1 show a reasonable agreement with the horizontal deflections physically measured in the dam. While the finite element model developed for this study is in its preliminary stages, this agreement of the measured and predicted deflections is encouraging. The refinement of the finite element model incorporating all the dam features is expected to make better predictions of deflections. Further analysis is being conducted to achieve a better correlation.

6. CONCLUSION

This paper presents the background of Burrinjuck Dam along with the history of its construction. A summary of instrumentation for various measurements is presented. A 3D finite element analysis was conducted to predict deflections of the dam for two seasonal temperature conditions. Owing to the preliminary nature of the finite element analysis, all the physical conditions of the dam structure were not modelled in the analysis. However, the results of the finite element analysis show reasonable agreement between the measured and predicted deflections. From the finite element analysis of the dam outlined in this paper, it may be concluded that it is possible to conduct stress analysis of dam structures and predict reasonably accurate thermal movements of dams.

ACKNOWLEDGEMENT

The authors are grateful to State Water, the bulk water delivery business of the Department of Land and Water Conservation, New South Wales for giving permission to publish the measured deflections and other details reported in this paper. The comments of Mr Clynt Sheehy and Mr Roger Mail in finalising this paper have been invaluable.

REFERENCES

Burrinjuck Dam Inspection Report 1995. NSW Department of Land & Water Conservation.

Geological Report on Northern Spillway Slope Stability Investigations 1977. NSW Department of Land & Water Conservation.

Munday S.M. 1955. Burrinjuck Dam - Strengthening and Enlargement, *Journal of the Institute of Engineers*, Australia, Vol. 27, No.6 June 1955.

Munday, S.M. 1955. Strengthening and Enlargement, Reprint from the *Journal of the Institution of Engineers*, Australia, June 1955, NSW Department of Land & Water Conservation.

Report on the Condition on Concrete at Burrinjuck Dam 1963. NSW Department of Land & Water Conservation, January 1963.

Savage J.L. 1942. *Report on Remedial measures for Burrinjuck Dam*, 1942.

ANSYS5.4 1998. *Users Manual*, ANSYS Inc., USA

Optimisation and reliability

Mechanics of Structures and Materials, Bradford, Bridge & Foster (eds)
© 1999 Balkema, Rotterdam, ISBN 90 5809 107 4

Real coded genetic algorithms for active control force optimisation

M.N.S.Hadi & Y.Arfiadi
Department of Civil, Mining and Environmental Engineering, University of Wollongong, N.S.W., Australia

ABSTRACT: Real coded genetic algorithms (GAs) that use vectors of real numbers are more appropriate in the active control system. This is because real coded GAs have the ability to explore the unknown domain as needed in designing the controller gain. In the optimisation, the Linear Quadratic Regulator (LQR) with output feedback is used as the objective function, where the feedback is not necessary to be a full state feedback. In this case we can select the number of sensors to be less than the state dimension. The numerical example of a six-storey building utilising active bracing systems is then performed to show the applicability of the proposed optimisation procedure.

1 INTRODUCTION

Genetic algorithms (GAs) are stochastic optimisation procedures that have received a remarkable attention from researchers due to their potential use in solving complex optimisation problems. Usually GAs use binary bit strings to represent design variables and have been successfully applied in various problems (Goldberg 1989, Holland 1992) including structural optimisation (Jenkins 1991), absorber optimisation for multi-degree of freedom systems (Hadi & Arfiadi 1998).

In active control structures, the domain of the controller gains, as design variables, is usually unknown. Although it is possible to predict the domain of the controller gain by using Routh Hurwitz criteria, for the MIMO (multi input multi output) system this approach is very cumbersome. Therefore, the approach that has the ability to explore the unknown domain is preferred. Real coded GAs (Michalewich 1996, Herrera et al. 1998) have the ability to cope this problem provided that the crossover and mutation operators are chosen appropriately. In addition, since the representation of design variables in real coded GAs uses vectors of real numbers it suggests that real coded GAs to be more natural.

In conjunction with the GA procedure, the performance index to be used in this paper is the Linear Quadratic Regulator (LQR). This performance index can be considered as a base line of modern control theory. The application of real coded GA associated with other control theory such as H_2 and H_∞ control can be found in Arfiadi & Hadi (1999). Notice also that the feedback control used in this paper is a static output feedback where not all states are needed to be measured.

2 FORMULATIONS OF THE PROBLEM

2.1 *Equations of motion*

Consider the equations of motion of actively controlled structure in a state space form as

$$\dot{\mathbf{Z}} = \mathbf{AZ} + \mathbf{Bu} + \mathbf{E}w \tag{1}$$

where

$$\mathbf{Z} = \begin{bmatrix}\mathbf{X}_s^T & \dot{\mathbf{X}}_s^T\end{bmatrix}^T \; \mathbf{A} = \begin{bmatrix}\mathbf{0}_{n\times n} & \mathbf{I}_{n\times n} \\ -\mathbf{M}_s^{-1}\mathbf{K}_s & -\mathbf{M}_s^{-1}\mathbf{C}_s\end{bmatrix} \; \mathbf{B} = \begin{bmatrix}\mathbf{0}_{n\times n_a} \\ \mathbf{M}_s^{-1}\mathbf{b}_s\end{bmatrix}, \mathbf{E} = \begin{Bmatrix}\mathbf{0}_{n\times 1} \\ \mathbf{M}_s^{-1}\mathbf{e}_s\end{Bmatrix}, \; w = \ddot{x}_g \tag{2a-e}$$

in which $\mathbf{X}_s$ = vector of displacement, $\mathbf{M}_s$ = mass matrix, $\mathbf{C}_s$ = damping matrix, $\mathbf{K}_s$ = stiffness matrix, $\mathbf{0}$ = null matrix with appropriate dimension, $\mathbf{I}$ = identity matrix, $\mathbf{b}_s$ = control force location matrix, $\mathbf{e}_s$ = vector induced ground acceleration, x_g = ground displacement, n = total degree of freedom of the structure, n_a = number of control forces, ^{T} and dot ($\dot{}$) represent transpose and time derivative, respectively.

2.2 *Feedback Control*

Generally control force $\mathbf{u}$ may be regulated by feeding back the displacements, velocities and absolute accelerations at certain locations as

$$\mathbf{u} = -\mathbf{G}_d\,\mathbf{X}_s - \mathbf{G}_v\,\dot{\mathbf{X}}_s - \mathbf{G}_a\,\ddot{\mathbf{X}}_{sa} \tag{3}$$

where $\mathbf{G}_d$, $\mathbf{G}_v$ and $\mathbf{G}_a$ are $n_a \times n$ displacement, velocity and absolute acceleration gains, respectively; and $\mathbf{X}_{sa}$ =absolute acceleration of the structure.

Absolute accelerations in eq. (3) can be written as

$$\ddot{\mathbf{X}}_{sa} = \mathbf{C}_{sa}\,\mathbf{Z} \tag{4}$$

where

$$\mathbf{C}_{sa} = \begin{bmatrix}-\mathbf{M}_{cl}^{-1}\mathbf{K}_{cl} & -\mathbf{M}_{cl}^{-1}\mathbf{C}_{cl}\end{bmatrix},\; \mathbf{M}_{cl} = \mathbf{M}_s + \mathbf{bG}_a,\; \mathbf{C}_{cl} = \mathbf{C}_s + \mathbf{bG}_v,\; \mathbf{K}_{cl} = \mathbf{K}_s + \mathbf{bG}_d \tag{5a-d}$$

such that the control force can be written as

$$\mathbf{u} = -\mathbf{G}_z\,\mathbf{Z} \tag{6}$$

where $\mathbf{G}_z$ = gain matrix which can be written as

$$\mathbf{G}_z = \begin{bmatrix}\mathbf{G}_d & \mathbf{G}_v\end{bmatrix} + \mathbf{G}_a\mathbf{C}_{sa} \tag{7}$$

Note that the elements of the gain matrices $\mathbf{G}_d$, $\mathbf{G}_v$, and $\mathbf{G}_a$ contain zero if there is no feedback from the corresponding measurement output.

2.3 *LQR with static output feedback*

To optimise the control force, the Linear Quadratic Regulator performance index (LQR) is used in this paper. In this case, the performance index can be written as

$$J = \frac{1}{2}\int_0^\infty \left(\mathbf{Z}^T\,\mathbf{Q}\,\mathbf{Z} + \mathbf{u}^T\,\mathbf{R}\,\mathbf{u}\right)dt \tag{8}$$

where **Q** and **R** are weighting matrices to impose the importance of each corresponding output.

In LQR usually the control force needs the feedback from the full state of the system, where all displacements and velocities of the structural system are needed to be available. In this case, the $\mathbf{G_d}$ and $\mathbf{G_v}$ matrices are full matrix and $\mathbf{G_a}$ is a null matrix. The feedback control gain can be obtained by solving the Ricatti equation and has been discussed elsewhere as this is a standard controller based on modern control theory (Soong 1990). As for economical reasons the number of sensors should be limited, the need of the output feedback, where not all state are available, is more pronounced. In this paper, the output feedback to be used is a static output feedback, where it is not necessary to construct the observer to estimate the state as in the case of dynamic output feedback.

The approach to obtain the static output feedback controller has been discussed by Levine & Athans (1970). According to Levine & Athans 1970 for the measurement vector

$$\mathbf{y} = \mathbf{C_y}\,\mathbf{Z} \tag{9}$$

and the control force

$$\mathbf{u} = -\mathbf{G_y}\,\mathbf{y} \tag{10}$$

the gain matrix $\mathbf{G_y}$ can be obtained by solving the following non-linear matrix equations

$$\mathbf{G_y} = \mathbf{R}^{-1}\mathbf{B}^{T}\tilde{\mathbf{K}}\,\mathbf{L}\mathbf{C}_y^{T}\left(\mathbf{C}_y\mathbf{L}\mathbf{C}_y^{T}\right)^{-1} \tag{11}$$

where

$$\tilde{\mathbf{M}}\,\mathbf{L} + \mathbf{L}\tilde{\mathbf{M}}^{T} + \mathbf{I} = \mathbf{0},\quad \tilde{\mathbf{K}}\,\tilde{\mathbf{M}} + \tilde{\mathbf{M}}^{T}\,\tilde{\mathbf{K}} + \mathbf{Q} + \mathbf{C}_y^{T}\,\mathbf{G}_y^{T}\,\mathbf{R}\mathbf{G}_y\,\mathbf{C}_y = \mathbf{0},\quad \tilde{\mathbf{M}} = \mathbf{A} - \mathbf{B}\mathbf{G}_y\,\mathbf{C}_y \tag{12-14}$$

In this paper the solution to the output feedback gain is solved by using real coded GA with the modified performance index. For a stable system the performance index can be written in the form

$$J = \mathbf{Z}_o^{T}\,\mathbf{P}\,\mathbf{Z}_o \tag{15}$$

where $\mathbf{Z_o}$ is the initial state condition and **P** is the solution of the Lyapunov equation

$$\mathbf{A}_{cl}^{T}\,\mathbf{P} + \mathbf{P}\,\mathbf{A}_{cl} = -\mathbf{Q}_o \tag{16}$$

in which

$$\mathbf{A}_{cl} = \mathbf{A} - \mathbf{B}\mathbf{G}_z \quad \text{and} \quad \mathbf{Q}_o = \mathbf{Q} + \mathbf{G}_z^{T}\mathbf{R}\mathbf{G}_Z \tag{17-18}$$

To make the performance index independent to the initial condition of the state vector, the performance index in eq. (15) can be averaged out as

$$J = tr(\mathbf{P}) \tag{19}$$

where *tr* (.) stands for trace of (.). This performance index is used by the GA to obtain the controller gain $\mathbf{G_z}$. Notice that directly minimising eq. (19) due to eqs. (16)-(18) by the use of GA is possible to be done.

3 REAL CODED GENETIC ALGORITHMS

3.1 *Representation of design variables*

In real coded GAs design variables are represented in a vector of real numbers. If we have *r* design variables of the gain $g_1\,g_2\ldots g_r$ then each individual (chromosome) is represented by

$$\mathbf{G} = [g_1 \quad g_2 \quad \dots \quad g_r] \tag{20}$$

where g_i = real numbers.

3.2 *Fitness evaluation*

The fitness of each individual can be obtained according to the defined objective function. Since in GA we seek the high positive fitness individual, the problem of minimisation is converted such that the fitness value is always positive. In this case the fitness f can be expressed as

$$f = \frac{\alpha}{J} \tag{21}$$

where α= positive constant to scale the fitness unction.

In addition, to prevent the exploitation of the individuals that cause unstable mode to the system, the fitness of the individuals that produces positive real part of the closed loop system eigenvalue is set to a very low value near to zero.

3.3 *Mechanics of real coded GAs*

The power of GAs lays on the ability of this procedure to adaptability manipulate individuals towards better individuals generation per generation according to the Darwinian's selection mechanisms. An initial population is first generated randomly according to the number of design variables. Each individual has its own fitness according to the defined objective function. From the initial population, we select the individuals based on their fitness. Not all individuals can be passed into the next generation as some individuals that have very low fitness might be died. On the other hand, individuals that have high value fitness might have more copy in the next generation. The number of population in every generation is preserved constant. After selection, individuals experience recombination and mutation through crossover and mutation operator. The process of selection, crossover and mutation are performed generation per generation until the convergent criteria are achieved. At the final generation, the best individual is chosen as the design point to be sought.

In order that real coded GAs have the ability to explore a larger domain the crossover and mutation procedures are performed as follows. For two parents $\tilde{\mathbf{G}}$ and $\overline{\mathbf{G}}$ the resulting offspring after crossover are

$$\tilde{\mathbf{G}}' = a(\tilde{\mathbf{G}} - \overline{\mathbf{G}}) + \tilde{\mathbf{G}} \tag{22a}$$

$$\overline{\mathbf{G}}' = a(\overline{\mathbf{G}} - \tilde{\mathbf{G}}) + \overline{\mathbf{G}} \tag{22b}$$

where a = random number between 0 and 1.

The mutation operator is chosen such that for a certain individual $\mathbf{G}$, the modified individual has the form

$$\hat{\mathbf{G}} = [\hat{g}_1 \quad \hat{g}_2 \quad \dots \quad \hat{g}_i' \quad \dots \quad \hat{g}_r] \tag{23}$$

where $\hat{g}_i' = \overline{\alpha}\, a\, \hat{g}_i$, $\overline{\alpha} > 1$ and a = random variable between 0 and 1.

By using operators defined above it is possible to thoroughly explore larger domains than the initial domain.

4 NUMERICAL EXAMPLE

The six-storey shear building as discussed in Jabbari et al. 1995 is taken as the numerical example. To reduce the response during the vibration, active bracing systems are installed at the first and fourth storey. The structural parameters are m_i=345.6 t, c_i=2937 kN-s/m, k_i= 3.404 $\times 10^5$, (i = 1, 2, . . ., 6). The equations of motion can be written as in eq. (1) where

$$\mathbf{M}_s = diag\begin{bmatrix} m_1 & m_2 & \dots & m_6 \end{bmatrix}, \mathbf{C}_s = \begin{bmatrix} c_1 + c_2 & -c_2 & 0 & \dots & 0 \\ -c_2 & c_2 + c_3 & -c_3 & \dots & 0 \\ 0 & -c_3 & c_3 + c_4 & \dots & 0 \\ \vdots & \vdots & \vdots & \ddots & \vdots \\ 0 & 0 & 0 & \dots & c_6 \end{bmatrix}$$

$$\mathbf{b}_s = \begin{bmatrix} 1 & 0 & 0 & 0 & 0 & 0 \\ 0 & 0 & -1 & 1 & 0 & 0 \end{bmatrix}^T \quad \mathbf{e}_s = \begin{bmatrix} -m_1 & -m_2 & \dots & -m_6 \end{bmatrix}^T$$

The stiffness matrix $\mathbf{K}_s$ can be obtained similar to the damping matrix $\mathbf{C}_s$ by replacing c_i with k_i. Using this formulation the displacement vector $\mathbf{X}_s$ contains relative displacements of floors with respect to the ground. Notice also that the active bracing is assumed to be installed such that control forces affect the floors above and below the location of the bracing.

To reduce the vibration, several feedback schemes are considered as follows: (a) each actuator utilises the feedback from absolute acceleration of the sixth floor and velocity of the first floor referred to as case A, (b) each actuator utilises the feedback from the absolute acceleration of the first floor and velocity of the first floor referred to as case B, (c) first storey actuator utilises the feedback from velocity of the first floor while the fourth storey actuator utilises the feedback from velocity of the third floor referred to as case C. For all cases the weighting matrix $\mathbf{R}$ is taken as unity and matrix $\mathbf{Q}$ are all zeroes except for $\mathbf{Q}(1,1) = q_1$. The values of q_1 are 1×10^{14}, 1×10^{15}, 1×10^{17} for cases A, B and C, respectively. The real coded GA described above is then used to obtain the controller gain $\mathbf{G}_z$. The GA parameters are taken as follows: population size = 20, maximum number of generation = 1000, probability of crossover = 0.8, probability of mutation = 0.01, the number of random fresh generation to be inserted = 10 % of the total population. In each case the real coded GA is run four times, and the history of the best individual is recorded. It can be said that the best individual typically evolves to the fitness to the high fitness value with the increase of the generation. For the illustration, the evolving best fitness of case A is depicted in Figure 1. The structure is then subjected to El Centro 1940 NS excitation. The response of the structure for each controller can be seen in Table 1. First storey displacements for each case are plotted in Figs.2-4.

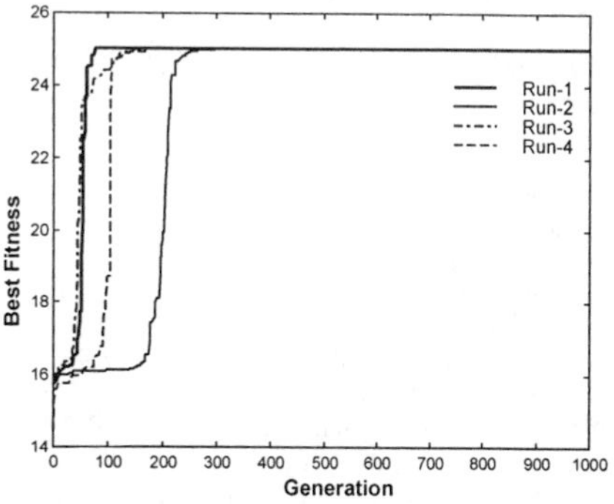

Figure 1. Evolving best fitness case A.

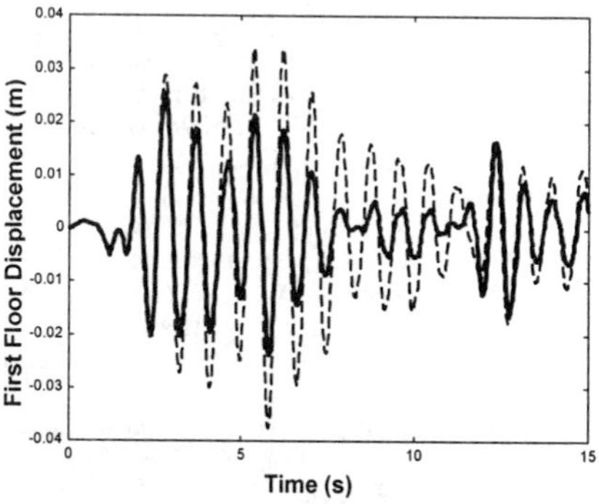

Figure 2. First storey displacement case A, dash = uncontrolled, solid = controlled.

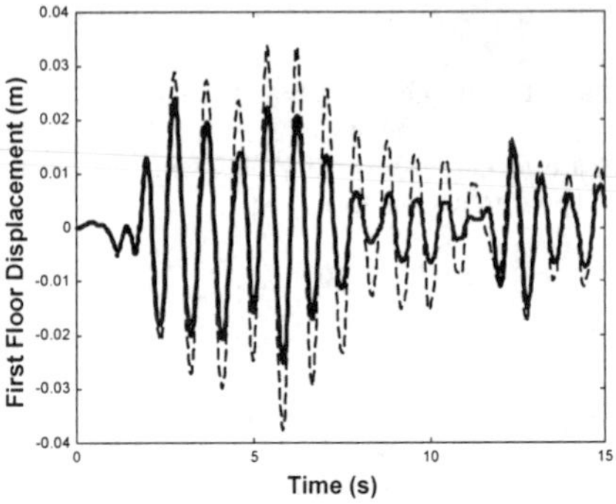

Figure 3. First storey displacement case B, dash = uncontrolled, solid = controlled.

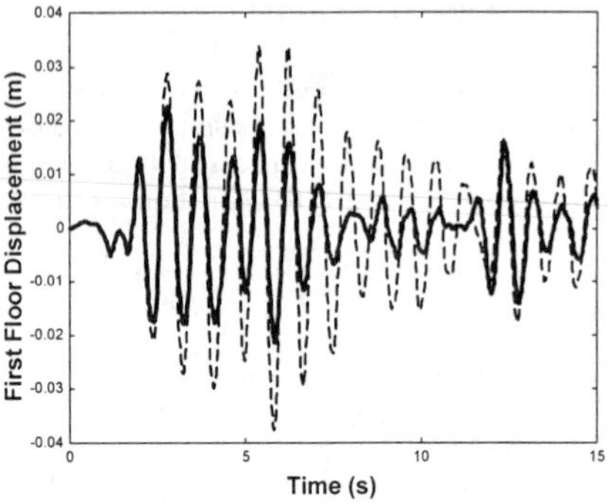

Figure 4. First storey displacement case C, dash = uncontrolled, solid = controlled.

Table 1. Peak responses subject to El Centro 1940 NS excitation

	Uncontrolled	Case A u_{max} =2060.8 kN	Case B u_{max} =1849.9 kN	Case C u_{max} =2037.6 kN
x_1 (cm)	3.76	2.53	2.52	2.27
x_2 (cm)	7.27	4.98	5.05	4.51
x_3 (cm)	10.35	7.14	7.30	6.49
x_4 (cm)	12.82	8.54	9.08	8.08
x_5 (cm)	14.53	9.68	10.37	9.20
x_6 (cm)	15.40	10.23	11.03	9.77

CONCLUSSION

Real coded GA has been used to obtain the controller gain where the LQR is used as the performance index. The feedback control used in this paper is a static output feedback where the number of sensors can be less than the state dimension. From the numerical simulation to a six-storey building utilising active bracing systems, it is found that the static output feedback controller can be obtained easily.

REFERENCES

Arfiadi, Y & Hadi, M. N. S. 1999. Active vibration control designs of civil structures using real coded genetic algorithms. *The seventh international conference on civil and structural engineering computing, 13-15 Sept. 1999.* Oxford, England: Civil-Comp Press (accepted).

Goldberg, D. E. 1989. *Genetic algorithm in search, optimization and machine learning*. Reading, Massachuset: Addison-Weshley Publishing Co., Inc.

Hadi, M. N. S. & Arfiadi, Y. 1998. Optimum design of absorbers for MDOF structures. *J. of Struct Engrg.*, ASCE, 124(11),1272-1280.

Herrera, F., Lozano, M., Verdegay, J. L. 1998. Tackling real-coded genetic algorithms: operator and tools for behavioural analysis. *Artificial Intelligence Rev.*,12,265-319.

Holland, J. H. 1992. *Adaptation in natural and artificial systems*. MIT Press.

Jabbari, F., Schmitendorf, W. E. & Yang, J. N. 1995. H_∞ control for seismic-excited buildings with acceleration feedback. *J. of Engrg. Mech.*, ASCE, 121(9), 994-1002.

Jenkins, W. M. 1991. Towards structural optimization via the genetic algorithm. *Computer and Struct.*, 40(5), 1321-1327.

Levine, W. S. & Athans, M. 1970. On the determination of the optimal output feedback gains. *IEEE Trans. Automatic Control*, AC-15,44-48.

Michalewics, Z. 1996. *Genetic algorithms + data structures=evolution programs.* Berlin: Springer.

Soong, T. T. 1990. *Active structural control: theory and practice.* New York: Longman Scientific and Technical.

Mechanics of Structures and Materials, Bradford, Bridge & Foster (eds)
© 1999 Balkema, Rotterdam, ISBN 90 5809 107 4

Single Input-Single Output (SISO) adaptive control in Gas Metal Arc Welding (GMAW) using Model Reference Adaptive Control (MRAC)

S-A. Moosavian & A. Basu
Department of Mechanical Engineering, University of Wollongong, N.S.W., Australia

ABSTRACT: A digital computer control system and the Model Reference Adaptive Control (MRAC) algorithm are proposed for improving the stability of the weld pool width shape in Metal Inert Gas (MIG) welding. Several cases where the system parameters such as welding process current, travel speed (welding table speed), electrode stickout (wire feed speed) and weld pool width are changed have been considered. An infrared sensor was used to sense the value of the weld pool width. At the instants where the welding process is carried out, the plant parameters of the welding process are unknown. As a consequence, their values were predicted using the well-known adaptive algorithm of the integration of product.

1 INTRODUCTION

A closed loop transfer function of the welding process will be explained and a general equation form of the welding process control will be obtained, in this section. Computer simulations are carried out to investigate the tracking characteristics of the MRAC system (Fiqure 1).

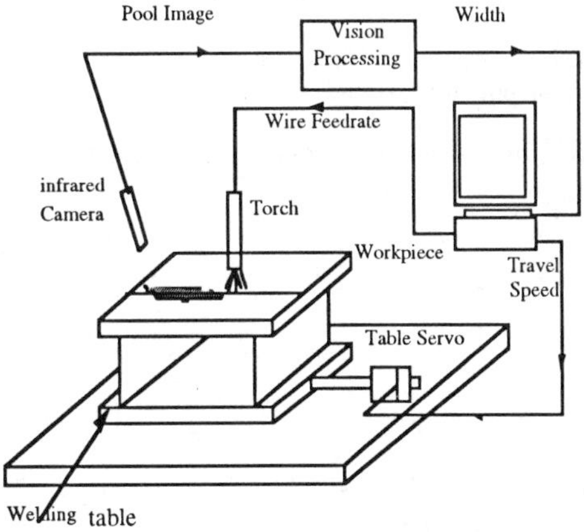

Figure1. Equipment of welding

2 APPLICATION OF ADAPTIVE CONTROL THEORY TO GAS METAL ARC WELDING SYSTEMS

A very simple block diagram for a closed-loop process control of the weld pool width can be defined in Figure 2

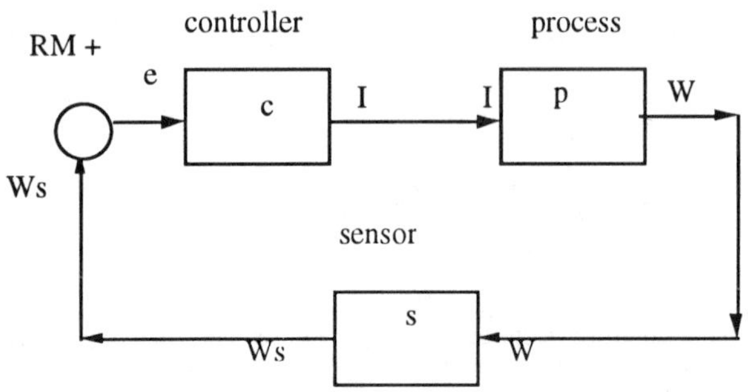

Figure 2. Block diagram for closed-loop process control of the weld pool width

S = Transfer function of the sensor; P = Transfer function of the process
C = Transfer function of the computer; RM = Reference Model value or desired value of the weld pool width (input to the system) (known).

"Process" means welding equipment such as robotics, welding torch, electrode, workpiece, etc.

Ws= Output of sensor (actual value of the width) to summing point located within the computer; e = Error (RM - Ws); W= Output from process to sensor; I = Current (output from computer and input to the process)

The relationship between input and output (value of width) of the system is defined as

$$Ws = b.RM + a \tag{1}$$

Where : a = Intercept of system (known) = The value of the width before the welding processstarts and usually its value is zero; b = Gain or transfer function of the whole system.

The main purpose is to find Ws, so that, its value becomes equal to the reference value or desired value of the weld pool width. In equation 1. the value of the "a" and RM parameters are known.

The transfer function of the whole process (b) can be found from equation 2 (Moosavian,1995):

$$b = \frac{Ws}{RM} = \frac{system\ output}{system\ input} \tag{2}$$

In regard to Figure 2

$$S = \frac{W_S}{W} = \frac{(output\ from\ sensor)}{(input\ to\ sensor)} \tag{3}$$

$$\mathrm{Ws} = \frac{\mathrm{S.C.P}}{1+\mathrm{S.C.P}}.\mathrm{RM} \tag{4}$$

To solve Equation 4 first find the C parameter. (C= I/ e).

To find I it is necessary to express the behaviour of the welding process..

The value of current (I) , the general form of the welding process control and the error of the system can be find from Equations. 5,6and 7

$$\overline{I(T)} = \frac{(Pm - (\overline{P(T)}))W(T) + qm.\mathrm{Im}(T)}{\overline{q(T)}} \tag{5}$$

$\overline{I(T)}$ is the input to the process using estimated parameters.

Because the process parameters P and q are unknown, they are replaced by their predicted quantities $\overline{P(T)}$ and $\overline{q(T)}$.

Pm and qm are the system parameters of the model reference adaptive control

W(T)=the output from the actual welding process.

Im(T) is the input parameter of the model reference adaptive control.

$$W(n) = P.W(n-1) + q.I(n-1) \tag{6}$$

w (n)= value of welding at the time of (n)

$$e^*(T) = \frac{Pm.e^*(T-1) + W(T-1)(P - \overline{P(T-1)})}{1 + Kp.(W(T-1))^2 + Kq.\overline{(I(T-1))}^2} + \frac{\overline{I(T-1)}.(q - \overline{q(T-1)})}{1 + Kp.(W(T-1))^2 + Kq.\overline{(I(T-1))}^2} \tag{7}$$

e^*(T-1)= Adaptive error in time of (T-1); and W(T-1)=Value of width in time of (T-1)

In the Equation 4 C,P and S parameters can be found, (Moosavian 1995) from:

$C = \frac{I(T)}{e^*(T)}$; P= $\frac{q.n^{-1}}{1 - P.n^{-1}}$ and S= x (108.97)

Now that all the parameters in equation 4 are known and it is possible to write a program to control the value of the weld pool width in a welding process.

3 DISCUSSIONS ON THE RESULTS AND FIGURES OF THE CONTROL PROGRAMS OF THE WELD POOL WIDTH.

Using the Matlab package, a program to controle weld pool width was written for seven cases. Tables 1 and 2 show all the information and differences of these cases.

Table 1 Cases of weld pool width control at periods of time when the welding parameters are constant

		number of sampling time	Total time sec.	Time of each sampling time (s)	Width error (mm)	Welding current (A)	p parameter	q parameter	(d)Wire diameter (mm)	(v) welding table speed (mm/s)	Wire feed speed (mm/s)	Desired value of width (mm)	Width obtained (mm)
Case (1,1)	non adaptive control	610	40.26	0.066	0.11	172.24	0.949	0.025	1.2	4.16	74.78	6	5.88
Case (1,2)	Adaptive control	610	40.26	0.066	*0	172.24	0.949	0.025	1.2	4.16	74.78	6	6
Case (2,1)	non adaptive control	610	40.26	0.066	0.08	119.87	0.965	0.016	0.9	3.33	84.66	4	3.920
Case (2,2)	Adaptive control	610	40.26	0.066	*0	119.87	0.965	0.016	0.9	3.33	84.66	4	4
Case (2,2,1)	Adaptive control	325	21.5	0.066	*0	119.87	0.965	0.016	0.9	3.33	84.66	4	4
Case (2,2,2)	Adaptive control	**135**	**8.91**	0.066	*0	119.87	0.965	0.016	0.9	3.33	84.66	4	4

* goes to zero

Table 2 Cases of weld pool width control ,when the welding parameters vary

		number of sampling time	Total time sec.	Time of each sampling time (s)	Width error (mm)	Welding current (A)	p parameter	q parameter	(d)Wire diameter (mm)	(v) welding table speed (mm/s)	Wire feed speed (mm/s)	Desired value of width (mm)	Width obtained (mm)
Case (1,1)	non adaptive control	3oo	19.8	0.066	0.10	172.24	0.895	0.052	1.6	5	42.53	6	5.891
Case (1,2)	Adaptive control	3oo	19.8	0.066	*0	84.18	0.895	0.052	1.6	5	42.53	6	6
Case (2,1)	non adaptive control	3oo	19.8	0.066	0.075	119.87	0.952	0.023	0.8	5.83	114.5	4	3.925
Case (2,2)	Adaptive control	3oo	19.8	0.066	*0	84.09	0.952	0.023	0.8	5.83	114.5	4	4
Case (2,2,1)	Adaptive control	175	11.55	0.066	*0	84.09	0.952	0.023	0.8	5.83	114.5	4	4
Case (2,2,2)	Adaptive control	90	5.94	0.066	*0	84.09	0.952	0.023	0.8	5.83	114.5	4	4
Case (3)	Adaptive control	910	60	0.066	varing	varing	varing	varing	varing	varing	varing	varing	varing

* goes to zero

Parameter q is dependent on the speed of the welding wire(VV), process current (A) and wire diameter (d). As a result of this, the input to the welding process depends on the value of parameter q and q is dependant on the speed of the welding wire(VV), process current (A) and wire diameter (d). As a result of this, the input to the welding process depends on the value of parameter q

4 DISCUSSIONS ON CASE (1)

In case (1), the reference value of the weld pool width was chosen as 6 (mm) and the time for the welding process was chosen as 60 (s). For more details see table 2
Case (1.1) was performed to find the value of the weld pool width without using adaptive control.

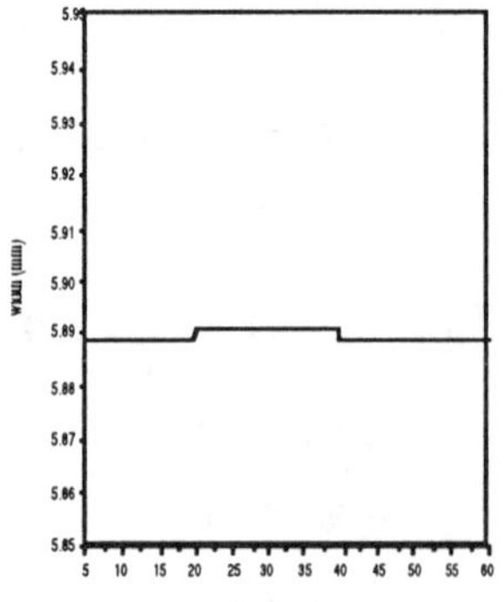

Figure 3: : Relationship between Time, Width and time in case (1,1)

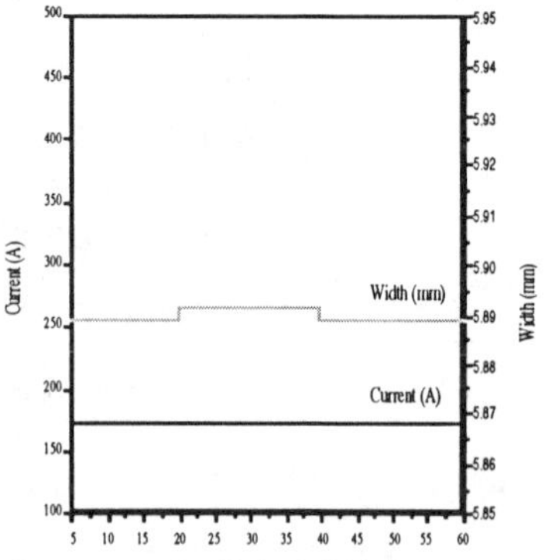

Figure 4:: Relationship between width and current in case (1,1)

In Figure3, since the welding process was not controlled, the value of the welding process current is constant, because of this, the welding process current does not play any role in changing the value of the weld pool width to its desired values.

Figure 4 shows that when the welding process parameters p and q, change after 20 seconds of the welding process, the value of weld pool width changes as well. It means, the value of the weld pool width has not been controlled. As a result of this, the workpiece will be returned as failed.

In Figure 4 $p = e^{-\frac{T}{\lambda}}$;T = sampling time of the process output
λ= Gain or transfer function which is depends on variations of the process welding parameters. $q= \frac{K}{\lambda} .(1-e^{-\frac{T}{\lambda}})$

Figure 5 shows that parameters p and q have an inverse relationship with each other. parameters P have a direct relationship with the weld pool width and an inverse relationship with wire diameter, speed of the welding table, value of each sampling time and parameters.q parameter q has a direct relationship with the weld pool width, wire diameter and speed of wire feed. However, it has an inverse relationship with the speed of the welding table, welding current, value of each sampling time and the parameter p.

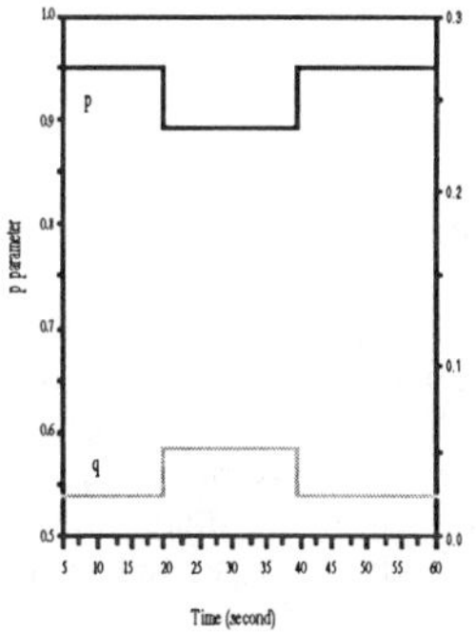

Figure 5: Relationship between p parameter q parameter and time in case (1,1)

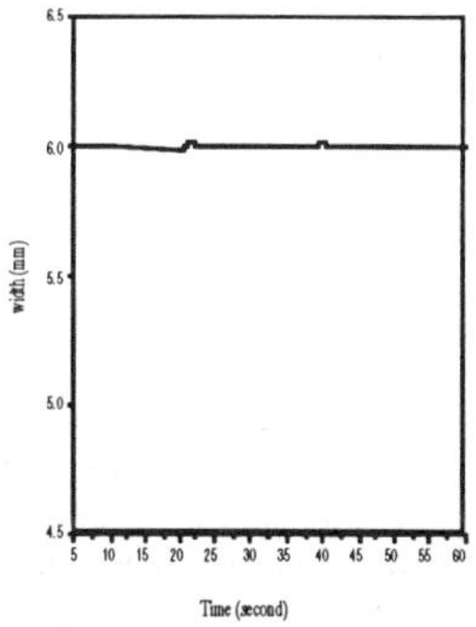

Figure 6 : Relationship between width-time in case (1,2)

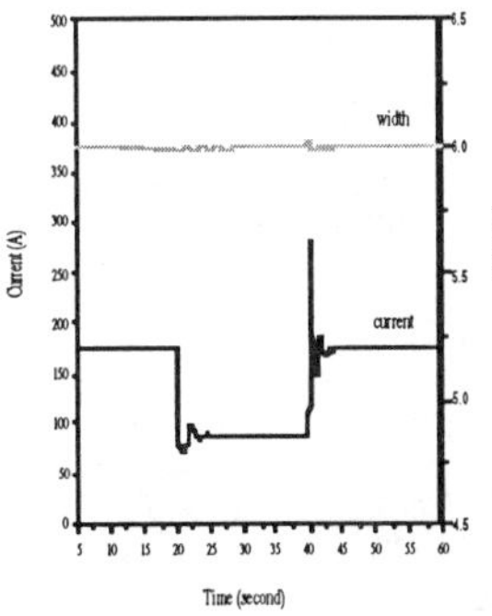

Figure 7: Relationship between current, width and time in case (1,2)

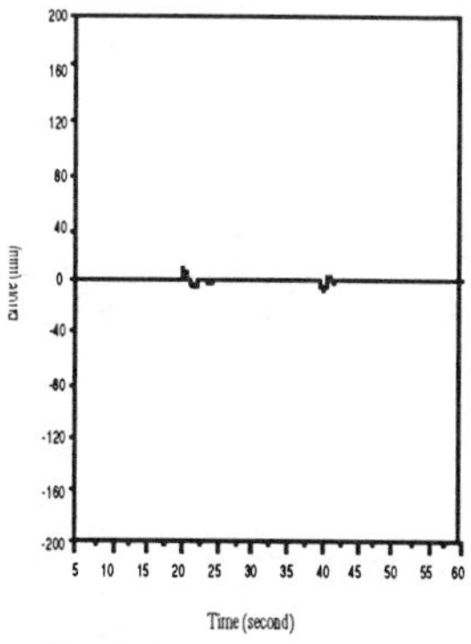

Figure 8: Relationship between time-error in case (1,2)

The effect of parameter q on the value of the weld pool width is higher than parameter.p Because of this, in the above figure, when p decreases, q increases.

In Figure 6,the value of the weld pool width has been controlled using model reference adaptive control. This figure shows that when welding parameters change in the period of time between 20 second and 40 second, the value of the weld pool width *does not change* , and it remains just at 6 (mm), (the desired value of width).

Figure 7 shows the role of the welding process current in the control of the weld pool width when the welding parameters change from a time of 20 second to 40 second. When a time of 20 seconds of welding process has passed, welding parameters change (parameter p decreases and parameter q increases). As a result of this, the value of weld pool width increases but the value of welding current immediately decreases (using the computer program). This action causes a return to the value of weld pool width to its previous value as quickly as possible. As is shown in the above figure, the value of weld pool width remains at 6 (mm) (the desired value). It means that adaptive control of welding process is feasible.

Figure 8 shows the situation of weld pool width error between the weld pool width and the desired value of the weld pool width. One of the basic aims of adaptive control is to make the value of weld pool width error zero. In this figure, the value of the error is precisely zero

When welding parameters change in the period of time from 20 second to 40 second of the welding process, the value of the error wants to change at the mentioned time but immediately the value of error is returned to its previous value (zero).(using the computer program). This action causes a desired value of weld pool width and a desired high quality of weld pool width control to be obtained.

5 CONCLUSION

Simulation results show that the MRAC system has a good tracking performance regardless of the difference between the parameters of the reference model and the controlled system. Simulations, carried out as part of this paper, show that during the duration of the welding process, the welding parameters change. As a result of this, the value of the weld pool width tends to change. However, the computer program, which has been written using the technique of model reference adaptive control, changes the value of the welding process current. This action causes the weld pool width error to converge to zero and finally, these actions result in the convergence of the value of the weld pool width to its desired value (reference value of weld pool width).

When welding parameters change in a simulation of the weld pool, without using an MRAC controller, the value of the weld pool width increases.

In conclusion, the above system demonstrates good tracking performance regardless of the difference in the welding process parameters.

6 REFERENCES

Moosavian, S.A. 1995 Single Input Single Output(SISO) Adaptive Control In Gas Metal Arc Welding(GMAW) Using Model Reference Adaptive Controle(MRAC), *Master thesis , Mech. Eng., University of Wollongong , Australia*

Mechanics of Structures and Materials, Bradford, Bridge & Foster (eds)
© 1999 Balkema, Rotterdam, ISBN 90 5809 107 4

Topological optimisation of simple trusses using a genetic algorithm

P. Dimmock & A. J. Deeks
Department of Civil and Resource Engineering, University of Western Australia, Nedlands, W.A., Australia

ABSTRACT: Genetic Algorithms are probabilistic optimisation methods with proven value in computer-aided structural design. They efficiently find global optima from vast, discontinuous search spaces. Furthermore, these methods easily handle discrete design variables, which are very common in structural engineering. However, most applications of Genetic Algorithms to structural design have involved the optimisation of set configurations of members by varying cross-sectional properties. Those research papers that do address the optimisation of structural shape approach the matter in a very controlled manner, limiting the search to a small number of possible configurations. The work reported in this paper is significant in showing that Genetic Algorithms can efficiently find global optima from vast topological design spaces. Specifically, the Genetic Algorithm is applied to the topological optimisation of a simply supported truss. The algorithm operates on a large set of possible configurations with minimal constraints on the design variables and with weight as the optimisation criteria.

1 INTRODUCTION

Operating as evolutionary processes, Genetic Algorithms have proven to be computationally efficient. They have found use in a variety of applications, and their use in structural optimisation is well established. However, most attention has been given to optimising cross-sectional properties of a given topology structure. This paper details an attempt to use a Genetic Alogorithm to optimise the structural topology of a simply supported truss. Existing research in truss topology optimisation has sought optima from a small pre-meditated set of topologies. In contrast, this study sought to extract optimum truss topologies with minimal human guidance, from a vast, relatively unconstrained, solution space. A secondary objective was to demonstrate that alternative (less contrived) representations of chromosomes, which are relatively inexpensive in terms of computer memory, could be used instead of a standard binary string.

2 STRUCTURAL OPTIMISATION AND GENETIC ALGORITHMS

Schmit (1960) heralded the modern era of computer-based design with a landmark paper. Following this paper, deterministic methods and practices were the most commonly used structural optimisation processes. The nature of these processes is described by Vanderplaats (1993). The classical optimisation methods were noted to be of limited practical application, due to their inability to deal with discontinuous solution spaces. They only operate with continuous design variables, and consequently provide the engineer with solutions containing non-standard section sizes, which are consequently neither practical nor economical. The drawbacks of classical methods led to a search for alternative algorithms. This stimulated the introduction of stochastic methods such as Genetic Algorithms.

Genetic Algorithms were first developed by John Holland in the 1970's at the University of Michigan (Holland, 1975). They are optimal search devices based on the proven evolutionary operations of natural selection and inherited characteristics. Goldberg (1989) highlighted the characteristics of these algorithms that distinguish them from contemporary optimisation schemes. Genetic Algorithms are outstanding performers in vast multi-variate solution spaces, and have been proven to be robust. Also, they are efficient, in that they encounter promising regions of the solution space within a few iterations (Goldberg, 1989).

Genetic Algorithms require operators governing parent selection and genetic data combination. The parents are selected on the basis of fitness for some purpose according to an objective function. Chromosomes are represented by strings, and contain all of the necessary information required to describe an individual. Manipulation of the chromosome strings of parents creates children. The encoding of solutions into chromosomes that the algorithm can manipulate is crucial to its correct functioning. After a new generation is created, the chromosome of each individual is decoded, and the fitness of the new individuals evaluated. Most Genetic Algorithms operate on chromosomes which are represented by a series of bit strings. It is necessary to have regularity in the order of the encoded parameters, and a unique representation for each solution. The performance of a Genetic Algorithm is effect by the population size, the number of generations, the probability of crossover, and the probability of mutation. Each may be varied to affect the rate of convergence and the probability of convergence to the optimum solution.

Various algorithms have been developed for the optimal topologies of trusses under static loading conditions. These are summarised in the review papers by Kirsch (1989) and Topping (1992). Amongst these algorithms, Shankar and Hajela (1991) proved that the genetic search is a viable strategy for the structural topology synthesis process. Despite the potential of Genetic Algorithms to efficiently search vast design spaces, most topology optimisation has been done with a very limited discrete range of structural geometries. For example, Jenkins (1991) used a Genetic Algorithm to find the minimum weight design of a trussed roof beam structure. Design variables governed the slope of the roof, the geometry of the tie, and maximum depth of the truss. Ohsaki (1995) provided another example of a Genetic Algorithm working with a very limited discrete range of structural topologies. A truss topology was optimised by deleting surplus struts or reducing the cross-sectional area of struts from a maximum network of members between nodes. The benefit of this research on limited numbers of possible geometric configurations are that the solution obtained could be compared with a known optimum. This process verified the accuracy of the technique.

Topological design variables are usually represented as fixed length binary strings in chromosomes. Ohsaki (1995) has used fixed length binary strings that mapped to section areas. Grierson and Pak (1993) used fixed length binary strings which mapped to a table of discrete column lengths. Other implementations should be considered, as the fixed length binary string implementations used by Ohsaki and Grierson and Pak are inefficient and require an excessive amount of computer memory when large populations are used.

3 PROBLEM DESIGN AND PROGRAM METHODOLOGY

The template for the simply supported truss considered in this work is two rows of nodes, where nodes are formed at grid line intersections (see Figure 1). Supports are applied at two bottom row nodes, and each node on the top row can assume variable height. The chromosome describing a structure records the position of all cross members and top node positions. Bottom and top chords of the truss are added automatically to each solution prior to analysis, and a screening module is applied to prevent structures with obvious mechanisms from being included, as these hinder convergence of the algorithm. The load is applied as a uniform load to the bottom members, and resolved by tributary length into point forces on bottom nodes that are connected by the truss.

There are two global degrees of freedom at each unrestrained node of the grid. A small spring is placed at each degree of freedom to prevent singularity of the stiffness matrix. If a structure does not have sufficient members to carry the load between the supports, then the deflections will be very large, from which the program can infer that the structure is a mechanism.

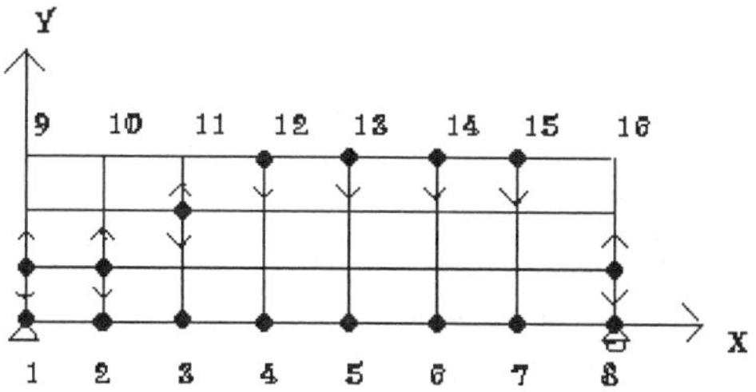

Figure 1. An example of a solution space for a simply supported truss.

2	11	7	13	8	15

8	15	2	11	7	13

Figure 2. Alternative chromosome representations of the same members.

The initial population of solutions are randomly generated. The program analyses each member of the population, using an embedded analysis module that utilises the matrix stiffness method. The analysis of each solution is required to provide data for comparison against the design constraints, which are strength and serviceability. Any violation of the design constraints is penalised. Mechanisms are given a high penalty to account for their invalid topology. Penalties for violation of the design constraints are superimposed on a penalty for the weight of the structure, and the fitness of each design is calculated as the inverse of the penalty.

This program uses the three genetic operators from the simple tripartite Genetic Algorithm used by Goldberg (1989). Each individual solution is represented by one chromosome. A weighted roulette wheel is used to select solutions from the present population to be parents. Crossover of chromosomes occurs according to a certain probability. A crossover point is randomly chosen as one of the vertical lines of nodes in the grid. The first child inherits members in the first parent chromosome, which are for the greater part to the left of this crossover line plus members from the second parent chromosome, which are predominantly to the right of this crossover line. The second child is formed using the same principle, but the other way around. Once the chromosomes of the children have been formed, some are mutated by random changes to some information within a chromosome. A group of parameters controls the rate of convergence of the algorithm. These parameters are the population size, the number of generations, the probability of crossover, and the probability of mutation. The affect of each of these parameters was investigated.

Each chromosome consists of a series of integers, each pair of integers in succession representing one member by a starting and ending node number in the grid. This approach obviates the need to decipher a bit string, and avoids the mapping problems that are associated with a binary string chromosome. Unlike the traditional binary string representation, the length of each chromosome is dependent on the number of members in the solution. In addition, since crossover is performed using a vertical line through the solution grid, the chromosome does not require order, unlike the traditional chromosome. The same member may be indicated at the end of one chromosome but at the beginning of another. Figure 2 shows alternative representations of the same members in two chromosome strings. Each pair of numbers in a chromosome represents the nodes connected by one element.

4 FINDINGS

The implementation of the methodology described above is illustrated using two trial examples. The first is a simply supported truss optimised on a grid of width 20m and height 4m. There are 10 grid points across the width and 4 grid points over the depth. The number of possible cross-members in this example is the number of bottom nodes (11) multiplied by the number of top nodes (11), ie. 121. In addition, there is the added variation of 4 different top node positions for each node. Example 2 consists of a simply supported truss optimised on a grid of width 24m and height 6m. There are 12 grid points across the width and 5 grid points over the depth. The number of possible cross-members is 169. In addition, each of the top nodes can assume one of five different heights. In accordance with the primary objective of this work, both examples are larger, less constrained topological solution spaces than have been tackled by previous workers. Ohsaki (1995) reported convergence for a plane truss with a maximum of 20 members. This is the largest topological solution space tackled using a Genetic Algorithm reported in the current literature.

In both examples, the procedure appeared to converge to an optimum. The plateau of average fitness in Figure 3 indicates convergence for Trial 1. The accompanying graphs of best fitness and average length of truss offer alternative perspectives on the process of convergence for Trial 1. The global optimum for this problem cannot be found analytically. Engineering judgement must be used to determine whether the designs achieved are likely to be global optima. Both solutions satisfied design criteria. In Trial 1, the algorithm appears to have converged on at least close to a global optimum (see Figure 4). However, Trial 2 appeared to converge to a structure that was only a local optimum. A larger population could solve this problem, but the computational time required is currently unrealistic.

Convergence of the program to a solution apparently close to a global optima in Trial 1 occurred from a solution space that was bounded by less stringent constraints than used in previous research. This research has shown that the Genetic Algorithm has the capability to find the optimum truss topology with fewer constraints placed on the design variables than has been attempted previously.

The performance of the alternative chromosome representation was compared with the expected performance of the traditional binary string. A binary string chromosome is longer than

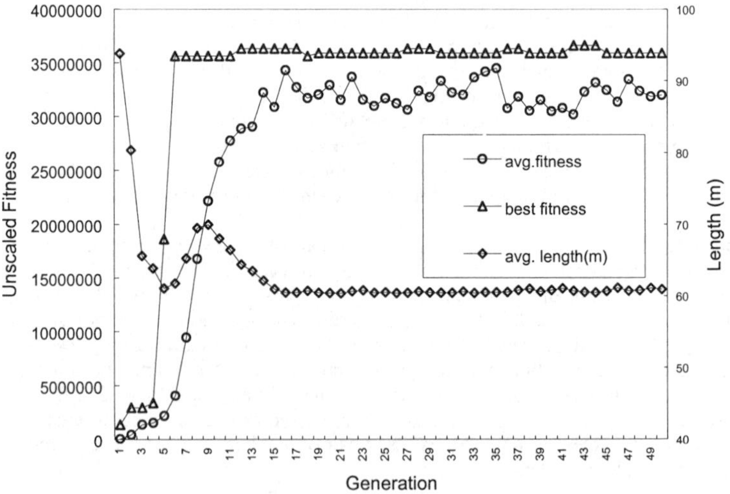

Figure 3. Trial 1, population fitness and average length versus generation number.

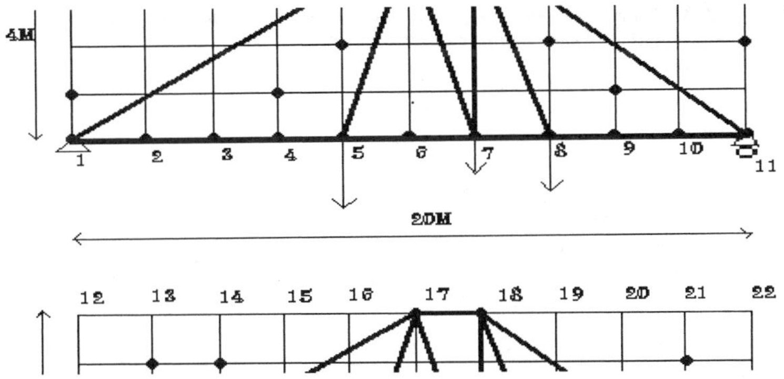

Figure 4. Trial 1, solution of highest fitness in the final generation.

the proposed representation, and it was found that the increased runtime of the program during Trial 2 was due largely to the processing of a longer chromosome string. On this basis, an implementation with the traditional bit string chromosomes would slow the program significantly. The traditional approach also consumes more computer memory. Hence, this alternative chromosome representation allowed the program to function more efficiently.

5 CONCLUSIONS

Genetic Algorithms have a future in the engineering profession as an initial search procedure for efficient designs. They can be useful in the scheming stage of the design process. However, they cannot be relied on to achieve the 'true' optimum unless a sufficiently large population is used. They are very efficient at searching large solution spaces for possible optima, but they do not use a proven mathematical method that guarantees at least a 'true' local optima, if not the global optimum.

The viability of Genetic Algorithms as a method for structural optimisation has been proven. However, more emphasis now needs to be placed on topological optimisation, and more specifically on the practicality of the design scenarios. The computer program that has been presented here, though having little practical design use in itself, is part of the process of proving that Genetic Algorithms can be utilised for the optimisation of geometry in practical design packages. The program has solved problems which show important characteristics of a practical design scenario, requiring a solution to be sought from a vast solution space with unknown optima.

REFERENCES

Goldberg, D.E. 1989. Genetic algorithms in search, optimisation and machine learning. Addison-Wesley.

Grierson, D.E. and W.H. Pak 1993. Optimal sizing, geometrical and topological design using a genetic algorithm. *Struct. Optimisation*. 6:151-159.

Holland, J.H. 1975. Adaptation in natural and artificial systems. Ann Arbor: University of Michigan Press.

Jenkins, W.M. 1991. Toward structural optimisation via the genetic algorithm. *Computers and Structures*. 40:1321-1327.

Kirsch, U. 1989. Optimal topologies of truss structures. *Appl. Mech. Rev*. 42:223-229.

Ohsaki, M. 1995. Genetic algorithm for topology optimisation of trusses. *Computers and Structures*. 57:219-225.

Schmit, L. A. 1960. Structural design by systematic synthesis. *Proc. 2nd Conf. on Electronic Computation*: 105-122. American Society of Civil Engineering, New York.

Shankar, N and P. Hajela 1991. Heuristics driven strategies for near-optimal structural topology development. *Artificial Intelligence and Structural Engineering*: 219-226.

Topping, B.H.V. 1992. Mathematical programming techniques for shape optimisation of skeletal structures. In G.I.N. Rozvany (ed.), *Shape and layout optimisation of structural systems and optimality criteria methods*. Vienna: Springer.
Vanderplaats, G.N. 1993. Thirty years of modern structural optimisation. *Advances In Engineering Software*. 16:81-88.

Mechanics of Structures and Materials, Bradford, Bridge & Foster (eds)

Finite element analysis in structural reliability

R.E. Melchers & X.L. Guan
Department of Civil, Surveying and Environmental Engineering, University of Newcastle, N.S.W., Australia

ABSTRACT: Complex structural and other problems, including fluid problems, increasingly use finite element analysis techniques. Probabilistic analyses are needed increasingly for a variety of purposes including analysis of existing systems. Various approaches have been proposed for handling the resulting computationally demanding problem. The 'Directional Simulation in the Load Process Space' approach is outlined both for time-invariant and the time-variant analysis and two example applications given for a flat plate containing a central hole and subject to in-plane stress. A comparison is made with the results provided by other solution techniques.

1 INTRODUCTION

The analysis of complex structural and other problems, including fluid problems, is increasingly utilizing various forms of finite element analysis techniques. These are expensive, generally, in terms of computer time particularly where fine discretization of the problem is required.

Increasingly there is a demand for probabilistic analysis of structures and solid mechanics systems. This arises from the need to calibrate design codes but increasingly also from the need to analyse and evaluate existing systems.

The essential techniques and tools for probabilistic structural analysis are well developed. Both simplified analytical and accurate but computationally very demanding approaches exist for such analysis when the system behaviour can be described in a relatively simple and computationally non-demanding way (Melchers, 1999a). When high-level system behaviour descriptions (such as provided through a finite element analysis) need to be employed, a significant computational problem arises.

In probabilistic structural analysis the approach to using finite element techniques has so far been restricted to two avenues - use of the simplified probabilistic estimation techniques such as First Order Second Moment (FOSM) methods (or refinements thereof) or simple Monte Carlo simulation. Experience indicates that the former may not be very accurate even for simple problems. The second can be very accurate but is extremely demanding of computer time since thousands of finite element analyses need to be carried out. Despite these draw-backs there are strong advocates and research activities for each approach.

To treat finite element problems in a probabilistic structural analysis framework, it is necessary to consider the structure to have uncertain properties and to be subject to a probabilistically defined loading system(s). In probabilistic analysis the uncertain properties are represented as random variables. Also the correlation structure of physical properties must be represented as a random field. This must be discretized in a manner which relates to the finite element discretization. Clearly there are a very large number of random variables involved. This makes FOSM-type methods problematical due to the extensive linearization involved. It also leads to the excessive computer demands for conventional Monte Carlo.

The problem can be made more tractable and accurate, with modest computational demands, by reformulating it in the load(-process) space and restricting simulation to particular operations. In its broadest perspective this approach has been termed Directional Simulation in the Load Process space. The basic concept has been elaborated previously but so far only a few examples have been given. The present paper provides results for two further example applications with the focus on estimating the probability of system failure in a pre-specified manner. The applications are both for a flat steel plate containing a central hole and loaded in-plane. The three solution approaches are compared with the other approaches for accuracy of failure probability estimation and for computation times.

2 BASIC FORMULATION

In the time-invariant domain, the probability of structural failure can be estimated from

$$p_f = P[G(\mathbf{X}) \le 0] = \int \ldots \int_{G(\mathbf{X}) \le 0} f_{\mathbf{x}}(\mathbf{x}) d\mathbf{x} \tag{1}$$

where $f_{\mathbf{X}}(\mathbf{x})$ is the joint probability density function for the n vector $\mathbf{X}$ of basic variables and $G(\mathbf{X}) \le 0$ defines the unacceptable or failure region. The usual resistance R and load effect S random variables are absorbed in $\mathbf{X}$ in this formulation. It is implicit that the load to cause failure is applied only once during the lifetime of the structure and that it is modelled through an extreme value distribution. When there is more than one load acting on the system the formulation needs to be modified to cater for 'load-combinations' or a time-variant approach must be adopted (see below). These matters are described in detail in various texts (e.g. Melchers, 1999a).

Except for $\mathbf{X}$ multi-normal or multi-lognormal (and some other less relevant special cases), the integration of (1) over the failure domain $G(\mathbf{X}) \le 0$ cannot be performed analytically. Simplification or numerical treatment (or both) of (i) the integration process, (ii) the integrand $f_{\mathbf{x}}(\,)$ and (iii) the definition of the failure domain $G(\mathbf{X}) \le 0$ is possible. The basic approaches are:

(a) numerical treatment such as simulation ('Monte Carlo') to perform the multi-dimensional integration in (1)

(b) transforming $f_{\mathbf{x}}(\mathbf{x})$ in (1) to a multi-normal probability density function (thereby sidestepping the integration process completely) and using well-known first and second moment properties together with a linearized failure domain to estimate an approximate probability of failure. This is the First Order Second Moment (FOSM) approach.

Because of the way FOSM was developed, it has become customary to transform the reliability problem to the space $(\mathbf{y})$ of the standard normal variables. The conversion of the problem from $(\mathbf{x})$ space to $(\mathbf{y})$ space can present a very interesting challenge in itself, requiring a Nataf, Rosenblatt or other transformation [Melchers, 1999a]. These transformations are convenient for some discussions but are not essential. Moreover, they are often approximate. For these reasons numerical methods still have a significant place in probabilistic systems analysis.

In most investigations integration of (1) is in the cartesian coordinate system. As will be seen in the next section, an alternative approach is to use the polar coordinate system.

3 DIRECTIONAL SIMULATION IN THE LOAD SPACE

For many structural systems the loads may be separated from the system resistances. This allows the reliability problem to be formulated in the m-dimensional load (process) space $\mathbf{q}$. Unlike the dimension of the $\mathbf{x}$ space, typically it will be of low order, since usually there are only a few load systems acting on a structure.

The possibility of solving structural reliability problems in the load space treated as a vector of random variables was explored also in various ways other than directional sampling [see Melchers 1999b for a review]. In most of these cases simplified methods of numerical multi-dimensional integration were proposed. In the following only directional simulation will be considered.

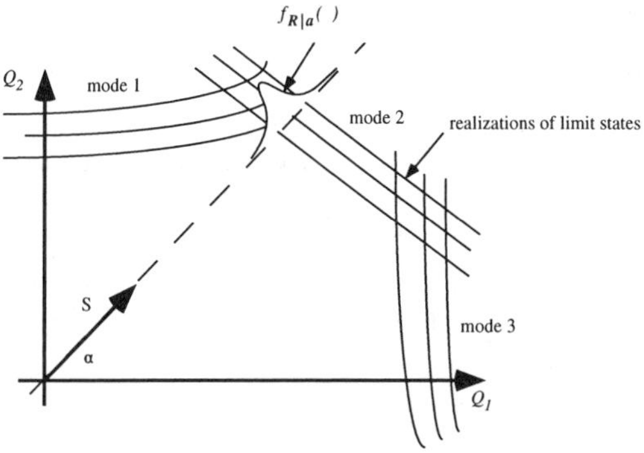

Figure 1. Two-dimensional load space showing a typical directional simulation and the probabilistic description of the limit state and some realizations.

For each component of $\mathbf{Q}$ let there be a corresponding resistance component $\mathbf{R} = \mathbf{R}(\mathbf{X})$ with pdf $f_{\mathbf{R}}(\ )$ and defined in terms of the n-dimensional vector $\mathbf{X}$ containing random material strengths, dimensions, etc. (generally in a similar manner to that used for collapse modes in the plastic analysis of rigid frames). It is evident that in the load space each conventional limit state function $G(\mathbf{x}) = 0$ becomes a probabilistic 'boundary' (see Figure 1).

For directional simulation, a point other than the load space origin may be selected as the basis point. Let this be defined arbitrarily. Then the (scalar) radial distance S representing system strength or resistance is defined as $\mathbf{R} = S.\mathbf{A} + \mathbf{c}$. It has a conditional pdf $f_{S|\mathbf{A}}(\ )$. It then follows that the probability of failure can be written as [Melchers, 1992]:

$$p_f = \int_{\substack{\text{unit}\\\text{sphere}}} f_{\mathbf{A}}(\mathbf{a}) \left[\int_S p_f(s|\mathbf{a}).f_{S|\mathbf{A}}(s|\mathbf{a})ds \right] d\mathbf{a} \tag{2}$$

with the conditional failure probability $p_f(s|\mathbf{a})$ being the probability that the structure will fail when the structural strength has the value $S(\mathbf{X}) = s$ for the direction $\mathbf{A} = \mathbf{a}$. This term can be evaluated for each radial direction using any method for evaluating a convolution integral, such as (i) direct integration, (ii) simulation and (iii) FOSM / FOR methods.

In (2), $f_{S|\mathbf{A}}(\ )$ represents the probability density function for the variation of structural strength $G(\ )|\alpha$ with distance along the given radial direction α (Figure 2). This can be evaluated in at least three different ways: (i) by multiple integration, (ii) by point-wise estimation and (iii) by sampling. Details are available elsewhere (Melchers, 1992, 1999b).

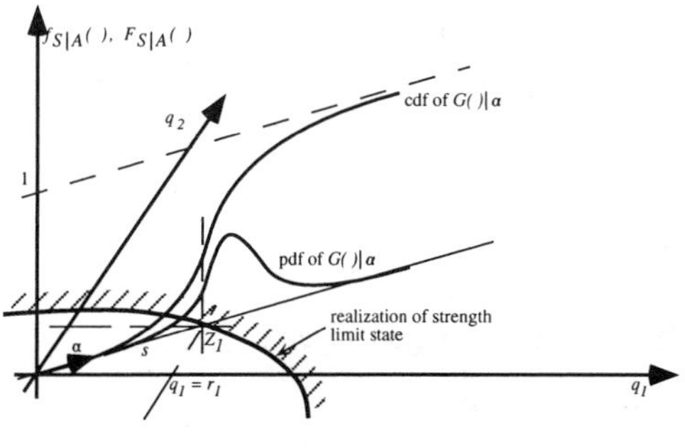

Figure 2. Schematic directional simulation in the load space showing variation of structural strength S along a sample radial direction $\mathbf{A} = \mathbf{a}$.

4 DIRECTIONAL SIMULATION IN THE LOAD PROCESS SPACE

Directional simulation in the load process space relies directly on well-known outcrossing ideas. The space ($\mathbf{X}$) is interpreted as the (m-dimensional) space of (stationary) load processes $\mathbf{Q}(t)$. Also, the domain boundary S_D can be interpreted as one realization of the structural resistance, denoted $\mathbf{R} = \mathbf{r}$. Further, the conventional limit state functions $G_i(\mathbf{q}, \mathbf{x}) = 0, i = 1, \ldots, k$ are again interpreted as probabilistic 'boundaries'. The only difference between the ideas of directional simulation presented above and that here is that the loads are now interpreted as load processes $\mathbf{Q}(t)$.

As before, the probability of failure for a structural system is given by (2). The conditional probability of failure $p_f(t_L|\mathbf{r})$ for a given resistance realization $\mathbf{R} = \mathbf{r}$ (and for the time period t_L of interest) can be estimated from the outcrossing rate v_D^+ and the initial failure probability $p_f(0)$ with the upper bound for high reliability systems (e.g.):

$$p_f(s|\mathbf{a}) \leq p_f(0, s|\mathbf{a}) + \left\{1 - \exp[-v_D^+(s|\mathbf{a})t_L]\right\} \tag{3}$$

where $p_f(0, s|\mathbf{a})$ is the failure probablity at time $t = 0$ and v_D^+ is the outcrossing rate of the vector process $\mathbf{Q}(t)$ out for the safe domain D. Directional simulation in the load process space may then be applied using (2).

The directional simulation approach in the load space has been applied for Gaussian processes and for Poisson pulse processes [Melchers, 1992; 1994; 1995]. Application to off-shore structural systems have been described by Moarefzedeh and Melchers (1996) and simple finite element applications by Guan and Melchers (1998).

5 EXAMPLE APPLICATION 1

A number of applications of the directional simulation technique are under study for structures modelled using finite elements (Guan and Melchers, 1998). Figure 3 shows one-quarter of a low-carbon steel plate with a central circular hole, subject to one-directional axial tension, modelled as uniformly applied and of uncertain magnitude. Because of the stochastic properties of the plate, symmetry of the plate physical and stochastic properties as implied in Figure 3 may not be strictly accurate. However, this assumption will be ignored in the following.

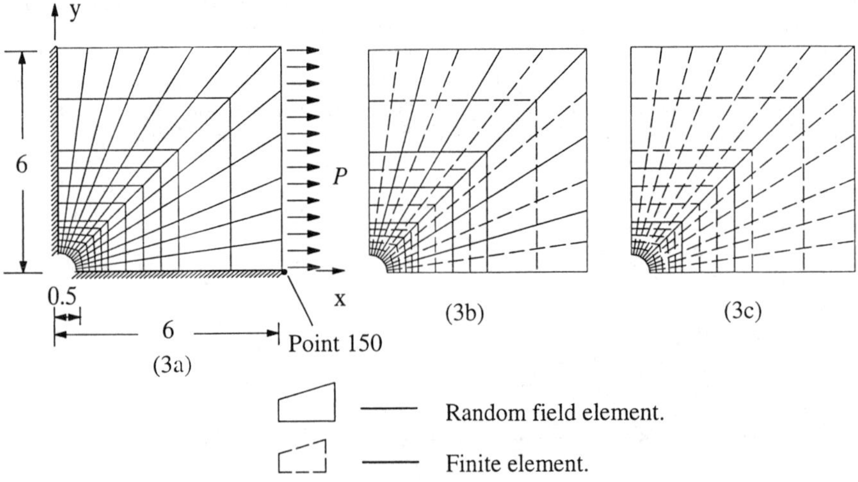

Figure 3. Finite element mesh with 132 elements and random field meshes with different stochastic discretization elements (dark boundaries) for one-quarter steel plate loaded in-plane.

Although the present problem degenerates to a one-dimensional load case it still requires a considerable number of finite elements and also stochastic discreterization for proper modelling of its stress state and its stochastic properties. The critical limit state condition $g = 0$ was taken as the condition $g(u) = 2.2\times10^{-5} - u_{150x}$ where u_{150x} denotes the horizontal displacement at point 150 (see Figure 3). Table 1 shows the assumed properties for various structural and stochastic idealizations, based on the discretization into the 132 finite elements. In order to examine the effect of the random field idealization on the distribution of s in eqn (2), three different discretizations were used for the random field E: 6, 26 and 132 elements.

Estimates for the probability of failure for a time-invariant case were obtained by Directional Simulation with 500 integration points along the radial direction s, Monte Carlo simulation with 100,000 sampling points and FOSM. The results are shown in Table 2. For the Monte Carlo approach it was found that this number of sampling points (i.e. calls to the finite element routine) was required to produce results which had sufficiently stabilized. Evidently a certain degree of judgement is implicit, therefore, in the CPU times reported for Monte Carlo. The output for the Monte Carlo approach was taken as the most accurate estimate of the probability of failure after careful consideration of the trend to convergence of the results.

It is clear that the directional simulation results are reasonably accurate, being generally close to the Monte Carlo results. It is evident also that for this example the FOSM results can be in considerable error, even though the computational times are much lower. Evidently, there may well be an appropriate trade-off between computation times and accuracy.

Table 1. Statistical properties for Example 1

Variable	Property	Mean	Standard Deviation
Uniformly distributed load intensity P	Gaussian	100 [force/unit length]	20 [force/unit length]
Elastic modulus E	Homogeneous Gaussian random field	3×10^7	4.5×10^6
Poissons ratio ν	Deterministic	0.3	
Autocorrelation coefficient for E	Isotropic	$\rho_{EE}(\Delta x) = \exp\left(-\dfrac{\Delta\mathbf{x}^T\Delta\mathbf{x}}{(\alpha L)^2}\right)$ $\Delta\mathbf{x}$ = position vector, L = plate side dimension, $\alpha = 0.125$ (dimensionless measure of correlation)	

Note: In this study all parameters are represented without specific units.

Table 2. Computational Results for Example 1

No. of Random Field Elements	Method	Estimated $p_f\times10^{-4}$	Lower Bound	Upper Bound	CPU time (minutes)
6	FOSM	13.4			0.55
	M-C	17.2	13.6	20.8	783
	DS	17.3	15.1	19.5	19.5
26	FOSM	4.07			1.8
	M-C	5.00	3.61	6.39	1616
	DS	5.01	5.01	5.01	40.6
132	FOSM	1.69			6.4
	M-C	2.20	1.28	3.12	1629
	DS	2.42	2.39	2.44	65

6 EXAMPLE APPLICATION 2

This example is similar to example 1 except that the loading is now a random variable along the edge of the plate. Let this be represented in the simple format of six uniformly distributed and independent random loads corresponding in discretization that of the main mesh. Each load has the properties shown in Table 1. Directional simulation is now in a six-dimensional space. Table 3 shows the results obtained.

Table 3. Computational Results for Example 2

No. of Random Field Elements	Method	Estimated $p_f \times 10^{-4}$	Lower Bound	Upper Bound	CPU time (minutes)
6	FOSM	7.42			0.82
	M-C	10.3	7.91	12.7	1101
	DS	8.53	7.93	9.13	70
26	FOSM	1.09			2.0
	M-C	1.63	0.74	2.51	1267
	DS	1.64	1.19	2.09	105
132	FOSM	2.82			8.34
	M-C	4.60	3.27	5.93	1631
	DS	4.64	4.36	4.92	172

7 DISCUSSION

An important assumption underlying the theory is that the system limit state function is 'load-path independent'. Thus it is not influenced by the exact nature of the realizations of the load (processes). This is also a limitation for FOSM and related techniques and also for elementary (or 'crude') Monte Carlo unless a complete loading history is traced out for each simulation.

While progress with incorporating finite element routines in structural reliability approaches has been rapid, this has been achieved mainly using simplified reliability calculation frameworks such as FOSM, together with reponse surface techniques. These have addressed time-invariant reliability or simplified time-variant reliability. As noted above, directional simulation in the load space appears to offer an intuitively simply way of joining reliability and finite element techniques, both for time-invariant and time variant situations. For the latter, no attempts appear yet to have been published - research in this area is currently under way. The aim of this work is to achieve full integration of the directional simulation approach with finite element structural (and other) analysis techniques, particularly commercially available packages, both for time-invariant and for time-variant reliability problems.

8 CONCLUSION

Directional simulation in the load space was reviewed in this paper. The technique has a natural affinity with the way structural engineers think about structural response under high load situations. The examples show that the technique can produce very good results compared with crude Monte Carlo analyses. The examples also indicate that results obtained by First Order Second Moment methods can be considerably in error even though the computation times are very low.

REFERENCES

Guan, X.L. and Melchers, R.E. (1998) A load space formulation for probabilistic finite element analysis of structural reliability, *Prob. Engineering Mechanics*, **14**, 73-81.

Melchers, R.E. (1992) Load space formulation of time dependent structural reliability, *J. Engineering Mechanics*, ASCE, **118** (5) 853-870.

Melchers, R.E. (1994) Structural System Reliability Assessment using Directional Simulation, *Structural Safety*, **16** (1+2) 23-39.

Melchers, R.E. (1995) Load space reliability formulation for Poisson pulse processes, *J. Engineering Mechanics*, ASCE, **121** (7) 779-784.

Melchers, R.E. (1999a) *Structural Reliability Analysis and Prediction* (Second Edn.) John Wiley & Sons, Chichester, England.

Melchers, R.E. (1999b) *Developments in Directional Simulation for Structural Reliability Estimation*, Proc APSSRA99, Taiwan, 51-62.

Moarefzadeh, M.R. and Melchers, R.E. (1996a) Sample-specific linearization in reliability analysis of off-shore structures, *Structural Safety*, **18** (2+3) 101-122.

Mechanics of Structures and Materials, Bradford, Bridge & Foster (eds)
© 1999 Balkema, Rotterdam, ISBN 90 5809 107 4

Use of Plant Design System (PDS) as a structural engineering tool

S.H.Teo
Barwood Parker Australia, Perth, W.A., Australia

A.K.Patnaik
School of Civil Engineering, Curtin University of Technology, Perth, W.A., Australia

D.Robinson
Plant Design System Centre, Advanced Manufacturing Technologies Centre

ABSTRACT: Plant Design System (PDS) is a powerful tool that can be used from project conception stage to plant commissioning and maintenance stage. PDS is currently very popular in many large consulting firms dealing with design of industrial plants. It offers a common ground for interaction in a multi-disciplinary environment of engineers from different disciplines. This paper outlines the successful attempt to link the plant design, and structural analysis and design process wherein the models developed by plant designers can be transferred to a structural analysis program (Space Gass) and vice versa. Also listed are a few limitations of the entire two-way transfer process. It is possible to use PDS system as a valuable structural engineering tool in future.

1 INTRODUCTION

Plant Design System (PDS) is now a popular tool to develop drawings from conception through to construction, commissioning and maintenance phases of major projects (Backert 1997). PDS is a common platform for engineers from different disciplines to interact, provide feedback and specify their disciplinary requirements in a multi-disciplinary project (Flowers 1998). The interaction between different disciplines can be represented as shown in Figure 1. Drawings are developed in different stages for any project. Initially, the plant layout is developed, and then the general arrangement drawings are developed. Structural engineers work from the preliminary drawings and come up with tentative structural layout drawings, and structural member shapes and sizes. In PDS environment, the structural engineers need to manually pass on the details of the members and structure geometry to the drafting designers who incorporate these details into their FrameWork Plus (Intergraph 1993) models. Structural engineers develop their own models to perform analysis and design of the structure using a structural analysis and design program, say Space Gass (Space Gass 1998).

In effect, two models are developed for the same plant structure, one in FrameWorks Plus by the drafting designers and one for structural analysis and design by the structural engineers. In a small project this type of duplication of efforts does not have major cost impact, but for a major project, it is possible to reduce the duplication by judiciously sharing the electronic models between all the parties involved in a project. In FrameWorks Plus model, geometry and properties of members are input to develop a 3D model. The same model can be used by structural engineers to analyse and design the structure. The structural engineer revises the member sizes if the members shown in the FrameWorks Plus model are structurally inadequate or overly conservative. It is possible to revise the member sizes in the analysis program and pass on the electronic file of the structure model to drafting designer who can read the file into a PDS program. This paper describes recent attempt to export/import electronic database from FrameWorks Plus to a structural analysis and design program (Space Gass), and vice versa.

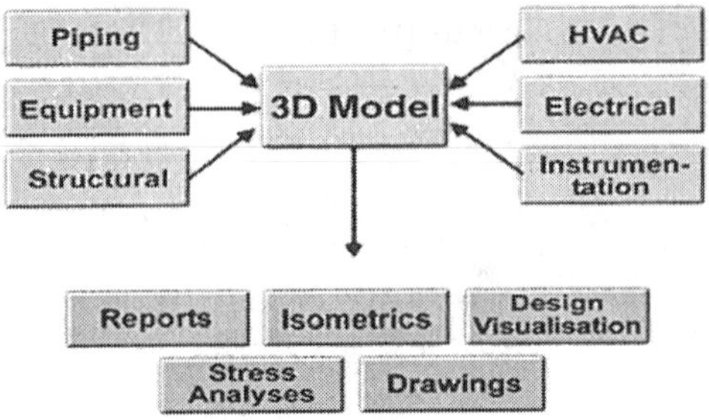

Figure 1 PDS Work Flow Diagram

2 AUTOMATING STRUCTURAL DRAFTING AND DESIGN

Automation of structural drafting and design was done in this project by integrating models developed in FrameWorks Plus and Space Gass. FrameWorks Plus provides a convenient platform for defining the geometry of a structure and placement of members into a model in the same way as they would be framed in the site. It automatically places node data at both ends of members and splits elements into finite segments during pre-processing, while maintaining the collective status of the physical elements during model revisions. This physical member data structure eliminates modelling errors by enabling the software to assign member releases at each end of the true physical member, while preventing release assignments from being made to interior nodes of continuous physical girders. FrameWorks Plus in the MicroStation (Bentley 1997) environment automates the design geometry with linear member layouts and modelling functions. However, a conversion program is necessary to make the information generated in MicroStation relevant to structural analysis and design programs. A "conversion" program (Lowe 1998) that is commercially available for transferring data from PDS format to Space Gass format was used in this project. This conversion program was made available for this study by ICF Kaiser Pty.Ltd (Nune 1997).

3 THE LINK

This link (a conversion program) is capable of reading data from both FrameWorks Plus files in a MicroStation environment and writing the data in format readable by Space Gass. Furthermore, the conversion program is also capable of reading Space Gass files and writing the data in a format that can be read by FrameWorks Plus. The link between MicroStation and Space Gass is essentially a translator utility program that allows the sharing of design and drafting data between FrameWorks Plus and Space Gass. By using an internally neutral format database and a collection of map files, the "converter" converts files between a range of popular modelling/analysis/design/drafting/detailing features in the structural engineering module.

A model of the structure can be generated within FrameWorks Plus and then imported into Space Gass to provide the engineer with a model that is dimensionally correct with section sizes, member orientation, member offsets, etc. Structural analysis and design can be performed in Space Gass and the member data updated as per structural requirements. The Space Gass file can then be transferred back to the FrameWorks Plus for development of construction drawings.

When a model is exported from FrameWorks Plus, additional files are generated that allow for a "changes file" that can be read back into FrameWorks Plus after analysis in Space Gass, thus updating the section properties of the members that needed revision. This feature is useful if a structure has prefixed layout and geometry, and only the section sizes can be changed by the structural engineer.

Conversely, a model can be generated within Space Gass to start with, and then its geometric data exported into FrameWorks Plus, by using the Space Gass text file facility.

4 HOW DOES THE STRUCTURAL CONVERTOR WORK ?

FrameWorks Plus assumes a simple concept that there are 4 elements in a steel model: beams, columns, horizontal braces and vertical braces. Space Gass does not identify such a difference. This is not a problem when exporting data from FrameWorks Plus to Space Gass. However, when a FrameWorks Plus model is created from Space Gass data, it can cause inconvenience. The converter does not recognise the difference in the element classification in FrameWorks Plus. By default, it classifies every element as a beam. This results in a model that requires extensive post-processing for use in FrameWorks Plus. To overcome this problem, the following classification to the element types was adopted and applied to the incoming elements:

Vertical	Column
Flat & parallel to the X-Y plane	Beam
Dead flat but not a beam	Horizontal brace
Others	Vertical brace

Space Gass allows the user to place a section mark (classification) against section designations. This feature is used when CAD drawings are produced from inside Space Gass. This information which is contained within the Space Gass text file can be used as a "marker" to indicate the element type. For example, if in the Space Gass model contains vertical braces, by placing the appropriate tag to the section mark in the Space Gass section table, this information is interpreted by the "converter" and creates the "tagged" sections as the correct FrameWorks Plus element type. The following tags were used, C for columns, B for beams, H for horizontal braces and V for vertical braces. If no tags are assigned, then the importation resorts to the "converter" recognising the untagged members as beam elements.

5 MAP FILES

Map files are utilised to configure a method of translating section designations between various packages. Two map files are used in a conversion. One map file for import and the other for the export file. During the import process, the section designations contained within the imported file are compared to the contents of the import map file. When a match is found the application stores the section designation into its own internal format.

When the export process occurs, the application's internal format section designation is compared with the export map file. When a match is found, the export format section designation is then written into the export file. If no matches are made in either the import or export phases, the "converter" will write a "BOGUS" designation into the export file, either a huge member or a member that is not used in the entire project.

The default structural information protocol of FrameWorks Plus is STAAD III. STAAD III is a structural analysis package that was developed by Research Engineers Incorporated and is widely used overseas. It's text file format is in an ASCII mode and has become a common basis for transferring data between structural engineering packages. The application compares the internal format section designation with the target file for a match. When a match is found, the target file section designation is written to the output file. Figure 3 shows the conversion process of element designation.

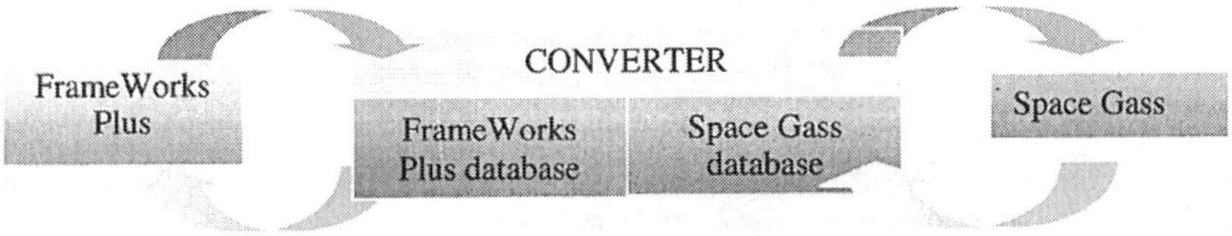

Figure 2 Flow between map file for conversion

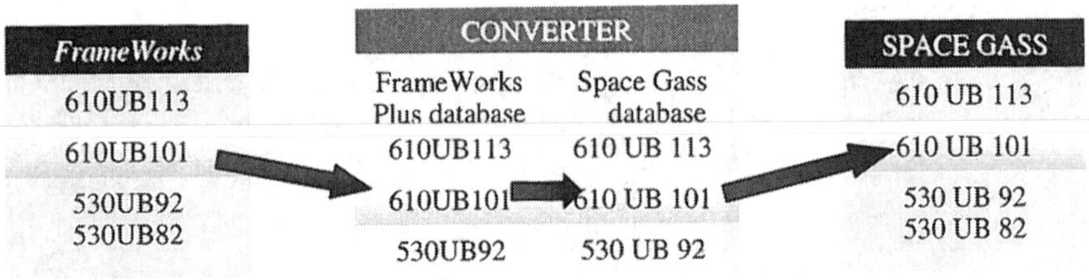

Figure 3 Conversion from STAAD III file designation to Space Gass file designation

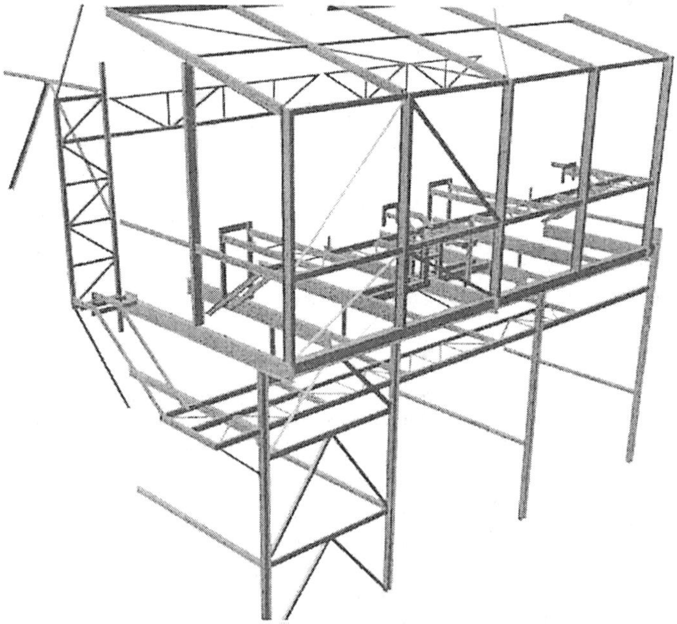

Figure 4 Example of approximate representation of structure

The map files are written in ASCII text file format and can be edited. Every option that is referenced in the section designation must contain this information within its map file.

Map files:
STAADIII (FrameWorks Plus) converts from/to STAAD.map
FrameWorks uses STAAD.map
Space Gass converts from/to Space Gass.map

6 WORK SEQUENCE

Following is an example of the work sequence with the conversion process between the FrameWorks Plus model and the analytical model in Space Gass.

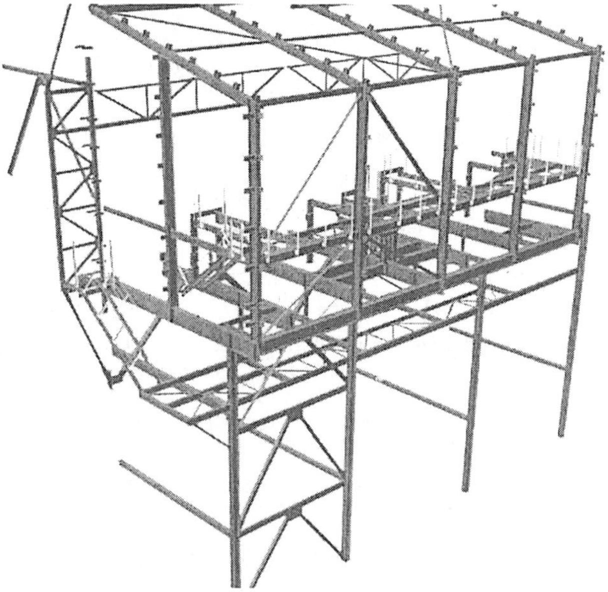

Figure 5 Shop detail model of the example structure of Figure 4

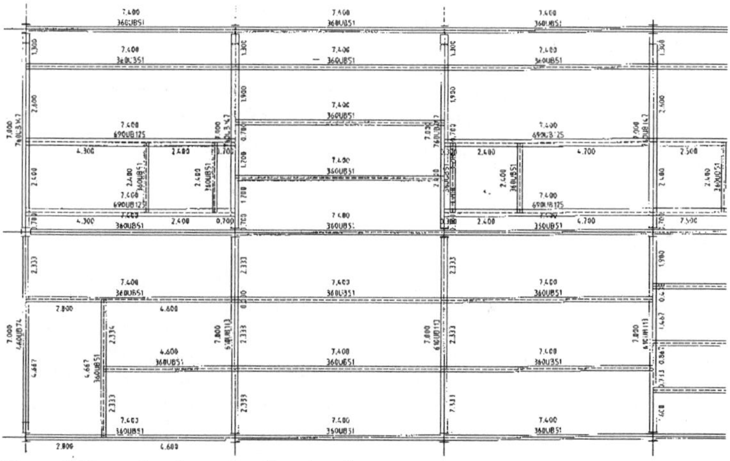

Figure 6 Example of construction drawing

6.1 Step 1

The modelling information of a structure (see Figure 4) is entered into the PDS model. The structural information at this stage is only geometrically correct, not necessarily structurally adequate. Geometry can be adopted from the owner's specifications and drawings and/or from the designers' intuition and experience. Even if the members are adequate, structural analysis and design check need to be conducted to prove structural integrity.

The model at this stage of the process is usually defined to conform to the requirements of the other engineering disciplines. The structure is developed to carry loads as per the design basis. The model at this stage is only geometrically correct.

6.2 *Step 2*

The conversion of FrameWorks Plus (in a STAAD III format) to Space Gass text file format is done in this step for analysis and design of the structure. After analysis and design, the model is converted back to STAAD III format in order to be imported into the 3D model. Any changes to the member sizes required from structural design are introduced in this step.

6.3 *Step 3*

Shop details of connections, floor coverings, purlins and handrails are added to the model in this step (see Figure 5).

6.4 *Step 4*

The structural model is now ready to be added to the global PDS model. The PDS model acts as a storage point of all the information developed from the project. This up-to-date information is now ready for extraction for other downstream applications. For example, Figure 6 is an extract drawing of a floor plan from a finished PDS model. Such geometric extractions can be sent to CNC steel fabricators and constructors.

7 LIMITATIONS

- Models cannot be directly updated by importing geometric data from Space Gass into the FrameWorks Plus model.
- Analytical outputs cannot be included in the transfer into the FrameWorks Plus.
- As PDS attaches an exclusive number in its database to every information, renumbering of nodes and members to suit convenience cannot be done in an external environment. PDS will reject such a change and deem the modified model illegal.
- The current way of sharing files, although convenient, has fundamental limitations that can impede its efficiency and usefulness as an engineering tool.
- The analytical output capabilities are indeed necessary in the global model. Remembering that the goal of PDS is to involve the processes from design to construction, the developed model has to have the capabilities of being extracted down to the steel detailers, the steel fabricators, and to produce construction drawings.

8 CONCLUDING COMMENTS

Multidisciplinary integration of drafting, design and analysis is the future of Computer Aided Engineering (CAE). PDS and information technology appear sure to play a major role in structural engineering in future. It is possible to automate design and drafting process, but the automation will lead to disasters if the person operating the programs does not have the expertise. However, judicious use of the transfer process briefly outlined in this paper will lead to reduced efforts in a major project. It is unlikely that such a transfer process can find day-to-day application in a design office in the near future, but, it may happen, one day !

REFERENCES

Beckert B.A. 1997. Making the move to 3D design, *Mechanical Solutions*, Fall, pp.12-14.

Bentley 1997. *MicroStation 95, (Windows NT),* [Computer program], Bentley Systems Inc.

Flowers, J. 1998. Multidisciplinary design is the future, *Computer-Aided Engineering*, March, pp.52-56.

Intergraph 1993. *FrameWorks Plus version 3.0*, [Computer program], Intergraph Corp., Huntsville, Alabama.

Lowe, A. 1998. *FrameWorks Plus Converter version 3.0*, (Windows NT), [Computer program], Wombat High Technology, Victoria.

Nune, R. 1997, Engineering Sequence CAE Implementation, *ICF Kaiser Pty.Ltd.*, Perth.

Space Gass Reference Manual 1998. V8.0, Integrated Technical Software Pty Ltd, Geelong, Victoria.

Mechanics of Structures and Materials, Bradford, Bridge & Foster (eds)
© 1999 Balkema, Rotterdam, ISBN 90 5809 107 4

Analysis, design and optimization of composite structures

A. L. Kalamkarov
Department of Mechanical Engineering, Dalhousie University, Halifax, N. S., Canada

ABSTRACT: Analysis and design problems for fibre-reinforced composite structures are based on the analytical solution for the effective moduli of high-stiffness fibre-reinforced composite material obtained using the asymptotic homogenization techniques. The set of prescribed effective stiffnesses for which the design problem is solvable, is described, and the method of the design parameters calculation based on the convex analysis is developed. Design problems are generalized on account of strength of composite material. The developed approach can be readily applied in practical design of highly effective and economical composite structures.

1 FIBRE-REINFORCED COMPOSITE MATERIAL

The effective properties of the composite material of a periodic structure can be calculated by means of the asymptotic homogenization method. This approach is mathematically justified, and it enables the calculation of both local and overall effective properties of the composite material, see Kalamkarov (1992) and Kalamkarov & Kolpakov (1997). In the present paper, these results are taken as basis to formulate and solve the design problems for the high-stiffness fibre-reinforced composite materials with the required set of effective moduli and the required strength.

Consider the fibre-reinforced composite material, and assume that the material of fibres is much stiffer than the matrix material, i.e., $E_F \gg E_M$. This assumption is typical for the polymer matrix fibre-reinforced composites. We further assume that all the fibres are parallel to coordinate plane $0x_1x_2$.

Application of the asymptotic homogenization method provides the following formulas for the effective moduli $\{a_{ijkl}\}$, (i,j,k,l,=1,2) in the plane of reinforcement, see Kalamkarov (1992):

$$a_{ijkl} = ES\sum_{\alpha=1}^{M} \gamma^{\alpha}{}_i\gamma^{\alpha}{}_j\gamma^{\alpha}{}_k\gamma^{\alpha}{}_l\mu_\alpha \qquad (1)$$

Here S is a fibre volume content, μ_α is a proportion of fibre content, and $\{\gamma^{\alpha}{}_i\}$, (i=1,2) are the direction cosines of the fibres within a reinforcing fibre family with a number α, $\gamma^{\alpha}{}_1 = \cos\varphi_\alpha$, $\gamma^{\alpha}{}_2 = \sin\varphi_\alpha$, φ_α, (α=1,2,...,M) are the fibre placement angles, and M is a total number of reinforcing fibre families.

The following natural limitations are imposed on μ_α and φ_α:

$$\sum_{\alpha=1}^{M} \mu_\alpha = 1, \quad \mu_\alpha \geq 0, \ \alpha = 1,2,\ldots,M \qquad (2)$$

$$\varphi_\alpha \in [0, \pi], \alpha = 1,2,...,M \tag{3}$$

The asymptotic homogenization method also provides the following formula for the micro stresses in the composite material, see Kalamkarov (1992):

$$\sigma_{ij} = E \gamma^{\alpha}{}_{i}\gamma^{\alpha}{}_{j} \sum_{k,l=1,2} \gamma^{\alpha}{}_{k} \gamma^{\alpha}{}_{l} \varepsilon_{kl} \tag{4}$$

Here $\{\varepsilon_{kl}\}$ are strains resulting from the solution of an averaged global problem formulated for the homogenized anisotropic material with the effective moduli $\{a_{ijkl}\}$, see Kalamkarov (1992). The axial stress in fibres of a family number α equals

$$\sigma_n{}^{\alpha} = E \sum_{k,l=1,2} \gamma^{\alpha}{}_{k} \gamma^{\alpha}{}_{l} \varepsilon_{kl} \tag{5}$$

2 DESIGN AND OPTIMIZATION PROBLEMS

Applying the above Eqs. (1) and (4), one can determine the effective elastic moduli and micro stresses in fibres. The strength of fibres can be then analyzed using the following strength criterion for the fibre material:

$$0 \le F(\sigma_{ij}) \le \sigma^* \tag{6}$$

That will constitute the problem of the strength analysis for the composite material. In the present paper, we will consider the design problem for the fibre-reinforced composite material that will represent an inverse problem in regards to the above formulated strength analysis problem. The design problem of interest will be concerned with the following question: what should be the composition and the reinforcment arrangement that will provide the composite material with the required (prescribed) set of the effective elastic moduli $\{a_{ijkl}\}$, and with the required strength in terms that this composite material will sustain the given averaged stresses

$$\hat{\sigma}_{ij} = a_{ijkl}\{\varepsilon_{kl}\} \tag{7}$$

The design parameters, i.e., the parameters that fully define the composition and structure of the composite material, are the following: S, M, $\{\mu_\alpha\}$ and $\{\varphi_\alpha\}$.

The mathematical formulation of the above design problems will be the following. According to Eq.(1), the effective moduli $\{a_{ijkl}\}$ are expressed in terms of the following four functions of the arguments $\varphi = (\varphi_1, \varphi_2, ..., \varphi_M)$, and $\mu = (\mu_1, \mu_2, ..., \mu_M)$:

$$y_1(\mu,\varphi) = \sum_{\alpha=1}^{M} \mu_\alpha \cos^4\varphi_\alpha,\ y_2(\mu,\varphi) = \sum_{\alpha=1}^{M} \mu_\alpha \sin^4\varphi_\alpha \tag{8}$$

$$y_3(\mu,\varphi) = \sum_{\alpha=1}^{M} \mu_\alpha \sin\varphi_\alpha \cos^3\varphi_\alpha,\ y_4(\mu,\varphi) = \sum_{\alpha=1}^{M} \mu_\alpha \sin^3\varphi_\alpha \cos\varphi_\alpha$$

Eq. (1) can be resolved explicitly, so that

$$y_1 = a_{1111}(ES)^{-1},\ y_2 = a_{2222}(ES)^{-1},\ y_3 = a_{1112}(ES)^{-1},\ y_4 = a_{2221}(ES)^{-1} \tag{9}$$

The following relation should be fulfilled:

$$a_{1122} = a_{1212} = 0.5\ (ES - a_{1111} - a_{2222}) \qquad (10)$$

Eq. (10) represents a solvability condition for the system (9). Eqs. (8) and (9) together with conditions (2) and (3) represent a problem of convex combinations for the fibre fractions $\{\mu_\alpha\}$. The solvability condition for this problem can be formulated as follows:

2.1 *Statement 1*

If $M \geq 5$, then the design problem (2), (3), (8) and (9) is solvable if and only if the point $\mathbf{y} = (y_1,y_2,y_3,y_4)$ belongs to a convex hull, *conv* Γ, (see Rockafellar, 1970) of the following curve Γ:

$$\Gamma = \{\mathbf{y} \in R^4: \mathbf{y} = (\cos^4\varphi,\ \sin^4\varphi,\ \sin\varphi\ \cos^3\varphi,\ \sin^3\varphi\ \cos\varphi),\ \varphi \in \Phi\} \qquad (11)$$

where Φ is a set of allowable fibre placement angles.

2.2 *Averaged strength criterion*

The averaged strains for the homogenized material with the effective moduli $\{a_{ijkl}\}$ are

$$\varepsilon_{kl} = \{a_{klmn}\}^{-1}\ \hat{\sigma}_{mn} \qquad (12)$$

Strength criterion (6) on account of Eqs. (4) and (12) yields

$$F_\alpha = F\Big(E\ \gamma^\alpha{}_i\gamma^\alpha{}_j \sum_{k,l=1,2} \gamma^\alpha{}_k\ \gamma^\alpha{}_l \sum_{m,n=1,2} \{a_{klmn}\}^{-1}\ \hat{\sigma}_{mn}\Big) \leq \sigma^* \qquad (13)$$

in the fibre family number α.

Eq. (13) represents an averaged strength criterion since, unlike the criterion (6), it is formulated in terms of the averaged stresses $\hat{\sigma}_{mn}$. Eq. (13) can be represented as follows, since $\{\gamma^\alpha{}_i\}$ could be expressed in terms of $\{\varphi_\alpha\}$, and $\{a_{ijkl}\}$ can be expresed in terms of $\mathbf{y}$:

$$F_\alpha(\varphi_\alpha,\ y,\ \hat{\sigma}_{mn}) \leq \sigma^* \qquad (14)$$

in the fibre family number α. Here function F_α is known since it can be derived from the left-hand side of the Eq. (13) by replacing $\{\gamma^\alpha{}_i\}$ and $\{a_{ijkl}\}$ by their expressions in terms of $\{\varphi_\alpha\}$ and $\mathbf{y}$.

To formulate a strength criterion for the all reinforcing families, we introduce the following function:

$$\mathfrak{M}(\hat{\sigma}_{mn},\ \{\varphi_\alpha\}) = \max_{\alpha=1,2,\ldots,M} F_\alpha(\varphi_\alpha,\ y,\ \hat{\sigma}_{mn}) \qquad (15)$$

where maximum is determined for the all placement angles, φ_α, of the all actually represented reinforcing families of the composite material.

The averaged strength criterion for the reinforcing fibres under the given averaged stresses $\hat{\sigma}_{mn}$ can be then formulated as follows:

$$\mathfrak{M}(\hat{\sigma}_{mn}, \{\varphi_\alpha\}) \leq \sigma^* \tag{16}$$

2.3 *Typical design problem formulations*

The following basic design problems can be distinguished:

Problem 1 consists in design of the fibre-reinforced composite material which will have the given set of the effective moduli $\{a_{ijkl}\}$. The problem given by the Eqs. (2), (3), (8) and (9) should be solved to calculate the design parameters S, M, $\{\mu_\alpha\}$ and $\{\varphi_\alpha\}$.

Problem 2 consists in design of the fibre-reinforced composite material with the required effective moduli $\{a_{ijkl}\}$ that will have a minimum fibre volume content. In this case, in addition to solving the problem (2), (3), (8) and (9), the fibre volume content, S, should be minimized:

$$S \to min \tag{17}$$

Since fibre material is commonly heavier than the matrix material, the overall weight of the composite material will be also minimized by the solution of this design problem.

Problem 3 consists in design of the fibre-reinforced composite material with the required effective moduli $\{a_{ijkl}\}$ and with the required strength: material should sustain the required average stresses. In this case, the problem (2), (3), (8) and (9) should be solved, and the condition (16) should be fulfilled for any of prescribed stresses $\hat{\sigma}_{mn}$.

Problem 4 consists in design of the maximum strength fibre-reinforced composite material with the required effective moduli $\{a_{ijkl}\}$. In this case, in addition to solving the problem formulated by the Equations (2), (3), (8) and (9), the left-hand side of the Eq. (16) should be minimized:

$$\mathfrak{M}(\hat{\sigma}_{mn}, \{\varphi_\alpha\}) \to min \tag{18}$$

and the value of $\sigma^*/[min\, \mathfrak{M}(\hat{\sigma}_{mn}, \{\varphi_\alpha\})]$ will define the safety factor of the composite material reinforcement.

The above design problems with the infinite number of the allowable fibre placement angles are usually rather complex. These problems can be reasonably simplified by discretization of magnitudes of the allowable placement angles. In the discrete design problem, the allowable placement angles are limited by a discrete set

$$\Phi_N = \{\varphi_\beta, \beta=1,2,...,N\} \tag{19}$$

3 SOLUTION OF THE DESIGN PROBLEMS

Let us denote, $\mathbf{z}_\alpha = (z_{\alpha 1}, z_{\alpha 2}, z_{\alpha 3}, z_{\alpha 4}) \in R^4$, $\alpha=1,2,...,M$, where

$$\mathbf{z}_{\alpha 1} = \cos^4\varphi_\alpha,\ \mathbf{z}_{\alpha 2} = \sin^4\varphi_\alpha,\ \mathbf{z}_{\alpha 3} = \sin\varphi_\alpha \cos^3\varphi_\alpha,\ \mathbf{z}_{\alpha 4} = \sin^3\varphi_\alpha \cos\varphi_\alpha \tag{20}$$

The problem (2) and (8) can be rewritten as follows:

$$\sum_{\alpha=1}^{M} z_{\alpha i}\, \mu_\alpha = y_i,\ (i=1,2,3,4), \qquad \sum_{\alpha=1}^{M} \mu_\alpha = 1,\ \mu_\alpha \geq 0,\ \alpha=1,2,...,M \tag{21}$$

We will solve the problem (21) by applying the method of convolution of the system of linear equation. This method is based on the possibility of explicit solution of the following one-dimensional problem

$$\sum_{\beta=1}^{m} z_\beta\, \mu_\beta = y, \quad \mu_\beta \geq 0, \quad (\beta=1,2,...,m), \qquad \sum_{\beta=1}^{m} \mu_\beta = 1 \tag{22}$$

3.1 *Statement 2*

Let us assume that z_β values in Eq. (22) are arranged in the order of growth, so that $z_1<z_2<...<z_m$. That does not limit the generality of the solution. The following statements are proved:
(i) Problem (22) is solvable if and only if the following condition is fulfilled:

$$z_1 \leq y \leq z_m \tag{23}$$

(ii) If the condition (23) is fulfilled, then point *y* can be represented as a convex combination of points $\{z_q\}$ and $\{z_r\}$, such that $z_q \leq y$, $z_r > y$, as follows:

$$y = \lambda_q z_q + \lambda_r z_r \tag{24}$$

For the given *q* and *r* values we introduce the *m*-dimensional vector

$$E_\eta = \{E_{\eta\beta} = 1, \text{ if } \beta=q, \text{ or } \beta=r; \text{ and } \beta= 0, \text{ if } \beta\neq q, \text{ and } \beta\neq r; \beta=1,2,...,m\} \tag{25}$$

The set of solutions of the problem (22) will be found in form of the following convolution:

$$\mu_\beta = \sum_{\eta=1}^{M_1} E_{\eta\beta}\, \lambda_\eta \tag{26}$$

Here M_1 is a total number of segments $[z_q,z_r]$, that contain the point y, $\eta=1,2,...,M_1$, and $\{\lambda_\eta\}$ are arbitrary numbers satisfying following condition:

$$\sum_{\eta=1}^{M_1} \lambda_\eta = 1, \quad \lambda_\eta \geq 0, \ (\eta=1,2,...,M_1) \tag{27}$$

Note that λ_q, λ_r values in Eq. (24) can be calculated as follows:

$$\lambda_q = \frac{(x - z_q)}{(z_q - z_r)}, \quad \lambda_r = 1 - \lambda_q \tag{28}$$

Let us return now to system (21). The convolution of this system can be performed in the following way. For *i* = *1* in Eq. (21) we get a first equation which is one-dimensional problem of complex combinations. If this problem is solvable, see Statement 2 (i), then its solution takes the form (26), (27). Substituting solution (26) into the remaining three equations of system (21), i.e. for *i=2,3,4*, and changing the order of summation, we obtain

$$\sum_{\eta_1=1}^{M_1} \lambda_{\eta_1} \sum_{\alpha=1}^{M} z_{\alpha i} E_{\eta_1 \alpha} = y_i, \ i=2,3,4 \tag{29}$$

First equation (for *i* =*2*) in the system (29) together with Eq. (27) is again a problem of complex combinations. This problem can be solved in the similar way, and the obtained solution can be substituted into two remaining equations (29), for *i=3,4*. As a result, by repeating this procedure four times, all the four equations (21) will be satisfied (if the condition (23) is fulfilled at each of these steps). The resulting solution can be represented as follows:

$$\mu_\alpha = \sum_{\eta_4=1}^{M_4} \lambda_{\eta_4} P_{\eta_4 \alpha}, \quad \lambda_{\eta_4} \geq 0, \quad \sum_{\eta_4=1}^{M_4} \lambda_{\eta_4} = 1 \tag{30}$$

where

$$P_{\eta_4} = \{P_{\eta_4 \alpha}\} = \{ \sum_{\eta_1=1}^{M_1} \sum_{\eta_2=1}^{M_2} \sum_{\eta_3=1}^{M_3} E_{\eta_4\eta_3} E_{\eta_3\eta_2} E_{\eta_2\eta_1} E_{\eta_1\alpha} \}, \quad \alpha = 1,2,\ldots,M \tag{31}$$

It is seen from Eqs. (30) and (31) that there is an infinite number of solutions of the problem (8) (or (21)), and these solutions are expressed in terms of a finite number of vectors P_{η_4}.

3.2 *Design example*

It is required to design a composite material with a general type of glass fibre reinforcement. The composite material should have the following effective moduli: a_{1111}= 25 GPa, and a_{2222} = 10 GPa. Young's modulus of glass fibres is E = 70 GPa, and the prescribed fibre volume content is S = 0.6. There are no limitations on the values of fibre placement angles, $\Phi = [-\pi/2, \pi/2]$. We discretize this interval with increments $\delta = \pi/15$. The solution (30), (31) is found numerically. We calculated M_4 = 281 different vectors P_{η_4}, and among them, the following two:

$$\begin{aligned} P_1 &= (0.1736, 0.0545, 0, 0, 0, 0, 0.4009, 0, 0, 0.2285, 0.1426, 0, 0, 0, 0) \\ P_2 &= (0, 0, 0.2116, 0, 0, 0, 0, 0.3946, 0.1819, 0, 0.0004, 0, 0.2115, 0, 0) \end{aligned} \tag{32}$$

The corresponding design projects are the following:

$$\begin{aligned} &\mu_1 = 0.1736\lambda_1, \mu_2 = 0.0545\lambda_1, \mu_3 = 0.2116\lambda_2, \mu_4 = \mu_5 = \mu_6 = 0, \mu_7 = 0.4009\lambda_1 \\ &\mu_8 = 0.3946\lambda_2, \mu_9 = 0.1819\lambda_2, \mu_{10} = 0.2285\lambda_1 \\ &\mu_{11} = 0.1426\lambda_1 + 0.0004\lambda_2, \mu_{12} = 0, \mu_{13} = 0.2115\lambda_2, \mu_{14} = \mu_{15} = 0 \end{aligned} \tag{33}$$

where $\lambda_1 + \lambda_2, \geq 0, \lambda_1 + \lambda_2 = 1$. In particular, we can choose $\lambda_1 = 0$, $\lambda_2 = 1$ to satisfy this condition. According to Eq. (33), in this case we obtain the following composite material design:

$$\begin{aligned} &\mu_3 = 0.2116, \varphi_3 = -\pi/2 + 3\pi/15, \quad \mu_8 = 0.3946, \varphi_8 = -\pi/2 + 8\pi/15, \quad \mu_9 = 0.1819 \\ &\varphi_9 = -\pi/2 + 9\pi/15, \mu_{11} = 0.0004, \varphi_{11} = -\pi/2 + 11\pi/15, \mu_{13} = 0.2115, \varphi_{13} = -\pi/2 + 13\pi/15 \end{aligned} \tag{34}$$

Other reinforcing fibre families are not used in the design (34).

REFERENCES

Kalamkarov, A.L. 1992. *Composite and Reinforced Elements of Construction*. Chichester, New York: Wiley.

Kalamkarov, A.L. & A.G. Kolpakov. (1997) *Analysis, Design and Optimization of Composite Structures*. Chichester, New York: Wiley.

Rockafellar, R.T. 1970. *Convex Analysis*. Princeton, N.J.: Princeton University Press.

Mechanics of Structures and Materials, Bradford, Bridge & Foster (eds)
© 1999 Balkema, Rotterdam, ISBN 90 5809 107 4

Optimum rigid road pavement design using genetic algorithm

M.N.S. Hadi & Y. Arfiadi
Department of Civil, Mining and Environmental Engineering, University of Wollongong, N.S.W., Australia

ABSTRACT: The design of rigid pavements according to AUSTROADS involves assuming a pavement structure then using a number of tables and figures to calculate the two governing design criteria, the flexural fatigue of the concrete base and the erosion of the subgrade/sub-base. Each of these two criteria needs to be less than 100%. The designer needs to ensure that both criteria are near 100% so that safe and economical designs are achieved. This paper presents a formulation for the problem of optimum rigid road pavement design by defining the objective function, which is the total cost of pavement materials, and all the constraints that influence the design. Genetic algorithm is used to find the optimum design. The results obtained from the genetic algorithm are compared with results obtained from a Newton-Raphson based optimisation solver. The latter being developed using spreadsheets.

1 INTRODUCTION

Rigid pavements have been used all over the world for several decades and different road authorities and organisations have developed design methods that suit their locale. To name a few, the American Association of State Highway and Transportation Officials, AASHTO, the Portland Cement Association, PCA, and the Corps of Engineers of the US Army. These methods and others are documented in several textbooks, for example Yoder and Witczak (1975) and Huang (1993). Most design methods are complex where the designer needs to refer to several tables and charts and use some rather lengthy design formulas. The iterative design methods invite the user to use several iterations in order to obtain a solution that satisfies the design requirements. Due to the lengthy design procedure most designers would stop after two or three iterations thus yielding designs that are safe but not necessarily economical. This over-design, which comes with a higher cost as a penalty, jeopardises the viability of using rigid pavements and reduces its effectiveness as a viable alternative for building road pavements. There has been a number of attempts to optimise the design of pavements (Hadi 1998), but none of these attempts are for the optimum design of rigid pavements and none by using Genetic Algorithm, GA.

In this paper, the design method of the Australian pavement design guide, AUSTROADS (1992) which is based on the recommendations of the PCA (1984) is presented and the optimisation problem is developed for this method. GA is chosen as a tool for the optimisation problem. GA was implemented using Matlab and a number of design examples are solved and compared with other optimisation techniques.

2 AUSTROADS DESIGN METHOD OF RIGID PAVEMENTS

The design method for rigid pavements presented by Fordyce and Yrjanson (1969) is the background of the current design method of Portland Cement Association. This method takes into consideration the stresses induced in the concrete slab due to the application of vehicular load and those induced due to concrete contraction and warping. This forms the basis of the design method of the PCA (1984) and hence of AUSTROADS (1992). The design method of AUSTROADS (1992) can be summarised as

(1) Decide on the type of pavement, ie PCP, JRCP, or CRCP.
(2) Decide whether concrete shoulders are to be provided.
(3) Estimate the daily number of commercial vehicular loading, C.
(4) Estimate the growth rate for the pavement, r.
(5) Decide on the design life of the pavement, N.
(6) Calculate the total number of commercial vehicles to be applied on the pavement throughout its design life, $CVAG$. This is done by applying equation 1.

$$CVAG = C * 365 * \frac{(1+r)^N - 1}{r} \tag{1}$$

where C is the daily number of commercial vehicles, r is the growth rate in percent and N is the pavement life in years.

(7) Estimate the CBR of the subgrade.
(8) Based on $CVAG$ and subgrade CBR and by using Figure 9.1 of AUSTROADS, determine the minimum sub-base requirement.
(9) Choose type and thickness of the sub-base and by using Figure 9.2 of AUSTROADS, determine the effective CBR.
(10) Choose the design compressive strength of concrete, f'_c which should have a minimum value of 32 MPa. From this and by applying equation 2, the flexural strength of concrete, f'_{cf} is to be calculated.

$$f'_{cf} = 0.75\sqrt{f'_c} \tag{2}$$

(11) Choose a trial thickness of the base, h which should be at least 150 mm.
(12) Based on h, the effective CBR, the existence of shoulders and whether the pavement is dowelled or undowlled, use Table 9.2 or 9.3 to calculate the stress factor and erosion factor for each type of axle.
(13) Choose a load safety factor, LSF. This depends on the locality and importance of the road.
(14) For each type of axle group and based on whether the pavement is located in a rural or urban area, determine the percentages of each of the axle groups and the distribution of the different axle loads in each group. This is given in Appendix I of AUSTROADS. Multiply this distribution by the total vehicular load, $CVAG$ and multiply it by the load safety factor, LSF. This will yield the expected load repetitions for each axle load within the axle groups.
(15) For each axle load of each axle group, calculate the allowable load repetitions based on limiting the fatigue of concrete and the erosion of the sub-base/subgrade. These can be calculated either by using the monographs shown in Figures 9.4, 9.5 and 9.6 of AUSTROADS or by using the following equations as given by Packard and Tayabji (1985). The equations for the fatigue of concrete are,

$$\log(N_f) = \left[\frac{0.9718 - S_r}{0.0828}\right] \qquad \text{when } S_r > 0.55 \tag{3}$$

$$N_f = \left[\frac{4.2577}{S_r - 0.4325}\right]^{3.268} \qquad \text{when } 0.45 \le S_r \le 0.55 \tag{4}$$

$$N_f \text{ is undefined} \qquad \text{when } S_r < 0.45$$

where

$$S_r = \frac{S_e}{f'_{cf}}\left[\frac{P \times LSF}{4.45F_1}\right]^{0.94} \tag{5}$$

and

N_f = allowable load repetitions based on fatigue
S_e = equivalent stress, MPa
f'_{cf} = design flexural strength of concrete, MPa

P = axle load, kN
LSF = load safety factor
F_1 = load adjustment factor
= 9 for single axle with single wheel
= 18 for single axle with dual wheel
= 36 for tandem axle with dual wheel
= 54 for triaxle with dual wheel

For erosion,

$$\log(F_2 N_e) = 14.524 - 6.777\left[\left(\frac{P \times LSF}{4.45F_1}\right)^2 \frac{10^{F_3}}{41.35} - 9.0\right]^{0.103} \tag{6}$$

where
P, LSF and F_1 are similar to the previous definitions
N_e = allowable load repetitions based on erosion
F_2 = adjustment factor for slab edge effects
= 0.06 for base with no shoulder
= 0.94 for base with shoulder
F_3 = erosion factor

(16) For each axle load of each axle group, divide the applied load repetitions by the allowable one obtained from fatigue and multiply by 100% to obtain the fatigue percentage. Add the fatigue percentages for all the axle loads to obtain the fatigue factor.
(17) Repeat step 16 for all axle loads to calculate the erosion factor by summing up the percentage damage for each axle load which is calculated by dividing the expected load repetitions by the allowable one to prevent erosion.
(18) If either the fatigue factor or the erosion factor is more than 100%, the designer has two options. Either to increase the base thickness, h thus to repeat steps 11 to 17 or to use a stronger sub-base thus to repeat steps 8 to 18.
(19) After the designer is satisfied with the structural design of the pavement, the final step of the design process is to calculate the amount of reinforcement required.

It is clear that the design of rigid pavements based on the recommendation of AUSTROADS is a lengthy process that requires several steps of elaborate calculations. Moreover, after calculating the fatigue and erosion factors, the designer is to make the decision whether a redesign is necessary or not. In the case when either of the factors is greater than 100% a redesign is necessary. However in the case when either or both of the factors is much less than 100%, in other words an over-design, it becomes the decision of the designer to whether to repeat the design by assuming a smaller base course or a weaker sub-base or accepting the over-design. The cost of the pavement is not explicitly specified in AUSTROADS as a design factor for both rigid and flexible pavements. It is the responsibility of the designer to ensure a safe and an economical design. It is to be noted here that the saving in the design of road pavements can be considerable due to the vast amount of materials involved in constructing the pavement structure.

This cost factor leads to the call to include the cost of initial construction and continuing maintenance of the pavement as an explicit design objective and to include the fatigue and erosion factors among other factors as design constraints. Using optimisation techniques as compared to "fully stressed designs" will yield an optimum design quickly using routine calculations. These designs satisfy the requirements of the design guide as well as being the most economical ones.

3 THE RIGID PAVEMENT OPTIMISATION PROBLEM

The first step in any optimisation problem is to define the problem, including the objective function and the constraints that control the solution. The formulation that is presented herein is based on the requirements of AUSTROADS (1992). However, other methods of design can be incorporated by extending the scope of the problem.

The objective function for which the minimum value is sought is presented as:

$$C_p = C_c \times h + C_{LMC} \times h_{LMC} + C_{bound} \times h_{bound} + C_s \times A_s \tag{7}$$

Subject to

$$h \geq 150 \ mm \tag{8}$$

$$0 \leq fatigue\ factor \leq 100 \tag{9}$$

$$0 \leq erosion\ factor \leq 100 \tag{10}$$

$$Type\ of\ sub-base \geq minimum\ recommended\ subbase \tag{11}$$

$$A_s \geq A_{s\,\min} \tag{12}$$

Where C_p is the cost of building the pavement per square metre, C_c is the cost of a cubic metre of concrete, h is the thickness of the base, C_{LMC} is the cost of a cubic metre of lean mixed concrete, h_{LMC} is the thickness of the lean mixed concrete layer, C_{bound} is the cost of a cubic metre of bound material, h_{bound} is the thickness of the bound material, C_s is the cost of one cubic metre of reinforcement and A_s is the area of reinforcement in a square metre of the base. The "*Type of sub-base*" variable is the type of sub-base that the designer can choose. AUSTROADS recommends these types: 100 mm, 125 mm and 150 mm of bound material and 100 mm, 125 mm and 150 mm of lean mix concrete, LMC. "minimum recommended sub-base" is the least sub-base type to satisfy Figure 9.1 of AUSTROADS (1992).

The design parameters that are input to the problem are type of pavement, location of the pavement, whether shoulders exist, daily commercial traffic, the growth rate, design life, subgrade CBR, strength of concrete, and strength of steel. The cost of concrete, lean mix concrete, LMC, and bound material, all expressed per cubic meter are also input. Based on these input parameters the optimisation process obtains the optimum solution, which includes the best type and thickness of sub-base, the thickness of the base and the amount of reinforcement.

4 GENETIC ALGORITHM

A new evolving branch of optimisation is genetic algorithm, GA that is based on nature's theory of evolution, survival of the fittest. GA has been developed by Holland (1975) as reported by Goldberg (1989). GA is an iterative process that involves Reproduction, Crossover and Mutation. Briefly, the steps involved in GA are:

(1) Represent all the variables in the optimisation problem in binary form. Hence all variables will be represented by a string of bits. All the variables will be in one block. For example, if we have three variables, *a*, *b*, and *c*; where *a* can have any value between 0 and 6, *b* can have any value between 0 and 32 and *c* can have any value between 0 and 4, then the following string will represent all three variables. In other words, one binary number of length 10 can represent all the three variables. See Figure 1.

(2) Randomly select two binary numbers of 10 bits length parents. Those two numbers need to satisfy all the constraints.

(3) Perform what is called crossover on the two selected numbers. Crossover between two variables is a simple process where the bits of the two variables after a randomly chosen point of crossover interchange values with the corresponding bit in the other variable. For example, if we have the two variables shown in Figure 2 (a), crossover will yield the variables shown in Figure 2 (b). In this example, bits 5 to 10 exchange values with corresponding bits in the other number.

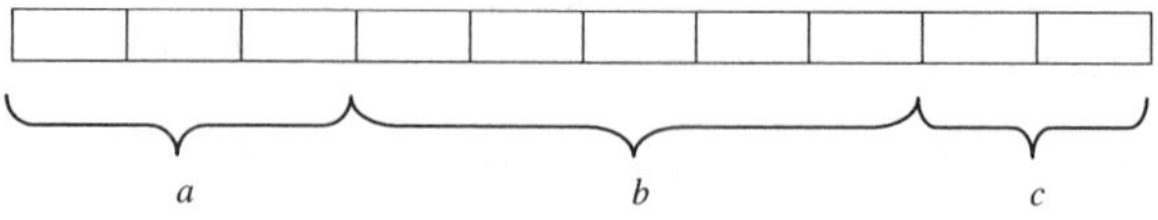

Figure 1. A Binary Representation of Three Variables

Parent A	1	1	0	0	0	1	0	0	0	0
Parent B	1	0	1	0	1	0	0	1	0	1
					↑ Point of Crossover					

(a) Before Crossover

Parent A	1	1	0	0	1	1	0	0	0	0
Parent B	1	0	1	0	0	0	0	1	0	1

(b) After Crossover

Figure 2. Crossover

Offspring A	1	1	0	0	1	0	0	1	0	1

(a) Before Mutation

Offspring A	1	1	0	0	0	0	0	1	0	1

(b) After Mutation

Figure 3. Mutation

(4) Perform mutation on the new offsprings. Mutation is simply to change the value in a randomly chosen bit from 0 to 1 or from 1 to 0, as shown in Figure 3.

If at any stage of mutation any of the variables violate the problem constraints, then such variables will be discarded and a new number is chosen randomly. Repeat steps 3 through to 4 for the pre-defined maximum number of iterations. The best solution will be achieved at the end of the iterations.

In our optimum design of rigid pavements, a string was chosen to represent the type of subbase and base thickness, those being the variables that their optimum value is sought. A computer program was written in Matlab comprising two modules, one for the analysis of the rigid pavement and the second to compute the optimum solution. Both modules are interconnected, as such the values of variables pass to either module during the GA process. Fitness of the design variables represents the value of the inverse of the objective function. The inverse was chosen as GA maximises the fitness and our problem is to minimise the cost. The GA operations of generating offsprings by selection, crossover and mutation is repeated a few hundred times in order to achieve the fittest (optimum) offspring (design).

Table 1. Design Examples

Pavement Number	1	2	3	4	5	6
Locality	Rural	Rural	Rural	Urban	Urban	Urban
Shoulders	Yes	Yes	Yes	Yes	Yes	Yes
Dowelled	Yes	Yes	Yes	Yes	Yes	Yes
CBR	5	5	5	5	5	5
LSF	1.2	1.2	1.2	1.2	1.2	1.2
CVAG	$1*10^6$	$3*10^6$	$5*10^6$	$1*10^6$	$3*10^6$	$5*10^6$

5 EXAMPLES

In this paper, a number of pavements were designed and compared with the corresponding optimum solutions presented by Hadi (1998). Table 1 summarises the pavements' design parameters and Table 2 summarises the results of both methods. Each of the GA runs was repeated four

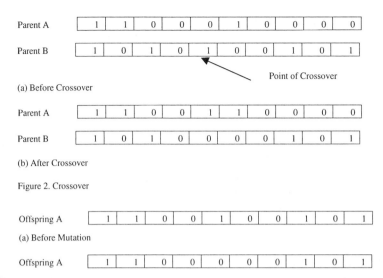

Table 2. Results of Optimisation

	GA				Optimisation (Hadi 1998)			
Pavement Number	h_{Bound} mm	h_{LMC} mm	h_{base} mm	Cost \$/m³	h_{Bound} mm	h_{LMC} Mm	h_{base} mm	Cost \$/m³
1	125	0	180	31.45	100	0	180	30.2
2	100	0	190	31.6	100	0	190	31.6
3	125	0	190	32.85	125	0	190	32.85
4	100	0	170	28.8	100	0	180	30.2
5	100	0	180	30.2	100	0	180	30.2
6	150	0	170	33.8	125	0	180	31.45

times and the number of generations were limited to 400. In each case, the same solution was achieved. Table 2 shows that the results of both methods are of comparable magnitude. The advantage of GA is making use of the discrete nature of GA and applying it to the discrete nature of the rigid pavement design.

6 CONCLUSIONS

Optimisation of rigid pavements yields pavements that are safe as well as economical. With advances of computing capabilities available to practicing engineers, it is herein proposed to include the cost of building the pavement as an explicit design criterion. This will produce pavements that are safe and economical. The current work is based on optimising the initial cost of pavement construction. As the maintenance cost of pavements is an important factor, this needs to be included in future development.

As mentioned above, optimisation of the structural design of pavements has been applied in a number of computer programs. But so far it has not been included in design guides, such as AUSTROADS (1992). This may be due to the fact that the design of rigid pavements used to be empirical and due to lack of computing power. Nowadays, most pavement design methods have some degree of determinism and there is a continuous upgrade of the design methods in order to make them more mechanistic. This is coupled with the availability of computer power of which design engineers can take advantage fairly easily.

Based on these facts, it appears that the time is ripe to develop equations for the optimum design of rigid pavements and apply these equations to obtain optimum design of rigid pavements.

In this paper, GA was easily adopted to the problem of rigid pavement design. The discrete nature of GA was utilised in solving the discrete rigid pavement design.

REFERENCES

AUSTROADS 1992. *A Guide to the Structural Design of Road Pavements*. AUSTROADS, Sydney.

Fordyce, P. and W.A. Yrjanson 1969. Modern Design of Concrete Pavements. Transportation Engineering Journal, ASCE, 95:TE3, August, pp. 407-438.

Goldberg, D.E. 1989. *Genetic Algorithms in Search, Optimizations, and Machine Learning*. Addison-Wesley.

Hadi, M.N.S. 1998. Cost Optimum Design of Rigid Road Pavements. 19th Australian Road Research Board Conference, (ARRB), Sydney, 6-11 December. pp. 295-309.

Holland, J.H. 1975. *Adaptation of Natural and Artificial Systems*. The University of Michigan Press.

Huang, Y.H. 1993. *Pavement Analysis and Design*. Prentice-Hall. New Jersey.

Packard, R.G., and S.D. Tayabji 1985. New PCA Design Procedure for Concrete Highway and Street Pavements. *Third International Conference on Concrete Pavement Design and Rehabilitation*, pp. 225-236. Purude University.

PCA 1984. *Thickness Design for Concrete Highway and Street Pavements*. Portland Cement Association, EB109P

Yoder, E.J., and Witczak, M.W. 1975. *Principles of Pavement Design*. John Wiley and Sons, Inc., New York, N.Y.

Mechanics of Structures and Materials, Bradford, Bridge & Foster (eds)
© 1999 Balkema, Rotterdam, ISBN 90 5809 107 4

Structural reliability of masonry walls in compression and flexure

Mark G. Stewart
Department of Civil, Surveying and Environmental Engineering, University of Newcastle, Callaghan, N.S.W., Australia

Bradley S. Wilkes
Lindsay and Dynan Pty Limited, Broadmeadow, N.S.W., Australia

ABSTRACT: Estimates of structural reliability were obtained for vertical loading of non-slender masonry walls and lateral loading of non-structural masonry panels. Structural reliabilities were most affected by variations in load eccentricity, workmanship and discretising of masonry unit thickness. It was found that safety indices for uninspected masonry is significantly lower than for inspected masonry. Considering that most masonry structures are uninspected, scope exists for the capacity reduction factor for masonry (vertical loading) to be increased from 0.45 to 0.75 if the masonry is subjected to thorough inspections. The results suggest also that poor quality mortar has a significant effect (by several orders of magnitude) on structural reliability for flexural masonry elements.

1 INTRODUCTION

Estimates of structural reliability may be used to assess the safety of various designs, of existing structures or of proposed changes in design codes. Reliability-based code-calibration is a method used in limit states design to achieve more uniform performance and reliability for different construction materials; it is a code optimisation procedure. To date, reliability-based methods have been applied to the development of structural steel, concrete and timber limit state design codes in Australia and elsewhere.

Very few studies have considered computational methods for calculating the structural reliability of masonry structures. Limit state codes in Australia, US, Canada and Europe have not been developed from reliability-based calibration methods, but rather simply from "past experience". Hence, it is unknown how structures designed the Masonry Structures Code [AS3700-1998] compare to other structural material codes in terms of reliability (or safety) and also if different masonry structural elements have similar levels of reliability.

The present paper develops a preliminary method to calculate the structural reliability (reliability index β) of typical masonry walls in compression and bending. First Order Reliability Methods (FORM) will be used as the computational tool that incorporates probabilistic information regarding element loads (dead, live and wind loads), material properties (compressive and flexural capacities), dimensions and modelling errors. The influence of workmanship is also considered. As such, reliability indices are obtained for vertical loading of non-slender single skin masonry walls and lateral wind loading of non-structural masonry panels considering variations of (i) live-to-dead load ratios, (ii) discretising of masonry unit thickness, (iii) unit compressive strength, (iv) mortar type, (v) load eccentricity, (vi) workmanship, (vii) tributary areas, and (viii) load intensity.

2 STRUCTURAL RELIABILITY

Failure of a structural element occurs when the load effect (S) exceeds the resistance (R). Reliability may then be expressed as a probability of failure (p_f) or "reliability index" (β):

Table 1. Statistical Parameters for Vertical Loading.

Parameter		Mean	Coefficient of Variation	Distribution	Reference
B (Inspection)		1.0	0.11	Normal	Ellingwood (1981)
B (No Inspection)		0.6	0.15	Normal	Ellingwood (1981)
a		1.2(e/t)+1	–	–	Ellingwood (1981)
f_m (MPa):	F'_{uc}=5MPa	4.25	0.15	Weibull	–
	F'_{uc}=20MPa	8.63	0.15	Weibull	–
	F'_{uc}=30MPa	10.33	0.15	Weibull	–
t (mm)		Nominal	0.02	Normal	Al-Harthy and Frangopol (1996)
L_p (N):	A_T=10m^2	0.68Q	0.42	Gumbel	Pham (1985)
	A_T=23m^2	0.70Q	0.26	Gumbel	Pham (1985)
	A_T=50m^2	0.74Q	0.25	Gumbel	Pham (1985)
D		1.0G	0.10	Normal	–

Table 2. Reliability Indices for Vertical Loading.

Varied Parameter	Inspected	F'_{uc} (MPa)	Mortar Type	A_T (m^2)	Q/G	Reliability index (β)
Workmanship	No	20	M3	23	1.1	3.96
	No	20	M3	23	1.1	3.96
	No	20	M3	23	1.1	3.96
F'_{uc} (MPa)	Yes	5	M3	23	1.1	4.92
	Yes	20	M3	23	1.1	4.92
	Yes	30	M3	23	1.1	4.92
Mortar Type	Yes	20	M2	23	1.1	4.92
	Yes	20	M3	23	1.1	4.92
	Yes	20	M4	23	1.1	4.91
A_T (m^2)	Yes	20	M3	10	1.1	4.80
	Yes	20	M3	23	1.1	4.92
	Yes	20	M3	50	1.1	4.88
Q/G	Yes	20	M3	23	0.4	4.72
	Yes	20	M3	23	1.1	4.92
	Yes	20	M3	23	2.1	4.98

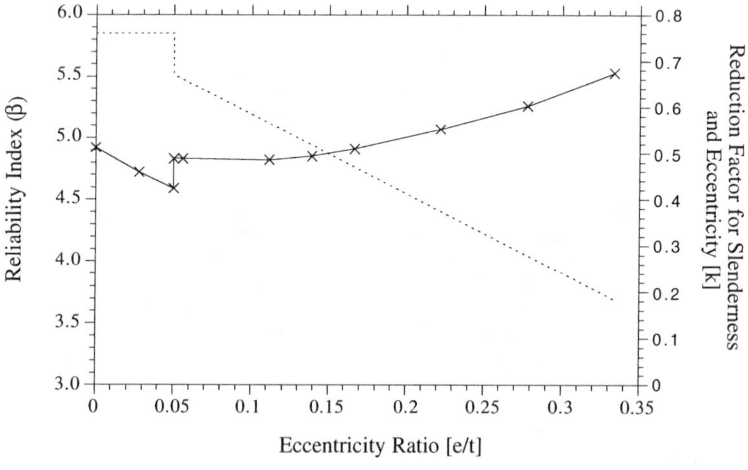

Figure 1. Effect of Load Eccentricity on Reliability Indices

$$p_f = \Pr(R - S \le 0) = \Pr(G(R,S) \le 0) = \int_0^{\infty} F_R(r) f_S(r) dr \qquad \beta = -\Phi^{-1}(p_f) \tag{1}$$

where Φ^{-1} is the inverse of the standard normal distribution function, G() is the "limit state function", $f_S(r)$ is the probability density function of the load effect and $F_R(r)$ is the cumulative probability density function of the resistance [e.g., Melchers, 1999].

3 MASONRY WALL - VERTICAL LOADING

The single storey wall is assumed to have a lateral support top and bottom, and with uniformly distributed dead and live load. There is no raking of bed joints and the masonry is fully bedded. It has been shown that the moment amplification may be neglected if the slenderness ratio is less than 14 [Ellingwood, 1981]; hence it reasonable to assume that a single storey wall is non-slender. According to AS3700-1998 the required wall thickness required for compressive capacity is calculated as

$$t = \frac{1.25G + 1.5Q}{k\phi f'_m b} \tag{2}$$

where G and Q are the design floor dead and live loads [AS1170.1-1989]; k is the reduction factor for slenderness and eccentricity; ϕ is the capacity reduction factor (0.45); f'_m is the characteristic strength of masonry and b is the run of wall (1000mm).

The limit state function for vertical loading is then

$$G(x) = Baf_m bt - \left[(D + L_P) + 6(e/t)(D + L_P)\right] \qquad 0 \le \frac{e}{t} \le \frac{1}{6}$$

$$G(x) = Baf_m bt - \left[\frac{4(D + L_P)}{3 - 6(e/t)}\right] \qquad \frac{1}{6} \le \frac{e}{t} \le \frac{1}{3} \tag{3}$$

where B is the workmanship factor, a is a model error (ratio of compressive strength when there is a strain gradient to the compressive strength under uniform compression), f_m is the compressive strength of masonry under uniform compression, D is the dead load, L_p is the peak live load (for a 50 year design life), and e/t represents the load eccentricity ratio.

The workmanship factor (B) accounts for the differences in fabrication and curing between the laboratory and the field since Ellingwood (1981) found a high correlation between B and workmanship quality. Ellingwood (1981) suggests that the strength of uninspected walls in the US tends to be approximately 60% (perhaps slightly conservatively) of the strength of inspected walls, based upon data obtained by Hendry (1976) and others. On the other hand, Turkstra (1989) suggests 70%. The effect of mortar mix, thickness of joints, environmental conditions and other factors would be expected to influence the workmanship factor, but are beyond the scope of the present study.

Mortar type and the unit compressive strength are the main factors that control the characteristic compressive strength of masonry (f'_m). Experimental data suggests that the compressive strength of masonry (f_m) follows a Weibull distribution with a coefficient of variation of 0.15. Since AS3700-1998 tabulates characteristic compressive strengths (5th percentiles) then standard statistical methods can be used to determine the Weibull distribution. The statistical parameters for the random variables are shown in Table 1.

For comparative purposes, the baseline structural configuration is concentric loading, 20MPa clay units, M3 mortar, inspected, slenderness ratio of 14, and tributary area of $23m^2$.

Table 2 shows that clay unit strength, mortar type, tributary area and load ratio have little influence on reliability indices. However, Figure 1 shows the effect of eccentricity on reliability indices, for the baseline case. Considering that eccentricities in excess of 1/6 are rare for most design applications the reliability indices are relatively consistent. It is observed also that the reliability index reduces from 4.92 to 3.96 when masonry is not inspected. This is a significant difference since in this case the probability of failure increases by a factor of eighty for uninspected masonry.

The reliability analyses conducted above assumed that the dimension of the constructed masonry is equal to the design value (t), see Eqn. (3). In practice, however, brick sizes are discretised (110mm, 220mm, etc.) and so designers are often conservative. The designer does have some control by changing the support spacing, masonry unit strength and various other factors that may bring the design capacity closer to the design actions. The effect of discretising unit thickness is examined by considering increasing design live loads while keeping the unit thickness (t) safely at 110mm, see Figure 2. Figure 2 shows some insight into how conservative (and safe) design can become if discretisation of masonry unit thickness is considered in a reliability analysis.

4 MASONRY WALL - LATERAL WIND LOADING

The wall is designed as a single skin in-fill panel simply supported top and bottom (one-way bending). It is assumed that the lateral load is transient, otherwise AS3700-1998 specifies that $f'_{mt}=0$. Since it is an in-fill panel there is no precompression, the only compressive stress is at the critical section (mid-height) due to masonry self-weight. As such, it is assumed that the wall will fail by exceeding the flexural tensile capacity of the mortar and not by local crushing. According to AS3700-1998 the required wall thickness (t) required is calculated as

$$t = \sqrt{\frac{3W_u bH^2}{4000\left(\phi f'_{mt} + \gamma H/2\right)}} \qquad (4)$$

where W_u is the design wind load, ϕ is the capacity reduction factor (0.60), f'_{mt} is the characteristic flexural tensile strength of masonry, H is the height of the wall and γ is the bulk density of the masonry. The limit state function becomes

$$G(x) = Bf_{mt}\left[\frac{3W_p H^2}{4t^2} - \gamma H/2\right] \qquad (5)$$

where f_{mt} is the flexural tensile strength of masonry and W_p is the 50-year maximum wind load.

For design the characteristic value of f'_{mt} is 0.2MPa [AS3700-1998] will be used. Experimental data suggests that the distribution of flexural tensile strength is a Weibull distribution with a coefficient of variation of 0.3. The mean can thus readily be obtained to give μ_{mt} =0.4MPa.

There appears to be no quantification of workmanship factors in the existing literature for lateral loading. If it is assumed that the workmanship factor considers only additional uncertainty for field-based conditions such as curing (and not workmanship) then statistical

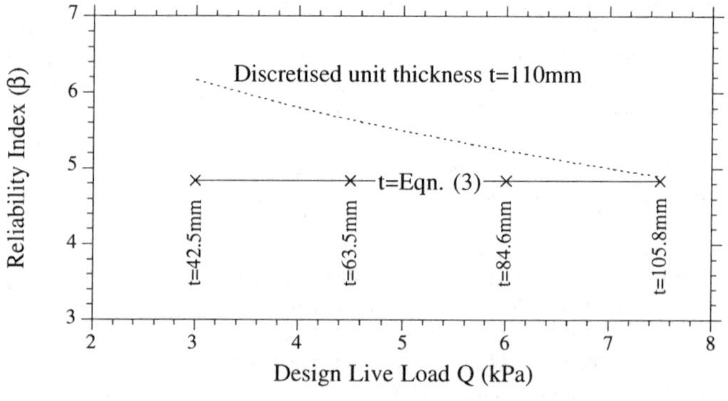

Figure 2. Effect of Discretisation of Masonry Unit Thickness on Reliability Indices

parameters for the inspected case (compression) are used. To investigate workmanship effects it is assumed that f_{mt} will be equal to $0.5\mu_{mt}$ and $0.1\mu_{mt}$ where μ_{mt} is the mean of the flexural tensile strength [AS3700-1998]. The reductions were based on results of laboratory work conducted at The University of Newcastle in which significant reductions in flexural tensile strength (poor bond) were observed for "typical" additives such as air entrainment and fireclay. The statistical parameters for the random variables are shown in Table 3.

Table 3. Statistical Parameters for Lateral Loading.

Parameter	Mean	Coefficient of Variation	Distribution	Reference
B	1.0	0.11	Normal	–
f_{mt} (MPa)	0.40	0.30	Weibull	–
t (mm)	Nominal	0.02	Normal	Al-Harthy and Frangopol (1996)
H (mm)	2400mm	–	–	–
W_p : (non-cyclonic)	0.30Wu	0.43	Lognormal	Pham (1985)
(cyclonic)	0.21Wu	0.76	Lognormal	Pham (1985)
γ	$19kN/m^3$	0.02	Normal	–

For comparative purposes, the baseline structural configuration is a wall height of 2.4m, non-cyclonic region [AS1170.2-1989] and bulk density of masonry of $19kN/m^3$. The reliability index for the baseline case was found to be 3.69. Load intensity (wind force regions) and bulk density of masonry had negligible effects on reliability indices. Table 4 shows the effects of workmanship on the reliability index. It is evident that the mis-use of mortar admixtures can significantly reduce the reliability index. In the present case, there is to an order of magnitude increase in the probability of failure (from 1.1×10^{-4} to 1.3×10^{-3}) for a 50% reduction in mortar flexural tensile strength, and a probability of failure of 0.16 (β=1.0) for a 90% reduction in mortar flexural tensile strength.

Table 4. Reliability Indices for Lateral Loading.

Workmanship	Reliability index (β)
Normal - μ_{mt}	3.69
Poor - $0.5\mu_{mt}$	3.00
Very Poor - $0.1\mu_{mt}$	1.00

5 DISCUSSION

The reliability indices estimated herein are generally consistent with those obtained for steel and concrete structural elements. Also, the present paper provides an indication of the sensitivity of structural reliabilities to all parameters.

Table 5 shows a comparison of typical reliability indices. AS3700-1998 does not discriminate between inspected and uninspected masonry; however, a possible mechanism to account for workmanship quality would be to use different capacity reduction factors. If the target reliability index is set at β_T=4.0 and considering that most masonry structures are uninspected, then further analysis reveals that the capacity reduction factor could be increased to ϕ=0.75 if the masonry was inspected for compression. This would provide for more uniform reliability and a possible incentive for designers to ensure that masonry construction is supervised and inspected. However, Turkstra (1989) notes that "designers are often reluctant to assume responsibility for field supervision and fear inspections will not be reliable. Contractors .. admit to only the best quality". This is clearly an area for further investigation. Finally, it should be noted also that there is considerable uncertainty about the statistical parameters for the workmanship factor and so observations regarding this change in capacity reduction factor are preliminary only.

Table 5. Comparison of Typical Reliability Indices

	ϕ	β
Vertical Loading (Inspected)	0.45	4.9
Vertical Loading (Not Inspected)	0.45	4.0
Lateral Wind Loading	0.60	3.7

Finally, there is an argument that the lateral loading case should be treated more conservatively than vertical loading since lateral loading will produce a sudden brittle failure with little warning. Hence, perhaps lateral loading structural figurations should have a reliability level higher than that obtained for vertical loading.

6 FURTHER WORK

The work reported herein are results from a Final Year Project Thesis [Wilkes, 1998]. Though some statistical and probabilistic analyses of the strength of masonry elements has been studied previously [e.g., Lawrence, 1991], there has been very little work conducted into the reliability analysis of masonry structures, either in Australia or elsewhere. There is clearly a need for more accurate probabilistic information regarding: (i) workmanship factors, (ii) model errors for other limit states, (iii) effect of mix proportions and admixtures on bond and (iv) correlations of strength between masonry units (system effects) for these and other structural configurations comprising two-way bending, slenderness effects, combined actions, shear, etc. As such, the present paper is very preliminary and more work is needed for the accurate reliability analysis of masonry structures.

7 CONCLUSIONS

Reliability indices were calculated for short masonry walls considering vertical and lateral loading. Reliability indices were most affected by variations in load eccentricity, workmanship and discretising of masonry unit thickness. It was found that reliability indices for uninspected masonry is significantly lower than for inspected masonry. Considering that most masonry structures are uninspected, scope exists for the capacity reduction factor for masonry (vertical loading) to be increased from 0.45 to 0.75 if the masonry is subjected to thorough inspections. The results suggest also that poor quality mortar (incorrect admixtures affecting flexural tensile strength) has a significant effect (by several orders of magnitude) on structural reliability for flexural masonry elements.

REFERENCES

Al-Harthy, A.S. and Frangopol, D.M., (1996), Reliability Analysis of Masonry Walls, *ASCE Specialty Conference on Probabilistic Mechanics and Structural Reliability*, Worchester, MA, pp.338-341.

AS 1170.1 (1989), *SAA Loading Code - Part 1: Dead and Live Loads and Load Combinations*, Standards Australia, Sydney.

AS 1170.2 (1989), *SAA Loading Code - Part 2: Wind Loads*, Standards Australia, Sydney.

AS3700 (1998), *Masonry Structures*, Standards Australia, Sydney.

Ellingwood, B.R. (1981), Analysis of Reliability of Masonry Structures, *Journal of Structural Engineering*, ASCE, Vol. 197, No. ST5, pp. 756-773.

Hendry, A.W. (1976), The Effect of Site Factors in Masonry Performance, *Proc. First Canadian Masonry Symposium,* E.L. Jessop and M.A. Ward (Eds.), Calgary, pp. 182-198.

Lawrence, S.J. (1991), Stochastic Analysis of Masonry Structures, *Computer Methods in Structural Masonry*, J. Middleton and G.N. Pande (Eds.), pp. 104-113.

Melchers, R.E. (1999), *Structural Reliability: Analysis and Prediction*, John Wiley, New York.

Pham, L. (1985), Load Combinations and Probabilistic Load Models for Limit States Codes, *Civil Engineering Transactions*, IEAust, Vol. CE27, No. 1, pp. 62-67.

Turkstra, C.J. (1989), Limit States in Masonry, *5th International Conference on Structural Safety and Reliability*, San Francisco, pp. 2043-2050.

Wilkes, B.S. (1998), Structural Reliability of Masonry Walls, Final Year Project, Dept. of Civil, Surveying and Environmental Engineering, The University of Newcastle, Australia.

Mechanics of Structures and Materials, Bradford, Bridge & Foster (eds)
© 1999 Balkema, Rotterdam, ISBN 90 5809 107 4

Author index